Climate Change and Infectious Fish Diseases

Climate Change and Infectious Fish Diseases

Edited by

Patrick T.K. Woo

University of Guelph, Guelph, Ontario, Canada

Jo-Ann Leong

University of Hawai'i at Manoa, Kane'ohe, Hawaii, USA

and

Kurt Buchmann

University of Copenhagen, Copenhagen, Denmark

CABI

CABI is a trading name of CAB International

CABI
Nosworthy Way
Wallingford
Oxfordshire OX10 8DE
UK

Tel: +44 (0)1491 832111
Fax: +44 (0)1491 833508
E-mail: info@cabi.org
Website: www.cabi.org

CABI
WeWork
One Lincoln Street
24th Floor
Boston, MA 02111
USA

Tel: +1 (617)682-9015
E-mail: cabi-nao@cabi.org

A catalogue record for this book is available from the British Library, London, UK.

Library of Congress Cataloging-in-Publication Data

Names: Woo, P. T. K., editor. | Leong, Jo-Ann, editor. | Buchmann, Kurt, editor.
Title: Climate change and infectious fish diseases / editors, Patrick T.K. Woo, University of Guelph, Guelph, Ontario, Canada, Jo-Ann Leong, University of Hawaii at Manoa, Kane'ohe, Hawaii, USA, Kurt Buchmann, University of Copenhagen, Copenhagen, Denmark.
Description: Wallingford, Oxfordshire ; Boston, MA: CAB International, [2020] | Includes bibliographical references and index. | Summary: "This definitive reference work explores the effects of current and expected climate change, taking place throughout the world, on selected bacterial, viral, fungal and parasitic infectious fish diseases of economically important fish in tropical and temperate waters"-- Provided by publisher.
Identifiers: LCCN 2020023959 (print) | LCCN 2020023960 (ebook) | ISBN 9781789243277 (hardback) | ISBN 9781789243284 (ebook) | ISBN 9781789243291 (epub)
Subjects: LCSH: Fishes--Diseases--Climatic factors. | Communicable disease in animals.
Classification: LCC SH172 .C54 2020 (print) | LCC SH172 (ebook) | DDC 333.95/6--dc23
LC record available at https://lccn.loc.gov/2020023959
LC ebook record available at https://lccn.loc.gov/2020023960

References to Internet websites (URLs) were accurate at the time of writing.

ISBN-13: 9781789243277 (hardback)
9781789243284 (ePDF)
9781789243291 (ePub)

Commissioning Editor: Caroline Makepeace
Editorial Assistant: Lauren Davies
Production Editor: Shankari Wilford

Typeset by SPi, Pondicherry, India

Contents

Contributors

Note: corresponding authors are indicated by an asterisk.

Carmen Amaro,* ERI BioTecMed, University of Valencia, Dr Moliner 50, 46100 Burjasot, Valencia, Spain. E-mail: carmen.amaro@uv.es

Julie D. Alexander,* Department of Microbiology, Oregon State University, 226 Nash Hall, Corvallis, OR 97330, USA. E-mail: alexanju@oregonstate.edu

Nurshuhada Ariff, Faculty of Fisheries and Food Sciences, Universiti Malaysia Terengganu, 21030 Kuala Nerus, Terengganu, Malaysia. E-mail: shu_look9092@yahoo.com

Brian Austin,* Institute of Aquaculture, School of Natural Sciences, University of Stirling, Stirling FK9 4LA, UK. E-mail: brian.austin@stir.ac.uk

Miguel Bao, Institute of Marine Research (IMR), Section of Contaminants and Biohazards, PO Box 1870 Nordnes, N-5817 Bergen, Norway. E-mail: miguel.bao-dominguez@hi.no

Jerri L. Bartholomew, Department of Microbiology, Oregon State University, 226 Nash Hall, Corvallis, OR 97330, USA. E-mail: barthoje@oregonstate.edu

Kurt Buchmann, Department of Veterinary and Animal Sciences, Faculty of Health and Medical Sciences, University of Copenhagen, Stigbøjlen 7, 1870 Frederiksberg C, Denmark. E-mail: kub@sund.ku.dk

Hsin-Yiu Chou,* Department of Aquaculture, National Taiwan Ocean University, 2 PeiNing Road, Keelung, Taiwan 20224, Republic of China. E-mail: hychou@email.ntou.edu.tw

Cindy Chu, Aquatic Research and Monitoring Section, Ontario Ministry of Natural Resources and Forestry, c/o Trent University, 2140 East Bank Drive, Peterborough, Ontario, Canada, K9L 1Z8. E-mail: cindy.chu@ontario.ca

Paolo Cipriani, Institute of Marine Research (IMR), Section of Contaminants and Biohazards, PO Box 1870 Nordnes, N-5817 Bergen, Norway. E-mail: paolo.cipriani@hi.no

Margaret Crumlish, Institute of Aquaculture, Faculty of Natural Sciences, University of Stirling, Stirling FK9 4LA, UK. E-mail: margaret.crumlish@stir.ac.uk

Sussie Dalvin, Institute of Marine Research, Postboks 1870 Nordnes, 5817 Bergen, Norway. E-mail: sussie.dalvin@hi.no

Derrick T. de Kerckhove,* Aquatic Research and Monitoring Section, Ontario Ministry of Natural Resources and Forestry, c/o Trent University, 2140 East Bank Drive, Peterborough, Ontario, Canada, K9L 1Z8. E-mail: dak.dekerckhove@ontario.ca

Carlos P. Dopazo,* Instituto de Acuicultura–Dpto Microbiología, University of Santiago de Compostela, E-15782 Santiago de Compostela, Spain. E-mail: carlos.pereira@usc.es

Mansour El-Matbouli,* Clinical Division of Fish Medicine, University of Veterinary Medicine, Veterinärplatz 1, 1210 Vienna, Austria. E-mail: Mansour.El-Matbouli@vetmeduni.ac.at

Mark D. Fast,* Department of Pathology & Microbiology, Atlantic Veterinary College, University of Prince Edward Island, 550 University Avenue, Charlottetown, Prince Edward Island, Canada, C1A 4P3. E-mail: mfast@upei.ca

Belén Fouz, ERI BioTecMed, University of Valencia, Dr Moliner 50, 46100 Burjasot, Valencia, Spain. E-mail: belen.fouz@uv.es

David T. Gauthier, Department of Biological Sciences, Old Dominion University, Norfolk, VA 23529, USA. E-mail: dgauthier@odu.edu

Lucilla Giulietti, Institute of Marine Research (IMR), Section of Contaminants and Biohazards, PO Box 1870 Nordnes, N-5817 Bergen, Norway. E-mail: lucilla.giulietti@hi.no

Matt J. Griffin,* Thad Cochran National Warmwater Aquaculture Center, College of Veterinary Medicine, Mississippi State University, Stoneville, MS 38776, USA. E-mail: matt.griffin@msstate.edu

Kollanoor Riji John, Faculty of Fisheries, Kerala University of Fisheries and Ocean Studies, Panangad – 682 506, Kochi, Kerala, India. E-mail: rijijohn@gmail.com

Louise von Gersdorff Jørgensen,* Department of Veterinary and Animal Sciences, Faculty of Health and Medical Sciences, University of Copenhagen, Stigbøjlen 7, 1870 Frederiksberg C, Denmark. E-mail: lvgj@sund.ku.dk

Dibyendu Kamilya,* Department of Aquatic Health and Environment, College of Fisheries, Central Agricultural University, Lembucherra, West Tripura – 799 210, Tripura, India. E-mail: dibyendu.kamilya@gmail.com

Anssi Karvonen,* University of Jyväskylä, Department of Biological and Environmental Science, PO Box 35, FI-40014 University of Jyväskylä, Finland. E-mail: anssi.t.karvonen@jyu.fi

Michael L. Kent, Departments of Microbiology and Biomedical Sciences, Oregon State University, Corvallis, OR 97331, USA. E-mail: Michael.kent@oregonstate.edu

Hidehiro Kondo, Graduate School of Marine Science and Technology, Tokyo University of Marine Science and Technology, 4-5-7 Konan, Minato, Tokyo, Japan 108-8477. E-mail: h-kondo@kaiyodai.ac.jp

Jesús Lamas,* Department of Functional Biology, Cell Biology, Institute of Aquaculture, CIBUS, Campus Vida, University of Santiago de Compostela, E-15782 Santiago de Compostela, Spain. E-mail: jesus.lamas@usc.es

Douglas W. Leaman, College of Science and Math Administration, Wright State University, Dayton, OH, 45435, USA. E-mail: douglas.leaman@wright.edu

José Manuel Leiro, Department of Microbiology and Parasitology, Laboratory of Parasitology, Institute of Research and Food Analysis, Campus Vida, University of Santiago de Compostela, E-15782 Santiago de Compostela, Spain. E-mail: josemanuel.leiro@usc.es

Jo-Ann Leong, Hawaii Institute of Marine Biology, University of Hawaii at Manoa, Manoa, HI 96744, USA. E-mail: joannleo@hawaii.edu

Arne Levsen,* Institute of Marine Research (IMR), Section of Contaminants and Biohazards, PO Box 1870 Nordnes, N-5817 Bergen, Norway. E-mail: arne.levsen@hi.no

Jia Huan Liew, Department of Biological Sciences, National University of Singapore, 14 Science Drive 4, Singapore 117543. Current address: School of Biological Sciences, The University of Hong Kong, Pokfulam, Hong Kong SAR, People's Republic of China. E-mail: jiahuan@u.nus.edu

Rayson Bock Hing Lim, Department of Biological Sciences, National University of Singapore, 14 Science Drive 4, Singapore 117543. E-mail: dbslbhr@nus.edu.sg

Bi Wei Low, Department of Biological Sciences, National University of Singapore, 14 Science Drive 4, Singapore 117543. E-mail: biweilow@gmail.com

Ken MacKenzie,* School of Biological Sciences (Zoology), The University of Aberdeen, Tillydrone Avenue, Aberdeen AB24 2TZ, UK. E-mail: k.mackenzie@abdn.ac.uk

David J. Marcogliese, Aquatic Contaminants Research Division, Water Science and Technology Directorate, Science and Technology Branch, Environment and Climate Change Canada, St. Lawrence Centre, 105 McGill, 7th floor, Montreal, Quebec, Canada, H2Y 2E7; and St. Andrews Biological Station, Fisheries and Oceans Canada, 125 Marine Science Drive, St. Andrews, New Brunswick, Canada, E5B 0E4. E-mail: david.marcogliese@canada.ca

Fernando O. Mardones, School of Veterinary Medicine, Pontifical Catholic University of Chile, Av. Vicuña Mackenna 4860, Macul, Santiago, Chile, 7820436. E-mail: femardones@uc.cl

Simonetta Mattiucci, Sapienza – University of Rome, Department of Public Health and Infectious Diseases, Piazzale Aldo Moro 5, 00185 Rome, Italy. E-mail: simonetta.mattiucci@uniroma1.it

Rajesh Mehrotra, Department of Biological Sciences, Birla Institute of Technology and Science, KK Birla Goa Campus, NH 17B, Bypass Road, Zuarinagar, Sancoale, Goa 403726, India. E-mail: rajeshm@goa.bits-pilani.ac.in

Sandhya Mehrotra,* Department of Biological Sciences, Birla Institute of Technology and Science, KK Birla Goa Campus, NH 17B, Bypass Road, Zuarinagar, Sancoale, Goa 403726, India. E-mail: sandhyam@goa.bits-pilani.ac.in

Tor Atle Mo,* Norwegian Institute for Nature Research – NINA, Sognsveien 68, 0855 Oslo, Norway. E-mail: tor.mo@nina.no

Maxine Allayne Darlene Mowe, Department of Biological Sciences, National University of Singapore, 14 Science Drive 4, Singapore 117543. E-mail: maxinemowe@nus.edu.sg

Namitha Nayak, Department of Biological Sciences, Birla Institute of Technology and Science, KK Birla Goa Campus, NH 17B, Bypass Road, Zuarinagar, Sancoale, Goa 403726, India. E-mail: nayaksnamitha@gmail.com

Ting Hui Ng, Department of Biological Sciences, National University of Singapore, 14 Science Drive 4, Singapore 117543. Current address: Lee Kong Chian Natural History Museum, National University of Singapore, 2 Conservatory Drive, Singapore 117377. E-mail: ngtinghui@nus.edu.sg

Megan D. Niner, Genetics and Genomics Group, Department of Environmental Sciences, University of Toledo, Toledo, OH 43606, USA. E-mail: megan.niner@rockets.utoledo.edu

Barbara F. Nowak,* IMAS, University of Tasmania, Locked Bag 1370, Launceston, Tasmania 7250, Australia. E-mail: b.nowak@utas.edu.au

Qi-Wei Qin, Guangdong Laboratory for Lingnan Modern Agriculture, College of Marine Sciences, South China Agricultural University, 484 Wushan Road, Guangzhou 510642, People's Republic of China. E-mail: qinqw@scau.edu.cn

Jesús L. Romalde, Department of Microbiology and Parasitology, CIBUS–Faculty of Biology, University of Santiago de Compostela, E-15782 Santiago de Compostela, Spain. E-mail: jesus.romalde@usc.es

Eva Sanjuán, ERI BioTecMed, University of Valencia, Dr Moliner 50, 46100 Burjasot, Valencia, Spain. E-mail: eva.sanjuan@uv.es

Kwanrawee Sirikanchana, Research Laboratory of Biotechnology, Chulabhorn Research Institute, Bangkok, Thailand 10210. E-mail: kwanrawee@cri.or.th

Pedro A. Smith,* Department of Animal Pathology, Faculty of Veterinary Sciences, University of Chile, Av. Santa Rosa 11735, La Pintana, Santiago, Chile, 8820808. E-mail: psmith@uchile.cl

Jadwiga Sokołowska, Institute of Aquaculture, Faculty of Natural Sciences, University of Stirling, Stirling FK9 4LA, UK. E-mail: j.p.sokolowska1@stir.ac.uk

Hatem Soliman, Clinical Division of Fish Medicine, University of Veterinary Medicine, Veterinärplatz 1, 1210 Vienna, Austria. E-mail: Hatem.soliman@vetmeduni.ac.at

Esteban Soto, Department of Medicine and Epidemiology, School of Veterinary Medicine, University of California – Davis, Davis, CA 95616, USA. E-mail: sotomartinez@ucdavis.edu

David J. Speare,* Atlantic Veterinary College, University of Prince Edward Island, Charlottetown, Prince Edward Island, Canada, C1A 4P3. E-mail: speare@upei.ca

Carol A. Stepien,* NOAA Pacific Marine Environmental Laboratory, Genetics and Genomics Group, 7600 Sand Point Way NE, Seattle, WA 98115, USA. E-mail: carol.stepien@noaa.gov and cstepien@uw.edu

Win Surachetpong,* Faculty of Veterinary Medicine, Kasetsart University, Bangkok, Thailand 10900. E-mail: fvetwsp@ku.ac.th

Christopher M. Whipps,* Departmental of Environmental and Forest Biology, State University of New York College of Environmental Science and Forestry (SUNY-ESF), Syracuse, NY 13210, USA. E-mail: cwhipps@esf.edu

David J. Wise, Thad Cochran National Warmwater Aquaculture Center, Mississippi Agriculture and Forestry Experiment Station, Mississippi State University, Stoneville, MS 38776, USA. E-mail: dwise@drec.msstate.edu

Patrick T.K. Woo,* Department of Integrative Biology, College of Biological Science, University of Guelph, Guelph, Ontario, Canada, N1G 2W1. E-mail: pwoo@uoguelph.ca

Darren Chong Jinn Yeo,* Department of Biological Sciences, National University of Singapore, 14 Science Drive 4, Singapore 117543; and Lee Kong Chian Natural History Museum, National University of Singapore, 2 Conservatory Drive, Singapore 117377. E-mail: dbsyeod@nus.edu.sg

Sandra C. Zainathan,* Faculty of Fisheries and Food Sciences, Universiti Malaysia Terengganu, 21030 Kuala Nerus, Terengganu, Malaysia. E-mail: sandra@umt.edu.my

Yiwen Zeng, Department of Biological Sciences, National University of Singapore, 14 Science Drive 4, Singapore 117543. E-mail: zengyiwen@nus.edu.sg

Preface

Climate change with global warming is not disputed by reputable scientists and the aquatic environment (e.g. rivers, lakes, oceans) is greatly affected. The two important greenhouse gases that contribute to climate change are carbon dioxide (CO_2; e.g. from extensive use of fossil fuels by industries and in transportation) and methane (CH_4; e.g. from the gas and oil industry, and agricultural activities such as large-scale breeding and raising of livestock for food) released into the atmosphere. These gases trap heat radiating from the earth and increase environmental temperatures. Of the two gases, atmospheric CO_2 is much more abundant and is the major factor contributing to climate change. Briefly, it also dissolves in water, acidifies aquatic ecosystems and decreases the amount of dissolved oxygen at higher temperatures, which may lead to hypoxic conditions for many aquatic organisms including fish. Climate change is also associated with more extreme and unusual weather conditions; e.g. it changes wind patterns, causes heavier rainfalls with high winds in some regions and prolongs droughts with widespread forest fires in other areas. Global warming also increases thawing of the permafrost, melting of glaciers and causes more rapid melting of the North and South Poles, which elevates sea levels and modifies salinities and water currents. These changes alter aquatic food webs and perhaps the compositions of aquatic animal communities. Many of these environmental changes may directly and/or indirectly affect the life history, biology and survival of aquatic organisms.

CH_4 (also commonly known as 'marsh gas') is the other important greenhouse gas, and the two main sources of atmospheric CH_4 are natural (e.g. due to decomposition of organic materials in wetlands, forest fires) and anthropogenic (e.g. from the oil and gas industry, agricultural activities). The amount in the atmosphere is significantly lower than that of CO_2; however, scientists are beginning to be concerned about it as a CH_4 molecule traps significantly more heat than a CO_2 molecule, and the amount of atmospheric CH_4 has continued to rise rapidly in recent years (e.g. the amount is estimated to have roughly doubled since the year 1800) due to increased anthropogenic activities and additional release of the gas from organic materials in areas previously covered by ice. It is tricky, but we can partially oxidize CH_4 to produce methanol or to produce CO_2 and water on complete oxidation of the gas.

Nearly 200 countries signed the United Nations 'Paris Agreement of 2015' to control climate change. An important component in the pledge was to reduce the current output of greenhouse gases as soon as possible so that global warming would be less than 2°C above pre-industrial levels. At present, it is about 1°C above and it is expected to rise to 3°C above or higher if countries continue with their current energy policies. The more recent '2018 UN Special Report' by the Intergovernmental Panel on Climate Change (IPCC; prepared by 91 authors and review editors from 40 countries) indicates the '2015 Paris Agreement' needs modifications. Many of the negative impacts due to global warming would be reduced at 1.5°C above pre-industrial levels compared with even 2°C. Limiting warming to 1.5°C is now considered possible if concerted efforts are made to integrate and implement most, if not all the recommendations. The slower rise in global temperature would provide us with more time: (i) to reduce the output of CO_2 by industries and transportation (e.g. use less-polluting fuels) and for national governments to impose a 'carbon tax' as well as to develop and implement sustainable energy policies (e.g. encourage the use of and accelerate the refinement of current technologies to harvest energy from the wind and waves); (ii) for governments (national/local) to provide additional incentives to develop new 'cleaner fuels'; (iii) to institute new programmes for reforestation, afforestation and the preservation of existing wetlands and forests; (iv) to develop more novel and practical strategies for the removal and storage of atmospheric CO_2; and (v) for organisms and ecosystems to adapt to ongoing changes in the aquatic environment.

Besides actions to be undertaken by national/local governments, there are a few simple lifestyle changes we all as individuals can make to reduce the size of our own carbon footprints. We suggest focusing on two changes because they are very cost-effective, well within our control and consequently are achievable. If we act collectively, these seemingly modest actions (also suggested by others; e.g. Seth Wynes, an environmental geographer at The University of British Columbia, Vancouver, Canada) will reduce emissions of the two most important greenhouse gases. First, we reduce unnecessary travel (especially by car and aeroplane), and second we modify our eating habits from an essentially red-meat diet to a more plant-based and/or fish-based diet.

Fish is an excellent and affordable source of animal protein for about 4.2 billion people, and with the growth in our population we expect the demand will continue to increase in the 21st century. According to the 2014 World Bank Report, the aquaculture industry will have to double its global production by the year 2030. The biology (e.g. development, physiology, behaviour, migration patterns) and well-being of fish are affected by current and expected changes to the aquatic environment. These are discussed in *Climate Change and Non-infectious Fish Disorders* (CCNFD; editors Patrick T.K. Woo and George K. Iwama), which has just been published (2020) by CABI, UK.

Our current book, entitled *Climate Change and Infectious Fish Diseases* (CCIFD), is the companion volume to CCNFD and it is also multidisciplinary. As in CCNFD, chapter contributors and topics in CCIFD are selected by the editors. We have more than 50 contributors from 18 countries and they include next-generation experts (e.g. Cindy Chu, Canada; Darren C.J. Yeo, Singapore; Matt J. Griffin, USA) as well as many well-established and highly respected experts (e.g. Brian Austin, UK; Tor Atle Mo, Norway; David J. Speare, Canada). The book has 25 chapters divided into three parts. Topics in Part I include changes (abiotic and biotic) in freshwater ecosystems in North America and South-East Asia, and biological sequestrations of atmospheric CO_2. The remaining 22 chapters are devoted to 12 microbial (Part II) and ten parasitic (Part III) infections. Although most are well-studied pathogens (e.g. viral haemorrhagic septicaemia virus; *Aeromonas salmonicida*; *Ichthyophthirius multifiliis*), CCIFD also includes emerging pathogens (e.g. tilapia lake virus; *Neoparamoeba perurans*). As in its companion volume (CCNFD), discussions in CCIFD address current and expected impacts due to environmental changes, point out gaps in our knowledge and articulate suggestions for future studies.

Briefly, Chapter 1 (Part I) focuses on biotic and abiotic changes (e.g. increases in water temperature, acidification of the aquatic environment, ice covers) in the Great Lakes Basin, while Chapter 2 is on changes due to anthropogenic activities, invasive fish species and changes in biodiversity in freshwater lakes and rivers in South-East Asia. Chapter 3 is on terrestrial and aquatic (biological) sequestrations of atmospheric CO_2 with suggestions to develop more novel strategies (e.g. use of molecular/genetic engineering to produce transgenic plants which will be more efficient) to enhance the sequestration process. Parts II and III are on infectious organisms with 12 chapters on microbial and ten chapters on parasitic diseases. Several criteria are used to select disease agents for discussions and they include well-studied and emerging pathogens (i) in economically important fish species, especially those that have wide host range and geographical distribution; (ii) that will adapt readily to new hosts and to environmental changes; (iii) with known reservoir or intermediate hosts; and (iv) that have or will have significant impact(s) on current or future fisheries and aquaculture.

All chapters on infectious organisms in CCIFD follow a relatively similar and logical format. This format was initially used in *Fish Diseases and Disorders: Protozoan and Metazoan Infections*, Volume 1 (1st edition; 1995), and it was well-received by many readers/users who found it to be quite 'user friendly'. The format has undergone some modifications and refinements. Briefly, each chapter in CCIFD is usually devoted to a specific pathogen and includes: (i) a brief description of the pathogen, its geographical distribution and its potential spread to new location(s) and/or novel host(s) (e.g. via changes in water currents, rise in water levels and temperatures, fish migration); (ii) clinical signs and diagnosis of the disease; (iii) pathology (gross and histological lesions) and pathobiology, which includes changes to fish immune responses and virulence of the pathogen; (iv) dynamics of the host–pathogen relationship, which may include transmission of the pathogen, acquisitions of new definitive reservoirs and/or intermediate hosts, etc.; and (v) control and prevention of the pathogen/disease with suggestions for future studies (e.g. novel vaccines and/or drugs; selective breeding of more adaptable fish including those with a more responsive immune system, higher tolerance to increases in water temperatures) to minimize the negative impacts climate change has/will have on fish health and longevity, and production of fish.

We hope our colleagues will find the two-volume set (CCNFD and CCIFD) useful and perhaps even thought-provoking. The primary audience of the books includes scientists in research institutes, universities, managers of aquaculture facilities, aquatic biologists and fish health consultants in government and private laboratories. The books may also be useful to environmentalists and ecologists who routinely monitor changes to the aquatic system. CCNFD and CCIFD are appropriate for the training of fish health specialists, and for specialized graduate students and senior undergraduates who study fish health and/or monitor changes to the aquatic environment.

Patrick T.K. Woo, Jo-Ann Leong and Kurt Buchmann

Previous titles by Patrick T.K. Woo

Climate Change and Non-infectious Fish Disorders
Edited by P.T.K. Woo and G.K. Iwama
2020 256 pages ISBN 978 1 78639 398 2

Fish Viruses and Bacteria: Pathobiology and Protection
Edited by P.T.K. Woo and R.C. Cipriano
2017 376 pages ISBN 978 1 78064 778 4

Diseases and Disorders of Finfish in Cage Culture, 2nd Edition
Edited by P.T.K. Woo and D.W. Bruno
2014 354 pages ISBN 978 1 78064 207 9

Fish Parasites: Pathobiology and Protection
Edited by P.T.K. Woo and K. Buchmann
2011 400 pages ISBN 978 1 84593 806 2

Fish Diseases and Disorders, Volume 3, 2nd Edition: Viral, Bacterial and Fungal Infections
Edited by P.T.K. Woo and D.W. Bruno
2011 944 pages ISBN 978 1 84593 554 2

Fish Diseases and Disorders, Volume 2, 2nd Edition: Non-infectious Disorders
Edited by J.F. Leatherland and P.T.K. Woo
2010 416 pages ISBN 978 1 84593 553 5

Fish Diseases and Disorders, Volume 1, 2nd Edition: Protozoan and Metazoan Infections
Edited by P.T.K. Woo
2006 800 pages ISBN 978 0 85199 015 6

1 Freshwater Ecosystems in North America with Reference to the Great Lakes Basin

Derrick T. de Kerckhove* and Cindy Chu

Ontario Ministry of Natural Resources and Forestry, Peterborough, Ontario, Canada

1.1 Introduction to Climate Change Models

The word 'climate' generally refers to a weather pattern over a long period of time (e.g. 30 years) and within a defined geographic area. Precipitation and wind data are often used to describe the mean, variability and possible year-to-year changes in climate (IPCC, 2007a,b). Weather is mainly determined by its energy budget; which is the balance between the incoming short-wave radiation from the Sun and the outgoing long-wave radiation from the Earth as influenced by absorption and scattering at the Earth's surface and the energy conversion processes (Trenberth *et al.*, 2009). From this budget, additional effects of ocean currents and topography, as well as negative and positive interactions among physical processes, determine the varied climates over space and time (Vecchi *et al.*, 2008; Andrews and Forster 2009; Stephens *et al.*, 2010). Greenhouse gases such as carbon dioxide (CO_2; e.g. from burning of fossil fuels by industries) and methane (CH_4; e.g. from large-scale breeding and raising of livestock) disrupt the energy budget by modifying the transfer of heat and radiation across the altitudinal zones.

Past climatic events will shed some light on the interrelationship between greenhouse gases and the physical processes to create regional variations in climate. The Medieval Climate Anomaly (800–1300 AD) followed by the 'Little Ice Age' (1350–1850 AD) demonstrated how volcanic activity had a profound influence on surface temperature, aridity, ocean weather and atmospheric concentrations of CO_2, leading to global warming in the former period and cooling in the latter (MacDonald, 2010). Over the last 300 years, anthropogenic activities have increased atmospheric concentrations of CO_2 by roughly one-third as a result of burning fossil fuels and deforestation (IPCC, 2007a,b). More recently, humans have begun to release smaller concentrations of more powerful greenhouse gases (e.g. sulfur hexafluoride and perfluoropentane) (Wuebbles *et al.*, 2019). The climate has changed in ways that cannot be explained through any natural physical processes (e.g. increased radiation from the Sun) nor natural variation in surface temperatures. Climate metrics including average global atmospheric temperature, surface seawater temperature, permafrost depth and melting of Arctic ice suggest that the Earth is warming (IPCC, 2013; USGCRP, 2017). In the last century, climatologists have noted an increase of 1°C in the global annual average air temperature over both land and oceans (USGCRP, 2017), and this rate of warming has increased since 2001 (NCEI, 2016). Also, the hottest years on record are the most recent (i.e. 2016, 2017, 2015 and 2018, in a descending ranked order) and they surpass the next warmest year (2014) by 0.16°C. Concurrent with these global patterns of air temperature, the intensity and frequency of heavy rain (i.e. storm events) have increased globally (Karl and Knight, 1998; O'Gorman and Schneider, 2009). These historical and recent changes along with long series of ocean, atmospheric and physical

*dak.dekerckhove@ontario.ca

data paint an alarming picture of an accelerating change in our climate. However, it allows climatologists to develop and test objective models to characterize past climate and predict future conditions (Giannini *et al.*, 2003; deMenocal, 2011).

The global circulation models (GCMs; a component of the global climate models) form the basis for climate predictions from the Intergovernmental Panel on Climate Change (IPCC) and as such have been widely used and vetted by climate scientists around the world. These models combine the energy budget models introduced earlier with other important physical processes (e.g. chemical transport) and superimpose their predictions on a spatially explicit surface (e.g. latitudinal and longitudinal grids). Some of the more common GCMs include data from the UK Meteorological Office's Hadley Centre's HadCM3 model and the US National Oceanic and Atmospheric Administration's (NOAA) Geophysical Fluid Dynamics Laboratory's (GFDL) CM2 series, which are both atmosphere–ocean models (Table 1.1). These models allow greenhouse gas concentrations to be included

Table 1.1. Global climate models and greenhouse gas (GHG) emission scenarios for future climate projections based on simulations conducted for the IPCC AR4 and IPCC AR5. (From Cherkauer and Sinha, 2010.)

Global climate model		
Model name	**Description**	**Sensitivity to GHGs**
GFDL	NOAA Geophysical Fluid Dynamics Laboratory (GFDL), version CM2.1.1	High
HadCM3	UK Met Office Hadley Centre Climate Model, version 3.1	Medium
AR4 emissions scenarios		
Scenario	**Description**	**Maximum CO_2 (ppm)**
B	Global population peaks in mid-century before declining Rapid changes in economic structures towards service and information economy Rapid introduction of clean and resource-efficient technologies	550
A1B	Very rapid economic growth Global population peaks in mid-century before declining Rapid introduction of new and more efficient technologies	720
A2	Gradual continuous increase in global population Regionally oriented economic growth Fragmented technological development	850
AR5 emissions scenarios		
Scenario	**Description**	**Radiative forcing (W/m^2)**
RCP2.6	Aggressive mitigation and policy for collective action Requires global participation in reducing GHGs Invention and application of CO_2 removal technologies	2.6
RCP4.5	Much like the B scenario above	4.5
RCP6.0	Slow efforts to curb emissions and apply mitigation Technologies capturing GHGs implemented by 2100	6.0
RCP8.5	Failure to curb warming through collective action GHGs are up to 7 times higher than pre-industrial levels	8.5

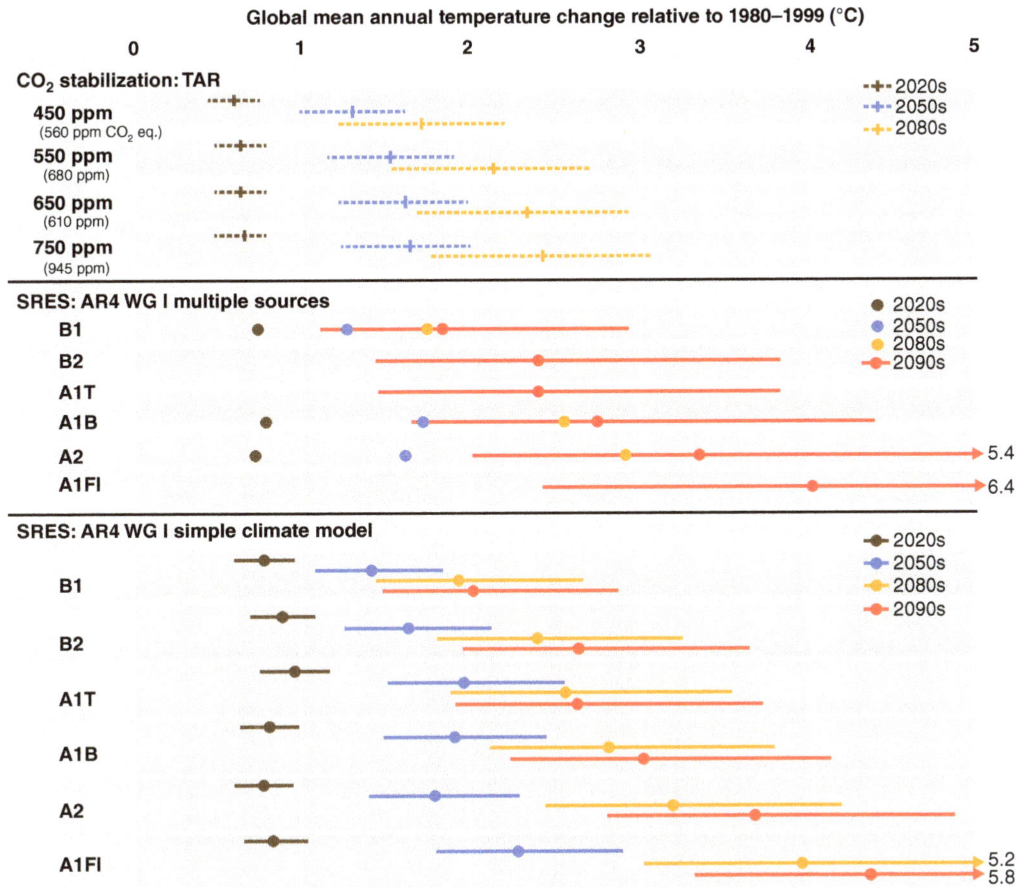

Fig. 1.1. Global mean temperature change projected for various emission scenarios. (From IPCC, 2007a,b.)

into projections to provide predictions of how the climate will change under different scenarios (Levy *et al.*, 2004). This is required because how humans respond to the growing concerns on the damaging effects of climate change and possibly reduce CO_2 emissions is the greatest source of uncertainty in these models. In response, the IPCC has developed 40 scenarios of human behaviour based on assumptions on future economic and regulatory conditions (Table 1.1 and Fig. 1.1), which have more recently been summarized as four representative concentration pathways (RCPs) which are identified by the watts of radiative forcing per square metre. Common scenarios include the extreme cases of A2 and RCP8.5 where the global dependence on fossil fuels and population growth remain high, the optimistic cases of B2 and RCP4.5 where fossil fuel use and population growth are curtailed, and a technological solutions scenario RCP2.6 in which reduced reliance on fossil fuels and effective carbon capture technologies are enhanced and new strategies implemented.

There is consensus among the experts that the global climate will continue to change significantly during the next century. The 2007 IPCC report concluded it is very likely that fossil fuel burning will be the dominant influence on atmospheric CO_2 concentrations in the 21st century. Model projections of atmospheric CO_2 concentrations estimated that by the end of the century, they could range from 490 to 1260 ppm, equivalent to between 75 and 350%

Table 1.2. Climate change and CO_2 concentration projections for the 21st century, if no climate policy interventions are made. (Modified from IPCC, 2001.)

Indicator	2025	2050	2100
CO_2 concentration (ppm)	415–460	460–625	475–1100
Global mean temperature change from 1990 (°C)	0.4–1.1	0.8–2.6	1.4–5.8
Global mean sea-level rise from 1990 (cm)	2–15	5–30	10–90

above estimated levels of CO_2 in the year 1750. Also, global air temperatures estimated by the models are expected to increase in the 21st century by 1.4 to 5.8°C, relative to 1990 temperatures (Table 1.2 and Fig. 1.1). These projected increases are considerably greater than those in the 20th century and are very likely to exceed any century-long trend in the past 10,000 years.

1.2 Climate Change Predictions for the Great Lakes Basin

The Great Lakes Basin in North America (Fig. 1.2) comprises the lakes, tributaries and associated watersheds of lakes Superior, Huron, Michigan, Erie and Ontario within the American states (Minnesota, Wisconsin, Illinois, Indiana, Michigan, Ohio, Pennsylvania and New York) and the Canadian province of Ontario. These large bodies of water have a strong influence on their surrounding climate by moderating air temperatures, as well as increasing winter precipitation downwind of the lakes while decreasing summer precipitation on the lakes (Notaro *et al.*, 2013). Given these influences on climate, and the wide geographic variability and latitude of the lakes, projecting long-term trends in the climate proved difficult for many years (Kling *et al.*, 2003). However, more recently it has become clear that global climate change is indeed influencing the local climate and the aquatic environment of the Great Lakes (Melillo *et al.*, 2014). Extremes in temperatures and precipitations are occurring more frequently across the Great Lakes Basin compared with the early half of the 20th century (Kunkel *et al.*, 1999; Winkler *et al.*, 2012). Further, over the last 30 years, the region has experienced increasing air temperatures (Zobel *et al.*, 2017), unpredictable lake-effect snowfalls (Suriano and Leathers, 2017), increased summer precipitations (Kunkel *et al.*, 2012) and increased runoff and shore erosions (Kelly *et al.*, 2017). These changes have led to shorter, warmer and wetter winters (Wuebbles and Hayhoe, 2004). Corresponding changes have been observed throughout the 20th century in winter processes (e.g. snowfall, snow melt, freezing and thawing of soil) including a significant reduction in mean snow cover area and earlier spring thaw from 1972 to 2000 in the northern hemisphere (Lemke *et al.*, 2007; Cherkauer and Sinha, 2010). Also, the occurrence of soil frost has been decreasing since the mid-1960s which indicates warmer temperatures (Cherkauer and Sinha, 2010).

The prediction of future climate for the Great Lakes Basin has a higher degree of uncertainty than global projections because it is inherently more difficult to model regional-scale phenomena. As such it is often best to view future projections as plausible scenarios rather than as forecasts (Jones *et al.*, 2006) and to use a range of models. Fortunately, there are a range of regional models available for future climate projections in the Great Lakes Basin, which are supplied by multiple agencies including the US Global Change Research Program and the Canadian Center for Climate Modelling and Analysis, as well as publications by academic institutions (Wuebbles *et al.*, 2019).

The consensus of the models is that under continuing greenhouse gas emissions, the Great Lakes Basin is likely to experience increases in annual and seasonal temperatures, a higher frequency of extreme heat (>32°C) and storms events, and seasonal changes to precipitation (Fig. 1.3). The models each project slightly different ranges of temperatures and focus on different areas, but a compelling image emerges among all of them. Under a moderate scenario (A1B) and in an averaging of 21 models, Christensen and Hewitson (2007) predicted increases of 3–5°C in the winter and 3.5–5°C in the summer across the Great Lakes Basin by the year 2100. According to Wuebbles and Hayhoe (2004), the regional temperature increases will be from 1 to 7°C in the winter and from 3 to 11°C in the summer over the same period. It is possible that the larger range of temperatures is more a reflection of what regions were chosen to analyse rather than the models themselves. For example, Jones *et al.* (2006) used two

Fig. 1.2. Laurentian Great Lakes and Basin.

GCMs to project air temperature increases of 3 to 8°C in the winter and 3 to 9°C in the summer over the same period while predicting that these seasonal increases would be less in the centre of the Great Lakes region due to the buffering effect of the lakes, with more extremes in the south and north. Mortsch and Quinn (1996), using four GCMs, suggested temperatures would increase most in the southern basins in Ohio and Indiana. This suggestion agrees with Wuebbles *et al.* (2010), who using the GFDL CM2.1, HadCM3 and US National Center for Atmospheric Research (NCAR) models demonstrated that under the extreme emission CO_2 scenario, mean annual air temperatures in Chicago would rise by 6°C by 2100. Their study not only demonstrated a relatively high increase in temperature for a local area, but also reinforced that the variability of the climate is almost a more important prediction than the rise. In that study the frequency of very hot summers was expected to increase, rather than a gradual rise in temperatures over time (also seen in Sousounis and Grover, 2002). Further, the frequency and intensity of extreme cold days and cold spells would likely decline, thus indicating an overall increase in mean temperature (Wuebbles *et al.*, 2010).

While average annual mean air temperatures can provide a good indicator of the rate of climate change, more detailed studies that look at seasonal trends can be more insightful into the nature of the temperature change that is expected across the basin. Recent analysis on the variation in both seasonal and spatial trends indicated that air temperature average maximums are expected to rise most in the southern Great Lakes in the summer, whereas winter average minimums will rise most in the northern Great Lakes (IPCC, 2013; USGCRP, 2017). The former finding reinforces the Wuebbles *et al.* (2010) study and suggests more extreme summer temperatures. However, the rise of the winter minimums in air temperature would continue the decreasing trend in soil frost that has been observed since the 1960s (Cherkauer and Sinha, 2010). Future projections suggest that the date of last frost will be about 30 and 20 days earlier by

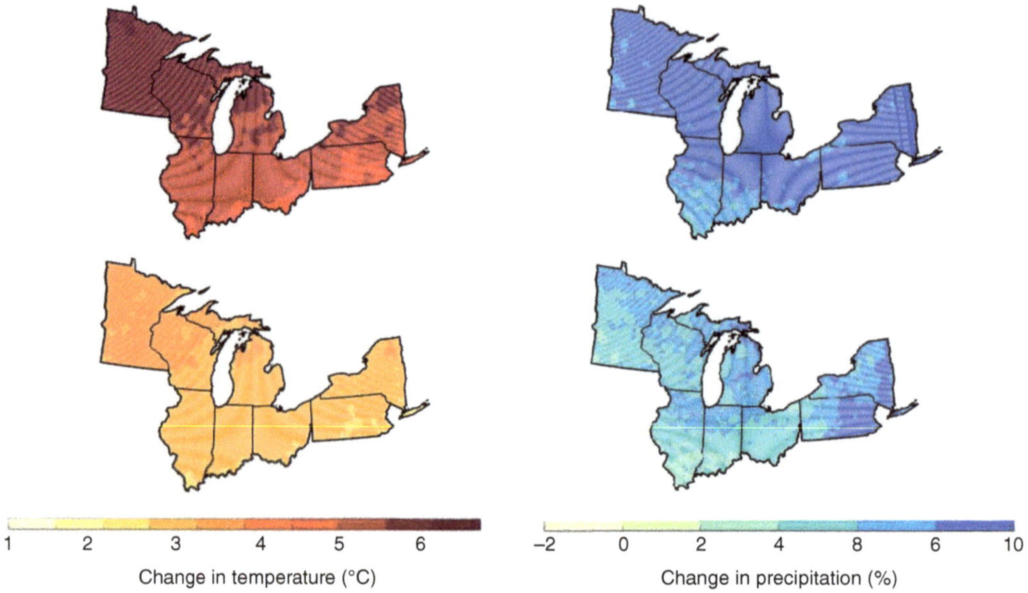

Fig. 1.3. Projected changes in annual temperature (left side, warm colours) and annual precipitation (right side, cool colours) for the RCP8.5 (upper row) and RCP4.5 (lower row) scenarios. (Modified from Wuebbles *et al.*, 2019 with permission from the Environmental Law and Policy Center.)

the end of the century under the extreme and optimistic greenhouse gas emissions scenarios, respectively (Lemmen and Warren, 2004; Erwin, 2009). Separate studies examining broad weather patterns using data from the Canadian Coupled Climate Model (CGCM1) and HadCM2 models confirm the predictions that there will be fewer cold air intrusions in winter and more heatwaves in summer by the end of this century (Sousounis and Grover, 2002; Polderman and Pryor, 2004). CGCM1 suggests a warmer climate scenario than HadCM2 yet both predict milder extremes in winter days as defined by an increased thickness of air mass of 10 to 20 decametres (dam) and slightly weaker winds, while extremely hot summers will be characterized by an increased air thickness of 10 dam and stronger winds. The main difference between the models lies in that HadCM2 predicts more moisture (southerly flow) at the surface while the CGCM1 predicts drier (more westerly) and shorter flow emanating from the influence of the Atlantic Bermuda High to the east (Sousounis and Grover, 2002).

The relationship between temperature change and precipitation is not necessarily intuitive because of feedbacks in physical processes discussed earlier. For example, the North American continent was divided in an east–west split in changes in precipitation over the 2000s (MacDonald, 2010). While the air temperatures either remained stable or increased in a relatively even pattern across the USA, the precipitation tended to decrease in the west and increase in the east, with an even greater increase surrounding the Great Lakes Basin. Seager and Vecchi (2010) determined that precipitation and evaporation are balanced by the divergence of a time-averaged, column-integrated moisture flux which ultimately leads to the conclusion that wet areas will generally become wetter while dry areas will become drier. Further, if topology and wind are included in the climate models, precipitation in one season should be highly influenced by the moisture content left behind from the previous season (Seager and Vecchi, 2010), which furthers the conclusion that shorter and drier winters can often contribute to drier summers.

These general conclusions seem to hold across North America; however, within the Great Lakes Basin most climate models do not predict a strong link between drier winters leading to drier summers. Instead, by the year 2100 increases in precipitation expected in the winter and spring across all Great Lakes are expected to be offset with a 5 to 15% decrease in precipitation in the summer (Byun and Hamlet, 2018). As such, the average annual precipitation is expected to have only minor changes with a projected increase between 0 and 10% (Wuebbles and Hayhoe, 2004; Christensen and Hewitson, 2007), although other studies have predicted as much as a 20% increase under some scenarios (Kling *et al.*, 2003). As with the spatial and seasonal patterns of temperature increases, the variation in precipitation emerges as a more important consideration than the averages. In general, higher precipitation will follow corresponding increases in extreme weather events such that the occurrence of 1-in-5-year storms (i.e. storms that have an intensity that now occurs only once in 5 years) could increase by 10.8 to 18.7% by 2085 depending on the emissions scenario (USGCRP, 2017). As these storms become more frequent, they are also expected to drop more precipitation individually (e.g. 7–8% more by the 2030s and 9–12% more by the 2050s) (Wuebbles *et al.*, 2019). In other studies, Wuebbles and Hayhoe (2004) and Cherkauer and Sinha (2010) used GCMs like HadCM3 and an ocean dynamics-linked model called the Parallel Climate Model (PCM; developed by NCAR) to predict precipitation under climate change for all emission scenarios. They concluded that an increase in thunderstorms and a doubling of 24-h and 7-day heavy rainfalls will occur by the end of the century. These more frequent storms will likely lead to greater risks of flooding and higher runoff into freshwater environments (Trapp *et al.*, 2007). Spatially, precipitation will decline most in the south-western part of the basin and increase in the more northerly areas which include western Ontario, northern Minnesota, Wisconsin and Michigan (Mortsch and Quinn, 1996).

Higher rates of evaporation and transpiration are expected in the next century. If precipitation falls most often in storm events leading to runoff, the recharging of soil moisture and groundwater resources will be affected. As such, soils are expected to be 30% drier in the Great Lakes Basin during the summer months and, correspondingly, stream flow and water levels will decline (Magnuson *et al.*, 1997; Kling *et al.*, 2003).

Wind speed and direction have important functions over the Great Lakes Basin in moving weather patterns and air masses across the landscape. Studies using the CGCM1 and HadCM2 models predict a decrease in surface wind speeds, more frequent easterly winds, and more frequent and intense warm fronts (Sousounis and Grover, 2002). Cyclones are rotating storms that are characterized by a low-pressure zone at the core and sustained localized winds. The severity of these storms is ranked on a scale. For example, in Tropical Cyclones a low ranking of 1 could refer to 40 m/s sustained winds, while 70 m/s winds would indicate a much higher ranking of 4 or 5. Over the Great Lakes, both climate models predict a decrease in low cyclone (sea-level pressure > 1005 hPa) numbers and a slight increase in strong cyclone (sea-level pressure < 1000 hPa) numbers (Sousounis and Grover, 2002). The main difference between the two models lies in their predictions for north-easterly and south-easterly winds in which the CGCM1 model predicts an increase in the former, while the HadCM2 predicts an increase in the latter as well as a decrease in winter south-westerly winds (Sousounis and Grover, 2002).

1.3 General Effects on Freshwater Ecosystems and Their Biota

Freshwater ecosystems occur from small ephemeral streams to lakes that are so large they share many qualities with the oceans. As such, the effect of a change in climate on a freshwater ecosystem will depend greatly on the type and location of that ecosystem. Further, through the lens of the biota that inhabit freshwater ecosystems, the effect of a warming climate will equally depend on the organism's own thermal preferences. As such, any discussion of climate change on freshwater systems is best partitioned into main ecosystem types (e.g. lakes, rivers and wetlands) and thermal guilds of the organisms (warm, cool and cold). In general, the predicted impacts of a warming climate will disturb all of these ecosystem types (Magnuson *et al.*, 1997; Schindler, 2001; Heino *et al.*, 2009) but not necessarily adversely affect all of the thermal guilds. The general threat that climate change imposes on freshwater ecosystems is dire considering that global freshwater biodiversity is the most threatened among the terrestrial and aquatic biomes of the world (Sala *et al.*, 2000).

The link between the climate and a changing freshwater environment lies mainly in the link between air temperature, precipitation patterns and extreme storms. These influence the quantity and quality of water in rivers, lakes and wetlands (Durance and Ormerod, 2007). While we present these effects for the specific environments below, a generally warming water temperature and declining water quality and quantity has some broad effects on aquatic organisms. Thermal guilds of aquatic organisms are often characterized by how broad a temperature range an organism will inhabit over its lifetime (e.g. stenotherms stay within very narrow temperature ranges) and its preferred ambient temperature. For example, for fish in the Great Lakes Basin, three main guilds exist for cold-water stenotherms, cool-water eurytherms (i.e. a broad temperature range) and warm-water eurytherms, with temperature preferences centred around 15, 24 and 28°C, respectively (Magnuson and DeStasio, 1996). While these guilds each have different temperature preferences, being forced to inhabit an environment that is warmer than the preferred temperature has some general effects on all guilds. Reproductive success in fish is highly dependent on completing all the requisite stages within a suitable temperature range. Gamete development, spawning, hatching and juvenile growth are moderated by temperature (Pankhurst and Munday, 2011). Further, the basic metabolic rate of any ectotherm is moderated by the ambient temperature. As such, the organism's life history must match its expected metabolic needs, which tend to increase exponentially with temperature. For example, if juveniles of a cold-water species (e.g. lake trout, *Salvelinus namaycush*, Walbaum 1792) experience a 3°C increase in ambient temperature, they may need to increase their consumption by 8.7 times to match their new higher metabolic costs (McDonald *et al.*, 1996). This scenario may be sustainable for the juvenile if its prey resources also increase in availability, but this is not always the case because warming water temperatures can also disrupt synchronized dynamics in predator–prey systems. For example, some studies found that warmer springs encouraged earlier phytoplankton and zooplankton population growth (Sommer *et al.*, 2007; Berger *et al.*, 2010; Shimoda *et al.*, 2011). If the fish did not also hatch earlier, they would be faced with having missed their predation opportunities while being laboured with greater metabolic costs leading to poorer body condition (Shuter *et al.*, 1980).

Consequently, it is perhaps not surprising that the main implication of changing water quality and quantity in freshwater environments is a shift in the spatial distributions of thermal guilds. As environments in turn become attractive or uninhabitable to a particular thermal guild, it will need to move across watersheds. This pattern is readily observed in North America with the distributions of warm-water fish (e.g. smallmouth bass, *Micropterus dolomieu*, Lacepède 1802) being predicted to shift 500 km northward with a doubling of greenhouse gas atmospheric concentrations (Eaton and Scheller, 1996; Heino *et al.*, 2009). From this same guild, 19 new species are expected to invade the lower Great Lakes Basin from the Mississippi or Atlantic Basins, and another eight lower Great Lake species are expected to migrate north (Mandrak, 1989; Rahel and Olden, 2008). While these predicted invaders are mainly minnows (e.g. Cyprinidae), sunfishes (e.g. Centrarchidae), suckers (e.g. Catostomidae) and topminnows (e.g. Fundulidae) (Mandrak, 1989), other models predict that 20% of lake trout (a cold-water species) lakes in Canada will be vulnerable to smallmouth bass (a warm-water species) invasions by 2050 (Sharma *et al.*, 2009). By the end of the century 9700 lakes would be vulnerable to smallmouth bass invasion, which would negatively impact cold-water species through competition and small-bodied forage fish species including northern redbelly dace (*Chrosomus eos*, Cope 1862), finescale dace (*Chrosomus neogaeus*, Cope 1867), fathead minnow (*Pimephales promelas*, Rafinesque 1820) and pearl dace (*Margariscus nachtriebi*, Cox 1896) through predation (Jackson and Mandrak, 2002; Sharma *et al.*, 2009). Using historical and contemporary species distribution data, Alofs *et al.* (2014) estimated that the range of smallmouth bass in Ontario, Canada has expanded northward at the rate of ~13 km/decade over a 30-year period. As will be discussed in greater detail in this book, the expansion of new species with their pathogens into new ecological communities may have detrimental effects on the native populations (Marcos-López *et al.*, 2010). However, it is equally possible that expanding populations also bring novel hosts for existing pathogens (Marcogliese, 2001; Heino *et al.*, 2009).

As freshwater ecosystems are typically already under environmental stress, the general effects of climate change must be considered under a multiple-stressor framework. For example, different sources of stress may have additive, multiplicative and even antagonistic interactions with each other, causing unpredictable reactions from freshwater biota. In a study of populations of a cool-water fish (walleye, *Sander vitreus*, Mitchill 1818) across Ontario, the effects of climate change were found to interact in unintuitive ways with the presence of invasive species, the recreational fishery and deforestation (Gutowsky *et al.*, 2019). Another example includes the interaction between warming temperatures and increasing water salinity in Saskatchewan as observed by Sereda *et al.* (2011). Together, these changes accounted for a 30% decrease in macroinvertebrate diversity and a decrease in algal primary productivity (Benoy *et al.*, 2007), even though warming water temperatures alone should improve diversity and production (Jackson *et al.*, 2007). Last, as climate change occurs in tandem with increasing human populations, the ability of freshwater environments to metabolize nutrients and contaminants from effluent could be compromised leading to an acceleration in water quality degradation (Tetreault *et al.*, 2012).

1.4 Lakes

An important classifying characteristic of lakes is whether they never mix layers within the water column (meromictic), mix once over the year (monomictic), twice over the year (dimictic) or remain well mixed year-round (unstratified). The stratification of water layers, even if temporary, has many great benefits for aquatic life, as does seasonal mixing. Stratified lakes are often deep and contain valuable cold-water refuges for thermally sensitive species (e.g. lake trout). However, under climate change, the stratification and mixing within lake types are expected to shift due to warming waters and less ice cover. For example, northern lakes that are currently monomictic would likely become dimictic if they began to stratify in the summer. Two main patterns of altered mixing are predicted: (i) strictly dimictic deep lakes migrating towards ice-free monomictic systems; and (ii) monomictic lakes switching to meromictic hydrodynamic regimes due to the increasing suppression of deeply penetrative mixing during mild winters (Magnuson *et al.*, 1997; Shimoda *et al.*, 2011). Further, large deep dimictic lakes (including the Great Lakes) are expected to be less likely to mix completely (McCormick, 1990).

The depth of the thermocline in stratified lakes is an important feature in defining the available warm-, cool- and cold-water fish habitats. A shallowing of the thermocline is expected in large temperate lakes as warmer surface-water temperatures create stronger thermal gradients which would generally be beneficial to all thermal guilds (Snucins and Gunn, 2000; Sharma *et al.*, 2011). Curiously, it is also possible that the trapping of more heat in the surface-mixed layer will allow less heat to transfer to the lower column, and deep waters could become cooler (Vincent, 2009). However, smaller lakes will be more sensitive to increased evaporation rates, increased wind and decreased dissolved organic carbon that would deepen the thermocline. This could put cold-water thermal guilds at considerable risk by squeezing their habitat and creating a stronger barrier to foraging opportunities in the littoral zone with suboptimal temperatures (Schindler *et al.*, 1996; Jackson, 2007; Plumb and Blanchfield, 2009). Further, warmer waters lose dissolved oxygen (DO) which presents challenges to organisms inhabiting shallow lakes if the full water column approaches anoxic conditions (Fang and Stefan, 1999; Ficke *et al.*, 2007). With the doubling of atmospheric concentrations of CO_2, Stefan *et al.* (1996) predicted declines in DO of 2 mg/l in the surface waters and as much as 8 mg/l in the deeper hypolimnetic waters in small lakes in Minnesota. The depletion in DO also was expected to occur for 2 months longer than the current period. These declines were most pervasive in eutrophic lakes. As mentioned above, from a metabolic perspective a corresponding reduction in DO will make it even more challenging for larger species to meet their metabolic needs (Willis and Magnuson, 2006).

All lake types are vulnerable to decreasing water levels under climate change in regions where evapotranspiration is expected to increase while discharges from groundwater and consistent runoff are expected to decrease. With this decrease in runoff, lakes may also receive less nutrients from the surrounding watershed. Phosphorus is an important nutrient for biotic growth and productivity in lakes and as such, under climate change, lakes could become increasingly oligotrophic (Schindler, 1998). Declining lake levels in general present challenges to species that use nearshore habitats, particularly for early- or late-season spawning. Many fish species (e.g. walleye, lake trout) tend to spawn over shallow cobble in the early spring or late autumn. If altered seasonal patterns in precipitation cause these habitats to be inundated while the fish are spawning but are generally dry for the rest of the year, many fish species could experience significant losses in population recruitment.

The duration of ice cover and the thickness of mid-winter ice are expected to be reduced for all lake types under climate change. In a study of 65 water bodies across the Great Lakes region, the average rates of change in freeze-up and break-up dates during 1975–2004 were 5.8 times and 3.3 times, respectively, more rapid than the average historical 1846–1995 rates (Jensen *et al.*, 2007). Similarly, Magnuson *et al.* (2000) showed that freeze-ups have a delay of up to 5.8 days and break-ups have been up to 6.5 days earlier, compared with the average 1846–1995 period in the northern hemisphere. Low ice cover is mainly expected to favour warm- and cool-water fish guilds by removing the competitive advantage that cold-water fish have accrued through their evolution under arctic conditions (Willis and Magnuson, 2006; Keller, 2007). For example, arctic char (*Salvelinus alpinus*, Linnaeus 1758) lose their competitive advantage over brown trout (*Salmo trutta*, Linnaeus 1758) in years with low ice coverage because they are better foragers at 1°C and begin foraging under the ice earlier in the spring (Finstad and Forseth, 2006; Helland *et al.*, 2011). However, ice cover also blocks the aeration of winter lake waters and so can contribute to large fish kills in the late winter. It is difficult to predict if there is an advantage to the oxygen concentrations for fish in lakes with reduced ice cover because the correspondingly earlier stratification could lead to earlier onset of anoxia in the hypolimnion and therefore negatively influence fish that inhabit deeper portions of the lake (De Stasio *et al.*, 1996). The relationship between ice cover and the productivity of lower trophic groups is not well known, although it has been suggested that low ice coverage in the Great Lakes Basin negatively impacts zooplankton abundances (Assel, 1991). Further, the timing of ice break-up influences the initial growth conditions for diatoms and the timing of the spring phytoplankton bloom (Blenckner *et al.*, 2007).

1.5 Rivers

The temperature of flowing water at any time of the year is strongly tied to the seasonal air temperature and groundwater discharge (Power *et al.*, 1999). Other natural modifiers of stream temperature include local patterns of rain and snowfall, the recharge rate for groundwater and the spring thaw (Ducharne, 2008). Anthropogenic impacts have a great influence on stream temperatures through the building of dams, removal of shade through deforestation and water withdrawals, which together tend to lead to warmer temperatures and a lower water quality (Poole and Berman, 2001; Chu *et al.*, 2010). Climate change is expected to modify ambient temperature, seasonal precipitation and soil moisture which will collectively lead to a shift in peak stream flows from spring to late winter (Rahel and Olden, 2008) and further disruptions to annual stream flow patterns (Mohseni *et al.*, 2003), including increased occurrences of short pulses of high discharge (i.e. flash floods; Heino *et al.*, 2009) and generally lower annual stream discharge (Xenopoulos *et al.*, 2005). Tisseuil *et al.* (2012) modelled rates of evaporation under climate change scenarios and found stream flow would decrease by 15% with a temperature increase of 1.2°C by the end of the 21st century. Similarly, Xenopoulos *et al.* (2005) conducted a global assessment under moderate and high climate change scenarios and predicted an 80% reduction in stream discharge over 133 watersheds. Future changes of this magnitude in water temperature and stream discharge will have significant impacts on stream biota. Chu *et al.* (2008) modelled how the predicted changes to air temperature and groundwater under climate change scenarios would affect the distribution of fish thermal guilds in a subset of Ontario watersheds. Under the optimistic scenario, warm-water species would be dominant in 66% of the watersheds tested by 2055, and under the extreme scenario all cold-water fish (e.g. brook trout, *Salvelinus fontinalis*, Mitchill 1814) would be lost. This study echoes Meisner's (1990) findings that an increase in air temperatures with a corresponding increase of 4.8°C in groundwater temperatures would decrease the length of suitable summer cold-water habitat in river headwaters by 30 to 40%. Lastly, Stefan *et al.* (1996) included riparian shading as a factor in their climate change predictions for Minnesota streams. They found that with a doubling of atmospheric CO_2 the stream temperatures would increase by 2.4 ± 4.7°C with an additional 6°C in the absence of shade.

In general, river ecosystems that have developed under repeated drying periods will fare best under climate change because they sustain themselves under low or stagnant discharge conditions (Meyer *et al.*, 1999). Rivers that are adapted to flooding may be able to handle the increased frequency of storms well but will be extremely vulnerable during the rest of the year. From the biotic perspective, warm-water fish (e.g. channel catfish, *Ictalurus punctatus*, Rafinesque 1818) that tend to inhabit the higher-order streams and mainstems of rivers would do better than headwater species (e.g. brook trout) under these conditions (Ostrand and Wilde, 2004; Wenger *et al.*, 2011). Finally, streams face the danger of acidification under extended drought conditions due to the oxidation of organic sulfur (Schindler *et al.*, 1997). Like with DO and temperature, cold-water fish species tend to be vulnerable to changes in pH and would fare worse under these conditions than species from the other thermal guilds.

1.6 Wetlands

Wetlands occur in a variety of forms, from fens to marshes to swamps to coastal wetlands, which are generally differentiated based upon their dominant source water. There is a wide range of general impacts that climate change will have on these systems including: change in base flows; altered hydrology in depth and period; increased heat stress on wildlife; extended range and activity of some pest and disease vectors; increased flooding, landslide, avalanche and mudslide damage; increased soil erosion; increased flood runoff resulting in a decrease in recharge of some floodplain aquifers; decreased water resource quantity and quality; increased risk of fires; and increased coastal erosion (Gitay *et al.*, 2011). Mid-continental wetlands are expected to be most vulnerable due to decreases in precipitation (Winter, 2000). The spatial distribution of semi-permanent and seasonal wetlands is expected to be reduced by increases in evapotranspiration and decreases in soil moisture. Once a wetland has dried up, the remaining plants are vulnerable to successional processes as terrestrial plants become established. Peatlands are broadly distributed in northern Canada, yet under a doubling of CO_2 concentrations the southern edge is expected to retreat 200 to 300 km northward (Anisimov and Fitzharris, 2001). This has implications for further greenhouse gas emissions as the drying of peatland releases CH_4 through an oxidation process. While the full destruction and loss of wetlands under climate change is the top concern, more nuanced changes to wetland types are also likely to be observed. For example, with increased variation in precipitation and more frequent storm events, marshes will begin to dominate over swamps, as trees lose dominance under fluctuating water levels (Mortsch, 1998).

1.7 Specific Effects on the Great Lakes Basin

Over the last 20,000 years the Great Lakes Basin experienced a change in mean annual air temperature of 5 to 6°C which caused major changes in the patterns of vegetation throughout the region, in some cases trading prairie for forest ecosystems (Kling *et al.*, 2003). Given that some models predict mean air temperatures will increase as much as 11°C, similar disruptions are likely to be observed under future scenarios. The Ontario government manages the aquatic resources of the Great Lakes Basin in Canada. In 2015 it released a vulnerability assessment for the inland aquatic resources (those within the basin's watersheds but not in the Great Lakes proper) divided by lakes, rivers and streams (Chu, 2015). As the northern portion of the basin covers a range of geographic and climatic clines, these assessments are useful for illustrating the spatial variation of the impacts to these freshwater systems. Under the extreme (A2) and optimistic (B1) scenarios maximum lake surface temperatures are expected to increase quite dramatically across the northern basin watersheds by 2080 (see Fig. 1.4). While maximum surface temperatures of inland lakes currently range between 18 and 26°C, they would shift to 24 to 30°C and 28 to 34°C under the optimistic and extreme scenarios, respectively (Chu, 2015). Under the optimistic scenario a small inland refuge for cool temperatures would only exist north of Lake Superior. The maximum weekly averaged temperatures for streams demonstrated a different pattern as both emission scenarios predicted 6 to 14°C increases across the northern watersheds (see Fig. 1.5). Wetlands were most vulnerable by 2080 in Lake Superior's eastern watersheds, around Georgian Bay and south-western Ontario in both scenarios (Fig. 1.6).

The Great Lakes have been monitored for decades and provide good long-term time series data for evaluating the early effects of climate change on large lake surface-water temperatures and lake-level fluctuations. Shimoda *et al.* (2011) demonstrated in north temperate deep lakes including the Great Lakes that an increase in overall lake and epilimnetic temperatures, an increase in thermal stability, a lengthening of the stratification period and a shortening of the ice cover period are the responses to our already warming climate (Table 1.3). However, long-term (25–87 years) observations of water temperatures at seven locations throughout the Great Lakes,

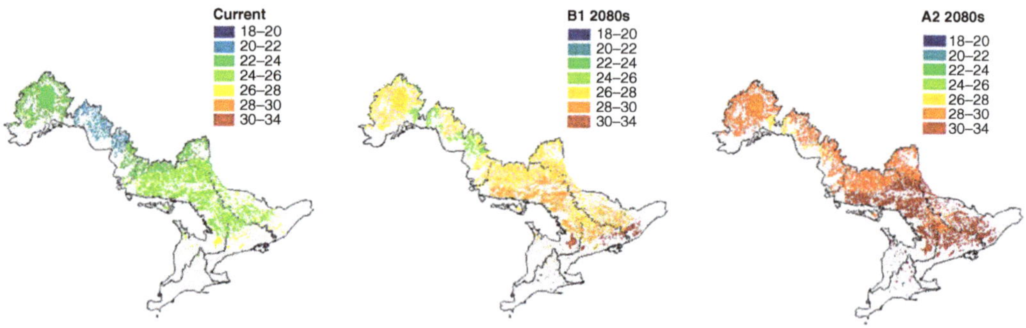

Fig. 1.4. Maximum predicted surface temperatures (°C) of lakes in Ontario's Great Lakes watersheds under two greenhouse gas emission scenarios, B1 and A2, by 2080. (From Chu, 2015.)

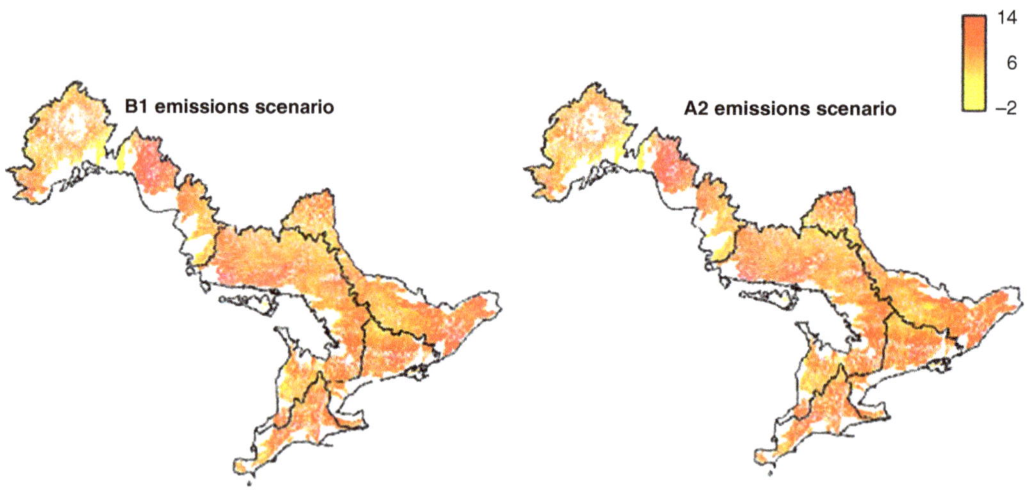

Fig. 1.5. Change in the predicted maximum weekly average temperature (°C) for streams in Ontario's Great Lakes watersheds under two greenhouse gas emission scenarios, B1 and A2, by 2080. (From Chu, 2015.)

i.e. Sault Ste Marie (Ontario), Green Bay (Wisconsin), St. Joseph (Michigan), Bay City (Michigan), Sandusky (Ohio), Put-In-Bay (Ohio) and Erie (Ohio), found strong trends towards increasing temperature at only two sites (Sault Ste Marie and Put-In-Bay) and weak trends at two other sites (Bay City and St. Joseph; McCormick and Fahnenstiel, 1999). In agreement with Shimoda *et al.* (2011), the stronger trends demonstrated a corresponding 4 and 6 h/year rate of increase in the maximum potential duration of summer stratification (14 and 18 days, respectively, over the time period). In this instance, the rate of increase in the duration data was mostly skewed towards earlier transitions to spring-like conditions, rather than a delayed onset as expected in smaller inland lakes.

While both McCormick and Fahnenstiel (1999) and Shimoda *et al.* (2011) did not study Lake Ontario, there is ample evidence of increasing water temperatures in other studies. A steady increase in lake water temperatures in the Bay of Quinte (Ontario) has been observed in summers (1950 to 2000) and winters (1980 to 2000; Casselman, 2002). Huang *et al.* (2012) also demonstrated increases in air (1.43 ± 0.39°C) and surface water (1.26 ± 0.32°C) temperatures at all seasonal and annual time scales during the last 40 years in

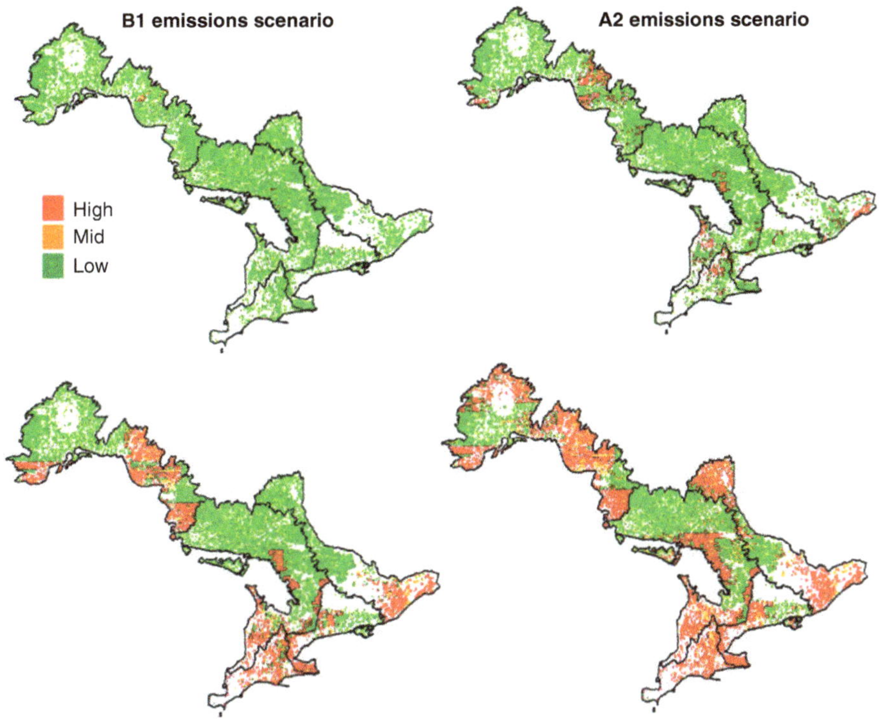

Fig. 1.6. Vulnerability of wetlands in Ontario's Great Lakes watersheds to the effects of climate change on groundwater inflows and air temperature following two scenarios, B1 (left column) and A2 (right column), for 2020 (top row) and 2080 (bottom row). (From Chu, 2015.)

Table 1.3. Summary of climate-induced changes in the thermal structure of the Great Lakes. (Modified from Shimoda *et al.*, 2011.)

	Increase in lake temperature (°C/year)		Increase in ice-free season (days/10 years)	
Lake	Observed change	Time period	Observed change	Time period
Lake Superior	0.01 (near-shore)	1906–1992	13	1973–2002
	0.110 (epilimnion)	1979–2006		
Lake Michigan	0.065 (epilimnion)	1979–2006	8.5	1973–2002
Lake Huron	0.086 (epilimnion)	1979–2006	2.3	1973–2002
Lake Erie	0.01 (nearshore)	1918–1992	5.9	1973–2002
Lake Ontario	NA	NA	10	1973–2002

NA, not available.

Lake Ontario. The rate of increase was higher for the air temperature than for surface-water temperature in winter and autumn. By contrast in spring and summer the surface water warmed faster than the air temperature. They also found that the length of the summer stratified season had increased by 12 ± 2 days since the early 1970s, most likely due to the increase in water temperature. Climate warming also had an effect on the wind speed over Lake Ontario, which declined and thus further enhanced the summer thermal stratification. This increasing rate is comparable to that of the duration of the summer stratification season of Lake Superior, which extended from 145 to 170 days over the last century (Austin and Colman, 2007).

The Great Lakes Environmental Research Laboratory (GLERL) through its Great Lakes Advanced Hydrologic Prediction System (AHPS) conducted a large-scale study of the influence of climate change on lake levels (Croley, 2005). Lake-level fluctuations were categorized in three distinct types: changes in long-term levels (annual), seasonal levels (following water-source variation trends) and short-period levels (mostly due to wind and storm surges).

Annual fluctuations accounted for most of the variability of the high and low lake levels. The overall range of the annual levels for most lakes was about 2 m, with precipitation influencing the major portions of long-term variations. Annual precipitation ranged from about 82 cm for Superior to 93 cm for Ontario, and these correlated very well with annual lake levels with a delay of 1 year. In addition, air temperature variations influenced lake-level fluctuations in multiple ways: (i) plants tended to use more water at higher temperatures; (ii) higher rates of evaporation were detected from the lake and ground surface; and (iii) greater humidity depletion in soils caused less runoff for the same amount of precipitation (Polderman and Pryor, 2004; Croley, 2005; Blanken *et al.*, 2011).

Seasonal variation in lake levels depends upon the individual water supplies. The seasonal range is about 30 cm on the upper lakes and about 38 cm on the lower lakes. When the net basin supplies diminish in the summer and autumn, the lakes begin their seasonal decline. Although the monthly precipitation is uniformly distributed throughout the year, the runoff has a peak during spring from the spring snow melt and a minimum in the late summer from higher evaporation rates. The higher evaporation period is due to colder dry air passing over warm lake surfaces (Croley, 2005; Blanken *et al.*, 2011). Croley's (2005) data confirmed observations of Lake Superior levels which between 1948 and 1999 decreased by 20% from 40 to 32 cm (Lenters, 2004). The seasonal decrease in water levels is explained by changes in runoff and over-lake precipitation rather than changes in evapotranspiration over the summer (Blanken *et al.*, 2011). Indeed, Lenters (2004) found that Lake Superior had begun to receive less influxes of water in the late spring (i.e. losing up to 1360 m^3/s) which would contribute to lower water levels and thus a smaller amplitude in the seasonal variations.

The evidence presented above suggests that the Great Lakes will undergo many of the same type of changes in physical and hydrological processes as many other lakes in North America. Similarly, the rivers that act as tributaries, and the coastal and protected wetlands, will be faced with the same type of challenges as further inland. As such, the biota within the Great Lakes Basin will need to deal with the same challenges in the form of changing thermal regimes, mismatched seasonal processes (e.g. the hatching of fish eggs versus the emergence of their invertebrate prey), the appearance of new species, and alterations in the quantity and quality of habitat. It is therefore important for scientists and managers to provide plausible scenarios now under climate change predictions, such that appropriate mitigation and adaptation plans can be implemented where possible.

References

Alofs, K.M., Jackson, D.A. and Lester, N.P. (2014) Ontario freshwater fishes demonstrate differing range-boundary shifts in a warming climate. *Diversity and Distributions* 20, 123–136.

Andrews, T. and Forster, P.M. (2009) Surface energy perspective on climate change. *Journal of Climate* 22, 2557–2570.

Anisimov, O. and Fitzharris, B. (2001) Polar regions (Arctic and Antarctic). In: White, K.S. (ed.) *Climate Change 2001: Impacts, Adaptation and Vulnerability. Contribution of Working Group II to the Third Assessment Report of the Intergovernmental Panel on Climate Change. Technical Report*. Cambridge University Press, Cambridge, UK, pp. 801–842.

Assel, R.A. (1991) Implications of CO_2 global warming on Great Lakes ice cover. *Climate Change* 18, 377–395.

Austin, J.A. and Colman, S.M. (2007) Lake Superior summer water temperatures are increasing more rapidly than regional air temperatures: a positive ice-albedo feedback. *Geophysical Research Letters* 34, L06604.

Benoy, G., Cash, K., McCauley, E. and Wrona, F. (2007) Carbon dynamics in lakes of the boreal forest under a changing climate. *Environmental Reviews* 15, 175–189.

Berger, S.A., Diehl, S., Stibor, H., Trommer, G. and Ruhenstroth, M. (2010) Water temperature and stratification depth independently shift cardinal events during plankton spring succession. *Global Change Biology* 16, 1954–1965.

Blanken, P.D., Spence, C., Hedstrom, N. and Lenters, J.D. (2011) Evaporation from Lake Superior: 1. Physical controls and processes. *Journal of Great Lakes Research* 37, 707–716.

Blenckner, T., Adrian, R., Livingstone, D.M., Jennings, E., Weyhenmeyer, G.A. *et al.* (2007) Large-scale climatic signatures in lakes across Europe: a meta-analysis. *Global Change Biology* 13, 1314–1326.

Byun, K. and Hamlet, A.F. (2018) Projected changes in future climate over the Midwest and Great Lakes region using downscaled CMIP5 ensembles. *International Journal of Climatology* 38, e531–e553.

Casselman, J. (2002) Effects of temperature, global extremes, and climate change on year-class production of warmwater, coolwater and coldwater fishes in the Great Lakes basin. In: *Fisheries in a Changing Climate. American Fisheries Society Symposium 32*. American Fisheries Society, Bethesda, Maryland, pp. 39–60.

Cherkauer, K.A. and Sinha, T. (2010) Hydrologic impacts of projected future climate change in the Lake Michigan region. *Journal of Great Lakes Research* 36, 33–50.

Christensen, J.H. and Hewitson, B. (2007) Regional climate projections. In: Solomon, S.D., Qin, D., Manning, M., Chen, Z., Marquis, M. *et al.* (eds) *Climate Change 2007: The Physical Science Basis. Contribution of Working Group I to the Fourth Assessment Report of the Intergovernmental Panel on Climate Change*. Cambridge University Press, New York, pp. 847–940.

Chu, C. (2015) Climate change vulnerability assessment for inland aquatic ecosystems in the Great Lakes basin, Ontario. *Climate Change Research Report No. CCRR-43*. Ontario Ministry of Natural Resources and Forestry, Science and Research Branch, Peterborough, Canada.

Chu, C., Jones, N.E., Mandrak, N.E., Piggott, A.R. and Minns, C.K. (2008) The influence of air temperature, ground water, discharge and climatic change on the thermal diversity of stream fishes in Southern Ontario Watersheds. *Canadian Journal of Fisheries and Aquatic Sciences* 65, 297–308.

Chu, C., Jones N.E. and Allin, L. (2010) Linking the thermal regimes of streams in the Great Lakes Basin, Ontario, to landscape and climate variables. *River Research Applications* 26, 221–241.

Croley, T.E. (2005) Using climate predictions in Great Lakes hydrologic forecasts. In: Garbrecht, J.D. and Piechota, T.C. (eds) *Climate Variations, Climate Change, and Water Resources Engineering*. Surface Water Hydrology Technical Committee and Environmental and Water Resources Institute, Virginia, pp. 166–187.

deMenocal, P.B. (2011) Climate and human evolution. *Science* 311, 540–541.

De Stasio, J., Hill, D.K., Kleinhaus, J.M., Nibblelink, N.P. and Magnuson, J.J. (1996) Potential effects of global climate change on small north-temperate lakes: physics, fish and plankton. *Limnology and Oceanography* 41, 1136–1149.

Ducharne, A. (2008) Importance of stream temperature to climate change impact on water quality. *Hydrology and Earth System Sciences* 12, 797–810.

Durance, I. and Ormerod, S.J. (2007) Climate change effects on upland stream macroinvertebrates over a 25-year period. *Global Change Biology* 13, 942–957.

Eaton, J.G. and Scheller, R.M. (1996) Effects of climate warming on fish thermal habitat in streams of the United States. *Limnology and Oceanography* 41, 1109–1115.

Erwin, K.L. (2009) Wetlands and global climate change: the role of wetland restoration in a changing world. *Wetland Ecology and Management* 17, 71–84.

Fang, X. and Stefan, H.G. (1999) Projections of climate change effects on water temperature characteristics of small lakes in the contiguous US. *Climate Change* 42, 377–412.

Ficke, A.D., Myrick, C.A. and Hansen, L.J. (2007) Potential impacts of global climate change on freshwater fisheries. *Reviews in Fish Biology and Fisheries* 17, 581–613.

Finstad, A.G. and Forseth, T. (2006) Adaptation to ice-cover conditions in Atlantic salmon, *Salmo salar* L. *Evolutionary Ecology Research* 8, 1249–1262.

Giannini, A., Saravanan, R. and Chang, P. (2003) Oceanic forcing of Sahel rainfall on interannual to interdecadal time scales. *Science* 302, 1027–1030.

Gitay, H., Finlayson, C.M. and Davidson, N. (2011) A framework for assessing the vulnerability of wetlands to climate change. *Ramsar Technical Report No. 05, CBD Technical Series 57*. Ramsar Convention Secretariat, Gland, Switzerland.

Gutowsky, L., Giacomini, H., de Kerckhove, D., Mackereth, R., McCormick, D. and Chu, C. (2019) Quantifying multiple pressure interactions affecting populations of a recreationally and commercially important freshwater fish. *Global Change Biology* 25, 1049–1062.

Heino, J., Virkkala, R. and Toivonen, H. (2009) Climate change and freshwater biodiversity: detected patterns, future trends and adaptations in northern regions. *Biological Reviews* 84, 39–54.

Helland, I.P., Finstad, A.G., Forseth, T., Hesthagen, T. and Ugedal, O. (2011) Ice-cover effects on competitive interactions between two fish species. *Journal of Animal Ecology* 80, 539–547.

Huang, A., Rao, Y.R. and Zhang, W. (2012) On recent trends in atmospheric and limnological variables in Lake Ontario. *Journal of Climate* 25, 5807–5816.

IPCC (Intergovernmental Panel on Climate Change) (2001) *Climate Change 2001: The Scientific Basis. Contribution of Working Group I to the Third Assessment Report of the Intergovernmental Panel on Climate Change* [Houghton, J.T., Ding, Y., Griggs, D.J., Noguer, M., van der Linden, P.J. *et al.* (eds)]. Cambridge University Press, Cambridge and New York.

IPCC (Intergovernmental Panel on Climate Change) (2007a) Summary for policymakers. In: Parry, M.L., Canziani, O.F., Palutikof, J.P., van der Linden, P.J. and Hanson, C.E. (eds) *Climate Change 2007: Impacts, Adaptation and Vulnerability. Contribution of Working Group II to the Fourth Assessment Report of the Intergovernmental Panel on Climate Change*. Cambridge University Press, Cambridge, UK, pp. 7–22.

IPCC (Intergovernmental Panel on Climate Change) (2007b) Technical summary. In: Parry, M.L., Canziani, O.F., Palutikof, J.P., van der Linden, P.J. and Hanson, C.E. (eds) *Climate Change 2007: Impacts, Adaptation and Vulnerability. Contribution of Working Group II to the Fourth Assessment Report of the Intergovernmental Panel on Climate Change*. Cambridge University Press, Cambridge, UK, pp. 23–78.

IPCC (Intergovernmental Panel on Climate Change) (2013) *Climate Change 2013: The Physical Science Basis. Contribution of Working Group I to the Fifth Assessment Report of the Intergovernmental Panel on Climate Change* [Stocker, T.F., Qin, D., Platner, G.K., Tignor, T., Allen, S.K. *et al.* (eds)]. Cambridge University Press, Cambridge and New York.

Jackson, B. (2007) Potential effects of climate change on lake trout in Atikokan Area. *Research Information Note No. 4*. Ontario Ministry of Natural Resources, Ontario, Canada.

Jackson, D.A. and Mandrak, N.E. (2002) Changing fish biodiversity: predicting the loss of cyprinid biodiversity due to global climate change. In: *Fisheries in a Changing Climate. American Fisheries Society Symposium 32*. American Fisheries Society, Bethesda, Maryland, pp. 89–98.

Jackson, L.J., Lauridsen, T.L., Sondergaard, M. and Jeppesen, E. (2007) A comparison of shallow Danish and Canadian lakes and implications of climate change. *Freshwater Biology* 52, 1782–1792.

Jensen, E.J., Ackerman, A.S. and Smith, J.A. (2007) Can overshooting convection dehydrate the tropical tropopause layer? *Journal of Geophysical Research* 112, D11209.

Jones, M.L., Shuter, B.J., Zhao, Y. and Stockwell, J.D. (2006) Forecasting effects of climate change on Great Lakes fisheries: models that link habitat supply to population dynamics can help. *Canadian Journal of Fisheries and Aquatic Sciences* 63, 457–468.

Karl, T.R. and Knight, R.W. (1998) Secular trends of precipitation amount, frequency, and intensity in the United States. *Bulletin of the American Meteorological Society* 79, 231–241.

Keller, W. (2007) Implications of climate warming for Boreal Shield lakes: a review and synthesis. *Environmental Reviews* 15, 99–112.

Kelly, S.A., Takbiri, Z., Belmont, P. and Foufoula-Georgiou, E. (2017) Human amplified changes in precipitation–runoff patterns in large river basins of the Midwestern United States. *Hydrology and Earth System Sciences* 21, 10.

Kling, G.W., Hayhoe, K., Johnson, L.B., Magnuson, J.J., Polasky, S. *et al.* (2003) *Confronting Climate Change in the Great Lakes Region: Impacts on our Communities and Ecosystems*. Union of Concerned Scientists, Cambridge, Massachusetts and Ecological Society of America, Washington, DC.

Kunkel, K.E., Andsager, K. and Easterling, D.R. (1999) Long-term trends in extreme precipitation events over the conterminous United States and Canada. *Journal of Climate* 12, 2515–2527.

Kunkel, K.E., Easterling, D.R., Kristovich, D.A., Gleason, B., Stoecker, L. and Smith, R. (2012) Meteorological causes of the secular variations in observed extreme precipitation events for the conterminous United States. *Journal of Hydrometeorology* 13, 1131–1141.

Lemke, P., Ren, J., Alley, R.B., Allison, I., Carrasco, J. *et al.* (2007) Observations: changes in snow, ice and frozen ground. In: Solomon, S., Qin, D., Manning, M., Chen, Z., Marquis, M. *et al.* (eds) *Climate Change 2007: The Physical Science Basis. Contribution of Working Group I to the Fourth Assessment Report of the Intergovernmental Panel on Climate Change*. Cambridge University Press, Cambridge and New York, pp. 337–383.

Lemmen, D.S. and Warren, F.J. (2004) *Climate Change Impacts and Adaptation: A Canadian Perspective*. Natural Resources Canada, Ottawa.

Lenters, J.D. (2004) Trends in the Lake Superior water budget since 1948: a weakening seasonal cycle. *Journal of Great Lakes Research* 30, 20–40.

Levy, P.E., Cannell, M.G.R. and Friend A.D. (2004) Modelling the impact of future changes in climate, CO_2 concentration and land use on natural ecosystems and the terrestrial carbon sink. *Global Environmental Change* 14, 21–30.

McCormick, M.J. (1990) Potential changes in thermal structure and cycle of Lake Michigan due to global warming. *Transactions of the American Fisheries Society* 119, 183–194.

McCormick, M.J. and Fahnenstiel, G.L. (1999) Recent climatic trends in nearshore water temperatures in the St. Lawrence Great Lakes. *Limnology and Oceanography* 44, 530–540.

MacDonald, G.M. (2010) Climate change and water in southwestern North America. Special Feature: Water, Climate Change, and Sustainability in the Southwest. *Proceedings of the National Academy of Sciences USA* 107, 21256–21262.

McDonald, M.E., Hershey, A.E. and Miller, M.C. (1996) Global warming impacts on Lake Trout in arctic lakes. *Limnology and Oceanography* 41, 1102–1108.

Magnuson, J.J. and DeStasio, B.T. (1996) Thermal niche of fishes and global warming. In: Wood, C.M. and McDonald, D.G. (eds) *Society for Experimental Biology Seminar Series 61: Global Warming: Implications for Freshwater and Marine Fish*. Cambridge University Press, Cambridge, UK, pp. 377–408.

Magnuson, J.J., Robertson, D.M., Benson, B.J., Wynne, R.H., Livingstone, D.M. *et al.* (2000) Historical trends in lake and river ice cover in the northern hemisphere. *Science* 289, 1743–1746.

Magnuson, J.J., Webster, K.E., Assel, R.A., Bowser, C.J., Dillon, P.J. *et al.* (1997) Potential effects of climate changes on aquatic systems: Laurentian Great Lakes and Precambrian Shield Region. *Hydrological Processes* 11, 825–871.

Mandrak, N.E. (1989) Potential invasion of the Great Lakes by fish species associated with climatic warming. *Journal of Great Lakes Research* 15, 306–316.

Marcogliese, D.J. (2001) Implications of climate change for parasitism of animals in the aquatic environment. *Canadian Journal of Zoology* 79, 1331–1352.

Marcos-López, M., Gale, P., Oidtmann, B.C. and Peeler, E.J. (2010) Assessing the impact of climate change on disease emergence in freshwater fish in the United Kingdom. *Transboundary Emergence of Disease* 57, 293–304.

Meisner, J.D. (1990) Potential loss of thermal habitat for brook trout, due to climatic warming, in two southern Ontario streams. *Transactions of the American Fisheries Society* 119, 282–291.

Melillo, J.M., Richmond, T.C. and Yohe, G.W. (eds) (2014) *Climate Change Impacts in the United States: The Third National Climate Assessment*. US Global Change Research Program, Washington, DC. https://doi.org/10.7930/J0Z31WJ2

Meyer, J.L., Sale, M.J., Mulholland, P.J. and Poff, N.L. (1999) Impacts of climate change on aquatic ecosystem functioning and health. *Journal of the American Water Research Association* 35, 1373–1386.

Mohseni, O., Stefan, H.G. and Eaton, J.G. (2003) Global warming and potential changes in fish habitat in US streams. *Climate Change* 59, 389–409.

Mortsch, L.D. (1998) Assessing the impact of climate change on the Great Lakes shoreline wetlands. *Climate Change* 40, 391–416.

Mortsch, L.D. and Quinn, F.H. (1996) Climate change scenarios for Great Lakes basin ecosystem studies. *Limnology and Oceanography* 41, 903–911.

NCEI (National Centers for Environmental Information) (2016) Climate at a Glance: Global Temperature Anomalies. Available at: http://www.ncdc.noaa.gov/cag/time-series/global/globe/land_ocean/ytd/12/1880-2015 (accessed June 2019).

Notaro, M., Zarrin, A., Vavrus, S. and Bennington, V. (2013) Simulation of heavy lake-effect snowstorms across the Great Lakes Basin by RegCM4: synoptic climatology and variability. *Monthly Weather Review* 141, 1990–2014.

O'Gorman, P.A. and Schneider, T. (2009) The physical basis for increase in precipitation extremes in simulations of 21st-century climate change. *Proceedings of the National Academy of Sciences USA* 106, 14773–14777.

Ostrand, K.G. and Wilde, G.R. (2004) Changes in prairie stream fish assemblages restricted to isolated streambed pools. *Transactions of the American Fisheries Society* 133, 1329–1338.

Pankhurst, N.W. and Munday, P.L. (2011) Effects of climate change on fish reproduction and early life history stages. *Marine and Freshwater Research* 62, 1015–1026.

Plumb, J.M. and Blanchfield, P.J. (2009) Performance of temperature and dissolved oxygen criteria to predict habitat use by lake trout (*Salvelinus namaycush*). *Canadian Journal of Fisheries and Aquatic Sciences* 66, 2011–2023.

Polderman, N.J. and Pryor, S.C. (2004) Linking synoptic-scale climate phenomena to lake-level variability in the Lake Michigan–Huron basin. *Journal of Great Lakes Research* 30, 419–434.

Poole, G.C. and Berman, C.H. (2001) An ecological perspective on in-stream temperature: natural heat dynamics and mechanisms of human-caused thermal degradation. *Environmental Management* 27, 787–802.

Power, G., Brown, R.S. and Imhof, J.G. (1999) Groundwater and fish – insights from northern North America. *Hydrological Processes* 13, 401–422.

Rahel, F.J. and Olden, J.D. (2008) Assessing the effects of climate change on aquatic invasive species. *Conservation Biology* 22, 521–533.

Sala, O.E., Chapin, F.S., Armesto, J.J., Berlow, E., Bloomfield, J. *et al.* (2000) Biodiversity – global biodiversity scenarios for the year 2100. *Science* 287, 1770–1774.

Schindler, D.W. (1998) A dim future for boreal waters and landscapes. *Bioscience* 48, 157–164.

Schindler, D.W. (2001) The cumulative effects of climate warming and other human stresses on Canadian freshwaters in the new millennium. *Canadian Journal of Fisheries and Aquatic Sciences* 58, 18–29.

Schindler, D.W., Bayley, S.E. and Parker, B.R. (1996) The effects of climatic warming on the properties of boreal lakes and streams at the Experimental Lakes Area, northwestern Ontario. *Limnology and Oceanography* 41, 1004–1017.

Schindler, D.W., Curtis, P.J., Bayley, S.E., Parker, B.R., Beaty, K.G. and Stainton, M.P. (1997) Climate induced changes in the dissolved organic carbon budgets of boreal lakes. *Biogeochemistry* 36, 9–28.

Seager, R. and Vecchi, G.A. (2010) Greenhouse warming and the 21st century hydroclimate of southwestern North America. *Proceedings of the National Academy of Sciences USA* 107, 21277–21282.

Sereda, J., Bogard, M., Hudson, J., Helps, D. and Dessouki, T. (2011) Climate warming and the onset of salinization: rapid changes in the limnology of two northern plains lakes. *Limnologica* 41, 1–9.

Sharma, S., Jackson, D.A. and Minns, C.K. (2009) Quantifying the potential effects of climate change and the invasion of smallmouth bass on native lake trout populations across Canadian lakes. *Ecography* 32, 517–525.

Sharma, S., Vander Zanden, M.J., Magnuson, J.J. and Lyons, J. (2011) Comparing climate change and species invasions as drivers of coldwater fish population extirpations. *PLoS One* 6, e22906. https://doi.org/10.1371/journal.pone.0022906

Shimoda, Y., Azim, M.E., Perhar, G., Ramin, M., Kenney, M.A. *et al.* (2011) Our current understanding of lake ecosystem response to climate change: what have we really learned from the north temperate deep lakes? *Journal of Great Lakes Research* 37, 173–193.

Shuter, B.J., MacLean, J.A., Fry, F.E.J. and Regier, H.A. (1980) Stochastic simulation of temperature effects on first-year survival of smallmouth bass. *Transactions of the American Fisheries Society* 109, 1–34.

Snucins, E. and Gunn, J. (2000) Interannual variation in the thermal structure of clear and colored lakes. *Limnology and Oceanography* 45, 1639–1646.

Sommer, U., Aberle, N., Engel, A., Hansen, T., Lengfellner, K. *et al.* (2007) An indoor mesocosm system to study the effect of climate change on the late winter and spring succession of Baltic Sea phyto- and zooplankton. *Oecologia* 150, 655–666.

Sousounis, P.J. and Grover, E.K. (2002) Potential future weather patterns over the Great Lakes Region. *Journal of Great Lakes Research* 28, 496–520.

Stefan, H.G., Hondzo, M., Fang, X., Eaton, J.G. and McCormick, J.H. (1996) Simulated long-term temperature and dissolved oxygen characteristics of lakes in the north-central US and associated habitat limits. *Limnology and Oceanography* 41, 1124–1135.

Stephens, G.L., L'Ecuyer, T., Forbes, R., Gettlemen, A., Golaz, J.C. *et al.* (2010) Dreary state of precipitation in global models. *Journal of Geophysical Research: Atmospheres* 115, D24211.

Suriano, Z.J. and Leathers, D.J. (2017) Synoptically classified lake-effect snowfall trends to the lee of Lakes Erie and Ontario. *Climate Research* 74, 1–13.

Tetreault, G.R., Bennett, C.J., Cheng, C., Servos, M.R. and McMaster, M.E. (2012) Reproductive and histopathological effects in wild fish inhabiting an effluent-dominated stream, Wascana Creek, SK, Canada. *Aquatic Toxicology* 110, 149–169.

Tisseuil, C., Vrac, M., Grenouillet, G., Wade, A.J., Gevrey, M. *et al.* (2012) Strengthening the link between climate, hydrological and species distribution modeling to assess the impacts of climate change on freshwater biodiversity. *Science of the Total Environment* 424, 123–201.

Trapp, R.J., Diffenbaugh, N.S., Brooks, H.E., Baldwin, M.E., Robinson, E.D. and Pal, J.S. (2007) Changes in severe thunderstorm environment frequency during the 21st century caused by anthropogenically enhanced global radiative forcing. *Proceedings of the National Academy of Sciences USA* 104, 19719–19723.

Trenberth, K.E., Fasullo, J.T. and Kiehl, J. (2009) Earth's global energy budget. *Bulletin of the American Meteorological Society* 90, 311–323.

USGCRP (US Global Change Research Program) (2017) *Climate Science Special Report: Fourth National Climate Assessment*, Vol. I [Wuebbles, D.J., Fahey, D.W., Hibbard, K.A., Dokken, D.J., Stewart, B.C. and Maycock, T.K. (eds)]. USGCRP, Washington, DC. https://doi.org/10.7930/J0J964J6

Vecchi, G.A., Swanson, K.L. and Soden, B.J. (2008) Climate change: whither hurricane activity? *Science* 322, 687–689.

Vincent, W.F. (2009) Effects of climate change on lakes. In: Likens, G.E. (ed.) *Encyclopedia of Inland Waters*. Vol. 3. *Pollution and Remediation*. Elsevier, Oxford, pp. 55–60.

Wenger, S.J., Isaak, D.J., Luce, C.H., Neville, H.M., Fausch, K.D. *et al.* (2011) Flow regime, temperature, and biotic interactions drive differential declines of trout species under climate change. *Proceedings of the National Academy of Sciences USA* 108, 14175–14180.

Willis, T.V. and Magnuson, J.J. (2006) Response of fish communities in five north temperate lakes to exotic species and climate. *Limnology and Oceanography* 51, 2808–2820.

Winkler, J.A., Arritt, R.W. and Pryor, S.C. (2012) Climate projections for the Midwest: availability, interpretation, and synthesis. In: Winkler, J., Andresen, J., Hatfield, J., Bidwell, D. and Brown, D. (coordinators) US National Climate Assessment Midwest Technical Input Report. Available at: http://glisa.umich.edu/media/files/NCA/MTIT_Future.pdf (accessed 7 April 2020).

Winter, T. (2000) The vulnerability of wetlands to climate change: a hydrologic landscape perspective. *Water Resources* 206, 50–57.

Wuebbles, D.J. and Hayhoe, K. (2004) Climate change projections for the United States Midwest. *Mitigation and Adaptation Strategies for Global Change* 9, 335–363.

Wuebbles, D.J., Hayhoe, K. and Parzen, J. (2010) Introduction: assessing the effects of climate change on Chicago and the Great Lakes. *Journal of Great Lakes Research* 36, 1–6.

Wuebbles, D., Cardinale, B., Cherkauer, K., Davidson-Arnott, R., Hellmann, J. *et al.* (2019) *An Assessment of the Impacts of Climate Change on the Great Lakes*. Environmental Law and Policy Center, Chicago, Illinois.

Xenopoulos, M.A., Lodge, D.M., Alcamo, J., Marker, M., Schilze, K. and Van Vuuren, D.P. (2005) Scenarios of freshwater fish extinctions from climate change and water withdrawal. *Global Change Biology* 11, 1557–1564.

Zobel, Z., Wang, J., Wuebbles, D.J. and Kotamarthi, V.R. (2017) High resolution dynamical downscaling ensemble projections of future extreme temperature distributions for the United States. *Earth's Future* 5, 1234–1251. https://doi.org/10.1002/2017EF000642

2 Tropical Freshwater Ecosystems, Biota, and Anthropogenic Activities with Reference to South-East Asia

JIA HUAN LIEW[1,2], RAYSON B.H. LIM[1], BI WEI LOW[1], MAXINE A.D. MOWE[1], TING HUI NG[1,3], YIWEN ZENG[1], AND DARREN C.J. YEO[1,3]*

[1]*Department of Biological Sciences, National University of Singapore, Singapore;* [2]*School of Biological Sciences, The University of Hong Kong, Hong Kong SAR, China;* [3]*Lee Kong Chian Natural History Museum, National University of Singapore, Singapore*

2.1 Introduction to Tropical Freshwater Ecosystems of South-East Asia

Tropical freshwater ecosystems differ from their temperate counterparts in the relative year-round stability of their thermal and ultraviolet radiation regimes (Payne, 1986). However, freshwater ecosystems in the tropics do experience some seasonality. Here, seasons are defined by fluctuations in water flow and/or water levels. These are driven by a wide range of factors including landscape morphology and rainfall patterns – so they are often less generalizable (Dudgeon, 1992).

Tropical fresh waters are highly diverse, but can be largely categorized into lentic (e.g. lakes) or lotic (e.g. rivers) systems. In South-East Asia (SE Asia), natural lentic systems are formed primarily by geomorphological processes including tectonics and vulcanism while the extent and distribution of river ecosystems are likely influenced by changing land masses since sea-level minima of the Pleistocene (Voris, 2000). For example, rivers along the west and east coasts of the Malay Peninsula and of Sumatra, respectively, are thought to be truncated tributaries of a now submerged and more extensive Malacca Straits River System. Smaller natural lakes may also be formed by river fluvial dynamics in conjunction with suitable landscape topography, such as when parts of a meandering river are cut off (e.g. oxbow lakes). Alternatively, lakes also form in seasonally inundated, low-lying regions that accumulate organic matter, as well as water (e.g. swamps). The largest natural lake in the SE Asian region, Tonlé Sap Lake in Cambodia, is one such example.

In general, lotic freshwater systems in tropical SE Asia resemble their Amazonian counterparts in that they are associated with mildly acidic, soft waters and are mostly nutrient-poor (Dudgeon, 1992). This is most evident in the upper reaches of river systems (e.g. low-order forest streams). Lake systems are more stochastic in their physical and chemical properties, possibly as a consequence of long water retention times which increase susceptibility to location-specific influence. Globally, the hydrology and physical properties of lakes are less well understood than those of rivers (Balek, 1983) – and SE Asia is no exception.

2.1.1 Human interaction with freshwater ecosystems

Human interaction with freshwater ecosystems vary widely in extent and intensity. Across SE Asia, freshwater resources provide potable water, food, transport and power, among others. Here, high densities of inland water bodies mean that human populations generally live closer to fresh water than

*dbsyeod@nus.edu.sg

in other parts of the world (Kummu *et al.*, 2011). Consequently, threats to human water security are thought to be relatively minor (Vorosmarty *et al.*, 2010). However, proximity to fresh waters may also mean that issues affecting freshwater ecosystems (e.g. water pollution) are more likely to impact humans.

One of the most ubiquitous types of human–freshwater interactions involves agriculture. Many SE Asian communities are historically agrarian and a large proportion still retain a close association with agricultural activities. Rice farming is particularly characteristic in SE Asia where it has a long cultivation history (Bellwood, 2004). Today, rice represents approximately 90% of all irrigated crops in the region (Facon, 2000). Rice agriculture and freshwater ecosystems are intricately linked because rice production is reliant on a steady water supply, often extracted or redirected from nearby river systems (Redfern *et al.*, 2012; Okazumi *et al.*, 2014). Although rice production and other agricultural activities are generally thought to impact freshwater diversity via pollutant runoff (Smith *et al.*, 1999; Zhao *et al.*, 2012), they can sometimes have unexpected ecological functions. For example, freshwater invertebrates found mainly in increasingly threatened swamp habitats are known to use irrigated rice fields as surrogate habitats (Yule and Yong, 2004).

Freshwater ecosystems are also closely linked to aquaculture. Here, aquacultural activities are largely focused on the production of commercially important non-native fish taxa, including hybrid clariid catfishes (*Clarias* spp.) and tilapia (*Oreochromis* spp.), with a few exceptions (De Silva *et al.*, 2005). The milkfish (*Chanos chanos*) is an example of a native species of significant regional importance that can be cultured in fresh waters (although the species can be bred only in saline environments) (Paclibare, 2005). In general, inland aquaculture production is more important in the continental parts of SE Asia, such as the Indochinese regions (e.g. Cambodia, Myanmar), than in insular SE Asia (e.g. Indonesia) (Hishamunda *et al.*, 2009). Besides commercial aquaculture, subsistence aquaculture is also practised in SE Asia. For example, rural Thai communities sometimes stock rice fields with aquaculture species (e.g. *Oreochromis niloticus*) to supplement their dietary protein (Little *et al.*, 1996).

Capture fisheries in SE Asian fresh waters are an important source of both nutrition (Youn *et al.*, 2014) and income (Dugan *et al.*, 2006) (Fig. 2.1). While inland fisheries are mostly artisanal and limited in scale (Nguyen and de Silva, 2006; Wilkinson *et al.*, 2018), commercial fishing is common in larger freshwater systems. For example, fishing lots in the Mekong River and the connected Tonlé Sap Lake are highly productive (Coates, 2002; Coates *et al.*, 2003). Collectively, fisheries in the Mekong River basin are more productive than other inland fisheries in the

Fig. 2.1. Freshwater fish caught from the Tonlé Sap Lake, Cambodia, on sale at a local market.

world (Baran and Myschowoda, 2009). Worryingly, inland fisheries and freshwater aquaculture are likely to be impacted by a rapidly changing climate that can impact the life cycles of commercially important species (see Section 2.3.4).

Another common interface between humans and freshwater ecosystems is via the ornamental pet trade (Ng and Tan, 1997). This is partly driven by both the region's rich aquatic biodiversity as well as the lucrative nature of the trade, especially when popular ornamental species are involved. One notable example is the Asian arowana (*Scleropages formosus*), a CITES (Convention on International Trade in Endangered Species of Wild Fauna and Flora) listed species known to fetch upwards of US$20,000 per individual (Yue *et al.*, 2004). Organisms in the trade are largely wild-caught because many popular species cannot be consistently or reliably bred in captivity (Ng and Tan, 1997). The continued exploitation of natural populations is also motivated by the higher prices commanded by wild-caught individuals (Sung and Fong, 2018).

2.2 Ecology of Freshwater Ecosystems

2.2.1 Seasonal dynamics

Seasonal variations in rainfall influence the hydrology, ecology and even the structure of freshwater ecosystems. Exemplifying this, the areal coverage of the Tonlé Sap Lake in Cambodia (Fig. 2.2) can fluctuate between 2500 and 10,000 km^2 across the dry and wet seasons (MRC, 2003). Precipitation patterns in the region are largely driven by monsoonal winds resulting from alternating low- and high-pressure zones in the landmasses north (Asia) and south (Australia) of the region. In general, Indochina experiences maximum rainfall during the Asian summer (~July) while insular SE Asia (e.g. Indonesia, Peninsular Malaysia) is wettest in the Australian summer (~November) (Chang and Wang, 2005).

Seasonal changes have a strong influence on aquatic biota (Lowe-McConnell, 1979) and the entire ecosystem through the ecological functions the biota provide (Lytle and Poff, 2004; Jardine *et al.*, 2015). This is because wet–dry season dynamics result in environmental fluctuations which serve as

Fig. 2.2. Tonlé Sap Lake, Cambodia.

cues for the life cycles of many freshwater taxa. For example, temperatures and oxygen levels in marginal habitats (e.g. lagoons and flood plain pools) of riverine systems increase significantly in wet seasons and this triggers spawning in many freshwater taxa (Meisner and Shuter, 1992). In addition, wet seasons are also important for facilitating the dispersal of some aquatic invertebrates (Bilton *et al.*, 2001).

2.2.2 Aquatic food webs

Food webs regulate a wide spectrum of ecological functions in freshwater ecosystems (Holt and Loreau, 2001; Thompson *et al.*, 2012). Unfortunately, food web data from SE Asian fresh waters are relatively scarce. In the absence of empirical data, our understanding of aquatic food webs is mainly inferred from existing literature on aquatic biodiversity (see Liew *et al.*, 2016a) or anecdotal observations. The limited empirical data available show that food webs in lentic water bodies are structurally conserved – even when comparisons are made across natural (e.g. lakes) and artificial (e.g. reservoir) systems (Liew *et al.*, 2016a). This suggests that aquatic food webs are mechanistically shaped, or at least constrained, by environmental conditions (Liew *et al.*, 2018a).

Tropical freshwater food webs are believed to be distinct from their temperate counterparts because of differences in the incorporation of primary productivity that are largely attributed to contrasting aquatic herbivore composition (Li and Dudgeon, 2009; Dudgeon *et al.*, 2010). As organic carbon is largely transferred up food webs through herbivorous primary consumers, the dominance of taxa associated with a specific dietary preference or feeding mechanism can determine the dominant organic carbon pathway. For example, a freshwater community comprising mainly shredding invertebrates (e.g. caddisflies) is likely to be supported by external organic carbon sources (e.g. leaf litter) while a primarily grazing invertebrate-dominated community (e.g. freshwater snails) would more likely rely on internal production (e.g. periphyton). Although our current understanding of the latitudinal patterns in the assimilation of primary production suggests that external carbon sources are more important in tropical fresh waters because of a higher capacity for processing detritus (Tiegs *et al.*, 2019), the rates at which organic matter are broken down may not necessarily correlate with greater uptake levels. This is because terrestrial detritus in fresh waters is processed primarily by microbiota (e.g. heterotrophic bacteria), especially in the tropics (Amado and Roland, 2017). Yet, bacteria are thought to be poor sources of nutrition (Taipale *et al.*, 2012; Galloway *et al.*, 2015), meaning that external (terrestrial) organic matter may not be taken up by aquatic food webs and is instead locked in the microbial biomass.

Aquatic food webs are also influenced by wet–dry season dynamics. One major mechanism through which seasonality regulates food web structure is via bottom-up forces (Liew *et al.*, 2018a) exerted through varying levels of terrestrial input. In river flood plains, high water level during wet seasons commonly improves connectivity between isolated water bodies (e.g. flood plain ponds), resulting in the influx of terrestrial resource subsidies. These are important for supporting aquatic biomasses which may exceed the carrying capacity of the system's internal primary productivity (Polis *et al.*, 1997; Jardine *et al.*, 2012). Terrestrial resource subsidies can also be brought in by runoff during periods of increased precipitation (Brett *et al.*, 2017). While there are some uncertainties surrounding the role of terrestrial production in freshwater systems (Brett *et al.*, 2017), species adapted to assimilating external resources (Batt *et al.*, 2012) are likely to benefit from the seasonal influx (McQueen *et al.*, 1989).

Another mechanism through which seasonal floods can influence aquatic food webs is by boosting internal primary production during high-water seasons, especially in ecosystems associated with longer inundation periods. Here, flood waters expand productive littoral zones which favours greater internal primary productivity (Junk *et al.*, 1989). This is ecologically significant because aquatic primary producers (e.g. algae) represent sources of fatty acids which are essential for sustaining the somatic growth of many freshwater organisms (Lau *et al.*, 2013; Brett *et al.*, 2017).

2.2.3 Key freshwater taxa

Phytoplankton

Phytoplankton are ubiquitous in freshwater bodies throughout SE Asia. In lentic water bodies, phytoplankton assemblages are determined by water stratification, flow and nutrient levels. Phytoplankton communities associated with natural lakes (e.g. Lake Taal, Lake Lanao, Laguna de Bay, Lake Inle, Tonlé

Sap Lake, Lake Toba) often differ significantly from those occurring in man-made reservoirs and ponds. While most natural lakes are oligotrophic to mesotrophic (Lewis, 1973, 1978; Green *et al.*, 1978; Papa and Mamaril, 2011) and are dominated by Chlorophyta (green algae) or Bacillariophyta (diatoms), man-made water bodies are more often eutrophic or hyper-eutrophic (Pongswat *et al.*, 2004; Khuantrairong and Traichaiyaporn, 2008; Te and Gin, 2011) and are, instead, dominated by Cyanophyta (blue-green algae). However, this distinction is increasingly lost as natural lakes are subjected to higher levels of anthropogenic nutrient enrichment (see Section 2.3.2), thus becoming more susceptible to reductions in phytoplankton diversity as well as to cyanobacterial blooms (Tamayo-Zafaralla *et al.*, 2002; Papa and Mamaril, 2011; Rahman *et al.*, 2016).

Unlike in lake systems, phytoplankton communities in SE Asian rivers are influenced by pH, nutrient levels and flow regimes. The most common phytoplankton group in lotic systems are Bacillariophyta (diatoms) and Chlorophyta (green algae). Abundance levels of these taxa are highly dependent on hydrological regimes (Sakset and Chankaew, 2012; Li *et al.*, 2013; Nursuhayati *et al.*, 2013). Another important subset of the region's lotic systems comprise low-pH rivers and these are mostly dominated by acid-tolerant diatoms (e.g. *Pinnularia acoricola*) (Lohr *et al.*, 2006).

Phytoplankton taxa have varying levels of sensitivity to environmental conditions, suggesting that they may be effective bio-indicators. For example, Bacillariophyta (diatoms) have been used to measure river pollution in Malaysia with a Saprobic index (Wan Maznah and Mansor, 2002; Wan Maznah, 2010). The index categorizes diatom species into three main groups: (i) pollution-sensitive species occurring only in oligotrophic rivers; (ii) moderately tolerant species resistant to polluted conditions; and (iii) highly tolerant species occurring in highly polluted rivers. Trials in the Pinang River basin (Malaysia) found species belonging to the third category to be dominant in most of the sites sampled (Wan Maznah and Mansor, 2002).

In addition to serving as indicators of water quality, phytoplankton composition has wider implications. In tropical Asia, blooms of cyanobacteria are mostly attributable to the genus *Microcystis*, among which more than 50% are producers of the hepatotoxin, microcystin (Mowe *et al.*, 2015a). *Cylindrospermopsis* has been found to be the second most prevalent genus in the region, causing blooms in Singapore and potentially in Thailand and Vietnam (Mowe *et al.*, 2015a). While *Microcystis* proliferates in high-nitrogen environments, *Cylindrospermopsis* blooms are better predicted by higher maximum water temperatures (Mowe *et al.*, 2015a). Importantly, there is a positive correlation between cyanobacterial cell counts and cyanotoxin concentrations in the water (Mowe *et al.*, 2015a).

Molluscs

Freshwater molluscs occur in habitats ranging from ephemeral ponds to large rivers. In SE Asia, molluscs are important sources of food (Fig. 2.3) and are sometimes kept in aquaria. However, freshwater molluscs are also of medical concern as some species are intermediate hosts of zoonotic parasites (Attwood, 2010; Sri-Aroon, 2010; Köhler *et al.*, 2012).

Freshwater snail diversity and endemism are thought to be highest in the tropics (Strong *et al.*, 2008). The ancient lakes of Sulawesi in Indonesia are host to ~75 species of endemic Pachychilidae (von Rintelen *et al.*, 2010), while more than 90 species of endemic Triculinae occur in the Mekong River basin alone (Davis, 1979). Unfortunately, the exact number of extant species in the region remains unclear (Strong *et al.*, 2008). Reliable data indicate a total of 325 species from the overall Indochinese region (Köhler *et al.*, 2012), 286 species from Thailand (Brandt, 1974; Nabhitabhata, 2009), 127 species from Vietnam (Do, 2015), 251 species from the Philippines (Pagulayan, 1995), 66 species from the island of Java in Indonesia (Marwoto *et al.*, 2011), about 60 species from

Fig. 2.3. Freshwater snails on sale at the local market in Cambodia.

Peninsular Malaysia (Maassen, 2001; Ng *et al.*, 2019) and more than 20 species from Singapore (Clements *et al.*, 2006; Tan *et al.*, 2012; Ng *et al.*, 2014a,b, 2015a, 2016a,b,c, 2017, 2019). However, baseline information of snails remains non-existent in many others parts of SE Asia. The need for additional surveys is evident in the discovery of new records (Adorable-Asis *et al.*, 2016) and possible cryptic species (Stelbrink *et al.*, 2019), even in relatively well-studied localities. For example, only about 30 freshwater gastropod species have been recorded from Borneo since the 1800s despite its size – suggesting that the figure is likely an underestimate (Ng *et al.*, 2015b, 2017).

SE Asia, as part of the larger Indo-Tropical or Oriental regions, has also been identified as a biodiversity hotspot for freshwater bivalves. The region is ranked second only to the much larger Palearctic region (Bogan, 2008; Graf, 2013; Zieritz *et al.*, 2018). A total of 135 freshwater mussel species (Unionida) have been recorded from SE Asia and the Indo-Burmese region (e.g. the Mekong River basin) is especially speciose with 51 species (Zieritz *et al.*, 2018). Vietnam has the highest number of freshwater mussels (59 species), followed by Thailand (52 species) and Myanmar (43 species) (Do *et al.*, 2018; Zieritz *et al.*, 2018). These numbers are likely to continue rising as new species are still being discovered (Bogan and Do, 2014; Kongim *et al.*, 2015; Bolotov *et al.*, 2017). In the rest of SE Asia, mussel species numbers vary from as many as 36 in Cambodia, to as few as seven in the Philippines. Like snails, information about SE Asian mussels is similarly in need of updating (Zieritz *et al.*, 2018).

Venerida clams are less speciose and only one or two *Corbicula* species are thought to be widely distributed across the region (Kijviriya *et al.*, 1991; Park and Kim, 2003) although much of the taxonomy of this genus remains unresolved. The Sulawesi great lakes of Indonesia host four species of endemic *Corbicula*, while one endemic species occurs in Sumatra (Glaubrecht *et al.*, 2003).

Crustaceans

Crustaceans represent one of the most diverse groups of freshwater taxa in SE Asia (e.g. Fernando, 1980; Yeo *et al.*, 2008; Korovchinsky, 2013; Rogers *et al.*, 2013), spanning a wide range of sizes from small planktonic Cladocera (Fernando, 1980; Forró *et al.*, 2008; Korovchinsky, 2013) to large and highly motile decapods (Crandall and Buhay, 2008; Yeo *et al.*, 2008). They fulfil many ecological roles vital to local ecosystems (e.g. Boxshall and Defaye, 2008; Forró *et al.*, 2008; Yeo *et al.*, 2008) and like molluscs, are associated with many aspects of human life. Crustaceans are food, pets, potential disease vectors and scientific model organisms (Yule and Yong, 2004; Forró *et al.*, 2008; Martens *et al.*, 2008; Zeng and Yeo, 2018). Some of the more diverse groups of crustaceans include, but are not limited to, Cladocera, Copepoda, Ostracoda, Amphipoda and Decapoda (Balian *et al.*, 2008).

Many crustacean groups constitute important components of the freshwater zooplankton community. For example, Cladocera and Copepoda occupy important functions in the planktonic food web (Fernando, 1980; Boxshall and Defaye, 2008; Korovchinsky, 2013), occurring in a wide range of habitat types including lotic systems (e.g. rivers and streams), reservoirs, rice fields and mining pools (Fernando, 1980; Rogers *et al.*, 2013). Despite their prevalence and importance, both groups are poorly studied (Korovchinsky, 2013; Rogers *et al.*, 2013). Estimates of Cladocera and Copepoda species richness vary and large knowledge gaps remain in our understanding of species delimitations and range boundaries (Korovchinsky, 2013; Rogers *et al.*, 2013).

Freshwater decapods in SE Asia, which include caridean shrimps, brachyuran crabs and parastacid crayfish, have been studied rather more extensively. Decapods maintain the stability of freshwater food webs (by feeding across multiple trophic levels), regulate water levels through their burrowing abilities (Ng, 1989; Stahl *et al.*, 2014; Zeng and Yeo, 2018), and are food and medicine for local human communities (Yeo and Ng, 1998; De Grave *et al.*, 2008; Yeo *et al.*, 2008; Zeng and Yeo, 2018). Freshwater crayfish are the least speciose of the main groups of decapods in SE Asia. All extant species belong to the genus *Cherax*, naturally occuring in the biogeographically Australasian part of the region (i.e. Indonesian West Papua), although one species, *Cherax quadricarinatus*, has been widely translocated throughout SE Asia (Zeng and Yeo, 2018; Zeng *et al.*, 2019). Conversely, Sundaic SE Asia is dominated by a high diversity of primary freshwater crabs (Gecarcinucidae and Potamidae) (Zeng and Yeo, 2018), with about 30% of the world's freshwater crab species (>400 species) found in an area making up only 4% of the global total (Zeng and Yeo, 2018). Freshwater shrimp assemblages, which include *Caridina* and *Macrobrachium* species, are similarly diverse in SE Asia (De Grave *et al.*, 2008).

Fishes

SE Asia comprises four ichthyofaunal diversity hotspots – coinciding with the global biodiversity hotspots of Indo-Burma, Sundaland, Wallacea and the Philippines (Myers *et al.*, 2000). There are at least 3108 species of freshwater fish from 137 families in the inland waters bounded by (and including) the Kaladan, Irrawaddy and Red River drainages, the Philippines, the Greater Sunda Islands and the Maluku Islands (Kottelat, 2013). Within this area, the dominant taxon is the family Cyprinidae, with at least 70 genera endemic to SE Asia, while many others have few known representatives outside the region (Zakaria-Ismail, 1994; Kottelat, 2013).

Freshwater fish richness and endemism are extraordinarily high in SE Asia. There are at least 1100 known species in the Mekong River alone, with some estimating close to 1700 species. This ranks the river third globally in total fish diversity, behind the Amazon and Zaire rivers (Dudgeon, 2005; Rainboth *et al.*, 2012). In Sundaic SE Asia, at least 100 species are stenotopic to highly acidic blackwater environments associated with ancient peat swamp forests (Giam *et al.*, 2012). Prominent examples of blackwater endemics are representatives from the family Osphronemidae including members of the genera *Betta*, *Parosphromenus* and *Sphaerichthys*. Species belonging to these genera are thought to be some of the most range-restricted fishes globally (Kottelat and Ng, 2005; Tan and Ng, 2005). They are also possibly some of the most threatened freshwater taxa in SE Asia (B.W. Low, 2019, unpublished results). Recent estimates suggest that up to 77% of stenotopic blackwater fishes in Sundaland could become extinct by 2050 owing to unabated peatland degradation (Giam *et al.*, 2012).

There is currently no consensus on classifications of freshwater fish into distinct zoogeograhic faunas (e.g. Zakaria-Ismail, 1994; Abell *et al.*, 2008). At present, the distribution patterns are broadly divided into seven overlapping, zoogeographic regions, defined by similarities in species assemblage and common ancestry. They include:

1. The Annam–Red River region in the north-east – closely allied with the Pearl and Yangtze River drainages of China (Yap, 2002);
2. The Irrawaddy–Salween–Mae Khlong region in the west – forming the eastern limit of several genera native to the Indian subcontinent (e.g. *Badis*, *Cabdio*, *Chagunius*) as well as the western limit of many SE Asian genera (e.g. *Albulichthys*, *Amblyrhynchichthys*, *Barbichthys*) (Zakaria-Ismail, 1994);
3. The Chao Phraya–Mekong region – the most speciose zoogeographic region in SE Asia (Dudgeon, 2005; Rainboth *et al.*, 2012);
4. The rivers of the Greater Sunda Islands (Java, Sumatra and Borneo) – centres of diversity and endemism for many prominent groups such as the Clariidae, Helostomatidae and Osphronemidae (Roberts, 1989; Kottelat, 2013) thought to have evolved late in the Pleistocene (Yap, 2002);
5. The Malay Peninsula – heavily influenced by elements from both the Indochinese Peninsula as well as the Greater Sunda Islands (Zakaria-Ismail, 1994);
6. The Philippines – distinguished by an explosive radiation of endemic cyprinid species flocks, especially in Lake Lanao (Ismail *et al.*, 2014); and
7. Wallacea (comprising Sulawesi and the Lesser Sunda Islands) – characterized by the presence of endemic species flocks from the family Telmatherinidae (von Rintelen *et al.*, 2012).

2.3 Anthropogenic Threats to Tropical Freshwater Ecosystems

Freshwater ecosystems are among the world's most threatened (Abell, 2002). The severity of impacts is even more pronounced in tropical east Asia, which includes SE Asia (Dudgeon, 2003). This is expected given the proximity of freshwater bodies to major urban population centres (Keppeler *et al.*, 2018). The impacts of anthropogenic activities can be direct (e.g. construction of physical barriers that modify flow) or indirect (e.g. climate change), with alarming consequences for associated biodiversity. As of 2012, freshwater vertebrate diversity had declined at a significantly higher rate than that of their marine or terrestrial counterparts (WWF, 2016).

2.3.1 Habitat modification/degradation

Modification and degradation of natural habitats incur widespread and persistent impacts on freshwater ecosystems globally (Dudgeon *et al.*, 2006; Reid *et al.*, 2019). Generally, habitat modification involves direct physical or hydrological alterations (e.g. impoundments convert running-water into still-water systems), while habitat degradation impacts physical and/or chemical properties of water bodies indirectly as a consequence of anthropogenic activity in the catchment (e.g. deforestation). In both instances, significant environmental change (Allan *et al.*, 2003; Dudgeon, 2014; Liew *et al.*, 2016b) drives the extirpation of sensitive taxa (Liew *et al.*, 2016a).

Freshwater ecosystems are often modified to manage water flow, e.g. dams minimize flow for water retention and canals regularize flow to attenuate flooding. While the consequences of flow modification are well documented (Melcher *et al.*, 2012; Quinn and Kwak, 2013; Liew *et al.*, 2016a), impacts are likely to be amplified in the tropics because fluctuations in flow and/or water levels are a major component of seasonality in inland waters, particularly in river systems (Payne, 1986). The artificial aseasonality in modified water bodies can have two main consequences to tropical freshwater biodiversity. First, the life cycles of many species would be disrupted in the absence of seasonal cues (Resh, 2007). In SE Asia, many species migrate to spawning areas in the wet season before returning to their primary habitats before the onset of the dry season (Welcomme *et al.*, 2016). Therefore, anthropogenic structures, such as dams, not only physically impede migrations, they may obfuscate this behaviour entirely. Second, the reduction in flow and/or water level heterogeneity represents a loss of ecological niches (Loke and Todd, 2016; Liew *et al.*, 2018b). The latter can impact overall species diversity by reducing species co-occurrence because taxa with similar biology may no longer be able to avoid competitive exclusion (Hardin, 1960) through habitat/resource partitioning that capitalizes on seasonal resource fluxes. For example, two separate species may occur sympatrically by mutually exclusive specialization on resources which become alternately abundant in the wet and dry seasons – but under artificially stable conditions, only one of the two is likely to persist.

Modified freshwater habitats are generally susceptible to invasion by non-native species which are capable of exploiting the drastically altered environmental conditions and ecological niches vacated by extirpated native taxa (Johnson *et al.*, 2008; Liew *et al.*, 2016a, 2018b). Moreover, invasion by non-native species will likely impact native freshwater biodiversity (see Section 2.3.3), possibly resulting in secondary extinctions (Liew *et al.*, 2016a). This creates a positive feedback loop which ultimately results in a progressively higher proportion of non-native taxa in modified water bodies.

Besides habitat modification, fresh waters are also threatened by environmental degradation, especially in denuded water catchments (Sodhi *et al.*, 2004; Carrasco *et al.*, 2016). In SE Asia, highly lucrative oil palm monocultures are thought to be the most important driver of land conversions in recent decades (Wilcove *et al.*, 2013). This commonly results in higher levels of pollutant runoff (see Section 2.3.2). Moreover, the proliferation of plantations near water bodies can also alter their physical properties (Sweeney and Newbold, 2014). For example, water temperatures are often higher when natural bank vegetation is replaced by monoculture (Sweeney and Newbold, 2014). In addition, removal of natural bank vegetation also reduces terrestrial subsidies which contribute both to resource availability and habitat complexity (Giam *et al.*, 2015), possibly impacting aquatic food webs via bottom-up forces (Liew *et al.*, 2018a).

2.3.2 Pollution

Eutrophication

Natural (e.g. Lake Toba, Lake Lanao and Lake Taal) and artificial lakes (e.g. reservoirs) in SE Asia have become increasingly eutrophic from runoffs associated with excessive aquaculture (Papa and Mamaril, 2011), agriculture (Tamayo-Zafaralla *et al.*, 2002; Meesukko *et al.*, 2007) and industry (Meesukko *et al.*, 2007; Huang *et al.*, 2015). The problem is most acute in watersheds where forests have been replaced with agricultural land (Meesukko *et al.*, 2007). For example, Laguna de Bay (the Philippines) recorded a drastic increase in nitrogen between 1975 and 2000, amounting to 7300 tonnes per year, with agricultural and domestic runoffs thought to account for 80% of total nutrient input (Tamayo-Zafaralla *et al.*, 2002). Apart from agricultural runoff, eutrophication is sometimes caused by direct chemical fertilizer application to floating vegetable gardens (e.g. Lake Inle, Myanmar) (Akaishi *et al.*, 2006). Regardless of the mechanism of entry of nutrients, eutrophication creates conditions which favour persistent cyanobacterial blooms (Tamayo-Zafaralla *et al.*, 2002; Meesukko *et al.*, 2007; Papa and Mamaril, 2011) (Fig. 2.4).

Lotic systems have also been impacted by eutrophication (Wan Maznah, 2010; Sakset and Chankaew, 2012; Huang *et al.*, 2015). The deterioration in water quality is especially evident in the lower Mekong River (Chea *et al.*, 2016), particularly across tributaries located downstream of densely populated cities (e.g. Vientiane, Phnom Penh, Chau Doc, Tan Chau). In addition, evidence suggests that

Fig. 2.4. Cyanobacterial bloom in a freshwater body.

agricultural activities account for more than 50% of nitrogen fluxes and more than 70% of phosphorus fluxes, despite the fact that most farms employ presumably less impactful traditional methods (Liljeström *et al.*, 2012). In general, eutrophication of river basins associated with agricultural activities is likely to intensify as commercial farming activities and the implicit application of fertilizers are expected to increase in the coming years (Liljeström *et al.*, 2012).

Metal pollution

Metal pollution is a long-standing environmental problem affecting freshwater ecosystems in SE Asia (Polprasert, 1982; Vicente-Beckett *et al.*, 1991; Vicente-Beckett, 1992; David, 2003). The two main sources of metal pollution are industrial and mining activities (David, 2003; Murphy *et al.*, 2009). While pollutants from industry commonly enter freshwater systems via effluents, mines pollute fresh waters through a combination of active effluent disposal as well as passive runoffs.

The Chao Phraya River is an example of an SE Asian water body with a history of metal pollution (e.g. copper, chromium) dating back to the 1970s (Menasveta, 1978; Polprasert, 1982). This coincided with the proliferation of non-registered industries along the river which were responsible for releasing untreated effluent containing more than 1 mg per litre of heavy metals (Polprasert, 1982). Metal pollution has persistent effects and can impact freshwater ecosystems even after the primary pollutant source is no longer active. For example, copper mining in the catchment of the Boac–Makulapnit River in the Philippines caused the influx of high concentrations of copper, even after the mine was closed (David, 2003).

Like most anthropogenic impacts, metal pollution acts synergistically with other forms of human disturbance. Deforestation of water catchments exemplifies this, and is often shown to exacerbate pollutant runoff (Benito *et al.*, 2003). The potent combination of deforestation and pollutant runoff is evident in the high concentrations of arsenic and mercury in groundwater around the largely denuded Mekong River delta (Murphy *et al.*, 2009).

Levels of metal pollutants may fluctuate between seasons, however, the exact mechanisms involved are not currently clear. In the Tonlé Sap and Bassac River of Cambodia where metal pollutants are mainly discharged as sewage, levels of heavy metals such as chromium, manganese and iron were higher in the wet season; while arsenic, selenium, barium and lead were higher in the dry season (Chanpiwat and Sthiannopkao, 2014). Conversely, in the Citarum

River of Indonesia, metal pollutants enter the water via bank erosions, with concentrations of titanium, zinc and silver elevated during the wet seasons while those of manganese and cobalt were elevated in the dry seasons (Chanpiwat and Sthiannopkao, 2014).

Organic pollution

Organic pollutants are present in relatively high concentrations in SE Asian rivers (Minh *et al.*, 2007; Kunacheva *et al.*, 2009; Huang *et al.*, 2015). In the Mekong River delta, organic pollutants, including DDT (dichlorodiphenyltrichoroethane), PCBs (polychlorinated biphenyls), CHLs (chlordanes), HCHs (hexachlorocyclohexanes) and HCB (hexachlorobenzene), are persistent problems, especially near urban complexes (Minh *et al.*, 2007). While our understanding of trends in organic pollutants across the region remains incomplete, a rare survey across Malaysian rivers suggested that BOD (biochemical oxygen demand) and SS (suspended solids) pollution were most prevalent (Huang *et al.*, 2015). The main source of SS loading was earthworks and land clearing, while sources of pollutants contributing to high BOD were agriculture-based industries and livestock farming (Huang *et al.*, 2015). Among the rivers surveyed, the Klang River recorded the highest levels of pollutant input at 142 tonnes of BOD and 360 tonnes of SS, daily.

Microplastic pollution

Microplastic pollution in fresh waters is an emerging environmental issue. However, there is little information on the impacts of microplastic on the ecology and biodiversity of the inland waters (Blettler *et al.*, 2018). In one of the few published studies, high concentrations of polyester fibres (172,000 to 519,000 items/m^3) and fragments (10 to 223 items/m^3) were recorded from the Saigon River in Vietnam (Lahens *et al.*, 2018). These were attributed to textile and plastic industries located along the river's catchment. Of the top 20 countries ranked by mass of mismanaged plastic, four (Indonesia, the Philippines, Vietnam and Malaysia) are in SE Asia (Jambeck *et al.*, 2015). Given that the ubiquitous use of plastic products is likely to continue in the foreseeable future, microplastic pollution represents a poorly understood threat to freshwater ecosystems which clearly requires more attention than it is currently receiving.

2.3.3 Biological invasion

Biological invasions are a major driver of species loss, contributing to about 54% of recorded extinctions (Clavero and Garcia-Berthou, 2005). Non-native and invasive taxa occuring in SE Asian fresh waters include amphibians (Ng and Yeo, 2012), molluscs (Ng *et al.*, 2016a, 2018), fishes (Liew *et al.*, 2014; Ng, H.H. *et al.*, 2014), crustaceans (Zeng and Yeo, 2018) and plants (Peh, 2010). These include several of the world's 100 most invasive species: water hyacinth (*Eichhornia crassipes*); golden apple snail (*Pomacea canaliculata*); American bullfrog (*Lithobates catesbeianus*); common carp (*Cyprinus carpio*); Mozambique tilapia (*Oreochromis mossambicus*); walking catfish (*Clarias batrachus*); western mosquitofish (*Gambusia affinis*); and the red-eared slider (*Trachemys scripta elegans*) (Lowe *et al.*, 2000; GISD, 2019). A significant percentage of alien taxa in the region is thought to originate from aquaculture (Arthur *et al.*, 2010) or the ornamental pet trade (Liew *et al.*, 2012, 2014; Ng *et al.*, 2016a).

Although species extirpations in SE Asia have not been directly attributed to biological invasions (Canonico *et al.*, 2005; Köhler *et al.*, 2012; Zieritz *et al.*, 2016), non-native species can nevertheless have substantial impacts on native biodiversity. One of the most direct mechanisms by which biological invasions impact freshwater biodiversity is through predation. The Nile perch (*Lates niloticus*) is a notable freshwater example of a highly predatory alien species. Since invading Lake Victoria in East Africa, the invasive fish drove nearly 200 species of haplochromine cichlids to extinction (Goldschmidt *et al.*, 1993). While SE Asia has so far been spared similarly destructive invasion events, several voracious alien predators are nevertheless already in the region's fresh waters. These include the assassin snail (Ng *et al.*, 2016b), peacock cichlids (Ng and Tan, 2010; Liew *et al.*, 2012), the flowerhorn cichlid (Herder *et al.*, 2012) and the African sharptooth catfish (Ng, H.H. *et al.*, 2014). Besides directly affecting population sizes of their native prey, predatory invasive species can also impact freshwater fauna indirectly – by altering food web structures (Vander Zanden *et al.*, 1999). This causes the effects of predation to cascade across trophic levels (Goldschmidt *et al.*, 1993; Walsh *et al.*, 2016), meaning that the predative influences of invasive species may be more extensive than expected.

Biological invasion can also impact native communities via competition, or more specifically, competitive displacement. This usually happens when two or more species have significantly overlapping ecological niches (DeBach, 1966). While competitive interactions are relatively common in many freshwater taxa, invasive taxa can sometimes drive native species to extirpation/extinction. For example, the non-native Chinese pond mussel (*Sinanodonta woodiana*) has largely replaced native unionids in Peninsular Malaysia and Borneo (Zieritz *et al.*, 2016, 2018), while globally invasive South American apple snails (*Pomacea* spp.) are rapidly displacing their native counterparts across Indonesia and the Malay Peninsula (Marwoto *et al.*, 2011; Ng *et al.*, 2019). There are also several examples of competitive displacement in freshwater vertebrates. Notably, introduced tilapia are displacing native cyprinid fishes in many parts of the Philippines (Canonico *et al.*, 2005) while in Singapore, the African sharptooth catfish (*Clarias gariepinus*) has driven the decline of at least one native congener from habitats where it was previously dominant (Ng, H.H. *et al.*, 2014). The competitive successes of invasive species are sometimes attributed to advantages conferred by life history traits. For instance, *C. gariepinus* grows faster, attains a larger maximum size and feeds more extensively across the water column compared with its native counterpart (Ng, H.H. *et al.*, 2014). Similarly, the introduced redclaw crayfish (*Cherax quadricarinatus*), which grows faster and larger than native freshwater crab analogues, likely outcompetes the latter for shelter (Zeng *et al.*, 2019). In yet another example, invasive South American apple snails, *Pomacea* spp., are more tolerant of a wide range of environmental conditions, are less selective of food resources and feed at higher rates relative to native apple snails (Chaichana and Sumpan, 2014, 2015; Ng, 2016; Ng *et al.*, 2019).

Hybridization is often overlooked when studying the impacts of biological invasions. While the process is thought to be an important contributor to the evolutionary process in natural systems (Cui *et al.*, 2013), hybridization between introduced and native species in invaded communities can have long-term ecological and genetic consequences on native populations and communities (Perry *et al.*, 2001a; Fitzpatrick *et al.*, 2010).

Alien–native hybridization has been shown to produce invaders that are superior to parental species in their survival, growth rates, reproductive capabilities or resource acquisition (Perry *et al.*, 2001a,b, 2002; Fitzpatrick *et al.*, 2010). This is because hybridization increases genetic variation and novel gene expressions, masks or removes deleterious loci and facilitates the transference of advantageous alleles (Lee, 2002). Consequences of invader–native hybridization are wide ranging and may include:

1. Creation of sterile hybrids which can compete with native communities for limited resources, causing a depletion of rare native gametes (Parker *et al.*, 1999);
2. Establishment of self-propagating native–invader hybrids capable of forming new biological entities as a result of post-invasion reproductive isolation (Mooney and Cleland, 2001; Lee, 2002); and
3. Formation of hybrid swarms capable of backcrossing with parental species, leading to the widespread contamination of native genotypes through genetic introgression and potentially the virtual extinction of native species (Parker *et al.*, 1999; Perry *et al.*, 2001a; D'Amato *et al.*, 2007).

Hybrid invasions have been documented in several key freshwater taxa, including fishes, decapod crustaceans and molluscs (Perry *et al.*, 2001a,b, 2002; Na-Nakorn *et al.*, 2004; D'Amato *et al.*, 2007). Notable examples include hybridization between introduced rainbow trout (*Oncorhynchus mykiss*) and native cutthroat trout (*Oncorhynchus clarkii*) in the western USA, which resulted in severe native genetic diversity losses as well as the extirpation of at least one genetically pure cutthroat trout subspecies. Similarly, the introduction of rusty crayfish (*Faxonius rusticus*) into northern Wisconsin, USA, led to the widespread genetic assimilation and displacement of native northern clearwater crayfish (*Faxonius propinquus*). This led to a subsequent decline in native populations, as well as multiple instances of localized extirpations (Perry *et al.*, 2001a,b, 2002).

In SE Asia, the extent of invader–native hybridization and introgression is likely to be severely underestimated because these processes are often difficult to detect without the use of advanced molecular tools (Perry *et al.*, 2002; Fitzpatrick *et al.*, 2010). Nevertheless, there is evidence to suggest that the introduction of artificial hybrids between the native broadhead walking catfish (*Clarias macrocephalus*) and introduced African sharptooth catfish (*C. gariepinus*) significantly altered the genetic composition of *C. macrocephalus* in Thailand through introgression – with alleles peculiar to *C. gariepinus* observed in nearly half (12 of 25) of wild *C. macrocephalus* populations (Na-Nakorn *et al.*, 2004). Further, recent work has suggested that hybrid *Clarias* catfish may have become firmly established in many parts of Indochina, possibly driving declines in populations of the native Asian common walking catfish (*C. batrachus*) (Welcomme and Vidthayanon, 2003; B.W. Low, 2019, unpublished results).

2.3.4 Climate change: general discussion

Climate change is one of the most globally pervasive anthropogenic impacts (Scheffers *et al.*, 2016). The abiotic regime shifts resulting from climate change have many biological consequences. In freshwater ecosystems, these include shifting spawning times (Parmesan and Yohe, 2003), changes in relative biomass (Elliot, 2012) and alteration of sex ratios (Schwanz and Janzen, 2008).

The biological impacts of climate change are likely to be particularly damaging for threatened or vulnerable species with limited distributions (Section 2.2.3). Moreover, many tropical species are thought to be at, or close to, their thermal optima (Franco *et al.*, 2006; Deutsch *et al.*, 2008). This is compounded by the fact that SE Asia is among the geographical regions expected to experience the greatest increase in water temperature (van Vliet *et al.*, 2013). While some tropical species may benefit from higher mean temperatures, many more are likely to experience population declines (Moritz and Agudo, 2013). Moreover, species expected to thrive under projected climate conditions may be harmful to humans and freshwater biodiversity alike. For example, rising temperatures may favour the proliferation of toxin-producing cyanobacteria (Brasil *et al.*, 2015; Mowe *et al.*, 2015a,b).

Geographical constraints also contribute to the vulnerability of SE Asian biodiversity to climate change impacts by precluding range shifts as an adaptation strategy (Chen *et al.*, 2011). In most parts of insular SE Asia, species under thermal stress will not be able to migrate to higher latitudes or altitudes (Bickford *et al.*, 2010). Fully aquatic freshwater organisms are especially limited in their dispersal abilities given the natural disconnect between inland water bodies. Moreover, anthropogenic barriers (e.g. dams) and expected decreases in stream flow (van Vliet *et al.*, 2013) can aggravate the existing disconnect, making movement across suitable habitats all but impossible for a large proportion of freshwater organisms.

In addition to increasing temperatures, climate change may alter precipitation patterns (Knapp *et al.*, 2008; Trenberth, 2011). Notably, the timings of monsoonal wind patterns are likely to shift, or become increasingly erratic (Arnell and Gosling, 2013; Loo *et al.*, 2015). Changes in the duration and intervality of precipitation patterns can have substantial impacts on the life cycles of aquatic organisms which rely on wet–dry season dynamics as environmental cues for biological processes (see Section 2.2.1). Moreover, extreme precipitation patterns – including floods (Hirobayashi *et al.*, 2013) and droughts (Dai, 2010; Wilkinson *et al.*, 2019) – may become common occurrences. For example, 52 out of 63 provinces in Vietnam were impacted by droughts in 2015 (FAO, 2016a) while floods in Myanmar in the same year affected at least 400,000 people (FAO, 2016b). Both phenomena can have substantial albeit contrasting consequences on freshwater ecosystems. While droughts can cause significant decreases in water levels and potential drying out of smaller water bodies, floods increase terrestrial pollutant input (Loeb *et al.*, 2007). Similarly, the intensity of anthropogenic activity is likely to exacerbate the effects of frequent droughts and floods. For example, high population densities may necessitate unsustainable levels of water extraction to sustain agricultural and potable water needs (Welcomme *et al.*, 2016). High densities of urban or agricultural infrastructure are also likely to aggravate pollution runoff in flood events (Loeb *et al.*, 2007). This contrasts with natural seasonal floods which are beneficial and possibly essential to some freshwater ecosystems (see Section 2.2.2).

Climate change can also impact freshwater ecosystems in less apparent ways. For instance, rising water temperatures are expected to increase microbial metabolism, thus elevating rates of organic matter decomposition (Hamilton, 2010). This may alter the relative composition of terrestrial and aquatic matter in the aquatic carbon pool at the bottom of aquatic food webs, influencing their overall structure (Liew *et al.*, 2018a). In addition to aerobic decomposition, anaerobic processes, including methanogenesis, are also likely to be amplified at higher temperatures (Fey *et al.*, 2004; Conrad *et al.*, 2009). A possible outcome of this change is an increase in the production of methane, a greenhouse gas. This is worrying because inland waters are a larger contributor to atmospheric carbon than previous estimates suggest (Raymond *et al.*, 2013). Fortunately, higher rates of methane production may not necessarily increase rates of emission as the latter is a function of interactions between several processes (e.g. fermentation, methanogenesis) (Conrad *et al.*, 2009; Hamilton, 2010).

Rising sea levels is another highly disruptive consequence of global climate change, albeit one that is more commonly discussed in relation to coastal (Nicholls and Cazenave, 2010) or terrestrial (Menon *et al.*, 2010) ecosystems. Despite that, low-lying fresh waters are vulnerable to being overtaken by encroaching shorelines. Prominent examples of ecosystems threatened by sea-level rise include the species-rich Irrawady, Salween and Mekong River deltas (Yusuf and Francisco, 2009; Salmivaara *et al.*, 2013). Even in less extreme scenarios, seawater ingress can still lead to the salinization of coastal rivers (Canedo-Arguelles *et al.*, 2013). This process may trigger a wide spectrum of biotic and abiotic changes (Hamilton, 2010), most likely involving the replacement of freshwater obligates with halophilic species. However, the fate of coastal rivers remains uncertain because saltwater intrusion could instead be balanced by increased river discharge resulting from intense precipitation events expected to occur with the changing climate (Day *et al.*, 2008; Loo *et al.*, 2015).

2.3.5 Climate change: peat swamps (a case study)

Few tropical freshwater ecosystems merit discussion in climate change science more than freshwater swamps and peat swamps. These are sometimes considered 'keystone' habitats (Dudgeon, 2000) because peat swamps especially are important stores of organic carbon (Rieley *et al.*, 1996). The significance of peat swamp carbon stores is exemplified by forest fires attributed to the 1997 El Niño event, during which the total amount of carbon released to the atmosphere from Indonesia alone was equivalent to 13–40% of mean annual emissions from fossil fuels (Page *et al.*, 2002).

About 62% of the world's peat swamps are in SE Asia – in lowlands across Peninsular Malaysia, Borneo, Sumatra and West Papua (Rieley *et al.*, 1996; Page *et al.*, 2009). Current estimates suggest that approximately 70 gigatonnes of carbon are locked in the region's peat swamps (Page *et al.*, 2011). In addition to their value as stores of organic carbon, freshwater swamps/peat swamps and the blackwater streams which drain them are also associated with a diverse and unique suite of freshwater taxa, including many stenotopic and endangered species (Ng *et al.*, 1994; Dudgeon, 2000) (Fig. 2.5).

Worryingly, peat swamps are disappearing at an alarming rate, with approximately 54,000 km^2 lost to deforestation between 1990 and 2010 (Miettinen *et al.*, 2011). Although deforestation has long been recognized as a contributor to global climate change (Bala *et al.*, 2007), the destruction of peat swamps is especially impactful because it is accompanied by a significant increase in greenhouse gas emissions from rapidly decomposing peat layers (Hooijer *et al.*, 2010). Moreover, deforestation also increases the susceptibility of peat material to 'deep fires' which are highly destructive to vegetation, difficult to extinguish

Fig. 2.5. Typical blackwater habitat in a peat swamp forest.

and often release significant amounts of carbon to the atmosphere (Page *et al.*, 2009).

The destruction of peat swamp forests is motivated primarily by timber extraction and agricultural expansion (Miettinen *et al.*, 2011). In SE Asia, lucrative crops, especially oil palm (*Elaeis guineensis*), drive large-scale conversions of peat swamp forests into plantations (Koh and Ghazoul, 2010). Unfortunately, there is little legal deterrence. Oil palm production remains an important driver of economic growth in the region, particularly among rural communities (Feintrenie *et al.*, 2010). Data suggest that only a small percentage of extant peat swamps occurs in protected areas (Posa *et al.*, 2011) and even then, many are subject to illegal logging or agricultural conversion (Yule, 2010).

Peat swamp conservation is critical for mitigating climate change (Miettinen *et al.*, 2011). Unfortunately, the persistence of peat swamps is, itself, threatened by climate change. This is because projected changes in precipitation patterns are likely to prolong droughts, possibly increasing the extraction rates of already over-taxed groundwater resources (Green *et al.*, 2011). In turn, depleted groundwater levels increase the risk of wildfires in degraded peat forests (Wosten *et al.*, 2008). Drought-induced low water tables also affect intact peat swamp ecosystems by exacerbating carbon dioxide emissions (Jauhiainen *et al.*, 2005). Overall, the sequence of events creates a potentially dangerous feedback loop in which the intensity of global climate change and its drivers are progressively escalated.

2.4 Future of Tropical Freshwater Ecosystems

Freshwater ecosystems in SE Asia are likely to be under increasing anthropogenic pressure in the coming decades. Human populations are expected to grow across most of the region (United Nations, 2019) and the pressure from increasing demands for freshwater resources may not be sufficiently attenuated by regional and national policies (Butchart *et al.*, 2010; Tittensor *et al.*, 2014). Notably, the rate at which new dams are constructed is likely to be sustained, if not accelarated, in the coming years (Andrews-Speed *et al.*, 2016; Stone, 2016). For example, more than 100 new hydropower dams are expected to be constructed along the Mekong River (Zarfl *et al.*, 2015).

Unfortunately, escalating demands for fresh water will likely be aggravated by a diminishing supply (Vörösmarty *et al.*, 2010). This may motivate more aggressive approaches to freshwater management across the region, possibly mirroring current water policies in Singapore. The island city state – the region's most densely populated nation – completed the damming of all its natural rivers in 2013 in what was widely thought to be a significant measure for reducing water stress. However, the extensive modification of freshwater ecosystems is associated with several environmental issues. For instance, Singapore's reservoirs are dominated by non-native species (e.g. South American cichlid fishes) (Liew *et al.*, 2012, 2018a), which is unsurprising when considering the evidence linking habitat modification with the establishment of alien taxa (Johnson *et al.*, 2008; Liew *et al.*, 2016a, 2018b). If this is replicated across the region, a likely outcome is the decline or extirpation of riverine specialists concomitant with an increased alien presence (Liew *et al.*, 2016b). Moreover, projections of climate change scenarios suggest that environmental conditions in freshwater ecosystems will become less favourable to native species and more conducive to biological invasion in the coming decades (Zeng and Yeo, 2018). This means that biotic homogenization of freshwater ecosystems (Lambdon *et al.*, 2008; Lososova *et al.*, 2012) may be inevitable without significant management interventions.

An escalation of anthropogenic threats on freshwater ecosystems (Section 2.3) is likely to also have societal impacts. A major example is transboundary conflict for water resources (Hansson *et al.*, 2011) – an issue that is highly relevant in the Mekong River basin which spans across the national borders of six countries (five of which are in SE Asia). Here, up-river dams fuel disputes (Biba, 2012) because they threaten the energy (e.g. hydropower), food (e.g. water for irrigation) and water security (e.g. potable water) of down-river communities. The impacts of extensive damming are arguably most severe for populations living in the low-lying regions of the Mekong delta where reduced sedimentation can amplify the threat of rising sea levels (Kondolf *et al.*, 2014).

Despite this, innovative management initiatives can be used to optimize trade-offs between increasing freshwater resource needs and the preservation of inland water bodies. One possible course of action is the maintenance of natural elements in and around freshwater ecosystems. For example, natural buffer zones (Giam *et al.*, 2015) and natural stream beds (Muotka *et al.*, 2002) can benefit species diversity. Other promising initiatives include efforts to optimize the utility of infrastructure aimed at resource extraction. A notable example includes Singapore's plans to extract solar power from its reservoirs using floating solar panels (Ang, 2019). While there are no guarantees that these will be successful in mitigating anthropogenic impacts on fresh waters, they may still be critical for preventing the collapse of key ecosystem functions until more effective measures are in place.

2.5 Conclusions

Freshwater ecosystems of SE Asia are highly diverse but remain relatively under-studied in comparison with temperate systems. The following areas of research, in particular, are in pressing need of current scientific information. First, baseline biodiversity information needs to be ascertained or updated in large parts of the region, without which conservation actions cannot be optimized. Second, quantitative ecological studies which incorporate empirical data (e.g. food webs, population dynamics) are also required for the development of more precise models to describe important ecosystem processes (e.g. organic carbon cycles). Without both sets of information, measures for mitigating impacts on freshwater systems will necessarily depend on generalized assumptions made from region-agnostic data. Consequently, the effectiveness of efforts to conserve freshwater ecosystems may be hampered by the lack of specificity offered by predictive models (e.g. Lim *et al.*, 2018). The adoption of modern research tools, e.g. eDNA metabarcoding (Deiner *et al.*, 2017) and compound-specific isotope analysis (Liew *et al.*, 2019), can be especially useful for increasing the temporal and spatial scale of future attempts to increase scientific data coverage.

SE Asia's fresh waters are expected to be under increasing pressure from the region's rapidly growing human population. This will likely be exacerbated by the synergy between various anthropogenic impacts. For example, the projected effects of global climate change are likely to drive the intensification of habitat modification, exacerbate influx of pollutants and aggravate risks of biological invasions, among others. In some cases, anthropogenic impacts on fresh waters may even trigger feedback mechanisms in which the causes and consequences are mutually reinforcing. Despite this, the importance of freshwater resources means that human exploitation of inland waters cannot be avoided entirely. More realistically, mitigative measures should instead focus on preserving or incorporating natural elements in freshwater ecosystems. Given sufficient motivation, trade-offs between human and biodiversity needs can be optimized for greater long-term sustainability.

References

Abell, R. (2002) Conservation biology for the biodiversity crisis: a freshwater follow-up. *Conservation Biology* 16, 1435–1437.

Abell, R., Thieme, M.L., Reyenga, C., Bryer, M., Kottelat, M. *et al.* (2008) Freshwater ecoregions of the world: a new map of biogeographic units for freshwater biodiversity conservation. *BioScience* 58, 403–414.

Adorable-Asis, A.-G.A., Cauyan, G.A., Pagulayan, R.C., Magbanua, F.S. and Papa, R.D.S. (2016) The macrogastropod communities of aquaculture-intensive lakes in the Philippines. *Molluscan Research* 36, 223–230.

Akaishi, F., Satake, M., Otaki, M. and Tominaga, N. (2006) Surface water quality and information about the environment surrounding Inle Lake in Myanmar. *Limnology* 7(1), 57–62.

Allan, D., Erickson, D. and Fay, J. (2003) The influence of catchment land use on stream integrity across multiple spatial scales. *Freshwater Biology* 37, 149–161.

Amado, A.M. and Roland, F. (2017) Editorial: Microbial role in the carbon cycle in tropical inland aquatic ecosystems. *Frontiers in Microbiology* 8, 20.

Andrews-Speed, P., Qiu, M. and Len, C. (2016) Chinese engagement in Southeast Asian energy and mineral resources: motivations and outlook. *Eurasian Geography and Economics* 57(3), 316–342.

Ang, P. (2019) Floating solar panel systems at Bedok, *Lower Seletar reservoirs operational middle of 2020. The Straits Times*, 30 October. Available at: https://www.straitstimes.com/singapore/environment/floating-solar-panel-systems-at-bedok-lower-seletar-reservoirs-operational (accessed 10 March 2020).

Arnell, N.W. and Gosling, S.N. (2013) The impacts of climate change on river flow regimes at the global scale. *Journal of Hydrology* 486, 351–364.

Arthur, R.I., Lorenzen, K., Homekingkeo, P., Sidavong, K., Sengvilaikham, B. *et al.* (2010) Assessing impacts of introduced aquaculture species on native fish communities: Nile tilapia and makor carps in SE Asian freshwaters. *Aquaculture* 299(1–4), 81–88.

Attwood, S.W. (2010) Studies on the parasitology, phylogeography and the evolution of host–parasite interactions for the snail intermediate hosts of medically important trematode genera in Southeast Asia. In: Zhou, X.N., Bergquist, R., Olveda, R. and Utzinger, J. (eds) *Important Helminth Infections in Southeast Asia: Diversity and Potential for Control and Elimination, Part B*, 1st edn. Elsevier, London, pp. 405–440.

Bala, G., Caldeira, K., Wickett, M., Phillips, T.J., Lobell, D.B. *et al.* (2007) Combined climate and carbon-cycle effects of large-scale deforestation. *Proceedings of the National Academy of Sciences USA* 104(16), 6550–6555.

Balek, J. (1983) *Hydrology and Water Resources in Tropical Regions*. Elsevier, New York.

Balian, E.V., Leveque, C., Segers, H. and Martens, K. (eds) (2008) *Freshwater Animal Diversity Assessment. Hydrobiologia* 595. Springer, Dordrecht, the Netherlands.

Baran, E. and Myschowoda, C. (2009) Dams and fisheries in the Mekong Basin. *Aquatic Ecosystem Health and Management* 12(3), 227–234.

Batt, R.D., Carpenter, S.R., Cole, J., Pace, M.L., Cline, T.J. *et al.* (2012) Resources supporting the food web of a naturally productive lake. *Limnology and Oceanography* 57(5), 1443–1452.

Bellwood, P. (2004) The origins and dispersals of agricultural communities in Southeast Asia. In: Glover, I. and Belwood, P. (eds) *Southeast Asia: From Prehistory to History*. Routledge Curzon, London, pp. 21–40.

Benito, E., Santiago, J.L., de Blas, E. and Varela, M.E. (2003) Deforestation of water-repellent soils in Galicia (NW Spain): effects on surface runoff and erosion under simulated rainfall. *Earth Surface Processes and Landforms* 28(2), 145–155.

Biba, S. (2012) China's continuous dam-building on the Mekong River. *Journal of Contemporary Asia* 42, 603–628.

Bickford, D., Howard, S.D., Ng, D.J.J. and Sheridan, J.A. (2010) Impacts of climate change on the amphibians and reptiles of Southeast Asia. *Biodiversity and Conservation* 19(4), 1043–1062.

Bilton, D.T., Freeland, J.R. and Okamura, B. (2001) Dispersal in freshwater invertebrates. *Annual Review of Ecology and Systematics* 32, 159–181.

Blettler, M.C., Abrial, E., Khan, F.R., Sivri, N. and Espinola, L.A. (2018) Freshwater plastic pollution: recognizing research biases and identifying knowledge gaps. *Water Research* 143, 416–424.

Bogan, A.E. (2008) Global diversity of freshwater mussels (Mollusca, Bivalvia) in freshwater. In: Balian, E.V., Leveque, C., Segers, H. and Martens, K. (eds) *Freshwater Animal Diversity Assessment. Hydrobiologia 595*. Springer, Dordrecht, the Netherlands, pp. 139–147.

Bogan, A.E. and Do, V.T. (2014) Two freshwater bivalve species new to the fauna of Vietnam (Mollusca: Bivalvia: Arcidae and Unionidae). *Tropical Natural History* 14, 113–116.

Bolotov, I.N., Vikhrev, I.V., Kondakov, A.V., Konopleva, E.S., Gofarov, M.Y. *et al.* (2017) New taxa of freshwater mussels (Unionidae) from a species-rich but overlooked evolutionary hotspot in Southeast Asia. *Scientific Reports* 7, 11573.

Boxshall, G.A. and Defaye, D. (2008) Global diversity of copepods (Crustacea: Copepoda) in freshwater. In: Balian, E.V., Leveque, C., Segers, H. and Martens, K. (eds) *Freshwater Animal Diversity Assessment. Hydrobiologia 595*. Springer, Dordrecht, the Netherlands, pp. 195–207.

Brandt, R.A.M. (1974) The non-marine aquatic Mollusca of Thailand. *Archiv für Molluskenkunde* 105, 1–423.

Brasil, J., Attayde, J.L., Vasconcelos, F.R., Dantas, D.D.F. and Huszar, V.L.M. (2016) Drought-induced water-level reduction favors cyanobacteria blooms in tropical shallow lakes. *Hydrobiologia* 770, 145–164.

Brett, M.T., Bunn, S.E., Chandra, S., Galloway, A.W.E., Guo, F. *et al.* (2017) How important are terrestrial organic carbon inputs for secondary production in freshwater ecosystems? *Freshwater Biology* 62, 833–853.

Butchart, S.H.M., Walpole, M., Collen, B., van Strien, A., Scharlemann, J.P.W. *et al.* (2010) Global biodiversity: indicators of recent declines. *Science* 328, 1164–1168.

Canedo-Arguelles, M., Kefford, B.J., Piscart, C., Prat, N., Schafer, R.B. *et al.* (2013) Salinisation of rivers: an urgent ecological issue. *Environmental Pollution* 173, 157–167.

Canonico, G.C., Arthington, A., McCrary, J.K. and Thieme, M.L. (2005) The effects of introduced tilapias on native biodiversity. *Aquatic Conservation: Marine and Freshwater Ecosystems* 15, 463–483.

Carrasco, L.R., Papworth, S.K., Reed, J., Symes, W.S., Ickowitz, A. *et al.* (2016) Five challenges to reconcile agricultural land use and forest ecosystem services in Southeast Asia. *Conservation Biology* 30(5), 962–971.

Chaichana, R. and Sumpan, T. (2014) The potential ecological impact of the exotic snail *Pomacea canaliculata* on the Thai native snail *Pila scutata*. *ScienceAsia* 40, 11–15.

Chaichana, R. and Sumpan, T. (2015) Environmental tolerance of invasive golden apple snails (*Pomacea canaliculata* (Lamarck, 1822)) and Thai native apple snails (*Pila scutata* (Mousson, 1848)). *Tropical Ecology* 56(3), 347–355.

Chang, C.P. and Wang, Z. (2005) Annual cycle of Southeast Asia–Maritime continent rainfall and the asymmetric monsoon transition. *Journal of Climate* 18, 287–301.

Chanpiwat, P. and Sthiannopkao, S. (2013) Status of metal levels and their potential sources of contamination in Southeast Asian rivers. *Environmental Science and Pollution Research* 21(1), 220–233.

Chea, R., Grenouillet, G. and Lek, S. (2016) Evidence of water quality degradation in lower Mekong basin revealed by self-organizing map. *PLoS ONE* 11(1), e0145527.

Chen, I.C., Hill, J.K., Ohlemuller, R., Roy, D.B. and Thomas, C.D. (2011) Rapid range shifts of species associated with high levels of climate warming. *Science* 333, 1024–1026.

Clavero, M. and Garcia-Berthou, E. (2005) Invasive species are a leading cause of animal extinctions. *Trends in Ecology and Evolution* 20, 110.

Clements, R., Koh, L.P., Lee, T.M., Meier, R. and Li, D. (2006) Importance of reservoirs for the conservation of freshwater molluscs in a tropical urban landscape. *Biological Conservation* 128, 136–146.

Coates, D. (2002) Inland capture fishery statistics of Southeast Asia: current status and information needs. *RAP Publication No. 2002(1)*. Asia-Pacific Fishery Commission, Bangkok.

Coates, D., Poeu, O., Suptornaratana, U., Tung, N.T. and Virayong, S. (2003) Biodiversity and fisheries in the Lower Mekong Basin. *Mekong Development Series 2*. Mekong River Commission, Phnom Penh.

Conrad, R., Close, M. and Noll, M. (2009) Functional and structural response of the methanogenic microbial community in a rice field soil to temperature change. *Environmental Microbiology* 11(7), 1844–1853.

Crandall, K.A. and Buhay, J.E. (2008) Global diversity of crayfish (Astacidae, Cambaridae, and Parastacidae – Decapoda) in freshwater. In: Balian, E.V., Leveque, C., Segers, H. and Martens, K. (eds) *Freshwater Animal Diversity Assessment. Hydrobiologia 595*. Springer, Dordrecht, the Netherlands, pp. 295–301.

Cui, R., Schumer, M., Kruesi, K., Walter, R., Andolfatto, P. *et al.* (2013) Phylogenomics reveals extensive reticulate evolution in Xiphophorus fishes. *Evolution* 67, 2166–2179.

Dai, A. (2010) Drought under global warming: a review. *Wiley Interdisciplinary Reviews: Climate Change* 2(1), 45–65.

D'Amato, M.E., Esterhuyse, M.M., van der Waal, B.C.W., Brink, D. and Volckaert, F.A.M. (2007) Hybridization and phylogeography of the Mozambique tilapia *Oreochromis mossambicus* in southern Africa evidenced by mitochondrial and microsatellite DNA genotyping. *Conservation Genetics* 8, 475–488.

David, C.P.C. (2003) Establishing the impact of acid mine drainage through metal bioaccumulation and taxa richness of benthic insects in a tropical Asian stream (The Philippines). *Environmental Toxicology and Chemistry* 22(12), 2952–2959.

Davis, G.M. (1979) *The Origin and Evolution of the Gastropod Family Pomatiopsidae, with Emphasis on the Mekong River Triculinae*. The Academy of Natural Sciences of Philadelphia, Philadelphia, Pennsylvania.

Day, J.W., Christian, R.R., Boesch, D.M., Yanez-Arancibia, A., Morris, J. *et al.* (2008) Consequences of climate change on the ecogeomorphology of coastal wetlands. *Estuaries and Coasts* 31, 477–491.

DeBach, P. (1966) The competitive displacement and coexistence principles. *Annual Review of Entomology* 11(1), 183–212.

De Grave, S., Cai, Y. and Anker, A. (2008) Global diversity of shrimps (Crustacea: Decapoda: Caridea) in freshwater. In: Balian, E.V., Leveque, C., Segers, H. and Martens, K. (eds) *Freshwater Animal Diversity Assessment. Hydrobiologia 595*. Springer, Dordrecht, the Netherlands, pp. 287–293.

Deiner, K., Bik, H.M., Mächler, E., Seymour, M., Lacoursière-Roussel, A. *et al.* (2017) Environmental DNA metabarcoding: transforming how we survey animal and plant communities. *Molecular Ecology* 26(21), 5872–5895.

De Silva, S.S., Nguyen, T.T.T., Abery, N.W. and Amarasinghe, U.S. (2005) An evaluation of the role and impacts of alien finfish in Asian inland aquaculture. *Aquaculture Research* 37(1), 1–17.

Deutsch, C.A., Tewksbury, J.J., Huey, R.B., Sheldon, K.S., Ghalambor, C.K. *et al.* (2008) Impacts of climate warming on terrestrial ectorherms across latitude. *Proceedings of the National Academy of Sciences USA* 105(18), 6668–6672.

Do, V.T. (2015) Freshwater snails of Vietnam: diversity and conservation status. In: *Proceedings of the Sixth National Scientific Conference on Ecology and Biological Resources*. Institute of Ecology and Biological Resources, Hanoi, pp. 977–986.

Do, V.T., Tuan, L.Q. and Bogan A.E. (2018) Freshwater mussels (Bivalvia: Unionida) of Vietnam: diversity, distribution, and conservation status. *Freshwater Mollusk Biology and Conservation* 21, 1–18.

Dudgeon, D. (1992) Endangered ecosystems: a review of the conservation status of tropical Asian rivers. *Hydrobiologia* 248, 167–191.

Dudgeon, D. (2000) Riverine biodiversity in Asia: a challenge for conservation biology. *Hydrobiologia* 418, 1–13.

Dudgeon, D. (2003) The contribution of scientific information to the conservation and management of freshwater biodiversity in tropical Asia. *Hydrobiologia* 500, 295–314.

Dudgeon, D. (2005) River rehabilitation for conservation of fish biodiversity in monsoonal Asia. *Ecology and Society* 10, 15.

Dudgeon, D. (2014) Threats to freshwater biodiversity in a changing world. In: Freedman, B. (ed.) *Global Environmental Change*. Springer, Dordrecht, the Netherlands, pp. 243–s.

Dudgeon, D., Arthington, A.H., Gessner, M.O., Kawabata, Z.I., Knowler, D.J. *et al.* (2006) Freshwater biodiversity: importance, threats, status, and conservation challenges. *Biological Reviews* 81, 163–182.

Dudgeon, D., Cheung, F.K.W. and Mantel, S.K. (2010) Food-web structure in small streams: do we need different models for the tropics? *Journal of the North American Benthological Society* 29, 395–412.

Dugan, P., Dey, M.M. and Sugunan, V.V. (2006) Fisheries and water productivity in tropical river basins: enhancing food security and livelihoods by managing water for fish. *Agricultural Water Management* 80, 262–275.

Elliot, J.A. (2012) Is the future blue-green? A review of the current model predictions of how climate change could affect pelagic freshwater cyanobacteria. *Water Research* 46(5), 1364–1371.

FAO (Food and Agriculture Organization of the United Nations) (2016a) *'El Niño' Event in Viet Nam – Agriculture, Food Security and Livelihood Needs Assessment in Response to Drought and Salt Water Intrusion*. FAO, Rome.

FAO (Food and Agriculture Organization of the United Nations) (2016b) *Southeast Asia Flood Situation: Food Security Concerns for Large Numbers of People in Several Asian Countries Following Severe Localized Monsoon Floods*. FAO, Rome.

Facon, T. (2000) Water management in rice in Asia: some issues for the future. In: Papademetriou, M.K., Dent, F.J. and Herath, E.M. (eds) *Bridging the Rice Yield Gap in the Asia-Pacific Region*. Food and Agriculture Organization of the United Nations, Rome, pp. 178–198.

Feintrenie, L., Chong, W.K. and Levang, P. (2010) Why do farmers prefer oil palm? Lessons learnt from Bungo district, Indonesia. *Small-Scale Forestry* 9(3), 379–396.

Fernando, C.H. (1980) The species and size composition of tropical freshwater zooplankton with special reference to the Oriental region (South East Asia). *International Review of Hydrobiology* 65, 411–426.

Fey, A., Claus, P. and Conrad, R. (2004) Temporal change of ^{13}C-isotope signatures and methanogenic pathways in rice field soil incubated anoxically at different temperatures. *Geochimica et Cosmochimica Acta* 68, 293–306.

Fitzpatrick, B.M., Johnson, J.R., Kump, D.K., Smith, J.J., Voss, S.R. *et al.* (2010) Rapid spread of invasive genes into a threatened native species. *Proceedings of the National Academy of Sciences USA* 107, 3606–3610.

Forró, L., Korovchinsky, N.M., Kotov, A.A. and Petrusek, A. (2008) Global diversity of cladocerans (Cladocera; Crustacea) in freshwater. In: Balian, E.V., Leveque, C., Segers, H. and Martens, K. (eds) *Freshwater Animal Diversity Assessment. Hydrobiologia 595*. Springer, Dordrecht, the Netherlands, pp. 117–184.

Franco, A.M.A., Hill, J.K., Kitschke, C., Collingham, Y.C., Roy, D.B. *et al.* (2006) Impacts of climate warming and habitat loss on extinctions at species' low-latitude range boundaries. *Global Change Biology* 12(8), 1545–1553.

Galloway, A.W.E., Brett, M.T., Holtgrieve, G.W., Ward, E.J., Ballantyne, A.J. *et al.* (2015) A fatty acid based Bayesian approach for inferring diet in aquatic consumers. *PLoS ONE* 10(6), e0129723.

Giam, X., Koh, L.P., Tan, H.H., Miettinen, J., Tan, H.T.W. *et al.* (2012) Global extinctions of freshwater fishes follow peatland conversion in Sundaland. *Frontiers in Ecology and the Environment* 10, 465–470.

Giam, X., Hadiaty, R.K., Tan, H.H., Parenti, L.R., Wowor, D. *et al.* (2015) Mitigating the impact of oil-palm monoculture on freshwater fishes in Southeast Asia. *Conservation Biology* 29(5), 1357–1367.

GISD (Global Invasive Species Database) (2019) 100 of the world's worst invasive alien species. Available at: http://www.iucngisd.org/gisd/100_worst.php (accessed 2 June 2019).

Glaubrecht, M., von Rintelen, T. and Korniushin, A.V. (2003) Toward a systematic revision of brooding freshwater Corbiculidae in Southeast Asia (Bivalvia, Veneroida): on shell morphology, anatomy and molecular phylogenetics of endemic taxa from islands in Indonesia. *Malacologia* 45, 1–40.

Goldschmidt, T., Witte, F. and Wanink, J. (1993) Cascading effects of the introduced Nile perch on the detritovorous/phytoplanktivorous species in the sublittoral areas of Lake Victoria. *Conservation Biology* 7(3), 686–700.

Graf, D.L. (2013) Patterns of freshwater bivalve global diversity and the state of phylogenetic studies on the Unionoida, Sphaeriidae, and Cyrenidae. *American Malacological Bulletin* 31, 135–153.

Green, J., Corbet, S.A., Watts, E. and Lan, O.B. (1978) Ecological studies on Indonesian lakes. The montane lakes of Bali. *Journal of Zoology* 186(1), 15–38.

Green, T.R., Taniguchi, M., Kooi, H., Gurdak, J.J., Allen, D.M. *et al.* (2011) *Journal of Hydrology* 405(3–4), 532–560.

Hamilton, S.K. (2010) Biogeochemical implications of climate change for tropical rivers and floodplains. *Hydrobiologia* 657(1), 19–35.

Hansson, S., Hellberg, S. and Öjendal, J. (2011) Politics and development in a transboundary watershed: the case of the Lower Mekong Basin. In: Öjendal, J., Hansson, S. and Hellberg, S. (eds) *Politics and Development in a Transboundary Watershed*. Springer, Dordrecht, the Netherlands, pp. 1–18.

Hardin, G. (1960) The competitive exclusion principle. *Science* 131, 1292–1297.

Herder, F., Schliewen, U.K., Geiger, M.F., Hadiaty, R.K., Gray, S.M. *et al.* (2012) Alien invasion in Wallace's dreamponds: records of the hybridogenic 'flowerhorn' cichlid in Lake Matano, with an annotated checklist of fish species introduced to the Malili Lakes system in Sulawesi. *Aquatic Invasions* 7(4), 521–535.

Hirobayashi, Y., Mahendran, R., Koirala, S., Konoshima, L., Yamazaki, D. *et al.* (2013) Global flood risk under climate change. *Nature Climate Change* 3, 816–821.

Hishamunda, N., Ridler, N.B., Bueno, P. and Yah, W.G. (2009) Commercial aquaculture in Southeast Asia: some policy lessons. *Food Policy* 34(1), 102–107.

Holt, R.D. and Loreau, M. (2001) Biodiversity and ecosystem functioning: the role of trophic interactions and the importance of system openness. In: Kinzig, A.P., Pacala, S.W. and Tilman, D. (eds) *The Functional Consequences of Biodiversity. Empirical Progress and Theoretical Extensions*. Princeton University Press, Princeton, New Jersey, pp. 246–262.

Hooijer, A., Page, S.E., Canadell, J.G., Silvius, M., Kwadijk, J. *et al.* (2010) Current and future CO_2 emissions from drained peatlands in Southeast Asia. *Biogeosciences* 7(5), 1505–1514.

Huang, Y.F., Ang, S.Y., Lee, K.M. and Lee, T.S. (2015) Quality of water resources in Malaysia. In: Lee, T.S. (ed.) *Research and Practices in Water Quality*. InTechOpen. https://doi.org/10.5772/58969

Ismail, G.B., Sampson, D.B. and Noakes, D.L.G. (2014) The status of Lake Lanao endemic cyprinids (*Puntius* species) and their conservation. *Environmental Biology of Fishes* 97, 425–434.

Jambeck, J.R., Geyer, R., Wilcox, C., Siegler, T.R., Perryman, M. *et al.* (2015) Plastic waste inputs from land into the ocean. *Science* 347(6223), 768–771.

Jardine, T.D., Pusey, B.J., Hamilton, S.K., Pettit, N.E., Davies, P.M. *et al.* (2012) Fish mediate high food web connectivity in the lower reaches of a tropical floodplain river. *Oecologia* 168(3), 829–838.

Jardine, T.D., Bond, N.R., Burford, M.A., Kennard, M.J., Ward, D.P. *et al.* (2015) Does flood rhythm drive ecosystem responses in tropical riverscapes? *Ecology* 96(3), 684–692.

Jauhiainen, J., Takahashi, H., Heikkinen, J.E.P., Martikainen, P.J. and Vasander, H. (2005) Carbon fluxes from a tropical peat swamp forest floor. *Global Change Biology* 11(10), 1788–1797.

Johnson, P.T., Olden, J.D. and Vander Zanden, M.J. (2008) Dam invaders: impoundments facilitate biological invasions into freshwaters. *Frontiers in Ecology and the Environment* 6, 357–363.

Junk, W.J., Bayley, P.B. and Sparks, R.E. (1989) The flood pulse concept in river–floodplain systems. In: Dodge, D.P. (ed.) *Proceedings of the International Large River Symposium (Canadian Special Publication of Fisheries and Aquatic Sciences)*. Canadian Government Publishing Centre, Ottawa, pp. 110–127.

Keppeler, F.W., de Souza, A.C., Halwass, G., Begossi, A., de Almeida, M.C. *et al.* (2018) Ecological influences of human population size and distance to urban centres on fish communities in tropical lakes. *Aquatic Conservation: Marine and Freshwater Ecosystems* 28(5), 1030–1043.

Khuantrairong, T. and Traichaiyaporn, S. (2008) Diversity and seasonal succession of the phytoplankton community in Doi Tao Lake, Chiang Mai province, northern Thailand. *Tropical Natural History* 8(2), 143–156.

Kijviriya, V., Upatham, E.S., Viyanant, V. and Woodruff, D.S. (1991) Genetic studies of Asian clams, *Corbicula*, in Thailand: allozymes of 21 nominal species are identical. *American Malacological Bulletin* 8, 97–106.

Knapp, A.K., Beier, C., Briske, D.D., Classen, A.T., Luo, Y. *et al.* (2008) Consequences of more extreme precipitation regimes for terrestrial ecosystems. *BioScience* 58(9), 811–821.

Koh, L.P. and Ghazoul, J. (2010) Spatially explicit scenario analysis for reconciling agricultural expansion, forest protection, and carbon conservation in Indonesia. *Proceedings of the National Academy of Sciences USA* 107, 11140–11144.

Köhler, F., Seddon, M., Bogan, A.E., Do, V.T., Sri-Aroon, P. *et al.* (2012) The status and distribution of freshwater molluscs of the Indo-Burma region. In: Allen, D.J., Smith, K.G. and Darwall, W.R.T. (eds) *The Status and Distribution of Freshwater Biodiversity in Indo-Burma*. IUCN, Cambridge and Gland, Switzerland, pp. 66–89.

Kondolf, G.M., Rubin, Z.K. and Minear, J.T. (2014) Dams on the Mekong: cumulative sediment starvation. *Water Resources Research* 50(6), 5158–5169.

Kongim, B., Sutcharit, C. and Panha, S. (2015) Cytotaxonomy of unionid freshwater mussels (Unionoida, Unionidae) from northeastern Thailand with description of a new species. *ZooKeys* 514, 93–110.

Korovchinsky, N.M. (2013) Cladocera (Crustacea: Branchiopoda) of South East Asia: history of exploration, taxon richness and notes on zoogeography. *Journal of Limnology* 72, 109–124.

Kottelat, M. (2013) The fishes of the inland waters of Southeast Asia: a catalogue and core bibliography of the fishes known to occur in freshwaters, mangroves and estuaries. *Raffles Bulletin of Zoology, Supplement* 27, 1–663.

Kottelat, M. and Ng, P.K.L. (2005) Diagnoses of six new species of Parosphromenus (Teleostei: Osphronemidae) from Malay Peninsula and Borneo, with notes on other species. *Raffles Bulletin of Zoology, Supplement* 13, 101–113.

Kummu, M., de Moel, H., Ward, P.J. and Varis, O. (2011) How close do we live to water? A global analysis of population distance to freshwater bodies. *PLoS ONE* 6(6), e20578.

Kunacheva, C., Boontanon, S.K., Fujii, S., Tanaka, S., Musirat, C. *et al.* (2009) Contamination of perfluorinated compounds (PFCs) in Chao Phraya River and Bangpakong River, Thailand. *Water Science and Technology* 60(4), 975–982.

Lahens, L., Strady, E., Kieu-Le, T.C., Dris, R., Boukerma, K. *et al.* (2018) Macroplastic and microplastic contamination assessment of a tropical river (Saigon River, Vietnam) transversed by a developing megacity. *Environmental Pollution* 236, 661–671.

Lambdon, P.W., Lloret, F. and Hulme, P.E. (2008) Do non-native species invasions lead to biotic homogenisation at small scales? The similarity and functional diversity of habitats compared for alien and native components of Mediterranean floras. *Diversity and Distributions* 14(5), 774–785.

Lau, D.C.P., Goedkopp, W. and Vrede, T. (2013) Cross-ecosystem differences in lipid composition and growth limitation of a benthic generalist consumer. *Limnology and Oceanography* 58, 1149–1164.

Lee, C.E. (2002) Evolutionary genetics of invasive species. *Trends in Ecology and Evolution* 17, 386–391.

Lewis, W.M. Jr (1973) A limnological survey of Lake Mainit, Philippines. *Internationale Revue der gesamten Hydrobiologie und Hydrographie* 58(6), 801–818.

Lewis, W.M. Jr (1978) Dynamics and succession of the phytoplankton in a tropical lake: Lake Lanao, Philippines. *Journal of Ecology* 66, 849–880.

Li, A.O.Y. and Dudgeon, D. (2009) Shredders: species richness, abundance, and role in litter breakdown in tropical Hong Kong streams. *Journal of the North American Benthological Society* 28, 167–180.

Li, J., Dong, S., Liu, S., Yang, Z., Peng, M. *et al.* (2013) Effects of cascading hydropower dams on the composition, biomass and biological integrity of phytoplankton assemblages in the middle Lancang–Mekong River. *Ecological Engineering* 60, 316–324.

Liew, J.H., Tan, H.H. and Yeo, D.C.J. (2012) Some cichlid fishes recorded in Singapore. *Nature in Singapore* 5, 229–236.

Liew, J.H., Tan, H.H., Yi, Y. and Yeo, D.C.J. (2014) Ecology and origin of the introduced cichlid *Acarichthys heckelii* in Singapore' fresh waters – first instance of development. *Environmental Biology of Fishes* 97(10), 1109–1118.

Liew, J.H., Carrasco, L.R., Tan, H.H. and Yeo, D.C.J. (2016a) Native richness and species level trophic traits predict establishment of alien freshwater fishes. *Biological Invasions* 18(12), 3495–3512.

Liew, J.H., Tan, H.H. and Yeo, D.C.J. (2016b) Dammed rivers: impoundments facilitate fish invasions. *Freshwater Biology* 61(9), 1421–1429.

Liew, J.H., Jardine, T.D., Lim, R.B.H., Kwik, J.T.B., Tan, H.H. *et al.* (2018a) Bottom-up influences on tropical freshwater food web structure support the 'environmental filtering' hypothesis. *Limnology and Oceanography* 63(5), 1877–1890.

Liew, J.H., Giam, X., Clews, E., Tan, K.Y.W., Tan, H.H. *et al.* (2018b) Contrasting changes in freshwater fish assemblages and food webs follow modification of tropical waterways. *Ecology of Freshwater Fish* 27(4), 1114–1125.

Liew, J.H., Chua, K.W.J., Arsenault, E.R., Thorp, J.H., Suvarnaraksha, A. *et al.* (2019) Quantifying terrestrial carbon in freshwater food webs using amino acid isotope analysis – case study with an endemic cave fish. *Methods in Ecology and Evolution* 10, 1594–1605. https://doi.org/10.1111/2041-210X.13230

Liljeström, I., Kummu, M. and Varis, O. (2012) Nutrient balance assessment in the Mekong Basin: nitrogen and phosphorus dynamics in a catchment scale. *International Journal of Water Resources Development* 28(2), 373-391.

Lim, B.H.R., Liew, J.H., Kwik, J.T.B. and Yeo, D.C.J. (2018) Predicting food web responses to biomanipulation using Bayesian Belief Network: assessment of accuracy and applicability using *in-situ* exclosure experiments. *Ecological Modelling* 384, 308–315.

Little, D.C., Surintaraseree, P. and Innes-Taylor, N. (1996) Fish culture in rainfed rice fields of northeast Thailand. *Aquaculture* 140(4), 295–321.

Loeb, R., van Daalen, E., Lamers, L.P.M. and Roelofs, J.G.M. (2007) How soil characteristics and water quality influence the biogeochemical response to flooding in riverine wetlands. *Biogeochemistry* 85(3), 289–302.

Lohr, A.J., Rutger, S., Olaveson, M.M., Ivorra, N., Van Gustel, C.A.M. *et al.* (2006) Macroinvertebrate and algal communities in an extremely acidic river and the Kawah Ijen crater lake (pH <0.3), Indonesia. *Archiv für Hydrologie* 165(1), 1–21.

Loke, L.H.L. and Todd, P.A. (2016) Structural complexity and component type increase intertidal biodiversity independently of area. *Ecology* 97, 383–393.

Loo, Y.Y., Billa, L. and Singh, A. (2015) Effect of climate change on seasonal monsoon in Asia and its impact on the variability of monsoon rainfall in Southeast Asia. *Geoscience Frontiers* 6(6), 817–823.

Lososova, Z., Chytry, M., Tichy, L., Danihelka, J., Fajmon, K. *et al.* (2012) Biotic homogenization of Central European urban floras depends on residence time of alien species and habitat types. *Biological Conservation* 145(1), 179–184.

Lowe, S., Browne, M., Boudjelas, S. and De Poorter, M. (2000) *100 of the World's Worst Invasive Alien Species: A Selection from the Global Invasive Database*. The Invasive Species Specialist Group (ISSG), a specialist group of the Species Survival Commission (SSC) of the World Conservation Union (IUCN), Gland, Switzerland.

Lowe-McConnell, R.A. (1979) Ecological aspects of seasonality in fishes of tropical waters. *Symposia of the Zoological Society of London* 44, 219–241.

Lytle, D.A. and Poff, N.L. (2004) Adaptation to natural flow regimes. *Trends in Ecology and Evolution* 19, 94–100.

Maassen, W.J.M. (2001) A preliminary checklist of the non-marine molluscs of West Malaysia: 'a hand list'. *De Kreukel, Extra Editie* 2001, 1–161.

McQueen, D.J., Johannes, M.R.S., Post, J.R., Stewart, T.J. and Lean, D.R. (1989) Bottom-up and top-down impacts on freshwater pelagic community structure. *Ecological Monographs* 59, 289–309.

Martens, K., Schon, I., Meisch, C. and Horne, D.J. (2008) Global diversity of ostracods (Ostracoda, Crustacea) in freshwater. In: Balian, E.V., Leveque, C., Segers, H. and Martens, K. (eds) *Freshwater Animal Diversity Assessment. Hydrobiologia 595*. Springer, Dordrecht, the Netherlands, pp. 185–193.

Marwoto, R.M., Isnaningsih, N.R., Mujiono, N., Heryanto, Alfiah and Riena (2011) *Freshwater Snails of Java*

Island (Mollusca, Gastropoda). Pusat Penelitian Biologi – LIPI, Bogor, Indonesia. (in Indonesian)

Meesukko, C., Gajaseni, N., Peerapornpisa, Y.L. and Voinov, A. (2007) Relationships between seasonal variation and phytoplankton dynamics in Kaeng Krachan Reservoir, Phetchaburi Province, Thailand. *Tropical Natural History* 7(2), 131–143.

Meisner, J.D. and Shuter, B.J. (1992) Assessing potential effects of global climate change on tropical freshwater fishes. *GeoJournal* 28(1), 21–27.

Melcher, A.H., Ouedraogo, R. and Schmutz, S. (2012) Spatial and seasonal fish community patterns in impacted and protected semi-arid rivers of Burkina Faso. *Ecological Engineering* 48, 117–129.

Menasveta, P. (1978) Distribution of heavy metals in the Chao Phraya River estuary. In: Lohani, B.W. and Thanh, N.C. (eds) *Water Pollution Control in Developing Countries*. Asian Institute of Technology, Bangkok, pp. 123–145.

Menon, S., Soberon, J., Li, X. and Townsend, P.A. (2010) Preliminary global assessment of terrestrial biodiversity consequences of sea-level rise mediated by climate change. *Biodiversity and Conservation* 19(6), 1599–1609.

Miettinen, J., Shi, C. and Liew, S.C. (2011) Two decades of destruction in Southeast Asia's peat swamp forests. *Frontiers in Ecology and the Environment* 10(3), 124–128.

Minh, N.H., Minh, T.B., Kajiwara, N., Kunisue, T., Iwata, H. *et al.* (2007) Pollution sources and occurrences of selected persistent organic pollutants (POPs) in sediments of the Mekong River delta, South Vietnam. *Chemosphere* 67(9), 1794–1801.

Mooney, H.A. and Cleland, E.E. (2001) The evolutionary impact of invasive species. *Proceedings of the National Academy of Sciences USA*, 98, 5446–5451.

Moritz, C. and Agudo, R. (2013) The future of species under climate change: resilience or decline. *Science* 341, 504–508.

Mowe, M.A.D., Mitrovic, S.M., Lim, R.P., Furey, A. and Yeo, D.C.J. (2015a) Tropical cyanobacterial blooms: a review of prevalence, problem taxa, toxins and influencing environmental factors. *Journal of Limnology* 74(2), 205–224.

Mowe, M.A.D, Porojan, C., Abbas, F., Mitrovic, S.M., Lim, R.P. *et al.* (2015b) Rising temperatures may increase growth rates and microcystin production in tropical *Microcystis* species. *Harmful Algae* 50, 88–98.

MRC (Mekong River Commission) (2003) *State of the Basin Report 2003*. MRC, Phnom Penh.

Muotka, T., Paavola, R., Haapala, A., Novikmec, M. and Laasonen, P. (2002) Long-term recovery of stream habitat structure and benthic invertebrate communities from in-stream restoration. *Biological Conservation* 105(2), 243–253.

Murphy, T.P., Irvine, K.N., Sampson, M., Guo, J. and Parr, T. (2009) Mercury contamination along the Mekong River, Cambodia. *Asian Journal of Water, Environment and Pollution* 6(1), 1–9.

Myers, N., Mittermeier, C.G., Mittermeier, G.A., da Fonseca, G.A.B. and Kent, J. (2000) Biodiversity hotspots for conservation priorities. *Nature* 403, 853–858.

Nabhitabhata, J. (2009) *Checklist of Mollusca Fauna in Thailand*. Office of Natural Resources and Environmental Policy and Planning, Bangkok.

Na-Nakorn, U., Kamonrat, W. and Ngamsin, T. (2004) Genetic diversity of walking catfish, *Clarias macrocephalus*, in Thailand and evidence of genetic introgression from introduced farmed *C. gariepinus*. *Aquaculture* 240, 145–163.

Ng, H.H. and Tan, H.H. (2010) An annotated checklist of the non-native freshwater fish species in the reservoirs of Singapore. *COSMOS* 6(1), 95–116.

Ng, H.H., Low, B.W., Kwik, J.T.B. and Yeo, D.C.J. (2014) The tables are turned: an invasive species under potential threat. *Biological Invasions* 16, 1567–1571.

Ng, P.K.L. (1989) *Terrathelphusa*, a new genus of semiterrestrial freshwater crabs from Borneo and Java (Crustacea: Decapoda: Brachyura: Sundathelphusidae). *Raffles Bulletin of Zoology* 37, 116–131.

Ng, P.K.L. and Tan, H.H. (1997) Freshwater fishes of Southeast Asia: potential for the aquarium fish trade and conservation issues. *Aquarium Sciences and Conservation* 1(2), 79–90.

Ng, P.K.L., Tay, J.B. and Lim, K.K.P. (1994) Diversity and conservation of blackwater fishes in Peninsular Malaysia, particularly in the north Selangor peat swamp forest. *Hydrobiologia* 285, 203–218.

Ng, T.H. (2016) Introduction and impacts of freshwater gastropods in Singapore. PhD thesis, National University of Singapore, Singapore.

Ng, T.H. and Yeo, D.C.J. (2012) Non-indigenous frogs in Singapore. *Nature in Singapore* 5, 95–102.

Ng, T.H., Tan, S.K. and Low, M.E.Y. (2014a) Singapore Mollusca: 7. The family Ampullariidae (Gastropoda: Caenogastropoda: Ampullarioidea). *Nature in Singapore* 7, 31–47.

Ng, T.H., Tan, S.K. and Yeo, D.C.J. (2014b) The taxonomy, distribution and introduction history of the earliest reported alien freshwater mollusc in Singapore *Sinotaia guangdungensis* (Gastropoda: Viviparidae). *Malacologia* 57, 401–408

Ng, T.H., Kahar, R.S. and Marshall, D.J. (2015a) Preliminary checklist of the freshwater Gastropoda of Brunei. *Occasional Molluscan Papers* 4, 1–5.

Ng, T.H., Tan, S.K., Yeo, D.C.J. (2015b) Clarifying the identity of the long-established, globally-invasive *Physa acuta* Draparnaud, 1805 (Gastropoda: Physidae) in Singapore. *BioInvasions Records* 4, 189–194.

Ng, T.H., Liew, J.H., Song, J.Z.E. and Yeo, D.C.J. (2016a) First record of the cryptic invader *Pyrgophorus platyrachis* Thompson 1968 (Gastropoda: Truncatelloidea:

Cochliopidae) outside the Americas. *BioInvasions Record* 5(2), 75–80.
Ng, T.H., Tan, S.K., Wong, W.H., Meier, R., Chan, S.Y. *et al.* (2016b) Molluscs for sale: assessment of freshwater gastropods and bivalves in the ornamental pet trade. *PLoS ONE* 11(8), e0161130.
Ng, T.H., Foon, J.K., Tan, S.K., Chan, M.K.K. and Yeo, D.C.J. (2016c) First non-native establishment of the carnivorous assassin snail, *Anentome helena* (von dem Busch in Philippi, 1847). *BioInvasions Records* 5(3), 143–148.
Ng, T.H., Dulipat, J., Foon, J.K., Lopes-Lima, M., Zieritz, A. *et al.* (2017) A preliminary checklist of the freshwater snails of Sabah (Malaysian Borneo) deposited in the BORNEENSIS collection, Universiti Malaysia Sabah. *ZooKeys* 673, 105–123.
Ng, T.H., Limpanont, Y., Chusongsang, Y., Chusongsang, P. and Panha, S. (2018) Correcting misidentifications and first confirmation of the globally invasive *Physa acuta* Draparnaud, 1805 (Gastropoda: Physidae) in Thailand and Laos. *BioInvasions Record* 7(1), 15–19.
Ng, T.H., Tan, S.K., Ahmad, A., Do, V.T., Joshi, R.C. *et al.* (2019) Not in the least concern: anthropogenic influences on a South-east Asian apple snail *Pila scutata* (Ampullariidae). *Oryx* 53, 230–238.
Nguyen, T.T.T. and de Silva, S.S. (2006) Freshwater finfish biodiversity and conservation: an Asian perspective. *Biodiversity and Conservation* 15, 3543–3568.
Nicholls, R.J. and Cazenave, A. (2010) Sea level rise and its impact on coastal zones. *Science* 328, 1517–1520.
Nursuhayati, A.S., Yusoff, F.M. and Shariff, M. (2013) Spatial and temporal distribution of phytoplankton in perak estuary, Malaysia, during monsoon season. *Journal of Fisheries and Aquatic Sciences* 8(4), 480.
Okazumi, T., Tanaka, S., Kwak, Y., Shrestha, B.B. and Sagiura, A. (2014) Flood vulnerability assessment in the light of rice cultivation characteristics in the Mekong River flood plain in Cambodia. *Paddy and Water Environment* 12, S275–S286.
Paclibare, J.O. (2005) *National Aquaculture Sector Overview, Philippines*. Food and Agriculture Organization of the United Nations, Fisheries and Aquaculture Department, Rome.
Page, S.E., Hoscilo, A., Langner, A., Tansey, K., Siegert, F. *et al.* (2009) Tropical peatland fires in Southeast Asia. In: Cochrane, M. (ed.) *Tropical Fire Ecology*. Springer, Berlin, pp. 263–287.
Page, S.E., Rieley, J.O. and Banks, C.J. (2011) Global and regional importance of the tropical peatland carbon pool. *Global Change Biology* 17, 798–818.
Page, S.E., Siegert, F., Rieley, J.O., van Boehm, H.-D., Jaya, A. *et al.* (2002) The amount of carbon released from peat and forest fires in Indonesia during 1997. *Nature* 420, 61–65.
Pagulayan, R.C. (1995) Status of taxonomic studies on Philippine molluscs: with an update on two freshwater taxa of medical significance. *Acta Medica Philippina* 31, 39–44.
Papa, R.D.S. and Mamaril, A.C. Sr (2011) History of the biodiversity and limno-ecological studies on Lake Taal with notes on the current state of Philippine limnology. *Philippine Science Letters* 4(1), 1–10.
Park, J.-K. and Kim, W. (2003) Two *Corbicula* (Corbiculidae: Bivalvia) mitochondrial lineages are widely distributed in Asian freshwater environment. *Molecular Phylogenetics and Evolution* 29, 529–539.
Parker, I.M., Simberloff, D., Lonsdale, W.M., Goodell, K., Wonham, M. *et al.* (1999) Impact: toward a framework for understanding the ecological effects of invaders. *Biological Invasions* 1, 3–19.
Parmesan, C. and Yohe, G. (2003) A globally coherent fingerprint of climate change impacts on natural systems. *Nature* 412, 37–42.
Payne, A.I. (1986) *The Ecology of Tropical Lakes and Rivers*. Wiley, Chichester, UK.
Peh, K.S.H. (2010) Invasive species in Southeast Asia: the knowledge so far. *Biodiversity and Conservation* 19(4), 1083–1099.
Perry, W.L., Feder, J.L., Dwyer, G. and Lodge, D.M. (2001a) Hybrid zone dynamics and species replacement between *Orconectes* crayfishes in a northern Wisconsin lake. *Evolution* 55, 1153–1166.
Perry, W.L., Feder, J.L. and Lodge, D.M. (2001b) Implications of hybridization between introduced and resident *Orconectes* crayfishes. *Conservation Biology* 15, 1656–1666.
Perry, W.L., Lodge, D.M. and Feder, J.L. (2002) Importance of hybridization between indigenous and nonindigenous freshwater species: an overlooked threat to North American biodiversity. *Systematic Biology* 51, 255–275.
Polis, G.A., Anderson, W.B. and Holt, R.D. (1997) Toward an integration of landscape and food web ecology: the dynamics of spatially subsidized food webs. *Annual Review of Ecology, Evolution, and Systematics* 28, 289–316.
Polprasert, C. (1982) Heavy metal pollution in the Chao Phraya River estuary, Thailand. *Water Research* 16(6), 775–784.
Pongswat, S., Thammathaworn, S., Peerapornpisal, Y., Thanee, N. and Somsiri, C. (2004) Diversity of phytoplankton in the Rama IX Lake, a man-made lake, Pathumthani Province, Thailand. *ScienceAsia* 30(3), 261–267.
Posa, M.R.C., Wijedasa, L.S. and Corlett, R.T. (2011) Biodiversity and conservation of tropical peat swamp forests. *BioScience* 61(1), 49–57.
Quinn, J.W. and Kwak, T.J. (2013) Fish assemblage change in an Ozark River after impoundment. A long-term perspective. *Transactions of the American Fisheries Society* 132, 110–119.
Rahman, A., Pratiwi, N.T.M. and Hariyadi, S. (2016) Struktur Komunitas Fitoplankton di Danau Toba,

Sumatera Utara. *Jurnal ilmu pertanian Indonesia* 21(2), 120–127.

Rainboth, W.J., Vidthayanon, C. and Mai, D.Y. (2012) Fishes of the greater Mekong ecosystem with species list and photographic atlas. *Miscellaneous Publications No. 201*. Museum of Zoology, University of Michigan, Ann Arbor, Michigan.

Raymond, P.A., Hartmann, J., Lauerwald, R., Sobek, S., McDonald, C. *et al.* (2013) Global carbon dioxide emissions from inland waters. *Nature* 503, 355–359.

Redfern, S.K., Azzu, N. and Binamira, J.S. (2012) Rice in Southeast Asia: facing risks and vulnerabilities to respond to climate change. In: *Building Resilience for Adaptation to Climate Change in the Agriculture Sector. Proceedings of a Joint FAO/OECD Workshop, Rome, Italy, 23–24 April 2012*. Food and Agriculture Organization of the United Nations, Rome, pp. 295–314.

Reid, A.J., Carlson, A.K., Creed, I.F., Eliason, E.J., Gell, P.A. *et al.* (2019) Emerging threats and persistent conservation challenges for freshwater biodiversity. *Biological Reviews* 94(3), 849–873. https://doi.org/10.1111/brv.12480

Resh, V.H. (2007) Multinational, freshwater biomonitoring programs in the developing world: lessons from African and Southeast Asian river surveys. *Environmental Management* 39(5), 737–748.

Rieley, J.O., Ahmad-Shah, A.A. and Brady, M.A. (1996) The extent and nature of tropical peat swamps. In: Maltby, E., Immirzi, C.P. and Safford, R.J. (eds) *Tropical Lowland Peatlands of Southeast Asia, Proceedings of the Workshop on Integrated Planning and Management of Tropical Lowland Peatlands*. IUCN, Gland, Switzerland, pp. 17–53.

Roberts, T.R. (1989) The freshwater fishes of Western Borneo (Kalimantan Barat, Indonesia). *Memoirs of the California Academy of Sciences* 14, 1–210.

Rogers, D.C., Thimuangphol, W., Saengphan, N. and Sanoamuang, L. (2013) Current knowledge of the South East Asian large branchiopod Crustacea (Anostraca, Notostraca, Laevicaudata, Spinicaudata, Cyclestherida). *Journal of Limnology* 72, 69–80.

Sakset, A. and Chankaew, W. (2013) Phytoplankton as a bio-indicator of water quality in the freshwater fishing area of Pak Phanang River Basin (Southern Thailand). *Chiang Mai Journal of Science* 40(3), 344–355.

Salmivaara, A., Kummu, M., Keskinen, M. and Varis, O. (2013) Using global datasets to create environmental profiles for data-poor regions: a case from the Irrawaddy and Salween river basins. *Environmental Management* 51(4), 897–911.

Scheffers, B.R., De Meester, L., Bridge, T.C.L., Hoffmann, A.A., Pandolfi, J.M. *et al.* (2016) The broad footprint of climate change from genes to biomes to people. *Science* 354, aaf7671.

Schwanz, L.E.and Janzen, F.J. (2008) Climate change and temperature-dependent sex determination: can individual plasticity in nesting phenology prevent extreme sex ratios? *Physiological and Biochemical Zoology* 81(6), 826–824.

Smith, V.H., Tilman, G.D. and Nekola, J.C. (1999) Eutrophication: impacts of excess nutrient unputs on freshwater, marine, and terrestrial ecosystems. *Environmental Pollution* 100(1–3), 179–196.

Sodhi, N.S., Koh, L.P., Brook, B.W. and Ng, P.K.L. (2004) Southeast Asian biodiversity: an impending disaster. *Trends in Ecology and Evolution* 19, 654–714.

Sri-Aroon, P. (2010) *Freshwater Snails of Medical Importance in Thailand*. Faculty of Tropical Medicine, Mahidol University, Bangkok.

Stahl, M.O., Tarek, M.H., Yeo, D.C.J., Badruzzaman, A.B.M. and Harvey, C.F. (2014) Crab burrows as conduits for groundwater–surface water exchange in Bangladesh. *Geophysical Research Letters* 41, 8342–8347.

Stelbrink, B., von Rintelen, T., Albrecht, C., Clewing, C. and Naga, P.O. (2019) Forgotten for decades: Lake Lanao and the genetic assessment of its mollusc diversity. *Hydrobiologia* 843, 31–49. https://doi.org/10.1007/s10750-018-3666-0

Stone, R. (2016) Dam-building threatens Mekong fisheries. *Science* 354, 1084–1085.

Strong, E.E., Gargominy, O., Ponder, W.F. and Bouchet, P. (2008) Global diversity of gastropods (Gastropoda; Mollusca) in freshwater. In: Balian, E.V., Leveque, C., Segers, H. and Martens, K. (2008) *Freshwater Animal Diversity Assessment. Hydrobiologia 595*. Springer, Dordrecht, the Netherlands, pp. 149–166.

Sung, Y.H. and Fong, J.J. (2018) Assessing consumer trends and illegal activity by monitoring the online wildlife trade. *Biological Conservation* 227, 219–225.

Sweeney, B.W. and Newbold, J.D. (2014) Streamside forest buffer width needed to protect stream water quality, habitat, and organisms: a literature review. *JAWRA Journal of the American Water Resource Association* 50(3), 560–584.

Taipale, S.J., Brett, M.T., Pulkkinen, K. and Kainz, M.J. (2012) The influence of bacteria-dominated diets on *Daphnia magna* somatic growth, reproduction, and lipid composition. *FEMS Microbiology Ecology* 82, 50–62.

Tamayo-Zafaralla, M., Santos, R.A.V., Orozco, R.P. and Elegado, G.C.P. (2002) The ecological status of lake Laguna de Bay, Philippines. *Aquatic Ecosystem Health and Management* 5(2), 127–138.

Tan, H.H. and Ng, P.K.L. (2005) The fighting fishes (Teleostei: Osphronemidae: Genus *Betta*) of Singapore, Malaysia and Brunei. *Raffles Bulletin of Zoology, Supplement* 13, 43–99.

Tan, S.K., Chan, S.Y. and Clements, R. (2012) *A Guide to Snails and Other Non-Marine Mollusca of Singapore*. Singapore Science Centre, Singapore.

Te, S.H. and Gin, K.Y.H. (2011) The dynamics of cyanobacteria and microcystin production in a tropical reservoir of Singapore. *Harmful Algae* 10(3), 319–329.

Thompson, R.M., Brose, U., Dunne, J.A., Hall, R.O. Jr, Hladyz, S. *et al.* (2012) Food webs: reconciling the structure and function of biodiversity. *Trends in Ecology and Evolution* 27(12), 689–697.

Tiegs, S.D., Costello, D.M., Isken, M.W., Woodward, G., McIntyre, P.B. *et al.* (2019) Global patterns and drivers of ecosystem functioning in rivers and riparian zones. *Science Advances* 5, eaav0486.

Tittensor, D.P., Walpole, M., Hill, S.L.L., Boyce, D.G., Britten, G.L. *et al.* (2014) A mid-term analysis of progress toward international biodiversity targets. *Science* 346, 241–244.

Trenberth, K.E. (2011) Changes in precipitation with climate change. *Climate Research* 47, 123–138.

United Nations, Department of Economic and Social Affairs, Population Division (2019) World Population Prospects: The 2019 Revision. Available at: https://population.un.org/wpp/Graphs/Probabilistic/POP/TOT/920 (accessed 10 March 2020).

Vander Zanden, M.J., Casselman, J.M. and Rasmussen, J.B. (1999) Stable isotope evidence for the food web consequences of species invasions in lakes. *Nature* 401, 464–467.

van Vliet, M.T.H., Franssen, W.H.P., Yearsley, J.R., Ludwig, F., Haddeland, I., Lettenmaier, D.P. *et al.* (2013) Global river discharge and water temperature under climate change. *Global Environmental Change* 23, 450–464.

Vicente-Beckett, V.A. (1992) Trace metal levels and speciation in sediments of some Philippine natural waters. *Science of the Total Environment* 125, 345–357.

Vicente-Beckett, V.A., Pascual, C.B., Kwan, C.S. and Beckett, R. (1991) Levels and distribution of trace metals in sediments of Laguna Lake (Philippines) and its tributary rivers. *International Journal of Environmental Analytical Chemistry* 45(2), 101–116.

von Rintelen, T., von Rintelen, K. and Glaubrecht, M. (2010) The species flocks of the viviparous freshwater gastropod *Tylomelania* (Mollusca: Cerithioidea: Pachychilidae) in the ancient lakes of Sulawesi, Indonesia: the role of geography, trophic morphology and color as driving forces in adaptive radiation. In: Glaubrecht, M. (ed.) *Evolution in Action*. Springer, Berlin/Heidelberg, pp. 485–512.

von Rintelen, T., von Rintelen, K., Glaubrecht, M., Schubart, C.D. and Herder, F. (2012) Aquatic biodiversity hotspots in Wallacea: the species flocks in the ancient lakes of Sulawesi, Indonesia. In: Gower, D.J., Johnson, K., Richardson, J., Rosen, B., Ruber, L. and Williams, S. (eds) *Biotic Evolution and Environmental Change in Southeast Asia*. Cambridge University Press, Cambridge, UK, pp. 290–315.

Voris, H.K. (2000) Maps of Pleistocene sea levels in Southeast Asia: shorelines, river systems and time durations. *Journal of Biogeography* 27(5), 1153–1167.

Vörösmarty, C.J., McIntyre, P.B., Gessner, M.O., Dudgeon, D., Prusevich, A. *et al.* (2010) Global threats to human water security and river biodiversity. *Nature* 467, 555–561.

Walsh, J.R., Carpenter, S.R. and Vander Zanden, M.J. (2016) Invasive species triggers a massive loss of ecosystem services through a trophic cascade. *Proceedings of the National Academy of Sciences USA* 113(15), 4081–4085.

Wan Maznah, W.O. (2010) Perspectives on the use of algae as biological indicators for monitoring and protecting aquatic environments, with special reference to Malaysian freshwater ecosystems. *Tropical Life Sciences Research* 21(2), 51.

Wan Maznah, W.O. and Mansor, M. (2002) Aquatic pollution assessment based on attached diatom communities in the Pinang River Basin, Malaysia. *Hydrobiologia* 487(1), 229–241.

Welcomme, R.L. and Vidthayanon, C. (2003) The impacts of introductions and stocking of exotic species in the Mekong Basin and policies for their control. *MRC Technical Paper No. 9*. Mekong River Commission, Phnom Penh.

Welcomme, R.L., Baird, I.G., Dudgeon, D., Halls, A., Lamberts, D. *et al.* (2016) Fisheries of the rivers of Southeast Asia. In: Craig, J.F. (ed.) *Freshwater Fisheries Ecology*. Wiley, Chichester, UK, pp. 363–376.

Wilcove, D.S., Giam, X., Edwards, D.P., Fisher, B. and Koh, L.P. (2013) Navjot's nightmare revisited: logging, agriculture, and biodiversity in Southeast Asia. *Trends in Ecology and Evolution* 28, 531–540.

Wilkinson, C.L., Yeo, D.C.J., Tan, H.H., Fikri, A.H. and Ewers, R.M. (2018) The availability of freshwater fish resources is maintained across a land-use gradient in Sabah, Borneo. *Aquatic Conservation: Marine and Freshwater Ecosystems* 28(5), 1044–1054. https://doi.org/10.1002/aqc.2920

Wilkinson, C.L., Yeo, D.C.J., Tan, H.H., Fikri, A.H. and Ewers, R.M. (2019) Resilience of tropical, freshwater fish (*Nematabramis everetti*) populations to severe drought over a land-use gradient in Borneo. *Environmental Research Letters* 14, 045008.

Wosten, J.H.M., Clymans, E., Page, S.E., Rieley, J.O. and Limon, S.H. (2008) Peat-water interrelationships in a tropical peatland ecosystem in Southeast Asia. *CATENA* 73(2), 212–224.

WWF (2016) *Living Planet Report 2016: Risk and Resilience in the New Era*. WWF International, Gland, Switzerland.

Yap, S.Y. (2002) On the distributional patterns of Southeast–East Asian freshwater fishes and their history. *Journal of Biogeography* 29, 1187–1199.

Yeo, D.C.J. and Ng, P.K.L. (1998) Freshwater crabs of the *Potamon tannanti* species group (Crustacea, Decapoda, Brachyura, Potamidae) of northern Indochina. *Raffles Bulletin of Zoology* 46, 627–650.

Yeo, D.C.J., Ng, P.K.L., Cumberlidge, N., Magalhães, C., Daniels, S.R. and Campos, M.R. (2008) A global assessment of freshwater crab diversity (Crustacea:

Decapoda: Brachyura). In: Balian, E.V., Leveque, C., Segers, H. and Martens, K. (eds) *Freshwater Animal Diversity Assessment. Hydrobiologia 595*. Springer, Dordrecht, the Netherlands, pp. 275–286.

Youn, S-J., Taylor, W.W., Lynch, A.A., Cowx, I.G., Beard, D.J. *et al.* (2014) Inland capture fishery contributions to global food security and threats to their future. *Global Food Security* 3, 142–148.

Yue, G.H., Li, Y., Lim, L.C. and Orban, L. (2004) Monitoring the genetic diversity of three Asian arowana (*Scleropages formosus*) captive stocks using AFLP and microsatellites. *Aquaculture* 237, 89–102.

Yule, C.M. (2010) Loss of biodiversity and ecosystem functioning in Indo-Malayan peat swamp forests. *Biodiversity and Conservation* 19(2), 393–409.

Yule, C.M. and Yong, H.S. (2004) *Freshwater Invertebrates of the Malaysian Region*. Academy of Sciences Malaysia, Kuala Lumpur.

Yusuf, A.A. and Francisco, H.A. (2009) *Climate change vulnerability mapping for Southeast Asia.* Economy and Environmental Program for Southeast Asia, Singapore. Available at: http://hdl.handle.net/10625/46380 (accessed 10 March 2020).

Zakaria-Ismail, M. (1994) Zoogeography and biodiversity of the freshwater fishes of Southeast Asia. *Hydrobiologia* 285, 41–48. https://doi.org/10.1007/BF00005652

Zarfl, C., Lumsdon, A.E., Berlekamp, J., Tydecks, L. and Tockner, K. (2015) A global boom in hydropower dam construction. *Aquatic Sciences* 77, 161–170.

Zeng, Y. and Yeo, D.C.J. (2018) Assessing the aggregated risk of invasive crayfish and climate change to freshwater crabs: a Southeast Asian case study. *Biological Convervation* 223, 58–67.

Zeng, Y., Shakir, K.K. and Yeo, D.C.J. (2019) Competition between a native freshwater crab and an invasive crayfish in tropical Southeast Asia. *Biological Invasions* 21, 2653–2663. https://doi.org/10.1007/s10530-019-02009-6

Zhao, X., Zhou, Y., Min, J., Wang, S., Shi, W. *et al.* (2012) Nitrogen runoff dominates water nitrogen pollution from rice-wheat rotation in the Taihu Lake region of China. *Agriculture, Ecosystems, and Environment* 156, 1–11.

Zieritz, A., Bogan, A.E., Froufe, E., Klishko, O., Kondo, T. *et al.* (2018) Diversity, biogeography and conservation of freshwater mussels (Bivalvia: Unionida) in East and Southeast Asia. *Hydrobiologia* 810, 29–44.

Zieritz, A., Lopes-Lima, M., Bogan, A.E., Sousa, R., Walton, S. *et al.* (2016) Factors driving changes in freshwater mussel (Bivalvia, Unionida) diversity and distribution in Peninsular Malaysia. *Science of the Total Environment* 571, 1069–1078.

3 Biological Sequestrations of Atmospheric Carbon Dioxide with Strategies to Enhance Storage of the Gas

NAMITHA NAYAK, RAJESH MEHROTRA AND SANDHYA MEHROTRA*

Birla Institute of Technology and Science, Pilani, India

3.1 Introduction

Global warming is a pressing issue and concerns over its disastrous consequences have sparked interest to reduce the release of greenhouse gases. Carbon dioxide (CO_2) is the primary greenhouse gas that contributes extensively to global warming. Its concentration in the atmosphere has increased drastically in recent times (Rodhe, 1990). The present level of atmospheric CO_2 is estimated to be 408 ppm, which corresponds to an approximate 43% increase since the Industrial Revolution. At this rate, it is expected it will be 60% by the year 2100 (Kumar *et al.*, 2016, 2018). Several strategies have been developed to mitigate CO_2 and include: (i) improve existing technologies to produce more efficient fuels; (ii) use of alternative fuels such as biodiesel, biohydrocarbon, etc.; and (iii) sequestration of CO_2 (Bharti *et al.*, 2014). Further reductions of emissions will not give immediate results since CO_2 will remain for a long time in the atmosphere (Archer *et al.*, 2009). Therefore, to achieve the goal of mitigation, sequestration is necessary.

Sequestration strategies can be broadly classified into two categories: biological and non-biological. Non-biological sequestration is attained through physical and chemical reactions and does not involve any organism (Lal, 2008a). It includes oceanic sequestration, geological sequestration and chemical sequestration by mineral carbonation. These methods, however, have limitations concerning their efficiency and the cost involved (Seibel and Walsh, 2001; Schrag, 2007). In the case of abiotic oceanic sequestration, the marine biota might also be adversely affected (Nogia *et al.*, 2016).

Biological sequestration refers to the capture and storage of CO_2 through biological processes. In comparison to non-biological sequestration, biotic techniques are natural, cost-effective and immediately applicable (Lal, 2008a). Autotrophic organisms naturally trap CO_2 during photosynthesis, facilitated by enzymes such as ribulose bisphosphate carboxylase/oxygenase (RuBisCO), carbonic anhydrase and phosphoribulokinase (Figueroa *et al.*, 2008; Kumar *et al.*, 2018). Photosynthetic uptake of atmospheric CO_2 is largely responsible for global carbon cycling (Jansson *et al.*, 2010). About 123×10^9 tonnes is the annual flux of carbon between the atmosphere and land, which represents the gross primary productivity of the global terrestrial system. Of this, a large portion of captured carbon is eventually lost to the atmosphere through plant respiration, heterotrophic nutrition and microbial respiration. About 10×10^9 tonnes of the captured carbon persists in the ecosystem and is the net ecosystem productivity (NEP). Depending on the nature of the preserved product, this carbon can potentially remain fixed for centuries. However, if biotic stresses and other disturbances are accounted for, then the long-term

*sandhyam@goa.bits-pilani.ac.in

sequestered carbon is brought down to only a fraction (approximately 30%) of the NEP (Tuskan and Walsh, 2001; Lal, 2004, 2008b). About 9 × 10^9 tonnes of carbon is released annually into the atmosphere via human activities, of which 5 × 10^9 tonnes is absorbed by the terrestrial and aquatic systems (Jansson *et al.*, 2010). By enhancing this natural process, biological sequestration can be achieved. The two processes involved in biological sequestration are oceanic sequestration through fertilization and terrestrial sequestration (Fig. 3.1).

3.2 Oceanic Sequestration

Phytoplankton, despite amounting to less than 1% of the photosynthetic biomass, fix about 50% of the atmospheric CO_2 (Falkowski *et al.*, 2000). The organic carbon they produce is mostly consumed by herbivores in the ocean surface and ultimately released as CO_2 through respiration. A portion of the organic carbon escapes this process and sinks to the bottom of the ocean to elevate carbon levels in the deep sea (Chisholm *et al.*, 2006). Photosynthetic phytoplankton therefore act as biological pumps of CO_2. The oceanic fertilization strategy aims to boost this 'pump' by increasing the concentration of certain limiting nutrients in the ocean (Yang *et al.*, 2008). A long-standing belief is that nitrogen and phosphorus are mainly responsible for limiting the primary productivity of phytoplankton. Yet, in the Southern, the equatorial Pacific and the subarctic north-east Pacific Oceans, despite unexhausted nitrogen and phosphorus, the phytoplanktonic biomass is not as high as expected. Martin and Fitzwater (1988) suggested the lack of biologically available iron is the cause behind making the utilization of nitrogen and phosphorus impossible in high-nutrient low-chlorophyll (HNLC) regions. Furthermore, atmospheric dust from land is recognized as an important source of iron for the sea. According to the 'iron hypothesis', when excess iron is available to the biological pump more CO_2 is delivered to the deep sea. This hypothesis is supported by ice-core records which showed the anti-correlation between atmospheric CO_2 and dust concentration over the past 180,000 years (Sigman and Boyle, 2000). Martin and Fitzwater (1998) extended the iron hypothesis to imply that deliberate addition of iron to the ocean surface has the potential to increase deep-sea carbon storage (Chisholm and Morel, 1991). Four small-scale experiments were conducted wherein small amounts (100–250 kg) of iron were added to the waters of the equatorial Pacific and southern oceans, to increase iron from ambient concentrations of 0.06 nM to ~3–4 nM (Martin *et al.*, 1994). Productivity and biomass of phytoplankton were determined for a few days to weeks (Martin *et al.*, 1994; Coale *et al.*, 1996; Boyd *et al.*, 2000). In the Southern Ocean iron fertilization experiment, an algal bloom of over 200 km was recorded. However,

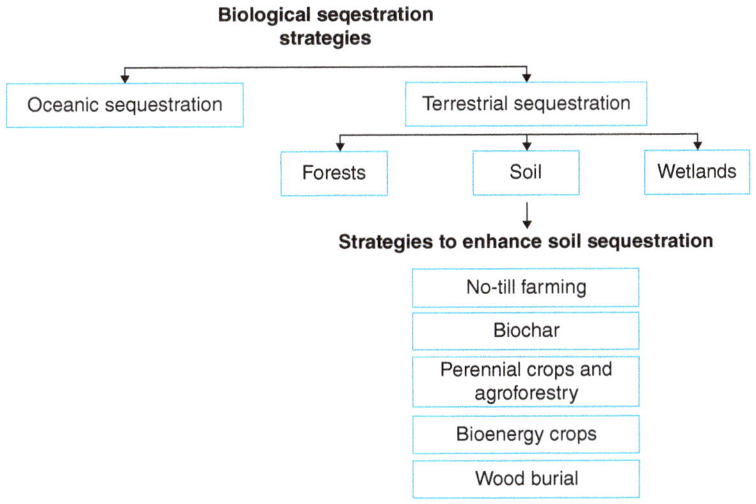

Fig. 3.1. Various strategies for biological carbon sequestration.

the process was not as efficient as predicted. According to the analysis, 1 tonne of iron forced about 1000 tonnes of carbon below 100 m. This was much less than the expected value of 100,000 tonnes (Dalton, 2002). These experiments could not verify CO_2 sequestration nor net transfer of CO_2 to the deep sea. If implemented on a larger scale, the ocean fertilization strategy may adversely affect ocean ecology. Models predict the occurrence of deep-ocean hypoxia or anoxia if sustained fertilization is carried out (Sarmiento and Orr, 1991). Phytoplankton bloom also increases the emission of other gases such as methyl bromide which depletes the ozone layer (Dalton, 2002). This strategy was not recommended because of less significant results and problems associated with its implementation.

3.3 Terrestrial Sequestration

Terrestrial sequestration is achieved through assimilation of atmospheric CO_2 into plant biomass (especially in regions of low turnover) as well as through allocation to deep roots for transfer into the soil organic carbon pool (Lal, 2008b). These are phytosequestration and soil carbon sequestration, respectively. Of the annual anthropogenically emitted CO_2, about 60% is removed by terrestrial carbon sinks through photosynthesis and storage in organic matter. Terrestrial sequestration also provides numerous ancillary benefits such as improved soil and water quality, increase in crop yield, restoration of degraded ecosystems, etc. Since it offers multiple benefits while effectively removing CO_2, this strategy is often regarded to be a 'win–win' strategy (Lal *et al.*, 2003). Terrestrial carbon sequestration involves three principal components: forests, soil and wetlands.

3.3.1 Forests

Carbon sequestered in forest ecosystems is harvestable timber, woody debris and wood products such as lignin and other resistant polymeric carbon compounds (Fan *et al.*, 1998; Wofsy, 2001). The net primary productivity (NPP) of terrestrial ecosystems is estimated to be saturated at about 800–1000 ppm of CO_2. Since the current CO_2 concentration has not reached the saturation point, the forest sink has the potential to increase by CO_2 fertilization (Falkowski *et al.*, 2000; Krishnamurthy and Machavaram, 2000). However, the NPP may still be limited by deficiency of nitrogen, phosphorus, water and other factors. Afforestation, the practice of planting trees on unforested land, is one of the viable options for carbon sequestration in terrestrial ecosystems. It has proven to be effective in different regions of the world; for example, in China, carbon sequestration in forests increased at an average rate of 21×10^{12} g/year through afforestation (Fang *et al.*, 2001). Yet large-scale afforestation is not recommended due to its negative impact on water resources and biodiversity of tropical forests (Jackson *et al.*, 2005).

3.3.2 Wetlands

Carbon sequestration in wetlands and peat soils has been estimated to be about 0.1×10^{15} g/year since the post-glaciation period, resulting in wetlands and associated soils constituting a large pedologic pool of carbon (Gorham, 1991; Kobak *et al.*, 1998). In fact, wetland soils may contain as much as 200 times more carbon than the associated vegetation (Milne and Brown, 1997; Garnett *et al.*, 2001). However, increasingly large areas of wetland (more than 50% in North America, Europe and Australia) have been drained due to anthropogenic activities such as agriculture, forestry, housing and transportation (Verhoeven and Setter, 2009). Drainage and subsequent cultivation of such ecosystems (e.g. agriculture, forestry) convert them into a net source of CO_2. To reverse this process, wetlands need to be restored and turned once more into a sink of atmospheric CO_2. A significant amount of time (a century or more) will be required before carbon is captured in a manner similar to that of natural wetlands (Lal, 2008a).

3.3.3 Soil

Plant roots translocate a portion of the total assimilated carbon (20–30% in cereals, 30–50% in pasture plants) to the soil, where it is stored in organic and inorganic forms (Kuzyakov and Domanski, 2000). Soil carbon sequestration involves restoring degraded or disturbed soil by enhancing pools of organic and inorganic carbon through adoption of management practices recommended for different kinds of ecosystems (Nogia *et al.*, 2016). The world's soil has the potential to re-sequester about 75–80% of the lost carbon provided ideal management practices are employed (Wojtkowski, 1998). The various methods proposed to enhance soil carbon sequestration are no-till farming, biochar, perennial crops and agroforestry, bioenergy crops and wood burial, described in the following sections.

No-till farming

Tillage refers to physical disturbance of the upper soil layers for seedbed preparation, weed control and other purposes. Depending on various factors such as climate, soil type, technology available, etc., tilling practices vary (Paustian *et al.*, 2000). The commonly used method is the most intensive type of tillage which includes inverting the soil (mouldboard plough) followed by a secondary tilling to homogenize the soil layer. This practice has a negative impact on the soil carbon pool. Tillage prevents formation of soil aggregates. While the dynamics between soil structure and soil organic matter (SOM) is not entirely known, it is widely accepted that soil aggregates increase the soil residence time of certain SOM fractions (Adu and Oades, 1978; Golchin *et al.*, 1994; Jastrow, 1996). Through tillage, new soil is continually exposed to freeze–thaw and wet–dry cycles at the surface which renders the soil aggregates vulnerable to disruption (Rovira and Greacen, 1957; Beare *et al.*, 1994). No-till or direct drilling is a tillage practice in which the seeds are directly sown into narrow slots made in the soil. This technique provides maximum protection to the soil profile while also reducing loss of soil carbon.

Biochar

Biochar is a black heterogeneous carbon product derived from the complete combustion of wood and other organic material. It contains both labile and recalcitrant aromatic compounds, depending on the ratio of which the biochar's half-life can vary from several hundred to several thousand years (Jansson *et al.*, 2010; Woolf *et al.*, 2010). This makes biochar an attractive option for long-term storage of carbon. In comparison to the carbon retained after burning (3%) or decomposition (10–20%) of biomass, conversion to biochar sequesters around 50% of the initial carbon. This yields more stable carbon in soil and reduces emission. Apart from sequestration, biochar also gives benefits such as improving soil fertility and crop production on application to soil (Jansson *et al.*, 2010).

Perennial crops and agroforestry

From the perspective of CO_2 mitigation, agricultural systems that include agroforestry and perennial cropping are desirable as they sequester CO_2 at higher rates (Toensmeier, 2018). The expanded root system of perennial crops helps store greater amounts of carbon in the root biomass thereby increasing the soil organic pool (Lal, 2008a). In addition to their sequestration potential, perennial crops also provide other benefits to the farm, farmer and the ecosystem in general. These benefits include reduced erosion, decreased nitrogen and nutrient leaching, increase in soil organic matter, etc. (Toensmeier, 2018). Perennial crops also eliminate the need for tillage which is again helpful in maintaining the soil organic carbon.

Bioenergy crops

A bioenergy crop is any crop (e.g. maize, wheat) that is grown for the purpose of producing energy but is generally characterized by its capability to produce large biomass and high energy potential. Of the world's total primary energy, 15% is supplied by bioenergy crops making them the fourth largest energy source in developing countries (Lemus and Lal, 2005). The advantage lies in decrease of atmospheric CO_2 enrichment due to cumulative effects of high biomass production. Bioenergy crops act as a link between sink (biomass and soil organic carbon) and source (fossil fuels). This is because the carbon incorporated into their biomass has a high possibility of entering the soil organic pool through the root system. The above-ground biomass is also utilized for energy production (Zan *et al.*, 2001). Since the CO_2 released during energy production was recently trapped from the atmosphere there will be no net increase in atmospheric CO_2 levels. Net gains in carbon sequestration, however, can only be achieved when bioenergy crops replace the annual crops (Lemus and Lal, 2005). The efficiency of bioenergy crops is determined by the species type considering factors such as productivity, fertilization and harvest management. At present, cereal crops such as maize, wheat, oil crops and sugarcane are being utilized as bioenergy crops.

Wood burial

In this technique, the CO_2 flux being captured from the atmosphere through photosynthesis and released back into the atmosphere through decomposition is hindered by cutting off the return pathway, thereby forming an effective sink (Zeng, 2008). It involves harvesting dead or live trees and stowing them in above-ground shelters or burying them in trenches so that decomposition is prevented because it is under anaerobic conditions. The burying of wood transfers it from a relatively fast carbon-decomposing pool (approximately 10 years) to a much slower one (100–1000 years or more). Wood burial is a sustainable strategy for CO_2 sequestration but theoretically to sequester all of the CO_2 emitted at present through wood burial and tree planting would require about 1 billion hectares of land area (Scholz and Hasse, 2008).

3.4 Plants as Carbon Sinks

As discussed in Section 3.3, plants are natural carbon sinks in two ways: by assimilation of atmospheric CO_2 and storage in above- and below-ground biomass, or by replacing fossil fuels through use of bioenergy crops (Fig. 3.2). Storage in above-ground biomass sequesters carbon for a relatively short period (decades to centuries), after which it is returned to the atmosphere upon plant decay. Long-term sequestration is achieved by transfer of above-ground biomass to roots and ultimately to the soil carbon pool. Fundamental to both these processes is photo-assimilation of atmospheric CO_2. Improvements in photosynthetic incorporation of CO_2 into plant biomass can effectively increase terrestrial sequestration. Of the several factors that affect biomass production in plants, the efficiency of solar energy conversion and light interception are the two most important (Jansson *et al.*, 2010). The Calvin cycle, in particular RuBisCO's carboxylation activity, plays a key role in conversion of solar energy to biomass.

3.4.1 Role of RuBisCO in biomass production

RuBisCO is the ubiquitous CO_2-fixing enzyme in plants and a key component of Calvin–Benson–Bassham (CBB) cycle. As the name suggests, RuBisCO has the capability of utilizing both CO_2 and O_2 as substrates because of the similarity in their molecular shapes. This phenomenon, however, reduces its catalytic efficiency (Whitney *et al.*, 2011b). Although CO_2 fixation remains the primary outcome, inadvertent fixing of O_2 (photorespiration) causes loss of fixed carbon and nitrogen (Price *et al.*, 2013). In evolutionary history, RuBisCO's oxygenation ability has not always been a problem. The enzyme evolved around 3.5 billion years ago when atmospheric oxygen (O_2) levels were very low and CO_2 concentration was many times higher than that observed at present (Badger and Price, 2003). Therefore, RuBisCO showed a much greater efficiency for carboxylation over oxygenation, until around 1.5 billion years ago when the O_2 in the atmosphere rose dramatically to the present levels. The decline in CO_2 led to the RuBisCO-catalysed step becoming rate limiting (Badger *et al.*, 1998; Badger and Price, 2003).

To adapt to the changed atmospheric conditions and enhance RuBisCO catalysis, various mechanisms evolved in different types of plants but can be broadly grouped as belonging to two kinds (Fig. 3.2). In the first approach the catalysis of RuBisCO is improved by increasing its affinity for CO_2 (Badger *et al.*, 1998). This strategy is employed by C3 plants. The improved RuBisCO can better discriminate between O_2 and CO_2, but the reaction rate decreases since the CO_2 level is much lower than that of O_2 (Tcherkez *et al.*, 2006). To achieve the same amount of CO_2 fixation, these plants require more molecules of RuBisCO. It has been estimated that in wheat plants up to 25% of the leaf nitrogen is invested in RuBisCO (Evans, 1989). In addition to improvements in RuBisCO, C3 plants have also adopted certain measures to decrease diffusive resistance to CO_2 which include pressing of chloroplasts against intracellular air spaces and increased chloroplast to leaf surface area (Evans and von Caemmerer, 1996). The second approach was developing a mechanism to concentrate inorganic carbon. Through biophysical or biochemical measures, a carbon-concentrating mechanism (CCM) accumulates inorganic carbon inside a photosynthesizing cell such that the RuBisCO enzyme attains saturation (Price *et al.*, 2013). This alternative strategy is used by algae, cyanobacteria as well as a subset of land plants, but the mechanism employed differs in each case (Badger *et al.*, 1998; Price *et al.*, 2013).

3.4.2 Cyanobacterial CCMs

Nearly 50% of the annual global primary productivity is contributed by cyanobacteria and other photosynthetic phytoplankton which act as major producers in the marine food web (Field *et al.*, 1998; Liu *et al.*, 1999). The diffusion of CO_2 in water is slower by 10^4 times. This puts great pressure on photosynthesis, leading to the development of CCMs in algal and cyanobacterial lineages (Price *et al.*, 2008). Cyanobacteria evolved an extremely efficient CCM that can bring about a 1000-fold increase in CO_2 concentration around RuBisCO (Badger and Price, 2003). Therefore, cyanobacterial

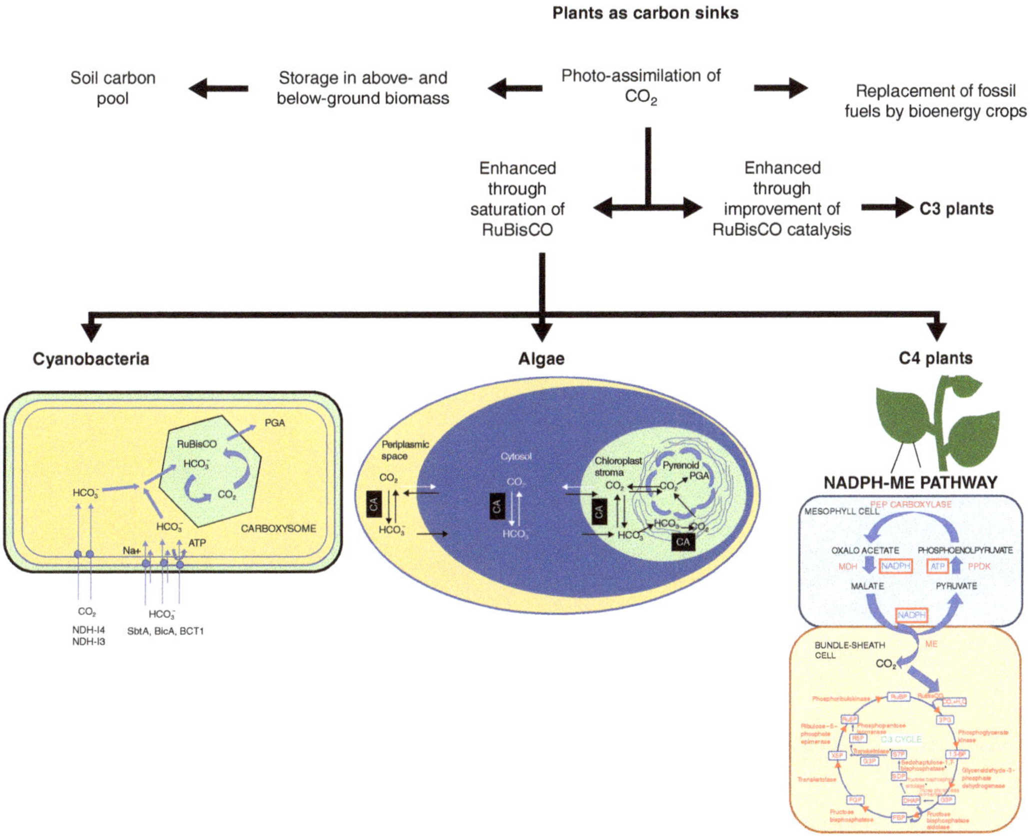

Fig. 3.2. Plants as natural carbon sinks: storage of photo-assimilated carbon into biomass and replacement of fossil fuels by bioenergy crops are the two ways in which plants can act as natural carbon sinks. Evolution has led to the development of two different strategies to enhance carbon assimilation in photosynthetic organisms: (i) saturation of RuBisCO enzyme with CO_2 achieved through carbon-concentrating mechanisms of cyanobacteria, algae and C4 plants; and (ii) improvement in catalysis of RuBisCO enzyme as observed in C3 plants. Abbreviation: PGA, phosphoglyceric acid; CA, carbonic anhydrase; PEP, phosphoenolpyruvate; MDH, malate dehydrogenase; PPDK, pyruvate phosphate dikinase; ME, malic enzyme; 3PG, 3-phosphoglycerate; 1,3-BP, 1,3-bisphosphoglycerate; G3P, glyceraldehyde-3-phosphate; FBP, fructose-1,6-bisphosphate; DHAP, dihydroxyacetone phosphate; SDP, sedoheptulose-1,7-bisphosphate; S7P, sedoheptulose-7-phosphate; F6P, fructose-6-phosphate; X5P, xylulose-5-phosphate; R5P, ribose-5-phosphate; Ru5P, ribulose-5-phosphate.

RuBisCO displays a high rate of carboxylation despite its lower CO_2 selective potential compared with C3 RuBisCO. Central to the functioning of cyanobacterial CCMs are carboxysomes, which are bacterial microcompartments enclosing the cyanobacterial RuBisCO and carbonic anhydrase enzymes. They are proteinaceous microbodies, 90–400 nm in diameter and icosahedral in shape, resembling viral capsids (Price *et al.*, 2013). The shell is composed of at least four different kinds of proteins which allow only the entry of polar molecules and restrict the passage of CO_2 and O_2. This property helps in prevention of CO_2 leakage as well as oxygenation activity of RuBisCO (Espie and Kimber, 2011). Cyanobacteria have two different kinds of carboxysomes (that emerged through parallel evolution) with two different types of RuBisCO. Alpha-carboxysomes are found in alpha-cyanobacteria and contain 1A type of RuBisCO, while beta-cyanobacteria contain beta-carboxysomes with 1B type of RuBisCO. Nevertheless, the two kinds of carboxysomes have the same function of sequestering and saturating RuBisCO with CO_2 produced by the action of carbonic anhydrases (Tabita, 1999; Badger and Price, 2003).

The inorganic carbon transporters act to selectively import inorganic carbon from the environment and pool it in the cytoplasm. So far, five major kinds of transporters have been characterized (Price *et al.*, 2008, 2011). BCT1, SbtA and BicA are bicarbonate transporters, located on the plasma membrane. BCT1 transporter belongs to the ATPase family of ATP-binding cassette (ABC) type of transporters (Omata *et al.*, 2002). It is a high-affinity uniporter that actively imports bicarbonate ions (HCO_3^-) by utilizing ATP and is inducible at low concentrations of CO_2 (Price *et al.*, 2008). BicA and SbtA, on the other hand, are Na^+-dependent, single-subunit symporters that utilize the periplasmic Na^+ gradient to pump in HCO_3^- ions (Shibata *et al.*, 2002; Price *et al.*, 2004). The CO_2 transporter NDH-I3 is an inducible high-affinity uptake system located on the thylakoid membrane (Herranen *et al.*, 2004; Price *et al.*, 2011). The other CO_2 transporter is NDH-I4 which has low affinity and is constitutively expressed (Price *et al.*, 2011). The collective action of these transporters generates a cytoplasmic pool of inorganic carbon. As its ionic nature prevents leakage through the plasma membrane, the inorganic carbon is retained as bicarbonate ions. To maintain the internal pH, an equivalent number of protons are either ejected from the cytoplasm or neutralized (Price *et al.*, 2008).

3.4.3 CCMs of eukaryotes

Like cyanobacteria, CCMs in eukaryotic photosynthetic microorganisms, such as algae, are based on CO_2 enrichment mechanisms. The inorganic carbon acquisition in algae, however, is generally lower than that of cyanobacteria. This is attributed to the nature of the algal RuBisCO which shows a comparatively higher affinity for CO_2 (Kaplan and Reinhold, 2002). Despite the low level of carbon accumulation, uptake is still mediated by active transporters (Badger and Price, 1992). In *Chlamydomonas*, inorganic carbon uptake transporters are present in the plasma membrane as well as the chloroplast (Kaplan and Reinhold, 2002). Plasma-membrane bicarbonate transporters include HLA3 and LCI1. HLA3 is an ABC type of transporter, overexpression of which increases acquisition especially at low CO_2 concentrations (Im and Grossman, 2001; Gao *et al.*, 2015). LCI1 transporter, on the other hand, is a novel protein that is predicted to have several transmembrane domains. The gene encoding LCI1 has no sequence-based homologues even in closely related organisms such as *Volvox* (Burow *et al.*, 1996). The chloroplast envelope is another membrane barrier to inorganic carbon uptake. LCIA is the only protein confirmed to function in active inorganic carbon uptake at the chloroplast envelope. Brueggeman *et al.* (2012) observed that on induction of CCM there was a 4000-fold increase in LCIA transcripts, making it the most responsive of all the CO_2-responsive CCM genes. An array of regulatory elements controls the functional expression of these uptake systems, enabling the organisms to shift between various acclimatization states (Meyer and Griffiths, 2013). Sixteen proteins have been found to be involved in the operation of *Chlamydomonas* CCM, which includes six membrane transporters (Wang *et al.*, 2011), two soluble proteins which are presumed to be involved in forming a barrier inside the chloroplast to recapture CO_2 (Yamano *et al.*, 2010) and two nuclear regulatory factors (Fukuzawa *et al.*, 2001; Xiang *et al.*, 2001; Kohinata *et al.*, 2008). Analogous to cyanobacterial carboxysomes, algae contain subcellular structures called pyrenoids as the site of RuBisCO localization. A pyrenoid or some form of it is found ubiquitously in most algal CCMs, exceptions being the genus *Chloromonas* and organisms having low carbon accumulation capacity (Morita *et al.*, 1998, 1999). It is the site of bicarbonate ion dehydration to form CO_2 catalysed by thylakoid-localized carbonic anhydrase, CAH3. The integrated functioning of these components results in a CCM that concentrates CO_2 around RuBisCO at ~40 times higher levels than ambient (Meyer and Griffiths, 2013).

Distinctly different from the CO_2-concentrating mechanisms discussed so far is the C4 photosynthesis observed in a subset of land plants. The single-cell-based CCMs of cyanobacteria and algae rely on multiple CCMs; C4 plants, on the other hand, have developed a combination of biochemical and morphological specializations to accumulate CO_2. C4 RuBisCO is localized in specialized cells known as bundle sheath cells adjacent to the vasculature (Dengler and Nelson, 1999). Mesophyll cells do not contain RuBisCO, but rather have a phosphoenolpyruvate (PEP) carboxylase enzyme that fixes atmospheric CO_2 producing a four-carbon compound.

This compound is transported to the bundle sheath cells and decarboxylated, releasing CO_2 that then enters the Calvin cycle. C4 photosynthesis has three subtypes depending on the decarboxylating enzyme: (i) nicotinamide adenine dinucleotide phosphate (NADP)-dependent malic enzyme (NADP-ME); (ii) nicotinamide adenine dinucleotide (NAD)-dependent malic enzyme (NAD-ME); or (iii) PEP carboxykinase (Sage, 2004). The PEP carboxylase does not have an oxygenase activity and as RuBisCO is kept out of reach of atmospheric air, oxygenation of ribulose-1,5-bisphosphate (RuBP) is avoided thereby eliminating photorespiration. The bundle sheath and mesophyll cells are kept in direct contact through a characteristic leaf anatomy wherein the bundle sheath cells surround the vascular bundles and are themselves enclosed by the mesophyll cells. Therefore, a repeating pattern of vein–bundle sheath–mesophyll is observed that forms a wreath-like structure and was named the Kranz anatomy (Sage, 2004). For optimal integration of C4 pathway, certain additional modifications are observed in C4 plants such as bundle sheath cells having lesser number of photosystem II complexes to reduce the production of O_2. At low temperatures, productivity can be higher in C3 plants, but theoretical models have shown that even at a temperature as low as 5°C, C4 photosynthesis is advantageous (Jansson *et al.*, 2010).

3.5 Genetic Engineering Approaches to Enhance Phytosequestration of Carbon Dioxide

One of the ways through which terrestrial sequestration can be enhanced is by improving the efficiency of plant photosynthesis. C3 plant photosynthesis is at a natural disadvantage because of the lack of a CCM. Despite having an improved RuBisCO with greater CO_2 specificity, photorespiration is not completely eliminated as in C4 plants. To increase the efficiency of C3 photosynthesis, several genetic engineering strategies have been suggested, including: (i) regeneration of RuBP; (ii) RuBisCO active site engineering; and (iii) introduction of CCM components from cyanobacteria or algae.

3.5.1 Regeneration of RuBP

CO_2 assimilation rate in high atmospheric CO_2 conditions will be limited by the rate of RuBP regeneration, which will in turn be limited by the capacity of the chloroplast electron transport chain and enzymes involved in RuBP regeneration (Stitt and Schulze, 1994; von Caemmerer and Evans, 2010). Plastid aldolase, ketolase and sedoheptulose-1,7-bisphosphatase are the enzymes responsible for the majority of the carbon flux in the C3 cycle (Fig. 3.3). It has been proposed that targeting these three enzymes can greatly enhance C3 photosynthesis (Raines, 2011). Transgenic tobacco plants that overexpressed SBPase from *Arabidopsis thaliana* exhibit greater rate of photosynthesis, carbon assimilation and electron transport compared with wild type at 585 ppm of CO_2 (Rosenthal *et al.*, 2011). Similar results were obtained in transgenic soybean overexpressing FBP/SBPase and transgenic *Arabidopsis* expressing plastidial aldolase, which further confirm these findings (Hay, 2012).

3.5.2 Engineering the active site of RuBisCO

A natural diversity exists in the catalytic property of RuBisCOs across different plant species. This diversity is influenced by varying environmental pressures as well as the changes in CO_2/O_2 ratio that occurred in the past (Galmes *et al.*, 2014). C3 plants adapted to the decrease in environmental CO_2 by increasing RuBisCO's CO_2 affinity (low K_c). Consequently, the velocity of the carboxylation reaction (K_{cat}) decreased which was recompensed by increasing the percentage of plant protein invested in RuBisCO. Yet, this trade-off is not universal, and deviations are found among existent species (Tcherkez *et al.*, 2006; Savir *et al.*, 2010; Whitney *et al.*, 2011b). For example, in plants adapted to certain specific environments such as carnivorous plants and C3 plants growing in arid or aquatic habitats, RuBisCO has evolved towards a higher K_{cat}/K_c ratio. Galmes *et al.* (2014) found the variability in RuBisCO catalysis to be related to the amino acid sequence of the large subunit (LS) of RuBisCO and identified 13 residues that were positively selected during evolution of RuBisCO kinetics. This information can be used to improve the low carboxylase activity as well as CO_2 specificity of C3 RuBisCO. Hybrid RuBisCOs have been generated successfully in tobacco plants wherein the LS of tobacco RuBisCO was replaced with that of bacterial (*Rhodospirillum rubrum*) RuBisCO (Whitney and Andrews, 2001; Alonso-Blanco *et al.*, 2009). Transplastomic tobacco lines have also been created with hybrid RuBisCO containing the LS

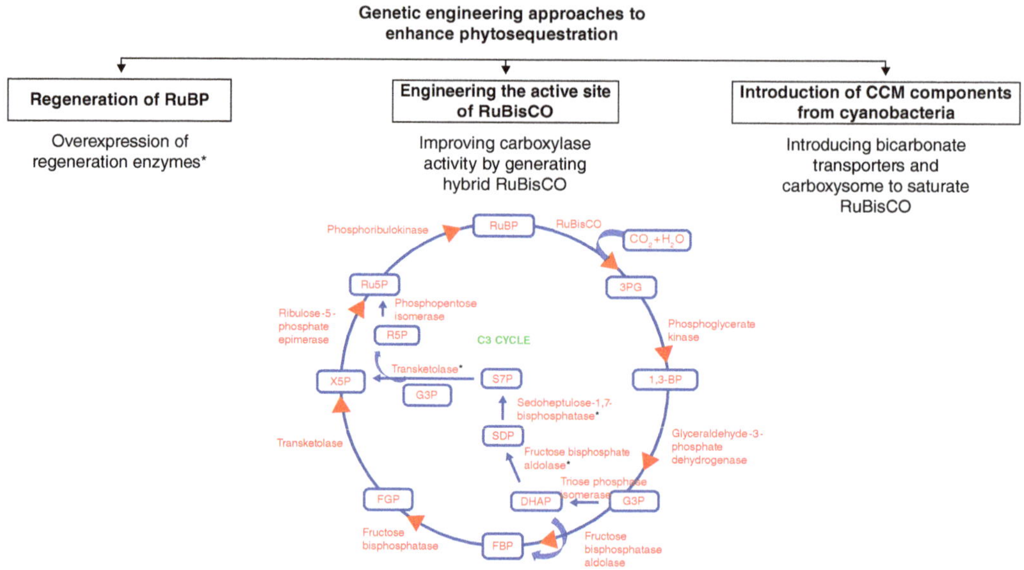

Fig. 3.3. Genetic engineering approaches to enhance phytosequestration in C3 plants: (i) regeneration of ribulose-1,5-bisphosphate (RuBP) by overexpressing enzymes (indicated by '*') sedoheptulose-1,7-bisphosphatase, transketolase and aldolase; (ii) engineering the active site of ribulose bisphosphate carboxylase/oxygenase (RuBisCO) to improve carboxylase activity; and (iii) introducing components of the carbon-capturing mechanism (CCM), bicarbonate transporters and carboxysomes, from cyanobacteria to elevate CO_2 around RuBisCO. Abbreviation: 3PG, 3-phosphoglycerate; 1,3-BP, 1,3-bisphosphoglycerate; G3P, glyceraldehyde-3-phosphate; FBP, fructose-1,6-bisphosphate; SDP, sedoheptulose-1,7-bisphosphate; DHAP, dihydroxyacetone phosphate; S7P, sedoheptulose-7-phosphate; F6P, fructose-6-phosphate; X5P, xylulose-5-phosphate; R5P, ribose-5-phosphate; Ru5P, ribulose-5-phosphate.

from different *Flavaria* species. Met-309-Ile was discovered to be the key substitution in the LS that is responsible for better carboxylation rates (Whitney *et al.*, 2011a). Chimeric rice RuBisCO containing the small subunit (SS) of RuBisCO from the C4 plant, *Sorghum bicolor*, also show an enhanced carboxylation rate. The drawback, however, was that the increased RuBisCO activity was not supported by increased CO_2 assimilation rate due to inadequate supply of electrons (Singh *et al.*, 2014). Liu and colleagues have recently gained success in forming active RuBisCO under laboratory conditions using chaperone proteins, which has helped pave the way for *in vitro* analysis of mutant and hybrid RuBisCO (Liu *et al.*, 2010).

3.5.3 Introduction of CCM components from cyanobacteria

An alternative approach to improving C3 photosynthesis is introducing a CCM from lower organisms into C3 plants and thereby saturating RuBisCO. As discussed in Section 3.4.2, cyanobacteria have evolved an extremely efficient CCM which is well characterized, making them the ideal candidate. Price *et al.* (2013) have developed the following pathway for engineering cyanobacterial CCM into C3 plants:

- Phase 1a – transferring active HCO_3^- pumps in the chloroplast envelope;
- Phase 1b – building a functional carboxysome in the chloroplast stroma;
- Phase 2 – combining traits of phases 1a and 1b;
- Phase 3 – eliminating stromal carbonic anhydrase; and
- Phase 4 – building a functional NDH1–CO_2 uptake complex in the thylakoid membrane.

The initial engineering goal is to introduce functional bicarbonate transporters in the chloroplast envelope of C3 plants for the purpose of reducing the deficit of CO_2 in chloroplasts with respect to leaf intracellular levels (Price *et al.*, 2013). It has been estimated that by using a single cyanobacterial bicarbonate transporter, CO_2 assimilation rates can be increased by 9%, while utilizing all transporters can improve assimilation rates by as much as 16% (McGrath and Long, 2014). BicA and SbtA are the simplest among the cyanobacterial bicarbonate transporters making them the most appropriate initial target for engineering. Moreover, modelling studies have shown that introducing SbtA or BicA into the chloroplast envelope can potentially achieve a 5–15% increase in CO_2 fixation rate at constant substomatal CO_2 levels (Price *et al.*, 2011). Recent developments regarding targeting of transporters to the chloroplast inner envelope and assembly of a functional carboxysome in *Escherichia coli* have laid the foundation for synergizing a functional CCM in higher plants (Singh *et al.*, 2014).

3.6 Sustainability of Plant Genetic Engineering

It has been estimated that the terrestrial pool has lost around 450 × 10^9 tonnes of carbon over the last 10,000 years (Lal, 2008b). If this amount is to be re-sequestered in the next 50 years, then sequestration should occur at a rate of 9 × 10^9 tonnes per year. Even if half this amount is recaptured it would constitute a major tap into the atmospheric carbon pool. Jansson *et al.* (2010) have speculated on the extent to which plant genetic engineering can help realize this goal. Agricultural cropland is assumed to be the ecosystem most likely to be affected by genetic engineering strategies whereas large untouched areas such as natural forests and grasslands are less likely to be benefited. Crops such as maize have shown nearly 100% increase in crop productivity over the last 50 years because of genetic improvements or improvements in management practices (Long *et al.*, 2006). It is estimated that the next 50 years will see a boost of at least 50% in biomass production of food and non-food crops (Jansson *et al.*, 2010). Eliminating photorespiration in C3 crop plants can theoretically increase the photosynthesis rate by 10–30% and yield by 6% (Metting *et al.*, 2001; Sinclair *et al.*, 2004). Collectively a 50% increase in productivity can be expected (Long *et al.*, 2006). If extrapolated to the entire cultivable land which has a current NPP of 6 × 10^9 tonnes per year, a 50% increase corresponds to an additional 3 × 10^9 tonnes NPP per year. In this case, the total sequestration potential of genetically engineered crops and plantations in biomass and soil carbon pool might reach 2–3 × 10^9 tonnes per year (Jansson *et al.*, 2010).

According to one study, bioenergy crops if cultivated in a land area of 750 million hectares have the potential to sequester about 1.6 × 10^9 tonnes of carbon in the soil per year (Lemus and Lal, 2005). Jansson *et al.* (2010) have speculated that if half this area is cultivated by the year 2050 using genetically improved crops having increased carbon allocation to roots, the same amount of sequestration can be achieved. Greenhouse gas emission equivalent to 5–8 × 10^9 tonnes per year is expected to be offset via bioenergy crops by 2050 and with the benefits of genetic engineering, an additional 4 × 10^9 tonnes per year can be sequestered.

However, these estimations are made disregarding several important issues, the foremost of which is societal resistance toward transgenic plants. Also, the effect of global warming and increased atmospheric CO_2 levels on carbon sequestration is unclear. The adaptive ability of plants to higher CO_2 levels, availability of water and nitrogen, and sensitivity of the soil carbon pool are uncertainties that need to be explored.

3.7 Conclusion

Biological sequestration strategies have the potential to tackle elevated carbon levels. But as each has its drawbacks, more than one strategy needs to be employed to achieve mitigation. At its heart, biotic sequestration is the photo-assimilation of carbon. Several examples of genetic engineering that target this process are discussed that not only can enhance carbon sequestration but also increase the yield of crops. Social and practical issues stand in the way of the widespread cultivation of transgenics, especially in developing countries. With time, the advantages of genetic engineering might overcome these problems. The ideas proposed with regard to genetic engineering are currently only at the initial stage of study. Further research is necessary to achieve optimal integration of transgenic proteins in higher plant systems.

References

Adu, J. and Oades, J. (1978) Physical factors influencing decomposition of organic materials in soil aggregates. *Soil Biology and Biochemistry* 10, 109–115.

Alonso-Blanco, C., Aarts, M.G.M., Bentsink, L., Keurentjes, J.J.B., Reymond, M. *et al.* (2009) What has natural variation taught us about plant development, physiology, and adaptation? *The Plant Cell* 21, 1877–1896.

Archer, D., Eby, M., Brovkin, V., Ridgwell, A., Cao, L. *et al.* (2009) Atmospheric lifetime of fossil fuel carbon dioxide. *Annual Review of Earth and Planetary Sciences* 37, 117–134.

Badger, M.R. and Price, G.D. (1992) The CO_2 concentrating mechanism in cyanobacteria and microalgae. *Physiologia Plantarum* 84(4), 606–615. https://doi.org/10.1111/j.1399-3054.1992.tb04711.x

Badger, M.R. and Price, G.D. (2003) CO_2 concentrating mechanisms in cyanobacteria: molecular components, their diversity and evolution. *Journal of Experimental Botany* 54(383), 609–622. https://doi.org/10.1093/jxb/erg076

Badger, M.R., Andrews, T.J., Whitney, S.M., Ludwig, M., Yellowlees, D.C. *et al.* (1998) The diversity and coevolution of Rubisco, plastids, pyrenoids, and chloroplast-based CO_2-concentrating mechanisms in algae. *Canadian Journal of Botany* 76(6), 1052–1071. https://doi.org/10.1139/b98-074

Beare, M., Hendrix, P. and Coleman, D. (1994) Water-stable aggregates and organic matter fractions in conventional- and no-tillage soils. *Soil Science Society of America Journal* 58, 777–786.

Bharti, R.K., Srivastava, S. and, Thakur, I.S. (2014) Production and characterization of biodiesel from carbon dioxide concentrating chemolithotrophic bacteria, *Serratia* sp. ISTD04. *Bioresource Technology* 153, 189–197.

Boyd, P., Watson, A.J., Law, C.S., Abraham, E.R., Trull, T. *et al.* (2000) A mesoscale phytoplankton bloom in the polar Southern Ocean stimulated by iron fertilization. *Nature* 407, 695–702.

Brueggeman, A.J., Gangadharaiah, D.S., Cserhati, M.F., Casero, D., Weeks, D.P. and Ladunga, I. (2012) Activation of the carbon concentrating mechanism by CO_2 deprivation coincides with massive transcriptional restructuring in *Chlamydomonas reinhardtii*. *The Plant Cell* 24(5), 1860–1875. https://doi.org/10.1105/tpc.111.093435

Burow, M.D., Chen, Z.-Y., Mouton, T.M. and Moroney, J.V. (1996) Isolation of cDNA clones of genes induced upon transfer of *Chlamydomonas reinhardtii* cells to low CO_2. *Plant Molecular Biology* 31, 443–448.

Chisholm, S.W. and Morel, F.M.M. (1991) What regulates phytoplankton production in nutrient-rich areas of the open sea? *Limnology and Oceanography Special Edition* 36(8), 493.

Chisholm, S.W., Falkowski, J.J.C. and Cullen, P.G. (2006) Dis-crediting ocean fertilization. *Science* 294, 309–310. https://doi.org/10.1126/science.1065349

Coale, K.H., Johnson, K.S., Fitzwater, S.E., Gordon, R.M., Tanner, S. *et al.* (1996) A massive phytoplankton bloom induced by an ecosystem-scale iron fertilization experiment in the equatorial Pacific Ocean. *Nature* 383, 495–501.

Dalton, R. (2002) Ocean tests raise doubts over use of algae as carbon sink. *Nature* 420, 722. https://doi.org/10.1038/420722a

Dengler, N. and Nelson, T. (1999) Leaf structure and development in C4 plants. In: Sage, R.F. and Monson, R.K. (eds) *C4 Plant Biology*. Academic Press, San Diego, California, pp. 133–172.

Espie, G.S. and Kimber, M.S. (2011) Carboxysomes: cyanobacterial RuBisCO comes in small packages. *Photosynthesis Research* 109, 7–20.

Evans, J.R. (1989) Photosynthesis and nitrogen relationships in leaves of C3 plants. *Oecologia* 78, 9–19.

Evans, J.R. and von Caemmerer, S. (1996) Carbon dioxide diffusion inside leaves. *Plant Physiology* 110, 339–346.

Falkowski, P., Scholes, R.J., Boyle, E., Canadell, J., Canfield, D. *et al.* (2000) The global carbon cycle: a test of our knowledge of earth as a system. *Science* 290, 291–296.

Fan, S., Gloor, M., Mahlman, J., Pacala, S., Sarmiento, J., Takahashi, T. and Tans, P. (1998) A large terrestrial carbon sink in North America implied by atmospheric and oceanic carbon dioxide data and models. *Science* 282, 442–446.

Fang, J., Chen, A., Peng, C., Zhao, S. and Ci, L. (2001) Changes in forest biomass carbon storage in China between 1949 and 1998. *Science* 292, 2320–2316.

Field, C.B., Behrenfeld, M.J., Randerson, J.T. and Falkowski, P. (1998) Primary production of the biosphere: integrating terrestrial and oceanic components. *Science* 281, 237–240.

Figueroa, D.J., Fout, T., Plasynski, S., McIlvried, H. and Srivastava, R.D. (2008) Advances in CO_2 capture technology – The US Department of Energy's Carbon Sequestration Program. *International Journal of Greenhouse Gas Control* 2, 9–20.

Fukuzawa, H., Miura, K., Ishizaki, K., Kucho, K.I., Saito, T. *et al.*, (2001) Ccm1, a regulatory gene controlling the induction of a carbon-concentrating mechanism in *Chlamydomonas reinhardtii* by sensing CO_2 availability. *Proceedings of the National Academy of Sciences USA* 98, 5347–5352.

Galmes, J., Kapralov, M.V., Andralojc, P.J., Conesa, M.À., Keys, A.J. *et al.* (2014) Expanding knowledge of the Rubisco kinetics variability in plant species: environmental and evolutionary trends. *Plant, Cell and Environment* 37, 1989–2001.

Gao, H., Wang, Y., Fei, X., Wright, D.A. and Spalding, M.H. (2015) Expression activation and functional analysis of HLA3, a putative inorganic carbon transporter in *Chlamydomonas reinhardtii*. *The Plant Journal* 82, 1–11.

Garnett, M.H., Ineson, P., Stevenson, A.C. and Howard, D.C. (2001) Terrestrial organic carbon in a British moorland. *Global Change Biology* 7, 375–388.

Golchin, A., Oades, J.M., Skjemstad, J.O. and Clarke, P. (1994) Soil structure and carbon cycling. *Australian Journal of Soil Research* 32, 1043–1068.

Gorham, E. (1991) Northern peatlands: role in the carbon cycle and probable responses to climate warming. *Ecological Applications* 1, 182–195.

Hay, W. (2012) Engineering cyanobacterial genes into *Glycine max* (soybean) leads to increased photosynthesis and productivity. Doctoral dissertation, University of Illinois at Urbana–Champaign, Urbana–Champaign, Illinois.

Herranen, M., Battchikova, N., Zhang, P., Graf, A., Sirpiö, S. *et al.* (2004) Towards functional proteomics of membrane protein complexes in *Synechocystis* sp. PCC 6803. *Plant Physiology* 134(1), 470–481. https://doi.org/10.1104/pp.103.032326

Im, C.S. and Grossman, A.R. (2001) Identification and regulation of high-light induced genes in *Chlamydomonas reinhardtii*. *The Plant Journal* 30, 301–313.

Jackson, R.B., Jobbágy, E.G., Avissar, R., Roy, S.B., Barrett, D.J. *et al.* (2005) Trading water for carbon with biological carbon sequestration. *Science* 310, 1944–1947.

Jansson, C., Wullschleger, S.D., Kalluri, U.C. and Tuskan, G.A. (2010) Phytosequestration: carbon biosequestration by plants and the prospects of genetic engineering. *BioScience* 60(9), 685–696. https://doi.org/10.1525/bio.2010.60.9.6

Jastrow, J. (1996) Soil aggregate formation and the accrual of particulate and mineral- associated organic matter. *Soil Biology and Biochemistry* 28, 656–676.

Kaplan, A. and Reinhold, L. (2002) CO_2 concentrating mechanisms in photosynthetic microorganisms. *Annual Review of Plant Physiology and Plant Molecular Biology* 50, 539–570. https://doi.org/10.1146/annurev.arplant.50.1.539

Kobak, K.I., Knodrasheva, N.Y. and Turchinovich, I.E. (1998) Changes in C pools of peatland and forests in north-western Russia during the Holocene. *Global Planet Change* 16/17, 75–84.

Kohinata, T., Nishino, H. and Fukuzawa, H. (2008) Significance of zinc in a regulatory protein, CCM1, which regulates the carbon-concentrating mechanism in *Chlamydomonas reinhardtii*. *Plant and Cell Physiology* 49, 273–283.

Krishnamurthy, R. and Machavaram, M. (2000) Is there a stable isotope evidence for the fertilization effect? *Proceedings of the Indian Academy of Sciences – Earth and Planetary Sciences* 109, 141–144.

Kumar, M., Gupta, A. and Thakur, I.S. (2016) Carbon dioxide sequestration by chemo-lithotrophic oleaginous bacteria for production and optimization of polyhydroxyalkanoate. *Bioresource Technology* 213, 249–256.

Kumar, M., Sundaram, S., Gnansounou, E., Larroche, C. and Thakur, I.S. (2018) Carbon dioxide capture, storage and production of biofuel and biomaterials by bacteria: a review. *Bioresource Technology* 247, 1059–1068. https://doi.org/10.1016/j.biortech.2017.09.050

Kuzyakov, Y. and Domanski, G. (2000) Carbon input by plants into the soil. Review. *Journal of Plant Nutrition and Soil Science* 163(4), 421–431.

Lal, R. (2004) Soil carbon sequestration impacts on global climate change and food security. *Science* 304, 1623–1627.

Lal, R. (2008a) Carbon sequestration. *Philosophical Transactions of the Royal Society B* 363, 815–830.

Lal, R. (2008b) Sequestration of atmospheric CO_2 in global carbon pools. *Energy and Environmental Science* 1, 86–100.

Lal, R., Follett, R.F. and Kimble, J.M. (2003) Achieving soil carbon sequestration in the United States: a challenge to the policy makers. *Soil Science* 168, 827–845.

Lemus, R. and Lal, R. (2005) Bioenergy crops and carbon sequestration. *Critical Reviews in Plant Sciences* 24(1), 1–21. https://doi.org/10.1080/07352680590910393

Liu, C., Young, A.L., Starling-Windhof, A., Bracher. A., Saschenbrecker, S. *et al.* (2010) Coupled chaperone action in folding and assembly of hexadecameric Rubisco. *Nature* 463, 197–202.

Liu, H., Landry, M.R., Vaulot, D. and Campbell, L. (1999) Prochlorococcus growth rates in the central equatorial Pacific: an application of the f_{max} approach. *Journal of Geophysics Research: Oceans* 104, 3391–3399.

Long, S.P., Zhu, X.G., Naidu, S.L. and Ort, D.R. (2006) Can improvement in photosynthesis increase crop yields? *Plant Cell and Environment* 29, 315–330. https://doi.org/10.1111/j.1365-3040.2005.01493.x

McGrath, J.M. and Long, S.P. (2014) Can the cyanobacterial carbon-concentrating mechanism increase photosynthesis in crop species? A theoretical analysis. *Plant Physiology* 164, 2247–2261.

Martin, J.H. and Fitzwater, S.E. (1988) Iron deficiency limits phytoplankton growth in the north-east Pacific subarctic. *Nature* 331, 341–343.

Martin, J.H., Coale, K.H., Johnson, K.S., Fitzwater, S.E., Gordon, R.M. *et al.* (1994) Testing the iron hypothesis in ecosystems of the equatorial Pacific Ocean. *Nature* 371, 123–129.

Metting, F.B., Smith, J.L., Amthor, J.S. and Izaurralde, R.C. (2001) Science needs and new technology for increasing soil carbon sequestration. *Climatic Change* 51, 11–34.

Meyer, M. and Griffiths, H. (2013) Origins and diversity of eukaryotic CO_2-concentrating mechanisms: lessons for the future. *Journal of Experimental Botany* 64(3), 769–786. https://doi.org/10.1093/jxb/ers390

Milne, R. and Brown, T.A. (1997) Carbon in the vegetation and soils of Great Britain. *Journal of Environmental Management* 49, 413–433.

Morita, E., Abe, T., Tsuzuki, M., Fujiwara, S., Sato, N. *et al.* (1998) Presence of the CO_2-concentrating mechanism in some species of the pyrenoid-less free-living algal genus *Chloromonas* (Volvocales, Chlorophyta). *Planta* 204, 269–276.

Morita, E., Abe, T., Tsuzuki, M., Fujiwara, S., Sato, N. *et al.* (1999) Role of pyrenoids in the CO_2-concentrating mechanism: comparative morphology, physiology and molecular phylogenetic analysis of closely related strains of *Chlamydomonas* and *Chloromonas* (Volvocales). *Planta* 208, 365–372. https://doi.org/10.1007/s004250050571

Nogia, P., Sidhu, G.K., Mehrotra, R. and Mehrotra, S. (2016) Capturing atmospheric carbon: biological and nonbiological methods. *International Journal of Low-Carbon Technologies* 11(2), 266–274. https://doi.org/10.1093/ijlct/ctt077

Omata, T., Price, G.D., Badger, M.R., Okamura, M., Gohta, S. and Ogawa, T. (2002) Identification of an ATP-binding cassette transporter involved in bicarbonate uptake in the cyanobacterium *Synechococcus* sp. strain PCC 7942. *Proceedings of the National Academy of Sciences USA* 96(23), 13571–13576. https://doi.org/10.1073/pnas.96.23.13571

Paustian, K., Six, J., Elliott, E.T. and Hunt, H.W. (2000) Management options for reducing CO_2 emissions from agricultural soils. *Biogeochemistry* 48, 147–163. https://doi.org/10.1023/A:1006271331703

Price, G.D., Woodger, F.J., Badger, M.R., Howitt, S.M. and Tucker, L. (2004) Identification of a SulP-type bicarbonate transporter in marine cyanobacteria. *Proceedings of the National Academy of Sciences USA* 101(52), 18228–18233. https://doi.org/10.1073/pnas.0405211101

Price, G.D., Badger, M.R., Woodger, F.J. and Long, B.M. (2008) Advances in understanding the cyanobacterial CO_2-concentrating-mechanism (CCM): functional components, Ci transporters, diversity, genetic regulation and prospects for engineering into plants. *Journal of Experimental Botany* 59(7), 1441–1461. https://doi.org/10.1093/jxb/erm112

Price, G.D., Badger, M.R. and von Caemmerer, S. (2011) The prospect of using cyanobacterial bicarbonate transporters to improve leaf photosynthesis in C_3 crop plants. *Plant Physiology* 155(1), 20–26. https://doi.org/10.1104/pp.110.164681

Price, G.D., Pengelly, J.J., Forster, B., Du, J., Whitney, S.M. *et al.* (2013) The cyanobacterial CCM as a source of genes for improving photosynthetic CO_2 fixation in crop species. *Journal of Experimental Botany* 64(3), 753–768.

Raines, C. (2011) Increasing photosynthetic carbon assimilation in C3 plants to improve crop yield: current and future strategies. *Plant Physiology* 155, 36–42.

Rodhe, H. (1990) A comparison of the contribution of various gases to the greenhouse effect. *Science* 248, 1217–1219.

Rosenthal, D.M., Locke, A.M., Khozaei, M., Raines, C.A., Long, S.P. and Or, D.R. (2011) Over-expressing the C_3 photosynthesis cycle enzyme sedoheptulose-1-7 bisphosphatase improves photosynthetic carbon gain and yield under fully open air CO_2 fumigation (FACE). *BMC Plant Biology* 11, 123.

Rovira, A. and Greacen, E. (1957) The effect of aggregate disruption on the activity of microorganisms in the soil. *Australian Journal of Agricultural Research* 8, 659–673.

Sage, R.F. (2004) The evolution of C_4 photosynthesis. *New Phytologist* 161, 341–370. https://doi.org/10.1111/j.1469-8137.2004.00974.x

Sarmiento, J.L. and Orr, J.C. (1991) Three-dimensional ocean model simulations of the impact of Southern Ocean nutrient depletion on atmospheric CO_2 and ocean chemistry. *Limnology and Oceanography Special Edition* 36, 1928–1950.

Savir, Y., Noor, E., Milo, R. and Tlusty, T. (2010) Cross-species analysis traces adaptation of RuBisCO toward optimality in a low-dimensional landscape. *Proceedings of the National Academy of Sciences USA* 107, 3475–3480.

Scholz, F. and Hasse, U. (2008) Permanent wood sequestration: the solution to the global carbon dioxide problem. *ChemSusChem* 1, 381–384.

Schrag, D.P. (2007) Preparing to capture carbon. *Science* 315, 812–813.

Seibel, B.A. and Walsh, P.J. (2001) Potential impacts of CO_2 injection on deep-sea biota. *Science* 294, 319–320.

Shibata, M., Katoh, H., Sonoda, M., Ohkawa, H., Shimoyama, M. *et al.* (2002) Genes essential to sodium-dependent bicarbonate transport in cyanobacteria: function and phylogenetic analysis. *Journal of Biological Chemistry* 277(21), 18658–18664. https://doi.org/10.1074/jbc.M112468200

Sigman, D.M. and Boyle, E.A. (2000) Glacial/interglacial variations in atmospheric carbon dioxide. *Nature* 407, 859–869.

Sinclair, T., Purcell, L. and Sneller, C. (2004) Crop transformation and the challenge to increase yield potential. *Trends in Plant Science* 9, 70–75.

Singh, J., Pandey, P., James, D., Chandrasekhar, K., Achary, V.M.M. *et al.* (2014) Enhancing C3 photosynthesis: an outlook on feasible interventions for crop improvement. *Plant Biotechnology Journal* 12(9), 1217–1230. https://doi.org/10.1111/pbi.12246

Stitt, M. and Schulze, D. (1994) Does Rubisco control the rate of photosynthesis and plant growth? An exercise in molecular ecophysiology. *Plant, Cell and Environment* 17, 465–487.

Tabita, F.R. (1999) Microbial ribulose 1,5-bisphosphate carboxylase/oxygenase: a different perspective. *Photosynthesis Research* 60, 1–28.

Tcherkez, G., Farquhar, G. and Andrews, T. (2006) Despite slow catalysis and confused substrate specificity, all ribulose bisphosphate carboxylases may be nearly perfectly optimized. *Proceedings of the National Academy of Sciences USA* 103, 7246–7251.

Toensmeier, E. (2018) Perennial staple crops and agroforestry for climate change mitigation. In: Montagnini, F. (ed.) *Advances in Agroforestry*. Vol. 12. *Integrating Landscapes: Agroforestry for Biodiversity Conservation and Food Sovereignty*. Springer, Cham, Switzerland, pp. 439–451. https://doi.org/10.1007/978-3-319-69371-2_18

Tuskan, G. and Walsh, M. (2001) Short-rotation woody crop systems, atmospheric carbon dioxide and carbon management: a US case study. *Forestry Chronicle* 77, 259–264.

Verhoeven, J.T.A. and Setter, T.L. (2010) Agricultural use of wetlands: opportunities and limitations. *Annals of Botany* 105, 155–163. https://doi.org/10.1093/aob/mcp172

von Caemmerer, S. and Evans, J. (2010) Enhancing C_3 photosynthesis. *Plant Physiology* 154, 589–592.

Wang, Y., Dunamu, D. and Spalding, M.H. (2011) Carbon dioxide concentrating mechanism in *Chlamydomonas reinhardtii*: inorganic carbon transport and CO_2 recapture. *Photosynthesis Research* 109, 115–122.

Whitney, S.M. and Andrews, T.J. (2001) The gene for the ribulose-1,5-bisphosphate carboxylase/oxygenase (Rubisco) small subunit relocated to the plastid genome of tobacco directs the synthesis of small subunits that assemble into Rubisco. *The Plant Cell* 13, 193–205.

Whitney, S.M., Sharwood, R.E., Orr, D., White, S.J., Alonso, H. and Galmés, J. (2011a) Isoleucine 309 acts as a C4 catalytic switch that increases ribulose-1,5-bisphosphate carboxylase/oxygenase (rubisco) carboxylation rate in *Flaveria*. *Proceedings of the National Academy of Sciences USA* 108, 14688–14693.

Whitney, S.M., Houtz, R.L. and Alonso, H. (2011b) Advancing our understanding and capacity to engineer nature's CO_2-sequestering enzyme, Rubisco. *Plant Physiology* 155(1), 27–35. https://doi.org/10.1104/pp.110.164814

Wofsy, S.C. (2001) Where has all the carbon gone? *Science* 292, 2261–2263.

Woolf, D., Amonette, J.E., Street-Perrott, F.A., Lehmann, J. and Joseph, S. (2010) Sustainable biochar to mitigate global climate change: supplementary information. *Nature Communications* 1, 56.

Wojtkowski, P.A. (1998) *The Theory and Practice of Agroforestry Design*. Science Publishers, Enfield, UK.

Xiang, Y., Zhang, J. and Weeks, D.P. (2001) The *Cia5* gene controls formation of the carbon concentrating mechanism in *Chlamydomonas reinhardtii*. *Proceedings of the National Academy of Sciences USA* 98, 5341–5346.

Yamano, T., Tsujikawa, T., Hatano, K., Ozawa, S., Takahashi, Y. and Fukuzawa, H. (2010) Light and low-CO_2-dependent LCIB–LCIC complex localization in the chloroplast supports the carbon-concentrating mechanism in *Chlamydomonas reinhardtii*. *Plant and Cell Physiology* 51, 1453–1468.

Yang, H., Xu, Z., Fan, M., Gupta, R., Slimane, R.B. *et al.* (2008) Progress in carbon dioxide separation and capture: a review. *Journal of Environmental Science* 20, 14–27.

Zan, C.S., Fyles, J.W., Girouard, P. and Samson, R.A. (2001) Carbon sequestration in perennial bioenergy, annual corn and uncultivated systems in southern Quebec. *Agriculture, Ecosystems and Environment* 86(2), 135–144. https://doi.org/10.1016/S0167-8809(00)00273-5

Zeng, N. (2008) Carbon sequestration via wood burial. *Carbon Balance and Management* 3, 1. https://doi.org/10.1186/1750-0680-3-1

4 Rhabdovirosis (Viral Haemorrhagic Septicaemia Virus)

CAROL A. STEPIEN[1]*, DOUGLAS W. LEAMAN[2] AND MEGAN D. NINER[3]

[1]*NOAA Pacific Marine Environmental Laboratory, Seattle, Washington, USA;* [2]*Wright State University, Dayton, Ohio, USA;* [3]*University of Toledo, Toledo, Ohio, USA*

4.1 Brief Introduction: Disease Description and Diversity, Geographical Distribution and Infected Fish Species

The family *Rhabdoviridae* comprises negative-sense single-stranded RNA viruses, which are in the order *Mononegavirales* (reviewed by Dietzgen and Kuzmin, 2012). The *Rhabdoviridae* contains several taxonomic groups of fish pathogens that infect economically important wild and aquacultured hosts. Some of these rhabdovirus species are generalists that infect wide ranges of fish host species. Some can extend across broad geographic areas, which vary in habitat and water conditions, spanning salinity and temperature regimes. Some live in anadromous hosts (such as salmons and trouts) and move among marine, brackish and freshwater ecosystems with their hosts. Such versatility likely facilitates the ecological successes of generalist rhabdovirus species, as well as their potential ability to further expand their geographic ranges and host diversities during climate change and other environmental opportunities. Other rhabdovirus species are more specialized on specific hosts and inhabit restricted geographic ranges, which may facilitate their long-term persistence. In recent years, descriptions and incidences of fish rhabdoviruses have been increasing due to their increased prevalence in aquaculture and global trade, accompanied by improved molecular diagnostics and surveillance (see ICTV, 2018; Walker *et al.*, 2018).

Rhabdovirions are encased in envelopes and most are bullet-shaped (Fig. 4.1), with widths of about 75 nm and lengths of approximately 180 nm (Dietzgen and Kuzmin, 2012). Their genomes are linear and approximately 11–15 kb in length, encoding five proteins: nucleoprotein (N), phosphoprotein (P), matrix protein (M), glycoprotein (G) and large protein (L). Functionally, the N protein encloses the RNA, the P protein is a cofactor for the RNA polymerase and a chaperone for N, the M protein is critical for virus assembly and budding, G is a transmembrane protein responsible for attachment to target cells and facilitates membrane fusion, and L is the viral RNA polymerase (Rieder and Conzelmann, 2011).

Each rhabdovirion contains an outer helix of matrix protein (M) and an inner helix of nucleoprotein (N) and RNA (Walker *et al.*, 2018; Fig. 4.1). Two layers of lipids, which are derived from the host cell membrane and the envelope glycoprotein (G), form the virion's outer layers and mediate interactions with cellular receptor(s) on the host's cells (see ICTV, 2018). G also plays a major antigenic role, eliciting strong innate and adaptive host immune responses (Kurath, 2012). M contains a hub domain having four contact sites that link to the neighbouring M and N subunits, providing rigidity and stability to maintain the bullet shape. Viral replication occurs in the host's cytoplasm by means of a combination of virally encoded and host-derived factors (ICTV, 2018).

The fish rhabdoviruses that have been well characterized belong to (at least) three different *Rhabdoviridae* lineages (ICTV, 2018), as depicted

*carol.stepien@noaa.gov and cstepien@uw.edu

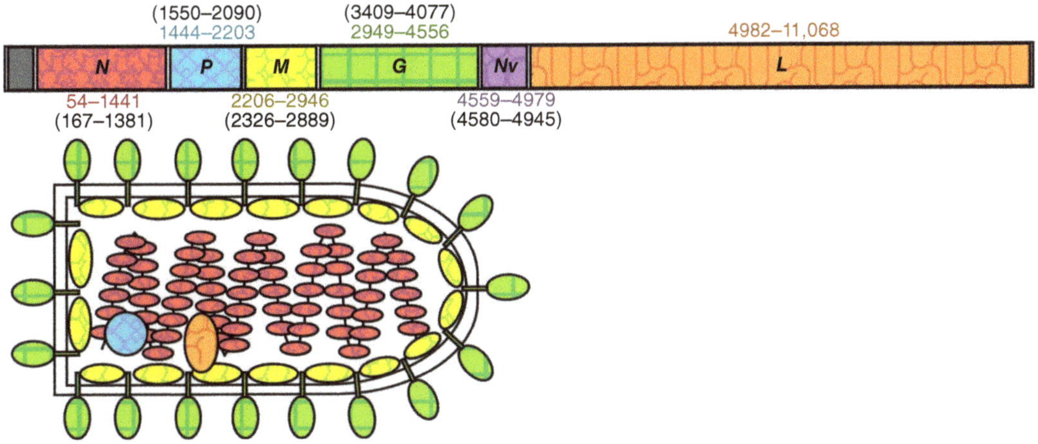

Fig. 4.1. *Piscine novirhabdovirus* VHSV-IVb structure and genome layout. Colours match the gene to the structure diagram. Numbers sharing the same colours as the gene refer to the nucleotide positions. Numbers in parentheses correspond to the region sequenced within each gene in the population genetic analyses. (Modified with permission from Pore, 2012.)

by their phylogenetic relationships among major groups. These relationships are based on our new analysis of their available whole-genome sequences (Fig. 4.2, fish rhabdoviruses are indicated with *). A literature search conducted uncovered about 21 fish rhabdoviruses that have been variously described and/or proposed. About half of these fish rhabdoviruses appear poorly described and many await verification. In this chapter, italics (with the first letter in upper case) designate the scientific names proposed by ICTV (2018), where available. The genera and higher taxa (with the first letter in upper case) are also in italics. The names that are not in italics are the 'traditional' virus names, which are used with their common abbreviations. Some of the latter have been grouped together into single species, based on what is known of their genetic relationships to date (ICTV, 2018). Many taxa need their genomes sequenced and also merit considerable future diagnostic work.

The disparate evolutionary lineages and origins of fish rhabdoviruses likely reflect a broad range of 'host jumping' from other taxa, which constitutes a principal mode of macroevolution (see Kitchen *et al.*, 2011). These incidents might become more acute with impending biogeographic distributional changes of their fish hosts, switches to new species and the myriad of other factors related to environmental alterations and stresses on host populations and their habitats. Some of the ongoing and growing stressors on fish host populations in marine and aquatic ecosystems include climate change, warming, hypoxia, acidification, introductions and spread of non-indigenous/invasive species, fishery exploitation, harmful algal blooms, pollutants, microplastics, food web alterations, sedimentation, habitat changes and losses, sea-level/lake-level changes and melting ice.

The three primary rhabdovirus genera that have been characterized in fish hosts are: *Novirhabdovirus*, *Sprivivirus* and *Perhabdovirus* (ICTV, 2018). The genus *Novirhabdovirus* has a more distant evolutionary relationship from the other two genera (Fig. 4.2) and possesses a unique non-virion (*Nv*) gene, which constitutes a synapomorphy that unites its species. The *Nv* gene is located between the *G* and *L* genes (see Pierce and Stepien, 2012; depicted in Fig. 4.1) and encodes a non-structural protein that is believed to play a role in blocking the host's innate immune response (Chinchilla and Gomez-Casado, 2017; ICTV, 2018). The genus *Sprivivirus* formerly was regarded as part of the genus *Vesiculovirus*, but now is recognized as a separate taxon that is the sister clade of *Vesiculovirus*, as shown in the new whole-genome phylogeny on Fig. 4.2. The genus *Perhabdovirus* appears as the sister group in the phylogeny to the ledanteviruses (which are ecologically associated with bats; see

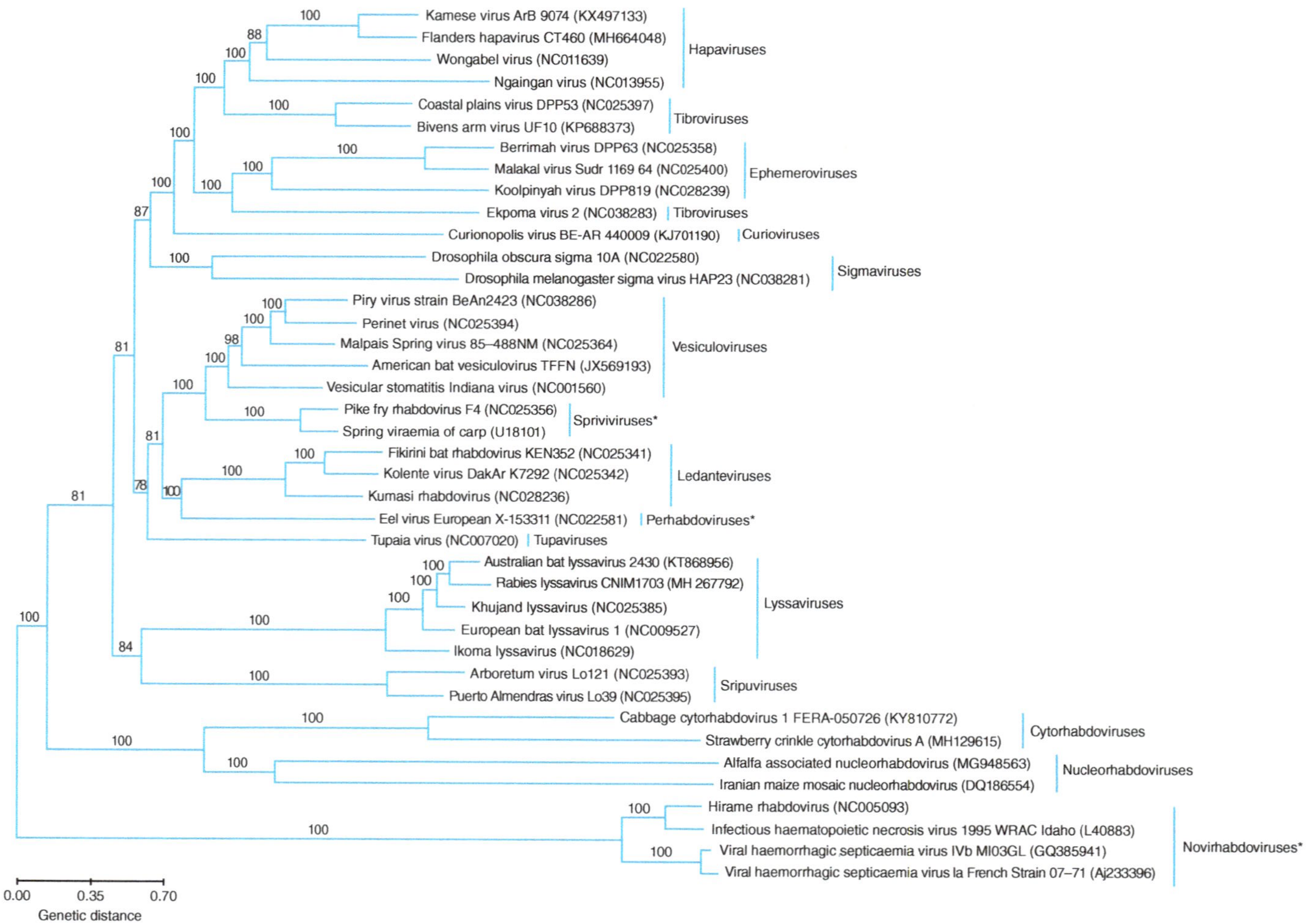

Fig. 4.2. Evolutionary relationships among rhabdoviruses, based on whole-genome sequences using Bayesian analysis (Mr. Bayes v3.2.7a; Ronquist *et al.*, 2012; program: Zhang *et al.*, 2019; http://nbisweden.github.io/MrBayes/authors.html (accessed 8 March 2020)). Bayesian posterior probabilities are indicated at nodes, GenBank accession numbers are in parentheses. Tree is rooted to the original viral haemorrhagic septicaemia virus (VHSV) isolate MI03GL from a muskellunge (GenBank: GQ385941).

Table 4.1. *Piscine novirhabdovirus* (VHSV) isolates used here in full genome analyses.

Isolate name	Genogroup	Host species	Location	Year	GenBank accession number	Reference	Coordinates	G gene haplotype
DK-Hededam	I	North Sea cod	North Sea, Denmark	1972	Z93412	Stone *et al.* (1997)	56.648, 9.271	N/A
French strain 07-71	Ia	Rainbow trout	Seine-Maritime, France	1971	AJ233396	Y. Yan Xing, P. Vende, S. Biacchesi, A. Lamoureux, M. Bearzotti and M. Bremont (1999, unpublished results)	49.9070, 0.7926	″
DK-3592B	″	″	North Sea, Denmark	1986	KC778774	Kim *et al.* (2015)	57.7376, 10.4160	″
De-Fil3	″	″	Baltic Sea, Germany	1999	Y18263	Schütze *et al.* (1999)	54.4212, 11.3928	″
Cod-Ulcus	Ib	North Sea cod	Denmark	1979	Z93414	Stone *et al.* (1997)	56.648, 9.271	″
SE-SVA-1033-3F	″	Rainbow trout	Kattegatt, Sweden	1998	AB839748	Ito *et al.* (2012)	56.9793, 12.2100	″
SE-SVA-1033-9C	″	″	″	″	AB839747	″	″	″
SE-SVA-14-3D	″	″	″	″	AB839745	″	″	″
SE-SVA-14-5G	″	″	″	″	AB839746	″	″	″
SE-SVA-1033	″	″	″	2000	FJ460591	Campbell *et al.* (2009)	″	″
KRRV9601	″	Olive flounder	Seto Inland Sea, Japan	1996	AB672614	Ito *et al.* (2012)	33.7252, 132.5249	″
DKp37	″	Blue whiting	North Sea, Denmark	1997	FJ460590	Campbell *et al.* (2009)	57.18766, 8.29058	″
DK-1p49	II	Atlantic herring	Baltic Sea	1996	KM244767	Lopez-Vazquez *et al.* (2015)	55.1418, 15.3020	″
23-75	III	Brown trout	Eure, France	1975	FN665788	Biacchesi *et al.* (2010)	49.2235, 1.2993	″
14-58	″	Rainbow trout	France	1990	AF143863	Betts and Stone (2000)	″	″
GH40	″	Greenland halibut	Flemish Cap, Newfoundland	1994	KM244768	Lopez-Vazquez *et al.* (2015)	47.0017, –44.9999	″
4p168	″	Atlantic herring	Skagerrack, Denmark	1996	AB672616	Ito *et al.* (2012)	57.7376, 10.4160	″
FA281107	″	Rainbow trout	Storfjorden, Norway	2007	EU481506	Duesund *et al.* (2010)	62.4056, 6.0446	″
BV060408-52	″	″	″	2008	FJ362510	″	″	″
KRRV9822	IVa	Olive flounder	Kagawa, Japan	1998	AB179621	J. Byon, I. Hirono and T. Aoki (2004, unpublished results)	34.3739, 133.9117	″
JF00Ehi1	″	Japanese flounder	Ehime, Japan	2000	AB490792	Ito *et al.* (2012)	33.7334, 132.6460	″
FYeosu05	″	″	South Korea	2005	KF477302	Kim *et al.* (2013)	34.7533, 127.6585	″
Paralichthys olivaceus rhabdovirus (VHSV)	″	″	China	″	KC685626	Zhu and Zhang (2014)	37.5297, 122.1231	″
KJ2008	″	″	Jeju, Korea	2008	JF792424.1	Kim *et al.* (2013)	33.4936, 126.5242	″

(*Continued*)

Table 4.1. Continued

Isolate name	Genogroup	Host species	Location	Year	GenBank accession number	Reference	Coordinates	G gene haplotype
JF-09	"	"	"	2009	KM926343	Kim *et al.* (2015)	33.4936, 126.5242	"
FP-VHS2010-1	"	"	Geoje, Korea	2010	KP334106	Hwang *et al.* (2016)	34.8965, 128.6190	"
ADC-VHS2012-10	"	"	Jeju, Korea	2012	KY979950	"	33.4936, 126.5242	"
ADC-VHS2012-11	"	"	"	"	KY979951	"	"	"
ADC-VHS2012-5	"	"	Gyeongbuk, Korea	"	KY979946	"	36.2894, 128.9379	"
ADC-VHS2012-6	"	"	Jeju, Korea	"	KY979947	"	33.4936, 126.5242	"
ADC-VHS2012-7	"	"	"	"	KY979948	"	"	"
ADC-VHS2012-9	"	"	"	"	KY979949	"	"	"
ADC-VHS2013-1	"	"	"	"	KY979952	"	"	"
ADC-VHS2013-2	"	"	"	"	KY979953	"	"	"
ADC-VHS2013-3	IVa	Olive flounder	Jeju, Korea	2013	KY979954	Hwang *et al.* (2018)	33.4936, 126.5242	N/A
ADC-VHS2013-4	"	"	"	"	KY979955	"	"	"
ADC-VHS2013-9	"	"	Gyeongbuk, Korea	"	KY979956	"	36.2894, 128.9379	"
ADC-VHS2014-2	"	"	Jeju, Korea	2014	KY979957	"	33.4936, 126.5242	"
ADC-VHS2014-4	"	"	"	"	KY979958	"	"	"
ADC-VHS2014-5	"	"	"	"	KY979959	"	"	"
ADC-VHS2015-2	"	"	"	2015	KY979960	"	"	"
ADC-VHS2015-5	"	"	"	"	KY979961	"	"	"
ADC-VHS2016-1	"	"	"	2016	KY979962	"	"	"
ADC-VHS2016-2	"	"	"	"	KY979963	"	"	"
C03MU*	IVb	Muskellunge	Lake St. Clair, USA	2003	GQ385941	Ammayappan and Vakharia (2009)	42.3908, –82.9114	a
E06FD	"	Freshwater drum	Lake Erie, USA	2006	MK783014	Present study	41.7559, –81.2868	"
E06WA	"	Walleye	"	"	MK782987	"	"	"
E06WBa	"	White bass	"	"	MK782986	"	"	"
E06YPa	"	Yellow perch	"	"	MK782985	"	"	"
E06SB	"	Smallmouth bass	"	"	MK782984	"	"	"
E06YPb	"	Yellow perch	"	"	MK782983	"	41.4922, –82.6670	"
E06YPc	"	"	"	"	MK782982	"	"	"
E06WBb	"	White bass	"	"	MK783013	"	"	"
O06SB	"	Smallmouth bass	Lake Ontario, USA	2006	KY359354	Getchell *et al.* (2017)	44.1167, –76.3333	b
C06NP	"	Northern pike	Lake St. Clair, USA	2006	MK782990	Present study	42.6348, –82.7779	a
C06GS	"	Gizzard shad	"	"	"	"	"	"
C06RB	"	Rock bass	"	"	"	"	"	"
C06SR	"	Shorthead redhorse	"	"	"	"	"	"

C06YP	″	Yellow perch	″	″	″	″	42.3430, –82.9020	″
C06FD	″	Freshwater drum	″	″	″	″	42.6348, –82.7779	″
B07BG	″	Bluegill	Budd Lake, Michigan, USA	2007	MK783006	″	44.0159, –84.7881	e
B07PS	″	Pumpkinseed	″	″	MK783008	″	″	f
E07CC	″	Common carp	Lake Erie, USA	″	MK783005	″	42.4906, –79.3381	c
E07YPa	″	Yellow perch	″	″	MK782989	″	41.8013, –81.3563	a
E07YPb	″	″	″	″	MK782988	″	41.7559, –81.2777	″
M07SB	″	Smallmouth bass	Lake Michigan, USA	″	MK783009	″	44.8846, –87.3889	h
M08RB	″	Rock bass	″	2008	MK783010	″	42.4854, –87.8000	i
E08ES	″	Emerald shiner	Lake Erie, USA	″	MK783012	″	41.7691, –81.2940	a
E08FDa	″	Freshwater drum	″	″	MK782993	″	41.7691, –81.3537	″
E08FDb	″	″	″	″	MK782992	″	″	″
M08AM	″	Amphipod	Lake Michigan, USA	2008	MK782990	Present study	43.6002, –86.9167	a
C08Lea	″	Leech	Lake St. Clair, USA	2008	″	Present study	42.6318, –82.7652	a
C08LEb	″	″	″	″	″	″	″	″
M08YP	″	Yellow perch	Lake Michigan, USA	″	MK783007	″	43.0397, –87.8024	j
C09MU	″	Muskellunge	Lake St. Clair, USA	2009	MK782990	″	42.6157, –82.7570	a
M11YP	″	Yellow perch	Lake Michigan, USA	2011	MK782991	″	43.0360, –87.8530	″
E12FD	″	Freshwater drum	Lake Erie, USA	2012	MK783004	″	41.4530, –82.7260	x
O13GS	″	Gizzard shad	Lake Ontario, USA	2013	KY359355	Getchell *et al.* (2017)	43.2362, -77.5345	bc
E14GS	″	″	Lake Erie, USA	2014	KY359356	″	42.4906, –79.3381	bd
E15RG	″	Round goby	″	2015	MK783003	Present study	41.7652, –81.2816	v
E16GSa	″	Gizzard shad	″	2016	MK783011	″	″	″
E16GSb	″	″	″	″	MK782997	″	″	″
E16GSc	″	″	″	″	MK782996	″	″	″
E16GSd	″	″	″	″	MK782994	″	″	″
E16GSe	″	″	″	″	MK782995	″	″	″
M16RGa	″	Round goby	Lake Michigan, USA	″	MK783001	″	42.9969, –87.8825	x
M16RGb	″	″	″	″	MK783000	″	″	″
CellC03	″	Muskellunge	Lake St. Clair, USA	2003	MK782981	″	Cell culture	N/A
Cell16a	″	Gizzard shad	Lake Erie	2016	MK782998	″	″	″
Cell16b	″	Largemouth bass	″	″	MK783002	″	″	″
Cell16c	″	″	″	″	MK782999	″	″	″

N/A, not available.

Table 4.2. Additional *Rhabdovirus* sequences used here in phylogenetic analyses.

Isolate name	Host	Location	Year	GenBank accession number	Reference
***Salmonid novirhabdovirus* = infectious haematopoietic necrosis virus (IHNV)**					
X89213	Rainbow trout	Oregon, USA	1969	X89213	Schütze *et al.* (1995)
WRAC	Chinook salmon	Idaho, USA	1994	L40883	Morzunov *et al.* (1995)
"	"	"	"	NC_001652	"
220-90	Rainbow trout	Idaho, USA	1990	GQ413939	Ammayappan *et al.* (2010)
HLJ-09	"	China	2009	JX649101	Wang *et al.* (2016)
Ch20101008	Brook trout	"	2010	KJ421216	Jia *et al.* (2014)
BjLL	Rainbow trout	"	2012	MF509592	Wang *et al.* (2016)
***Snakehead novirhabdovirus* = snakehead rhabdovirus (SHRV)**					
NC_000903	Snakehead murrel	Thailand	1988	NC_000903	Johnson *et al.* (1999)
AF147498	"	"	"	AF147498	Johnson *et al.* (2000)
***Hirame novirhabdovirus* = hirame rhabdovirus (HIRRV)**					
CA 9703	Japanese flounder	Japan	1984	NC_005093	Kim *et al.* (2005)
"	"	"	"	AF104985	"
80113	Stone flounder	China	2008	FJ376982	Yingjie *et al.* (2011)

Blasdell *et al.*, 2015), whose clade (*Perhabdovirus* + Ledanteviruses) together comprises the sister group to the (*Sprivivirus* + *Vesiculovirus*) clade (Fig. 4.2). These three rhabdovirus genera which infect fish hosts are detailed here.

The genus *Novirhabdovirus* solely infects fishes, of which four have been well described and are recognized as valid species, including: *Piscine novirhabdovirus* (scientific name per ICTV, 2018) = viral haemorrhagic septicaemia virus (VHSV), *Salmonid novirhabdovirus* = infectious haematopoietic necrosis virus (IHNV), *Hirame novirhabdovirus* = hirame rhabdovirus (HIRRV) and *Snakehead novirhabdovirus* = snakehead rhabdovirus (SHRV) (see ICTV, 2018). Of these, *Piscine novirhabdovirus* and *Salmonid novirhabdovirus* are the best-studied novirhabdoviruses, with both possessing broad geographic ranges and exerting pronounced economic impacts on aquaculture (Kurath *et al.*, 2003; Kurath, 2012). Occurrences of *Piscine novirhabdovirus* and *Salmonid novirhabdovirus* are required to be reported to the World Organization for Animal Health and are detailed in Section 4.1.1.

The genus *Sprivivirus* is most closely related to *Vesiculovirus* (Fig. 4.2), and primarily infects cypriniform fishes (i.e. carps, minnows, loaches, etc.; Order Cypriniformes) (ICTV, 2018). Spriviviruses include *Carp sprivivirus* = spring viraemia of carp virus (SVCV), which is the type species, and *Pike fry sprivivirus*, a species that encompasses three closely related described variants (ICTV, 2018), as follows:

1. pike fry rhabdovirus (PFRV);
2. grass carp rhabdovirus (GCRV); and
3. tench rhabdovirus (TRV).

Carp sprivivirus (SVCV) is the most well-known and widespread sprivivirus, being reportable to the World Organization for Animal Health. *Carp sprivivirus* (SVCV) and *Pike fry sprivivirus* (PFRV) are detailed in Section 4.1.2.

The third fish rhabdovirus genus, *Perhabdovirus*, comprises a monophyletic clade according to a phylogeny based on its complete *L* gene sequences (Walker *et al.*, 2018) and in the new whole-genome phylogeny here (Fig. 4.2). *Perhabdovirus* is the sister taxon to the ledanteviruses, which clade (*Perhabdovirus* + Ledanteviruses) then constitutes the sister group to the clade containing the spriviviruses and the vesiculoviruses (Spriviviruses + Vesiculoviruses) (Fig. 4.2). Perhabdoviruses cause diseases that mostly occur in farmed perciform fishes. *Perhabdovirus* has three recognized species (ICTV, 2018):

1. *Perch perhabdovirus* = perch rhabdovirus (PRV) (which is the virus genus' type species), infecting the European pike-perch (*Sander lucioperca*), and;
2. *Anguillid perhabdovirus*, which includes two variants:
   - **2a.** eel virus European X (EVEX) in the European eel (*Anguilla anguilla*) (isolated in 1977) and occurring in a wide variety of wild and cultured populations, and
   - **2b.** eel virus American (EVA) in the American eel (*Anguilla rostrata*); and
3. *Sea trout perhabdovirus*, containing two putative lineages:
   - **3a.** lake trout rhabdovirus (LTRV) in brown trout (*Salmo trutta lacustris*) (see Gubala, 2012) and
   - **3b.** Swedish sea trout virus (SSTV) in sea trout (*Salmo trutta trutta*) in Europe (isolated in 1996) (see explanatory table in ICTV, 2018).

Perch perhabdovirus (PRV) undergoes both horizontal and vertical transmission and has resulted in mass mortality of pike-perch larvae (Bigarré *et al.*, 2017).

The *Anguillid perhabdovirus* genogroups EVEX and EVA are very closely related (Gadd, 2013). The two eel host species, one from North America (*A. rostrata*) and the other from Europe (*A. anguilla*), each migrate to the Sargasso Sea in the North Atlantic Ocean for spawning and in that habitat, they may co-mingle and exchange the virus. The eel larvae, termed leptocephali, then respectively migrate to the fresh waters of North America or Europe, where they maintain their genetic distinctions (see van Ginneken *et al.*, 2005; Bellec *et al.*, 2014). The two virus populations also slightly genetically diverge, according to phylogenetic trees from the *G*, *N* and *P* genes by Bellec *et al.* (2014), with their respective *P* gene sequences exhibiting the greatest genetic difference. This virus is hypothesized to have contributed to the marked declines of wild freshwater eel populations in both Europe and America over the past decade (Bellec *et al.*, 2014) and further study is merited to establish its possible role in these losses.

Another group of fish rhabdoviruses is the hypothesized 'Sinistar group', which appears related to the perhabdoviruses (ICTV, 2018). This name is derived from a combination of *Siniperca* and starry flounder (Gubala, 2012). These include:

1. *Siniperca chuatsi* rhabdovirus (SCRV), occurring in mandarin fish (*Siniperca chuatsi*) in China (isolated in 1999); and
2. starry flounder rhabdovirus (SFRV) from the starry flounder (*Platichthys stellatus*) in US North Pacific waters (isolated in 2000).

Siniperca chuatsi rhabdovirus (SCRV) possesses a gene located between the *M* and *G* genes, which encodes a small protein that is absent from the perhabdoviruses (including *Perch perhabdovirus* and *Anguillid perhabdovirus*; Gadd, 2013). SCRV also infected juvenile largemouth bass (*Micropterus salmoides*) on a fish farm in China, which resulted in lethargy, irregular swimming, distended abdomens and 40% mortality (Ma *et al.*, 2013). SFRV was described from a wild starry flounder in Puget Sound, Washington (Mork *et al.*, 2004). Other members of this potential Sinistar group are:

3. eelpout rhabdovirus (EPRV), which caused mass mortality of eelpout (*Zoarces viviparous*) along the Swedish Baltic Sea coast in 2014 (Axén *et al.*, 2017); and
4. *Scophthalmus maximus* rhabdovirus (SMRV) in cultured turbot (*Scophthalmus maximus*) in China, which also displayed ability to infect grass carp (*Ctenopharyngodon idella*) (Zhang *et al.*, 2007).

Little else appears known about these Sinistar group viruses, their identities or their relationships (ICTV, 2018).

Other poorly known fish rhabdoviruses, which may be related to *Vesiculovirus* and *Sprivivirus*, and whose relationships and veracity appear unresolved, include:

1. *Monopterus albus* rhabdovirus (MoARV);
2. ulcerative disease rhabdovirus (UDRV); and
3. snakehead vesiculovirus (SHVV).

Among these, MoARV infects the farmed rice field eel (*Monopterus albus*) in China, causing skin haemorrhages, loss of equilibrium and disorganized swimming (Ou *et al.*, 2013). UDRV infects the striped snakehead (*Channa striata*) and other freshwater fishes in South-East Asia, and is distinct from SHVV (Frerichs *et al.*, 1989). SHVV was isolated in 2014 from diseased cultured hybrid snakehead fish in Guangdong Province, China, and causes high mortality (Liu *et al.*, 2019). Whole-genome analyses, accompanied by gene expression results, likely will resolve these identities and relationships, and uncover their evolutionary patterns.

4.1.1 The novirhabdoviruses

Members of the rhabdovirus genus *Novirhabdovirus* possess a unique sixth gene (Fig. 4.1) that is located between the *G* and the *L* genes (*5′-N–P–M–G–Nv–L-3′*) and encodes a 370–425 nt long non-structural 'non-virion' (Nv) protein, which plays a role in pathogenesis (ICTV, 2018). The Nv protein is believed to enhance viral replication and augment persistence of the virus in the fish host, prolonging shedding and thereby increasing spread of the virions to other hosts and over geographic areas (Ammayappan *et al.*, 2011).

The *Nv* gene exhibited more substitutions in *Piscine novirhabdovirus* genogroups than were found in other genes (4.1%; Pierce and Stepien, 2012). Similarly, research by He *et al.* (2013, 2014) detected more variation in the *Nv* gene of *Salmonid novirhabdovirus* than in its *N* and *G* genes. *Nv* gene sequences are highly divergent among the different novirhabdoviruses, displaying little sequence conservation (Kurath, 2012). In a study of challenged yellow perch (*Perca flavescens*), fish infected with *Nv*-knockout mutants of *Piscine novirhabdovirus* (VHSV-IVb) experienced less mortality than did wild-type challenged fish (Ammayappan *et al.*, 2011).

Novirhabdovirus infections cause acute haemorrhaging in multiple organs of infected fishes, which exhibit common signs of darkened colour, external and internal lesions, and abdominal distension (Kurath, 2012). Disease most commonly occurs in juveniles, with resistance increasing with age and maturity of the hosts. Notably, surviving fishes can continue to shed *Piscine novirhabdovirus* (VHSV-IVb) in challenge experiments for up to 3 months following infection at 12°C (Kim and Faisal, 2010). *Salmonid novirhabdovirus*, *Piscine novirhabdovirus* and *Hirame novirhabdovirus* optimally replicate at 12–15°C and are inactivated at water temperatures of 18–20°C. In contrast, *Snakehead novirhabdovirus* occurs in warm waters, optimally replicating at 15–25°C and is inactivated at 30–37°C (Kurath, 2012).

Of the novirhabdoviruses, *Salmonid novirhabdovirus* and *Hirame novirhabdovirus* are the most closely related, and form the sister clade to *Piscine novirhabdovirus*, with *Snakehead novirhabdovirus* being more distantly related. Their phylogenetic relationships, based on the whole-genome analyses, are shown in Fig. 4.3.

Piscine novirhabdovirus = *viral haemorrhagic septicaemia virus (VHSV)*

Piscine novirhabdovirus (= VHSV) infects over 140 species of fishes in marine, estuarine and freshwater environments across the northern hemisphere, making it one of the world's most serious fish infectious diseases (Escobar *et al.*, 2018). *Piscine novirhabdovirus* is transmitted most readily when fish congregate during the spring spawning season in temperatures of 9–12°C, with infection leading to erratic swimming behaviour, exophthalmia (bulging eyes), bloated abdomens, and extensive external and internal bleeding that leads to liver and kidney damage (Winton and Einer-Jensen, 2002; Daniels and Watanabe, 2010). Fish-to-fish transmission occurs via shed mucus and urine (Winton and Einer-Jensen, 2002). *Piscine novirhabdovirus* (VHSV-IVb) viral particles can live for up to 14 days in the water at 12°C and are transported via diverse vectors, including boating, ballast water, fishing tackle and animals (e.g. amphipod crustaceans, leeches, turtles and birds) (Faisal and Schulz, 2009; Bain *et al.*, 2010; Faisal and Winters, 2011; Goodwin and Merry, 2011).

Piscine novirhabdovirus (e.g. VHSV-IVa) first was described from European cultured salmonid fishes by Schäperclaus (1938), and later was isolated in 1962 from infected rainbow trout (VHSV genogroup I, isolate DK-F1; Einer-Jensen *et al.*, 2004). *Piscine novirhabdovirus*, unlike *Salmonid novirhabdovirus*, has low virulence in rainbow trout (Kurath, 2012). Four *Piscine novirhabdovirus* genogroups (also known as 'strains') and several subgenogroups ('substrains') have been described; comparative analyses of the isolates indicate that there is considerable genetic diversity within the genogroups (Pierce and Stepien, 2012; Stepien *et al.*, 2015). The common evolutionary ancestry of the four *Piscine novirhabdovirus* genogroups (VHSV-I–IV) has been traced to a hypothesized marine origin in the North Atlantic Ocean (Figs 4.4 and 4.5; Pierce and Stepien, 2012).

Evolutionary trees indicate that the *Piscine novirhabdovirus* evolutionary lineage then diverged into two primary clades, one comprising genogroup VHSV-IV in North America and Asia (along with a recent discovery in lumpfish in Iceland; Guðmundsdóttir *et al.*, 2019) and the other containing genogroups I–III that diversified in the North-Eastern Atlantic region (i.e. Europe; including the North, Baltic and Mediterranean seas) (Pierce and Stepien, 2012; see Figs 4.3 and 4.4). Across all *Piscine*

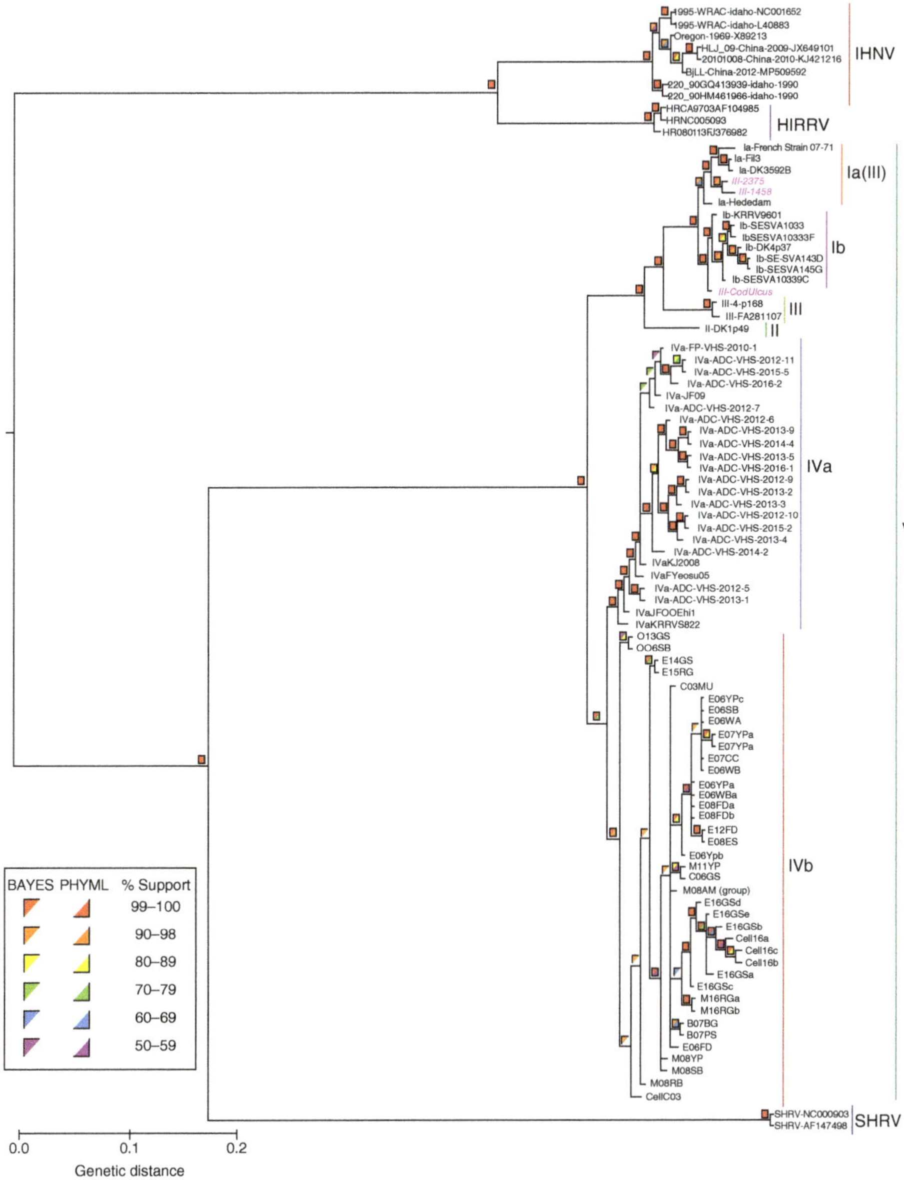

Fig. 4.3. Phylogenetic tree of *Novirhabdovirus* relationships, based on available full-genome sequences (see Tables 4.1 and 4.2) and maximum likelihood (PHYML v3.0; Guindon and Gascuel, 2003; program: https://ngphylogeny.fr/workflows/oneclick/ (accessed 8 March 2020)) and Bayesian analyses (Mr. Bayes v3.2.7a; Ronquist *et al.*, 2012; program: Zhang *et al.*, 2019; http://nbisweden.github.io/MrBayes/authors.html (accessed 8 March 2020)). Coloured squares designate support values, top left half = Bayesian posterior probabilities, bottom right = 500 bootstrap pseudoreplicates for maximum likelihood. Tree is rooted to the *Snakehead novirhabdovirus* (SHRV, GenBank: AF147498). Note that some isolates in subgenogroup III (italics) are paraphyletic, rendering them incorrectly placed in III. They need to be placed into I, to correct their taxonomy.

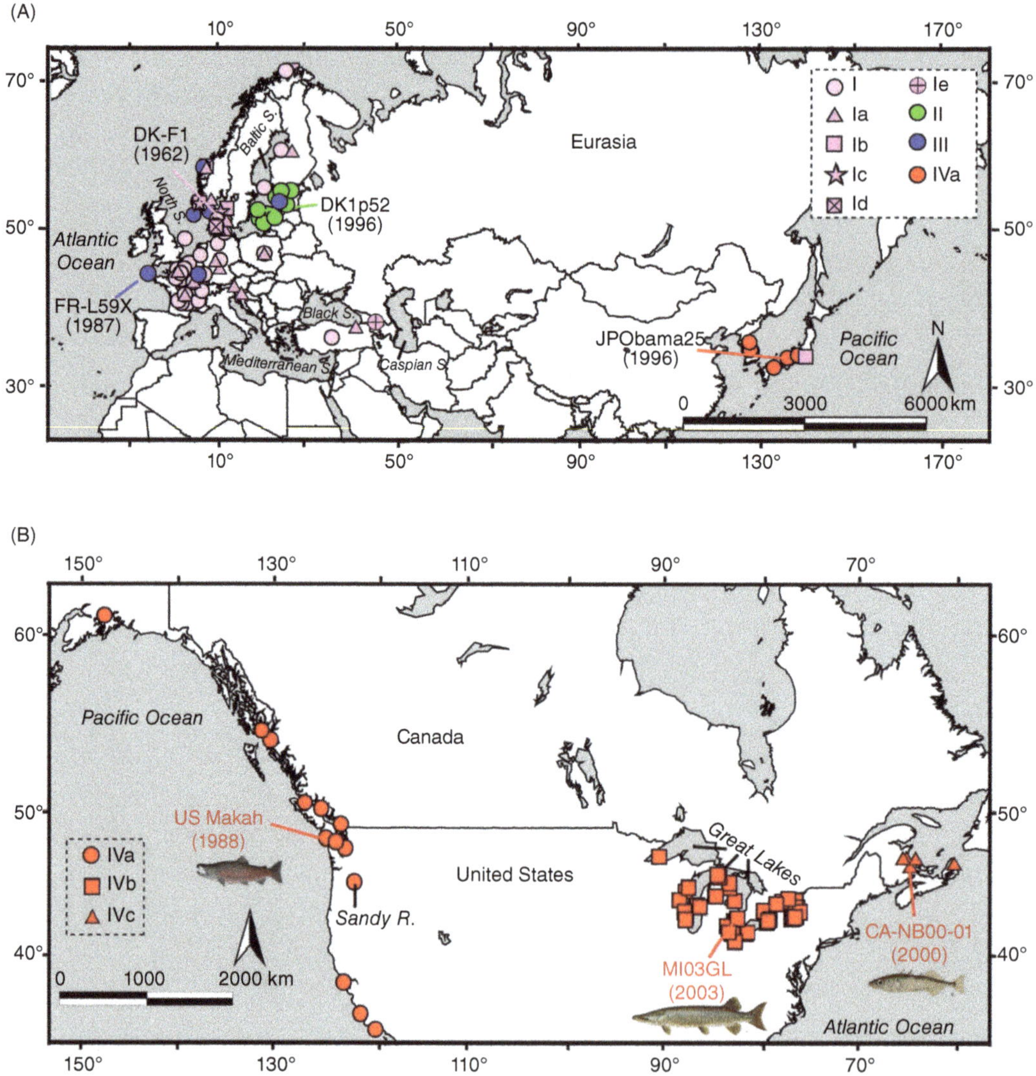

Fig. 4.4. Maps showing distributions of *Piscine novirhabdovirus* VHSV genogroups in (A) Eurasia (genogroups I–IV) and (B) North America (genogroup IV).

novirhabdovirus genogroups (VHSV-I–IV), an average of three additional host fish species have been identified each year since 1962 (Escobar *et al.*, 2018). Thus, recognition of this fish viral pathogen and its impacts on fisheries and aquaculture are growing.

Piscine novirhabdovirus genogroup (also termed 'strain') VHSV-II differentiated in the estuarine waters of the Baltic Sea region and comprises the sister group to genogroups I and III, with I mostly occurring in freshwater and III in marine and estuarine waters (Figs 4.3 and 4.4). Genogroup I possesses a wide and diverse geographic range across western Europe, having several subgenogroups and the greatest number of fish host species (Kurath, 2012; Pierce and Stepien, 2012), and exerting significant aquaculture impacts (Abbadi *et al.*, 2016;

Ghorani *et al.*, 2016). Genogroup III spread from wild marine fishes to nearby turbot farms during the 1990s (Stone *et al.*, 1997). Our whole-genome phylogeny (Fig. 4.3) indicates that genogroup III needs to be redefined, as some III sequences belong to the Ia subgenogroup clade.

Genogroup VHSV-IV comprises the sister group to the VHSV-I–III clade (Fig. 4.3) and occurs as three allopatrically distributed subgenogroups in North America, as IV a–c (Pierce and Stepien, 2012; Stepien *et al.*, 2015). Subgenogroup VHSV-IVa infects North-Eastern Pacific salmonids and many marine fishes, whereas VHSV-IVb is endemic to the freshwater Great Lakes and IVc occurs in marine/estuarine North Atlantic waters (Fig. 4.4B). A recent discovery of a divergent form of IVa from lumpfish in Greenland (Guðmundsdóttir *et al.*, 2019) lends support to the hypothesis of Pierce and Stepien (2012) that genogroup IV originated in the North Atlantic Ocean. Moreover, the appearance of IVa in the North-Eastern Pacific coastal region (in the 1980s) may have resulted from inadvertent transport in aquaculture to the North American west coast from the Atlantic coast, perhaps during Atlantic salmon (*Salmo salar*) introductions.

Subgenogroup (also termed 'substrain') VHSV-IVa has a marine reservoir and is prevalent in the keystone foraging species of Pacific herring (*Clupea pallasii*) and Pacific sardine (*Sardinops sagax*) (see Garver *et al.*, 2013). VHSV-IVa also commonly occurs in Pacific cod (*Gadus macrocephalus*), chinook salmon (*Oncorhynchus tshawytscha*) and coho salmon (*Oncorhynchus kisutch*). IVa leads to periodic disease outbreaks, which variously are triggered by unusual temperatures, high fish host densities and/or pollution (Hershberger *et al.*, 1999). Two separate VHSV subgenogroup lineages are believed to have independently appeared in the North-Western Pacific region (Asia) in 1996, with subgenogroup Ib originating from the west and IVa from the east (Pierce and Stepien, 2012).

The Great Lakes' endemic *Piscine novirhabdovirus* subgenogroup – VHSV-IVb – was back-traced to a 2003 muskellunge (*Esox masquinongy*) from Lake St. Clair (Ammayappan and Vakharia, 2009). Its first outbreaks occurred during the 2005 and 2006 spring months, manifested in massive fish kills across the Great Lakes (Groocock *et al.*, 2007; Lumsden *et al.*, 2007; Thompson *et al.*, 2011). Significant spatial patterns of genetic divergence have characterized the virus' populations among the Upper, Central and Lower Great Lakes, along with significant differentiation and diversification over time (Stepien *et al.*, 2015; Niner, 2019; see Figs 4.5 and 4.6). VHSV-IVb has genetically diversified during its less than two-decade history in the Great Lakes, accompanied by declining outbreaks and virulence (Figs 4.5 and 4.6).

Among *Piscine novirhabdovirus* genogroups, VHSV-IVb infects the broadest fish host range, suggesting strong ability to spread to naïve species (Escobar *et al.*, 2018). Continuing evolutionary diversification likely has allowed the virus to persist at low levels in resident fish populations, and potentially may facilitate successful spread to new habitats and unacclimatized hosts (Stepien *et al.*, 2015). It is possible that *Piscine novirhabdovirus* genogroups may become more successful if they move northward into new habitats and if they infect additional species, over the coming course of climatic change. Additionally, some genogroups could be transported to the southern hemisphere and possibly infect fish hosts there.

Salmonid novirhabdovirus = *infectious haematopoietic necrosis virus (IHNV)*

Salmonid novirhabdovirus (ICTV, 2018) (= IHNV) originally was known to infect salmon and trout along the Pacific coast of the USA and Canada, and later spread via aquaculture to Asia and Europe (Kurath, 2012). Epidemics of *Salmonid novirhabdovirus* first were reported from Washington and the Oregon fish hatcheries during the 1950s (Rucker *et al.*, 1953). Today, most wild Pacific salmonid populations and most watersheds house some *Salmonid novirhabdovirus* (Kurath, 2012). The virus is transmitted by shedding in waste products, sexual fluids and mucus, and through the surrounding water, entering other fish through the gills and the vent. It also spreads through transport of virus-contaminated fish eggs in aquaculture and in the wild (Dixon *et al.*, 2016). *Salmonid novirhabdovirus* can survive in fresh water for up to 1 month at 9°C (Bergmann *et al.*, 2003). Vertical transmission via eggs has been described (OIE, 2018).

Clinical signs of *Salmonid novirhabdovirus* infection include abdominal distension, bulging eyes, darkened skin, anaemia, pale gills, and haemorrhages located in the mouth, behind the head, pectoral fins and near the vent (OIE, 2018). Infected fishes typically are lethargic, which may be punctuated by periods of frenzied swimming (OIE, 2018).

Salmonid novirhabdovirus has a smaller host range than does *Piscine novirhabdovirus* (Kurath,

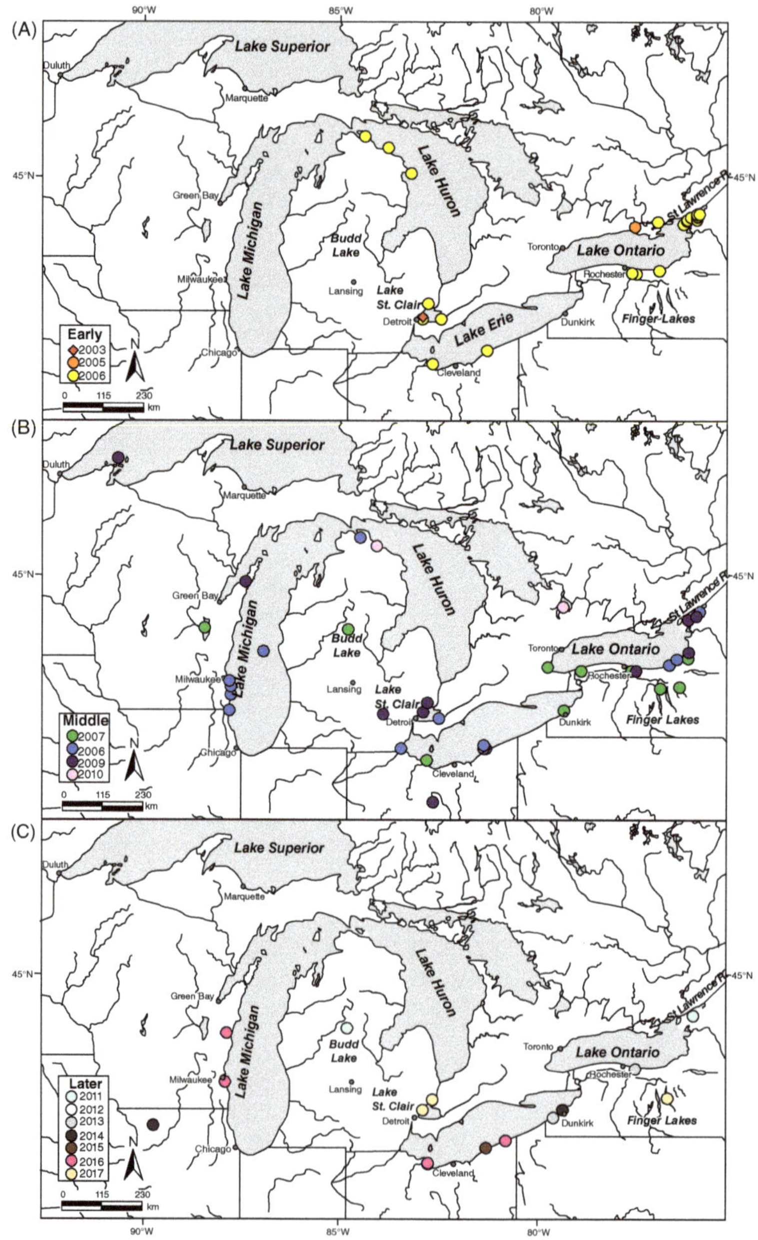

Fig. 4.5. Maps showing locations (circles, coloured by year) of *Piscine novirhabdovirus* VHSV-IVb isolates in the Great Lakes, analysed per time period: (A) early (2003–2006), (B) middle (2007–2010) and (C) later (2011–2019).

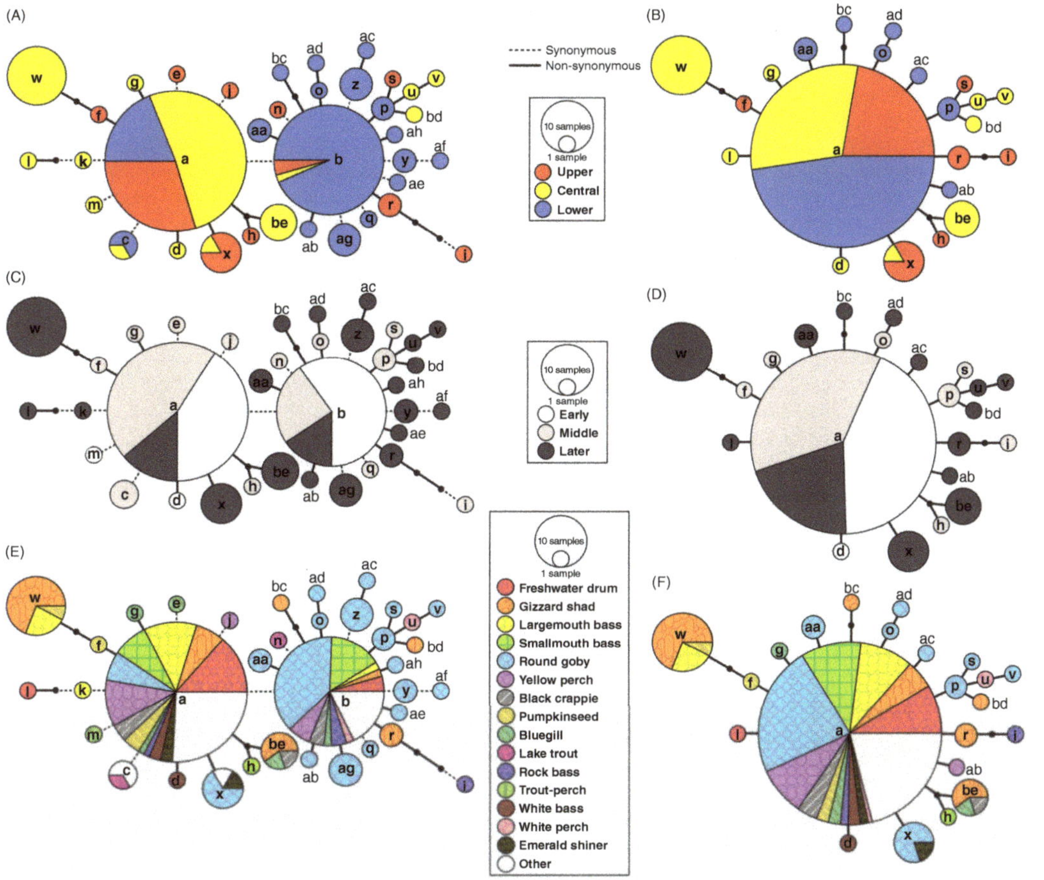

Fig. 4.6. *G* gene haplotype networks for *Piscine novirhabdovirus* VHSV-IVb in the Laurentian Great Lakes, constructed from partial *G* gene sequences (669 nt) of 176 isolates using POPART (https://popart.otago.ac.nz (accessed 3 March 2020)) and TCS (Clement *et al.*, 2000), shown grouped among: (A, B) Great Lakes regions (Upper, Central, Lower), (C, D) time periods (Early, Middle, Later) and (E, F) host species. (A), (C) and (E) are based on nucleotide substitutions and (B), (D) and (F) on amino acid changes. Circles are sized according to frequency of the haplotype in the population. Lines denote a single substitution step between haplotypes, with dashed lines for synonymous changes and solid lines for non-synonymous changes. Small, unlabelled black circles represent hypothesized haplotypes. The 'Other' category in (E) and (F) contains all host species in which three or fewer non-unique isolates were detected: alewife (*Alosa pseudoharengus*), amphipod (*Diporeia* spp.), brown bullhead (*Ameiurus nebulosus*), burbot (*Lota lota*), channel catfish (*Ictalurus punctatus*), chinook salmon (*Oncorhynchus tshawytscha*), common carp (*Cyprinus carpio*), cisco (*Coregonus artedi*), lake whitefish (*Coregonus clupeaformis*), leech (*Myzobdella lugubris*), muskellunge (*Esox masquinongy*), northern pike (*Esox lucius*), rainbow trout (*Oncorhynchus mykiss*), sea lamprey (*Petromyzon marinus*), shorthead redhorse (*Moxostoma macrolepidotum*) and walleye (*Sander vitreus*).

2012). Five major genetic groups have been described, based on gene sequences. Genogroups 'U', 'M' and 'L' originated in the North American Pacific coastal salmonid fishes and were named for their coastal geographic distribution patterns (Upper, Middle, Lower coastal groups), which also differ in their host specificities (Kurath, 2012). Genogroup 'U' primarily infects *Oncorhynchus nerka* (sockeye and kokanee salmon), 'M' occurs in *Oncorhynchus mykiss* (rainbow and steelhead trout) and 'L' in *O. tshawytscha* (chinook salmon, primarily in California and southern Oregon). 'U' is postulated to be the ancestral *Salmonid novirhabdovirus* genogroup, whereas 'M' possesses the greatest genetic diversity and has been found to be the fastest evolving genogroup (Kurath, 2012). Epidemics of the 'U' genotype have occurred in aquaculture sea pens of Atlantic salmon *S. salar* reared along the Pacific coast of Canada (Saksida, 2006). Within the 'U' genogroup in North America, subgenogroup 'UP' is more common in sockeye salmon (*O. nerka*), whereas 'UC' infects chinook salmon and steelhead trout, revealing some geographic regional differences (Black *et al.*, 2016).

Salmonid novirhabdovirus genogroup 'U' was introduced to Asia (Japan in 1971) and 'M' to Europe, both via fish eggs used for aquaculture (Enzmann *et al.*, 2010). *Salmonid novirhabdovirus* first was reported in Europe in 1987 in France and Italy, and then in 1992 in Germany. European 'M' descendants have diverged to become the 'E' genogroup, which infects rainbow trout in aquaculture, and there is evidence for other regionally differentiated clades (Enzmann *et al.*, 2010). First reports of *Salmonid novirhabdovirus* in China occurred in 1985 and in Korea in 1991, which likely originated from Japan. Asian *Salmonid novirhabdovirus* now constitutes a distinct and genetically diverse 'J' lineage (descendent from 'U'), which primarily is found in rainbow trout (Xu *et al.*, 2018). In recent years, *Salmonid novirhabdovirus* has been spreading in Asia and the Middle East (ICTV, 2018). The evolutionary history of *Salmonid novirhabdovirus* thus has undergone distinctive geographic and temporal patterns of rapid differentiation.

Hirame novirhabdovirus = *hirame rhabdovirus* (HIRRV)

Hirame novirhabdovirus (ICTV, 2018) (= HIRRV) predominantly infects cultured fishes in Asia, causing acute haemorrhages in Japanese flounder (*Paralichthys olivaceus*), black sea bream (*Acanthopagrus schlegeli*), stone flounder (*Kareius bicoloratus*) and sea bass (*Lateolabrax maculatus*) (Kurath, 2012). It is not known from wild fishes. Infected fishes usually have mouth haemorrhages, enlarged spleens, reddened fins and distended abdomens (Kurath, 2012). *Hirame novirhabdovirus* also has been identified in cultured grayling (*Thymallus thymallus*) from Poland (Borzym *et al.*, 2014). *Hirame novirhabdovirus* is the sister group to *Salmonid novirhabdovirus* as discerned from whole-genome phylogenetic analysis (Fig. 4.2).

Snakehead novirhabdovirus = *snakehead rhabdovirus* (SHRV)

Snakehead novirhabdovirus (ICTV, 2018) (= SHRV) infects warm-water freshwater and estuarine fishes in the wild and in pond culture, from Asia and the Pacific region (see Alonso *et al.*, 2004). It originally was isolated from a diseased cultured snakehead fish (*C. striata*) during an epizootic outbreak in Thailand (Alonso *et al.*, 2004). Cultured snakehead, walking (= Philippine) catfish (*Clarius batrachus*) and sand (= marble) goby (*Oxyeleotris marmorata*) have been the most affected hosts (Hedrick, 1986). *Snakehead novirhabdovirus* disease is characterized by severe ulcerative dermal necrosis, at an optimal temperature range between 28 and 31°C. *Snakehead novirhabdovirus* is more distantly related to the other novirhabdoviruses (i.e. *Hirame novirhabdovius*, *Piscine novirhabdovirus* and *Salmonid novirhabdovirus* are more closely related to one another) (Fig. 4.3; Pierce and Stepien, 2012). Knockout gene analyses have shown that the *Nv* gene of SHRV is not required for pathogenesis (Alonso *et al.*, 2004).

4.1.2 The spriviviruses

The genus *Sprivivirus* is believed to be closely related to vesiculoviruses, which primarily cause diseases in farm animals, including horses, cattle and pigs (see Fig. 4.2 phylogeny). Spriviviruses infect fishes, including *Carp sprivivirus* and the closely related *Pike sprivivirus*, along with the latter's grass carp rhabdovirus (GCRV) and tench rhabdovirus (TRV) variants (Rodriguez and Pauszek, 2012). Antibodies of PFRV, GCRV and TRV cross-react, and their sequences are closely related in a monophyletic clade, denoting a common evolutionary origin (Rodriguez and Pauszek, 2012), of which three have been grouped together as the *Pike sprivivirus* species (ICTV, 2018).

Carp sprivivirus = *spring viraemia of carp virus (SVCV)*

Carp sprivivirus (ICTV, 2018) (= SVCV), which also is called 'infectious dropsy', occurs in the wild in a variety of cyprinids and other freshwater fish groups (Kurath, 2012). Infected cyprinid species include common carp (*Cyprinus carpio carpio*), koi carp (*Cyprinus carpio koi*), Crucian carp (*Carassius carassius*), silver carp (*Hypophthalmichthys molitrix*), bighead carp (*Hypophthalmichthys nobilis*), grass carp (also known as white amur) (*Ctenopharyngodon idella*), goldfish (*Carassius auratus*), orfe (*Leuciscus idus*), tench (*Tinca tinca*), bream (*Abramis brama*), emerald shiner (*Notropis atherinoides*) and roach (*Rutilus rutilus*) (ICTV, 2018). The varieties of its non-cyprinid fish hosts include pike (*Esox lucius*), largemouth bass (*M. salmoides*), rainbow trout, bluegill sunfish (*Lepomis macrochirus*), Siberian sturgeon (*Acipenser baerii*) and sheatfish (also known as European catfish or wels) (*Silurus glanis*). *Carp sprivivirus* is reportable to the World Organization for Animal Health (ICTV, 2018; OIE, 2018).

Carp sprivivirus is prevalent in ornamental species, especially goldfish and koi carp, which regularly are transported around the world, along with the virus (OIE, 2018). The disease first was noted in 1930, and the virus originally was isolated in 1971 from common carp in a Yugoslavian fish farm. It occurs in a variety of European countries and in the USA, Canada, China, the Middle East and South America. Phylogenetic analyses based on the G gene have revealed four subgenogroups across the world, which appear to have evolved independently in different geographic regions (OIE, 2018). These include genogroup Ia from Asia, the UK and North America; genogroups Ib and Ic from Eastern Europe; and Id from the UK and some other European countries (summarized in ICTV, 2018).

Carp sprivivirus is shed into the water via faeces and urine (OIE, 2018). The virus can be transmitted directly through the water, as well as through ectoparasites, including carp lice (*Argulus foliaceus*) and leech (*Pisicola geometra*) (ICTV, 2018). It has been shown to remain viable outside the host for 5 weeks in river water at 10°C, for more than 6 weeks in pond mud at 4°C and for 4 days in pond mud at 10°C (Ahne, 1976). *Carp sprivivirus* enters its fish hosts through the gills, and the virus then spreads to the liver, kidney, spleen and digestive tract (OIE, 2018). Fishes with *Carp sprivivirus* infections often appear darker in colour, and may exhibit bulging eyes, pale gills, haemorrhages on the skin, base of the fins and the vent, distended abdomens and a protruding vent (anus), often with trailing mucus/faeces. Behaviourally, infected fishes become lethargic, separate from others and may exhibit loss of equilibrium. *Carp sprivivirus* most commonly affects young individuals, less than 1 year of age (OIE, 2018). The virus is believed to have been spread internationally via aquaculture (ICTV, 2018) and likely will spread further in the future.

Pike fry sprivivirus = *pike fry rhabdovirus (PFRV)*

Pike fry sprivivirus (ICTV, 2018) (= PFRV) is a member of the genus *Sprivivirus*, which causes a disease similar to *Carp sprivivirus* in hatchery-reared pike fry (*E. lucius*) in Europe, and first appeared in the Netherlands in 1956 (Ahne *et al.*, 1998). It also has been reported from barbel (*Barbus barbus*) in the Czech Republic (Vicenova *et al.*, 2011). Challenge experiments showed that *Pike sprivivirus* was capable of infecting a variety of cyprinid species that were tested, and thus should be considered a threat to cyprinid fish farms (Haenin and Davidse, 1993). Its clinical signs include lethargy, bulging eyes and haemorrhages (usually around the pelvic fins). Control methods include isolation of infected individuals and iodophor treatment of eggs to remove surface virus contamination (OIE, 2018).

4.1.3 The perhabdoviruses

The genus *Perhabdovirus* is most closely related to the lenteviruses (Fig. 4.2), which clade then comprises the sister group to the (Spriviviruses + Vesiculoviruses) clade. Perhabdoviruses cause diseases in farmed perciform fishes and some wild species (ICTV, 2018). Three viral species are recognized in the genus (according to ICTV, 2018), including: *Perch perhabdovirus*; *Anguillid perhabdovirus* comprising both the eel virus European X and eel virus American; and *Sea trout perhabdovirus*, which contains both the lake trout rhabdovirus (LTRV) and the Swedish sea trout virus (SSTV). Clinical disease occurs at 5–18°C, appearing maximal at temperatures <10°C, when the host immune system appears suppressed (ICTV, 2018).

Perch perhabdovirus = *perch rhabdovirus (PRV)*

Perch perhabdovirus (ICTV, 2018) (= PRV) infects a variety of European freshwater percid fishes, including the European pike-perch (*S. lucioperca*), cultured European perch (*Perca fluviatilis*), grayling and largemouth bass (ICTV, 2018). *Perch perhabdovirus* is both horizontally and vertically transmitted and has resulted in mass mortality of pike-perch larvae (Bigarré *et al.*, 2017).

Sea trout perhabdovirus

The Swedish sea trout (SSTV) variant of *Sea trout perhabdovirus* has occurred in the closely related brown trout (*S. t. lacustris*) in Finland and sea trout (*S. t. trutta*) from Sweden (ICTV, 2018).

4.2 Diagnosis of the Fish Rhabdoviruses and Their Hosts' Immune Responses

Teleost fish hosts possess the vertebrate components of innate and adaptive immunity, which viruses must evade or suppress during viral replication (Purcell *et al.*, 2012). Neutralizing antibodies provide hosts with long-term protection, and several investigations also have described cell-mediated immunity. Innate immunity, especially the interferon (IFN) system, is induced rapidly in response to infection. IFN induces an antiviral state in which host cells block mRNA transcription and translation to prevent viral replication in infected cells (Katze, 2002). It also upregulates expression of hundreds of early response genes, known collectively as IFN-stimulated genes (ISGs), which impact a variety of additional cellular processes to slow the replication of viruses and prime the adaptive immune response (Schneider *et al.*, 2014).

To combat this coordinated innate immune response, RNA viruses that replicate entirely within cytoplasm frequently shut down host transcription globally, as this frees up cellular translational machinery for viral gene products and inhibits host antiviral responses by preventing the synthesis of antiviral proteins (Purcell *et al.*, 2012; OIE, 2018). Rhabdoviruses commonly utilize the matrix (M) protein to suppress IFN and other innate immune responses. The M protein has been implicated in cellular apoptosis and transcriptional suppression in *Piscine novirhabdovirus* and *Salmonid novirhabdovirus* (Choi *et al.*, 2011; Ke *et al.*, 2017). Research studies have indicated that the Nv protein from the *Piscine novirhabdovirus* VHSV-IVb genogroup suppressed apoptosis (Ammayappan and Vakharia, 2011) and Nv from the Ia genogroup suppressed innate immune responses (Choi *et al.*, 2011; Kim and Kim, 2013). To date, however, the Nv mechanism of action remains unknown, and these conflicting reports on the impact on cellular effects suggest that much remains to be learned about Nv function. Despite the reported anti-host functions of M and Nv, primarily aimed at suppressing host antiviral responses, survivors of *Piscine novirhabdovirus* and *Salmonid novirhabdovirus* have exhibited broad protective immunity against reinfection (Kurath, 2012). Subsequent adaptive immune responses remain intact in fish that clear the primary infection, thereby establishing lifelong immunity.

The currently accepted method for detection and identification of fish viruses is isolation of the virus in cell culture, followed by its immunological or molecular identification (OIE, 2018). Since there is considerable variation in the strength and duration of the serological responses of fish to virus infections, detection of fish antibodies to viruses is not currently accepted for routine diagnostics. However, positive serological response is considered presumptive evidence of past exposure to a virus (Jorgensen *et al.*, 1991; OIE, 2018).

Carp sprivivirus is diagnosed by clinical signs, isolation of virus in cell culture and molecular methods (Ahne *et al.*, 2002). Antibodies against *Carp sprivivirus* are effective, but also cross-react with the *Pike sprivivirus*, indicating that the two viruses are closely related. *Carp sprivivirus* and *Pike sprivivirus* can be distinguished by the ribonuclease protection assay (Ahne *et al.*, 2002), although a newer reverse transcription-polymerase chain reaction (RT-PCR) strategy may allow for more rapid detection of *Carp sprivivirus* that is less technologically challenging than ribonuclease protection (Shimahara *et al.*, 2016).

Most fish viruses are diagnosed by various real-time quantitative RT-PCR (RT-qPCR) assays. For example, for *Salmonid novirhabdovirus*, an accurate qPCR assay developed by Pierce *et al.* (2013a,b) incorporates synthetic competitive template internal standards for quality control to circumvent false negative results. Results demonstrated high signal-to-analyte response and a linear dynamic range that spanned seven orders of magnitude (R^2 = 0.99), ranging from 6 to 6 × 10^6 molecules. Infected fishes were found to harbour levels of virus that ranged to 1.2 × 10^6 *Piscine novirhabdovirus* (VHSV-IVb) molecules/10^6 actb1 (β-actin) molecules, with 1000 viral molecules as a cut-off point for clinical signs of disease (Fig. 4.7). This assay is rapid, inexpensive, and possesses significantly greater accuracy than other published RT-qPCR tests and traditional cell culture diagnostic tests.

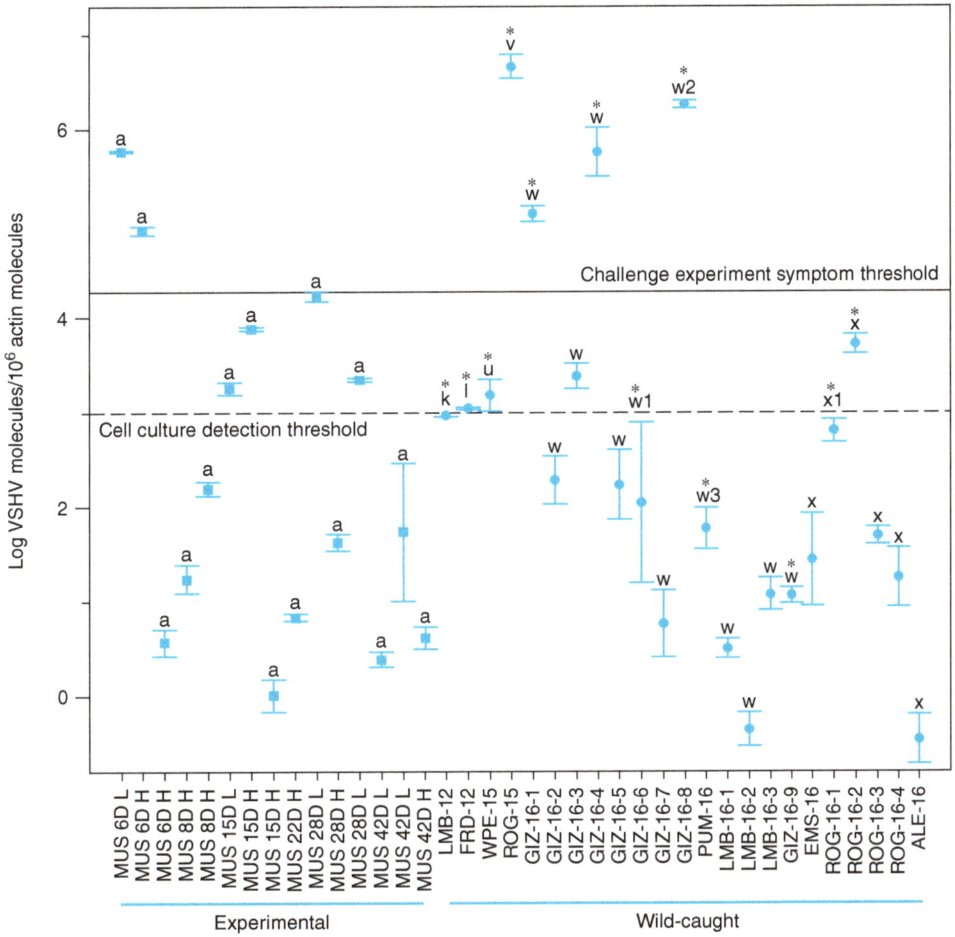

Fig. 4.7. Concentrations of *Piscine novirhabdovirus* VHSV-IVb (with their standard errors represented by vertical bars) in wild-caught fish tissues, compared with results from experimental laboratory haplotype 'a' challenged muskellunge, determined with our laboratory qPCR assay using internal standards (method published by our research team (led by C.A.S. and D.W.L.) in Pierce *et al.*, 2013a,b). Laboratory samples (squares) are named by the number of days (6–42D) after VHSV-IVb inoculation, H = high virus dosage (1 × 10^5 plaque-forming units (pfu)/ml) and L = low dosage (100 pfu/ml). Haplotype of each sample is listed above its standard error bars. * above the haplotype indicates that the sample was sequenced for both the *G* gene and the whole genome analyses. Solid line denotes the experimental symptom threshold and dashed line the cell culture detection threshold (adapted from Pierce *et al.*, 2013b). Wild-caught samples (circles) are designated by abbreviated common name, followed by collection year and sample number. Fish species names: MUS = muskellunge, LMB = largemouth bass, FRD = freshwater drum, WPE = white perch, ROG = round goby, GIZ = gizzard shad, PUM = pumpkinseed, EMS = emerald shiner, ALE = alewife.

4.3 Expected/Potential Spread of the Pathogen or Its Hosts to New Geographical Area(s)

Virus sources for *Salmonid novirhabdovirus* and *Piscine novirhabdovirus*, along with other fish rhabdoviruses, include wild fish reservoirs, as well as fish and fish eggs used in aquaculture. Movements of fish eggs spread *Salmonid novirhabdovirus* to Japan and Europe (see Kurath, 2012). It appears possible that aquaculture transport spread *Piscine novirhabdovirus*

VHSV-IV from the North American east (Atlantic) coast to the west (Pacific Coast; see Pierce and Stepien, 2012). An additional spread source is the use of raw fish products in fish feed, which was a mechanism that spread *Piscine novirhabdovirus* in Europe (Einer-Jensen *et al.*, 2004) and *Salmonid novirhabdovirus* in Washington sockeye hatcheries (Kurath, 2012). The adoption of pasteurized fish feeds eliminated or decreased that avenue of spread. Other transmission mechanisms have included housing aquacultured species in close location to wild populations (or related host species) and using common water supplies or net pens. Introduction of exotic species into areas where the virus is endemic constitutes another infection route, as occurred when Atlantic salmon was cultured in marine net pens (Saksida, 2006).

Host switching by *Salmonid novirhabovirus* and *Piscine novirhabdovirus* among cultured salmonid species has been common and may alter virulence (Kurath, 2012). For example, *Salmonid novirhabovirus* 'M' and 'J' genogroups and *Piscine novirhabdovirus* VHSV-Ia are believed to have differentiated with evolutionary host switches to cultured rainbow trout, which then were accompanied by higher virulence and rapid evolution (Kurath, 2012; Einer-Jensen *et al.*, 2014). Such adaptation to new hosts and unique conditions may readily occur in aquaculture farms. These factors also are likely to influence wild populations.

Rhabdoviruses often diversify into a multidirectional 'cloud-like' burst of closely related variants over time, from one or more central ancestral types, termed a 'quasispecies' pattern (Belshaw *et al.*, 2008; Pereira and Amorim, 2013; Andino and Domingo, 2015). The resultant pool of similar variants may serve as a genetic reservoir to facilitate adaptation to new hosts and environments (Quer *et al.*, 1996; Lauring and Andino, 2010; Andino and Domingo, 2015), as has been proposed for *Piscine novirhabdovirus* (Pierce and Stepien, 2012; Stepien *et al.*, 2015; see Fig. 4.6). Furthermore, a genetically diverse host pool, whether in terms of population or species, can increase diversity of viral sequences (Ojosnegros and Beerenwinkel, 2010); this also appears to be the case for *Piscine novirhabdovirus*. Such diversification may facilitate evading and overcoming host immune system responses and may be enhanced by environmental change.

4.4 Increase in Pathogen-induced Mortality

Disease patterns are influenced by water temperature, fish physiological condition, population density and stress factors. For example, *Carp sprivivirus* typically manifests at water temperatures of 5–18°C and its clinical signs of disease appear greatest at <10°C, at which the host immune response may be suppressed or delayed. *Piscine novirhabdovirus* is most readily transmitted when fish congregate during the spring spawning season in temperatures of 9–12°C (Winton and Einer-Jensen, 2002; Daniels and Watanabe, 2010). Its viral particles have been documented for up to 14 days at 12°C in the water and are transported via diverse vectors, including boating, ballast water, fishing tackle and animals (e.g. amphipod crustaceans, leeches, turtles and birds) (Faisal and Schulz, 2009; Bain *et al.*, 2010; Faisal and Winters, 2011; Goodwin and Merry, 2011). Fish-to-fish transmission occurs mainly through infected mucus and urine that are shed into the environment, and this likely increases during spawning aggregations (Winton and Einer-Jensen, 2002).

Temperature plays a major role in fish–rhabdovirus interactions and responses since fish are poikilothermic vertebrates and highly sensitive to temperature changes. The severity of disease often decreases at higher temperatures, which may reflect greater fish immune response at the higher temperatures (see Purcell *et al.*, 2012). Individual fish surviving rhabdovirus infections develop a robust immune response and clear the virus. In contrast, fish held at lower temperatures (e.g. 4–10°C) may rely more on innate immunity and display reduced specific immune responses (e.g. Lorenzen *et al.*, 2009). Low or cold water temperatures are also linked to persistent rhabdoviral infections, with infected rainbow trout at 4°C retaining *Piscine novirhabdovirus* in the brain for over 400 days post-infection, yet without clinical signs or detectable serum neutralizing antibodies. This observation may indicate the suppression of the adaptive immune response at colder temperatures (Neukirch, 1986). As discussed in Section 4.6.3, climate change also may impact the spread of the disease, if waters warm earlier in the season, thereby altering fish spawning and available food for developing young. Resistance to *Piscine novirhabdovirus* and other fish rhabdovirus infections also has been shown to be modulated by diet (Beaulaurier, 2012). Other environmental factors, including age, diet, seasonality and reproductive status, likely influence fish immune response to rhabdoviruses (Purcell *et al.*, 2012).

Ke *et al.* (2017) screened each of the six *Piscine novirhabdovirus* structural and non-structural genes, identifying M as the most potent anti-host protein, regulating cytopathicity and host transcriptional suppression. M of *Piscine novirhabdovirus* (VHSV-IVb) suppressed the host mitochondrial antiviral signalling protein (MAVS) and type I IFN-induced gene expression in a dose-dependent manner. M of *Piscine novirhabdovirus* VHSV-Ia subgenogroup F1 was significantly less potent than the VHS-IVb M at inhibiting general transcription in EPC (endothelial progenitor) cells despite differing by just four amino acids (Ke *et al.*, 2017). Ultimately, viral determinants of pathogenicity are more complex *in vivo*, with *Piscine novirhabdovirus* VHSV-Ia and VHSV-IVb gene swaps among multiple genes demonstrating a non-restrictive role for G, L or Nv proteins in promoting virulence in trout (Yusuff *et al.*, 2019). More recently, the combined actions of the N and P proteins have been implicated in VHSV-IVb versus Ia host-specific virulence (Vakharia *et al.*, 2019; Yusuff *et al.*, 2019). Future work thus should focus on the relative roles of these genes and, ultimately, analyses of the mechanisms by which these gene products impact infectivity, host response or cellular function in directing host-dependent effects.

4.5 Control and/or Prevention

For all novirhabdoviruses, DNA vaccines with the viral glycoprotein gene have been developed, including *Salmonid novirhabdovirus*, *Piscine novirhabdovirus*, *Hirame novirhabdovirus* and *Snakehead novirhabdovirus* (Kurath, 2012). For *Salmonid novirhabdovirus*, killed vaccines and a DNA vaccine are licensed for commercial use in Atlantic salmon net-pen aquaculture on the west coast of North America, which is delivered by injection (OIE, 2018). Vaccines developed to date necessitate individual fish handling and typically are injected (Purcell *et al.*, 2012). The major barrier that remains to commercial vaccination is the ability to mass vaccinate small fishes.

For *Piscine novirhabdovirus*, a vaccine containing a DNA plasmid, with a cytomegalovirus (CMV) promoter and the VHSV-IVb glycoprotein (*G*) gene insert was developed by Standish *et al.* (2016) and found to provide protection (95–100%) in muskellunge (*E. masquinongy*), which is a very susceptible species in the wild and in fish challenge studies (Kim and Faisal, 2010; Millard *et al.*, 2014). The vaccine also conferred protection against VHSV-IVb for less susceptible salmonids, including rainbow trout, brown trout and lake trout (*Salvelinus namaycush*). The vaccine induced significant levels of circulating *Piscine novirhabdovirus*-binding antibodies in muskellunge, as measured by indirect enzyme-linked immunosorbent assay (ELISA), reaching peak levels 6–7 weeks post-vaccination. Viral shedding in vaccinated survivors was minimal and did not persist (Standish *et al.*, 2016). A highly similar vaccine also produced a response against *Salmonid novirhabdovirus* in inoculated rainbow trout (Millard *et al.*, 2017). In both cases, however, widespread utilization of DNA vaccines in fish targeted for human consumption is limited by regulatory restrictions in the USA (Adams, 2019).

A variety of attenuated viral vaccine candidates have been developed for *Piscine novirhabdovirus*, involving rearrangement of the standard viral gene order or *Nv* gene deletion (Kim *et al.*, 2011; Rouxel *et al.*, 2016). These have the advantage of ease of immunization through submersion challenge and targeting of the mucosal membranes for optimal effectiveness (Munang'andu *et al.*, 2015). Ke *et al.* (2017) suggested that targeting the *Piscine novirhabdovirus* VHSV-IVb *M* gene for induced mutation also has the potential to undermine its anti-host function and suppress replication, which both are desirable characteristics of an effective attenuated virus. However, since recombinant viruses are considered genetically modified organisms, regulatory approval would be required to allow their use on potential food for human consumption.

Recommended virus disease control practices in aquaculture include regular equipment disinfection (OIE, 2018). Reducing fish density, particularly during winter and early spring, has been advocated to reduce the spread of *Carp sprivivirus* (OIE, 2018), and should also be practised against *Piscine novirhabdovirus* and others. In rearing facilities with temperature-controlled environments, elevation of water temperature above 19–20°C may stop or prevent *Carp sprivivirus* outbreaks. A safe and effective vaccine for *Carp sprivivirus* is not currently available (OIE, 2018).

A study by Balmer *et al.* (2017) indicated that the broad-spectrum small-molecule antiviral rhodanine derivative, LJ001, is useful as a preventive and/or therapeutic agent for infections by enveloped viruses in aquaculture. Laboratory tests demonstrated that *Salmonid novirhabdovirus* infectivity was significantly inhibited *in vitro* in EPC fish cell lines and *in vivo* in rainbow trout fry, in a dose-dependent and time-dependent manner. LJ001 appeared to stimulate the innate immune response in the rainbow trout host (Balmer *et al.*, 2017).

4.6 Conclusions with Suggestions for Future Studies

The number and variety of fish rhabdoviruses capable of causing significant disease are extensive and growing, due to a variety of activities that include increased diagnostic capability and routine monitoring of the disease in aquaculture and net-pen rearing of susceptible fish species. Within a single viral type, the array of fish species affected is large and, in some cases, extends across marine, estuarine and freshwater environments. Human influences are expanding the impact of many rhabdoviruses beyond former barriers as shipping practices, transportation of bait fish and climate change alter the landscape of natural controls. Among the research areas that remain under-explored are the three outlined in Sections 4.6.1 to 4.6.3 below. We believe that epidemiological assessment of rhabdovirus spread and evolution will constitute important areas of new research.

4.6.1 Monitoring viral genotypes in wild populations

Little is known currently about the molecular determinants that underlie massive fish kills, for example with *Piscine novirhabdovirus*, which occur sporadically in ecosystems around the globe. Aside from novel co-infections that might alter host susceptibility, the most likely explanation centres on viral genetic changes that alter immunogenicity, augment cytopathicity, increase suppression of host responses and/or enhance evasion of detection. Studies to assess these changes are underway, utilizing high-throughput sequencing strategies that allow rapid whole-genome sequencing and analysis. These studies can identify quickly hotspots for mutation and search for corresponding changes in virulence. Follow-up laboratory studies on these isolates will be necessary to identify which changes confer the most significant changes in anti-host activities, including cytopathicity, host transcriptional inhibition and innate immune response regulation. The availability of cell-based model systems for many viruses, and the advent of reverse genetic systems for *Piscine novirhabdovirus* and *Salmonid novirhabdovirus*, allow hypothesis-driven, fully controlled studies to examine isolates or synthetic mutants in a controlled system to compare changes in replication and anti-host functions.

4.6.2 Vaccine development for aquaculture and stocking programmes

Nearly all vaccines currently in commercial use involve intraperitoneal injection. Injection delivery drives up labour costs and limits vaccine use to those facilities set up to handle large-scale manipulation of fish. As discussed earlier, development of new strategies that would allow immersion vaccination with attenuated viruses, or establishment of small-molecule therapeutics that are bioavailable through mucosal membrane absorption, would significantly facilitate high-throughput prevention/treatment. Indeed, mucosal infection routes with attenuated viruses or encapsulated inactivated viruses generally are more effective at promoting lasting immunity (Munang'andu *et al.*, 2015). However, additional work on immune priming and boost schedules, as well as adjuvants and infection doses, is needed to determine the optimal conditions to produce sustained and effective responses.

4.6.3 Effects of climate change on outbreaks and adaptations

The role of water temperature in determining pathogenicity has been alluded to throughout the present narrative. For most fish rhabdoviruses, low temperatures are correlated with enhanced infectivity, with rapid loss in pathogenicity occurring at higher temperatures. While this might seem to imply that warming waters will become less hospitable to cold-water viral diseases, equally likely scenarios are that: (i) the virus will adapt and, in the process, take on new characteristics that may be more harmful; (ii) fish hosts will be stressed due to temperature fluctuations and more susceptible to infections at a range of temperatures; or (iii) emergence of new invasive pathogens will lead to enhanced likelihood of co-infection, leading in some cases to more severe disease. Further spread northward into cold temperate and polar waters is likely, with possible spread into melt waters and infections of new host fish species and populations. Although hypothetical at this stage, these scenarios provide a variety of testable possibilities to consider as the spread of pathogenic viruses is monitored. *Piscine novirhabdovirus* isolates northern climates display greater temperature sensitivity than do isolates from more southern locations, indicating that rhabdoviruses can adapt to temperature change (Arkush *et al.*, 2006). However, it is the impact of the environment on the physiological and immunological status of affected fishes, coupled with changes in microbiomes as a result of warming waters, which will almost certainly provide the most fodder for investigation into the factors affecting the spread of rhabdoviral diseases into new regions and/or their periodic re-emergence in existing locales.

Acknowledgements

This is contribution #5002 from the NOAA Pacific Marine Environmental Lab. This work was supported by National Science Foundation (NSF) IOS award #DBI-135406 to D.W.L. and C.A.S.

References

Abbadi, M., Fusaro, A., Ceolin, C., Casarotto, C., Quartesan, R. *et al.* (2016) Molecular evolution and phylogeography of co-circulating IHNV and VHSV in Italy. *Frontiers in Microbiology* 7, 1306. https://doi.org/10.3389/fmicb.2016.01306

Adams, A. (2019) Progress, challenges and opportunities in fish vaccine development. *Fish and Shellfish Immunology* 90, 210–214. https://doi.org/10.1016/j.fsi.2019.04.066

Ahne, W. (1976) Uptake and multiplication of spring viraemia of carp virus in carp, *Cyprinus carpio* L. *Journal of Fish Diseases* 1(3), 265–268. https://doi.org/10.1111/j.1365-2761.1978.tb00029.x

Ahne, W., Kurath, G. and Winton, J. (1998) A ribonuclease protection assay can distinguish spring viremia of carp virus from pike fry rhabdovirus. *Bulletin of European Association of Fish Pathology* 18, 220–224. Available at: https://eafp.org/download/1998-Volume18/Issue%206/18_6%20220.pdf (accessed 30 December 2019)

Ahne, W., Bjorklund, H.V., Essbauer, S., Fijan, N., Kurath, G. and Winton, J.R. (2002) Spring viremia of carp (SVC). *Diseases of Aquatic Organisms* 52, 261–272. https://doi.org/10.3354/dao052261

Alonso, M., Kim, C.H., Johnson, M.C., Pressley, M. and Leong, J.A. (2004) The Nv gene of snakehead rhabdovirus (SHRV) is not required for pathogenesis, and a heterologous glycoprotein can be incorporated into the SHRV envelope. *Journal of Virology* 78(11), 5875–5882. http://doi.org/10.1128/JVI.78.11.5875-5882.2004

Ammayappan, A. and Vakharia, V.N. (2009) Molecular characterization of the Great Lakes viral hemorrhagic septicemia virus (VHSV) isolate from USA. *Virology Journal* 6(1), 171. https://doi.org/10.1186/1743-422X-6-171

Ammayappan, A. and Vakharia, V.N. (2011) Nonvirion protein of *Novirhabdovirus* suppresses apoptosis at the early stage of virus infection. *Journal of Virology* 18(16),8393–8402.https://doi.org/10.1128/JVI.00597-11

Ammayappan, A., LaPatra, S.E. and Vakharia, V.N. (2010) Molecular characterization of the virulent infectious hematopoietic necrosis virus (IHNV) strain 220-90. *Virology Journal* 7, 10. https://doi.org/10.1186/1743-422X-7-10

Ammayappan, A., Kurath, G., Thompson, T.M. and Vakharia, V.N. (2011) A reverse genetics system for the Great Lakes strain of viral hemorrhagic septicemia virus: the Nv gene is required for pathogenicity. *Marine Biotechnology* 13(4), 672–683. https://doi.org/10.1007/s10126-010-9329-4

Andino, R. and Domingo, E. (2015) Viral quasispecies. *Virology* 479, 46–51. https://doi.org/10.1016/j.virol.2015.03.022

Arkush, K.D., Mendonca, H.L., McBride, A.M., Yun, S., McDowell, T.S. and Hedrick, R.P. (2006) Effects of temperature on infectivity and of commercial freezing on survival of the North American strain of viral hemorrhagic septicemia virus (VHSV). *Diseases of Aquatic Organisms* 69(2–3), 145–151. https://doi.org/10.3354/dao069145

Axén C., Hakhverdyan, M., Boutrup, T.S., Blomkvist, E., Ljunghager, F. *et al.* (2017) Emergence of a new rhabdovirus associated with mass mortalities in eelpout (*Zoarces viviparus*) in the Baltic Sea. *Journal of Fish Diseases* 40(2), 219–229. https://doi.org/10.1111/jfd.12506

Bain, M.B., Cornwell, E.R., Hope, K.M., Eckerlin, G.E., Casey, R.N. *et al.* (2010) Distribution of an invasive aquatic pathogen (viral hemorrhagic septicemia virus) in the Great Lakes and its relationship to shipping. *PLoS ONE* 5, e10156. https://doi.org/10.1371/journal.pone.0010156

Balmer, B.F., Powers, R.L., Zhang, T.H., Lee, J., Vigant, F. *et al.* (2017) Inhibition of an aquatic rhabdovirus demonstrates promise of a broad-spectrum antiviral for use in aquaculture. *Journal of Virology* 91(4), e02181-16. https://doi.org/10.1128/JVI.02181-16

Beaulaurier, J., Bickford, N., Gregg, J.L., Grady, C.A., Gannam, A. *et al.* (2012) Susceptibility of Pacific herring to viral hemorrhagic septicemia is influenced by diet. *Journal of Aquatic Animal Health* 24(1), 43–48. https://doi.org/10.1080/08997659.2012.668511

Bellec, L., Cabon, J., Bergmann, S., de Boisséson, C., Engelsma, M. *et al.* (2014) Evolutionary dynamics and genetic diversity from three genes of *Anguillid rhabdovirus*. *Journal of General Virology* 95(11), 23905-401. https://doi.org/10.1099/vir.0.069443-0

Belshaw, R., Gardner, A., Rambaut, A. and Pybus, O.G. (2008) Pacing a small cage: mutation and RNA viruses. *Trends in Ecology and Evolution* 23(4), 188–193. https://doi.org/10.1016/j.tree.2007.11.010

Bergmann, S.M., Fichtner, D., Skall, H.F., Schlotfeldt, H.-J. and Olesen, N.J. (2003) Age- and weight-dependent susceptibility of rainbow trout *Oncorhynchus mykiss* to isolates of infectious haematopoietic necrosis virus (IHNV) of varying virulence. *Diseases of Aquatic*

Organisms 55(3), 205–210. https://doi.org/10.3354/dao055205

Betts, A.M. and Stone, D.M. (2000) Nucleotide sequence analysis of the entire coding regions of virulent and avirulent strains of viral haemorrhagic septicaemia virus. *Virus Genes* 20(3), 259–262. https://doi.org/10.1023/A:1008148813746

Biacchesi, S., Lamoureux, A., Mérour, E., Bernard, J. and Brémont, M. (2010) Limited interference at the early stage of infection between two recombinant novirhabdoviruses: viral hemorrhagic septicemia virus and infectious hematopoietic necrosis virus. *Journal of Virology* 84(19), 10038–10050. https://doi.org/10.1128/JVI.00343-10

Bigarré, L., Plassiart, G., Boisséson, C., Pallandre, L., Pozet, F. *et al.* (2017) Molecular investigations of outbreaks of *Perch perhabdovirus* infections in pikeperch. *Diseases of Aquatic Organisms* 127(1), 19–27. https://doi.org/10.3354/dao03177

Black, A., Breyta, R., Bedford, T. and Kurath, G. (2016) Geography and host species shape the evolutionary dynamics of U genogroup infectious hematopoietic necrosis virus. *Virus Evolution* 2(2), vew034. https://doi.org/10.1093/ve/vew034

Blasdell, K.R., Guzman, H., Widen, S.G., Firth, C., Wood, T.G. *et al.* (2015) *Ledantevirus*: a proposed new genus in the *Rhabdoviridae* has a strong ecological association with bats. *American Journal of Tropical Medical Hygiene* 92(2), 405–410. https://doi.org/10.4269/ajtmh.14-0606

Borzym, E., Matras, M., Maj-Paluch, J., Baud, M., De Boisséson, C. *et al.* (2014) First isolation of hirame rhabdovirus from freshwater fish in Europe. *Journal of Fish Diseases 2014*, 37(5), 423–430. https://doi.org/10.1111/jfd.12119

Campbell, S., Collet, B., Einer-Jensen, K., Secombes, C.J. and Snow, M. (2009) Identifying potential virulence determinants in viral haemorrhagic septicaemia virus (VHSV) for rainbow trout. *Diseases of Aquatic Organisms* 86(3), 205–212. https://doi.org/10.3354/dao02127

Chinchilla, B. and Gomez-Casado, E. (2017) Identification of the functional regions of the viral haemorrhagic septicaemia virus (VHSV) Nv protein: variants that improve function. *Fish and Shellfish Immunology* 70, 343–350. https://doi.org/10.1016/j.fsi.2017.09.021

Choi, M.K., Moon, C.H., Ko, M.S., Lee, U.-H., Cho, W.J. *et al.* (2011) A nuclear localization of the infectious haematopoietic necrosis virus Nv protein is necessary for optimal viral growth. *PLoS ONE* 6(7), e22362. https://doi.org/10.1371/journal.pone.0022362

Clement, M., Posada, D.C. and Crandall, K.A. (2000) TCS: a computer program to estimate gene genealogies. *Molecular Ecology* 9(10), 1657–1659. https://doi.org/10.1046/j.1365-294x.2000.01020.x

Daniels, H.V. and Watanabe, W.O. (2010) *Practical Flatfish Culture and Stock Enhancement.* Wiley-Blackwell, Ames, Iowa. https://doi.org/10.1002/9780813810997

Dietzgen, R.G. and Kuzmin I.V. (2012) Taxonomy of rhabdoviruses. In: Dietzgen, R.G. and Kuzmin I.V. (eds) *Rhabdoviruses: Molecular Taxonomy, Evolution, Genomics, Ecology, Host–Vector Interactions, Cytopathology and Control.* Caister Academic Press, Norfolk, UK, pp. 5–12.

Dixon, P., Paley, R., Alegria-Moran, R. and Oidtmann, B. (2016) Epidemiological characteristics of infectious hematopoietic necrosis virus (IHNV): a review. *Veterinary Research* 47(1), 63. https://doi.org/10.1186/s13567-016-0341-1

Duesund, H., Nylund, S., Watanabe, K., Ottem, K.F. and Nylund, A. (2010) Characterization of a VHS virus genotype III isolated from rainbow trout (*Oncorhychus mykiss*) at a marine site on the west coast of Norway. *Journal of Virology* 7, 19. https://doi.org/10.1186/1743-422X-7-19

Einer-Jensen, K., Ahrens, P., Forsberg, R. and Lorenzen, N. (2004) Evolution of the fish rhabdovirus viral haemorrhagic septicaemia virus. *Journal of General Virology* 85(5), 1167–1179. http://doi.org/10.1099/vir.0.79820-0

Einer-Jensen, K., Harmache, A., Biacchesi, S., Bremont, M., Stegmann, A. and Lorenzen, N. (2014) High virulence differences between phylogenetically distinct isolates of the fish rhabdovirus VHSV is not associated with variability of the surface glycoprotein G nor the nonvirion protein Nv. *Journal of General Virology* 95(2), 307–316. https://doi.org/10.1099/vir.0.057448-0

Enzmann, P.J., Castric, J., Bovo, G., Thiery, R., Fichtner, D. *et al.* (2010) Evolution of infectious hematopoietic necrosis virus (IHNV), a fish rhabdovirus, in Europe over 20 years: implications for control. *Diseases of Aquatic Organisms* 89(1), 9–15. http://doi.org/10.3354/dao02182

Escobar, L.E., Escobar-Dodero, J. and Phelps, N.B. (2018) Infectious disease in fish: global risk of viral hemorrhagic septicemia virus. *Reviews in Fish Biology and Fisheries* 28(3), 637–655. https://doi.org/10.1007/s11160-018-9524-3

Faisal, M. and Schulz, C.A. (2009) Detection of viral hemorrhagic septicemia virus (VHSV) from the leech *Myzobdella lugubris* Leidy, 1851. *Parasites and Vectors* 2(1), 45. https://doi.org/10.1186/1756-3305-2-45

Faisal, M. and Winters, A.D. (2011) Detection of viral hemorrhagic septicemia virus (VHSV) from *Diporeia* spp. (Pontoporeiidae, Amphipoda) in the Laurentian Great Lakes, USA. *Parasites and Vectors* 4(1), 2. https://doi.org/10.1186/1756-3305-4-2

Frerichs, G.N., Hill, B.J. and Way, K. (1989) Ulcerative disease rhabdovirus: cell-line susceptibility and serological comparison with other fish rhabdoviruses. *Journal of Fish Diseases* 12(1), 51–56. https://doi.org/10.1111/j.1365-2761.1989.tb01290.x

Gadd, T. (2013) Fish rhabdoviruses. Viral hemorrhagic septicemia virus (VHSV) and perch rhabdovirus (PRV): study of viral strains and the disease epidemiology in Finland. PhD thesis, University of Helsinki, Helsinki. Available at: https://helda.helsinki.fi/

bitstream/handle/10138/40955/gadd_dissertation.pdf?sequence=1 (accessed 4 March 2020).

Garver, K.A., Traxler, G.S., Hawley, L.M., Richard, J., Ross, J.P. and Lovy, J. (2013) Molecular epidemiology of viral haemorrhagic septicaemia virus (VHSV) in British Columbia, Canada, reveals transmission from wild to farmed fish. *Diseases of Aquatic Organisms* 104(2), 93–104. https://doi.org/10.3354/dao02588

Getchell, R.G., Cornwell, E.R., Bogdanowicz, S., Andrés, J., Batts, W.N. *et al.* (2017) Complete sequences of 4 viral hemorrhagic septicemia virus IVb isolates and their virulence in northern pike fry. *Diseases of Aquatic Organisms* 126(3), 211–227. https://doi.org/10.3354/dao03171

Ghorani, M., Adel, M., Dadar, M., Langeroudi, A.G., Kamyabi, R. *et al.* (2016) Phylogenetic analysis of the glycoprotein gene of viral hemorrhagic septicemia virus from Iranian trout farms points towards a common European origin. *Veterinary Microbiology* 186, 97–101. https://doi.org/10.1016/j.vetmic.2016.02.019

Goodwin, A.E. and Merry, G.E. (2011) Replication and persistence of VHSV-IVb in freshwater turtles. *Diseases of Aquatic Organisms* 94(3), 173–177. https://doi.org/10.3354/dao02328

Groocock, G.H., Getchell, R.G., Wooster, G.A., Britt, K.L., Batts, W.N. *et al.* (2007) Detection of viral hemorrhagic septicemia in round gobies in New York State (USA) waters of Lake Ontario and the St. Lawrence River. *Diseases of Aquatic Organisms* 76(3), 187–192. https://doi.org/10.3354/dao076187

Gubala, A. (2012) Recent advances in the characterization of animal rhabdoviruses. In: Dietzgen, R.G. and Kuzmin, I.V. (eds) *Rhabdoviruses: Molecular Taxonomy, Evolution, Genomics, Ecology, Host–Vector Interactions, Cytopathology, and Control*. Caister Academic Press, Norfolk, UK, pp. 165–203.

Guðmundsdóttir, S., Vendramin, N., Cuenca, A., Sigurðardóttir, H., Kristmundsson, A. *et al.* (2019) Outbreak of viral haemorrhagic septicaemia (VHS) in lumpfish (*Cyclopterus lumpus*) in Iceland caused by VHS virus genotype IV. *Journal of Fish Diseases* 42(1), 47–62. http://doi.org/10.1111/jfd.12910

Guindon, S. and Gascuel, O. (2003) A simple, fast, and accurate algorithm to estimate large phylogenies by maximum likelihood. *Systematic Biology* 52(5), 696–704. https://doi.org/10.1080/10635150390235520

Haenin, O.L.M. and Davidse, A. (1993) Comparative pathogenicity of two strains of pike fry rhabdovirus and spring viremia of carp virus for young roach, common carp, grass carp, and rainbow trout. *Diseases of Aquatic Organisms* 15, 87–92. Available at: https://www.int-res.com/articles/dao/15/d015p087.pdf (accessed 4 March 2020).

He, M., Ding, N.Z., He, C.Q., Yan, X.C. and Teng, C.B. (2013) Dating the divergence of the infectious hematopoietic necrosis virus. *Infection, Genetics and Evolution* 18, 145–150. https://doi.org/10.1016/j.meegid.2013.05.014

He, M., Yan, X.C., Liang, Y., Sun, X.W. and Teng, C.B. (2014) Evolution of the viral hemorrhagic septicemia virus: divergence, selection and origin. *Molecular Phylogenetics and Evolution* 77, 34–40. https://doi.org/10.1016/j.ympev.2014.04.002

Hedrick, R.P., Eaton, W.D., Fryer, J.L., Groberg, W.G. and Boonyaratpalin, S. (1986) Characteristics of a birnavirus isolated from cultured sand goby *Oxyeleotris marmoratus*. *Diseases of Aquatic Organisms* 1, 219–225. Available at: https://www.int-res.com/articles/dao/1/d001p219.pdf (accessed 30 December 2019).

Hershberger, P.K., Kocan, R.M., Elder, N.E., Meyers, T.R. and Winton, J.R. (1999) Epizootiology of viral hemorrhagic septicermia virus in Pacific herring from the spawn-on-kelp fishery in Prince William Sound, Alaska, USA. *Diseases of Aquatic Organisms* 37(1), 23–31. https://doi.org/10.3354/dao037023

Hwang, J.Y., Kwon, M.G., Seo, J.S., Do, J.W., Park, M.A. *et al.* (2016) Differentially expressed genes after viral haemorrhagic septicaemia virus infection in olive flounder (*Paralichthys olivaceus*). *Veterinary Microbiology* 193, 72–82. https://doi.org/10.1016/j.vetmic.2016.05.024

Hwang, J.Y., Lee, S., Priyathilaka, T.T., Yang, H., Kwon, H. *et al.* (2018) Phylogenetic analysis and duplex RT-PCR detection of viral hemorrhagic septicemia virus in olive flounder (*Paralichthys olivaceus*) from Korea. *Aquaculture* 484, 242–249. https://doi.org/10.1016/j.aquaculture.2017.11.038

ICTV (International Committee on the Taxonomy of Viruses) (2018) *Rhabdoviridae*. In: Virus Taxonomy: The Classification and Nomenclature of Viruses. 10th ICTV Report. Available at: https://talk.ictvonline.org/ictv-reports/ictv_online_report/negative-sense-rna-viruses/mononegavirales/w/rhabdoviridae/ (accessed 30 December 2019).

Ito, T., Kurita, J., Sano, M., Skall, H.F., Lorenzen, N. *et al.* (2012) Typing of viral hemorrhagic septicemia virus by monoclonal antibodies. *Journal of General Virology* 93(12), 2546–2557. https://doi.org/10.1099/vir.0.043091-0

Jia, P., Zheng, X.C., Shi, X.J., Kan, S.F., Wang, J.J. *et al.* (2014) Determination of the complete genome sequence of infectious hematopoietic necrosis virus (IHNV) Ch20101008 and viral molecular evolution in China. *Infection, Genetics and Evolution* 27, 418–431. https://doi.org/10.1016/j.meegid.2014.08.013

Johnson, M.C., Maxwell, J.M., Loh, P.C. and Leong, J.A.C. (1999) Molecular characterization of the glycoproteins from two warm water rhabdoviruses: snakehead rhabdovirus (SHRV) and rhabdovirus of penaeid shrimp (RPS)/spring viremia of carp virus (SVCV). *Virus Research* 64(2), 95–106. https://doi.org/10.1016/S0168-1702(99)00071-4

Johnson, M.C., Simon, B.E., Kim, C.H. and Leong, J.A.C. (2000) Production of recombinant *Snakehead rhabdovirus*: the Nv protein is not required for viral replication. *Journal of Virology* 74(5), 2343–2350. https://doi.org/10.1128/JVI.74.5.2343-2350.2000

Jorgensen, P.E.V., Olesen, N.J., Lorenzen, N., Winton, J.R. and Ristow, S.S. (1991) Infectious hematopoietic necrosis (IHN) and viral hemorrhagic septicaemia (VHS): detection of trout antibodies to the causative viruses by means of plaque neutralization, immunofluorescence, and enzyme-linked immunosorbent assay. *Journal of Aquatic Animal Health* 3(2), 100–108. https://doi.org/10.1577/1548-8667(1991)003<0100:IHNIAV>2.3.CO;2

Katze, M.G., He, Y. and Gale, M. Jr (2002) Viruses and interferon: a fight for supremacy. *Nature Reviews in Immunology* 2, 675–687. https://doi.org/10.1038/nri888

Ke, Q., Weaver, W., Pore, A., Gorgoglione, B., Wildschutte, J.H. *et al.* (2017) Role of viral hemorrhagic septicemia virus matrix (M) protein in suppressing host transcription. *Journal of Virology* 91(19), e00279-17. http://doi.org/10.1128/JVI.00279-17

Kim, D.H., Oh, H.K., Eou, J.I., Seo, H.J., Kim, S.K. *et al.* (2005) Complete nucleotide sequence of the hirame rhabdovirus, a pathogen of marine fish. *Virus Research* 107(1), 1–9. https://doi.org/10.1016/j.virusres.2004.06.004

Kim, J.O., Kim, W.S., Nishizawa, T. and Oh, M.J. (2013) Complete genome sequence of viral hemorrhagic septicemia virus isolated from an olive flounder in South Korea. *Genome Announcements* 1(5), e00681-13. https://doi.org/10.1128/genomeA.00681-13

Kim, M.S. and Kim, K.H. (2013) The role of viral hemorrhagic septicemia virus (VHSV) Nv gene in TNF-α- and VHSV infection-mediated NF-κB activation. *Fish and Shellfish Immunology* 34(5), 1315–1319. https://doi.org/10.1016/j.fsi.2013.02.026

Kim, M.S., Kim, D.S. and Kim, K.H. (2011) Oral immunization of olive flounder (*Paralichthys olivaceus*) with recombinant live viral hemorrhagic septicemia virus (VHSV) induces protection against VHSV infection. *Fish and Shellfish Immunology* 31(2), 212–216. https://doi.org/10.1016/j.fsi.2011.05.003

Kim, R. and Faisal, M. (2010) Experimental studies confirm the wide host range of the Great Lakes viral haemorrhagic septicaemia virus genotype IVb. *Journal of Fish Diseases* 33(1), 83–88. https://doi.org/10.1111/j.1365-2761.2009.01093.x

Kim, S.H., Yusuff, S., Vakharia, V.N. and Evensen, Ø. (2015) Interchange of L polymerase protein between two strains of viral haemorrhagic septicemia virus (VHSV) genotype IV alters temperature sensitivities *in vitro*. *Virus Research* 195, 203–206. https://doi.org/10.1016/j.virusres.2014.10.013

Kitchen, A., Shackelton, L.A. and Holmes, E.C. (2011) Family level phylogenies reveal modes of macroevolution in RNA viruses. *Proceedings of the National Academy of Sciences USA* 108(1), 238–243. https://doi.org/10.1073/pnas.1011090108

Kurath, G. (2012) Molecular epidemiology and evolution of fish novirhabdoviruses. In: Dietzgen, R.G. and Kuzmin, I.V. (eds) *Rhabdoviruses: Molecular Taxonomy, Evolution, Genomics, Ecology, Host–Vector Interactions, Cytopathology, and Control.* Caister Academic Press, Norfolk, UK, pp. 89–116.

Kurath, G., Garver, K.A., Troyer, R.M., Emmenegger, E.J., Einer-Jensen, K. and Anderson, E.D. (2003) Phylogeography of infectious haematopoietic necrosis virus in North America. *Journal of General Virology* 84(4), 803–814. https://doi.org/10.1099/vir.0.18771-0

Lauring, A.S. and Andino, R. (2010) Quasispecies theory and the behavior of RNA viruses. *PLoS Pathogens* 6(7), e1001005. https://doi.org/10.1371/journal.ppat.1001005

Liu, X., Qin, Z., Babu, S., Zhao, L., Li, J. *et al.* (2019) Transcriptomic profiles of striped snakehead cells (SSN-1) infected with snakehead vesiculovirus (SHVV) identifying IFI35 as a positive factor for SHVV replication. *Fish and Shellfish Immunology* 86, 46–52. https://doi.org/10.1016/j.fsi.2018.11.031

Lopez-Vazquez, C., Bandín, I. and Dopazo, C.P. (2015) Real-time RT-PCR for detection, identification and absolute quantification of viral haemorrhagic septicaemia virus using different types of standards. *Diseases of Aquatic Organisms* 114(2), 99–116. https://doi.org/10.3354/dao02840

Lorenzen, E., Einer-Jensen, K., Rasmussen, J.S., Kjaer, T.E., Collet, B. *et al.* (2009) The protective mechanisms induced by a fish rhabdovirus DNA vaccine depend on temperature. *Vaccine* 27(29), 3870–3880. https://doi.org/10.1016/j.vaccine.2009.04.012

Lumsden, J.S., Morrison, B., Yason, C., Russell, S., Young, K. *et al.* (2007) Mortality event in freshwater drum *Aplodinotus grunniens* from Lake Ontario, Canada, associated with viral haemorrhagic septicemia virus, Type IV. *Diseases of Aquatic Organisms* 76(2), 99–111. https://doi.org/10.3354/dao076099

Ma, D., Deng, G., Bai, J., Li, S., Yu, L. *et al.* (2013) A strain of *Siniperca chuatsi* rhabdovirus causes high mortality among cultured largemouth bass in South China. *Journal of Aquatic Animal Health* 25(3), 197–204. https://doi.org/10.1080/08997659.2013.799613

Millard, E.V., Brenden, T.O., LaPatra, S.E., Marcquenski, S. and Faisal, M. (2014) Detection of viral hemorrhagic septicemia virus-IVb antibodies in sera of muskellunge *Esox masquinongy* using competitive ELISA. *Diseases of Aquatic Organisms* 108(3), 187–199. https://doi.org/10.3354/dao02712

Millard, E.V., Bourke, A.M., LaPatra, S.E., Brenden, T.O., Fitzgerald, S.D. and Faisal, M. (2017) DNA vaccination partially protects muskellunge against viral hemorrhagic septicemia virus (VHSV-IVb). *Journal of Aquatic Animal Health* 29(1), 50–56. https://doi.org/10.1080/08997659.2016.1238413

Mork, D., Hershberger, P., Kocan, R., Batts, W. and Winton J. (2004) Isolation and characterization of a rhabdovirus from starry flounder (*Platichthys stellatus*) collected from the northern portion of Puget Sound, Washington, USA. *Journal of General Virology* 85(2), 495–505. https://doi.org/10.1099/vir.0.19459-0

Morzunov, S.P., Winton, J.R. and Nichol, S.T. (1995) The complete genome structure and phylogenetic relationship of infectious hematopoietic necrosis virus. *Virus Research* 38(2–3), 175–192. https://doi.org/10.1016/0168-1702(95)00056-V

Munang'andu, H.M., Mutoloki, S. and Evensen, Ø. (2015) An overview of challenges limiting the design of protective mucosal vaccines for finfish. *Frontiers in Immunology* 6, 542. https://doi.org/10.3389/fimmu.2015.00542

Neukirch, M. (1986) Demonstration of persistent viral haemorrhagic septicaemia (VHS) virus in rainbow trout after experimental waterborne infection. *Journal of Veterinary Medicine* 33(1–10), 471–476. https://doi.org/10.1111/j.1439-0450.1986.tb00058.x

Niner, M.D. (2019) Evolutionary patterns and occurrences of the fish viral hemorrhagic septicemia virus in the Laurentian Great Lakes. PhD thesis, University of Toledo, Toledo, Ohio.

OIE (World Organization for Animal Health) (2018) Manual of Diagnostic Tests for Aquatic Animals, 12/10/2018. Available at: http://www.oie.int/standard-setting/aquatic-manual/access-online/ (accessed 30 December 2019).

Ojosnegros, S. and Beerenwinkel, N. (2010) Models of RNA virus evolution and their roles in vaccine design. *Immunome Research* 6(2), S5. https://doi.org/10.1186/1745-7580-6-S2-S5

Ou, T., Zhu, R.L, Chen, Z.Y. and Zhang, Q. (2013) Isolation and identification of a lethal rhabdovirus from farmed rice field eels *Monopterus albus*. *Diseases of Aquatic Organisms* 106(3), 197–206. https://doi.org/10.3354/dao02660

Pereira, F. and Amorim, A. (2013) Evolution: viruses. In: Maloy, S. and Hughes, K. (eds) *Brenner's Encyclopedia of Genetics*. Elsevier, New York, pp. 566–568. https://doi.org/10.1016/B978-0-12-374984-0.00499-X

Pierce, L.R. (2013) The evolution and detection of the fish viral hemorrhagic septicemia virus (VHSV). PhD thesis, University of Toledo, Toledo, Ohio.

Pierce, L.R. and Stepien, C.A. (2012) Evolution and biogeography of an emerging quasispecies: diversity patterns of the fish viral hemorrhagic septicemia virus (VHSV). *Molecular Phylogenetics and Evolution* 63(2), 327–341. https://doi.org/10.1016/j.ympev.2011.12.024

Pierce, L.R., Willey, J.C., Crawford, E.L., Palsule, V.V., Leaman, D.W. *et al.* (2013a) A new StaRT-PCR approach to detect and quantify fish viral hemorrhagic septicemia virus (VHSV): enhanced quality control with internal standards. *Journal of Virological Methods* 189(1), 129–142. https://doi.org/10.1016/j.jviromet.2013.01.006

Pierce, L.R., Willey, J.C., Palsule, V.V., Yeo, J., Shepherd, B.S. *et al.* (2013b) Accurate detection and quantification of the fish viral hemorrhagic septicemia virus (VHSV) with a two-color fluorometric real-time PCR assay. *PLoS ONE* 8(8), e71851. https://doi.org/10.1371/journal.pone.0071851

Pore, A. (2012) Studies on host–virus interaction for viral hemorrhagic septicemia virus (VHSV). Master's thesis, University of Toledo, Toledo, Ohio. Available at: https://etd.ohiolink.edu/pg_10?0::NO:10:P10_ACCESSION_NUM:toledo1336766667 (accessed 4 March 2020).

Purcell, M.K., Laing, K.J. and Winton, J.R. (2012) Immunity to fish rhabdoviruses. *Viruses* 4(1), 140–166. https://doi.org/10.3390/v4010140

Quer, J., Huerta, R., Novella, I.S., Tsimring, L., Domingo, E. and Holland, J.J. (1996) Reproducible nonlinear population dynamics and critical points during replicative competitions of RNA virus quasispecies. *Journal of Molecular Biology* 264(3), 465–471. https://doi.org/10.1006/jmbi.1996.0654

Rieder, M. and Conzelmann, K.-K. (2011) Interferon in rabies virus infection. *Advances in Virus Research* 79, 91–114. https://doi.org/10.1016/B978-0-12-387040-7.00006-8

Rodriguez, L.L. and Pauszek, S.J. (2012) Genus *Vesiculovirus*. In: Dietzgen, R.G. and Kuzmin, I.V. (eds) *Rhabdoviruses: Molecular Taxonomy, Evolution, Genomics, Ecology, Host–Vector Interactions, Cytopathology and Control*. Caister Academic Press, Norfolk, UK, pp. 23–35.

Ronquist, F., Teslenko, M., van der Mark, P., Ayres, D.L., Darling, A. *et al.* (2012) MRBAYES 3.2: efficient Bayesian phylogenetic inference and model selection across a large model space. *Systematic Biology* 61(3), 539–542. https://doi.org/10.1093/sysbio/sys029

Rouxel, R.N., Tafalla, C., Mérour, E., Leal, E., Biacchesi, S. and Brémont, M. (2016) Attenuated infectious hematopoietic necrosis virus with rearranged gene order as potential vaccine. *Journal of Virology* 90(23), 10857–10866. https://doi.org/10.1128/JVI.01024-16

Rucker, R.R., Whipple, W.J., Parvin, J.R. and Evans, C.A. (1953) A contagious disease of salmon possibly of virus origin. *Fishery Bulletin* 76, 35–46. Available at: https://www.st.nmfs.noaa.gov/spo/FishBull/54-1/rucker.pdf (accessed 4 March 2020).

Saksida, S. (2006) Infectious haematopoietic necrosis epidemic (2001 to 2003) in farmed Atlantic salmon *Salmo salar* in British Columbia. *Diseases of Aquatic Organisms* 72(3), 213–223. https://doi.org/10.3354/dao072213

Schäperclaus, W. (1938) Die Immunisierung von karlen gegen Bauchwa sucht auf natürilchem und künstlichem Wege. *Fisherei-zeitung (Neudamm)* 41, 193–196.

Schneider, W.M., Chevillotte, M.D. and Rice, C.M (2014) Interferon-stimulated genes: complex web of host defenses. *Annual Review of Immunology* 32, 513–545. https://doi.org/10.1146/annurev-immunol-032713-120231

Schütze, H., Enzmann, P.J., Kuchling, R., Mundt, E., Niemann, H. and Mettenleiter, T.C. (1995) Complete genomic sequence of the fish rhabdovirus infectious haematopoietic necrosis virus. *Journal of General Virology* 76(10), 2519-27. https://doi.org/10.1099/0022-1317-76-10-2519

Schütze, H., Mundt, E. and Mettenleiter, T.C. (1999) Complete genomic sequence of viral hemorrhagic septicemia virus, a fish rhabdovirus. *Virus Genes* 19, 59–65. https://doi.org/10.1023/a:1008140707132

Shimahara, Y., Kurita, J., Nishioka, T., Kiryu, I., Yuasa, K. *et al.* (2016) Development of an improved RT-PCR for specific detection of spring viraemia of carp virus. *Journal of Fish Diseases* 39(3), 269–275. https://doi.org/10.1111/jfd.12357

Standish, I.F., Millard, E.V., Brenden, T.O. and Faisal, M. (2016) A DNA vaccine encoding the viral hemorrhagic septicemia virus genotype IVb glycoprotein confers protection in muskellunge (*Esox masquinongy*), rainbow trout (*Oncorhynchus mykiss*), brown trout (*Salmo trutta*), and lake trout (*Salvelinus namaycush*). *Virology Journal* 13(1), 203. https://doi.org/10.1186/s12985-016-0662-8

Stepien, C.A., Pierce, L.R., Leaman, D.W., Niner, M.D. and Shepherd, B.S. (2015) Gene diversification of an emerging pathogen: a decade of mutation in a novel fish viral hemorrhagic septicemia (VHS) substrain since its first appearance in the Laurentian Great Lakes. *PLoS ONE* 10(8), e0135146. https://doi.org/10.1371/journal.pone.0135146

Stone, D.M., Way, K. and Dixon, P.F. (1997) Nucleotide sequence of the glycoprotein gene of viral haemorrhagic septicaemia (VHS) viruses from different geographical areas: a link between VHS in farmed fish species and viruses isolated from North Sea cod (*Gadus morhua* L.). *Journal of General Virology* 78(6), 1319–1326. http://doi.org/10.1099/0022-1317-78-6-1319

Thompson, T.M., Batts, W.N., Faisal, M., Bowser, P., Casey, J.W. *et al.* (2011) Emergence of viral hemorrhagic septicemia virus in the North American Great Lakes region is associated with low viral genetic diversity. *Diseases of Aquatic Organisms* 96(1), 29–43. https://doi.org/10.3354/dao02362

Vakharia, V.N., Li, J., McKenney, D.G. and Kurath, G. (2019) The nucleoprotein and phosphoprotein are major determinants of virulence of viral hemorrhagic septicemia virus in rainbow trout. *Journal of Virology* 93(18), e00382-19. https://doi.org/10.1128/JVI.00382-19

Van Ginneken, V., Ballieux, B., Willemze, R., Coldenhoff, K., Lentjes, E. *et al.* (2005) Hematology patterns of migrating European eels and the role of EVEX virus. *Comparative Biochemistry and Physiology Part C: Toxicology and Pharmacology* 140(1), 97–102. https://doi.org/10.1016/j.cca.2005.01.011

Vicenova, M., Reschova, S., Pokorová, D., Hulova, J. and Vesely, T. (2011) First detection of pike fry-like rhabdovirus in barbel and spring viraemia of carp virus in sturgeon and pike in aquaculture in the Czech Republic. *Diseases of Aquatic Organisms* 95(2), 87–95. https://doi.org/10.3354/dao02340

Walker, P.J., Blasdell, K.R., Calisher, C.H., Dietzgen, R.G., Kondo, H. *et al.* (2018) ICTV virus taxonomy profile: *Rhabdoviridae*. *Journal of General Virology* 99(4), 447–448. https://doi.org/10.1099/jgv.0.001020

Wang, C., Lian, G.H., Zhao, L.L., Wu, Y., Li, Y.J., Tang, L.J. *et al.* (2016) Virulence and serological studies of recombinant infectious hematopoietic necrosis virus (IHNV) in rainbow trout. *Virus Research* 220, 193–202. https://doi.org/10.1016/j.virusres.2016.04.015

Winton J.R. and Einer-Jensen K. (2002) Molecular diagnosis of infectious hematopoietic necrosis and viral hemorrhagic septicemia. In: Cunningham, C.O. (ed.) *Reviews: Methods and Technologies in Fish Biology and Fisheries. Vol. 3. Molecular Diagnosis of Salmonid Diseases*. Springer, Dordrecht, the Netherlands, pp. 49–79. https://doi.org/10.1007/978-94-017-2315-2_3

Xu, L., Zhao, J., Liu, M., Kurath, G., Breyta, R.B. *et al.* (2018) Phylogeography and evolution of infectious hematopoietic necrosis virus in China. *Molecular Phylogenetics and Evolution* 131, 19–28. https://doi.org/10.1016/j.ympev.2018.10.030

Yingjie, S., Min, Z., Hong, L., Zhiqin, Y., Xiaocong, Z. and Zhe, W. (2011) Analysis and characterization of the complete genomic sequence of the Chinese strain of *hirame* rhabdovirus. *Journal of Fish Diseases* 34(2), 167–171. https://doi.org/10.1111/j.1365-2761.2010.01218.x

Yusuff, S., Kurath, G., Kim, M.S., Tesfaye, T.M., Li, J. *et al.* (2019) The glycoprotein, non-virion protein, and polymerase of viral hemorrhagic septicemia virus are not determinants of host-specific virulence in rainbow trout. *Virology Journal* 16, 31. https://doi.org/10.1186/s12985-019-1139-3

Zhang, C., Huelsenbeck, J., van der Mark, P., Ronquist, F. and Taslenko, M. (2019) Mr. Bayes: Bayesian Inference of Phylogeny. 3.2.7a. Available at: https://github.com/NBISweden/MrBayes/tree/v3.2.7a (accessed 30 December 2019).

Zhang, Q.Y., Tao, J.J., Gui, L., Zhou, G.Z., Ruan, H.M., Li, Z.Q. and Gui, J.F. (2007) Isolation and characterization of *Scophthalmus maximus* rhabdovirus. *Diseases of Aquatic Organisms* 74(2), 95–105. https://doi.org/10.3354/dao074095

Zhu, R.L. and Zhang, Q.Y. (2014) Determination and analysis of the complete genome sequence of *Paralichthys olivaceus* rhabdovirus (PORV). *Archives in Virology* 159(4), 817–820. https://doi.org/10.1007/s00705-013-1716-5

5 Nodavirosis (Striped Jack Nervous Necrosis Virus)

SANDRA C. ZAINATHAN* AND NURSHUHADA ARIFF

Universiti Malaysia Terengganu, Terengganu, Malaysia

5.1 Introduction

The *Nodaviridae* family consists of two genera (*Alphanodavirus* and *Betanodavirus*). Alphanodaviruses are known to infect insect species such as adult common fruit flies (*Drosophila melanogaster*), tsetse flies (*Glossina morsitans morsitans*), reduviid bugs (*Rhodnius prolixus*) and several species of mosquito (*Aedes aegypti*, *Culex pipiens*, *Armigeres subalbatus* and *Anopheles gambiae*), while betanodaviruses mostly cause disease in fish species (Thiéry *et al.*, 2004; Viral Zone, 2018). The current taxonomy classifies the causative agent of viral nervous necrosis (VNN) into the genus *Betanodavirus* within the family *Nodaviridae* (Schneemann *et al.*, 2005; OIE, 2018). VNN disease was formerly known as sea bass viral encephalitis (SVE) (Bellance and Gallet de Saint-Aurin, 1988), fish virus encephalitis (FVE) (Comps *et al.*, 1994), viral encephalopathy and retinopathy (VER) (Munday *et al.*, 2002) and viral encephalitis and retinitis (Tanaka *et al.*, 2004) due to the neurological disorder that resulted from virus infection. VNN caused by betanodavirus is recognized as a very important viral disease in teleost fish, resulting in severe mortality in more than 120 cultured marine fish, wild fish and invertebrate species worldwide (Munday *et al.*, 2002; Kokawa *et al.*, 2008). The first case was discovered in 1994 and there have been continuous reports of cases mainly in South-East Asian countries (China, Chinese Taipei, India, Indonesia, Iran, Japan, Korea, Malaysia, Philippines, Thailand, Vietnam), the Mediterranean (France, Greece, Israel, Italy, Malta, Portugal, Spain, Tunisia), Oceania (Australia, Tahiti), the UK, Norway, the Caribbean, Canada and the USA (Munday *et al.*, 2002; Costa and Thompson, 2016; OIE, 2017). The causative agent of VNN causes severe infection in the central nervous system and affects the larvae and juvenile stage of susceptible fish (Munday *et al.*, 2002; Panzarin *et al.*, 2012). The causative agent of VNN was first identified as a new member of the family *Nodaviridae* following virus purification from the brain tissues of affected striped jack larvae (*Pseudocaranx dentex*), and the name *Striped jack nervous necrosis virus* (SJNNV) was adopted (Mori *et al.*, 1992; OIE, 2018).

The virion of the VNN causative agent is small, non-enveloped, with a spherical shape, measuring 23 to 25 nm in diameter. The viral genome is comprised of two single-stranded positive-sense RNA molecules, RNA1 and RNA2. Both lack a poly(A) sequence at the 3′ terminus. RNA1 (3.1 kb) contains a single open reading frame (ORF) which encodes 'protein A', the viral RNA-dependent RNA polymerase (RdRp) and the only enzyme known to be encoded by the virus (Mori *et al.*, 1992; Chi *et al.*, 2001; Iwamoto *et al.*, 2001; Tan *et al.*, 2001; Panzarin *et al.*, 2012; Costa and Thompson, 2016). The RdRp encoded by RNA1 has been shown to be responsible for temperature sensitivity of virus replication (Hata *et al.*, 2010; Panzarin *et al.*, 2014). Similar to RNA1, RNA2 (1.4 kb) consists of a single ORF that encodes the capsid protein whose precursor is cleaved to form capsid protein α (Nishizawa *et al.*, 1995; Panzarin *et al.*, 2012; Costa and Thompson, 2016). RNA3 is a subgenomic transcript formed by RdRp during the replication of RNA1 (Iwamoto *et al.*, 2004) and is co-terminal with the 3′ terminus of RNA1; RNA1 accumulates within the host cell during the initial virus transcription (Iwamoto *et al.*, 2005; He and Teng, 2015).

*sandra@umt.edu.my

RNA3 encodes one or two small non-structural proteins (B1 and B2). Previous study suggests that the RNA3 within all four betanodavirus genotypes is missing the B1 ORF which is in the same reading frame as protein A (ORF A) and B2 is a +1 ORF relative to ORF A (Okinaka and Nakai, 2008; Costa and Thompson, 2016). The level of B2 expression is based on whether the stage of infection is either acute or chronic as B2 is absent in chronically infected fish (Mezeth *et al.*, 2009).

Sequence divergences have been found among the betanodavirus isolates from different geographical locations (Nishizawa *et al.*, 1997; OIE, 2014). RNA2 contains a highly variable region identified from the early phylogenetic analyses that classified betanodavirus into four genotypes based on the T4 region of RNA2 SJNNV: (i) *Redspotted grouper nervous necrosis virus* (RGNNV) (*Epinephelus akaara*); (ii) *Striped jack nervous necrosis virus* (SJNNV); (iii) *Tiger puffer nervous necrosis virus* (TPNNV) (*Takifugu rubripes*); and (iv) *Barfin flounder nervous necrosis virus* (BFNNV) (*Verasper moseri*) (Nishizawa *et al.*, 1995, 1997; Iwamoto *et al.*, 2000; Dalla Valle *et al.*, 2001; Skliris *et al.*, 2001; Tanaka *et al.*, 2001; Chi *et al.*, 2003; Gagné *et al.*, 2004; Thiéry *et al.*, 2004; Panzarin *et al.*, 2012; OIE, 2014; Costa and Thompson, 2016). In 2014, two genotypes of betanodavirus were discovered in two species including turbot (*Scophthalmus maximus*) and Atlantic cod (*Gadus morhua*). Atlantic cod nervous necrosis virus (ACNNV) was also found in haddock (*Melanogrammus aeglefinus*) and winter flounder (*Pseudopleuronectes americanus*) (Gagné *et al.*, 2004). However, only turbot nervous necrosis virus (TNNV) was accepted as the new and fifth genotype; meanwhile ACNNV has been classified as a subgroup of the BFFNV genotype. Later, to enable the subgroups within genotypes based on specific genotypes rather than host specificity, the new nomenclature system for betanodavirus genotypes was introduced as I, II, III and IV for RGNNV, BFNNV, TPNNV and SJNNV, respectively) and TNNV (genotype V) (Thiéry *et al.*, 2004; Cutrín *et al.*, 2007; Moody *et al.*, 2009).

The SJNNV genotype was commonly described in farmed fish particularly in striped jack and red sea bream (*Pagrus major*) in Japan (Nishizawa *et al.*, 1994) before it was reported in the Iberian Peninsula countries (Spain and Portugal) in 2007 (Cutrín *et al.*, 2007; García-Rosado *et al.*, 2007). SJNNV is highly virulent to larval striped jack in natural and experimental conditions (Arimoto *et al.*, 1993, 1994). Panzarin *et al.* (2012) examined 120 viral strains isolated between 2000 and 2009 in six different countries in Southern Europe and found six isolates genotyped as RGNNV and one isolate genotyped as SJNNV; 23/120 samples were classified as RGNNV/SJNNV reassortants. The SJNNV genotype has been described to cause disease in cultured Senegalese sole (*Solea senegalensis*) in Spain within the Mediterranean Sea (Thiéry *et al.*, 2004; Cutrín *et al.*, 2007). However, virus sequences of the Spanish isolates were different from the isolates from Japan, indicating that the sequence variation shown by both isolates was high although clustered within the same genotype. The emergence of SJNNV in Spain and Portugal was reported in the isolates of Senegalese sole and gilthead sea bream (*Sparus aurta*) (Olveira *et al.*, 2009). However, these isolates were distinguished from the Japanese and Iberian strains based on geographical origin as the subtypes were distinctly clustered. SJNNV reassortants exist as either SJNNV/RGNNV or RGNNV/SJNNV. These data indicated that both combinations of genomic segments of SJNNV and RGNNV genotypes are successful in producing disease. Interestingly, from the results of Cherif *et al.* (2011) and Olveira *et al.* (2009), a certain relationship between the type of reassortant and the susceptible host species seems to exist: SJNNV/RGNNV affecting European sea bass (*Dicentrarchus labrax* (L.)) and RGNNV/SJNNV affecting Senegalese sole and gilthead sea bream. Recently, the VNN causative agent detected in hybrid groupers from Malaysia showed high similarities (88–100%) to SJNNV and the reassortant strain RGNNV/SJNNV isolate 430.2004 (GenBank accession number JN189932.1) (Ariff *et al.*, 2019).

In striped jack, mortalities most frequently occur within 10 days after hatching (Mori *et al.*, 1998). Total losses (100% mortality) of the Japanese flounder juveniles were seen 2 weeks after the onset of SJNNV disease at two farms in Japan. The disease initiated 35 days post-hatching and mass mortalities occurred when the fishes were at 25 mm (Nguyen *et al.*, 1994). The virus was purified from diseased larvae and the viral aetiology was established by experimental infection. The virion was non-enveloped and spherical, about 25 nm in diameter, with two structural proteins and contained two single-stranded, positive-sense RNA molecules. From these morphological and biochemical properties, this virus was identified as a member of the *Nodaviridae* and designated as SJNNV.

5.2 Diagnosis

5.2.1 Morphological features

Nodavirus virions are non-enveloped, roughly spherical, 25–33 nm in diameter and have icosahedral symmetry ($T = 3$) (Fig. 5.1). Electron microscopy of negatively stained betanodaviruses shows surface projections; these are not observed in alphanodaviruses. Virion buoyant density in CsCl ranges from 1.30 to 1.36 g/cm^3. Virions are stable to pH values ranging from 2 to 9 and are resistant to heating at 56°C for 30 min (Sahul Hameed *et al.*, 2019). SJNNV contains two single-stranded, positive-sense RNA molecules of 1.01×10^6 Da (RNA1) and 0.49×10^6 Da (RNA2). RNA2 encodes a structural protein of 42 kDa of the virus (Mori *et al.*, 1992). Nishizawa *et al.* (1995) found that the SJNNV coat protein gene was 1410 bases in length and contained a single ORF of 1023 bases encoding a protein of 340 amino acids (Munday *et al.*, 2002).

5.2.2 Clinical signs

The appearance of clinical signs caused by VNN infection varies with fish species, the biological stage of the fish, the virulence of the viral strain and the temperature of the water (Souto *et al.*, 2015a). VNN causative agent multiplies within the eye, brain and spinal cord as well as in the internal organs including gonad, liver, stomach and intestine. The virus infection results in severe tissue vacuolation (OIE, 2017). Commonly, larvae with hyperacute infection suffer from anorexia that leads to death. Affected species will develop several neurological abnormalities due to the presence of vacuolation and cell necrosis in the retina and central nervous system (Fig. 5.2) leading to mortality rates of up to 100% (Munday *et al.*, 2002; Iwamoto *et al.*, 2004). As a result, fish affected by VNN infection will display abnormal swimming behaviour such as corkscrew motion, looping swimming pattern, darting and upside-down orientation due to neurological disorders (Kokawa *et al.*, 2008). Affected species will also exhibit skin colour variation (pale or dark), loss of appetite, lethargy and hyperinflation of the swim bladder (Ransangan and Manin, 2010; Costa and Thompson, 2016). In fact, it was reported that the fish with bright silver coloration tend to survive better than fish with dark to grey skin pigmentation (Ransangan and Manin, 2010). The authors of that report proposed that the bright silver fish may have a stronger defence system against the VNN agent and could possibly serve as virus carriers (Gomez *et al.*, 2004; Xylouri *et al.*, 2007).

The first detection of VNN in larval striped jack described fish with central nervous systems containing infected nerve cells with numerous virus particles in the cytoplasm (Nishizawa *et al.*, 1994). The outbreak in larval striped jack in Japan affected 2- to 20-day-old larvae (Arimoto *et al.*, 1993). The affected larvae demonstrated loss of appetite, skinny body and mortality after 2 to 4 days in larvae less than 10 days old; in older larvae (>11 days), the clinical signs were swim bladder enlargement and vertebral deformity (Arimoto *et al.*, 1993). Japanese flounder (*Paralichthys olivaceus*) juveniles infected with SJNNV had abnormal swimming behaviour such as whirling and there was congestion in the brains and gills in some fish. No histopathological changes were observed in the gills, liver, kidney and spleen; heavy vacuolation was observed in the brain, spinal cord and retina (Nguyen *et al.*, 1994). The progression of infection in naturally and experimentally infected striped jack larvae with SJNNV was followed histopathologically by Nguyen *et al.* (1996). These investigators found necrosis and vacuolation

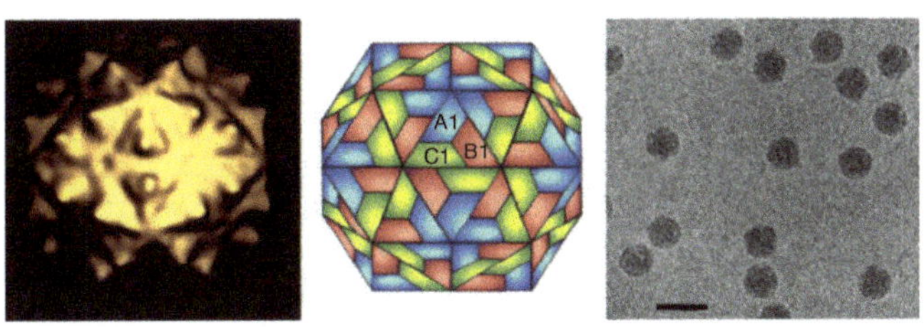

Fig. 5.1. *Flock House virus* is an alphanodavirus that infects insects and represents the general structure of nodaviruses. (a) Image reconstruction. (b) Schematic representation of a $T = 3$ icosahedral lattice; A1, B1 and C1 indicate three different quasi-equivalent copies of the capsid protein. (c) Cryo-electron micrograph (scale bar = 50 nm). (Courtesy of N. Olson and T. Baker.) (Reproduced with permission of *Journal of General Virology* from Sahul Hameed *et al.*, 2019.)

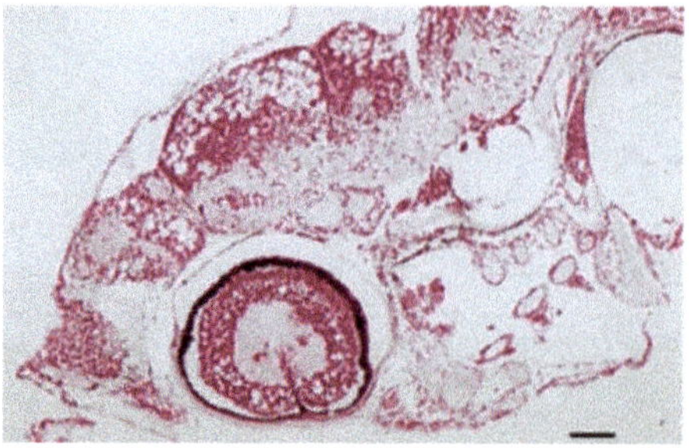

Fig. 5.2. Barramundi larva with betanodavirus infection. Note severe vacuolation of the brain and retina. Stained with haematoxylin and eosin; scale bar = 100 μm. (Reproduced with permission of *Journal of Fish Diseases* from Munday *et al*., 2002.)

first in the nerve cells in the spinal cord, above the swim bladder, followed by the lesions in the brain and then the retina. Mortalities occurred at 1 to 2 days after the commencement of lytic degeneration of the cells with heavy vacuolation in the nervous tissues (Nguyen *et al*., 1996). Mori *et al*. (1998) observed no specific external signs, except enlargement of the swim bladder in striped jack juveniles during mortalities. The clinical signs of infection observed in the Malaysian hybrid grouper that were found positive for the VNN causative agent (RGNNV/SJNNV) included darkened skin, deformation of the backbone, abdominal distension, skin lesions and fin erosion (Ariff *et al*., 2019). The most typical histopathological change was extended vacuolation in the retinal and central nervous tissues. Non-enveloped spherical virus particles were observed in the cytoplasm of the degenerated nerve cells by electron microscopic observations (Mori *et al*., 1998). Swim bladder hyperinflation has been reported in barramundi (*Lates calcarifer*), European sea bass and striped jack (Breuil *et al*. 1991; Mori *et al*., 1992; Munday *et al*., 2002). Sea bass affected by SJNNV/RGNNV demonstrated dark skin pigmentation, exophthalmia and nervous swim behaviour (Cherif *et al*., 2011). Although SJNNV causes no mortality or very low mortality in sea bass, several authors demonstrated that SJNNV replicates in sea bass brain without development of any histopathological lesions characteristic of SJNNV infection (Souto *et al*., 2015a; Carballo *et al*., 2016; Moreno *et al*., 2018).

5.2.3 Methods

Several immunoassays and DNA-based diagnostic methods have been developed for betanodavirus pathogen screening and disease diagnosis such as immunohistochemistry (IHC) (Le Breton *et al*., 1997), indirect enzyme-linked immunosorbent assay (ELISA) (Shieh and Chi, 2005), *in situ* hybridization (Comps *et al*., 1996), fish cell-line cytopathic effect (CPE) (Chi *et al*., 1999) and reverse transcription-polymerase chain reaction (RT-PCR) (Nishizawa *et al*., 1995; Dalla Valle *et al*., 2005).

Cell culture

For VNN, cell culture is one of the diagnostic assays developed for detection of the causative agent from supernatants of brain or eye homogenates (OIE, 2000). The SSN-1 cell line was derived from striped snakehead (*Channa striata*) (Frerichs *et al*., 1996) and a cloned cell line (E-11) derived from SSN-1 itself (Iwamoto *et al*., 2000). Both these cell lines are useful for qualitative and quantitative analyses of all betanodaviruses. Other susceptible cell cultures, including GF-1, have been developed and described (Chi *et al*., 1999), and may be used for research and diagnostic purposes provided cell lines' sensitivity to virus infection is regularly monitored (OIE, 2018). To detect virus in carriers, the PCR method is preferred and more sensitive (Gomez *et al*., 2004). According to Zorriehzahra *et al*. (2019), the SSN-1 cell line has been shown to be able to support replication for 17 isolates of fish nodaviruses, including the RGNNV, SJNNV, TPNNV and BFNNV types (Iwamoto *et al*., 2000, 2001; Dalla Valle *et al*., 2001; Chi *et al*., 2003). CPE is evident by the dark, granular and contracted cells which tend to detach from the surface of the flask. Vacuoles inside the cytoplasm of infected cells are observed. CPE further evolves into extended necrotic foci, resulting in the complete disruption of the cell monolayer (Panzarin, 2016).

Immunological methods

SJNNV was first detected using indirect ELISA in eggs, larvae and broodstock of hatchery-reared and captured striped jack (Arimoto *et al*., 1992). Indirect ELISA using purified SJNNV, rabbit anti-SJNNV serum and enzyme-conjugated goat anti-rabbit IgG antibody was employed for antibody detection from the plasma of the striped jack broodstock (Mushiake *et al*., 1992). Antibody to SJNNV was detected at high frequency (65%) in plasma samples collected from broodstock reared at various facilities regardless of their sex or origin (wild or domestic). Mouse monoclonal antibodies (MAbs) to SJNNV were produced by using a homogenate of infected larval striped jack. Among eight MAbs reacting with the homogenate of infected larvae in ELISA, three MAbs (SJ-102B, SJ-204D, SJ-207C) were found to recognize the 42 kDa coat protein of SJNNV by Western blot analysis (Nishizawa *et al.*, 1995).

The first study addressing the sea bass anti-SJNNV immune response was conducted in SJNNV-infected European sea bass (Moreno *et al.*, 2018). In that study, the transcription of IFN-I (interferon type 1), ISG-12 (interferon-stimulated gene 12), ISG-15 (interferon-stimulated gene 15) and MxA (Myxovirus resistance gene A) genes was analysed in the brain and head kidney. The findings demonstrated that SJNNV genotypes induced less intense response of the IFN-I system but higher IgM (immunoglobulin M) titres in sera than the RGNNV genotype (Moreno *et al.*, 2018). Labella *et al.* (2018) showed that a highly virulent RGNNV/SJNNV reassortant, named wild-type (wt) isolate (wSs160.03) in Senegalese sole, upregulated the immune genes such as type I IFN expression (DHX58, IRF3, IRF7) and IFN-stimulated genes (ISG15, Mx, PKR (protein kinase RNA activated), Gig1 (IFN-stimulated genes), ISG12, IFI44 (interferon-induced protein 44), IFIT-1 (interferon-induced proteins with tetratricopeptide repeats)) in infected fish.

Molecular methods

The conventional RT-PCR method is the current accepted method to detect the agent that causes VNN. There are some limitations to this method including complex processing of the sample and low-efficiency replication during amplification (Nishizawa *et al.*, 1995; Starkey *et al.*, 2004; Harikrishnan *et al.*, 2010). However, these limitations have been addressed by the use and design of more efficient PCR primers for RT-PCR and quantitative PCR (qPCR). PCR primers were designed based on the nucleotide sequence of SJNNV RNA2 (Nishizawa *et al.*, 1994). Various types of primers have been designed to optimize the detection of the virus corresponding to different genotypes and geographical areas (Nishizawa *et al.*, 1994; Gagné *et al.*, 2004; Gomez *et al.*, 2004; Ucko *et al.*, 2004; Manin and Ransangan, 2011). The PCR primers used to detect VNN causative agent were synthesized from the ORF of SJNNV covering the 1410 bp RNA2 coat protein gene between nt 17 and 1036. The ORF consisted of five potential target regions (T1, T2, T3, T4, T5) used for PCR amplification with a combination of five different pairs of PCR primers (F1–R1 (T1), F1–R3 (T2), F2–R1 (T3), F2–R3 (T4) and F1–R2 (T5)). All the target regions were expected to amplify the product size of 1147 bp (T1), 875 bp (T2), 698 bp (T3), 426 bp (T4) and 175 bp (T5) (Fig. 5.3). Interestingly, the PCR amplification of betanodavirus genotypes was observed only on T2 and T4 regions of SJNNV (Nishizawa *et al.*, 1995). The detection of VNN agent in fish that survived the outbreaks or infection can be confirmed by RT-PCR amplification. RT-PCR has been widely used for molecular detection of VNN disease in maricultured striped jack juveniles in South Asian countries (Nishizawa *et al.*, 1994). It was reported that the number of VNN viral copies detected by real-time RT-PCR in fish with clinical signs was higher than in those survivor fish that tested positive for the virus (Toffan *et al.*, 2017). However, studies conducted by Toffan *et al.* (2017) have shown that sea bream that survived after several months of infection tested positive for VNN agent in the absence of VNN disease.

A relative quantification RT-PCR assay was developed for the detection of SJNNV and RGNNV in sea bass, Senegalese sole and gilthead sea bream (Cherif *et al.*, 2011). A combined RT-PCR and dot-blot hybridization method was developed to detect the coexistence of SJNNV and RGNNV betanodavirus genotypes in wild meagre (*Argyrosomus regius*) (Lopez-Jimena *et al.*, 2010). The combination of RT-PCR and blot hybridization increases the detection rate up to 90.62% from asymptomatic carriers (Lopez-Jimena *et al.*, 2010). A new one-step real-time RT-PCR targeting RNA1 of most genotypes of betanodaviruses including SJNNV was developed by Baud *et al.* (2015) and this assay detected reliably 50–100 copies of five plasmids each containing a genomic portion of RNA1 from different genotypes. The sensitivity of real-time PCR is capable of detecting subclinical VNN infection in carrier fish and screening out infected broodstock to reduce or prevent the vertical transmission of the virus (Costa and Thompson, 2016). Zorriehzahra *et al.* (2019) also pointed out another method developed by Deiman *et al.* (2002). This method, NASBA, consists of an isothermal method for nucleic acid amplification that is particularly suited to RNA targets. Toubanaki *et al.* (2015) developed a novel allele-specific PCR method for the detection of RGNNV and SJNNV suitable for farming facilities

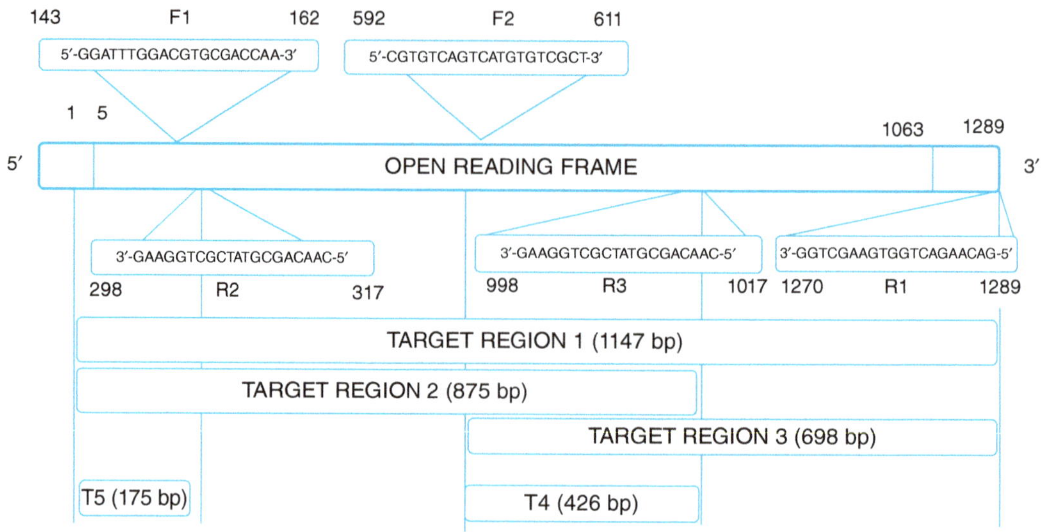

Fig. 5.3. Schematic illustration of the physical map of the SJNNV RNA2 composed from two forward primers and three reverse primers with five target regions (T1, T2, T3, T4 and T5) for PCR amplification of SJNNV gene (re-illustrated and similar to Nishizawa *et al.*, 1994). (Reproduced with permission of *Diseases of Aquatic Organisms* from Nishizawa *et al.*, 1994.)

and medium-scale research laboratories. This method can be conducted within 4 h and costs less than 2€ per sample. Phase 1 of the PCR confirms the existence of betanodavirus in a sample of either genotype using degenerate primers, whereas the product of phase 2 indicates the specific genotype (Toubanaki *et al.*, 2015). Most recently, a dual lateral flow biosensor for simultaneous detection of the most prevalent betanodavirus genotypes (RGNNV and SJNNV) was developed and optimized; the biosensor is still awaiting further evaluation with clinical samples before it can be used in the development of a diagnostic kit (Toubanaki and Karagouni, 2018).

5.3 Potential Spread of the Pathogen/Intermediate Host/Reservoir Host (Transmission)

5.3.1 Transmission pathways

Biological and environmental stress factors are involved in the development of latent to clinical infections via vertical and horizontal transmission (Breuil *et al.*, 2002; Azad *et al.*, 2006; Nerland *et al.*, 2007; Hick *et al.*, 2011; Krishnan *et al.*, 2016). Vertical transmission of the agent that causes VNN can occur from the broodstock via eggs or sperm, or it may enter the aquaculture facility through water sources and/or feed as the basis for horizontal transmission (Castric *et al.*, 2001; Breuil *et al.*, 2002; Gomez *et al.*, 2008a; Manin and Ransangan, 2011; Patel and Nerland, 2014). The occurrence of horizontal transmission has been reported previously in infected halibut larvae where high concentrations of VNN agent have been detected (Nerland *et al.*, 2007). However, the authors speculated that the virus is transmitted vertically from subclinically infected broodstock to fingerlings and these larvae may spread the disease in water horizontally.

A prior study by Mushiake *et al.* (1994) showed that the detection of SJNNV using RT-PCR amplification in broodstock gonads corresponded to the presence of VNN causative agent in their offspring. This finding demonstrated that removing the VNN-positive broodstock from the spawning batch reduced the risk of mortalities associated with VNN in larvae. Mushiake and Arimoto (2000) investigated the effects of multiple spawnings and rearing conditions for spawning (hormone injection and rearing density) on the multiplication of SJNNV and its effect on the disease in broodstock of striped jack. The results suggested that rearing conditions such as multiple spawnings (more than ten), HCG (human chorionic gonadotropin) injection and higher rearing density would lead to higher SJNNV multiplication in the spawners by reducing their immune resistance, which would increase the vertical transmission of SJNNV from spawners to their offspring (Mushiake, 2000). A probable vertical mode of piscine nodavirus transmission was reported for a case of nodavirus-associated larval mortalities in hatchery-produced Asian sea bass/barramundi (*L. calcarifer*) (Azad *et al.*, 2006). Polyclonal rabbit anti-SJNNV antibodies (SGWak97) detected the viral antigens in the tissue sections from the eggs and the larvae at different time intervals from 1 to 42 days post-hatch (Azad *et al.*, 2006). A vertical transmission of VNN causative agent was postulated and confirmed in outbreaks caused by RGNNV/SJNNV reassortants in gilthead sea bream (Toffan *et al.*, 2017). Younger infected sea bream larvae (10–15 days post-hatch) showed positive results for the detection of the reassortants and it was speculated that horizontal transmission played a vital role in spreading the disease via water (Toffan *et al.*, 2017).

Cutrín *et al.* (2007) concluded that trade in live fish and eggs is a key factor in the dissemination of nodaviruses to different geographical locations (Nishizawa *et al.*, 1997; Aspehaug *et al.*, 1999) and commercial exchange is regarded as mainly responsible for the spread of nodaviruses within the natural range of affected fish species (Munday *et al.*, 2002). Striped jack is a widely distributed circumtropical species and it could be argued that individuals of these species carrying SJNNV might have spread the virus to other fish species inhabiting other regions. Although the existence of highly migratory fish serving as carriers and transporting the virus to distant geographical locations has been proposed (Curtis *et al.*, 2001), Cutrín *et al.* (2007) speculated that no migration of fish species from Japanese to southern European waters occurred at the time. The epidemiological and phylogenetic analysis revealed that the Iberian strains may be derived from a common ancestor originally found in Senegalese sole followed by sea bream and European sea bass. Thus, extensive aquaculture activities in Spain and Portugal may trigger the virus in latently infected wild Senegalese sole that remain in the area for a period of time (Cutrín *et al.*, 2007). The presence of asymptomatic or subclinically infected wild fish may lead to the occurrence of viral disease in cultured fish (Gomez *et al.*, 2008b). SJNNV has been isolated from asymptomatic wild Japanese jack mackerel (*Trachurus japonicas*) from Japan (Nishioka *et al.*, 2016). That study revealed that SJNNV in subclinically infected wild fish was pathogenic and suggested that wild fish can be a potential source for disease transmission via water or as untreated food sources (Nishioka *et al.*, 2016).

5.3.2 Water temperature

Survivability of the agent that causes VNN under extreme environmental conditions has been reported previously (Frerichs *et al.*, 2000). In fact, betanodavirus genotypes are not highly associated with species specificity but tend to correlate more with geographical area in which water temperature plays the biggest role (Hata *et al.*, 2010). The relationship between the occurrence of VNN disease and environmental factors has been described in several studies (Le Breton *et al.*, 1997; Vendramin *et al.*, 2013). Clinical signs of VNN can also be induced under optimum water temperature (OIE, 2017). It has also been reported that optimum growth temperature can induce the appearance of cytopathic effects under *in vitro* conditions within 3–5 days post-incubation (Iwamoto *et al.*, 2000; OIE, 2017).

According to Arimoto *et al.* (1993), the agent that causes VNN occurred at water temperatures from 20 to 26°C, which are the temperatures used for seed production of striped jack, and the replication of SJNNV increased most rapidly at 24°C. Quantitative analysis conducted by Iwamoto *et al.* (2000) using the cloned E-11 cell line for virus culture has shown that the optimum growth temperatures for the four genotypes – RGNNV, SJNNV, TPNNV and BFNNV – were 25–30, 20–25, 20 and 15–20°C, respectively. The adaptation of the genotype clusters to certain water temperature correlated with the environmental water temperatures in different geographical regions and host species adapted to those temperatures (Thiéry *et al.*, 2004; Cherif *et al.*, 2009).

Experimental trials have also demonstrated the effect of temperature, infectious dose and viral multiplication rate on betanodavirus pathogenicity and disease course (Panzarin *et al.*, 2014). Panzarin *et al.* (2014) demonstrated that the mortality rate for a given strain increased with increasing water temperature; including RGNNV/SJNNV at 25°C in experimentally infected European sea bass. The analysis indicated that at 15°C, viral titres of strains RGNNV, SJNNV and SJNNV/RGNNV increased significantly more than RGNNV/SJNNV titres over time whereas SJNNV showed a reduced fitness at 25°C. At 20°C, no significant difference was observed between the increase over time of viral titres of strains RGNNV and RGNNV/SJNNV, and strains SJNNV and SJNNV/RGNNV, while isolates SJNNV and SJNNV/RGNNV showed a significantly higher replication efficiency compared with strain RGNNV and strain RGNNV/SJNNV. At 25 and 30°C, SJNNV viral titres were significantly lower than RGNNV, RGNNV/SJNNV and SJNNV/RGNNV titres over time. However, at both temperatures, no significant differences were noticeable among the replication efficiencies of viruses RGNNV, RGNNV/SJNNV and SJNNV/RGNNV (Panzarin *et al.*, 2014). More recently, there have been increasing reports of the susceptibility of fish species to reassortant betanodavirus (RGNNV/SJNNV). Experimental challenges using the reassortant betanodavirus (RGNNV-RNA1 and SJNNV-RNA2) on juvenile Senegalese sole have shown different virulence levels at three different temperatures: 22, 18 and 16°C (Souto *et al.*, 2015c). The results revealed that the virulence level of the betanodavirus decreased with the reduction of temperature where the highest mortalities (100%) were recorded at 22°C and the lowest were at 16°C (8%). Souto *et al.* (2015a) demonstrated that the RGNNV/SJNNV reassortant isolate obtained from diseased Senegalese sole exhibited a slightly modified SJNNV capsid protein compared with the parental genotype. The results suggested that changes in the SJNNV RNA2 segment of the reassortant isolate may involve host specificity and/or virulence determinants for European sea bass. Souto *et al.* (2019) suggested that RNA1 encoding the RdRP plays a key role in controlling viral replication at different temperatures and that the region comprising amino acid positions 1–445 of the protein A is involved in thermotolerance of the agent that causes VNN (Hata *et al.*, 2010; Panzarin *et al.*, 2014). They compared the replication of a recombinant virus with point mutations (r1_445) with the recombinant with no mutations, the wild-type strain and strains belonging to the parental genotypes RGNNV and SJNNV at 15, 20, 25 and 30°C. These findings indicate that r1_445 showed a temperature sensitivity (ts) phenotype and confirmed that the 1–445 region of RNA polymerase is involved in the temperature adaptation of VNN causative agent as previously reported. At high temperature (25°C), r1_445 demonstrated poor adaptation to temperature due to the mutations in the polymerase N region (Souto *et al.*, 2019).

5.4 Increased Prevalence of Pathogen and Disease Severity due to Reassortants

In recent years, the occurrence of reassortant strains between the RGNNV and SJNNV genotypes has been increasingly reported in sole, sea bream and sea bass species mainly in Southern Europe (Toffolo *et al.*, 2007; Olveira *et al.*, 2009; Panzarin *et al.*, 2012; Souto *et al.*, 2015b). Olveira *et al.* (2009) showed that almost all the isolates from RNA2 analyses belong to SJNNV strain. However, the analysis of RNA1 provided the opposite result where the isolates were classified within RGNNV genotypes. Toffolo *et al.* (2007) reported that the betanodavirus isolates were comprised of RNA1 of SJNNV and RNA2 of RGNNV. This finding revealed the occurrence of natural reassortment between SJNNV and RGNNV within the same isolates and these reassortants can produce disease. In fact, the chimeric betanodavirus developed previously using SJNNV and RGNNV indicated that reassortment between two genotypes was possible (Iwamoto *et al.*, 2004). Interestingly, the other type of reassortment event between RGNNV and SJNNV was also found with RNA1-RGNNV genotype and RNA2-SJNNV genotype (RGNNV/SJNNV) within European sea bass and Senegalese sole (Souto *et al.*, 2015b). According to Souto *et al.* (2015b), the reassortant of RGNNV/SJNNV is the dominant cause of the clinical signs and mortalities in European sea bass and Senegalensis sole as these isolates have altered genome sequences of RNA2 SJNNV.

A VNN outbreak was reported at low temperature in France where the BFNNV was isolated from European sea bass (Thiéry *et al.*, 2004). In a later study by Panzain *et al.* (2014), RGNNV, SJNNV, RGNNV/SJNNV and SJNNV/RGNNV field isolates were fully sequenced and growth curves generated *in vitro* at four different temperatures (15, 20, 25, 30°C) were developed for each isolate. The experimental data demonstrated that viral titres varied significantly with incubation temperature, particularly at 30°C. Viruses containing the RNA1 of the RGNNV genotype showed the best replication efficiency at 30°C compared with the other isolates. Vendramin *et al.* (2014) examined the pathogenicity of ten virus isolates in European sea bass. Their study showed no strong correlation between genotype, water temperature, host of origin and pathogenicity. To date, the occurrence of reassortment events between RGNNV and SJNNV has been described in several studies (Table 5.1). Until 2015,

Table 5.1. The occurrence of reassortment between RGNNV and SJNNV exhibited by both RNA1 and RNA2. (From Ariff *et al.*, 2019.)

Isolate	Source	RNA1	RNA2	Reference
SpSa-IAusc156.03	Gilthead sea bream (*Sparus aurata*)	RGNNV	SJNNV	Olveira *et al.* (2009)
SpSs-IAusc160.03	Senegalese sole (*Solea senegalensis*)	RGNNV	SJNNV	Olveira *et al.* (2009)
VNNV/S.aurata/Farm1/127-1/ Mar2015	Gilthead sea bream (*S. aurata*)	RGNNV	SJNNV	Toffan *et al.* (2017)
VNNV/S.aurata/Farm1/461-3/ Nov2014	Gilthead sea bream (*S. aurata*)	RGNNV	SJNNV	Toffan *et al.* (2017)
292.7.8.2009 D. labrax Greece F	European sea bass (*Dicentrarchus labrax*)	RGNNV	SJNNV	Panzarin *et al.* (2012)
250.2.2009 S. aurata Cyprus F	Gilthead sea bream (*S. aurata*)	RGNNV	SJNNV	Panzarin *et al.* (2012)
250.3.2009 A. salina Cyprus F	Brine shrimp (*Artemia salina*)	RGNNV	SJNNV	Panzarin *et al.* (2012)
477.2004 S. solea Italy (VE) F	Senegalese sole (*S. senegalensis*)	RGNNV	SJNNV	Panzarin *et al.* (2012)
367.2.2005	European sea bass (*D. labrax*)	RGNNV	SJNNV	Panzarin *et al.* (2012)
389/196	European sea bass (*D. labrax*)	SJNNV	RGNNV	Vendramin *et al.* (2014)
DI-HR-96	European sea bass (*D. labrax*)	SJNNV	RGNNV	Toffolo *et al.* (2007)
DI-1-96b	European sea bass (*D. labrax*)	SJNNV	RGNNV	Toffolo *et al.* (2007)
82/107	Gilthead sea bream (*S. aurata*)	RGNNV	SJNNV	Beraldo *et al.* (2011)

the occurrence of inter-genotype reassortments between RGNNV and SJNNV demonstrated the ability of these isolates to co-infect fish and the finding of reassortants was solely restricted to South European countries (He and Teng, 2015).

He and Teng (2015) found that viral genotypes clustered genetically within subgroups despite different geographical origins. All the viruses identified in the study conducted by Toffan *et al.* (2017) also indicated genetic similarities despite different geographic origins and date of isolation. Among the factors that contributed to this phenomenon included trade activity of fish stocks, transportation of asymptomatic carriers and convergent evolution (Dalla Valle *et al.*, 2001; Toffolo *et al.*, 2007).

5.5 Pathogen Control Methods

5.5.1 Chemical agents

The early study by Arimoto *et al.* (1996) examined the effects of chemical disinfectants, organic solvents, hydrogen ions, heat, ultraviolet (UV) irradiation and ozone on the inactivation of SJNNV. The following inactivated SJNNV successfully: (i) sodium hypochlorite, calcium hypochlorite, benzalkonium chloride and iodine, 50 ppm, for 10 min at 20°C; (ii) cresol at concentrations >10,000 ppm; (iii) 60% (v/v) ethanol and 50% (v/v) methanol; (iv) pH 12 for 10 min at 20°C; (v) heat treatment at 60°C for 30 min; (vi) UV irradiation at 410 $\mu W/cm^2$ for 244 s; and (vii) ozone at 0.1 pg/ml as a total residual oxidant for 2.5 min.

5.5.2 Temperature

Senegalese sole is extremely susceptible to the reassortant strain RGNNV/SJNNV, with 100% mortality at a water temperature of 22°C. Mortality can be reduced dramatically to 8% when water temperature is lowered to 16°C. Notably, the rise of water temperature caused a significant increase in the viral load in fish challenged at 16°C, with a consequent escalation of mortalities. This suggests that the agent that causes VNN can produce a persistent infection in Senegalese sole at low temperatures, and the virus can easily be reactivated by an increase of water temperature even a long time, 66 days, after the infection (Souto *et al.*, 2015a,b).

5.5.3 Vaccination

Several vaccines against VNN disease have been developed. These include DNA vaccines, killed vaccines, attenuated vaccines, recombinant technology vaccine and synthetic peptide vaccines (Assefa and Abunna, 2018). Previously, an experimental challenge was conducted on adult turbot and Atlantic halibut (*Hippoglossus hippoglossus*) using intraperitoneal injection of an oil-emulsified recombinant partial capsid protein (rT2) derived from the capsid gene of SJNNV (Húsgağ *et al.*, 2001). The results indicated that a specific humoral immune response was triggered, and the level of rT2-specific antibodies was increased significantly during the 20 weeks after a single and a double immunization trial. In addition, the efficacy of the vaccine was determined using the juvenile turbot that were vaccinated with an oil-emulsified rT2 as a test vaccine and challenged by the intramuscular inoculation of SJNNV at 5 and 10 weeks post-vaccination. The test vaccine showed that the rT2 protein could induce full protection in turbot at 10 weeks post-vaccination with SJNNV. The mortality and severity of clinical signs were reduced markedly in vaccinated fish. Similar vaccination protocols have been applied on the juvenile turbot using 10 and 50 µg of recombinant capsid protein from the Atlantic halibut betanodavirus (recAHNV-C) (Sommerset and Nerland, 2004). The findings indicate that the Atlantic halibut vaccinated with 50 µg demonstrated a higher survival rate than those vaccinated with 10 µg recombinant capsid protein. However, the plasmid vaccine (pAHNV-C) developed in the study was unsuccessful in providing full protection against betanodavirus challenge. The lack of efficient stimulation of innate immunity exhibited by vaccinated fish might hinder the development of the specific immunity of pAHNV-C in vaccinated fish. Turbot vaccinated with a recombinant partial capsid protein of SJNNV (Húsgağ *et al.*, 2011) is protected significantly against betanodavirus challenge (reviewed by Shetty *et al.*, 2012). Nishizawa *et al.* (2012) demonstrated the potential of a live VNN vaccine for sevenband grouper (*Epinephelus septemfasciatus*) at a low rearing temperature (17°C). The fish mortality was reduced at 17°C.

5.6 Control and Prevention

Biosecurity in aquaculture involves the practices, procedures and policies used to prevent the introduction of infectious diseases and their spread to the other cultured animals within the facility (Yanong and Erlacher-Reid, 2012). Biosecurity measures recommended by the World Organization for Animal Health (OIE, 2018) for VNN disease in general are: UV treatment of water entering the hatchery; the adoption of sanitary barriers; regular fallowing and disinfection of tanks and biological filters; disinfection of facilities and utensils; avoidance of raw fish for feeding; improvement of the spawning-induction method, which includes providing adequate food for broodstock; and decreasing the stocking density of larvae and juveniles (Mushiake *et al.*, 1994). Watanabe *et al.* (2000) proposed the use of ELISA to test the level of specific antibody in each broodstock. Unfortunately, these measures have not been sufficiently implemented. For Japanese flounder, screening for the agent that causes VNN using ELISA is done at 3 months before spawning and only negative fish are reared as broodstock for egg production (Yoshimizu, 2009). In the sea bass industry, it has been suggested that restocking of on-growing facilities located in infected areas should be performed during autumn when the number of clinical outbreaks is decreasing (OIE, 2018). Common disinfectants such as sodium hypochlorite, iodine, hydrogen peroxide and benzalkonium chloride have been shown to inactivate betanodaviruses; ozone has also been used to avoid or reduce virus contamination on eggshell surfaces (Grotmol and Totland, 2000). Chlorine at 25–100 ppm for 5–30 min has been used to inactivate VNN causative agent (reviewed by Munang'andu, 2016). Water supplies from unreliable sources can be filtered and sterilized using UV light, chlorine and ozonation. Vertical transmission from the broodstock to the offspring can be reduced by washing the fertilized eggs with ozonated seawater (Mori *et al.*, 1998; Hick *et al.*, 2011; OIE, 2017). The ozonated seawater was shown to disinfect Atlantic halibut and haddock (*M. aeglefinus*) eggs against betanodavirus by viral inactivation (Arimoto *et al.*, 1996; Grotmol and Totland, 2000; Buchan *et al.*, 2006). Experimental challenge studies have shown that the Atlantic halibut eggs were free from the infection after being washed with seawater ozonated at 0.3 mg/l for 30 s (Grotmol and Totland, 2000). In fact, increased survival rates were observed in striped jack larvae from diseased broodstock when the eggs were exposed to a total residual oxidant concentration of 0.5 µg/l for 1 min (Arimoto *et al.*, 1996). These results proved that ozonated seawater can be used to disinfect eggs against betanodavirus without affecting the hatching rates of the eggs (Ben-Atia *et al.*, 2001; Buchan *et al.*, 2006). The disinfection of eggs before the incubation period is important to reduce the disease transmission from surface contamination with bacteria and viruses (Buchan *et al.*, 2006). However, the optimum levels of dissolved ozone will vary depending on the fish species, so the authors recommended evaluation of the tolerance of eggs to ozonated seawater before establishing a disinfection protocol (Grotmol *et al.*, 2003).

5.6.1 Potential effect of climate change

According to Pickering *et al.* (2011), climate change may increase the risk posed by disease through alterations in the distribution, prevalence and virulence of viruses and changes in the susceptibility of the host species. A delicate balance exists between the host, the pathogen and the environment; disturbing this balance can create opportunities for pathogens to proliferate. Climate changes such as temperature fluctuations, salinity changes, low pH, low dissolved oxygen, habitat alterations and harmful algal blooms can stress the host and suppress its immune system (Pickering *et al.*, 2011). The sequence of disease development also depends to a large extent on environmental factors. Virulence of the pathogen, disease resistance mechanisms of the host and the prevailing environmental factors determine the pathology in the host.

The study of potential effects of climate change on VNN disease is limited. The agent that causes VNN has a wide host range and the ability to affect both freshwater and marine species. RNA viruses, including the VNN causative agent, have relatively high mutation rates, and are extremely heterogeneous, which allows for greater adaptability and the rapid evolution of their RNA genomes (Sanjuan *et al.*, 2007). This genetic diversity allows a viral population to rapidly adapt to dynamic environments and evolve resistance to vaccines and antiviral drugs (Lauring and Andino, 2010). Host shifts are defined as the ability of a pathogen to jump between different host species (Roberts *et al.*, 2018). Pathogen evolution lowers the species barrier so new strains are more likely to extend their host range (Kuiken *et al.*, 2006). Thereby, the evolution and emergence of reassortant strain are more likely to extend the host range of the VNN agent.

A qualitative screening-level risk assessment was developed to evaluate relative levels of risk from climate change to aquaculture industries (Doubleday *et al.*, 2013). Temperature was the most frequently cited climate change driver based on that study. Water temperatures above or below the physiological optimum of the fish species may cause stress, increasing susceptibility to infection and the likelihood of disease and mortality (Marcos-López *et al.*, 2010). As explained in Section 5.3.2, SJNNV displayed high virulence at 20 to 25°C and reduced fitness at 25 and 30°C (Panzarin *et al.*, 2014). At low temperature (15°C), SJNNV/RGNNV showed higher virulence and mortality, in contrast to RGNNV/SJNNV which demonstrated low mortality (8%) at 16°C (Souto *et al.*, 2105c). These findings may be explained by the fact that RGNNV is mostly found in warm-water species and SJNNV in cold-water species. RGNNV is able to display higher virulence at high temperature and SJNNV at low temperature. Thus, depending on the strain (SJNNV, RGNNV, SJNNV/RGNNV, RGNNV/SJNNV), the effect of temperature due to climate change could be altered. SJNNV and SJNNV/RGNNV would survive under low temperature, whereas RGNNV and RGNNV/SJNNV would survive under high temperature. Increasing water temperature has been shown to shift the balance of host–pathogen relationships and the frequency and distribution of disease (Marcos-López *et al.*, 2010).

The occurrence of SJNNV and reassortant strains (RGNNV/SJNNV and SJNNV/RGNNV) were previously found in European countries and Japan. However, a recent study showed that the VNN causative agent found in Malaysian groupers demonstrated high similarities to reassortant strains (Ariff *et al.*, 2109). Could the spread of VNN agent from European countries/Japan to Malaysia be due to the effect of climate change which enables a cold-water species strain to survive under warmer temperature? The ability of a pathogen to mutate will enable it to respond rapidly to novel opportunities created by climate change, such as the establishment of new host species (Gale *et al.*, 2009). Similarly, RGNNV/SJNNV adapts poorly to high temperature due to the mutation points in the polymerase N region (Souto *et al.*, 2019). Six substitutions in the RNA1 segment encoding amino acids 1–445 has reduced the viral replication *in vitro* compared with that of the wild-type strain at 25 and 30°C. Therefore, it is speculated that the effect of climate change could either increase or decrease the effect of VNN causative agents in susceptible species. Roberts *et al.* (2018) investigated whether species' susceptibilities towards RNA virus infection changed with temperature due to climate change and asked if susceptibility is greatest at different temperatures in different species. They concluded that as temperature increases, the most susceptible species become more susceptible, and the least susceptible less so to a novel pathogen. These results are similar to other segmented RNA viruses, such as the family *Orthomyxoviridae*, that have higher mutation rates compared with other viruses (Holland *et al.*, 1982) and thus are more likely to emerge as new diseases or in new hosts. Infectious salmon anaemia virus (ISAV), an orthomyxovirus, appears to have evolved from a wild avirulent ancestor on at least two occasions (Cunningham and Snow, 2000; Nylund *et al.*, 2003).

5.7 Conclusions with Suggestions for Future Studies

To date, many reports have highlighted risk factors associated with the development of the viral disease of RGNNV rather than SJNNV. Limited studies have been conducted on the epidemiology of SJNNV disease, the associated factors which influence SJNNV as well as treatment or vaccine development for SJNNV specifically. The emergence of reassortants such as RGNNV/SJNNV and SJNNV/RGNNV, which have been shown to cause disease, mortality and higher virulence than RGNNV or SJNNV alone, is a forewarning to the aquaculture industry. Future work should be focused on studies of the reassortants, the epidemiology of SJNNV disease, treatment and vaccination to prevent and control SJNNV.

References

Ariff, N., Abdullah, A., Azmai, M.N.A., Musa, N. and Zainathan, S.C. (2019) Risk factors associated with viral nervous necrosis in hybrid groupers in Malaysia and the high similarity of its causative agent nervous necrosis virus to reassortant red-spotted grouper nervous necrosis virus/striped jack nervous necrosis virus strains. *Veterinary World* 12(8), 1272–1284.

Arimoto, M., Mushiake, K., Mizuta, Y., Nakai, T., Muroga, K. and Furusawa, I. (1992) Detection of striped jack nervous necrosis virus (SJNNV) by enzyme-linked immunosorbent assay (ELISA). *Fish Pathology* 27, 191–195.

Arimoto, M., Mori, K., Nakai, T., Muroga, K. and Furusawa, I. (1993) Pathogenicity of the causative agent of viral necrosis disease in striped jack, *Pseudocaranx dentex* (Block & Schneider). *Journal of Fish Diseases* 16, 461–469.

Arimoto, M., Maruyama, K. and Furusawa I. (1994) Epizootiology of viral nervous necrosis (VNN) in striped jack. *Gyobyo Kenkyu* 29, 19–24. (in Japanese)

Arimoto, M., Sato, J., Maruyama, K., Mimura, G. and Furusawa, I. (1996) Effect of chemical and physical treatments on the inactivation of striped jack nervous necrosis virus (SJNNV). *Aquaculture* 143, 15–22.

Aspehaug, V., Devold, M. and Nylund, A. (1999) The phylogenetic relationship of nervous necrosis virus from halibut (*Hippoglossus hippoglossus*). *Bulletin of the European Association of Fish Pathologists* 19, 196–202.

Assefa, A. and Abunna, F. (2018) Maintenance of fish health in aquaculture: review of epidemiological approaches for prevention and control of infectious disease of fish. *Veterinary Medicine International* 2018, 5432497.

Azad, I.S., Shekhar, M.S., Thirunavukkarasu, A.R. and Jithendran, K.P. (2006) Viral nerve necrosis in hatchery-produced fry of Asian sea bass *Lates calcarifer*: sequential microscopic analysis of histopathology. *Diseases of Aquatic Organisms* 73, 123–130.

Baud, M., Cabon, J., Salomoni, A., Toffan, A., Panzarin, V. and Bigarré, L. (2015) First generic one step real-time Taqman RT-PCR targeting the RNA1 of betanodaviruses. *Journal of Virological Methods* 211, 1–7.

Bellance, R. and Gallet de Saint-Aurin. D. (1988) L'encéphalite virale du loup de mer. *Caraibes Médical* 2, 105–114.

Ben-Atia, I., Lutzky, S., Barr, Y., Gamsiz, K., Shtupler, Y. *et al.* (2001) The effect of ozone treatment on egg and larvae performance in the gilt-head sea bream (*Sparus aurata*) and other marine species. *European Aquaculture Society Specific Publications* 30, 58–59.

Beraldo, P., Panzarin, V., Galeotti, M. and Bovo, G. (2011) Isolation and molecular characterization of viral encephalopathy and retinopathy virus from gilthead sea bream larvae (*Sparus aurata*) showing mass mortalities. In: *15th EAFP Conference Abstract Book*. European Association of Fish Pathologists, p. 351.

Breuil, G., Bonami, J.R., Pepin, J.F. and Pichot, Y. (1991) Viral infection (picorna-like virus) associated with mass mortalities in hatchery-reared sea bass (*Dicentrarchus labrax*) larvae and juveniles. *Aquaculture* 97, 109–116.

Breuil, G., Pépin, J.F.P., Boscher, S. and Thiéry, R. (2002) Experimental vertical transmission of nodavirus from broodfish to eggs and larvae of the sea bass, *Dicentrarchus labrax* (L.). *Journal of Fish Diseases* 25, 697–702.

Buchan, K.A.H., Martin-Robichaud, D.J., Benfey, T.J., MacKinnon, A.M. and Boston, L. (2006) The efficacy of ozonated seawater for surface disinfection of haddock (*Melanogrammus aeglefinus*) eggs against piscine nodavirus. *Aquaculture Engineering* 35, 102–107.

Carballo, C., Garcia-Rosado, E., Borrego, J.J. and Carmen Alonso, M. (2016) SJNNV down-regulates RGNNV replication in European sea bass by the induction of the type I interferon system viruses infecting fish. *Veterinary Research* 47, 6.

Castric, J., Thiéry, R., Jeffroy, J., de Kinkelin, P. and Raymond, J.C. (2001) Sea bass *Sparus aurata*, an asymptomatic contagious fish host for nodavirus. *Diseases of Aquatic Organisms* 47, 33–38.

Cherif, N., Lopez-Jimena, B., Garcia-Rosado, E., Cano, I., Castro, D. *et al.* (2011) Detection of SJNNV and RGNNV genotypes using a relative quantification RT-PCR assay. *Journal of Applied Ichthyology* 27, 805–812.

Chi, S.C., Hu, W.W. and Lo, B.J. (1999) Establishment and characterization of a continuous cell line (GF-1) derived from grouper, *Epinephelus coicoides* (Hamilton): a cell line susceptible to grouper nervous necrosis virus (GNNV). *Journal of Fish Diseases* 22, 173–182.

Chi, S.C., Lee, K.W. and Hwang, S.J. (2001) Investigation of host range of fish nodavirus in Taiwan. In: *Proceedings of the 10th International Conference of the European Association of Fish Pathologists, Dublin, 9–14 September 2001*. Albion Press, Aberdeen, UK, p. 40, abstract 49.

Chi, S.C., Shieh, J.R. and Lin, S.J. (2003) Genetic and antigenic analysis of betanodaviruses isolated from aquatic organisms in Taiwan. *Diseases of Aquatic Organisms* 55, 221–228.

Comps, M., Pépin, J.F. and Bonami, J.R. (1994) Purification and characterization of two fish encephalitis viruses (FEV) infecting *Lates calcarifer* and *Dicentrarchus labrax*. *Aquaculture* 123, 1–10.

Comps, M., Trindade, M. and Delsert, C. (1996) Investigation of fish encephalitis viruses (FEV) expression in marine fishes using DIG-labelled probes. *Aquaculture* 143, 113–121.

Costa, J.Z. and Thompson, K.D. (2016) Understanding the interaction between Betanodavirus and its host for the development of prophylactic measures for viral encephalopathy and retinopathy. *Fish and Shellfish Immunology* 53, 35–49.

Cunningham, C.O. and Snow, M. (2000) Genetic analysis of infectious salmon anaemia virus (ISAV) from Scotland. *Diseases of Aquatic Organisms* 41, 1–8.

Curtis, P.A., Drawbridge, M., Iwamoto, T., Nakai, T., Hedrick, R.P. and Gendron, A.P. (2001) Nodavirus infection of juvenile whitebass cultured in southern California: first record of viral nervous necrosis (VNN) in North America. *Journal of Fish Diseases* 24, 263–271.

Cutrín, J.M., Dopazo, C.P., Thiéry, R., Leao, P., Olveira, J.G. *et al.* (2007) Emergence of pathogenic betanodaviruses belonging to the SJNNV genogroup in farmed fish species from the Iberian Peninsula. *Journal of Fish Diseases* 30, 225–232.

Dalla Valle, L., Negrisolo, E., Patarnello, P., Zanella, L., Maltese, C. *et al.* (2001) Sequence comparison and phylogenetic analysis of fish nodaviruses based on the coat protein gene. *Archives of Virology* 146, 1125–1137.

Dalla Valle, L., Toffolo, V., Lamprecht, M., Maltese, C., Bovo, G. *et al.* (2005) Development of a sensitive and quantitative diagnostic assay for fish nervous necrosis virus based on two-target real-time PCR. *Veterinary Microbiology* 110, 167–179.

Deiman, B., Van Aarle, P. and Sillekens, P. (2002) Characteristics and applications of nucleic acid sequence-based amplification (NASBA). *Molecular Biotechnology* 20, 163–179.

Doubleday, Z., Clarke, S., Li, X., Pecl, G., Ward, T. *et al.* (2013) Assessing the risk of climate change to aquaculture: a case study from south-eastern Australia. *Aquaculture Environment Interactions* 3, 163–175.

Frerichs, G.N., Rodger, H.D. and Peric, Z. (1996) Cell culture isolation of piscine neuropathy nodavirus from juvenile sea bass, *Dicentrarchus labrax*. *Journal of General Virology* 77, 2067–2071.

Frerichs, G.N., Tweedie, A., Starkey, W.G. and Richards R.H. (2000) Temperature, pH and electrolyte sensitivity, and heat, UV and disinfectant inactivation of sea bass (*Dicentrarchus labrax*) neuropathy nodavirus. *Aquaculture* 185, 13–24.

Gagné, N., Johnson, S.C., Cook-Versloot, M., MacKinnon, A.M. and Olivier, G. (2004) Molecular detection and characterization of nodavirus in several marine fish species from the northeastern Atlantic. *Diseases of Aquatic Organisms* 62, 181–189.

Gale, P., Drew, T., Phipps, L.P., David, G. and Wooldridge, M. (2009) The effect of climate change on the occurrence and prevalence of livestock disease in Great Britain: a review. *Journal of Applied Microbiology* 106, 1409–1423.

García-Rosado, E., Cano, I., Martín-Antonio, B., Labella, A., Manchado, M. *et al.* (2007) Co-occurrence of viral and bacterial pathogens in disease outbreaks affecting newly cultured sparid fish. *International Microbiology* 10, 193–199.

Gomez, D.K., Sato, J., Mushiake, K., Isshiki, T., Okinaka, Y. and Nakai, T. (2004) PCR-based detection of betanodaviruses from cultured and wild marine fish with no clinical signs. *Journal of Fish Diseases* 27, 603–608.

Gomez, D.K., Baeck, G.W., Kim, J.H., Choresca, C.H. Jr and Park, S.C. (2008a) Molecular detection of betanodavirus in wild marine fish populations in Korea. *Journal of Veterinary Diagnostic Investigation* 20, 38–44.

Gomez, D.K., Baeck, G.W., Kim, J.H., Choresca, C.H. Jr and Park, S.C. (2008b) Genetic analysis of betanodaviruses in subclinically infected aquarium fish and invertebrates. *Current Microbiology* 56, 499–504.

Grotmol, S. and Totland, G.K. (2000) Surface disinfection of Atlantic halibut *Hippoglossus hippoglossus* eggs with ozonated seawater inactivates nodavirus and increases survival of the larvae. *Diseases of Aquatic Organisms* 39, 89–96.

Grotmol, S., Dahl-Paulsen, E. and Totland, G.K. (2003) Hatchability of eggs from Atlantic cod, turbot and Atlantic halibut after disinfection with ozonated seawater. *Aquaculture* 221, 245–254.

Harikrishnan, R., Balasundaram, C. and Heo, M.S. (2010) Molecular studies, disease status and prophylactic measures in grouper aquaculture: economic importance, diseases and immunology. *Aquaculture* 309, 1–14.

Hata, N., Okinaka, Y., Iwamoto, T., Kawato, Y., Mori, K. and Nakai, T. (2010) Identification of RNA regions that determine temperature sensitivities in Betanodaviruses. *Archives of Virology* 155, 1597–1606.

He, M. and Teng, C.B. (2015) Divergence and codon usage bias of Betanodavirus, a neurotropic pathogen in fish. *Molecular Phylogenetics and Evolution* 83, 137–142.

Hick, P., Schipp, G., Bosmans, J., Humphrey, J. and Whittington, R. (2011) Recurrent outbreaks of viral nervous necrosis in intensively cultured barramundi

(*Lates calcarifer*) due to horizontal transmission of betanodavirus and recommendations for disease control. *Aquaculture* 319, 41–52.

Holland, J., Spindler, K. and Horodyski, F. (1982) Rapid evolution of RNA genomes. *Science* 215, 1577–1585.

Húsgağ, S., Grotmol, S., Hjeltnes, B.K., Rodseth, O.M. and Biering, E. (2001) Immune response to a recombinant capsid protein of striped jack nervous necrosis virus (SJNNV) in turbot *Scophthalmus maximus* and Atlantic halibut *Hippoglossus hippoglossus*, and evaluation of a vaccine against SJNNV. *Diseases of Aquatic Organisms* 45, 33–44.

Iwamoto, T., Nakai, T., Mori, K., Arimoto, M. and Furusawa, I. (2000) Cloning of the fish cell line SSN-1 for piscine nodaviruses. *Diseases of Aquatic Organisms* 43, 81–89.

Iwamoto, T., Mise, K., Mori, K., Arimoto, M., Nakai, T. and Okuno, T. (2001) Establishment of an infectious RNA transcription system for striped jack nervous necrosis virus, the type species of the betanodavirus. *Journal of General Virology* 82, 2653–2662.

Iwamoto, T., Okinaka, Y., Mise, K., Mori, K., Arimoto, M. *et al.* (2004) Identification of host-specificity determinants in betanodaviruses by using reassortants between striped jack nervous necrosis virus and sevenband grouper nervous necrosis virus. *Journal of Virology* 78, 1256–1262.

Iwamoto, T., Mise, K., Takeda, A., Okinaka, Y., Mori, K.I. *et al.* (2005) Characterization of striped jack nervous necrosis virus sub genomic RNA3 and biological activities of its encoded protein B2. *Journal of General Virology* 86, 2807–2816.

Kokawa, Y., Takami, I., Nishizawa, T. and Yoshimizu, M. (2008) A mixed infection in sevenband grouper *Epinephelus septemfasciatus* affected with viral nervous necrosis (VNN). *Aquaculture* 284, 41–45.

Krishnan, K., Gopiesh Khanna, V. and Sahul Hameed, A.S. (2010) Antiviral activity of dasyscyphin C extracted from *Eclipta prostrata* against fish nodavirus. *Journal of Antivirals and Antiretrovirals* 2, 29–32.

Kuiken, T., Holmes, E.C., McCauley, J., Rimmelzwaan, G.F., Williams, C.S. and Grenfell, B.T. (2006) Host species barriers to influenza virus infections. *Science* 312, 394–397.

Labella, A.M., Garcia-Rosado, E., Bandín, I., Dopazo, C.P., Castro, D. *et al.* (2018) Transcriptomic profiles of Senegalese sole infected with nervous necrosis virus reassortants presenting different degree of virulence. *Frontiers in Immunology* 9, 1626.

Lauring, A.S. and Andino, R. (2010) Quasispecies theory and the behavior of RNA viruses. *PLoS Pathogens* 6, e1001005.

Le Breton, A., Grisez, L., Sweetman, J. and Ollevier, F. (1997) Viral nervous necrosis (VNN) associated with mass mortalities in cage-reared sea bass, *Dicentrarchus labrax* (L.). *Journal of Fish Diseases* 20, 1145–1151.

Lopez-Jimena, B., Cherif, N., Garcia-Rosado, E., Infante, C., Cano, I. *et al.* (2010) A combined RT-PCR and dot-blot hybridization method reveals the coexistence of SJNNV and RGNNV betanodavirus genotypes in wild meagre (*Argyrosomus regius*). *Journal of Applied Microbiology* 109, 1361–1369.

Manin, B.O. and Ransangan, J. (2011) Experimental evidence of horizontal transmission of betanodavirus in hatchery-produced Asian sea bass, *Lates calcarifer* and brown-marbled grouper, *Epinephelus fuscoguttatus* fingerling. *Aquaculture* 321, 157–165.

Marcos-López, M., Gale, P., Oidtmann, B.C. and Peeler, E.J. (2010) Assessing the impact of climate change on disease emergence in freshwater fish in the United Kingdom. *Transboundary and Emerging Diseases* 57, 293–304.

Mezeth, K.B., Patel, S., Henriksen, H., Szilvay, A.M. and Nerland, A.H. (2009) B2 protein from betanodavirus is expressed in recently infected but not in chronically infected fish. *Diseases of Aquatic Organisms* 83, 97–103.

Moody, N.J.G., Horwood, P.F., Reynolds, A., Mahony, T.J., Anderson, I.G. and Oakey, H.J. (2009) Phylogenetic analysis of betanodavirus isolates from Australian finfish. *Diseases of Aquatic Organisms* 87, 151–160.

Moreno, P., Lopez-Jimena, B., Randelli, E., Scapigliati, G., Buonocore, F. *et al.* (2018) Immuno-related gene transcription and antibody response in nodavirus (RGNNV and SJNNV)-infected European sea bass (*Dicentrarchus labrax* L.). *Fish and Shellfish Immunology* 78, 270–278.

Mori, K., Nakai, T., Muroga, K., Arimoto, M., Mushiake, K. and Furusawa, I. (1992) Properties of a new virus belonging to nodaviridae found in larval striped jack (*Pseudocaranx dentex*) with nervous necrosis. *Virology* 187, 368–371.

Mori, K., Mushiaka K. and Arimoto M. (1998) Control measures for viral nervous necrosis in striped jack. *Fish Pathology* 33, 443–444.

Munang'andu, H.M. (2016) Environmental viral metagenomics analysis in aquaculture: applications in epidemiology and disease control. *Frontiers in Microbiology* 7, 1986.

Munday, B.L., Kwang, J. and Moody, N. (2002) Betanodavirus infections of teleost fish: a review. *Journal of Fish Diseases* 25, 127–142.

Mushiake, K. (2000) Effects of spawning conditions on multiplication of the causative virus (SJNNV) of viral nervous necrosis (VNN) in broodstocks of striped jack, *Pseudocaranx dentex*. *Aquaculture Science* 48, 109–115.

Mushiake, K. and Arimoto, M. (2000) Control of viral nervous necrosis (VNN) of striped jack in hatcheries. *Saibai Giken* 28, 47–55. (in Japanese with English abstract)

Mushiake, K., Arimoto, M., Furusawa, T., Furusawa, I., Nakai, T. and Muroga, K. (1992) Detection of antibodies against striped jack nervous necrosis virus (SJNNV) from broodstock of striped jack. *Nippon Suisan Gakkaishi* 58, 2351–2356.

Mushiake, K., Nishizawa, T., Nakai, T., Furusawa, I. and Muroga, K. (1994) Control of VNN in striped jack: selection of spawners based on the detection of SJNNV gene by polymerase chain reaction (PCR). *Journal of Fish Pathology* 29, 177–182.

Nerland, A.H., Skaar, C., Eriksen, T.B. and Bleie, H. (2007) Detection of nodavirus in seawater from rearing facilities for Atlantic halibut *Hippoglossus hippoglossus* larvae. *Journal of Fish Diseases* 73, 201–205.

Nguyen, H.D., Mekuchi, T., Imura, K., Nakai, T., Nishizawa, T. and Muroga, K. (1994) Occurrence of viral nervous necrosis (VNN) in hatchery-reared juvenile Japanese flounder *Paralichthys olivaceus*. *Fisheries Science* 60, 551–554.

Nguyen, H.D., Nakai, T. and Muroga, K. (1996) Progression of striped jack nervous necrosis virus (SJNNV) infection in naturally and experimentally infected striped jack *Pseudocaranx dentex* larvae. *Diseases of Aquatic Organisms* 24, pp. 99–105.

Nishioka, T., Sugaya, T., Kawato, Y., Koh, M. and Nakai, T. (2016) Pathogenicity of striped jack nervous necrosis virus (SJNNV) isolated from asymptomatic wild Japanese jack mackerel *Trachurus japonicus*. *Journal of Fish Pathology* 51, 176–183.

Nishizawa, T., Mori, K., Nakai, T., Furusawa, I. and Muroga, K. (1994) Polymerase chain reaction (PCR) amplification of RNA of striped jack nervous necrosis virus (SJNNV). *Diseases of Aquatic Organisms* 18, 103–107.

Nishizawa, T., Mori, K., Furuhashi, M., Nakai, T., Furusawa, I. and Muroga, K. (1995) Comparison of the coat protein genes of five fish nodaviruses, the causative agents of viral nervous necrosis in marine fish. *Journal of General Virology* 76, 1563–1569.

Nishizawa, T., Furuhashi, M., Nagai, T., Nakai, T. and Muroga, K. (1997) Genomic classification of fish nodaviruses by molecular phylogenetic analysis of the coat protein gene. *Applied and Environmental Microbiology* 63, 1633–1636.

Nishizawa, T., Gye, H.J., Takami, I. and Oh, M.J. (2012) Potentiality of a live vaccine with nervous necrosis virus (NNV) for sevendband grouper *Epinephelus septemfasciatus* at a low rearing temperature. *Vaccine* 30, 1056–1063.

Nylund, A., Devold, M., Plarre, H., Idsal, E. and Aarseth, M. (2003) Emergence and maintenance of infectious salmon anaemia virus (ISAV) in Europe: a new hypothesis. *Diseases of Aquatic Organisms* 56, 11–24.

OIE (World Organization for Animal Health) (2000) *Manual of Diagnostic Tests for Aquatic Animals*. OIE, Paris.

OIE (World Organization for Animal Health) (2014) *Manual of Diagnostic Tests for Aquatic Animals*. OIE, Paris.

OIE (World Organization for Animal Health) (2017) *Manual of Diagnostic Tests for Aquatic Animals*. OIE, Paris.

OIE (World Organization for Animal Health) (2018) *Manual of Diagnostic Tests for Aquatic Animals*. OIE, Paris.

Okinaka, Y. and Nakai T. (2008) Comparisons among the complete genomes of four betanodavirus genotypes. *Diseases of Aquatic Organisms* 80, 113–121.

Olveira, J.G., Souto, S., Dopazo, C.P., Thiéry, R., Barja, J.L. and Bandín, I. (2009) Comparative analysis of both genomic segments of betanodaviruses isolated from epizootic outbreaks in farmed fish species provides evidence for genetic reassortment. *Journal of General Virology* 90, 2940–2951.

Panzarin, V. (2016) A multidisciplinary approach to the study of Betanodaviruses in the Mediterranean basin. PhD thesis, Università degli Studi di Padova, Padova, Italy.

Panzarin, V., Fusaro, A., Monne, I., Cappellozza, E., Patarnello, P. *et al.* (2012) Molecular epidemiology and evolutionary dynamics of betanodavirus in southern Europe. *Infection, Genetics and Evolution* 12, 63–70.

Panzarin, V., Cappellozza, E., Mancin, M., Milani, A., Toffan, A. *et al.* (2014) *In vitro* study of the replication capacity of the RGNNV and the SJNNV betanodavirus genotypes and their natural reassortants in response to temperature. *Veterinary Research* 45, 56.

Patel, S. and Nerland, A.H. (2014) Vaccination against diseases caused by Betanodavirus. In: Gudding, R., Lillehaug, A. and Evensen, Ø. (eds) *Fish Vaccination*. Wiley, Chichester, UK, pp. 341–351.

Pickering, T.D., Ponia, B., Hair, C.A., Southgate, P.C., Poloczanska, E.S. *et al.* (2011) Vulnerability of aquaculture in the tropical Pacific to climate change. In: Bell, J.D., Johnson, J.E. and Hobday, A.J. (eds) *Vulnerability of Tropical Pacific Fisheries and Aquaculture to Climate Change*. Secretariat of the Pacific Community, Noumea, New Caledonia, pp. 647–732.

Ransangan, J. and Manin, B.O. (2010) Mass mortality of hatchery-produced larvae of Asian sea bass, *Lates calcarifer* (Bloch), associated with nervous necrosis in Sabah, Malaysia. *Veterinary Microbiology* 145, 153–157.

Roberts, K.E., Hadfield, J.D., Sharma, M.D. and Longdon, B. (2018) Changes in temperature alter the potential outcomes of virus host shifts. *PLoS Pathogens* 14(10), e1007185.

Sahul Hameed, A.S., Ninawe, A.S., Nakai, T., Chi, S.C., Johnson, K.L. and ICTV Report Consortium (2019) ICTV Virus Taxonomy Profile: *Nodaviridae*. *Journal of General Virology* 100, 3–4.

Sanjuan, R., Cuevas, J.M., Furio, V., Holmes, E.C. and Moya, A. (2007) Selection for robustness in mutagenized RNA viruses. *PLoS Genetics* 3, e93.

Schneemann, A., Ball, L.A., Delsert, C., Johnson, J.E. and Nishizawa, T. (2005) Family *Nodaviridae*. In: Fauquet, C.M., Mayo, M.A., Maniloff, J., Desselberger, U. and Ball, L.A. (eds) *Virus Taxonomy Eighth Report of the International Committee on Taxonomy of*

Viruses. Elsevier Academic Press, San Diego, California, pp. 865–872.

Shetty, M., Maiti, B., Santhosh, K.S., Venugopal, M.N. and Karunasagar, I. (2012) Betanodavirus of marine and freshwater fish: distribution, genomic organization, diagnosis and control measures. *Indian Journal of Virology* 23, 114–123.

Shieh, J.R. and Chi, S.C. (2005) Production of monoclonal antibodies against grouper nervous necrosis virus (GNNV) and development of an antigen capture ELISA. *Diseases of Aquatic Organisms* 63, 53–60.

Skliris, G.P., Krondiris, J.V., Sideris, D.C., Shinn, A.P., Starkey, W.G. and Richard, R.H. (2001) Phylogenetic and antigenic characterization of newfish nodavirus isolates from Europe and Asia. *Virus Research* 75, 59–67.

Sommerset, I. and Nerland, A.H. (2004) Complete sequence of RNA1 and subgenomic RNA3 of Atlantic halibut nodavirus (AHNV). *Diseases of Aquatic Organisms* 58, 117–125.

Souto, S., Lopez-Jimena, B., Alonso, M.C., Garcia-Rosado, E. and Bandin, I. (2015a) Experimental susceptibility of European sea bass and Senegalese sole to different betanodavirus isolates. *Veterinary Microbiology* 177, 53–61.

Souto, S., Merour, E., Biacchesi, S., Bremont, M., Olveira, J.G. and Bandin, I. (2015b) *In vitro* and *in vivo* characterization of molecular determinants of virulence in reassortant betanodavirus. *Journal of General Virology* 96, 1287–1296.

Souto, S., Olveira, J.G. and Bandin, I. (2015c) Influence of temperature on Betanodavirus infection in Senegalese sole (*Solea senegalensis*). *Veterinary Microbiology* 179, 162–167.

Souto, S., Olveira, J.G., García-Rosado, E., Dopazo, C.P. and Bandín, I. (2019) Amino acid changes in the capsid protein of a reassortant betanodavirus strain: effect on viral replication *in vitro* and *in vivo*. *Journal of Fish Diseases* 42, 221–227.

Starkey, W.G., Millar, R.M., Jenkins, M.E., Ireland, J.H., Muir, K.F. and Richards, R.H. (2004) Detection of piscine nodaviruses by real-time nucleic acid sequence-based amplification (NASBA). *Diseases of Aquatic Organisms* 59, 93–100.

Tan, C., Huang, B., Chang, S.F., Ngoh, G.H., Munday, B.L. *et al.* (2001) Determination of the complete nucleotide sequence of RNA1 and RNA2 from greasy grouper (*Epinephelus tauvina*) nervous necrosis virus, Singapore strain. *Journal of General Virology* 82, 647–653.

Tanaka, S., Mori, K., Arimoto, M., Iwamoto, T. and Nakai, T. (2001) Protective immunity of sevenband grouper, *Epinephelus septemfasciatus* Thunberg, against experimental viral nervous necrosis. *Journal of Fish Diseases* 24, 15–22

Tanaka, S., Takagi, M. and Miyazaki, T. (2004) Histopathological studies on viral nervous necrosis of sevenband grouper, *Epinephelus septemfasciatus* Thunberg, at the grow-out stage. *Journal of Fish Diseases* 27, 385–399.

Thiéry, R., Cozien, J., De Boisseson, C., Kerbart-Boscher, S. and Nevarez, L. (2004) Genomic classification of new *Betanodavirus* isolates by phylogenetic analysis of the coat protein gene suggests a low host-fish species specificity. *Journal of General Virology* 85, 3079–3087.

Toffan, A., Pascoli, F., Pretto, T., Panzarin, V., Abbadi, M. *et al.* (2017) Viral nervous necrosis in gilt-head sea bream (*Sparus aurata*) caused by reassortant betanodavirus RGNNV/SJNNV: an emerging threat for Mediterranean aquaculture. *Scientific Reports* 7, 46755.

Toffolo, V., Negrisolo, E., Maltese, C., Bovo, G., Belvedere, P. *et al.* (2007) Phylogeny of betanodaviruses and molecular evolution of their RNA polymerase and coat proteins. *Molecular Phylogenetics and Evolution* 43, 298–308.

Toubanaki, D.K. and Karagouni, E. (2018) Towards a dual lateral flow nanobiosensor for simultaneous detection of virus genotype-specific PCR products. *Journal of Analytical Methods in Chemistry* 2018, 11.

Toubanaki, D.K., Margaroni, M. and Karagouni, E. (2015) Development of a novel allele-specific PCR method for rapid assessment of nervous necrosis virus genotypes. *Current Microbiology* 71, 529–539.

Ucko, M., Colorni, A. and Diamant A. (2004) Nodavirus infections in Israeli mariculture. *Journal of Fish Diseases* 27, 459–469.

Vendramin, N., Patarnello, P., Toffan, A., Panzarin, V., Cappellozza, E. *et al.* (2013) Viral Encephalopathy and Retinopathy in groupers (*Epinephelus* spp.) in southern Italy: a threat for wild endangered species? *BMC Veterinary Research* 9, 20.

Vendramin, N., Toffan, A., Mancin, M., Cappellozza, E., Panzarin, V. *et al.* (2014) Comparative pathogenicity study of ten different betanodavirus strains in experimentally infected European sea bass, *Dicentrarchus labrax* (L). *Journal of Fish Diseases* 37, 371–383.

Viral Zone (2018) ExPASy SIB Bioinformatics Resource Portal – Home. Available at: https://www.expasy.org/ (accessed 28 April 2019).

Watanabe, K., Nishizawa, T. and Yoshimizu, M. (2000) Selection of broodstock candidates of barfin flounder using an ELISA system with recombinant protein of barfin flounder nervous necrosis virus. *Diseases of Aquatic Organisms* 41, 219–223.

Xylouri, E., Kotzamanis, Y.P., Athanassopoulou, F., Dong, L., Pappas, I. *et al.* (2007) Isolation, characterization, and sequencing of Nodavirus in sturgeon (*Acipenser gueldestaedi* L.) reared in freshwater facilities. *Israeli Journal of Aquaculture – Bamidgeh* 59, 37–42.

Yanong, R.P.E. and Erlacher-Reid, C. (2012) Biosecurity in aquaculture, part 1: an overview. *SRAC Publication*

No. 4707. USDA Southern Regional Aquaculture Center, Stoneville, Mississippi.

Yoshimizu, M. (2009) Control strategy for viral diseases of salmonid fish, flounders and shrimp at hatchery and seed production facility in Japan. *Fish Pathology* 44, 9–13.

Zorriehzahra, M.J., Adel, M., Dadar, M., Ullah, S. and Ghasemi, M. (2019) Viral nervous necrosis (VNN) an emerging disease caused by *Nodaviridae* in aquatic hosts: diagnosis, control and prevention: a review. *Iranian Journal of Fisheries Sciences* 18, 30–47.

6 Aquatic Birnavirosis (Infectious Pancreatic Necrosis Virus)

Carlos P. Dopazo*

University of Santiago de Compostela, Santiago de Compostela, Spain

6.1 Introduction

The infectious pancreatic necrosis (IPN) is a lethal viral disease mainly affecting cultured salmonid fish. The disease was originally called 'acute catarrhal enteritis', as reviewed by Munro and Midtlyn (2011), and produces high mortalities (frequently over 90%) in different species of trout and salmon, mainly in very young fry (immediately after absorption of the yolk sack). IPN is caused by a virus, the *Infectious pancreatic necrosis virus* (IPNV), which is the type species of the genus *Aquabirnavirus* within the family *Birnaviridae*. Aquabirnaviruses include a wide variety of very closely related viruses (often named 'IPNV-like viruses'), which have been isolated from a wide range of fish and shellfish species (see below) where they can produce specific differentiated symptoms (OIE, 2003; Munro and Midtlyn, 2011) or no symptoms at all (Moreno *et al.*, 2014). However, the term 'IPNV' is strictly used for those aquabirnaviruses which affect salmonid fish and develop the characteristic symptoms, whereas the general term 'aquatic birnaviruses' is employed for all other cases. Just like the wide range of the host fish species, the virus is distributed worldwide and has been isolated from diseased and carrier animals from the five continents, as discussed below.

An infectious disease is a complete set of *circumstances* leading to an alteration of the physiological status of the affected individual. Such circumstances include factors related to three groups of participants: the host, the infectious agent and the environment. The interaction between the three of them determines the development – or not – of the disease, as well as its intensity.

In the case of fish, their environment is an aquatic ecosystem. In addition, because they are poikilothermic animals, even small changes in the environment may provoke substantial effects on their physiological status and, therefore, on their susceptibility to an infection. Global warming is undoubtedly affecting oceans and other aquatic ecosystems, as well as other types of animal ecosystems and human society worldwide, and it is well documented and accepted that the changes experienced by those ecosystems will – and already are – causing changes in the epidemiology of infectious diseases, both in humans and animals, in the wild and under intensive culture, including aquaculture (de Silva and Soto, 2009; Slenning, 2010; Bett *et al.*, 2017).

Climate change is a consequence of the exponential accumulation of greenhouse gases due to anthropogenic action since the beginning of the industrial era. In fact, although not the only one, the most apparent effect of climate change – the one which is most talked about – is global warming. According to the Surface Temperature Analysis (SISS) application available at the National Aeronautics and Space Administration (NASA, 2019) website, the anomalies in surface temperature over the last 10 years, with respect to the last decade of the 20th century, have been between +0.2 and +1°C (0.2–0.5°C in most oceans), reaching +2°C in northern areas of the world (Fig. 6.1). Predictions for the end of the present century by the 5th Assessment Report of the Intergovernmental Panel on Climate Change (IPCC, 2013) indicate that, in the best scenario, the average world temperature will increase by 2.5°C; without additional mitigation strategies the increase would be in the range of 3.7–4.8°C and, in the worst-case scenario, the increase could reach 7.8°C. Another phenomenon associated with global warming is the increase of precipitations which will

*carlos.pereira@usc.es

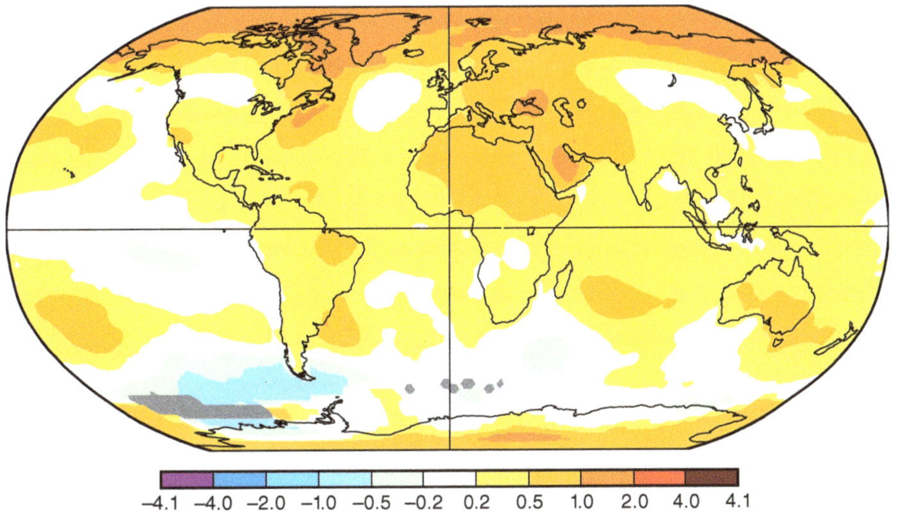

Fig. 6.1. Temperature anomalies (land-ocean temperature index, LOTI) by 2008–2018, versus 1990–2000, as determined by NASA's Surface Temperature Analysis online application (https://data.giss.nasa.gov/gistemp/maps/index_v3.html (accessed 11 March 2019)).

produce – are already producing – changes in the chemistry of aquatic systems, affecting salinity and pH. Furthermore, both water temperature and chemistry changes will affect ocean currents.

All these processes will influence wild and cultured fish worldwide, and their diseases. In this respect, a list of the impacts on aquaculture, related to each climate change element, was presented as early as 2008 in a report from a workshop on 'Climate Change Implications for Fisheries and Agriculture', organized in Rome by the Food and Agriculture Organization of the United Nations (FAO, 2008). For instance, it claimed that global warming could raise water temperature over the tolerance limit of farmed species and increase the virulence of certain pathogens and the expansion of others; acidification could have an adverse impact on shellfish, and extreme weather events could affect farming facilities.

In the present chapter we focus on the analysis of all the data available on IPN and its aetiological agent to predict what might happen with this disease in the future in the present scenario of climate change, looking into the capacity of the virus to adapt to new conditions, the effect of increasing stress factors affecting the fish, the influence of all those changes in the aquatic environment on the transmission of the disease, and the options the industry has to mitigate the effect on production.

6.2 The Pathogen, the Disease and Its Diagnosis

6.2.1 Characteristics of the pathogen

Structural and molecular features

The aquatic birnaviruses are unenveloped, icosahedral, double-stranded and bisegmented RNA viruses of diameter about 65 nm, although other sizes have been reported (Lago *et al.*, 2016). The genome encodes three structural proteins: segment B has a single open reading frame (ORF) corresponding to the viral polymerase (RNA polymerase-RNA dependent or RpRd), whereas segment A has two ORFs. The largest one encodes a polyprotein (NH_2–pVP2–VP4–VP3–COOH) which is cleaved – by the internal protease activity corresponding to the VP4– into the non-structural VP4 and two structural viral proteins: VP2 (after modification of pVP2) and VP3 (see review by Munro and Midtlyn, 2011). A second small ORF in segment A encodes VP5, a non-structural protein, detected only in infected cells, which is supposed to be involved in the suppression of apoptosis at the beginning of the infective process (Liu and Vakharia, 2006).

VP2 is an external protein containing the domain to recognize the cell receptors and, therefore, is related to cell susceptibility and host range. In addition, since it is the major outer capsid protein, it is highly antigenic and contains the neutralization epitopes. VP3 is an internal protein – although some units might be also on the surface (Nicholson, 1993). This protein interacts with the RNA polymerase and is involved in the control of viral replication and morphogenesis.

Replication takes place between 16 and 20 h in cell culture and starts by adsorption and the introduction of uncoated particles in vesicles into the cytoplasm. Just 2–4 and 4–6 h after infection, transcription and replication, respectively, are detected. Although the production ratio is relatively low (1000–2000 progeny particles per each parental one) as reported by Malsberger and Cerini (1965), its capacity to infect new cells makes it a very productive virus.

The capacity of the virus to infect the cells in a host is the first condition for defining its virulence and depends on specific adsorption domains coded in the VP2 sequence region. In this same region, Santi *et al.* (2004) reported the presence of certain residues related to virulence, so that changes in a single amino acid in these positions could be responsible for the high or low virulence of the IPNV strains. However, discrepancies in field samples have been found by some authors (Dadar *et al.*, 2013; Salgado-Miranda *et al.*, 2014) which indicates that other factors, including the host, must be involved. In this regard, reassortment, a phenomenon exclusive for segmented viruses like this one and discovered in aquatic birnaviruses by Romero-Brey *et al.* (2009), and polyploidy, also recently discovered in these viruses (Lago *et al.*, 2016), are mechanisms used by aquabirnaviruses to modulate virulence.

Viral stability and survival

Probably due to the structural features of the virus, IPNV is a very stable virus, surviving for long periods of time – even months – in water at 4–10°C (see review by Munro and Midtlyn, 2011). The virus was also tested at temperatures up to 20°C, showing that it maintains infectivity for at least 2 weeks (Toranzo and Hetrick, 1982; Yoshimizu *et al.*, 2005). In addition, it is necessary to point out that the virus has been isolated from a water environment at even higher temperatures: 25, 27 and even 30°C (Chou *et al.*, 1993; Kitamura and Suzuki, 2000; Munro and Midtlyn, 2011). Mortensen *et al.* (1998) demonstrated that the virus was able to maintain its infectivity for 4 months at temperatures ranging from below 0 to 20°C, and under salinity values of 0 to 40‰. In their study, they also reported a reduction of 5 logarithms of infectivity in 2 months at 30°C; unfortunately, they did not test intermediate temperatures. Furthermore, IPNV is remarkably stable under extreme conditions, since it was demonstrated to persist at 60°C for 24 hours with a reduction of between 2 and 5 logarithms of infectivity, depending on the strain, and for 28 days at pH 4 with no significant reduction of titre (Dixon *et al.*, 2012). This high stability, which can be increased in the presence of suspended particulate matter and sediments, makes the virus very efficient insofar as horizontal transmission is concerned.

Diversity and adaptation

Aquatic birnaviruses are a very diverse group of viruses. In 1995, Hill and Way organized the many strains of IPNV and IPNV-like viruses known at that time by applying cross-neutralization to a large panel of strains and established the existence of two serogroups. Most of the viruses were assigned to one of them – serogroup A – containing nine serotypes which they named from A1 to A9. A reference strain was assigned to each of those serotypes: WB (West Buxton), Sp (Spjarup), Ab (Abildt), He (Hecht), Te (Tellina), C1, C2 and C3 (Canada 1 to 3, respectively) and Ja (Jasper), from A1 to A9, respectively. Serogroup B would only include serotype B1 (reference strain TV-1) (Hill and Way, 1995). However, there was still some confusion as it was sometimes difficult to serotype some new isolates. The application of molecular techniques for genotyping – mainly genome sequencing – was able to simplify the classification of these viruses (Blake *et al.*, 2001; Nishizawa *et al.*, 2005) into seven recognized genotypes which mostly correspond to the previously defined serotypes.

The diversity of aquabirnaviruses is now known to be even higher than previously thought because of the discovery of reassortment in these viruses (Romero-Brey *et al.*, 2009). Reassortment is one of the main resources that segmented genome viruses can use to generate diversity and it provides them with a high capacity to adapt to new hosts and environmental conditions. Before the knowledge on the existence of this phenomenon in aquatic birnaviruses, their typing was based on genome A sequencing, where the capsid VP2 protein is encoded. Since then, few IPNV-like epidemiological studies have typed both segments, but those that have, have demonstrated a higher diversity of the virus in nature. In 2014, Bandín *et al.* (2014) and Moreno *et al.* (2014) published the results of epidemiology studies performed on wild eels from the Albufera lake (Spain) and on the wild marine species in the Gulf of Cádiz (south of the Iberian Peninsula), respectively, reporting similar results in both cases: about 85% of strains were WB/WB (segments A/B) and 15% were reassortants of the WB/Ab type. More recently, Panzarin *et al.* (2018) and Ulrich *et al.* (2018) reported phylogenetic analysis of aquabirnaviruses isolated from Italy and Scotland, respectively, over a three- to four-decade period. In Scotland, among 57 strains, 8.9% were reassortants of the type C1/Ab, 1.8% were C1/Sp and the remaining were wild-type European Sp (78.5%), American WB or Canadian C1 types (5.9% in both cases). In Italy, all 75 tested strains corresponded to reassortants type Sp/C1 or Sp/Te. Over the last two decades, our team at the Instituto de Acuicultura, Universidad de Santiago de Compostela (IA-USC), Spain, has continued to perform epidemiological studies on wild marine populations in coastal waters in the north-west of Spain, where we have detected the wild types Ab/Ab (21%) and Sp/Sp (7%) and three types of reassortants: WB/Ab (8%), WB/C1 (61%) and C1/Ab (3%). In the Flemish Cap we have detected reassortants Sp/Ab (12.5%) and WB/Ab (12.5%) and wild types WB/WB and Ab/Ab (37.5% in both cases) (C.P. Dopazo, J.G. Olveira and I. Bandín, 2020, unpublished results). Finally, in an unpublished study performed in collaboration with Ruben Avendaño on IPNV isolates from Chile, 3% of the strains turned out to be reassortants of the type WB/Sp and the remaining were wild types Sp/Sp and WB/WB (64.5 and 32.5%, respectively) (C.P. Dopazo, J.G. Olveira, R. Avendaño and I. Bandín, 2020, unpublished results).

6.2.2 Characteristics of the disease

Clinical signs

The IPN, the disease caused by IPNV, mainly affects fry of salmonid species and its virulence is progressively reduced with the age of the fish, which become refractory after the sixth month. The effect of the virus on fish also depends on water temperature, the optimum being around 10°C; mortality is delayed below 6°C and suppressed over 15°C (Dorson and Torchy, 1981). Another critical factor is the concentration of the virus in the environment, establishing a minimum level of about 10^4 plague-forming units (pfu)/l to produce mortality in a population (see review by Munro and Midtlyn, 2011). It is therefore understandable to consider the extreme fish density of tanks in aquaculture facilities as an important risk factor of the disease, as well as the continuous surveillance of tanks to remove symptomatic fish, in order to reduce the infected sources from spreading viruses.

The IPN is mainly characterized by not being associated with specific external symptoms (Munro and Midtlyn, 2011; McAllister, 2016). Perhaps the only one specifically associated with the disease is the abnormal swimming behaviour, with erratic corkscrew and violent whirling; but even this symptom is characteristic only in very young salmon and trout fry, which can also show pseudofaecal tails, composed of a mixture of mucus and intestinal epithelium extruded through – and linked to – the anus (Fig. 6.2).

Other external symptoms, like exophthalmia, skin darkening sometimes accompanied by petechial haemorrhages, pale gills or abdominal swelling due to the accumulation of ascitic fluid, do not differentiate IPN from other fish diseases. Anorexia is another

Fig. 6.2. Trout fry infected with IPNV showing the typical skin darkening and pseudofaecal tails.

behavioural change frequently observed in both Atlantic salmon post-smolt and affected non-salmonid fish.

The virus enters via the gills and intestine and quickly spreads to other tissues and organs. Internally, the most frequent sign is a pale and yellowish liver; heart, spleen and kidney can also appear abnormally pale, and internal petechial haemorrhages in the muscle and accumulation of visceral ascites are also seen. As the name of the disease indicates, pancreatic necrosis should be the most characteristic histopathological sign, but this is exclusive to salmonid fish and not always observed; for non-salmonid species, other histopathological and internal signs lend their names to different diseases: branchionephritis in eels, viral ascites in yellowtail (*Seriola quinqueradiata*) or gill necrosis in clams.

Tropism and shedding of the virus

As expected, the tropism of the virus explains the external and internal symptoms, and the effect on the tissues explains the evolution of the disease and the spread of the virus. After the infection of the fish, and the entrance of the virus through gills and intestine, the virus takes a few days to reach the internal organs, probably transported by phagocytic cells, from the intestine, which is the first replication site (Swanson *et al.*, 1982). The intense replication in the intestinal tissues causes the severe necrosis of the intestinal epithelium – and the pseudofaecal tails – and indirectly the anorexia. The high production of virus in the intestinal cells and the extrusion of that tissue through the anus strongly contribute to the spread of the virus into the environment, which is confirmed by the high concentration of infective particles found in faeces (around 10^7 $TCID_{50}$/ml; see review by Munro and Midtlyn, 2011).

Viraemia explains the rapid distribution of the virus throughout the internal organs, and exocrine pancreas and hepatic tissue necrosis quickly develop; the capillary walls are also affected, explaining the internal petechias. Finally, the virus also reaches the brain, where extensive focal vacuolization is produced, causing the abnormal swimming behaviour.

Transmission and spread of the virus

The most commonly known ways of transmission of the IPNV are vertical and horizontal, via faeces. Although the presence of the virus in ovarian fluids was demonstrated as early as 1963 (Wolf *et al.*, 1963), intra-ovarian transmission could not be demonstrated until 1981 (Dorson and Torchy, 1981), although the real mechanism remains unclear. Despite this, vertical transmission is widely accepted and is of special concern for fish farmers who must control the sanitary status of their broodstock, mainly knowing that the survivors of the disease remain infected in an asymptomatic carrier state. Regarding the second transmission route, since the infected fish shed high quantities of virus into the environmental water via faeces and urine, the levels of viral concentration in tanks with an infected population can be high, reaching values of 10^5 $TCID_{50}$/ml (Desautels and MacKelvie, 1997). Therefore, healthy susceptible fish sharing the environment of acutely infected individuals are extremely likely to be infected and to suffer the disease. In fact, a single infected individual can be responsible for an epizootic if it is introduced into a healthy population, given that as little as 10^{-1} $TCID_{50}$/ml is enough to produce infection in bath immersion (Urquhart *et al.*, 2008). On the other hand, other transmission routes, such as rotifers, molluscs and even sediments, are less documented. In this regard, Comps *et al.* (1991) reported the visualization of viral particles in rotifers massively cultured for feeding fish; their size and morphological characteristics resembled those of IPNV, and they speculated on the risk of transmission through infected rotifers.

Further evidence points to the risk of transmission by molluscs, since the presence of these viruses in species of scallops (Mortensen *et al.*, 1992) and even diseased oysters (Kitamura *et al.*, 2000) has been identified since the 1990s. However, it was more important to discover the existence of IPNV-like viruses in molluscs in the proximities of fish culture facilities and to demonstrate that in some cases there was a certain molecular similarity between the strains of both origins (Rivas *et al.*, 1993; Gregory *et al.*, 2007). More recently, the possibility of transmission from infected mussels to cohabitating fish was demonstrated experimentally (Molloy *et al.*, 2013), making epidemiologists consider this means of transmission to be a real risk.

In addition, as already indicated, due to the high stability of the viral particle, the virus can survive for long periods of time in the aquatic environment, and it has been demonstrated that it is even more stable in the presence of particulate matter and sediments both in marine and fresh waters (Gregory *et al.*, 2007; Labrana *et al.*, 2008). Finally, the virus has also been detected in piscivorous birds, which makes this a particularly concerning mode of transmission (McAllister and Owens, 1992).

6.2.3 Diagnosis

Isolation in cell culture

In spite of the many serological and molecular methods of diagnosis of this – and any – virus, the reference technique is still isolation in cell culture. This is simply because it is the only method which, along with the detection of the virus, demonstrates that the virus is active and, therefore, represents a real risk. Although the World Organization for Animal Health does not include IPNV in its latest edition of the online Manual of Diagnostic Tests for Aquatic Animals, it is still available in the 2003 edition (OIE, 2003), where they indicate the use of BF-2, CHSE-214 or RTG-2 for the isolation of the virus and the procedure to apply the technique to infected tissues. The procedure can also be found in chapter 4.1 by McAllister (2016), in the Fish Health Section (FHS) Blue Book (FHS Blue Book, 2016), where the use of only CHSE-214 is advised. In both cases, incubation temperature after viral inoculation is 15°C, for 7 or 14 days (depending on the source). As soon as cytopathic effect (CPE) is obtained, or after the incubation time if no CPE is visualized, a re-inoculation must be performed on new monolayers and the same incubation time applied. Two consecutive results are considered the definitive result. Then, the diagnosis must be confirmed by serological or molecular techniques.

Serological techniques

Among all the serological techniques, serum neutralization (SNT), the enzyme-linked immunosorbent assay (ELISA) and the indirect fluorescent antibody test (IFAT) are the most frequently used. Mainly used is SNT, which is the only serological method advised by the FHS Blue Book (2016). The procedure, described step-by-step in the OIE (2003) manual, is quite simple and does not need any special equipment; the main disadvantage is that it requires an extended time to complete a diagnosis. Kits for diagnosis of IPNV by ELISA are commercially available, both using polyclonal or monoclonal antibodies. Their limit of detection is around 10^3 pfu/ml (Dopazo and Bandín, 2011), which is admissible to identify the virus after isolation in cell culture, but not for direct detection in infected fish tissues in most cases.

IFAT is probably more used than ELISA for the diagnosis of this virus and several commercial kits are available which employ monoclonal antibodies, the advantage being the simplicity of the procedure (OIE, 2003). However, the use of monoclonal antibodies represents a disadvantage because they can fail (or produce lower signals) with certain strains. Recently, Vázquez *et al.* (2016a) reported a new technology adapted from traditional flow cytometry (FC), namely the microflow cytometry (μFC), which reduces the high costs by avoiding the very expensive equipment needed in FC, and ensures detection and quantification of the virus as early as 18 h after its inoculation in cell culture.

Molecular techniques

Although a list of molecular IPNV diagnostic techniques has been reviewed (Dopazo and Barja, 2002; Munro and Midtlyn, 2011), including nucleic acid hybridization (NAH), *in situ* hybridization and loop-mediated isothermal amplification (LAMP), the fact is that nowadays polymerase chain reaction (PCR)-based techniques have displaced any others, except cell culture isolation. The clear advantage of PCR is that even using its basic procedure (which means developing the results by visualizing the amplicon in agarose gels), it provides a limit of detection of 15 fg, which can even be improved ten-or 100-fold by complementing it with Southern blotting and NAH, or nested PCR, respectively (Dopazo and Bandín, 2011).

A very simple procedure of reverse transcription-PCR (RT-PCR) is described in the FHS Blue Book (2016) based on one previously described by Blake *et al.* (1995), using primers located in the intergene NS–VP3 region and producing an amplicon of 174 bp. Although the visualization of a band corresponding to the right size of the amplicon is reported to be 'confirmatory of IPNV', we advise the use of any of the confirmatory strategies indicated above, or even a simple Sanger sequencing of the purified amplicon.

Real-time quantitative PCR (qPCR) is nowadays displacing traditional PCR, not only because it provides higher sensitivities, reaching a limit of detection as low as 31 $TCID_{50}$/ml (Vázquez *et al.*, 2016b), but also because it confirms the identification at the same time as performing the detection, dramatically reducing the time for diagnosis. The first report of the application of this technology for

IPNV diagnosis was by McBeath *et al.* (2007), using the Sp strains as reference. Unfortunately, this procedure was designed for studying the gene expression in IPNV-infected fish, and therefore it was not actually validated (meaning no sensitivity and specificity data are available). More recently Vázquez *et al.* (2016b) reported the design and validation of a diagnosis and quantification RT-qPCR method for IPNV, which showed a sensitivity of 31 $TCID_{50}$/ml or 50 pfu/ml or 66 RNA copies/ml (with a dynamic range between 1 ng and 10 ag), which was tested on 30 IPNV strains from different types and showed high repeatability and reproducibility.

6.3 Potential Spread of the Pathogen

6.3.1 Geographic distribution of the virus

The IPNV virus – and the disease, as well – has been detected all around the world (Fig 6.3). Except in Australia, where the disease is considered exotic and only one isolation from diseased rainbow trout has been reported (McCowan *et al.*, 2015), and in Africa, where little effort has been made regarding the epidemiology of this virus and also just one isolate – from Kenya – has been documented (Mulei *et al.*, 2018), in all the other continents many scientists have reported the disease in a variety of cultured species and/or performed surveillance on wild fish populations and reservoirs (see epidemiology maps published by the author on the IA-USC website; https://www.usc.gal/gl/institutos/acuicultura/difusion/aportacions-cientificas.html (accessed 10 March 2020)). For instance, in Europe the disease and the virus have been detected in most countries – except in Eastern ones – mainly in outbreaks in cultured rainbow trout (*Oncorhynchus mykiss*) and Atlantic salmon (*Salmo salar*), but also in pike (*Esox lucius*), eel (*Anguilla* spp.), turbot (*Scophthalmus maximus*) and halibut (*Hippoglossus* spp.). It has also been isolated from asymptomatic cultured fish (sole, *Solea* spp., salmonids, eel and carp), as well as from a large variety of wild fish.

Lots of efforts have been made in Asia, namely in Turkey and Iran, where most isolations are from rainbow trout, both diseased and asymptomatic; also in South Korea and Japan, isolations have been performed from diseased eels, rainbow trout, flounder (*Paralichthys olivaceus*), yellowtail (*Seriola* spp.), ayu (*Plecoglossus altivelis*), as well as from diseased pearl oysters (*Pinctada fucata*) and a variety of wild fish species. Finally, regarding North and South America, isolations are well documented in Canada, the USA, Mexico and Argentina, most of them from salmonids.

6.3.2 The effect of climate change on the virus

'Viruses are the most successful inhabitants of the biosphere.' This is the main conclusion in a review by Wasik and Turner (2013) on the capacity of viruses to survive in new environmental conditions. Due to their parasitic way of life, viruses must adapt to their host and environment. Insect viruses must adapt to particularly harsh soil conditions; human and in general homeotherm animal viruses must adapt to a narrow range of temperatures to replicate, and fish viruses have adapted to a wide variety of environmental conditions and range of optimal temperatures.

Fish viruses have chosen marine/estuarine or freshwater fish species as their host; some infect warm-water (>28°C) and others cool-water (20–28°C) and cold-water (≤20°C) fish. But, among fish viruses, IPNV has been able to take advantage of almost any type of environment. In fact, although the disease is especially virulent to freshwater salmonid fry, it also produces mortality in salmonids in the marine environment and has been isolated from a variety of marine fish species. Moreover, although IPNV has been reported to have an optimum replication temperature of around 10°C and a range between 6 and 16°C (Dorson and Torchy, 1981), this is true only for some strains and not for all the wide variety of types known. In fact, it has been isolated from fish and shellfish at temperatures as high as 23°C, from eels in the Netherlands (Haenen *et al.*, 2002) and at 25°C from asymptomatic cultured loach (*Misgurnus anguillicaudatus*) in Taiwan (Chou *et al.*, 1993) and wild pearl oysters in Japan (Kitamura *et al.*, 2002), or as low as 5°C from diseased rainbow trout in Iran (Büyükekiz *et al.*, 2018). Additionally, and even more surprisingly, Lo *et al.* (1988) demonstrated that an aquabirnavirus of the Ab type, isolated from clams (*Meretrix lusoria*), was able to replicate at a maximum temperature of 30°C.

What makes the virus so adaptable? First, it has extremely high stability at different pH and salinity conditions and, mainly, at extremely high environmental temperatures, as reviewed above. Normally, unenveloped viruses are much more stable in physicochemical conditions than enveloped ones. Therefore, the global warming scenario will affect different viruses at different levels; and, among all these, enveloped viruses will see their – at present – natural environment reduced and will be progressively detected in more northern areas, where temperatures might still be within their permissive range. Therefore, by the end of this century, diseases such as the infectious hematopoietic necrosis (IHN) and the viral hematopoietic necrosis (VHS) will probably disappear from where their causative viral agents (IHNV and VHSV, respectively) are now present, just because they will not withstand the expected rise in water temperatures (Yoshimizu *et al.*, 2005; Marcos-López *et al.*, 2010).

What will happen to IPNV? Will it survive global warming? If water temperatures continue to rise – as seen over the last two to three decades, based on the anomalies map obtained from the NASA surface temperatures analysis (Hansen *et al.*, 2010; GISTEMP Team, 2019) and shown in Fig. 6.1 – in approximately 2050 the average water temperature in most areas where the virus has been detected so far would increase by 0.5°C, except in Northern European countries and North Canada, where the increase would be almost 1°C and even 1.5–2°C in some parts of Turkey and Iran. As mentioned above, the virus is in fact adapted to a wide range of temperatures and probably would survive those temperatures in most cases. The question about its virulence in those conditions is more difficult to answer.

The large diversity of aquabirnaviruses can be understood as a reflection of their ability to adapt to any situation. As reviewed by Wasik and Turner (2013), this ability is based on three conditions: (i) large progeny production; (ii) adaptation via natural selection; and (iii) recombination, and IPNV-like viruses are experts in all the three. First, although the virus is not especially highly productive in terms of ratio of progeny per parental particle, as already reviewed above, its high infectivity to new cells makes it very efficient and highly productive. Second, being an RNA virus, there is a high frequency of mutation per round of genome replication. These mutations are the very reason behind this genetic variability, providing the virus with the ability to adapt, and they are present in the quasispecies of the progeny from an infection. One of them – or a set of them – might allow the virus to adapt to a new host which it accidentally has encountered; some can allow the virus to infect and replicate up to a higher – or lower – temperature; one might allow the VP2 to adsorb at a lower pH. Or perhaps none of those new conditions will be present in the virus environment, and then none of those mutations will succeed. Third, IPNV is a genome-segmented virus, and co-infection is possible; therefore, reassortment is a demonstrated phenomenon naturally occurring in aquabirnaviruses and since reassortment is a source of diversity, it provides aquatic birnaviruses with a high capacity to adapt (Lago *et al.*, 2017).

The virus has a high capacity of adaptation to changing water chemistry (pH and salinity) and temperature conditions and we now know the strategies it uses. But which are the critical steps involved? Salinity, pH and temperature are parameters affecting the stability of protein, and the virus is mostly protein. Therefore, viral stability depends, to a large extent, on the capacity of those proteins to maintain their activity – for instance, capsid stability – under changing conditions. Obviously, the protein function is sequence dependent and, therefore, the capacity of the virus to survive in a new environment in a global warming scenario depends on the ability of the viral genome to evolve and code more adapted proteins. Unfortunately, no studies have been performed to discover the effect of those parameters on coat proteins.

Viral survival in an aquatic environment depends on the opportunity of the virus to encounter a host in which to replicate. Otherwise, it must find any kind of binding particle (or a living reservoir) as an intermediate step to maintain itself 'alive' before reaching the right host. For IPNV and IPNV-like viruses, sediments are an effective inanimate reservoir, where the virus can remain stable and adsorbed to the particles for long periods of time. However, this adsorption is highly influenced by the water chemistry and its fluctuations (Harvey and Ryan, 2004).

Salinity and pH also affect the capacity of the virus to infect the cell; viral adaptation to cell receptors follows the same physicochemical process as to any particle, at least in the first steps. Unfortunately, although some knowledge exists regarding the effect of these parameters on the adaptation capacity of some viruses (Grant *et al.*, 1993), no data are available for IPNV. However, some authors have speculated on the effect of temperature on the first steps of 'viral life' (Hata *et al.*, 2007). For the remaining replication steps, only segment A, the one coding the capsid protein, has been involved in the adaptation of the virus to new hosts and conditions such as temperature. In this regard, Inaba *et al.* (2009), studying marine aquabirnavirus strains isolated from Japan's southern islands 'where water temperature is higher than 25°C throughout the year', detected a 5-nucleotide deletion in the VP2/Ns inter-region. However, for other viruses, such as the nervous necrosis virus (NNV), also a bisegmented virus, Hata *et al.* (2007) demonstrated that both segments were involved in the adaptation of the virus to temperature. The implication of the polymerase gene in adaptation to temperature has also been demonstrated for the influenza virus (Hayashi *et al.*, 2015).

In a very extensive report by Chursov *et al.* (2013), the authors describe a web server to analyse RNA sequences to predict regions of secondary structure which could be affected by changes in temperature. Unfortunately, the server is not available and therefore we could not analyse the IPNV genome segments to predict which genes might be affected in a scenario of rising temperatures.

Much remains to be studied regarding aquabirnaviruses to have a more complete vision of the ways that climate change can affect the stability, replication and infectivity of the virus.

6.3.3 The risks of the pathogen spreading

The pathogen has already spread

The fact that aquatic birnaviruses have already spread worldwide was made clear in a previous section: a wide range of types of IPNV-like viruses has been isolated in all five continents, but not only from cultured fish; insofar as the risks of the pathogen spreading are concerned, it seems even more suggestive that the virus has been detected in a wide variety of asymptomatic carrier fish and shellfish (McAllister *et al.*, 1984; Suzuki *et al.*, 1997; Romero-Brey *et al.*, 1999; Inaba *et al.*, 2007; Labrana *et al.*, 2008; Wallace *et al.*, 2008; Jeon *et al.*, 2011; Moreno *et al.*, 2014; Kim *et al.*, 2016). Considering that most isolations in salmonid outbreaks are from types WB and Sp, and that in the wild there are many other types and reassortants, there is greater diversity in the wild. Typing is just a way of 'giving name' to the differences; but, as discussed above, the high mutation rates of these kinds of viruses mean that additional molecular differences exist among them.

Carrier fish from the wild are mostly asymptomatic, which implies that they have reached a virus–host balance, and such a balance depends not only on the fish themselves but also on environmental conditions. Climate warming is producing – and will produce – changes in the aquatic environment including at least changes in pH, salinity and temperature. These environmental changes will break the balance by generating stress for the host and/or for a selection of certain viral molecular variants developing a disease, as demonstrated for other fish species and viruses (Jiravanichpaisal *et al.*, 2004). In the future, could a certain strain of aquabirnavirus be responsible for an outbreak in the wild because of climate warming? This is very speculative and it is tempting to apply the rule that 'over 16°C the disease is not possible' (Dorson and Torchy, 1981), but this is true only for salmonids and, moreover, it has been demonstrated that some of the carrier species in the wild (such as turbot, sole or flounder) are susceptible to disease under farming conditions.

Regarding cultured salmonids, a high number of studies have defended the principle of the higher the environmental temperature (within the range of tolerance of the salmonid host), the lower the virulence (of the virus) and the lower and more delayed the fish mortality (Dorson and Torchy, 1981; Okamoto *et al.*, 1987). Therefore, it seems possible that a rise in temperature will reduce the frequency of outbreaks and the level of mortality. However, we should not underestimate the virus's ability to adapt, and the possible presence of some variants which may already be adapted to produce mortality at higher temperatures. In the following section, we approach another issue introducing uncertainty: the rising temperature also affects the host, making it more susceptible.

Furthermore, just considering the aquabirnaviruses that have already spread, the problem is not only limited to cultured and carrier species known so far, based on the surveillance studies performed to date: climate change will also bring about the colonization by naïve invasive fish species from areas where the virus is already present. In the pre-existing molecular diversity, a viral particle adapted to the new species might already exist and produce new outbreaks. All these arguments, of course, are speculative but nevertheless cannot be dismissed, since they are based on scientific facts. Therefore, they should be taken into consideration in future studies on IPNV–host interaction and climate change scenarios.

The risk of the virus spreading

The spread of viruses can be via anthropogenic and/or natural mechanisms, and climate change may be involved in the process itself and/or in its final effect.

ANTHROPOGENIC SPREAD AND CLIMATE WARMING. Farmed fish are a source of pathogens when an outbreak is produced and it has been reported that viruses can be detected in water downstream from an infected fish farm (McAllister and Bebak, 1997). More recently, Mulei *et al.* (2018) described the isolation of IPNV from a rainbow trout farm in Kenya located in a mountain area with water temperatures ranging from 8 to 18°C; the same virus was detected in tilapia farms downstream, at much higher temperatures (24–28°C), but fish were asymptomatic. In the marine environment, transmission from farms has also been reported. Wallace *et al.* (2008), studying the prevalence of IPNV in wild fish in the vicinity of salmonid farms, discovered significantly higher viral prevalence at distances below 5 km. Another source of pathogens to the environment is fish escaping from cages, which not only represents a thread to wild fish, but also an additional source of viral diversity. Furthermore, it is well known that commercial movements of fish stocks also represent an efficient mechanism for spreading viruses.

The contribution of new variants to the environment, the high mutation rate and the reassortment ability of aquabirnavirus, as discussed above, all promote higher diversity in the environment; the pressure exerted by climate warming can do the rest. Under stable conditions, the balance between non-salmonid species and the virus makes those become carriers; under external pressure, the right viral variant might appear to encounter the right cell receptor in those species, which might transform them from poor to highly susceptible hosts (Wasik and Turner, 2013).

CLIMATE CHANGE AND VIRAL SPREAD. Global warming will not only be the direct cause of increasing the risk for aquatic animals, but also the cause of other phenomena that will, at the same time, be the cause of additional risks. Global warming is producing ever more important alterations in hydrological cycles (FAO, 2008; Danovaro *et al.*, 2010; Bett *et al.*, 2017): (i) the modification of water circulation patterns, altering oceanic currents; (ii) flooding events due to increased precipitation in some areas and regions, and, on the contrary, lower levels of water and flow rates in rivers in others; and (iii) a higher frequency of extreme events like stronger-than-usual storms, among others. As we discuss later, just the warming of the aquatic environment will be responsible for a redistribution of fish species, worldwide; this, together with the fact that the modification of currents will alter the migration routes of plankton and fish, helps us understand that the global distribution of viral types will also be altered. Moreover, flooding events will favour the escape of fish from farming facilities, contributing to the virus spreading (Marcos-López *et al.*, 2010). Harsh storm events will also contribute to increasing escapes by producing significant damage to farming facilities, as reported recently by Fish Information and Services (FIS, 2015): 'On 27 January 2015 more than 51,000 farmed Atlantic salmon were reported to have escaped from an open-net pen sited in... The escape occurred near... due to extreme weather...'. This escape was caused by the important damage to the cages, as seen in the photograph provided in the news.

6.4 Potential Spread Through Natural and Intermediate Hosts

6.4.1 Effect of climate change on IPNV natural hosts

Climate change and global warming will directly affect the average temperature of aquatic ecosystems, but indirectly will also modify important water parameters like pH, salinity, dissolved oxygen, turbidity and ammonium concentration (Alborali, 2006). All of them affect the fitness of fish in important processes and stages, such as reproduction, development and growth, migration and disease survival; even fish behaviour may be influenced by external conditions. Each fish species has an optimum value and a range of tolerance for each of those parameters and they can change with the development stage. In the case of temperature, which is perhaps the most important parameter affecting fish, for salmonid species (those affected by the disease of reference: the IPN) the optimum temperature is below 20°C, i.e. they are cold-water species. Additionally, they are stenothermal, i.e. they have a narrow tolerance to water temperature (Table 6.1). Cold-water species, and moreover stenothermal species, will be those to more dramatically suffer the effects of climate change. To this regard, Marcos-López *et al.* (2010) suggested that climate change will convert the UK into a non-appropriate environment for Atlantic salmon.

Within the range of tolerance, and with a permanent source of food, an increase in temperature might represent a benefit for fish growth. However, such an advantage would be higher at younger ages;

Table 6.1. Fish tolerance and optimum temperatures[a].

	Optimum	Tolerance	
		Range	Maximum
Brown trout[b]	13	3.5–19.5	25
Atlantic salmon[b]	11	6–22.5	28
Pacific salmonid species[c]	15	10–16	25

[a]Temperatures (°C) are given for juvenile growth; data may change with the stage.
[b]From UK Environment Agency (2008).
[c]From Richter and Kolmes (2017).

for larger fish, the metabolic cost of such enhanced growth would be too high. This would lead to truncated populations, lacking the older fish due to fish surviving for a shorter time (Collingsworth *et al.*, 2017). The warming and the reduction in water levels and flow rates (in rivers) will also have an effect of decreasing dissolved oxygen, at the same time that fish would have an increase in their demand for oxygen; with a reduced capacity of the involved circulatory and ventilatory systems to provide enough oxygen, the survival of the fish will be seriously compromised.

The variation, within certain limits, of any of these parameters produces stress, which is a physiological condition affecting the immune response of fish (Ellis, 2001; Bett *et al.*, 2017; Kim *et al.*, 2017; Wang *et al.*, 2018). Fish, like any animal, depend on their immune system to defend themselves from pathogens and under environmental parameters compromising this immune response, they will be more vulnerable to infections, even from opportunistic and low-virulent agents (Alborali, 2006; Marcos-López *et al.*, 2010).

Therefore, cold-water fish, such as salmonids, will have their competitive efficiency reduced compared with other species more adapted to warmer water, and they will have no choice but to change to a new environment. If they are living in the wild! But what will happen with those under intensive culture?

6.4.2 Effect of climate change on cultured IPNV-like-susceptible fish species

Although salmonid fish are the most commonly considered IPNV-susceptible hosts, other cultured species have been reported to be affected by aquatic birnaviruses, mainly turbot, sole, yellowtail, flounder and eels (Takano *et al.*, 2001; Hirayama *et al.*, 2007; Varvarigos *et al.*, 2011; and see review by Munro and Midtlyn, 2011).

Reviewing Fig. 6.3 and considering sites of isolation as locations of intensive culture, the farming of non-salmonids seems to be restricted to the northern hemisphere (turbot and sole in Southern Europe; flounder and yellowtail in Eastern Asia; eels in Northern and Southern Europe and Eastern Asia), whereas salmonid fish are also intensively cultured in the southern hemisphere (Europe, North America and Mexico, Chile and Western Asia). However, there is no doubt that this is just part of the picture, since globalization not only applies to global warming, but also to introducing into new areas the culture of species already cultured in other areas, or even introducing fish species for the first time into culture. The effect of climate warming on fish will be different depending on their temperature tolerance: cold-water fish will suffer more – and their culture will be more affected – than warm-water species; and eurythermal fish species – those with a wide range of temperature tolerance – will be more likely to adapt to the new conditions, mainly considering that the change will not be sudden, giving fish time to adapt, both physiologically and genetically.

Reviewing Fig. 6.1, we have additional information to try to predict what will happen to the intensive culture level. Over the last 30 years, the rise in temperature has been higher in the northern hemisphere, and even more pronounced as we get closer to the North Pole. Following this progression, by 2050 the temperature will increase on average by 0.5°C in North and South America, Central and Southern Europe, and Eastern Asia, and by between 1.0 and 1.5°C in North and Northwest Canada, Eastern and Northern Europe, and Western Asia (Iran); and this prediction is even worse by the end of the century (IPCC, 2013).

These data are bad enough for cultured salmonids, given that a rising temperature, in the best scenario – remaining within the tolerance range – will affect the production of larger fish sizes and breeders, as already discussed (Collingsworth *et al.*, 2017), and in the worst case, salmonid culture would no longer be successful in areas where it is quite productive nowadays (Marcos-López *et al.*, 2010). Even if in a certain area the average temperature rise is still within tolerance, its daily evolution must be of concern since high and low daily peaks also provoke stress in the fish. And, as we already know, stress is a process which prevents an adequate response to other factors, such as infections. Therefore, cultured fish, under global warming, would a priori be more susceptible to infections. In addition, the effect of reduced river flow rates due to warming has been demonstrated to be associated with fish mortalities due to endemic pathogens (Winton, 2016). However, this is true just for certain pathogens and fish species. For instance, climate warming is expected to eradicate VHSV from salmonid culture in some European areas because the negative effect of higher temperatures on the virus would combine with an enhanced fish immune response – of course before the water temperature gets so high that the immune system would also be affected (Hershberger *et al.*, 2013). Can we predict what will happen with virus–fish interaction in the case of IPNV and salmonid culture? And with no salmonids?

As we have shown before, if the fish are in their optimal temperature and under no significant stress, they may be refractory at least to low-virulent strains (Miller *et al.*, 2014; Kim *et al.*, 2017). Furthermore, the narrow temperature tolerance of salmonids might compromise their immune response to infection if the environmental temperature is near the higher tolerance temperature, making it possible for even low-virulent strains to produce mortality in a fish stock (Chou *et al.*, 1999).

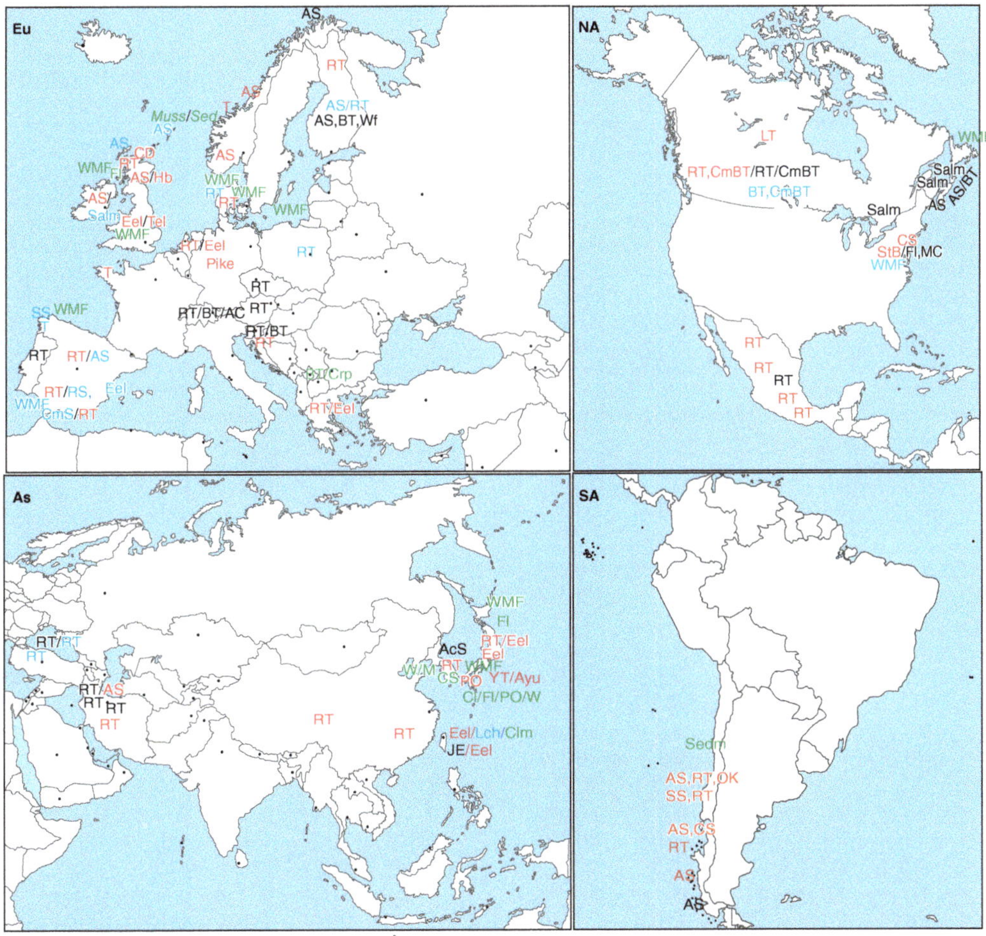

Fig. 6.3. Continued

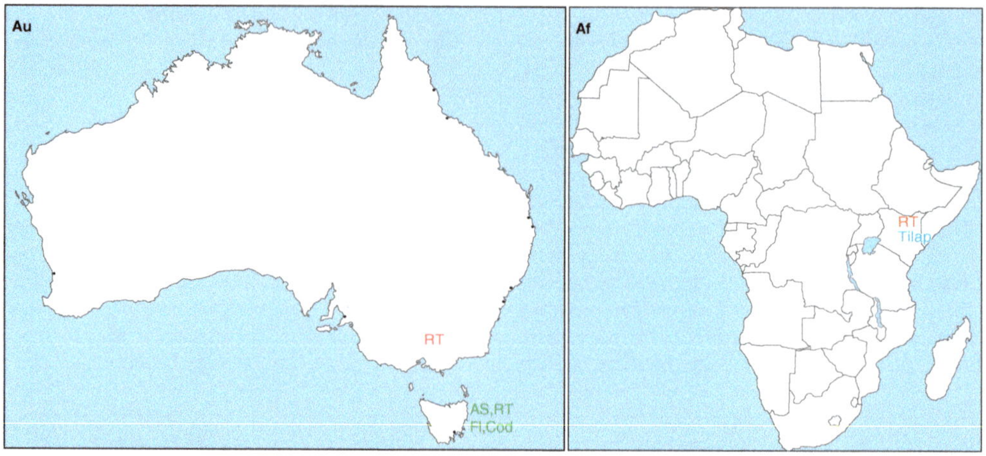

Fig. 6.3. Detection of aquatic birnaviruses worldwide. Six maps (from Europe (Eu), North America (NA), Asia (As), South America (SA), Australia (Au) and Africa (Af)) show the locations of isolation and detection of aquabirnaviruses worldwide, and indicate the species (see below), the type of population (from cultured or wild animals) and the type of case (asymptomatic/diseased animals). Species: AC, Arctic char (*Salvelinus alpinus*); AcS, various aquaria species; AS, Atlantic salmon (*Salmo salar*); Ayu (*Plecoglossus altivelis*); BT, brown trout (*Salmo trutta*); CD, common dab (Limanda limanda); ChS, chum salmon (*Oncorhynchus keta*); Cl, clams (different species); CmBT, common brook trout (*Salvelinus fontinalis*); CmS, common sea bream (*Pagus pagrus*); Crp, carp (*Cyprinus carpio*); CS, coho salmon (*Oncorhynchus kisutch*); Eel, eels (*Anguilla* spp.); Fl, flounder (*Paralichthys olivaceus*); Hb, halibut (*Hippoglossus* spp.); JE, Japanese eel (*Anguilla japonica*); Lch, loach (*Misgurnus anguillicaudatus*); LT, lake trout (*Salvelinus namaycush*); M, mussels; MC, mummichog (*Fundulus heteroclitus*); Pike (*Esox lucius*); PO, pearl oyster (*Pinctada fucata*); RS, red sea bream (*Pagellus bogaraveo*); RT, rainbow trout (*Oncorhynchus mykiss*); Salm, salmonids (various species); Sed, sediments; SS, Senegalese sole (*Solea senegalensis*); StB, striped bass (*Morone saxatilis*); T, turbot (*Scophthalmus maximus*); Tel, tellina (*Tellina* spp.); Tilap, tilapia (*Oreochromis niloticus*); W, water; Wf, whitefish (*Coregonus clupeaformis*); WMF, wild marine fish (different species); YT, yellowtail (*Seriola lalandi*). Type of populations: green colour indicates wild populations; the rest are for cultured ones. Type of case: diseased (red); asymptomatic (blue); data not available (black).

However, up to certain levels, warmer temperatures could enhance the immune response of fish (Le Morvan *et al.*, 1998; Hershberger *et al.*, 2013) and make them refractory to certain strains of the virus.

We present Fig. 6.4 as a practical example, where two scenarios are proposed: A, for those areas where temperature is expected to increase by at least 0.5°C until 2050; and B, for those with a minimum increment of 1°C from now. Hypothesizing that cultured fish were at their optimum temperature nowadays, but that an increase up to +0.5°C would not affect fish survival and would even improve their immune response, mortalities would be reduced until 2050 in scenario A, whereas in scenario B, mortalities would be reduced until around 2035 but would then increase. However, in some areas nowadays temperatures are already so high that scenario C or D would be more appropriate and would mean that salmonid culture would not succeed due to pathologies. However, this is very speculative and does not take into consideration how the rising water temperature will affect viral virulence. Unfortunately, as already discussed, the available data only tell us that nowadays the IPNV salmonid isolates were isolated at temperatures ranging from 6 to 15°C, and other IPNV-like viruses have been isolated from fish of different species at higher temperatures. However, due to the high capacity of the virus to adapt to new situations, the appearance of new strains adapted to higher temperatures and infective to salmonids cannot be ruled out.

A similar argument can be used for non-salmonid species with similar tolerance temperatures, like turbot, with an optimum between 14 and 18°C and maximum tolerance at 23°C; but not for others

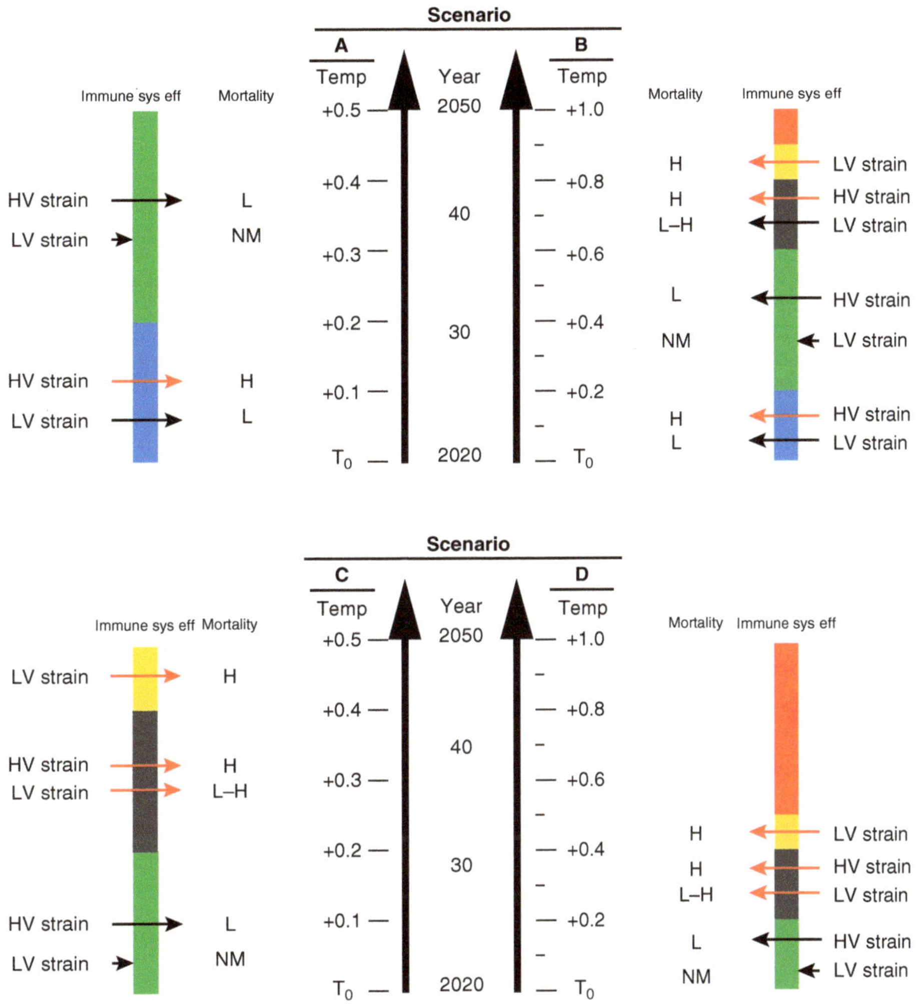

Fig. 6.4. A practical example of what hypothetically might happen under global warming with fish susceptibility to viral infection in different scenarios. The four scenarios are proposed from 2020 (baseline, T_0) to 2050, assuming temperature increases of 0.5 or 1.0°C (scenarios A and C, and B and D, respectively). Additionally, scenarios C and D are proposed for fish species living in environments where their temperature tolerance rates are closer to the initial water temperature in comparison to fish in scenarios A and B. For each scenario, the column 'immune system efficiency' (Immune sys eff) uses colours to show how the temperature increase might affect the fish immune response capacity: blue, the immune response is not affected by the temperature at present; green, the immune system is favoured by the environmental temperature; black, it begins to be negatively affected by temperature; yellow, immune response is affected drastically; red, it has been completely cancelled (the fish is lethally damaged). The infection by low-virulent (LV) strains would trigger high (H) mortality only if the fish immune system is affected to some extent; otherwise, they would develop low (L) or no mortality (NM) at all. The high-virulent (HV) strains would develop low mortality to fish with their immune system activated; otherwise, they would produce high mortality.

with higher optimum temperatures, such as sole (15–21°C, optimum at 18–20°C), or for cool-water species (20–28°C) such as flounder (optimum 20–25°C, maximum 30°C).

6.4.3 Fish migration and virus spread

Global warming is provoking the evolution of aquatic ecosystems, making them less appropriate to native populations and more to others in the vicinity. This is producing movements of cold-water fish species to northern latitudes – or elevations – to find cooler environments, more appropriate to their physiological characteristics, and northern populations to colonize those areas now also more adjusted to them (Holbrook *et al.*, 1997; Alborali, 2006). Temperate regions, as reported by Britton *et al.* (2010), will be more vulnerable to global warming, meaning cold-water native species will be displaced by non-native populations.

This process is not occurring in separated events, but as a single progressive and simultaneous one: as water temperature rises, native species that have long been adapted to previous environmental conditions move to cooler waters, to find more appropriate environments. This provokes a permanent movement of populations and, consequently, changes in species' distributions (Cheung *et al.*, 2009). In the interphase, both populations compete for feeding and space. While environmental conditions still favour the native population, the non-native is just an 'explorer'. As the temperature rises, the explorer starts to be better adapted; it turns into a 'colonizer' and the former native species starts to be displaced.

In the whole process, native and non-native populations are carriers each of their own set of pathogens (Winton, 2016). The native population might be a carrier of – and adapted to – a specific aquatic birnavirus strain (let us call it sA) whereas the invader species might carry a different strain (sB). Native and non-native species would now be naïve for the viral strains new to each of them. Disease is quite probable in either population and, therefore, although we can blame immigrating species for introducing new IPNV strains – and new pathologies, in general – we should not forget that they can also suffer from the already present pathogens. As interaction continues, both viral strains will adapt to both fish populations, either by mutation or by reassortment, or both, giving rise to new viral types, either more virulent to either species or becoming an avirulent parental strain making the fish asymptomatic carriers.

The most important summary of the whole process is that it is not a discrete one with separate steps: (i) adaptation of sA to the native populations; (ii) invasion of the non-native population and introduction of sB; (iii) infection of sB to the native population and vice versa; and (iv) mutation/reassortment provoking adaptation and/or worsening of the epidemic situation. Instead, it is a progressive evolution of the complete colonization process, involving three levels of interactions: (i) between two (or more) fish populations; (ii) between two (or more) fish viral strains; and (iii) between fish and virus. And such a process simply introduces more viral diversity and, therefore, more uncertainty as to what could happen in the future due to climate warming.

Furthermore, to make the prediction even more imprecise and inexact, we must consider that the modifications of aquatic ecosystems and oceanic currents will imply changes in fish migrations worldwide. It is well known that those long-distance routes expose fish to a variety of habitats, contributing to the spread of pathogens (Altizer *et al.*, 2001). The modification of their routes will mean they are exposed to new and different environments and populations, spreading the viral strain they carry to naïve populations and being infected by new strains they will spread to the successive aquatic ecosystems they go through. This will progressively modify the epidemiology of IPNV, implying a higher diversity than we now know, and making the appearance of new strains via reassortment more likely.

6.5 Control and Prevention

Control of fish viral diseases is much more limited, compared with bacterial diseases. The development of chemotherapeutic agents has not been successful and, due to IPNV's special resistance, disinfection is not an option either. Therefore, control of the disease is limited to specific strategies of prevention. In this section, it is not our intention to review methods of prevention and control of IPNV; for that purpose, there are other sources (Munro and Midtlyn, 2011; Sommerville, 2012). Our aim is simply to start a discussion on what we should do to prevent and, as far as possible, avoid the risks of the effects of climate change on the pathologies that aquatic birnaviruses can cause in fish and shellfish. To our understanding, any efforts made in this direction must focus on reducing the spread of the virus, selection of the application of breeding programmes, selecting fish strains more resistant to the virus, and on the development of a deep epidemiological knowledge to enable the design of predictive vaccines.

6.5.1 Reducing the risk of horizontal transmission

Reducing horizontal transmission risk is easier said than done, and even utopian, if the objective is to control transmission between fish. It is impossible in the wild; if an IPNV-carrier invader species enters a naïve population area, it will introduce carried IPNV strains and there is nothing that can be done. Or is there? In the case of wild populations, as we will see below, the only thing – but of special importance – that we can do is intensify surveillance studies to know 'what there is' and 'what is coming' and to promote molecular (and challenge) research studies to 'predict' what will happen in the future. But, additionally, knowing that molluscs and sediments can act as reservoirs in the wild (Rivas *et al.*, 1993; Gregory *et al.*, 2007; Labrana *et al.*, 2008; Morley, 2010), this surveillance must not be limited to fish. Regarding cultured fish, the tools available for reducing the horizontal spread of the virus are even more directly efficient and include reducing to a minimum the risk of escapes, implementing retention systems to avoid (or reduce) releasing infectious agents into the environment during outbreaks, as well as monitoring critical points in the facilities which could act as reservoirs for the virus, like biological and sand filters.

6.5.2 Reducing the risk of vertical transmission

The vertical route is perhaps the most important way of transmission under culture conditions. Therefore, the control of breeders at any facility should be a standard obligation. As reviewed by Munro and Midtlyn (2011), the systematic testing of breeders was used for the first time as early as in the late 1970s, and applied for decades; but the recommended method for diagnosis was isolation in cell culture from internal organs, which made it possible only to determine the probability of breeder stocks being free of the virus; it was also applied to eggs and sperm, but in both cases with the limitations of the diagnostic procedure itself. In that review, a method by RT-PCR was also referenced, which, according to the authors, should be discarded because it was inefficient and not very repetitive (Storset *et al.*, 2006). However, in the same review the authors missed another reference, by Cutrín *et al.* (2005), where a nested RT-PCR procedure was reported to detect IPNV in fish blood samples. In fact, this method has been employed over the last decade by that team for the selection of salmon breeders – captured from the wild – free of IPNV (and other viruses) to be used in a restocking programme by the Galician Government (north-west Spain) (Bandín and Dopazo, 2006), which has proven to be a reliable and efficient procedure, the use of which should be generalized.

6.5.3 Prediction of IPNV types evolution, and vaccines

The relevance of vaccination for the control of viral diseases is unquestionable, and there have been many researchers working on the design of different types of vaccines for this virus; but the focus has been aimed at a few specific strains traditionally known to represent the highest risk for cultured salmon. Here, we propose something much more ambitious: to develop an international research programme to provide, in the medium term, the necessary epidemiological knowledge to be able to design a prediction method for the viral strain that will circulate in a specific location and time; something similar to what human medicine has developed to predict the influenza virus strain that should be used to design the following season's vaccine.

Such a programme should include: (i) permanent monitoring of IPNV viral types affecting salmonid (and non-salmonid) farms worldwide, including next-generation sequencing of those viruses to analyse their evolution associated with level of virulence, water temperature, species and fish characteristics; (ii) intensive surveillance of wild fish and shellfish in the vicinity of those farms (and in other locations) to know what is present in the wild and how it evolves under the influence of temperature and nearby outbreaks; and (iii) molecular research studies to understand how this virus can adapt to temperature, how certain types can evolve changing their sequence, and how reassortment between the strains that are present in a certain location can evolve to new types adapted to higher temperatures, to new species and/or to different levels of virulence.

But because it is so ambitious, such a programme should be approached from an international perspective, including worldwide locations, and certainly including other pathogens of interest and concern in aquaculture.

6.6 Conclusions

Climate change is a worldwide accepted fact; and it is such a fact that even fulfilling the international treaties and applying the strategies designed to reduce to a minimum the greenhouse gas emissions, global warming would be slowed down but not completely stopped in this century. Therefore, designing measures to reduce its effect on certain activities and to adapt them to the inevitable consequences are important issues of concern.

Regarding the issue we have approached in this chapter, the IPNV and the diseases it causes, the main conclusion is the enormous lack of knowledge on the effect that climate change might have on them. In fact, much of the discussion has been speculative; based in scientific arguments, indeed, but with too many uncertainties. What is the temperature limit for this virus? Is it strain dependent? What happens with virus stability as water temperature rises? How does it affect viral replication, and in which step(s)? What happens to the viral genome sequence as temperature rises, i.e. which are the molecular determinants of adaptation to temperature? Can climate change be responsible for the appearance and spread of new viral types? Might the viral molecular mechanisms be responsible for the appearance of a new super-virulent strain? With the knowledge available at present, we can just discuss all the possibilities. But we should promote research studies to accumulate as much knowledge as needed to be able to predict how climate warming will affect the virus and the disease.

Nowadays, knowledge on the effect of climate change on fish is deeper. For many species, we know the tolerance ranges and much about the effects that high temperatures have on the physiological and immunological state of the individuals. We need to have more information on the effect of warming on the population dynamics in nature, because it will suggest to us which species will be more appropriate for culture in specific areas, but also how the viral agent can be spread in the future.

For the industry, there are several lines of action for resilience and reduction of economic losses. First, regarding the adaptation to the changing climate, the best to do for farmers is selection of the right species for a specific location, considering the predicted evolution of temperatures in such a location and the changing latitude or elevation for those species cultured in locations where the temperature is reaching – or will reach in a short future – the limit of tolerance. In addition, the industry and administrations should invest more in breeding programmes to obtain fish strains with a wider temperature tolerance range (or with a higher limit), to delay the substitution of species or locations. However, those programmes must be combined with the selection of fish families also resistant to the disease.

Finally, the three social actors involved – industry, administration and research – should invest more in reducing the risk of spreading the virus and on the surveillance of its evolution worldwide. Research should be applied to design methods to retain viruses during an outbreak, at least in hatcheries (where the water flow is significantly lower); and the industries should be concerned of the need to apply those systems, when available, to reduce the release of virus to the environment. They should also implement programmes to select IPNV-free breeders (by the non-lethal molecular techniques available) to avoid vertical transmission and viral spread.

Administrations and the research institutions should intensify surveillance and monitoring programmes in farms and in wild populations, to have an epidemiological and evolutive map of the IPNV strains and types that are circulating at any time and in any location. This will let us predict 'what is coming next' and design control strategies to minimize their effect on aquaculture and the natural environment.

References

Alborali, L. (2006) Climatic variations related to fish diseases and production. *Veterinary Research Communications* 30(Suppl. 1), 93–97. https://doi.org/10.1007/s11259-006-0019-7

Altizer, S., Bartel, R. and Han, B.A. (2001) Animal migration and infectious disease risk. *Science* 331, 296–302. https://doi.org/10.1126/science.1194694

Bandín, I. and Dopazo, C.P. (2006) Restocking of salmon in Galician rivers: a health management program to reduce risk of introduction of certain fish viruses. *DIPNET (Disease interactions and pathogen exchange between farmed and wild aquatic animal populations – a European network) Newsletter No. 35.*

Bandín, I., Souto, S., Cutrín, J.M., López-Vázquez, C., Olveira, J.G. *et al.* (2014) Presence of viruses in wild eels *Anguilla anguilla* L, from the Albufera Lake (Spain). *Journal of Fish Diseases* 37, 597–607. https://doi.org/10.1111/jfd.1392

Bett, B., Kiunga, P., Gachochi, J., Sindato, C., Mbotha, D. *et al.* (2017) Effects of climate change on the occurrence and distribution of livestock diseases. *Preventive Veterinary Medicine* 137, 119–129. https://doi.org/10.1016/j.prevetmed.2016.11.019

Blake, S.L., Schill, W.B., McAllister, P.E., Lee, M.K., Singer, J.T. and Nicholson, B.L. (1995) Detection and identification of aquatic birnaviruses by PCR assay. *Journal of Clinical Microbiology* 33, 835–839.

Blake, S.L., Ma, J.Y., Caporale, D.A., Jairath, S. and Nicholson, B.L. (2001) Phylogenetic relationships of aquatic birnaviruses based on deduced amino acid sequences of genome segment A cDNA. *Diseases of Aquatic Organisms* 45, 89–102. https://doi.org/10.3354/dao045089

Britton, J.R., Cucherousset, J., Davies, G.D., Godard, M.J. and Copp, G.H. (2010) Non-native fishes and climate change: predicting species responses to warming temperatures in a temperate region. *Freshwater Biology* 55, 1130–1141. https://doi.org/10.1111/j.1365-2427.2010.02396.x

Büyükekiz, A.G., Altun, S., Hansen, E.F., Saticioglu, I.B., Duman, M. *et al.* (2018) Infectious pancreatic necrosis virus (IPNV) serotype Sp is prevalent in Turkish rainbow trout farms. *Journal of Fish Diseases* 41, 95–104. https://doi.org/10.1111/jfd.12675

Cheung, W.W.L., Lam, V.W.Y., Sarmiento, J.L., Kearney, K., Watson, R. and Pauly, D. (2009) Projecting global marine biodiversity impacts under climate change scenarios. *Fish and Fisheries* 10, 235–251. https://doi.org/10.1111/j.1467-2979.2008.00315.x

Chou, H.-Y., Lo, C.-F., Tung, M.-C., Wang, C.-H., Fukuda, H. and Sano, T. (1993) The general characteristics of a birnavirus isolated from cultured loach (*Misgurnus anguillicaudatus*) in Taiwan. *Fish Pathology* 28, 1–7. https://doi.org/10.3147/jsfp.28.1

Chou, H.Y., Peng, T.Y., Chang, S.J., Hsu, Y.L. and Wu, J.L. (1999) Effect of heavy metal stressors and salinity shock on the susceptibility of grouper (*Epinephelus* sp.) to infectious pancreatic necrosis virus. *Virus Research* 63, 121–129. https://doi.org/10.1016/S0168-1702(99)00065-9

Chursov, A., Kopetzky, S.J., Bocharov, G., Frishman, D. and Shneider, A. (2013) RNAtips: analysis of temperature-induced changes of RNA secondary structure. *Nucleic Acids Research* 41, W486–W491. https://doi.org/10.1093/nar/gkt486

Collingsworth, P., Bunnell, D.B., Murray, M.V., Kao, Y.-C., Feiner, Z.S. *et al.* (2017) Climate change as a long-term stressor for the fisheries of the Laurentian Great Lakes of North America. *Reviews in Fish Biology and Fisheries* 27,363–391.https://doi.org/10.1007/s11160-017-9480-3

Comps, M., Menu, B., Breuil, G. and Bonami, J.R. (1991) Viral-infection associated with rotifer mortalities in mass-culture. *Aquaculture* 93, 1–7. https://doi.org/10.1016/0044-8486(91)90200-Q

Cutrín, J.M., López-Vázquez, C., Olveira, J.G., Castro, S., Dopazo, C.P. and Bandín, I. (2005) Isolation in cell culture and detection by PCR-based technology of IPNV-like virus from leucocytes of carrier turbot, *Scophthalmus maximus* (L.). *Journal of Fish Diseases* 28, 713–722. https://doi.org/10.1111/j.1365-2761.2005.00675.x

Dadar, M., Peyghan, R., Memari, H.R., Shapouri, M.R.S.A., Hasanzadeh, R. *et al.* (2013) Sequence analysis of infectious pancreatic necrosis virus isolated from Iranian reared rainbow trout (*Oncorhynchus mykiss*) in 2012. *Virus Genes* 47, 574–578. https://doi.org/10.1007/s11262-013-0981-4

Danovaro, R., Corinaldesi, C., Dell'Anno, A., Fuhrman, J.A., Middelburg, J.J. *et al.* (2010) Marine viruses and global climate change. *FEMS Microbiology Reviews* 35, 993–1034. https://doi.org/10.1111/j.1574-6976.2010.00258.x

Das, S. and Sahoo, P.K. (2014) Markers for selection of disease resistance in fish: a review. *Aquaculture International* 6, 1793–1812. https://doi.org/10.1007/s10499-014-9783-5

Desautels, D. and MacKelvie, R.M. (1975) Practical aspects of survival and destruction of infectious pancreatic necrosis virus. *Journal of the Fisheries Research Board of Canada* 32, 523–531. https://doi.org/10.1139/f75-064

De Silva, S.S. and Soto, D. (2009) Climate change and aquaculture: potential impacts, adaptation and migration. In: Cochrane, K., de Young, C., Soto, D. and Bahri, T. (eds) Climate change implications for fisheries and aquaculture: overview of current scientific knowledge. *FAO Fisheries and Aquaculture Technical Paper No. 530*. Food and Agriculture Organization of the United Nations, Rome, pp. 151–212.

Dixon, P.F., Smail, D.A., Algoët, M., Hastings, T.S., Bayley, A. *et al.* (2012) Studies on the effect of temperature and pH on the inactivation of fish and bacterial pathogens. *Journal of Fish Diseases* 35, 51–64. https://doi.org/10.1111/j.1365-2761.2011.01324.x

Dopazo, C.P. and Bandín, I. (2011) Techniques of diagnosis of fish and shellfish virus and viral diseases. In: Nollet, L. and Toldra, F. (eds) *Safety Analysis of Foods of Animal Origin*. CRC Press, Boca Raton, Florida, pp. 531–576.

Dopazo, C.P. and Barja, J.L. (2002) Diagnosis and identification of IPNV in salmonids by molecular methods. In: Cunningan, C. (ed.) *Molecular Diagnosis of Salmonid Diseases*, Vol. 3. Kluwer Academic, London, pp. 23–48.

Dorson, M. and Torchy, C. (1981) The influence of fish age and water temperature on mortalities of rainbow-trout, *Salmo gairdneri* Richardson, caused by a European strain of infectious pancreatic necrosis virus. *Journal of Fish Diseases* 4, 213-221. https://doi.org/10.1111/j.1365-2761.1981.tb01128.x

Ellis, A.E. (2001) Innate host defense mechanisms of fish against viruses and bacteria. *Developmental and Comparative Immunology* 25, 827–839. https://doi.org/10.1016/S0145-305X(01)00038-6

FAO (Food and Agriculture Organization of the United Nations) (2008) Expert workshop on climate change implications for fisheries and aquaculture, Rome, Italy, 7–9 April 2008. *FAO Fisheries Report No. 870*. FAO, Rome.

FHS Blue Book (2016) *Suggested Procedures for the Detection and Identification of Certain Finfish and Shellfish Pathogens*. AFS Fish Health Section. Available at: https://units.fisheries.org/fhs/fish-health-section-blue-book-2016 (accessed 6 March 2020).

FIS (Fish Information and Services) (2015) NGO calls for more transparence about salmon escapes, ISA. Available at: https://www.fis.com/fis/worldnews/worldnews.asp?monthyear=&day=7&id=75620&l=e&special=&ndb=1%20target= (accessed 3 April 2019).

GISTEMP Team (2019) *GISS Surface Temperature Analysis (GISTEMP)*. NASA Goddard Institute for Space Studies. Available at: https://data.giss.nasa.gov/gistemp/maps/index_v3.html (accessed 30 March 2019).

Grant, S.B., List, E.J. and Lidstrom, M.E. (1993) Kinetic analysis of virus adsorption and inactivation in batch experiments. *Water Resources Research* 29, 2067–2085. https://doi.org/10.1029/93WR00757

Gregory, A. Munro, L.A., Wallace, I.S., Bain, N. and Raynard, R.S. (2007) Detection of infectious pancreatic necrosis virus (IPNV) from the environment in the vicinity of IPNV-infected Atlantic salmon farms in Scotland. *Journal of Fish Diseases* 30, 621–630. https://doi.org/10.1111/j.1365-2761.2007.00844.x

Haenen, O.L.M., Dijkstra, S.G., Tulden, P.W., Davidse, A., van Nieuwstadt, A.P. *et al.* (2002) *Herpesvirus anguillae* (HVA) isolations from disease outbreaks in cultured European eel, *Anguilla anguilla* in the Netherlands since 1996. *Bulletin of the European Association of Fish Pathologists* 22, 247–257.

Hansen, J., Ruedy, R., Sato, M. and Lo, K. (2010) Global surface temperature change. *Reviews in Geophysics* 48, RG4004. https://doi.org/10.1029/2010RG000345

Harvey, R.W. and Ryan, J.N. (2004) Use of PRD1 bacteriophage in groundwater viral transport, inactivation, and attachment studies. *FEMS Microbiology Ecology* 49, 3–16. https://doi.org/10.1016/j.femsec.2003.09.015

Hata, N., Okinaka, Y., Sakamoto, T., Iwamoto, T. and Nakai, T. (2007) Upper temperature limits for the multiplication of betanodaviruses. *Fish Pathology* 42, 225–228. https://doi.org/10.3147/jsfp.42.225

Hayashi, T., Wills, S., Bussey, K.A. and Takimoto, T. (2015) Identification of influenza A virus PB2 residues involved in enhanced polymerase activity and virus growth in mammalian cells at low temperatures. *Journal of Virology* 89, 8042–8049. https://doi.org/10.1128/JVI.00901-15

Hershberger, P.K., Purcell, M.K., Hart, L.M., Gregg, J.L., Thomson, R.L. *et al.* (2013) Influence of temperature on viral hemorrhagic septicemia (Gebogroup IVa) in Pacific herring, *Clupea pallasis* Valenciennes. *Journal of Experimental Marine Biology and Ecology* 444, 81–86. https://doi.org/10.1016/j.jembe.2013.03.006

Hill, B.J. and Way, K. (1995) Serological classification of infectious pancreatic necrosis (IPN) virus and other aquatic birnaviruses. *Annual Review of Fish Diseases* 5, 55–77.

Hirayama, T., Nagano, I., Shinmoto, H., Yagyu, K.I. and Oshima, S.I. (2007) Isolation and characterization of virulent yellowtail ascites virus. *Microbiology and Immunology* 51, 397–406. https://doi.org/10.1111/j.1348-0421.2007.tb03927.x

Holbrook, S.J., Schmitt, R.J. and Stephens, J. (1997) Changes in an assemblage of temperate reef fishes associated with a climate shift. *Ecological Applications* 7, 1299–1310. https://www.jstor.org/stable/2641215

Inaba, M., Kimura, T., Kikukawa, R., Iwasaki, M., Nose, M. and Suzuki, S. (2007) Annual dynamics of marine birnavirus (MABV) in cultured Japanese flounder *Paralichthys olivaceus* and sea water. *Fisheries Science* 73, 615–622.

Inaba, M., Suzuki, S., Kitamura S.-I., Kumazawa, N. and Kodama, H. (2009) Distribution of marine birnavirus (MABV) in marine organisms from Okinawa, Japan, and a unique sequence variation of the VP2/NS region. *Journal of Microbiology* 47, 76–84. https://doi.org/10.1007/s12275-008-0250-8

IPCC (Intergovernmental Panel on Climate Change) (2013) *Climate Change 2013: The Physical Science Basis. Contribution of Working Group I to the Fifth Assessment Report of the Intergovernmental Panel on Climate Change* [Stocker, T.F., Qin, D., Plattner, G.-K., Tignor, G.-K., Allen, S.K. *et al.* (eds)]. Cambridge University Press, Cambridge and New York.

Jeon, C.-H., Kim, S.-R., Kim, W.-S., Lee, C.-H., Seong, K.-B. *et al.* (2011) Monitoring of viruses in chum salmon (*Oncorhynchus keta*) migrating to Korea. *Archives of Virology* 156, 1025–1030. https://doi.org/10.1007/s00705-011-0944-9

Jiravanichpaisal, P., Soderhall, K. and Soderhall, I. (2004) Effect of water temperature on the immune response and infectivity pattern of white spot syndrome virus (WSSV) in freshwater crayfish. *Fish and Shellfish Immunology* 17, 265–275. https://doi.org/10.1016/j.fsi.2004.03.010

Kim, J.H., Park, H.J., Kim, K.W., Hwang, I.K., Kim, D.H. *et al.* (2017) Growth performance, oxidative stress, and non-specific immune responses in juvenile sablefish, *Anoplopoma fimbria*, by changes of water temperature and salinity. *Fish Physiology and Biochemistry* 43, 1421–1431. https://doi.org/10.1007/s10695-017-0382-z

Kim, K.I., Kwon, W.J., Kim, Y.C., Kim, M.S., Hong, S. and Do Jeong, H. (2016) Surveillance of aquatic animal viruses in seawater and shellfish in Korea.

Aquaculture 461, 17–24. https://doi.org/10.1016/j.aquaculture.2016.03.053
Kitamura, S.I. and Suzuki, S. (2000) Occurrence of marine birnavirus through the year in coastal seawater in the Uwa Sea. *Marine Biotechnolology* 2, 188-194.
Kitamura, S.I., Jug, S.J. and Suzuki, S. (2000) Seasonal change of infective state of marine birnavirus in Japanese pearl oyster *Pinctada fucata*. *Archives of Virology* 145, 2003–2014. https://doi.org/10.1007/s007050070036
Kitamura, S.I., Tomaro, Y., Kawabata, Z. and Suzuki, S. (2002) Detection of marine birnavirus in the Japanese pearl oyster *Pinctada fucata* and seawater from different depths. *Diseases of Aquatic Organisms* 50, 211–217. https://doi.org/10.3354/dao050211
Labrana, R., Espinoza, J.C. and Kuznar, J. (2008) Detection of infectious pancreatic necrosis virus (IPNV) in freshwater sediments. *Archivos de Medicina Veterinaria* 40, 203–205. https://doi.org/10.4067/S0301-732X2008000200014
Lago, L., Rodríguez, J.F., Bandín, I. and Dopazo, C.P. (2016) Aquabirnavirus polyploidy: a new strategy to modulate virulence? *Journal of General Virology* 97, 1168–1177. https://doi.org/10.1099/jgv.0.000434
Lago, M., Bandín, I., Olveira, J.G. and Dopazo, C.P. (2017) *In vitro* reassortment between infectious pancreatic necrosis virus (IPNV) strains: the mechanisms involved and its effect on virulence. *Virology* 501, 1–11. https://doi.org/10.1016/j.virol.2016.11.003
Liu, M. and Vakharia, V.N. (2006) Non-structural protein of infectious bursal disease virus inhibits apoptosis at the early stage of virus infection. *Journal of Virology* 80, 3369–3377. https://doi.org/10.1128/JVI.80.7.3369-3377.2006
Lo, C.F., Hong, Y.W., Huang, S.Y. and Wang, C.H. (1988) The characteristics of a virus isolated from the gill of clam, *Meretrix lusoria*. *Fish Pathology* 23, 147–154. https://doi.org/10.3147/jsfp.23.147
Malsberger, R.G. and Cerini, C.P. (1965) Multiplication of infectious pancreatic necrosis virus. *Annals of the New York Academy of Sciences* 126, 320–327. https://doi.org/10.1111/j.1749-6632.1965.tb14283.x
Marcos-López, M., Gale, P., Oidtmann, B.C. and Peeler, E.J. (2010) Assessing the impact of climate change on disease emergence in freshwater fish in the United Kingdom. *Transboundary and Emerging Diseases* 57, 293–304. https://doi.org/10.1111/j.1865-1682.2010.01150.x
McAllister, P.E. (2016) Infectious pancreatic necrosis. In: *Fish Health Section Blue Book: Suggested Procedures for the Detection and Identification of Certain Finfish and Shellfish Pathogens. AFS Fish Health Section*. Available at: https://units.fisheries.org/fhs/fish-health-section-blue-book-2016/ (accessed 3 April 2019).
McAllister, P.E. and Bebak, J. (1997) Infectious pancreatic necrosis virus in the environment: relationship to effluent from aquaculture facilities. *Journal of Fish Diseases* 20, 201–207.https://doi.org/10.1046/j.1365-2761.1997.00297.x
McAllister, P.E. and Owens, W.J. (1992) Recovery of infectious pancreatic necrosis virus from the faeces of wild piscivorous birds. *Aquaculture* 106, 227–232. https://doi.org/10.1016/0044-8486(92)90254-I
McAllister, P.E., Newman, M.W., Sauber, J.H. and Owens, W.J. (1984) Isolation of infectious pancreatic necrosis virus (serotype-ab) from diverse species of estuarine fish. *Helgoländer Meeresuntersuchungen* 37, 317–328. https://doi.org/10.1007/BF01989314
McBeath, A.J.A., Snow, M., Secombes, C.J., Ellis, A.E. and Collet, B. (2007) Expression kinetics of interferon and interferon-induced genes in Atlantic salmon (*Salmo salar*) following infection with infectious pancreatic necrosis virus and infectious salmon anaemia virus. *Fish and Shellfish Immunology* 22, 210–241. https://doi.org/10.1016/j.fsi.2006.05.004
McCowan, C., Motha, J., Crane, M.St.J., Moody, N.J.G., Crameri, S. *et al.* (2015) Isolation of a novel aquatic birnavirus from rainbow trout *Onchorhynchus mykiss* in Australia. *Diseases of Aquatic Organisms* 114, 117–125. https://doi.org/10.3354/dao02858
Miller, K.M., Teffer, A., Tucker, S., Li, S., Schulze, A.D. *et al.* (2014) Infectious disease, shifting climates, and opportunistic predators: cumulative factors potentially impacting wild salmon declines. *Evolutionary Applications* 7, 812–855. https://doi.org/10.1111/eva.12164
Molloy, S.D., Pietrak, M.R., Bricknell, I. and Bouchard, D.A. (2013) Experimental transmission of infectious pancreatic necrosis virus from the blue mussel, *Mytilus edulis*, to cohabitating Atlantic salmon (*Salmo salar*) smolts. *Applied and Environmental Microbiology* 79, 5882–5890. https://doi.org/10.1128/AEM.01142-13
Moreno, P., Olveira, J.G., Labella, A., Cutrín, J.M., Baro, J.C., Borrego, J.J. and Dopazo, C.P. (2014) Surveillance of viruses in wild fish populations in areas around the Gulf of Cadiz (South Atlantic Iberian Peninsula). *Applied and Environmental Microbiology* 80, 6560–6571. https://doi.org/10.1128/AEM.02090-14
Morley, N.J. (2010) Interactive effects of infectious diseases and pollution in aquatic molluscs. *Aquatic Toxicology* 96, 27–36. https://doi.org/10.1016/j.aquatox.2009.09.017
Mortensen, S.H., Bachere, E., Legall, G. and Mialhe, E. (1992) Persistence of infectious pancreatic necrosis virus (IPNV) in scallops *Pecten maximus*. *Diseases of Aquatic Organisms* 12, 221–227. https://doi.org/10.3354/dao012221
Mortensen, S.H., Nilsen, R.K. and Hjeltnes, B. (1998) Stability of an infectious pancreatic necrosis virus (IPNV) isolate stored under different laboratory conditions. *Diseases of Aquatic Organisms* 33, 67–71. https://doi.org/10.3354/dao033067
Morvan, C.L., Troutaud, D. and Deschaux, P. (1998) Differential effects of temperature on specific and nonspecific immune defences in fish. *Journal of Experimental Biology* 201, 165–168.

Mulei, I.R., Nyaga, P.N., Mbuthia, P.G., Waruiru, R.M., Njagi, L.W. *et al.* (2018) Infectious pancreatic necrosis virus isolated from farmed rainbow trout and tilapia in Kenya is identical to European isolates. *Journal of Fish Diseases* 41, 1191–1200. https://doi.org/10.1111/jfd.12807

Munang'andu, H.M., Galindo-Villegas, J. and David, L. (2018) Teleosts genomics: progress and prospects in disease prevention and control. *International Journal of Molecular Sciences* 19, 1083. https://doi.org/10.3390/ijms19041083

Munro, E. and Midtlyn, P.J. (2011) Infectious pancreatic necrosis virus and associated aquatic birnaviruses. In: Woo, P.T.K. and Bruno, D.W. (eds) *Fish and Shellfish Disorders*. Vol. 3. *Viral, Bacterial and Fungal Infections*, 2nd edn. CAB International, Wallingford, UK, pp. 1–65.

NASA (National Aeronautics and Space Administration) (2019) SISS Surface Temperature Analysis. Available at: https://data.giss.nasa.gov/gistemp/maps/index_v3.html (accessed 11 March 2019).

Nicholson, B.L. (1993) Use of monoclonal antibodies in identification and characterization of fish viruses. *Annual Review of Fish Diseases* 3, 241–257.

Nishizawa, T., Kinoshita, S. and Yoshimizu, M. (2005) A approach for genogrouping of Japanese isolates of aquabirnaviruses in a new genogroup, VII, based on the VP2/NS junction region. *Journal of General Virology* 86, 1973–1978. https://doi.org/10.1099/vir.0.80438-0

OIE (World Organization for Animal Health) (2003) *Manual of Diagnostic Tests for Aquatic Animals*. OIE, Paris.

Okamoto, N., Yasutomi, R., Shibazaki, H., Hanzawa, S. and Sano, T. (1987) The influence of immersing temperature for inoculation with IPNV and/or rearing temperature on mortality of rainbow trout fry postinfection. *Nippon Suisan Gakkaishi* 53, 1125–1128. (in Japanese) https://doi.org/10.2331/suisan.53.1125

Ozaki, A., Khoo, S., Yoshiura, Y., Ototake, M., Sakamoto, T. *et al.* (2007) Identification of additional quantitative trait loci (QTL) responsible for susceptibility to infectious pancreatic necrosis virus in rainbow trout. *Fish Pathology* 42, 131–140. https://doi.org/10.3147/jsfp.42.131

Panzarin, P., Holmes, E.C., Abbadi, M., Zamperin, G., Quartesan, R. *et al.* (2018) Low evolutionary rate of infectious pancreatic necrosis virus (IPNV) in Italy is associated with reduced virulence in trout. *Virus Evolution* 4(2), vey09. https://doi.org/10.1093/ve/vey019

Richter, A. and Kolmes, S.A. (2017) Maximum temperature limits for chinook, coho, and chum salmon, and steelhead trout in the Pacific Northwest. *Reviews in Fisheries Science* 13, 23–49. https://doi.org/10.1080/10641260590885861

Rivas, C., Cepeda, C., Dopazo, C.P., Novoa, B., Noya, M. and Barja, J.L. (1993) Marine environment as reservoir of birnaviruses from poikilothermic animals. *Aquaculture* 115, 183–194. https://doi.org/10.1016/0044-8486(93)90135-L

Romero-Brey, I., Bandín, I., Dopazo, C.P. and Barja, J.L. (1999) Isolation of marine birnavirus from new species of wild fishes. *FHS Newsletters* 31, 21–23.

Romero-Brey, I., Bandín, I., Cutrín, J.M., Vakharia, V.N. and Dopazo, C.P. (2009) Genetic analysis of aquabirnaviruses isolated from wild fish reveals occurrence of natural reassortment of infectious pancreatic necrosis virus. *Journal of Fish Diseases* 32, 585–595. https://doi.org/10.1111/j.1365-2761.2009.01020.x

Salgado-Miranda, C., Rojas-Anaya, E., García-Espinosa, G. and Loza-Rubio, E. (2014) Molecular characterization of the VP2 gene of infectious pancreatic necrosis virus (IPNV) isolates from Mexico. *Journal of Aquatic Animal Health* 6, 43–51. https://doi.org/10.1080/08997659.2013.860060

Santi, N., Vakharia, V.N. and Evensen, Ø. (2004) Identification of putative motifs involved in the virulence of infectious pancreatic necrosis virus. *Virology* 322, 31–40. https://doi.org/10.1016/j.virol.2003.12.016

Slenning, B.D. (2010) Global climate change and implications for disease emergence. *Veterinary Pathology* 47, 28–33. https://doi.org/10.1177/0300985809354465

Sommerville, C. (2012) Advances in non-chemical methods for parasite prevention and control in fish. In: Austin, E. (ed.) *Infectious Disease in Aquaculture: Prevention and Control*. Woodhead Publishing, Cambridge, UK, pp 480–512.

Storset, A., Evensen, Ø. and Midtlyng, P.J. (2006) A user's inter-laboratory comparison of broodfish screening for infectious pancreatic necrosis virus using molecular and conventional diagnostic methods. *Developments in Biologicals* 126, 101–105.

Suzuki, S., Nakata, T., Kamakura, M., Yoshimoto, M., Furukawa, Y. *et al.* (1997) Isolation of birnavirus from agemaki (jack knife clam) *Sinonovacura constricta* and survey of the virus using PCR technique. *Fisheries Science* 63, 563–566. https://doi.org/10.2331/fishsci.63.563

Swanson, R.N., Carlisle, J.C. and Gillespie, J.H. (1982) Pathogenesis of infectious pancreatic necrosis virus-infection in brook trout, *Salvelinus fontinalis* (Mitchill), following intraperitoneal injection. *Journal of Fish Diseases* 5, 449–460. https://doi.org/10.1111/j.1365-2761.1982.tb00504.x

Takano, R., Mori, K.-I., Nishizawa, T., Arimoto, M. and Muro, K. (2001) Isolation of viruses from wild Japanese flounder *Paralichthys olivaceus*. *Fish Pathology* 36, 153–160. https://doi.org/10.3147/jsfp.36.153

Toranzo, A.E. and Hetrick, F.M. (1982) Comparative stability of two salmonid viruses and poliovirus in fresh, estuarine and marine waters. *Journal of Fish Diseases* 5, 223–231. https://doi.org/10.1111/j.1365-2761.1982.tb00477.x

UK Environment Agency (2008) *The thermal biology of brown trout and Atlantic salmon: science summary*. Report No. SCHO0808BOLV-E-P. Available at: https://assets.publishing.service.gov.uk/government/

uploads/system/uploads/attachment_data/file/291742/scho1008boue-e-e.pdf (accessed 18 March 2019).

Ulrich, K., Wehner, S., Bekaert, M., Di Paola, N., Dilcher, M. *et al.* (2018) Molecular epidemiological study on infectious pancreatic necrosis virus isolates from aquafarms in Scotland over three decades. *Journal of General Virology* 99, 1567–1581. https://doi.org/10.1099/jgv.0.001155

Urquhart, K., Murray, A.G., Gregory, A., O'Dea, M., Munro, L.A. *et al.* (2008) Estimation of infectious dose and viral shedding rates for infectious pancreatic necrosis virus in Atlantic salmon, *Salmo salar* L., post-smolts. *Journal of Fish Diseases* 31, 879–887. https://doi.org/10.1111/j.1365-2761.2008.00989.x

Varvarigos, P., Vendramin, N., Cappellozza, E., Bovo, G. (2011) First confirmation of herpes virus anguillae (HVA) and infectious pancreatic necrosis (IPN) virus infecting European eel *Anguilla anguilla* farmed in Greece. *Bulletin of the European Association of Fish Pathologists* 31, 101–111.

Vázquez, D., López-Vázquez, C., Cutrín, J.M. and Dopazo, C.P. (2016a) A novel procedure of quantitation of virus based on microflow cytometry analysis. *Applied Microbiology and Biotechnology* 100, 2347–2354. https://doi.org/10.1007/s00253-015-7228-3

Vázquez, D., Cutrín, J.M., Olveira, J.G. and Dopazo, C.P. (2016b) Design and validation of a RT-qPCR procedure for diagnosis and quantification of most types of infectious pancreatic necrosis virus using a single pair of degenerated primers. *Journal of Fish Diseases* 40, 1155–1167. https://doi.org/10.1111/jfd.12590

Wallace, I.S., Gregory, A., Murray, A.G., Munro, E.S. and Raynard, R.S. (2008) Distribution of infectious pancreatic necrosis virus (IPNV) in wild marine fish from Scottish waters with respect to clinically infected aquaculture sites producing Atlantic salmon, *Salmo salar* L. *Journal of Fish Diseases* 31, 177–186. https://doi.org/10.1111/j.1365-2761.2007.00886.x

Wang, L., Liu, P., Wan, Z.Y., Huang, S.Q., Wen, Y.F. *et al.* (2018) RNA-Seq revealed the impairment of immune defence of tilapia against the infection of *Streptococcus agalactiae* with simulated climate warming. *Fish and Shellfish Immunology* 55, 679–689. https://doi.org/10.1016/j.fsi.2016.06.058

Wasik, B.R. and Turner, P.E. (2013) On the biological success of viruses. *Annual Review of Microbiology* 67, 519–541. https://doi.org/10.1146/annurev-micro-090110-102833

Winton, J.R. (2016) Anthropogenic drivers of emerging viruses in fish. *Bulletin of the European Association of Fish Pathologists* 36, 164–168.

Wolf, K., Quimby, M.C. and Bradford, A.D. (1963) Egg-associated transmission of IPN virus of trout. *Virology* 21, 317–321. https://doi.org/10.1016/0042-6822(63)90192-2

Yoshimizu, M., Yoshinaka, T., Hatori, S. and Kasai, H. (2005) Survivability of fish pathogenic viruses in envi-

7 Herpesvirosis (Koi Herpesvirus)

Hatem Soliman and Mansour El-Matbouli*

University of Veterinary Medicine, Vienna, Austria

7.1 Introduction

Herpesviruses comprise a group of linear, double-stranded DNA-containing, large and complex viruses that are widespread among the vertebrates and certain invertebrates (Minson *et al.*, 2000; Davison *et al.*, 2005). These viruses have a common virion structure and biological properties and are host-specific pathogens (Davison, 2002; McGeoch *et al.*, 2008; Mettenleiter *et al.*, 2008).

Three families of herpesviruses constitute the order *Herpesvirales*: the family *Herpesviridae* contains the mammalian, avian and reptilian viruses; the family *Alloherpesviridae* includes the fish and amphibian viruses; and the family *Malacoherpesviridae* comprises the bivalve viruses (Davison *et al.*, 2009). In addition to the virion structure, the members of the order *Herpesvirales* exhibit the presence of other conserved characteristics as well, such as the replication process, the ability to establish long-term latency and manipulation of the host immune response (Hanson *et al.*, 2011). Currently, family *Alloherpesviridae* comprises 12 species distributed among the following four genera: *Cyprinivirus*, *Salmonivirus*, *Ictalurivirus* and *Batrachovirus*. These 12 species form a highly diverse group, which includes the largest (295 kbp) and the smallest (134 kbp) known sequenced genomes among the herpesviruses, namely *Cyprinid herpesvirus 3* and channel catfish virus, respectively (McGeoch *et al.*, 2006; Davison *et al.*, 2009; Davison, 2010).

The genus *Cyprinivirus* consists of phylogenetically related viruses, including members that are able to cause devastating diseases in fish resulting in severe economic losses to the aquaculture industry. This genus comprises four species, one of which is *Anguillid herpesvirus 1* (AngHV1), which is the economically most relevant virus present in wild and cultured freshwater eels belonging to the genus *Anguilla* (Sano *et al.*, 1990). The other three species of genus *Cyprinivirus* constitute the group of herpesviruses of cyprinids (family *Cyprinidae*), which include *Cyprinid herpesvirus 1* (CyHV-1), *Cyprinid herpesvirus 2* (CyHV-2) and *Cyprinid herpesvirus 3* (CyHV-3). These three herpesviruses are closely related, although they cause distinctly different diseases that result in significant economic losses in the cyprinid aquaculture.

Cyprinid herpesvirus 1 is the causative agent for the oldest-known fish disease, i.e. carp pox, a disease that dates back to the Middle Ages (Gessner, 1558; Hofer, 1904; Sano *et al.*, 1985). Carp pox disease is characterized by mucoid to waxy epidermal growths developing on the skin of common carp (*Cyprinus carpio carpio*), koi carp (*Cyprinus carpio koi*) and the other cyprinids (Mawdesley-Thomas and Bucke, 1967; Sano *et al.*, 1991; Davison *et al.*, 2013). Although mortalities may occur in carp fry (Sano *et al.*, 1991), the effect of the carp pox disease on its host is limited and is primarily a problem for the cultivation of ornamental carp and fish trade because of the occurrence of focal epidermal hyperplasia (Engelsma *et al.*, 2013).

Cyprinid herpesvirus 2, also known as the goldfish haematopoietic necrosis virus, is the causative agent for the herpesviral haematopoietic necrosis (HVHN) disease that affects goldfish (*Carassius auratus*) and Prussian carp (*Carassius gibelio*) (Jung and Miyazaki, 1995; Danek *et al.*, 2012). The disease results in severe epizootics among all ages of goldfish during the spring and autumn seasons.

Cyprinid herpesvirus 3 is the aetiological agent for the highly contagious viral disease known as the koi herpesvirus disease (KHVD), which causes massive damage to the production of koi and common carp worldwide (Hedrick *et al.*, 2000).

*mansour.el-matbouli@vetmeduni.ac.at

Considering the importance of CyHV-3 as a threat to the koi and common carp industry, this chapter describes the CyHV-3 virus in detail.

7.2 Cyprinid herpesvirus 3 (Koi Herpesvirus)

7.2.1 Brief introduction

Description of disease

CyHV-3 is the causative agent of a severe epizootic disease, known as the koi herpesvirus disease (KHVD), which causes significant losses to the common carp and koi carp populations, and impacts fish breeders, retailers and hobbyists due to cumulative mortalities in the fish populations (Hedrick *et al.*, 2000; Gilad *et al.*, 2003; Ronen *et al.*, 2003). The disease is characterized by severe gill necrosis and high mortality that may reach 80–90% of the infected fish, leading to serious economic losses in the common carp and koi industries throughout the world (Gilad *et al.*, 2002; Perelberg *et al.*, 2003). Prior to taxonomic classification, the aetiological agent of the KHVD was designated as koi herpesvirus (KHV) based on the affected host (Hedrick *et al.*, 2000), and also as interstitial nephritis and gill necrosis virus (CNGV) on the basis of the pathological observations in the affected host (Ilouze *et al.*, 2011). Afterwards, evidence was obtained in support of the classification of the virus as a herpesvirus, and it was named as CyHV-3 (Waltzek *et al.*, 2005). The comparison of complete genome sequences of the three CyHV-3 strains isolated from Israel (I), USA (U) and Japan (J) revealed the presence of two lineages (J and U/I) (Aoki *et al.*, 2007). Afterwards, a third lineage intermediate between the J and U/I lineages was also identified (Bigarré *et al.*, 2009). Subsequently, a fourth intermediate lineage emerged in Indonesia (Sunarto *et al.*, 2011). Certain CyHV-3 variants did not produce clinical symptoms consistent with the usual CyHV-3 infection; as a result, such strains were referred to as low- or non-pathogenic variants of CyHV-3 (Engelsma *et al.*, 2013). Latency was established for the CyHV-3 infection, and the main site for the latency was identified to be the IgM^+ white blood cells, B lymphocytes specifically in koi (Eide *et al.*, 2011b; Reed *et al.*, 2014). Reactivation of the CyHV-3 latent infection has been reported to cause reinfection in the surviving populations, which may spread to naïve fish as well (St-Hilaire *et al.*, 2005).

Current prevalence and geographical distribution

Since its first description in the 1990s, CyHV-3 has been reported from different continents. There is evidence for the presence of CyHV-3 in carp in the UK in 1996 (Aoki *et al.*, 2007). However, the first description of the disease was reported in Germany in 1997 (Bretzinger *et al.*, 1999), followed by Israel and the USA in 1998, where the virus was isolated for the first time (Hedrick *et al.*, 2000). Since then, the virus has spread to several countries worldwide (Table 7.1).

Table 7.1. Geographical distribution of KHVD.

Country/region	Reference
Europe	
UK	Walster (1999)
Germany	Bretzinger *et al.* (1999)
Israel	Hedrick *et al.* (2000)
Denmark	Haenen *et al.* (2004)
Luxembourg	"
Italy	"
France	"
Netherlands	"
Belgium	"
Austria	"
Switzerland	"
Poland	Bergmann *et al.* (2006)
Czech Republic	Novotny *et al.* (2010)
Ireland	Gotesman *et al.* (2013)
Lithuania	"
Hungary	Láng *et al.* (2014)
Romania	OIE (2018)
Slovenia	"
Spain	"
Sweden	"
Asia	
South Korea	Oh *et al.* (2001)
Japan	Sano *et al.* (2004)
Taiwan	Tu *et al.* (2004)
Thailand	Pikulkaew *et al.* (2009)
Philippines	Somga *et al.* (2010)
Indonesia	Sunarto *et al.* (2011)
Iran	Rahmati-Holasoo *et al.* (2016)
China	OIE (2018)
Hong Kong	"
Chinese Taipei	"
Malaysia	"
Singapore	"
North America	
USA	Hedrick *et al.* (2000)
Canada	Garver *et al.* (2010)
Africa	
South Africa	McVeigh (2004)

In addition to impacting the koi and common carp industries, CyHV-3 has also exerted a negative impact on the environment by affecting the wild carp populations (Rakus *et al.*, 2013). Mass mortalities in wild carp have been reported in the Yoshi River and Lake Biwa in Japan, in angling water in the UK, in New York and South Carolina, USA, and in the Kawartha Lakes region, Ontario, Canada (Denham, 2003; Terhune *et al.*, 2004; Grimmett *et al.*, 2006; Uchii *et al.*, 2009; Garver *et al.*, 2010). It has been implicated that CyHV-3 is able to persist in the wild carp population and may subsequently be transmitted to naïve fish (Minamoto *et al.*, 2009, 2012; Uchii *et al.*, 2009).

The number of reports available in the literature on the prevalence of CyHV-3 in the farmed or wild populations of carp is limited. A minimum of 10–25% of the total common carp population displayed high antibody titres and an immunological response was detectable for several months after experimental infection with CyHV-3 when the carp populations were maintained at a temperature lower than the permissive temperature (St-Hilaire *et al.*, 2009). Furthermore, even in the absence of clinical symptoms of the disease, the CyHV-3 DNA was detected through polymerase chain reaction (PCR) assays in carp populations that were maintained at 13°C, which implied that the infected fish surviving at low temperatures could serve as reservoirs for the virus (Gilad *et al.*, 2004). In the wild carp population, an analysis of those carp that survived the KHVD outbreak in Lake Biwa, the largest freshwater lake in Japan, in 2006 revealed that 54% of the older carp were seropositive and 31% were PCR-positive (Uchii *et al.*, 2009). In a study on the prevalence of CyHV-3 in common carp inhabiting the three rivers in Kochi Prefecture, Shikoku Island, Japan in 2013, CyHV-3 DNA (strain J, genotype A1) was detected in 16.7% of the brain samples from Kagami River, in 3.9% each of brain and gill samples from the Monobe River, and in 5.1% of brain samples and 1.3% of gill samples from the Wajiki River. The results of the CyHV-3 distribution survey conducted in England and Wales in 2007, which included sites that experienced clinical outbreaks of KHVD in 2006 and received no further introductions of fish since that time, revealed that three out of four investigated sites produced positive results and exhibited 85–93% seroprevalence in the samples of the surviving carp population, while the fourth site produced negative results (Taylor *et al.*, 2010).

Susceptible host species and reservoir hosts

Common carp and its variety koi carp are both known to be naturally susceptible hosts to CyHV-3 (Hedrick *et al.*, 2000; Perelberg *et al.*, 2003; Michel *et al.*, 2010a). All age groups of these fish appear to be susceptible to the CyHV-3 infection (Sano *et al.*, 2004). Carp larvae have been reported to be resistant to CyHV-3 infection; however, on maturation, the same carp fish were observed to be susceptible to CyHV-3 infection (Ito *et al.*, 2007).

Hybrids of common carp and koi have been observed to be susceptible to CyHV-3 to different degrees. Among all the hybrids, the ghost carp (koi carp × common carp) hybrid has been reported to be the most susceptible one to the CyHV-3 infection (OIE, 2018). Experimental infection with CyHV-3 in (koi carp × crucian carp) and (koi carp × goldfish) hybrids through bath immersion resulted in 91–100% and 35–42% mortality rate, respectively (Bergmann *et al.*, 2010a). On the other hand, the hybrid (common carp × goldfish) exhibited only a little susceptibility to CyHV-3 infection, with low (5%) mortality rate (Hedrick *et al.*, 2006).

Detection of CyHV-3 DNA in certain cyprinid and non-cyprinid fish species, and also in other aquatic organisms (Table 7.2) without the observation of any symptoms of infection in these species, suggested that these species could serve as reservoirs for CyHV-3 and consequently transmit the virus to naïve carp (El-Matbouli *et al.*, 2007a; Kempter and Bergmann, 2007; Sadler *et al.*, 2008; Kempter *et al.*, 2009, 2012; Kielpinski *et al.*, 2010; El-Matbouli and Soliman, 2011; Minamoto *et al.*, 2011; Radosavljević *et al.*, 2012; Fabian *et al.*, 2013; Rakus *et al.*, 2013).

7.2.2 Diagnosis of KHVD

Morphological features of the pathogen

CyHV-3 is a spherical virion (Fig. 7.1) that contains a core, capsid, tegument and envelope in its structure (Davison *et al.*, 2009). The core of the virion consists of viral genome packaged as a single copy of a large, linear, double-stranded DNA molecule inside a preformed icosahedral capsid (Mettenleiter *et al.*, 2009). A host-derived lipid envelope carries the viral glycoproteins and an amorphous layer of proteins termed the tegument which resides between the capsid and the envelope (Miyazaki *et al.*, 2008). The diameter of the CyHV-3 virion has been reported to range between 167 and 230 nm

Table 7.2. List of fish species and other aquatic organisms that act as reservoir for CyHV-3 without clinical signs.

Aquatic organism	Scientific name	Reference
Goldfish	*Carassius auratus*	El-Matbouli *et al.* (2007a); Sadler *et al.* (2008)
Grass carp	*Ctenopharyngodon idella*	Kempter *et al.* (2012); Radosavljević *et al.* (2012)
Silver carp	*Hypophthalmichthys molitrix*	"
Prussian carp	*Carassius gibelio*	Kempter *et al.* (2008); Radosavljević *et al.* (2012)
Tench	*Tinca tinca*	"
Vimba bream	*Vimba vimba*	Kempter and Bergmann (2007)
Gudgeon	*Gobio gobio*	Kempter *et al.* (2008); Fabian *et al.* (2013)
Common roach	*Rutilus rutilus*	"
Northern pike	*Esox lucius*	"
Common bream	*Abramis brama*	"
European perch	*Perca fluviatilis*	"
Crucian carp	*Carassius carassius*	Kempter *et al.* (2008)
European chub	*Leuciscus cephalus*	"
Common barbel	*Barbus barbus*	"
European bullhead	*Cottus gobio*	Kempter *et al.* (2008); Rakus *et al.* (2013)
Spined loach	*Cobitis taenia*	Kempter *et al.* (2008); Rakus *et al.* (2013)
Belica	*Leucaspius delineatus*	Kempter *et al.* (2008)
Common nase	*Chondrostoma nasus*	"
Ide	*Leuciscus idus*	"
Eurasian ruffe	*Gymnocephalus cernua*	Kempter *et al.* (2012)
Common dace	*Leuciscus leuciscus*	Kempter and Bergmann (2007); Fabian *et al.* (2013)
Ornamental catfish	*Ancistrus* sp.	Bergmann *et al.* (2009)
Russian sturgeon	*Acipenser gueldenstaedtii*	Kempter *et al.* (2009)
Atlantic sturgeon	*Acipenser oxyrinchus*	"
Brown bullhead	*Ameiurus nebulosus*	Fabian *et al.* (2013)
Three-spined stickleback	*Gasterosteus aculeatus*	"
Rudd	*Scardinius erythrophthalmus*	"
Swan mussels	*Anodonta cygnea*	Kielpinski *et al.* (2010)
Scuds (crustacea)	*Gammarus pulex*	"
Planktons	Phylum *Rotifera*	Minamoto *et al.* (2011)

according to the cell types infected both *in vitro* and *in vivo* (Hedrick *et al.*, 2000; Miwa *et al.*, 2007; Miyazaki *et al.*, 2008).

Behavioural changes and clinical symptoms

BEHAVIOURAL CHANGES The CyHV-3-infected fish exhibit behavioural changes such as gasping at the water surface, anorexia, increased respiratory frequency, lethargy, separation from the shoal and gathering close to the water inlet or to the sides of the pond, remaining at the bottom of the tank with folded dorsal fins, erratic swimming, loss of equilibrium and disorientation; on the other hand, the infected fish may exhibit signs of hyperactivity (Hedrick *et al.*, 2000; Perelberg *et al.*, 2003; Haenen *et al.*, 2004).

KOI HERPESVIRUS DISEASE: CLINICAL SYMPTOMS Mass mortality occurs in the koi and common carp populations during a KHVD outbreak. Loss of osmoregulation in the kidney, gut and gills of the fish contributes to mortality during acute infection with CyHV-3 (Gilad *et al.*, 2004). The course of infection and the clinical symptoms observed during the disease vary with the individual fish. The infected fish may exhibit one or more symptoms of the KHVD; however, the pathogen may be present even in the absence of any symptoms. The affected fish may exhibit gill necrosis, pale discoloration of the gills, increased (sometimes decreased) mucus secretion (Fig. 7.2), changes in skin including hyperaemia at the base of the fins, irregular pale patches or blisters on the skin, skin ulcers, sandpaper-like texture of the skin, sloughing of scales,

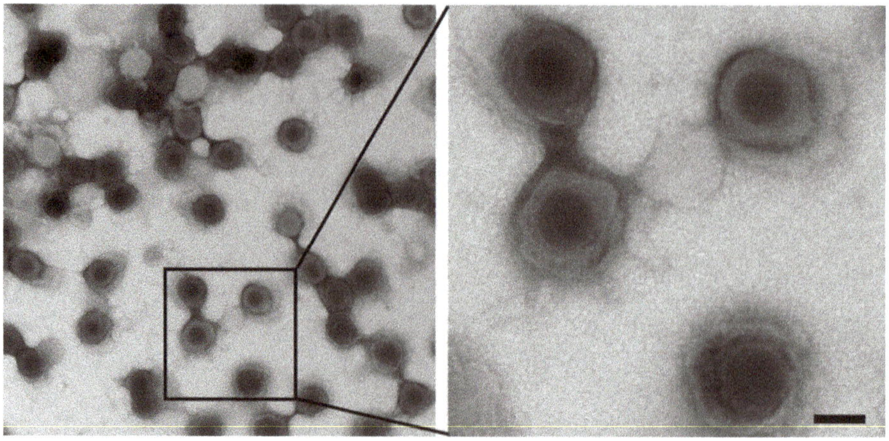

Fig. 7.1. Demonstration of purified CyHV-3 virions by transmission electron microscopy (TEM). Purified virions were negatively stained with uranyl acetate and observed by TEM; scale bar = 50 nm. (Adapted from Michel *et al.*, 2010b with permission from the Microbiology Society via the Copyright Clearance Center.)

Fig. 7.2. *Cyprinus carpio* with gross lesions associated with KHVD. The operculum is removed, demonstrating gill necrosis and anaemia. Increased mucus production is marked on the dorsal fin. (Image from the archive of the Clinical Division of Fish Medicine.)

fin erosion and bilateral exophthalmia (Walster, 1999; Hedrick and Yun, 2005; Hara *et al.*, 2006; McDermott and Palmeiro, 2013).

Macroscopic and microscopic lesions

Certain CyHV-3-infected fish demonstrate inconsistent changes in necropsy including abdominal adhesions and accumulation of abdominal fluid, and darkening, enlargement and petechial haemorrhages in certain internal organs (Bretzinger *et al.*, 1999; Walster, 1999; Hedrick *et al.*, 2000).

The most prominent histopathological alterations that occur as a result of the disease have been observed in the skin, gills, kidneys, liver, gut, spleen, heart and brain of CyHV-3-infected fish (Hedrick *et al.*, 2000; Miyazaki *et al.*, 2008).

The histopathological changes have been observed in both gill lamellae and gill rakers. It has been reported that the affected gill lamellae exhibit hypertrophy, hyperplasia, nuclear degeneration of the branchial epithelium and complete loss of the gill lamellae architecture due to erosion, necrosis and fusion of the primary lamellae (Fig. 7.3) (Pikarsky *et al.*, 2004; Miyazaki *et al.*, 2008; Ouyang *et al.*, 2013). In gill rakers, increased subepithelial inflammation, infiltration of the inflammatory cells and congestion of the gill arch blood vessels, accompanied by a reduction in the length of the rakers, have been reported (Pikarsky *et al.*, 2004).

Obvious histopathological changes have also been observed in the kidney, beginning with weak peritubular inflammatory infiltrate, which, within a few days, changes to heavy interstitial inflammatory infiltrate, accompanied by congestion of blood vessels and degeneration of the tubular epithelium in several nephrons (Pikarsky *et al.*, 2004).

It has been reported that intranuclear inclusion bodies are evident mainly in the haematopoietic cells (Miyazaki *et al.*, 2008; Miwa *et al.*, 2015). Similarly, mild inflammatory infiltrates have been observed in the liver parenchyma, while certain

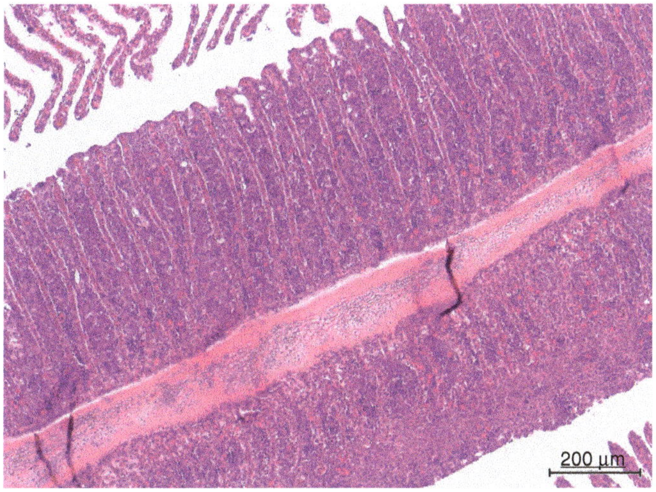

Fig. 7.3. *Cyprinus carpio* gill section stained with haematoxylin and eosin showing fusion and clubbing of the gill lamellae, interstitial oedema and necrosis as a result of KHVD. (Image from the archive of the Clinical Division of Fish Medicine.)

cases have exhibited large numbers of necrotic splenocytes along with haemorrhages in the spleen (Pikarsky *et al.*, 2004; Miyazaki *et al.*, 2008).

Alterations in the myofibril bundles, the disappearance of cross-striation and nuclear degeneration are the most prominent histopathological changes reported in the myocardial cells of the heart of CyHV3-infected fish (Miyazaki *et al.*, 2008).

CyHV-3 has been reported to cause hyperplasia of the stomach and intestinal epithelium, which consequently expresses intranuclear inclusion bodies that may detach from mucosa into the lumen of the organ (Mohi Eldin, 2011).

Furthermore, the brains of the infected fish demonstrating neurological symptoms have been reported to exhibit meningeal and parameningeal focal inflammation, oedematous dissociation of the nerve fibres, and congestion of the small veins and blood capillaries in the medulla oblongata and valvula cerebella (Miyazaki *et al.*, 2008; Miwa *et al.*, 2015).

Histopathological investigations of the skin of the infected fish revealed degeneration and necrosis in the epidermis as well as the presence of intranuclear inclusion bodies. Furthermore, goblet cells appeared slim and slender, empty of mucus, and their number was observed to be reduced by 50% in the infected fish in comparison to the control fish (Adamek *et al.*, 2013; Miwa *et al.*, 2015).

Confirmation of clinical diagnosis

Even in the modern era of molecular analyses, virus isolation remains the 'gold standard' for the detection of viruses in the important aquaculture species. The viruses exhibit cytopathic effects (CPEs) on tissue cultured cells and the virus identification may be confirmed by immunoassays and/or amplification of a specific gene fragment of the particular virus using PCR assays (Leland and Ginocchio, 2007; Crane and Hyatt, 2011; Munang'andu *et al.*, 2017). It is possible to isolate CyHV-3 using the common carp brain (CCB) or koi fin (KF-1) cell lines, as described in several previous studies (Neukirch *et al.*, 1999; Hedrick *et al.*, 2000; Neukirch and Kunz, 2001). Recently, novel cell lines derived from koi head kidney and brain have been developed and utilized for the isolation of CyHV-3 (Wang *et al.*, 2018; Eckart *et al.*, 2019). A variety of tissues (gill, skin, kidney, liver, spleen, intestine and encephalon) in different combinations have been utilized for the inoculation of cell cultures (Hedrick *et al.*, 2000; Neukirch and Kunz, 2001; Gilad *et al.*, 2002, 2003; Sano *et al.*, 2004). However, since the cell culture isolation of CyHV-3 is not as sensitive as the PCR-based methods, cell culture is not considered a reliable diagnostic method for the detection and isolation of CyHV-3 (Haenen *et al.*, 2004; OIE, 2018).

IMMUNOASSAYS FOR DETECTION OF CYHV-3. Immunoassay is one of the epidemiological tools employed for the screening of target populations of fish with the aim of prevention and control of the disease caused by CyHV-3. Serological assays are considered reliable screening methods as these assays allow non-lethal sampling, which is particularly beneficial in the case of expensive fish breeds (Haenen *et al.*, 2004). Enzyme-linked immunosorbent assays (ELISAs) have been utilized extensively for the detection of biological molecules such as antigens or antibodies in samples. In a previous study, the results of ELISA revealed an elevated titre of CyHV-3 antibodies at 14 days post-infection (dpi), which remained elevated up to 51 dpi, with peak titre observed at 21 dpi, in the fish that survived the CyHV-3 infection (Ronen *et al.*, 2003). Several ELISAs were developed that were able to detect the CyHV-3 antibodies for up to 1 year after the natural infection as well as experimental infection (Adkison *et al.*, 2005; St-Hilaire *et al.*, 2005, 2009). The disadvantage of the ELISAs previously used for CyHV-3 antibodies detection was the occurrence of cross-reactivity with CyHV-1, which could be overcome by the dilution of the serum (Haenen and Hedrick, 2006; St-Hilaire *et al.*, 2009). Another optimized CyHV-3 antibody detection ELISA with high diagnostic and analytical sensitivity and high specificity was developed to serve as a practical tool for indirect detection of CyHV-3 even in the absence of the clinical symptoms (Bergmann *et al.*, 2017a). Although the CyHV-3 antibodies detection ELISAs serve as valuable tools for establishing a previous exposure to CyHV-3, these assays are not able to determine whether the virus persisted in the infected fish. Therefore, an antigen-capture ELISA was developed for detection of the CyHV-3 antigen, with a detection limit below 10^4 virus particles/ml (Bergmann *et al.*, 2017b). Another CyHV-3 antigen detection ELISA was developed which could detect CyHV-3 in the fish droppings (Dishon *et al.*, 2005). As a result of insufficient knowledge regarding the serological responses of fish to viral infections, it has been recommended not to use immunoassays as the primary diagnostic tools or as routine screening methods for assessment of the status of CyHV-3 in fish populations.

MOLECULAR DETECTION OF CYHV-3 DNA. Various molecular assays have been developed for the amplification of CyHV-3 DNA directly from the infected tissues or the cell culture supernatants. The most prominent assay used for this purpose is the PCR assay. Several conventional and nested PCR assays have been developed and evaluated for the detection of the CyHV-3 DNA (Gilad *et al.*, 2002; Gray *et al.*, 2002; Bercovier *et al.*, 2005; Ishioka *et al.*, 2005; Yuasa *et al.*, 2005; Bergmann *et al.*, 2006, 2010b; El-Matbouli *et al.*, 2007b). Certain studies have compared the sensitivity of the PCR assays developed for the detection of CyHV-3 (Bergmann *et al.*, 2010b; Monaghan *et al.*, 2015). The conventional PCR that was based on the amplification of the thymidine kinase gene of CyHV-3 has been reported to be more sensitive in comparison to the other assays as it is able to detect CyHV-3 DNA quantities as little as 10 fg, which corresponds to just 30 CyHV-3 virions (Bercovier *et al.*, 2005). Moreover, a real-time PCR assay has also been developed for the quantitative detection of the CyHV-3 DNA (Gilad *et al.*, 2004). This real-time assay possesses the ability to detect considerably low copy numbers of the CyHV-3 DNA. Furthermore, loop-mediated isothermal amplification (LAMP) and recombinase polymerase amplification (RPA) assays were developed for the isothermal detection of the CyHV-3 DNA. These assays provided a sensitivity equivalent to that provided by certain conventional PCR assays (Gunimaladevi *et al.*, 2004; Soliman and El-Matbouli, 2005, 2009, 2010, 2018; Yoshino *et al.*, 2006, 2009; Prescott *et al.*, 2016). It has been suggested that the diagnosis of the KHV disease should not rely on just one test; instead, a combination of two or three tests must be applied (Haenen *et al.*, 2004). Based on recommendations provided by OIE (2018), the most reliable PCR assays for the detection of CyHV-3 are the conventional one-round assays developed by Bercovier *et al.* (2005) and Yuasa *et al.* (2005), followed by a sequence analysis of the PCR products and the quantitative real-time PCR assay developed by Gilad *et al.* (2004).

7.3 Potential Spread of CyHV-3

7.3.1 Spread of CyHV-3 to new geographical areas

The rapid worldwide spread of CyHV-3 has been attributed to intensive fish culture, domestic and international trading, and koi exhibitions prior to the current knowledge of the disease and the means to detect it (Gilad *et al.*, 2003). Only the horizontal transmission of CyHV-3 has been reported so far, while the vertical transmission of the virus has not been reported yet (Rakus *et al*., 2013). Horizontal transmission may occur directly through fish-to-fish contact, or indirectly through vector-based transmission. Skin-to-skin contact allows CyHV-3 to transmit through the skin, while cannibalistic and necrophagous behaviours of carp allow CyHV-3 to be transmitted from the infected or carrier fish to naïve fish through pharyngeal periodontal mucosa (Raj *et al*., 2011; Fournier *et al*., 2012). Virulent CyHV-3 strains may be shed through gills, skin mucus, urine and faeces into the water, which contains abiotic vectors that transmit CyHV-3 (Minamoto *et al*., 2009). Several fomites and animate vectors including piscivorous birds, fish droppings, aquatic invertebrates, planktons and sediments may be involved in the transmission of CyHV-3 (Dishon *et al*., 2005; Minamoto *et al*., 2009, 2011; Kielpinski *et al*., 2010; Ilouze *et al*., 2011; Honjo *et al*., 2012).

The disposal of infected fish through selling or through release of the infected fish into the ponds of public parks by certain persons, as happened in Indonesia and Taiwan, respectively, may also serve as a means for the spread of CyHV-3 (Tu *et al*., 2004; Sunarto *et al*., 2005).

7.3.2 Effect of water temperature on the spread of CyHV-3

Warm water temperatures have been reported to induce the CyHV-3-caused clinical disease more rapidly in comparison to low temperatures, which supports the fact that KHVD is temperature dependent. The lowest temperature associated with a CyHV-3 outbreak (15.5°C) was recorded in Japan; however; KHVD has been reported to occur at water temperatures ranging from 18 to 28°C in natural conditions, and at temperatures ranging from 16 to 28°C in experimental conditions (Gilad *et al*., 2003, 2004; Hara *et al*., 2006; Ilouze *et al*., 2006). Similarly, water temperature has been reported to affect the onset of mortality caused by CyHV-3 infection. When the infected fish were maintained in a temperature range of 23–28°C, it required 5–8 days post-exposure for the mortality to occur, while 14–21 days post-exposure were required for the mortality to occur when the fish were maintained at a temperature range of 16–18°C (Gilad *et al*., 2003; Yuasa *et al*., 2008). No mortalities were recorded at the temperature of 29 or 30°C. At 13°C, although no mortalities were recorded, CyHV-3 DNA was detected in the fish through PCR (Gilad *et al*., 2004; Ilouze *et al*., 2011). Similarly, 4 months prior to a CyHV-3 outbreak in a Japanese river, viral DNA was detected in the river water sample at a temperature ranging from 9 to 11°C (Haramoto *et al*., 2007). Additionally, the expression of certain CyHV-3 genes encoding structural and non-structural proteins was detected in fish maintained at low temperatures, although no infectious particles could be isolated (Gilad *et al*., 2004; Eide *et al*., 2011a; Sunarto *et al*., 2012, 2014; Baumer *et al*., 2013). Studies have also demonstrated that the transfer of CyHV-3-infected fish, which were earlier maintained at 13°C, to permissive temperatures led to the occurrence of KHVD, while the transfer of the CyHV-3-infected fish to 13 or 30°C reduced the mortality rate significantly (Gilad *et al*., 2003; Ronen *et al*., 2003; St-Hilaire *et al*., 2005, 2009; Eide *et al*., 2011b; Sunarto *et al*., 2014). These findings support the hypothesis that CyHV-3 is able to replicate at low temperatures without inducing mortalities, and that the infected fish which had survived at low temperatures may be serving as reservoirs for CyHV-3 (Gilad *et al*., 2004; Sunarto *et al*., 2012, 2014). CyHV-3 is able to remain infectious in water at temperatures ranging from 23 to 25°C for 4 h, but the virus is non-infectious after 21 h at these temperatures (Perelberg *et al*., 2003). The infectivity of CyHV-3 was reported to be significantly reduced in environmental water or sediment at 15°C within 3 days in the absence of a host (Shimizu *et al*., 2006). Temperature fluctuation, for example ±3°C/day, was reported to induce stress in fish due to increased release of cortisol and to increase the susceptibility and severity of the CyHV-3 infection (Takahara *et al*., 2014). Moreover, the water temperature has been reported to control the conversion from latency to lytic infection of CyHV-3, which consequently permits the virus to persist in the host for several seasons even at non-permissive temperatures for productive viral replication (Boutier *et al*., 2015a).

7.4 Control and Prevention

The control and prevention of KHVD are imperative as the disease raises a challenge to the sustainable development of koi and common carp industries. Different approaches have been reported to be utilized in order to prevent and control

KHVD and to provide disease resistance, such as biosecurity measures, environmental manipulation, chemotherapy, immunization, gene therapy and cross-breeding.

7.4.1 Biosecurity

CyHV-3 pathogen may spread to farms from multiple sources such as water, wild fish, newly introduced fish, contaminated equipment, predators and human visitors. Therefore, biosecurity measures should be undertaken ensuring implementation of stringent rules to prevent the spread of the CyHV-3 pathogen to farms or the transfer of the pathogen to an uninfected area. Farm-level biosecurity measures include egg disinfection, traffic control, water treatment, regular disinfection of the ponds, clean feed, hygienic disposal of the mortalities and strict quarantine measures. If new fish must be introduced, the source of such fish must be KHVD-free, and the new fish should be maintained with sentinel fish at a permissive temperature for KHVD, followed by conducting quarantine measures for 4 weeks to 2 months prior to mixing with the naïve fish. It has also been recommended to destroy the stocks of fish already infected with CyHV-3 and disinfect the farm and restock with fish that are free of CyHV-3.

7.4.2 Environmental manipulation

Environmental manipulation measures such as elevation of the water temperature have been utilized to control KHVD (Ronen *et al.*, 2003), resulting in a reduction in mortality rate of up to approximately 39% when healthy carp were exposed to CyHV-3 through cohabitation with the diseased fish at 22–23°C for 3–5 days followed by transferring to and maintaining the fish in ponds with water at a temperature of 30°C for 30 days. This approach has several disadvantages, the most important one being that these fish have been exposed to wild-type virulent CyHV-3 and would presumably serve as carrier fish, due to latent infection, if they encounter naïve carp (Michel *et al.*, 2010a).

7.4.3 Antiviral substances

The main obstacle in effective drug treatment of viral diseases is the strict parasitic relationship between the virus and the host which causes most antiviral agents to be toxic to the host as well. Targeting processes unique to virus replication, the ones which are not essential for host cell metabolism, has been suggested as the best strategy for developing antiviral drugs (Kibenge *et al.*, 2012). Antiviral drugs have been studied in order to identify *in vitro* activity against CyHV-3, although such drugs have not yet been commercially applied in aquaculture.

The antiviral activities of acyclovir (ACV) and acyclovir monophosphate (ACV-MP) against CyHV-3 have been studied *in vitro* using CCB and KF-1 cell lines (Troszok *et al.*, 2018). Using 66.67 μM of ACV and 66.67 μM of ACV-MP separately resulted in inhibition of the CPE of CyHV-3 by 66 and 58%, respectively, in CCB and by 25 and 37%, respectively, in KF-1. Moreover, the expression of viral genes (ORF149, ORF3, ORF134 and ORF78) in the CyHV-3-infected CCB cells was observed to be strongly downregulated (78–91%) in response to both ACV and ACV-MP, demonstrating the ability of these antiviral agents to inhibit the replication of CyHV-3 *in vitro*.

Other antiviral agents that were investigated in previous studies also demonstrated promising effects. The exopolysaccharides from *Arthrospira platensis* (>18 μg/ml) were reported to suppress viral replication in CCB cells relative to the commercially available antiviral drug Ganciclovir (Reichert *et al.*, 2017).

The crude extract of *Clinacanthus nutans* demonstrated effective antiviral activity against CyHV-3 both pre- and post-infection in koi carp (Haetrakul *et al.*, 2018). Similarly, koi carp fed with food containing 1.62% dry extract of lemon balm (*Melissa officinalis*) and subsequently challenged with CyHV-3 exhibited the lowest cumulative mortality (Haselmeyer *et al.*, 2018).

Since the clinical signs of KHVD appear after the occurrence of viral replication and pathology, utilizing antiviral agents against the infection in fish may be of value only as a prophylaxis or in anticipation of a stress that might reactivate the latent infection of the herpesvirus or depress the immune system and consequently increase the susceptibility to the viral infection (Goris *et al.*, 2008). So far, antiviral drugs have not been applied in aquaculture.

7.4.4 Gene therapy

Gene therapy is a promising treatment that utilizes genes to treat or prevent diseases. Several approaches to gene therapy have been investigated in different studies, including the inactivation or knockout of gene(s) that are important for replication or virulence of a pathogen. Gene silencing using RNA interference (RNAi) that utilizes small interfering RNAs (siRNAs) has been investigated as a potential antiviral tool. These siRNAs have been used for specifically targeting a homologous sequence for cleavage by cellular ribonucleases, through which they trigger a sequence-specific post-transcriptional gene silencing process (Hannon, 2002). The siRNAs were also used to inhibit the *in vitro* replication of CyHV-3 in CCB cells. Individually targeting the thymidine kinase or DNA polymerase genes of CyHV-3 using gene-specific siRNAs reduced the number of CyHV-3 particles released from the CCB cells; the siRNA targeting DNA polymerase was observed to be the most effective in reducing the viral release according to the measurements of quantitative real-time PCR (Gotesman *et al.*, 2014). In contrast, another study demonstrated that the treatment of CyHV-3-infected CCB cells with a mixture of siRNAs specific to the genes encoding CyHV-3 capsid triplex protein, major capsid protein, DNA helicase and DNA polymerase caused a significant reduction in the number of CCB cells that were killed during the process (survival: 79% of the host cells); however, the treatment of the cells with a single target siRNA did not cause a significant reduction in the number of cells killed (survival: 48–57% of the CCB cells) (Adamek *et al.*, 2014). All the siRNA investigations available in the literature were *in vitro* trials, and no *in vivo* investigations have been reported to date.

Unlike RNAi, which requires the continued presence of effector moieties to maintain gene silencing, just a single treatment with clustered regularly interspaced short palindromic repeats (CRISPR)/Cas9 is required for permanent disruption of the targeted genes (Manjunath *et al.*, 2013). The potential of CRISPR/Cas9 to inhibit CyHV-3 replication has been investigated by Zhao *et al.* (2016), in which two guide RNAs targeting the thymidine kinase (TK) and DNA polymerase (DP) genes of CyHV-3 were designed and used to establish stable KF-1 cell lines expressing Cas9 along with one of the following: TK-guide RNA, DP-guide RNA or both of these RNAs together. CyHV-3 proliferation was observed to be reduced in all three CRISPR/Cas9-positive KF-1 cells by 65–80% in comparison to the control cells. Moreover, the antiviral effect of the CRISPR/Cas9-positive KF-1 cells was stable and significant. No investigations using CRISPR/Cas9 *in vivo* have been reported to date. The progress in gene silencing and gene editing technologies may lead the way for developing effective gene therapies against CyHV-3 in the future.

7.4.5 Vaccination

Sustainable development in aquaculture is considered equivalent to disease prevention, and vaccination has become the single most important tool for achieving this (Evensen, 2009; Plant and LaPatra, 2011). With the expansive development of aquaculture, much effort has been dedicated to vaccine development for stimulation of the fish immune system to assist in fighting against diseases. Various immunization strategies have been applied to prevent CyHV-3 infection. Subsequent to the development of attenuated CyHV-3 through a 20-serial passage in KF-1 cells along with ultraviolet irradiation, fish were immunized through immersion in water containing the attenuated CyHV-3 at 10–100 plague-forming units/ml for 40 min, followed by incubation at permissive temperatures for 48–72 h (Ronen *et al.*, 2003; Perelberg *et al.*, 2005). The titre of CyHV-3-specific antibodies increased 7 days post-immunization, its peak was observed at 21 days post-immunization and this remained at high levels for 56 days (Ronen *et al.*, 2003). The use of this attenuated vaccine was restricted as it raised a risk of residual virulence in fish that weighed less than 50 g, in addition to a reversion to pathogenic wild type which could not be excluded (Meeusen *et al.*, 2007; Zak *et al.*, 2007; Weber *et al.*, 2014).

A recombinant attenuated vaccine against CyHV-3 was developed by deleting ORF56 and ORF57. This vaccine induced a protective mucosal immune response capable of preventing entry of the virulent CyHV-3 pathogen (Boutier *et al.*, 2015b). In addition, some DNA vaccines based on ORF25 and ORF81 were developed and reported to demonstrate different degrees of protection against CyHV-3 (Zhou *et al.*, 2014a,b; Liu *et al.*, 2018; Embregts *et al.*, 2019). Furthermore, oral immunization of carp with formalin-inactivated CyHV-3 trapped within liposomes and mixed with food caused a reduction of 70% in the mortality induced by viral challenge (Yasumoto *et al.*, 2006).

7.4.6 Prospects for production of CyHV-3-resistant carp

So far, vaccines against CyHV-3 have not performed well, and antiviral drugs have not yet been applied successfully in aquaculture. It is anticipated that developing fish strains that are genetically resistant to CyHV-3 would serve as a sustainable solution to this problem.

Cross-breeding of the sensitive domesticated carp strains with resistant wild carp strains, followed by exposure of the progeny to CyHV-3, resulted in mortalities ranging from 8.9 to 89.7% and from 8.6 to 68.5% in the field trial and laboratory trial of CyHV-3 exposures, respectively (Shapira *et al.*, 2005). Furthermore, the survival rates post challenge with CyHV-3 in the progeny produced from breeding three hybrids and three pure crosses from two Hungarian strains and one Israeli strain of carp were 4–20% (Zak *et al.*, 2007). Similarly, investigations involving di-allelic cross-breeding of 96 carp families derived from two wild carp strains native to the Amur and Danube rivers and two domesticated strains from Tat and Szarvas, Hungary, demonstrated that the most resistant carp families were obtained from the wild-type strains (Dixon *et al.*, 2009; Ødegård *et al.*, 2010). In addition, most of the Czech cross-breeds and strains that are genetically related to the wild Amur carp were reported to be significantly more resistant to the CyHV-3 infection compared with the other strains that were not related to the Amur carp (Piackova *et al.*, 2013). CyHV-3 DNA was detected by PCR in the surviving hybrids of male goldfish and female common carp which were observed to be resistant to KHVD, suggesting that these hybrids were potential CyHV-3 carriers (Hedrick *et al.*, 2006). Similarly, hybrids of koi carp and goldfish exhibited partial resistance to KHVD, marked by reduced mortality (35–42%) following the infection with CyHV-3 by an immersion process (Bergmann *et al.*, 2010a). Furthermore, CyHV-3 resistance was introgressed from the CyHV-3-resistant strain (Amur Sassn) into two susceptible cultured strains, with the variation in the resistance of families from F1 and first backcross generations being significantly greater than that among the families of any of the susceptible parent lines (Tadmor-Levi *et al.*, 2017).

7.5 Conclusions

Alloherpesviridae is one among three families of herpesviruses that constitute the order *Herpesvirales*. This family contains 12 species distributed into four genera. Genus *Cyprinivirus* contains four species, three of which, namely *Cyprinid herpesvirus 1* (CyHV-1), *Cyprinid herpesvirus 2* (CyHV-2) and *Cyprinid herpesvirus 3* (CyHV-3), cause significant economic losses in the rearing of cyprinids. The most important species among these three species is CyHV-3, which is the causative agent of KHVD, is a contagious viral disease that has exerted a severe economic impact on the koi carp and common carp industries. The isolation of CyHV-3 and complete sequencing of its genome has assisted in the accurate characterization and taxonomic identification of this virus. The mortality and morbidity of CyHV-3 are restricted to common carp and koi carp. Although the DNA of CyHV-3 has been detected in other fish species as well, such as goldfish, no clinical symptoms or mortalities have been reported in these species; as a result, such species may be considered carriers of CyHV-3, which may disseminate this virus everywhere. After the isolation and identification of CyHV-3 in 2000, the virus has spread rapidly and been reported from many countries. Generally, KHVD has been reported after the introduction of new fish into a naïve population without prior quarantine and the occurrence of sudden death in the fish population with or without apparent clinical symptoms such as gill necrosis, excessive mucus and other septicaemic conditions. Various sensitive diagnostic assays have been developed for the detection and screening of CyHV-3. While no antiviral drug has been applied in the field to control KHVD, there is also no effective vaccine available that could be used to prevent the disease. Several trials have been performed for the selection and breeding of a CyHV-3-resistant carp strain, although no such carp strain that has genetic resistance to CyHV-3 has been successfully selected to date.

Disease is the result of interaction among host, pathogen and environment. Climate change may alter the host–pathogen relationship by changing pathogen transmission rates, increasing pathogen survival in the environment, extending carrier host species' range or changing host susceptibility to disease. Climate change may affect fish, as poikilothermic animals that are directly affected by the ambient temperature, more severely than other homeothermic animals. Similarly, CyHV-3 also has an optimal temperature range for replication. Temperature is the key determinant shifting the outcome of CyHV-3 infection towards disease and mortality or immunity and recovery. Changes in weather and climate can change the distribution, by elongating the periods of transmission, and timing of KHVD outbreaks, as they can start earlier and end later in the year. Accordingly, KHVD is likely to become more widely spread and outbreaks will be observed over a greater period of the year. Climate change can also enable the CyHV-3 to establish new host species. Likewise, climate change might affect disease transmission through increases or decreases in host susceptibility. Fish immune system is optimal at normal summer temperature for each species. Water temperature above or below the physiological optimum may cause stress, increasing susceptibility to infection and the likelihood of disease and mortality. It can be assumed that warming will decrease immune function because cellular and humoral immune defences are difficult to maintain and can collapse under thermal stress, accordingly fish become highly susceptible to KHVD.

Further investigations on CyHV-3 are required in order to identify the biological functions of the virus ORFs, to determine the virulence genes, to evaluate the virogenesis of this virus in the carrier non-susceptible hosts (such as goldfish) and to understand the mechanisms involved in the regulation of latency of the virus.

Additionally, antiviral agents and gene therapy approaches should be investigated to develop safe and effective control of CyHV-3. The developed agents should be assessed *in vitro* and *in vivo* to demonstrate their effectiveness in reducing mortalities and controlling KHVD.

Moreover, advancing vaccines is one of the most important approaches to prevent infectious aquaculture diseases. More research should be directed to develop next-generation vaccines against CyHV-3 that can induce a specific long-term protection and should have the advantage of safety, ease of production and administration. New adjuvants should be investigated to enhance immunogenicity of the CyHV-3 vaccines.

It is speculated that the sustainable way to prevent and control KHVD is the development of CyHV-3-resistant carp strains. Accordingly, new biotechnological approaches should be applied to understand the genetic resistance of carp to diseases and elucidate suitable genetic criteria that help in the selection of CyHV-3-resistant carp strains.

References

Adamek, M., Syakuri, H., Harris, S., Rakus, K.L., Brogden, G. *et al.* (2013) Cyprinid herpesvirus 3 infection disrupts the skin barrier of common carp (*Cyprinus carpio* L.). *Veterinary Microbiology* 162, 456–470.

Adamek, M., Rauch, G., Brogden, G. and Steinhagen, D. (2014) Small interfering RNA treatment can inhibit Cyprinid herpesvirus 3 associated cell death *in vitro*. *Polish Journal of Veterinary Sciences* 17, 733–735.

Adkison, M.A., Gilad, O. and Hedrick, R.P. (2005) An enzyme linked immunosorbent assay (ELISA) for detection of antibodies to the koi herpesvirus (KHV) in the serum of koi *Cyprinus carpio*. *Fish Pathology* 40, 53–62.

Aoki, T., Hirono, I., Kurokawa, K., Fukuda, H., Nahary, R. *et al.* (2007) Genome sequences of three koi herpesvirus isolates representing the expanding distribution of an emerging disease threatening koi and common carp worldwide. *Journal of Virology* 81, 5058–5065.

Baumer, A., Fabian, M., Wilkens, M.R., Steinhagen, D. and Runge, M. (2013) Epidemiology of cyprinid herpesvirus-3 infection in latently infected carp from aquaculture. *Diseases of Aquatic Organisms* 105, 101–108.

Bercovier, H., Fishman, Y., Nahary, R., Sinai, S., Zlotkin, A. *et al.* (2005) Cloning of the koi herpesvirus (KHV) gene encoding thymidine kinase and its use for a highly sensitive PCR based diagnosis. *BMC Microbiology* 5, 13.

Bergmann, S.M., Kempter, J., Sadowski, J. and Fichtner, D. (2006) First detection, confirmation and isolation of koi herpesvirus (KHV) in cultured common carp (*Cyprinus carpio* L.) in Poland. *Bulletin of the European Association of Fish Pathologists* 26, 97–104.

Bergmann, S.M., Schütze, H., Fischer, U., Fichtner, D., Riechardt, M. *et al.* (2009) Detection KHV genome in apparently healthy fish. *Bulletin of the European Association of Fish Pathologists* 29, 145–152.

Bergmann, S.M., Sadowski, J., Kielpinski, M., Bartlomiejczyk, M., Fichtner, D. *et al.* (2010a) Susceptibility of koi × crucian carp and koi × goldfish hybrids to koi herpesvirus (KHV) and the development of KHV disease (KHVD). *Journal of Fish Diseases* 33, 267–272.

Bergmann, S.M., Riechardt, M., Fichtner, D., Lee, P. and Kempter, J. (2010b) Investigation on the diagnostic sensitivity of molecular tools used for detection of koi herpesvirus. *Journal of Virological Methods* 163, 229–233.

Bergmann, S.M., Wang, Q., Zeng, W., Li, Y., Wang, Y. *et al.* (2017a) Validation of a KHV antibody enzyme-linked immunosorbent assay (ELISA). *Journal of Fish Diseases* 40, 1511–1527.

Bergmann, S.M., Engler, Ch., Wang, Q., Zeng, W., Li, Y. *et al.* (2017b) Investigation on antigen ELISA for detection of envelope glycoprotein coded by ORF 149 of different koi herpesvirus isolates obtained from cell culture. *Journal of Veterinary Science and Medicine* 5, 7.

Bigarré, L., Baud, M., Cabon, J., Antychowicz, J., Bergmann, S.M. *et al.* (2009) Differentiation between cyprinid herpesvirus type-3 lineages using duplex PCR. *Journal of Virological Methods* 158, 51–57.

Boutier, M., Ronsmans, M., Rakus, K., Jazowiecka-Rakus, J., Vancsok, C. *et al.* (2015a) Cyprinid herpesvirus 3: an archetype of fish alloherpesviruses. *Advances in Virus Research* 93, 161–256.

Boutier, M., Ronsmans, M., Ouyang, P., Fournier, G., Reschner, A. *et al.* (2015b) Rational development of an attenuated recombinant cyprinid herpesvirus 3 vaccine using prokaryotic mutagenesis and *in vivo* bioluminescent imaging. *PLoS Pathogens* 11, e1004690.

Bretzinger, A., Fischer-Scherl, T., Oumouna, M., Hoffmann, R. and Truyen, U. (1999) Mass mortalities in koi carp, *Cyprinus carpio*, associated with gill and skin disease. *Bulletin of the European Association of Fish Pathologists* 19, 182–185.

Crane, M. and Hyatt A. (2011) Viruses of fish: an overview of significant pathogens. *Viruses* 3, 2025–2046.

Danek, T., Kalous, L., Vesely, T., Krasova, E., Reschova, S. *et al.* (2012) Massive mortality of Prussian carp *Carassius gibelio* in the upper Elbe basin associated with herpesviral hematopoietic necrosis (CyHV-2). *Diseases of Aquatic Organisms* 102, 87–95.

Davison, A.J. (2002) Evolution of the herpesviruses. *Veterinary Microbiology* 86, 69–88.

Davison, A.J. (2010) Herpesvirus systematics. *Veterinary Microbiology* 143, 52–69.

Davison, A.J., Eberle, R., Hayward, G.S., McGeoch, D.J., Minson, A.C. *et al.* (2005) *Herpesviridae*. In: Fauquet, C.M., Mayo, M.A., Maniloff, J., Desselberger, U. and Ball, L.A. (eds) *Virus Taxonomy: VIIIth Report of the International Committee on Taxonomy of Viruses*. Elsevier Academic Press, Amsterdam, pp. 193–212.

Davison, A.J., Eberle, R., Ehlers, B., Hayward, G.S., McGeoch, D.J. *et al.* (2009) The order *Herpesvirales*. *Archives of Virology* 154, 171–177.

Davison, A., Kurobe, T., Gatherer, D., Cunningham, C., Korf, I. *et al.* (2013) Comparative genomics of carp herpesviruses. *Journal of Virology* 87, 2908–2922.

Denham, K.L. (2003) Koi herpesvirus in wild fish. *Veterinary Record* 153, 507.

Dishon, A., Perelberg, A., Bishara-Shieban, J., Ilouze, M., Davidovich, M. *et al.* (2005) Detection of carp interstitial nephritis and gill necrosis virus in fish droppings. *Applied and Environmental Microbiology* 71, 7285–7291.

Dixon, P.F., Joiner, C.L., Way, K., Reese, R.A., Jeney, G. and Jeney, Z. (2009) Comparison of the resistance of selected families of common carp, *Cyprinus carpio* L., to koi herpesvirus: preliminary study. *Journal of Fish Diseases* 32, 1035–1039.

Eckart, V., Yamaguchi, T., Franzke, K., Bergmann, S.M., Boudinot, P. *et al.* (2019) New cell lines for efficient propagation of koi herpesvirus and infectious salmon anaemia virus. *Journal of Fish Diseases* 42, 181–187.

Eide, K.E., Miller-Morgan, T., Heidel, J.R., Bildfell, R.J. and Jin, L. (2011a) Results of total DNA measurement in koi tissue by koi herpes virus real-time PCR. *Journal of Virological Methods* 172, 81–84.

Eide, K.E., Miller-Morgan, T., Heidel, J.R., Kent, M.L., Bildfell, R.J. *et al.* (2011b) Investigation of koi herpesvirus latency in koi. *Journal of Virology* 85, 4954–4962.

El-Matbouli, M. and Soliman, H. (2011) Transmission of cyprinid herpesvirus-3 (CyHV-3) from goldfish to naive common carp by cohabitation. *Research in Veterinary Science* 90, 536–539.

El-Matbouli, M., Saleh, M. and Soliman, H. (2007a) Detection of cyprinid herpesvirus type 3 in goldfish cohabiting with CyHV-3-infected koi carp (*Cyprinus carpio koi*). *Veterinary Record* 161, 792–793.

El-Matbouli, M., Rucker, U. and Soliman, H. (2007b) Detection of cyprinid herpesvirus-3 (CyHV-3) DNA in infected fish tissues by nested polymerase chain reaction. *Diseases of Aquatic Organisms* 78, 23–28.

Embregts, C.W.E., Tadmor-Levi, R., Vesely, T., Pokorova, D., David, L. *et al.* (2019) Intra-muscular and oral vaccination using a koi herpesvirus ORF25 DNA vaccine does not confer protection in common carp (*Cyprinus carpio* L.). *Fish and Shellfish Immunology* 85, 90–98.

Engelsma, M.Y., Way, K., Dodge, M.J., Voorbergen-Laarman, M., Panzarin, V. *et al.* (2013) Detection of novel strains of cyprinid herpesvirus closely related to koi herpesvirus. *Diseases of Aquatic Organisms* 107, 113–120.

Evensen, O. (2009) Development in fish vaccinology with focus on delivery methodologies, adjuvants and formulations. In: Rogers, C. and Basurco, B. (eds) The use of veterinary drugs and vaccines in Mediterranean aquaculture. *Options Méditerranéennes: Série A. Séminaires Méditerranéens 86*. CIHEAM, Zaragoza, Spain, pp. 177–186.

Fabian, M., Baumer, A., and Steinhagen, D. (2013) Do wild fish species contribute to the transmission of koi herpesvirus to carp in hatchery ponds? *Journal of Fish Diseases* 36, 505–514.

Fournier, G., Boutier, M., Stalin Raj, V., Mast, J., Parmentier, E. *et al.* (2012) Feeding *Cyprinus carpio* with infectious materials mediates cyprinid herpesvirus 3 entry through infection of pharyngeal periodontal mucosa. *Veterinary Research* 43, 6.

Garver, K.A., Al-Hussinee, L., Hawley, L.M., Schroeder, T., Edes, S. *et al.* (2010) Mass mortality associated with koi herpesvirus in wild common carp in Canada. *Journal of Wildlife Diseases* 46, 1242–1251.

Gessner, C. (1558) *Historiae Animalium*, Vol. IV. C. Froschauer, Zurich, Switzerland.

Gilad, O., Yun, S., Andree, K.B., Adkison, M.A., Zlotkin, A. *et al.* (2002) Initial characteristics of koi herpesvirus and development of a polymerase chain reaction assay to detect the virus in koi, *Cyprinus carpio* koi. *Diseases of Aquatic Organisms* 48, 101–108.

Gilad, O., Yun, S., Andree, K.B., Adkison, M.A., Way, K. *et al.* (2003) Molecular comparison of isolates of an emerging fish pathogen, the koi herpesvirus, and the effect of water temperature on mortality of experimentally infected koi. *Journal of General Virology* 84, 1–8.

Gilad, O., Yun, S., Zagmutt-Vergara, F.J., Leutenegger, C.M., Bercovier, H. and Hedrick, R.P. (2004) Concentrations of a koi herpesvirus (KHV) in tissues of experimentally-infected *Cyprinus carpio* koi as assessed by real-time TaqMan PCR. *Diseases of Aquatic Organisms* 60, 179–187.

Goris, N., Vandenbussche, F. and De Clercq, K. (2008) Potential of antiviral therapy and prophylaxis for controlling RNA viral infections of livestock. *Antiviral Research* 78, 170–178.

Gotesman, M., Kattlun, J., Bergmann, S.B. and El-Matbouli, M. (2013) CyHV-3: the third cyprinid herpesvirus. *Diseases of Aquatic Organisms* 105, 163–174.

Gotesman, M., Soliman, H., Besch, R. and El-Matbouli, M. (2014) *In vitro* inhibition of cyprinid herpesvirus-3 replication by RNAi. *Journal of Virological Methods* 206, 63–66.

Gray, W., Mullis, L., Lapatra, S., Groff, J. and Goodwin, A. (2002) Detection of koi herpesvirus DNA in tissue of infected fish. *Journal of Fish Diseases* 25, 171–178.

Grimmett, S.G., Warg, J.V., Getchell, R.G., Johnson, D.J. and Bowser, P.R. (2006) An unusual koi herpesvirus associated with a mortality event of common carp (*Cyprinus carpio*) in New York State, USA. *Journal of Wildlife Diseases* 42, 658–662.

Gunimaladevi, I., Kono, T., Venugopal, M.N. and Sakai, M. (2004) Detection of koi herpesvirus in common carp, *Cyprinus carpio* L., by loop-mediated isothermal amplification. *Journal of Fish Diseases* 27, 583–589.

Haenen, O. and Hedrick, R. (2006) Koi herpesvirus workshop. *Bulletin of the European Association of Fish Pathologists* 26, 26–37

Haenen, O.L.M., Way, K., Bergmann, S.M. and Ariel, E. (2004) The emergence of koi herpesvirus and its significance to European to aquaculture. *Bulletin of the European Association of Fish Pathologists* 24, 293–307.

Haetrakul, T., Dunbar, S.G. and Chansue, N. (2018) Antiviral activities of *Clinacanthus nutans* (Burm.f.) Lindau extract against cyprinid herpesvirus 3 in koi (*Cyprinus carpio koi*). *Journal of Fish Diseases* 41, 581–587.

Hannon, G.J. (2002) RNA interference. *Nature* 418, 244–251.

Hanson, H., Dishon, A. and Kotler, M. (2011) Herpesviruses that infect fish. *Viruses* 3, 2160–2191.

Hara, H., Aikawa, H., Usui, K. and Nakanishi, T. (2006) Outbreaks of koi herpesvirus disease in rivers of Kanagawa Prefecture. *Fish Pathology* 41, 81–83.

Haramoto, E., Kitajima, M., Katayama, H. and Ohgaki, S. (2007) Detection of koi herpesvirus DNA in river water in Japan. *Journal of Fish Diseases* 30, 59–61.

Haselmeyer, A., Nowotny, N., Heistinger, H., Kolodziejek, J., Homola, J. *et al.* (2018) *Melissa officinalis* L. extract and its main phenolic compound rosmarinic acid as phytoprophylactic feed additives against koi herpesvirus infection in a pilot study. *Wiener Tierärztliche Monatsshrift* 105, 175–183.

Hedrick, R.P. and Yun, S. (2005) Initial isolation and characterization of a herpes-like virus (KHV) from koi and common carp. *Special issue: International symposium on koi herpesvirus disease. Bulletin of Fish Research Agency* 2, 1–7.

Hedrick, R.P., Gilad, O., Yun, S., Spangenberg, J.V., Marty, G.D. *et al.* (2000) A herpesvirus associated with mass mortality of juvenile and adult koi, a strain of a common carp. *Journal of Aquatic Animal Health* 12, 44–57.

Hedrick, R.P., Waltzek, T.B. and McDowell, T.S. (2006) Susceptibility of koi carp, common carp, goldfish and goldfish × common carp hybrids to cyprinid herpesvirus-2 and herpesvirus-3. *Journal of Aquatic Animal Health* 18, 26–34.

Hofer, B. (1904) *Handbuch der Fischkrankheiten*. Verlag der Allgemeinen Fischerei-Zeitung, B. Heller, Munich, Germany.

Honjo, M.N., Minamoto, T. and Kawabata, Z. (2012) Reservoirs of cyprinid herpesvirus 3 (CyHV-3) DNA in sediments of natural lakes and ponds. *Veterinary Microbiology* 155, 183–190.

Ilouze, M., Dishon, A., Kahan, T. and Kotler, M. (2006) Cyprinid herpes virus-3 (CyHV-3) bears genes of genetically distant large DNA viruses. *FEBS Letters* 580, 4473–4478.

Ilouze, M., Davidovich, M., Diamant, A., Kotler, M. and Dishon, A. (2011) The outbreak of carp disease caused by CyHV-3 as a model for new emerging viral diseases in aquaculture: a review. *Ecological Research* 26, 885–892.

Ishioka, T., Yoshizumi, M., Izumi, S., Suzuki, K., Suzuki, H. *et al.* (2005) Detection and sequence analysis of DNA polymerase and major envelope protein genes in koi herpesviruses derived from *Cyprinus carpio* in Gunma Prefecture, Japan. *Veterinary Microbiology* 110, 27–33.

Ito, T., Sano, M., Kurita, J., Yuasa, K. and Iida, T. (2007) Carp larvae are not susceptible to koi herpesvirus. *Fish Pathology* 42, 107–109.

Jung, S.J., and Miyazaki, T. (1995) Herpesviral haematopoietic necrosis of goldfish, *Carassius auratus* (L.). *Journal of Fish Diseases* 18, 211–220.

Kempter, J. and Bergmann, S.M. (2007) Detection of koi herpesvirus (KHV) genome in wild and farmed fish from northern Poland. *Aquaculture* 272, S275–S275.

Kempter, J., Kielpinski, M., Panicz, R. and Sadowski, J. (2008) *Determination of the Carrier Species and the Infection Susceptibility for Koi Herpes Virus Infections among Selected Cyprinid Fish Species and Their Hybrids Acquired from Open Waters and Culture Sites of the Odra River Drainage*. Publishers of Scientific Agricultural University in Szczecin, Szczecin, Poland.

Kempter, J., Sadowski, J., Schütze, H., Fischer, U., Dauber, M. *et al.* (2009) Koi herpes virus: do acipenserid restitution programs pose a threat to carp farms in the disease-free zones? *Acta Ichthyologica et Piscatoria* 39, 119–126.

Kempter, J., Kielpinski, M., Panicz, R., Sadowski, J., Mysłowski, B. and Bergmann, S.M. (2012) Horizontal transmission of koi herpes virus (KHV) from potential vector species to common carp. *Bulletin of the European Association of Fish Pathologists* 32, 212–219.

Kibenge, F.S.B., Godoy, M.G., Fast, M., Workenhe, S. and Kibenge, M.J.T. (2012) Countermeasures against viral diseases of farmed fish. *Antiviral Research* 95, 257–281.

Kielpinski, M., Kempter, J., Panicz, R., Sadowski, J., Schütze, H. *et al.* (2010) Detection of KHV in freshwater mussels and crustaceans from ponds with KHV history in common carp (*Cyprinus carpio*). *Israeli Journal of Aquaculture – Bamidgeh* 62, 28–37.

Láng, M., Glávits, R., Papp, M., Paulus, P., György, T.Á. and Dán, A. (2014) First detection of koi herpesvirus (KHV) disease in Hungary. *Magyar Allatorvosok Lapja* 136, 721–727. (in Hungarian)

Leland, D.S. and Ginocchio, C.C. (2007) Role of cell culture for virus detection in the age of technology. *Clinical Microbiology Reviews* 20, 49–78.

Liu, L., Gao, S., Luan, W., Zhou, J. and Wang, H. (2018) Generation and functional evaluation of a DNA vaccine co-expressing cyprinid herpesvirus-3 envelope protein and carp interleukin-1β. *Fish and Shellfish Immunology* 80, 223–231.

Manjunath, N., Yi, G., Dang, Y. and Shankar, P. (2013) Newer gene editing technologies toward HIV gene therapy. *Viruses* 5, 2748–2766.

Mawdesley-Thomas, L. and Bucke, D. (1967) Fish pox in the roach (*Rutilus rutilus* L.). *Veterinary Record* 81, 56–57.

McDermott, C. and Palmeiro, B. (2013) Selected emerging infectious diseases of ornamental fish. *Veterinary Clinics of North America: Exotic Animal Practice* 16, 261–282.

McGeoch, D.J., Rixon, F.J. and Davison, A.J. (2006) Topics in herpesviruses genomics and evolution. *Virus Research* 117, 90–104.

McGeoch, D.J., Davison, A.J., Dolan, A., Gatherer, D. and Sevilla-Reyes, E. (2008) Molecular evolution of the *Herpesvirales*. In: Domingo, E., Parrish, C.R. and Holland, J.J. (eds) *Origin and Evolution of Viruses*, 2nd edn. Elsevier, Oxford, pp. 447–475.

McVeigh, S. (2004) KHV identified in South Africa. *Fish Farming International* 31, 27.

Meeusen, E.N., Walker, J., Peters, A., Pastoret, P.P. and Jungersen, G. (2007) Current status of veterinary vaccines. *Clinical Microbiology Reviews* 20, 489–510.

Mettenleiter, T.C., Keil, G.M. and Fuchs, W. (2008) Molecular biology of animal herpesviruses. In: Mettenleiter, T.C. and Sobrino, F. (eds) *Animal Viruses: Molecular Biology*. Caister Academic Press, Norfolk, UK, pp. 375–456.

Mettenleiter, T.C., Klupp, B.G. and Granzow, H. (2009) Herpesvirus assembly: an update. *Virus Research* 143, 222–234.

Michel, B., Fournier, G., Lieffrig, F., Costes, B. and Vanderplasschen, A. (2010a) Cyprinid herpesvirus 3. *Emerging Infectious Diseases* 16, 1835–1843.

Michel, B., Leroy, B., Stalin Raj, V., Lieffrig, F., Mast, J. *et al.* (2010) The genome of cyprinid herpesvirus 3 encodes 40 proteins incorporated in mature virions. *Journal of General Virology* 91, 452–462.

Minamoto, T., Honjo, M.N., Uchii, K., Yamanaka, H., Suzuki, A.A. *et al.* (2009) Detection of cyprinid herpesvirus 3 DNA in river water during and after an outbreak. *Veterinary Microbiology* 135, 261–266.

Minamoto, T., Honjo, M.N., Yamanaka, H., Tanaka, N., Itayama, T. and Kawabata, Z. (2011) Detection of cyprinid herpesvirus-3 DNA in lake plankton. *Research in Veterinary Science* 90, 530–532.

Minamoto, T., Honjo, M.N., Yamanaka, H., Uchii, K. and Kawabata, Z. (2012) Nationwide cyprinid herpesvirus 3 contamination in natural rivers of Japan. *Research in Veterinary Science* 93, 508–514.

Minson, A.C., Davison, A., Eberle, R., Desrosiers, R.C., Fleckenstein, B. *et al.* (2000) Family *Herpesviridae*. In: Van Regenmortel, M.H.V., Fauquet, C.M., Bishop, D.H.L., Carstens, E.B., Estes, M.K. *et al.* (eds) *Virus Taxonomy*. Academic Press, San Diego, California, pp. 203–225.

Miwa, S., Ito, T. and Sano, M. (2007) Morphogenesis of koi herpesvirus observed by electron microscopy. *Journal of Fish Diseases* 30, 715–722.

Miwa, S., Kiryu, I., Yuasa, K., Ito, T. and Kaneko, T. (2015) Pathogenesis of acute and chronic diseases caused by cyprinid herpesvirus-3. *Journal of Fish Diseases* 38, 695–712.

Miyazaki, T., Kuzuya, Y., Yasumoto, S., Yasuda, M. and Kobayashi, T. (2008) Histopathological and ultrastructural features of koi herpesvirus (KHV)-infected carp *Cyprinus carpio*, and the morphology and morphogenesis of KHV. *Diseases of Aquatic Organisms* 80, 1–11.

Mohi Eldin, M.M. (2011) Histopathological studies in experimentally infected koi carp (*Cyprinus carpio koi*) with koi herpesvirus in Japan. *World Journal of Fish and Marine Science* 3, 252–259.

Monaghan, S.J., Thompson, K.D., Adams, A. and Bergmann, S.M. (2015) Sensitivity of seven PCRs for early detection of koi herpesvirus in experimentally infected carp, *Cyprinus carpio* L., by lethal and non-lethal sampling methods. *Journal of Fish Diseases* 38, 303–319.

Munang'andu, H.M., Mugimba, K.K., Byarugaba, D.K., Mutoloki, S. and Evensen, Ø. (2017) Current advances on virus discovery and diagnostic role of viral metagenomics in aquatic organisms. *Frontiers in Microbiology* 8, 406.

Neukirch, M. and Kunz, U. (2001) Isolation and preliminary characterization of several viruses from koi (*Cyprinus carpio*) suffering gill necrosis and mortality. *Bulletin of the European Association of Fish Pathologists* 21, 125–135.

Neukirch, M., Böttcher, K. and Bunnajrakul, S. (1999) Isolation of a virus from koi with altered gills. *Bulletin of the European Association of Fish Pathologists* 19, 221–224.

Novotny, L., Pokorova, D., Reschova, S., Vicenova, M., Axmann, R. *et al.* (2010) First clinically apparent koi herpesvirus infection in the Czech Republic. *Bulletin of the European Association of Fish Pathologists* 30, 85–91.

Ødegård, J., Olesen, I., Dixon, P., Jeney, Z., Nielsen, H.-M. *et al.* (2010) Genetic analysis of common carp (*Cyprinus carpio*) strains II: Resistance to koi herpesvirus and *Aeromonas hydrophila* and their relationship with pond survival. *Aquaculture* 304, 7–13.

OIE (World Organization for Animal Health) (2018) Koi herpesvirus disease. In: Manual of Diagnostic Tests for Aquatic Animals. Available at: http://www.oie.int/index.php?id=2439&L=0&htmfile=chapitre_koi_herpesvirus.htm (accessed 9 March 2020).

Oh, M.J., Jung, S.J., Choi, T.J., Kim, H.R., Rajendran, K.V. *et al.* (2001) A viral disease occurring in cultured carp *Cyprinus carpio* in Korea. *Fish Pathology* 36, 147–151.

Ouyang, P., Rakus, K., Boutier, M., Reschner, A., Leroy, B. *et al.* (2013) The IL-10 homologue encoded by cyprinid herpesvirus 3 is essential neither for viral replication *in vitro* nor for virulence *in vivo*. *Veterinary Research* 44, 53.

Perelberg, A., Smirnov, M., Hutoran, M., Diamant, A., Bejerano, I. and Kotler, M. (2003) Epidemiological description of a new viral disease afflicting cultured *Cyprinus carpio* in Israel. *Israeli Journal of Aquaculture – Bamidgeh* 55, 5–12.

Perelberg, A., Ronen, A., Hutoran, M., Smith, Y. and Kotler, M. (2005) Protection of cultured *Cyprinus carpio* against a lethal viral disease by an attenuated virus vaccine. *Vaccine* 23, 3396–3403.

Piackova, V., Flajshans, M., Pokorova, D., Reschova, S., Gela, D. *et al.* (2013) Sensitivity of common carp, *Cyprinus carpio* L., strains and crossbreeds reared in the Czech Republic to infection by cyprinid herpesvirus 3 (CyHV-3; KHV). *Journal of Fish Diseases* 36, 75–80.

Pikarsky, E., Ronen, A., Abramowitz, J., Levavi-Sivan, B., Hutoran, M. *et al.* (2004) Pathogenesis of acute viral disease induced in fish by carp interstitial nephritis and gill necrosis virus. *Journal of Virology* 78, 9544–9551.

Pikulkaew, S., Meeyam, T. and Banlunara, W. (2009) The outbreak of koi herpesvirus (KHV) in koi (*Cyprinus carpio* koi) from Chiang Mai Province, Thailand. *Thai Journal of Veterinary Medicine* 39, 53–8.

Plant, K.P. and LaPatra, S.E. (2011) Advances in fish vaccine delivery. *Developmental and Comparative Immunology* 35, 1256–1262.

Prescott, M.A., Reed, A.N. and Jin, L. (2016) Rapid detection of cyprinid herpesvirus 3 in latently infected koi by recombinase polymerase amplification. *Journal of Aquatic Animal Health* 28, 173–180.

Radosavljević, V., Jeremić, S., ćirković, M., Lako, B., Milićević, V. *et al.* (2012) Common fish species in polyculture with carp as cyprinid herpes virus 3 carriers. *Acta Veterinaria (Beograd)* 62, 675–681.

Rahmati-Holasoo, H., Zargar, A., Ahmadivand, S., Shokrpoor, S., Ezhari, S. and Ebrahimzadeh Mousavi, H.A. (2016) First detection of koi herpesvirus from koi, *Cyprinus carpio* L. experiencing mass mortalities in Iran: clinical, histopathological and molecular study. *Journal of Fish Diseases* 39, 1153–1163.

Raj, V.S., Fournier, G., Rakus, K., Ronsmans, M., Ouyang, P. *et al.* (2011) Skin mucus of *Cyprinus carpio* inhibits cyprinid herpesvirus 3 binding to epidermal cells. *Veterinary Research* 42, 92.

Rakus, K., Ouyang, P., Boutier, M., Ronsmans, M., Reschner, A. *et al.* (2013) Cyprinid herpesvirus 3: an interesting virus for applied and fundamental research. *Veterinary Research* 44, 85.

Reichert, M., Bergmann, S.M., Hwang, J., Buchholz, R. and Lindenberger, C. (2017) Antiviral activity of exopolysaccharides from *Arthrospira platensis* against koi herpesvirus. *Journal of Fish Diseases* 40, 1441–1450.

Reed, A.N., Izume, S., Dolan, B.P., LaPatra, S., Kent, M. *et al.* (2014) Identification of B cells as a major site for cyprinid herpesvirus 3 latency. *Journal of Virology* 88, 9297–9309.

Ronen, A., Perelberg, A., Abramowitz, J., Hutoran, M., Tinman, S. *et al.* (2003) Efficient vaccine against the virus causing a lethal disease in cultured *Cyprinus carpio*. *Vaccine* 21, 4677– 4684.

Sadler, J., Marecaux, E. and Goodwin, A.E. (2008) Detection of koi herpes virus (CyHV-3) in goldfish, *Carassius auratus* (L.), exposed to infected koi. *Journal of Fish Diseases* 31, 71–72.

Sano, T., Fukuda, H., Furukawa, M., Hosoya, H. and Moriya, Y. (1985) A herpesvirus isolated from carp papilloma in Japan. In: Ellis, A.E. (ed.) *Fish and Shellfish Pathology*. Academic Press, London, pp. 307–311.

Sano, M., Fukuda, H. and Sano, T. (1990) Isolation and characterization of a new herpesvirus from eel. In: Perkins, F.O. and Cheng, T.C. (eds) *Pathology in Marine Sciences*, 1st edn. Academic Press, San Diego, California, pp. 15–31.

Sano, T., Morita, N., Shima, N. and Akimoto, M. (1991) Herpesvirus cyprini: lethality and oncogenicity. *Journal of Fish Diseases* 14, 533–543.

Sano, M., Ito, T., Kurita, J., Yanai, T., Watanabe, N. *et al.* (2004) First detection of koi herpesvirus in cultured common carp *Cyprinus carpio* in Japan. *Fish Pathology* 39, 165–167.

Shapira, Y., Magen, Y., Zak, T., Kotler, M., Hulata, G. and Levavi-Sivan, B. (2005) Differential resistance to koi herpes virus (KHV9)/carp interstitial nephritis and gill necrosis virus (CNGV) among common carp (*Cyprinus carpio* L.) strains and crossbreds. *Aquaculture* 245, 1–11.

Shimizu, T., Yoshida, N., Kasai, H. and Yoshimizu, M. (2006) Survival of koi herpesvirus (KHV) in environmental water. *Fish Pathology* 41, 153–157.

Soliman, H. and El-Matbouli, M. (2005) An inexpensive and rapid diagnostic method of the koi herpesvirus (KHV) infection by a loop-mediated isothermal amplification method. *Virology Journal* 2, 83.

Soliman, H., and El-Matbouli, M. (2009) Immunocapture and direct binding loop mediated isothermal amplification simplify molecular diagnosis of cyprinid herpesvirus-3. *Journal of Virological Methods* 162, 91–95.

Soliman, H., and El-Matbouli, M. (2010) Loop mediated isothermal amplification combined with nucleic acid lateral flow strip for diagnosis of cyprinid herpes virus-3. *Molecular and Cellular Probes* 24, 38–43.

Soliman, H. and El-Matbouli, M. (2018) Rapid detection and differentiation of carp oedema virus and cyprinid herpes virus-3 in koi and common carp. *Journal of Fish Diseases* 41, 761–772.

Somga, J.R., de la Pena, L.D., Sombito, C.D., Paner, M.G., Suarnaba, V.S. *et al.* (2010) Koi herpesvirus-associated mortalities in quarantined koi carp in the Philippines. *Bulletin of the European Association of Fish Pathologists* 30, 2–7.

Sunarto, A., Rukyani, A. and Itami, T. (2005) Indonesian experience on the outbreak of koi herpesvirus in koi and carp (*Cyprinus carpio*). *Bulletin of the European Association of Fish Pathologists* 2, 15–21.

Sunarto, A., McColl, K.A., Crane, M.S., Sumiati, T., Hyatt, A.D. *et al.* (2011) Isolation and characterization of koi herpesvirus (KHV) from Indonesia: identification of a new genetic lineage. *Journal of Fish Diseases* 34, 87–101.

Sunarto, A., Liongue, C., McColl, K.A., Adams, M.M., Bulach, D. *et al.* (2012) Koi herpesvirus encodes and expresses a functional interleukin-10. *Journal of Virology* 86, 11512–11520.

Sunarto, A., McColl, K.A., Crane, M.S., Schat, K.A., Slobedman, B. *et al.* (2014) Characteristics of cyprinid herpesvirus 3 in different phases of infection: implications for disease transmission and control. *Virus Research* 188, 45–53.

St-Hilaire, S., Beevers, N., Way, K., Le Deuff, R.M., Martin, P. and Joiner, C. (2005) Reactivation of koi herpesvirus infections in common carp *Cyprinus carpio*. *Diseases of Aquatic Organisms*, 67, 15–23.

St-Hilaire, S., Beevers, N., Joiner, C., Hedrick, R.P. and Way, K. (2009) Antibody response of two populations of common carp, *Cyprinus carpio* L., exposed to koi herpesvirus. *Journal of Fish Diseases* 32, 311–320.

Tadmor-Levi, R., Asoulin, E., Hulata, G. and David, L. (2017) Studying the genetics of resistance to CyHV-3 diseases using introgression from feral to cultured common carp strains. *Frontiers in Genetics* 8, 24.

Takahara, T., Honjo, M.N., Uchii, K., Minamoto, T., Doi, H. *et al.* (2014) Effects of daily temperature fluctuation on the survival of carp infected with cyprinid herpesvirus 3. *Aquaculture* 433, 208–213.

Taylor, N.G.H., Dixon, P.F., Jeffery, K.R., Peeler, E.J., Denway, K.L. and Way, K. (2010) Koi herpesvirus: distribution and prospects for control in England and Wales. *Journal of Fish Diseases* 33, 221–230.

Terhune, J.S., Grizzle, J.M., Hayden, K., McClenahan, S.D., Lamprecht, S.D. and White, M.G. (2004) First report of koi herpesvirus in wild common carp in the western hemisphere. *Fish Health Newsletter* 32, 8–9.

Troszok, A., Kolek, L., Szczygiet, J., Wawrzeczko, J., Borzym, E. *et al.* (2018) Acyclovir inhibits cyprinid herpesvirus 3 multiplication *in vitro*. *Journal of Fish Diseases* 41, 1709–1718.

Tu, C., Weng, M.C., Shiau, J.R. and Lin, S.Y. (2004) Detection of koi herpesvirus in koi *Cyprinus carpio* in Taiwan. *Fish Pathology* 39, 109–110.

Uchii, K., Matsui, K., Iida, T. and Kawabata, Z. (2009) Distribution of the introduced cyprinid herpesvirus 3 in a wild population of common carp, *Cyprinus carpio* L. *Journal of Fish Diseases* 32, 857–864.

Walster, C.I. (1999) Clinical observations of severe mortalities in koi carp, *Cyprinus carpio*, with gill disease. *Fish Veterinary Journal* 3, 54–58.

Waltzek, T.B., Kelley, G.O., Stone, D.M., Way, K., Hanson, L. *et al.* (2005) Koi herpesvirus represents a third cyprinid herpesvirus (CyHV-3) in the family *Herpesviridae*. *Journal of General Virology* 86, 1659–1667.

Wang, Y., Zeng, W., Wang, Q., Li, Y., Bergmann, S.M. *et al.* (2018) Establishment and characterization of a new cell line from koi brain (*Cyprinus carpio* L.). *Journal of Fish Diseases* 41, 357–364.

Weber, E.P., Malm, K.V., Yun, S.C., Campbell, L.A., Kass, P.H. *et al.* (2014) Efficacy and safety of a modified-live cyprinid herpesvirus 3 vaccine in koi (*Cyprinus carpio* koi) for prevention of koi herpesvirus disease. *American Journal of Veterinary Research* 75, 899–904.

Yasumoto, S., Kuzuya, Y., Yasuda, M., Yoshimura, T. and Miyazaki, T. (2006) Oral immunization of common carp with a liposome vaccine fusing koi herpesvirus antigen. *Fish Pathology* 41, 141–145.

Yoshino, M., Watari, H., Kojima, T. and Ikedo, M. (2006) Sensitive and rapid detection of koi herpesvirus by LAMP method. *Fish Pathology* 41, 19–27.

Yoshino, M., Watari, H., Kojima, T., Ikedo, M. and Kurita, J. (2009) Rapid, sensitive and simple detection method for koi herpesvirus using loop-mediated isothermal amplification. *Microbiology and Immunology* 53, 375–383.

Yuasa, K., Sano, M., Kurita, J., Ito, T. and Iida, T. (2005) Improvement of a PCR method with the SphI-5 primer set for the detection of koi herpesvirus (KHV). *Journal of Fish Pathology* 40, 37–39.

Yuasa, K., Ito, T. and Sano, M. (2008) Effect of water temperature on mortality and virus shedding in carp experimentally infected with koi herpesvirus. *Fish Pathology* 43, 83–85.

Zak, T., Perelberg, A., Magen, I., Milstein, A. and Joseph, D. (2007) Heterosis in the growth rate of Hungarian–Israeli common carp crossbreeds and evaluation of their sensitivity to koi herpes virus (KHV) disease. *Israeli Journal of Aquaculture – Bamidgeh* 59, 63–72.

Zhao, Y., Wang, T., Yu, Z, Wang, H., Liu, B. *et al.* (2016) Inhibiting cyprinid herpesvirus-3 replication with CRISPR/Cas9. *Biotechnology Letters* 38, 573–578.

Zhou, J., Xue, J., Wang, Q., Zhu, X., Li, X. *et al.* (2014a) Vaccination of plasmid DNA encoding ORF81 gene of CJ strains of KHV provides protection to immunized carp. *In vitro Cellular and Developmental Biology – Animal* 50, 489-495.

Zhou, J., Wang, H., Li, X.W., Zhu, X., Lu, W.L. and Zhang, D.M. (2014b) Construction of KHV-CJ ORF25 DNA vaccine and immune challenge test. *Journal of Fish Diseases* 37, 319–325.

8 Orthomyxovirosis (Tilapia Lake Virus)

Win Surachetpong[1]* and Kwanrawee Sirikanchana[2]

[1]Kasetsart University, Bangkok, Thailand; [2]Chulabhorn Research Institute, Bangkok, Thailand

8.1 Introduction

Tilapia is the second most important farmed food fish with an annual global production of 6.4 million tonnes (FAO, 2017a). It is considered a disease-resistant fish with few pathogens, particularly bacterial pathogens which cause high mortality (Mian *et al*., 2009; Sirimanapong *et al*., 2018). However, the recent emergence of a new viral disease that causes massive mortality in both wild and farm-raised tilapia in many countries has changed this perception. The disease called tilapia lake virus disease (TiLVD) is caused by a novel pathogen, tilapia lake virus (TiLV), also referred to as syncytial hepatitis of tilapia (SHT)-associated virions, and was first identified in Israel and Ecuador in 2014 (Eyngor *et al*., 2014; Ferguson *et al*., 2014). The virus may have existed before 2014 in wild-caught tilapia in Lake Galilee, Israel, where tilapia numbers had declined dramatically since 2007 due to a disease of unknown aetiology (Eyngor *et al*., 2014). Following the first detection in Israel (Eyngor *et al*., 2014) and Ecuador (Ferguson *et al*., 2014; Del-Pozo *et al*., 2017), TiLV was found in moribund fish in Colombia (Kembou Tsofack *et al*., 2017), Egypt (Fathi *et al*., 2017; Nicholson *et al*., 2017), Chinese Taipei (OIE, 2017a), Malaysia (Amal *et al*., 2018), Indonesia (Koesharyani *et al*., 2018), India (Behera *et al*., 2018), Thailand (Dong *et al*., 2017a; Surachetpong *et al*., 2017), Uganda and Tanzania (Mugimba *et al*., 2018), the Philippines (OIE, 2017b), Mexico (OIE, 2018a), Peru (OIE, 2018b) and the USA (OIE, 2019). In these reports, morbidity and mortality ranging from 9 to 100% were described in different tilapia strains. Further, different life stages of tilapia from fry, juvenile, fingerling, adult and bloodstock were shown to be susceptible to TiLVD (Ferguson *et al*., 2014; Dong *et al*., 2017b; Fathi *et al*., 2017; Surachetpong *et al*., 2017; Amal *et al*., 2018; Behera *et al*., 2018). While the socio-economic impacts of TiLVD worldwide have not been fully investigated, it has been reported that annual production losses of tilapia in Egypt alone associated with TiLV infection was 98,000 tonnes, accounting for $US100 million (Fathi *et al*., 2017). Nevertheless, awareness of the spread of this emerging viral disease has been raised in different countries, especially in countries where most food resources are still limited (FAO, 2017b; Hounmanou *et al*., 2018).

TiLV is an icosahedral, enveloped, negative-sense, single-stranded RNA virus with genome size of 10,323 bp (Bacharach *et al*., 2016). The TiLV genomic RNA is comprised of ten segments of which segment 1 contains an open reading frame with approximately 17% amino acid identity to the influenza C virus PB1 subunit (Bacharach *et al*., 2016). While the rest of the genome of TiLV shares no homology to other known viruses (Eyngor *et al*., 2014; Bacharach *et al*., 2016), the virus does possess comparable conserved complementary sequences at the 5′ and 3′ termini to that of the genome found for viruses in the family *Orthomyxoviridae* (Weber *et al*., 1999; Sandvik *et al*., 2000; Bacharach *et al*., 2016). Initially, TiLV was classified as an orthomyxo-like virus (Bacharach *et al*., 2016); however, a recent publication assigned a new taxonomy to the International Committee on Taxonomy of Viruses (ICTV) for a novel genus *Tilapinevirus*, and species *Tilapia tilapinevirus*, in the family *Amnoonviridae* (Kuhn, 2018). Four complete genome sequences of TiLV isolates from Thailand, Ecuador and Israel are available in the GenBank database (Bacharach *et al*., 2016; Surachetpong *et al*., 2017; Al-Hussinee *et al*., 2018; Subramaniam *et al*., 2019). Comparison of

*fvetwsp@ku.ac.th

 Climate Change and Infectious Fish Diseases (P.T.K. Woo, J.-A. Leong and K. Buchmann)

isolates from Thai Nile tilapia (WVL18053-01A) indicated that the segment 1 of this virus shared high sequence identity (99.1%; 1546/1560 nucleotides) with the virus (GenBank accession no. KY615742) previously isolated in Thailand (Al-Hussinee *et al.*, 2018). In addition to segment 1, analysis of the nucleotide sequences of other segments from different geographic locations showed that the viruses had sequence identity of 93 to 100% (Dong *et al.*, 2017a; Surachetpong *et al.*, 2017; Behera *et al.*, 2018; Mugimba *et al.*, 2018).

8.2 Diagnosis of TiLVD

8.2.1 Morphological features of the pathogen

TiLV or SHT-associated virion particles are enveloped and have a highly pleomorphic structure with a diameter ranging between 55 and 100 nm under transmission electron microscopy (TEM) (Fig. 8.1A) (Eyngor *et al.*, 2014; Del-Pozo *et al.*, 2017; Tattiyapong *et al.*, 2017). Different morphologies of the virus including round- to oval-shaped and filamentous structures (Fig. 8.1B) were reported in infected fish (Ferguson *et al.*, 2014; Del-Pozo *et al.*, 2017; Tattiyapong *et al.*, 2017). Ferguson *et al.* (2014) reported the presence of virus-like particles in the cytoplasm of hepatocytes and space of Disse (perisinusoidal space) in diseased tilapia with pathology in the liver and gastrointestinal tract. A single or small group of virus-like structures were described as capsid-like trilaminar structures with thin bands of electron-dense particles in association with the surface of the endoplasmic reticulum (Ferguson *et al.*, 2014; Del-Pozo *et al.*, 2017). TEM studies did identify multinucleated hepatocytes in infected fish (Ferguson *et al.*, 2014) as well as the intracellular location of the virions within intracytoplasmic structures and perinuclear areas of the infected cells (Del-Pozo *et al.*, 2017).

Although the function(s) of receptors, structural proteins and spikes of TiLV remain(s) to be determined, the morphology of the virus is very close to other piscine orthomyxoviruses such as infectious salmon anaemia virus (ISAV). ISAV is enveloped and highly pleomorphic with a diameter ranging from 80 to 120 nm in its spherical form; its filamentous form is longer than 300 nm and both forms have glycoprotein spikes (Kibenge and Kibenge, 2016). A further study showed that ISAV spikes contained haemagglutinin-esterase which hydrolysed Neu4,5Ac_2 sialic acids (Hellebø *et al.*, 2004) and the roles of the TiLV proteins are crucial to understanding how the virus infects the host and its cells.

8.2.2 Clinical signs

During disease outbreaks, mass morbidity and mortality ranging from 20 to 100% (Fig. 8.2A) were recorded 7–10 days after the first fish mortality

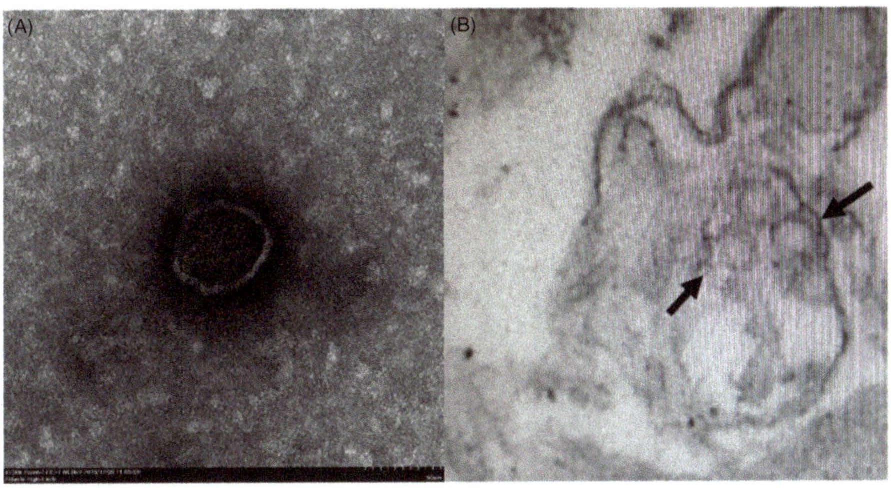

Fig. 8.1. Transmission electron micrographs of: (A) a viral particle from infected E-11 cells showing a round, enveloped virion of size 55–80 nm; and (B) an infected brain showing multiple virus particles (arrows). (Original figures.)

Fig. 8.2. Mass mortality of tilapia and gross signs of TiLV infection. (A) Mass mortality of red hybrid tilapia (*Oreochromis* spp.) in a floating cage. (B) Nile tilapia (*Oreochromis niloticus*) showing corneal opacity and skin erosions. (C) Mozambique tilapia (*Oreochromis mossambicus*) showing skin erosion and discoloration. (D) Red hybrid tilapia showing skin haemorrhage, fin congestion, eye exophthalmos, scale protrusion and abdominal swelling. (Original figures.)

(Eyngor *et al.*, 2014; Surachetpong *et al.*, 2017; Behera *et al.*, 2018). Moribund fish associated with TiLV infection have been described as displaying loss of appetite, emaciation, swimming at the water surface and erratic swimming (Eyngor *et al.*, 2014; Surachetpong *et al.*, 2017; Tattiyapong *et al.*, 2017). Other clinical signs of infection include multifocal areas of haemorrhages and congestion of the skin, skin erosions, skin darkening, gill pallor, pale skin, ocular opacity, eye protrusion and abdominal distension (Eyngor *et al.*, 2014; Ferguson *et al.*, 2014; Surachetpong *et al.*, 2017; Tattiyapong *et al.*, 2017). Similar to the natural infection, the clinical signs of fish in experimental challenge studies (Tattiyapong *et al.*, 2017; Behera *et al.*, 2018) include skin erosion, exophthalmia and abdominal distension (Fig. 8.2B–D). In a study on clinically healthy fish, investigators have detected virus using highly sensitive molecular methods (see Section 8.2.4) and also found histopathological features resembling syncytial hepatitis in the liver (Senapin *et al.*, 2018). These results indicate that routine monitoring for TiLV even in apparently healthy stocks of tilapia is warranted.

8.2.3 Histopathology and *in situ* hybridization

Although the pathology of TiLVD is not well characterized, dramatic histopathological changes were found in various organs such as the liver, spleen, brain and kidney. Initially, the liver and brain were the most studied organs for pathological changes (Eyngor *et al.*, 2014; Ferguson *et al.*, 2014; Del-Pozo *et al.*, 2017; Surachetpong *et al.*, 2017). In the liver, the main histopathology includes hepatic necrosis and syncytial cell formations (Fig. 8.3) (Dong *et al.*, 2017a; Kembou Tsofack *et al.*, 2017; Tattiyapong *et al.*, 2017; Amal *et al.*, 2018; Behera *et al.*, 2018; Jaemwimol *et al.*, 2018), accumulation of eosinophilic lipoproteinaceous material and eosinophilic intracytoplasmic inclusion bodies in the hepatocytes (Fig. 8.3). Conversely, Behera *et al.* (2018) reported no eosinophilic inclusion bodies in TiLV-infected fish in India. In Israel, lesions in the infected brain were oedema, focal haemorrhages in the leptomeninges and capillary congestion (Eyngor *et al.*, 2014). Moreover, multiple foci of gliosis, encephalitis, degeneration of neurons that exhibit less cytoplasmic density, vacuolation and central chromatolysis, and

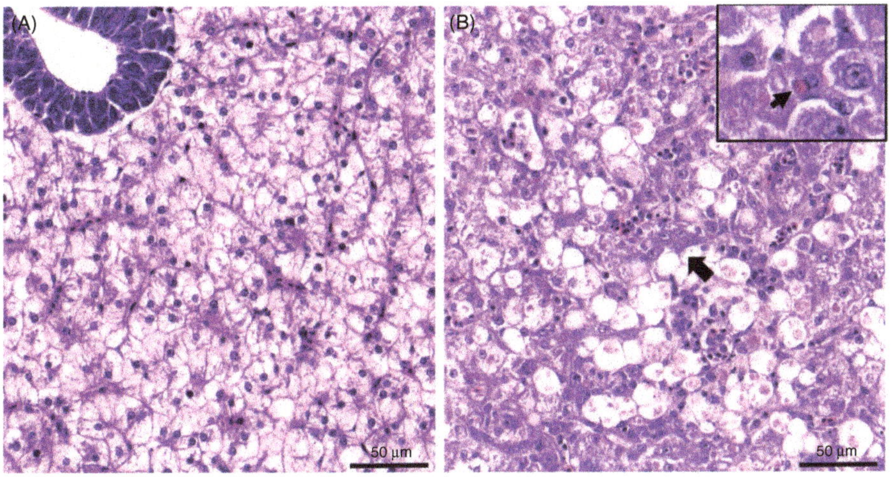

Fig. 8.3. Liver of normal and TiLV-infected red hybrid tilapia (*Oreochromis* spp.) showing the histopathological changes associated with TiLVD. (A) Normal liver: hepatocytes, the exocrine pancreas surrounds the hepatic portal vein, glycogen storage. (B) TiLV-infected liver: multiple necrosis and degeneration of hepatocytes, presence of syncytial hepatic cells (arrow) and eosinophilic intracytoplasmic inclusion bodies in hepatocytes (insert, arrow). Haematoxylin and eosin stain; scale bars = 50 µm. (Original figures.)

mild periodic perivascular cuffing of lymphocytes were found in the brain of infected fish (Eyngor *et al.*, 2014; Tattiyapong *et al.*, 2017; Amal *et al.*, 2018). In other studies, no apparent lesions in the brain were reported (Ferguson *et al.*, 2014; Behera *et al.*, 2018). Ferguson *et al.* (2014) also reported pathological changes in the gastrointestinal tract, with the characteristic pervasive necrosis of gastric glands with inflammatory cells' infiltration into the submucosa, accumulation of proteinaceous material in the lumen of the intestine and varying eosinophilic necrotic debris existing within the glands. *In situ* hybridization using digoxigenin-labelled DNA probes against segment 3 of the virus showed a positive signal in the liver, gills, kidney and brain of naturally infected tilapia (Dong *et al.*, 2017a). Specifically, the hybridization signals were clearly observed in the nuclei and cytoplasm of multinucleated liver cells of infected fish (Bacharach *et al.*, 2016). Using an oligonucleotide probe against segment 1 of TiLV, the hybridization signal showed that the virus was located only near the blood vessels of leptomeninges (Bacharach *et al.*, 2016). Besides tilapia, a positive hybridization was demonstrated in the brain and liver of TiLV-infected giant gourami (*Osphronemus goramy*) (Jaemwimol *et al.*, 2018).

8.2.4 Molecular techniques

Generally, the internal organs of individual fish including the liver, spleen, brain, heart and kidney are collected for TiLV examination. For resource-limited situations, pooling of samples from three to five fish does produce reliable molecular diagnostic results (Fathi *et al.*, 2017; Kembou Tsofack *et al.*, 2017; Surachetpong *et al.*, 2017). Non-lethal samples using mucus have been found to contain virus using reverse transcription-quantitative polymerase chain reaction (RT-qPCR) assays and virus isolation in susceptible cell culture (Liamnimitr *et al.*, 2018). Several TiLV detection protocols using reverse transcription-polymerase chain reaction (RT-PCR) techniques have been developed (Eyngor *et al.*, 2014; Dong *et al.*, 2017a; Kembou Tsofack *et al.*, 2017). The first work involved development of one specific primer pair against segment 3 of TiLV to amplify isolates in Israel (Eyngor *et al.*, 2014). Later, a nested RT-PCR protocol with higher sensitivity and lower detection limit (7 copies per reaction) was applied for TiLV detection in samples originating from Colombia, Israel and Ecuador (Kembou Tsofack *et al.*, 2017). Thereafter, the semi-nested RT-PCR method was developed and shown to amplify TiLV in archived samples (Dong *et al.*, 2017b) as well as in apparently normal juvenile and adult tilapia (Senapin *et al.*, 2018). In addition to RT-PCR and semi-nested RT-PCR, two SYBR Green I-based RT-qPCR methods using specific primers against segment 3 of the virus were developed (Kembou Tsofack *et al.*, 2017; Tattiyapong *et al.*, 2018). These assays are highly specific and sensitive for TiLV with the detection limit of 70 copies and 2 copies, respectively (Kembou Tsofack *et al.*, 2017; Tattiyapong *et al.*, 2018). The RT-qPCR assay is 100 and 10,000 times more sensitive than the conventional RT-PCR and virus isolation in cell culture, respectively (Tattiyapong *et al.*, 2018). Additionally, a RT-qPCR method based on the TaqMan probe assay against segment 3 has been developed to detect TiLV from different geographic regions (Waiyamitra *et al.*, 2018). The TaqMan probe-based RT-qPCR protocol is more specific due to the binding of probe and primers, and it offers high sensitivity compared with the other qPCR assays (Waiyamitra *et al.*, 2018). Most current PCR assays rely on the submission of processed samples to the laboratory; however, a molecular assay that could be performed at the farm site (with rapid result) would be very useful and facilitate disease control. For example, simple and rapid reverse transcription loop-mediated isothermal amplification (RT-LAMP) assays based on colorimetric change were recently developed to detect TiLV in fish tissues (Phusantisampan *et al.*, 2019; Yin *et al.*, 2019). Although all of the PCR methods have been developed in different laboratories, a standard PCR protocol for TiLV detection has not yet been completely validated. Preliminary comparison of conventional RT-PCR and SYBR Green I-based RT-qPCR by two laboratories to detect TiLV in fish tissues revealed that both assays provide high sensitivity and specificity for TiLV detection (Nicholson *et al.*, 2018).

8.2.5 Cell cultures

The gold standard for diagnosis of viral disease is the isolation of the virus using susceptible cell lines. Nine primary and continuous cell lines – CHSE-214, BF-2, BB, EPC, KF-1, RTG-2, FHM, E-11 and primary brain cells – were tested for TiLV propagation (Eyngor *et al.*, 2014). Apparently, only primary tilapia brain cells and E-11 cells from snakehead fish (*Channa striata*) allowed TiLV replication, with development of cytopathic effect (CPE) at 5–12 days post-inoculation (dpi) (Eyngor *et al.*, 2014). The infected E-11 cells were round and swollen, and detached from the culture flask with complete CPE formation at 7 dpi (Fig. 8.4). A recent study suggested that two additional cell lines from Mozambique tilapia (*Oreochromis mossambicus*) brain (OmB) and bulbus arteriosus (TmB) were quite sensitive to TiLV infection; however, the CPE was more difficult to monitor than in E-11 cells. The OmB cells seemed suitable for endpoint dilution

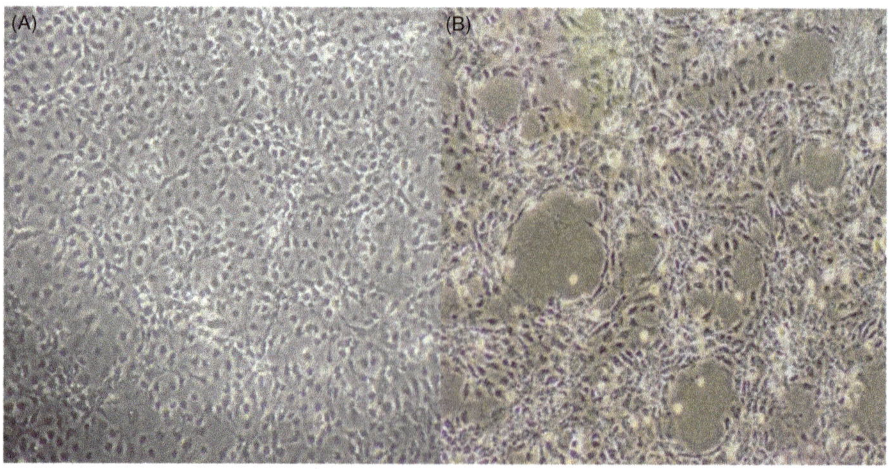

Fig. 8.4. Transmission electron micrographs showing CPE formation in TiLV-infected E-11 cells and morphology of viral particles. (A) Normal E-11 cells; (B) TiLV-infected cells at 5 days post-infection, showing CPE formation, cell vacuolation and detachment. (Original figures.)

assay and production of purified TiLV (Kembou Tsofack *et al.*, 2017). In the CFF cell line derived from Malayan leaffish (*Pristolepis fasciatus*), TiLV caused CPE formation as early as 3 dpi with >50% cell detachment at 7 dpi (Behera *et al.*, 2018). Subsequently, two tilapia (*Oreochromis niloticus*) cell lines, OnB (brain cells) and OnL (liver cells), showed syncytial formation, cell shrinkage, increased granularity and destruction of the monolayer appearance at 5 to 8 dpi with viral titre at $10^{7.0-7.3}$ $TCID_{50}$/ml (Thangaraj *et al.*, 2018). The supernatant of infected cells was confirmed positive for TiLV using RT-PCR. Most recently, it was shown that TiLV could be propagated in primary tilapia brain (TiB) cells with the viral titre of $10^{7.43}$ $TCID_{50}$/ml at 10 dpi (Wang, Y. *et al.*, 2018).

There are no immunological assays for TiLV. For other piscine viruses, immunological techniques such as enzyme-linked immunosorbent assay (ELISA) have been developed to detect fish that were previously exposed to the virus (Jaramillo *et al.*, 2016). According to our observation (W. Surachetpong, Bangkok, 2019, personal communication) and the previous report by Eyngor *et al.* (2014), fish that survived TiLV infection in the field or under laboratory challenge were resistant to subsequent infections, indicating that the ELISA-based assay would be a potent method to detect previously TiLV-exposed fish.

8.3 Potential Spread of TiLV to Other Fish Species

There are several reports of TiLV associated with high mortality in various species of tilapia, mainly in the tribe Tilapiini and genera *Sarotherodon* and *Oreochromis* (Eyngor *et al.*, 2014; FAO, 2017a; Jansen and Mohan, 2017). In Israel, morbidity and mortality of TiLV were reported in commercial hybrid tilapia *O. niloticus* × *Oreochromis aureus*, wild tilapia *Sarotherodon galilaeus* (St. Peter's fish), *Tilapia zilli* (common tilapia), *O. aureus* (Jordan tilapia) and *Tristamella simonis intermedia* (Eyngor *et al.*, 2014). However, the role that wild tilapia has in the disease epizootiology and spread of the virus is not well understood. In Ecuador, a similar disease caused by an unknown virus was reported in the fingerlings of farmed Nile tilapia (*O. niloticus*) strain Chitralada (Ferguson *et al.*, 2014). In Asia, outbreaks of TiLV have been reported in Nile tilapia in Thailand (Dong *et al.*, 2017a; Surachetpong *et al.*, 2017) and India (Behera *et al.*, 2018), in red hybrid tilapia (*Oreochromis* spp.) in Thailand (Dong *et al.*, 2017a; Surachetpong *et al.*, 2017) and in red tilapia (*O. niloticus* × *O. mossambicus*) in Malaysia (Amal *et al.*, 2018). In addition, a recent study from Lake Victoria, Africa, detected the virus in 14.66% (28 out of 191) of tilapia with no clinical signs, which suggests that healthy Nile tilapia could serve as a potential carrier for other susceptible fish (Mugimba *et al.*, 2018).

While most fish species are resistant to TiLV infection, two recent studies suggested the susceptibility of other warm-water fish to the virus. In Malaysia, TiLV genomic RNA was detected in wild river barb (*Barbonymus schwanenfeldii*) and tilapia in a man-made pond (Abdullah *et al.*, 2018). When intraperitoneally injected, giant gourami (*O. goramy*) showed rapid progress of the disease, with clinical signs of infection within 2–3 days and mortality reached 100% within 7 days (Jaemwimol *et al.*, 2018). However, a similar study demonstrated resilience of the following warm-water species to TiLV infection: snakeskin gourami (*Trichogaster pectoralis*), iridescent shark (*Pangasianodon hypophthalmus*), walking catfish (*Clarias macrocephalus*), striped snakehead fish (*C. striata*), climbing perch (*Anabas testudineus*), common carp (*Cyprinus carpio*), silver barb (*Barbodes gonionotus*) and Asian sea bass (*Lates calcarifer*) (Jaemwimol *et al.*, 2018). In addition to these species, wild fishes including carp (*C. carpio*) and grey mullet (*Mugil cephalus*) (Eyngor *et al.*, 2014), thin-lipped mullet (*Liza ramada*) (Fathi *et al.*, 2017), catfish and peacock bass (*Cichla* spp.) (Abdullah *et al.*, 2018), milkfish (*Chanos chanos*), pearl spot (*Etroplus suratensis*) and the Indian major carps rohu (*Labeo rohita*), catla (*Catla catla*) and mrigal (*Cirrhinus mrigala*) (Behera *et al.*, 2018) that share water resources with infected tilapia were not affected by the virus. Thus, further studies are needed to understand the susceptibility and pathogenesis as well as identify the specific receptors required for TiLV infection. These findings indicate that other fish species could be potential carriers for the virus. Therefore, strict biosecurity measures must be emphasized in polyculture systems or in open environments where other fish species exist. Additionally, future studies should investigate the susceptibility to TiLV of other fish species, including those that could serve as a virus carrier.

Cohabitation of fish is one factor that could promote TiLV transmission. Recent experiments showed that TiLV infection occurred after the cohabitation of healthy tilapia with moribund fish (Eyngor *et al.*, 2014; Liamnimitr *et al.*, 2018). Healthy tilapia developed clinical signs of infection within 3 days of cohabitation and cumulative mortality was 80% (Eyngor *et al.*, 2014). Eyngor *et al.* (2014) reported a delay of 2 to 3 days in the progression of the infection via cohabitation challenge compared with intraperitoneal challenge. In another study, TiLV genomic RNA was detected in fish mucus up to 12 days post-infection (dpi) while the virus was present in liver and intestine for 14 and 12 dpi, respectively (Liamnimitr *et al.*, 2018). In addition to cohabitation, the disease could be induced via intragastric challenge, suggesting that the gastrointestinal tract is another route of virus entry and spread in fish populations (Pierezan *et al.*, 2019). Although vertical transmission of the virus from broodstock to progeny was not confirmed, TiLV genomic RNA was detected in eggs, juveniles and broodstock (Jansen *et al.*, 2018). Our recent study revealed that infectious virus could be isolated from the gonads of male and female broodstock from naturally infected and experimentally challenged fish (Yamkasem *et al.*, 2019). That study supports the hypothesis that TiLV can be transferred from infected broodstock to their progeny.

8.4 Potential Spread of TiLV Across Carriers Other Than Fish

The presence and spread of TiLV through non-piscine intermediate and/or reservoir hosts have not been fully investigated. It is possible that the transfer of TiLV from local intermediate or reservoir hosts to susceptible fish species may exist and contribute to the distribution of this viral disease in new geographical regions. Such an example of the cross-infection of a viral pathogen between different aquatic phyla was demonstrated in a recent study of a novel shrimp alphanodavirus that can cause disease in farmed Japanese flounder (*Paralichthys olivaceus*) (Wang, C. *et al.*, 2018). Further, the genomic RNA of the betanodavirus that causes viral nervous necrosis (VNN) or viral encephalopathy and retinopathy (VER) was detected in marine invertebrates (Gomez *et al.*, 2008). However, the role of invertebrates in TiLV transmission has not been elucidated. Our preliminary attempts to amplify TiLV using RT-PCR from molluscs and fish parasites (*Argulus* spp.) collected from TiLV-infected fish and ponds produced negative results (W. Surachetpong, Bangkok, 2019, personal communication). Aquatic birds also could serve as carriers to spread the virus from infected ponds/farms to neighbouring farms or other areas. We often see aquatic birds near TiLV-infected ponds or cages, and they feed on moribund tilapia and carcasses. The role of invertebrates and other animals in the spread of TiLV requires further studies. Importantly, identification of possible reservoir hosts will facilitate the development of control measures such as setting up biosecurity or eliminating the intermediate host and carriers.

8.5 Impact of Environment and Risk Factors Associated with Disease Outcome

Changes in fish-rearing practices such as aquaculture intensification, movement of live fish and distribution of potential carriers could drive the genetic variation and adaptation of the virus to the host, facilitating the emergence of novel viruses and increased disease severity (Retel *et al.*, 2019). In addition, abiotic factors including dissolved oxygen level, pH, salinity, nutrient availability and temperature have been shown to affect viral persistence; for example, the fluctuation of pH and salinity can promote survival by increasing virus adsorption to particles (Danovaro *et al.*, 2011; Horas *et al.*, 2018). In a recent study, multiple production factors were shown to increase the incidence and severity of SHT that occurred from 2011 to 2013 in Ecuador (Kabuusu *et al.*, 2018). The risk factors that were analysed included water temperature, number of production cycles per year, dissolved oxygen, weight and strain of the fish, and stocking density. Specifically, tilapia strain Chitralada had a higher risk of SHT (relative risk = 2.1; 95% confidence interval 1.8–2.4) than the genetically male tilapia (GMT) or genetically improved farm tilapia (GIFT) strains. Increase of stocking density, dissolved oxygen and pond production cycle at 1 fish/m^2, 1 mg/l and 1 cycle, respectively, was associated with higher mortality in the pond with SHT, while raising the stocking weight and water temperature by 1 g and 1°C respectively reduced the presence and severity of SHT. The resistance of large fish to infection is possibly due to the more mature, complete immune function of larger fish (Breine *et al.*, 2015). Other factors, including season and stocking year, were not significantly associated with SHT emergence (Kabuusu *et al.*, 2018). Similarly, the presence and severity of TiLV outbreaks in Thailand seemed to be associated with increased stocking density, poor water quality, seasonal change and increased stress of fish (W. Surachetpong, Bangkok, 2019, personal communication). During TiLV outbreaks, if control measures were not properly implemented, the disease caused high mortalities in infected fish with severe economic losses. A study by Yang *et al.* (2018) reported a deterministic susceptible infectious mortality model to derive R_0 (basic reproduction number) which quantitatively characterizes the average number of new fish infected by a fish with TiLV in an entirely susceptible population. The estimated R_0 of the disease was 2.59 in a cohabitation challenge model of 2.6×10^5 $TCID_{50}$/fish. For the disease control model, the control reproduction number R_c (average number of new fish infected by a TiLV-infected fish given control measures) was estimated. It was suggested that appropriate control measures such as controlling the contact rate, population density and infection time reduced R_c to <1 to limit the virus transmission and disease impact (Yang *et al.*, 2018). Furthermore, the control measures may focus on secondary or opportunistic infections of bacteria and parasites. Co-infection of *Aeromonas* and TiLV has been commonly detected in moribund fish (Ferguson *et al.*, 2014; Surachetpong *et al.*, 2017; Amal *et al.*, 2018). Our recent study on the effects of co-infections revealed that the concurrent infection of *Aeromonas hydrophila* and TiLV resulted in higher mortality and worsening of the disease outcome than the single bacterial or viral infection (Nicholson *et al.*, 2020).

8.6 Effect of Climate Change on TiLV Pathogenicity

Pathogens of emerging diseases have been suspected to be a result of environmental changes driven by both climate change and human activities (Harvell *et al.*, 1999). One of the major effects of climate change is an increase in water temperature (Hoegh-Guldberg *et al.*, 2018). Although the impacts of environmental change on TiLV infection have not been fully elucidated, these changes could enhance the disease emergence and severity in several ways. The outbreaks in aquaculture systems showed more cumulative mortality with shorter outbreak durations in the tropical countries compared with other regions (Leung and Bates, 2013). Temperature at 28°C or above appears to favour TiLV outbreaks (Eyngor *et al.*, 2014; Nicholson *et al.*, 2017; Surachetpong *et al.*, 2017). Likewise, high mortality of tilapia was reported during warm seasons in Israel, Ecuador and Egypt (Eyngor *et al.*, 2014; Ferguson *et al.*, 2014; Nicholson *et al.*, 2017). The higher mortality rate during TiLV infection may be caused by reduced immune capability in fish due to heat stress, making them more susceptible to infection (Harvell *et al.*, 1999). Although the high temperature appeared to favour TiLVD progression, it could also promote TiLVD control by rendering higher efficacy to certain disinfectants (e.g. iodine, hydrogen peroxide, quaternary ammonium compounds, monochloramine) widely used as disease control measures in aquaculture farms (Sirikanchana *et al.*, 2008; Mon-on *et al.*, 2018; Jaemwimol *et al.*, 2019). Consequently, applications of disinfectants to reduce viral load in the environment or farm equipment should be done routinely.

In addition to high temperature, extreme weather events, such as storms and unexpected higher precipitation, contribute to more disease outbreaks in aquaculture. Notably, higher nutrient loads being carried with water runoff could impair hosts' immune systems, making them more susceptible to various pathogens (Leung and Bates, 2013). Extreme storms increase suspended solids in water by both transport from land and resuspension of riverbed sediments. Disturbance of riverbed sediments could reintroduce viruses from sediment reservoirs into overlying water (Mackowiak *et al.*, 2018), which increases risk of exposure to aquaculture animals as viruses appear to remain infectious longer in riverbed sediments. Furthermore, solid particles could serve as a physical shield to protect viruses from natural inactivation by ultraviolet (UV) radiation. Together with more cloud cover due to climate change that leads to reduced solar radiation to ground and water, solar inactivation of viruses is expected to decrease with the effect of climate change. In contrast, suspended solids in water could remove freely available viruses by adsorption and also by inactivation (US EPA, 2015; Booncharoen *et al.*, 2018). Although no information about TiLV inactivation with solar or UV disinfection is available, high efficacy of UV inactivation of RNA viruses has been reported (Hijnen *et al.*, 2006).

Higher water salinity, as driven by higher evaporation rate, could impact inactivation rates of viruses; however, higher salinity might also promote more efficient infection as shown in studies that found inorganic salts such as magnesium salts at optimal ranges promote bacteriophage–host binding, leading to more efficient infection (Moldovan *et al.*, 2007).

Additionally, water acidification could change water pH and affect the efficacy of disinfectants (Sirikanchana *et al.*, 2008). Water pH also affects the surface charge of viral particles, depending on the isoelectric points of each virus type, and can induce adsorption among viral particles to form clumps or promote adsorption of the virus particles to hosts (Ellis, 2001; Michen and Graule, 2010). These examples highlight the effect of extreme weather conditions on disease outbreaks, while the impact of these factors on TiLVD outbreaks awaits further investigation.

8.7 Control and Prevention

Since there are no effective therapies and vaccines are not available against TiLVD, the control and prevention of the disease rely on biosecurity measures and good management practices (FAO, 2018). Active surveillance in wild and farmed tilapia is needed to identify early disease outbreaks and to limit the spread of the virus. The movement of live fish within a region or across countries should be screened for the virus. Continued efforts on the validation of diagnostic assays for the detection of all TiLV isolates are also important. Such reliable assays will help in the screening of infected fish and will help the control programme and eventually eliminate the disease. Additionally, the development of a specific pathogen-free stocking broodstock or vaccine will serve as a tool for TiLV disease prevention (FAO, 2017a; Jansen *et al.*, 2018).

Although TiLVD is not on the World Organization for Animal Health's list of diseases, the emergence of the virus and its potential economic impact have raised the attention of managers and investors in global tilapia aquaculture (FAO, 2017a,b, 2018). To reduce spread of the virus, tight regulations should govern the movement of live fish and products. Current legislation to limit the introduction and spread of disease within a region or across countries is one of the most important approaches to prevent the further spread of TiLV. Such legislation from local governments includes regulation of farm registration, promoting good aquaculture practice, and implementing hazard analysis critical control point (HACCP) and code of conduct procedures at fish farms and distribution sites (Chinabut and Puttinaowarat, 2005). In addition, the potential risk for the introduction of TiLV into TiLV-free countries should be determined, and proper risk management measures should be in place based on well-conducted scientific studies (FAO, 2018). For example, a recent study suggested that frozen tilapia fillets from healthy and subclinically infected fish had no risk of TiLV spread (Thammatorn *et al.*, 2019).

The application of vaccines to prevent and control infectious disease in fish is widely used. For TiLV, fish that survive the infection develop protective immunity as re-exposure to TiLV does not lead to clinical signs or mortality (Eyngor *et al.*, 2014; W. Surachetpong, Bangkok, 2019, personal communication). These observations indicate that fish are immune to the infection and a TiLV vaccine has the potential to control the disease. Currently, both attenuated and inactivated TiLV vaccines based on cell culture preparations are being developed. A live attenuated TiLV immersion vaccine has been developed by Kovax Co., Ltd in Israel. In our laboratory, a formalin-inactivated TiLV vaccine is being tested. Remarkably, the vaccine improved the survival rate above 60–70% compared with the control group. However, the mechanism behind this protection remains to be determined. One potential mechanism is the development of protective antibody such as immunoglobulin M (IgM). Our preliminary investigation revealed that the vaccinated fish have significant anti-TiLV IgM levels compared with the unvaccinated fish. Besides vaccination, the application of therapeutic agents such as antiviral compounds or general immunostimulants to control TiLVD remains largely unexplored.

The selective breeding of disease resistance in fish is of major interest to the aquaculture industry (Ferguson *et al.*, 2014; Janssen *et al.*, 2017). Ferguson *et al.* (2014) indicated that the mortality rate from SHT in Chitralada strain is very high (80–90%) while the mortality rate of GMT strains of tilapia was less than 20%. Yet, there is no comprehensive breeding programme for the selection of TiLV-resistant tilapia. Another strategy that could be applied to reduce the spread of virus is to establish TiLV-free broodstock. The selection of broodstock using extensive disease screening and the removal of infected stock will produce tilapia fingerlings that are free of TiLV. Intensive biosecurity must be implemented for the aquaculture facilities to maintain the TiLV-free environment. Such biosecurity strategies include routine disease monitoring, tight regulations for the introduction of new fish, and disinfecting water and vehicles (Fig. 8.5A). When TiLVD outbreaks occur in a fish farm, immediate actions should be implemented to limit the economic losses from TiLVD. Such approaches

Fig. 8.5. Proper biosecurity measures and good management practices. (A) After transporting live fish, the vehicle should be properly cleaned and disinfected to reduce the potential spread of the virus. (B) Avoid continuous production of fish and apply the 'all in–all out' practice. (Original figures.)

include the immediate removal of severely moribund and dead fish, avoiding the movement of infected fish to another pond or location, maintaining good water quality and reducing the stress of the fish. To avoid virus spread in the farm or environment, fish should be cultured from the same source or at the same age (applying the 'all in–all out' concept) (Fig. 8.5B).

To reduce viral outbreaks, cleaning procedures using disinfectants should be implemented to reduce the spread of virus in fish farms. This would include disinfecting the water reservoir to kill the virus or dipping and immersing the equipment in disinfectant before use. The disinfectants applied should be environmentally friendly with minimal negative impacts to aquatic organisms in the environment (Méndez-Vilas, 2013). Until now, there has been no scientific report on the persistence of TiLV in the natural environment. For other piscine orthomyxoviruses, iodophors, chloramine T and mixtures of peracetic acid, hydrogen peroxide and acetic acid showed different efficacy in the inhibition of ISAV (Smail *et al.*, 2004). Likewise, a study on mixtures of quaternary ammonium compounds and other inorganic and organic solvents reduced the amount of ISAV by 100,000-fold (García *et al.*, 2013). A recent study by Tidbury *et al.* (2018) indicated that Virkon® (pentapotassium bis-peroxymonosulfate bis-sulfate) effectively inactivates infectious pancreatic necrosis virus following exposure to 0.5 to 1% v/v for 2 to 5 min under no organic load condition. We used common disinfectants such as 10 ppm NaOCl and 5000 ppm of Virkon® in our study and the combination reduced the amount of TiLV by 100,000-fold $TCID_{50}$/ml after exposure at 28°C for 1 min under laboratory conditions (Jaemwimol *et al.*, 2019).

8.8 Conclusions with Suggestions for Future Studies

TiLVD is an infectious disease caused by the TiLV. It has been a global threat to world tilapia production since it was first reported in 2014. TiLV has been associated with massive mortalities in tilapia and their hybrid species. Further, recent reports showed that the virus could cause clinical infections in other freshwater fish species such as giant gourami and wild river barb. Although the impact of this disease on tilapia production has been well documented, little is known regarding the biology of the virus. The disease is not listed as being reportable to the World Organization for Animal Health. Consequently, legislation for control and prevention programmes has not been instituted and implemented. More extensive studies are needed especially on strategies to control the virus during tilapia production. These studies should include:

1. Geographic distribution and host range of TiLV. A detailed investigation into the prevalence and/or re-emergence of TiLV and the spread of different TiLV strains in tilapia-growing regions is needed.
2. Pathophysiology and immune response of the host. Little is known with regard to the mechanisms

of pathogenesis in fish and their immune response to the virus, including the kinetics of virus infection, the functions of the viral proteins and the genes that regulate host immune response. Further detailed analysis of the structure and function of viral proteins should be undertaken. Moreover, a complete study of pathogenesis such as mechanisms of viral entry, tissue tropism, port of viral entry, potential for genetic reassortment of the virus by environmental and host factors, and distribution of the virus at the cellular level should be conducted.

3. Development of diagnostic assays for rapid and sensitive detection of the virus. A number of diagnostic assays has been developed for TiLV detection, but the sensitivity and accuracy of these assays should be further validated by different laboratories. Importantly, a rapid, accurate and inexpensive on-farm assay should be developed and made available as a tool for TiLV control. The application of molecular diagnostic techniques is important to identify potential carriers and sources of the pathogen, determine the route(s) of transmission and understand the mechanisms of disease outbreaks.

4. Controls and prevention measures. Further studies are required to develop control strategies including an effective vaccine, improved biosecurity measures and identification of intermediate hosts and carriers. For example, identification of additional fish species or potential vectors, as well as intermediate hosts, remains to be done. Screening of carriers and other hosts that are susceptible to TiLV will help to reduce trans-species infection. Restriction of live fish movement through national and international trade is urgently needed to limit the further spread of the virus. Currently, there is no information on viral persistence outside the host in the water column. Such information is important for the implementation of control strategies. Procedures for the management of secondary or co-infection of bacteria and parasites through proper antibiotic usage, for the application of immunostimulants to reduce morbidity and mortality, and to limit the economic impact of the disease should be developed. As no commercial vaccine is available for TiLV, the need for continuous effort to develop an effective and affordable vaccine is urgent. The vaccine should be easy to administer/deliver, especially to small fish e.g. by immersion, while able to stimulate protective immunity in fish throughout the culture period.

Overall, future research should be evaluated to better understand all the parameters that impact on fish associated with TiLV and SHT, especially environmental effects due to climate change. Additional needs are the implementation of biosecurity at fish farms, control measures, and the development of an effective vaccine which will limit further spread of the virus and minimize the economic impact of the disease.

References

Abdullah, A., Ramly, R., Ridzwan, M.S.M., Sudirwan, F., Abas, A. *et al.* (2018) First detection of tilapia lake virus (TiLV) in wild river carp (*Barbonymus schwanenfeldii*) at Timah Tasoh Lake, Malaysia. *Journal of Fish Diseases* 41(9), 1459–462. https://doi.org/10.1111/jfd.12843

Al-Hussinee, L., Subramaniam, K., Ahasan, M.S., Keleher, B. and Waltzek, T.B. (2018) Complete genome sequence of a tilapia lake virus isolate obtained from Nile tilapia (*Oreochromis niloticus*). *Genome Announcement* 6(26), e00580-18. https://doi.org/10.1128/genomeA.00580-18

Amal, M.N.A., Koh, C.B., Nurliyana, M., Suhaiba, M., Nor-Amalina, Z. *et al.* (2018) A case of natural co-infection of tilapia lake virus and *Aeromonas veronii* in a Malaysian red hybrid tilapia (*Oreochromis niloticus* × *O. mossambicus*) farm experiencing high mortality. *Aquaculture* 485, 12–16. https://doi.org/10.1016/j.aquaculture.2017.11.019

Bacharach, E., Mishra, N., Briese, T., Zody, M.C., Kembou Tsofack, J.E. *et al.* (2016) Characterization of a novel orthomyxo-like virus causing mass die-offs of tilapia. *mBio* 7(2), e00431-16. https://doi.org/10.1128/mBio.00431-16

Behera, B.K., Pradhan, P.K., Swaminathan, T.R., Sood, N., Paria, P. *et al.* (2018) Emergence of tilapia lake virus associated with mortalities of farmed Nile tilapia *Oreochromis niloticus* (Linnaeus 1758) in India. *Aquaculture* 484, 168–174. https://doi.org/10.1016/j.aquaculture.2017.11.025

Booncharoen, N., Mongkolsuk, S. and Sirikanchana, K. (2018) Comparative persistence of human sewage-specific enterococcal bacteriophages in freshwater and seawater. *Applied Microbiology and Biotechnology* 102(14), 6235–6246. https://doi.org/10.1007/s00253-018-9079-1

Breine, J., Van Thuyne, G. and De Bruyn, L. (2015) Development of a fish-based index combining data from different types of fishing gear. A case study of reservoirs in Flanders (Belgium). *Belgian Journal of Zoology* 145(1), 17–39. Available at: https://repository.uantwerpen.be/docman/irua/9888a3/127000.pdf (accessed 22 July 2019).

Chinabut, S. and Puttinaowarat, S. (2005) The choice of disease control strategies to secure international market access for aquaculture products. *Developments* in *Biologicals* 121, 255–261. Available at: https://europepmc.org/abstract/med/15962488 (accessed 22 July 2019).

Danovaro, R., Corinaldesi, C., Dell'Anno, A., Fuhrman, J.A., Middelburg, J.J. *et al.* (2011) Marine viruses and global climate change. *FEMS Microbiology Reviews* 35(6), 993–1034. https://doi.org/10.1111/j.1574-6976.2010.00258.x

Del-Pozo, J., Mishra, N., Kabuusu, R., Cheetham, S., Eldar, A. *et al.* (2017) Syncytial hepatitis of tilapia (*Oreochromis niloticus* L.) is associated with orthomyxovirus-like virions in hepatocytes. *Veterinary Pathology* 54(1), 164–170. https://doi.org/10.1177/0300985816658100

Dong, H.T., Siriroob, S., Meemetta, W., Santimanawong, W., Gangnonngiw, W. *et al.* (2017a) Emergence of tilapia lake virus in Thailand and an alternative semi-nested RT-PCR for detection. *Aquaculture* 476, 111–118. https://doi.org/10.1016/j.aquaculture.2017.04.019

Dong, H.T., Ataguba, G.A., Khunrae, P., Rattanarojpong, T., and Senapin, S. (2017b) Evidence of TiLV infection in tilapia hatcheries from 2012 to 2017 reveals probable global spread of the disease. *Aquaculture* 479, 579–583. https://doi.org/10.1016/j.aquaculture.2017.06.035

Ellis, A.E. (2001) Innate host defense mechanisms of fish against viruses and bacteria. *Developmental and Comparative Immunology*, 25(8–9), 827–839. https://doi.org/10.1016/S0145-305X(01)00038-6

Eyngor, M., Zamostiano, R., Kembou Tsofack, J.E., Berkowitz, A., Bercovier, H. *et al.* (2014) Identification of a novel RNA virus lethal to tilapia. *Journal of Clinical Microbiology* 52(12), 4137–4146. https://doi.org/10.1128/JCM.00827-14

FAO (Food and Agricultural Organization of the United Nations) (2017a) Outbreaks of Tilapia lake virus (TiLV) threaten the livelihoods and food security of millions of people dependent on tilapia farming. Global Information and Early Warning System on Food and Agriculture (GIEWS), Special Alert No. 338. Available at: http://www.fao.org/3/a-i7326e.pdf (accessed 22 July 2019).

FAO (Food and Agricultural Organization of the United Nations) (2017b) Social and economic performance of tilapia farming in Africa. *FAO Fisheries and Aquaculture Circular No.* 1130. Available at: http://www.fao.org/3/a-i7258e.pdf (accessed 22 July 2019).

FAO (Food and Agricultural Organization of the United Nations) (2018) Tilapia lake virus: expert knowledge elicitation risk assessment. *Animal Health Risk Analysis Assessment No.* 7. Available at: http://www.fao.org/3/CA2864EN/ca2864en.pdf (accessed 22 July 2019).

Fathi, M., Dickson, C., Dickson, M., Leschen, W., Baily, J. *et al.* (2017) Identification of tilapia lake virus in Egypt in Nile tilapia affected by 'summer mortality' syndrome. *Aquaculture* 473, 430–432. https://doi.org/10.1016/j.aquaculture.2017.03.014

Ferguson, H.W., Kabuusu, R., Beltran, S., Reyes, E., Lince, J.A. and del Pozo, J. (2014) Syncytial hepatitis of farmed tilapia, *Oreochromis niloticus* (L.): a case report. *Journal of Fish Diseases*, 37(6), 583–589. https://doi.org/10.1111/jfd.12142

García, K., Díaz, A., Navarrete, A., Higuera, G., Guiñez, E. and Romero, J. (2013) New strategies for control, prevention and treatment of ISA virus in aquaculture. In: Méndez-Vilas, V. (ed.) *The Microbial Pathogens and Strategies for Combating Them: Science, Technology and Education*. Formatex Research Center, Badajoz, Spain, pp. 587–597. Available at: https://pdfs.semanticscholar.org/1e85/9c3137a0e672df772f6dc883f23e0fb15004.pdf (accessed 9 March 2020).

Gomez, D.K., Baeck, G.W., Kim, J.H., Choresca, C.H. and Park, S.C. (2008) Genetic analysis of betanodaviruses in subclinically infected aquarium fish and invertebrates. *Current Microbiology* 56(5), 499–504. https://doi.org/10.1007/s00284-008-9116-x

Harvell, C., Kim, K., Burkholder, J., Colwell, R., Epstein, P.R. *et al.* (1999) Emerging marine diseases – climate links and anthropogenic factors. *Science* 285(5433), 1505–1510. https://doi.org/10.1126/science.285.5433.1505

Hellebø, A., Vilas, U., Falk, K. and Vlasak, R. (2004) Infectious salmon anemia virus specifically binds to and hydrolyzes 4-*O*-acetylated sialic acids. *Journal of Virology* 78(6), 3055–3062. https://doi.org/10.1128/JVI.78.6.3055-3062.2004

Hijnen, W., Beerendonk, E. and Medema, G.J. (2006) Inactivation credit of UV radiation for viruses, bacteria and protozoan (oo)cysts in water: a review. *Water Research* 40(1), 3–22. https://doi.org/10.1016/j.watres.2005.10.030

Hoegh-Guldberg, O., Jacob, D., Taylor, M., Bindi, M., Brown, S. *et al.* (2018) Impacts of 1.5ºC global warming on natural and human systems. In: Masson-Delmotte, V., Zhai, P., Pörtner, H.-O., Roberts, D., Skea, J. *et al.* (eds) *Global Warming of 1.5°C. An IPCC Special Report on the impacts of global warming of 1.5°C above pre-industrial levels and related global greenhouse gas emission pathways, in the context of strengthening the global response to the threat of climate change, sustainable development, and efforts to eradicate poverty*. World Meteorological Organization, Geneva, Switzerland.

Horas, E., Theodosiou, L. and Becks, L. (2018) Why are algal viruses not always successful? *Viruses* 10(9), 474. https://doi.org/10.3390/v10090474

Hounmanou, Y.M.G., Mdegela, R.H., Dougnon, T.V., Achoh, M.E., Mhongole, O.J. *et al.* (2018) Tilapia lake virus threatens tilapiines farming and food security: socioeconomic challenges and preventive measures in Sub-Saharan Africa. *Aquaculture* 493, 123–129. https://doi.org/10.1016/j.aquaculture.2018.05.001

Jaemwimol, P., Rawiwan, P., Tattiyapong, P., Saengnual, P., Kamlangdee, A. and Surachetpong, W. (2018) Susceptibility of important warm water fish species to tilapia lake virus (TiLV) infection. *Aquaculture*

497, 462–468. https://doi.org/10.1016/j.aquaculture.2018.08.028

Jaemwimol, P., Sirikanchan, K., Tattiyapong, P., Mongkulsuk, S. and Surachetpong, W. (2019) Virucidal effects of common disinfectants against tilapia lake virus. *Journal of Fish Diseases* 42(10), 1383–1389. https://doi.org/10.1111/jfd.13060

Jansen, M.D. and Mohan, C.V. (2017) Tilapia lake virus (TiLV): literature review. Working Paper FISH-2017-2004. CGIAR Research Program on Fish Agri-Food Systems, Penang, Malaysia. Available at: https:/// fish.cgiar.org/publications/tilapia-lake-virus-tilv-literature-review (accessed 22 July 2019).

Jansen, M.D., Dong, H.T. and Mohan, C.V. (2018) Tilapia lake virus: a threat to the global tilapia industry? *Reviews in Aquaculture* 11(3), 725–739. https://doi.org/10.1111/raq.12254

Janssen, K., Chavanne, H., Berentsen, P. and Komen, H. (2017) Impact of selective breeding on European aquaculture. *Aquaculture* 472(Suppl. 1), 8–16. https://doi.org/10.1016/j.aquaculture.2016.03.012

Jaramillo, D., Hick, P., Deece, K., Tweedie, A., Kirkland, P. *et al.* (2016) Comparison of ELISA formats for detection of antibodies specific for nervous necrosis virus (Betanodavirus) in the serum of immunized barramundi *Lates calcarifer* and Australian bass *Macquaria novemaculeata*. *Aquaculture* 451, 33–38. https://doi.org/10.1016/j.aquaculture.2015.08.015

Kabuusu, R.M., Aire, A.T., Stroup, D.F., Macpherson, C.N.L. and Ferguson, H.W. (2018) Production-level risk factors for syncytial hepatitis in farmed tilapia (*Oreochromis niloticus* L). *Journal of Fish Diseases* 41(1), 61–66. https://doi.org/10.1111/jfd.12672

Kembou Tsofack, J.E., Zamostiano, R., Watted, S., Berkowitz, A., Rosenbluth, E. *et al.* (2017) Detection of tilapia lake virus in clinical samples by culturing and nested reverse transcription-PCR. *Journal of Clinical Microbiology* 55(3), 759–767. https://doi.org/10.1128/JCM.01808-16

Kibenge, F.S.B. and Kibenge, F.J.T. (2016) *Orthomyxoviruses of fish*. In: Kibenge, F.S. and Godoy, M. (eds) *Aquaculture Virology*. Academic Press, Cambridge, Massachusetts, pp. 299–326.

Koesharyani, I., Gardenia, L., Widowati, Z., Khumaira, K. and Rustianti, D. (2018) Studi kasus infeksi tilapia lake virus (TiLV) pada ikan nila (*Oreochromis niloticus*). *Jurnal Riset Akuakultur* 13(1), 85–92. https://doi.org/10.15578/jra.13.1.2018.85-92

Kuhn, J.H. (2018) Megataxonomy of negative-sense RNA viruses: phylum *Negarnaviricota*. Available at: https:///www.researchgate.net/publication/325934040_MEGATAXONOMY_OF_NEGATIVE-SENSE_RNA_VIRUSES_PHYLUM_Negarnaviricota (accessed 22 July 2019).

Leung, T.L. and Bates, A.E. (2013) More rapid and severe disease outbreaks for aquaculture at the tropics: implications for food security. *Journal of Applied Ecology* 50(1), 215–222. https://doi.org/10.1111/1365-2644.12017

Liamnimitr, P., Thammatorn, W., U-thoomporn, S., Tattiyapong, P. and Surachetpong, W. (2018) Non-lethal sampling for tilapia lake virus detection by RT-qPCR and cell culture. *Aquaculture* 486, 75–80. https://doi.org/10.1016/j.aquaculture.2017.12.015

Mackowiak, M., Leifels, M., Hamza, I.A., Jurzik, L. and Wingender, J. (2018) Distribution of *Escherichia coli*, coliphages and enteric viruses in water, epilithic biofilms and sediments of an urban river in Germany. *Science of the Total Environment* 626, 650–659. https://doi.org/10.1016/j.scitotenv.2018.01.114

Méndez-Vilas, A. (ed.) (2013) *Microbial Pathogens and Strategies for Combating Them: Science, Technology and Education*, 1st edn. Formatex Research Center, Badajoz, Spain.

Mian, G., Godoy, D., Leal, C., Yuhara, T., Costa, G. and Figueiredo, H.C.P. (2009) Aspects of the natural history and virulence of *S. agalactiae* infection in Nile tilapia. *Veterinary Microbiology* 136(1–2), 180–183. https://doi.org/10.1016/j.vetmic.2008.10.016

Michen, B, and Graule, T. (2010) Isoelectric points of viruses. *Journal of Applied Microbiology* 109(2), 388–397. https://doi.org/10.1111/j.1365-2672.2010.04663.x

Moldovan, R., Chapman-McQuiston, E. and Wu, X. (2007) On kinetics of phage adsorption. *Biophysical Journal* 93(1), 303–315. https://doi.org/10.1529/biophysj.106.102962

Mon-on, N., Surachetpong, W., Mongkolsuk, S. and Sirikanchana, K. (2018) Roles of water quality and disinfectant application on inactivation of fish pathogenic *Streptococcus agalactiae* with povidone iodine, quaternary ammonium compounds and glutaraldehyde. *Journal of Fish Diseases* 41(5), 783–789. https://doi.org/10.1111/jfd.12776

Mugimba, K.K., Chengula, A.A., Wamala, S., Mwega, E.D., Kasanga, C.J. *et al.* (2018) Detection of tilapia lake virus (TiLV) infection by PCR in farmed and wild Nile tilapia (*Oreochromis niloticus*) from Lake Victoria. *Journal of Fish Diseases* 41(8), 1181–1189. https://doi.org/10.1111/jfd.12790

Nicholson, P., Fathi, M.A., Fischer, A., Mohan, C., Schieck, E. *et al.* (2017) Detection of tilapia lake virus in Egyptian fish farms experiencing high mortalities in 2015. *Journal of Fish Diseases* 40(12), 1925–1928. https://doi.org/org/10.1111/jfd.12650

Nicholson, P., Rawiwan, P. and Surachetpong, W. (2018) Detection of tilapia lake virus using conventional RT-PCR and SYBR Green RT-qPCR. *Journal of Visualized Experiments* (141), e58596. https://doi.org/10.3791/58596

Nicholson, P., Mon-on, N., Jaemwimol, P., Tattiyapong, P. and Surachetpong, W. (2020) Coinfection of tilapia lake virus and *Aeromonas hydrophila* synergistically increased mortality and worsened the disease severity in tilapia (*Oreochromis* spp.). *Aquaculture*

520, 734746. https://doi.org/10.1016/j.aquaculture.2019.734746

OIE (World Organization for Animal Health) (2017a) Tilapia lake virus disease, Chinese Taipei. Available at: http://www.oie.int/wahis_2/public/wahid.php/Reviewreport/Review?reportid=24033 (accessed 22 July 2019).

OIE (World Organization for Animal Health) (2017b) Tilapia lake virus (TiLV), Philippines. Available at: http://www.oie.int/wahis_2/public/wahid.php/Reviewreport/Review?page_refer=MapFullEventReport&reportid=25278 (accessed 22 July 2019).

OIE (World Organization for Animal Health) (2018a) Tilapia lake virus, Mexico. Available at: https:///www.oie.int/wahis_2/public/wahid.php/Reviewreport/Review?page_refer=MapFullEventReport&reportid=27650 (accessed 22 July 2019).

OIE (World Organization for Animal Health) (2018b) Tilapia lake virus, Peru. Available at: http://www.oie.int/wahis_2/public/wahid.php/Reviewreport/Review?page_refer=MapFullEventReport&reportid=26027 (accessed 22 July 2019).

OIE (World Organization for Animal Health) (2019) Tilapia lake virus, United States of America. Available at: http://www.oie.int/wahis_2/public/wahid.php/Reviewreport/Review?page_refer=MapFullEventReport&reportid=30412 (accessed 22 July 2019).

Phusantisampan, T., Tattiyapong, P., Mutrakulcharoen, P., Sriariyanun, M. and Surachetpong, W. (2019) Rapid detection of tilapia lake virus using a one-step reverse transcription loop-mediated isothermal amplification assay. *Aquaculture* 507, 35–39. https://doi.org/10.1016/j.aquaculture.2019.04.015

Pierezan, F., Yun, S., Surachetpong, W. and Soto, E. (2019) Intragastric and intracoelomic injection challenge models of tilapia lake virus infection in Nile tilapia (*Oreochromis niloticus* L.) fingerlings. *Journal of Fish Diseases* 42(9), 1301–1307. https://doi.org/10.1111/jfd.13052

Retel, C., Märkle, H., Becks, L. and Feulner, P.G. (2019) Ecological and evolutionary processes shaping viral genetic diversity. *Viruses* 11(3), 220. https://doi.org/10.3390/v11030220

Sandvik, T., Rimstad, E. and Mjaaland, S. (2000) The viral RNA 3′- and 5′-end structure and mRNA transcription of infectious salmon anaemia virus resemble those of influenza viruses. *Archives of Virology* 145(8), 1659–1669. https://doi.org/10.1007/s007050070082

Senapin, S., Shyam, K.U., Meemetta, W., Rattanarojpong, T. and Dong, H.T. (2018) Inapparent infection cases of tilapia lake virus (TiLV) in farmed tilapia. *Aquaculture* 487, 51–55. https://doi.org/10.1016/j.aquaculture.2018.01.007

Sirikanchana, K., Shisler, J.L. and Mariñas, B.J. (2008) Inactivation kinetics of adenovirus serotype 2 with monochloramine. *Water Research* 42(6–7), 1467–1474. https://doi.org/10.1016/j.watres.2007.10.024

Sirimanapong, W., Thompson, K.D., Shinn, A.P., Adams, A. and Withyachumnarnkul, B. (2018) *Streptococcus agalactiae* infection kills red tilapia with chronic *Francisella noatunensis* infection more rapidly than the fish without the infection. *Fish and Shellfish Immunology* 81, 221–232. https://doi.org/10.1016/j.fsi.2018.07.022

Smail, D.A., Grant, R., Simpson, D.A., Bain, N. and Hastings, T.S. (2004) Disinfectants against cultured infectious salmon anaemia (ISA) virus: the virucidal effect of three iodophors, chloramine T, chlorine dioxide and peracetic acid/hydrogen peroxide/acetic acid mixture. *Aquaculture* 240(1–4), 29–38. https://doi.org/10.1016/j.aquaculture.2004.05.045

Subramaniam, K., Ferguson, H.W., Kabuusu, R. and Waltzek, T.B. (2019) Genome sequence of tilapia lake virus associated with syncytial hepatitis of tilapia in an Ecuadorian aquaculture facility. *Microbiology Resource Announcements* 8(18), e00084-19. https://doi.org/10.1128/MRA.00084-19

Surachetpong, W., Janetanakit, T., Nonthabenjawan, N., Tattiyapong, P., Sirikanchana, K. and Amonsin, A. (2017) Outbreaks of tilapia lake virus infection, Thailand, 2015–2016. *Emerging Infectious Diseases* 23(6), 1031–1033. https://doi.org/10.3201/eid2306.161278

Tattiyapong, P., Dachavichitlead, W. and Surachetpong, W. (2017) Experimental infection of tilapia lake virus (TiLV) in Nile tilapia (*Oreochromis niloticus*) and red tilapia (*Oreochromis* spp.). *Veterinary Microbiology* 207, 170–177. https://doi.org/10.1016/j.vetmic.2017.06.014

Tattiyapong, P., Sirikanchana, K. and Surachetpong, W. (2018) Development and validation of a reverse transcription quantitative polymerase chain reaction for tilapia lake virus detection in clinical samples and experimentally challenged fish. *Journal of Fish Diseases* 41(2), 255–261. https://doi.org/10.1111/jfd.12708

Thammatorn, W., Rawiwan, P. and Surachetpong, W. (2019) Minimal risk of tilapia lake virus transmission via frozen tilapia fillets. *Journal of Fish Diseases* 42(1), 3–9. https://doi.org/10.1111/jfd.12924

Thangaraj, R.S., Ravi, C., Kumar, R., Dharmaratnam, A., Saidmuhammed, B.V. *et al*. (2018) Derivation of two tilapia (*Oreochromis niloticus*) cell lines for efficient propagation of tilapia lake virus (TiLV). *Aquaculture* 492, 206–214. https://doi.org/10.1016/j.aquaculture.2018.04.012

Tidbury, H.J., Joiner, C.L., Rimmer, G.S.E., Potter, H.V. and Taylor, N.G.H. (2018) The effectiveness of fishery net dips: advice for the improvement of biosecurity measures. *Journal of Fish Diseases* 41(11), 1625–1630. https://doi.org/10.1111/jfd.12868

Waiyamitra, P., Tattiyapong, P., Sirikanchana, K., Mongkolsuk, S., Nicholson, P. and Surachetpong, W. (2018) A TaqMan RT–qPCR assay for tilapia lake virus (TiLV) detection in tilapia. *Aquaculture* 497,

184–188. https://doi.org/10.1016/j.aquaculture.2018.07.060

Wang, C., Liu, S., Li, X., Hao, J., Tang, K.F.J. and Zhang, Q. (2018a) Infection of covert mortality nodavirus in Japanese flounder reveals host jump of the emerging alphanodavirus. *Journal of General Virology* 100(2), 166–175. https://doi.org/10.1099/jgv.0.001177

Wang, Y., Wang, Q., Zeng, W., Yin, J., Li, Y. *et al.* (2018b) Establishment and characterization of a cell line from tilapia brain for detection of tilapia lake virus. *Journal of Fish Diseases* 41(12), 1803–1809. https://doi.org/10.1111/jfd.12889

Weber, F., Gruber, S., Haller, O. and Kochs, G. (1999) PB2 polymerase subunit of Thogoto virus (*Orthomyxoviridae* family). *Archives of Virology* 144(8), 1601–1609. https://doi.org/10.1007/s007050050613

Yamkasem, J., Tattiyapong, P., Kamlungdee, A. and Surachetpong, W. (2019) Evidence of potential vertical transmission of tilapia lake virus. *Journal of Fish Diseases* 42(9), 1293–1300. https://doi.org/10.1111/jfd.13050

Yang, Y.F., Lu, T.H., Lin, H.C., Chen, C.Y. and Liao, C.M. (2018) Assessing the population transmission dynamics of tilapia lake virus in farmed tilapia. *Journal of Fish Diseases*, 41(9), 1439–1448. https://doi.org/10.1111/jfd.12845

Yin, J., Wang, Q., Wang, Y., Li, Y., Zeng, W. *et al.* (2019) Development of a simple and rapid reverse transcription-loop mediated isothermal amplification (RT-LAMP) assay for sensitive detection of tilapia lake virus. *Journal of Fish Diseases* 42(6), 817–824. https://doi.org/10.1111/jfd.12983

US EPA (Environmental Protection Agency) (2015) Review of coliphages as possible indicators of fecal contamination for ambient water quality. Available at: https://www.epa.gov/sites/production/files/2015-10/documents/coliphages-literature-review-report-2015.pdf (accessed 22 July 2019).

9 Iridovirosis

Hsin-Yiu Chou[1]*, Hidehiro Kondo[2] and Qi-Wei Qin[3]

[1]*National Taiwan Ocean University, Taiwan, Republic of China;* [2]*Tokyo University of Marine Science and Technology, Tokyo, Japan;* [3]*South China Agricultural University, Guangzhou, People's Republic of China*

9.1 Brief Introduction

9.1.1 History

Fish iridovirosis is a highly contagious disease which causes mass mortality in wild and cultured fish populations, from juvenile to market-sized fish. Primarily, the disease is represented by red sea bream iridoviral disease (RSIVD), the first identified fish iridovirosis. The outbreak of RSIVD was first reported in farmed red sea bream (*Pagrus major*) in Shikoku Island, Japan in 1990 (Inouye *et al.*, 1992). To date, more than 30 other species of cultured marine fish have been recorded with iridovirosis, and its prevalence in cultured fish populations has been well documented not only in Japan but also widely in East and South-East Asian countries, including China, Korea, Singapore, Taiwan, Thailand and Vietnam (Jung *et al.*, 1997; Miyata *et al.*, 1997; Chou *et al.*, 1998; Jung and Oh, 2000; Kim *et al.*, 2002; Sudthongkong *et al.*, 2002b; Jeong *et al.*, 2003, 2006; Do *et al.*, 2004, 2005; Gibson-Kueh *et al.*, 2004; Kurita *et al.*, 2004; Oseko *et al.*, 2004; Nakajima and Kunita, 2005; Dong *et al.*, 2017).

9.1.2 Aetiology

The causal agent, red sea bream iridovirus (RSIV), is a large, icosahedral, cytoplasmic DNA virus and is a member of the family *Iridoviridae*. The disease is characterized by basophilic enlarged cells in the spleen, heart, kidney, liver and gill. Hexagonal virions, measuring 200–240 nm in diameter, were found in the cytoplasm of infected cells (Inouye *et al.*, 1992; Nakajima and Sorimachi, 1994). The epidemiology and diagnosis have been well described in previous reviews (Whittington *et al.*, 2010; Kurita and Nakajima, 2012; Hick *et al.*, 2016; Kawato *et al.*, 2017a).

9.1.3 Taxonomy

Iridoviruses are large, icosahedral, double-stranded DNA-containing viruses that can infect various poikilothermic vertebrates and invertebrates. The family *Iridoviridae* is divided into two subfamilies, *Alphairidovirinae* and *Betairidovirinae*. The former is comprised of iridoviruses that infect a wide range of bony fish, amphibians and reptiles, and is classified into three genera: *Lymphocystivirus*, *Ranavirus* and *Megalocytivirus*. The *Betairidovirinae* contains two genera (*Iridovirus* and *Chloriridovirus*) that infect mainly invertebrates such as insects and crustaceans (Chinchar *et al.*, 2017). Piscine iridoviruses in the genus *Lymphocystivirus* generally cause non-fatal, superficial dermal infections, while the ranaviruses and megalocytiviruses are notoriously known for causing high mortality in many economically important fish species, including RSIV of Japan, *Singapore grouper iridovirus* (SGIV) and *Infectious spleen and kidney necrosis virus* (ISKNV) of China (Inouye *et al.*, 1992; He *et al.*, 1998, 2000; Qin *et al.*, 2003). In Taiwan, the incidence of systemic infections of marine fish caused by iridoviruses can be traced back to 1998. The infectious agents have been isolated and designated as grouper iridovirus of Taiwan (TGIV) and grouper iridovirus (GIV), which belong to the genus *Megalocytivirus* and genus *Ranavirus*, respectively (Chou *et al.*, 1998; Chao *et al.*, 2002; Murali *et al.*, 2002).

*hychou@email.ntou.edu.tw

9.2 Diagnosis of the Disease

9.2.1 Clinical signs

Megalocytivirus infects more than 50 species of marine and freshwater fish, causing systemic disease characterized by cellular hypertrophy in multiple internal organs, commonly the kidney and spleen (Gias *et al.*, 2011; Chinchar *et al.*, 2017). However, the three types of megalocytivirus infections showed different clinical signs. RSIV infection in red sea bream (*P. major*) results in enlarged spleen, erratic swimming and lethargy (Wang *et al.*, 2003). Spontaneously diseased grouper (*Epinephelus* spp.) due to TGIV have similar signs with anorexia and weight loss, lethargy, an overall darkening, pale gills and enlarged spleen (Fig. 9.1). Clinical signs of ISKNV infection in mandarin fish (*Siniperca chuatsi*) include anorexia, abnormal swimming, faded body pigmentation and pale gills (He *et al.*, 2002; Subramaniam *et al.*, 2012). In addition, petechial haemorrhages appear widely on the body. The kidney and spleen are swollen, whereas the heart and liver tend to be pale (He *et al.*, 2002; Subramaniam *et al.*, 2012). Fish infected with turbot reddish body iridovirus (TRBIV) are anorexic, lethargic, with dark pigmentation of the tail and fins, and display abnormal swimming. The gills are pale and are more susceptible to TRBIV infection than kidney and liver (Subramaniam *et al.*, 2012). A 'reddish body syndrome' (RBS) was first found in both juveniles and adults of farmed turbot (*Scophthalmus maximus*) in China. The gross signs of diseased turbot were pale gills with local haemorrhages, petechial haemorrhages in fins and fin bases, and especially haemorrhages in muscle and skin (Shi *et al.*, 2004). The turbot iridovirus (TBIV)-infected fish in Go-Chang, Korea, have pale body colour, an enlarged abdomen, protruding eyes, an enlarged spleen and kidney, and pale gills and/or liver (Kim *et al.*, 2005).

Fig. 9.1. Appearance and internal symptoms of diseased grouper (*Epinephelus* sp.) due to TGIV collected from southern Taiwan in 1995. (A) Diseased fish show underweight condition, an overall darkening and enlarged spleen. (B) When compared with the spleen of normal fish, 87 of 107 diseased fish show spleen enlargement.

Ranaviruses cause severe systemic disease involving multiple internal organs that often culminates in organ necrosis and massive haemorrhaging (Miller *et al.*, 2011). The gross signatures of infected amphibians include swelling of the legs and body, erythema and ecchymosis on the body, and irregular skin discoloration (Miller *et al.*, 2011). In the USA, ranaviruses primarily affect the larvae of amphibians, such as tiger salamander (*Ambystoma tigrinum*) and northern leopard frog (*Lithobates pipiens*), with distinct clinical signs including erratic swimming, haemorrhaging and swellings. However, in Europe, infected adult amphibians have different signs presented by: (i) a chronic disease accompanied with skin ulceration but no obvious internal gross lesions; and (ii) a peracute disease characterized by systemic haemorrhages (Lesbarrères *et al.*, 2012). Not just in amphibians, haemorrhages, fluid accumulation and skin ulceration also occur in fish (e.g. rainbow trout) and reptiles (e.g. green tree pythons), and necrosis of the oral cavity can be especially severe in chelonians (Lesbarrères *et al.*, 2012). Redfin perch (*Perca fluviatilis*) infected with *Epizootic haematopoietic necrosis virus* (EHNV) have petechial haemorrhages at the base of fins and excessive amounts of serosanguinous peritoneal fluid, together with swelling of the kidney and spleen (Hick *et al.*, 2016). Infection with largemouth bass virus (LMBV) in China causes skin and muscle ulcers and splenomegaly (Deng *et al.*, 2011). SGIV infection is characterized by haemorrhages and enlargement of the spleen (Qin *et al.*, 2003). Other clinical signs of SGIV-infected fish include lethargy, absence of response to stimulation, inappetence and hanging at the surface or bottom of the net (Hick *et al.*, 2016).

9.2.2 External/internal macroscopic and microscopic lesions

The most characteristic pathology of megalocytivirus infection is hypertrophied, basophilic cells in a variety of tissues, with or without degeneration and necrosis. Histopathological observations identify the hypertrophic cells present in the spleen, haematopoietic tissue in kidney, cranial connective tissues, liver, intestine, gills and endocardium (He *et al.*, 2000). The hypertrophied cells caused by lymphocystis are usually named lymphocystis cells. These cells show a thick hyaline capsule, an enlarged nucleus and cytoplasmic DNA inclusions (Kitamura *et al.*, 2006; Hossain and Oh, 2011).

A common histological lesion in ranavirus infection of amphibians, reptiles and fishes is necrosis of endothelial cells that results in destruction of many organs and necrosis of haematopoietic tissue. The severity of these lesions varies with host species and ranavirus isolate (Lesbarrères *et al.*, 2012). Moreover, intracytoplasmic inclusion bodies are in multiple cell types. In iridovirus-infected fish, this may include enlargement of cells and necrosis of the renal and splenic haematopoietic tissues (Qin *et al.*, 2003). The affected tissues of RBS-infected turbot were the endothelial and connective tissues of the spleen, kidney, gill and intestine. The infected cells showed hypertrophic appearances (Shi *et al.*, 2004). A most common feature of the histopathology in EHNV-infected fish is the presence of many scattered, individual necrotic cells within blood vessels (Fig. 9.2) (Hick *et al.*, 2016).

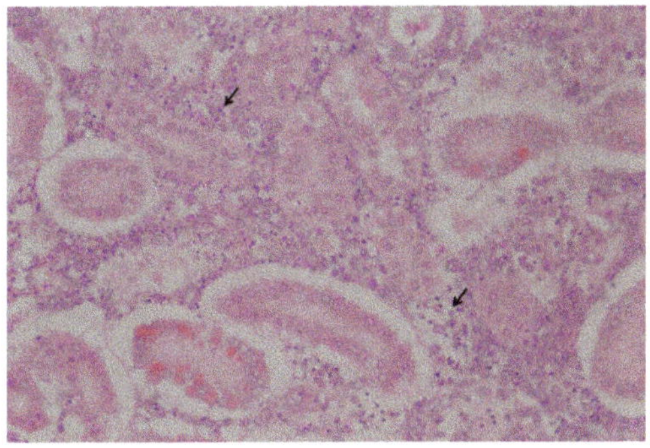

Fig. 9.2. Histopathology of EHNV disease in the kidney of redfin perch (*Perca fluviatilis*). There is interstitial (haematopoietic) necrosis characterized by karyorrhexis and karyolysis of individual cells throughout the section, creating a 'starry sky' appearance (arrows). Formalin-fixed, paraffin-embedded section, haematoxylin and eosin stain.

9.2.3 Biochemical/immunological/ molecular diagnosis

To confirm the clinical diagnosis of iridovirus, assays including biochemical, immunological and molecular techniques have been developed. Polymerase chain reaction (PCR) assay is currently one of most common and widely used technologies. Nearly all reported iridoviruses have detailed procedures for PCR assays (Oshima *et al.*, 1998; Chen *et al.*, 2003; Grizzle *et al.*, 2003). PCR assays usually target the gene encoding major capsid protein (MCP), ATPase or polymerase. Compared with conventional PCR, real-time PCR assay has increased sensitivity and high-throughput testing for certification testing. Wang *et al.* (2006) developed a real-time PCR using a molecular beacon for quantitative detection of large yellow croaker iridovirus (LYCIV) infection. Loop-mediated isothermal amplification (LAMP) also has the advantage of better sensitivity and high specificity and efficiency for detection of iridoviruses such as SGIV and RSIV (Caipang *et al.*, 2004; Mao *et al.*, 2008; Subramaniam *et al.*, 2012). In addition, based on a non-radioactive digoxigenin-labelled DNA probe, an *in situ* hybridization method can be an important tool (Huang *et al.*, 2004). Furthermore, aptamer, an oligonucleotide fragment, has been selected to specifically target SGIV and may be developed as a potential probe in diagnostics and pathogen detection (Li *et al.*, 2014, 2015).

Further isolation and detection of virus can be performed in suitable fish cell lines, which is one of the standard assays to ascertain the presence of the live virus. *Frog virus 3* (FV3), bohle iridovirus (BIV), pike-perch iridovirus (PPIV), European catfish virus (ECV), European sheatfish virus (ESV), EHNV, doctor fish virus (DFV), guppy virus 6 (GF6), short-finned eel virus (SERV) and *Rana esculenta* virus Italy 282/102 (REV 282/102) will grow well in bluegill fry (BF-2), epithelioma papulosum cyprini (EPC) and chinook salmon embryo (CHSE-214) cells; in fathead minnow (FHM) and rainbow trout gonad (RTG-2) cells the virus titres are lower (Ariel *et al.*, 2009). SGIV grows well in a series of cell lines from grouper, such as grouper embryonic (GP) and grouper spleen (GS) cells. Mandarin fish fry (MFF-1) displayed high sensitivity to ISKNV. Flounder embryonic cell line (FEC) is susceptible to lymphocystis virus (LCV) and turbot reddish body iridovirus (TRBIV), while bluegill fry-2 (BF-2), Grunt fit (GF) and CRF-1 (a new red bream *P. major* cell line) can be used to carry out diagnosis for RSIV. An antigen capture enzyme-linked immunosorbent assay (ELISA) is also used in diagnosis. The same antibody regents can be used in immunohistochemical staining procedures. A rapid diagnosis method using specific monoclonal antibodies has been developed for RSIV disease (Ito *et al.*, 2013).

9.3 Expected/Potential Spread of the Pathogen

9.3.1 Pathogenesis of iridovirosis

Pathogenesis is a process that describes how the pathogen induces disease. Herein, SGIV is taken as an example to illustrate the pathogenesis mechanism of iridovirosis (Fig. 9.3) (Wang *et al.*, 2014).

SGIV enters host cells via clathrin-mediated endocytosis and macropinocytosis in a dynamin-, pH-dependent manner, but not via caveola-dependent endocytosis (Wang *et al.*, 2014). This process differs from the caveola-mediated endocytosis in tiger frog virus (TFV) and ISKNV, or the clathrin-mediated endocytosis in FV3 (Braunwald *et al.*, 1985; Guo *et al.*, 2011, 2012a). Moreover, it has been demonstrated for the first time that micropinocytosis is involved in iridovirus entry (Wang *et al.*, 2014). Micropinocytosis is also essential for infection of the soft-shell turtle iridovirus (Huang, Y.H. *et al.*, 2018). Thus, the authors proposed that micropinocytosis might be an important entry pathway for iridovirus. By tracking individual SGIV particles in real time, SGIV was observed to travel along actin-rich protrusions to enter cells. Once taken up by cells, SGIV could be transported along actin or microtubules. When microtubules or actin filaments were disrupted by nocodazole or CytoD, respectively, the motility of SGIV was remarkably impaired. Following internalization, SGIV could be sorted to acidic endosomes, such as Rab5- or Rab7-positive endosomes (Wang *et al.*, 2014).

Furthermore, using a novel ultrafast tracking technique based on atomic force microscopy, the force tracing technique, the entry of single viral particle-induced invagination could be monitored and quantified. The maximum velocity of SGIV entry into the host cell is approximately 200 nm/s and the endocytic force is approximately 60.8 ± 18.5 pN (Pan *et al.*, 2015). In addition, the binding energy density becomes larger with increasing engulfment depth, indicating that additional binding events between viral ligands and receptors gradually occurred to provide enough energy for accomplishing viral invagination (Pan *et al.*, 2015).

Virus assembly was also studied using electron microscopy. Large quantities of electron-dense core-like materials, tubular membrane-like structures, empty capsids and partly formed empty capsids by sealing tubular membranes were within the viral matrix. The fully matured nucleocapsids were probably formed by insertion of the electron-dense nucleoprotein core into a partly formed empty capsid just before the capsid is completely sealed. The budding and release of the viral particle was also observed (Qin *et al.*, 2001). Besides the well-known budding mechanism through the cell membrane, a novel budding process in which viral particles bud into a tubular-like structure within vacuoles was observed (Liu *et al.*, 2016).

9.3.2. Signalling events in virus infection

Mitogen-activated protein kinase (MAPK) cascades are evolutionarily conserved intracellular signalling networks that regulate cell growth, apoptosis and cellular responses to stress (Liu *et al.*, 2007). In SGIV-infected cells, MAPK cascades are involved in SGIV-induced parapoptosis and viral replication, including extracellular signal-regulated kinase (ERK), p38 mitogen-activated protein kinase (p38 MAPK) and c-Jun N-terminal kinase (JNK) signalling (Huang *et al.*, 2011a,b,c). Furthermore, the expression of grouper immune genes including interferon regulatory factor 1 (IRF1), interleukin-8 (IL-8) and tumour necrosis factor alpha (TNF-α) is regulated by JNK, while only TNF-α is regulated by p38 MAPK. Therefore, it is proposed that the JNK pathway is essential for SGIV replication and the modulation of inflammatory responses during virus infection (Huang *et al.*, 2011b).

The ubiquitin–proteasome system (UPS) is the major intracellular protein degradation pathway and plays crucial roles in a variety of fundamental cellular processes including regulation of gene transcription, cell cycle progression, autophagy, and modulation of the immune and inflammatory responses (Gustin *et al.*, 2011; Kleiger and Mayor, 2014). During SGIV infection, the expression of 65 genes within the UPS pathway including ubiquitin encoding, ubiquitination, deubiquitination and proteasome formation are up- or downregulated.

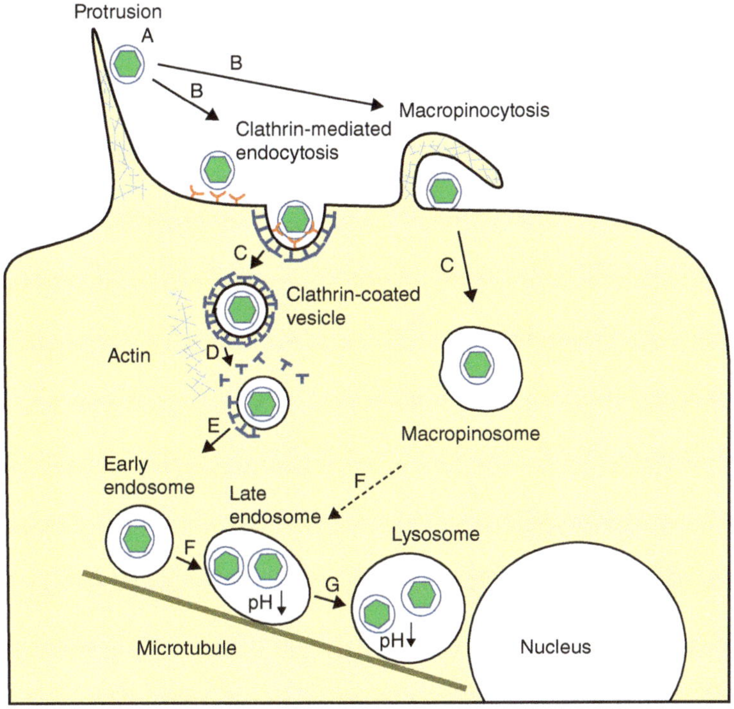

Fig. 9.3. Model entry route of SGIV into GS cells. A, SGIV particles may transport along actin-rich protrusions to reach the cell surface; B, internalization of SGIV particles by clathrin-mediated endocytosis and micropinocytosis; C, internalized particles contained within clathrin-coated vesicles and macropinosomes; D, clathrin-coated vesicles containing individual SGIV particles are rapidly uncoated; E, the virus is transported to the early endosome; F, an early endosome matures into a late endosome by decreasing its pH, and the macropinosome may also undergo acidification; G, further acidification brings the late endosome to the lysosome. C and D may represent actin-dependent movement, while E, F and G may represent microtubule-dependent movement.

Inhibition of proteasome formation by different proteasome inhibitors significantly decreases SGIV replication *in vitro*, accompanied by inhibition of virus assembly site formation, and viral gene transcription and protein transportation. Moreover, ubiquitin overexpression partly rescues the inhibitory effect of ubiquitin inhibitor on SGIV replication, suggesting that UPS plays a crucial role in SGIV replication *in vitro* (Huang, X.H. *et al.*, 2018).

9.4. Expected/Potential Spread of Intermediate and/or Reservoir Host

9.4.1. Host description

Megalocytivirus infects a wide range of fish species. The susceptible species include: red sea bream, *P. major* (Inouye *et al.*, 1992); groupers, *Epinephelus* spp. (Chua *et al.*, 1994; Danayadol *et al.*, 1997; Lu *et al.*, 2005; Wang *et al.*, 2007); mandarin fish, *S. chuatsi* (Wu *et al.*, 1997; He *et al.*, 1998, 2000); sea bass, *Lateolabrax* spp. (Jung *et al.*, 1997); angelfish, *Pterophyllum scalare* (Rodger *et al.*, 1997); tilapia, *Oreochromis niloticus* (McGrogan *et al.*, 1998); rock bream, *Oplegnathus fasciatus* (Jung and Oh, 2000); dwarf gourami, *Colisa lalia* (Sudthongkong *et al.*, 2002a); African lampeye, *Aplocheilichthys normani* (Sudthongkong *et al.*, 2002b); red drum, *Sciaenops ocellata* (Weng *et al.*, 2002); large yellow croaker, *Larimichthys crocea* (Chen *et al.*, 2003); turbot, *S. maximus* (Shi *et al.*, 2004); Murray cod, *Maccullochella peelii peelii* (Go *et al.*, 2006); and spotted knifejaw, *Oplegnathus punctatus* (Dong *et al.*, 2010).

However, the susceptibility to different types of megalocytivirus may vary by fish species. For example, red sea bream is resistant to ISKNV and rock bream is resistant to TBIV, but both fish species are susceptible to RSIV. A more complete list is available online at the World Organization for Animal Health's website (http://www.oie.int/fileadmin/Home/eng/Internationa_Standard_Setting/docs/pdf/2.3.07_RSIVD.pdf (accessed 10 March 2020)).

Members of the genus *Ranavirus* are a group of viruses capable of infecting a wide variety of vertebrate hosts including fish, amphibians and reptiles; at least 175 species across 52 families were reported (Marschang, 2011; Miller *et al.*, 2011; Gray *et al.*, 2015).

9.4.2 Potential spread of reservoir host

Megalocytivirus is classified into three subgroups, RSIV, ISKNV and TRBIV types, according to the nucleotide sequences of their MCP and ATPase genes (Kurita and Nakajima, 2012). This is confirmed by a preliminary whole-genome comparison (Fig. 9.4). The RSIV-type viruses are in marine fish of East and South-East Asia, whereas the ISKNV-type viruses are in marine and freshwater fish in South-East Asia, China and Taiwan (Kurita and Nakajima, 2012). TRBIV-type viruses are believed to mainly affect flatfishes, such as flounder and turbot in China and South Korea (Shuang *et al.* 2013), although some isolates have also been reported from barred knifejaw (Kurita and Nakajima, 2012). Recently, *Scale drop disease virus* (SSDV), showing low homology of nucleotide sequences to the three types of megalocytivirus (De Groof *et al.*, 2015), has also been classified as a megalocytivirus (Chinchar *et al.*, 2017). Because of their higher nucleotide sequence identities among the RSIV-type viruses (more than 93%), the isolates were designated in various names such as giant sea perch iridovirus (GSIV), LYCIV, orange-spotted grouper iridovirus (OSGIV) and rock bream iridovirus (RBIV).

Virus strains closely related to the ISKNV subgroup were reported to cause disease in more than 50 marine and freshwater fish species (orders Perciformes, Pleuronectiformes, Clupeiformes, Tetraodontiformes, Myctophiformes and Mugiliformes) in China (Wang *et al.*, 2007), as well

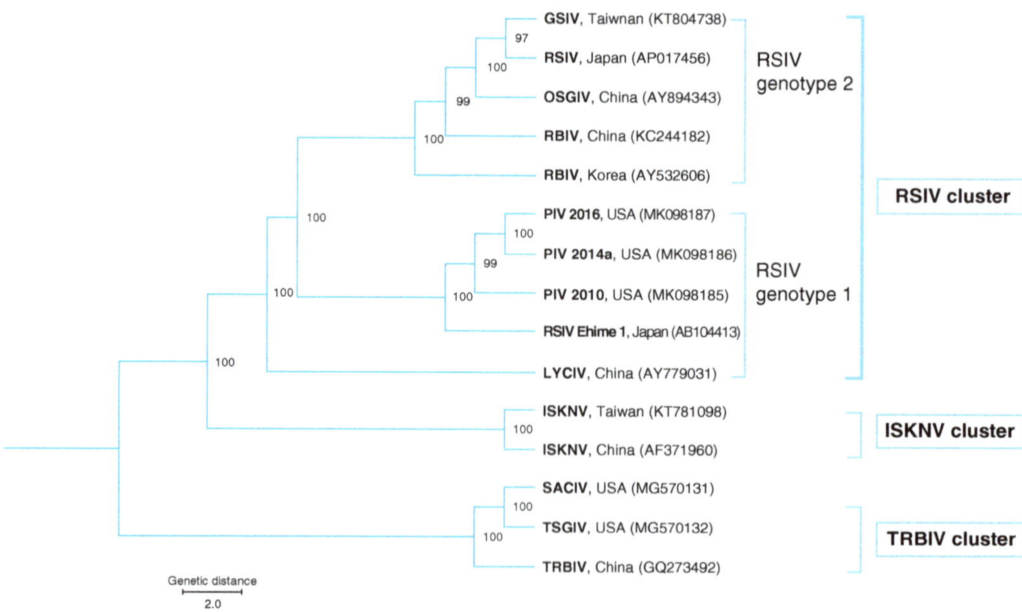

Fig. 9.4. Phylogenetic tree based on single-nucleotide polymorphisms analysis of 15 megalocytiviruses (whole genome). Bootstrap values are represented by percentages at nodes and GenBank accession numbers are in parentheses.

as ornamental fish (Yanong and Waltzek, 2010). Fu *et al.* (2011) compared the complete major capsid protein genes of nine ISKNV isolates with other known megalocytiviruses. They found there were some significant molecular characters and divided 33 megalocytiviruses into three genotypes. Further, based on the phylogenetic tree, there was a strong host species signal in three genotypes: genotype I (mainly from marine fish); genotype II (from freshwater fish; belonging to the same order, Perciformes); and genotype III (mainly from marine flatfish belonging to Pleuronectiformes). Herein, nine ISKNV isolates belong to genotype I or genotype II. It was suggested that hosts of ISKNV consisted of both freshwater and marine fishes. This finding is consistent with previous reports by Nakajima and Kunita (2005) and Wang *et al.* (2007). Since mandarin fish is the host of genotype I and II viruses, it may be a mixing vessel host of ISKNV and other megalocytiviruses from marine fish.

Previously, TRBIV-like megalocytiviruses had not been documented in ornamental fish. In a recent study, Go *et al.* (2016) analysed two formalin-fixed, paraffin-embedded materials of archival ornamental fish cases from 1986 and 1988 in conjunction with data for a range of genes from five cases of fresh frozen tissues obtained from 1991 through to 2010. This was the first evidence of TRBIV-like megalocytiviruses in ornamental fish species and it was suggested that the TRBIV genotype might have been the major genotype affecting ornamental fish from the time when megalocytiviruses appeared to emerge in ornamental fish in the late 1980s until the early 1990s.

Among fish ranavirus isolates, two genetically distinct but related ranaviruses, SGIV and GIV, have negative impacts on grouper mariculture in Asia (Murali *et al.*, 2002; Qin *et al.*, 2002; Peng *et al.*, 2015). Notably, SGIV and GIV appear to be the most related viruses among the current isolates of the genus *Ranavirus*. Grouper iridoviruses possess few regions of collinearity in genome sequence with other ranaviruses (Jancovich *et al.*, 2015). Therefore, GIV/SGIV was recognized as a novel species. For a better understanding of the relationships among ranavirus species, the genomic sequence of EHNV was determined and compared with FV3, TFV, *Ambystoma tigrinum* virus (ATV), soft-shelled turtle ranavirus (STIV) and two grouper iridoviruses (GIV/SGIV) (Jancovich *et al.*, 2010). The data suggest several recent hosts shifts have occurred. These findings indicate that pathogens of cold-blooded vertebrates may cross species barriers easily and led to recent host jumping of ranaviruses. The authors proposed that the ancestral ranavirus was a fish virus and postulated that there must have been at least three species jumps, from fish to frogs, from fish to salamanders and from frogs to reptiles, and perhaps as many as four species jumps, including a jump from tetrapod amphibians back to fish (Jancovich *et al.*, 2010).

9.5 Role of Climate in Host Distribution and Susceptibility

The first outbreak of RSIV was recorded in red sea bream (*P. major*) during August to September in Ehime Prefecture, Japan (Inouye *et al.*, 1992), where the water temperature is around 25°C. Likewise, the optimal temperature for RSIV *in vitro* replication is 20 or 25°C (Nakajima and Sorimachi, 1994). In Korea, outbreaks of RSIV disease in cultured rock bream (*O. fasciatus*) occurred from August to September, when water temperature was 23–27°C (Jung and Oh, 2000). Choi *et al.* (2006) sampled the yearling rock bream in May, and suggested that RSIV establishes a latent infection when water temperature is below 18°C.

In 1994–1998, the natural outbreaks of ISKNV in mandarin fish occurred from March to November at water temperatures ranging from 20 to 32°C. Nevertheless, no outbreak of the disease was reported from December to April when the water temperature ranges from 9 to 20°C in Guangdong Province, China. He *et al.* (2002) indicated experimentally that infected mandarin fish show no clinical signs or mortality when held at 20 and 15°C; however, at temperatures above 25°C, clinical signs with high mortality took place which is consistent with the natural outbreaks of ISKNV.

The same epidemic situation was reported in juvenile turbot (*S. maximus*) mariculture facilities at Go-Chang, South Korea, in 2003. The mortality of cultured turbot due to TBIV increased rapidly when the water temperature rose from 17–18 to 20–23°C (Kim *et al.*, 2005). Further, Oh *et al.* (2006) confirmed that TBIV causes up to 60% mortality in turbot kept at 22°C, but no mortality occurred at 20 and 17°C. Their results of PCR detection and the occurrence of enlarged spleen cells in TBIV-injected fish clearly indicated an optimum growth temperature for TBIV higher than 22°C, which is consistent with field results.

Temperature can affect the replication and thus also the viral virulence of megalocytiviruses, including ISKNV, RSIV and TGIV, *in vitro* and *in vivo* (Inouye *et al.*, 1992; Kusuda *et al.*, 1994; Chou *et al.*, 1998; He *et al.*, 2002; Nakajima *et al.*, 2002; Sano *et al.*, 2002; Jun *et al.*, 2009; Liu *et al.*, 2016). In Korea, RSIV causes recurrent outbreaks in marine farms year after year. Jun *et al.* (2009) investigated the influence of temperature shifts on the onset of RSIVD in rock bream and the viral concentration in the spleen of infected fish. Their results revealed that iridovirus may persist in rock bream at low water temperatures for extended periods. Then, when summer arrives and the water warms, viral replication accelerates and eventually leads to high mortalities in farm fish. Therefore, water temperature appears to be a major factor that influences outbreaks of megalocytiviruses. Climate change will increase fish mortality and cause a change in the distribution of marine resources. Since fish iridoviruses have a very broad host range, the probable spread of iridovirosis to new regions will need continuous attention.

9.6 Increase in Pathogen Mortality Due to Temperature Changes

Environmental temperature often plays an important role in the outbreak of diseases in aquatic animals by altering viral replication directly and the host immune response indirectly (Snieszko, 1974; Bly and Clem, 1992; Alcorn *et al.*, 2002). Outbreaks of RSIVD have resulted in high mortalities in cultures of red sea bream and other marine fish in south-west Japan, primarily in the summer. Temperature-associated outbreaks of diseases have been reported in several piscine iridoviruses as well.

For the ISKNV in China, the outbreaks occurred from March to November in 1994–1998 (He *et al.*, 1998, 2000). ISKNV caused no clinical signs or mortality in mandarin fish held at 20 and 15°C; however, significant mortality was observed when the water temperature rose to 25°C (He *et al.*, 2002). This observation is consistent with the natural outbreaks of ISKNV observed in the field. In addition, a similar observation was reported for ENHV infection. Results of viral transmission studies with EHNV in redfin perch showed that while a longer period of incubation (up to 28 days) at colder temperatures was required for the disease outbreak, a shorter incubation period (about 11 days) for the same virus was achieved at a water temperature of 19–21°C. No occurrence of disease was observed when the temperature was below 12°C (Whittington and Reddacliff, 1995). These two viral agents have proved to be highly virulent in different teleosts in aquaculture. Natural outbreaks of diseases caused by these viruses occur most frequently in spring and summer at higher temperatures (Wang *et al.*, 2003; Li *et al.*, 2011).

Liu *et al.* (2016) determined the impact of different temperatures (18, 25 and 32°C) on TGIV infection in grouper. As the temperature rose from 25 to 32°C, the presence of pyknotic nuclei and chromatin margination became more prominent in the infected cells, indicating an apoptotic death. The morphological feature of apoptotic cells was confirmed using electron microscopy. In addition, the apoptosis of TGIV-infected cells promoted by hyperthermia (32°C) was confirmed using DNA laddering, DNA content and annexin V staining. Also, cumulative mortalities due to TGIV reached 100% in the fish held at 25°C on day 10 post-infection, whereas it was only 37.5% at 32°C at 2 weeks post-infection. The mortality in the infected fish held at 32–25°C skyrocketed to 97.5% by day 8. Altogether, the data suggest while the lower temperature (18°C) adversely affects the propagation of TGIV, the hyperthermic temperature (32°C) promotes apoptosis to prevent the spreading of the virus, resulting in higher resistance of the host against the virus (Liu *et al.*, 2016). This result suggests that projecting the effect of climate change due to increases in water temperatures is difficult.

Likewise, temperature strongly influenced the incidence of RSIVD in rock bream (*O. fasciatus*) exposed to Sachun (IVS-1) and RSIV. Rock bream is more susceptible to RSIV-type virus than other fish species (OIE, 2018). The disease progressed more rapidly at higher water temperatures (e.g. 25, 21 or 18°C) but no mortality was observed at ≤18°C. Oh *et al.* (2014) reported that fish survivors from previous viral infection under 18°C were resistant to further virus challenge. When the water temperature rose from 13 to 25°C, the cumulative mortality reached 100%. Their experiments also suggested that lower viral numbers can cause iridoviral disease in rock bream at lower water temperatures, possibly due to the suppressed defence system of the fish (Jun *et al.*, 2009). Furthermore, neutralizing antibody against RSIV was in the sera. When fish received sera from the virus-resistant fish, the injected fish showed resistance against virus challenge; however, the protective activity was low (Zenke *et al.*, 2014).

Type I interferon (IFN-I) is an important cytokine for host protection against various viral infections. IFN-I induces the expression of interferon-stimulated genes (ISGs) leading to inhibition of viral proliferation in cells. Polyinosinic-polycytidylic acid (poly(I:C)), which is a double-stranded RNA analogue, causes the induction of IFN-I and ISGs which leads to an increase of antiviral status in the fish. Although Kim *et al.* (2012) showed that the administration of poly(I:C) did not protect rock bream from the virus, Jung and Jung (2017) demonstrated that poly(I:C) administration increased the mRNA level of Mx gene (an ISG) and resulted in slight protection of fish against viral challenge. Furthermore, administration of a certain CpG oligodeoxynucleotide (ODN), which is also known to induce IFN-I, protected fish slightly against viral challenge at a lower viral concentration (Jung *et al.*, 2017). The protective effect of IFN-I on viral infection is affected by the rearing temperatures of the fish (Dios *et al.*, 2010; Hori *et al.*, 2012). For example, the mRNA levels of ISGs in Japanese flounder (*Paralichthys olivaceus*) spleen at 3 and 24 h after poly(I:C) injection at 15°C were significantly higher than those at 25°C. These results indicate that the type I IFN gene expressions of *P. olivaceus* treated with poly(I:C) can be regulated in a temperature-dependent manner (Thanasaksiri *et al.*, 2015). In the case of RSIV infection, the optimal temperature for the viral replication in the cultured cells was 25°C (Nakajima and Sorimachi, 1994). Since the levels of induction for fish IFN-I and ISG mRNAs are transient at warmer water temperature (such as 25°C), the effects of the IFN-I system against RSIV might be restricted at warmer temperatures.

9.7 Control and/or Prevention

Fish iridovirosis has attracted more attention recently because of its economic impacts on the aquaculture industry. A growing number of strategies have been developed for prevention and control of the disease.

9.7.1 Inactivated vaccine

The formalin-inactivated RSIV vaccine showed high protection in red sea bream under laboratory conditions (Nakajima *et al.*, 1997) and in field trial tests (Nakajima *et al.*, 1999). The vaccine also showed efficacy in various marine fish species (Nakajima *et al.*, 2002) and has been approved as a commercial vaccine for red sea bream, *Seriola* spp., groupers, etc. The vaccine is prepared in GF (grouper fin) cells (Clem *et al.*, 1961) because of their susceptibility to the virus (Nakajima *et al.*, 1997). Various cultured cells, such as BF-2, CHSE-214, FHM, JSKG (a cell line from Japanese striped knifejaw), KRE-3 (a cell line from kelp and red-spotted groupers), RTG-2 and YTF (a cell line from yellowtail), are susceptible to the virus (Nakajima and Sorimachi, 1994), but serial passages of the virus in BF-2 and KRE-3 caused a decrease of the viral infectivity (Nakajima and Sorimachi, 1995). To maintain the infectivity after serial passages, certain cell lines have been developed (Imajoh *et al.*, 2007; Wen *et al.*, 2008; Oh and Nishizawa, 2016; Kawato *et al.*, 2017b).

It is generally accepted that use of vaccines would be a pivotal strategy to achieve effective control of diseases and a significant decrease of economic losses in the aquaculture industry. The formalin- or β-propiolactone (BPL)-inactivated SGIV vaccine was highly protective against SGIV challenge in the laboratory tests (Ou-yang *et al.*, 2012a). This inactivated SGIV vaccine can significantly upregulate the expression of ISGs and the major histocompatibility complex (MHC) class I gene, and produces a substantial amount of specific serum antibody; indicating that effective vaccination in groupers induced an early, non-specific antiviral immunity and a specific immune response involving both humoral and cell-mediated immunity. The inactivated vaccine was prepared using GS cells because of their susceptibility to SGIV (Huang *et al.*, 2009). Moreover, various cultured cells including grouper brain (GB), grouper liver (GL), grouper kidney (GK), *Epinephelus lanceolatus* grouper heart (ELGH), *Epinephelus lanceolatus* grouper snout (ELGSN) and *Epinephelus moara* brain (EMB) are also susceptible to SGIV (Huang *et al.*, 2011b; Ou-yang *et al.*, 2010; Gong *et al.*, 2011; Huang *et al.*, 2014; Guo *et al.*, 2015; Liu *et al.*, 2018).

9.7.2 Component (subunit) vaccine (recombinant protein and DNA vaccine)

In contrast to the whole pathogen vaccine, such as the formalin-inactivated vaccine, a vaccine composed of only certain antigen(s) is a component (subunit) vaccine. Although this type of vaccine has not been approved for use in aquaculture, it has been tested against many pathogens in laboratory studies. Studies on component vaccines of megalocytivirus are listed in Table 9.1. These component vaccines are in the form of recombinant vaccines or DNA vaccines. A recombinant vaccine is prepared in *Escherichia coli* or other microorganisms by transformation and administrated as a purified protein or the genetically modified organism. A DNA vaccine is composed of a plasmid harbouring the antigenic protein gene(s) under the control of a eukaryotic gene promoter. By intramuscular injection, the antigenic protein gene(s) of the pathogen in the DNA vaccine is (are) transcribed and translated in the host cells. The produced antigen(s) will be recognized by the host immune system and induce specific immunity.

MCP, which is identified as a major component of the megalocytivirus capsid, is an antigen that has been well studied. Intraperitoneal injection of recombinant MCP produced in *E. coli* cells protected fish against the challenge with megalocytiviruses in rock bream (Kim *et al.*, 2008). However, Shimmoto *et al.* (2010) showed that intraperitoneal injection of formalin-killed *E. coli* cells producing recombinant MCP did not protect red sea bream in the challenge studies. Yet, oral administration of recombinant MCP expressed in cells of the transformed rice callus (Shin *et al.*, 2013) or yeast (Seo *et al.*, 2013) showed high protection in challenge tests. Through repeated studies, Liu *et al.* (2015) demonstrated that MCP is a suitable antigen for anti-megalocytivirus vaccines. Nevertheless, it is still uncertain why different protection efficacies of the recombinant MCP were observed among these experiments. Matsuyama *et al.* (2018) produced 72 different recombinant proteins, including predicted membrane or virion protein of RSIV and showed that 24 proteins reacted against convalescent fish sera.

DNA vaccines against megalocytivirus have been developed for many fish species including red sea bream (Caipang *et al.*, 2006b), rock bream (Jung *et al.*, 2018), Chinese perch (*S. chuatsi*) (Fu *et al.*, 2014) and turbot (*S. maximus*) (Zhang *et al.*, 2012; Zhang and Li, 2015; Zhang *et al.*, 2015; Zheng *et al.*, 2016a,b). The protection efficiency of these vaccines depends on the antigen gene used since certain genes showed higher protection in fish than others. In red sea bream injected with RSIV MCP DNA vaccine, mRNA levels of MHC class I genes were increased in the vaccinated fish (Caipang *et al.*, 2006a). The mRNA levels of immune-related genes in turbot immunized with DNA vaccines were increased on day 3 after injection (Zhang *et al.*, 2015; Zheng *et al.*, 2016b). Since Caipang *et al.* (2006a) showed that the DNase-treated formalin-inactivated RSIV vaccine lacked protective activity in the challenge test, it may suggest that the DNA in the formalin-inactivated vaccine might be incorporated by cells and then transcribed and translated as in the case of DNA vaccine. Moreover, seven DNA vaccines based on seven genes of RBIV isolate 1 from China (RBIV-C1) were constructed, and their protective potentials were examined in a turbot model. A vaccine candidate, pCN86, which is a plasmid that expresses an 86-residue viral protein, showed the highest protection and immunological responses against iridovirosis (Zhang *et al.*, 2012).

In addition, DNA vaccine has been shown to be highly efficient against fish viruses such as infectious hematopoietic necrosis virus (IHNV) (Lorenzen and LaPatra, 2005), RBIV (Jung *et al.*, 2018) and viral haemorrhagic septicaemia virus (VHSV) (Tonheim *et al.*, 2008). DNA vaccine against SGIV has also been reported in orange-spotted grouper, *Epinephelus coioides* (Ou-yang *et al.*, 2012a). Thirteen vaccine candidate genes from SGIV were cloned into a eukaryotic expression vector individually. Fish vaccinated with plasmid DNA encoding viral major capsid protein, VP39 and VP36, exhibited a relative percentage survival rate of more than 58%. Furthermore, a similar protective effect was obtained after fish were vaccinated with recombinant protein prepared from ORF072 (Ou-yang *et al.*, 2012b). Taken together, these results suggest that an effective vaccine may be used to control the outbreak of iridovirosis in aquaculture.

9.7.3 Fish immunity against iridoviral infection

The majority of serious diseases in fish farming, causing important economic losses worldwide, have a viral aetiology (OIE, 2012). In most cases, viral infection leads to a significant fish loss. Identifying components of the immune system which limit viral replication and boost host recovery are critical in developing adequate prophylactic control strategies (Collet, 2014). In cold-blooded vertebrates, establishment of the rapid induced innate immune response is more critical for prevention of viral replication than the late but more pathogen-specific adaptive immune response.

Table 9.1. List for the component antigens and vaccine experiments.

Antigen	Nucleotide accession no.	GenBank accession no.	Virus	Type of vaccine	Vaccinated fish	Route of injection	Effect (%)[a]	Note	Reference
MCP			RSIV	Recombinant (10 μg protein)	*Oplegnathus fasciatus*		94		Kim *et al.* (2008)
18R	AB104413.1	BAK14220	RSIV	400 mg wet weight of FKC expressing the recombinant protein	*Pagrus major*		19		Shimmoto *et al.* (2010)
351R	"	BAK14272	"	400 μg wet weight of FKC expressing the recombinant protein	"		29		"
MCP	"	BAK14277	"	400 μg wet weight of FKC expressing the recombinant protein	"		ND		"
MCP			ISKNV	Recombinant (20 μg protein)	*Siniperca chuatsi*		89		Fu *et al.* (2012)
"			"	Recombinant (50 μg protein)	"		57		"
"			"	Recombinant (100 μg protein)	"		38		"
MCP			RBIV	Yeast (1 × 10^8 cells in 25 g diet) expressing the recombinant protein	*O. fasciatus*	Oral	92		Seo *et al.* (2013)
MCP			RBIV	Rice (10 μg/10 g fish) expressing the recombinant protein	*O. fasciatus*	Oral	80		Shin *et al.* (2013)
"			"	Rice (30 μg/10 g fish) expressing the recombinant protein	"	"	90		"
ORF004		BAZ95618	RSIV	FKC expressing the recombinant protein	*Seriola quinqueradiata*		33		Matsuyama *et al.* (2018)
"		"	"	"	*Seriola dumerili*		18		"
ORF017		BAZ95631	"	FKC expressing the recombinant protein	*S. quinqueradiata*		17		"
"		"	"	"	*S. dumerili*		36		"
ORF020		BAZ95634	"	FKC expressing the recombinant protein	*S. quinqueradiata*		16		"
ORF057		BAZ95671	"	FKC expressing the recombinant protein	*S. quinqueradiata*		83		"
"		"	"	"	*S. dumerili*		54		"
ORF076		BAZ95690	"	FKC expressing the recombinant protein	*S. quinqueradiata*		83		"
"		"	"	"	*S. dumerili*		72		"
MCP	AB104413.1	BAK14277	RSIV	DNA vaccine (25 μg plasmid)	*P. major*		57		Caipang *et al.* (2006b)
ORF569	"	BAK14313	"	"	"		48		"
ORF374	"	BAK14276	"	"	"		30		"
ORF575	"	BAK14314	"	"	"		19		"
ORF018	"	BAK14265	"	"	"		5		"
ORF291	"	BAK14261	"	"	"		0		"
ORF086	JX134501		RBIV	DNA vaccine (20 μg plasmid)	*Scophthalmus maximus*		72		Zhang *et al.* (2012)

continued

Table 9.1. Continued.

Antigen	Nucleotide accession no.	GenBank accession no.	Virus	Type of vaccine	Vaccinated fish	Route of injection	Effect (%)[a]	Note	Reference
ORF137	JX134505		"	"	"		28		"
ORF142	JX134504		"	"	"		19		"
ORF224	JX134503		"	"	"		38		"
ORF336	HQ105007		"	"	"		34		"
ORF453	HQ105005		"	"	"		11		"
ORF454	JX134502		"	"	"		4		"
MCP			ISKNV	DNA vaccine	*S. chuatsi*		80	Immunized with QCDC adjuvant	Fu *et al.* (2014)
P247		AGG37986	RBIV	DNA vaccine	*S. maximus*		75		Zhang *et al.* (2015)
P523		AGG37965	"	"	"		63		"
ORF75	KC244182		RBIV	DNA vaccine	*S. maximus*		73		Zhang and Li (2015)
MCP	GQ273492		TRBIV	DNA vaccine (5 μg plasmid)	*S. maximus*		66		Zheng *et al.* (2016b)
MMP	AAT71823		RBIV	DNA vaccine (0.1 μg plasmid)	*O. fasciatus*		73		Jung *et al.* (2018)

FKC, formalin-killed cell; ND, not determined; QCDC, quilA (20 μg/ml), cholesterol (20 μg/ml), dimethyl dioctadecyl ammonium bromide (10 μg/ml) and carbopol (0.05%, v/v).
[a] Grey shading indicates a high survival rate of more than 60%.

Groupers are a commercial marine fish which are widely cultured on the south-east coast of China and other Asian countries (Wei *et al.*, 2010). Innate immunity is induced rapidly after infection, characterized by the absence of immunological memory, not specific to antigens of pathogens, and is modulated by molecules encoded by germ-line genes. Results of a recent study on paired grouper cell lines suggest that innate immune factors contribute a great deal of the overall resistance to viral infection (Verrier *et al.*, 2012). Studies by Takeuchi and Akira (2009) revealed that innate antiviral defence mechanisms have humoral and cellular components that can directly, and on their own, exclusively eradicate the viral pathogens in an infected animal population. Contributions of innate and adaptive immune responses to the iridovirus infections are summarized briefly below.

Specific antibodies induced by formalin-inactivated RSIV vaccine are a central player to provide protection in the vaccinated fish. The specific antibodies in the sera of the vaccinated fish were not detectable using an ELISA that detects specific antibodies in fish that survived an RSIV challenge (Kwon *et al.*, 2010). Matsuyama *et al.* (2016) reported the effects of serum antibodies from convalescent and immunized red sea bream, Japanese amberjack (*Seriola quinqueradiata*), yellowtail amberjack (*Seriola lalandi*), greater amberjack (*Seriola dumerili*) and rock bream (*O. fasciatus*) on the protection against the virus. Except for rock bream, fish injected with the convalescent serum showed high protection in challenge tests, and thus the specific antibody in the treated fish sera might be important for the protection.

Cellular adaptive immunity, which mainly depends on the activation of cytotoxic T cells, is also considered to be important for protection of fish from viral infections. The injection of formalin-inactivated RSIV vaccine induced the mRNA 1 of MHC class I gene, which is important for the activation of cellular adaptive immunity (Caipang *et al.*, 2006a). However, there are still no suitable methods to estimate the activity of fish cellular adaptive immunity, whereas the antibody titres can be evaluated using ELISA or a neutralizing assay. Therefore, it is still uncertain whether the cellular adaptive immunity is involved in the protection against RSIV.

Strains of red sea bream resistant against iridoviruses are not available. Once the resistant fish strains are developed, these resistant fish strains may help to better understand the host–pathogen relationship of iridoviruses (Sawayama and Tagaki, 2017; Sawayama *et al.*, 2017).

9.7.4 Antimicrobial peptide responses to iridovirus infection

Antimicrobial peptides (AMPs) serve as the first line of innate immune response against pathogens in the host (Zasloff, 2002). AMPs are known to be involved in defending against iridoviruses. Defensins are a group of AMPs that play an important role in innate host defence (White *et al.*, 1995). Guo *et al.* (2012b) showed that β-defensin from grouper exhibited a key role in inhibiting the infection and replication of SGIV. The expression profiles of EcDefensin (*E. coioides* defensin) were significantly ($P < 0.001$) upregulated after challenging with SGIV *in vivo*. EcDefensin was found to possess dual antiviral activity, inhibiting the infection and replication of SGIV *in vitro* (Guo *et al.*, 2012b). Hepcidin is also an important innate immune component in fish (Krause *et al.*, 2000). Two novel types of hepcidin genes (designated EC-hepcidin1 and EC-hepcidin2) were cloned from *E. coioides* (Zhou *et al.*, 2011). The expressions of both EC-hepcidins were upregulated by viral challenges. The replication of SGIV was inhibited by EC-hepcidin1 and EC-hepcidin2. Tachyplesin I is a potent AMP isolated from the haemocytes of the horseshoe crab, *Tachypleus tridentatus* (Nakamura *et al.*, 1988). Tachyplesin I exhibits a wide spectrum of antimicrobial activities against Gram-negative and Gram-positive bacteria (e.g. *Staphylococcus* species and *Salmonella* strains), fungi (e.g. *Candida albicans* M9), protozoa (e.g. *Perkinsus marinus*) and viruses (e.g. SGIV and *Red-spotted grouper nervous necrosis virus*, RGNNV) (Miyata *et al.*, 1989; Morvan *et al.*, 1997; Xie *et al.*, 2016). It has been shown recently by Xie *et al.* (2016) through *in vivo* studies that treatment with tachyplesin I drastically decreased the SGIV titre and viral gene expression.

9.7.5 Interferon response to iridovirus infection

The IFN response provides a significant contribution to antiviral immunity. IFN responses generally arise as the result of recognition of viral products through an array of host pathogen recognition receptors (PRRs), including toll-like receptors

(TLRs), retinoic acid-inducible gene 1 (RIG-I)-like receptors (RLRs) and cytosolic DNA sensors (Sadler and Williams, 2008; Baum and Garcia-Sastre, 2010). This branch of antiviral immunity consists of three classes of cytokines: type I, type II and type III IFNs (Sadler and Williams, 2008). Presently, only the type I IFN systems of bony fish have been explored in detail. Fish type I IFNs fall into two groups (group 1 and 2) based on the number of cysteine residues they contain, either two or four, respectively (Zhang and Gui, 2012). The IFN response provides a powerful cellular defence against viral infection in primitive vertebrates (e.g. teleosts) by inducing the expression of hundreds of ISGs in responsive cells. These ISGs play various functions, including antiviral activity, antiproliferation, immunomodulation, antigen presentation and antigen processing.

Vig-1, a gene induced by VHSV infection, was first isolated in rainbow trout (Boudinot *et al.*, 1999; O'Farrell *et al.*, 2002) and was later found in humans and renamed as Viperin (virus inhibitory protein, endoplasmic reticulum-associated, IFN-inducible) (Seo *et al.*, 2011). A functional domain of Viperin capable of interacting with lipid rafts was recently identified in humans (Fitzgerald, 2011). A novel grouper immune gene, EcVig, was identified in orange-spotted grouper (*E. coioides*) (Yeh *et al.*, 2016), and levels of expression of EcVig and other ISGs were significantly increased following infection by RSIV and GIV. Furthermore, application of recombinant type I IFN to fish also increased EcVig expression. Therefore, EcVig may be a novel ISG that exhibits an antiviral immune response.

IFN-stimulated gene 15 (ISG15) is a ubiquitin homologue that has been shown to be significantly induced by type I IFNs or viral infections (Zhang and Zhang, 2011). An ISG15 homologue (EcISG15) was cloned from grouper, *E. coioides* (Huang *et al.*, 2013). Expression analysis showed that EcISG15 was dramatically induced by grouper nervous necrosis virus (GNNV) infection or treatment with poly(I:C) or poly(dA-dT), but no obvious change was observed following SGIV infection. Immunofluorescence assay showed that EcISG15 localized mainly in the cytoplasm of grouper cells in response to poly(I:C) stimulation or GNNV infection, but not in mock- or SGIV-infected cells. Western blot analysis indicated that the ISGylation (covalent addition of ISG15 to cytoplasmic and nuclear proteins similar to ubiquitination) was absent in SGIV-infected cells, but significantly enhanced in GNNV-infected or poly(I:C)-transfected cells, suggesting that EcISG15 might play different roles in SGIV and GNNV infection.

Members of the newly discovered virus-induced tripartite motif (TRIM)-containing protein families have been identified in grouper and found to be strongly induced during SGIV infection. Initially discovered as induced by viral infection, some members of the antiviral TRIM family in grouper have been shown to be induced by IFN (Everett and Chelbi-Alix, 2007). A novel TRIM32 gene from orange-spotted grouper (EcTRIM32) was cloned and characterized (Yu *et al.*, 2017a). After injection with SGIV, the relative expression of EcTRIM32 in grouper spleen was differentially regulated. EcTRIM32 significantly inhibited the replication of SGIV. Moreover, EcTRIM32 positively upregulated the IFN immune response, evidenced by the significant increase in the expression level of IFN-related signalling molecules, including IFN regulatory factor 3 (IRF3), IRF7, ISG15, IFN-induced 35-kDa protein (IFP35), MXI, TIR-domain-containing adaptor-inducing IFN-β (TRIF) and melanoma differentiation-associated protein 5 (MDA5). After incubation with SGIV, the ectopic expression of EcTRIM16L significantly enhanced the viral replication as demonstrated by an increase of cytopathic effect severity and viral gene transcriptions. Moreover, the ectopic expression of EcTRIM16L significantly decreased both MDA5- and mediator of IRF3 activation (MITA)-induced interferon immune responses (Yu *et al.*, 2017b). After incubation with SGIV, EcTRIM25 significantly decreased the viral gene transcription of SGIV *in vitro*. EcTRIM25 significantly increased the expression level of IFN-related signalling molecules, including IRF3, IFP35, MXI, IRF7 and myeloid differentiation factor 88 (MyD88) (Yang *et al.*, 2016).

9.7.6 Apoptosis in iridovirus infection

Apoptosis is essential for the maintenance of homeostasis in the immune system and apoptotic cell death occurs in a wide range of viral infections (Clarke and Tyler, 2009; Kinpara *et al.*, 2013). To ensure their own survival and propagation, viruses modulate the crucial aspects of host homeostasis through influencing the cell cycle and regulating the apoptotic machinery of host cells (Tortorella *et al.*, 2000; Gougeon and Piacentini, 2009).

Virus-induced apoptosis is related to activation of JNK (Hrincius *et al.*, 2010; Shi *et al.*, 2012), nuclear factor kappa B (NF-κB) and p53 pathways (Myskiw *et al.*, 2009; Kinpara *et al.*, 2013). As described previously, SGIV can induce apoptosis in individual cells (Huang *et al.*, 2011a). To better understand the mechanism of SGIV-induced apoptosis, many molecules have been identified in apoptosis in grouper. Ec-JNK1 was identified from orange-spotted grouper, *E. coioides* (Guo *et al.*, 2016a). Ec-JNK1 is involved in the immune response to pathogen *in vivo* challenges and the SGIV-induced apoptosis under *in vitro* conditions. Furthermore, Ec-JNK2 is involved in the immune response to pathogen *in vivo* challenges and the infection by SGIV *in vitro* (Guo *et al.*, 2016b). Fas-associated protein with death domain (FADD) is the key adaptor protein that transmits apoptotic signals mediated by the main death receptors (DRs). FADD is an important pro-apoptotic adaptor in DR-induced apoptosis (Tourneur and Chiocchia, 2010). A FADD homologue (EcFADD) from the orange-spotted grouper was cloned (Zhang *et al.*, 2018). The expression of EcFADD in the spleen of the orange-spotted grouper was differentially upregulated when the fish was challenged with SGIV. EcFADD inhibited SGIV infection and replication and SGIV-induced apoptosis. TNF receptor-associated factors (TRAFs) and TNF receptor-associated proteins (TTRAPs) are the latest identified cytosolic proteins that serve as negative regulators in the TNF signalling pathway. TTRAPs are also proved to serve as a switch of signal transduction which inhibits the activation of NF-κB and turns the TNF signalling pathway to JNK activation, finally leading to DNA fragmentation and apoptosis. TTRAP from grouper (EcTTRAP) was identified as a remarkably upregulated protein responding to SGIV infection and inhibited apoptosis induced by SGIV. EcTTRAP also inhibited the replication of SGIV (Wei *et al.*, 2015).

9.7.7 Oligonucleotide antiviral therapeutics: small interfering RNAs and antisense morpholino oligonucleotides

RNA interference (RNAi) is a mechanism that degrades exogenous and endogenous RNA (Hannon, 2002; Leung and Whittaker, 2005). By incorporating small molecules of double-stranded RNA consisting of 19–23 bases – or small interfering RNAs (siRNAs) – into cells, the intracellular RNAi mechanism can be triggered, leading to the degradation of mRNAs in a sequence-specific manner (Leung and Whittaker, 2005). By using this technique, therapeutic effects are expected for various viruses in humans (Levanova and Poranen, 2018), and it has also been studied to control aquatic viral infections (Reshi *et al.*, 2014).

Dang *et al.* (2008b) synthesized siRNA against the MCP of RSIV and introduced it into cultured cells that were constantly expressing MCP. They observed a marked decrease in MCP mRNA in the cells. Since introduction of this siRNA into RSIV-infected cells led to a marked decrease of viral DNA copies, the viral gene-specific siRNA suppresses infection by RSIV. Zenke and Kim (2010) also developed an siRNA expression vector and showed the reduction of viral replication in the transfected cells. Direct injection of long-chain, double-stranded RNA encoding MCP into fish muscle induced the increase of the mRNA level of the Mx gene similarly to those injected with non-specific double-stranded RNA. Furthermore, the same result was obtained an *in vivo* challenge test (Zenke and Kim, 2010). Since it is costly to synthesize siRNAs, the effect of microRNA (miRNA) on controlling viruses has also been studied. In viral infected cells transfected with a plasmid containing an miRNA sequence, the transcribed pre-miRNA molecule in the host cells is converted to mature miRNA and that leads to a reduction in the copies of the viral DNA in the cells (Dang *et al.*, 2008a). However, since the miRNA expression vectors without virus-specific sequences also increased mRNA levels of the Mx gene, it is unclear whether the reduction was caused in a sequence-specific manner.

Antisense morpholino oligonucleotides (asMOs) have been used extensively to knock-down levels of gene expression in developing embryos and inhibit gene expression in cultured cells (Deas *et al.*, 2005). During SGIV infection, knock-down of ORF018R expression resulted in a reduction in the expression of viral late genes. This led to distortion of viral particle assembly and inhibition of SGIV infection in grouper embryonic cells. The result suggests that ORF018R plays an important role in expression of viral late genes and virion assembly (Wang *et al.*, 2008). In addition, asMOs have also successfully been used to inhibit translation of viral genes during FV3 infection (Sample *et al.*, 2007). Thus, asMOs could be chosen as an alternative tool to investigate the function of iridoviral genes *in vitro*.

9.7.8 *In silico* linear epitope prediction for potential vaccine candidates

Epitopes (antigenic determinants) are clusters of amino acid segments located on the surfaces of an antigen which induce B-cell antibody production and stimulate T-cell activation. Basically, B-cell epitopes are categorized into two types: linear epitopes (LEs) and conformational epitopes (CEs) (Barlow *et al.*, 1986). Although B-cell LEs occupy only a small part of the entire epitope group, they are important in vaccine development (Yadav *et al.*, 2011). In recent years, based on the characteristics of amino acids by their hydrophobicity, surface accessibility, mobility, protrusion area, physicochemical properties, antigenicity and pocket characteristics (Alix, 1999; Saha and Raghava, 2004; Andersen *et al.*, 2006; Chang *et al.*, 2006, 2008; Pai *et al.*, 2006), bioinformatics can enable rapid prediction of potential epitopes (Davies and Cohen, 1996). However, there is room for improvement in the identification process that needs to include retrieving the unpredictable epitopes and enhancing the antigenicity score *in silico*.

Wang *et al.* (2011) designed a novel B-cell LE prediction system named Linear Epitope Prediction by Propensities and Support Vector Machine (LEPS) that combined physicochemical propensity identification and support vector machine (SVM) classification. Furthermore, taking the *Iridoviridae* family as an example, Shih *et al.* (2019) applied a voting-mechanism-based LE prediction system to analyse any two differently clustered pathogen groups, allowing both conserved and exclusive LEs to be identified simultaneously. In addition, to increase the successful results of vaccine design, they emphasized the surface structure characteristics of the predicted epitopes. The predicted LE candidates were further validated using ELISAs for evaluating the strength of antigenicity and cross-antigenicity. The conserved LEs for *Iridoviridae* family reflected high antigenicity responses. Therefore, the proposed system may provide an effective approach for *in silico* LE prediction of epitopes for vaccine development. The designed voting-mechanism-based LE prediction system is presented in Fig. 9.5.

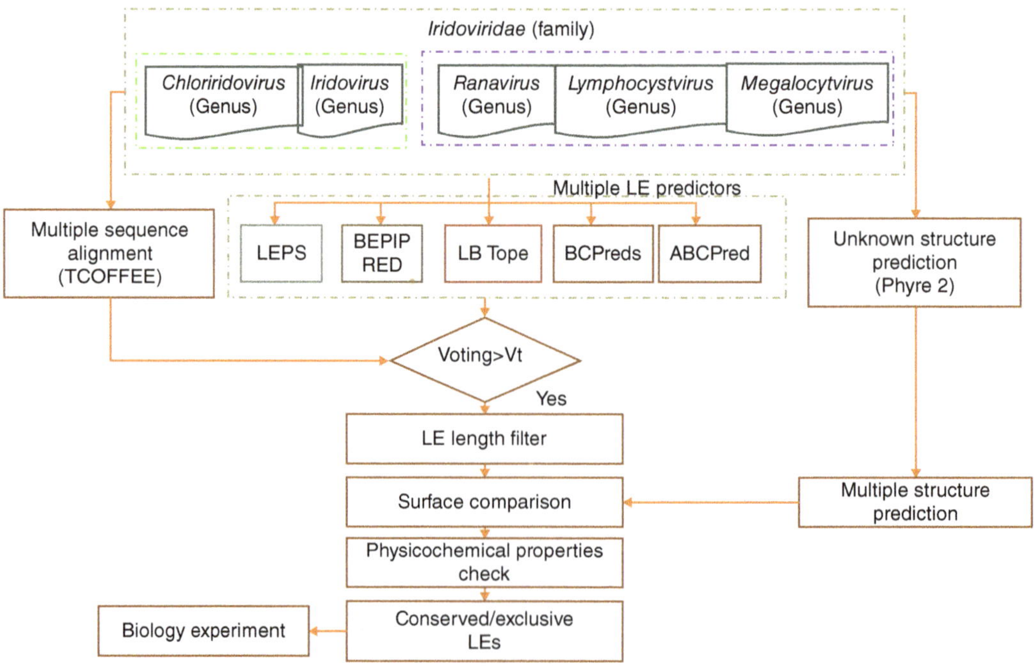

Fig. 9.5. The designed voting-mechanism-based LE prediction system. The different coloured dotted lines separate two subfamilies of *Iridoviridae*. Vt, voting threshold (three of the five predictors).

9.8 Conclusions with Suggestions for Future Studies

Climate change is ongoing and inevitable. For the aquaculture industry, the main impacts are water temperature and extreme weather events such as heavy rainfalls, droughts, etc. The immune response of fish and the replication of pathogenic viruses correlate with water temperature. According to the optimal multiplication temperature of the three groups of megalocytivirus, water temperature may be the factor that affects the geographical distribution of the RSIV-type, ISKNV-type and TRBIV-type viruses. Climate change may cause the virus to spread to new geographical locations. Moreover, novel strains can be generated by genetic recombination and cause outbreaks through cross-species transmission. Therefore, each country should establish a risk framework soon to evaluate the influence of climate change on disease emergence in its aquaculture facilities.

For improving the sustainability of aquaculture, the possibility of raising varieties of fish with traits that include disease resistance, rapid growth and environment tolerance (temperature, salinity) is an attractive prospect in breeding programmes. Disease resistance is one of the most desirable traits, because infectious disease is a major constraint for the success of aquaculture. Genetic approaches including classical selection, genetic marker-assisted selection (MAS), intraspecific/interspecific cross-breeding and transgenesis have been used successfully in reducing disease incidence in aquaculture. The RSIVD-resistant trait of broodstock of red sea bream has been evaluated in commercial production based on DNA parentage analysis (Sawayama and Takagi, 2017). From two groups of the same production lot, one population was without outbreak of disease (farm A) and the other population suffered high mortalities after an RSIVD outbreak (farm B). The survival of offspring from farm B males was 82.3% and this was much higher than that from farm A males (estimated survival 2.5%). The result indicates that some broodstock have resistance traits against RSIVD and shows the potential for developing an RSIVD-resistant strain of red sea bream. Since disease resistance cannot be selected by the appearance of fish, marker-assisted breeding using DNA markers genetically linked to disease resistance is the best approach for breed improvement. Quantitative trait loci (QTL) can be useful in MAS of elites resistant to specific pathogens. Wang *et al.* (2017) conducted a genome scan for QTL associated with SGIV resistance in an Asian sea bass (*Lates calcarifer*) family, and the results could be useful both for MAS and advanced genomic selection in the selective breeding programmes of Asian sea bass.

Recently, genome editing technology has opened avenues for modifying disease resistance in farmed animals, using techniques such as zinc finger nucleases (ZFNs), transcription activator-like effector nucleases (TALENs) and clustered regularly interspaced short palindromic repeats (CRISPR)/CRISPR-associated protein 9 (Cas9). It has great potential to address the threats of disease problems in aquaculture. Most aquaculture species are unimproved stocks. Consequently, genomic selection in breeding for disease resistance, temperature tolerance and other sought-after traits is worth serious consideration in future studies.

References

Alcorn, S.W., Murray, A.L. and Pascho, R.J. (2002) Effects of rearing temperature on immune functions in sockeye salmon (*Oncorhynchus nerka*). *Fish and Shellfish Immunology* 12, 303–334.

Alix, A.J.P. (1999) Predictive estimation of protein linear epitopes by using the program PEOPLE. *Vaccine* 18, 311–314.

Andersen, P.H., Nielsen, M. and Lund, O. (2006) Prediction of residues in discontinuous B-cell epitopes using protein 3D structures. *Protein Science* 15, 2558–2567.

Ariel, E., Nicolajsen, N., Christophersen, M.B., Holopainen, R., Tapiovaara, H. and Jensen, B.B. (2009) Propagation and isolation of ranaviruses in cell culture. *Aquaculture* 294, 159–164.

Barlow, D.J., Edwards, M.S. and Thornton, J.M. (1986) Continuous and discontinuous protein antigenic determinants. *Nature* 322, 747–748.

Baum, A. and Garcia-Sastre, A. (2010) Induction of type I interferon by RNA viruses: cellular receptors and their substrates. *Amino Acids* 38, 1283–1299.

Bly, J.E. and Clem, L.W. (1992) Temperature and teleost immune functions. *Fish and Shellfish Immunology* 2, 159–171.

Boudinot, P., Massin, P., Blanco, M., Riffault, S. and Benmansour, A. (1999) *vig-1*, a new fish gene induced by the rhabdovirus glycoprotein, has a virus-induced homologue in humans and shares conserved motifs with the MoaA family. *Journal of Virology* 73, 1846–1852.

Braunwald, J., Nonnenmacher, H. and Tripier-Darcy, F. (1985) Ultrastructural and biochemical study of frog virus 3 uptake by BHK-21 cells. *Journal of General Virology* 66, 283–293.

Caipang, C.M.A., Haraguchi, I., Ohira, T., Hirono, I. and Aoki, T. (2004) Rapid detection of a fish iridovirus using loop-mediated isothermal amplification (LAMP). *Journal of Virological Methods* 121, 155–161.

Caipang, C.M.A., Hirono, I. and Aoki, T. (2006a) Immunogenicity, retention and protective effects of the protein derivatives of formalin-inactivated red seabream iridovirus (RSIV) vaccine in red seabream, *Pagrus major*. *Fish and Shellfish Immunology* 20, 597–609.

Caipang, C.M.A., Takano, T., Hirono, T. and Aoki, T. (2006b) Genetic vaccines protect red seabream, *Pagrus major*, upon challenge with red seabream iridovirus (RSIV). *Fish and Shellfish Immunology* 21, 130–138.

Chang, H.T., Liu, C.H. and Pai, T.W. (2008) Estimation and extraction of B-cell linear epitopes predicted by mathematical morphology approaches. *Journal of Molecular Recognition* 21, 431–441.

Chang, H.T., Pai, T.W., Fan, T.C., Su, B.H., Wu, P.C. *et al.* (2006) A reinforced merging methodology for mapping unique peptide motifs in members of protein families. *BMC Bioinformatics* 7, 38.

Chao, C.B., Yang, S.C., Tsai, H.Y., Chen, C.Y., Lin, C.S. and Huang, H.T. (2002) A nested PCR for the detection of grouper iridovirus in Taiwan (TGIV) in cultured hybrid grouper, giant seaperch, and largemouth bass. *Journal of Aquatic Animal Health* 14, 104–113.

Chen, X.H., Lin, K.B. and Wang, X.W. (2003) Outbreaks of an iridovirus disease in maricultured large yellow croaker, *Larimichthys crocea* (Richardson). *Journal of Fish Diseases* 26, 615–619.

Chinchar, V.G., Hick, P., Ince, I.A., Jancovich, J.K., Marschang, R. *et al.* (2017) ICTV Virus Taxonomy Profile: *Iridoviridae*. *Journal of General Virology*, 98, 890–891.

Choi, S.K., Kwon, S.R., Nam, Y.K., Kim, S.K. and Kim, K.H. (2006) Organ distribution of red sea bream iridovirus (RSIV) DNA in asymptomatic yearling and fingerling rock bream (*Oplegnathus fasciatus*) and effects of water temperature on transition of RSIV into acute phase. *Aquaculture* 256, 23–26.

Chou, H.Y., Hsu, C.C. and Peng, T.Y. (1998) Isolation and characterization of a pathogenic iridovirus from cultured grouper (*Epinephelus* sp.) in Taiwan. *Fish Pathology* 33, 201–206.

Chua, F.H.C., Ng, M.L., Ng, K.L., Loo, J.J. and Wee, J.Y. (1994) Investigation of outbreaks of a novel disease, 'sleepy grouper disease', affecting the brown-spotted grouper, *Epinephelus tauvina* Forskal. *Journal of Fish Diseases* 17, 417–427.

Clarke, P. and Tyler, K.L. (2009) Apoptosis in animal models of virus-induced disease. *Nature Reviews Microbiology* 7, 144–155.

Clem, L.W., Moewus, L. and Michael Sigel, M. (1961) Studies with cells from marine fish in tissue culture. *Proceedings of the Society for Experimental Biology and Medicine* 108, 762–766.

Collet, B. (2014) Innate immune responses of salmonid fish to viral infections. *Developmental and Comparative Immunology* 43, 160–173.

Danayadol, Y., Direkbusarakom, S., Boonyaratpalin, S., Miyazaki, T. and Miyata, M. (1997) Iridovirus infection in brown-spotted grouper (*Epinephelus malabaricus*) cultured in Thailand. In: Flegal, T.W. and MacRae, I.H. (eds) *Disease in Asian Aquaculture*, III. Fish Health Section, Asian Fisheries Society, Manila, pp. 67–72.

Dang, L.T., Kondo, H., Aoki, T. and Hirono, I. (2008a) Engineered virus-encoded pre-microRNA (pre-miRNA) induces sequence-specific antiviral response in addition to nonspecific immunity in a fish cell line: convergence of RNAi-related pathways and IFN-related pathways in antiviral response. *Antiviral Research* 80, 316–323.

Dang, L.T., Kondo, H., Hirono, I. and Aoki, T. (2008b) Inhibition of red seabream iridovirus (RSIV) replication by small interfering RNA (siRNA) in a cell culture system. *Antiviral Research* 77, 142–149.

Davies, D.R. and Cohen, G.H. (1996) Interactions of protein antigens with antibodies. *Proceedings of the National Academy of Sciences USA* 93, 7–12.

De Groof, A., Guelen, L., Deijs, M., Van Der Wal, Y., Miyata, M. *et al.* (2015) A novel virus causes scale drop disease in *Lates calcarifer*. *PLoS Pathogens* 11, e1005074.

Deas, T.S., Binduga-Gajewska, I., Tilgner, M., Ren, P., Stein, D.A. *et al.* (2005) Inhibition of flavivirus infections by antisense oligomers specifically suppressing viral translation and RNA replication. *Journal of Virology* 79, 4599–4609.

Deng, G.C., Li, S.J., Xie, J., Bai, J.J., Chen, K.C. *et al.* (2011) Characterization of a ranavirus isolated from cultured largemouth bass (*Micropterus salmoides*) in China. *Aquaculture* 312, 198–204.

Dios, S., Romero, A., Chamorro, R., Figueras, A. and Novoa, B. (2010) Effect of the temperature during antiviral immune response ontogeny in teleosts. *Fish and Shellfish Immunology* 29, 1019–1027.

Do, J.W., Moon, C.H., Kim, H.J., Ko, M.S., Kim, S.B. *et al.* (2004) Complete genomic DNA sequence of rock bream iridovirus. *Virology* 325, 351–363.

Do, J.W., Cha, S.J., Kim, J.S., An, E.J., Lee, N.S. *et al.* (2005) Phylogenetic analysis of the major capsid protein gene of iridovirus isolates from cultured flounders *Paralichthys olivaceus* in Korea. *Diseases of Aquatic Organisms* 64, 193–200.

Dong, C.F., Weng, S.P., Luo, Y.W., Huang, M.M., Ai, H.S. *et al.* (2010) A new marine megalocytivirus from spotted knifejaw, *Oplegnathus punctatus*, and its pathogenicity to freshwater mandarinfish, *Siniperca chuatsi*. *Virus Research* 147, 98–106.

Dong, H.T., Jitrakorn, S., Kayansamruaj, P., Pirarat, N., Rodkhum, C. *et al.* (2017) Infectious spleen and kidney necrosis disease (ISKND) outbreaks in farmed barramundi (*Lates calcarifer*) in Vietnam. *Fish and Shellfish Immunology* 68, 65–73.

Everett, R.D. and Chelbi-Alix, M.K. (2007) PML and PML nuclear bodies: implications in antiviral defence. *Biochimie* 89, 819–830.

Fitzgerald, K.A. (2011) The interferon inducible gene: Viperin. *Journal of Interferon and Cytokine Research* 31, 131–135.

Fu, X.Z., Li, N.Q., Liu, L.H., Lin, Q., Wang, F. *et al.* (2011) Genotype and host range analysis of infectious spleen and kidney necrosis virus (ISKNV). *Virus Genes* 42, 97–109.

Fu, X., Li, N., Lai, Y., Liu, L., Lin, Q. *et al.* (2012) Protective immunity against iridovirus disease in mandarin fish, induced by recombinant major capsid protein of infectious spleen and kidney necrosis virus. *Fish and Shellfish Immunology* 33, 880–885.

Fu, X.Z., Li, N.Q., Lin, Q., Guo, H.Z., Zhang, D.F. *et al.* (2014) Protective immunity against infectious spleen and kidney necrosis virus induced by immunization with DNA plasmid containing *mcp* gene in Chinese perch *Siniperca chuatsi*. *Fish and Shellfish Immunology* 40, 259–266.

Gias, E., Johnston, C., Keeling, S., Spence, R.P. and McDonald, W.L. (2011) Development of real-time PCR assays for detection of megalocytiviruses in imported ornamental fish. *Journal of Fish Diseases* 34, 609–618.

Gibson-Kueh, S., Ngoh-Lim, G.H., Netto, P., Kurita, J., Nakajima, K. and Ng, M.L. (2004) A systemic iridoviral disease in mullet, *Mugil cephalus* L., and tiger grouper, *Epinephelus fuscoguttatus* Forsskal: a first report and study. *Journal of Fish Diseases* 27, 693–699.

Go, J., Lancaster, M., Deece, K., Dhungyel, O. and Whittington, R. (2006) The molecular epidemiology of iridovirus in Murray cod (*Maccullochella peelii peelii*) and dwarf gourami (*Colisa lalia*) from distant biogeographical regions suggests a link between trade in ornamental fish and emerging iridoviral diseases. *Molecular and Cellular Probes* 20, 212–222.

Go, J., Waltzek, T.B., Subramaniam, K., Yun, S.C., Groff, J.M. *et al.* (2016) Detection of infectious spleen and kidney necrosis virus (ISKNV) and turbot reddish body iridovirus (TRBIV) from archival ornamental fish samples. *Diseases of Aquatic Organisms* 122, 105–123.

Gong, J., Huang, Y., Huang, X., Ouyang, Z., Guo, M. and Qin, Q. (2011) Establishment and characterization of a new cell line derived from kidney of grouper, *Epinephelus akaara* (Temminck & Schlegel), susceptible to Singapore grouper iridovirus (SGIV). *Journal of Fish Diseases* 34, 677–686.

Gougeon, M.L. and Piacentini, M. (2009) New insights on the role of apoptosis and autophagy in HIV pathogenesis. *Apoptosis* 14, 501–508.

Gray, M.J., Brunner, J.L., Earl, J.E. and Ariel, E. (2015) Design and analysis of ranavirus studies: surveillance and assessing risk. In: Gray, M. and Chinchar, V. (eds) *Ranaviruses: Lethal Pathogens of Ectothermic Vertebrates*. Springer, Cham, Switzerland, pp. 209–240.

Grizzle, J.M., Altinok, I. and Noyes, A.D. (2003) PCR method for detection of largemouth bass virus. *Diseases of Aquatic Organisms* 54, 29–33.

Guo, C.J., Liu, D., Wu, Y.Y., Yang, X.B., Yang, L.S. *et al.* (2011) Entry of tiger frog virus (an iridovirus) into HepG2 cells via a pH-dependent, atypical, caveola-mediated endocytosis pathway. *Journal of Virology* 85, 6416–6426.

Guo, C.J., Wu, Y.Y., Yang, L.S., Yang, X.B., He, J. *et al.* (2012a) Infectious spleen and kidney necrosis virus (a fish iridovirus) enters mandarin fish fry cells via caveola-dependent endocytosis. *Journal of Virology* 86, 2621–2631.

Guo, C.Y., Huang, Y.H., Wei, S.N., Ouyang, Z.L., Yan, Y. *et al.* (2015) Establishment of a new cell line from the heart of giant grouper, *Epinephelus lanceolatus* (Bloch), and its application in toxicology and virus susceptibility. *Journal of Fish Diseases* 38, 175–186.

Guo, M.L., Wei, J.G., Huang, X.H., Huang, Y.H. and Qin, Q.W. (2012b) Antiviral effects of beta-defensin derived from orange-spotted grouper (*Epinephelus coioides*). *Fish and Shellfish Immunology* 32, 828–838.

Guo, M.L., Wei, J.G., Huang, X.H., Zhou, Y.C., Yan, Y. and Qin, Q.W. (2016a) JNK1 derived from orange-spotted grouper, *Epinephelus coioides*, involving in the evasion and infection of Singapore grouper iridovirus (SGIV). *Frontiers in Microbiology* 7, 121.

Guo, M.L., Wei, J.G., Zhou, Y.C. and Qin, Q.W. (2016b) Molecular clone and characterization of c-Jun N-terminal kinases 2 from orange-spotted grouper, *Epinephelus coioides*. *Fish and Shellfish Immunology* 49, 355–363.

Gustin, J.K., Moses, A.V., Fruh, K. and Douglas, J.L. (2011) Viral takeover of the host ubiquitin system. *Frontiers in Microbiology* 2, 161.

Hannon, G.J. (2002) RNA interference. *Nature* 418, 244–251.

He, J.G., Weng, S.P., Huang, Z.J. and Zeng, K. (1998) Identification of outbreak and infectious diseases pathogen of *Siniperca chuatsi*. *Acta Scientiarum Naturalium Universitatis Sunyatseni* 5, 74–77.

He, J.G., Wang, S.P., Zeng, K., Huang, Z.J. and Chan, S.-M. (2000) Systemic disease caused by an iridovirus-like agent in cultured mandarinfish, *Siniperca chuatsi* (Basilewsky), in China. *Journal of Fish Diseases* 23, 219–222.

He, J.G., Zeng, K., Weng, S.P. and Chan, S.M. (2002) Experimental transmission, pathogenicity and physical-chemical properties of infectious spleen and kidney necrosis virus (ISKNV). *Aquaculture* 204, 11–24.

Hick, P., Becker, J. and Whittington, R. (2016) Iridoviruses of fish. In: Kibenge, F.S.B. and Godoy, M.G. (eds) *Aquaculture Virology*. Academic Press, San Diego, California, pp. 127–152.

Hori, T.S., Gamperl, A.K., Booman, M., Nash, G.W. and Rise, M.L. (2012) A moderate increase in ambient temperature modulates the Atlantic cod (*Gadus morhua*) spleen transcriptome response to intraperitoneal viral mimic injection. *BMC Genomics* 13, 431.

Hossain, M. and Oh, M.J. (2011) Histopathology of marine and freshwater fish lymphocystis disease virus (LCDV). *Sains Malaysiana* 40, 1049–1052.

Hrincius, E.R., Wixler, V., Wolff, T., Wagner, R., Ludwig, S. and Ehrhardt, C. (2010) CRK adaptor protein expression is required for efficient replication of avian influenza A viruses and controls JNK-mediated apoptotic responses. *Cellular Microbiology* 12, 831–843.

Huang, C.H., Zhang, X.B., Gin, K.Y.H. and Qin, Q.W. (2004) *In situ* hybridization of a marine fish virus, Singapore grouper iridovirus with a nucleic acid probe of major capsid protein. *Journal of Virological Methods* 117, 123–128.

Huang, X.H., Huang, Y.H., Sun, J.J., Han, X. and Qin, Q. (2009) Characterization of two grouper *Epinephelus akaara* cell lines: application to studies of Singapore grouper iridovirus (SGIV) propagation and virus–host interaction. *Aquaculture* 292, 172–179.

Huang, X.H., Huang, Y.H., Ouyang, Z.L., Cai, J., Yan, Y. and Qin, Q.W. (2011a) Roles of stress-activated protein kinases in the replication of Singapore grouper iridovirus and regulation of the inflammatory responses in grouper cells. *Journal of General Virology* 92, 1292–1301.

Huang, X.H., Huang, Y.H., Ouyang, Z.L. and Qin, Q.W. (2011b) Establishment of a cell line from the brain of grouper (*Epinephelus akaara*) for cytotoxicity testing and virus pathogenesis. *Aquaculture* 311, 65–73.

Huang, X.H., Huang, Y.H., Ouyang, Z.L., Xu, L.X., Yan, Y. *et al.* (2011c) Singapore grouper iridovirus, a large DNA virus, induces nonapoptotic cell death by a cell type dependent fashion and evokes ERK signaling. *Apoptosis* 16, 831–845.

Huang, X.H., Huang, Y.H., Cai, J., Wei, S.N., Ouyang, Z.L. and Qin, Q.W. (2013) Molecular cloning, expression and functional analysis of ISG15 in orange-spotted grouper, *Epinephelus coioides*. *Fish and Shellfish Immunology* 34, 1094–1102.

Huang, X.H., Wei, S.N., Ni, S.W., Huang, Y.H. and Qin, Q.W. (2018a) Ubiquitin–proteasome system is required for efficient replication of Singapore grouper iridovirus. *Frontiers in Microbiology* 9, 2798.

Huang, Y.H., Huang, X.H., Ouyang, Z.L., Wei, S.N., Guo, C.Y. and Qin, Q.W. (2014) Development of a new cell line from the snout of giant grouper, *Epinephelus lanceolatus* (Bloch), and its application in iridovirus and nodavirus pathogenesis. *Aquaculture* 432, 265–272.

Huang, Y.H., Huang, X.H., Wang, S.W., Yu, Y.P., Ni, S.W. and Qin, Q.W. (2018b) Soft-shelled turtle iridovirus enters cells via cholesterol-dependent, clathrin-mediated endocytosis as well as macropinocytosis. *Archives of Virology* 163, 3023–3033.

Imajoh, M., Kawa, T. and Oshima, S.I. (2007) Characterization of a new fibroblast cell line from a tail fin of red sea bream, *Pagrus major*, and phylogenetic relationships of a recent RSIV isolate in Japan. *Virus Research* 126, 45–52.

Inouye, K., Yamano, K., Maeno, Y., Nakajima, K., Matsuoka, M. *et al.* (1992) Iridovirus infection of cultured red sea bream, *Pagrus major*. *Fish Pathology* 27, 19–27.

Ito, T., Yoshiura, Y., Kamaishi, T., Yoshida, K. and Nakajima, K. (2013) Prevalence of red sea bream iridovirus among organs of Japanese amberjack (*Seriola quinqueradiata*) exposed to cultured red sea bream iridovirus. *Journal of General Virology* 94, 2094–2101.

Jancovich, J.K., Bremont, M., Touchman, J.W. and Jacobs, B.L. (2010) Evidence for multiple recent host species shifts among the ranaviruses (family *Iridoviridae*). *Journal of Virology* 84, 2636–2647.

Jancovich, J.K., Qin, Q., Zhang, Q.-Y. and Chinchar, V.G. (2015) Ranavirus replication: molecular, cellular, and immunological events. In: Gray, M.J. and Chinchar, V.G. (eds) *Ranaviruses: Lethal Pathogens of Ectothermic Vertebrates*. Springer, Cham, Switzerland, pp. 105–139.

Jeong, J.B., Jun, L.J., Yoo, M.H., Kim, M.S., Komisar, J.L. and Jeong, H.D. (2003) Characterization of the DNA nucleotide sequences in the genome of red sea bream iridoviruses isolated in Korea. *Aquaculture* 220, 119–133.

Jeong, J.B., Jun, L.J., Park, K.H., Kim, K.H., Chung, J.-K. *et al.* (2006) Asymptomatic iridovirus infection in various marine fishes detected by a 2-step PCR method. *Aquaculture* 255, 30–38.

Jun, L.J., Jeong, J.B., Kim, J.H., Nam, J.H., Shin, K.W. *et al.* (2009) Influence of temperature shifts on the onset and development of red sea bream iridoviral disease in rock bream *Oplegnathus fasciatus*. *Diseases of Aquatic Organisms* 84, 201–208.

Jung, M.H. and Jung, S.J. (2017) Innate immune responses against rock bream iridovirus (RBIV) infection in rock bream (*Oplegnathus fasciatus*) following poly (I:C) administration. *Fish and Shellfish Immunology* 71, 171–176.

Jung, M.H., Lee, J., Ortega-Villaizan, M., Perez, L. and Jung, S.J. (2017) Protective immunity against Megalocytivirus infection in rock bream (*Oplegnathus fasciatus*) following CpG ODN administration. *Vaccine* 35, 3691–3699.

Jung, M.H., Nikapitiya, C. and Jung, S.J. (2018) DNA vaccine encoding myristoylated membrane protein (MMP) of rock bream iridovirus (RBIV) induces protective immunity in rock bream (*Oplegnathus fasciatus*). *Vaccine* 36, 802–810.

Jung, S., Miyazaki, T., Miyata, M., Danayadol, Y. and Tanaka, S. (1997) Pathogenicity of iridovirus from Japan and Thailand for the red sea bream *Pagrus major* in Japan, and histopathology of experimentally infected fish. *Fisheries Science* 63, 735–740.

Jung, S.J. and Oh, M.J. (2000) Iridovirus-like infection associated with high mortalities of striped beakperch, *Oplegnathus fasciatus* (Temminck & Schlegel), in southern coastal areas of the Korean peninsula. *Journal of Fish Diseases* 23, 223–226.

Kawato, Y., Subramaniam, K., Nakajima, K., Waltzek, T. and Whittington, R. (2017a) Iridoviral diseases: red sea bream iridovirus and white sturgeon iridovirus. In: Woo, P.T.K. and Cipriano, R.C. (eds) *Fish Viruses and Bacteria: Pathobiology and Protection*. CAB International, Wallingford, UK, pp. 147–159.

Kawato, Y., Yamashita, H., Yuasa, K., Miwa, S. and Nakajima, K. (2017b) Development of a highly permissive cell line from spotted knifejaw (*Oplegnathus punctatus*) for red sea bream iridovirus. *Aquaculture* 473, 291–298.

Kim, J.O., Oh, S.Y., Matsui, T., Oh, M.J. and Nishizawa, T. (2012) RSIV is probably insensitive to the transient innate immune response induced by administration of poly(I:C), a synthetic double-stranded RNA. *Fish Pathology* 47, 137–142.

Kim, T.J., Jang, E.J. and Lee, J.I. (2008) Vaccination of rock bream, *Oplegnathus fasciatus* (Temminck & Schlegel), using a recombinant major capsid protein of fish iridovirus. *Journal of Fish Diseases* 31, 547–551.

Kim, W.S., Oh, M.J., Jung, S.J., Kim, Y.J. and Kitamura, S.I. (2005) Characterization of an iridovirus detected from cultured turbot *Scophthalmus maximus* in Korea. *Diseases of Aquatic Organisms* 64, 175–180.

Kim, Y.J., Jung, S.J., Choi, T.J., Kim, H.R., Rajendran, K.V. and Oh, M.J. (2002) PCR amplification and sequence analysis of irido-like virus infecting fish in Korea. *Journal of Fish Diseases* 25, 121–124.

Kinpara, S., Kijiyama, M., Takamori, A., Hasegawa, A., Sasada, A. *et al.* (2013) Interferon-α (IFN-α) suppresses HTLV-1 gene expression and cell cycling, while IFN-α combined with zidovudin induces p53 signaling and apoptosis in HTLV-1-infected cells. *Retrovirology* 10, 52.

Kitamura, S.I., Jung, S.J., Kim, W.S., Nishizawa, T., Yoshimizu, M. and Oh, M.J. (2006) A new genotype of lymphocystivirus, LCDV-RF, from lymphocystis diseased rockfish. *Archives of Virology*, 151, 607–615.

Kleiger, G. and Mayor, T. (2014) Perilous journey: a tour of the ubiquitin–proteasome system. *Trends in Cell Biology* 24, 352–359.

Krause, A., Neitz, S., Magert, H.J., Schulz, A., Forssmann, W.G. *et al.* (2000) LEAP-1, a novel highly disulfide-bonded human peptide, exhibits antimicrobial activity. *FEBS Letters* 480, 147–150.

Kurita, J. and Nakajima, K. (2012) Megalocytiviruses. *Viruses* 4, 521–538.

Kurita, J., Ngoh-Lim, G.H., Gibson-Kueh, S., De La Pena, L., Chuah, T.T. *et al.* (2004) Phylogenetic analysis of red sea bream iridovirus-like viruses in Southeast Asia. In: *7th Asian Fisheries Forum 04 Abstracts*. Asian Fisheries Society, Penang, Malaysia, p. 381.

Kusuda, R., Nagato, K. and Kawai, K. (1994) Characteristics of an iridovirus isolated from red sea bream, *Pagrus major*. *Suisanzoshoku* 42, 151–156.

Kwon, S.R., Nishizawa, T., Takami, I. and Yoshimizu, M. (2010) Antibody detection against red sea bream iridovirus (RSIV) in yellowtail *Seriola quinqueradiata* using ELISA. *Fish Pathology* 45, 73–76.

Lesbarrères, D., Balseiro, A., Brunner, J., Chinchar, V.G., Duffus, A. *et al.* (2012) Ranavirus: past, present and future. *Biology Letters* 8, 481–483.

Leung, R.K.M. and Whittaker, P.A. (2005) RNA interference: from gene silencing to gene-specific therapeutics. *Pharmacology and Therapeutics* 107, 222–239.

Levanova, A. and Poranen, M.M. (2018) RNA interference as a prospective tool for the control of human viral infections. *Frontiers in Microbiology* 9, 422–429.

Li, H., Sun, Z.P., Li, Q. and Jiang, Y.L. (2011) Characterization of an iridovirus detected in rock bream (*Oplegnathus fasciatus*; Temminck & Schlegel). *Chinese Journal of Virology* 27, 158–164.

Li, P.F., Yan, Y., Wei, S.N., Wei, J.G., Gao, R. *et al.* (2014) Isolation and characterization of a new class of DNA aptamers specific binding to Singapore grouper iridovirus (SGIV) with antiviral activities. *Virus Research* 188, 146–154.

Li, P.F., Wei, S.N., Zhou, L.L., Yang, M., Yu, Y.P. *et al.* (2015) Selection and characterization of novel DNA aptamers specifically recognized by Singapore grouper iridovirus-infected fish cells. *Journal of General Virology* 96, 3348–3359.

Liu, H.I., Chiou, P.P., Gong, H.Y. and Chou, H.Y. (2015) Cloning of the major capsid protein (MCP) of grouper iridovirus of Taiwan (TGIV) and preliminary evaluation of a recombinant MCP vaccine against TGIV. *International Journal of Molecular Sciences* 16, 28647–28656.

Liu, X.F., Wu, Y.H., Wei, S.N., Wang, N., Li, Y.Z. *et al.* (2018) Establishment and characterization of a brain-cell line from kelp grouper *Epinephelus moara*. *Journal of Fish Biology* 92, 298–307.

Liu, Y., Tran, B.N., Wang, F., Ounjai, P., Wu, J.L. and Hew, C.L. (2016) Visualization of assembly intermediates and budding vacuoles of Singapore grouper iridovirus in grouper embryonic cells. *Scientific Reports* 6, 18696.

Liu, Y.S., Shepherd, E.G. and Nelin, L.D. (2007) MAPK phosphatases – regulating the immune response. *Nature Reviews Immunology* 7, 202–212.

Lorenzen, N. and LaPatra, S.E. (2005) DNA vaccines for aquacultured fish. *Revue Scientifique et Technique (International Office of Epizootics)* 24, 201.

Lu, L., Zhou, S.Y., Chen, C., Weng, S.P., Chan, S.M. and He, J.G. (2005) Complete genome sequence analysis of an iridovirus isolated from the orange-spotted grouper, *Epinephelus coioides*. *Virology* 339, 81–100.

McGrogan, D.G., Ostland, V.E., Byrne, P.J. and Ferguson, H.W. (1998) Systemic disease involving an iridovirus-like agent in cultured tilapia, *Oreochromis niloticus* L. – a case report. *Journal of Fish Diseases* 21, 149–152.

Mao, X.L., Zhou, S., Xu, D., Gong, J., Cui, H.C. and Qin, Q.W. (2008) Rapid and sensitive detection of Singapore grouper iridovirus by loop-mediated isothermal amplification. *Journal of Applied Microbiology* 105, 389–397.

Marschang, R.E. (2011) Viruses infecting reptiles. *Viruses* 3, 2087–2126.

Matsuyama, T., Minami, T., Fukuda, Y., Sano, N., Sakai, T. *et al.* (2016) Passive immunization against red sea bream iridoviral disease in five marine fish species. *Fish Pathology* 51, 32–35.

Matsuyama, T., Sano, N., Takano, T., Sakai, T., Yasuike, M. *et al.* (2018) Antibody profiling using a recombinant protein-based multiplex ELISA array accelerates recombinant vaccine development: case study on red sea bream iridovirus as a reverse vaccinology model. *Vaccine* 36, 2643–2649.

Miller, D., Gray, M. and Storfer, A. (2011) Ecopathology of ranaviruses infecting amphibians. *Viruses* 3, 2351–2373.

Miyata, M., Matsuno, K., Jung, S.J., Danayadol, Y. and Miyazaki, T. (1997) Genetic similarity of iridoviruses from Japan and Thailand. *Journal of Fish Diseases* 20, 127–134.

Miyata, T., Tokunaga, F., Yoneya, T., Yoshikawa, K., Iwanaga, S. *et al.* (1989) Antimicrobial peptides, isolated from horseshoe crab hemocytes, tachyplesin II, and polyphemusins I and II: chemical structures and biological activity. *Journal of Biochemistry* 106, 663–668.

Morvan, A., Iwanaga, S., Comps, M. and Bachere, E. (1997) *In vitro* activity of the limulus antimicrobial peptide tachyplesin I on marine bivalve pathogens. *Journal of Invertebrate Pathology* 69, 177–182.

Murali, S., Wu, M.F., Guo, I.C., Chen, S.C., Yang, H.W. and Chang, C.Y. (2002) Molecular characterization and pathogenicity of a grouper iridovirus (GIV) isolated from yellow grouper, *Epinephelus awoara* (Temminck & Schlegel). *Journal of Fish Diseases* 25, 91–100.

Myskiw, C., Arsenio, J., van Bruggen, R., Deschambault, Y. and Cao, J. (2009) Vaccinia virus E3 suppresses expression of diverse cytokines through inhibition of the PKR, NF-κB, and IRF3 pathways. *Journal of Virology* 83, 6757–6768.

Nakajima, K. and Kunita, J. (2005) Red sea bream iridoviral disease. *Uirusu* 55, 115–125.

Nakajima, K. and Sorimachi, M. (1994) Biological and physico-chemical properties of the iridovirus isolated from cultured red sea bream, *Pagrus major*. *Fish Pathology* 29, 29–33.

Nakajima, K. and Sorimachi, M. (1995) Production of monoclonal-antibodies against red-sea bream iridovirus. *Fish Pathology* 30, 47–52.

Nakajima, K., Maeno, Y., Kurita, J. and Inui, Y. (1997) Vaccination against red sea bream iridoviral disease in red sea bream. *Fish Pathology* 32, 205–209.

Nakajima, K., Maeno, Y., Honda, A., Yokoyama, K., Tooriyama, T. and Manabe, S. (1999) Effectiveness of a vaccine against red sea bream iridoviral disease in a field trial test. *Diseases of Aquatic Organisms* 36, 73–75.

Nakajima, K., Ito, T., Kurita, J., Kawakami, H., Itano, T. *et al.* (2002) Effectiveness of a vaccine against red sea bream iridoviral disease in various cultured marine fish under laboratory conditions. *Fish Pathology* 37, 90–91.

Nakamura, T., Furunaka, H., Miyata, T., Tokunaga, F., Muta, T. *et al.* (1988) Tachyplesin, a class of antimicrobial peptide from the hemocytes of the horseshoe crab (*Tachypleus tridentatus*). *Isolation and Chemical Structure* 263, 16709–16713.

O'Farrell, C., Vaghefi, N., Cantonnet, M., Buteau, B., Boudinot, P. and Benmansour, A. (2002) Survey of transcript expression in rainbow trout leukocytes reveals a major contribution of interferon-responsive genes in the early response to a rhabdovirus infection. *Journal of Virology* 76, 8040–8049.

Oh, M.J., Kitamura, S.I., Kim, W.S., Park, M.K., Jung, S.J. *et al.* (2006) Susceptibility of marine fish species to a megalocytivirus, turbot iridovirus, isolated from turbot, *Psetta maximus* (L.). *Journal of Fish Diseases* 29, 415–421.

Oh, S.Y. and Nishizawa, T. (2016) Establishment of rock bream *Oplegnathus fasciatus* embryo (RoBE-4) cells with cytolytic infection of red seabream iridovirus (RSIV). *Journal of Virological Methods* 238, 1–5.

Oh, S.Y., Oh, M.J. and Nishizawa, T. (2014) Potential for a live red seabream iridovirus (RSIV) vaccine in rock bream *Oplegnathus fasciatus* at a low rearing temperature. *Vaccine* 32, 363–368.

OIE (World Organization for Animal Health) (2012) General introduction. In: *Manual of Diagnostic Tests for Aquatic Animals*, 6th edn. OIE, Paris.

OIE (World Organization for Animal Health) (2018) *Red sea bream iridoviral disease*. In: Manual of Diagnostic Tests for Aquatic Animals. OIE, Paris. Available at: http://www.oie.int/fileadmin/Home/eng/Health_standards/aahm/current/chapitre_rsbid.pdf (accessed 31 March 2020).

Oseko, N., Chuah, T.T., Palamisamy, V., Maeno, Y. and Kurita, J. (2004) Iridovirus isolated from diseased sea bass *Lates calcarifer* and red drum *Sciaenops ocellatus* causing mass mortality in Malaysia. In: *7th Asian Fisheries Forum 04 Abstracts*. Asian Fisheries Society, Penang, Malaysia, p. 127.

Oshima, S., Hata, J., Hirasawa, N., Ohtaka, T., Hirono, I. *et al.* (1998) Rapid diagnosis of red sea bream iridovirus infection using the polymerase chain reaction. *Diseases of Aquatic Organisms* 32, 87–90.

Ou-yang, Z.L., Huang, X.H., Huang, E.Y., Huang, Y.H., Gong, J. *et al.* (2010) Establishment and characterization of a new marine fish cell line derived from red-spotted grouper *Epinephelus akaara*. *Journal of Fish Biology* 77, 1083–1095.

Ou-yang, Z.L., Wang, P.R., Huang, Y.H., Huang, X.H., Wan, Q.J. *et al.* (2012a) Selection and identification of Singapore grouper iridovirus vaccine candidate antigens using bioinformatics and DNA vaccination. *Veterinary Immunology and Immunopathology* 149, 38–45.

Ou-yang, Z.L., Wang, P.R., Huang, X.H., Cai, J., Huang, Y.H. *et al.* (2012b) Immunogenicity and protective effects of inactivated Singapore grouper iridovirus (SGIV) vaccines in orange-spotted grouper, *Epinephelus coioides*. *Developmental and Comparative Immunology* 38, 254–261.

Pai, T.W., Chang, M.D., Tzou, W.S., Su, B.H., Wu, P.C. *et al.* (2006) REMUS: a tool for identification of unique peptide segments as epitopes. *Nucleic Acids Research* 34, W198–W201.

Pan, Y.G., Wang, S.W., Shan, Y.P., Zhang, D.L., Gao, J. *et al.* (2015) Ultrafast tracking of a single live virion during the invagination of a cell membrane. *Small* 11, 2782–2788.

Peng, C., Ma, H.L., Su, Y.L., Wen, W.G., Feng, J. *et al.* (2015) Susceptibility of farmed juvenile giant grouper *Epinephelus lanceolatus* to a newly isolated grouper iridovirus (genus *Ranavirus*). *Veterinary Microbiology* 177, 270–279.

Qin, Q.W., Lam, T.J., Sin, Y.M., Shen, H., Chang, S.F. *et al.* (2001) Electron microscopic observations of a marine fish iridovirus isolated from brown-spotted grouper, *Epinephelus tauvina*. *Journal of Virological Methods* 98, 17–24.

Qin, Q.W., Shi, C.Y., Gin, K.Y.H. and Lam, T.J. (2002) Antigenic characterization of a marine fish iridovirus from grouper, *Epinephelus* spp. *Journal of Virological Methods* 106, 89–96.

Qin, Q.W., Chang, S.F., Ngoh-Lim, G.H., Gibson-Kueh, S., Shi, C. and Lam, T.J. (2003) Characterization of a novel ranavirus isolated from grouper *Epinephelus tauvina*. *Diseases of Aquatic Organisms* 53, 1–9.

Reshi, M.L., Wu, J.-L., Wang, H.-V. and Hong, J.-R. (2014) RNA interference technology used for the study of aquatic virus infections. *Fish and Shellfish Immunology* 40, 14–23.

Rodger, H.D., Kobs, M., Macartney, A. and Frerichs, G.N. (1997) Systemic iridovirus infection in freshwater angelfish, *Pterophyllum scalare* (Lichtenstein). *Journal of Fish Diseases* 20, 69–72.

Sadler, A.J. and Williams, B.R.G. (2008) Interferon-inducible antiviral effectors. *Nature Reviews Immunology* 8, 559–568.

Saha, S. and Raghava, G.P.S. (2004) BcePred: prediction of continuous B-cell epitopes in antigenic sequences using physico-chemical properties. In: Nicosia, G., Cutello, V., Bentley, P.J. and Timmis, J. (eds) *Artificial Immune Systems*. Springer, Berlin/Heidelberg, Germany, pp. 197–204.

Sample, R., Bryan, L., Long, S., Majji, S., Hoskins, G. *et al.* (2007) Inhibition of iridovirus protein synthesis and virus replication by antisense morpholino oligonucleotides targeted to the major capsid protein, the 18 kDa immediate-early protein, and a viral homolog of RNA polymerase II. *Virology* 358, 311–320.

Sano, M., Minagawa, M. and Nakajima, K. (2002) Multiplication of red sea bream iridovirus (RSIV) in the experimentally infected grouper *Epinephelus malabaricus*. *Fish Pathology* 37, 163–168.

Sawayama, E. and Takagi, M. (2017) Evaluation of an RSIVD-resistant trait of red sea bream *Pagrus major* broodstock using DNA-based pedigree tracings: a field study. *Fish Pathology* 52, 23–30.

Sawayama, E., Tanizawa, S., Kitamura, S.I., Nakayama, K., Ohta, K. *et al.* (2017) Identification of quantitative trait loci for resistance to RSIVD in red sea bream (*Pagrus major*). *Marine Biotechnology* 19, 601–613.

Seo, J.Y., Yaneva, R. and Cresswell, P. (2011) Viperin: a multifunctional, interferon-inducible protein that regulates virus replication. *Cell Host and Microbe* 10, 534–539.

Seo, J.Y., Chung, H.J. and Kim, T.J. (2013) Codon-optimized expression of fish iridovirus capsid protein in yeast and its application as an oral vaccine candidate. *Journal of Fish Diseases* 36, 763–768.

Shi, C.Y., Wang, Y.G., Yang, S.L., Huang, J. and Wang, Q.Y. (2004) The first report of an iridovirus-like agent infection in fanned turbot, *Scophthalmus maximus*, in China. *Aquaculture* 236, 11–25.

Shi, W.F., Li, X., Hou, X.L., Peng, H.J., Jiang, Q.B. *et al.* (2012) Differential apoptosis gene expressions of rhabdomyosarcoma cells in response to enterovirus 71 infection. *BMC Infectious Diseases* 12, 327.

Shih, T.-C., Ho, L.-P., Wu, J.-L., Chou, H.-Y. and Pai, T.-W. (2019) A voting mechanism-based linear epitope prediction system for the host-specific *Iridoviridae* family. *BMC Bioinformatics* 20(Suppl. 7), 192.

Shimmoto, H., Kawai, K., Ikawa, T. and Oshima, S. (2010) Protection of red sea bream *Pagrus major* against red sea bream iridovirus infection by vaccination with a recombinant viral protein. *Microbiology and Immunology* 54, 135–142.

Shin, Y.J., Kwon, T.H., Seo, J.Y. and Kim, T.J. (2013) Oral immunization of fish against iridovirus infection using recombinant antigen produced from rice callus. *Vaccine* 31, 5210–5215.

Shuang, F., Luo, Y., Xiong, X.-P., Weng, S., Li, Y. *et al.* (2013) Virions proteins of an RSIV-type megalocytivirus from spotted knifejaw *Oplegnathus punctatus* (SKIV-ZJ07). *Virology* 437, 89–99.

Snieszko, S.F. (1974) The effects of environmental stress on outbreaks of infectious diseases of fishes. *Journal of Fish Biology* 6, 197–208.

Subramaniam, K., Shariff, M., Omar, A.R. and Hair-Bejo, M. (2012) Megalocytivirus infection in fish. *Reviews in Aquaculture* 4, 221–233.

Sudthongkong, C., Miyata, M. and Miyazaki, T. (2002a) Iridovirus disease in two ornamental tropical freshwater fishes: African lampeye and dwarf gourami. *Diseases of Aquatic Organisms* 48, 163–173.

Sudthongkong, C., Miyata, M. and Miyazaki, T. (2002b) Viral DNA sequences of genes encoding the ATPase and the major capsid protein of tropical iridovirus isolates which are pathogenic to fishes in Japan, South China Sea and Southeast Asian countries. *Archives of Virology* 147, 2089–2109.

Takeuchi, O. and Akira, S. (2009) Innate immunity to virus infection. *Immunological Reviews* 227, 75–86.

Thanasaksiri, K., Hirono, I. and Kondo, H. (2015) Temperature-dependent regulation of gene expression in poly (I:C)-treated Japanese flounder, *Paralichthys olivaceus*. *Fish and Shellfish Immunology* 45, 835–840.

Tonheim, T.C., Bogwald, J. and Dalmo, R.A. (2008) What happens to the DNA vaccine in fish? A review of current knowledge. *Fish and Shellfish Immunology* 25, 1–18.

Tortorella, D., Gewurz, B.E., Furman, M.H., Schust, D.J. and Ploegh, H.L. (2000) Viral subversion of the immune system. *Annual Review of Immunology* 18, 861–926.

Tourneur, L. and Chiocchia, G. (2010) FADD: a regulator of life and death. *Trends in Immunology* 31, 260–269.

Verrier, E.R., Langevin, C., Tohry, C., Houel, A., Ducrocq, V. *et al.* (2012) Genetic resistance to rhabdovirus infection in teleost fish is paralleled to the derived cell resistance status. *PLoS ONE* 7, e33935.

Wang, C.S., Shih, H.H., Ku, C.C. and Chen, S.N. (2003) Studies on epizootic iridovirus infection among red sea bream, *Pagrus major* (Temminck & Schlegel), cultured in Taiwan. *Journal of Fish Diseases* 26, 127–133.

Wang, F., Bi, X., Chen, L.M. and Hew, C.L. (2008) ORF018R, a highly abundant virion protein from Singapore grouper iridovirus, is involved in serine/threonine phosphorylation and virion assembly. *Journal of General Virology* 89, 1169–1178.

Wang, H.W., Lin, Y.C., Pai, T.W. and Chang, H.T. (2011) Prediction of B-cell linear epitopes with a combination of support vector machine classification and amino acid propensity identification. *Journal of Biomedicine and Biotechnology* 2011, 432830.

Wang, L., Bai, B., Huang, S., Liu, P., Wan, Z.Y. *et al.* (2017) QTL mapping for resistance to iridovirus in Asian seabass using genotyping-by-sequencing. *Marine Biotechnology* 19, 517–527.

Wang, S.W., Huang, X.H., Huang, Y.H., Hao, X., Xu, H.J. *et al.* (2014) Entry of a novel marine DNA virus, Singapore grouper iridovirus, into host cells occurs via clathrin-mediated endocytosis and macropinocytosis in a pH-dependent manner. *Journal of Virology* 88, 13047–13063.

Wang, X.-W., Ao, J.-Q., Li, Q.-G. and Chen, X.-H. (2006) Quantitative detection of a marine fish iridovirus isolated from large yellow croaker, *Pseudosciaena crocea*, using a molecular beacon. *Journal of Virological Methods* 133, 76–81.

Wang, Y.Q., Lu, L., Weng, S.P., Huang, J.N., Chan, S.M. and He, J.G. (2007) Molecular epidemiology and phylogenetic analysis of a marine fish infectious spleen and kidney necrosis virus-like (ISKNV-like) virus. *Archives of Virology* 152, 763–773.

Wei, J.G., Xu, D., Zhou, J.G., Cui, H.C., Yan, Y. *et al.* (2010) Molecular cloning, characterization and expression analysis of a C-type lectin (Ec-CTL) in orange-spotted grouper, *Epinephelus coioides*. *Fish and Shellfish Immunology* 28, 178–186.

Wei, J., Zhang, P., Guo, M., Xu, M., Li, P. *et al.* (2015) TTRAP is a critical factor in grouper immune response to virus infection. *Fish and Shellfish Immunology* 46, 274–284.

Wen, C.M., Lee, C.W., Wang, C.S., Cheng, Y.H. and Huang, H.Y. (2008) Development of two cell lines from *Epinephelus coioides* brain tissue for characterization of betanodavirus and megalocytivirus infectivity and propagation. *Aquaculture* 278, 14–21.

Weng, S.P., Wang, Y.Q., He, J.G., Deng, M., Lu, L. *et al.* (2002) Outbreaks of an iridovirus in red drum, *Sciaenops ocellata*, cultured in southern China. *Journal of Fish Diseases* 25, 681–685.

White, S.H., Wimley, W.C. and Selsted, M.E. (1995) Structure, function, and membrane integration of defensins. *Current Opinion in Structural Biology* 5, 521–527.

Whittington, R.J. and Reddacliff, G.L. (1995) Influence of environmental temperature on experimental infection of redfin perch (*Perca fluviatilis*) and rainbow trout (*Oncorhynchus mykiss*) with epizootic haematopoietic necrosis virus, an Australian iridovirus. *Australian Veterinary Journal* 72, 421–424.

Whittington, R.J., Becker, J.A. and Dennis, M.M. (2010) Iridovirus infections in finfish – critical review with emphasis on ranaviruses. *Journal of Fish Diseases* 33, 95–122.

Wu, S.Q., Li, X.H., Pan, H.J. and Huang, Z.B. (1997) Research on the pathogen of the outbreak-infective disease of *Siniperca chuatsi*. *Journal of Fisheries China* 21(Suppl.), 56–60. (in Chinese with English abstract)

Xie, H.W., Wei, J.G. and Qin, Q.W. (2016) Antiviral function of tachyplesin I against iridovirus and nodavirus. *Fish and Shellfish Immunology* 58, 96–102.

Yadav, M., Liebau, E., Haldar, C. and Rathaur, S. (2011) Identification of major antigenic peptide of filarial glutathione-*S*-transferase. *Vaccine* 29, 1297–1303.

Yang, Y., Huang, Y.H., Yu, Y.P., Yang, M., Zhou, S. *et al.* (2016) RING domain is essential for the antiviral activity of TRIM25 from orange spotted grouper. *Fish and Shellfish Immunology* 55, 304–314.

Yanong, R.P.E. and Waltzek. T.B. (2010) Megalocytivirus infections in fish, with emphasis on ornamental species. *University of Florida IFAS Extension Bulletin No. FA182*. Program in Fisheries and Aquatic Sciences, School of Forest Resources and Conservation, UF/IFAS Extension, University of Florida, Gainesville, Florida.

Yeh, Y.C., Wang, T.Y., Chou, H.Y., Lin, H.Y., Chen, T.Y. *et al.* (2016) A member of the immunoglobulin superfamily, orange-spotted grouper novel immune gene EcVig, is induced by immune stimulants and type I interferon. *Fish and Shellfish Immunology* 58, 415–422.

Yu, Y.P., Huang, X.H., Liu, J.X., Zhang, J.C., Hu, Y. *et al.* (2017a) Fish TRIM32 functions as a critical antiviral molecule against iridovirus and nodavirus. *Fish and Shellfish Immunology* 60, 33–43.

Yu, Y.P., Huang, Y.H., Ni, S.W., Zhou, L.L., Liu, J.X. *et al.* (2017b) Singapore grouper iridovirus (SGIV) TNFR homolog VP51 functions as a virulence factor via modulating host inflammation response. *Virology* 511, 280–289.

Zasloff, M. (2002) Antimicrobial peptides of multicellular organisms. *Nature* 415, 389–395.

Zenke, K. and Kim, K.H. (2010) Effects of long double-stranded RNAs on the resistance of rock bream *Oplegnathus fasciatus* fingerling against rock bream iridovirus (RBIV) challenge. *Journal of Fish Pathology* 23, 273–280.

Zenke, K., Yoon, K.J., Kim, M.S., Choi, S.H. and Kim, K.H. (2014) Acquired resistance of rock bream (*Oplegnathus fasciatus*) against rock bream iridovirus (RBIV) through undergoing low water temperature period. *Journal of Fish Pathology* 27, 85–89.

Zhang, D.X. and Zhang, D.E. (2011) Interferon-stimulated gene 15 and the protein ISGylation system. *Journal of Interferon and Cytokine Research* 31, 119–130.

Zhang, J. and Li, M.F. (2015) ORF75 of megalocytivirus RBIV-C1: a global transcription regulator and an effective vaccine candidate. *Fish and Shellfish Immunology* 45, 486–494.

Zhang, J., Zhang, B.C. and Sun, L. (2015) P247 and p523: two *in vivo*-expressed megalocytivirus proteins that induce protective immunity and are essential to viral infection. *PLoS ONE* 10(3), e0121282.

Zhang, M., Hu, Y.H., Xiao, Z.Z., Sun, Y. and Sun, L. (2012) Construction and analysis of experimental DNA vaccines against megalocytivirus. *Fish and Shellfish Immunology* 33, 1192–1198.

Zhang, X., Zang, S., Li, C., Wei, J. and Qin, Q. (2018) Molecular cloning and characterization of FADD from the orange-spotted grouper (*Epinephelus coioides*). *Fish and Shellfish Immunology* 74, 517–529.

Zhang, Y.B. and Gui, J.F. (2012) Molecular regulation of interferon antiviral response in fish. *Developmental and Comparative Immunology* 38, 193–202.

Zheng, F., Liu, H., Sun, X., Zhang, Y., Zhang, B. *et al.* (2016a) Development of oral DNA vaccine based on chitosan nanoparticles for the immunization against reddish body iridovirus in turbots (*Scophthalmus maximus*). *Aquaculture* 452, 263–271.

Zheng, F., Liu, H., Sun, X., Qin, X., Xu, Z. and Wang, B. (2016b) Construction and expression of DNA vaccine against reddish body iridovirus and evaluation of immune efficacy in turbot (*Scophthalmus maximus*). *Aquaculture Research* 48, 4174–4183.

Zhou, J.G., Wei, J.G., Xu, D., Cui, H.C., Yan, Y. *et al.* (2011) Molecular cloning and characterization of two novel hepcidins from orange-spotted grouper, *Epinephelus coioides*. *Fish and Shellfish Immunology* 30, 559–568.

10 Vibriosis

Carmen Amaro[1]*, Belén Fouz[1], Eva Sanjuán[1] and Jesús L. Romalde[2]

[1]University of Valencia, Burjasot, Spain; [2]University of Santiago de Compostela, Santiago de Compostela, Spain

10.1 General Introduction

Vibrios are ubiquitous in the aquatic environment, although they colonize preferentially habitats located in temperate, tropical and subtropical areas with few exceptions. In these environments, vibrios survive either as planktonic free-living forms in the water column or as sessile living forms attached to organic and inorganic surfaces (Gómez-Gil *et al.*, 2014). The life cycle of vibrios in water depends on temperature and salinity as well as on the presence of an available source of nutrients. Regarding nutrients and metabolism, vibrios constitute a very versatile group of organotrophs that can ferment or respire, either aerobically or anaerobically, a wide variety of organic compounds. Regarding temperature and salinity, vibrios generally prefer waters of medium salinities (1–2% NaCl) and temperatures above 15°C, with variable optimal values depending on the species.

Within the genus *Vibrio* there are approximately 140 species (http://www.bacterio.net/vibrio.html (accessed 12 March 2020)), including some recognized as important pathogens for animals and/or humans, causing multiple diseases known globally as vibriosis. Most pathogenic vibrios are pathogens of multiple hosts, and at least one of them, *Vibro vulnificus*, is considered a true zoonotic agent (Veenstra *et al.*, 1992; Dalsgaard *et al.*, 1996). In addition, vibrios are unusual because they have more than one entrance portal in their respective hosts; they infect fish by colonizing the gills and/or the anus after contact with water containing the pathogen, or by colonizing the intestine after ingestion of contaminated food (Toranzo *et al.*, 2005, 2017; Amaro *et al.*, 2015). Among the different animal vibrioses, the most studied is fish vibriosis, which has multiple clinical manifestations depending on the specific host and the pathogenic species involved in the interaction. However, they all have in common that the acute form of the disease is a septicaemia that can lead to death of the host, especially if the host is immunocompromised.

Pathogenic *Vibrio* species include virulent and avirulent strains that differ in the presence and combination of virulence genes. Many virulence genes are present in mobile genetic elements that are exchanged by horizontal gene transfer, mainly in biofilms where bacteria coexist proximally. In addition, many vibrios produce a type VI secretion system to kill neighbouring cells that are not siblings, together with a natural transformation machinery that allows them to take DNA and recombine it with their own DNA (Borgeaud *et al.*, 2015). All these processes of DNA exchange, when they affect the virulence genes, can contribute to the appearance of new pathogenic variants.

Vibrios have a marked seasonal distribution in temperate regions where the greatest number of bacteria and the maximum occurrence of vibriosis in humans and animals generally occur during periods of warmer temperatures. Consequently, the increase in water temperature caused by global warming would increase vibrio populations to which humans and animals would be exposed. In addition, weather-mediated physiological stress may also compromise the disease resistance of fish and increase their susceptibility to vibriosis.

This chapter focuses on the most relevant *Vibrio* species, the vibrioses that they cause in fish and how, the main methodologies for their diagnosis and control, as well as the biology and culture of their main hosts. Variations in the incidence and

*carmen.amaro@uv.es

severity of these vibrioses in relation to global warming are also discussed.

10.2 *Vibrio anguillarum* and the Classical Vibriosis

Vibrio anguillarum was first described in 1909 as the aetiological agent responsible for a haemorrhagic septicaemia in eels called 'red eel pest'. That disease, now designated 'classical vibriosis' (*Va*-vibriosis), affects a wide variety of warm- and cold-water fish species of economic importance (Table 10.1). Recently, a case of a putative fatal *Va*-vibriosis was reported in an immunocompromised patient in Maine (USA), although it was not possible to ascertain whether the pathogen was the only agent responsible for the patient's death (Sinatra and Colby, 2018).

10.2.1 Taxonomy and phylogeny of the pathogen

V. anguillarum's taxonomy has been subject to continuous revisions. Originally described as *Bacterium anguillarum*, a few years later the current name was coined (Austin, 2012). Then, two biotypes were described, and the old biotype II was subsequently reclassified as *Vibrio ordalii* (Schiewe *et al.*, 1981).

MacDonell and Colwell (1985) proposed that the *Listonella* genus should encompass *V. anguillarum*, *Vibrio damsela* and *Vibrio pelagius*, based on a phylogenetic study of the 5S rRNA gene. Later, Thompson *et al.* (2011) showed that *Listonella* was a heterotopic synonym of the genus *Vibrio* and proposed to return to the original taxonomic assignment.

In addition to its phenotypic heterogeneity, *V. anguillarum* also presents a great antigenic diversity. The European classification comprises a total of 23 serotypes, described on the basis of the variation of the O-antigen. Only strains of serotype O1, O2 and, to a lesser extent, O3 have been associated with outbreaks of diseases, while the rest of the serotypes are considered innocuous strains (Toranzo *et al.*, 2005). In addition, serotypes O2 and O3 are antigenically heterogeneous, each subdivided into two subgroups (Toranzo *et al.*, 2005).

As other vibrios, *V. anguillarum* presents two circular chromosomes of around 3.0 and 1.2 Mbp with a G + C content of 43–46%. The presence of two chromosomes could represent a survival strategy, since it could allow the bacterium a rapid adaption to environmental changes or different niches. Studies of intraspecific genetic diversity have demonstrated the existence of different clonal lineages with epidemiological significance among the major pathogenic serotypes (Toranzo *et al.*, 2017).

10.2.2 Ecology of the pathogen

V. anguillarum is considered a major threat to marine aquaculture due to the high frequency of disease outbreaks, its wide geographical distribution and the high number of species of fish, molluscs and crustaceans affected by *Va*-vibriosis (Table 10.1).

Table 10.1. Host range of the most relevant *Vibrio* species in the context of global warming[a].

Species	Human	Fish	Others
Vibrio alginolyticus	+	+ (croaker, cobia, grouper, sea bream, snapper)	+ (clams, shrimps, prawns)
Vibrio anguillarum	(+)[b]	+ (ayu, cod, eel, rainbow trout, salmon, sea bass, sea bream, sole, striped bass, turbot)	
Vibrio harveyi	(+)[c]	+ (amberjack, groupers, horse mackerel, salmonids, sea bass, sea bream, sharks, sole)	+ (shrimp, abalone, coral)
Vibrio ichthyoenteri		+ (flounder, sea bream, snapper, salmon, sole, turbot, wrasse)	
Vibrio ordalii		+ (ayu, rockfish, salmonids)	
Vibrio splendidus		+ (flounder, rainbow trout, salmon, sea bream, turbot, wrasse)	+ (oyster, scallop)
Vibrio vulnificus	+	+ (eel, grouper, sea bass, red snapper, tilapia)	+ (shrimp)

[a]Data from: Toranzo *et al.* (2005), Austin and Zhang (2006), Haldar *et al.* (2010), Actis *et al.* (2011), Frans *et al.* (2011), Austin (2012), Ruwandeepika *et al.* (2012), Angelidis (2014), Gómez-Gil *et al.* (2014), Amaro *et al.* (2015), Toranzo *et al.* (2017) and Ceccarelli *et al.* (2019).

[b]A case of fatal infection has been reported recently in an immunocompromised patient, although the unique aetiology of *V. anguillarum* could not be demonstrated.

[c]Associated with wound infections mostly in patients with impaired immunity.

The pathogen is part of the natural microbiota of fish and is also associated with planktonic rotifers, which are the main source of fish feed in aquaculture. Consequently, rotifers can play an important role as vectors for *Va*-vibriosis. High environmental levels of organic material, caused by an excess of fish feed, could promote the growth and persistence of *V. anguillarum* in aquaculture facilities. Therefore, the bacterium could persist in a fish farm for years, leading to recurrent infections and outbreaks.

V. anguillarum is a moderate halophilic bacterium (0.5–2% NaCl) that is adapted to withstand natural changes in salinity, temperature and nutrient availability. In fact, studies on the detection of this pathogen in water show that temperature and salinity play an important role in its cultivability and distribution (Hickey and Lee, 2018). Therefore, it seems to be more abundant in low-salinity than in high-salinity waters (salinities above 6% are lethal) in cold areas. *V. anguillarum* survives under starvation in water microcosms at salinities between 1 and 3.5% for more than 50 weeks (Hickey and Lee, 2018). This resistance to nutrient starvation would facilitate its dissemination in the water column, which could act as a reservoir for the pathogen. Water temperature and salinity influence the chemotaxis of *V. anguillarum* towards the mucus of fish and, consequently, the colonization of fish (Larsen *et al.*, 2004). Interestingly, the pathogen can also cause diseases in fish grown in fresh water in subtropical countries (ayu in Japan). In this case, it seems that its survival strategy is to form a biofilm to create a protective microenvironment but with easy access to essential nutrients (Fujiwara-Nagata and Eguchi, 2003).

10.2.3 Biology and culture of the main fish hosts

As mentioned above, *Va*-vibriosis affects a wide variety of fish species in both natural and artificial conditions (Table 10.1). From an economic point of view, the most important host species are Atlantic salmon, turbot, sea bass, sea bream and cod.

Atlantic salmon

Atlantic salmon, *Salmo salar*, occur naturally along both east and west coasts of the North Atlantic Ocean, occupying a variety of freshwater environments (Jones, 2004).

All the main production areas are within latitudes 40–70° in the northern hemisphere and 40–50° in the southern hemisphere (Jones, 2004). The culture lasts about 2 years and has two phases, one in fresh water (eggs to juveniles 40–120 g) and the other in seawater, after the induction of early smoltification (adaptation of juvenile fish from living in fresh water to living in seawater). The commercial size is 2–3 kg. Atlantic salmon grows best in places where the water temperature is in the range of 6 to 16°C, and the salinities are close to ocean levels (33–34‰) (Table 10.2).

Turbot

The turbot, *Scophthalmus maximus*, is widely distributed in European waters from Iceland and Norway (68°N) to Morocco (30°N) (Nelson, 2006). Turbot is a flatfish that lives on sandy and rocky bottoms up to 70–80 m deep, and feeds mainly on fish, crustaceans and bivalves (Nelson, 2006). This fish has been a target species in marine aquaculture due to its high economic value. As a result of this activity, turbot is currently present in other geographical areas, such as Chile and China. The age of maturity is estimated at 2 years, and the reproduction period is usually from the end of March to August (Nelson, 2006).

The growth rate of juvenile turbot is significantly influenced by the interaction of temperature and fish size (Imsland *et al.*, 1996). Thus, the optimal temperature for the growth of juveniles of 25–75 g is between 16 and 19°C, while for fish of 100 g it is between 13 and 16°C (Table 10.2). The turbot tolerates overcrowding (the density of the population can reach 100 kg/m^2) and its productivity depends on the quality of the offspring, reproduction temperatures and control of the main pathologies.

Sea bass

The European (or common) sea bass (ESB), *Dicentrarchus labrax* spends most of its life in coastal lagoons and estuaries. This species is euryhaline and eurythermic, which determines its wide geographical distributions from southern Norway to Western Sahara, along the Mediterranean Sea and the Black Sea (Haffray *et al.*, 2007; Pérez-Ruzafa and Marcos, 2014). Temperature is the main environmental condition that affects its distribution and behaviour. Optimum and tolerable growth

Table 10.2. Type of culture and physicochemical conditions at which the most important hosts for *Vibrio vulnificus*, *Vibrio anguillarum* and *Vibrio harveyi* are grown in fish farms[a].

		Range of				
		Temperature (°C)		Salinity (%)		
Pathogen	Host	Growth	Optimal	Growth	Optimal	Type of culture (density in tanks)
V. vulnificus	Eels	23–28	26–28 (*Anguilla japonica*) 24–26 (*Anguilla anguilla*)	0–3.5	0–0.5	Extensive pond systems, intensive culture in recirculation systems (densities 100–150 kg/m²) or valliculture (elvers of 15–35 g stocked at rate of 4–15 kg/ha)
	Nile tilapia	25–30	28–30	0–1.6	8	Semi-intensive pond systems (up to 3 fishes/m²) or intensive in tanks (up to 185 kg/m³)
V. anguillarum	Atlantic salmon	4–18	12–13	0–3.5	2.2–2.8	Intensive in sea cages (up to 20 kg/m³)
	Turbot	8–26	16–18	1.2–4.0	2.5–3.0	Intensive in circular concrete tanks (variable, stocking density can reach 100 kg/m²)
	Sole	12–22	18–20	0.5–5.5	3.9	Intensive (up to 30 kg/m²)
	Gilthead sea bream	14–30	18–24	0.5–4	3.0-3.7	Intensive in sea cages (10–15 kg/m³) but also in concrete tanks (15–45 kg/m³)
	Cod	4–20	8–10	1.4–2.8	1.4	Intensive in sea cages (up to 40 kg/m³)
V. harveyi	European sea bass	18–27	18 or 25[b]	0.5–3.7	1.5 or 3[c]	Intensive in sea cages (10–20 kg/m³) but also in concrete raceways and ponds (<2 kg/m³)

[a]Data from: R. Barrera (Valencia, 1990, personal communication), Bœuf and Payan (2001), Haffray *et al.* (2007), Person-Le Ruyet (2010), Morais *et al.* (2016), FAO (2018) and respective fact sheets of FAO-CASIP (Food and Agriculture Organization of the United Nations, Cultured Aquatic Species Information Programme) (http://www.fao.org/tempref/FI/DOCUMENT/aquaculture/CulturedSpecies/index.htm (accessed 25 March 2020)). In the case of eels, the data refer to fresh-water aquaculture; brackish-water aquaculture has been abandoned in Europe precisely due to *V. vulnificus*.
[b]Atlantic or Mediterranean populations, respectively (Pérez-Ruzafa and Marcos, 2014).
[c]Larval rearing and nursery or adults, respectively (Pérez-Ruzafa and Marcos, 2014).

temperatures and salinities for ESB populations of the Atlantic and Mediterranean areas are indicated in Table 10.2. Temperature also regulates migratory movements to deeper areas in the open coast for reproduction and the duration of the spawning season. ESB is a gonochoristic (separation of the sexes in different individuals) species that, under culture conditions, has an undifferentiated sexual period throughout the first year of life. The females spawn in winter in the Mediterranean Sea (December to March) and until June in the Atlantic Ocean. They have a high fecundity (on average 200,000 eggs/kg of female), begin to reproduce at more than 2 kg and can live 6–7 years in the wild (Haffray *et al.*, 2007). Eggs and larvae have a large dispersion during the first 3 months of life and adults migrate for several hundred kilometres.

The ESB is not an endangered species since it is listed as Least Concern by the International Union for the Conservation of Nature. It is widespread and represents one of the most abundant and productive fish species in European coasts (Pérez-Ruzafa and Marcos, 2014). ESB culture production in the EU is ten times higher than the yields of fishing and has also been introduced in Israel, Oman and the United Arab Emirates. Although the ESB is mainly bred in marine cages, it can also be reared in concrete raceways and ponds (Table 10.2).

Gilthead sea bream

The gilthead sea bream, *Sparus aurata*, is common in the Mediterranean Sea and present along the Eastern Atlantic coasts from Great Britain to Senegal (http://www.fao.org/tempref/FI/DOCUMENT/aquaculture/CulturedSpecies/file/en/en_giltheadseabr.htm (accessed 25 March 2020)). It is mainly carnivorous and inhabits seagrass beds and sandy bottoms, commonly to depths of about 30 m. The sea bream is a protandrous hermaphrodite. Very sensitive to low temperatures (lower lethal limit is 4°C), the optimum conditions for its culture range from 18 to 26°C, in which they reach commercial size (350–400 g) in about 1 year (Table 10.2). At present, the main producers in the Mediterranean are Greece, Turkey, Spain and Italy (FAO, 2018).

Cod

The cod, *Gadus morhua*, is a species of cold-water fish that is distributed in the continental shelves and in the coastal waters of the North Atlantic Ocean (Otterå, 2009). The cod is demersal, living usually on or close to the sea bottom. Adult cod prefers water temperatures of 2 to 8°C but can also be found at temperatures up to 20°C (Table 10.2). The adult cod is omnivorous, also presenting a cannibalistic behaviour towards the smaller cod. The cod is a batch spawner that generates between ten and 20 batches during a period of 2 to 3 months, usually between December and June (Otterå, 2009). Spawning can easily be manipulated to occur at other times of the year using a photoperiod setting.

Although the culture of Atlantic cod has a long history, it was around 1980 when the use of seawater enclosures for the production of juvenile cod in Norway began, obtaining the first significant production of juvenile cod in 1983 (Otterå, 2009). Aquaculture production reached a maximum of 21,000 tonnes in 2009–2010 and has suffered a significant decrease in recent years (FAO, 2018).

10.2.4 The classical vibriosis

Fish affected by *Va*-vibriosis generally show typical signs of septicaemia with haemorrhages at the base of the fins, exophthalmia and corneal opacity (Fig. 10.1). Moribund fish are often anorexic with pale gills that reflect severe anaemia. Oedematous lesions are frequently observed, predominantly centred in the hypodermis (Toranzo *et al.*, 2017). In the case of acute disease, darkening of the skin and ulcerations are observed, while internally there is a dilatation and liquefaction of the spleen and kidney. In the case of chronic disease, the cutaneous lesions can progress to granulomatosis, the gills lose colour and the haemorrhages of the abdominal cavity end with fibrinous adhesions between the viscera.

Outbreaks of *Va*-vibriosis usually occur when the water temperature rises (spring in the northern hemisphere) or falls (autumn in the northern hemisphere) rapidly, being more frequent at temperatures below 10–11°C in salmonids or between 15 and 16°C in flatfish. The incubation period is related to the temperature of the water, the virulence of the strain and the degree of stress of the host.

10.2.5 Host–pathogen interaction

Although the mechanisms of pathogenicity of *V. anguillarum* are still not well understood, some factors and genes related to virulence have been identified, which include those related to chemotaxis, motility, iron acquisition, and production of exoenzymes, haemolysins, proteases, exopolysaccharides (EPS) and lipopolysaccharides (LPS).

The portals of entry of *V. anguillarum* into the host seem to be the intestine, gills and some areas of the skin (Hickey and Lee, 2018). In rainbow trout, *V. anguillarum* is able to grow faster and more abundantly in the skin mucus than in the intestinal mucus, suggesting that primary colonization may be the skin (Weber *et al.*, 2010). In contrast, in transparent zebrafish, the invasion occurs after adherence and colonization of the intestine (O'Toole *et al.*, 2004).

Whatever the point of entry into the host, motility and chemotaxis play an essential role in the pathogenesis (Toranzo *et al.*, 2017). In particular, mutants defective in chemotaxis through a deletion in *cheR* are also deficient in virulence by immersion but not by intraperitoneal (IP) injection, which suggests that chemotaxis is important for attachment and colonization but not for invasion (Toranzo *et al.*, 2017).

V. anguillarum is able to adhere to different types of cell lines using adhesins, such as pili and fimbriae, and also polysaccharides (EPS and LPS) and glycoproteins. In addition, the bacterium utilizes the O-antigen to mask the molecular patterns located on the surface that are recognized by

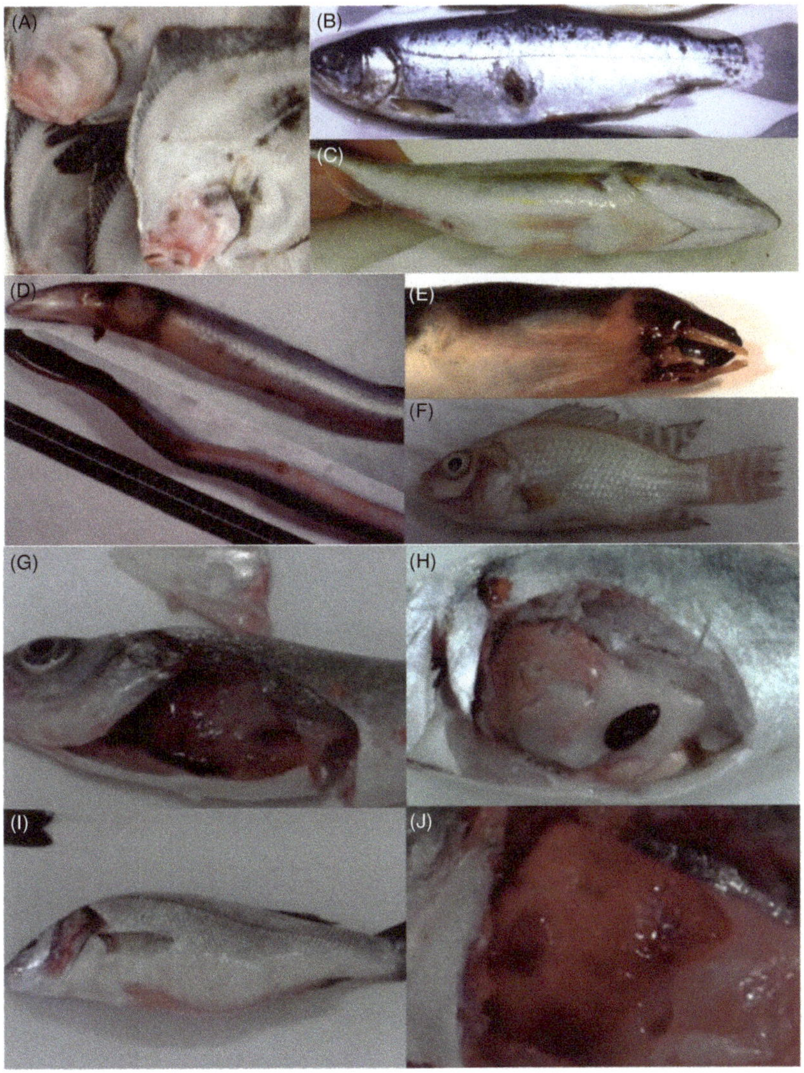

Fig. 10.1. Clinical signs of fish vibriosis caused by *Vibrio anguillarum* in turbot (*Scophthalmus maximus*) (A), Atlantic salmon (*Salmo salar*) (B) and gilthead sea bream (*Sparus aurata*) (C), by *Vibrio vulnificus* in eel (*Anguilla anguilla*) (D, E) and tilapia (*Oreochromis niloticus*) (F), and by *Vibrio harveyi* in sea bass (*Dicentratus labrax*) (G–J).

the receptors on the phagocytic skin cells and, therefore, is able to evade phagocytosis (Lindell *et al.*, 2012).

The growth in the mucus induces in *V. anguillarum* the expression of a different pattern of outer membrane proteins and exoenzymes that may be involved in the resistance to the mucosal antimicrobial components. Among these exoenzymes, the most important is the zinc metalloprotease encoded by the *empA* gene (homologues *vvpE* in *V. vulnificus*), which degrades host tissues and produces visible lesions (Hickey and Lee, 2018). *V. anguillarum* also secretes several haemolysins (VAH 1–5) (Rodkhum *et al.*, 2005) as well as produces an RTX toxin (repeats-in-toxin) with a proven role, but not fully understood, in virulence (Li *et al.*, 2008).

The ability to obtain iron is considered one of the main virulence factors in *V. anguillarum*. The species produces two catechol-type siderophores: anguibactin and vanchrobactin. Anguibactin biosynthesis is encoded in a 65 kb plasmid, named pJM1, which it is present only in O1 strains, although its synthesis also requires chromosomal genes (Li and Ma, 2017). The genes for ferric-anguibactin transport through the outer and inner membrane (Fat proteins and others) are also found in pJM1 (Li and Ma, 2017). The biosynthesis of vanchrobactin is encoded on chromosome I (Li and Ma, 2017) and appears to be widespread and ancestral in the species. It has been proposed that the vanchrobactin-biosynthesis pathway is inactivated by transposon insertion when the bacterium acquires pJM1 (or a similar plasmid) (Toranzo *et al.*, 2017). The true role of the vanchrobactin system in virulence remains unknown.

Balado *et al.* (2018) described a novel genomic island present in many virulent strains that encodes a third siderophore, named piscibactin, with a strong homology to the piscibactin encoded in pPHDP70 from *Photobacterium damselae* subsp. *piscicida*. This siderophore is produced simultaneously with vanchrobactin and seems to be a key virulence factor for *Va*-vibriosis. The synthesis of piscibactin is favoured at low temperatures, when many outbreaks of *Va*-vibriosis occur.

Three quorum sensing (QS) systems have been described in *V. anguillarum*: two involving the signal synthase/receptor pair VanI/VanR, and another known as the three-channel system that is also present in other *Vibrio* species (Frans *et al.*, 2011). The *in vivo* production of the auto-inducer AHL (acylhomoserine lactone) has been demonstrated in *V. anguillarum* but the results obtained to date indicate that QS is not involved in the regulation of virulence (Frans *et al.*, 2011). Alternative sigma factors have also been involved in the virulence of *V. anguillarum*. Among them, RpoN seems to have a role in motility, as an *rpoN* mutant was not motile due to a deficiency in one of the flagellins, and RpoS seems to have a role in toxin and exoenzyme production (phospholipase, diastase, lipase, caseinase, haemolysin, catalase and protease) as an *rpoS* mutant produced significantly less exoenzymes and toxins than the wild-type strain (Ma *et al.*, 2009).

10.2.6 Preventive measures

Manipulate physicochemical parameters

Good management practices at fish farms, including the control of the physicochemical parameters of water, the density of fish, the elimination of dead fish and excess food, etc., are basic practices to avoid the occurrence of outbreaks of *Va*-vibriosis.

Temperature can be a key factor in the increase in mortalities caused by *V. anguillarum* although this influence depends on the fish species (or even the genetic family) or the geographic region. Hoare *et al.* (2002) reported that Canadian and Icelandic halibut were more susceptible to *Va*-vibriosis and, consequently, suffered a higher mortality, when cultured at 18°C than 12°C. There have been reports of *Va*-vibriosis associated with the increase in water temperature in both turbot and halibut (Toranzo *et al.*, 2017), which suggests that the high temperature allowed a more rapid bacterial growth and spread. These studies also suggest that control of water temperature in some aquaculture facilities, specifically those consisting of concrete tanks on the coast, may prevent outbreaks.

Monitoring V. anguillarum

Several protocols for rapid detection of *V. anguillarum* from environmental samples have been proposed as a preventive measure to avoid outbreaks of *Va*-vibriosis. Most of them are based on the polymerase chain reaction (PCR) assay using different target genes such as those encoding haemolysins, sigma factors, metalloproteases or chaperones (Toranzo *et al.*, 2017; Hickey and Lee, 2018) as well as on multiplex real-time PCR procedures (Hickey *et al.*, 2015). These protocols, which show small differences in sensitivity and specificity, are useful for differentiating *V. anguillarum* and *V. ordalii*, two vibrio species with up to 99% similarity in the 16S rRNAS gene. In addition, some multiplex-PCR approaches based on *rpoN*, *empA* and the haemolysin genes have been developed to quickly detect *V. anguillarum* in water. Finally, loop-mediated isothermal amplification (LAMP) procedures based on the *empA* or *amiB* gene have also been designed as cost-effective alternatives to PCR protocols (Toranzo *et al.*, 2017). We proposed the protocol shown in Fig. 10.2 that involves isolation of the causative agent and identification by PCR and/or colony hybridization with species-specific probes.

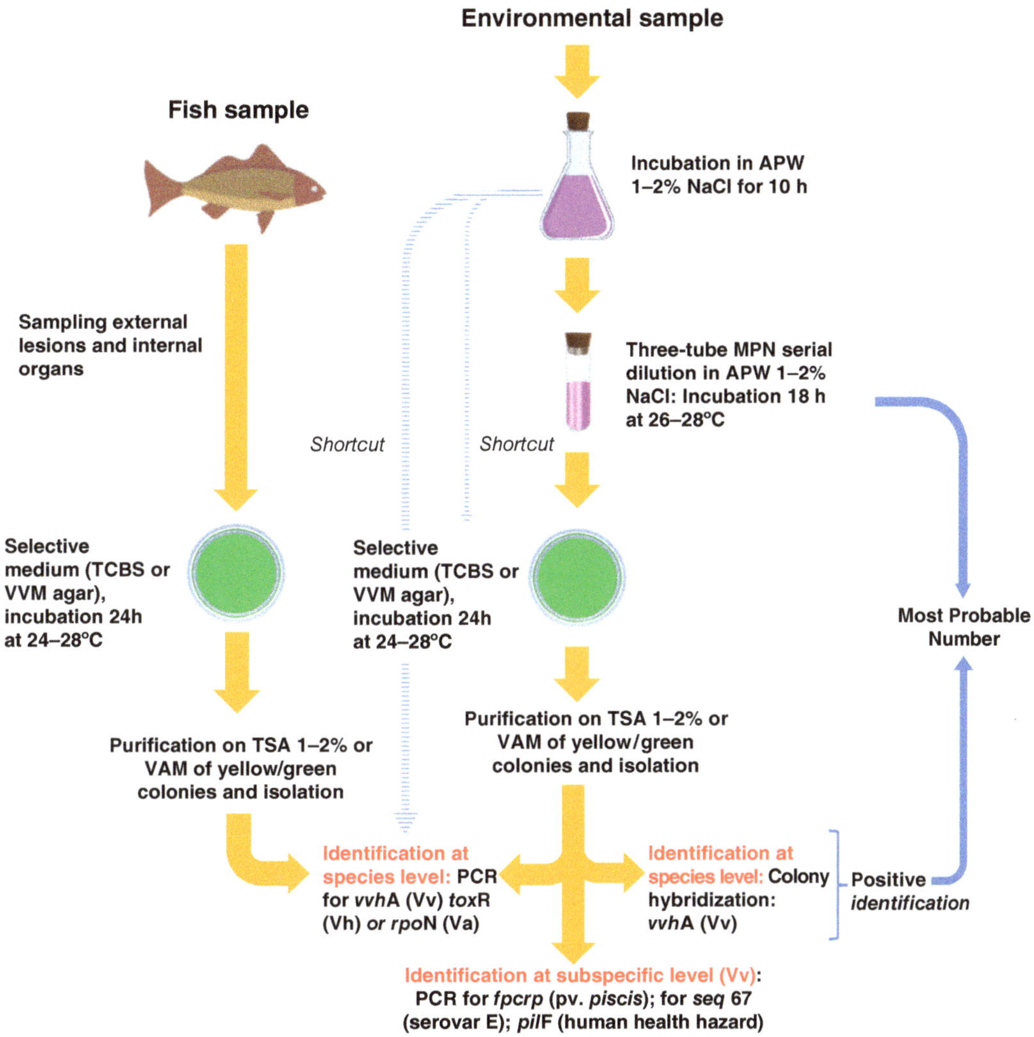

Fig. 10.2. Scheme for the isolation and identification of *Vibrio anguillarum* (*Va*), *Vibrio vulnificus* (*Vv*) and *Vibrio harveyi* (*Vh*) from environmental and fish samples. The scheme is based on the protocol recommended by the US Food and Drug Administration's Bacteriological Analytical Manual (https://www.fda.gov/Food/FoodScienceResearch/LaboratoryMethods/ucm070830.htm (accessed 13 March 2010)) with modifications of Sanjuán and Amaro (2004). Each *Vibrio* species can be identified by PCR by targeting *empA* (*Va*) (Xiao *et al.*, 2009), *vvhA* (*Vv*) (Wright *et al.*, 1985) and *toxR* (*Vh*) (Pang *et al.*, 2006). For identification at subspecific level within *Vv*, the following genes are recommended to be targeted for PCR: *fpcrp* (fish phagocytosis and complement resistance protein; formerly *vep07*) (Sanjuán and Amaro, 2007); *seq67*, a positive result identifies the zoonotic serovar (SerE) (Sanjuán and Amaro, 2007); $pilF_{variant}$, a positive result indicates public health hazard (Roig *et al.*, 2010). Once the vibrio has been isolated and identified, the pure culture should be maintained at –80°C in LB-1 (Luria-Bertani broth supplemented with 0.5% NaCl) or TSB-1 (tryptic soy broth, 1% NaCl) plus 15–20% glycerol. MPN, most probable number; APW 1–2%, alkaline peptone water containing 1–2% NaCl; TCBS, thiosulfate citrate bile salts sucrose; VVM, *V. vulnificus* medium (Sanjuán and Amaro, 2004); TSA 1–2%, tryptic soy agar, 1–2% NaCl; VAM, *V. anguillarum* medium (Toranzo *et al.*, 2017).

Probiotics

Probiotic bacteria (*Phaeobacter*, *Roseobacter*, *Ruegeria*, *Pseudomonas*, *Lactobacillus*, etc.) show inhibitory activity against the growth of *V. anguillarum* and, therefore, could prevent vibriosis (Sorroza *et al.*, 2012). Different routes of administration have been tried, including its addition to the water or to the food, although one of the most promising would be the bioencapsulation of probiotics in rotifers (Planas *et al.*, 2006).

Bacteriophages

The use of bacteriophages has been proposed as an alternative preventive measure in aquaculture, especially during the larval stages of the host when the immune system is immature. Promising results have been obtained both *in vitro* and *in vivo* against *V. anguillarum* and other fish pathogens, using suspensions of individual phages or phage cocktails (Mateus *et al.*, 2014; Rørbo *et al.*, 2018). In general, phage treatment improves the survival of fish larvae and eliminates the pathogen from the system, especially when a large multiplicity of infection of phage is administered.

The application of bacteriophages in aquaculture can be hampered by the emergence of resistant bacteria. In fact, different mechanisms of resistance to *V. anguillarum* phages have been described which include cell surface modification, cell aggregation, phage inactivation through clustered regularly interspaced short palindromic repeats (CRISPR)/CRISPR-associated protein (Cas) systems or the regulation of receptor expression in QS (Castillo *et al.*, 2019).

Immunostimulants and inhibition of quorum sensing

Non-specific immunostimulants administered in water and feed have also been suggested as a strategy for preventing *Va*-vibriosis in fish farms (Toranzo *et al.*, 2017). Recently, the antibacterial activity of some leaf extracts has been tested *in vitro* (Beltrán *et al.*, 2018), with contradictory results. Therefore, although the study of plant extracts may be an area of research in the future, their current use is not reliable.

Another alternative approach is the use of QS disruptors (Defoirdt *et al.*, 2004). These disruptors caused a reduction in rainbow trout mortality with decreased biofilm formation and protease production. Again, it is necessary to optimize the administration procedures and calculate their cost effectiveness.

Vaccination

Multiple commercial vaccines have been developed to prevent outbreaks of *Va*-vibriosis in different fish species consisting of whole-cell bacterins, containing or not adjuvants, to be administered by bath or injection (Frans *et al.*, 2011; Toranzo *et al.*, 2017). Vaccines include different serotypes in their formulations, as well as non-mineral oil adjuvants to avoid side-effects caused by oil adjuvants (Poppe and Koppang, 2014).

The most recent developments include subunit, recombinant, attenuated and DNA vaccines (Toranzo *et al.*, 2017; Hickey and Lee, 2018; Bao *et al.*, 2019; Xu *et al.*, 2019), as well as attenuated vaccines that can express the full range of protective antigens and may be administered by the natural route of infection (Toranzo *et al.*, 2017; Bao *et al.*, 2019). However, evidence on their safety, duration of protection and immunoprotection mechanisms are still areas of additional study.

10.2.7 Diagnosis and treatment

The traditional diagnosis includes the analysis of the clinical signs (Fig. 10.1), the isolation of the pathogen using general (TSA-1, tryptic soy agar at 1% salt concentration) or selective media (TCBS, thiosulfate citrate bile salts sucrose; or VAM, *V. anguillarum* medium) (Fig. 10.2) and the phenotypical identification of the strains with a final serological confirmation (Toranzo *et al.*, 2017). However, the phenotypic methods, including the metabolic fingerprinting, are not always conclusive due to their limited accuracy and reliability (Frans *et al.*, 2011). Figure 10.2 summarizes the selected entire protocol for *Va*-vibriosis diagnosis after *V. anguillarum* identification.

Once the disease is diagnosed, multiple antibiotics can be used as therapeutic agents against *V. anguillarum* (Frans *et al.*, 2011; Angelidis, 2014). However, such widespread use not only caused environmental contamination but also led to the development of antibiotic resistance among strains. Since antibiotics are still used to control outbreaks, dosing regimens should be based on pharmacokinetic and pharmacological evidence and should be used under strict control to avoid the appearance of multi-resistant strains.

10.3. *Vibrio vulnificus* and the Warm-Water Vibriosis

V. vulnificus is a multi-host pathogen able to cause disease (*Vv*-vibriosis) in aquatic vertebrate (fish; the eel is the most susceptible host) and invertebrate (shrimps) animals as well as humans (Oliver, 2015; Table 10.1). Clinical signs associated with *Vv*-vibriosis depend on both the transmission route (ingestion or contact) and the host, but all these vibrioses (even the human ones) commonly result in a septicaemia with a high probability of death by sepsis (Amaro *et al.*, 2015; Oliver, 2015). Unlike other vibrios, this pathogen is a zoonotic agent as *Vv*-vibriosis can be transmitted directly from diseased fish to humans by contact (Veenstra *et al.*, 1992; Dalsgaard *et al.*, 1996).

10.3.1 Taxonomy and phylogeny of the pathogen

Classically, *V. vulnificus* has been subdivided into three biotypes according to a few phenotypic and host-range differences (Oliver, 2015). All three biotypes can cause human *Vv*-vibriosis but only biotype 2 can cause disease in fish thanks to a virulence plasmid (pVvBt2) transferable by conjugation (Amaro *et al.*, 2015). A recent phylogenomic analysis on the core genome of 80 strains showed that the species is subdivided into five well-supported phylogenetic lineages that do not correspond to the biotypes (Roig *et al.*, 2018). That study also concluded that all the strains should be considered as potentially virulent to humans since most of the genes involved in virulence for humans were present in the core genome of the species. In contrast, only the strains possessing pVvBt2 should be considered as potentially virulent for fish. The lack of congruence between the phylogenetic trees of plasmids and chromosomes also suggests that pVvBt2 was acquired independently by different clones, probably in fish farms (Roig *et al.*, 2018). The authors propose a new updated classification of the species based on phylogenetic lineages and the inclusion of all biotype 2 strains in a pathovar (named *piscis*) with the specific ability to cause fish *Vv*-vibriosis. This pathovar is further subdivided into three serotype-related sublineages, one of which (SerE) constitutes a zoonotic clonal complex that is distributed throughout the world.

10.3.2 Ecology of the pathogen

V. vulnificus inhabits coastal brackish-water ecosystems in temperate, subtropical and tropical areas. In temperate regions, the pathogen is especially abundant in the warm season and seems to disappear from the water column in the cold season, although it is present in a non-recoverable state known as VBNC (viable but not cultivable) (Oliver, 2015). Epidemiological and ecological data suggest that the presence and abundance of the pathogen in these ecosystems is controlled by the temperature and salinity and, more importantly, by the interaction between both parameters (Takemura *et al.*, 2014). Thus, the true temperature limits for *V. vulnificus* survival are narrower (22–33°C) at higher salinities (>1% NaCl) and wider (10–33°C) at low salinities (0.5–1.0% NaCl). Consequently, this pathogen inhabits water bodies of higher salinity in warm areas and of lower salinity in temperate areas. *V. vulnificus*, like the rest of the *Vibrio* species, switches between free-swimming and sessile life forms throughout its life cycle in the water (Oliver, 2015). The sessile *V. vulnificus* are associated to abiotic (i.e. sediment) and biotic (i.e. the internal/external mucosa of fish) surfaces, forming biofilms. In addition, free-living and microparticulate-associated *V. vulnificus* can be concentrated by more than 100 times the concentration found in the surrounding waters by filter feeders such as small crustaceans and bivalves (Oliver, 2015). Bivalves, shrimp and fish are considered the main environmental reservoirs for *V. vulnificus* (Oliver, 2015). However, *V. vulnificus* pv. *piscis* has been isolated from fish but not from bivalves (Amaro *et al.*, 2015), suggesting a niche adaptation for this pathovar and explaining why it has never caused a human infection acquired by ingestion of seafood. Recently, it has been proposed that that temperature is one of the main factors that determines the life strategy of the pathogen (Hernández-Cabanyero *et al.*, 2020). Figure 10.3 shows the life cycle proposed for this bacterium in the environment.

10.3.3 Biology and culture of the main fish hosts

The eel and the Nile tilapia are the main hosts for *Vv*-vibriosis (Amaro *et al.*, 2015; B. Fouz and C. Amaro, 2020, unpublished results).

Eels

Eels are teleost, euryhaline and catadromous fish that preferentially inhabit temperate and subtropical areas (Aida *et al.*, 2003). The most studied eel

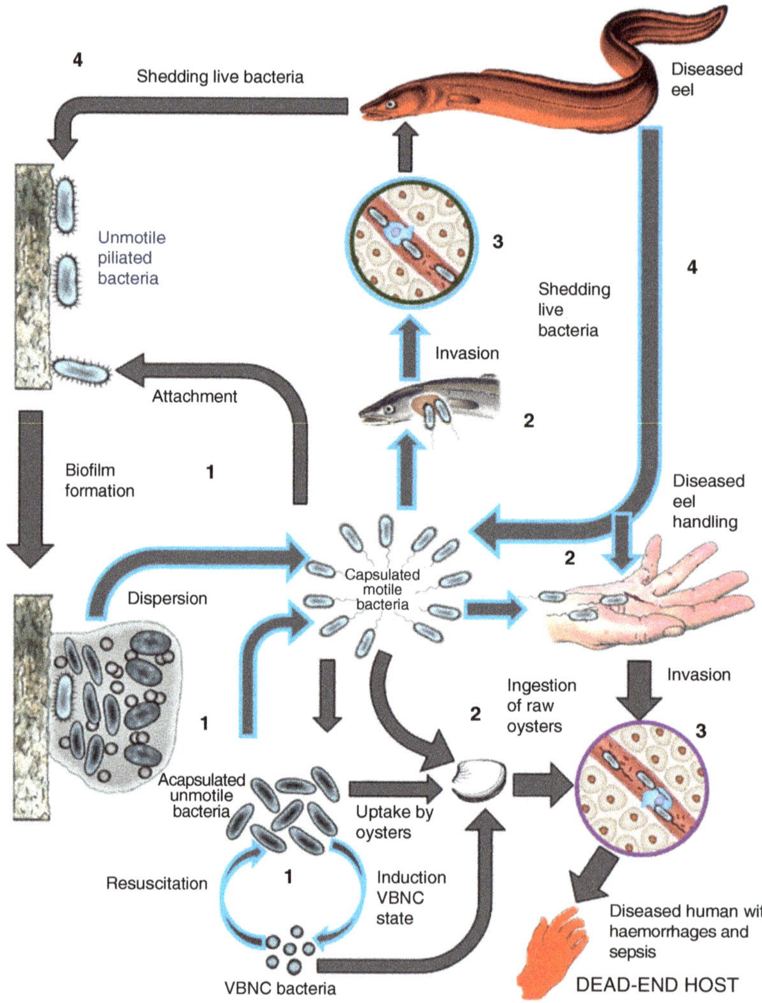

Fig. 10.3. Life cycle of *Vibrio vulnificus*: role of temperature (Pajuelo *et al.*, 2016; Hernández-Cabanyero *et al.*, 2019, 2020). Processes controlled by temperature are surrounded by a blue line. 1, Survival in water. Low temperatures induce the bacterium to enter a 'dormant' state known as VBNC (viable but not culturable), while warm temperatures activate resuscitation and biofilm dispersion. 2, Host colonization: bacteria are attracted by blood and/or mucus from their susceptible hosts and colonize them, a process that increases at warm temperatures. Bacteria can also be taken up by filtering organisms and these can be ingested by humans. 3, Invasion and sepsis: from the colonized tissue, the pathogen invades the bloodstream of the host. Only the cells that resist the innate immunity multiply and secrete the toxins RtxA1 and VvhA that will cause the death of the host by a toxic sepsis (Murciano *et al.*, 2017). To resist the innate immunity in human blood, the pathogen produces a capsule whose synthesis is increased under iron excess conditions in risk patients. To resist innate immunity in fish blood, the pathogen produces an outer membrane enriched in O-antigen plus two proteins, Fpcrp (fish phagocytosis complement resistance protein) and Ftbp (fish transferrin binding protein), whose synthesis is increased under iron starvation and at warm temperatures. 4, Transmission to new hosts and the environment: diseased fish can infect humans (zoonosis) by contact and shed live bacteria in the water as well. Transmission of the pathogen is increased at warm temperatures.

species are *Anguilla anguilla* (European eel) and *Anguilla japonica* (Japanese eel). The spawning of these two species takes place several hundred metres deep, either in the Sargasso Sea (European eel) or in the Pacific Ocean west of the Mariana Islands (Japanese eel). The young larvae (leptocephali) move to the nearest continental coast with the ocean currents and arrive 0.5–3 years later. When approaching the coast, the larvae metamorphose into the 'glass eel' stage, enter the estuaries and begin to migrate upstream. During this second migration, the glass eels metamorphose into elvers (young eels) and colonize ponds, lagoons, lakes, etc. The elver grows and turns yellow ('yellow eel' or adult eel) and, after 5–20 years, some of the eels metamorphose again, become sexually mature ('silver eels') and migrate back to the ocean to spawn and die. The life cycle of the eel is so complex and poorly understood that no successful methodology for its reproduction in captivity has been developed. Therefore, the production of eels is based on the capture of wild glass eels or elvers and their growth in farms under semi-intensive or intensive conditions (Table 10.2).

Nile tilapia

Nile tilapia, *Oreochromis niloticus*, is a freshwater tropical species that preferentially inhabits shallow waters. It is an omnivorous fish that lives at temperatures between 11 and 42°C with an optimum between 31 and 36°C (Webster and Lim, 2010). Its life cycle in nature is well known. Sexual maturity is reached at 5–6 months of age and spawning is controlled by water temperature (over 24°C). The male is territorial, digs a nest and the female spawns in the nest. Immediately after fertilization, the female collects the eggs in the mouth and incubates them for 1 to 2 weeks, depending on the temperature. If there is no cold period, the female may spawn continuously. Nile tilapia can live for about 10 years and reach a weight of more than 3 kg. Tilapia is cultured in ponds or floating cages in large lakes in many tropical countries (Webster and Lim, 2010). In temperate regions, recirculation systems have been developed to culture tilapia throughout the year under controlled conditions (Table 10.2).

10.3.4 The warm-water vibriosis

The *Vv*-vibriosis caused by *V. vulnificus* pv. *piscis* in fish is known as warm-water vibriosis (WWV) because the most serious outbreaks always occur at more than 25°C. Epidemiological data from studies in eels and tilapia suggest that WWV is an acute haemorrhagic septicaemia that occurs in two modalities: the brackish-water modality, caused by SerE; and the freshwater modality, caused by the rest of the serotypes. The first occurs as outbreaks of high mortality in eel farms, while the second one occurs as outbreaks of low mortality in both eel and tilapia farms (Chen *et al.*, 2006; Amaro *et al.*, 2015). The main clinical signs of WWV are shown in Fig. 10.1. The common external lesions are haemorrhages at the base of the anal and dorsal fins along with a reddening in the region of the operculum and, occasionally, petechiae on the abdomen. The belly can be swollen due to the accumulation of fluid. Internally, the liver can appear pale and/or haemorrhagic, and the kidney and the intestine haemorrhagic. In the case of diseased eels, two clinical signs specific to the modality have been described in the largest sized specimens: ulcers on the head (brackish-water modality) and degradation of the jaw (fresh-water modality). These signs are presented by very few specimens (Amaro *et al.*, 2015) (Fig. 10.1).

Experiments with eels and tilapia artificially infected by different routes revealed that immersion in water followed by ingestion are the prime vehicles for the transmission of WWV (Amaro *et al.*, 2015; B. Fouz and C. Amaro, 2020, unpublished results). Further, there is a link between serology and the preferential infection route. SerE preferably infects through water while the other serotypes infect preferentially by the oral route (Amaro *et al.*, 2015). In any case, the virulence of the strain is strongly dependent on water salinity (maximum at 1.5% NaCl for SerE and 0.5–1% NaCl for the other serotypes) and temperature (maximum at 28°C for all serotypes, the highest tested temperature), and this dependence is mainly explained by the survival and persistence of *V. vulnificus* in water.

Experiments of eel infection by immersion with a SerE strain show that *V. vulnificus* colonizes the gills and multiplies on them, forming biofilms. Subsequently, it invades the bloodstream and spreads to internal organs where it reaches population sizes markedly lower than those associated with other vibriosis (Amaro *et al.*, 2015). Parallel to bacterial spread, extensive haemorrhages are produced in all organs and the animal dies in less than 72 h. Granulocytes are the main cell type that shows clear signs of damage, which is evidenced by the release of cytoplasmic content, including granules. Very few bacteria are present in the tissues of infected eels, and they are mostly near or inside the lumen of the capillaries. The rapid death of eels is congruent with the hypothesis that eels die from an acute septic shock. No similar study has been performed in orally infected eels or tilapia.

10.3.5 Host–pathogen interaction: role of temperature

In this section, the main pathogen and host factors involved in WWV are described and classified according to their role in colonization, invasion and sepsis (Fig. 10.4). Finally, the role of temperature on the infectivity and virulence of the pathogen is also considered.

Colonization

The pathogen is attracted by the gill mucus and colonizes this organ probably via an indirect interaction between mucus-coated bacteria and the cellular receptors for mucus (Amaro *et al.*, 2015). The protease VvpE and the capsule are two of the virulence factors involved in this process; genetic variants deficient in either the capsule or VvpE are significantly less efficient in branchial colonization than the wild-type strain and are completely avirulent for eels by immersion (Amaro *et al.*, 2015).

Invasion

Once the gills are colonized, the pathogen probably expresses an important toxin that is involved in the destruction of phagocytes and tissue damage, the toxin RtxA1, a toxin of the MARTX family (multifunctional autoprocessing RTX) (Satchell, 2011). This toxin is an early expressed protein that is secreted when the bacteria come into contact with multiple types of eukaryotic cells. MARTX toxins are modular proteins of high molecular weight with conserved external modules (in particular, N-terminal and C-terminal regions) that form pores in the eukaryotic membrane, through which the central module of the toxin is translocated. Then, the effector domains in the central module are released after activation of the cysteine protease domain (CPD) in the cytosol and attack the eukaryotic cells by different domain-specific mechanisms (Fig. 10.5). The comparison of the immune response against the wild-type strain and the toxin-deficient mutant suggests that the toxin causes a local inflammation (Callol *et al.*, 2015). Additional evidence of inflammation comes from the observation of cells (dendritic cells?) expressing immune receptors (toll-like receptors (TLRs) 2 and 5) and effectors (interleukin (IL)-8) that are located in the second lamella and are relocated during the infection (Callol *et al.*, 2015). This local damage would favour the entrance of the bacteria into the bloodstream.

Sepsis

The bacterium is able to survive and proliferate in the blood and, therefore, induce the septicaemia that ends with the death of the animal. When the bacterium reaches the blood, it finds important defences such as: (i) transferrin, an iron chelator that inhibits the growth of pathogens; (ii) serum complement, a series of proteins that once activated destroy the pathogen or mark it for later destruction by phagocytosis; and (iii) different types of phagocytic cells that ingest the pathogen or secrete products for its destruction, such as nitric oxide (Austin, 2012). Recent transcriptomic studies performed in serum revealed that *V. vulnificus* is able to detect iron restriction in the blood and organize a response that activates a virulent phenotype responsible for the death of the eel (Hernández-Cabanyero *et al.*, 2019). This virulent phenotype is defined by: (i) an anaerobic metabolism of amino compounds based on nitrite/nitrate respiration that activates, concomitantly, a protective mechanism against nitric oxide, one of the by-products of this type of anaerobic respiration; (ii) a bacterial envelope especially rich in the high-molecular-weight part of the LPS, which confers partial protection against complement, microcidal peptides and phagocytosis (Amaro *et al.*, 2015); and (iii) the overexpression of two important toxins VvhA and RtxA1, which probably act cooperatively, as revealed by experiments carried out with mutants deficient in each of the toxins or in both toxins. The results obtained by Hernández-Cabanyero *et al.* (2019) also suggest that VvhA, a potent haemolysin, would be expressed before RtxA1, lyse erythrocytes and create a haem-rich microenvironment that would activate the transcription of RtxA1. This toxin would then cause the release of cytokines in the eel similarly to that found in mice, causing the death of the eel (Murciano *et al.*, 2017). Preliminary transcriptomic studies performed in eels infected by immersion with a mutant deficient in *rtxA1* support this hypothesis

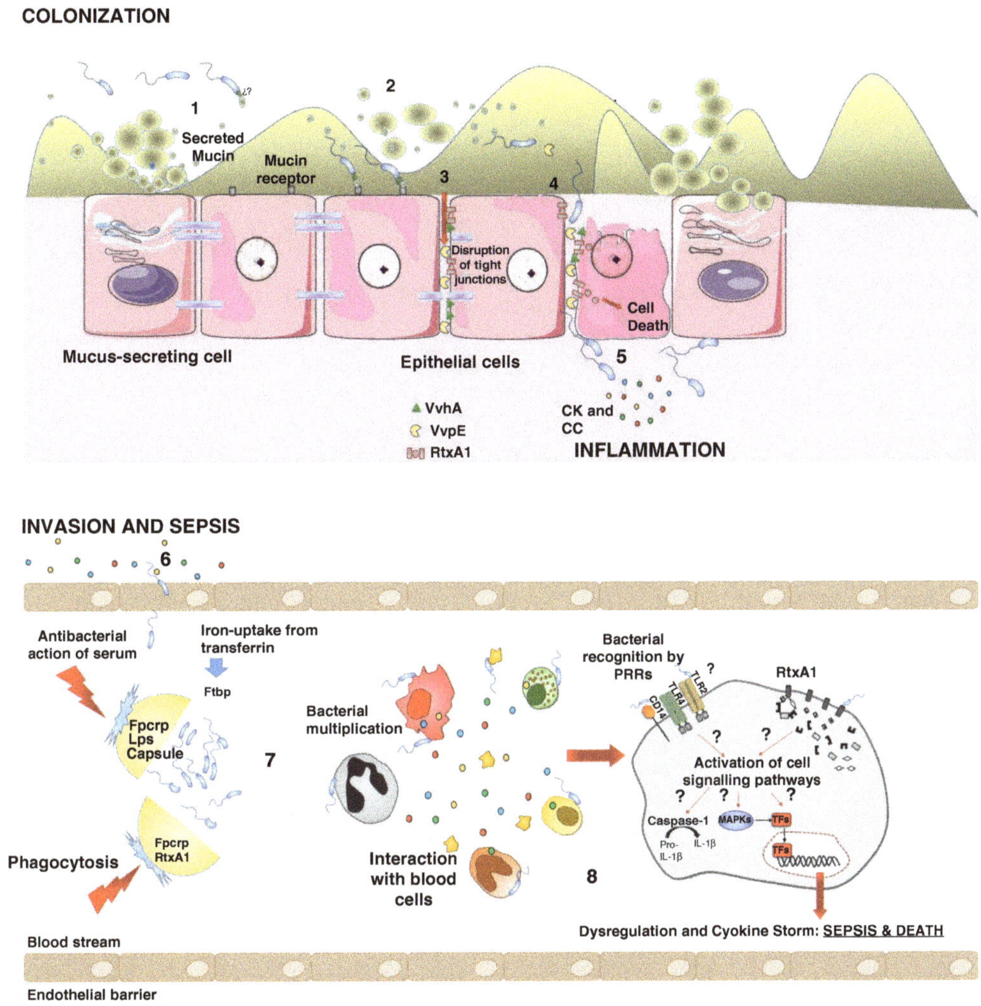

Fig. 10.4. Steps in colonization, invasion and sepsis caused by *Vibrio vulnificus* (pv. *piscis*) in the eel: a holistic model. Colonization: motile bacteria are attracted by gill mucus, bind mucin on their surface by unknown receptors (1) and the mucin-coated cell binds to mucin receptors on epithelial cells (2). Attached cells will produce protease VvpE and toxins VvhA and RtxA1, whose joint activity (mainly RtxA1) will result in increased permeability and tight junction disruption (3). Cells will pass through the epithelium and will continue producing VvpE, VvhA and RtxA1, which will be able to cause cell death by different mechanisms (4). Attacked cells will secrete cytokines (CK) and chemokines (CK), which will trigger a local inflammation (5). Established bacteria will multiply on the epithelium. Invasion and sepsis: inflammation alters endothelial cells and bacteria will be able to cross the endothelial barrier (6). Bacteria multiply in blood thanks to the combination of a series of protective mechanisms in which Ftbp (fish transferrin binding protein), Fpcrp (fish phagocytosis complement resistance protein), LPS (lipopolysaccharides) and RtxA1 will be involved (7) (see text for details). Bacteria will interact with different immune cells and, finally, will cause sepsis and death probably by inducing a cytokine storm (8). PPR, pattern recognition receptor; TLR, toll-like receptor; CD, cluster of differentiation; IL, interleukin; MAPK, mitogen-activated protein kinase; TF, transcription factor. Figure not to scale.

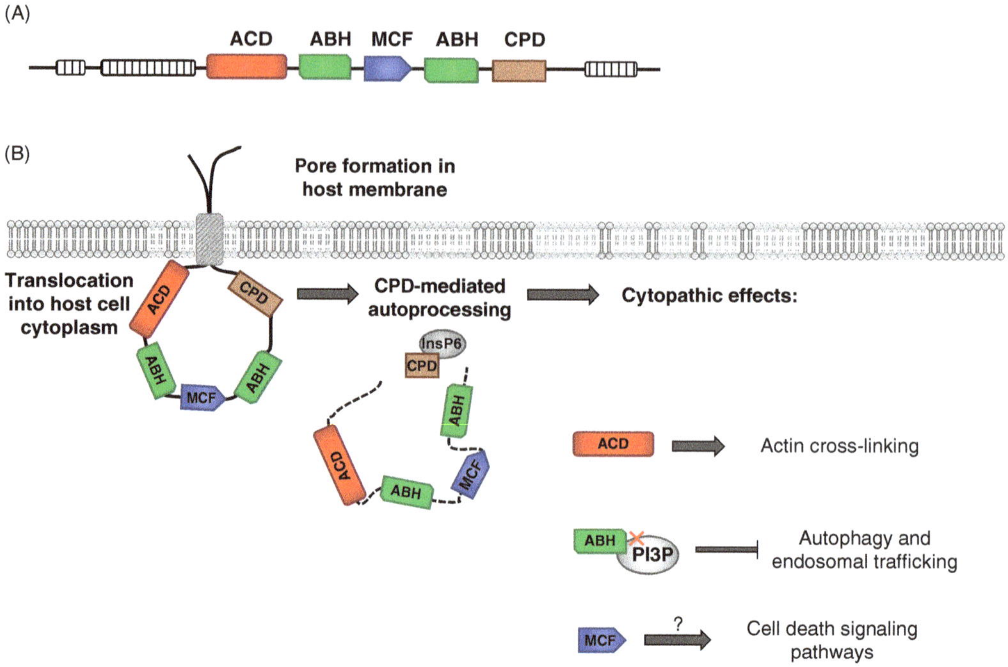

Fig. 10.5. RtxA1 toxin of *Vibrio vulnificus* (pv. *piscis*): scheme of RtxA1 structure (A) and mode of action at cellular level (B). The scheme shows the conserved external modules and the internal module, containing the five effector domains represented with different colours. The toxin is secreted and the external module is associated with the target cell membrane by forming a pore that allows the central module to be exposed to the cytosol. CPD, cysteine protease domain (this domain catalyses the release of the rest of the internal domains after being activated by binding inositol hexakisphosphate (InsP6)); ACD, actin cross-linking domain (this domain activates actin cross-linking); ABH, α/β hydrolase domain (this domain binds inositol-3-phosphate (PI3P) and inhibits autophagy and endosomal trafficking); MCF, 'makes caterpillars floppy'-like domain (this domain induces depolarization of the mitochondrial membrane potential, which causes activation of cell death) (Satchell *et al.*, 2011; Murciano *et al.*, 2017).

(C. Hernández-Cabanyero and C. Amaro, 2020, unpublished results).

The study of Hernández-Cabanyero *et al.* (2019) and a previous one (Pajuelo *et al.*, 2015) found that *V. vulnificus* overexpressed in serum two plasmid iron-regulated outer membrane proteins: Ftrp (fish transferrin binding protein), a recently described receptor for eel transferrin (Pajuelo *et al.*, 2015); and Fpcrp (fish phagocytosis complement resistance protein) (formerly *vep07*, the gene used for the *V. vulnificus* pv. *piscis* identification by PCR (Fig. 10.2)), encoding a multifunctional protein that collaborates in specific resistance to eel complement and phagocytosis. The pathogen could use both proteins to resist innate immunity in the blood and multiply, which would explain why this bacterium is so virulent for eels (LD_{50} per IP injection is less than 100 bacteria per fish). Homologous genes with an identity and coverage of almost 100% are present in the plasmids of all *V. vulnificus* pv. *piscis* strains sequenced to date. It should be noted that both genes are also present in the plasmid pVH1 of *V. harveyi* (https://www.ncbi.nlm.nih.gov/nuccore/HM752267 (accessed 25 March 2020)) and, unexpectedly, in other *V. vulnificus* strains belonging to other lineages, not previously associated with fish diseases but with human infection after ingestion of raw seafood (previously, strains of biotype 1). This result suggests that fish farming is probably favouring genetic exchanges and, consequently, the emergence of new groups potentially pathogenic for fish. In addition, the phylogenetic analysis of both proteins divided the strains into groups supposedly related to the specificity of the host, indicating that a change in a few amino acids could adapt the protein to the immune system of a new host.

Role of temperature in infectivity and virulence

The severity of the vibriosis in the eel depends largely on the water temperature (highly virulent at 28°C, avirulent at 20°C or below). According to Hernández-Cabanyero *et al.* (2020), the reason for this is that warm temperatures activate adaptative traits that prepare the bacteria for host colonization (metabolism, motility, chemotaxis and the protease activity) and fish septicaemia (iron uptake from transferrin and production of O-antigen of high molecular weight) (Fig. 10.3).

10.3.6 Human vibriosis, an important health concern for fish farmers and consumers

There have been worldwide reports of human *Vv*-vibriosis associated with handling fish. The most important cases occurred among fish farmers and consumers of tilapia cultivated in Israel in the 1990s (Bisharat *et al.*, 1999). These infections were caused by a new variant of *V. vulnificus* with the ability to produce outbreaks of *Vv*-vibriosis in humans but not in tilapia. This variant was called biotype 3 and corresponds to lineage 3 (Roig *et al.*, 2018). Other studies showed that strains of biotype 3 were present in the water and mucosal surfaces of tilapia, a prickly fish that produces lesions easily if not handled with care (Bisharat *et al.*, 1999). The only documented zoonotic cases caused by *V. vulnificus* were recorded among eel farmers after the management of diseased eels in several farms located in north-west Europe (Veenstra *et al.*, 1992; Dalsgaard *et al.*, 1996). All strains isolated belong to the SerE of *V. vulnificus* pv. *piscis*. Fortunately, there is a PCR for the detection of the isolates that constitute a public health hazard regardless of lineage or biotype. The PCR primer targets a polymorphism in the *pilF* gene (a gene that has a role in the biogenesis of pili type IV) that is present in all the human clinical isolates, as well as in all strains belonging to the zoonotic clonal complex (Roig *et al.*, 2010).

Transcriptomic studies performed with human serum demonstrated that the zoonotic SerE strains can only multiply in human serum if it contains an iron concentration similar to that presented in a patient with haemochromatosis (Hernández-Cabanyero *et al.*, 2019). In this serum with iron overload, the bacterium develops an anaerobic metabolism based on glycans/sugars and produces a large amount of capsule that protects it against human complement and phagocytosis. Interestingly, the bacterium also produces the same toxins as in eel blood, VvhA and RtxA1, which suggests that the regulatory process involved in the production of toxins is a complex process that involves other factors in addition to iron. The bacterium is probably attracted to the blood of a wound and colonizes the skin, invades the bloodstream, produces the capsule and the toxins, and triggers a cytokine storm that causes death if the patient does not receive treatment in time (Murciano *et al.*, 2017) (Fig. 10.3). In conclusion, special care must be taken when handling diseased fish suspected of suffering from WWV.

10.3.7 Preventive measures

Manipulate physicochemical parameters

One of the preventive measures to control WWV in farms is to keep the physicochemical parameters outside the limits that favour the proliferation of *V. vulnificus* and the WWV transmission. Since eels can be grown in fresh and brackish water, it is recommended to use fresh water to prevent *Vv*-vibriosis. In the case of eels and tilapia grown under intensive conditions by using recirculating water, we recommend maintaining the temperature below 25°C, although production will be lower. Another parameter that could be controlled to reduce the presence of *V. vulnificus* in water is pH. *V. vulnificus* prefers pH values higher than 6, so, if possible, the pH of the water should be less than 6.

Monitoring V. vulnificus

Another preventive measure that can be applied on farms is to monitor the presence of *V. vulnificus* in water and animals by periodic sampling. Figure 10.2 represents a scheme of a general isolation and identification procedure. The protocol for the identification of *V. vulnificus* both at species and subspecies level allows to distinguish innocuous from potentially virulent strains for humans and/or fish, including those that constitute a danger to public health (Fig. 10.2). If *V. vulnificus* is detected, we recommend modifying the physicochemical parameters, according to the instructions explained above, and if the *V. vulnificus* pv. *piscis* is detected, it is recommended to vaccinate the animals.

Vaccination

The ideal preventive measure is to vaccinate animals, especially because any *V. vulnificus* strain should be considered as potentially pathogenic for humans. There is a patented vaccine called Vulnivaccine that has been shown to be effective in farms (Fouz *et al.*, 2001). The vaccine is a bacterin supplemented with toxoids and is administered after dilution at 1:1000–1:10,000 by prolonged immersion in three doses separated by 14 days. The vaccine protects the animals during the entire period that they are in the farm (Fouz *et al.*, 2001). A series of laboratory-scale studies showed that the vaccine is effective because it stimulates active immunity in the gills, intestine and blood, regardless of the route of administration (Esteve-Gassent *et al.*, 2004) (Fig. 10.6). Especially interesting is that this vaccine can stimulate a protective mucosal immunity in the gills, the main portal of entry for the zoonotic serovar (Esteve-Gassent *et al.*, 2003), thus protecting the eels against WWV in its first step, the process of colonization.

10.3.8 Diagnosis and treatment of WWV

Specimens with clinical signs, especially those showing signs of septicaemia (Fig. 10.1), should be analysed microbiologically according to Fig. 10.2. In this case, a general medium such as TSA-1 could be used instead of TCBS or VVM (*V. vulnificus* medium) agar since *V. vulnificus* is recovered as a pure culture from diseased eels and tilapia. *V. vulnificus* is generally sensitive to all antimicrobials that can be used on farms in the EU and the USA. Therefore, the treatment should be started as soon as possible. In parallel, an antibiogram should be performed to determine if an alternative antibiotic should be used.

10.4 *Vibrio harveyi* and the Emerging Marine Vibriosis

V. harveyi (synonymous with *Vibrio carchariae* and *Vibrio trachuri*) is a widespread marine pathogen capable of causing disease (*Vh*-vibriosis) in multiple fish, mollusc and crustacean species of economic importance (Table 10.1). *V. harveyi* has been recognized as an opportunistic (or secondary) pathogen, since the most severe diseases occur in immunosuppressed or environmentally stressed hosts, but its role as a primary pathogen has been gaining attention in recent years. Clinical signs vary depending on the infected species: luminescent vibriosis, Bolitas negricans in shrimp, white spots on the foot in the abalone, bleaching or 'white syndrome' in tropical corals, and gastroenteritis/necrotizing enteritis, skin ulcers or haemorrhagic septicaemia in fish (Fig. 10.1) (Austin and Zhang, 2006; Travers *et al.*, 2009; Luna *et al.*, 2010; Ruwandeepika *et al.*, 2012; Zhu *et al.*, 2018).

10.4.1 Taxonomy and phylogeny of the pathogen

V. harveyi belongs to the *Harveyi* clade (or *Vibrio* core group) that also includes the species *Vibrio campbellii*, *Vibrio parahaemolyticus*, *Vibrio alginolyticus*, *Vibrio natriegens*, *Vibrio rotiferianus* and *Vibrio mytili* (Sawabe *et al.*, 2007). All these species present a value of similarity for 16S rRNA gene higher than 97.6% and a value for DNA–DNA association close to 70% (Pascual *et al.*, 2010). *V. harveyi* is genetically variable and includes avirulent and virulent strains, the last ones clustered in groups related to infected species and epizootic events in fish farms (Austin and Zhang, 2006; Haldar *et al.*, 2010; Ruwandeepika *et al.*, 2012). Recent studies suggest that the ability of *V. harveyi* to cause disease in fish and abalone could be related to virulence plasmids acquired in the fish farm environment. Thus, virulent strains isolated from diseased sea bass contain a putative virulence plasmid (pVH1) with a high homology to pVvBt2 of *V. vulnificus* pv. *piscis* (Pajuelo *et al.*, 2015; B. Fouz and C. Amaro, 2020, unpublished results) and the strains from diseased abalone harbour another putative virulence plasmid (pVCR1) (Travers *et al.*, 2009). Among the genes encoded in pVH1 there are two with a similarity close to 100% to *ftbp* and *fpcrp*, genes of *V. vulnificus* pv. *piscis* that encode two proteins related to resistance and growth in fish blood (Pajuelo *et al.*, 2015; Hernández-Cabanyero *et al.*, 2019).

10.4.2 Ecology of the pathogen

V. harveyi is an aquatic bacterium that inhabits marine and estuarine environments located in temperate, subtropical (optimal temperature range between 18 and 25°C) and tropical zones (around 28–30°C) (Austin and Zhang, 2006; Zhu *et al.*, 2018). *V. harveyi* can be present in these ecosystems either as a free-living form in the water column or as a living form associated with biotic (skin and intestinal surface of marine animals) or abiotic (sediment) surfaces, forming biofilms (Karunasagar *et al.*, 1996).

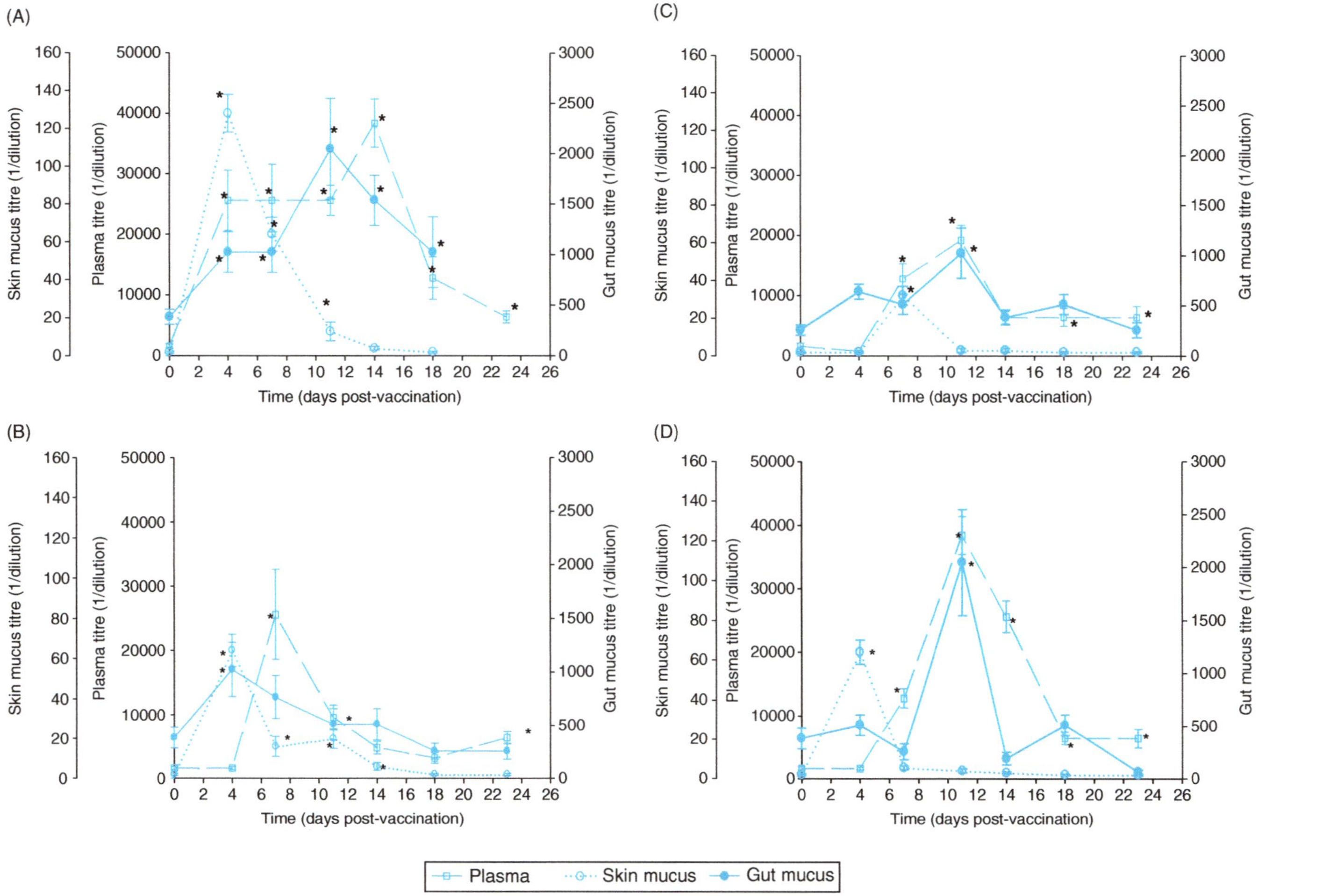

Fig. 10.6. Antibody production after immunization of eels with a vaccine against *Vibrio vulnificus* (pv. *piscis*) by using different administration routes. ELISA values are given as the reciprocal of the highest dilution giving a positive response. Positive reaction was set up as 2 × OD of the zero sample; values are means with their standard deviations represented by vertical bars. (A) Intraperitoneal route; (B) immersion route; (C) oral route; (D) anal route. * Denotes significant differences ($P < 0.05$) with respect to the control group by a Student *t*-test. (From Esteve-Gassent *et al.*, 2004.)

V. harveyi has been isolated from multiple species of vertebrate and invertebrate animals that co-inhabit the same environments.

Low temperatures and/or salinities under nutrient starvation conditions can induce the VBNC state in *V. harveyi* (Ramaiah *et al.*, 2002; Sun *et al.*, 2008). Interestingly, the bacterium seems to retain its pathogenic potential after a long-term incubation under starvation conditions (Sun *et al.*, 2008). The abundance of the pathogen in natural ecosystems is probably controlled by variations in salinity, temperature and nutrient availability. In temperate regions, high temperatures in the warm season favour multiplication of the bacterium while the pathogen could survive in the cold season in the VBNC state.

10.4.3 Biology and culture of the main fish hosts

Vh-vibriosis mainly affects grouper aquaculture in tropical areas (Zhu *et al.*, 2018) and sea bass aquaculture in temperate areas (B. Fouz and C. Amaro, 2020, unpublished results). Precisely, the sea bass aquaculture (see Section 10.2.3) will be one of the most affected by global warming.

10.4.4 The emerging marine vibriosis

There have been numerous reports of diseases caused by *V. harveyi* in marine fish species throughout the world since 1990 (Austin and Zhang, 2006; Ruwandeepika *et al.*, 2012). In all of them, *V. harveyi* seems to act as an opportunistic pathogen, causing moderate mortality. Recently, an increased number of *Vh*-infections has been reported in groupers and juvenile ESB (Mohamad *et al.*, 2018; Zhu *et al.*, 2018; B. Fouz and C. Amaro, 2020, unpublished results).

The disease in groupers is mainly characterized by scale drop, a swollen intestine that contains yellow fluid and deep necrotic dermal lesions (Zhu *et al.*, 2018). This pathology may not be caused exclusively by *V. harveyi* but by a consortium of bacteria. In fact, some authors recently demonstrated the impact of concurrent infection involving *V. alginolyticus* and *V. harveyi* in marine fish cultured in tropical areas (Mohamad *et al.*, 2018).

On the contrary, the vibriosis in juvenile ESB could be a new emerging marine disease caused by a primary pathogen since *V. harveyi* is recovered in pure culture from internal organs and the isolates reproduce the natural disease in experimental challenges (B. Fouz and C. Amaro, 2020, unpublished results). This vibriosis could be related to global warming since the outbreaks of the highest mortalities occur at temperatures over 20°C, temperatures that are adequate for the survival and persistence of *V. harveyi* in seawater, as indicated above.

The diseased juvenile ESB show signs of haemorrhagic septicaemia (Fig. 10.1). External clinical signs include haemorrhages in the fins together with redness in the operculum region and petechiae on the abdomen, mild abdominal distension and skin lesions that may ulcerate and bleed. Internally, the viscera may appear haemorrhagic or pale (mainly liver) and splenomegaly and enteritis (producing yellow exudate) are commonly observed.

As reported for other vibrios that cause septicaemia, *V. harveyi* probably colonizes the external surface of susceptible fish (gills, skin lesions, oral mucosa, etc.), invades the bloodstream and spreads to internal organs, causing haemorrhagic lesions and death in less than 2 to 3 days. Experimental challenges by IP injection in juvenile ESB reveal that the isolated strains present values of LD_{50} around 10^5 colony-forming units/fish. The pathogen can also infect fish by immersion or ingestion (B. Fouz and C. Amaro, 2020, unpublished results).

In conclusion, epidemiological data suggest that the emergent *Vh*-vibriosis is a haemorrhagic septicaemia affecting juvenile ESB cultured in temperate areas that could be associated to global warming.

10.4.5 Host–pathogen interaction

Different virulence factors have been described in *V. harveyi* but the specific role in the pathogenesis of most of them remains to be determined. Probably, the pathogen attaches to the gill mucosa, then penetrates in the bloodstream, proliferates without visible signs and, finally, causes clinical manifestations and death, as it has been described for abalone (Travers *et al.*, 2009).

Colonization

V. harveyi produces an EPS that probably improves the survival and persistence of the pathogen in fish farming systems (as reported in shrimp farms) by providing protection against predators and antimicrobial compounds, and also by contributing to the colonization of fish surfaces (Bramhachari and Dubey, 2006).

Sepsis

To proliferate in the blood and arrive at the internal organs, the pathogen needs to overcome the innate immunity. Probably, LPS together with the plasmid proteins Fpcrp and Ftbp (see Section 10.3.5 on *V. vulnificus*) confer both resistance to innate immunity and ability to grow by sequestering iron from fish transferrin (B. Fouz and C. Amaro, 2020, unpublished results).

Tissue lesions

V. harveyi secretes lytic enzymes/toxins that are likely involved in the destruction of immune cells and tissue damage (invasion and sepsis steps) and the production seems to be regulated by QS (Natrah *et al.*, 2011). Haemolysins, proteases and lipases have been widely documented in *V. harveyi* strains from different origins, with cysteine protease and haemolysins being the main exotoxins for shrimps and fish, respectively (Austin and Zhang, 2006; Ruwandeepika *et al.*, 2012). The haemolysin genes are duplicated in the most virulent strains and there is a correlation between the degree of virulence for salmonids and the degree of haemolytic activity (Austin and Zhang, 2006). Interestingly, environmental strains can also be haemolytic and proteolytic and contain virulence genes in their genomes, suggesting that potential virulence for fish extends widely within the species (Ruwandeepika *et al.*, 2012).

10.4.6 Human *Vh*-vibriosis

V. harveyi has been associated with human infections in wounds after exposure to seawater (Hundenborn *et al.*, 2013; Del Gigia-Aguirre *et al.*, 2017) or shark bite (Pavia *et al.*, 1989) and even with a secondary bacteraemia related to catheters (Wilkins *et al.*, 2008) in immunocompromised patients. Recent studies have shown that fish virulent strains cannot resist the bactericidal effect of human serum and are not virulent for mice, which supports the hypothesis that *V. harveyi* can infect humans only when the immune system is impaired (B. Fouz and C. Amaro, 2020, unpublished results).

10.4.7 Preventive measures

Environmental adjustments

As in the case of other fish vibrioses, modifying the physicochemical parameters to disadvantage the proliferation and transmission of *V. harveyi* would help to prevent the disease. In intensive conditions, the first recommendation is to keep the temperature below 18–20°C.

V. harveyi *monitoring*

Periodic sampling is recommended to control the presence of *V. harveyi* in water and fish. The conventional identification procedure (isolation on TCBS agar, Gram staining and biochemical tests) is not suitable for identifying *V. harveyi* since the species is phenotypically very heterogeneous and cannot be distinguished from other members of the clade, mainly from *V. alginolyticus*. Even 16S rRNA gene sequencing often leads to the misidentification of species within the clade. Several PCR techniques have been developed that amplify sequences of the *vhh* haemolysin or the *toxR* genes to identify *V. harveyi* (Conejero and Hedreyda, 2004; Pang *et al.*, 2006). Among them, *toxR* is an effective taxonomic marker for the species-specific identification within the genus *Vibrio*. Other multiplex PCR/real-time PCR strategies have been reported in the literature for the detection of the main pathogenic *Vibrio* species, including *V. harveyi* (Pinto *et al.*, 2017; Nishiki *et al.*, 2018). We recommend a protocol for the isolation and identification of *V. harveyi* from environmental and fish samples that is based on the US Food and Drug Administration's protocol for *V. vulnificus* (Fig. 10.2). Once the species is identified, we recommend an additional multiplex PCR that targets the plasmid-encoded virulence genes *ftbp* and *fpcrp*. The strains possessing these virulence markers should be considered as potentially virulent for fish (B. Fouz and C. Amaro, 2020, unpublished results).

Serological techniques using specific polyclonal antisera would be desirable for epidemiological purposes, since they would allow the identification at serotype level. In fact, although the *V. harveyi* isolates recently recovered from diseased ESB are serologically heterogeneous, three serotypes seem to be dominant, which includes the most virulent strains for ESB and other fish species (B. Fouz and C. Amaro, 2020, unpublished results). Therefore, the precise identification of *V. harveyi* is an important issue in the investigation of aquatic animal diseases.

If potentially virulent strains are detected, fish vaccination is recommended.

Vaccination

Different types of experimental vaccines against *V. harveyi* have been developed to confer protection: killed (bacterins) (Nguyen *et al*., 2017), attenuated (Hu *et al*., 2012), recombinant (Nguyen *et al*., 2018) and DNA vaccines (Hu and Sun, 2011). Currently, the administration of bacterins by immersion and/or injection seems to be the most affordable strategy to prevent *Vh*-vibriosis. As an example, good protection can be achieved in ESB after administration of a bacterin in three doses (two by immersion and one by injection) to fish (F. Unzué, 2018, Spain, personal communication). Finally, since *Vh*-vibriosis can occur in vaccinated ESB against *V. anguillarum*, the formulation of polyvalent bacterins may be the best option, as long as simultaneous exposure to different antigens does not affect the protective response against any of them.

Non-specific immunostimulation

Dietary administration of lactic acid bacteria and *Pseudoalteromonas*, among others, has shown good results in the reduction of mortality in experimental challenges in abalone and shrimp with virulent *V. harveyi* strains (Offret *et al*., 2018; Quiroz-Guzmán *et al*., 2018a). Similar results have been observed after the inclusion of fermented soybean products in shrimp feed (Yatip *et al*., 2018). These findings support the benefits of functional foods in preventing *Vh*-vibriosis in fish.

Phage therapy

The use of bacteriophages (individual phages and cocktails) has been shown to be effective in improving survival in shrimp larval rearing systems by controlling *V. harveyi* populations (Quiroz-Guzmán *et al*., 2018b).

10.4.8 Diagnosis and treatment

Individual ESB with the clinical signs of septicaemia (Fig. 10.1) should be analysed microbiologically as soon as possible. The recommended procedure for the isolation and identification of *V. harveyi* from fish is shown in Fig. 10.2. *V. harveyi* is highly sensitive to all authorized antimicrobials in aquaculture in the EU (B. Fouz and C. Amaro, 2020, unpublished results) and treating diseased fish quickly is recommended. A laboratory test (antibiogram) should be performed to verify the efficacy of the applied antibiotic and to identify an alternative treatment.

Although it is necessary to control outbreaks, therapy with antibiotics should not be used extensively in the treatment of *Vh*-vibriosis. On the contrary, preventive measures should be implemented in order to minimize the contamination of the environment with antibiotics as well as the emergence of multi-resistant strains. In fact, *Harveyi* clade strains isolated from diseased groupers were resistant to multiple antibiotics (Zhu *et al*., 2018).

10.5 Other Vibrios

10.5.1 *Vibrio alginolyticus*

The pathogenicity of *V. alginolyticus* has been demonstrated for finfish, shellfish and humans (Table 10.1). In humans, the more frequent clinical signs caused by *V. alginolyticus* include otitis, diarrhoea and wound infections, although endophthalmitis and conjunctivitis have also been reported (Ceccarelli *et al*., 2019).

The ability to adhere to host tissues and to form biofilms, to secrete siderophores, exotoxins and proteases, as well as to produce a type III secretion system (T3SS) have been described as essential for the virulence of this species (Ceccarelli *et al*., 2019). In addition, regulators for some of these virulence factors have been detected and examined *in vitro* and *in vivo*, such as the phosphatase PppA encoded in the T6SS gene cluster, the *toxR* gene, the signal transduction RstB–RstA system (transcriptional regulators involved in QS) and the senor kinases LuxR–LuxO from the QS regulatory system (Gao *et al*., 2018; Yang *et al*., 2018).

Among the detection methods, various PCR assays including conventional, multiplex and real time-procedures, LAMP assays, matrix-assisted laser desorption/ionization–time-of-flight (MALDI-TOF) mass spectrometry, aptameric technologies and the use of multi-walled carbon nanotubes, functioning as immune, magnetic or fluorescent sensors, are now available for the detection and discrimination of *V. alginolyticus* in fish and environmental samples (Ceccarelli *et al*., 2019).

Antibiotics have been used to control vibriosis in fish and shellfish farms and, although they are efficient agents, their excessive use or misuse resulted in the appearance and transmission of resistance. The use of bacteriophages has been reported to inhibit the growth and biofilm formation capacity of the bacterium (Kokkari *et al.*, 2018). Bacterins enriched with extracellular products, to more recent formulations based on attenuated, subunit or DNA vaccines have been tested, mainly in grouper and gilthead sea bream, and some of them have produced promising results (Pang *et al.*, 2018).

Some studies have evaluated the effect of changes in temperature and salinity on the virulence of *V. alginolyticus*. In general, abrupt changes in temperature or salinity from the optimal values of orange-spotted grouper suppress immunity and increase the virulence of *V. alginolyticus* (Cheng *et al.*, 2009; Chen *et al.*, 2018).

10.5.2 *Vibrio ichthyoenteri*

Vibrio ichthyoenteri was described in 1996 after the characterization of six isolates obtained from diseased larvae of Japanese flounder (*Paralichthys olivaceous*) showing intestinal opacity and necrosis with high mortality rates (Ishimaru *et al.*, 1996). *V. ichthyoenteri* strains are inhibited by low pH, which can indicate that the differentiation of the stomach during the physiological development from flounder larvae to the juvenile stage provides the host a non-immunological defence mechanism.

V. ichthyoenteri has been associated with different pathologies in the fish species shown in Table 10.1. Pathogenicity for turbot, with clinical signs similar to those in Japanese flounder, was demonstrated by Lv *et al.* (2009).

Currently, the disease is treated with antibiotics that, although they can be partially useful, may result in microbial resistance and environmental contamination. Some efforts have been made to develop an effective vaccine based on the recombinant outer membrane protein T (OmpT) expressed in *Escherichia coli*. This vaccine induces strong innate and humoral immune responses in flounder and confers up to 80% protection in experimental trials (Tang *et al.*, 2019).

10.5.3 *Vibrio ordalii*

Biotype II (or phenon II) strains of *V. anguillarum* were assigned to the new species *V. ordalii* in 1981 (Schiewe *et al.*, 1981). Later, different techniques demonstrated that *V. ordalii* is genetically homogeneous although shows certain serological variability related to the geographic origin of the strain, sharing antigens with the serotype O2a of *V. anguillarum* (Toranzo *et al.*, 2017). *V. ordalii* causes vibriosis in the fish species shown in Table 10.1.

V. anguillarum and *V. ordalii* are indistinguishable on the basis of 16S rRNA gene variability but can be differentiated by PCR targeting different genes such as the *V. ordalii* haemolysin (*vohB*) (Avendaño-Herrera *et al.*, 2014).

The presence of the bacterium on the fish surface and gastrointestinal tract suggests that *V. ordalii* invades the host through the invasion of the skin and intestinal epithelium (Toranzo *et al.*, 2017). Related to this process, several putative virulence determinants have been described in *V. ordalii*, including the factors involved in adhesion, colonization and invasion, exotoxins, iron-uptake systems or cell surface components (Toranzo *et al.*, 2017). Among these, the best characterized were the iron-uptake systems, one mediated by siderophores and the other by direct binding to haemin and haemoglobin. The relationship between pathogenicity and iron uptake was demonstrated through *in vivo* challenges using rainbow trout as a fish model (Toranzo *et al.*, 2017).

There is limited information on the behaviour of *V. ordalii* in aquatic environments, apart from some studies on its survival in microcosms. Therefore, the role of seawater as a reservoir or route of transmission is not clear (Toranzo *et al.*, 2017).

10.5.4 *Vibrio splendidus*

Different studies during the 1980s and 1990s described a number of cases of disease outbreaks presumptively associated with *V. splendidus* in the fish species shown in Table 10.1. Most of these isolates were identified as *Vs*-related strains on the sole basis of phenotypic characterization and, therefore, it is hard to determine whether any of those isolates were bona fide *V. splendidus*, especially since species differentiation within the clade *splendidus* is reliable only by molecular methods (Austin, 2012).

In 2015, Gulla *et al.* (2015) demonstrated the virulence capacities of *V. splendidus sensu stricto* to cleaner fish, lumpsucker (*Cyclopterus lumpus*) and ballan wrasse (*Labrus bergylta*), employed as a biocontrol measure against salmon louse in Norwegian farms. A considerable degree of genetic diversity was observed among the *V. splendidus* isolates in these studies. The authors attributed the infections to contact with seawater where this bacterial species is abundant.

10.6 Impact of Global Warming on Vibriosis

There is strong scientific evidence that global warming is increasing the sea surface temperature, which is about 1°C higher than 100 years ago (https://www.epa.gov/climate-indicators/climate-change-indicators-sea-surface-temperature (accessed 12 March 2020)). It is also predicted that the melting of the ice at the poles would also produce a decrease in the salinity of seawater around the world. Since vibrios, with very few exceptions, grow preferentially in warm brackish water (>15°C, 1% salinity), the warming of marine waters would probably support larger populations of these bacteria and, therefore, would increase the risk of *Vibrio* infections (Le Roux *et al.*, 2015).

Most studies on global warming and vibrios have focused on human-pathogenic species. These studies found evidence that global warming is increasing the number of human *Vibrio* infections worldwide and propose vibrios as microbial barometer of climate change (Baker-Austin *et al.*, 2017). The data are especially alarming in Northern Europe and North America, where a marked increment in cases of human infections has been detected in recent summers.

Although there are no similar data in the case of fish vibriosis, it is probable that what is happening in humans will also occur in fish. The hot and long-lasting summers of recent years (especially the one recorded in 2018 in north-west Europe) represented a great challenge for fish species that live naturally in coastal ecosystems due to their exposure to higher concentrations of vibrios. For example, the growth and rate of propagation of *V. anguillarum* increase with temperature, as do the mortalities caused by this pathogen in fish (Hickey and Lee, 2018). Therefore, the frequency and intensity of *Va*-vibriosis outbreaks can be expected to increase, especially in the current cold regions. For *V. harveyi*, the warming of seawater can lead to a wider geographical distribution of the pathogen, since its optimum temperature range is 18 to 28–30°C. *V. vulnificus* causes serious outbreaks of WWV in eel and tilapia only when the water temperature exceeds 25–26°C, which means that the species could also spread if the physicochemical conditions favour its persistence and infectivity. Temperature also affects the infective capability of this pathogen, making it more infectious and invasive at warm temperatures (Hernández-Cabanyero *et al.*, 2020). This fact should be taken into account to predict the real risk of *V. vulnificus* infection caused by global warming.

There is some current evidence of how global warming plus genetic exchange in fish farms could accelerate the emergence and spreading of novel vibrioses and new pathogenic vibrios. One of these examples refers to *V. vulnificus* and *V. harveyi*. *V. vulnificus* pv. *piscis* has acquired the virulence plasmid pVvBt2, which encodes the ability to resist the innate immunity of the fish, a resistance probably adapted to the fish host. This plasmid has already been transmitted to *V. harveyi*, resulting in new *Vh* strains capable of acting as primary pathogens for sea bass that are the responsible ones for the emergent vibriosis in juvenile ESB (B. Fouz and C. Amaro, 2020, unpublished results).

Global warming also affects fish species, which are all poikilothermic. Therefore, any temperature variation in their aquatic habitats will have a significant influence on their metabolism and their susceptibility/resistance to diseases. Determining how fish pathogens will respond to climate change, and how they will interact with their farm hosts in modified scenarios, is essential. Among all the host species for fish vibriosis, the eel will probably be most affected by global warming. In fact, the eel is in decline in nature, especially in the northern hemisphere, and is included in the Convention on International Trade in Endangered Species of Wild Fauna and Flora (https://www.cites.org/sites/default/files/common/cop/14/inf/E14i-21.pdf (accessed 25 March 2020)). There is a growing number of reports that relate global warming to this decline. Since temperature determines the final sex in the eel, a bias of the sex ratio was observed with respect to the females (van Ginneken and Niemantsverdriet, 2017). This finding, together with a general decrease in marine production because of reduced survival of eel larvae during the first stages of life, provides for a bleak outlook for the future of the European eel (Bonhommeau *et al.*, 2008).

From an ecological point of view, the effects of global warming on coastal habitats can also affect the *Vibrio* population associated with biotic and abiotic surfaces. For instance, it has been shown that the attachment of vibrios to chitin increases with temperature. Elevated water temperatures are known to trigger algal blooms that, in turn, promote the increase in zooplankton grazers feeding on algae. In such warm environments, the increased number of copepods and related chitinous zooplankton species is expected to lead to an increase in the number of vibrios (Vezzulli *et al.*, 2015). Other important direct impacts of warmer water temperature include changes in the structure of the communities, earlier times of important events in the life cycle and movements towards the polar regions in the distribution of zooplankton.

10.7 Conclusions and Suggestions for Future Research

Climate change will affect a variety of biotic and abiotic parameters, from modifications in physicochemical conditions (temperature, pH, oxygen, etc.) to alterations in the ecological distribution and fitness of different organisms (from phyto- and zooplankton to shellfish and fish) (Barange *et al.*, 2018). From an aquatic animal health management perspective, the development of geographically specific best management practices to address specific fish health risks for farming systems subjected to climate change impacts will enable these aquaculture settings to become resilient. For instance, the establishment of a breeding programme for thermotolerance in areas in which increases in temperature are expected would be prudent. Some of the key areas of future research will include climate projections for aquaculture hotspots, the effects of climate change on fish production and, of course, the influence of climate change on fish diseases.

Vibrios, for their ubiquity in marine environments, metabolic versatility and thermal dependence, constitute a perfect model to study the association between ocean warming and persistence and spread of pathogens in seawater (Vezzulli *et al.*, 2015). However, the biocomplexity of interactions between these bacteria and their natural environment makes these studies difficult. Future studies should include comparative field observations on the occurrence of these bacteria on long temporal and spatial scales and the interactions of vibrios with their aquatic reservoirs in different environmental and climatic scenarios, as well as the influence of climate change on host susceptibility and exposure patterns. The integration of the results obtained from these multidisciplinary studies would allow the development of models and measures to predict, prevent and control *Vibrio*-related fish diseases.

Acknowledgements

Carmen Amaro and Belén Fouz acknowledges the support of MINECO (Ministerio de Economía Industria y Competitividad, Spain) (Projects AGL2017-87723, AGL2014-58933 and AGL2011-29639) and GVA (Generalitat Valenciana) (Project AICO2018-123) that funded the research that generated part of the *V. vulnificus* and *V. harveyi* results presented here. Jesús L. Romalde also thanks MINECO (AGL2016-77539-R) for support.

References

Actis, L.A., Tolmasky, M.E. and Crosa, J.H. (2011) Vibriosis. In: Woo, P.T.K. and Bruno, D.W. (eds) *Fish Diseases and Disorders*. Vol. 3. *Viral, Bacterial and Fungal Infections*, 2nd edn. CAB International, Wallingford, UK, pp. 570–605.

Aida, K., Tsukamoto, K. and Yamauchi, K. (eds) (2003) *Eel Biology*. Springer, Tokyo. https://doi.org/10.1007/978-4-431-65907-5

Amaro, C., Sanjuán, E., Fouz, B., Pajuelo, D., Lee, C.T. *et al.* (2015) The fish pathogen *Vibrio vulnificus* biotype 2: epidemiology, phylogeny, and virulence factors involved in warm-water vibriosis. *Microbiology Spectrum* 3(3). https://doi.org/10.1128/microbiolspec.VE-0005-2014

Angelidis, P. (2014) *Vibrio anguillarum*-associated vibriosis in the Mediterranean aquaculture. In: Angelidis, P. (ed) *Aspects of Mediterranean Marine Aquaculture*. Blue Crab PC, Thessaloniki, Greece, pp. 243–264.

Austin, B. (ed) (2012) *Infectious Disease in Aquaculture: Prevention and Control*, 1st edn. *Woodhead Publishing Series in Food Science, Technology and Nutrition No.* 231. Woodhead Publishing, Cambridge, UK.

Austin, B. and Zhang, X.-H. (2006) *Vibrio harveyi*: a significant pathogen of marine vertebrates and invertebrates. *Letters in Applied Microbiology* 43, 119–124. https://doi.org/10.1111/j.1472-765X.2006.01989.x

Avendaño-Herrera, R., Maldonado, J.P., Tapia-Cammas, D., Feijoó, C.G., Calleja, F. and Toranzo, A.E. (2014) PCR protocol for the detection of *Vibrio ordalii* by amplification of the *vohB* (hemolysin) gene. *Diseases of Aquatic Organisms* 107, 223–234. https://doi.org/10.3354/dao02684

Baker-Austin, C., Trinanes, J., Gonzalez-Escalona, N. and Martinez-Urtaza, J. (2017) Non-cholera vibrios: the microbial barometer of climate change. *Trends in Microbiology* 25, 76–84. https://doi.org/10.1016/j.tim.2016.09.008

Balado, M., Lages, M.A., Fuentes-Monteverde, J.C., Martínez-Matamoros, D., Rodríguez, J. *et al.* (2018) The siderophore piscibactin is a relevant virulence factor for *Vibrio anguillarum* favored at low temperatures. *Frontiers in Microbiology* 9, 1766. https://doi.org/10.3389/fmicb.2018.01766

Bao, P., Sun, X., Liu, Q., Zhang, Y. and Liu, X. (2019) Synergistic effect of a combined live *Vibrio anguillarum* and *Edwardsiella* piscicida vaccine in turbot.

Fish and Shellfish Immunology 88, 84–90. https://doi.org/10.1016/j.fsi.2019.02.014

Barange, M., Fernandes J.A., Kay, S., Hossain, M.A.R., Ahmed, M. and Lauria, V. (2018) Marine ecosystems and fisheries: trends and prospects. In: Nicholls, R., Hutton, C., Adger, W., Hanson, S., Rahman, M. and Salehin M. (eds) *Ecosystem Services for Well-Being in Deltas*. Palgrave Macmillan, Cham, Switzerland, pp. 469–488.

Beltrán, J.M.G., Espinosa, C., Guardiola, F.A. and Esteban, M.A. (2018) *In vitro* effects of *Origanum vulgare* leaf extracts on gilthead seabream (*Sparus aurata* L.) leucocytes, cytotoxic, bactericidal and antioxidant activities. *Fish and Shellfish Immunology* 79, 1–10. https://doi.org/10.1016/j.fsi.2018.05.005

Bisharat, N., Agmon, V., Finkelstein, R., Raz, R., Ben-Dror, G. *et al.* (1999) Clinical, epidemiological, and microbiological features of *Vibrio vulnificus* biogroup 3 causing outbreaks of wound infection and bacteraemia in Israel. *Lancet* 354, 1421–1424. https://doi.org/10.1016/S0140-6736(99)02471-X

Bœuf, G. and Payan, P. (2001) How should salinity influence fish growth? *Comparative Biochemistry and Physiology Part C: Toxicology & Pharmacology* 130, 411–423. https://doi.org/10.1016/S1532-0456(01)00268-X

Bonhommeau, S., Chassot, E., Planque, B., Rivot E., Knap, A.H. and Le Pape, O. (2008) Impact of climate on eel populations of the northern hemisphere. *Marine Ecology Progress Series* 373, 71–80. https://doi.org/10.3354/meps07696

Borgeaud, S., Metzger, L.C., Scrignari, T. and Blokesch, M. (2015) The type VI secretion system of *Vibrio cholerae* fosters horizontal gene transfer. *Science* 347, 63–67. https://doi.org/10.1126/science.1260064

Bramhachari, P.V. and Dubey, S.K. (2006) Isolation and characterization of exopolysaccharide produced by *Vibrio harveyi* strain VB23. *Letters in Applied Microbiology* 43, 571–577. https://doi.org/10.1111/j.1472-765X.2006.01967.x

Callol, A., Pajuelo, D., Ebbesson, L., Teles, M., MacKenzie, S. and Amaro, C. (2015) Early steps in the European eel (*Anguilla anguilla*)–*Vibrio vulnificus* interaction in the gills: role of the $RtxA1_3$ toxin. *Fish and Shellfish Immunology* 43, 502–509. https://doi.org/10.1016/j.fsi.2015.01.009

Castillo, D., Rørbo, N., Jørgensen, J., Lange, J., Tan, D. *et al.* (2019) Phage defense mechanisms and their genomic and phenotypic implications in the fish pathogen *Vibrio anguillarum*. *FEMS Microbiology Ecology* 95, *fiz004*. https://doi.org/10.1093/femsec/fiz004

Ceccarelli, D., Amaro, C., Romalde, J.L., Suffredini, E. and Vezzulli, L. (2019) *Vibrio* species. In: Doyle, M.P., Diez-Gonzalez, F. and Hill, C. (eds) *Food Microbiology: Fundamentals and Frontiers*, 5th edn. American Society for Microbiology, Washington, DC, pp. 347–388. https://doi.org/10.1128/9781555819972.ch13

Chen, C.Y., Chao, C.B. and Bowser, P.R. (2006) Infection of tilapia *Oreochromis* sp. by *Vibrio vulnificus* in freshwater and low-salinity environments. *Journal of the World Aquaculture Society* 37, 82–88. https://doi.org/10.1111/j.1749-7345.2006.00010.x

Chen, Y.-Y., Cheng, A.-C., Cheng, S.-A. and Chen, J.C. (2018) Orange-spotted grouper *Epinephelus coioides* that have encountered low salinity stress have decreased cellular and humoral immune response reactions and increased susceptibility to *Vibrio alginolyticus*. *Fish and Shellfish Immunology* 80, 392–396. https://doi.org/10.1016/j.fsi.2018.06.028

Cheng, A.-C., Cheng, S.-A., Chen, Y.-Y. and Chen, J.-C. (2009) Effects of temperature change on the innate cellular and humoral immune responses of orange-spotted grouper *Epinephelus coioides* and its susceptibility to *Vibrio alginolyticus*. *Fish and Shellfish Immunology* 26, 768–772. https://doi.org/10.1016/j.fsi.2009.03.011

Conejero, M.J.U. and Hedreyda, C.T. (2004) PCR detection of hemolysin (*vvh*) gene in *Vibrio harveyi*. *Journal of General and Applied Microbiology* 50, 137–142. https://doi.org/10.2323/jgam.50.137

Dalsgaard, A., Frimodt-Møller, N., Bruun, B., Høi, L. and Larsen, J.L. (1996) Clinical manifestations and molecular epidemiology of *Vibrio vulnificus* infections in Denmark. *European Journal of Clinical Microbiology and Infectious Diseases* 15, 227–232. https://doi.org/10.1007/BF01591359

Defoirdt, T., Boon, N., Bossier, P. and Verstraete, W. (2004) Disruption of bacterial quorum sensing: an unexplored strategy to fight infections in aquaculture. *Aquaculture* 240, 69–88. https://doi.org/10.1016/j.aquaculture.2004.06.031

Del Gigia-Aguirre, L., Sánchez-Yebra-Romera, W., García-Muñoz, S. and Rodríguez-Maresca, M. (2017) First description of wound infection with *Vibrio harveyi* in Spain. *New Microbes and New Infections* 19, 15–16. https://doi.org/10.1016/j.nmni.2017.05.004

Esteve-Gassent, M.D., Nielsen, M.E. and Amaro, C. (2003) The kinetics of antibody production in mucus and serum of European eel (*Anguilla anguilla* L.) after vaccination against *Vibrio vulnificus*: development of a new method for antibody quantification in skin mucus. *Fish and Shellfish Immunology* 15, 51–61. https://doi.org/10.1016/S1050-4648(02)00138-9

Esteve-Gassent, M.D., Fouz, B. and Amaro, C. (2004) Efficacy of a bivalent vaccine against eel diseases caused by *Vibrio vulnificus* after its administration by four different routes. *Fish and Shellfish Immunology* 16, 93–105. https://doi.org/10.1016/S1050-4648(03)00036-6

FAO (Food and Agriculture Organization of the United Nations) (2018) *Fishery and Aquaculture Statistics 2016*. FAO Yearbook. FAO Fisheries and Aquaculture Department, Rome. Available at: http://www.fao.org/3/i9942t/I9942T.pdf (accessed 12 March 2020).

Fouz, B., Esteve-Gassent, M.D., Barrera, R., Larsen, J.L., Nielsen, M.E. and Amaro, C. (2001) Field testing of a vaccine against eel diseases caused by *Vibrio vulnificus*. *Diseases of Aquatic Organisms* 45, 183–189. https://doi.org/10.3354/dao045183

Frans, I., Michiels, C.W., Bossier. P., Willems, K.A., Lievens, B. and Rediers, H. (2011) *Vibrio anguillarum* as a fish pathogen: virulence factors, diagnosis and prevention. *Journal of Fish Diseases* 34, 643–661. https://doi.org/10.1111/j.1365-2761.2011.01279.x

Fujiwara-Nagata, E. and Eguchi, M. (2003) Survival of *Vibrio anguillarum*, a fish pathogen, in freshwater by forming biofilms. *Microbes and Environments* 18, 196–202. https://doi.org/10.1264/jsme2.18.196

Gao, X., Wang, X., Mao, Q., Xu, R., Zhou, X. *et al.* (2018) VqsA, a novel LysR-type transcriptional regulator, coordinates quorum sensing (QS) and is controlled by QS to regulate virulence in the pathogen *Vibrio alginolyticus*. *Applied and Environmental Microbiology* 84, e00444-18. https://doi.org/10.1128/AEM.00444-18

Gomez-Gil, B., Thompson, C.C., Matsumura, Y., Sawabe, T., Iida, T. *et al.* (2014) The Family *Vibrionaceae*. In: Rosenberg, E., DeLong, E.F., Lory, S., Stackebrandt, E. and Thompson, F. (eds) *The Prokaryotes*. Springer, Berlin/Heidelberg, Germany, pp. 659–747. https://doi.org/10.1007/978-3-642-38922-1_225

Gulla, S., Sørum, H., Vågnes, Ø. and Colquhoun, D.J. (2015) Phylogenetic analysis and serotyping of *Vibrio splendidus*-related bacteria isolated from salmon farm cleaner fish. *Diseases of Aquatic Organisms* 117, 121–131. https://doi.org/10.3354/dao02938

Haffray, P., Tsigenopoulos, C.-S., Bonhome, F., Chatain, B., Magoulas, A. *et al.* (2007) European sea bass – *Dicentrarchus labrax*. GENIMPACT (Evaluation of genetic impact of aquaculture activities on native populations – A European network) Final Scientific Report, pp. 40–46. Available at: https://www.researchgate.net/publication/252768351_European_sea_bass_-_Dicentrarchus_labrax (accessed 25 March 2020).

Haldar, S., Maharajan, A., Chatterjee, S., Hunter, S.A., Chowdhury, N. *et al.* (2010) Identification of *Vibrio harveyi* as a causative bacterium for a tail rot disease of sea bream *Sparus aurata* from research hatchery in Malta. *Microbiological Research* 165, 639–648. https://doi.org/10.1016/j.micres.2009.12.001

Hernández-Cabanyero, C., Lee, C.T., Tolosa-Enguis, V., Sanjuán, E., Pajuelo, D. *et al.* (2019) Adaptation to host in *Vibrio vulnificus*, a zoonotic pathogen that causes septicemia in fish and humans. *Environmental Microbiology* 21(8), 3118–3139. https://doi.org/10.1111/1462-2920.14714

Hernández-Cabanyero, C., Sanjuán, E., Fouz, B., Pajuelo, D., Vallejos-Vidal, E. *et al.* (2020) The effect of the environmental temperature on the adaptation to host in the zoonotic pathogen *Vibrio vulnificus*. *Frontiers in Microbiology*, 11, 489. https://doi.org/10.3389/fmicb.2020.00489.

Hickey, M.E. and Lee, J.-L. (2018) A comprehensive review of *Vibrio* (*Listonella*) *anguillarum*: ecology, pathology and prevention. *Reviews in Aquaculture* 10, 585–610. https://doi.org/10.1111/raq.12188

Hickey, M.E., Richards, G.P. and Lee, J.-L. (2015) Development of a two-step, non-probed multiplex real-time PCR for surveilling *Vibrio anguillarum* in seawater. *Journal of Fish Diseases* 38, 551–559. https://doi.org/10.1111/jfd.12264

Hoare, R., Hivland, H., Langston, A.L., Imsland, A., Stefansson, S.O. *et al.* (2002) Susceptibility of three different strains of juvenile Atlantic halibut (*Hippoglossus hippoglossus* L.) cultured at two different temperatures to *Vibrio anguillarum* and temperature effect on antibiotic response. *Fish and Shellfish Immunology* 13, 111–123. https://doi.org/10.1006/fsim.2001.0385

Hu, Y.H. and Sun, L. (2011) A bivalent *Vibrio harveyi* DNA vaccine induces strong protection in Japanese flounder (*Paralichthys olivaceus*). *Vaccine* 29, 4328–4333. https://doi.org/10.1016/j.vaccine.2011.04.021

Hu, Y.-H., Deng, T., Sun, B.-G. and Sun. L. (2012) Development and efficacy of an attenuated *Vibrio harveyi* vaccine candidate with cross protectivity against *Vibrio alginolyticus*. *Fish and Shellfish Immunology* 32, 1155–1161. https://doi.org/10.1016/j.fsi.2012.03.032

Hundenborn, J., Thurig, S., Kommerell, M., Haag, H. and Nolte, O. (2013) Severe wound infection with *Photobacterium damselae* ssp. *damselae* and *Vibrio harveyi*, following a laceration injury in marine environment: a case report and review of the literature. *Case Reports in Medicine* 2013, 610632. https://doi.org/10.1155/2013/610632

Imsland, A.K., Sunde, L.M., Folkvord, A. and Stefansson, S.O. (1996) The interaction of temperature and fish size on growth of juvenile turbot. *Journal of Fish Biology* 49, 926–940. https://doi.org/10.1111/j.1095-8649.1996.tb00090.x

Ishimaru, K., Akagawa-Matsushita, M. and Muroga, K. (1996) *Vibrio ichthyoenteri* sp. nov., a pathogen of Japanese flounder (*Paralichthys olivaceus*) larvae. *International Journal of Systematic Bacteriology* 46, 155–159. https://doi.org/10.1099/00207713-46-1-155

Jones, M. (2004) *Cultured Aquatic Species Information Programme, Salmo salar*. Food and Agriculture Organization of the United Nations, Fisheries and Aquaculture Department, Rome. Available at: http://www.fao.org/fishery/culturedspecies/Salmo_salar/en (accessed 13 March 2020).

Karunasagar, I., Otta, S.K. and Karunasagar, I. (1996) Biofilm formation of *Vibrio harveyi* on surfaces. *Aquaculture* 140, 241–245. https://doi.org/10.1016/0044-8486(95)01180-3

Kokkari, C., Sarropoulou, E., Bastias, R., Mandalakis, M. and Katharios, P. (2018) Isolation and characterization of a novel bacteriophage infecting *Vibrio alginolyticus*. *Archives of Microbiology* 200, 707–718. https://doi.org/10.1007/s00203-018-1480-8

Larsen, M.H., Blackburn, N., Larsen, J.L. and Olsen, J.E. (2004) Influences of temperature, salinity and starvation on the motility and chemotactic response of *Vibrio anguillarum*. *Microbiology* 150, 1283–1290. https://doi.org/10.1099/mic.0.26379-0

Le Roux, F., Wegner, K.M., Baker-Austin, C., Vezzulli, L., Osorio, C.R. *et al.* (2015) The emergence of *Vibrio* pathogens in Europe: ecology, evolution, and pathogenesis (Paris, 11–12th March 2015). *Frontiers in Microbiology* 6, 830. https://doi.org/10.3389/fmicb.2015.00830

Li, L., Rock, J.L. and Nelson, D.R. (2008) Identification and characterization of a repeat-in-toxin gene cluster in *Vibrio anguillarum*. *Infection and Immunity* 76, 2620–2632. https://doi.org/10.1128/IAI.01308-07

Li, Y. and Ma, Q. (2017) Iron acquisition strategies of *Vibrio anguillarum*. *Frontiers in Cellular and Infection Microbiology* 7, 342. https://doi.org/10.3389/fcimb.2017.00342

Lindell, K., Fahlgren, A., Hjerde, E., Willassen, N.-P., Fällman, M. and Milton, D.L. (2012) Lipopolysaccharide O-antigen prevents phagocytosis of *Vibrio anguillarum* by rainbow trout (*Oncorhynchus mykiss*) skin epithelial cells. *PLoS ONE* 7, e37678. https://doi.org/10.1371/journal.pone.0037678

Luna, G.M., Bongiorni, L., Gili, C., Biavasco, F. and Danovaro, R. (2010) *Vibrio harveyi* as a causative agent of the white syndrome in tropical stony corals. *Environmental Microbiology Reports* 2, 120–127. https://doi.org/10.1111/j.1758-2229.2009.00114.x

Lv, J., Li, X., Han, Y., Chen, J.X. and Zhang, X.H. (2009) Isolation and identification of *Vibrio ichthyoenteri* in farmed turbot and histopathology study of the diseased fish. *Journal of Fisheries China* 33, 311–317.

Ma, L., Chen, J., Liu, R., Zhang, X.-H. and Jiang, Y.-A. (2009) Mutation of *rpoS* gene decreased resistance to environmental stresses, synthesis of extracellular products and virulence of *Vibrio anguillarum*. *FEMS Microbiology Ecology* 70, 286–292. https://doi.org/10.1111/j.1574-6941.2009.00713.x

MacDonell, M.T. and Colwell, R.R. (1985) Phylogeny of the Vibrionaceae, and recommendation for two new genera, *Listonella* and *Shewanella*. *Systematic and Applied Microbiology* 6, 171–182. https://doi.org/10.1016/S0723-2020(85)80051-5

Mateus, L., Costa, L., Silva, Y.J., Pereira, C., Cunha, A. and Almeida, A. (2014) Efficacy of phage cocktails in the inactivation of *Vibrio* in aquaculture. *Aquaculture* 424, 167–173. https://doi.org/10.1016/j.aquaculture.2014.01.001

Mohamad, N., Mohd Roseli, F.A., Azmai, M.N.A., Saad, M.Z., Md Yasin, I.S. *et al.* (2018) Natural concurrent infection of *Vibrio harveyi* and *V. alginolyticus* in cultured hybrid groupers in Malaysia. *Journal of Aquatic Animal Health* 31, 88–96. https://doi.org/10.1002/aah.10055

Morais, S., Aragão, C., Cabrita, E., Conceição, L.E., Constenla, M. *et al.* (2016) New developments and biological insights into the farming of *Solea senegalensis* reinforcing its aquaculture potential. *Reviews in Aquaculture* 8, 227–263. https://doi.org/10.1111/raq.12091

Murciano, C., Lee, C.T., Fernandez-Bravo, A., Hsieh, T.H., Fouz, B. *et al.* (2017) MARTX toxin in the zoonotic serovar of *Vibrio vulnificus* triggers an early cytokine storm in mice. *Frontiers in Cellular and Infection Microbiology* 7, 332. https://doi.org/10.3389/fcimb.2017.00332

Natrah, F.M., Ruwandeepika, H.A., Pawar, S., Karunasagar, I., Sorgeloos, P. *et al.* (2011) Regulation of virulence factors by quorum sensing in *Vibrio harveyi*. *Veterinary Microbiology* 154, 124–129. https://doi.org/10.1016/j.vetmic.2011.06.024

Nelson, J.S. (2006) *Fishes of the World*, 4th edn. Wiley, New York.

Nguyen, H.T., Nguyen, T.T.T., Wang, Y.-T., Wang, P.C. and Chen, S.C. (2017) Effectiveness of formalin-killed vaccines containing CpG oligodeoxynucleotide 1668 adjuvants against *Vibrio harveyi* in orange-spotted grouper. *Fish and Shellfish Immunology* 68, 124–131. https://doi.org/10.1016/j.fsi.2017.07.018

Nguyen, H.T., Nguyen, T.T.T., Chen, Y.-C., Vu-Khac, H., Wang, P.C. and Chen, S.C. (2018) Enhanced immune responses and effectiveness of refined outer membrane protein vaccines against *Vibrio harveyi* in orange-spotted grouper (*Epinephelus coioides*). *Journal of Fish Diseases* 41, 1349–1358. https://doi.org/10.1111/jfd.12828

Nishiki, I., Minami, T., Murakami, A., Hoai, T.D. and Fujiwara, A. (2018) Multilocus sequence analysis of Vibrionaceae isolated from farmed amberjack and the development of a multiplex PCR assay for the detection of pathogenic species. *Journal of Fish Diseases* 41, 1295–1301. https://doi.org/10.1111/jfd.12823

Offret, C., Rochard, V., Laguerre, H., Mounier, J., Huchette, S. *et al.* (2018) Protective efficacy of a *Pseudoalteromonas* strain in European abalone, *Haliotis tuberculata*, infected with *Vibrio harveyi* ORM4. *Probiotics and Antimicrobial Proteins* 11, 239–247. https://doi.org/10.1007/s12602-018-9389-8

Oliver, J.D. (2015) The biology of *Vibrio vulnificus*. *Microbiology Spectrum* 3(3). https://doi.org/10.1128/microbiolspec.VE-0001-2014

O'Toole, R., Von Hofsten, J., Rosqvist, R., Olsson, P.E. and Wolf-Watz, H. (2004) Visualisation of zebrafish infection by GFP-labelled *Vibrio anguillarum*. *Microbial Pathogenesis* 37, 41–46. https://doi.org/10.1016/j.micpath.2004.03.001

Otterå, H. (2009) *Cultured Aquatic Species Information Programme, Gadus morhua*. Food and Agriculture Organization of the United Nations, Fisheries and Aquaculture Department, Rome. Available at: http://www.fao.org/fishery/culturedspecies/Gadus_morhua/en (accessed 13 March 2020).

Pajuelo, D., Lee, C.T., Roig, F.J., Hor, L.I. and Amaro, C. (2015) Novel host-specific iron acquisition system in the zoonotic pathogen *Vibrio vulnificus*. *Environmental Microbiology* 17, 2076–2089. https://doi.org/10.1111/1462-2920.12782

Pajuelo, D., Hernandez-Cabanyero, C., Sanjuan, E., Lee, C.T., Silva-Hernandez, F.X. *et al.* (2016) Iron and Fur in the life cycle of the zoonotic pathogen *Vibrio vulnificus*. *Environmental Microbiology* 18, 4005–4022. https://doi.org/10.1111/1462-2920.13424

Pang, H., Qiu, M., Zhao, J., Hoare, R., Monaghan, S.J. *et al.* (2018) Construction of a *Vibrio alginolyticus hopPmaJ* (*hop*) mutant and evaluation of its potential as a live attenuated vaccine in orange-spotted grouper (*Epinephelus coioides*). *Fish and Shellfish Immunology* 76, 93–100. https://doi.org/10.1016/j.fsi.2018.02.012

Pang, L., Zhang, X.-H., Zhong, Y., Chen, J., Li, Y. and Austin, B. (2006) Identification of *Vibrio harveyi* using PCR amplification of the *toxR* gene. *Letters in Applied Microbiology* 43, 249–255. https://doi.org/10.1111/j.1472-765X.2006.01962.x

Pascual, J., Macián, M.C., Arahal, D.R., Garay, E. and Pujalte, M.J. (2010) Multilocus sequence analysis of the central clade of the genus *Vibrio* by using the 16S rRNA, *recA*, *pyrH*, *rpoD*, *gyrB*, *rctB* and *toxR* genes. *International Journal of Systematic and Evolutionary Microbiology* 60, 154–165. https://doi.org/10.1099/ijs.0.010702-0

Pavia, A.T., Bryan, J.A., Maher, K.L., Hester, T.R. and Farmer, J.J. (1989) *Vibrio carchariae* infection after a shark bite. *Annals of Internal Medicine* 111, 85–86. https://doi.org/10.7326/0003-4819-111-1-85

Pérez-Ruzafa, A. and Marcos, C. (2014) Ecology and distribution of *Dicentrarchus labrax* (Linnaeus 1758). In: Sánchez Vásquez, F.J. and Muñoz-Cueto, J.A. (eds) *Biology of European Sea Bass*. CRC Press, Boca Raton, Florida, pp. 3–33.

Person-Le Ruyet, J. (2010) Turbot culture. In: Daniels, H.V. and Watanabe, W.O. (eds) *Practical Flatfish Culture and Stock Enhancement*. Wiley-Blackwell, Ames, Iowa, pp. 123–139. https://doi.org/10.1002/9780813810997.ch7

Pinto, M.F., Baptista, T. and Afonso, C.C.N. (2017) Development of a new multiplex-PCR tool for the simultaneous detection of the fish pathogens *Vibrio alginolyticus*, *Vibrio anguillarum*, *Vibrio harveyi* and *Edwardsiella tarda*. *Aquatic Living Resources* 30, 4. https://doi.org/10.1051/alr/2017005

Planas, M., Pérez-Lorenzo, M., Hjelm, M., Gram, L. Fisksdal, I.U. *et al.* (2006) Probiotic effect *in vivo* of *Roseobacter* strain 27-4 against *Vibrio* (*Listonella*) *anguillarum* infections in turbot (*Scophthalmus maximus* L.) larvae. *Aquaculture* 255, 323–333. https://doi.org/10.1016/j.aquaculture.2005.11.039

Poppe, T.T. and Koppang, E. (2014) Side-effects of vaccination. In: Gudding, R., Lillehaug, A. and Evensen, Ø. (eds) *Fish Vaccination*. Wiley-Blackwell, Chichester, UK, pp. 153–161.

Quiroz-Guzmán, E., Vázquez-Juárez, R., Luna-González, A., Balcázar, J.L., Barajas-Sandoval, D.R. and Martínez-Díaz, S.F. (2018a) Administration of probiotics improves the brine shrimp production and prevents detrimental effects of pathogenic *Vibrio* species. *Marine Biotechnology* 20, 512–519. https://doi.org/10.1007/s10126-018-9822-8

Quiroz-Guzmán, E., Peña-Rodríguez, A., Vázquez-Juárez, R., Barajas-Sandoval, D.R., Balcázar, J.L. and Martínez-Díaz, S.F. (2018b) Bacteriophage cocktails as an environmentally-friendly approach to prevent *Vibrio parahaemolyticus* and *Vibrio harveyi* infections in brine shrimp (*Artemia franciscana*) production. *Aquaculture* 492, 273–279. https://doi.org/10.1016/j.aquaculture.2018.04.025

Ramaiah, N., Ravel, J., Straube, W.L., Hill, R.T. and Colwell, R.R. (2002) Entry of *Vibrio harveyi* and *Vibrio fischeri* into the viable but nonculturable state. *Journal of Applied Microbiology* 93, 108–116. https://doi.org/10.1046/j.1365-2672.2002.01666.x

Rodkhum, C., Hirono, I., Crosa, J.H. and Aoki, T. (2005) Four novel hemolysin genes of *Vibrio anguillarum* and their virulence to rainbow trout. *Microbial Pathogenesis* 39, 109–119. https://doi.org/10.1016/J.MICPATH.2005.06.004

Roig, F.J., Sanjuan, E., Llorens, A. and Amaro, C. (2010) *pilF* polymorphism-based PCR to distinguish *Vibrio vulnificus* strains potentially dangerous to public health. *Applied and Environmental Microbiology* 76, 1328–1333. https://doi.org/10.1128/AEM.01042-09

Roig, F.J., González-Candelas, F., Sanjuán, E., Fouz, B., Feil, E.J. *et al.* (2018) Phylogeny of *Vibrio vulnificus* from the analysis of the core-genome: implications for intra-species taxonomy. *Frontiers in Microbiology* 8, 2613. https://doi.org/10.3389/fmicb.2017.02613

Rørbo, N., Rønneseth, A., Kalatzis, P.G., Rasmussen, B.B., Engell-Sørensen, K. *et al.* (2018) Exploring the effect of phage therapy in preventing *Vibrio anguillarum* infections in cod and turbot larvae. *Antibiotics* 7, 42. https://doi.org/10.3390/antibiotics7020042

Ruwandeepika, H.A.D., Jayaweera, T.S.P., Bhowmick, P.P., Karunasagar, I., Bossier, P. and Defoirdt, T. (2012) Pathogenesis, virulence factors and virulence regulation of vibrios belonging to the *Harveyi* clade. *Reviews in Aquaculture* 4, 59–74. https://doi.org/10.1111/j.1753-5131.2012.01061.x

Sanjuán, E. and Amaro, C. (2004) Protocol for specific isolation of virulent strains of *Vibrio vulnificus* serovar E (biotype 2) from environmental samples. *Applied and Environmental Microbiology* 70, 7024–7032. https://doi.org/10.1128/AEM.70.12.7024-7032.2004

Sanjuán, E. and Amaro, C. (2007) Multiplex PCR assay for detection of *Vibrio vulnificus* biotype 2 and simultaneous discrimination of serovar E strains. *Applied and Environmental Microbiology* 73, 2029–2032. https://doi.org/10.1128/AEM.02320-06

Satchell, K.J. (2011) Structure and function of MARTX toxins and other large repetitive RTX proteins. *Annual Review of Microbiology* 65, 71–90. https://doi.org/10.1146/annurev-micro-090110-102943

Sawabe, T., Kita-Tsukamoto, K. and Thompson, F.L. (2007) Inferring the evolutionary history of vibrios by means of multilocus sequence analysis. *Journal of Bacteriology* 189, 7932–7936. https://doi.org/10.1128/JB.00693-07

Schiewe, M.H., Trust, T.J. and Crosa, J.H. (1981) *Vibrio ordalii* sp. nov.: a causative agent of vibriosis in fish. *Current Microbiology* 6, 343–348. https://doi.org/10.1007/BF01567009

Sinatra, J.A. and Colby, K. (2018) Fatal *Vibrio anguillarum* infection in an immunocompromised patient – Maine, *2017. Morbidity and Mortality Weekly Report* 67, 962. https://doi.org/10.15585/mmwr.mm6734a5

Sorroza, L., Padilla, D., Acosta, F., Román, L., Grasso, V. *et al.* (2012) Characterization of the probiotic strain *Vagococcus fluvialis* in the protection of European sea bass (*Dicentrarchus labrax*) against vibriosis by *Vibrio anguillarum*. *Veterinary Microbiology* 155, 369–373. https://doi.org/10.1016/j.vetmic.2011.09.013

Sun, F., Chen, J., Zhong, L., Zhang, X.H., Wang, R. *et al.* (2008) Characterization and virulence retention of viable but nonculturable *Vibrio harveyi*. *FEMS Microbiology Ecology* 64, 37–44. https://doi.org/10.1111/j.1574-6941.2008.00442.x

Takemura, A.F., Chien, D.M. and Polz, M.F. (2014) Associations and dynamics of Vibrionaceae in the environment, from the genus to the population level. *Frontiers in Microbiology* 5, 38. https://doi.org/10.3389/fmicb.2014.00038

Tang, X., Wang, H., Liu, F., Sheng, X., Xing, J. and Zhan, W. (2019) Recombinant outer membrane protein T (OmpT) of *Vibrio ichthyioenteri*, a potential vaccine candidate for flounder (*Paralichthys olivaceus*). *Microbial Pathogenesis* 126, 185–192. https://doi.org/10.1016/j.micpath.2018.11.001

Thompson, F.L., Thompson, C.C., Dias, G.M., Naka, H., Dubay, C. and Crosa, J.H. (2011) The genus *Listonella* MacDonell and Colwell 1986 is a later heterotopic synonym of the genus *Vibrio* Pacini 1854 (Approved lists 1980) – a taxonomic opinion. *International Journal of Systematic and Evolutionary Microbiology* 61, 3023–3027. https://doi.org/10.1099/ijs.0.030015-0

Toranzo, A.E., Magariños, B. and Romalde, J.L. (2005) A review of the main bacterial fish diseases in mariculture systems. *Aquaculture* 246, 37–61. https://doi.org/10.1016/j.aquaculture.2005.01.002

Toranzo, A.E., Magariños, B. and Avendaño-Herrera, R. (2017) Vibriosis: *Vibrio anguillarum*, *V. ordalii* and *Aliivibrio salmonicida*. In: Woo, P.T.K. and Cipriano, R.C. (eds) *Fish Viruses and Bacteria: Pathobiology and Protection*. CAB International, Wallingford, UK, pp. 314–333.

Travers, M.-A., Basuyaux, O., Le Goïc, N., Huchette, S., Nicolas, J.-L. *et al.* (2009) Influence of temperature and spawning effort on *Haliotis tuberculata* mortalities caused by *Vibrio harveyi*: an example of emerging vibriosis linked to global warming. *Global Change Biology* 15, 1365–1376. https://doi.org/10.1111/j.1365-2486.2008.01764.x

van Ginneken, V. and Niemantsverdriet, P. (2017) Is global warming the cause for the dwindling European eel population? *Ocean & Fish Open Access Journal* 2(5), 555597. https://doi.org/10.19080/OFOAJ.2017.02.555597

Veenstra, J., Rietra, P.J., Stoutenbeek, C.P., Coster, J.M., de Gier H.H. and Dirks-Go, S. (1992) Infection by an indole-negative variant of *Vibrio vulnificus* transmitted by eels. *Journal of Infectious Diseases* 166, 209–210. https://doi.org/10.1093/infdis/166.1.209

Vezzulli, L., Pezzati, E., Brettar, I., Höfle, M. and Pruzzon, C. (2015) Effects of global warming on *Vibrio* ecology. *Microbiology Spectrum* 3(3). https://doi.org/10.1128/microbiolspec.VE-0004-2014

Weber, B., Chen, C. and Milton, D.L. (2010) Colonization of fish skin is vital for *Vibrio anguillarum* to cause disease. *Environmental Microbiology Reports* 2, 133–139. https://doi.org/10.1111/j.1758-2229.2009.00120.x

Webster, C.D. and Lim, C. (2010) *Tilapia: Biology, Culture and Nutrition, 1st edn*. CRC Press, Boca Raton, Florida.

Wilkins, S., Millar, M., Hemsworth, S., Johnson, G., Warwick, S. and Pizer, B. (2008) *Vibrio harveyi* sepsis in a child with cancer. *Pediatric Blood and Cancer* 50, 891–892. https://doi.org/10.1002/pbc.21356

Wright, A.C., Morris, J.G. Jr, Maneval, D.R. Jr, Richardson, K. and Kaper, J.B. (1985) Cloning of the cytotoxin-hemolysin gene of *Vibrio vulnificus*. *Infection and Immunity* 50, 922–924.

Xiao, P., Mo, Z.L., Mao, Y.X., Wang, C.L., Zou, Y.X. and Li, J. (2009) Detection of *Vibrio anguillarum* by PCR amplification of the *empA* gene. *Journal of Fish Diseases* 32, 293–296. https://doi.org/10.1111/j.1365-2761.2008.00984.x

Xu, H., Xing, J., Tang, X., Sheng, X. and Zhan, W. (2019) Intramuscular administration of a DNA vaccine encoding OmpK antigen induces humoral and cellular immune responses in flounder (*Paralichthys olivaceus*) and improves protection against *Vibrio anguillarum*. *Fish and Shellfish Immunology* 86, 618–626. https://doi.org/10.1016/j.fsi.2018.11.073

Yang, Z., Wang, X., Xu, W., Zhou, M., Zhang, Y. *et al.* (2018) Phosphorylation of PppA at threonine 253 controls T6SS2 expression and bacterial killing capacity in the marine pathogen *Vibrio alginolyticus*. *Microbiological Research* 209, 70–78. https://doi.org/10.1016/j.micres.2018.02.004

Yatip, P., Nitin Chandra Teja, D., Flegel, T.W. and Soowannayan, C. (2018) Extract from the fermented soybean product natto inhibits *Vibrio* biofilm formation and reduces shrimp mortality from *Vibrio harveyi* infection. *Fish and Shellfish Immunology* 72, 348–355. https://doi.org/10.1016/j.fsi.2017.11.008

Zhu, Z.M., Dong, C.F., Weng, S.P. and He, J.G. (2018) The high prevalence of pathogenic *Vibrio harveyi* with multiple antibiotic resistance in scale drop and muscle necrosis disease of the hybrid grouper, *Epinephelus fuscoguttatus* (♀) × *E. lanceolatus* (♂), in China. *Journal of Fish Diseases* 41, 589–601. https://doi.org/10.1111/jfd.12758

11 Aeromoniosis (*Aeromonas salmonicida*)

Margaret Crumlish and Brian Austin*

University of Stirling, Stirling, UK

11.1 Introduction

Aeromonas species have been associated with fish diseases, with pathogens loosely linked with the motile aeromonads (e.g. *Aeromonas hydrophila*) and the non-motile *Aeromonas salmonicida* (Table 11.1). Motile aeromonads are ubiquitous inhabitants of freshwater ecosystems and infect a wide range of (freshwater) fish with virtually a worldwide distribution, causing generalized septicaemias and fin/tail rot. However, it is often difficult to determine if these bacteria are primary or opportunistic pathogens (Austin and Austin, 2016). In contrast, *A. salmonicida* is recognized as a primary pathogen. It was originally associated with a haemorrhagic septicaemia, known as furunculosis, which was named because of boil-like lesions, termed furuncles, in the musculature of salmonids (Fig. 11.1; McCarthy, 1975). Infected older fish are lethargic, display slight exophthalmia, haemorrhaged fins, bloody discharges from the nares and vent, and haemorrhages particularly in the musculature. There may be haemorrhaging in the liver, swelling of the spleen and kidney necrosis (Fig. 11.2; Snieszko, 1958; McCarthy and Roberts, 1980). This form of furunculosis results typically in low levels of mortality, and the fish may survive (McCarthy, 1975). Actively growing fish and adults display generalized septicaemia with sudden onset in which disease signs also include melanosis, inappetence, lethargy, and small haemorrhages particularly at the base of the fins. Mortality is high, with death occurring 2–3 days after appearance of clinical signs. Co-infections have been established experimentally between *A. salmonicida* and largemouth bass virus leading to high mortalities in juvenile smallmouth bass (Boonthai *et al.*, 2018). The pathogen is regarded as an obligate fish pathogen as recovery occurs only from within infected specimens, but this may have reflected the lack of suitable methods rather than evidence of absence from habitats other than fish (Austin and Austin, 2016). Certainly, the organism may be in the blood and widely disseminated throughout the tissues in clinically diseased fish (McCarthy, 1975). Furunculosis is widespread wherever there is salmonid farming, namely in Europe, Japan and South and North America. The organism has recently been linked to human infections (Vincent *et al.*, 2018).

The organism was first described by Emmerich and Weibel (1894), who recovered it from diseased brown trout (*Salmo trutta*) in a German hatchery. The organism went through a series of names until it acquired its current name of *Aeromonas salmonicida* as a result of the detailed characterization by Griffin *et al.* (1953). They published the data necessary to formulate the species description. Thus, the 7th edition of *Bergey's Manual of Determinative Bacteriology* (published in 1957) included the organism as *A. salmonicida*. Isolates from salmonids are regarded as homogeneous and are classified as *A. salmonicida* subsp. *salmonicida*, which are also referred to as 'typical' *A. salmonicida*. The organism is characterized by its ability to produce brown diffusible melanin pigment around colonies (Fig. 11.3) when growing on protein-containing media, such as tryptone soy agar (TSA) at ≤22°C (Qiao *et al.*, 2019), and the dissociation into three distinct colony types, referred to as 'rough', 'smooth' and 'G-phase' (intermediate) colonies. These colony types have relevance to pathology,

*brian.austin@stir.ac.uk

Table 11.1. *Aeromonas* fish pathogens. (Modified from Austin and Austin, 2016.)

Pathogen	Name of disease	Host	Reported in
Aeromonas allosaccharophila	–	Elvers	Spain
Aeromonas bestiarum	–	Cyprinids	UK, USA
Aeromonas caviae	Septicaemia	Atlantic salmon (*Salmo salar*)	Turkey
Aeromonas dhakensis	Generalized septicaemia	Nile tilapia (*Oreochromis niloticus*)	Mexico
Aeromonas hydrophila	Haemorrhagic septicaemia, motile aeromonas septicaemia, redsore disease, fin rot	Many freshwater fish species	Wide distribution
Aeromonas jandaei	–	Eels (*Anguilla* spp.)	Spain
Aeromonas piscicola	–	?	Spain
Aeromonas salmonicida (subsp. *achromogenes*, *masoucida*, *salmonicida* and *smithia*) (= *Haemophilus piscium*)	Furunculosis, carp erythrodermatitis, ulcer disease	Salmonids, cyprinids and marine fish species	Wide distribution
Aeromonas sobria	–	Garra rufa (*Garra rufa*), perch (*Perca fluvialitis*), gizzard shad (*Dorosoma cepedianum*), stone loach (*Triplophysa siluroides*), tilapia	China, Slovakia, Switzerland, USA
Aeromonas schubertii	Septicaemia	Snakehead (*Ophiocephalus argus*)	China
Aeromonas veronii biovar sobria	Epizootic ulcerative syndrome, infectious dropsy	African catfish (*Clarias gariepinus*), rajputi (*Puntius gonionotus*), rui (*Labeo rohita*), catla (*Catla catla*), striped snakehead (*Channa striata*), oscar (*Astronotus ocellatus*)	Bangladesh, India
Aeromonas veronii biovar veronii		Chinese longsnout (*Leiocassis longirostris*)	China

Fig. 11.1. Burst furuncle on a rainbow trout revealing the underlying liquefying muscle, which is oozing over the surface. (Original photograph.)

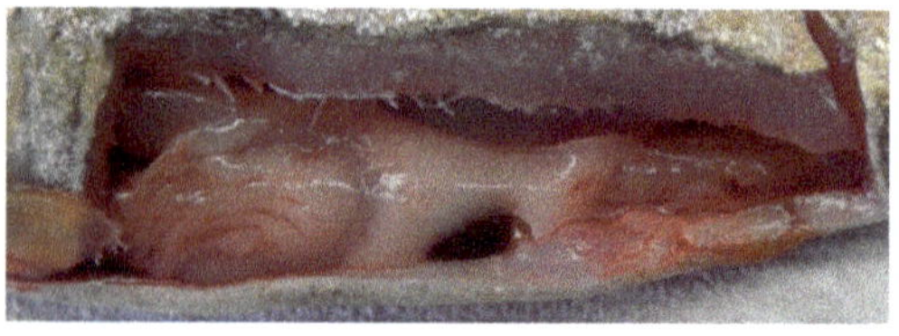

Fig. 11.2. Haemorrhagic septicaemia in a rainbow trout. Note the presence of bloody ascites. (Original photograph.)

with cells from the rough and smooth colonies being the most and least virulent, respectively. By electron microscopy, cells from the rough and smooth colony types have the presence or absence of an extracellular 49 kDa proteinaceous layer (= the A-layer), respectively. In addition to isolates

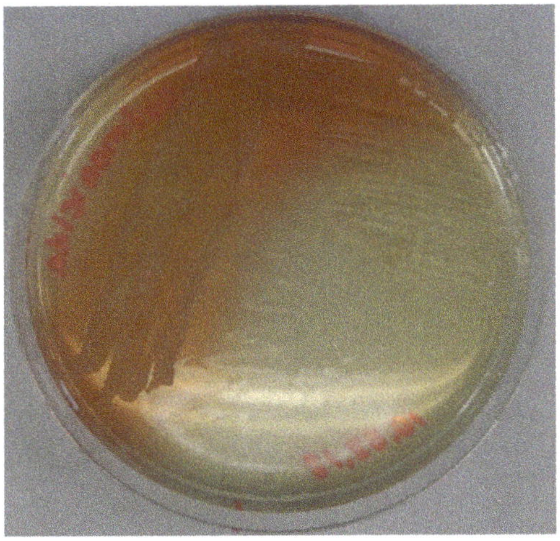

Fig. 11.3. *Aeromonas salmonicida* subsp. *salmonicida* on TSA after incubation at 18°C for 48 h. Note the brown diffusible pigment around the rough-looking colonies. (Original photograph.)

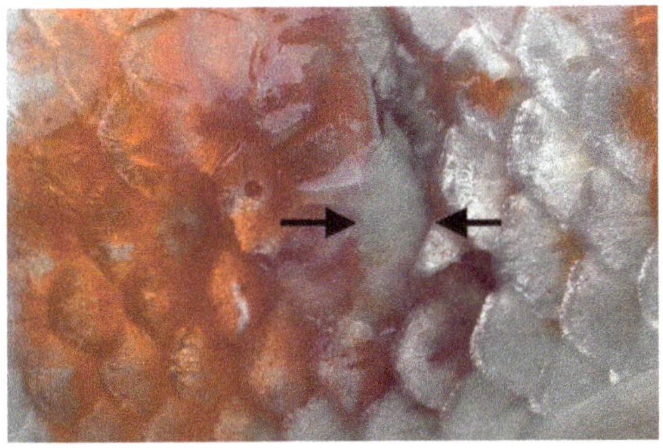

Fig. 11.4. Early stages of ulceration (arrowed) in koi carp. (Original photograph.)

Fig. 11.5. Ulcer (arrowed) in koi carp. (Original photograph.)

from salmonids, it was realized that the pathogen could be recovered from non-salmonids principally displaying ulcerations (Figs 11.4 and 11.5; e.g. Han *et al.*, 2011). All these manifestations in non-salmonids have been linked to other subspecies (*achromogenes*, *masoucida* and *smithia*) and to so-called 'atypical' isolates (Yamamoto, 2017).

Apart from the salmonid isolates that are recovered in subsp. *salmonicida*, some cultures may be classified in other subspecies of *A. salmonicida* (*achromogenes*, *masoucida* and *smithia*) but many others are unique and are not accommodated by the current subspecies descriptions; these are labelled as 'atypical'. In comparison to the typical counterparts, atypical isolates may show a lack of, weak or slow brown-pigment production (Koppang *et al.*, 2000), catalase and/or oxidase negativity (Kaku *et al.*, 1999), nutritional fastidiousness, namely for blood, serum or blood products (Austin and Austin, 2016), slow growth (Kaku *et al.*, 1999) and recovery from hosts apart from salmonids. Thus, isolates grouped in subsp. *achromogenes*, *masoucida* and *smithia* and the unclustered atypical isolates have been recovered from a wide range of fish species including cyprinids in freshwater habitats, particularly in Central and Eastern Europe (Austin and Austin, 2016), and in an increasing range of marine fish notably in the Baltic Sea and North Sea (Austin and Austin, 2016; Vercauteren *et al.*, 2018), including Atlantic lumpfish (*Cyclopterus lumpus*) (Rouleau *et al.*, 2018), common dab (*Limanda limanda*) (Vercauteren *et al.*, 2018), common wolffish (*Anarhichas lupus*) (Hellberg *et al.*, 1996), plaice and flounder (*Platichthys flesus*) (Wiklund and Dalsgaard, 1995), marbled sole (*Pleuronectes yokohamae*) (Kumagai *et al.*, 2006), sea lamprey (*Petromyzon marinus*) (Diamanka *et al.*, 2014), Senegalese sole (*Solea senegalensis*) (Silva *et al.*, 2016), turbot (*Scophthalmus maximus*) (Pedersen *et al.*, 1994) and wrasse (Treasurer and Cox, 1991).

There has been discussion over *Haemophilus piscium*, which is the causal agent of ulcer disease, with the species name coined by Snieszko *et al.* (1950). However, Kilian (1976) showed that the organism did not belong in the genus *Haemophilus* based on DNA, biochemical, serological and bacteriophage sensitivity data. Moreover, Austin *et al.* (1989) suggested that *H. piscium* should probably be grouped with *A. salmonicida*.

Infectious disease inevitably reflects the interaction of a susceptible host, the pathogen and a stressor, which may include unsuitable changes to the environment and poor hygiene leading to an accumulation of harmful microorganisms and chemicals, such as by-products of metabolism. Climate change involves enhanced greenhouse effects that trap more of the sun's energy (= heat) leading to warming of the environment, including aquatic systems, i.e. rivers, lakes, seas and oceans. Carbon dioxide, much of which is emissions from industry and agriculture, dissolves in the water leading to increasing acidification, and higher temperatures decrease the oxygen content in the water resulting in hypoxia. Global warming also changes precipitation and wind patterns and increases the melting of frozen structures such as glaciers and ice caps at the poles, leading to elevated sea levels, modified ocean currents and salinities, and altered food webs. Droughts and increasing extreme weather conditions, i.e. storms/floods, may occur. These are all likely stressors for fish and therefore impact on the development of disease. Simply put, stressed fish are more susceptible to disease. Wild fish would move to more favourable habitats; however, fish farmed in ponds and lakes would not be so fortunate. As *A. salmonicida* subsp. *salmonicida* prefers some of the abiotic factors associated with climate change, such as a preference for temperatures of >16°C (Austin and Austin, 2016), then the events associated with global warming are likely to favour this pathogen.

In summary, *A. salmonicida* is recognized to have greater biological potential than garnered from its initial association with furunculosis in salmonids. Not only has the host range expanded and will undoubtedly continue to expand to include both freshwater and marine fish species, but the pathogen also demonstrates the ability to adjust to changing environmental situations from increasing temperature, pH fluctuations to the salinities found in estuarine waters and seawater. With increasing stresses to potential hosts caused by climate change, this pathogen is ideally suited to benefit and is likely to become more troublesome in the future.

11.2 Diagnosis of the Pathogen/Disease

The traditional approach to diagnosis involved culturing the pathogen on Coomassie brilliant blue agar (CBB; Markwardt *et al.*, 1989), TSA or brain heart infusion agar (BHIA), when *A. salmonicida* subsp. *salmonicida* produces colonies after incubation at 25°C for 48 h. The pathogen may be identified from phenotypic tests, notably the Gram-stain (appears as small Gram-negative rods), motility (non-motile), fermentative metabolism, growth at 37°C (usually negative), arginine dihydrolase, catalase and oxidase production (all positive) but not ornithine decarboxylase, gluconate oxidation (negative), degradation of gelatin, starch (both positive) and urea (negative), and acid production from xylose (negative) (Wiklund *et al.*, 1992). Matrix-assisted laser desorption/ionization–time-of-flight (MALDI-TOF) mass spectrometry was regarded as effective at identifying *Aeromonas* isolates to the genus level but there were some issues with correctly equating to the species (Jung-Schroers *et al.*, 2018).

It is uncertain what proportion of cells are actually capable of forming colonies on laboratory media. Moreover, culturing was not suitable for the recovery of colony-forming units (cfu) from asymptomatic carrier fish and from the aquatic environment. Some of these drawbacks could be resolved by the developments in serology, some methods of which could be used directly on fish tissues without the need for culturing.

Serodiagnosis offered the possibility of rapid diagnoses directly from infected fish tissues. Initially, it was reported that whole-cell agglutination using polyclonal antisera was effective in recognizing smooth cells of *A. salmonicida* (Rabb *et al.*, 1964), but not the auto-agglutinating rough, virulent cells. The latex agglutination technique, in which antisera/antibodies were absorbed on to latex particles, proved effective for recognizing the pathogen and opened up the possibility of rapid, field-based diagnoses although sensitivity was an issue (McCarthy, 1975; Sakai *et al.*, 1986). The fluorescent antibody technique permitted the observation of bacterial cells directly in tissues and was considered better than culturing for the diagnosis of atypical *A. salmonicida* infections (Kawahara and Kusuda, 1987). Nevertheless, latex agglutination and co-agglutination (where antibodies are attached to bacterial cells) were determined to be more sensitive than the indirect fluorescent antibody technique. A subsequent development was the sensitive and specific enzyme-linked immunosorbent assay (ELISA) that really did open up the possibility of quick and reliable field-based diagnoses (Austin *et al.*, 1986); ELISA was more sensitive than culturing (Hiney *et al.*, 1994). A later refinement involved the use of polyclonal antibody-coated gold nanoparticles that were used in an immunoassay which enabled the specific, sensitive and rapid detection of *A. salmonicida* in tissues within 45 min (Saleh *et al.*, 2011).

Scientists were quick to embrace molecular tools for the diagnosis of *A. salmonicida* infections and the recognition of the pathogen in the natural environment (e.g. Høie *et al.*, 1997; Keeling *et al.*, 2013). The sensitivity was quickly established, with Hiney *et al.* (1992) reporting a polymerase chain reaction (PCR) that was capable of detecting ~2 cells of *A. salmonicida*. Høie *et al.* (1997) detected 10^3 and 10^4 cfu in 100 ml of kidney suspension by use of 16S rRNA and plasmid primers, respectively. It was realized that the PCR was capable of detecting the pathogen more so than culturing (O'Brien *et al.*, 1994). Developments continued:

- Terminal restriction fragment length polymorphism (RFLP) detected ~30 cfu/mg of artificially infected kidney tissue (Nilsson and Strom, 2002).
- PCR-RFLP, which targeted the *rpoD* gene, permitted the differentiation of *Aeromonas* cultures, including *A. salmonicida*. It was considered that the approach could be used for identification and epizootiology (Puah *et al.*, 2018).
- Multiplex PCR permitted the simultaneous detection (the detection limit was equivalent to 5.33×10^4 cfu/ml) of *A. salmonicida*, *Piscirickettsia salmonis*, *Streptococcus phocae* and *Vibrio anguillarum* (Tapia-Cammas *et al.*, 2011).
- Real-time PCR was highly specific and sensitive, detecting 5 fg of DNA, 2.2×10^4 cfu/g of kidney tissue without enrichment, and 40 cfu/g with enrichment (Keeling *et al.*, 2013). According to Fernandez-Alvarez *et al.* (2016), the sensitivity was reported as 1–2 bacterial cells and 6–60 bacteria per reaction for seeded kidney and blood. Quantitative real-time PCR was considered more sensitive than conventional PCR with a lower detection limit of 5.6 copies of the positive plasmids and was suggested for use in aquaculture (Du *et al.*, 2017). A multiplex real-time PCR permitted the early detection (the limit was 10^4 cfu/ml = 2×10^2 cfu/tube) of infection by *A. salmonicida*, *Tenacibaculum maritimum* and/or *V. anguillarum* in diseased fish and was considered again to be suitable for use in aquaculture (Chapela *et al.*, 2018).
- Nested PCR with species-specific primers rather than universal eubacterial primers improved detection from 1.4×10^4 cfu/reaction to <14 cfu/sample (Taylor and Winton, 2002).
- Reverse transcription-multiplex PCR enabled Rattanachaikunsopon and Phumkhachorn (2012) to detect 10 cfu in pure culture and 30 cfu in tissue, with the system distinguishing viable from non-viable cells, and typical from atypical *A. salmonicida*.
- Denaturing gradient gel electrophoresis (DGGE) of 16S rDNA was evaluated for the non-lethal detection of *A. salmonicida* from mucus leading to a distinctive and reproducible four-band pattern. Again, DGGE was more sensitive than culturing, recognizing 36/52 coho salmon mucus samples as positive for *A. salmonicida* compared with 31 positives by culturing (Quinn and Stevenson, 2012).
- Multilocus sequence typing (MLST) is used extensively for diagnosis of human diseases (e.g. Piao *et al.*, 2018) but the uptake for aquatic diseases has been slower. Although the technique has been used with *A. salmonicida*, the focus was on taxonomic and epizootiological studies. Carnahan *et al.* (2001) included *A. salmonicida* in an MLST examination of 94 aeromonad cultures. Subsequently, Beaz-Hidalgo *et al.* (2009) isolated four cultures from moribund fish that revealed an RFLP pattern similar to *A. salmonicida* and *Aeromonas bestiarum*. However, multilocus sequencing of housekeeping genes *rpoD*, *gyrB*, *recA* and *dnaJ* confirmed that the cultures belonged to an unknown genetic lineage (for which the name of *Aeromonas piscicola* was proposed) with *A. salmonicida* as one of the phylogenetically nearest neighbours. MLST, based on *dnaK*, *gltA*, *gyrB*, *radA*, *rpoB*, *tsf* and *zipA* genes, confirmed the current taxonomy in evolutionary terms of the 195 *Aeromonas* cultures studied, and in particular *A. salmonicida*. The fish pathogen formed a clonal complex (Roger *et al.*, 2012).

Since the development of serological and then molecular-based methods, diagnosis of disease and detection of pathogens have become highly specific and ultra-sensitive. Rapid field-based diagnoses have become achievable. However, when methods are capable of detecting only a few cells, questions must be raised about the meaning of 'positivity'. The presence of a potential pathogen does not necessarily infer a disease event but could reflect natural background populations of the organism.

In summary, diagnosis has progressed from the comparatively insensitive culture-dependent to the ultra-sensitive culture-independent methods, which are capable of detecting with a high degree of accuracy low numbers of cells. Thus, diagnostic techniques are keeping well abreast of potential changes in the biological potential of *A. salmonicida*, as well as those that may develop as a response to climate change. However, to overemphasize a previously made point, detection of low numbers of bacterial cells does not infer the occurrence of a clinical disease cycle.

11.3 Potential Spread of the Pathogen

The ecology of the organism is uncertain as isolation is usually only possible from clinically diseased fish. Consequently, the absence of colony-forming units has been inferred as evidence of absence. Therefore, it is generally accepted that the organism is restricted to fish and does not occur freely in the aquatic environment (Austin and Austin, 2016). Using the differential growth medium CBB, the pathogen was recovered from 56% of mucus samples and 6% of kidney tissue from salmonids (Cipriano *et al.*, 1992). Subsequently, *A. salmonicida* was cultured from 15% of gills and 19% of mucus samples, which pointed to external carriage on fish (Cipriano *et al.*, 1996). Indeed, the organism was found in higher numbers in mucus rather than the kidneys of six salmon (Cipriano *et al.*, 1996). Hiney *et al.* (1994) supported the view that the pathogen colonizes mucus and gills, and additionally fins and intestine, the latter of which may be the primary location of *A. salmonicida* in asymptomatic Atlantic salmon. A culture of *A. salmonicida* subsp. *masoucida*, AS-C4, was detected on the skin and in the gills and intestine of Atlantic salmon. Using quantitative real-time PCR right after immersion challenge, these sites were regarded as likely portals of entry; the pathogen was detected within 3 h in the blood and thereafter colonization occurred in the kidney, liver, muscle and spleen (Du *et al.*, 2018). This leads to the question about how the organism enters fish. The gills, skin/mucus, lateral line, mouth, anus or a surface injury have been identified as possible sites of entry into the host (e.g. Effendi and Austin, 1995b; Du *et al.*, 2018). Moreover, McCarthy (1980) reported that rainbow trout that resisted furunculosis subsequently died from furunculosis after their flanks were abraded with sandpaper. Translocation across the intestinal epithelia has been described (Jutfelt *et al.*, 2006). Thus, intestinal segments of rainbow trout were exposed for 90 min to isothiocyanate-labelled cells of virulent *A. salmonicida* leading to translocation.

For many years, contact with overtly and covertly (= carrier) infected fish or contaminated water and fish farm materials, and transovarian transmission were regarded as the most likely routes of infection albeit when presence or absence reflected the ability to culture the pathogen on laboratory media (McCarthy, 1980). Certainly, the international trade in fish eggs poses a risk of spreading the pathogen particularly if the health of the broodstock is not certified or disinfection of the eggs is not carried out or badly done. Carrier fish have been blamed for constituting a reservoir of the pathogen in fish populations. Carrier fish would develop clinical disease if the water temperature was increased from 5 to 18°C (Blake and Clark, 1931). Subsequently, Bullock and Stuckey (1975) reported that increasing the water temperature to 18°C and injecting corticosteroids would activate the carrier state with the development of furunculosis. If heat stress was used without corticosteroid, mortalities ensued but it was difficult to recover the pathogen on bacteriological media. Thus, it is expected that warmer temperatures, associated with climate change, would trigger carrier fish to develop clinical disease.

There is a risk associated with the transfer of infected salmon or trout from fresh water to seawater. Lund (1967) recovered the pathogen from 4/234 smolts in the River Coquet, England. When in the sea such fish could develop clinical disease and serve as a reservoir for transmission to other susceptible species. *A. salmonicida* causes mortalities in anadromous fish in seawater (e.g. Evelyn, 1971; Smith *et al.*, 1982). Smith *et al.* (1982) reported that mortalities in Atlantic salmon from two marine fish farms in Ireland were attributable to lateral transmission of the pathogen in seawater. These workers suggested that subsequent to stocking during spring 1978 and removal of carrier fish the following year, the pathogen became established and persisted in the fish farm for at least 6 months after the removal of the carrier fish.

There has been an association with a wider range of macroorganisms than those previously linked with the pathogen and this may suggest how the disease spreads. For example, sea lampreys have been found to harbour typical *A. salmonicida* and may well be a source of infection for salmonids (El Morabit *et al.*, 2004). Trash fish used in the production of feed have been implicated in the spread of furunculosis. Thus, Kim *et al.* (2013) deduced that outbreaks of disease in Korean rockfish (*Sebastes schlegelii*) were attributable to the fish used to prepare feed, i.e. big head croaker, Japanese anchovy and Konoshiro gizzard shad.

The problems with conventional culturing techniques that are not ideal for detecting *A. salmonicida* may have led to underestimation of its presence in the absence of clinical disease. For example, using the indirect fluorescent antibody technique (iFAT), epifluorescence microscopy and the direct viable count techniques (after Kogure *et al.*, 1979), Effendi and Austin (1994) observed that *A. salmonicida* remained in the environment after plate counts revealed the absence of colony-forming units. Thus, maximal survival of the pathogen was recorded in brackish water (i.e. salinity = 25‰), conditions on wood and in sediment rather than in water (Effendi and Austin, 1994). Furthermore, Wiklund (1995) reported that better survival occurred at 4°C than 15°C, but again in brackish rather than sea- or fresh water, and in the presence of particulates, specifically sand. This suggests that survival of the organism in the environment may be adversely influenced by climate change if waters become warmer. As the pathogen is facultatively anaerobic (= fermentative metabolism), then it is likely to do well in environments with reduced levels of oxygen and in more acidic conditions that could occur with global warming. Yet, it is unclear how long the pathogen may survive in the environment as studies have mostly relied on culturing methods, which have undoubtedly underestimated the longevity of viable cells (e.g. McCarthy, 1980). Using whole bacterial cells and DNA released into lake water microcosms, with culturing and PCR for detection, Deere *et al.* (1996a) recovered viable *A. salmonicida* cells for <4 weeks but determined that the DNA remained intact for >13 weeks. This raises the question about why DNA was found for over 2 months after culturing showed that the pathogen had effectively disappeared. An examination of sterile seeded river sediments led Michel and Dubois-Darnaudpeys (1980) to conclude that *A. salmonicida* could survive and multiply for over 10 months. Pathogenicity was retained for up to 8–9 months. These workers deduced that such a time period would enable aeromonad cells to be released from sediment into the overlying water, with bottom-feeding fish facilitating the direct contamination of fish, possibly allowing them to become carriers. Thus, sediments could provide a reservoir of infection. According to Sakai (1986), virulent cultures of *A. salmonicida* could survive for extended periods, i.e. >15 weeks, if dilute humic acid at 10 µg/ml was present. He reasoned that humic acid and amino acid–humic acid complexes were absorbed on to sand, which led to a build-up of amino acids on the surface of the sand particles. The electrostatic interrelationships led to the attachment and perhaps colonization/attachment of virulent *A. salmonicida*, resulting in their increased survival. Furthermore, virulent, agglutinating cells and avirulent non-agglutinating strains expressed net negative and positive charges, respectively. The suggestion was made that the negatively charged virulent cells were able to survive under starvation conditions in river sediments. These free-living cells may enter a dormant phase with viability declining because of lack of suitable nutrients. The suggestion was that free-living cells were transitional with a loss of viability eventually occurring (Sakai, 1986). In short, there was a recognition that culturability of *A. salmonicida* cells could reflect the physiological state of the cells, specifically the onset of the so-called non-culturable but viable (NCBV) state (Allen-Austin *et al.*, 1984). This was confirmed by Rose *et al.* (1990), who observed cells microscopically in water after colony counts reached zero. Also, Morgan *et al.* (1991) examined survival of the pathogen in lake water by using techniques which included cell culture, cell revival, epifluorescence microscopy, flow cytometry, membrane fatty acid analysis, plasmid maintenance and respiration. Cells became unculturable in sterile lake water after 8 days at 10°C when microscopic and flow cytometric methods revealed their continued presence. These uncultured cells possessed genomic and plasmid DNA, and RNA. Flow cytometry using rhodamine 123 indicated that cellular properties associated with viability were lost soon after culturability disappeared in distilled water but not so in lake water (Morgan *et al.*, 1993; Deere *et al.*, 1996b). The NCBV state could be postponed by the addition of 125 µM of arginine and methionine when *A. salmonicida* decreased in size and became rounder, but were still culturable (Pickup *et al.*, 1996). Also, the NCBV state occurred in seawater at 4°C (Ferguson *et al.*, 1995). However, the relevance of these NCBV cells to fish pathology may be questioned because of the lack of ability to establish any clinical disease with them (Stanley *et al.*, 2002). Effendi and Austin (1995a) examined the characteristics of the NCBV cells in a marine microcosm when it was observed that the cells decreased in size becoming coccoid in shape but retaining respiratory activity. There was not any alteration in the lipopolysaccharide (LPS) composition, but there was a reduction in 15, 17, 22, 30 and 70 kDa proteins and an increase in a 49 kDa protein. There was a loss of DNA. However, by means of the addition of nalidixic acid and yeast extract (after Kogure *et al.*, 1979), the cells developed large bizarre shapes, indicative of viability.

There is a current awareness of the widespread presence of microplastics in the aquatic – notably marine – environment (Virsek *et al.*, 2017). Moreover, there is concern that these particles could serve as vectors for the long-distance transfer of microorganisms. Thus, a study in the North Adriatic Sea sought DNA from microplastic particles using PCR amplification of the 16S rDNA with the result that *A. salmonicida* was one of 28 bacterial taxa identified. Consequently, there is a risk that the pathogen could be transferred over wider geographical areas on particulates, including microplastics, due to changes in oceanic currents resulting from climate change (Virsek *et al.*, 2017).

King and Shotts (1988) recovered *A. salmonicida* from the digestive tract of the ciliated protozoan, *Tetrahymena pyriformis*, determining that the pathogen had undergone limited multiplication, i.e. twofold. *A. salmonicida* was recovered from the liver and kidney and skin lesions in goldsinny (*Ctenolabrus rupestris*), rock cook (*Ctenolabrus exoletus*) and cuckoo wrasse (*Labrus mixtus*), which are used for the biological control of sea lice (*Lepeophtheirus salmonis*) in farmed Atlantic salmon (Treasurer and Cox, 1991). These data raised the concern that sea lice and/or wrasse could transmit the pathogen between salmonid populations. This concern was addressed by means of culturing and immunomagnetic beads coated with monoclonal antibodies to the LPS when *A. salmonicida* was found in low numbers, i.e. ~10 bacterial cells, in sea lice and in marine plankton (populations = 600 bacterial cells/g of homogenized plankton) (Nese and Enger, 1993). In addition, Frerichs *et al.* (1992) had recovered atypical *A. salmonicida* from healthy wild wrasse, although the isolates were not pathogenic to Atlantic salmon smolts.

In summary, the ecology of the pathogen is incomplete, and it remains difficult to locate *A. salmonicida* away from diseased fish in a form that is relevant to epizootiology and pathology. It would appear likely that cells are capable of spreading in the aquatic environment, but it is not proven that infection with these cells – if it occurs at all – leads to clinical disease. In terms of the risks associated with climate change, stressors including increased temperatures and acidification, hypoxic conditions and changes in salinity may well trigger outbreaks of clinical disease in fish that are harbouring *A. salmonicida*. The pathogen may adapt to the conditions associated with climate change including increasing temperatures and salinities; already cultures have been recognized to be capable of growing at 37°C, which is much higher than the normal growth range, and in seawater. The risk of increased spread could reflect changes in the movement of fish associated with environmental changes, including higher water temperatures, where fish are able to swim to more conducive habitats.

11.4 Potential Spread to New Geographical Areas

Historically, the geographical distribution of *A. salmonicida* has encompassed salmonids and cyprinids in fresh water, and subsequently appeared in marine fish species with some cultures having an obligate requirement for sodium chloride, i.e. marine characteristics (Austin and Austin, 2016). The spread is not well understood and will undoubtedly reflect the movement of healthy carriers and infected eggs. An unanswered questioned is whether or not salmonids farmed in seawater pose a risk to native marine or, for that matter, freshwater fish. Conversely, could atypical *A. salmonicida*, which is recovered with increasing frequency in wild fish, pose a threat to cultured salmonids? Certainly, the potential role of waterborne transmission is unclear. However, it is clear that *A. salmonicida* becomes metabolically more active as temperatures increase beyond 15°C. For example, a 70 kDa serine protease was produced in far higher quantities at high temperatures, i.e. 25°C, compared with 10°C (Fyfe *et al.*, 1987). The corollary is that as temperatures increase, cells may become attenuated/non-virulent (Ishiguro *et al.*, 1981).

Thus, *A. salmonicida* may be recovered from new geographical areas. Isolates from a range of healthy wild and farmed fish and invertebrate species, including red spot emperor, king soldier bream, white-spotted rabbit fish, tilapia and abalone, in Oman during 2011–2012 were studied. Some virulence-associated genes were found widely in the isolates, and they included the cytotoxic enterotoxin *ast* gene and aerolysin-like protein (*act*), but there was not any evidence of disease or of the ability to cause disease in laboratory-based pathogenicity experiments. It remains speculative if *A. salmonicida* causes a realistic risk to fish in this geographical location if situations change to favour the growth of the pathogen, expression of virulence factors and the availability of susceptible fish species (Alghabshi *et al.*, 2018).

In summary, it may be anticipated that spread of the pathogen will reflect the transportation of infected undisinfected eggs and asymptomatic carriers, and possibly the movement of the bacterial cells on water currents, on particulates and invertebrates/vertebrates. Also, wild populations may well move to new locations according to the changing environmental conditions associated with climate change and transfer the pathogen accordingly. Such movement could constitute reservoirs of infection to other resident fish species. Overall, there is a knowledge gap concerning the actual spread of the pathogen between remote fish populations. The level of risk associated with waterborne transfer needs elucidation; for example, could the pathogen actually be transferred in water or on inanimate or animate particulates, invertebrates or vertebrates? It is unknown if the pathogen is already in habitats not associated with clinical disease and is merely awaiting appropriate conditions and the availability of susceptible hosts to trigger infections.

11.5 Population Dynamics of Intermediate Host and Fish

The dogma has been that *A. salmonicida* is a fish-only organism that spends its existence in the host (see Austin and Austin, 2016). Then came the realization that the organism could exist in other animal species that were not considered as the primary hosts, but was *Aeromonas* merely a contaminant/chance invader or a resident? For example, *A. salmonicida* was recovered from sablefish (*Anoplopoma fimbria*; Evelyn, 1971) and was identified in wrasse (Treasurer and Cox, 1991), which function for biological control of sea lice populations in Atlantic salmon. The goldsinny, cuckoo and rock cook wrasse developed clinical furunculosis, from which the pathogen was recovered from kidney and liver (Treasurer and Cox, 1991). Was the wrasse a natural host, or had the pathogen been transmitted from Atlantic salmon or sea lice? Could sea lice be a host or a vector? Using immunomagnetic beads coated with monoclonal antibodies to LPS with culturing methods, the pathogen was recovered from sea lice and marine plankton (Nese and Enger, 1993). Also, atypical *A. salmonicida* was recovered from healthy wild wrasse captured in the open sea, although there was no evidence of pathogenicity to Atlantic salmon smolts (Frerichs *et al.*, 1992). The ability of *A. salmonicida* to multiply in invertebrates was determined by King and Shotts (1988), who reported twofold multiplication in the digestive tract of the ciliated protozoan, *T. pyriformis*. However, it is unclear if sablefish, wrasse or invertebrates could be considered as alternative or intermediate hosts or whether the presence of the pathogen reflected ingestion of contaminated tissue/lateral transmission (Klontz and Wood, 1972). Certainly, lateral transmission was considered to be responsible for furunculosis developing and persisting in two marine Atlantic salmon farms in Ireland (Smith *et al.*, 1982). If the pathogen becomes established in feral fish in the vicinity of cages or in the underlying sediment, then these sources will serve as a constant threat to aquaculture.

Could more aggressive forms of the pathogen develop as a result of environmental pressures associated with climate change? There is evidence of varying levels of pathogenicity among existing strains, but this is the norm in microbial pathology. During 1992, there was a substantial dip in production of Atlantic salmon in Scotland due to furunculosis with subsequent recovery in harvest levels due to the availability of improved disease control, namely effective commercial vaccines (Munro and Gauld, 1996). *A. salmonicida* subsp. *masoucida* has caused substantial economic losses to salmonid production in China (Du *et al.*, 2018). Was the pathogen already present in China or transferred with the importation of salmonid stocks from other countries? The situation could be repeated if the conditions become favourable to the pathogen at the expense of the host; increasing water temperature, changing salinities and hypoxia are among the predisposing factors (Austin and Austin, 2016).

In summary, the role of *A. salmonicida* in non-salmonid fish and invertebrates is unclear, but these organisms undoubtedly represent a potential reservoir of infection. As seen in China and Scotland, the pathogen is capable of impacting salmon production. The problems may well become exacerbated if fish become further disadvantaged as a result of impacts due to climate change.

11.6 Increase in Pathogen Mortality Due to Environmental Change

Could the biology of *A. salmonicida* offer an insight into possible effects of climate change? The answer is complicated because of the diversity of pathogenicity factors that have been described among isolates. However, to date, the dogma has been that in terms of culturability *A. salmonicida* prefers lower temperatures with maxima of ~25–30°C; thus, there has not been much incentive to check the ability to grow at >30°C (Austin and Austin, 2016). As a few isolates can grow at 37°C (McIntosh and Austin, 1991), there is even a link to human infections (Vincent *et al.*, 2018). An interpretation of the data is that the organism is more adaptable than the current species definition. Certainly, some pathogenicity traits appear to be adversely influenced by increasing temperature (to 30°C). For example, tenfold more of an unidentified extracellularly secreted 100 kDa protein was produced at lower temperatures (Fyfe *et al.*, 1987); and haemolysin was relatively heat labile (Nomura and Saito, 1982). This could infer that the standard isolates would not fare well with increasing water temperatures. Conversely, expression of the type III secretion system in one isolate, A449, was definitely temperature dependent, being active within 30 min at 28°C especially when followed by exposure to low levels of calcium, but not 17°C. However, expression was induced at 16°C in the presence of 0.19 to 0.38 M NaCl (Ebanks *et al.*, 2006). Moreover, the 70 kDa serine protease was produced in greater amounts after incubation for 18 h at 25°C compared with 125 h at 10°C (Fyfe *et al.*, 1987). It may be anticipated that the resilience of the pathogen will permit its modification to suit the prevailing conditions, including increasing environmental temperatures. Together with likely increasing stress on the host fish, it is predicted that *A. salmonicida* will remain a serious pathogen in times of increasing environmental stress.

In summary, there are aggressive cultures of *A. salmonicida* that maintain virulence even after prolonged periods in the laboratory. More than likely, the pathogen will respond to the challenges of climate change with the emergence of new virulent strains with the ability to cause disease in hosts that may be compromised as a result of environmental stressors; these could include combinations of increasing temperature, changes in salinity and/or hypoxia. Already, it is recognized that the pathogen has diverse biological activity and is capable of infecting a wide range of freshwater and marine fish. It is predicted that the organism will spread to new locations, and affect an ever-increasing range of fish species, particularly if these are badly stressed and therefore vulnerable to infection. Undoubtedly, *A. salmonicida* will continue to be a serious fish pathogen with the events associated with climate change.

11.7 Control and Prevention

Chemotherapy will continue to be discouraged amid concerns over the development and spread of drug resistance, and residues in tissues. Prophylaxis will continue to be highlighted for disease control. Here, it is envisaged that non-specific approaches, e.g. non-specific immunostimulants, will be more successful than their specific counterparts, namely vaccines. The reason is that with weakening due to increasing stress resulting from climate change, fish will be vulnerable to disease. Whereas protection may result from the strains included in the vaccine, weakened/vulnerable fish would be likely to succumb to other pathogens, including new or different strains of *A. salmonicida*. Where immunostimulation results in herd immunity, the overall health of the fish is likely to be improved.

11.7.1 Antimicrobial compounds

Since the end of World War II until the 1970s, chemicals were the major if not the only means used to control disease. Sulfonamides, notably sulfamerazine, are effective for chemotherapy at a daily dose of 22 g/100 kg of fish (Snieszko, 1958). The increased value of potentiated sulfonamides was established widely (McCarthy *et al.*, 1974). Interest continued to the antibiotics, including chloramphenicol, furazolidone, oxytetracycline and polymyxin (Snieszko, 1958; McCashion and Lynch, 1987), and to synthetic compounds, notably the quinolone flumequine (Michel *et al.*, 1980) and oxolinic acid (Endo *et al.*, 1973; Barnes *et al.*, 1991a). Subsequently, the value of newer compounds has been documented including amoxicillin (Barnes *et al.*, 1994) and florfenicol (Samuelsen *et al.*, 1998). Attention has been directed towards 4-quinolones/fluoroquinolones, which have inhibitory activity against *A. salmonicida* (Barnes *et al.*, 1991b; Elston *et al.*, 1995). The outcome of the research led to difloxacin, enrofloxacin and sarofloxacin that were more effective than oxolinic acid, in terms of minimal inhibitory concentration (MIC) (e.g. Elston *et al.*, 1995). Enrofloxacin, which was dosed orally daily at 10 mg/kg body weight of fish for 10 days, was used successfully in field trials with lake trout (*Salvelinus namaycush*) (Hsu *et al.*, 1995). Enrofloxacin has demonstrated effectiveness in controlling atypical *A. salmonicida* in tomcod, with a single injection with 5 mg enrofloxacin/kg fish halting the development of disease (Williams *et al.*, 1997). Alas, use of many antimicrobial compounds has led to the emergence and spread of drug-resistant strains (Barnes *et al.*, 1994; Samuelsen *et al.*, 1998) including plasmid-mediated resistance for antibiotics (Mitoma *et al.*, 1984; Hedges *et al.*, 1985). Du *et al.* (2019) monitored the real-time evolution of a clone of *A. salmonicida* in response to chemotherapy and determined its responsibility for a 4-year outbreak of furunculosis in a recirculating Atlantic salmon farm in China. The resistance profile of the clone provided by acquired mobile genetic elements closely reflected the antibiotics used for chemotherapy. Also, there are concerns about the fate of bioactive residues in fish tissues (McCarthy and Roberts, 1980) and the aquatic environment. Such concerns negate the value of the compounds for disease control in aquaculture. The caveat to the use of chemotherapeutants is that a suitable period of time must lapse, following the conclusion of treatment, before the fish may be sold for human consumption. This should allow for the purging from the fish of all traces of the active compound and the metabolites. It is worth remembering the opinions of Snieszko (1958), who wrote that chemotherapy should only be a stopgap measure until the sources of infection by *A. salmonicida* could be eliminated, or disease-resistant strains of fish developed.

A novel approach involved the use of rainbow trout that were exposed to silver nanoparticles (100 µg/l) by immersion for 3 h or by intraperitoneal (IP) injection (17 µg/ml) followed by challenge with *A. salmonicida*. The outcome was the absence of clinical disease and the inability to recover the pathogen after 35 days (Shaalan *et al.*, 2018). In addition, *trans*-cinnamic acid, which is a naturally occurring aromatic acid, was moderately inhibitory to *A. salmonicida* albeit *in vitro* (Yilmaz *et al.*, 2018).

11.7.2 Disease-resistant fish

There has been some interest in the concept of genetically disease-resistant fish starting with the pioneering work of Embody and Hayford (1925), who used selective breeding to increase resistance of brook trout to furunculosis. Efforts continued with Wolf (1954), who attempted to develop furunculosis- and ulcer disease-resistant brook trout and brown trout, and Snieszko *et al.* (1959), who concluded disease resistance was genetically determined. This was followed by Ehlinger (1977), who showed furunculosis resistance in the progeny of brook trout. Cipriano (1983) described varying levels of resistance, which was correlated with serum neutralization titre, to furunculosis in 11 strains of rainbow trout. The McConnaughy strain was the most susceptible, with 83% of the fish dying within 14 days of challenge by a 1 min bath in 1.2×10^9 cells of *A. salmonicida*. The serum neutralization titre was 1:80 against one of the extracellular fractions of the pathogen. Conversely, there was no mortality in the Wytheville strain, which had a serum neutralization titre of 1:2560. Cipriano (1983) highlighted that serum (which enabled the neutralization of toxic components of the pathogen) from naturally resistant rainbow trout could protect brook trout by passive immunization from challenge with virulent cells. However, administration of serum from susceptible Atlantic salmon did not confer resistance to brook trout. A decade later, Gjedrem and Gjoen (1995) described genetic variation in the susceptibility to furunculosis of 1-year-old Atlantic salmon. Then, Rodriguez-Ramilo *et al.* (2011) indicated differential resistance of four turbot families to *A. salmonicida* and noted the significance of the research for breeding disease-resistant fish. The theme has relevance for the future of disease control strategies in aquaculture, although a genetic engineering route would likely be an anathema in some countries.

11.7.3 Vaccine development

A. salmonicida was the first fish pathogen to be targeted for vaccine development, starting with the work of Duff (1942), who developed a chloroform-inactivated whole-cell preparation, which was administered orally. However, with the discovery of antibiotics, chemotherapy became the principal means of disease control. Antibiotic resistance developed quickly, leading to a resurgence of interest in vaccines with formulations including: inactivated whole cells; iron-regulated outer membrane proteins (IROMP) of *A. salmonicida* with effectiveness being concentration dependent (Marana *et al.*, 2017a); subcellular components/soluble extracts; and genetically engineered live attenuated cells (including cells without the A-layer and O-antigen; Thornton *et al.*, 1991, 1994) of *A. salmonicida* (Marana *et al.*, 2017b) or other taxa, notably *Aliivibrio salmonicida* (Norqvist *et al.*, 1989) or *A. hydrophila* (Vivas *et al.*, 2004), which have been administered by IP injection (with or without adjuvant;

Midtlyng *et al.*, 1996; Noor *et al.*, 2017), bathing (Villumsen and Raida, 2013) with or without ultrasound (Navot *et al.*, 2011), or via food (e.g. Irie *et al.*, 2005). Of these, injection is often regarded as the most successful in protection, whereas the oral route is often not as protective (Midtlyng *et al.*, 1996). The problems with oral vaccines may be that the antigens become degraded during passage through the stomach. To overcome the possibility of antigenic degradation, liposome-entrapped antigens of atypical *A. salmonicida* were fed to carp with the aim of controlling ulcer disease with the result that there were fewer mortalities, a reduced incidence of ulceration and stimulation of the immune response with antibodies recorded in bile, intestinal mucus and serum (Irie *et al.*, 2005). Ultrasound has been used to administer vaccines to goldfish when soluble A-layer protein was applied by immersion (100 µg A-protein/ml for 10 min) after ultrasound (1 MHz frequency of ultrasound/ 1 min) pre-treatment, and led to promising results against challenge (Navot *et al.*, 2011). Also, low-frequency sonophoresis at 37 kHz has been used successfully to increase the uptake of antigens across rainbow trout skin (Cobo *et al.*, 2014).

The water temperature during vaccination may be crucial to the development of protection. With Atlantic lumpfish (*C. lumpus*), vaccination at low temperatures, i.e. 5°C compared with 10 and 15°C, led to a lower antibody response (Erkinharju *et al.*, 2018). Therefore, the increased temperatures associated with global warming could have a beneficial effect of stimulating the development of protection. The caveat is that higher temperatures lead to increased stress and weakening of the fish, leading it to be more susceptible to diseases.

Adjuvants are beneficial in injectable vaccines (Midtlyng, 1996) with 16S rRNA and LPS being detected in the head kidney and spleen at 2 weeks (and in the head kidney at 12 weeks) after injection with a commercial oil-adjuvanted, formalin-inactivated vaccine (Grove *et al.*, 2003). Certainly, Midtlyng (1996) determined that IP administration of furunculosis vaccine in a mineral oil adjuvant gave the best protection in Atlantic salmon. Apart from Freund's complete adjuvant (FCA; Olivier *et al.*, 1985b), the use of β-1,3-glucan (Vita-Stim-Taito), lentinan and formalin-killed cells of *Renibacterium salmoninarum* have enhanced the effectiveness of formalized whole-cell vaccines (Nikl *et al.*, 1991).

Products may be monovalent, i.e. just based on *A. salmonicida*, or bi- or polyvalent in which case other antigens are present, such as *Aliivibrio salmonicida* (Hoel *et al.*, 1998) and/or *V. anguillarum* (Marana *et al.*, 2017a). Vaccine components are taken up into the body of the fish, via the head kidney and spleen (Høie *et al.*, 1996). It is not known if there is cross- protection between typical and atypical *A. salmonicida*; the benefit of autogenous products has been documented (Gudmundsdóttir and Gudmundsdóttir, 1997). Severe side-effects have been reported following IP injection of oil-adjuvanted vaccines (Gudmundsdóttir *et al.*, 2003; Villumsen *et al.*, 2015), in which the extracellular products (ECP) component contributed to an inflammatory response (Mutoloki *et al.*, 2006). Also, an initial and temporary immunosuppressive effect has been described, with the problem resolved following the addition of antibiotics, notably amoxicillin (Inglis *et al.*, 1996). Initially, vaccines were not noted for their success insofar as protection was not especially good except for passive immunization and attenuated live preparations (Cipriano and Starliper, 1982; Ellis *et al.*, 1988; Vaughan *et al.*, 1993).

Various antigens have been identified as conferring immunoprotection against challenge with virulent isolates. Use of the 28 kDa outer membrane pore-forming porin led to protection in rainbow trout (Lutwyche *et al.*, 1995). Furthermore, an inactivated whole-cell vaccine based on the metalloendopeptidase AsaP1(Y309F)-toxoid mutant protected Arctic char (Schwenteit *et al.*, 2015). Similarly, fish have responded to LPS O-antigen and protease components of the ECP (Hastings and Ellis, 1985; Ellis *et al.*, 1988). However, some workers have been sceptical about whether vaccines based on virulent cells expressing the A-layer protein are better or worse than avirulent cells without A-layer at conferring protection (McCarthy *et al.*, 1983; Olivier *et al.*, 1985a). The A-layer protein is an important protective antigen in non-oily Montanide-adjuvanted injectable whole-cell inactivated vaccines (relative percentage survival (RPS) = 51–78%). However, there was no correlation between protection and humoral antibody response (Lund *et al.*, 2003). Moreover, Lund *et al.* (2008a,b) discussed the importance of A-layer in vaccine preparations designed to protect Atlantic cod against atypical isolates. Yet with atypical *A. salmonicida* and Atlantic cod, there was a correlation reported between protection and cross-reacting LPS-specific antibodies (Lund *et al.*, 2008a,b). Furthermore, other researchers have linked protection with a humoral antibody response (e.g. Romstad *et al.*,

2013; Ronneseth *et al.*, 2017). Perhaps it is timely to consider other aspects of the immune response, notably cell-mediated immunity, which has been suggested previously by McCarthy and Roberts (1980).

There is an interest to use vaccines developed for salmonids with cyprinids and marine fish species. The ongoing dilemma is whether vaccines to *A. salmonicida* subsp. *salmonicida* are applicable for controlling disease caused by atypical isolates. A commercial polyvalent product for salmon failed to protect turbot from experimental challenge with *A. salmonicida* subsp. *achromogenes* (Björnsdóttir *et al.*, 2005), although Santos *et al.* (2005) achieved better success with turbot. Here, the commercial furunculosis vaccine and an autogenous vaccine resulted in RPS of 72–99% when challenged 120 days after IP vaccination. Reasonable protection, i.e. RPS = 50–52%, was recorded after 6 months. Conversely, immersion vaccination did not lead to significant protection, nor did an oral booster dose improve survival (Santos *et al.*, 2005). Furthermore, the efficiency of vaccination in disease management has been demonstrated with regard to the positive effect on antibiotic resistance (Du *et al.*, 2019). Certainly, vaccines are important as a primary means of disease control, but their specificity may work against them if the recipient fish is otherwise stressed by climate change and therefore vulnerable to infection.

11.7.4 Non-specific immunostimulants

Apart from vaccines, a wide range of compounds have been reported to stimulate immunity and enhance resistance to *A. salmonicida*. The list includes Baypamum (Ortega *et al.*, 1996), dimerized lysozyme (Siwicki *et al.*, 1998), β-1,3-glucan (Ji *et al.*, 2017; Libran-Perez *et al.*, 2018) and synthetic peptides (Kitao and Yoshida, 1986). For example, rainbow trout fed for 42 days with food supplemented with 0.1 and 0.2% β-glucan improved growth and weight gain, leading to significantly enhanced survival after challenge. Furthermore, there was a stimulation of catalase, lysozyme, peroxidase and superoxide dismutase activities, but lower serum glutamic oxalacetic transaminase and glutamic pyruvic transaminase levels (Ji *et al.*, 2017). Also, food supplemented with natural plant products, notably 0.5–1.0% (w/w) of garlic (*Allium sativum*), was successful in controlling the pathogen (Breyer *et al.*, 2015).

Pure bacterial cultures, including *Bacillus licheniformis*, *Bacillus subtilis*, *Bacillus velezensis*, *Carnobacterium* and *Vibrio alginolyticus*, are immunostimulatory and confer protection when administered orally (e.g. Irianto and Austin, 2002; Park *et al.*, 2017; Yi *et al.*, 2018). Probiotics normally lead to weight gain and enhanced cellular and innate immune parameters, including lysozyme, phagocytic and superoxide dismutase activities (Irianto and Austin, 2002; Park *et al.*, 2017).

We believe herd immunity may have better success in the long term for controlling diseases in fish that have been compromised by climate change.

11.7.5 Biological control

A Myoviridae bacteriophage, from rainbow trout farm sediment in Korea, lysed *A. salmonicida* (Kim *et al.*, 2012). Similarly, bacteriophage PAS-1 and AS-A protected rainbow trout and Senegalese sole (*S. senegalensis*), respectively, against mortalities caused by the pathogen (Kim *et al.*, 2015). In artificially infected Senegalese sole, bacteriophage AS-A inhibited the multiplication of the pathogen after 6 h, and there was no mortality at 72 h compared with 36% deaths in control fish (Silva *et al.*, 2016). Subsequently, two better bacteriophage cultures of the Myoviridae were recovered, one of which, AS-D, led to a higher rate of pathogen reduction and less resistance. Cocktails including the two new bacteriophage cultures led to better control of the pathogen in less time with reduced development of resistant mutants (Duarte *et al.*, 2018). Moreover, five T4 bacteriophages of the Caudovirales with heterogeneous lytic capacities to *A. salmonicida* were used as cocktails, with some demonstrating significantly higher inhibitory activity than others or individual bacteriophage preparations *in vitro* (Chen *et al.*, 2018).

Oysters (*Crassostrea gigas*) have been used as an eco-friendly method to control populations of *A. salmonicida* in wastewater from Atlantic salmon farms. Using green fluorescent protein tags, the pathogen removal efficiency and ingestion rate were studied in oyster larvae and adults. There were substantial differences, as larvae removed 88–95% of *A. salmonicida* compared with 79–92% removed by adult oysters (Ma *et al.*, 2017). Given the changes in aquaculture practices and the reintroduction to integrated farming systems, a fish/invertebrate farm system could be a conceivable development.

11.7.6 Novel approaches

Common carp (*Cyprinus carpio*) exposed to humic-rich compounds had significant reduction in infection rates when challenged with atypical *A. salmonicida*, with a commensurate rise in antibody titre but not innate immune parameters. Moreover, the pathogen did not grow as well in liquid medium supplemented with humic-rich water, Leonardite humic-rich extract or synthetic humic acid (Yamin *et al.*, 2017).

Research into developing effective disease control strategies will continue, with emphasis on genetic resistance to diseases and (non-specific) immunomodulation.

In summary, some of the current approaches to disease control were initiated for *A. salmonicida*, including the first fish vaccine. The momentum has continued with research morphing from therapeutic (e.g. antibiotics) to prophylactic measures. Non-specific immunostimulation may have better success than vaccines for controlling disease in fish that have been compromised by climate change. The impetus will continue into novel areas, including genetics and biological control, reflecting the need for effective disease control in the rapidly developing aquaculture industry.

11.8 Conclusions and Suggestions for Further Study

The effects anticipated with climate change, namely temperature increases, salinity and pH (= acidification) changes, hypoxia and modifications in oceanic currents, are conducive to the survival and spread of *A. salmonicida*. The pathogen has been found already in habitats not previously associated with furunculosis or ulcerative conditions (e.g. Oman; Alghabshi *et al.*, 2018). However, it is not known if this represents the spread of the organism to new geographical areas or if the organism is part of the normal microflora in the region. Clearly, environmental shifts associated with climate change are stressors that will weaken fish, leaving them more susceptible to infection. Certainly, adaptations in farming practices due to climate change may also lead to alterations in incidence of some diseases due to warming conditions, which are not suitable to some pathogens such as *Flavobacterium* spp., whereas *A. salmonicida* not only does better under warmer conditions but may also change in virulence and host species. Therefore, increases in the incidence and severity of diseases are predicted; *A. salmonicida* not only survives in freshwater and marine habitats, it causes large-scale fish mortalities. It is likely the pathogen will demonstrate increasing biological potential and will cause more severe disease in an increasing range of fish species, especially those that are stressed. Could *A. salmonicida* act in concert synergistically with other pathogens and/or parasites to exacerbate disease conditions? Time will tell, and fish disease specialists need to be aware of this possibility.

In aquaculture, non-specific immunomodulatory agents are likely to have better success in combating diseases than narrowly focused vaccines that will protect against the strains within the products but leave weakened/stressed fish liable to infection by other strains of the same pathogen or different diseases. Combinations of specific and non-specific immunological products may be the appropriate way forward in fish production.

Certainly, climate change is already impacting the development and spread of furunculosis with the disease now appearing in the Ouje-Bougoumou region of northern Quebec, Canada. In particular, regression analysis demonstrated a significant, positive temporal trend in average air temperature, the range of which is conducive to the survival of *A. salmonicida*. Indeed, furunculosis occurred initially in this temperature range, and was considered to likely persist throughout the 21st century. Although climate change was one of the factors causing furunculosis in the Ouje-Bougoumou region, other stressors such as the effects of previous mining activities and the release of contaminants that may have adversely affected the host's immune system may be contributing factors (Tam *et al.*, 2011). Elevated temperatures (e.g. 19–20°C rather than 12°C) in the presence of the organophosphate pesticide malathion led to enhanced mortalities of chinook salmon following challenge with *A. salmonicida* (Dietrich *et al.*, 2014). In addition, movement/introduction of alien fish species led to increased risk of spreading pathogens, including *A. salmonicida*, to farmed and wild fish in Swedish lakes and rivers. These problems were predicted to increase with disturbances in the ecosystems and climate change (Josefsson, 1999).

Could climate change impact fish immunity? It has been observed that a moderate rise in water temperatures as experienced during spring and summer, i.e. from 10 to 16°C, accelerated the spleen transcriptome response to the IP injection of polyriboinosinic polyribocytidylic acid (albeit a viral mimic) in Atlantic cod (*Gadus morhua*) reared in sea cages in Newfoundland, Canada (Hori *et al.*, 2012). The response of Atlantic cod to IP injection with formalized *A. salmonicida* cells at moderately elevated temperatures had less effect on the fish's spleen transcriptome response (= antibacterial response) to the bacterial pathogen compared with the antiviral response to polyriboinosinic polyribocytidylic acid. This suggests that the impact of higher temperatures on the Atlantic cod's immune response may well be pathogen dependent (Hori *et al.*, 2013). Furthermore, there was a positive correlation between water temperature and the level of mucosal antibodies in olive flounder (*Paralichthys olivaceus*). In contrast, there was an inverse relationship between the titre of mucosal antibodies and the activity of mucosal haemagglutinin and proteases. Moreover, there was not any relationship between lysozyme activity and other innate immune parameters in olive flounder (Jung *et al.*, 2012). These relationships may be compensation for the fish to protect itself against pathogens and could indicate possible changes with increasing water temperatures as would occur with climate change. Therefore, more profound modifications in the host's immune system could be anticipated with climate change leading to even higher temperatures. However, fish will seek to increase their body temperature by moving to warmer water in response to infection, i.e. a behavioural fever response which limits the development of the pathogen leading to mitigation of the disease (Rakus *et al.*, 2017; Rey *et al.*, 2017). In this situation, the increasing water temperatures associated with climate change could have a positive impact on fish health.

Even immunoprophylaxis is not without problems as, in north Spanish salmonid hatcheries, there has been a change in the occurrence of disease following vaccination and increasing temperatures. Thus, the fish have been protected against the antigens in the preparation, but succumb to different diseases instead (Marquez *et al.*, 2014).

When fish become stressed, such as with climate change, and weakened by pathogens, e.g. *A. salmonicida*, this instigates clinical disease cycles. Unfortunately, there is evidence that if efforts are made to control some specific pathogens, such as by vaccination, but the hosts are still in a weakened/stressed state, other pathogens not associated with the vaccine will become involved to establish clinical disease. This situation appears to have happened already in the case of *Yersinia ruckeri* in England, whereby rainbow trout were vaccinated against classical strains, but a new and aggressive biotype became established in the weakened stock (Austin and Austin, 2016). How long will it take for the scenario to be repeated with *A. salmonicida*?

In summary, *A. salmonicida* is one of the best studied fish pathogens that is recognized as a major constraint on farmed salmonids and cyprinids in fresh water. Moreover, the disease has been recognized in wild marine fish and salmonids in sea cages. The pathogen has diverse biological potential; fresh water-tolerating and salt water-tolerating, psychrophilic and mesophilic cultures have been recovered. Therefore, the organism is likely to adapt to the environmental conditions resulting from climate change, especially as the prospective hosts become even more stressed. This could lead to the heightened severity of disease with increased numbers of epidemics.

11.8.1 Suggestions for further work

- Fundamental gaps in our knowledge of *A. salmonicida* include the precise location and function of cells within carrier fish. Are they intracellular in a form analogous to mitochondria or stable L-forms? Certainly, the role of these bacteria is unclear as they form an obvious reservoir of infection when the host becomes stressed, such as by climate change, allowing multiplication of the pathogen and progression to a pathogenic state.
- How widespread is *A. salmonicida* in the aquatic environment? It is unclear if the pathogen occurs as 'free-living' cells in the aquatic environment when there are no fish or if the spread of the disease reflects the movement of fish populations. Could the organism be spread in ocean currents or in rivers?
- Do cells exist in an active state in the aquatic environment away from fish? It is unknown if these cells are active metabolically and actually pose a realistic threat to fish. The cells could be inactive, dormant or senescent. Answers to this could help to understand the risk of the disease surviving and spreading in the aquatic environment.
- Although much work has been done to explain the pathogenicity of *A. salmonicida*, the definitive answer is still unclear because of the multiplicity

of mechanisms, which may reflect strain differences or the inability of *in vitro* methodologies to adequately model disease in a real host in the natural environment. Will the organism become more aggressive with climate change?

- How do the increasing water temperatures associated with climate change relate to the 'fever response' whereby fish prefer to become warmer to negate the impact of infection?
- Research needs to focus on the possibility of synergistic effects of two or more pathogens on a susceptible host, i.e. could disease be a manifestation of multiple organisms working sequentially or simultaneously, such as *A. salmonicida* with *A. hydrophila* or other motile aeromonads? Current diagnostic procedures are biased towards the notion that disease results from single species of pathogens, but this need not always be the case. Moreover, could synergies become more important in fish that are stressed as a result of climate change?
- With ever more sensitive and specific diagnostic systems, attention needs to be on determining the relevance of such positive outcomes to the pathological process. Are low numbers of cells indicative of the likely development of clinical disease or do they reflect background populations levels of the pathogen that occur in the environment?
- With current and anticipated advances in genetics and molecular biology, could science develop/select fish strains more suited to climate change, e.g. disease-resistant fish that have greater temperature tolerance?
- The impetus to develop novel and effective means of disease control strategies needs to continue with the overriding aim for use in aquaculture. The emphasis needs to be on disease prevention rather than control, and more needs to be done to exploit immunomodulatory agents that seem to offer so much potential for controlling disease. Combinations of non-specific (= conferring herd immunity) and specific (= vaccines) immunological products deserve further exploration particularly if vaccines fail to control clinical disease in severely stressed fish.

References

Alghabshi, A., Austin, B. and Crumlish, M. (2018) *Aeromonas salmonicida* isolated from wild and farmed fish and invertebrates in Oman. *International Aquatic Research* 10, 145–152. https://doi.org/10.1007/s40071-018-0195-4

Allen-Austin, D., Austin, B. and Colwell, R.R. (1984) Survival of *Aeromonas salmonicida* in river water. *FEMS Microbiology Letters* 21, 143–146. https://doi.org/10.1111/j.1574-6968.1984.tb00200.x

Austin, B. and Austin, D.A. (2016) *Bacterial Fish Pathogens, Disease of Farmed and Wild Fish, 6th edn*. Springer, Dordrecht, the Netherlands.

Austin, B., Bishop, I., Gray, C., Watt, B. and Dawes, J. (1986) Monoclonal antibody-based enzyme-linked immunosorbent assays for the rapid diagnosis of clinical cases of enteric redmouth and furunculosis in fish farms. *Journal of Fish Diseases* 9, 469–474. https://doi.org/10.1111/j.1365-2761.1986.tb01042.x

Austin, D.A., McIntosh, D. and Austin, B. (1989) Taxonomy of fish associated *Aeromonas* spp., with the description of *Aeromonas salmonicida* subsp. *smithia* subsp. nov. *Systematic and Applied Microbiology* 11, 277–290. https://doi.org/10.1016/S0723-2020(89)80026-8

Barnes, A.C., Lewin, C.S., Hastings, T.S. and Amyes, S.G.B. (1991a) *In vitro* susceptibility of the fish pathogen *Aeromonas salmonicida* to flumequine. *Antimicrobial Agents and Chemotherapy* 35, 2634–2635. https://doi.org/10.1128/AAC.35.12.2634

Barnes, A.C., Amyes, S.G.B., Hastings, T.S. and Lewin, C.S. (1991b) Fluoroquinolones display rapid bactericidal activity and low mutation frequencies against *Aeromonas salmonicida*. *Journal of Fish Diseases* 14, 661–667. https://doi.org/10.1111/j.1365-2761.1991.tb00624.x

Barnes, A.C., Hastings, T.S. and Amyes, S.G. (1994) Amoxycillin resistance in Scottish isolates of *Aeromonas salmonicida*. *Journal of Fish Diseases* 17, 357–363. https://doi.org/10.1111/j.1365-2761.1994.tb00231.x

Beaz-Hidalgo, R., Alperi, A., Figueras, M.J. and Romalde, J.L. (2009) *Aeromonas piscicola* sp. nov., isolated from diseased fish. *Systematic and Applied Microbiology* 32, 471–479. https://doi.org/10.1016/j.syapm.2009.06.004

Björnsdóttir, B., Gudmundsdóttir, S., Bambir, S.H. and Gudmundsdóttir, B.K. (2005) Experimental infection of turbot, *Scophthalmus maximus* (L.), by *Aeromonas salmonicida* subsp. *achromogenes* and evaluation of cross protection induced by a furunculosis vaccine. *Journal of Fish Diseases* 28, 181–188. https://doi.org/10.1111/j.1365-2761.2005.00617.x

Blake, I. and Clark, J.C. (1931) Observations on experimental infection of trout by *B. salmonicida*, with particular reference to 'carriers' of furunculosis and to certain factors influencing susceptibility. *Fisheries Board of Scotland, Salmon Fisheries No. 7*. HMSO, Edinburgh, pp. 1–13.

Boonthai, T., Loch, T.P., Yamashita, C.J., Smith, G.D., Winters, A.D. *et al.* (2018) Laboratory investigation into the role of largemouth bass virus (*Ranavirus*, *Iridoviridae*) in smallmouth bass mortality events in

Pennsylvania rivers. *BMC Veterinary Research* 14, 62. https://doi.org/10.1186/s12917-018-1371-x

Breyer, K.E., Getchell, R.G., Cornwell, E.R., Wooster, A.A., Ketola, H.G. and Bowser, P.R. (2015) Efficacy of an extract from garlic, *Allium sativum*, against infection with the furunculosis bacterium, *Aeromonas salmonicida*, in rainbow trout, *Oncorhynchus mykiss*. *Journal of the World Aquaculture Society* 46, 273–282. https://doi.org/10.1111/jwas.12195

Bullock, G.L. and Stuckey, H.M. (1975) *Aeromonas salmonicida*: detection of asymptomatically infected trout. *The Progressive Fish-Culturist* 37, 237–239. https://doi.org/10.1577/1548-8659(1975)37[237:AS]2.0.CO;2

Carnahan, A.M., Stine, O.C., Morris, J.G., Waddington, M. and Joseph, S.W. (2001) MLST (multilocus strain typing) of a diverse group of *Aeromonas* strains. *Abstracts of the General Meeting of the 101st American Society for Microbiology* 101, 681–682.

Chapela, M.-J., Ferreira, M., Ruiz-Cruz, A., Martin-Varela, I., Fernandez-Casal, J. and Garrido-Maestu, A. (2018) Application of real-time PCR for early diagnosis of diseases caused by *Aeromonas salmonicida*, *Vibrio anguillarum*, and *Tenacibaculum maritimum* in turbot: a field study. *Journal of Applied Aquaculture* 30, 76–89. https://doi.org/10.1080/10454438.2017.1406419

Chen, L., Yuan, S.-J., Liu, Q., Mai, G.-Q., Yang, J.-F. *et al.* (2018) *In vitro* design and evaluation of phage cocktails against *Aeromonas salmonicida*. *Frontiers in Microbiology* 9, 1476. https://doi.org/10.3389/fmicb.2018.01476

Cipriano, R.C. (1983) Resistance of salmonids to *Aeromonas salmonicida*: relation between agglutinins and neutralizing activities. *Transactions of the American Fisheries Society* 112, 95–99. https://doi.org/10.1577/1548-8659(1983)112<95:ROSTAS>2.0.CO;2

Cipriano, R.C. and Starliper, C.E. (1982) Immersion and injection vaccination of salmonids against furunculosis with an avirulent strain of *Aeromonas salmonicida*. *The Progressive Fish-Culturist* 44, 167–169.

Cipriano, R.C., Ford, L.A., Teska, J.D. and Hale, L.E. (1992) Detection of *Aeromonas salmonicida* in the mucus of salmonid fishes. *Journal of Aquatic Animal Health* 4, 114–118. https://doi.org/10.1577/1548-8667(1992)004<0114:DOASIT>2.3.CO;2

Cipriano, R.C., Ford, L.A., Teska, J.D., Schachte, J.H., Petrie, C. *et al.* (1996) Use of non-lethal procedures to detect and monitor *Aeromonas salmonicida* in potentially endangered or threatened populations of migrating and post-spawning salmon. *Diseases of Aquatic Organisms* 27, 233–236. https://doi.org/10.3354/dao027233

Cobo, C., Makosch, K., Jung, R., Kohlmann, K. and Knopf, K. (2014) Enhanced *Aeromonas salmonicida* bacterin uptake and side effects caused by low frequency sonophoresis in rainbow trout (*Oncorhynchus mykiss*). *Fish and Shellfish Immunology* 36, 444–452. https://doi.org/10.1016/j.fsi.2013.12.010

Deere, D., Porter, J., Pickup, R.W. and Edwards, C. (1996a) Survival of cells and DNA of *Aeromonas salmonicida* released into aquatic microcosms. *Journal of Applied Bacteriology* 81, 309–318. https://doi.org/10.1111/j.1365-2672.1996.tb04333.x

Deere, D., Porter, J., Pickup, R. and Edwards, C. (1996b) Direct analysis of starved *Aeromonas salmonicida*. *Journal of Fish Diseases* 19, 459–467. https://doi.org/10.1046/j.1365-2761.1996.d01-98.x

Diamanka, A., Loch, T.P., Cipriano, R.C., Winters, A.D. and Faisal, M. (2014) Infection of sea lamprey with an unusual strain of *Aeromonas salmonicida*. *Journal of Wildlife Diseases* 50, 159–170. https://doi.org/10.7589/2013-01-026

Dietrich, J.P., Van Gaest, A.L., Strickland, S.A. and Arkoosh, M.R. (2014) The impact of temperature stress and pesticide exposure on mortality and disease susceptibility of endangered Pacific salmon. *Chemosphere* 108,353–359. https://doi.org/10.1016/j.chemosphere.2014.01.079

Du, Y.S., Liu, Y., Xiao, P., Meng, L.J. and Liu, P.F. (2017) Development and application of a quantitative real-time polymerase chain reaction assay for the detection of *Aeromonas salmonicida*. *Journal of the World Aquaculture Society* 48, 574–582. https://doi.org/10.1111/jwas.12395

Du, Y.S., Liu, P.F., Meng, L.J., Sharawy, Z. and Liu, Y. (2018) Colonization of *Aeromonas salmonicida* subsp. *masoucida* strains in Atlantic salmon (*Salmo salar* L.) during infection. *Aquaculture Research* 49, 1826–1833. https://doi.org/10.1111/are.13637

Du, X.-C., Bayliss, S.C., Edward, J., Liu, Y., Wang, C. *et al.* (2019) Real time monitoring of *Aeromonas salmonicida* evolution in response to successive antibiotic therapies in a commercial fish farm. *Environmental Microbiology* 21, 1113–1123. https://doi.org/10.1111/1462-2920.14531

Duarte, J., Pereira, C., Moreirinha, C., Salvio, R., Lopes, A. *et al.* (2018) New insights on phage efficacy to control *Aeromonas salmonicida* in aquaculture systems: an *in vitro* preliminary study. *Aquaculture* 495, 970–982. https://doi.org/10.1016/j.aquaculture.2018.07.002

Duff, D.C.B. (1942) The oral immunization of trout against *Bacterium salmonicida*. *Journal of Immunology* 44, 87–94.

Ebanks, R.O., Knickle, L.C., Goguen, M., Boyd, J.M., Pinto, D.M. *et al.* (2006) Expression of and secretion through the *Aeromonas salmonicida* type III secretion system. *Microbiology* 152, 1275–1286. https://doi.org/10.1099/mic.0.28485-0

Effendi, I. and Austin, B. (1994) Survival of the fish pathogen *Aeromonas salmonicida* in the marine environment. *Journal of Fish Diseases* 17, 375–385. https://doi.org/10.1111/j.1365-2761.1994.tb00233.x

Effendi, I. and Austin, B. (1995a) Dormant/unculturable cells of the fish pathogen *Aeromonas salmonicida*. *Microbial Ecology* 30, 183–192. https://doi.org/10.1007/BF00172573

Effendi, I. and Austin, B. (1995b) Uptake of *Aeromonas salmonicida* by Atlantic salmon (*Salmo salar* L.). *Bulletin of the European Association of Fish Pathologists* 15, 115–118.

Ehlinger, N.F. (1977) Selective breeding of trout for resistance to furunculosis. *New York Fish and Game Journal* 24, 25–36.

Ellis, A.E., Burrows, A.S., Hastings, T.S. and Stapleton, K.J. (1988) Identification of *Aeromonas salmonicida* extracellular protease as a protective antigen against furunculosis by passive immunization. *Aquaculture* 70, 207–218. https://doi.org/10.1016/0044-8486(88)90097-X

El Morabit, A., García-Márquez, S. and Santos, Y. (2004) Is sea lamprey a potential source of infection with *Aeromonas salmonicida* for wild and farmed fish? *Bulletin of the European Association of Fish Pathologists* 24, 100–103.

Elston, R., Drum, A.S. and Bunnell, P.R. (1995) Efficacy of orally administered difloxacin for the treatment of Atlantic salmon held in seawater. *Journal of Aquatic Animal Health* 7, 22–28. https://doi.org/10.1577/1548-8667(1995)007<0022:EOOADF>2.3.CO;2

Embody, G.C. and Hayford, C.O. (1925) The advantage of rearing brook trout fingerlings from selected breeders. *Transactions of the American Fisheries Society* 55, 135–142.

Emmerich, R. and Weibel, E. (1894) Über eine durch Bakterien erzengte Seuche unter den Forellen. *Archives für Hygiene und Bakteriologie* 21, 1–21.

Endo, T., Ogishima, K., Hayasaki, H., Kaneko, S. and Ohshima, S. (1973) Application of oxolinic acid as a chemotherapeutic agent for treating infectious diseases in fish. I. Antibacterial activity, chemotherapeutic effect and pharmacokinetic effect of oxolinic acid in fish. *Bulletin of the Japanese Society of Scientific Fisheries* 3, 165–171.

Erkinharju, T., Dalmo, R.A., Vagsnes, O., Hordvik, I. and Seternes, T. (2018) Vaccination of Atlantic lumpfish (*Cyclopterus lumpus* L.) at a low temperature leads to a low antibody response against *Aeromonas salmonicida*. *Journal of Fish Diseases* 41, 613–623. https://doi.org/10.1111/jfd.12760

Evelyn, T.P.T. (1971) An aberrant strain of the bacterial fish pathogen *Aeromonas salmonicida* isolated from a marine host, the sablefish (*Anoplopoma fimbria*) and from two species of cultured Pacific salmon. *Journal of the Fisheries Research Board of Canada* 28, 1629–1634. https://doi.org/10.1139/f71-243

Ferguson, Y., Glover, L.A., McGillivray, D.M. and Prosser, J.I. (1995) Survival and activity of lux-marked *Aeromonas salmonicida* in seawater. *Applied and Environmental Microbiology* 61, 3494–3498.

Fernandez-Alvarez, C., Gonzalez, S.F. and Santos, Y. (2016) Development of a SYBR green I real-time PCR assay for specific identification of the fish pathogen *Aeromonas salmonicida* subspecies *salmonicida*. *Applied Microbiology and Biotechnology* 100, 10585–10595. https://doi.org/10.1007/s00253-016-7929-2

Frerichs, G.N., Millar, S.D. and McManus, C. (1992) Atypical *Aeromonas salmonicida* isolated from healthy wrasse (*Ctenolabrus rupestris*). *Bulletin of the European Association of Fish Pathologists* 12, 48–49.

Fyfe, L., Coleman, G. and Munro, A.L.S. (1987) A comparative study of the formation of extracellular proteins by *Aeromonas salmonicida* at two different temperatures. *Journal of Applied Bacteriology* 62, 367–370. https://doi.org/10.1111/j.1365-2672.1987.tb04932.x

Gjedrem, T. and Gjoen, H.M. (1995) Genetic variation in susceptibility of Atlantic salmon, *Salmo salar* L., to furunculosis, BKD and cold water vibriosis. *Aquaculture Research* 26, 129–134. https://doi.org/10.1111/j.1365-2109.1995.tb00892.x

Griffin, P.J., Snieszko, S.F. and Friddle, S.B. (1953) A more comprehensive description of *Bacterium salmonicida*. *Transactions of the American Fisheries Society* 82, 129–138. https://doi.org/10.1577/1548-8659(1952)82[129:AMCDOB]2.0.CO;2

Grove, S., Høie, S. and Evensen, Ø. (2003) Distribution and retention of antigens of *Aeromonas salmonicida* in Atlantic salmon (*Salmo salar* L.) vaccinated with a Δ*aroA* mutant or formalin-inactivated bacteria in oil-adjuvant. *Fish and Shellfish Immunology* 15, 349–358. https://doi.org/10.1016/S1050-4648(02)00184-5

Gudmundsdóttir, B.K. and Gudmundsdóttir, S. (1997) Evaluation of cross protection by vaccines against atypical and typical furunculosis in Atlantic salmon, *Salmo salar* L. *Journal of Fish Diseases* 20, 343–350. https://doi.org/10.1046/j.1365-2761.1997.00307.x

Gudmundsdóttir, S., Lange, S., Magnadóttir, B. and Gudmundsdóttir, B.K. (2003) Protection against atypical furunculosis in Atlantic halibut, *Hippoglossus hippoglossus* (L.); a comparison of a commercial furunculosis vaccine and an autogenous vaccine. *Journal of Fish Diseases* 26, 331–338. https://doi.org/10.1046/j.1365-2761.2003.00462.x

Han, H.-J., Kim, D.-Y., Kim, W.-S., Kim, C.-S., Jung, S.-J. *et al.* (2011) Atypical *Aeromonas salmonicida* infection in the black rockfish, *Sebastes schlegeli* Hilgendorf, in Korea. *Journal of Fish Diseases* 34, 47–55. https://doi.org/10.1111/j.1365-2761.2010.01217.x

Hastings, T.S. and Ellis, A.E. (1985) Differences in the production of hemolytic and proteolytic activities by various isolates of *Aeromonas salmonicida*. In: Ellis, A.E. (ed.) *Fish and Shellfish Pathology; First International Conference of the European Association of Fish Pathologists, Plymouth, England, Sept. 20–23, 1983*. Academic Press, London/Orlando, Florida, pp. 69–78.

Hedges, R.W., Smith, P. and Brazil, G. (1985) Resistance plasmids of aeromonads. *Journal of General Microbiology* 131, 2091–2095.

Hellberg, H., Moksness, E. and Høie, S. (1996) Infection with atypical *Aeromonas salmonicida* in farmed common wolffish, *Anarhichas lupus*. *Journal of Fish Diseases* 19, 329–332. https://doi.org/10.1111/j.1365-2761.1996.tb00711.x

Hiney, M., Dawson, M.T., Heery, D.M., Smith, P.R., Gannon, F. and Powell, R. (1992) DNA probe for *Aeromonas salmonicida*. *Applied and Environmental Microbiology* 58, 1039–1042.

Hiney, M.P., Kilmartin, J.J. and Smith, P.R. (1994) Detection of *Aeromonas salmonicida* in Atlantic salmon with asymptomatic furunculosis infections. *Diseases of Aquatic Organisms* 19, 161–167. https://doi.org/10.3354/dao019161

Hoel, K., Reitan, L.J. and Lillehaug, A. (1998) Immunological cross reactions between *Aeromonas salmonicida* and *Vibrio salmonicida* in Atlantic salmon (*Salmo salar* L.) and rabbit. *Fish and Shellfish Immunology* 8, 171–182. https://doi.org/10.1006/fsim.1997.0135

Høie, S., Heum, M. and Thoresen, O.F. (1996) Detection of *Aeromonas salmonicida* by polymerase chain reaction in Atlantic salmon vaccinated against furunculosis. *Fish and Shellfish Immunology* 6, 199–206. https://doi.org/10.1006/fsim.1996.0020

Høie, S., Heum, M. and Thoresen, O.F. (1997) Evaluation of a polymerase chain reaction-based assay for the detection of *Aeromonas salmonicida* ss. *salmonicida* in Atlantic salmon *Salmo salar*. *Diseases of Aquatic Organisms* 30, 27–35. https://doi.org/10.3354/dao030027

Hori, T.S., Gamperl, A.K., Booman, M., Nash, G.W. and Rise, M.L. (2012) A moderate increase in ambient temperature modulates the Atlantic cod (*Gadus morhua*) spleen transcriptome response to intraperitoneal viral mimic injection. *BMC Genomics* 13, 431. https://doi.org/10.1186/1471-2164-13-431

Hori, T.S., Gamperl, A.K., Nash, G., Booman, M., Barat, A. and Rise, M.L. (2013) The impact of a moderate chronic temperature increase on spleen immune-relevant gene transcription depends on whether Atlantic cod (*Gadus morhua*) are stimulated with bacterial versus viral antigens. *Genome* 56, 567–576. https://doi.org/10.1139/gen-2013-0090

Hsu, H.-M., Bowser, P.R., Schachte, J.H. Jr, Scarlett, J.M. and Babish, J.G. (1995) Winter field trials of enrofloxacin for the control of *Aeromonas salmonicida* infection in salmonids. *Journal of the World Aquaculture Society* 26, 307–314. https://doi.org/10.1111/j.1749-7345.1995.tb00259.x

Inglis, V., Robertson, D., Miller, K., Thompson, K.D. and Richards, R.H. (1996) Antibiotic protection against recrudescence of latent *Aeromonas salmonicida* during furunculosis vaccination. *Journal of Fish Diseases* 19, 341–348. https://doi.org/10.1046/j.1365-2761.1996.d01-86.x

Irianto, A. and Austin, B. (2002) Use of probiotics to control furunculosis in rainbow trout, *Oncorhynchus mykiss* (Walbaum). *Journal of Fish Diseases* 25, 332–342. https://doi.org/10.1046/j.1365-2761.2002.00375.x

Irie, T., Watarai, S., Iwasaki, T. and Kodama, H. (2005) Protection against experimental *Aeromonas salmonicida* infection in carp by oral immunisation with bacterial antigen entrapped liposomes. *Fish and Shellfish Immunology* 18, 235–242. https://doi.org/10.1016/j.fsi.2004.07.006

Ishiguro, E.E., Kay, W.W., Ainsworth, T., Chamberlain, J.B., Austen, R.A. *et al.* (1981) Loss of virulence during culture of *Aeromonas salmonicida* at high temperature. *Journal of Bacteriology* 148, 333–340.

Ji, L.Q., Sun, G.X., Li, J., Wang, Y., Du, Y.S. *et al.* (2017) Effect of dietary β-glucan on growth, survival and regulation of immune processes in rainbow trout (*Oncorhynchus mykiss*) infected with *Aeromonas salmonicida*. *Fish and Shellfish Immunology* 64, 56–67. https://doi.org/10.1016/j.fsi.2017.03.015

Josefsson, M. (1999) Introduced organisms in Swedish freshwater environment. *Fauna och Flora (Stockholm)* 94, 75–84.

Jung, T.S., del Castillo, C.S., Javaregowda, P.K., Dalvi, R.S., Nho, S.W. *et al.* (2012) Seasonal variation and comparative analysis of non-specific humoral immune substances in the skin mucus of olive flounder (*Paralichthys olivaceus*). *Developmental and Comparative Immunology* 38, 295–301. https://doi.org/10.1016/j.dci.2012.06.005

Jung-Schroers, V., Jung, A., Ryll, M., Bauer, J., Teitge, F. and Steinhagen, D. (2018) Diagnostic methods for identifying different *Aeromonas* species and examining their pathogenicity factors, their correlation to cytotoxicity and adherence to fish mucus. *Journal of Fish Diseases* 42, 189–219. https://doi.org/10.1111/jfd.12917

Jutfelt, F., Olsen, R.E., Glette, J., Ringø, E. and Sundell, K. (2006) Translocation of viable *Aeromonas salmonicida* across the intestine of rainbow trout, *Oncorhynchus mykiss* (Walbaum). *Journal of Fish Diseases* 29, 255–262. https://doi.org/10.1111/j.1365-2761.2006.00715.x

Kaku, Y., Yamada, Y. and Wakabayashi, H. (1999) Characterization of atypical *Aeromonas salmonicida* isolated from an epizootic ulcerative disease in carp (*Cyprinus carpio*). *Fish Pathology* 34, 155–162. https://doi.org/10.3147/jsfp.34.155

Kawahara, E. and Kusuda, R. (1987) Direct fluorescent antibody technique for diagnosis of bacterial disease in eel. *Nippon Suisan Gakkaishi* 53, 395–399.

Keeling, S.E., Brosnahan, C.L., Johnston, C., Wallis, R., Gudkovs, N. and McDonald, W.L. (2013) Development and validation of a real-time PCR assay for the detection of *Aeromonas salmonicida*. *Journal of Fish Diseases* 36, 495–503. https://doi.org/10.1111/jfd.12014

Kilian, M. (1976) Taxonomic study of genus *Haemophilus*, with the proposal of a new species. *Journal of General*

Microbiology 93, 9–62. https://doi.org/10.1099/00221287-93-1-9
Kim, D.-H., Choi, S.-Y., Kim, C.S., Oh, M.-J. and Jeong, H.-D. (2013) Low-value fish used in aquaculture were a source of furunculosis caused by atypical *Aeromonas salmonicida*. *Aquaculture* 408–409, 113–117. https://doi.org/10.1016/j.aquaculture.2013.05.014
Kim, J.H., Son, J.S., Choi, Y.J., Choresca, C.H., Shin, S.P. *et al.* (2012) Isolation and characterization of lytic *Myoviridae* bacteriophage PAS-1 with broad infectivity in *Aeromonas salmonicida*. *Current Microbiology* 64, 418–426. https://doi.org/10.1007/s00284-012-0091-x
Kim, J.H., Choresca, C.H., Shin, S.P., Han, J.E., Jun, J.W. and Park, S.C. (2015) Biological control of *Aeromonas salmonicida* subsp. *salmonicida* infection in rainbow trout (*Oncorhynchus mykiss*) using *Aeromonas* phage PAS-1. *Transboundary and Emerging Diseases* 62, 81–86. https://doi.org/10.1111/tbed.12088
King, C.H. and Shotts, E.B. (1988) Enhancement of *Edwardsiella tarda* and *Aeromonas salmonicida* through ingestion by the ciliated protozoan *Tetrahymena pyriformis*. *FEMS Microbiology Letters* 51, 95–100. https://doi.org/10.1111/j.1574-6968.1988.tb02976.x
Kitao, T. and Yoshida, Y. (1986) Effect of an immunopotentiator on *Aeromonas salmonicida* infection in rainbow trout (*Salmo gairdneri*). *Veterinary Immunology and Immunopathology* 12, 287–296. https://doi.org/10.1016/0165-2427(86)90132-7
Klontz, G.W. and Wood, J.W. (1972) Observations on the epidemiology of furunculosis disease in juvenile coho salmon (*Oncorhynchus kisutch*). *FI:EIFAC 72/SC II Symposium* 27, 1–8.
Kogure, K., Simidu, U. and Taga, N. (1979) Tentative direct microscopic method for counting living marine bacteria. *Canadian Journal of Microbiology* 25, 415–420. https://doi.org/10.1139/m79-063
Koppang, E.O., Fjølstad, M., Melgård, B., Vigerust, M. and Sørum, H. (2000) Non-pigment-producing isolates of *Aeromonas salmonicida* subspecies *salmonicida*: isolation, identification, transmission and pathogenicity in Atlantic salmon, *Salmo salar* L. *Journal of Fish Diseases* 23, 39–48. https://doi.org/10.1046/j.1365-2761.2000.00205.x
Kumagai, A., Sugimoto, K., Itou, D., Kamaishi, T., Miwa, S. and Iida, T. (2006) Atypical *Aeromonas salmonicida* in cultured marbled sole *Pleuronectes yokohamae*. *Fish Pathology* 41, 7–12. https://doi.org/10.3147/jsfp.41.7
Libran-Perez, M., Costs, M.M., Figueras, A. and Novoa, B. (2018) β-Glucan administration induces metabolic changes and differential survival rates after bacterial or viral infection in turbot (*Scophthalmus maximus*). *Fish and Shellfish Immunology* 82, 173–182. https://doi.org/10.1016/j.fsi.2018.08.005
Lund, M. (1967) A study of the biology of Aeromonas salmonicida (Lehmann and Neumann 1896) Griffin 1954. MSc thesis, University of Newcastle upon Tyne, Newcastle upon Tyne, UK.
Lund, V., Arnesen, J.A., Coucheron, D., Modalsli, K. and Syvertsen, C. (2003) The *Aeromonas salmonicida* A-layer protein is an important protective antigen in oil-adjuvanted vaccines. *Fish and Shellfish Immunology* 15, 367–372. https://doi.org/10.1016/S1050-4648(02)00180-8
Lund, V., Arnesen, J.A., Mikkelsen, H., Gravningen, K., Brown, L. and Schrøder, M.B. (2008a) Atypical furunculosis vaccines for Atlantic cod (*Gadus morhua*); vaccine efficacy and antibody responses, *Vaccine* 26, 6791–6799. https://doi.org/10.1016/j.vaccine.2008.10.012
Lund, V., Mikkelsen, H. and Schrøder, M.B. (2008b) Comparison of atypical furunculosis vaccines in spotted wolffish (*Anarhichas minor* O.) and Atlantic halibut (*Hippoglossus hippoglossus* L.). *Vaccine* 26, 2833–2840. https://doi.org/10.1016/j.vaccine.2008.03.049
Lutwyche, P., Exner, M.M., Hancock, R.E.W. and Trust, T.J. (1995) A conserved *Aeromonas salmonicida* porin provides protective immunity to rainbow trout. *Infection and Immunity* 63, 3137–3142.
Ma, X.N., Li, X., Sun, G.X., Sharawy, Z.Z., Qiu, T.L. *et al.* (2017) The biofilteration ability of oysters (*Crassostrea gigas*) to reduce *Aeromonas salmonicida* in salmon culture. *Applied Microbiology and Biotechnology* 101, 5869–5880. https://doi.org/10.1007/s00253-017-8296-3
McCarthy, D.H. (1975) Detection of *Aeromonas salmonicida* antigen in diseased fish tissue. *Journal of General Microbiology* 88, 384–386. https://doi.org/10.1099/00221287-88-2-384
McCarthy, D.H. (1980) Some ecological aspects of the bacterial fish pathogen – *Aeromonas salmonicida*. In: *Aquatic Microbiology. Symposium of the Society of Applied Bacteriology No. 6*, pp. 299–324.
McCarthy, D.H. and Roberts, R.J. (1980) Furunculosis of fish – the present state of our knowledge. In: Droop, M.A. and Jannasch, H.W. (eds) *Advances in Aquatic Microbiology*. Academic Press, London, pp. 293–341.
McCarthy, D.H., Stevenson, J.P. and Salsbury, A.W. (1974) Therapeutic efficacy of a potentiated sulphonamide in experimental furunculosis. *Aquaculture* 4, 407–410. https://doi.org/10.1016/0044-8486(74)90069-6
McCarthy, D.H., Amend, D.F., Johnson, K.A. and Bloom, J.V. (1983) *Aeromonas salmonicida*: determination of an antigen associated with protective immunity and evaluation of an experimental bacterin. *Journal of Fish Diseases* 6, 155–174. https://doi.org/10.1111/j.1365–2761.1983.tb00063.x
McCashion, R.N. and Lynch, W.H. (1987) Effects of polymyxin B nonapeptide on *Aeromonas salmonicida*. *Antimicrobial Agents and Chemotherapy* 31, 1414–1419. https://doi.org/10.1128/AAC.31.9.1414
McIntosh, D. and Austin, B. (1991) Atypical characteristics of the salmonid pathogen *Aeromonas salmonicida*.

Journal of General Microbiology 137, 1341–1343. https://doi.org/10.1099/00221287-137-6-1341

Marana, M.H., Skov, J., Chettri, J.K., Krossoy, B., Dalsgaard, I. *et al.* (2017a) Positive correlation between *Aeromonas salmonicida* vaccine antigen concentration and protection in vaccinated rainbow trout *Oncorhynchus mykiss* evaluated by a tail fin infection model. *Journal of Fish Diseases* 40, 507–516. https://doi.org/10.1111/jfd.12527

Marana, M.H., Jorgensen, L.V., Skov, J., Chettri, J.K., Mattsson, A.H. *et al.* (2017b) Subunit vaccine candidates against *Aeromonas salmonicida* in rainbow trout *Oncorhynchus mykiss*. *PLoS ONE* 12, e0171944. https://doi.org/10.1371/journal.pone.0171944

Markwardt, N.M., Gocha, Y.M. and Klontz, G.W. (1989) A new application for Coomassie Brilliant Blue agar: detection of *Aeromonas salmonicida* in clinical samples. *Diseases of Aquatic Organisms* 6, 231–233. https://doi.org/10.3354/dao006231

Marquez, I., Garcia-Vazquez, E. and Borrell, Y.J. (2014) Possible effects of vaccination and environmental changes on the presence of disease in northern Spanish fish farms. *Aquaculture* 431, 118–123. https://doi.org/10.1016/j.aquaculture.2013.12.030

Michel, C. and Dubois-Darnaudpeys, A. (1980) Persistence of the virulence of *Aeromonas salmonicida* kept in river sediments. *Annales de Recherche Vétérinaire* 11, 375–386.

Michel, C., Gerard, J.-P., Fourbet, B., Collas, R. and Chevalier, R. (1980) Emploi de la flumequine contre la furunculose des salmonides; essais therapeutiques et perspectives pratiques. *Bulletin Français de Pisciculture* 52, 154–162.

Midtlyng, P.J. (1996) A field study on intraperitoneal vaccination of Atlantic salmon (*Salmo salar* L.) against furunculosis. *Fish and Shellfish Immunology* 6, 553–565. https://doi.org/10.1006/fsim.1996.0052

Midtlyng, P.J., Reitan, L.J., Lillehaug, A. and Ramstad, A. (1996) Protection, immune responses and side effects in Atlantic salmon (*Salmo salar* L.) vaccinated against furunculosis by different procedures. *Fish and Shellfish Immunology* 6, 599–613. https://doi.org/10.1006/fsim.1996.0055

Mitoma, Y., Aoki, T. and Crosa, J.H. (1984) Phylogenetic relationships among *Vibrio anguillarum* plasmids. *Plasmid* 12, 143–148. https://doi.org/10.1016/0147-619X(84)90038-6

Morgan, J.A.W., Cranwell, P.A. and Pickup, R.W. (1991) Survival of *Aeromonas salmonicida* in lake water. *Applied and Environmental Microbiology* 57, 1777–1782.

Morgan, J.A.W., Rhodes, G. and Pickup, R.W. (1993) Survival of nonculturable *Aeromonas salmonicida* in lake water. *Applied and Environmental Microbiology* 59, 874-880.

Munro, A.L.S. and Gauld, J.A. (1996) *Scottish Fish Farms Annual Production Survey 1995*. Marine Laboratory, Aberdeen, UK.

Mutoloki, S., Brudeseth, B., Reite, O.B. and Evensen, Ø. (2006) The contribution of *Aeromonas salmonicida* extracellular products to the induction of inflammation in Atlantic salmon (*Salmo salar* L.) following vaccination with oil-based vaccines. *Fish and Shellfish Immunology* 20, 1–11. https://doi.org/10.1016/j.fsi.2005.01.005

Navot, N., Sinyakov, S. and Avtalion, R.R. (2011) Application of ultrasound in vaccination against goldfish ulcer disease: a pilot study. *Vaccine* 29, 1382–1389. https://doi.org/10.1016/j.vaccine.2010.12.069

Nese, L. and Enger, Ø. (1993) Isolation of *Aeromonas salmonicida* from salmon lice *Lepeophtheirus salmonis* and marine plankton. *Diseases of Aquatic Organisms* 16, 79–81. https://doi.org/10.3354/dao016079

Nikl, L., Albright, L.J. and Evelyn, T.P.T. (1991) Influence of seven immunostimulants on the immune response of coho salmon to *Aeromonas salmonicida*. *Diseases of Aquatic Organisms* 12, 7–12. https://doi.org/10.3354/dao012007

Nilsson, W.B. and Strom, M.S. (2002) Detection and identification of bacterial pathogens of fish in kidney tissue using terminal restriction fragment length polymorphism (T-RFLP) analysis of 16S rRNA genes. *Diseases of Aquatic Organisms* 48, 175–185. https://doi.org/10.3354/dao048175

Nomura, S. and Saito, H. (1982) Production of the extracellular hemolytic toxin by an isolated strain of *Aeromonas salmonicida*. *Bulletin of the Japanese Society of Scientific Fisheries* 48, 1589–1597.

Noor, H.F., Yanuhar, U. and Maftuch (2017) Application of inactivated *Aeromonas salmonicida* vaccine towards *Cyprinus carpio* koi fish immunogenicity. *Research Journal of Pharmaceutical, Biological and Chemical Sciences* 8, 2349–2355.

Norqvist, A., Hagström, Å. and Wolf-Watz, H. (1989) Protection of rainbow trout against vibriosis and furunculosis by the use of attenuated strains of *Vibrio salmonicida*. *Applied and Environmental Microbiology* 55, 1400–1405.

O'Brien, D., Mooney, J., Ryan, D., Powell, E., Hiney, M. *et al.* (1994) Detection of *Aeromonas salmonicida*, causal agent of furunculosis in salmonid fish, from the tank effluent of hatchery-reared Atlantic salmon smolts. *Applied and Environmental Microbiology* 60, 3874–3877.

Olivier, G., Evelyn, T.P.T. and Lallier, R. (1985a) Immunogenicity of vaccines from a virulent and an avirulent strain of *Aeromonas salmonicida*. *Journal of Fish Diseases* 8, 43–55. https://doi.org/10.1111/j.1365-2761.1985.tb01186.x

Olivier, G., Evelyn, T.P.T. and Lallier, R. (1985b) Immunity to *Aeromonas salmonicida* in coho salmon (*Oncorhynchus kisutch*) induced by modified Freund's complete adjuvant: its non-specific nature and the probable role of macrophages in the phenomenon. *Developmental and Comparative Immunology* 9, 419–432. https://doi.org/10.1016/0145-305X(85)90005-9

Ortega, C., Ruiz, I., De Blas, I., Muzquiz, J.L., Fernandez, A. and Alonso, J.L. (1996) Furunculosis control using a paraimmunization stimulant (Baypamun) in rainbow trout. *Veterinary Research (Paris)* 27, 561–568.

Park, Y., Lee, S., Hong, J., Kim, D., Moniruzzaman, M. and Bai, S.C. (2017) Use of probiotics to enhance growth, stimulate immunity and confer disease resistance to *Aeromonas salmonicida* in rainbow trout (*Oncorhynchus mykiss*). *Aquaculture Research* 48, 2672–2682. https://doi.org/10.1111/are.13099

Pedersen, K., Kofod, H., Dalsgaard, I. and Larsen, J.L. (1994) Isolation of oxidase-negative *Aeromonas salmonicida* from diseased turbot *Scophthalmus maximus*. *Diseases of Aquatic Organisms* 18, 149–154. https://doi.org/10.3354/dao018149

Piao, D.-R., Liu, X., Di, D.-D., Xiao, P., Zhao, Z.-Z. *et al.* (2018) Genetic polymorphisms identify in species/biovars of *Brucella* isolated in China between 1953 and 2013 by MLST. *BMC Microbiology* 18, 7. https://doi.org/10.1186/s12866-018-1149-0

Pickup, R.W., Rhodes, G., Cobban, R.J. and Clarke, K.J. (1996) The postponement of non-culturability in *Aeromonas salmonicida*. *Journal of Fish Diseases* 19, 65–74. https://doi.org/10.1111/j.1365-2761.1996.tb00121.x

Puah, S.M., Khor, W.C., Kee, B.P., Tan, J.A.M.A., Puthucheary, S.D. and Chua, K.H. (2018) Development of a species-specific PCR-RFLP targeting *rpoD* gene fragment for discrimination of *Aeromonas* species. *Journal of Medical Microbiology* 67, 1271–1278. https://doi.org/10.1099/jmm.0.000796

Qiao, Y.-Q., Wang, J., Wang, H., Chai, B.-Z., Rao, C.-F. *et al.* (2019) 4-Hydroxyphenylpyruvate dioxygenase thermolability is responsible for temperature-dependent melanogenesis in *Aeromonas salmonicida* subsp. *salmonicida*. *Applied and Environmental Microbiology* 85, e01926-18. https://doi.org/10.1128/AEM.01926-18

Quinn, R.A. and Stevenson, R.M.W. (2012) Denaturing gradient gel electrophoresis for nonlethal detection of *Aeromonas salmonicida* in salmonid mucus and its potential for other bacterial fish pathogens. *Canadian Journal of Microbiology* 58, 563–571. https://doi.org/10.1139/W2012-024

Rabb, L., Cornick, J.W. and MacDermott, L.A. (1964) A macroscopic slide agglutination test for the presumptive diagnosis of furunculosis in fish. *The Progressive Fish-Culturist* 26, 118–120. https://doi.org/10.1577/1548-8640(1964)26[118:AMATFT]2.0.CO;2

Rakus, K., Ronsmans, M., Forlenza, M., Boutier, M., Piazzon, M.C. *et al.* (2017) Conserved fever pathways across vertebrates: a herpesvirus expressed decoy TNF-α receptor delays behavioural fever in fish. *Cell Host and Microbe* 21, 244–253. https://doi.org/10.1016/j.chom.2017.01.010

Rattanachaikunsopon, P. and Phumkhachorn, P. (2012) Detection of *Aeromonas salmonicida* by reverse transcription-multiplex polymerase chain reaction. *Bioscience, Biotechnology, and Biochemistry* 76, 665–670. https://doi.org/10.1271/bbb.110744

Rey, S., Moiche, V., Boltana, S., Teles, M. and MacKenzie, S. (2017) Behavioural fever in zebrafish larvae. *Developmental and Comparative Immunology* 67, 287–292. https://doi.org/10.1016/j.dci.2016.09.008

Rodriguez-Ramilo, S.T., Toro, M.A., Bouza, C., Hermida, M., Pardo, B.G. *et al.* (2011) QTL detection for *Aeromonas salmonicida* resistance related traits in turbot (*Scophthalmus maximus*). *BMC Genomics* 12, 541. https://doi.org/10.1186/1471-2164-12-541

Roger, F., Marchandin, H., Jumas-Bilak, E., Kodjo, A. and Lamy, B. (2012) Multilocus genetics to reconstruct aeromonad evolution. *BMC Microbiology* 12, 62. https://doi.org/10.1186/1471-2180-12-62

Romstad, A.B., Reitan, L.J., Midtlyng, P., Gravningen, K. and Evensen, Ø. (2013) Antibody responses correlate with antigen dose and *in vivo* protection for oil-adjuvanted, experimental furunculosis (*Aeromonas salmonicida* subsp. *salmonicida*) vaccines in Atlantic salmon (*Salmo salar* L.) and can be used for batch potency testing of vaccines. *Vaccine* 31, 791–796. https://doi.org/10.1016/j.vaccine.2012.11.069

Ronneseth, A., Haugland, G.T., Colquhoun, D.J., Brudal, E. and Wergeland, H.I. (2017) Protection and antibody reactivity following vaccination of lumpfish (*Cyclopterus lumpus* L.) against atypical *Aeromonas salmonicida*. *Fish and Shellfish Immunology* 64, 383–391. https://doi.org/10.1016/j.fsi.2017.03.040

Rose, A.S., Ellis, A.E. and Munro, A.L.S. (1990) Evidence against dormancy in the bacterial fish pathogen *Aeromonas salmonicida* subsp. *salmonicida*. *FEMS Microbiology Letters* 68, 105–108. https://doi.org/10.1111/j.1574-6968.1990.tb04131.x

Rouleau, F.D., Vincent, A.T. and Charette, S.J. (2018) Genomic and phenotypic characterization of an atypical *Aeromonas salmonicida* strain isolated from a lumpfish and producing unusual granular structures. *Journal of Fish Diseases* 41, 673–681. https://doi.org/10.1111/jfd.12769

Sakai, D.K. (1986) Electrostatic mechanism of survival of virulent *Aeromonas salmonicida* strains in river water. *Applied and Environmental Microbiology* 51, 1343–1349.

Sakai, M., Atsuta, S. and Kobayashi, M. (1986) A streptococcal disease of cultured jacopever, *Sebastes schlegeli*. *Suisanzoshuku (Aquiculture)* 34, 171–177.

Saleh, M., Soliman, H., Haenen, O. and El-Matbouli, M. (2011) Antibody-coated gold nanoparticles immunoassay for direct detection of *Aeromonas salmonicida* in fish tissues. *Journal of Fish Diseases* 34, 845–852. https://doi.org/10.1111/j.1365-2761.2011.01302.x

Samuelsen, O.B., Hjeltnes, B. and Glette, J. (1998) Efficacy of orally administered florfenicol in the treatment of furunculosis in Atlantic salmon. *Journal of Aquatic Animal Health* 10, 56–61. https://doi.org/10.1577/1548-8667(1998)010<0056:EOOAFI>2.0.CO;2

Santos, Y., García-Marquez, S., Pereira, P.G., Pazos, F., Riaza, A. *et al.* (2005) Efficacy of furunculosis vaccines in turbot, *Scophthalmus maximus* (L.): evaluation of immersion, oral and injection delivery. *Journal of Fish Diseases* 28, 165–172. https://doi.org/10.1111/j.1365-2761.2005.00610.x

Schwenteit, J.M., Weber, B., Milton, D.L., Bornscheuer, U.T. and Gudmundsdottir, B.K. (2015) Construction of *Aeromonas salmonicida* subsp. *achromogenes* AsaP1-toxoid strains and study their ability to induce immunity in Arctic char, *Salvelinus alpinus* L. *Journal of Fish Diseases* 38, 891–900. https://doi.org/10.1111/jfd.12303

Shaalan, M., El-Mandy, M., Theiner, S., Dinhopl, N., El-Matbouli, M. and Saleh, M. (2018) Silver nanoparticles: their role as antibacterial agent against *Aeromonas salmonicida* subsp. *salmonicida* in rainbow trout (*Oncorhynchus mykiss*). *Research in Veterinary Science* 119, 196–204. https://doi.org/10.1016/j.rvsc.2018.06.019

Silva, Y.J., Moeeirinha, C., Pereira, C., Costa, L., Rocha, R.J.M. *et al.* (2016) Biological control of *Aeromonas salmonicida* infection in juvenile Senegalese sole (*Solea senegalensis*) with Phage AS-A. *Aquaculture* 450, 225–233. https://doi.org/10.1016/j.aquaculture.2015.07.025

Siwicki, A.K., Klein, P., Morand, M., Kiczka, W. and Studnicka, M. (1998) Immunostimulatory effects of dimerized lysozyme (KLP-602) on the nonspecific defense mechanisms and protection against furunculosis in salmonids). *Veterinary Immunology and Immunopathology* 61, 369–378. https://doi.org/10.1016/S0165-2427(97)00140-2

Smith, P.R., Brazil, G.M., Drinan, E.M., O'Kelly, J., Palmer, R. and Scallan, A. (1982) Lateral transmission of furunculosis in sea water. *Bulletin of the European Association of Fish Pathologists* 3, 41–42.

Snieszko, S.F. (1958) Fish furunculosis. *Fishery Leaflet No. 467.* US Fish and Wildlife Service, Washington, DC.

Snieszko, S.F., Griffin, P. and Friddle, S.B. (1950) A new bacterium (*Haemophilus piscium* n. sp.) from ulcer disease of trout. *Journal of Bacteriology* 59, 699–710.

Snieszko, S.F., Dunbar, C.E. and Bullock, G.L. (1959) Resistance to ulcer disease and furunculosis in eastern brook trout, *Salvelinus fontinalis*. *The Progressive Fish-Culturist* 21, 111–116. https://doi.org/10.1577/1548-8659(1959)21[111:RTUDAF]2.0.CO;2

Stanley, C., Hiney, M., Gilroy, D., Powell, R., Padley, D. and Smith, P. (2002) Furunculosis-inducing ability of microcosms seeded with *Aeromonas salmonicida* correlates with colony-forming ability but not with PCR and ELISA data. *Aquaculture* 210, 35–48. https://doi.org/10.1016/S0044-8486(01)00856-0

Tam, B., Gough, W.A. and Tsuji, L. (2011) The impact of warming on the appearance of furunculosis in fish of the James Bay region, Quebec, Canada. *Regional Environmental Change* 11, 123–132. https://doi.org/10.1007/s10113-010-0122-8

Tapia-Cammas, D., Yanez, A., Arancibia, G., Toranzo, A.E. and Avendano-Herrera, R. (2011) Multiplex PCR for the detection of *Piscirickettsia salmonis*, *Vibrio anguillarum*, *Aeromonas salmonicida* and *Streptococcus phocae* in Chilean fish farms. *Diseases of Aquatic Organisms* 97, 135–142. https://doi.org/10.3354/dao02395

Taylor, P.W. and Winton, J.R. (2002) Optimization of nested polymerase chain reaction assay for identification of *Aeromonas salmonicida*, *Yersinia ruckeri*, and *Flavobacterium psychrophilum*. *Journal of Aquatic Animal Health* 14, 216–224. https://doi.org/10.1577/1548-8667(2002)014<0216:OONPCR>2.0.CO;2

Thornton, J.C., Garduño, R.A., Newman, S.G. and Kay, W.W. (1991) Surface-disorganized, attenuated mutants of *Aeromonas salmonicida* as furunculosis live vaccines. *Microbial Pathogenesis* 11, 85–99. https://doi.org/10.1016/0882-4010(91)90002-R

Thornton, J.C., Garduño, R.A. and Kay, W.W. (1994) The development of live vaccines for furunculosis lacking the A-layer and O-antigen of *Aeromonas salmonicida*. *Journal of Fish Diseases* 17, 195–204. https://doi.org/10.1111/j.1365-2761.1994.tb00215.x

Treasurer, J. and Cox, D. (1991) The occurrence of *Aeromonas salmonicida* in wrasse (*Labridae*) and implications for Atlantic salmon farming. *Bulletin of the European Association of Fish Pathologists* 11, 208–210.

Vaughan, L.M., Smith, P.R. and Foster, T.J. (1993) An aromatic-dependent mutant of the fish pathogen *Aeromonas salmonicida* is attenuated in fish and is effective as a live vaccine against the salmonid disease furunculosis. *Infection and Immunity* 61, 2172–2181.

Vercauteren, M., De Swaef, E., Declercq, A., Bosseler, L., Gulla, S. *et al.* (2018) First isolation of *Vibrio tapetis* and an atypical strain of *Aeromonas salmonicida* from skin ulceration in common dab (*Limanda limanda*) in the North Sea. *Journal of Fish Diseases* 41, 329–335. https://doi.org/10.1111/jfd.12729

Villumsen, K.R. and Raida, M.K. (2013) Long-lasting protection by bath vaccination against *Aeromonas salmonicida* subsp. *salmonicida* in rainbow trout. *Fish and Shellfish Immunology* 35, 1649–1653. https://doi.org/10.1016/j.fsi.2013.09.013

Villumsen, K.R., Koppang, E.O. and Raida, M.K. (2015) Adverse and long-term protective effects following oil-adjuvanted vaccination against *Aeromonas salmonicida* in rainbow trout. *Fish and Shellfish Immunology* 42, 193–203. https://doi.org/10.1016/j.fsi.2014.09.024

Vincent, A.T., Fernandez-Bravo, A., Sanchis, M., Mayayo, E., Figueras, M.J. and Charette, S.J. (2018) Investigation of the virulence and genomics of *Aeromonas salmonicida* strains isolated from human patients. *Infection, Genetics and Evolution* 68, 1–9. https://doi.org/10.1016/j.meegid.2018.11.019

Virsek, M.K., Manca, K., Lovsin, M.N., Koren, S., Krzan, A. and Peterlin, M. (2017) Microplastics as a vector for the transport of the bacterial fish pathogen species *Aeromonas salmonicida*. *Marine Pollution Bulletin* 125, 301–309. https://doi.org/10.1016/j.marpolbul.2017.08.024

Vivas, J., Riaño, J., Carracedo, B., Razquin, B.E., López-Fierro, P. *et al.* (2004) The auxotrophic *aroA* mutant of *Aeromonas hydrophila* as a live attenuated vaccine against *A. salmonicida* infections in rainbow trout (*Oncorhynchus mykiss*). *Fish and Shellfish Immunology* 16, 193–206. https://doi.org/10.1016/S1050-4648(03)00078-0

Wiklund, T. (1995) Survival of 'atypical' *Aeromonas salmonicida* in water and sediment microcosms of different salinities and temperatures. *Diseases of Aquatic Organisms* 21, 137–143. https://doi.org/10.3354/dao021137

Wiklund, T. and Dalsgaard, I. (1995) Atypical *Aeromonas salmonicida* associated with ulcerated flatfish species in the Baltic sea and the North Sea. *Journal of Aquatic Animal Health* 7, 218–224. https://doi.org/10.1577/1548-8667(1995)007<0218:AASAWU>2.3.CO;2

Wiklund, T., Sazonov, A., L'Iniova, G.P., Pugaewwa, V.P., Zoobaha, S.V. and Bylund, G. (1992) Characteristics of *Aeromonas salmonicida* subsp. *salmonicida* isolated from wild Pacific salmonids in Kamchatka, *Russia. Bulletin of the European Association of Fish Pathologists* 12, 76–79.

Williams, P.J., Courtenay, S.C. and Vardy, C. (1997) Use of enrofloxacin to control atypical *Aeromonas salmonicida* in Atlantic tomcod. *Journal of Aquatic Animal Health* 9, 216–222. https://doi.org/10.1577/1548-8667(1997)009<0216:UOETCA>2.3.CO;2

Wolf, L.E. (1954) Development of disease resistant strains of fish. *Transactions of the American Fisheries Society* 83, 342–349. https://doi.org/10.1577/1548-8659(1953)83[342:DODSOF]2.0.CO;2

Yamamoto, A. (2017) Typical and atypical *Aeromonas salmonicida* infections in fish. *Fish Pathology* 52, 126–130. https://doi.org/10.3147/jsfp.52.126

Yamin, G., Falk, R., Avtalion, R.R., Shoshana, N., Ofek, T. *et al.* (2017) The protective effect of humic-rich substances on atypical *Aeromonas salmonicida* subsp. *salmonicida* infection in common carp (*Cyprinus carpio* L.). *Journal of Fish Diseases* 40, 1783–1790. https://doi.org/10.1111/jfd.12645

Yi, Y.-L., Zhang, Z.-H., Zhao, F., Li, H., Yu, L.-J. *et al.* (2018) Probiotic potential of *Bacillus velezensis* JW: antimicrobial activity against fish pathogenic bacteria and immune enhancement effects on *Carassius auratus*. *Fish and Shellfish Immunology* 78, 322–330. https://doi.org/10.1016/j.fsi.2018.04.055

Yilmaz, S. Soya, M. and Ergun, S. (2018) Antimicrobial activity of *trans*-cinnamic acid and commonly used antibiotics against important fish pathogens and non-pathogenic isolates. *Journal of Applied Microbiology* 125, 1714–1727. https://doi.org/10.1111/jam.14097

12 Edwardsiellosis

MATT J. GRIFFIN[1]*, ESTEBAN SOTO[2] AND DAVID J. WISE[3]

[1]*College of Veterinary Medicine, Mississippi State University, Stoneville, Mississippi, USA;* [2]*University of California – Davis, Davis, California, USA;* [3]*Mississippi Agriculture and Forestry Experiment Station, Stoneville, Mississippi, USA*

12.1 Introduction

Global climate change will have a range of direct and indirect impacts on marine and freshwater capture fisheries and aquaculture. Regional fisheries are already under strain from overfishing, loss of habitat, pollution and introduced species stemming from increased industrialization and rampant globalization. In addition to these pressures, shifting weather patterns and putative increases in global temperatures have the potential to dramatically change aquatic ecosystems both biotically and chemically, conceivably modifying fish population structures as environmental suitability for different aquatic animal species is altered. Moreover, spans of extreme temperatures or drought may result in prolonged periods exceeding thermal tolerances of endemic fish species, causing undue stress, with resultant compromised states increasing susceptibility to both primary and opportunistic infectious agents. Aquaculture industries will need to expand efforts to meet increasing global demand for high-quality fish protein and other fish products and accommodate precipitous declines in wild catch harvests from marine and freshwater fisheries that may be exacerbated by changes in regional weather patterns.

Current predictions suggest global aquaculture productivity and yields will need to double by 2030 to meet worldwide demand (World Bank, 2014). In addition to this expansion, there is increased risk for the spread of aquatic pathogens through international trade and trans-boundary translocation of fish and fish products to new geographic regions for aquaculture purposes. Furthermore, infectious disease poses greater risk to aquaculture than wild capture fisheries given the propensity for proliferation of disease in intensive systems with high stocking densities and heavy organic loads. At present, there are limited studies evincing the direct effects of climate change on aquatic animal disease and elements of climate change are likely to affect discrete geographic locales variably. Impacts of a changing climate will be stochastic and, depending on locality, may be beneficial, detrimental or negligible. Regardless, shifting environmental conditions have the potential to alter the distribution, prevalence and virulence of infectious agents and may increase the susceptibility of respective host species. Consequently, geographic ranges of many aquatic animal pathogens could expand in the foreseeable future. Still, predictions regarding the effects of climate change on aquatic animal health are complicated as responses of endemic, emergent and exotic aquatic pathogens to major environmental changes are unfixed and likely depend on a combination of factors which include pathogen, host, geography and ecological niche.

The *Edwardsiella* are Gram-negative enteric pathogens mostly known for the diseases they cause in wild and cultured cool-, temperate- and warm-water fish (Mohanty and Sahoo, 2007; Griffin *et al.*, 2017). While historically a problem for temperate- and warm-water fish species, increased global temperatures and incidence of extreme temperature events could expand the geographic range of *Edwardsiella* spp. and disease outbreaks in cold-water fish may become more frequent, as prolonged exposure to temperatures that exceed thermal tolerances increase susceptibility to disease agents while simultaneously providing environments conducive to the proliferation of these enigmatic pathogens.

The *Edwardsiella* genus was first proposed in 1965 to accommodate a group of organisms isolated from humans and other terrestrial animals in the USA,

*matt.griffin@msstate.edu

Brazil, Ecuador, Israel and Japan that did not fit within any other Enterobacteriaceae genera. The type species *Edwardsiella tarda* represented organisms deemed 'bacterium 1483-59' which were phenotypically consistent with isolates previously categorized as the 'Bartholomew' and 'Asakusa' groups from humans and snakes (Ewing, 1965). In 1980 the second member of the genus, *Edwardsiella hoshinae*, was described as a commensal organism from birds and reptiles (Grimont *et al*., 1980). While often mentioned in the literature as a pathogen of fish, reports of *E. hoshinae* causing disease in fish, birds, reptiles or humans are notional. In 1981, Hawke *et al*. (1981) described the third member of the genus, *Edwardsiella ictaluri*, as the aetiological agent of enteric septicaemia of catfish (*Ictalurus punctatus*) (ESC) which is one of the most important pathogens in global catfish aquaculture.

Recently the *Edwardsiella* underwent a significant reorganization. Genetic studies in the late 1990s revealed isolates previously categorized as typical fish pathogenic *E. tarda* and atypical fish pathogenic *E. tarda* were more similar to *E. ictaluri* than non-fish pathogenic *E. tarda* (Yamada and Wakabayashi, 1999). Further work demonstrated isolates representing typical, atypical and non-fish pathogenic *E. tarda* carried variable pathogenicity in different fish hosts (Matsuyama *et al*., 1995). Throughout the 2000s and early 2010s, there was mounting evidence that fish pathogenic (typical and atypical) and non-fish pathogenic *E. tarda* formed polyphyletic groups, representing multiple genetically discrete taxa (Castro *et al*., 2006, 2011; Wang *et al*., 2011; Abayneh *et al*., 2012; Yang *et al*., 2012; Griffin *et al*., 2013). This culminated in the recognition of *Edwardsiella piscicida* (Abayneh *et al*., 2013) and *Edwardsiella anguillarum* (Shao *et al*., 2015) to represent isolates previously deemed typical and atypical fish pathogenic *E. tarda*, respectively (Yamada and Wakabayashi, 1999; Sakai *et al*., 2009a,b; Griffin *et al*., 2014; Reichley *et al*., 2017; Buján *et al*., 2018a). Comparably, bona fide *E. tarda* is synonymous with what was previously classified as non-fish pathogenic *E. tarda*, consistent with the original type strain from humans (ATCC 15947) (Griffin *et al*., 2014; Reichley *et al*., 2017). As a result, much of the historical literature regarding *E. tarda* must be viewed in consideration of current *Edwardsiella* systematics, expanding host records for some *Edwardsiella* spp. while deleting host records for others (Griffin *et al*., 2017).

The *Edwardsiella* are arguably one of the most important groups of fish pathogens with extremely wide host and geographic ranges (Table 12.1; Fig. 12.1). The impacts of *E. ictaluri* and *E. piscicida* on catfish aquaculture in the USA are well known (Wise *et al*., 2004; Griffin *et al*., 2014, 2019a). Similarly, outbreaks of *Edwardsiella* spp. in European aquaculture have been well documented (Castro *et al*., 2006; Padrós *et al*., 2006; Lan *et al*., 2008; Buján *et al*., 2018b) and there are numerous reports of *Edwardsiella* spp. from cultured and wild fish species in Asia, India, Latin America, the Caribbean, Africa and Australia (Bragg, 1988, 1991; Clavijo *et al*., 2002; Akinbowale *et al*., 2006; Sakai *et al*., 2008; Abraham, 2011; Ibrahem *et al*., 2011; Wei *et al*., 2011; Chenia and Vietze, 2012; Hassan *et al*., 2012; Soto *et al*., 2012; Walakira *et al*., 2014; Abraham *et al*., 2015; Mo *et al*., 2015; Lymbery *et al*., 2016; Phillips *et al*., 2017; Reichley *et al*., 2017; Kelly *et al*., 2018; Yang *et al*., 2018; Buján *et al*., 2018a,b; Dubey *et al*., 2019).

In addition to reports of infections in warm-water fishes, *Edwardsiella* spp. are considered a significant pathogen of cultured cool- and cold-water flatfishes such as turbot (*Psetta maxima*) (Castro *et al*., 2006, 2011) and flounder (*Paralichthys olivaceus*) (Rashid *et al*., 1994) in Europe and Asia. Similarly, *E. piscicida* is increasingly isolated from largemouth bass (*Micropterus salmoides*) in the USA (Fogelson *et al*., 2016; Camus *et al*., 2019) and *Edwardsiella* spp. have been isolated from disease outbreaks in salmonids, namely sea trout (*Salmo trutta*) (Reichley *et al*., 2017), rainbow trout (*Oncorhynchus mykiss*) (Reddacliff *et al*., 1996; Keskin *et al*., 2004; Řehulka *et al*., 2012), chinook salmon (*Oncorhynchus tshawytscha*) and brook trout (*Salvelinus fontinalis*) (Uhland *et al*., 2000). More recently *E. piscicida* has been isolated from mortality events in Atlantic salmon (*Salmo salar*) cultured in North America (M.J. Griffin and E. Soto, 2019, unpublished results). The host and geographic ranges of *Edwardsiella* appear to be expanding, although at the time of printing it is unknown if these new reports are indicative of future trends or represent isolated, unrelated incidents.

Established and nascent aquaculture industries in the tropics and subtropics have the potential for increased losses due to climate change-mediated disease mortality. It is expected that deterioration of the aquatic environment associated with increased global temperatures and extreme weather events will be more severe in these regions, especially for inland water bodies and pond culture, leading to thermal stress and increased incidence of disease. Tilapia are a popular cultured fish in the expanding

Table 12.1. Aquatic animal hosts of *Edwardsiella* spp.

Species	Host	Reference
Edwardsiella anguillarum	Blackspot sea bream (*Pagellus bogaraveo*)	Buján *et al.* (2018a)
	Blue striped grunt (*Haemulon sciurus*)	Reichley *et al.* (2017)
	Climbing perch (*Anabas testudineus*)	Dubey *et al.* (2019)
	Dusky grouper (*Epinephelus marginatus*)	Ucko *et al.* (2016)[b]
	European eel (*Anguilla anguilla*)	Shao *et al.* (2015)
	Gilthead bream (*Sparus aurata*)	Buján *et al.* (2018b)
	Japanese eel (*Anguilla japonica*)	Shao *et al.* (2015)
	Marbled eel (*Anguilla marmorata*)	Shao *et al.* (2015)
	Red sea bream (*Pagrus major*)	Matsuyama *et al.* (2005)[a]; Shao *et al.* (2015)
	Rohu (*Labeo rohita*)	Dubey *et al.* (2019)
	Sharpsnout sea bream (*Diplodus puntazzo*)	Katharios *et al.* (2019)
	Silver surfperch (*Hyperprosopon ellipticum*)	Dubey *et al.* (2019)
	Spotted snakehead (*Ophiocephalus punctatus*)	Dubey *et al.* (2019)
	Striped bass (*Morone saxatilis*)	Reichley *et al.* (2017)
	Striped catfish (*Pangasianodon hypophthalmus*)	Dubey *et al.* (2019)
	Taiwanese worm eel (*Sympenchelys taiwanensis*)	Dubey *et al.* (2019)
	Tilapia (*Oreochromis* sp.)	Griffin *et al.* (2014)[b]; Armwood *et al.* (2019); López-Porras *et al.* (2019)
	Walking catfish (*Clarias batrachus*)	Dubey *et al.* (2019)
	White grouper (*Epinephelus aeneus*)	Ucko *et al.* (2016)[b]; Reichley *et al.* (2017)
Edwardsiella ictalurid	Ayu (*Plecoglossus altivelis*)	Nagai *et al.* (2008)
	Bagrid catfish (*Pelteobagrus nudiceps*)	Sakai *et al.* (2009b)
	Brown bullhead (*Ameiurus nebulosus*)	Iwanowicz *et al.* (2006)
	Channel catfish (*Ictalurus punctatus*)	Hawke *et al.* (1981)
	Danio (*Danio devario*)	Waltman *et al.* (1985)
	Green knife fish (*Eigemannia virescens*)	Kent and Lyons (1982)
	Rainbow trout (*Oncorhynchus mykiss*)	Keskin *et al.* (2004)
	Rosy barb (*Pethia conchonius*)	Humphrey *et al.* (1986)
	Striped catfish (*Pangasius hypophthalmus*)	Crumlish *et al.* (2002)
	Tadpole madtom (*Noturus gyrinus*)	Klesius *et al.* (2003)
	Tilapia (*Oreochromis niloticus*)	Soto *et al.* (2012)
	Walking catfish (*C. batrachus*)	Kasornchandra *et al.* (1987)
	Wet tropics tandan (*Tandanus tropicanus*)	Lymbery *et al.* (2016); Kelly *et al.* (2018)
	Yellow catfish (*Pelteobagrus fulvidraco*)	Ye *et al.* (2009)
	Zebrafish (*Danio rerio*)	Hawke *et al.* (2013)
Edwardsiella piscicida	African catfish (*Clarias gariepinus*)	Shao *et al.* (2015)
	Ayu (*P. altivelis*)	Yamada and Wakabayashi (1999)[a]; Buján *et al.* (2018b)
	Barramundi (*Lates calcarifer*)	Loch *et al.* (2017)
	Black crappie (*Pomoxis nigridis*)	Griffin *et al.* (2019b)
	Blackspot sea bream (*P. bogaraveo*)	Castro *et al.* (2006)[a]; Buján *et al.* (2018b)
	Blotched fantail stingray (*Taeniura meyeni*)	Camus *et al.* (2016)
	Blue catfish (*Ictalurus furcatus*)	Griffin *et al.* (2014)
	Channel catfish (*I. punctatus*)	Griffin *et al.* (2014)
	European eel (*A. anguilla*)	Abayneh *et al.* (2013)
	Hybrid catfish (*I. punctatus* × *I. furcatus*)	Griffin *et al.* (2014, 2019a)

Continued

Table 12.1. Continued.

Species	Host	Reference
	Japanese eel (*A. japonica*)	Yamada and Wakabayashi (1999)[a]; Buján *et al.* (2018b)
	Japanese flounder (*Paralichthys olivaceus*)	Nakatsugawa (1983)[a]; Buján *et al.* (2018b); Dubey *et al.* (2019)
	Koi (*Cyprinus carpio*)	Reichley *et al.* (2017)
	Korean catfish (*Silurus asotus*)	Yu *et al.* (2009)[a]; Abayneh *et al.* (2013); Shao *et al.* (2015); Buján *et al.* (2018b)
	Largemouth bass (*Micropterus salmoides*)	Fogelson *et al.* (2016); Camus *et al.* (2019)
	Marbled eel (*A. marmorata*)	Yang *et al.* (2013)[a]; Buján *et al.* (2018b)
	Olive flounder (*P. olivaceus*)	Oguro *et al.* (2014)
	Rainbow trout (*O. mykiss*)	Reichley *et al.* (2017)
	Sea trout (*Salmo trutta* morpha *trutta*)	Reichley *et al.* (2017)
	Serpae tetra (*Hyphessobrycon eques*)	Ling *et al.* (2000)[a]; Shao *et al.* (2015)
	Sharpsnout bream (*D. puntazzo*)	Katharios *et al.* (2015)[a]; Buján *et al.* (2018b)
	Smallmouth bass (*Micropterus dolomieu*)	Reichley *et al.* (2017)
	Sole (*Solea senegalensis*)	Castro *et al.* (2012)[a]; Buján *et al.* (2018b)
	Striped bass (*M. saxatilis*)	Baya *et al.* (1997)[a]; Reichley *et al.* (2017); Buján *et al.* (2018b)
	Striped catfish (*P. hypophthalmus*)	Shetty *et al.* (2014)[a]; Buján *et al.* (2018b)
	Tilapia (*Oreochromis* sp.)	Yamada and Wakabayashi (1999)[a]; Griffin *et al.* (2014)
	Turbot (*Scophthalmus maximus*)	Castro *et al.* (2006)[a]; Abayneh *et al.* (2013)
	Whitefish (*Coregonus lavaretus*)	Shafiei *et al.* (2016)
Edwardsiella tarda	Asian catfish (*C. gariepinus*)	Abraham *et al.* (2015)
	Asian swamp eel (*Monopterus albus*)	Shao *et al.* (2016)
	Ayu (*P. altivelis*)	Yamada and Wakabayashi (1999)
	Barcoo grunter (*Scortum barcoo*)	Ye *et al.* (2010)
	Barramundi (*L. calcarifer*)	Humphrey and Langdon (1986); Loch *et al.* (2017)
	Brook trout (*Salvelinus fontinalis*)	Uhland *et al.* (2000)
	California sea lion (*Zalophus californianus*)	Coles *et al.* (1978)
	Channel catfish (*I. punctatus*)	Meyer and Bullock (1973)
	Chinese soft-shell turtle (*Trionyx sinensis*)	Pan *et al.* (2010)
	Chinook salmon (*Oncorhynchus tshawytscha*)	Amandi *et al.* (1982)
	Coloured carp (*C. carpio*)	Sae-oui *et al.* (1984)
	Crimson sea bream (*Evynnis japonica*)	Kusuda *et al.* (1977)
	Dabry's sturgeon (*Acipenser dabryanus*)	Yang *et al.* (2018)
	European eel (*A. anguilla*)	Alcaide *et al.* (2006)
	Golden tiger barb (*Puntius tetrazona*)	Akinbowale *et al.* (2006)
	Harbour porpoise (*Phocena phocena*)	Coles *et al.* (1978)
	Hybrid striped bass (*Morone chrysops* × *M. saxatilis*)	Griffin *et al.* (2014)
	Japanese eel (*A. japonica*)	Yamada and Wakabayashi (1999); Shao *et al.* (2015)
	Japanese flounder (*P. olivaceus*)	Yamada and Wakabayashi (1999)
	Korean catfish (*S. asotus*)	Yu *et al.* (2009)
	Largemouth bass (*M. salmoides*)	White *et al.* (1973)
	Mullet (*Mugil cephalus*)	Kusuda *et al.* (1976)

Continued

Table 12.1. Continued.

Species	Host	Reference
	Oscar fish (*Astronotus ocellatus*)	Wang *et al.* (2011)
	Pangas catfish (*Pangasius pangasius*)	Nakhro *et al.* (2013)
	Red sea bream (*Chrysophrys major*)	Yamada and Wakabayashi (1999)
	Sea bass (*Dicentrarchus labrax*)	Blanch *et al.* (1990)
	Siamese fighting fish (*Betta splendens*)	Humphrey *et al.* (1986)
	Silver carp (*Hypophthalmichthys molitrix*)	Xu and Zhang (2014)
	Sperm whale (*Physeter macrocephalus*)	Cools *et al.* (2013)
	Striped bass (*M. saxatilis*)	Herman and Bullock (1986)
	Steller's sea lion (*Eumetopias jubata*)	Coles *et al.* (1978)
	Tilapia (*O. niloticus*)	Yamada and Wakabayashi (1999); Reichley *et al.* (2017)
	Toadfish	Reichley *et al.* (2017)
	Western African lungfish (*Protopterus annectens*)	Rousselet *et al.* (2018)
	False killer whale (*Pseudorca crassidens*)	Lee *et al.* (2018)

[a]Reported as *E. tarda*.
[b]Reported as *E. piscicida*-like.

Fig. 12.1. Range map of *Edwardsiella* spp. Several reports of *Edwardsiella tarda* occurred prior to the recognition of *Edwardsiella piscicida* and *Edwardsiella anguillarum*. As such, the geographic range of *E. piscicida* and *E. anguillarum* may be under-represented.

aquaculture industries of Latin America, Africa and the Caribbean (Fitzsimmons, 2000; Hernández-Rodríguez *et al.*, 2001) and *E. ictaluri*, *E. piscicida*, *E. anguillarum* and *E. tarda* have all been reported from disease outbreaks in cultured tilapia (Iregui *et al.*, 2012; Soto *et al.*, 2012; Griffin *et al.*, 2014; Buján *et al.*, 2018b; Armwood *et al.*, 2019; Dong *et al.*, 2019; López-Porras *et al.*, 2019). If current predictions hold true, it is likely *Edwardsiella* spp. will continue to burden tilapia cage and pond aquaculture if effective remedial strategies are not identified.

In contrast to climate change dogma, shifting weather patterns could also lead to unseasonable periods of anomalous low temperatures in some localities. For example, atypical mid-summer cold fronts have occurred in the catfish-farming region of the south-eastern USA over the past several production cycles, resulting in atypical ESC outbreaks during phases of peak production. In catfish aquaculture, the temperature range for ESC is typically between 22 and 28°C (Francis-Floyd *et al.*, 1987). Pond temperatures commonly exit this 'ESC window' by mid-June, with warm temperatures (>30°C) extending through the rest of the growing season until the autumn when temperatures decline with the onset of shorter days and cooler nights. However, in recent years anomalous cold fronts in late June and early July have expanded the 'ESC season', as cool night-time temperatures and cold front-associated precipitation have led to periods of unseasonably low pond temperatures. Furthermore, heavy precipitation events (>50 mm) associated with these frontal systems can cause rapid temperature changes and dramatic pH shifts in shallower ponds with poor buffering capacity, and 'turnover' in deeper ponds can result in brief periods of hypoxic stress for fish. These changes can have significant impacts on the health status and feeding activity of fish, which can impede treatment options as disease-induced inappetence and environmental disruptions to feed intake can reduce the efficacy of medicated feeds. Furthermore, aberrant weather events can complicate immunization practices, especially for orally delivered or live attenuated vaccines, as feeding activity is reduced and pond conditions may be conducive for disease outbreaks. The impacts of recent non-typical weather patterns in US catfish aquaculture in relation to the *Edwardsiella* are discussed further in Section 12.5.3.

Regardless, management practices to minimize impacts of *Edwardsiella* spp. in aquaculture will predictably stay the current course, with reliance on antibiotic feeds and vaccination. While bacteriophage therapies have been proposed as alternatives to antibiotics in aquaculture (Defoirdt *et al.*, 2011; Oliveira *et al.*, 2012; Richards, 2014) and *Edwardsiella*-specific bacteriophages have been identified (Walakira *et al.*, 2008; Carrias *et al.*, 2011; Yasuike *et al.*, 2013), anti-*Edwardsiella* spp. phage therapies have yet to garner widespread adoption in aquaculture settings.

Despite predictions of climatic shifts and extreme weather events leading to prolonged heatwaves, atypical cold snaps and anomalous precipitation, the most effective measures to manage edwardsiellosis in global aquaculture will assuredly be prophylactic approaches stemming from the development of effective vaccines. Efficacious vaccines can increase survival, growth and feed efficiency while reducing dependency on medicated feeds, leading to increased farm profitability (Thorarinsson and Powell, 2006; Wise *et al.*, 2020). Reducing dependence on antibiotic feeds is another important benefit, particularly given the reluctance of regulatory bodies to approve additional medications for aquaculture use, the growing recognition of antimicrobial resistance reservoirs in agriculture systems and increasing consumer demand for antibiotic-free foods.

Vaccination practices in multiple aquaculture industries across the globe have delivered promising results. Moreover, with the recent recognition that isolates previously designated as typical, atypical and fish-non-pathogenic *E. tarda* actually represented three genetically and serologically discrete taxa (Abayneh *et al.*, 2013; Shao *et al.*, 2015; Reichley *et al.*, 2015b, 2017), previous hurdles preventing the development of effective *Edwardsiella* vaccines have been removed (Griffin *et al.*, 2017). Regardless of meteorological conditions, antibiotic intervention will continue to be a course of action in response to disease outbreaks. As efficacious and practical vaccine candidates are identified and delivery strategies are optimized, reliance on medicated feeds to treat *Edwardsiella* spp. should be reduced. Research investigating measures to reduce impact of *Edwardsiella* spp. on global fish health, climate and meteorological dynamics notwithstanding, is summarized in Sections 12.2, 12.3, 12.4 and 12.5.

12.2 *Edwardsiella tarda*

12.2.1 Background

E. tarda has been one of the most economically and environmentally important pathogens of fish, with reports from all seven continents (Meyer and Bullock, 1973; Van Damme and Vandepitte, 1980; Clavijo *et al.*, 2002; Akinbowale *et al.*, 2006; Alcaide *et al.*, 2006; Leotta *et al.*, 2009; Joh *et al.*, 2011; Xu and Zhang, 2014). In 1973, Meyer and Bullock reported *E. tarda* as the causative agent of emphysematous putrefactive disease of channel catfish (*I. punctatus*) from an aquaculture operation in Arkansas. Genetic studies have since identified these original case isolates as bona fide *E. tarda* (Reichley *et al.*, 2017). While the segregation of *E. tarda* into multiple discrete taxa has minimized its status as a significant fish pathogen, *E. tarda* is still regarded in many circles as a globally important disease agent of temperate- and warm-water fish (Abayneh *et al.*, 2012; Griffin *et al.*, 2013, 2014, 2017, 2019a; Shao *et al.*, 2015; Reichley *et al.*, 2017, 2018; Buján *et al.*, 2018b).

Although much of the historical data describing fish pathogenic *E. tarda* infections is likely referencing *E. piscicida* or *E. anguillarum*, *E. tarda* remains an important pathogen of cultured fish, predominantly associated with opportunistic infections in warm-water environments which could be exacerbated by current predictions towards a warmer climate. Environmental variables such as high temperature, poor water quality and high organic content contribute to the severity of infections (Meyer and Bullock, 1973; Hawke and Khoo, 2004; Mohanty and Sahoo, 2007; Park *et al.*, 2012). Seasonal outbreaks in temperate- and cold-water cultured and wild fish typically occur in the late summer/early autumn, when water temperatures are at maximum and approach the upper ranges of thermal tolerance for cool- and temperate-water fish. If global temperatures continue to rise, incidence and prevalence of *E. tarda* may increase in cool-, temperate- and warm-water fish as periods of abnormally high temperatures and diminished water quality are prolonged.

In channel catfish, the disease begins as small cutaneous ulcers, progressing into large gas-filled abscesses in the musculature, typically when water temperatures exceed 30°C. In chronic infections, abscesses fill with malodorous gas and necrotic tissue, although a generalized septicaemia is more common. While rare, outbreaks occur in large fish (>450 g) from July through October, during periods of high water temperatures and peak production, when ponds have elevated levels of organic material (Meyer and Bullock, 1973; Hawke and Khoo, 2004; Griffin *et al.*, 2019a).

Similarly, *E. tarda* was implicated in a largemouth bass kill in a Florida lake late in the summer of 1991. These die-offs were similar to other seasonal fish kills in eutrophic and mesotrophic Florida lakes attributed to oxygen deficits during periods of high water temperatures. The mortality event persisted for 6 weeks, involving approximately 1500 fish. While mixed populations of aerobic bacteria were recovered from fish, pure cultures of *E. tarda* were consistently recovered from viscera (Francis-Floyd *et al.*, 1993). This finding is consistent with previous reports of *E. tarda* recovered from reptiles, birds, fish and surface waters of Florida lakes (White *et al.*, 1973) during periods of high water temperatures. Similarly, Wyatt *et al.* (1979) reported that isolation of *E. tarda* from fish, reptiles and invertebrates correlated with increased temperatures in catfish ponds in the south-eastern USA and Miniero Davies *et al.* (2018) surmised poor water quality and high temperatures were contributing factors to *E. tarda* outbreaks in fish and aquatic birds in Brazil.

There are also reports of *E. tarda* from cold-water fishes. Fish kills occurred in pre-spawning wild adult chinook salmon in Oregon in 1979, 1980 and 1981. Isolates biochemically and serologically consistent with *E. tarda* were isolated from moribund and dead fish (Amandi *et al.*, 1982). Amandi *et al.* (1982) and Uhland *et al.* (2000) also reported summer outbreaks of *E. tarda* in brook trout in Canada, attributing outbreaks to stress-induced immunosuppression brought on by high water temperatures and drought. Congruous to these reports, *E. tarda* has been isolated from outbreaks in intensively reared rainbow trout in the Czech Republic (Řehulka *et al.*, 2012), although water temperatures (~14°C) were deemed within acceptable limits, suggesting heat stress was not a contributing factor.

E. tarda is also considered a zoonotic pathogen, although the evidence is conflicting. Research performed prior to the recognition of *E. piscicida* and *E. anguillarum* demonstrated that isolates phenotypically characterized as *E. tarda* carried varying degrees of pathogenicity in mice, with median lethal doses ranging from $<10^7$ to $>10^8$ colony-forming units (cfu) (Janda *et al.*, 1991; Baya *et al.*, 1997; Castro *et al.*, 2011). Furthermore, *E. tarda* FL95-01, *E. anguillarum* LADL 05-105 and *E. piscicida* S11-285 were found avirulent in mice with intraperitoneal (IP) doses exceeding 10^8 cfu (M.J. Griffin, 2016, unpublished results). Similar to other pathogens (e.g. aeromonads, vibrios and salmonellas), *E. tarda* has been implicated in severe extraintestinal infections in compromised individuals. However, human infections with *E. tarda* are extremely rare and primarily limited to sporadic cases of gastroenteritis (Clarridge *et al.*, 1980; Janda and Abbott, 1993; Hirai *et al.*, 2015). While noteworthy, most reports of zoonotic edwardsiellosis occurred prior to the recognition of *E. piscicida* and *E. anguillarum*. It is presently unknown which of the three taxa (*E. anguillarum*, *E. piscicida* or *E. tarda*) are typically associated with disease in humans as all are capable of growth at 37°C (Abayneh *et al.*, 2013; Griffin *et al.*, 2013; Shao *et al.*, 2015). Still, the role of the *Edwardsiella* in human disease is limited and they likely pose no greater zoonotic threat than other Gram-negative bacteria frequently encountered in aquatic environments. While environmental changes could increase the incidence of edwardsiellosis in temperate- and cold-water fish, *Edwardsiella* spp. are unlikely to become a more substantial threat to consumers, fish handlers or sportsman given the opportunistic nature of these reported zoonoses.

Still, as with any food-borne Gram-negative bacterium, proper sanitation techniques should be employed when handling raw fish and raw fish products to reduce exposure to *Edwardsiella* spp. (Haenen *et al.*, 2013).

12.2.2 Treatment

Independent of changing weather patterns, antibiotic treatments remain one of the few reliable tools aquaculturists have to combat bacterial infections. The effect of common aquaculture chemicals topically applied to surfaces against *E. tarda* (isolate USFWS 9.36) found that ethyl alcohol (30, 50 or 70%), benzyl-4-chlorophenol/phenylphenol (1%), sodium hypochlorite (50, 100, 200 or 50,000 mg/l), *N*-alkyl dimethyl benzyl ammonium chloride (1:256), povidone iodine (50 or 100 mg/l), glutaraldehyde (2%) and potassium peroxymonosulfate/sodium chloride (1%) were effective disinfectants, while chloramine-T (15 mg/l) and formalin (250 mg/l) were not (Mainous *et al.*, 2010).

A survey in the 1980s of isolates from the USA and Taiwan found *E. tarda* susceptible to aminoglycosides, cephalosporins, penicillins, nitrofurantoin, sulfamethoxazole/trimethoprim and quinolones. Meanwhile, resistance was observed to penicillin G, sulfadiazine, colistin, novobiocin, spectinomycin, ampicillin, tetracycline and chloramphenicol (Waltman and Shotts, 1986a). Reichley *et al.* (2017) examined the antibiotic susceptibility profiles of several molecularly confirmed *E. tarda* isolates, including the original isolates from channel catfish (Meyer and Bullock, 1973), and found them susceptible to a range of antibiotics, including florfenicol and oxytetracycline, both of which are approved for use in US food-fish aquaculture.

In the original report of *E. tarda* from fish, Meyer and Bullock (1973) reported that oxytetracycline administered with feed (55 mg/kg for 10 days) reduced mortality in channel catfish within 3 days. Similarly, an *E. tarda* outbreak in brook trout in Quebec, Canada was treated using oxytetracycline mixed with the feed and vegetable oil (dosage 100 mg/kg live weight) (Uhland *et al.*, 2000). Moreover, oxytetracycline, florfenicol and oxolinic acid have been reported efficacious against *E. tarda* in fish (Kusuda and Kawai, 1998), with reports of tetracycline being used to treat *E. tarda* in Asia (Lo *et al.*, 2014).

As can be expected with prolonged or repeated use, antibiotic resistance in *E. tarda* has been documented. Terramycin (oxytetracycline) resistance was reported in *E. tarda* from channel catfish polycultured with tilapia (Hilton and Wilson, 1980) and R plasmid-mediated antibiotic resistance to oxytetracycline in *E. tarda* has been reported (Aoki and Takahashi, 1987). Isolates from eels in Taiwan carried resistance to oxytetracycline and doxycycline (Lo *et al.*, 2014). A survey of *E. tarda* from fish, water and sediments in a freshwater culture system revealed 78% of isolates carried multi-drug resistance (Acharya *et al.*, 2007) and *E. tarda* from golden tiger barb (*Puntius tetrazona*) in Australia was resistant to ampicillin, amoxycillin, erythromycin, chloramphenicol, tetracycline and oxytetracycline (Akinbowale *et al.*, 2006).

Similar findings were reported from Korea, as tetracycline resistance determinants were detected on both mobile and non-mobile genetic elements of *E. tarda* isolates from epizootics on fish farms occurring from 1993 to 2002 (Jun *et al.*, 2004). Likewise, Loch *et al.* (2017) reported plasmid-mediated tetracyline resistance in *E. tarda* from barramundi (*Lates calcarifer*) cultured in the mid-western USA and multi-drug resistant *E. tarda* isolates have been reported from China (Sun *et al.*, 2009). In fact, the emergence of antibiotic resistance has spurred the development of new therapeutics in Japan. Eels infected with drug-resistant strains of *E. tarda* were successfully treated with a 1:3 combination of ormetoprim/sulfamonomethoxine (25 mg/kg daily), oxolinic acid (12.5 mg/kg daily) and miloxacin (6.2 mg/kg daily) (Aoki *et al.*, 1989). If increases in global temperatures result in increased prevalence of bacterial infections in aquaculture, it is likely antibiotic use will also increase if alternative measures are not identified. These studies further emphasize the importance of appropriate and responsible antibiotic use to prolong the effectiveness of the limited antimicrobial agents currently available to manage bacterial infections in fish.

12.2.3 Prevention

The economic importance of the *Edwardsiella*, coupled with concerted efforts to mitigate antibiotic use in global agriculture, have spurred tremendous research investment in vaccine development. However, much of the historical literature is clouded by the recent reorganization of the genus. Still, these recent developments offer an explanation as to why effective *E. tarda* vaccines have been so elusive (Kawai *et al.*, 2004; Mohanty and Sahoo, 2007), largely because researchers, aquaculturists and fish health professionals were unknowingly targeting multiple intraspecific *Edwardsiella* spp. (Griffin *et al.*, 2017).

Despite these obstacles, researchers have demonstrated limited success in immunizing fish against *E. tarda* in experimental trials. In rohu (*Labeo rohita*), strong protection against *E. tarda* infection was obtained via bath immersions with an *E. tarda* bacterin, although efficacy was dependent on the age of fish and duration of the bath. Fish less than 3 weeks old did not respond to immunization (Swain *et al.*, 2002). In eels (*Anguilla japonica*), intramuscular immunization seemingly enhanced survival when challenged via intramuscular (IM) injections with virulent field isolates (Salati and Kusuda, 1985). Japanese flounder (*P. olivaceus*) fed formalin-killed *E. tarda* cells, the immunostimulant Curdlan (bacterial β-1,3-glucan) and a quillaja saponin suspension daily for 3 weeks survived better than control fish (Ashida *et al.*, 1999). Also, in flounder, vaccination with an *E. tarda* outer membrane protein via IP injection evoked strong protection against multiple *E. tarda* serotypes (Kawai *et al.*, 2004). Likewise, vaccination of flounder with a double-knockout *E. tarda* strain (NH1) conferred protection against challenge when exposed to wild-type NH1 (Choi and Kim, 2011) and injection of turbot with live and formalin-killed *E. tarda* resulted in significant improvements in survival in laboratory trials (Castro *et al.*, 2008). As is the case for most fish vaccines, there is no shortage of viable candidates. However, many are delivered as injectable vaccines and/or recombinant, genetically modified organisms, which carry logistic, regulatory and market complications that have precluded widespread adoption in most aquaculture industries.

12.3 *Edwardsiella anguillarum*

12.3.1 Background

E. anguillarum (syn. *E. piscicida*-like sp.) represents a group of genetically distinct isolates phenotypically indistinguishable from *E. tarda* and *E. piscicida*. As such, confirmatory identification of *E. anguillarum* requires species-specific polymerase chain reaction (PCR) assays (Griffin *et al.*, 2014; Reichley *et al.*, 2015b, 2017) or sequencing of relevant, high-resolution genetic markers (Yamada and Wakabayashi, 1999; Griffin *et al.*, 2014, 2016; Reichley *et al.*, 2017). *E. anguillarum* is genetically synonymous with isolates previously described as atypical, non-motile *E. tarda* (Yamada and Wakabayashi, 1999; Matsuyama *et al.*, 2005; Sakai *et al.*, 2007, 2009c; Griffin *et al.*, 2017; Reichley *et al.*, 2017). In the description of *E. anguillarum*, Shao *et al.* (2015) cited motility by peritrichous flagella and Griffin *et al.* (2013) observed motility at both 25 and 37°C for *E. anguillarum* isolate LADL 05-105 (Reichley *et al.*, 2015a). Conversely, Katharios *et al.* (2019) reported that *E. anguillarum* isolate EA011113 associated with disease outbreaks in aquaculture operations in Greece lacked flagella, suggesting motility is variable for this group.

Originally described as the cause of haemorrhagic septicaemia in eels (Shao *et al.*, 2015), recent reports have expanded the host range of *E. anguillarum* (Reichley *et al.*, 2017) and new host and geographic records continue to emerge (Dubey *et al.*, 2019; López-Porras *et al.*, 2019). In Taiwan, *E. anguillarum* has been isolated from the Taiwanese worm eel (*Sympenchelys taiwanensis*) and the silver surfperch (*Hyperprosopon ellipticum*) (Dubey *et al.*, 2019). Similarly, *E. anguillarum* has been reported from spotted snakehead (*Ophiocephalus punctatus*), climbing perch (*Anabas testudineus*), walking catfish (*Clarias batrachus*), striped catfish (*Pangasianodon hypophthalmus*) and rohu (*L. rohita*) in India (Dubey *et al.*, 2019). While the name would suggest *E. anguillarum* is notably a pest of *Anguilla* spp., it has been suggested this nomenclature is misleading (Ucko *et al.*, 2016), as *E. anguillarum* is also recognized as a pathogen of several marine fish (Matsuyama *et al.*, 2005; Griffin *et al.*, 2017; Reichley *et al.*, 2017;). Furthermore, recent reports suggest *E. anguillarum* is of significant concern in cultured tilapia (Armwood *et al.*, 2019), grouper (*Epinephelus aeneus* and *Epinephelus marginatus*) (Ucko *et al.*, 2016) and sharpsnout sea bream (*Diplodus puntazzo*) (Katharios *et al.*, 2019).

In tilapia, outbreaks of *E. anguillarum* result in erratic swimming, exophthalmia and what has been described as a progressive lethargy. Histologically, *E. anguillarum* causes an acute necrotizing to chronic granulomatous septicaemia, with acute lesions consisting of variably sized foci of coagulative necrosis, while chronic lesions consist of discrete granulomas (Fig. 12.2). Bacteria can be observed within macrophage cytoplasm, consistent with the facultative intracellular nature of *Edwardsiella* spp. In some instances, granulomatous inflammation lines the spinal meninges, sometimes infiltrating the spinal cord, and granulomatous infiltrates have been observed enveloping the olfactory nerves. While tilapia lack a cranial fontanelle necessary for development of

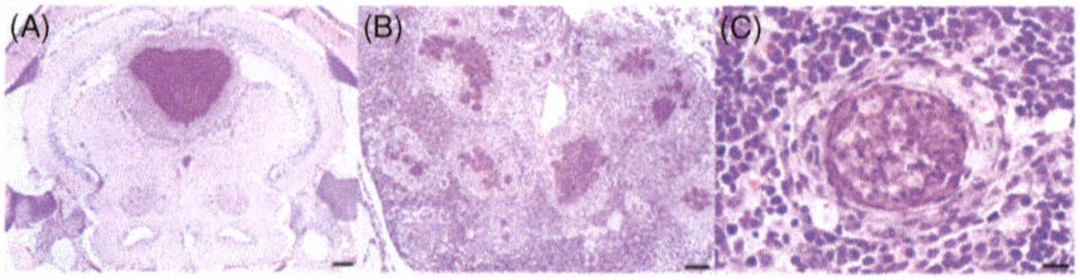

Fig. 12.2. Photomicrographs of naturally occurring *Edwardsiella anguillarum*-induced lesions in cultured Nile tilapia, *Oreochromis niloticus*. (A) A large granuloma in the brain, composed of a broad outer layer of epithelioid macrophages surrounding an inner core of necrotic debris and bacterial bacilli, fills the third ventricle, with granulomatous inflammation extending over the dorsal surface of the optic tectum. Scale bar = 200 μm. (B) Marked splenomegaly is caused by widespread, multifocal to coalescing, variably demarcated granulomas, composed of epithelioid macrophages, which often surround extensive areas of central necrosis when more chronic. Spleen and kidney are frequently severely affected. Scale bar = 100 μm. (C) *E. anguillarum* in the head kidney of a Nile tilapia. Inflammation can vary from acute, necrotizing lesions to chronic granulomas. Subacute granulomas are characterized by thin outer mantles of attenuated epithelioid macrophages, surrounding an inner core of necrotic debris and numerous intracellular bacterial bacilli. Scale bar = 10 μm. All sections stained with haematoxylin and eosin. (Photographs courtesy of Dr Abigail Armwood, University of Georgia.)

'hole-in-the head' as seen in catfish, a recent histological description of an *E. anguillarum* outbreak in cultured tilapia in Central America reported meningoencephalitis and/or ventriculitis extending into the cranial cavity. This would suggest a possible route of infection to the central nervous system through the olfactory epithelium (Armwood *et al.*, 2019).

E. anguillarum has also been reported from cultured grouper (*E. aeneus* and *E. marginatus*) (Ucko *et al.*, 2016). Behavioural changes and clinical signs were non-descript; affected fish showed loss of equilibrium and were observed lying at the bottom of the tank or floating at the surface as a result of hyperinflated swim bladders. Grossly, skin ecchymoses and ulcerations, fin and tail erosion and, occasionally, exophthalmia and cataracts were also observed. Infections are also associated with branchitis and pale gills. Internally, there was accumulation of ascitic fluid in the abdominal cavity, with congestion, haemorrhage and oedema present in the kidney, heart, spleen and liver. Whitish watery abscesses were present on the surface of these organs. Occasionally, large purulent abscesses spread into the surrounding musculature. Similarly, *E. anguillarum* has been reported from sharpsnout sea bream raised in floating cage culture in Greece. Fish were moribund, and upon necropsy presented with nodules and abscesses in the spleen and kidney, with cumulative mortality of around 5% (Katharios *et al.*, 2015, 2019).

Tantamount to isolates previously deemed 'atypical' fish pathogenic *E. tarda*, the *E. anguillarum* phylogroup has been long recognized as an important pathogen of cultured, marine fishes (Yamada and Wakabayashi 1999; Matsuyama *et al.*, 2005; Sakai *et al.*, 2009a), with an expansive and increasing host range (Shao *et al.*, 2015; Griffin *et al.*, 2017; Reichley *et al.*, 2017; Dubey *et al.*, 2019). Factors contributing to epizootics appear to be associated with suboptimal culture conditions, including overcrowding, increased organic loads and high water temperatures. Indeed, warming trends could be associated with the ostensible spread of this pathogen, although it is more likely attributed to expansion of aquaculture and new culture species in developing regions like Latin America and Africa.

12.3.2 Treatment/prevention

Limited information exists regarding the prevention and treatment of *E. anguillarum* as it was first recognized as a distinct species in 2015 (Shao *et al.*, 2015). Armwood *et al.* (2019) evaluated minimal inhibitory concentrations of a panel of compounds against *E. anguillarum* isolated from diseased tilapia in Central America, revealing sensitivity to tetracycline, oxytetracycline and florfenicol, among others. Similarly, Reichley *et al.* (2017) found *E. anguillarum* from multiple fish hosts to be susceptible to a range of antibiotic agents, including oxytetracycline and florfenicol, suggesting these agents would be effective in response to disease outbreaks.

Vaccine research focusing on *E. anguillarum* is still in its infancy. Isolate FPC503, deemed *E. anguillarum* by molecular methods, demonstrated protective efficacy (relative percentage survival (RPS) = 85–100%) as a formalin-killed injectable vaccine against *E. anguillarum* infection in red sea bream (Takano *et al*., 2011). A recombinant *E. anguillarum* outer membrane protein A (OmpA) was more effective than formalin-killed whole-cell *E. anguillarum* in preventing haemorrhagic septicaemia in Japanese eel (LiHua *et al*., 2019). Similarly, Costa *et al*. (1998) demonstrated atypical, non-motile *E. tarda* (syn. *E. anguillarum*) from sea bream (*Pagrus major* and *Evynnis japonica*) and typical *E. tarda* (syn. *E. piscicida*) from Japanese eel and flounder had the same O-serotype and similar surface antigens. While molecular data are lacking, these findings suggest the potential for cross-protective efficacy of an *E. piscicida* and *E. anguillarum* vaccine.

The recent recognition of *E. anguillarum* as a pathogen of concern in nascent aquaculture, most notably tilapia culture in Latin America, suggests *E. anguillarum* will likely receive significant research investment moving forward. Contemporary research suggests antibiotics currently used in global aquaculture would be effective against *E. anguillarum*. Still, with increasing recognition as a pathogen of cultured tilapia, effective vaccine candidates will likely be identified in the near future, providing fish culturists additional tools to manage this emerging pathogen.

12.4 *Edwardsiella piscicida*

12.4.1 Background

As with *E. anguillarum* and *E. tarda*, a review of current trends regarding *E. piscicida* is complicated by the large number of erroneous classifications perpetuated in the literature. Despite the recognition of *E. piscicida* in 2013, there continue to be published scientific reports focusing on *E. tarda* that discuss work using an erroneously classified *E. piscicida* isolate (Griffin *et al*., 2017), most notably *E. piscicida* isolate EIB202 (Xiao *et al*., 2008; Griffin *et al*., 2014, 2017; Shao *et al*., 2015). As such, the true geographic and host ranges of *E. piscicida* remain unclear, although recent accounts using proper molecular methods suggest increasing occurrence and an expanding host range (Abayneh *et al*., 2013; Griffin *et al*., 2014, 2017; Reichley *et al*., 2017; Buján *et al*., 2018a; Dubey *et al*., 2019).

First described in 2013, *E. piscicida* is synonymous with what is deemed in the literature as 'typical, fish pathogenic *E. tarda*' (Yamada and Wakabayashi, 1999; Matsuyama *et al*., 2005; Sakai *et al*., 2007, 2009a,b; Griffin *et al*., 2014, 2017; Reichley *et al*., 2017). Similar to *E. anguillarum*, *E. piscicida* is identified by most commercial phenotypic diagnostic kits as *E. tarda*. In clinical settings, matrix-assisted laser desorption/ionization–time-of-flight (MALDI-TOF) mass spectrometry has gained popularity as a rapid, precise and cost-effective method for bacterial identification (Carbonnelle *et al*., 2011). However, since most institutional MALDI-TOF databases lag behind current *Edwardsiella* systematics, accurate identification of *Edwardsiella* spp. by MALDI-TOF requires analysis of individual spectral profiles and comparisons with molecularly confirmed controls (Reichley *et al*., 2017). Confirmatory diagnosis can also be made using species-specific PCR (Griffin *et al*., 2014; Reichley *et al*., 2015b, 2017) or sequence comparisons of appropriate genetic markers (Griffin *et al*., 2014, 2016, 2017; Reichley *et al*., 2017).

Since its recognition in 2013, retrospective studies have identified *E. piscicida* as a significant global pathogen with reports from the USA, Norway, Greece, France, Spain, Holland, Portugal, China, Japan, South Korea, Taiwan, Australia and India (Griffin *et al*., 2017; Reichley *et al*., 2017; Buján *et al*., 2018a; Dubey *et al*., 2019). At present, *E. piscicida* is considered an emergent pathogen in catfish aquaculture in the south-eastern USA, suggesting an increased prevalence of *E. piscicida* associated with the culture of channel (♀) × blue (♂) (*Ictalurus furcatus*) hybrid catfish, particularly in larger market-sized fish (Griffin *et al*., 2019a) (Fig. 12.3). Experimental data support these reports, as hybrid catfish were shown more susceptible to *E. piscicida* infection than channel catfish (Reichley *et al*., 2018).

There have also been recent reports of *E. piscicida* causing disease in important temperate-water sportfish in the USA. Fogelson *et al*. (2016) described outbreaks of *E. piscicida* in largemouth bass from display aquariums and recirculating systems in the USA. Similarly, Camus *et al*. (2019) reported spontaneous mortality events in pond-cultured largemouth bass in the US state of California (Figs 12.4 and 12.5). More recently, Griffin *et al*. (2019b) described opportunistic infections of *E. piscicida* in a black crappie (*Pomoxis nigridis*) from a freshwater temperate gamefish display in the mid-western USA. While the environments wherein these outbreaks occurred do not necessarily reflect the environments these fish inhabit in the wild, these reports expand the known host range of *E. piscicida* and suggest that *E. piscicida* should be considered when investigating fish kills involving these important recreational species.

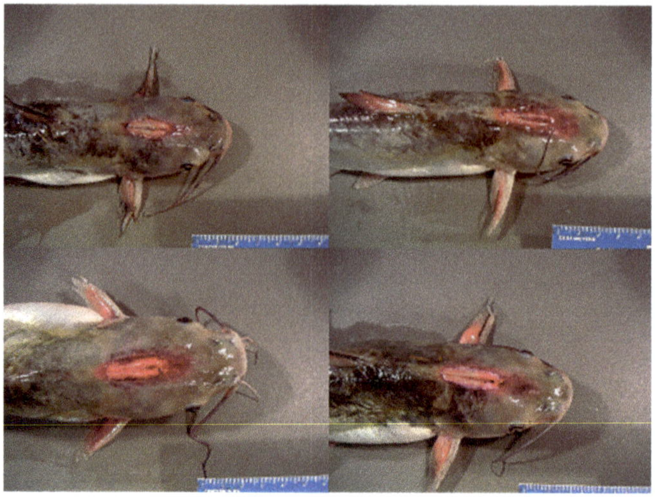

Fig. 12.3. Cranial midline lesions associated with *Edwardsiella piscicida* infection in catfish, with haemorrhagic ulcers exposing frontal bones and the open fontanelle. (Photographs courtesy of Dr Lester Khoo, Mississippi State University.)

E. piscicida has also been reported from diseased whitefish (*Coregonus lavaretus*) in Finland (Shafiei *et al.*, 2016), farmed barramundi in the mid-western USA (Loch *et al.*, 2017), a septic blotched fantail stingray (*Taeniura meyeni*) from a large display aquarium in the south-eastern USA (Camus *et al.*, 2016), wild European eels (*Anguilla anguilla*) in Spain (Esteve and Alcaide, 2018) and striped catfish (*P. hypophthalmus*) and olive flounder (*P. olivaceus*) in India (Dubey *et al.*, 2019). Typically, fish present with clinical signs indicative of a generalized septicaemia (erratic swimming, ecchymosis, exophthalmia, ascites, etc.). The number of new reports of *E. piscicida* from different hosts and geographic regions appears to outpace the occurrence of *E. tarda* in the literature. It is difficult to ascertain whether this putative expansion of host species and geographic range is indicative of an emergent pathogen associated with a dynamic climate, global aquaculture expansion or merely better recognition of the disease agent by improved detection methods.

Similar to *E. anguillarum*, the putative increases in *E. piscicida* outbreaks could be a function of changing meteorological conditions, or simply representative of aquaculture development in new geographic regions, changes in production practices or adoption of new culture species. In US farm-raised catfish, *E. piscicida* infections were typically associated with warm water and high organic loads occurring late in the production cycle, consistent with the original report of *E. tarda* (Meyer and Bullock, 1973). However, with increased adoption of hybrid catfish as a culture fish, recent outbreaks have been reported throughout the growing season. Furthermore, outbreaks have occurred at pond water temperatures (22–28°C) previously associated with outbreaks of ESC, suggesting warm temperatures (>30°C) are not the only predisposing factor for *E. piscicida* outbreaks in US farmed catfish and that host specificity plays a critical role. Similarly, evidence of *E. piscicida* in cool- and cold-water flatfishes (turbot and flounder) in Europe and Asia, as well as salmonids in North America and Europe, suggests the temperature and host ranges of *E. piscicida* are wider than previously thought and not limited to temperate- and warm-water fish. If cool- and cold-water fish are forced to endure periods approaching their thermal tolerances due to environmental warming trends, it is conceivable that outbreaks of *E. piscicida* and other temperate- and warm-water pathogens will become more commonplace in cool- and cold-water species.

12.4.2 Treatment

Congruous to *E. anguillarum*, much of the literature documenting treatment strategies to combat *E. tarda* can be similarly applied to *E. piscicida*, as the majority of work was done prior to the systematic reorganization of the *Edwardsiella*. Recent research

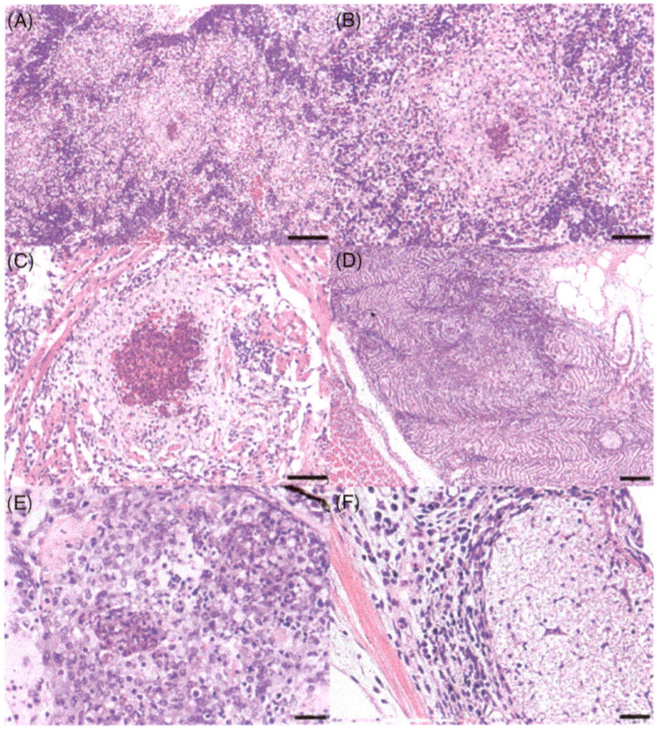

Fig. 12.4. Photomicrographs of naturally occurring *Edwardsiella piscicida*-induced lesions in largemouth bass, *Micropterus salmoides*. (A) Low-magnification image of head kidney with extensive, coalescing, pale-staining areas of necrosis, granulomatous inflammation and granuloma formation. Scale bar = 100 µm. (B) Higher-magnification image of organizing granuloma with central, intensely eosinophilic areas of necrosis surrounded by a broad mantle of epithelioid macrophages in the head kidney. Scale bar = 50 µm. (C, D) Widespread lesions predominantly affect vascular tissues, including the heart (C) and pseudobranch (D). Scale bars = 50 µm and 200 µm, respectively. (E) Granuloma formation in the meninges. Free and phagocytized bacteria are common in areas of necrosis but decrease in number as lesions become more chronic. Scale bar = 20 µm. (F) Granulomatous inflammatory infiltrates surrounding the olfactory tract, suggesting possible ascending infection from the olfactory epithelium. Scale bar = 20 µm. All sections stained with haematoxylin and eosin. (Photographs courtesy of Dr Alvin Camus, University of Georgia.)

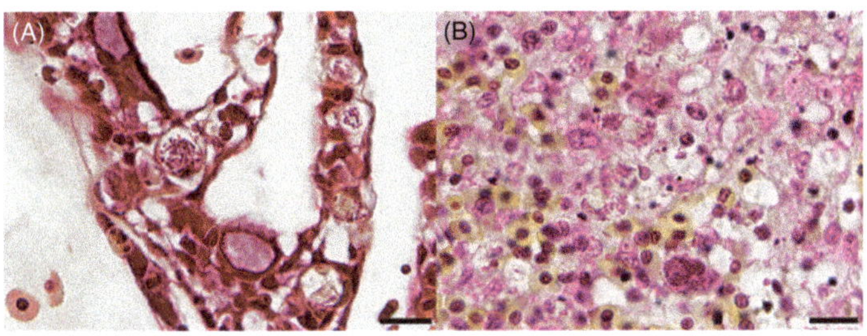

Fig. 12.5. Photomicrographs of Gram-stained histological sections of naturally occurring *Edwardsiella piscicida* infection in largemouth bass, *Micropterus salmoides*. (A) Section of gill demonstrating Gram-negative bacterial rods free and within the cytoplasm of macrophages circulating in lamellar capillaries. Scale bar = 100 µm. (B) Head kidney with numerous, predominantly phagocytized, Gram-negative bacterial rods within a developing area of necrosis. Scale bar = 100 µm. (Photographs courtesy of Dr Alvin Camus, University of Georgia.)

using molecularly confirmed isolates suggests *E. piscicida* is susceptible to a panel of antimicrobials, including florfenicol and oxytetracycline (Reichley *et al.*, 2017). Similarly, a contemporary survey of *E. piscicida* from catfish aquaculture in the south-eastern USA suggests the majority of *E. piscicida* isolates are susceptible to florfenicol, oxytetracycline and Romet® (Hoffman LaRoche), a potentiated sulfonamide formulated as a 5:1 combination of sulfadimethoxine and ormetoprim (Griffin *et al.*, 2019a).

While susceptible to a range of antimicrobial agents, *E. piscicida* also has the potential to acquire and disseminate antibiotic resistance genes to other bacteria in the environment. Researchers in China and South Korea identified conjugative plasmids for isolates TX01 and CK41, both members of the *E. piscicida* phylogroup. The pTX01 harboured genes encoding resistance to chloramphenicol and tetracycline, while pCK41 was able to confer resistance to kanamycin, tetracycline and streptomycin by conjugal transfer to *Escherichia coli* (Sun *et al.*, 2009; Yu *et al.*, 2012). The presence of these antibiotic resistance genes, as well as conjugal machinery, suggests *E. piscicida* has high potential to disseminate antibiotic resistance to other environmental microbes (Leung *et al.*, 2019). These reports further emphasize the need for effective antibiotic alternatives to combat this global fish pathogen, particularly if the expanding host and geographic ranges of *E. piscicida* are being driven by current warming trends.

12.4.3 Prevention

The most effective measures to combat edwardsiellosis associated with *E. piscicida* will conceivably be through improved immunization strategies as regulatory bodies and consumer interest steer aquaculture away from reliance on antibiotic therapies. One benefit to the wide host range and global distribution of *E. piscicida* is the large number of government and private entities with a vested interest in controlling this ubiquitous pathogen. As a result, a sizeable number of viable vaccine candidates against *E. piscicida* are available, although given the recent reorganization of the *Edwardsiella*, pathogens of interest require molecular confirmation to ensure vaccine candidates are targeting the appropriate bacterial species. Still, similar to *E. ictaluri*, there is no shortage of *Edwardsiella* vaccines that have been shown effective in laboratory-controlled studies (Cheng *et al.*, 2018; Bao *et al.*, 2019). A selection of these studies is summarized herein.

In 2010, the efficacy of five avirulent *Edwardsiella* spp. strains as live vaccines against edwardsiellosis was investigated in Japanese flounder (Takano *et al.*, 2010). Molecular data places four of the isolates within the *E. tarda* phylogroup, while one (E22) clustered with *E. piscicida* (Griffin *et al.*, 2017). Efficacy was evaluated by challenge with *E. piscicida* strain NUF806, with protection provided only by immunization with *E. piscicida* E22. The *E. tarda* isolates did not protect against *E. piscicida*. Conversely, exposure of Japanese flounder to the bona fide *E. tarda* type strain from humans (ATCC 15947) resulted in resistance against subsequent challenge with fish pathogenic isolate TX1 (Cheng *et al.*, 2010), later identified as a member of the *E. piscicida* cluster (Yang *et al.*, 2012; Shao *et al.*, 2015). This would suggest at least some degree of shared epitopes between *E. piscicida* and *E. tarda*, potentially reducing the need to develop numerous vaccines for each bacterium or fish culture species.

The surface antigen Esa1 from *E. piscicida* isolate TX1 has demonstrated tremendous potential as a recombinant subunit vaccine. In turbot (*Scophthalmus maximus*), the purified recombinant subunit vaccine (expressed in a *Pseudomonas* sp.) delivered orally or injected led to 52 and 79% RPS, respectively (Sun *et al.*, 2010). Esa1 was also efficacious as a DNA vaccine, resulting in 57% RPS in fish injected with plasmid constructs carrying the Esa1 inserts (Sun *et al.*, 2011a). Similar to the Esa1 work, purified recombinant subunit and DNA vaccines exploited the antigenicity of surface antigen Eta2, also from *E. piscicida* isolate TX1. Injections of the Eta2 purified recombinant subunit or the DNA vaccine in flounder resulted in RPS values of 83 and 67%, respectively. Furthermore, passive immunization with sera from fish vaccinated with the recombinant Eta2 vaccine resulted in greater survival (RPS = 57%) than immunization with sera from fish immunized with the DNA vaccine (RPS = 29%; Sun *et al.*, 2011b).

Moreover, bacterial 'ghosts' have been evaluated as vaccine candidates against *Edwardsiella* infections in fish. 'Ghosts' were created from *Edwardsiella* isolate FSW910410 cultured from a moribund olive flounder on a fish farm in South Korea. In tilapia, two IP immunizations with ~10^6 FSW910410 ghost cells, administered 2 weeks apart, significantly improved survival over cohorts receiving formalin-killed cells or non-vaccinated controls when exposed to wild-type FSW910410. Furthermore, oral delivery of FSW910410 bacterial ghosts to flounder was more effective than formalin-killed cells *per os* (Kwon *et al.*, 2006, 2007). Analysis of *sodB* sequences identifies isolate FSW910410 as *E. piscicida* (GenBank AB232158; Han *et al.*, 2006).

Along these lines, a large body of work exists regarding *Edwardsiella* isolate EIB202, originally isolated from a mortality event on a turbot mariculture farm in Yantai, China (Xiao *et al.*, 2008; Wang *et al.*, 2009; GenBank CP001135). Since the recognition of *E. piscicida* in 2013, there have been numerous reports evincing isolate EIB202 is in fact *E. piscicida* (Griffin *et al.*, 2014, 2017; Shao *et al.*, 2015). Despite these reports, there continue to be published studies investigating immunization against *E. tarda* focusing on isolate EIB202 as a genomic template for recombinant targets or a candidate for an attenuated or killed vaccine. While the subject organism has been misidentified, these studies still have merit and application in the march towards an effective *E. piscicida* vaccine.

Immersion immunization with a live EIB202 *esrB* mutant (EsrB positively regulates the type III and VI secretion systems) protected turbot against subsequent *E. piscicida* challenge (Yang *et al.*, 2015). Similarly, IP injection using an EIB202 mutant lacking UDP-glucose dehydrogenase improved protection (RPS = 43.3–76.7%) in turbot against parental wild-type EIB202 (Lv *et al.*, 2012). Similarly, turbot injected with a recombinant EIB202 glyceraldehyde-3-phosphate dehydrogenase (GAPDH) had lower cumulative mortality when compared with identically challenged control fish (Liang *et al.*, 2012) and turbot immunized with formalin-killed EIB202 in the presence of flagellin (FlgD) as an adjuvant resulted in significantly improved survival over formalin-killed cells alone (Liu *et al.*, 2017). Similar success was achieved vaccinating flounder with a DNA vaccine encoding molecular chaperone GroEL derived from EIB202 (Liu *et al.*, 2016). Likewise, recombinant EIB202 outer membrane proteins OmpC, OmpI and OmpX are all promising vaccine candidates to protect flounder against edwardsiellosis (Liu *et al.*, 2016; Liu *et al.*, 2017). Furthermore, polyvalent DNA vaccines containing recombinant OmpA from EIB202 have been shown effective in protecting zebrafish (*Danio rerio*) against the wild-type EIB202 strain (Cheng *et al.*, 2018).

Lastly, there is evidence that a recently developed, orally delivered, live attenuated *E. ictaluri* vaccine (Wise *et al.*, 2015) offers protection in channel and hybrid catfish against both *E. ictaluri* and *E. piscicida* infection. While researchers did not set out to develop a multivalent vaccine, the attenuated strain was found effective against *E. piscicida* isolate S11-285 (Reichley *et al.*, 2016) in controlled laboratory trials (Griffin *et al.*, 2020). Similarly, in an attempt to develop a multivalent vaccine against edwardsiellosis in tilapia (*Oreochromis niloticus*), a recombinant GAPDH derived from the outer membrane protein of *E. ictaluri* and expressed in *E. coli* was delivered concurrently with injections of formalin-inactivated *E. ictaluri* cells. The combination of recombinant protein and formalin-killed cells resulted in 71.4% RPS following challenge with *Edwardsiella* sp. strain OT9805-27 (Trung Cao *et al.*, 2014), which based on genetic sequence data is actually *E. piscicida* (Griffin *et al.*, 2017). These findings hold promise that a multivalent vaccine against both *E. piscicida* and *E. ictaluri* is achievable through a single vaccine construct. If increasing global temperatures result in increased incidence and prevalence of *E. piscicida* and *E. ictaluri* in cultured and wild fish stocks, the potential for efficacious polyvalent vaccines is a promising avenue of research and could potentially mitigate losses attributed to these two important, globally distributed pathogens.

12.5 *Edwardsiella ictaluri*

12.5.1 Background

E. ictaluri is the causative agent of ESC and a significant pathogen of commercially cultured catfish and other temperate freshwater fishes (Hawke, 1979; Hawke *et al.*, 1981). The disease is endemic on most commercial catfish operations in the southeastern USA (Plumb and Vinitnantharat, 1993) and can cause catastrophic losses, particularly in first-year catfish fingerlings (Wise *et al.*, 2004). Initial studies suggested isolates from different geographic locations were biochemically, biophysically and serologically homogenous (Newton *et al.*, 1988; Plumb and Vinitnantharat, 1989; Hawke and Khoo, 2004). Genetically, *E. ictaluri* is mostly clonal in US catfish aquaculture (Griffin *et al.*, 2011) with some exceptions (Bader *et al.*, 1998). However, subtypes have been identified from wild and farmed fish outside the USA and *E. ictaluri* is now recognized as a pathogen of significance in aquaculture industries in Latin America and Asia (Ferguson *et al.*, 2001; Crumlish *et al.*, 2002; Nagai *et al.*, 2008; Sakai *et al.*, 2008; Xu *et al.*, 2009; Ye *et al.*, 2009;

Liu *et al.*, 2010; Bartie *et al.*, 2012; Soto *et al.*, 2012; Geng *et al.*, 2013; Rogge *et al.*, 2013; Dong *et al.*, 2015).

Stress, poor nutrition, co-infection and suboptimal water quality increase ESC-related mortality (Wise *et al.*, 1993, 2004). As such, environmental degradation as a result of climate change could exacerbate ESC-related losses in US farm-raised catfish. Outbreaks of ESC in the catfish-farming region of the south-eastern USA mostly occur in late spring and early autumn, when water temperatures range between 22 and 28°C (Francis-Floyd *et al.*, 1987; Wise *et al.*, 2004), although ESC related die-offs can occur outside this range (Plumb and Shoemaker, 1995). Generally, ESC outbreaks do not occur when water temperatures exceed 30°C. As a result, increased average temperatures associated with putative warming trends may not have measurable effects on incidence of *E. ictaluri* in catfish aquaculture in the USA and Asia, especially if resultant temperatures lead to prolonged periods exceeding temperatures conducive to ESC. In fact, warmer temperatures may be beneficial to catfish aquaculture, particularly in the USA, as milder winters could reduce incidence of winter kill syndrome caused by *Saprolegnia* spp. Similarly, warmer spring and summer temperatures may lead to earlier spawns, longer growing seasons, better feed conversion and a reduced time from hatch to harvest.

It should be noted that climate change is not restricted to warming and climactic shifts may result in unseasonably cool weather in certain geographical regions, in addition to abnormal precipitation patterns which can have cooling effects on pond-based culture. This has been observed in the catfish-farming region of the USA, where atypical summer cold fronts have occurred over the past several years. This summer cooling pattern has extended the temperature window conducive for *E. ictaluri* infections and has had a significant economic impact on catfish culture. Historically, outbreaks were limited to the autumn of the year as pond water temperatures began to cool. Most recently, *E. ictaluri* infections have become problematic throughout much of the growing season due to these unseasonable cool weather patterns occurring in this region of the country.

Epizootics of ESC in catfish aquaculture are characterized by disease-induced inappetence, with complete cessation of feeding at the height of severe outbreaks. Fish become listless near the surface, accumulating along the leeward pond bank. Individuals are often observed swimming in circles. Presumptive diagnosis is based on clinical signs ranging from multifocal, pinpoint red and white ulcers and petechial haemorrhages on the skin, to distended abdomens (Fig. 12.6), exophthalmia and, in more chronic forms, meningoencephalitis. This meningoencephalitis leads to erosion of the skin covering the fontanelle of the frontal bones, producing the characteristic 'hole-in-the-head' lesion (Fig. 12.3), a common colloquialism for the disease. Internally the abdomen can fill with clear, yellow or bloody ascites accompanied by renomegaly and pale, necrotic, mottled liver (Thune *et al.*, 1993). Similar pathology was seen in a spontaneous *E. ictaluri* outbreak in brown bullhead catfish (*Ameiurus nebulosa*) (Iwanowicz *et al.*, 2006).

In Vietnam, China and other parts of Asia, *E. ictaluri* is the causative agent of bacillary necrosis (BN) in cultured *Pangasius* spp. (Ferguson *et al.*, 2001; Crumlish *et al.*, 2002), which has also been reported from several other cultured catfish species in Asia (Yuasa *et al.*, 2003; Dung *et al.*, 2008; Xu *et al.*, 2009; Ye *et al.*, 2009; Liu *et al.*, 2010; Geng *et al.*, 2013; Dong *et al.*, 2015). Similar to pathology in channel and hybrid catfish in the south-eastern USA, clinical signs associated with BN involve emaciation, swollen abdomens and petechial haemorrhages on the skin. Internally, small (1–3 mm), multifocal white miliary lesions of varying sizes are

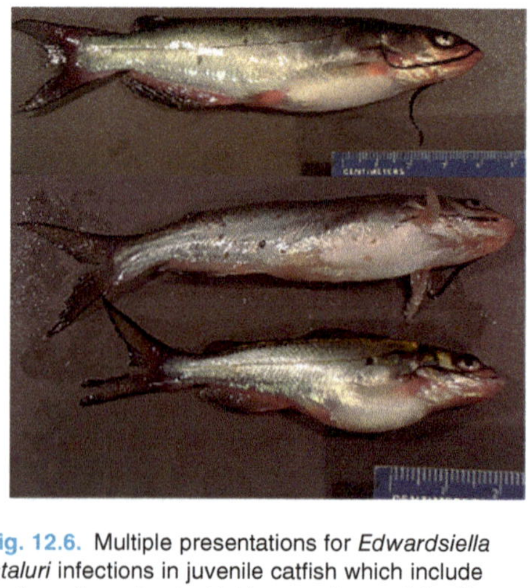

Fig. 12.6. Multiple presentations for *Edwardsiella ictaluri* infections in juvenile catfish which include ecchymotic and petechial haemorrhage of the fins and skin, dermal ulcerations and distended fluid-filled abdomens. (Photographs courtesy of Dr Lester Khoo, Mississippi State University.)

present on several organs including liver, spleen and kidney (Ferguson *et al.*, 2001; Crumlish *et al.*, 2002).

Edwardsiella has also emerged in non-ictalurid fishes (e.g. zebrafish, tilapia, ayu (*Plecoglossus altivelis*)). Outbreaks have been reported in laboratory populations of zebrafish, seemingly triggered by handling (Hawke *et al.*, 2013). Clinical signs were similar to those in catfish, with systemic disease characterized by necrotic spleen, kidney, liver, gut and brain with large numbers of bacteria observed, often within macrophages. Likewise, *E. ictaluri* has also been linked to mortality in cultured tilapia (Soto *et al.*, 2012; Dong *et al.*, 2019). Fingerlings had consistent multifocal nodules in the spleen and head kidney, with hepatomegaly and reduced fat in the liver and peritoneum. Histologically, sheets of granulomatous inflammation, discrete granulomas and multifocal areas of necrosis were seen in the spleen, head kidney and liver (Soto *et al.*, 2012). Moreover, late summer–early autumn mortality associated with *E. ictaluri* has been observed among wild ayu in Japanese rivers (Sakai *et al.*, 2008). Diseased fish had haemorrhagic ascites and exophthalmia with reddening of the body surface, anus or bases of fins and research has demonstrated *E. ictaluri* was comparatively more virulent to ayu at 28°C than at 20°C (Nagai and Nakai, 2014). Lastly, *E. ictaluri* has recently emerged as a pathogen of significance in ornamental fish aquaculture in the south-eastern USA (Shelley *et al.*, 2018). Again, the expanding host range of *E. ictaluri* could be indicative of environmental shifts, a function of improved diagnostics or merely a result of aquaculture expansion into new geographic locales.

Regardless, *E. ictaluri* continues to be one of the most well-studied fish pathogens, largely due to the significant economic impact it has had on catfish aquaculture in the south-eastern USA. With the recent emergence of *E. ictaluri* in Asia and tilapia culture in Latin America, *E. ictaluri* has received considerable attention from government and private institutions investigating methodologies for treatment and prevention.

12.5.2 Treatment

Control of ESC in US catfish aquaculture has typically relied on feed restriction to slow the spread of infection and use of antibiotic-medicated feeds. Feed restriction when water temperatures are permissive for ESC is a common management practice in the south-eastern USA. In catfish, restricting feed to every other day or every third day can improve survival during an ESC outbreak by limiting ingestion of the bacterium, in turn precluding the faecal/oral route of infection (Wise and Johnson, 1998; Wise *et al.*, 2008). In 2009, nearly 30% of fingerling operations reported withdrawing feed to control ESC (USDA/APHIS/NAHMS, 2010); however, prophylactic feed restriction results in reduced growth due to lost feed days and can delay time to harvest (Wise *et al.*, 2004). Historically, these practices were not implemented until the end of the production season, although with recent anomalous weather patterns resulting in unseasonably cool summer temperatures, these practices must be implemented on and off throughout the growing season, forcing US catfish producers to deviate from typical production practices.

Disinfection with ethyl alcohol (30, 50 or 70%), benzyl-4-chlorophenol/phenylphenol (1%), sodium hypochlorite (50, 100, 200 or 50,000 mg/l), *N*-alkyl dimethyl benzyl ammonium chloride (1:256), povidone iodine (50 or 100 mg/l), glutaraldehyde (2%) and potassium peroxymonosulfate/sodium chloride (1%) reduced or eliminated the number of detectable organisms within 1 min of contact time. Comparatively, chloramineT (15 mg/l) and formalin (250 mg/l) did not reduce or eliminate *E. ictaluri* (Mainous *et al.*, 2010). Moreover, *E. ictaluri* is susceptible to aminoglycosides, cephalosporins, penicillins, quinolones, tetracyclines, chloramphenicol, nitrofurantoin and potentiated sulfonamides (Waltman and Shotts, 1986b). In the USA, Romet® and Aquaflor® are approved for the control of *E. ictaluri* in catfish. As mentioned above, Romet® (Hoffman LaRoche) is a potentiated sulfonamide formulated as a 5:1 combination of sulfadimethoxine and ormetoprim, which effectively reduces ESC mortality in channel catfish fingerlings at a recommended dose of 50 mg/kg for five consecutive days (Plumb *et al.*, 1987). Aquaflor® (Merck Animal Health) is a broad-spectrum antibiotic of the phenicol class that is approved for use in a variety of fish species worldwide. It is delivered as a feed premix (50% w/w florfenicol) (Gaunt *et al.*, 2004, 2006, 2015) at 10 mg/kg body weight daily for 10 days (Gaikowski *et al.*, 2003). In the USA, the decision to treat with antibiotics must be weighed against the ability to harvest and market fish. Romet® has a mandated withdrawal period of only 3 days, while fish treated with Aquaflor® require a 15-day withdrawal period before they can be sent to slaughter.

While effective, proper antimicrobial stewardship must be performed when treating *E. ictaluri* with medicated feeds as plasmid-mediated antimicrobial resistance to tetracycline, sulfadimethoxine/ormetoprim and florfenicol (Waltman *et al.*, 1989; Welch *et al.*, 2008) has been documented for *E. ictaluri* and field isolates resistant to tetracycline, sulfadimethoxine/ormetoprim and florfenicol have been reported (Starliper *et al.*, 1993; Welch *et al.*, 2008). Similarly, investigations of *E. ictaluri* isolates from BN outbreaks in the Mekong Delta, Vietnam from 2002 to 2005 revealed acquired resistance to oxytetracycline, trimethoprim and spectomycin, with >70% of isolates examined demonstrating multi-drug resistance (Dung *et al.*, 2008). Furthermore, *incK* plasmid-mediated tetracycline resistance has been observed among *E. ictaluri* isolates from diseased freshwater catfish in Vietnam (Dung *et al.*, 2009). Multi-drug resistance has also been reported for *E. ictaluri* isolate MS-17-156 from the USA, with resistance to tetracycline, oxytetracycline, doxycycline, florfenicol, erythromycin, chloramphenicol, streptomycin, penicillin, novobiocin, azithromycin and spectinomycin (Abdelhamed *et al.*, 2018b). As with *E. tarda* and *E. piscicida*, the potential for *E. ictaluri* to acquire antibiotic resistance further emphasizes the importance of proper antibiotic use and stresses the need for antibiotic alternatives to manage bacterial fish pathogens in the wake of current warming trends and putative increases in *Edwardsiella* prevalence in global aquaculture.

12.5.3 Prevention

Environmental conditions aside, there is high potential for controlling *E. ictaluri* through vaccination as *E. ictaluri* is regionally genetically and antigenically homogeneous (Bertolini *et al.*, 1990; Griffin *et al.*, 2011), although differences have been observed among geographically discrete isolates (Bader *et al.*, 1998; Bartie *et al.*, 2012; Rogge *et al.*, 2013; Griffin *et al.*, 2016; Phillips *et al.*, 2017). Fish that survive *E. ictaluri* infection develop high levels of protective immunity (Shoemaker and Klesius, 1997; Wise *et al.*, 2000) and there is evidence of effective immunization of catfish using a range of vaccine strategies, including formalin-inactivated *E. ictaluri*, recombinant protein expression and bacterial ghosts (Wang *et al.*, 2016; Yang *et al.*, 2016; Zhu *et al.*, 2019).

Early attempts at vaccinating catfish against ESC in the USA focused on bacterin preparations delivered as an immersion bath, by incorporation in feed, or a combination of both. These efforts met with mixed results (Saeed and Plumb, 1986; Plumb and Vinitnantharat, 1993; Plumb *et al.*, 1994; Thune *et al.*, 1994). It was later suggested that killed vaccines do not elicit a sufficient protective immune response against *E. ictaluri* in channel catfish (Thune *et al.*, 1997). Lack of efficacy is attributed to the inability of killed vaccines to stimulate the strong, cell-mediated immune responses necessary to protect against facultative-intracellular pathogens (Nahm *et al.*, 1999; Thune *et al.*, 1999). Conversely, live attenuated vaccines stimulate the robust cell-mediated immunity necessary to immunize catfish against *E. ictaluri* infection (Klesius and Sealey, 1995; Seder and Hill, 2000; Shoemaker *et al.*, 2009). As a result, a number of live attenuated vaccine candidates against *E. ictaluri* have been developed and there is no shortage of vaccine candidates proven effective in controlled laboratory studies (Lawrence *et al.*, 1997; Thune *et al.*, 1999; Pridgeon and Klesius, 2011; Santander *et al.*, 2012; Abdelhamed *et al.*, 2013, 2016a,b, 2018a; Dahal *et al.*, 2013, 2014; Nho *et al.*, 2017). Limitations associated with vaccine efficacy in the commercial setting are dictated by catfish production practices and logistical challenges associated with delivering vaccines to immunocompetent fish (Wise *et al.*, 2015).

In the USA, catfish reproductive characteristics limit egg collection to the spring of the year, when spawns can be collected from broodstock ponds and transferred to a hatchery. Approximately 7–10 days post-hatch (dph), swim-up fry are transported to nursery ponds where they are maintained through the autumn (Tucker and Robinson, 1990) when, as first-year fingerlings, they typically encounter *E. ictaluri* for the first time. This initial encounter often leads to considerable losses (Wise *et al.*, 2004). The most opportune time to vaccinate would be during transfer of 7- to 10-day-old fish from the hatchery to the nursery pond, when fish are confined in hauling tanks. Regrettably, strategies targeting this stage of production have had negligible success and as a result, previous commercially available immersion vaccines have struggled to gain widespread industry adoption (Bebak and Wagner, 2012; Wise *et al.*, 2015). This limited success is likely related to an immature immune system at the time of vaccination. Studies evaluating immune system development indicate catfish are not fully immunocompetent until >21 dph (Petrie-Hanson and Ainsworth, 1999, 2001), which is considerably later than the typical transfer of hatchery fry to nursery ponds.

Once fish are stocked into ponds, catfish industry dogma suggests vaccination is no longer economically or logistically feasible. As such, immunization practices in catfish production have been limited and recent attempts to develop new vaccine candidates have met a fair amount of scepticism from catfish producers (Bebak and Wagner, 2012).

Perhaps the most successful effort to date to immunize fish in aquaculture settings against *E. ictaluri* has been reported by researchers at the Thad Cochran National Warmwater Aquaculture Center in Stoneville, Mississippi, USA. Scientists developed an effective *E. ictaluri* management practice using a live attenuated ESC vaccine (340X2) coupled with an oral delivery platform. Oral administration facilitates in-pond delivery, which has been shown to provide high levels of protection in laboratory and small-scale experimental pond trials (Wise *et al.*, 2015, 2020; Peterson *et al.*, 2016; Greenway *et al.*, 2017; Chatakondi *et al.*, 2018).

A key benefit of oral delivery is it facilitates delivery of the vaccine to older, immunocompetent fish. This circumvents limitations associated with vaccinating very young fish in the hatchery or during transfer from the hatchery to the pond prior to maturation of the catfish immune arsenal. As proof of concept, Wise *et al.* (2020) evaluated the efficacy of the oral delivery platform and the live attenuated *E. ictaluri* vaccine under simulated commercial conditions. Channel catfish fingerlings (~60–70 dph) were vaccinated approximately 40–50 days post-stocking by mixing the attenuated vaccine with feed using an in-line mechanized delivery system. Across three different years, oral vaccination using the live attenuated *E. ictaluri* vaccine resulted in marked improvements in survival, feed conversion ratio, feed fed and total yield. Furthermore, economic analysis of experimental data suggested a significant positive net economic benefit from vaccination, evincing that oral immunization of channel catfish fingerlings against ESC is an effective strategy to mitigate ESC-related losses. Furthermore, there is evidence of a cross-protective effect of the live attenuated *E. ictaluri* vaccine against at least one strain of *E. piscicida* (S11-285) in both channel and hybrid catfish (Griffin *et al.*, 2020). Recent industry-scale vaccine trials in hybrid catfish have demonstrated significant net economic benefits from vaccinating hybrid catfish with the live attenuated *E. ictaluri* vaccine and it is hypothesized this benefit is multifactorial and associated with minimizing economic losses to both *E. ictaluri* and *E. piscicida* in hybrid production (Kumar *et al.*, 2019). The ability to minimize losses related to both *E. ictaluri* and *E. piscicida* has the potential to significantly improve production efficiency and profitability on commercial catfish operations in the south-eastern USA.

One limitation to using live vaccines is that effectiveness is related to the elicitation of an underlying infection. This can result in unintended mortality if the target population harbours ongoing infections. Moreover, caution should be exercised when vaccinating fish during periods when environmental conditions predispose fish to disease, such as episodes of stress or when environmental temperatures are permissive for epizootics. However, in the absence of sick fish there are limited practical methods of assessing population health and predicting potential adverse vaccine reactions is difficult. In the catfish-farming region of the south-eastern USA, observed summer temperatures and precipitation, including extreme precipitation events, have remained largely consistent with long-term averages and the region has not experienced the warming or climatic shifts reported for other locales (NOAA, 2019). Still, the catfish-farming industry in Mississippi, USA, has endured unseasonable cold spells in June and July over the past several growing seasons. These cold spells extended the 'ESC window' into mid-summer, which can increase the risk of adverse reactions to vaccination as conditions are conducive to underlying disease throughout the summer grow-out. In US catfish aquaculture, these atypical weather events may be anomalous or indicative of future trends which will complicate current vaccination protocols that adhere to production schedules targeting vaccination of fish 40–60 days post-stocking (Wise *et al.*, 2020). Considering these and other climactic shifts, similar considerations will have to be made for other culture species and industries which are geographically, facility and situationally dependent.

12.6 Conclusion

While the root cause is a source of debate in the political arena, there is a consensus among climatologists that the average global temperature will be 4°C above pre-industrial levels by 2100 (World Bank, 2014). Thermal stress associated with either low or high temperature extremes is a serious concern to fish health. Prolonged exposure to extreme temperatures can lead to reduced growth, poor feed efficiency, decreased immune function and, in some instances, death. Although thermal stress is not uncommon and occurs in both cold- and warm-water fishes, widely fluctuating temperatures can be of concern in the crowded, high-stress environments typically associated with aquaculture. Furthermore, periods of thermal extremes not only impact fish physiology but can also trigger disease occurrence, as environmental conditions may be conducive to rapid proliferation of pathogens within the host and environment.

It is difficult to predict the effects these climate dynamics will have on global fish health, although it is expected that *Edwardsiella* will remain a significant threat to aquaculture species in the years ahead. We assert that research investment should continue to focus on management practices that will mitigate *Edwardsiella*-related losses, particularly the development of effective, logistically practical vaccines. Given their ability to grow at higher temperatures, elevated global temperatures could increase incidence and prevalence of *E. piscicida*, *E. anguillarum* and even *E. tarda*, as temperature-driven eutrophication may lead to increased organic loads and diminished water quality in intensive systems. Comparably, the permissive growth range of *E. ictaluri* is much lower. As a result, warming may reduce incidence of *E. ictaluri* in land-based pond aquaculture as water temperatures during grow-out may exceed the permissive 'ESC window' typically associated with disease outbreaks, particularly in catfish culture in the USA and Asia.

Responsive treatments will continue to be reliant on medicated feeds until effectual preventive measures can be identified and effective vaccines made commercially available. It is disappointing that the majority of vaccine research funded by government agencies largely focuses on recombinant and other forms of genetically manipulated organisms. Although effective in controlled laboratory trials, these genetically modified organisms carry significant political baggage and face formidable obstacles in terms of regulatory approval and consumer sentiment. As such, commercialization and mass distribution of recombinant and other genetically modified organisms remains complicated. Still, in the wake of putative increases in global temperatures, incidence of edwardsiellosis could very well increase in cold- and cool-water fish hosts as temperature extremes and resultant heatwaves stress cold-water species like trout and salmon in the Pacific North-Western USA and Canada, and whitefish, flounder and turbot in Northern Europe. Comparably, fledgling and established temperate- and warm-water fish aquaculture industries across the Americas, Africa, Europe and Asia, which have been battling *Edwardsiella* spp. outbreaks for decades, will likely see little change in terms of research investment or management strategies regardless of climactic conditions.

References

Abayneh, T., Colquhoun, D. and Sørum, H. (2012) Multilocus sequence analysis (MLSA) of *Edwardsiella tarda* isolates from fish. *Veterinary Microbiology* 158, 367–375.

Abayneh, T., Colquhoun, D. and Sørum, H. (2013) *Edwardsiella piscicida* sp. nov., a novel species pathogenic to fish. *Journal of Applied Microbiology* 114, 644–654.

Abdelhamed, H., Lu, J., Shaheen, A., Abbass, A., Lawrence, M. and Karsi, A. (2013) Construction and evaluation of an *Edwardsiella ictaluri fhuC* mutant. *Veterinary Microbiology* 162, 858–865.

Abdelhamed, H., Lu, J., Lawrence, M.L. and Karsi, A. (2016a) Involvement of *tolQ* and *tolR* genes in *Edwardsiella ictaluri* virulence. *Microbial Pathogenesis* 100, 90–94.

Abdelhamed, H., Lu, J., Lawrence, M.L. and Karsi, A. (2016b) Ferric hydroxamate uptake system contributes to *Edwardsiella ictaluri* virulence. *Microbial Pathogenesis* 100, 195–200.

Abdelhamed, H., Ibrahim, I., Baumgartner, W., Lawrence, M.L. and Karsi, A. (2018a) The virulence and immune protection of *Edwardsiella ictaluri* HemR mutants in catfish. *Fish and Shellfish Immunology* 72, 153–160.

Abdelhamed, H., Tekedar, H.C., Ozdemir, O., Hsu, C.Y., Arick, M.A. *et al.* (2018b) Complete genome sequence of multidrug-resistant *Edwardsiella ictaluri* strain MS-17-156. *Genome Announcements* 6, e00477-18.

Abraham, T.J. (2011) Food safety hazards related to emerging antibiotic resistant bacteria in cultured freshwater fishes of Kolkata, India. *Advance Journal of Food Science and Technology* 3, 69–72.

Abraham, T., Mallick, P., Adikesavalu, H. and Banerjee, S. (2015) Pathology of *Edwardsiella tarda* infection in African catfish, *Clarias gariepinus* (Burchell 1822), fingerlings. *Archives of Polish Fisheries* 23, 141–148.

Acharya, M., Maiti, N.K., Mohanty, S., Mishra, P. and Samanta, M. (2007) Genotyping of *Edwardsiella tarda* isolated from freshwater fish culture system. *Comparative Immunology, Microbiology and Infectious Diseases* 30, 33–40.

Akinbowale, O., Peng, H. and Barton, M. (2006) Antimicrobial resistance in bacteria isolated from aquaculture sources in Australia. *Journal of Applied Microbiology* 100, 1103–1113.

Alcaide, E., Herraiz, S. and Esteve, C. (2006) Occurrence of *Edwardsiella tarda* in wild European eels *Anguilla anguilla* from Mediterranean Spain. *Diseases of Aquatic Organisms* 73, 77–81.

Amandi, A., Hiu, S., Rohovec, J. and Fryer, J. (1982) Isolation and characterization of *Edwardsiella tarda* from fall Chinook salmon (*Oncorhynchus tshawytscha*). *Applied and Environmental Microbiology* 43, 1380–1384.

Aoki, T. and Takahashi, A. (1987) Class D tetracycline resistance determinants of R plasmids from the fish pathogens *Aeromonas hydrophila*, *Edwardsiella tarda*, and *Pasteurella piscicida*. *Antimicrobial Agents and Chemotherapy* 31, 1278–1280.

Aoki, T., Kitao, T. and Fukudome, M. (1989) Chemotherapy against infection with multiple drug resistant strains of *Edwardsiella tarda* in cultured eels. *Fish Pathology* 24, 161–168.

Armwood, A.R., Camus, A.C., López-Porras, A., Ware, C., Griffin, M.J. and Soto, E. (2019) Pathologic changes in cultured Nile tilapia (*Oreochromis niloticus*) associated with an outbreak of *Edwardsiella anguillarum*. *Journal of Fish Diseases* 42, 1463–1469.

Ashida, T., Okimasu, E., Ui, M., Heguri, M., Oyama, Y. and Amemura, A. (1999) Protection of Japanese flounder *Paralichthys olivaceus* against experimental edwardsiellosis by formalin-killed *Edwardsiella tarda* in combination with oral administration of immunostimulants. *Fisheries Science* 65, 527–530.

Bader, J., Shoemaker, C., Klesius, P., Connolly, M. and Barbaree, J. (1998) Genomic subtyping of *Edwardsiella ictaluri* isolated from diseased channel catfish by arbitrarily primed polymerase chain reaction. *Journal of Aquatic Animal Health* 10, 22–27.

Bao, P., Sun, X., Liu, Q., Zhang, Y. and Liu, X. (2019) Synergistic effect of a combined live *Vibrio anguillarum* and *Edwardsiella piscicida* vaccine in turbot. *Fish and Shellfish Immunology* 88, 84–90.

Bartie, K.L., Austin, F.W., Diab, A., Dickson, C., Dung, T.T. *et al.* (2012) Intraspecific diversity of *Edwardsiella ictaluri* isolates from diseased freshwater catfish, *Pangasianodon hypophthalmus* (Sauvage), cultured in the Mekong Delta, Vietnam. *Journal of Fish Diseases* 35, 671–682.

Baya, A.M., Romalde, J.L., Green, D.E., Navarro, R.B., Evans, J. *et al.* (1997) Edwardsiellosis in wild striped bass from the Chesapeake Bay. *Journal of Wildlife Diseases* 33, 517–525.

Bebak, J. and Wagner, B. (2012) Use of vaccination against enteric septicemia of catfish and columnaris disease by the US catfish industry. *Journal of Aquatic Animal Health* 24, 30–36.

Bertolini, J.M., Cipriano, R.C., Pyle, S.W. and McLaughlin, J.J. (1990) Serological investigation of the fish pathogen *Edwardsiella ictaluri*, cause of enteric septicemia of catfish. *Journal of Wildlife Diseases* 26, 246–252.

Blanch, A., Pintó, R. and Jofre, J. (1990) Isolation and characterization of an *Edwardsiella* sp. strain, causative agent of mortalities in sea bass (*Dicentrarchus labrax*). *Aquaculture* 88, 213–222.

Bragg, R.R. (1988) First isolation of *Edwardsiella tarda* from fish in South Africa. *Bulletin of the European Association of Fish Pathologists* 8, 87–88.

Bragg, R.R. (1991) Health status of salmonids in river systems in Natal. III. Isolation and identification of bacteria. *The Onderstepoort Journal of Veterinary Research* 58, 67–70.

Buján, N., Mohammed, H., Balboa, S., Romalde, J.L., Toranzo, A.E. *et al.* (2018a) Genetic studies to re-affiliate *Edwardsiella tarda* fish isolates to *Edwardsiella piscicida* and *Edwardsiella anguillarum* species. *Systematic and Applied Microbiology* 41, 30–37.

Buján, N., Toranzo, A.E. and Magariños, B. (2018b) *Edwardsiella piscicida*: a significant bacterial pathogen of cultured fish. *Diseases of Aquatic Organisms* 131, 59–71.

Camus, A., Dill, J., McDermott, A., Hatcher, N. and Griffin, M. (2016) *Edwardsiella piscicida*-associated septicaemia in a blotched fantail stingray *Taeniura meyeni* (Müeller & Henle). *Journal of Fish Diseases* 39, 1125–1131.

Camus, A., Griffin, M., Armwood, A. and Soto, E. (2019) A spontaneous outbreak of systemic *Edwardsiella piscicida* infection in largemouth bass *Micropterus salmoides* (Lacépède, 1802) in California, USA. *Journal of Fish Diseases* 42, 759–763

Carbonnelle, E., Mesquita, C., Bille, E., Day, N., Dauphin, B. *et al.* (2011) MALDI-TOF mass spectrometry tools for bacterial identification in clinical microbiology laboratory. *Clinical Biochemistry* 44, 104–109.

Carrias, A., Welch, T.J., Waldbieser, G.C., Mead, D.A., Terhune, J.S. and Liles, M.R. (2011) Comparative genomic analysis of bacteriophages specific to the channel catfish pathogen *Edwardsiella ictaluri*. *Virology Journal* 8, 6.

Castro, N., Toranzo, A.E., Barja, J.L., Nunez, S. and Magarinos, B. (2006) Characterization of *Edwardsiella tarda* strains isolated from turbot, *Psetta maxima* (L.). *Journal of Fish Diseases* 29, 541–547.

Castro, N., Toranzo, A., Nunez, S. and Magariños, B. (2008) Development of an effective *Edwardsiella tarda* vaccine for cultured turbot (*Scophthalmus maximus*). *Fish and Shellfish Immunology* 25, 208–212.

Castro, N., Toranzo, A., Bastardo, A., Barja, J.L. and Magariños, B. (2011) Intraspecific genetic variability of *Edwardsiella tarda* strains from cultured turbot. *Diseases of Aquatic Organisms* 95, 253–258.

Castro, N., Toranzo, A.E., Devesa, S., González, A., Nuñez, S. and Magariños, B. (2012) First description of *Edwardsiella tarda* in Senegalese sole, *Solea senegalensis* (Kaup). *Journal of Fish Diseases* 35, 79–82.

Chatakondi, N., Peterson, B.C., Greenway, T.E., Byars, T.S. and Wise, D.J. (2018) Efficacy of a live-attenuated *Edwardsiella ictaluri* oral vaccine in channel and hybrid catfish. *Journal of the World Aquaculture Society* 49, 686–691.

Chenia, H.Y. and Vietze, C. (2012) Tetracycline resistance determinants of heterotrophic bacteria isolated from a South African tilapia aquaculture system. *African Journal of Microbiology Research* 6, 6761–6768.

Cheng, S., Hu, Y., Zhang, M. and Sun, L. (2010) Analysis of the vaccine potential of a natural avirulent *Edwardsiella tarda* isolate. *Vaccine* 28, 2716–2721.

Cheng, Z.X., Chu, X., Wang, S.N., Peng, X.X. and Li, H. (2018) Six genes of *ompA* family shuffling for development of polyvalent vaccines against *Vibrio alginolyticus* and *Edwardsiella tarda*. *Fish and Shellfish Immunology* 75, 308–315.

Choi, S. and Kim, K. (2011) Generation of two auxotrophic genes knock-out *Edwardsiella tarda* and assessment of its potential as a combined vaccine in olive flounder (*Paralichthys olivaceus*). *Fish and Shellfish Immunology* 31, 58–65.

Clarridge, J., Musher, D., Fainstein, V. and Wallace, R. (1980) Extraintestinal human infection caused by *Edwardsiella tarda*. *Journal of Clinical Microbiology* 11, 511–514.

Clavijo, A., Conroy, G., Conroy, D., Santander, J. and Aponte, F. (2002) First report of *Edwardsiella tarda* from tilapias in Venezuela. *Bulletin of the European Association of Fish Pathologists* 22, 280–282.

Coles, B.M., Stroud, R.K. and Sheggeby, S. (1978) Isolation of *Edwardsiella tarda* from three Oregon sea mammals. *Journal of Wildlife Diseases* 14, 339–341.

Cools, P., Haelters, J., dos Santos Santiago, G.L., Claeys, G., Boelens, J. *et al.* (2013) *Edwardsiella tarda* sepsis in a live-stranded sperm whale (*Physeter macrocephalus*). *Veterinary Microbiology* 166, 311–315.

Costa, A., Kanai, K. and Yoshikoshi, K. (1998) Serological characterization of atypical strains of *Edwardsiella tarda* isolated from sea breams. *Fish Pathology* 33, 265–274.

Crumlish, M., Dung, T., Turnbull, J., Ngoc, N. and Ferguson, H. (2002) Identification of *Edwardsiella ictaluri* from diseased freshwater catfish, *Pangasius hypophthalmus* (Sauvage), cultured in the Mekong Delta, Vietnam. *Journal of Fish Diseases* 25, 733–736.

Dahal, N., Abdelhamed, H., Lu, J., Karsi, A. and Lawrence, M.L. (2013) Tricarboxylic acid cycle and one-carbon metabolism pathways are important in *Edwardsiella ictaluri* virulence. *PLoS ONE* 8, e65973.

Dahal, N., Abdelhamed, H., Karsi, A. and Lawrence, M.L. (2014) Tissue persistence and vaccine efficacy of tricarboxylic acid cycle and one-carbon metabolism mutant strains of *Edwardsiella ictaluri*. *Vaccine* 32, 3971–3976.

Defoirdt, T., Sorgeloos, P. and Bossier, P. (2011) Alternatives to antibiotics for the control of bacterial disease in aquaculture. *Current Opinion in Microbiology* 14, 251–258.

Dong, H.T., Nguyen, V.V., Phiwsaiya, K., Gangnonngiw, W., Withyachumnarnkul, B. *et al.* (2015) Concurrent infections of *Flavobacterium columnare* and *Edwardsiella ictaluri* in striped catfish, *Pangasianodon hypophthalmus* in Thailand. *Aquaculture* 448, 142–150.

Dong, H.T., Senapin, S., Jeamkunakorn, C., Nguyen, V.V., Nguyen, N.T. *et al.* (2019) Natural occurrence of edwardsiellosis caused by *Edwardsiella ictaluri* in farmed hybrid red tilapia (*Oreochromis* sp.) in Southeast Asia. *Aquaculture* 499, 17–23.

Dubey, S., Maiti, B., Kim, S.H., Sivadasan, S.M., Kannimuthu, D. *et al.* (2019) Genotypic and phenotypic characterization of *Edwardsiella* isolates from different fish species and geographical areas in Asia: implications for vaccine development. *Journal of Fish Diseases* 42, 835–850.

Dung, T.T., Haesebrouck, F., Tuan, N.A., Sorgeloos, P., Baele, M. and Decostere, A. (2008) Antimicrobial susceptibility pattern of *Edwardsiella ictaluri* isolates from natural outbreaks of bacillary necrosis of *Pangasianodon hypophthalmus* in Vietnam. *Microbial Drug Resistance* 14, 311–316.

Dung, T.T., Haesebrouck, F., Sorgeloos, P., Tuan, N.A., Pasmans, F. *et al.* (2009) IncK plasmid-mediated tetracycline resistance in *Edwardsiella ictaluri* isolates from diseased freshwater catfish in Vietnam. *Aquaculture* 295, 157–159.

Esteve, C. and Alcaide, E. (2018) Seasonal recovery of *Edwardsiella piscicida* from wild European eels and natural waters: isolation methods, virulence and reservoirs. *Journal of Fish Diseases* 41, 1613–1623.

Ewing, W., McWhorter, A., Escobar, M. and Lubin, A. (1965) *Edwardsiella*, a new genus of Enterobacteriaceae based on a new species, *E. tarda*. *International Bulletin of Bacteriological Nomenclature and Taxonomy* 15, 33–38.

Ferguson, H.W., Turnbull, J.F., Shinn, A., Thompson, K., Dung, T.T. and Crumlish, M. (2001) Bacillary necrosis in farmed *Pangasius hypophthalmus* (Sauvage) from the Mekong Delta, Vietnam. *Journal of Fish Diseases* 24, 509–513.

Fitzsimmons, K. (2000) Future trends of tilapia aquaculture in the Americas. *Tilapia Aquaculture in the Americas* 2, 252–264.

Fogelson, S., Petty, B., Reichley, S., Ware, C., Bowser, P. *et al.* (2016) Histologic and molecular characterization of *Edwardsiella piscicida* infection in largemouth bass (*Micropterus salmoides*). *Journal of Veterinary Diagnostic Investigation* 28, 338–344.

Francis-Floyd, R., Beleau, M., Waterstrat, P. and Bowser, P. (1987) Effect of water temperature on the clinical outcome of infection with *Edwardsiella ictaluri* in channel catfish. *Journal of the American Veterinary Medical Association* 191, 1413–1416.

Francis-Floyd, R., Reed, P., Bolon, B., Estes, J. and McKinney, S. (1993) An epizootic of *Edwardsiella tarda* in largemouth bass (*Micropterus salmoides*). *Journal of Wildlife Diseases* 29, 334–336.

Gaikowski, M., Wolf, J., Endris, R. and Gingerich, W. (2003) Safety of Aquaflor (florfenicol, 50% type A medicated article), administered in feed to channel catfish, *Ictalurus punctatus*. *Toxicologic Pathology* 31, 689–697.

Gaunt, P., Endris, R., Khoo, L., Howard, R., McGinnis, A., Santucci, T.D. and Katz, T. (2004) Determination of dose rate of florfenicol in feed for control of mortality in channel catfish *Ictalurus punctatus* (Rafinesque) infected with *Edwardsiella ictaluri*, etiological agent of enteric septicemia. *Journal of the World Aquaculture Society* 35, 257–267.

Gaunt, P., McGinnis, A., Santucci, T., Cao, J., Waeger, P. and Endris, R. (2006) Field efficacy of florfenicol for control of mortality in channel catfish, *Ictalurus punctatus* (Rafinesque), caused by infection with *Edwardsiella ictaluri*. *Journal of the World Aquaculture Society* 37, 1–11.

Gaunt, P.S., Chatakondi, N., Gao, D. and Endris, R. (2015) Efficacy of florfenicol for control of mortality associated with *Edwardsiella ictaluri* in three species of catfish. *Journal of Aquatic Animal Health* 27, 45–49.

Geng, Y., Wang, K.Y., Li, C.W., Ren, S.Y., Zhou, Z.Y. *et al.* (2013) Isolation and characterization of *Edwardsiella ictaluri* from southern catfish, *Silurus soldatovi meridionalis*, (Chen) cultured in China. *Journal of the World Aquaculture Society* 44, 273–281.

Greenway, T.E., Byars, T.S., Elliot, R.B., Jin, X., Griffin, M.J. and Wise, D.J. (2017) Validation of fermentation and processing procedures for the commercial-scale production of a live, attenuated *Edwardsiella ictaluri* vaccine for use in channel catfish aquaculture. *Journal of Aquatic Animal Health* 29, 83–88.

Griffin, M., Mauel, M., Greenway, T., Khoo, L. and Wise, D. (2011) A real-time polymerase chain reaction assay for quantification of *Edwardsiella ictaluri* in catfish pond water and genetic homogeneity of diagnostic case isolates from Mississippi. *Journal of Aquatic Animal Health* 23, 178–188.

Griffin, M., Quiniou, S., Cody, T., Tabuchi, M., Ware, C. *et al.* (2013) Comparative analysis of *Edwardsiella* isolates from fish in the eastern United States identifies two distinct genetic taxa amongst organisms phenotypically classified as *E. tarda*. *Veterinary Microbiology* 165, 358–372.

Griffin, M., Ware, C., Quiniou, S., Steadman, J., Gaunt, P. *et al.* (2014) *Edwardsiella piscicida* identified in the southeastern USA by *gyrB* sequence, species-specific and repetitive sequence-mediated PCR. *Diseases of Aquatic Organisms* 108, 23–35.

Griffin, M.J., Reichley, S.R., Greenway, T.E., Quiniou, S.M., Ware, C. *et al.* (2016) Comparison of *Edwardsiella ictaluri* isolates from different hosts and geographic origins. *Journal of Fish Diseases* 39, 947–969.

Griffin, M.J., Greenway, T.E. and Wise, D.J. (2017) *Edwardsiella* spp. In: Woo, P.T.K. and Cipriano, R.C. (eds) *Fish Viruses and Bacteria: Pathobiology and Protection*. CAB International, Wallingford, UK, pp. 190–210.

Griffin, M.J., Reichley, S.R., Baumgartner, W.A., Aarattuthodiyil, S., Ware, C. *et al.* (2019a) Emergence of *Edwardsiella piscicida* in farmed channel ♀, *Ictalurus punctatus* × blue ♂, *Ictalurus furcatus*, hybrid catfish cultured in Mississippi. *Journal of the World Aquaculture Society* 50, 420–432.

Griffin, M., Petty, B.D., Ware, C. and Fogelson, S.B. (2019b) Recovery and confirmation of *Edwardsiella piscicida* from a black crappie *Pomoxis nigromaculatus* (Lesueur, 1829). *Journal of Fish Diseases* 42, 1457–1461.

Griffin, M.J., Greenway, T.E., Byars, T.S., Ware, C., Aarattuthodiyil, S. *et al.* (2020) Cross-protective potential of a live, attenuated *Edwardsiella ictaluri* vaccine against *Edwardsiella piscicida* in channel (*Ictalurus punctatus*) and channel × blue (*Ictalurus furcatus*) hybrid catfish. *Journal of the World Aquaculture Society* (in press).

Grimont, P., Grimont, F., Richard, C. and Sakazaki, R. (1980) *Edwardsiella hoshinae*, a new species of Enterobacteriaceae. *Current Microbiology* 4, 347–351.

Haenen, O., Evans, J. and Berthe, F. (2013) Bacterial infections from aquatic species: potential for and prevention of contact zoonoses. *Revue Scientifique et Technique (International Office of Epizootics)* 32, 497–507.

Han, H.J., Kim, D.H., Lee, D.C., Kim, S.M. and Park, S.I. (2006) Pathogenicity of *Edwardsiella tarda* to olive flounder, *Paralichthys olivaceus* (Temminck & Schlegel). *Journal of Fish Diseases* 29, 601–609.

Hassan, E.S., Mahmoud, M.M., Kawato, Y., Nagai, T., Kawaguchi, O. *et al.* (2012) Subclinical *Edwardsiella ictaluri* infection of wild ayu *Plecoglossus altivelis*. *Fish Pathology* 47, 64–73.

Hawke, J.P. (1979) A bacterium associated with disease of pond cultured channel catfish, *Ictalurus punctatus*. *Journal of the Fisheries Board of Canada* 36, 1508–1512.

Hawke, J. and Khoo, L. (2004) Infectious diseases. In: Tucker, C. and Hargreaves, J. (eds) *Biology and Culture of the Channel Catfish*, 1st edn. Elsevier, Amsterdam, pp. 347–443.

Hawke, J., McWhorter, A., Steigerwalt, A. and Brenner, D. (1981) *Edwardsiella ictaluri* sp. nov., the causative agent of enteric septicemia of catfish. *International Journal of Systematic Bacteriology* 31, 396–400.

Hawke, J.P., Kent, M., Rogge, M., Baumgartner, W., Wiles, J. *et al.* (2013) Edwardsiellosis caused by

Edwardsiella ictaluri in laboratory populations of zebrafish *Danio rerio*. *Journal of Aquatic Animal Health* 25, 171–183.

Herman, R. and Bullock, G. (1986) Pathology caused by the bacterium *Edwardsiella tarda* in striped bass. *Transactions of the American Fisheries Society* 115, 232–235.

Hernández-Rodríguez, A., Alceste-Oliviero, C., Sanchez, R., Jory, D., Vidal, L. and Constain-Franco, L.F. (2001) Aquaculture development trends in Latin America and the Caribbean. In: Subasinghe, R., Bueno, P.B., Phillips, M.J., Hough, C., McGladdery, S.E. and Arthur, J.R. (eds) *Technical Proceedings of the Conference on Aquaculture in the Third Millennium*. Network of Aquaculture Centres in Asia-Pacific, Bangkok and Food and Agriculture Organization of the United Nations, Rome, pp. 317–340.

Hilton, L. and Wilson, J. (1980) Terramycin-resistant *Edwardsiella tarda* in channel catfish. *The Progressive Fish-Culturist* 42, 159.

Hirai, Y., Ashata-Tago, S., Ainoda, Y., Fujita, T. and Kikuchi, K. (2015) *Edwardsiella tarda* bacteremia. A rare but fatal water- and foodborne infection: review of the literature and clinical cases from a single centre. *Canadian Journal of Infectious Disease and Medical Microbiology* 26, 313–318.

Humphrey, J. and Langdon, J. (1986) Pathological anatomy and diseases of barramundi (*Lates calcarifer*). In: Copland, J.W. and Grey, D.L. (eds) *Management of wild cultured sea bass/barramundi (Lates calcarifer): proceedings of an international workshop held at Darwin, NT, Australia, 24–30 September 1986. ACIAR Proceedings No.* 20. Australian Centre for International Agricultural Research, Canberra, pp. 198–203.

Humphrey, J.D., Lancaster, C., Gudkovs, N. and McDonald, W. (1986) Exotic bacterial pathogens *Edwardsiella tarda* and *Edwardsiella ictaluri* from imported ornamental fish *Betta splendens* and *Puntius conchonius*, respectively: isolation and quarantine significance. *Australian Veterinary Journal* 63, 369–371.

Ibrahem, M.D., Shaheed, B., Yazeed, H.A.E. and Korani, H. (2011) Assessment of the susceptibility of polyculture reared African catfish and Nile tilapia to *Edwardsiella tarda*. *Journal of American Science* 7, 779–786.

Iregui, C.A., Guarín, M., Tibatá, V.M. and Ferguson, H.W. (2012) Novel brain lesions caused by *Edwardsiella tarda* in a red tilapia (*Oreochromis* spp.). *Journal of Veterinary Diagnostic Investigation*, 24, 446–449.

Iwanowicz, L., Griffin, A., Cartwright, D. and Blazer, V. (2006) Mortality and pathology in brown bullheads *Amieurus nebulosus* associated with a spontaneous *Edwardsiella ictaluri* outbreak under tank culture conditions. *Diseases of Aquatic Organisms* 70, 219–225.

Janda, J. and Abbott, S. (1993) Infections associated with the genus *Edwardsiella*: the role of *Edwardsiella tarda* in human disease. *Clinical Infectious Diseases* 17, 742–748.

Janda, J., Abbott, S., Kroske-Bystrom, S., Cheung, W., Powers, C. *et al.* (1991) Pathogenic properties of *Edwardsiella* species. *Journal of Clinical Microbiology* 29, 1997–2001.

Joh, S., Kim, M., Kwon, H., Ahn, E., Jang, H. and Kwon, J. (2011) Characterization of *Edwardsiella tarda* isolated from farm-cultured eels, *Anguilla japonica*, in the Republic of Korea. *Journal of Veterinary Medical Science* 73, 7–11.

Jun, L.J., Jeong, J.B., Huh, M.D., Chung, J.K., Choi, D.L. *et al.* (2004) Detection of tetracycline-resistance determinants by multiplex polymerase chain reaction in *Edwardsiella tarda* isolated from fish farms in Korea. *Aquaculture* 240, 89–100.

Kasornchandra, J., Rogers, W. and Plumb, J. (1987) *Edwardsiella ictaluri* from walking catfish, *Clarias batrachus* L., in Thailand. *Journal of Fish Diseases* 10, 137–138.

Katharios, P., Kokkari, C., Dourala, N. and Smyrli, M. (2015) First report of edwardsiellosis in cage-cultured sharpsnout sea bream, *Diplodus puntazzo* from the Mediterranean. *BMC Veterinary Research* 11, 155.

Katharios, P., Kalatzis, P.G., Kokkari, C., Pavlidis, M. and Wang, Q. (2019) Characterization of a highly virulent *Edwardsiella anguillarum* strain isolated from Greek aquaculture, and a spontaneously induced prophage therein. *Frontiers in Microbiology* 10, 141.

Kawai, K., Liu, Y., Ohnishi, K. and Oshima, S. (2004) A conserved 37 kDa outer membrane protein of *Edwardsiella tarda* is an effective vaccine candidate. *Vaccine* 22, 3411–3418.

Kelly, E., Martin, P.A.J., Gibson-Kueh, S., Morgan, D.L., Ebner, B.C. *et al.* (2018) First detection of *Edwardsiella ictaluri* (Proteobacteria: Enterobacteriaceae) in wild Australian catfish. *Journal of Fish Diseases* 41, 199–208.

Kent, M. and Lyons, J. (1982) *Edwardsiella ictaluri in the green knife fish, Eigemannia virescens. Fish Health News 2, 2*. Eastern Fish Disease Laboratory, US Department of the Interior, Fish and Wildlife Service, Kearneysville, West Virginia.

Keskin, O., Seçer, S., İzgür, M., Türkyilmaz, S. and Mkakosya, R.S. (2004) *Edwardsiella ictaluri* infection in rainbow trout (*Oncorhynchus mykiss*). *Turkish Journal of Veterinary and Animal Sciences* 28, 649–653.

Klesius, P.H. and Sealey, W.M. (1995) Characteristics of serum antibody in enteric septicemia of catfish. *Journal of Aquatic Animal Health* 7, 205–210.

Klesius, P., Lovy, J., Evans, J., Washuta, E. and Arias, C. (2003) Isolation of *Edwardsiella ictaluri* from tadpole madtom in a southwestern New Jersey River. *Journal of Aquatic Animal Health* 15, 295–301.

Kumar, G., Byars, T.S., Greenway, T.E., Aarattuthodiyil, S., Khoo, L.H. *et al.* (2019) Economic assessment of

commercial-scale *Edwardsiella ictaluri* vaccine trials in US catfish industry. *Aquaculture Economics & Management* 23, 254–275.
Kusuda, R. and Kawai, K. (1998) Bacterial diseases of cultured marine fish in Japan. *Fish Pathology* 33, 221–227.
Kusuda, R., Toyoshima, T., Iwamura, Y. and Sako, H. (1976) *Edwardsiella tarda* from an epizootic of mullets (*Mugil cephalus*) in Okitsu Bay. *Nippon Suisan Gakkaishi* 42, 271–275.
Kusuda, R., Itami, T., Munekiyo, M. and Nakajima, H. (1977) Characteristics of an *Edwardsiella* sp. from an epizootic of cultured sea breams. *Nippon Suisan Gakkaishi* 43, 129–134.
Kwon, S., Nam, Y., Kim, S. and Kim, K. (2006) Protection of tilapia (*Oreochromis mosambicus*) from edwardsiellosis by vaccination with *Edwardsiella tarda* ghosts. *Fish and Shellfish Immunology* 20, 621–626.
Kwon, S., Lee, E., Nam, Y., Kim, S. and Kim, K. (2007) Efficacy of oral immunization with *Edwardsiella tarda* ghosts against edwardsiellosis in olive flounder (*Paralichthys olivaceus*). *Aquaculture* 269, 84–88.
Lan, J., Zhang, X., Wang, Y., Chen, J. and Han, Y. (2008) Isolation of an unusual strain of *Edwardsiella tarda* from turbot and establish a PCR detection technique with the *gyrB* gene. *Journal of Applied Microbiology* 105, 644–651.
Lawrence, M., Cooper, R. and Thune, R. (1997) Attenuation, persistence, and vaccine potential of an *Edwardsiella ictaluri purA* mutant. *Infection and Immunity* 65, 4642–4651.
Lee, K., Kim, H.K., Park, S.K., Sohn, H., Cho, Y. *et al.* (2018) First report of the occurrence and whole-genome characterization of *Edwardsiella tarda* in the false killer whale (*Pseudorca crassidens*). *Journal of Veterinary Medical Science* 80, 1041–1046.
Leotta, G., Piñeyro, P., Serena, S. and Vigo, G. (2009) Prevalence of *Edwardsiella tarda* in Antarctic wildlife. *Polar Biology* 32, 809–812.
Leung, K.Y., Wang, Q., Yang, Z. and Siame, B. (2019) *Edwardsiella piscicida*: a versatile emerging pathogen of fish. *Virulence* 10, 555–567.
Liang, S., Wu, H., Liu, B., Xiao, J., Wang, Q. and Zhang, Y. (2012) Immune response of turbot (*Scophthalmus maximus* L.) to a broad spectrum vaccine candidate, recombinant glyceraldehyde-3-phosphate dehydrogenase of *Edwardsiella tarda*. *Veterinary Immunology and Immunopathology* 150, 198–205.
LiHua, D., JianJun, F., Peng, L., SongLin, G., Le, H. and YiQun, X. (2019) Evaluation of an outer membrane protein as a vaccine against *Edwardsiella anguillarum* in Japanese eels (*Anguilla japonica*). *Aquaculture* 498, 143–150.
Ling, S.H.M., Wang, X.H., Xie, L., Lim, T.M. and Leung, K.Y. (2000) Use of green fluorescent protein (GFP) to study the invasion pathways of *Edwardsiella tarda* in *in vivo* and *in vitro* fish models. *Microbiology* 146, 7–19.
Liu, F., Tang, X., Sheng, X., Xing, J. and Zhan, W. (2016) DNA vaccine encoding molecular chaperone GroEL of *Edwardsiella tarda* confers protective efficacy against edwardsiellosis. *Molecular Immunology* 79, 55–65.
Liu, J.Y., Li, A.H., Zhou, D.R., Wen, Z.R. and Ye, X.P. (2010) Isolation and characterization of *Edwardsiella ictaluri* strains as pathogens from diseased yellow catfish *Pelteobagrus fulvidraco* (Richardson) cultured in China. *Aquaculture Research* 41, 1835–1844.
Liu, X., Zhang, H., Jiao, C., Liu, Q., Zhang, Y. and Xiao, J. (2017) Flagellin enhances the immunoprotection of formalin-inactivated *Edwardsiella tarda* vaccine in turbot. *Vaccine* 35, 369–374.
Lo, D., Lee, Y., Wang, J. and Kuo, H. (2014) Antimicrobial susceptibility and genetic characterisation of oxytetracycline-resistant *Edwardsiella tarda* isolated from diseased eels. *Veterinary Record* 175, 203–203.
Loch, T.P., Hawke, J.P., Reichley, S.R., Faisal, M., Del Piero, F. and Griffin, M.J. (2017) Outbreaks of edwardsiellosis caused by *Edwardsiella piscicida* and *Edwardsiella tarda* in farmed barramundi (*Lates calcarifer*). *Aquaculture* 481, 202–210.
López-Porras, A., Elizondo, C., Chaves, A.J., Camus, A.C., Griffin, M.J. *et al.* (2019) Application of multiplex quantitative Polymerase chain reaction methods to detect common bacterial fish pathogens in Nile tilapia, *Oreochromis niloticus*, hatcheries in Costa Rica. *Journal of the World Aquaculture Society* 50, 645–658.
Lv, Y., Zheng, J., Yang, M., Wang, Q. and Zhang, Y. (2012) An *Edwardsiella tarda* mutant lacking UDP-glucose dehydrogenase shows pleiotropic phenotypes, attenuated virulence, and potential as a vaccine candidate. *Veterinary Microbiology* 160, 506–512.
Lymbery, A., Kueh, S., Kelly, E., Morgan, D., Buller, N. *et al.* (2016) *A Survey of Edwardsiella ictaluri in Wild Catfish Populations in Australia*. Fisheries Research and Development Corporation and Murdoch University, Perth, Australia.
Mainous, M., Smith, S. and Kuhn, D. (2010) Effect of common aquaculture chemicals against *Edwardsiella ictaluri* and *E. tarda*. *Journal of Aquatic Animal Health* 22, 224–228.
Matsuyama, T., Kamaishi, T., Ooseko, N., Kurohara, K. and Iida, T. (2005) Pathogenicity of motile and non-motile *Edwardsiella tarda* to some marine fish. *Fish Pathology* 40, 133–135.
Meyer, F. and Bullock, G. (1973) *Edwardsiella tarda*, a new pathogen of channel catfish (*Ictalurus punctatus*). *Applied Microbiology* 25, 155–156.
Miniero Davies, Y., Xavier de Oliveira, M.G., Paulo Vieira Cunha, M., Soares Franco, L., Pulecio Santos, S.L. *et al.* (2018) *Edwardsiella tarda* outbreak affecting fishes and aquatic birds in Brazil. *Veterinary Quarterly* 38, 99–105.

Mo, Z.Q., Zhou, L., Zhang, X., Gan, L., Liu, L. and Dan, X.M. (2015) Outbreak of *Edwardsiella tarda* infection in farm-cultured giant mottled eel *Anguilla marmorata* in China. *Fisheries Science* 81, 899–905.

Mohanty, B. and Sahoo, P. (2007) Edwardsiellosis in fish: a brief review. *Journal of Biosciences* 32, 1331–1344.

Nagai, T. and Nakai, T. (2014) Water temperature effect on *Edwardsiella ictaluri* infection of ayu *Plecoglossus altivelis*. *Fish Pathology* 49, 61–63.

Nagai, T., Iwamoto, E., Sakai, T., Arima, T., Tensha, K. *et al.* (2008) Characterization of *Edwardsiella ictaluri* isolated from wild ayu *Plecoglossus altivelis* in Japan. *Fish Pathology* 43, 158–163.

Nahm, M.H., Apicella, M.A. and Briles, D.E. (1999) Immunity to extracellular bacteria. In: Paul, W.E. (ed.) *Fundamental Immunology*. Lippincott-Raven, Philadelphia, Pennsylvania, pp. 1373–1386.

Nakatsugawa, T. (1983) *Edwardsiella tarda* isolated from cultured young flounder. *Fish Pathology* 18, 99–101.

Nakhro, K., Devi, T. and Kamilya, D. (2013) *In vitro* immunopathogenesis of *Edwardsiella tarda* in catla *Catla catla* (Hamilton). *Fish and Shellfish Immunology* 35, 175–179.

Newton, J.C., Bird, R.C., Blevins, W.T., Wilt, G.R. and Wolfe, L.G. (1988) Isolation, characterization, and molecular cloning of cryptic plasmids isolated from *Edwardsiella ictaluri*. *American Journal of Veterinary Research* 49, 1856–1860.

Nho, S.W., Abdelhamed, H., Karsi, A. and Lawrence, M.L. (2017) Improving safety of a live attenuated *Edwardsiella ictaluri* vaccine against enteric septicemia of catfish and evaluation of efficacy. *Veterinary Microbiology* 210, 83–90.

NOAA (National Ocean and Atmosphere Administration), National Centers for Environmental Information (2019) *State Climate Summaries* 2019. Available at: https://statesummaries.ncics.org/ (accessed 22 August 2019).

Oguro, K., Tamura, K., Yamane, J., Shimizu, M., Yamamoto, T. *et al.* (2014) Draft genome sequences of two genetic variant strains of *Edwardsiella piscicida*, JF1305 and RSB1309, isolated from olive flounder (*Paralichythys olivaceus*) and red sea bream (*Pagrus major*) cultured in Japan, respectively. *Genome Announcements* 2, e00546-14.

Oliveira, J., Castilho, F., Cunha, A. and Pereira, M.J. (2012) Bacteriophage therapy as a bacterial control strategy in aquaculture. *Aquaculture International* 20, 879–910.

Padrós, F., Zarza, C., Dopazo, L., Cuadrado, M. and Crespo, S. (2006) Pathology of *Edwardsiella tarda* infection in turbot, *Scophthalmus maximus* (L.). *Journal of Fish Diseases* 29, 87–94.

Pan, X., Hao, G., Yao, J., Xu, Y., Shen, J. and Yin, W. (2010) Identification and pathogenic facts studying for *Edwardsiella tarda* from edwardsiellosis of *Trionyx sinensis* [J.]. *Freshwater Fisheries* 6, 40–45.

Park, S., Aoki, T. and Jung, T. (2012) Pathogenesis of and strategies for preventing *Edwardsiella tarda* infection in fish. *Veterinary Research* 43, 67.

Peterson, B.C., Flora, C., Wood, M., Bosworth, B.G., Quiniou, S.M. *et al.* (2016) Vaccination of full-sib channel catfish families against enteric septicemia of catfish with an oral live attenuated *Edwardsiella ictaluri* vaccine. *Journal of the World Aquaculture Society* 47, 207–211.

Petrie-Hanson, L. and Ainsworth, A.J. (1999) Humoral immune responses of channel catfish (*Ictalurus punctatus*) fry and fingerlings exposed to *Edwardsiella ictaluri*. *Fish and Shellfish Immunology* 9, 579–589.

Petrie-Hanson, L. and Ainsworth, A.J. (2001) Ontogeny of channel catfish lymphoid organs. *Veterinary Immunology and Immunopathology* 81, 113–127.

Phillips, A.C.N., Reichley, S.R., Ware, C. and Griffin, M.J. (2017) *Edwardsiella ictaluri* infection in Pangasius catfish imported from West Bengal into the Southern Caribbean. *Journal of Fish Diseases* 40, 743–756.

Plumb, J. and Shoemaker, C. (1995) Effects of temperature and salt concentration on latent *Edwardsiella ictaluri* infections in channel catfish. *Diseases of Aquatic Organisms* 21, 171–175.

Plumb, J. and Vinitnantharat, S. (1989) Biochemical, biophysical, and serological homogeneity of *Edwardsiella ictaluri*. *Journal of Aquatic Animal Health* 1, 51–56.

Plumb, J. and Vinitnantharat, S. (1993) Vaccination of channel catfish, *Ictalurus punctatus* (Rafinesque), by immersion and oral booster against *Edwardsiella ictaluri*. *Journal of Fish Diseases* 16, 65–71.

Plumb, J.A., Maestrone, G. and Quinlan, E. (1987) Use of a potentiated sulfonamide to control *Edwardsiella ictaluri* infection in channel catfish (*Ictalurus punctatus*). *Aquaculture* 62, 187–194.

Plumb, J., Vinitnantharat, S. and Paterson, W. (1994) Optimum concentration of *Edwardsiella ictaluri* vaccine in feed for oral vaccination of channel catfish. *Journal of Aquatic Animal Health* 6, 118–121.

Pridgeon, J.W. and Klesius, P.H. (2011) Development of a novobiocin-resistant *Edwardsiella ictaluri* as a novel vaccine in channel catfish (*Ictalurus punctatus*). *Vaccine* 29, 5631–5637.

Rashid, M.M., Honda, K., Nakai, T. and Muroga, K. (1994) An ecological study on *Edwardsiella tarda* in flounder farms. *Fish Pathology* 29, 221–227.

Reddacliff, G.L., Hornitzky, M. and Whittington, R.J. (1996) *Edwardsiella tarda* septicaemia in rainbow trout (*Oncorhynchus mykiss*). *Australian Veterinary Journal* 73, 30.

Řehulka, J., Marejková, M. and Petráš, P. (2012) Edwardsiellosis in farmed rainbow trout (*Oncorhynchus mykiss*). *Aquaculture Research* 43, 1628–1634.

Reichley, S., Waldbieser, G., Lawrence, M. and Griffin, M. (2015a) Complete genome sequence of an

Edwardsiella piscicida-like species, recovered from tilapia in the United States. *Genome Announcements* 3, e01004-15.

Reichley, S., Ware, C., Greenway, T., Wise, D. and Griffin, M. (2015b) Real-time polymerase chain reaction assays for the detection and quantification of *Edwardsiella tarda*, *Edwardsiella piscicida*, and *Edwardsiella piscicida*-like species in catfish tissues and pond water. *Journal of Veterinary Diagnostic Investigation* 27, 130–139.

Reichley, S.R., Waldbieser, G.C., Tekedar, H.C., Lawrence, M.L. and Griffin, M.J. (2016) Complete genome sequence of *Edwardsiella piscicida* isolate S11-285 recovered from channel catfish (*Ictalurus punctatus*) in Mississippi, USA. *Genome Announcements* 4, e01259-16.

Reichley, S.R., Ware, C., Steadman, J., Gaunt, P.S., García, J.C. *et al.* (2017) Comparative phenotypic and genotypic analysis of *Edwardsiella* isolates from different hosts and geographic origins, with emphasis on isolates formerly classified as *E. tarda*, and evaluation of diagnostic methods. *Journal of Clinical Microbiology* 55, 3466–3491.

Reichley, S.R., Ware, C., Khoo, L.H., Greenway, T.E., Wise, D.J. *et al.* (2018) Comparative susceptibility of channel catfish, *Ictalurus punctatus*; blue catfish, *Ictalurus furcatus*; and channel (♀) × blue (♂) hybrid catfish to *Edwardsiella piscicida*, *Edwardsiella tarda*, and *Edwardsiella anguillarum*. *Journal of the World Aquaculture Society* 49, 197–204.

Richards, G.P. (2014) Bacteriophage remediation of bacterial pathogens in aquaculture: a review of the technology. *Bacteriophage* 4, e975540.

Rogge, M., Dubytska, L., Jung, T., Wiles, J., Elkamel, A. *et al.* (2013) Comparison of Vietnamese and US isolates of *Edwardsiella ictaluri*. *Diseases of Aquatic Organisms* 106, 17–29.

Rousselet, E., Stacy, N.I., Rotstein, D.S., Waltzek, T.B., Griffin, M.J. and Francis-Floyd, R. (2018) Systemic *Edwardsiella tarda* infection in a Western African lungfish (*Protopterus annectens*) with cytologic observation of heterophil projections. *Journal of Fish Diseases* 41, 1453–1458.

Saeed, M.O. and Plumb, J.A. (1986) Immune response of channel catfish to lipopolysaccharide and whole cell *Edwardsiella ictaluri* vaccines. *Diseases of Aquatic Organisms* 2, 21–26.

Sae-oui, D., Muroga, K. and Nakai, T. (1984) A case of *Edwardsiella tarda* infection in cultured colored carp *Cyprinus carpio*. *Fish Pathology* 19, 197–199.

Sakai, T., Iida, T., Osatomi, K. and Kanai, K. (2007) Detection of Type 1 fimbrial genes in fish pathogenic and non-pathogenic *Edwardsiella tarda* strains by PCR. *Fish Pathology* 42, 115–117.

Sakai, T., Kamaishi, T., Sano, M., Tensha, K., Arima, T. *et al.* (2008) Outbreaks of *Edwardsiella ictaluri* infection in ayu *Plecoglossus altivelis* in Japanese rivers. *Fish Pathology* 43, 152–157.

Sakai, T., Matsuyama, T., Sano, M. and Iida, T. (2009a) Identification of novel putative virulence factors, adhesin AIDA and type VI secretion system, in atypical strains of fish pathogenic *Edwardsiella tarda* by genomic subtractive hybridization. *Microbiology and Immunology* 53, 131–139.

Sakai, T., Yuasa, K., Sano, M. and Iida, T. (2009b) Identification of *Edwardsiella ictaluri* and *E. tarda* by species-specific polymerase chain reaction targeted to the upstream region of the fimbrial gene. *Journal of Aquatic Animal Health* 21, 124–132.

Salati, F. and Kusuda, R. (1985) *Vaccine* preparations used for immunization of eel *Anguilla japonica* against *Edwardsiella tarda* infection. *Nippon Suisan Gakkaishi* 51, 1233–1237.

Santander, J., Golden, G., Wanda, S. and Curtiss, R. (2012) Fur-regulated iron uptake system of *Edwardsiella ictaluri* and its influence on pathogenesis and immunogenicity in the catfish host. *Infection and Immunity* 80, 2689–2703.

Seder, R.A. and Hill, A.V.S. (2000) *Vaccines* against intracellular infections requiring cellular immunity. *Nature* 406, 793–797.

Shafiei, S., Viljamaa-Dirks, S., Sundell, K., Heinikainen, S., Abayneh, T. and Wiklund, T. (2016) Recovery of *Edwardsiella piscicida* from farmed whitefish, *Coregonus lavaretus* (L.), in Finland. *Aquaculture* 454, 19–26.

Shao, J., Yuan, J., Shen, Y., Hu, R. and Gu, Z. (2016) First isolation and characterization of *Edwardsiella tarda* from diseased Asian swamp eel, *Monopterus albus* (Zuiew). *Aquaculture Research* 47, 3684–3688.

Shao, S., Lai, Q., Liu, Q., Wu, H., Xiao, J. *et al.* (2015) Phylogenomics characterization of a highly virulent *Edwardsiella* strain ET080813T encoding two distinct T3SS and three T6SS gene clusters: propose a novel species as *Edwardsiella anguillarum* sp. nov. *Systematic and Applied Microbiology* 38, 36–47.

Shelley, J.P., Yanong, R.P.E., Hawke, J.P. and Griffin, M. (2018) Edwardsiellosis in ornamental fish. In: *Proceedings of the 8th International Symposium on Aquatic Animal Health, Prince Edward Island, Canada, 2–6 September* 2018, p.102.

Shetty, M., Maiti, B., Venugopal, M., Karunasagar, I. and Karunasagar, I. (2014) First isolation and characterization of *Edwardsiella tarda* from diseased striped catfish, *Pangasianodon hypophthalmus* (Sauvage). *Journal of Fish Diseases* 37, 265–271.

Shoemaker, C. and Klesius, P. (1997) Protective immunity against enteric septicaemia in channel catfish, *Ictalurus punctatus* (Rafinesque), following controlled exposure to *Edwardsiella ictaluri*. *Journal of Fish Diseases* 20, 361–368.

Shoemaker, C.A., Klesius, P.H., Evans, J.J. and Arias, C.R. (2009) Use of modified live vaccines in aquaculture.

Journal of the World Aquaculture Society 40, 573–585.

Soto, E., Griffin, M., Arauz, M., Riofrio, A., Martinez, A. and Cabrejos, M. (2012) *Edwardsiella ictaluri* as the causative agent of mortality in cultured Nile tilapia. *Journal of Aquatic Animal Health* 24, 81–90.

Starliper, C., Cooper, R., Shotts, E. and Taylor, P. (1993) Plasmid-mediated Romet resistance of *Edwardsiella ictaluri*. *Journal of Aquatic Animal Health* 5, 1–8.

Sun, K., Wang, H.L., Zhang, M., Xiao, Z.Z. and Sun, L. (2009) Genetic mechanisms of multi-antimicrobial resistance in a pathogenic *Edwardsiella tarda* strain. *Aquaculture* 289, 134–139.

Sun, Y., Liu, C.-S. and Sun, L. (2010) Identification of an *Edwardsiella tarda* surface antigen and analysis of its immunoprotective potential as a purified recombinant subunit vaccine and a surface-anchored subunit vaccine expressed by a fish commensal strain. *Vaccine* 28, 6603–6608.

Sun, Y., Liu, C.-S. and Sun, L. (2011a) Construction and analysis of the immune effect of an *Edwardsiella tarda* DNA vaccine encoding a D15-like surface antigen. *Fish and Shellfish Immunology* 30, 273–279.

Sun, Y., Liu, C.-S. and Sun, L. (2011b) Comparative study of the immune effect of an *Edwardsiella tarda* antigen in two forms: subunit vaccine vs DNA vaccine. *Vaccine* 29, 2051–2057.

Swain, P., Nayak, S., Sahu, A., Mohapatra, B. and Meher, P. (2002) Bath immunisation of spawn, fry and fingerlings of Indian major carps using a particulate bacterial antigen. *Fish and Shellfish Immunology* 13, 133–140.

Takano, T., Matsuyama, T., Oseko, N., Sakai, T., Kamaishi, T. *et al.*, (2010) The efficacy of five avirulent *Edwardsiella tarda* strains in a live vaccine against edwardsiellosis in Japanese flounder, *Paralichthys olivaceus*. *Fish and Shellfish Immunology* 29, 687–693.

Takano, T., Matsuyama, T., Sakai, T. and Nakayasu, C. (2011) Protective efficacy of a formalin-killed vaccine against atypical *Edwardsiella tarda* infection in red sea bream *Pagrus major*. *Fish Pathology* 46, 120–122.

Thorarinsson, R. and Powell, D.B. (2006) Effects of disease risk, vaccine efficacy, and market price on the economics of fish vaccination. *Aquaculture* 256, 42–49.

Thune, R., Stanley, L. and Cooper, R. (1993) Pathogenesis of Gram-negative bacterial infections in warmwater fish. *Annual Review of Fish Diseases* 3, 37–68.

Thune, R., Hawke, J. and Johnson, M. (1994) Studies on vaccination of channel catfish, *Ictalurus punctatus*, against *Edwardsiella ictaluri*. *Journal of Applied Aquaculture* 3, 11–24.

Thune, R.L., Collins, L.A. and Penta, M.P. (1997) A comparison of immersion, immersion/oral combination and injection methods for the vaccination of channel catfish *Ictalurus punctatus* against *Edwardsiella ictaluri*. *Journal of the World Aquaculture Society* 28, 193–201.

Thune, R., Fernandez, D. and Battista, J. (1999) An *aroA* mutant of *Edwardsiella ictaluri* is safe and efficacious as a live, attenuated vaccine. *Journal of Aquatic Animal Health* 11, 358–372.

Trung Cao, T., Tsai, M., Yang, C., Wang, P., Kuo, T. *et al.* (2014) Vaccine efficacy of glyceraldehyde-3-phosphate dehydrogenase (GAPDH) from *Edwardsiella ictaluri* against *E. tarda* in tilapia. *Journal of General and Applied Microbiology* 60, 241–250.

Tucker, C.C. and Robinson, E.H. (1990) *Channel Catfish Farming Handbook*. Springer Science & Business Media, New York.

Ucko, M., Colomi, A., Dubytska, L. and Thune, R. (2016) *Edwardsiella piscicida*-like pathogen in cultured grouper. *Diseases of Aquatic Organisms* 121, 141–148.

Uhland, F., Hélie, P. and Higgins, R. (2000) Infections of *Edwardsiella tarda* among brook trout in Quebec. *Journal of Aquatic Animal Health* 12, 74–77.

USDA/APHIS/NAHMS (US Department of Agriculture/Animal and Plant Health Inspection Service, National Animal Health Monitoring System) (2010) *Catfish 2010. Part I: Reference of Catfish Health and Production Practices in the United States, 2009*. USDA/APHIS/NAHMS, Fort Collins, Colorado.

Van Damme, L. and Vandepitte, J. (1980) Frequent isolation of *Edwardsiella tarda* and *Pleisiomonas shigelloides* from healthy Zairese freshwater fish: a possible source of sporadic diarrhea in the tropics. *Applied and Environmental Microbiology* 39, 475–479.

Walakira, J.K., Carrias, A.A., Hossain, M.J., Jones, E., Terhune, J.S. and Liles, M.R. (2008) Identification and characterization of bacteriophages specific to the catfish pathogen, *Edwardsiella ictaluri*. *Journal of Applied Microbiology* 105, 2133–2142.

Walakira, J., Akoll, P., Engole, M., Sserwadda, M., Nkambo, M. *et al.* (2014) Common fish diseases and parasites affecting wild and farmed tilapia and catfish in Central and Western Uganda. *Uganda Journal of Agricultural Sciences* 15, 113–125.

Waltman, W. and Shotts, E. (1986a) Antimicrobial susceptibility of *Edwardsiella tarda* from the United States and Taiwan. *Veterinary Microbiology* 12, 277–282.

Waltman, W. and Shotts, E. (1986b) Antimicrobial susceptibility of *Edwardsiella ictaluri*. *Journal of Wildlife Diseases* 22, 173–177.

Waltman, W., Shotts, E. and Blazer, V. (1985) Recovery of *Edwardsiella ictaluri* from danio (*Danio devario*). *Aquaculture* 46, 63–66.

Waltman, W. II, Shotts, E. and Wooley, R. (1989) Development and transfer of plasmid-mediated antimicrobial resistance in *Edwardsiella ictaluri*. *Canadian Journal of Fisheries and Aquatic Sciences* 46, 1114–1117.

Wang, Q., Yang, M., Xiao, J., Wu, H., Wang, X. *et al.* (2009) Genome sequence of the versatile fish pathogen *Edwardsiella tarda* provides insights into its

adaptation to broad host ranges and intracellular niches. *PLoS ONE* 4, e7646.
Wang, R.H., Xiao, T.Y., Zeng, L.B., Liu, X.Y., Zhou, Y. and Ma, J. (2016) Generation and use of *Edwardsiella ictaluri* ghosts as a vaccine against enteric septicemia of catfish (ESC). *Aquaculture* 456, 9–15.
Wang, Y., Wang, Q., Xiao, J., Liu, Q., Wu, H. and Zhang, Y. (2011) Genetic relationships of *Edwardsiella* strains isolated in China aquaculture revealed by rep-PCR genomic fingerprinting and investigation of *Edwardsiella* virulence genes. *Journal of Applied Microbiology* 111, 1337–1348.
Wei, L.S., Musa, N., Seng, C.T., Mohd, N.A., Wee, W. *et al.* (2011) Antibiogram and plasmid profiling from *Edwardsiella tarda* isolated from freshwater fish in east coast Malaysia. *Journal of Sustainability Science and Management* 6, 19–27.
Welch, T., Evenhuis, J., White, D., McDermott, P., Harbottle, H. *et al.* (2008) IncA/C plasmid-mediated florfenicol resistance in the catfish pathogen *Edwardsiella ictaluri*. *Antimicrobial Agents and Chemotherapy* 53, 845–846.
White, F., Simpson, C. and Williams, L. (1973) Isolation of *Edwardsiella tarda* from aquatic animal species and surface waters in Florida. *Journal of Wildlife Diseases* 9, 204–208.
Wise, D. and Johnson, M. (1998) Effect of feeding frequency and Romet-medicated feed on survival, antibody response, and weight gain of fingerling channel catfish *Ictalurus punctatus* after natural exposure to *Edwardsiella ictaluri*. *Journal of the World Aquaculture Society* 29, 169–175.
Wise, D., Schwedler, T. and Otis, D. (1993) Effects of stress on susceptibility of naive channel catfish in immersion challenge with *Edwardsiella ictaluri*. *Journal of Aquatic Animal Health* 5, 92–97.
Wise, D.J., Klesius, P.H., Shoemaker, C.A. and Wolters, W.R. (2000) Vaccination of mixed and full-sib families of channel catfish *Ictalurus punctatus* against enteric septicemia of catfish with a live attenuated *Edwardsiella ictaluri* isolate (RE-33). *Journal of the World Aquaculture Society* 31, 206–212.
Wise, D., Camus, A., Schwedler, T. and Terhune, J. (2004) Health management. In: Tucker, C. and Hargreaves, J. (eds) *Biology and Culture of the Channel Catfish*, 1st edn. Elsevier, Amsterdam, pp. 444–503.
Wise, D., Greenway, T., Li, M., Camus, A. and Robinson, E. (2008) Effects of variable periods of food deprivation on the development of enteric septicemia in channel catfish. *Journal of Aquatic Animal Health* 20, 39–44.
Wise, D., Greenway, T., Byars, T., Griffin, M. and Khoo, L. (2015) Oral vaccination of channel catfish against enteric septicemia of catfish using a live attenuated *Edwardsiella ictaluri* isolate. *Journal of Aquatic Animal Health* 27, 135–143.
Wise, D.J., Greenway, T.E., Byars, T.S., Kumar, G., Griffin, M.J. *et al.* (2020) Validation of an *Edwardsiella ictaluri* oral vaccination platform in experimental pond trials. *Journal of the World Aquaculture Society* 51(2), 346–363.
World Bank (2014) Fish to 2030: prospects for fisheries and aquaculture. *World Bank Report No. 83177-GLB*. World Bank, Washington, DC.
Wyatt, L., Nickelson, R. and Vanderzant, C. (1979) *Edwardsiella tarda* in freshwater catfish and their environment. *Applied and Environmental Microbiology* 38, 710–714.
Xiao, J., Wang, Q., Liu, Q., Wang, X., Liu, H. and Zhang, Y. (2008) Isolation and identification of fish pathogen *Edwardsiella tarda* from mariculture in China. *Aquaculture Research* 40, 13–17.
Xu, J., Luo, X.S. and Zeng, L.B. (2009) Isolation, identification and pathogenicity of *Edwardsiella ictaluri* from Chinese yellow catfish (*Pelteobagrus fulvidraco*). *Freshwater Fisheries* 39, 47–53.
Xu, T. and Zhang, X. (2014) *Edwardsiella tarda*: an intriguing problem in aquaculture. *Aquaculture* 431, 129–135.
Yamada, Y. and Wakabayashi, H. (1999) Identification of fish-pathogenic strains belonging to the genus *Edwardsiella* by sequence analysis of *sodB*. *Fish Pathology* 34, 145–150.
Yang, M., Lv, Y., Xiao, J., Wu, H., Zheng, H. *et al.*, (2012) *Edwardsiella* comparative phylogenomics reveal the new intra/inter-species taxonomic relationships, virulence evolution and niche adaptation mechanisms. *PLoS ONE* 7, e36987.
Yang, M., Shao, S., Xiao, J., Wang, Q. and Zhang, Y. (2013) Phylogenetic investigation of *Edwardsiella tarda* with multilocus sequence typing (MLST) and pulsed field gel electrophoresis (PFGE) typing methods. *Aquaculture* 410, 79–85.
Yang, Q., Pan, Y.L., Wang, K.Y., Wang, J., He, Y. *et al.* (2016) OmpN, outer membrane proteins of *Edwardsiella ictaluri* are potential vaccine candidates for channel catfish (*Ictalurus punctatus*). *Molecular Immunology* 78, 1–8.
Yang, R., Liu, Y., Wang, Y., Lei, M., Pan, G. *et al.* (2018) Pathogenesis and pathological analysis of *Edwardsiella tarda* from Dabry's sturgeon (*Acipenser dabryanus*) in China. *Aquaculture* 495, 637–642.
Yang, W., Wang, L., Zhang, L., Qu, J., Wang, Q. and Zhang, Y. (2015) An invasive and low virulent *Edwardsiella tarda esrB* mutant promising as live attenuated vaccine in aquaculture. *Applied Microbiology and Biotechnology* 99, 1765–1777.
Yasuike, M., Sugaya, E., Nakamura, Y., Shigenobu, Y., Kawato, Y. *et al.* (2013) Complete genome sequences of *Edwardsiella tarda*-lytic bacteriophages KF-1 and IW-1. *Genome Announcements* 1, e00089-12.
Yuasa, K., Kholidin, E.B., Panigoro, N. and Hatai, K. (2003) First isolation of *Edwardsiella ictaluri* from cultured striped catfish *Pangasius hypophthalmus* in Indonesia. *Fish Pathology* 38, 181–183.

Ye, S., Li, H., Qiao, G. and Li, Z. (2009) First case of *Edwardsiella ictaluri* infection in China farmed yellow catfish *Pelteobagrus fulvidraco*. *Aquaculture* 292, 6–10.

Ye, X., Lin, X. and Wang, Y. (2010) Identification and detection of virulence gene of the pathogenic bacteria *Edwardsiella tarda* in cultured *Scortum barcoo*. *Freshwater Fisheries* 40, 50–54.

Yu, J., Han, J., Park, K., Park, K. and Park, S. (2009) *Edwardsiella tarda* infection in Korean catfish, *Silurus asotus*, in a Korean fish farm. *Aquaculture Research* 41, 19–26.

Yu, J.E., Cho, M.Y, Kim, J. and Kang, H.Y. (2012) Large antibiotic-resistance plasmid of *Edwardsiella tarda* contributes to virulence in fish. *Microbial Pathogenesis* 52, 259–266.

Zhu, W., Zhang, Y., Zhang, J., Yuan, G., Liu, X. *et al.* (2019) Astragalus polysaccharides, chitosan and poly(I:C) obviously enhance inactivated *Edwardsiella ictaluri* vaccine potency in yellow catfish *Pelteobagrus fulvidraco*. *Fish and Shellfish Immunology* 87, 379–385.

13 Fish Mycobacteriosis

Christopher M. Whipps[1]*, David T. Gauthier[2] and Michael L. Kent[3]

[1]*State University of New York College of Environmental Science and Forestry, Syracuse, New York, USA;* [2]*Old Dominion University, Norfolk, Virginia, USA;* [3]*Oregon State University, Oregon, USA*

13.1 Background

Mycobacteriosis is typically a chronic disease caused by bacteria in the genus *Mycobacterium*, occurring in a wide range of freshwater and marine fish (Decostere *et al*., 2004; Gauthier and Rhodes, 2009). A challenge to the study and control of this disease is that it is not caused by a single aetiological agent, but many different *Mycobacterium* species. Transmission can occur between individual fish, between fish and other species, and to and from the environment. Treatment is often not attempted because infections are notoriously difficult to treat (Chang *et al*., 2017) and disinfecting equipment can be challenging due to mycobacterial resistance to some disinfectants (Mainous and Smith, 2005; Chang *et al*., 2015). Historically, mycobacteriosis in fish was usually ascribed to one of three species, *Mycobacterium chelonae*, *Mycobacterium fortuitum* or *Mycobacterium marinum* (Belas *et al*., 1995; Chinabut, 1999). These species continue to be found in fish, but as research on mycobacteriosis in wild and captive fish has increased, the genetic tools for diagnosis and discrimination have improved (Telenti *et al*., 1993; Rhodes *et al*., 2003; Whipps *et al*., 2003; Kent *et al*., 2004; McNabb *et al*., 2006). Consequently, many *Mycobacterium* species and strains are now being reported from fish.

For example, in recent years there has been thorough investigation of the mycobacteria found in fish from Chesapeake Bay in the USA (Rhodes *et al*., 2003, 2005; Stine *et al*., 2010; Gauthier *et al*., 2011). Several related species have been identified and described, including *M. marinum*, *Mycobacterium shottsii* and *Mycobacterium pseudoshottsii*. Another well-studied fish species is the laboratory zebrafish (*Danio rerio*), where several *Mycobacterium* species have been described (Whipps *et al*., 2012; Nogueira *et al*., 2015; Whipps and Kent, 2020). The list includes the traditional *M. chelonae*, *M. fortuitum* and *M. marinum*, but also many others (e.g. *Mycobacterium haemophilum*, *Mycobacterium abscessus* and *Mycobacterium saopaulense*). *Mycobacterium salmoniphilum* (formerly classified as *M. chelonae* or *M. fortuitum*) has been reported in aquaculture fish in the western USA, eastern Canada, Australia and Europe (Whipps *et al*., 2007a; Zerihun *et al*., 2011a,b; Righetti *et al*., 2014). In addition to the species listed above, other new species from fish have been described using genetic techniques, such as *Mycobacterium syngnathidarum* from syngnathid fish (Fogelson *et al*., 2018), *Mycobacterium montefiorense* from moray eels (Muranidae spp.) (Levi *et al*., 2003) and *Mycobacterium stephanolepidis* from filefish (Monacanthidae) (Fukano *et al*., 2017), to name a few. Even when a species cannot be isolated or ascribed to a particular species based on gene sequences, it will be catalogued as '*marinum*-like' or '*fortuitum*-like', etc. and DNA sequences deposited in GenBank for subsequent studies.

Mycobacterium species appear to occupy a wide range of habitats at a range of temperatures. Mycobacteriosis is common in tropical fish but is also reported in temperate fish, suggesting there is no one set of conditions that limits their growth, environmental persistence and ability to infect fish. Species descriptions of mycobacteria often include data on *in vitro* growth at different temperatures. For example, *M. montefiorense* was described as growing at 25°C, but not at 30 or 37°C (Levi *et al*.,

*cwhipps@esf.edu

2003). *M. shottsii* was described as growing at 23°C, weakly at 30°C and not at 37°C (Rhodes *et al.*, 2003). Few studies attempt to determine the lower limit of growth, but mycobacteria are readily frozen for storage (at −70°C) in the laboratory (Kim and Kubica, 1972) and likely tolerate cold or freezing conditions in the environment. In a study of *Mycobacterium avium* biofilms in drinking-water systems, Torvinen *et al.* (2007) were able to culture *M. avium* from biofilms held at 7 and 20°C (over several weeks), but those from 20°C produced significantly higher numbers of bacteria. Given the great diversity of fish mycobacteria, environmental changes with an average increase or decrease in temperature will likely enable some species and strains to grow more readily and limit the growth of others. *In vivo*, mycobacteria may be able to grow at a wider range of temperatures. For example, Kent *et al.* (2006) found that although a strain of *M. marinum* from fish did not grow well in media culture at 37°C, it was able to grow within a mouse host and at 37°C in macrophage cell lines. Furthermore, *M. marinum* infections in people are often associated with handling fish or contact with aquatic systems (Gauthier, 2015; Aubry *et al.*, 2017). Here, the infections are typically localized in the extremities (e.g. hand infections), where temperatures are more permissible for growth.

13.2 Geographic Distribution

Mycobacteria are ubiquitous in aquatic ecosystems (Falkinham, 1996) and have been identified globally in a wide range of fish (Gauthier and Rhodes, 2009). They readily form surface biofilms and therefore can persist without a host. Many likely live only in the environment or are opportunistic pathogens. Many reports come from tropical fish (Beran *et al.* 2006; Gauthier and Rhodes, 2009), but we have also documented a species similar to *M. montefiorense* causing granulomatous lesions in rock fish (*Sebastes* spp.) that were collected in deep waters off Oregon, USA (Whipps *et al.*, 2003). A few species like *M. salmoniphilum* have been reported in many places including eastern and western North America, Europe and Australia (Whipps *et al.*, 2007a; Zerihun *et al.*, 2011a). Although there are some genetic differences between strains of *M. salmoniphilum*, it is likely that these bacteria have been spread by the introduction of salmonid fish to new geographic areas. Similarly, there is a global pet trade in tropical fish and mycobacteriosis is a common finding in ornamental fish (Lansdell *et al.*, 1993; Smith, 1997; Beran *et al.*, 2006; Zanoni *et al.*, 2008; Kušar *et al.*, 2017; Puk and Guz, 2019). The likely consequence of intentional fish transplantations is that pathogens will be spread to new geographic areas, decreasing isolation of certain fish populations and exposing them to previously unencountered pathogens.

Mycobacteria have been studied at length in human drinking-water systems as they are potential human pathogens (Falkinham *et al.*, 2001). Thus, there is an appreciation of the potential diversity of mycobacteria in any aquatic system, and specific studies into aquarium or aquaculture systems have similarly revealed many different mycobacteria in surface biofilms (Beran *et al.*, 2006; Whipps *et al.*, 2008). Mycobacteria readily adhere to surfaces because they are among the most hydrophobic bacteria (Falkinham *et al.*, 2001; Falkinham, 2009). They are also oligotrophic and therefore well adapted to closed aquaculture or aquarium systems where nutrient loads and competition with other prokaryotes are low (Whipps *et al.*, 2012). In studies investigating mycobacteria in both fish and the environment, it is important to note that many more species are found in the environment than appear to be able to infect fish (Beran *et al.*, 2006; Whipps *et al.*, 2007a, 2008, 2012). However, some are found in both the environment and in fish. Genetic analysis of mycobacteria from fish and surface biofilms from the same systems have found identical strains of *M. chelonae* in zebrafish (Whipps *et al.*, 2008) and *M. marinum* in pompano (*Trachinotus carolinus*) (Yanong *et al.*, 2010), demonstrating the epidemiological link between fish and biofilms. Through bacterial shedding, infected fish can be a source of environmental bacteria and, likewise, surface biofilms can be a source of infection for fish. Chang *et al.* (2019a) found that fish can become infected when housed in tanks with surface biofilms of *M. chelonae* in as little as 2 weeks, with increasing prevalence over time. In addition, the same study reported that *M. chelonae* was detectable in as little as 1 week from tank biofilms containing infected zebrafish. The latter point has important implications for monitoring, because environmental samples can be used to screen aquatic systems (Crim *et al.*, 2017). Testing of environmental samples should be as specific as possible because mycobacteria are so common in aquatic systems and many different species and strains exist, with differences in virulence (Gauthier, 2015; Whipps and Kent,

2020). Lack of strain-level discrimination can lead to mistaken conclusions about zoonotic potential. For example, *Mycobacterium ulcerans*, a significant human pathogen, belongs to a larger group of ecovars that contains the fish and frog pathogens *M. pseudoshottsii* and *Mycobacterium liflandii*, respectively (Doig *et al*., 2012). While these ecovars share common genetic markers (e.g. insertion sequence IS*2404*), there is no evidence for zoonotic potential of the poikilotherm-pathogenic members of this group.

13.3 Diagnosis

Mycobacteriosis is often diagnosed by the presence of acid-fast bacteria in tissue sections or imprints (Fig. 13.1) because specialized media are typically required for isolation (Decostere *et al*., 2004; Kent *et al*., 2004; Whipps *et al*., 2008). Routine histological screening may not include acid-fast staining of sections, but it should be carried out if mycobacteriosis is suspected or if granulomatous inflammation is observed in haematoxylin and eosin-stained sections (Fig. 13.2). There may be few granulomas present in early or acute disseminated infections (Whipps *et al*., 2007b) and in these cases, acid-fast staining of tissue sections is recommended. Well-developed granulomas in experimentally infected fish may also be acid-fast negative; therefore, the absence of acid-fast bacteria cannot be taken as evidence of the absence of mycobacteriosis (Gauthier *et al*., 2003). Autofluorescence has been reported for mycobacteria (Whipps *et al*., 2014), thus the presence of fluorescent foci corresponding to locations of granulomas may be suggestive, but not diagnostic, of mycobacteriosis. Histology cannot determine the species of *Mycobacterium* present in the specimen but allows for characterization of the severity of infection and overall pathology. In some cases, DNA can be extracted from tissues preserved for histology, or the tissues fixed in blocks or sections, then the presence of mycobacteria confirmed by polymerase chain reaction (PCR) assay and the species identified by sequencing (Zerihun *et al*., 2011c; Peterson *et al*., 2013a; Meritet *et al*., 2017). The success of these methods depends on fixative, time in fixative, histological processing and recovery methods. Preferably, fresh, frozen or ethanol-preserved tissues would be used for DNA analysis and ideally mycobacteria would be isolated in culture.

Mycobacterial culture and a wide range of biochemical tests for identification of species are well documented (Kent and Kubica, 1985). Most notably, mycobacteria are slow-growing relative to many other fish pathogens and require specialized media. Most species grow on Middlebrook 7H10 agar plates or Löwenstein–Jensen slants, which contain supplements known to facilitate mycobacterial growth and inhibit growth of other bacteria and fungi. This inhibition of non-specific growth is critical because other organisms can overgrow the medium before mycobacterial growth is visible. A notable example from zebrafish is *M. haemophilum*, which takes approximately 4–8 weeks for observable growth on Middlebrook or Löwenstein–Jensen supplemented with iron (Whipps *et al*., 2007b; Lindeboom *et al*., 2011). In wild fish, *M. shottsii* was originally reported to take 4–6 weeks to grow (Rhodes *et al*., 2003), but in our experience growth can take 8–12 weeks. Mycobacteria can generally be divided into slow-growers and rapid-growers, with the former taking 2–8 weeks to grow and the latter typically

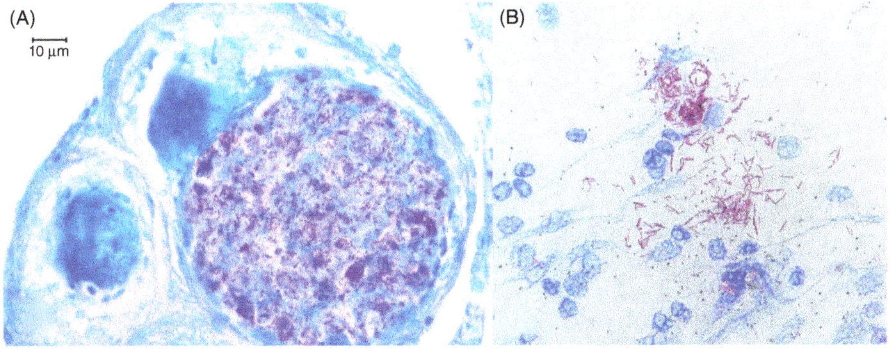

Fig. 13.1. Acid-fast mycobacteria in granuloma in (A) a histological section and (B) tissue imprint.

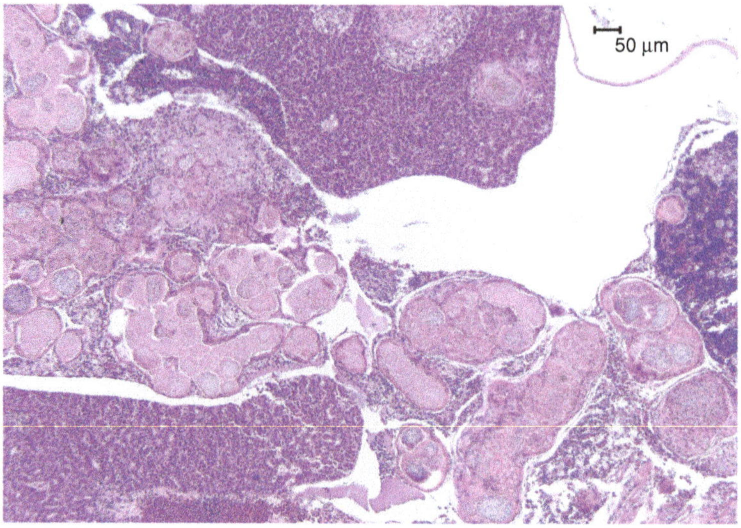

Fig. 13.2. Mycobacteriosis-related granulomatous inflammation observed in a zebrafish (*Danio rerio*) histological section stained with haematoxylin and eosin.

5–7 days. Isolates from fish tend to grow at a broad range of temperatures from 15 to 30°C, with optima varying by species and strain (Arakawa and Fryer, 1984; Levi *et al.*, 2003; Rhodes *et al.*, 2005; Whipps *et al.*, 2007a). Most do not grow well at 37°C, so an incubator set at this temperature to culture human pathogens would not be appropriate. Lacking an incubator, cultures may grow sufficiently at room temperature (20°C). Pure cultures can be used for a variety of strain-typing methods (Ostland *et al.*, 2008; Whipps *et al.*, 2008) and for whole-genome sequencing (Das *et al.*, 2018), or for subsequent virulence testing (Watral and Kent, 2007; Ostland *et al.*, 2008). Furthermore, pure cultures can be lyophilized or frozen in glycerol at −80°C, for long-term storage.

Testing using PCR has become commonplace in fish disease diagnostics. Whipps *et al.* (2008) reported that PCR testing on zebrafish tissues compares well with both culture and histology tests for the pathogen. Quantitative or real-time PCR assays for mycobacteria in fish are likely even more sensitive than conventional PCR (Gauthier *et al.*, 2010; Zerihun *et al.*, 2011c; Meritet *et al.*, 2017). Furthermore, such assays can be adapted to test environmental samples for mycobacteria (Whipps *et al.*, 2008, 2012; Gauthier *et al.*, 2010; Crim *et al.*, 2017). Because mycobacteria are common in aquatic systems, a PCR test that is specific to mycobacteria in general is likely to detect these organisms in the environment. When screening such samples, the assay should be specific to certain species or a subset of species known to infect fish. In general, PCR is more rapid than other diagnostic tests and can be used on preserved tissues or tissues fixed in blocks as mentioned above, directly on infected tissues or on environmental samples. A disadvantage of not culturing the bacteria is that many strain-typing techniques require pure culture. In addition, PCR positives alone do not indicate the clinical significance of the result that could otherwise be determined by histology.

Analysis of DNA sequences from PCR positives is often recommended to confirm the identity of the bacterial species present. This can be accomplished in a variety of ways. A common genomic target for species identification is the *hsp65* gene, which is analysed using DNA sequencing or restriction fragment length polymorphism analysis (Ringuet *et al.*, 1999; Kim *et al.*, 2005; McNabb *et al.*, 2006; Whipps *et al.*, 2008). In some cases, the DNA sequencing results will allow for assignment to a species, and other times only to a species complex. A phylogenetic analysis or other methods (biochemical tests) may need to be performed for very closely related species (Kent *et al.*, 2004; Whipps *et al.*, 2007a; Nogueira *et al.*, 2015). It is important to note from a practical standpoint that *Mycobacterium* spp. that are highly similar at various housekeeping genes may be clinically and ecologically different. As previously mentioned, *M. ulcerans* ecovars are a prime example; *M. ulcerans* from humans and *M. pseudoshottsii*, which has never been isolated from humans, differ from one another in several housekeeping genes at a similar level to the dissimilarity between human-isolated *M. ulcerans* strains. Consequently, caution must be exercised in claiming isolation of human pathogens from fish sources, as in many cases no true zoonotic may exist (Gauthier, 2015).

13.4 Clinical Signs and Pathobiology

The manifestation of mycobacteriosis in fish ranges from a chronic disease with few clinical signs, to more acute infections (Kent *et al.*, 2004; Gauthier and Rhodes, 2009). When clinical signs are apparent, these too range broadly. In some cases, fish may present with dermal lesions (Fig. 13.3), have raised scales or have swollen abdomens. In zebrafish, *M. haemophilum* infections are often associated with a more acute and severe disease where emaciation is commonly reported (Whipps *et al.*, 2007b). Fish may show other signs such as erratic swimming or lethargy. The complete lack of any clinical signs is not uncommon (Whipps *et al.*, 2008; Gauthier and Rhodes, 2009) and confounds recognition that fish are infected, decreasing aquaculture production, impacting research using fish, and presenting a risk for consumption and to those handling infected fish.

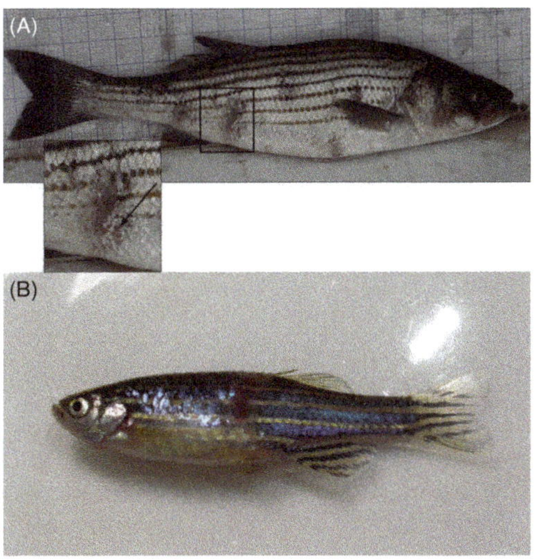

Fig. 13.3. (A) Striped bass (*Morone saxatilis*) with external mycobacteriosis. Inset shows ulcerative dermal lesion with scale loss. Characteristic pigmented foci (arrow) are present and are typically the first manifestations observed externally. (B) Non-specific haemorrhagic skin lesion associated with *Mycobacterium marinum* infection in zebrafish (*Danio rerio*).

Internally, granulomas are often visible as distinct white spots throughout all tissues (Fig. 13.4), but are typically observed in the spleen, kidney and liver. Granulomas may not always be present in early infections, or where diffuse systemic infections have been reported in the cases of *M. haemophilum* (Whipps *et al.*, 2007b) and *M. marinum* (Ramsay *et al.*, 2009a). In zebrafish, colonization of the swim bladder is common, with lesions extending into the body cavity (Whipps and Kent, 2020). Involvement of the central nervous system has been reported for *M. haemophilum* (Whipps *et al.*, 2007b).

Transmission of mycobacteria to naïve fish is most likely primarily through ingestion. Ross (1970) reported transmission in salmon hatcheries when young fish were fed raw and infected carcasses of adult fish. When feed was pasteurized, prevalence dropped precipitously (Ross reported 55% prevalence in one group of fish, and that infections were 'practically eliminated' following the change in feed). Harriff *et al.* (2007) specifically identified the gastrointestinal tract as the site of infection in laboratory zebrafish, and oral transmission has been reported multiple times in the zebrafish (Harriff *et al.*, 2007; Peterson *et al.*, 2013b; Chang *et al.*, 2019b). The zebrafish swim bladder, which is connected to the gastrointestinal tract by the pneumatic duct, has also been noted as a potential point of invasion because aerocystitis (inflammation of

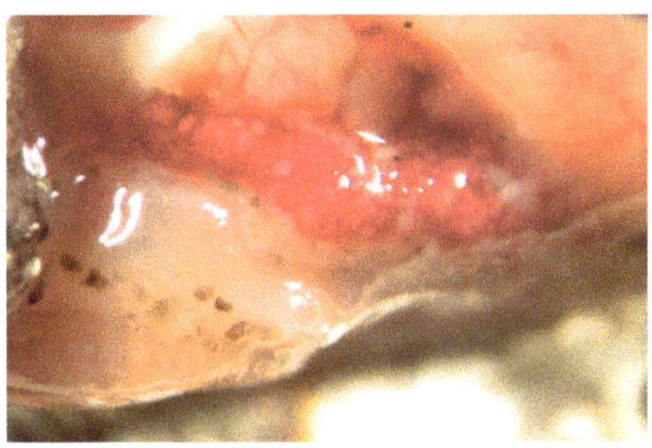

Fig. 13.4. Mycobacteriosis-related visible granulomas in the spleen of a minnow (Cyprinidae).

the swim bladder) is often reported in infected fish (Whipps *et al.*, 2008; Whipps and Kent, 2020).

Sources for mycobacteria in general are many, given that they are considered ubiquitous in aquatic systems. Exposure by ingestion of contaminated food, detritus or other fish are the most likely sources. In laboratory zebrafish, Chang *et al.* (2019b) determined that fish can be reliably infected by feeding them live feeds (paramecia, brine shrimp (*Artemia*) and rotifers) containing mycobacteria. Fish can also be infected by feeding mycobacteria directly, with similar severity resulting from all exposure methods. The extent that live feeds may be contaminated with mycobacteria is unknown, but Beran *et al.* (2006) did report finding mycobacteria in brine shrimp, which are often used in larval or tropical fish aquaculture. Pasteurization of feeds greatly reduces or eliminates exposure (Ross, 1970). Biofilms or other environmental sources also present a risk for infection. Whipps *et al.* (2012) reported a case of infected zebrafish found in the system sump, where they had only detritus and surface biofilms to feed on. More directly, Chang *et al.* (2019a) placed zebrafish in tanks after allowing mycobacterial biofilms to form on the tanks and found that fish could become infected in as little as 2 weeks. In aquaculture, this highlights the potential utility of regular tank cleaning or tank exchanges where feasible.

In addition to the mycobacteria present in surface biofilms, protozoans also occur in high abundances (Arndt *et al.*, 2003). Many of these aquatic protozoans are phagocytic, actively consuming bacteria, and in some cases, this enhances the survival, distribution and virulence of these bacteria (Barker and Brown, 1994). Uptake and persistence of mycobacteria within such free-living phagocytic organisms like amoeba have been reported for many species (Steinert *et al.*, 1998; Winiecka-Krusnell and Linder, 2001; Gupta *et al.*, 2013). Some of the adaptations that mycobacteria possess for infection of vertebrate animals, such as intracellular replication within the host and blockage of phagolysosomal fusion, may have evolved first in these environmental interactions (Cosson and Soldati, 2008). As such, when mycobacteria are ingested by any phagocytic cells, they upregulate virulence genes (Harriff *et al.*, 2007). When fish then consume mycobacteria that are primed in this way, greater virulence may be observed, as has been reported for exposures with mycobacteria in paramecium and mosquito larvae (Mutoji, 2011; Peterson *et al.*, 2013b). However, a study by Chang *et al.* (2019b) investigating uptake and virulence of mycobacteria in paramecia, brine shrimp and rotifers found no significant differences in virulence compared with controls. More studies will be necessary to resolve this question of whether organisms in surface biofilms enhance virulence for fish mycobacteria.

Once a fish is infected, bacteria may be shed in faeces or from skin lesions (Noga, 2010). In histological sections of zebrafish, bacteria are in the lumen of the intestine and associated with the intestinal epithelia (Whipps *et al.*, 2007b). When zebrafish are infected with *M. chelonae* by injection or by feeding, this same bacterium can be detected in the surface biofilms and tank detritus within 1 to 2 weeks (Chang *et al.*, 2019a). Furthermore, Chang *et al.* (2019a) observed this mycobacterial shedding throughout this 4-month experiment. This highlights the importance of removing sick fish from any captive population because they are potential sources of infection for other individuals. The shedding of bacteria does provide an opportunity for diagnostic testing in a closed aquaculture system. For example, Crim *et al.* (2017) tested tank detritus, faeces and water from tanks containing infected zebrafish, and reliably obtained positive PCR results from these environmental samples.

Mycobacterium species are potentially zoonotic, which may present a risk for personnel handling wild fish at capture, cultured fish at harvest or handling fish for research (Gauthier, 2015). Human skin infections caused by *M. marinum* are known to be associated with aquaria or handling fish, with the majority of cases associated with hobbyists' home aquaria (Ang *et al.*, 2000; Lewis *et al.*, 2003; Slany *et al.*, 2011; Aubry *et al.*, 2017). Overall, *M. marinum* appears to be a genetically diverse taxon (Das *et al.*, 2018). In an evaluation of strain differences between epidemiologically unrelated isolates of *M. marinum* from humans and fish, Ucko and Colorni (2005) found that strains tended to cluster by host rather than finding the same strains in fish and humans. Similar clustering by host was reported by Broutin *et al.* (2012), but with many more isolates, some intermixing of fish and human isolates was observed. This suggests that there is some host specificity for *M. marinum*, but some strains can opportunistically infect both humans and fish. For a case in point, Ostland *et al.* (2008) found identical strains of *M. marinum* in striped bass and from a human (skin infection) based on pulsed-field gel electrophoresis. However, whole-genome sequencing identified some differences between these same isolates (Das *et al.*, 2018), again pointing towards the great diversity even between very similar strains. Mason *et al.* (2016) reported a hand infection in a member of the husbandry staff, concurrent with a *M. marinum* outbreak in the fish in a zebrafish facility. The weight of evidence does suggest a linkage between fish and human infections in some cases, but detailed strain discrimination methods are likely necessary to draw any definitive conclusions. Human infections with other non-tuberculosis mycobacteria (e.g. *M. chelonae*, *M. fortuitum*, *M. abscessus*) are not typically associated with handling fish or home aquaria and are generally considered to be caused by opportunistic human pathogens from a variety of other sources (Brown-Elliott and Wallace, 2002; Whipps *et al.*, 2012).

13.5 Expected/Potential Spread of the Pathogen

Because mycobacteria tend to adhere to surfaces and can persist in the environment without a host, their capacity to spread is likely high. Fish migration to new areas and intentional and unintentional transplantation of fish are possible drivers of pathogen spread. As discussed earlier (Section 13.2), this might explain the wide distribution of *M. salmoniphilum* and is an important concern in the global tropical fish trade. Much of this risk is inferred, but we also have direct evidence of spread by transplantation of fish. One example of likely spread of mycobacteria with human movement of fish was with a unique *M. marinum* isolate that caused severe disease in hybrid striped bass (*Morone chrysops* × *Morone saxatilis*) in a specific fish farm (Ostland *et al.*, 2008). Genetic characterization of several isolates showed that the infection over several years was caused by an essentially clonal organism, and the same strain was found in hybrid striped bass in a different state that was the source of this fish for this farm. In laboratory zebrafish, genetically identical strains of *M. chelonae* have been reported from two research facilities on the same university campus, which likely spread by the sharing of fish between facilities (Whipps *et al.*, 2008).

13.6 Dynamics of Host–Pathogen Relationships

The interactions of *Mycobacterium* species and their hosts have been well characterized because multiple species are important human pathogens (Puissegur *et al.*, 2004; Houben *et al.*, 2006). Although these interactions are not understood as well in fish, there is increasing interest in such studies because *M. marinum* in zebrafish has been proposed as a model for *Mycobacterium tuberculosis* infections (van Der Sar *et al.*, 2004; Lesley and Ramakrishnan, 2008; Tobin and Ramakrishnan, 2008; Meijer and Spaink, 2011). Within a vertebrate host, mycobacteria are typically intracellular within phagocytic cells such as macrophages (Schorey *et al.*, 1997). It is thought that mycobacteria inhibit phagosome and lysosome fusion, and delay or block maturation and acidification of the phagosome (Armstrong and Hart, 1971; Sturgill-Koszycki *et al.*, 1994; Clemens and Horwitz, 1995). Most of these observations have been made on mammalian hosts, but the lack of phagolysosomal fusion has also been observed in carp cell cultures with *Mycobacterium smegmatis* (El-Etr *et al.*, 2001). Nevertheless, Gauthier and Rhodes (2009) point out that different observations have been made with fish and other poikilotherms. Specifically, phagosomes containing *M. marinum* have been observed to undergo fusion with lysosomes, both *in vitro* (Gauthier and Vogelbein, 2003) and *in vivo* (Gauthier *et al.*, 2004). A granulomatous response is typical for mycobacterial infections, whereby the immune system attempts to contain the infection (Saunders and Cooper, 2000). *M. haemophilum* in zebrafish causes somewhat chronic but ultimately deadly infections (Whipps *et al.*, 2007b). In addition to extensive granulomas replete with mycobacteria, massive numbers of bacterial colonies not contained within the granulomas are observed throughout the fish, including in the central nervous system.

Using a study system of goldfish gene expression and *M. marinum* infections, Hodgkinson *et al.* (2012) reported that early infections stimulate expression of inflammatory cytokines, followed by an increase in anti-inflammatory cytokines and overall stabilization in gene expression after a month. Extracellular products from mycobacteria are immunostimulatory and associated with a strong innate immune response (Chen *et al.*, 1996; Choi and Oh, 2000). Swaim *et al.* (2006) used different genetic knockouts of zebrafish and *M. marinum* to conclude that adaptive immunity plays a role in moderating mycobacterial infections. The testing of different vaccines, where decreased severity and mortality are reported in vaccinated fish (see discussion in Section 13.9 below on control using vaccines), also supports an adaptive immune response to mycobacteria in fish.

13.7 Population Dynamics of Host Fish

Population-level effects of fish mycobacteriosis are often difficult to determine, due to the generally chronic nature of the disease and the resultant cryptic mortality. In the absence of directly observable fish kills, other approaches may be employed, such as mathematical modelling of prevalence data using approaches modified from terrestrial disease epidemiology. In one such approach, Gauthier *et al.* (2008) demonstrated evidence for mycobacteriosis-associated mortality in Chesapeake Bay striped bass, which have high endemic levels of the disease. A subsequent tag-recapture study in the Rappahannock River, a tributary of Chesapeake Bay, demonstrated that externally diseased fish had lower recapture rates (i.e. annual survival) and that relative survival decreased markedly with increasing skin lesion severity (Hoenig *et al.*, 2017). These studies provide strong evidence that mycobacteriosis is involved with mortality of this fish in a natural system; however, the population-level impacts are more difficult to determine. Striped bass are found along the US and Canadian Atlantic coast, and fish from different areas may be susceptible to infection and disease depending on local conditions such as temperature and salinity. Spawning areas along the North American Atlantic coast contribute to varying degrees to an overall coastal migratory stock, so the level and severity of disease in different areas may affect the population.

Several studies have linked mycobacteriosis to environmental conditions. In a series of studies, Coutant (1985) presented the 'thermal-oxygen squeeze' hypothesis to explain observed poor health due to an unspecified disease agent in striped bass inhabiting inland reservoirs. Under this hypothesis, deep-water hypoxia due to seasonal stratification and eutrophic nutrient input forces striped bass from cool-water summer refugia into water at a higher-than-optimal temperature (Coutant, 1985, 1990; Coutant and Benson, 1990). Chesapeake Bay is a naturally stratifying estuary and anthropogenic nutrient input has likely exacerbated hypoxic deep-water 'dead zones' in the last half-century. Lapointe *et al.* (2014) demonstrated that elevated temperature, hypoxia and existing mycobacteriosis interact synergistically to reduce the metabolic scope of striped bass, and a follow-on study to the Rappahannock tag-recapture work demonstrated that annual survival was negatively correlated with sea-surface water temperature (Groner *et al.*, 2018). These studies collectively suggest that mycobacteriosis is linked with elevated water temperature and increased estuarine hypoxia, although the causal relationship between disease and environment is still unclear. Future projections of this scenario are complicated by the apparent upper thermal limit of the main striped bass mycobacterial pathogens, *M. shottsii* and *M. pseudoshottsii*, which may be inhibited by temperatures at the upper limits of striped bass thermal tolerance (30°C). Both pathogens are present in the Delaware River, north of Chesapeake Bay, and a scenario is plausible in which disease pressure from these pathogens spreads north in the face of warming water temperatures.

Sublethal effects likely contribute to the overall epidemiological picture of piscine mycobacteriosis in wild fish. Striped bass with mycobacteriosis exhibit reduced size-at-length in Chesapeake Bay (Latour *et al.*, 2012), consistent with reduced growth that is frequently cited for infected fish in aquaculture. Mycobacteriosis in these fish has also been associated with decreased age-at-maturity, suggesting disease may play a role in reduced reproductive potential in the population (Gervasi *et al.*, 2019). In laboratory zebrafish, Ramsay *et al.* (2009a,b) reported that fish stressed by repeated handling had elevated body cortisol and greater severity of disease and mortality due to mycobacteriosis relative to controls. Other environmental conditions that may increase physiological stress could have similar effects in other wild or aquaculture species, where underlying infections become more severe in less optimal conditions.

13.8 Mycobacteria and Emerging Aquaculture

There has been a dramatic increase in finfish aquaculture around the globe, and this will continue. Indeed, Froehlich *et al.* (2018) concluded that an increase in aquaculture will be required to provide the additional animal protein needed in future decades as the human population continues to increase. Starting over a decade ago, concurrent with the rapid growth of aquaculture around the world, there have been increased reports on outbreaks of mycobacteriosis in farmed fish (Seng and Colorni, 2004). For example, high mortalities caused by *M. marinum* and *Mycobacterium frederiksbergense* have been documented in pen-reared turbot (*Scophthalmus maximus*), sea bream (*Sparus aurata*) and sea bass (*Dicentrachus labrax*) in various Mediterranean countries (dos Santos *et al.*, 2002; Ucko and Colorni, 2005; Urku *et al.*, 2018). Marine net-pen farming of Atlantic salmon (*Salmo salar*) is the largest net-pen industry in the world, and *M. salmoniphilum* has caused outbreaks in fish farms in western Norway (Zerihun *et al.*, 2011a) and British Columbia (Brocklebank *et al.*, 2003).

Mycobacteriosis has extended into land-based aquaculture. Mycobacteriosis, caused by *M. marinum* and *M. chelonae*, has also caused outbreaks in turbot (*Scophthalmus maximus*) in Portugal (dos Santos *et al.*, 2002). Recirculating land-based systems are also becoming popular, and outbreaks of *M. marinum* over several years have been severe in a recirculating facility with hybrid striped bass (*M. chrysops* × *M. saxatilis*) (Ostland *et al.*, 2008). *M. marinum* infections have also been reported in pompano reared in a recirculating system in Florida (Yanong *et al.*, 2010). Mycobacteria are transmitted directly from fish to fish, or through biofilms and invertebrate vectors in the fish tanks (Chang *et al.*, 2019b). Therefore, the high density of fish and sharing of water between tanks in recirculating systems provides a strong opportunity for the mycobacteria to spread in a facility, particularly when disinfection of recirculating effluent is not employed.

13.9 Future Strategies for Control

Controlling mycobacteriosis in aquatic species is challenging. These bacteria are resistant to many antibiotics and treatment regimens can be lengthy and expensive. Depending on the regulations in a country, there may be few antibiotics approved for fish destined for human consumption (Whipps *et al.*, 2012). Treatment has been attempted on occasion, often opportunistically, and with mixed results (reviewed in Chang and Whipps, 2015). Briefly, different antibiotics have been used, exposure methods and duration of treatment have varied, and the outcomes ranged widely from elimination of infections to no apparent effect. Kawakami and Kasuda (1990) evaluated three different antibiotics (rifampicin, streptomycin and erythromycin) delivered orally in the feed to yellowtail (*Seriola quinqueradiata*) and found a greater survival in the treated group (56%) versus controls (14%), but fish did not clear the infections. In another study on striped bass, Hedrick *et al.* (1987) described the ineffective treatment (rifampicin) of fish infected with *M. marinum*, ultimately requiring the fish to be destroyed.

In laboratory zebrafish, Chang and Whipps (2015) outlined a method for identifying appropriate antibiotics; then in a subsequent study (Chang *et al.*, 2017) evaluated tolerance of antibiotics (tigecycline at 1 μg/g per day and clarithromycin at 4 μg/g per day) in zebrafish and carried out treatments for 2 to 4 weeks. Chang *et al.* (2017) found a reduction in severity of infection, but not complete elimination of infection, a finding similar to earlier reports. This highlights the idea that potentially long treatments are necessary and, even then, such regimens may only reduce severity. Nevertheless, detailed studies are still needed, and ideally each species and strain would be tested. It is important to note that due to the potentially prolonged treatment regimens and zoonotic potential of mycobacteria, there is concern that overuse of antibiotics could lead to genetic resistance to antibiotics. Thus, we recommend that if antibiotics are used, they be used at a small scale, targeting valuable fish or broodstock, to clean up parental fish prior to breeding.

Another approach is the development of a vaccine. Vaccination for mycobacteria has had some success in mammalian systems. The BCG vaccine prepared from *Mycobacterium bovis* bacillus Calmette-Guérin is the most widely used vaccine worldwide for human and bovine tuberculosis (Martin, 2006; Liu *et al.*, 2009). Cui *et al.* (2010) showed some success for reducing *M. marinum* infections in zebrafish with an attenuated live vaccine. Similarly, Kato *et al.* (2010, 2011) reported that BCG stimulated an adaptive immune response in flounder and subsequent testing as a vaccine showed some reduction in mortality. A DNA vaccine based on the *M. marinum* gene encoding Ag85A, developed by Pasnik and Smith (2005, 2006), also stimulated an adaptive immune response and showed protection against *M. marinum* infections in hybrid striped bass. These vaccine trials tend to show reductions in severity and mortality in infected fish, but not complete protection. Further testing and development may improve these outcomes even more, and even a partially protective vaccine may useful.

Due to the difficulties with treatment of mycobacterial infections, it is recommended that prevention by avoidance using quarantine measures be the focus of control in captive fish. As with other fish pathogens, screening brood fish and progeny is recommended to avoid potential maternal transmission and introduction of the bacteria with transfer of embryos. With zebrafish, *M. chelonae* is very widespread in fish and biofilms, including those that lack fish (Whipps *et al.*, 2012). Hence, it would be very difficult to avoid this species. However, particularly virulent strains (e.g. H1E2) (Whipps *et al.*, 2008) or other species (Whipps and Kent, 2020) could be avoided by screening fish before transport or screening broodstock fish to avoid potential maternal transmission.

Selective breeding for resistance to mycobacterial infections has not been reported in fish but has been explored in other animals such as cattle infected with *M. bovis* (Tsairidou *et al.*, 2018). A genetic link to mycobacterial resistance has been explored at length, given the importance of these pathogens in humans and livestock (Orme *et al.*, 1986; Skamene, 1989; Bellamy, 2003), and presumably some of these same mechanisms occur in fish. In zebrafish, Whipps *et al.* (2008) found that prevalence of mycobacteriosis was statistically different in fish with different genetic backgrounds. Specifically, Tübingen ('TU') zebrafish had higher prevalence than other lines of zebrafish. Similar findings were reported by Murray *et al.* (2011), suggesting a link between genetics and susceptibility in zebrafish.

13.10 Concluding Remarks

Considering the future for mycobacteriosis in fish in the context of climate change and global warming, it is important to realize that many *Mycobacterium* species can cause this disease, and the number of potential host and pathogen interactions are many. Most *Mycobacterium* spp. grow better as temperatures increase, with optimal growth at about 30–32°C. Given that many fish have subclinical chronic infections, physiological stress leading to immunosuppression will increase the severity of mycobacteriosis. Studies on zebrafish showing that crowding stress increases the severity and prevalence of *M. marinum* and *M. chelonae* infections support this concern of a connection between stress (whatever the source) and severity of mycobacteriosis. In conclusion, it is very likely that diseases caused by *Mycobacterium* spp. will continue to expand in both wild and captive fish considering global climate change, particularly warming of water, increase in global aquaculture, increased fish transplantation and trade, and degradation of certain aquatic environments.

References

Ang, P., Rattana-Apiromyakij, N. and Goh, C.L. (2000) Retrospective study of *Mycobacterium marinum* skin infections. *International Journal of Dermatology* 39(5), 343–347.

Arakawa, C.K. and Fryer, J.L. (1984) Isolation and characterization of a new subspecies of *Mycobacterium chelonei* infectious for salmonid fish. *Helgoländer Meeresuntersuchungen* 37(1), 329–342.

Armstrong, J.A. and Hart, P.A. (1971) Response of cultured macrophages to *Mycobacterium tuberculosis*, with observations on fusion of lysosomes with phagosomes. *Journal of Experimental Medicine* 134(3), 713–740.

Arndt, H., Schmidt-Denter, K., Auer, B. and Weitere, M. (2003) Protozoans and biofilms. In: Krumbein, W.E., Paterson, D.M. and Zavarzin, G.A. (eds) *Fossil and Recent Biofilms: A Natural History of Life on Earth*. Springer, Dordrecht, the Netherlands, pp. 161–179.

Aubry, A., Mougari, F., Reibel, F. and Cambau, E. (2017) *Mycobacterium marinum*. *Microbiology Spectrum* 5(2), TNMI700382016.

Barker, J. and Brown, M.R.W. (1994) Trojan horses of the microbial world: protozoa and the survival of bacterial pathogens in the environment. *Microbiology* 140(6), 1253–1259.

Belas, R., Faloon, P. and Hannaford, A. (1995) Potential applications of molecular biology to the study of fish mycobacteriosis. *Annual Review of Fish Diseases* 5, 133–173.

Bellamy, R. (2003) Susceptibility to mycobacterial infections: the importance of host genetics. *Genes and Immunity* 4(1), 4–11.

Beran, V., Matlova, L., Dvorska, L., Svastova, P. and Pavlik, I. (2006) Distribution of mycobacteria in clinically healthy ornamental fish and their aquarium environment. *Journal of Fish Diseases* 29(7), 383–393.

Brocklebank, J., Raverty, S. and Robinson, J. (2003) Mycobacteriosis in Atlantic salmon farmed in British Colombia. *Canadian Veterinary Journal* 44(6), 486–489.

Broutin, V., Bañuls, A.L., Aubry, A., Keck, N., Choisy, M. *et al.* (2012) Genetic diversity and population structure of *Mycobacterium marinum*: new insights into host and environmental specificities. *Journal of Clinical Microbiology* 50(11), 3627–3634.

Brown-Elliott, B.A. and Wallace, R.J. (2002) Clinical and taxonomic status of pathogenic nonpigmented or late-pigmenting rapidly growing mycobacteria. *Clinical Microbiology Reviews* 15(4), 716–746.

Chang, C.T. and Whipps, C.M. (2015) Activity of antibiotics against *Mycobacterium* species commonly found in laboratory zebrafish. *Journal of Aquatic Animal Health* 27(2), 88–95.

Chang, C.T., Colicino, E.G., DiPaola, E.J., Al-Hasnawi, H.J. and Whipps, C.M. (2015) Evaluating the effectiveness of common disinfectants at preventing the propagation of *Mycobacterium* spp. isolated from zebrafish. *Comparative Biochemistry and Physiology Part C: Toxicology and Pharmacology* 178, 45–50.

Chang, C.T., Doerr, K.M. and Whipps, C.M. (2017) Antibiotic treatment of zebrafish mycobacteriosis: tolerance and efficacy of treatments with tigecycline and clarithromycin. *Journal of Fish Diseases* 40(10), 1473–1485.

Chang, C.T., Lewis, J.A. and Whipps, C.M. (2019a) Source or sink: examining the role of biofilms in transmission of *Mycobacterium* spp. in laboratory zebrafish. *Zebrafish* 16(2), 197–206.

Chang, C.T., Benedict, S. and Whipps, C.M. (2019b) Transmission of *Mycobacterium chelonae* and *Mycobacterium marinum* in laboratory zebrafish through live feeds. *Journal of Fish Diseases* 42(10), 1425–1431.

Chen, S.C., Yoshida, T., Adams, A., Thompson, K.D. and Richards, R.H. (1996) Immune response of rainbow trout to extracellular products of *Mycobacterium* spp. *Journal of Aquatic Animal Health* 8(3), 216–222.

Chinabut, S. (1999) Mycobacteriosis and nocardiosis. In: Woo, P.T.K. and Bruno, D.W. (eds) *Fish Diseases and Disorders*, Vol. 3. *Viral, Bacterial and Fungal Infections*. CAB International, Wallingford, UK, pp. 319–340.

Choi, S.H. and Oh, C.H. (2000) Stimulatory effects of extracellular products of mycobacterium spp. and various adjuvants on non-specific immune response of Nile tilapia, *Oreochromis nilotica*. *Korean Journal of Biological Sciences* 4(3), 299–304.

Clemens, D.L. and Horwitz, M.A. (1995) Characterization of the *Mycobacterium* tuberculosis phagosome and evidence that phagosomal maturation is inhibited. *Journal of Experimental Medicine* 181(1), 257–270.

Cosson, P. and Soldati, T. (2008) Eat, kill or die: when amoeba meets bacteria. *Current Opinion in Microbiology* 11(3), 271–276.

Coutant, C.C. (1985) Striped bass, temperature, and dissolved oxygen: a speculative hypothesis for environmental risk. *Transactions of the American Fisheries Society* 114(1), 31–61.

Coutant, C.C. (1990) Temperature–oxygen habitat for freshwater and coastal striped bass in a changing climate. *Transactions of the American Fisheries Society* 119(2), 240–253.

Coutant, C.C. and Benson, D.L. (1990) Summer habitat suitability for striped bass in Chesapeake Bay: reflections on a population decline. *Transactions of the American Fisheries Society* 119(4), 757–778.

Crim, M.J., Lawrence, C., Livingston, R.S., Rakitin, A., Hurley, S.J. and Riley, L.K. (2017) Comparison of antemortem and environmental samples for zebrafish health monitoring and quarantine. *Journal of the American Association for Laboratory Animal Science* 56(4), 412–424.

Cui, Z., Samuel-Shaker, D., Watral, V. and Kent, M.L. (2010) Attenuated *Mycobacterium marinum* protects zebrafish against mycobacteriosis. *Journal of Fish Diseases* 33(4), 371–375.

Das, S., Pettersson, B.F., Behra, P.R.K., Mallick, A., Cheramie, M. *et al.* (2018) Extensive genomic diversity among *Mycobacterium marinum* strains revealed by whole genome sequencing. *Scientific Reports* 8(1), 12040.

Decostere, A., Hermans, K. and Haesebrouck, F. (2004) Piscine mycobacteriosis: a literature review covering the agent and the disease it causes in fish and humans. *Veterinary Microbiology* 99(3–4), 159–166.

Doig, K.D., Holt, K.E., Fyfe, J.A.M., Lavender, C.J., Eddyani, M. *et al.* (2012) On the origin of *Mycobacterium ulcerans*, the causative agent of Buruli ulcer. *BMC Genomics* 13, 258.

dos Santos, N.M.S., do Vale, A., Sousa, M.J. and Silva, M.T. (2002) Mycobacterial infection in farmed turbot *Scophthalmus maximus*. *Diseases of Aquatic Organisms* 52(1), 87–91.

El-Etr, S.H., Yan, L. and Cirillo, J.D. (2001) Fish monocytes as a model for mycobacterial host–pathogen interactions. *Infection and Immunity* 69(12), 7310–7317.

Falkinham, J.O. 3rd (1996) Epidemiology of infection by nontuberculous mycobacteria. *Clinical Microbiology Reviews* 9(2), 177–215.

Falkinham, J.O. (2009) Surrounded by mycobacteria: nontuberculous mycobacteria in the human environment. *Journal of Applied Microbiology* 107(2), 356–367.

Falkinham, J.O. 3rd, Norton, C.D. and LeChevallier, M.W. (2001) Factors influencing numbers of *Mycobacterium avium*, *Mycobacterium intracellulare*, and other mycobacteria in drinking water distribution systems. *Applied and Environmental Microbiology* 67(3), 1225–1231.

Fogelson, S.B., Camus, A.C., Lorenz, W., Phillips, A., Bartlett, P. and Sanchez, S. (2018) *Mycobacterium syngnathidarum* sp. nov., a rapidly growing mycobacterium identified in syngnathid fish. *International Journal of Systematic and Evolutionary Microbiology* 68(12), 3696–3700.

Froehlich, H.E., Runge, C.A., Gentry, R.R., Gaines, S.D. and Halpern, B.S. (2018) Comparative terrestrial feed and land use of an aquaculture-dominant world. *Proceedings of the National Academy of Sciences USA* 115(20), 5295–5300.

Fukano, H., Wada, S., Kurata, O., Katayama, K., Fujiwara, N. and Hoshino, Y. (2017) *Mycobacterium stephanolepidis* sp. nov., a rapidly growing species related to *Mycobacterium chelonae*, isolated from marine teleost fish, *Stephanolepis cirrhifer*. *International Journal of Systematic and Evolutionary Microbiology* 67(8), 2811–2817.

Gauthier, D.T. (2015) Bacterial zoonoses of fishes: a review and appraisal of evidence for linkages between fish and human infections. *Veterinary Journal* 203(1), 27–35.

Gauthier, D.T. and Rhodes, M.W. (2009) Mycobacteriosis in fishes: a review. *Veterinary Journal* 180(1), 33–47.

Gauthier, D.T. and Vogelbein, W.K. (2003) Phagolysosomal fusion in striped bass (*Morone saxatilis*) macrophages infected with *Mycobacterium marinum*. In: *Proceedings of the 28th Annual Eastern Fish Health Workshop, Gettysburg, Pennsylvania*, p. 84.

Gauthier, D.T., Rhodes, M.W., Vogelbein, W.K., Kator, H. and Ottinger, C.A. (2003) Experimental mycobacteriosis in striped bass *Morone saxatilis*. *Diseases of Aquatic Organisms* 54(2), 105–117.

Gauthier, D.T., Vogelbein, W.K. and Ottinger, C.A. (2004) Ultrastructure of *Mycobacterium marinum* granuloma in striped bass *Morone saxatilis*. *Diseases of Aquatic Organisms* 62(1–2), 121–132.

Gauthier, D.T., Latour, R.J., Heisey, D.M., Bonzek, C.F., Gartland, J. *et al*. (2008) Mycobacteriosis-associated mortality in wild striped bass (*Morone saxatilis*) from Chesapeake Bay, USA. *Ecological Applications* 18(7), 1718–1727.

Gauthier, D.T., Reece, K.S., Xiao, J., Rhodes, M.W., Kator, H.I. *et al*. (2010) Quantitative PCR assay for *Mycobacterium pseudoshottsii* and *Mycobacterium shottsii* and application to environmental samples and fishes from the Chesapeake Bay. *Applied Environmental Microbiology* 76(18), 6171–6179.

Gauthier, D.T., Helenthal, A.M., Rhodes, M.W., Vogelbein, W.K. and Kator, H.I. (2011) Characterization of photochromogenic *Mycobacterium* spp. from Chesapeake Bay striped bass *Morone saxatilis*. *Diseases of Aquatic Organisms* 95(2), 113–124.

Gervasi, C.L., Lowerre-Barbieri, S.K., Vogelbein, W.K., Gartland, J. and Latour, R.J. (2019) The reproductive biology of Chesapeake Bay striped bass with consideration of the effects of mycobacteriosis. *Bulletin of Marine Science* 95(2), 117–137.

Groner, M.L., Hoenig, J.M., Pradel, R., Choquet, R., Vogelbein, W.K. *et al*. (2018) Dermal mycobacteriosis and warming sea surface temperatures are associated with elevated mortality of striped bass in Chesapeake Bay. *Ecology and Evolution* 8(18), 9384–9397.

Gupta, T., Fine-Coulson, K., Karls, R., Gauthier, D. and Quinn, F. (2013) Internalization of *Mycobacterium shottsii* and *Mycobacterium pseudoshottsii* by *Acanthamoeba polyphaga*. *Canadian Journal of Microbiology* 59(8), 570–576.

Harriff, M.J., Bermudez, L.E. and Kent, M.L. (2007) Experimental exposure of zebrafish, *Danio rerio* (Hamilton), to *Mycobacterium marinum* and *Mycobacterium peregrinum* reveals the gastrointestinal tract as the primary route of infection: a potential model for environmental mycobacterial infection. *Journal of Fish Diseases* 30(10), 587–600.

Hedrick, R.P., McDowell, T. and Groff, J. (1987) Mycobacteriosis in cultured striped bass from California. *Journal of Wildlife Diseases* 23(3), 391–395.

Hodgkinson, J.W., Ge, J.Q., Grayfer, L., Stafford, J. and Belosevic, M. (2012) Analysis of the immune response in infections of the goldfish (*Carassius auratus* L.) with *Mycobacterium marinum*. *Developmental and Comparative Immunology* 38(3), 456–465.

Hoenig, J.M., Groner, M.L., Smith, M.W., Vogelbein, W.K., Taylor, D.M. *et al*. (2017) Impact of disease on the survival of three commercially fished species. *Ecological Applications* 27(7), 2116–2127.

Houben, E.N., Nguyen, L. and Pieters, J. (2006) Interaction of pathogenic mycobacteria with the host immune system. *Current Opinion in Microbiology* 9(1), 76–85.

Kato, G., Kondo, H., Aoki, T. and Hirono, I. (2010) BCG vaccine confers adaptive immunity against *Mycobacterium* sp. infection in fish. *Developmental and Comparative Immunology* 34(2), 133–140.

Kato, G., Kato, K., Saito, K., Pe, Y., Kondo, H. *et al*. (2011) Vaccine efficacy of *Mycobacterium bovis* BCG against *Mycobacterium* sp. infection in amberjack *Seriola dumerili*. *Fish and Shellfish Immunology* 30(2), 467–472.

Kawakami, K. and Kusuda, R. (1990) Efficacy of rifampicin, streptomycin and erythromycin against experimental *Mycobacterium* infection in cultured yellowtail. *Nippon Suisan Gakkaishi* 56(1), 51–53.

Kent, P.T. and Kubica, G.P. (1985) *Public Health Mycobacteriology: A Guide for the Level III Laboratory*. US Department of Health and Human Services, Centers for Disease Control and Prevention, Atlanta, Georgia.

Kent, M.L., Matthews, J.L., Bishop-Stewart, J.K., Whipps, C.M., Watral, V. *et al*. (2004) Mycobacteriosis in zebrafish (*Danio rerio*) research facilities. *Comparative Biochemistry and Physiology Part C: Toxicology and Pharmacology* 138(3), 383–390.

Kent, M.L., Watral, V., Wu, M. and Bermudez, L.E. (2006) *In vivo* and *in vitro* growth of *Mycobacterium marinum* at homoeothermic temperatures. *FEMS Microbiology Letters* 257(1), 69–75.

Kim, H., Kim, S.H., Shim, T.S., Kim, M.N., Bai, G.H. *et al*. (2005) Differentiation of *Mycobacterium* species by analysis of the heat-shock protein 65 gene (*hsp65*). *International Journal of Systematic and Evolutionary Microbiology* 55(4), 1649–1656.

Kim, T.H. and Kubica, G.P. (1972) Long-term preservation and storage of mycobacteria. *Applied and Environmental Microbiology* 24(3), 311–317.

Kušar, D., Zajc, U., Jenčič, V., Ocepek, M., Higgins, J. *et al*. (2017) Mycobacteria in aquarium fish: results of a 3-year survey indicate caution required in handling pet-shop fish. *Journal of Fish Diseases* 40(6), 773–784.

Lansdell, W., Dixon, B., Smith, N. and Benjamin, L. (1993) Communications: Isolation of several *Mycobacterium* species from fish. *Journal of Aquatic Animal Health* 5(1), 73–76.

Lapointe, D., Vogelbein, W.K., Fabrizio, M.C., Gauthier, D.T. and Brill, R.W. (2014) Temperature, hypoxia, and mycobacteriosis: effects on adult striped bass *Morone saxatilis* metabolic performance. *Diseases of Aquatic Organisms* 108(2), 113–127.

Latour, R.J., Gauthier, D.T., Gartland, J., Bonzek, C.F., McNamee, K.A. and Vogelbein, W.K. (2012) Impacts of mycobacteriosis on the growth of striped bass (*Morone saxatilis*) in Chesapeake Bay. *Canadian Journal of Fisheries and Aquatic Sciences* 69(2), 247–258.

Lesley, R. and Ramakrishnan, L. (2008) Insights into early mycobacterial pathogenesis from the zebrafish. *Current Opinion in Microbiology* 11(3), 277–283.

Levi, M.H., Bartell, J., Gandolfo, L., Smole, S.C., Costa, S.F. *et al.* (2003) Characterization of *Mycobacterium montefiorense* sp. nov., a novel pathogenic mycobacterium from moray eels that is related to *Mycobacterium* triplex. *Journal of Clinical Microbiology* 41(5), 2147–2152.

Lewis, F.M., Marsh, B.J. and von Reyn, C.F. (2003) Fish tank exposure and cutaneous infections due to *Mycobacterium marinum*: tuberculin skin testing, treatment, and prevention. *Clinical Infectious Diseases* 37(3), 390–397.

Lindeboom, J.A., van Coppenraet, L.E.B., van Soolingen, D., Prins, J.M. and Kuijper, E.J. (2011) Clinical manifestations, diagnosis, and treatment of *Mycobacterium haemophilum* infections. *Clinical Microbiology Reviews* 24(4), 701–717.

Liu, J., Tran, V., Leung, A.S., Alexander, D.C. and Zhu, B. (2009) BCG vaccines: their mechanisms of attenuation and impact on safety and protective efficacy. *Human Vaccines* 5(2), 70–78.

McNabb, A., Adie, K., Rodrigues, M., Black, W.A. and Isaac-Renton, J. (2006) Direct identification of mycobacteria in primary liquid detection media by partial sequencing of the 65-kilodalton heat shock protein gene. *Journal of Clinical Microbiology* 44(1), 60–66.

Mainous, M.E. and Smith, S.A. (2005) Efficacy of common disinfectants against *Mycobacterium marinum*. *Journal of Aquatic Animal Health* 17(3), 284–288.

Martin, C. (2006) Tuberculosis vaccines: past, present and future. *Current Opinion in Pulmonary Medicine* 12(3), 186–191.

Mason, T., Snell, K., Mittge, E., Melancon, E., Montgomery, R. *et al.* (2016) Strategies to mitigate a *Mycobacterium marinum* outbreak in a zebrafish research facility. *Zebrafish* 13(Suppl. 1), S77–S87.

Meijer, A.H. and Spaink, H.P. (2011) Host–pathogen interactions made transparent with the zebrafish model. *Current Drug Targets* 12(7), 1000–1017.

Meritet, D.M., Mulrooney, D.M., Kent, M.L. and Löhr, C.V. (2017) Development of quantitative real-time PCR assays for postmortem detection of *Mycobacterium* spp. common in zebrafish (*Danio rerio*) research colonies. *Journal of the American Association for Laboratory Animal Science* 56(2), 131–141.

Murray, K.N., Bauer, J., Tallen, A., Matthews, J.L., Westerfield, M. and Varga, Z.M. (2011) Characterization and management of asymptomatic *Mycobacterium* infections at the Zebrafish International Resource Center. *Journal of the American Association for Laboratory Animal Science* 50(5), 675–679.

Mutoji, K.N. (2011) Investigation into mechanisms of mycobacterial transmission between fish. PhD thesis, University of Louisiana at Lafayette, Lafayette, Louisiana.

Noga, E. (2010) *Fish Disease: Diagnosis and Treatment, 2nd edn*. John Wiley and Sons, Ames, Iowa.

Nogueira, C.L., Whipps, C.M., Matsumoto, C.K., Chimara, E., Droz, S. *et al.* (2015) *Mycobacterium saopaulense* sp. nov., a rapidly growing mycobacterium closely related to members of the *Mycobacterium chelonae–Mycobacterium* abscessus group. *International Journal of Systematic and Evolutionary Microbiology* 65(12), 4403–4409.

Orme, I.M., Stokes, R.W. and Collins, F.M. (1986) Genetic control of natural resistance to nontuberculous mycobacterial infections in mice. *Infection and Immunity* 54(1), 56–62.

Ostland, V.E., Watral, V., Whipps, C.M., Austin, F.W., St-Hilaire, S. *et al.* (2008) Biochemical, molecular, and virulence characteristics of select *Mycobacterium marinum* isolates in hybrid striped bass *Morone chrysops* × *M. saxatilis* and zebrafish *Danio rerio*. *Diseases of Aquatic Organisms* 79(2), 107–118.

Pasnik, D.J. and Smith, S.A. (2005) Immunogenic and protective effects of a DNA vaccine for *Mycobacterium marinum* in fish. *Veterinary Immunology and Immunopathology* 103(3–4), 195–206.

Pasnik, D.J. and Smith, S.A. (2006) Immune and histopathologic responses of DNA-vaccinated hybrid striped bass *Morone saxatilis* × *M. chrysops* after acute *Mycobacterium marinum* infection. *Diseases of Aquatic Organisms* 73(1), 33–41.

Peterson, T.S., Kent, M.L., Ferguson, J.A., Watral, V.G. and Whipps, C.M. (2013a) Comparison of fixatives and fixation time for PCR detection of *Mycobacterium* in zebrafish *Danio rerio*. *Diseases of Aquatic Organisms* 104(2), 113–120.

Peterson, T.S., Ferguson, J.A., Watral, V.G., Mutoji, K.N., Ennis, D.G. and Kent, M.L. (2013b) *Paramecium caudatum* enhances transmission and infectivity of *Mycobacterium marinum* and *M. chelonae* in zebrafish *Danio rerio*. *Diseases of Aquatic Organisms* 106(3), 229–239.

Puk, K. and Guz, L. (2019) Occurrence of *Mycobacterium* spp. in ornamental fish. *Annals of Agricultural and Environmental Medicine*. https://doi.org/10.26444/aaem/114913

Puissegur, M.P., Botanch, C., Duteyrat, J.L., Delsol, G., Caratero, C. and Altare, F. (2004) An *in vitro* dual model of mycobacterial granulomas to investigate the molecular interactions between mycobacteria and human host cells. *Cellular Microbiology* 6(5), 423–433.

Ramsay, J.M., Watral, V., Schreck, C.B. and Kent, M.L. (2009a) Husbandry stress exacerbates mycobacterial

infections in adult zebrafish, *Danio rerio* (Hamilton) *Journal of Fish Diseases* 32(11), 931–941.
Ramsay, J.M., Feist, G.W., Varga, Z.M., Westerfield, M., Kent, M.L. and Schreck, C.B. (2009b) Whole-body cortisol response of zebrafish to acute net handling stress. *Aquaculture* 297(1–4), 157–162.
Rhodes, M.W., Kator, H., Kotob, S., van Berkum, P., Kaattari, I. *et al.* (2003) *Mycobacterium shottsii* sp. nov., a slowly growing species isolated from Chesapeake Bay striped bass (*Morone saxatilis*). *International Journal of Systematic and Evolutionary Microbiology* 53(Pt 2), 421–424.
Rhodes, M.W., Kator, H., McNabb, A., Deshayes, C., Reyrat, J.M. *et al.* (2005) *Mycobacterium pseudoshottsii* sp. nov., a slowly growing chromogenic species isolated from Chesapeake Bay striped bass (*Morone saxatilis*) *International Journal of Systematic and Evolutionary Microbiology* 55(3), 1139–1147.
Righetti, M., Favaro, L., Antuofermo, E., Caffara, M., Nuvoli, S. *et al.* (2014) *Mycobacterium* salmoniphilum infection in a farmed Russian sturgeon, *Acipenser gueldenstaedtii* (Brandt and Ratzeburg). *Journal of Fish Diseases* 37(7), 671–674.
Ringuet, H., Akoua-Koffi, C., Honore, S., Varnerot, A., Vincent, V. *et al.* (1999) *hsp65* sequencing for identification of rapidly growing mycobacteria. *Journal of Clinical Microbiology* 37(3), 852–857.
Ross, A.J. (1970) Mycobacteriosis among Pacific salmonid fishes. In: Sniesko, S.F. (ed.) *A Symposium on Diseases of Fishes and Shellfishes*. American Fisheries Society, Washington, DC, pp. 279–283.
Saunders, B.M. and Cooper, A.M. (2000) Restraining mycobacteria: role of granulomas in mycobacterial infections. *Immunology and Cell Biology* 78(4), 334–341.
Schorey, J.S., Carroll, M.C. and Brown, E.J. (1997) A macrophage invasion mechanism of pathogenic mycobacteria. *Science* 277(5329), 1091–1093.
Seng, L.T. and Colorni, A. (2004) Infectious diseases of warmwater fish in marine and brackish waters. In: Woo, P.T.K., Bruno, D.W. and Lim, L.H.S. (eds) *Diseases and Disorders of Finfish in Cage Culture*. CAB International, Wallingford, UK, pp. 193–230.
Skamene, E. (1989) Genetic control of susceptibility to mycobacterial infections. *Reviews of Infectious Diseases* 11(Suppl. 2), S394–S399.
Slany, M., Jezek, P., Fiserova, V., Bodnarova, M., Stork, J. *et al.* (2011) *Mycobacterium marinum* infections in humans and tracing of its possible environmental sources. *Canadian Journal of Microbiology* 58(1), 39–44.
Smith, S.A. (1997) Mycobacterial infections in pet fish. *Seminars in Avian and Exotic Pet Medicine* 6(1), 40–45.
Steinert, M., Birkness, K., White, E., Fields, B. and Quinn, F. (1998) *Mycobacterium avium* bacilli grow saprozoically in coculture with *Acanthamoeba polyphaga* and survive within cyst walls. *Applied and Environmental Microbiology* 64(6), 2256–2261.
Stine, C.B., Kane, A.S. and Baya, A.M. (2010) Mycobacteria isolated from Chesapeake Bay fish. *Journal of Fish Diseases* 33(1), 39–46.
Sturgill-Koszycki, S., Schlesinger, P.H., Chakraborty, P., Haddix, P.L., Collins, H.L. *et al.* (1994) Lack of acidification in *Mycobacterium* phagosomes produced by exclusion of the vesicular proton-ATPase. *Science* 263(5147), 678–681.
Swaim, L.E., Connolly, L.E., Volkman, H.E., Humbert, O., Born, D.E. and Ramakrishnan, L. (2006) *Mycobacterium marinum* infection of adult zebrafish causes caseating granulomatous tuberculosis and is moderated by adaptive immunity. *Infection and Immunity* 74(11), 6108–6117.
Telenti, A., Marchesi, F., Balz, M., Bally, F., Böttger, E.C. and Bodmer, T. (1993) Rapid identification of mycobacteria to the species level by polymerase chain reaction and restriction enzyme analysis. *Journal of Clinical Microbiology* 31(2), 175–178.
Tobin, D.M. and Ramakrishnan, L. (2008) Comparative pathogenesis of *Mycobacterium marinum* and *Mycobacterium* tuberculosis. *Cellular Microbiology* 10(5), 1027–1039.
Torvinen, E., Lehtola, M.J., Martikainen, P.J. and Miettinen, I.T. (2007) Survival of *Mycobacterium* avium in drinking water biofilms as affected by water flow velocity, availability of phosphorus, and temperature. *Applied and Environmental Microbiology* 73(19), 6201–6207.
Tsairidou, S., Allen, A., Banos, G., Coffey, M., Anacleto, O. *et al.* (2018) Can we breed cattle for lower bovine TB infectivity? *Frontiers in Veterinary Science* 5, 310.
Ucko, M. and Colorni, A. (2005) *Mycobacterium marinum* infections in fish and humans in Israel. *Journal of Clinical Microbiology* 43(2), 892–895.
Urku, C., Genc, G.E., Wittwer, F., Erturan, Z. and Pfyffer, G. (2018) Mycobacteriosis in farmed sea bream (*Sparus aurata*) caused by *Mycobacterium frederiksbergense* in Turkey. *Acta Veterinaria* 68(4), 391–400.
van der Sar, A.M., Appelmelk, B.J., Vandenbroucke-Grauls, C.M. and Bitter, W. (2004) A star with stripes: zebrafish as an infection model. *Trends in Microbiology* 12(10), 451–457.
Watral, V. and Kent, M.L. (2007) Pathogenesis of *Mycobacterium* spp. in zebrafish (*Danio rerio*) from research facilities. *Comparative Biochemistry and Physiology Part C: Toxicology and Pharmacology* 145(1), 55–60.
Whipps, C.M. and Kent, M.L. (2020) Bacterial and fungal diseases of zebrafish. In: Cartner, S., Eisen, J., Farmer, S., Guillemin, K., Kent, M. and Sanders, G. (eds) *The Zebrafish in Biomedical Research: Biology, Husbandry, Diseases, and Research Applications*, 1st edn. Academic Press, London, pp. 495–508.
Whipps, C.M., Watral, V.G. and Kent, M.L. (2003) Characterization of a *Mycobacterium* sp. in rockfish, *Sebastes alutus* (Gilbert) and *Sebastes reedi*

(Westrheim and Tsuyuki), using rDNA sequences. *Journal of Fish Diseases* 26(4), 241–245.

Whipps, C.M., Butler, W.R., Pourahmad, F., Watral, V.G. and Kent, M.L. (2007a) Molecular systematics support the revival of *Mycobacterium salmoniphilum* (ex Ross 1960) sp. nov., nom. rev., a species closely related to *Mycobacterium chelonae*. *International Journal of Systematic and Evolutionary Microbiology* 57(11), 2525–2531.

Whipps, C.M., Dougan, S.T. and Kent, M.L. (2007b) *Mycobacterium haemophilum* infections of zebrafish (*Danio rerio*) in research facilities. *FEMS Microbiology Letters* 270(1), 21–26.

Whipps, C.M., Matthews, J.L. and Kent, M.L. (2008) Distribution and genetic characterization of *Mycobacterium chelonae* in laboratory zebrafish *Danio rerio*. *Diseases of Aquatic Organisms* 82(1), 45–54.

Whipps, C.M., Lieggi, C. and Wagner, R. (2012) Mycobacteriosis in zebrafish colonies. *ILAR Journal* 53(2), 95–105.

Whipps, C.M., Moss, L.G., Sisk, D.M., Murray, K.N., Tobin, D.M. and Moss, J.B. (2014) Detection of autofluorescent *Mycobacterium chelonae* in living zebrafish. *Zebrafish* 11(1), 76–82.

Winiecka-Krusnell, J. and Linder, E. (2001) Bacterial infections of free-living amoebae. *Research in Microbiology* 152(7), 613–619.

Yanong, R.P., Pouder, D.B. and Falkinham, J.O. III (2010) Association of mycobacteria in recirculating aquaculture systems and mycobacterial disease in fish. *Journal of Aquatic Animal Health* 22(4), 219–223.

Zanoni, R.G., Florio, D., Fioravanti, M.L., Rossi, M. and Prearo, M. (2008) Occurrence of *Mycobacterium* spp. in ornamental fish in Italy. *Journal of Fish Diseases* 31(6), 433–441.

Zerihun, M.A., Nilsen, H., Hodneland, S. and Colquhoun, D.J. (2011a) *Mycobacterium salmoniphilum* infection in farmed Atlantic salmon, *Salmo salar* L. *Journal of Fish Diseases* 34(10), 769–781.

Zerihun, M.A., Berg, V., Lyche, J.L., Colquhoun, D.J. and Poppe, T.T. (2011b) *Mycobacterium salmoniphilum* infection in burbot *Lota lota*. *Diseases of Aquatic Organisms* 95(1), 57–64.

Zerihun, M.A., Hjortaas, M.J., Falk, K. and Colquhoun, D.J. (2011c) Immunohistochemical and Taqman real-time PCR detection of mycobacterial infections in fish. *Journal of Fish Diseases* 34(3), 235–246.

14 Piscirickettsiosis (*Piscirickettsia salmonis*)

Pedro A. Smith[1]* and Fernando O. Mardones[2]

[1]*University of Chile, Santiago, Chile;* [2]*Pontifical Catholic University of Chile, Santiago, Chile*

14.1 Introduction/Description of the Disease

Piscirickettsiosis is a contagious infectious disease, usually with an acute presentation, that has been reported in several teleost species, predominantly in salmonid fish, reared in seawater or brackish waters. This condition is caused by *Piscirickettsia salmonis*, a Gram-negative, non-motile and non-sporulating pleomorphic bacterium. *In vivo* this organism replicates in the cytoplasm of different kinds of host cells, but it is considered a facultative intracellular bacterium because it can also grow extracellularly *in vitro* in special axenic culture media. Initial characterization of this bacterium indicated it shared several phenotypical features with prokaryotes of the Rickettsiales order and for this reason it was preliminarily identified as a rickettsiales-like organism (RLO) by several researchers (Fryer *et al.*, 1990; Cvitanich *et al.*, 1991). However, further analyses of the 16S rRNA gene demonstrated that *P. salmonis* belonged to the Gammaproteobacteria (Fryer *et al.*, 1992) and therefore is phylogenetically distant to the rickettsiales that belong to the Alphaproteobacteria class (Fryer and Hedrick, 2003; Fryer and Lannan, 2005). Although *P. salmonis* is not a rickettsial agent and its susceptible hosts are not only salmonid fish, the name 'salmonid rickettsial septicaemia' (SRS), proposed by Cvitanich *et al.* (1991), is still used as a piscirickettsiosis synonym (Smith *et al.*, 2019). The first well-documented reports of this disease occurred from coho salmon (*Oncorhynchus kisutch*) dying in seawater net-pens in southern Chile in the year 1989 (Fryer *et al.*, 1990; Branson and Nieto Diaz-Munoz, 1991; Cvitanich *et al.*, 1991). Although initially piscirickettsiosis affected only coho salmon, in the following years all the other salmonid species cultured in the early 1990s in Chile, including Atlantic salmon (*Salmo salar*), rainbow trout (*Oncorhynchus mykiss*), chinook salmon (*Oncorhynchus tshawytscha*) (Garcés *et al.*, 1991) and masou salmon (*Oncorhynchus masou*) (Bravo, 1994), suffered mortalities associated with this disease (Mauel and Fryer, 1997; Fryer and Hedrick, 2003; Fryer and Lannan, 2005). Since then, piscirickettsiosis became endemic in Chile and at present the infection is highly prevalent (58.1% of the sea sites), affecting all salmonid species reared in the last years in this country which are Atlantic salmon, coho salmon and rainbow trout (Gaete-Carrasco *et al.*, 2019; Smith *et al.*, 2019). In seawater-cultured salmonid fish, after the first epidemic outbreaks of piscirickettsiosis in Chile, the disease was subsequently reported in North America (eastern and western Canada) and northern Europe (Ireland, Scotland and Norway). Atlantic salmon having piscirickettsiosis was documented in the Pacific and the Atlantic coast of Canada (Brocklebank *et al.*, 1993; Jones *et al.*, 1998), Ireland (Palmer *et al.*, 1996), Scotland (Grant *et al.*, 1996) and Norway (Olsen *et al.*, 1997; Karatas *et al.*, 2008). In addition, in western Canada this disease was verified in chinook salmon (Brocklebank *et al.*, 1993). Retrospectively, it is suspected that this infectious condition caused losses in pink salmon (*Oncorhynchus gorbuscha*) as early as the year 1970 in this geographical area (Evelyn *et al.*, 1998). Also, in the southern hemisphere it was reported that RLOs antigenically and

*psmith@uchile.cl

genetically related to *P. salmonis* caused disease in Atlantic salmon reared in sea cages in Australia (Corbeil *et al.*, 2005) and chinook salmon in New Zealand (Gias *et al.*, 2018). In non-salmonid fish this disease has been detected in captive white sea bass (*Atractoscion nobilis*) from California, USA (Chen *et al.*, 2000; Arkush *et al.*, 2006), in European sea bass (*Dicentrarchus labrax*) cultured in the Mediterranean Sea (McCarthy *et al.*, 2005; Zrnčić *et al.*, 2015) and in lumpfish (*Cyclopterus lumpus*) reared in seawater tanks in Ireland (Marcos-López *et al.*, 2017). Besides, piscirickettsiosis has been experimentally induced in zebrafish (*Danio rerio*) as an infection model for studying this disease (Berger, 2014; Tandberg *et al.*, 2017).

Piscirickettsiosis has not been described in wild fish nor has *P. salmonis* been isolated from them. Nevertheless, the bacterium has been detected, using either immunofluorescence or polymerase chain reaction (PCR) assay, from wild fish (naturalized salmonids and native species) in southern Chile, including coho salmon migrating to spawning areas in the Claro River (Pérez *et al.*, 1998), chinook salmon in the Toltén River estuary (Irgang *et al.*, 2019) and in the native fish species Patagonian blenny (*Eleginops maclovinus*), Cape redfish (*Sebastes capensis*), Chilean silverside (*Odontesthes regia*) and tadpole codling (*Saliota australis*) captured off the coast in the Pacific Ocean (Contreras-Lynch *et al.*, 2015). Rickettsia-like organisms, most likely *P. salmonis*, have been detected using PCR in wild sockeye salmon (*Oncorhynchus nerka*) from British Columbia, Canada. Prevalence was low in returning adults in the Fraser River (Thakur *et al.*, 2019) and negligible in post-smolt fish in the sea (Nekouei *et al.*, 2018).

Interestingly, it was found that bacterioplankton sampled in coastal waters of Oregon contained 16S rDNA with up to 99.5% similarity to the DNA sequence of the same genes of *P. salmonis*. This indicates that a *P. salmonis*-like organism (PLO) was present in that aquatic environment (Mauel and Fryer, 2001). The source of this PLO is unknown, but it could be from salmonids (wild or hatchery stocks) and/or from other teleost species that share this habitat acting as susceptible hosts or animal reservoirs of this bacterium. Although piscirickettsiosis has not been reported from fish in Oregon waters, this disease has occurred in the vicinity of this geographical area (i.e. British Columbia, Canada and California, USA).

P. salmonis survives for only a limited time in the aquatic environment outside its fish hosts and, thus, it is considered an obligate pathogen (Smith *et al.*, 2015a). An *in vitro* study about the extracellular survival of *P. salmonis* showed that this bacterium was inactivated virtually immediately in fresh water, but a few of these organisms were still viable after 14 days in seawater at 5, 10 and 15°C. Infectious titres declined with time in every case, but this decrease was more pronounced as the temperature increased (Lannan and Fryer, 1994). Consistently, in seawater surrounding cages that had had salmonid fish with a piscirickettsiosis outbreak, putative *P. salmonis* DNA was detected up to 30 days (and not at 40 days) in samples collected every 10 days after the initiation of the fallowing period (Olivares and Marshall, 2010). The evidence strongly suggests that *P. salmonis* requires reservoir host(s) to remain viable in the aquatic environment. Probably some wild anadromous (e.g. salmonids), estuarine and/or marine fishes are the natural reservoirs, but this has not yet been clearly determined. *P. salmonis* has been detected in some arthropod ectoparasites of salmonid fish such as *Ceratothoa gaudichaudii* (Garcés *et al.*, 1994) and *Caligus rogercresseyi* (Maquera, 2017), but it is not known if these invertebrates play any role in the ecology of piscirickettsiosis. It is well documented that this disease can be transmitted directly from fish to fish by cohabitation in the absence of any vectors (Cvitanich *et al.*, 1991; Almendras *et al.*, 1997; Rozas-Serri *et al.*, 2017; Meza *et al.*, 2019). The epidemiological status of the disease in the last years is globally characterized by an endemic presentation in salmonid fish in the southern hemisphere (South America) and scattered outbreaks in salmonid (North America and Europe) and non-salmonid fish (Europe) in the northern hemisphere. In South America, piscirickettsiosis severely affects salmonids reared in the South-East Pacific Ocean along the southern coastline of Chile from approximately 39 to 52°S. In this regard, although there was a slight decline in the severity of the 2018 outbreaks, piscirickettsiosis remains the most important cause for the overall disease-specific mortalities in the salmonid fish reared in the sea in Chile (Sernapesca, 2019). In North America there were few (one to six) piscirickettsiosis outbreaks per year, generally with low or moderate mortality, in Atlantic salmon farmed in the sea off the west coast of Canada from year 2002 to 2017, except the years 2010 to 2012 in which the disease was not observed (ICES, 2018; Jones, 2019). Besides, a high-mortality piscirickettsiosis outbreak did occur in 2018 in Atlantic salmon in the same area of Canada (Jones *et al.*, 2020). In Europe, the epidemiology of piscirickettsiosis in salmonid fish in recent years has been characterized by sporadic episodes with low mortality in Atlantic salmon farmed in Ireland, Scotland (H. Rodger, Ireland, 2019, personal communication) and Norway (A.B. Olsen, Norway, 2019, personal communication). In non-salmonid fish the disease has been observed once in lumpfish reared in a research station on the west coast of Ireland in 2015 (Marcos-López *et al.*, 2017) and sporadically in European sea bass cultured in the Mediterranean Sea (S. Zrnčić, Croatia, 2019, personal communication). Mortalities were substantial (up to 35% cumulative mortality) in European sea bass (S. Zrnčić, Croatia, 2019, personal communication), but very low in lumpfish (Marcos-López *et al.*, 2017). The best-characterized outbreak in European sea bass occurred on the Croatian coast in the year 2016, but other episodes of the disease have been observed in the same fish species cultured in the Mediterranean Sea close to Italy and France in more recent years (S. Zrnčić, Croatia, 2019, personal communication). Significantly, all these outbreaks of piscirickettsiosis in the Mediterranean Sea have occurred in winter when the water temperature was below 16°C (S. Zrnčić, Croatia, 2019, personal communication).

14.2 Diagnosis of the Disease

Diagnosis is based on the detection of *P. salmonis* in the tissues of one or more individuals from a fish population having mortalities with clinical signs and/or pathological manifestations compatible with piscirickettsiosis. Presumptive identification of the bacterium is via microscopic examinations of smears, imprints or histological sections of infected tissues stained with non-specific methods including Gram, Giemsa, methylene blue, toluidine blue, orange acridine, Giménez or Pinkerton's procedures. Definitive recognition of this pathogen is done using either specific antibody-mediated tests – such as immunofluorescence (Lannan *et al.*, 1991), immunohistochemistry (Alday-Sanz *et al.*, 1994), enzyme-linked immunosorbent assay (ELISA) (Aguayo *et al.*, 2002) or dot–blot (Larenas *et al.*, 2012) – or via the identification of particular genomic fragments of the bacterium using *in situ* hybridization (Venegas *et al.*, 2004) or, more commonly, a variety of PCR tests (Mauel *et al.*, 1996; Marshall *et al.*, 1998; Karatas *et al.*, 2008). Isolation of the bacterium provides the most definitive diagnosis although it is not routinely used given its technical difficulty, high cost and the fact that it takes a longer time, compared with its direct detection from tissues. If required, *P. salmonis* can be isolated using either different fish cell lines (Fryer *et al.*, 1990; Birkbeck *et al.*, 2004; Smith *et al.*, 2015b) or enriched agars (Mauel *et al.*, 2008; Mikalsen *et al.*, 2008; Yañez *et al.*, 2013; Makrinos and Bowden, 2017). For optimal growth, cultures should be incubated at 15–18°C (Fryer *et al.*, 1990; OIE, 2003). Clinicopathological manifestations exhibited by fish with piscirickettsiosis are multiple, but they are not specific to the disease and they vary between fish species and from one individual to another. In salmonids, clinical signs include lethargy, anorexia, skin darkening in the dorsal area of the fish, abdominal swelling and abnormal swimming (Smith *et al.*, 2019). Typical external gross lesions comprise gill paleness (Cvitanich *et al.*, 1991) along with lifted scales as well as small nodules in the skin that progress to shallow ulcerations (Branson and Nieto Diaz-Munoz, 1991). Extensive and deep skin ulcerations have been commonly observed in the piscirickettsiosis outbreaks in more recent years (Fig. 14.1). Usual internal macroscopic abnormalities are enlargement of the liver, kidney and spleen along with some degree of ascites (Branson and Nieto Diaz-Munoz, 1991) and the presence of haemorrhagic foci in one or more organs (Cvitanich *et al.*, 1991). The most diagnostic lesions (Fig. 14.2) – although not pathognomonic and found only in a few fish – are mottled livers with white to yellowish, circular foci that appear solid or as ring-shaped formations (Branson and Nieto Diaz-Munoz, 1991;

Fig. 14.1. Atlantic salmon (*Salmo salar*) affected by piscirickettsiosis and showing a number of skin ulcers. (Image courtesy of P. Bustos, ADL Diagnostic Chile.)

Cvitanich *et al.*, 1991; Fryer and Lannan, 1993). Meningeal congestion, increase of the cerebrospinal fluid volume and hydro- or haemopericardium are sometimes present. Thickening, opacity, congestion and/or haemorrhage of the swim bladder wall are also frequently found in salmonids with piscirickettsiosis (Fig. 14.3). Microscopically, inflammation, haemorrhage, degeneration and necrosis are typically observed in the affected tissues (Branson and Nieto Diaz-Munoz, 1991; Bartholomew *et al.*, 2017). These histological changes are accompanied by the presence of basophilic or amphophilic organisms (haematoxylin and eosin staining) located

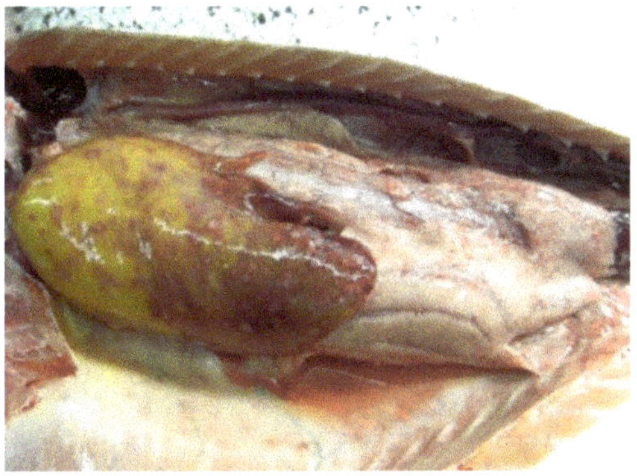

Fig. 14.2. Atlantic salmon (*Salmo salar*) with piscirickettsiosis showing a mottled liver with solid or ring-shaped subcapsular nodules along with congestion and opacity of the swim bladder wall. (Image courtesy of P. Bustos, ADL Diagnostic Chile.)

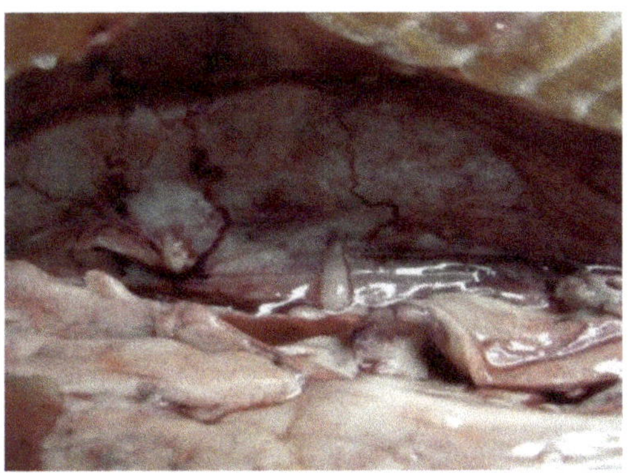

Fig. 14.3. Atlantic salmon (*Salmo salar*) with piscirickettsiosis showing thickening, opacity, congestion and haemorrhage of the swim bladder wall. (Image courtesy of P. Bustos, ADL Diagnostic Chile.)

in cytoplasmic vacuoles of host cells (OIE, 2003) and sometimes dispersed extracellularly (Branson and Nieto Diaz-Munoz, 1991). These bacteria are 0.5–1.5 μm in diameter. They are pleomorphic, predominantly coccoid, and pairs of curved rods are frequently seen in tissue smears (Garcés *et al.*, 1991; Lannan and Fryer, 1991). In salmonids, the most prominent microscopic lesions are seen in the liver, kidney, spleen and intestine, but pathological changes in the brain, heart, ovary and gill can also be observed (OIE, 2003) as well as in visceral fat, stomach and body musculature (Fryer and Hedrick, 2003). Main histopathological findings in European sea bass are multifocal areas of necrosis and extensive inflammation of the brain. These lesions are consistent with the nervous clinical signs, mainly whirling movements, observed in this fish species affected with piscirickettsiosis (Zrnčić *et al.*, 2015).

14.3 Spread of the Pathogen

Since the available evidence shows that *P. salmonis* survives for only a limited time outside its hosts (i.e. susceptible teleost fishes), it is reasonable to think that the geographical distribution of the pathogen will be mainly associated with the aquatic habitats used by its hosts or the water bodies where these animals are held or transported for aquaculture, research, exhibition or other purposes. Salmonid fish, either farmed or free-living, are likely to be the most significant hosts of *P. salmonis* at least in temperate- or cold-water environments. Significant water temperature changes would cause free-living salmonid populations, both natives and naturalized, to have plastic and genetically driven modifications in their life histories (Crozier *et al.*, 2008). Potential response of these fish to climate change would include modifications of their migratory patterns to avoid water with higher temperatures, lower dissolved oxygen and lesser food availability (Crozier *et al.*, 2008). There are, of course, particular situations dependent on specific currents and other oceanographic conditions, but in general it would be expected that the salmonid habitat will be displaced towards the polar regions in both hemispheres and that the same situation will occur subsequently with the potential distribution of *P. salmonis*. Along with this expansion towards polar locations and unless the pathogen and its hosts become adapted to higher water temperatures, an absence or a decreased presence of the bacterium is likely to occur in oceanic or fresh waters at present in the lower geographical latitudes of its distribution. Predicted life history adaptations of salmonids, which include phenological adjustments, not only imply the potential modification of the migratory routes, but especially temporal migration changes as already observed in some wild populations (Kovach *et al.*, 2015). These spatio-temporal variations in the presence or abundance of the hosts of *P. salmonis* in specific habitats should have the same associated effect on the distribution of this pathogen. Salmonids reared in sea cages, given the significant biomass and high fish loading density of these aquaculture operations, are probably the main source of this bacterial shedding to the environment. In this respect, it could be anticipated that salmonid sea farms will be increasingly located poleward in areas near the coast or at offshore facilities. Most likely, *P. salmonis* will move along with these fish. An example of such displacement is the Chilean case, where in the last decade salmon culture has expanded rapidly further south to colder waters of the Magallanes Region at freshwater and seawater farming sites. During the first years after the beginning of this aquaculture expansion, *P. salmonis* was not detected in the salmonid populations, but lately these fish, while reared in sea cages, have tested positive for this bacterium (Gaete-Carrasco *et al.*, 2019) and have suffered some piscirickettsiosis outbreaks as well (Leal, 2017).

Conversely, it is probable that climate change will cause farmers to remove cages with salmonids located at present in lower latitudes as water temperatures exceed the physiological requirements of these fish. Therefore, *P. salmonis* would not have these potential hosts for its survival and multiplication in these places.

Although the geographical distribution of *P. salmonis* is most likely linked to its host location, the bacterium can also be spread through the water after being shed by infected fish. This spreading area has not been thoroughly determined and probably depends on the water flow velocity, currents and tides among other factors. Since the bacterium can remain viable for at least 2 weeks in seawater, becoming non-infective almost immediately in fresh water (Lannan and Fryer, 1994), this physical dissemination should have more importance in seawater. In this respect, based on epidemiological models it was reported that transmission of *P. salmonis* can occur up to 10 km from its source in Atlantic salmon sea farms (Rees *et al.*, 2014).

Non-salmonid species, such as native Chilean marine fish (see Section 14.1) in which genomic DNA of *P. salmonis* has been found (Contreras-Lynch *et al.*, 2015), may also play a role as susceptible hosts or reservoirs of the pathogen, contributing to its spread in the environment. Due to climate change, these fish populations would also move to new habitats, which would result in modifications in the location and abundance of these hosts. Such adaptations could have effects on the pathogen distribution, although we have no data to foresee the outcome of this process.

In marine environments warmer than the typical temperature range used by salmonids, the only place where piscirickettsiosis has been observed is in the Mediterranean Sea. Accordingly, it can be presumed that there are one or more natural reservoirs, such as the wild European sea bass, in this aquatic environment. The pathogen distribution in the Mediterranean Sea is largely unknown, but since this water body is vast and spatio-temporally heterogeneous in terms of temperature and salinity (Pastor *et al.*, 2018), a difference in the bacterium spreading would be expected there. Since the extracellular survival time of *P. salmonis* decreases with increasing seawater temperature (Lannan and Fryer, 1994) and as the south-eastern regions of the Mediterranean Sea are expected to have the highest surface water temperatures (Pastor *et al.*, 2018), the spread of the pathogen will probably decline in these aquatic environments.

14.4 Spread of Reservoir Hosts

Reservoir hosts of *P. salmonis* have not been clearly identified although several marine invertebrates or most likely wild fish species – including marine, estuarine and diadromous teleosts – could have this biological function. Wild salmonid and non-salmonid fish species in which *P. salmonis* has been detected (Pérez *et al.*, 1998; Contreras-Lynch *et al.*, 2015; Irgang *et al.*, 2019) may be some of the reservoirs of this pathogen, but this hypothesis remains to be demonstrated. Since it is postulated here that the pathogen distribution is closely related to the presence of its susceptible hosts and/or reservoirs, the possible modifications in the spread of the potential reservoirs of *P. salmonis* caused by the climate change can be found above (Section 14.3).

14.5 Disease Presentation Features

Piscirickettsiosis may occur in wild fish but, in the scarce research carried out so far about this subject, it has never been reported in those fish populations and there are no data available to predict the way the disease can vary under the influence of climate change.

Regarding salmonid fish reared in sea cages and based on field data, it is expected that there would be an increased prevalence and severity of piscirickettsiosis. Water temperature rise is a risk factor associated with a greater incidence of and mortality from piscirickettsiosis outbreaks. In this regard and after an epidemiological study in salmonids reared in seawater in southern Chile, Leal (2017) found significantly higher mortality rates in the outbreaks of this disease in summer when the water temperature was elevated. In addition, in other research (Rees *et al.*, 2014) where field data were obtained from the same location, it was found that water temperature is a predictor of the presentation of piscirickettsiosis outbreaks for Atlantic salmon, rainbow trout and coho salmon. In that study it was observed that the probability of an outbreak appearance was reduced as the temperature decreased.

The empirical trends observed in the works of Leal (2017) and Rees *et al.* (2014) are robust because they analysed data obtained from several million cultured salmonid fish. However, there may be variations in the virulence of *P. salmonis* strains in their salmonid hosts when held at different water temperatures. In this respect, it is worth mentioning that Larenas *et al.* (1997), in a controlled experiment infecting rainbow trout with the LF strain of *P. salmonis*, reported cumulative mortalities significantly higher in these fish when held at 14°C compared with 8 or 18°C.

In relation to the future presentation of piscirickettsiosis in European sea bass farmed in the Mediterranean Sea, any prediction is uncertain since data availability is scarce. However, based on the fact that the outbreaks have occurred when water temperature was low (below 16°C), it could be expected that the incidence of the disease and its associated mortalities – in contrast to the situation of cultured salmonid fish – should diminish under the effect of climate change.

14.6 Control and/or Prevention

Besides using novel approaches in the control and prevention of this disease under the new conditions posed by the climate change, it would be important to rigorously reinforce the procedures that are currently recommended to attain these goals. The strategy that is presently recommended to decrease the risk of having outbreaks of piscirickettsiosis and/or their severity starts with the execution of good husbandry procedures that include the use of an appropriate aquatic environment (water quality and quantity), the right fish biomass and, in general, actions to minimize fish stress. Biosecurity measures such as the use of disinfection systems, sanitary barriers, frequent mortality removal, procedures to avoid moving infected fish and fallowing periods are advised to prevent or control the disease (Smith *et al.*, 2019). Co-infections with other agents such as the bacterium *Renibacterium salmoninarum* (Larenas, 1999) or the ectoparasite *C. rogercresseyi* (Lhorente *et al.*, 2014; Arriagada *et al.*, 2019) increase the severity of piscirickettsiosis and therefore their avoidance, wherever possible, is beneficial. Periodic surveillance of fish health for an early detection and timely treatment of the disease is helpful for its efficient control. In this regard, opportune diagnosis and necropsy training for the recognition of piscirickettsiosis were considered the most important protective factors to control this disease by fish health experts (Estévez *et al.*, 2019). Considering the findings of Rees *et al.* (2014), the use of distances longer than 10 km between different sea farms would be advisable to decrease the risk of *P. salmonis* spread into them.

As for vaccines against *P. salmonis*, they have been extensively used, particularly in Chile, but they have not provided significant protection in the field (Maisey *et al.*, 2017). Thus, until effective vaccines against this pathogen are available, the employment of this prophylactic measure is questionable. Oxytetracycline, florfenicol and other antimicrobials are used for the disease treatment, even though the presence of bacterial resistance has been demonstrated (Smith *et al.*, 1996; Miranda *et al.*, 2018; Saavedra *et al.*, 2018) and a poor response to these therapies has been reported in some cases (Price *et al.*, 2016). In order to get an effective prevention and/or control of this disease under the expected environmental change, it will be necessary to evaluate the adequacy of the current strategies and adjust them to the new circumstances as well as to explore new approaches. To successfully face these future tasks, it would be convenient to have more tools available such as new vaccines and antimicrobials as well as genetically selected fish. Regarding vaccines against *P. salmonis*, the development of cost-effective products against this pathogen that confer long-lasting protective immunity in salmonids and other teleost fishes at different water temperatures would be highly desirable. As to antimicrobials, at present only a few of them are available to treat fish with piscirickettsiosis. Therefore, to minimize the generation of bacterial resistance and improve the therapeutic efficacy of treatments, new drugs with bacteriostatic or bactericidal activity to *P. salmonis* should be developed. With respect to genetic selection, this can be directed to develop fish with greater resistance to piscirickettsiosis – ideally also with greater resistance to the usual co-infecting agents – and with better adaptability to aquaculture systems with higher temperatures. Genetic selection of fish for resistance to piscirickettsiosis might be successful since significant susceptibility/resistance variation to this disease among families of salmonid fish has been found in controlled experiments (Smith *et al.*, 2009; Dettleff *et al.*, 2015), with the estimated heritability of this trait ranging from 0.11 to 0.41 (Yáñez *et al.*, 2013). However, for a selective breeding programme it should be considered that no genetic correlation was found between the resistance to piscirickettsiosis as a single infection and that with a co-infection with *C. rogercresseyi* in Atlantic salmon, which indicates that different genes control these processes (Lhorente *et al.*, 2014). Genetic selection to improve fitness and productivity of fish reared at higher temperatures also seems to be feasible since unpublished research has shown significant variation in the growth rate among rainbow trout families farmed under thermal stress (water temperature 20–22°C) along with a 0.19 heritability value for this trait (J.M. Yáñez, 2019, unpublished results). To decrease the risks of having piscirickettsiosis outbreaks and their clinical severity, as well as to improve animal welfare issues, it would be necessary to adapt some aquaculture systems using lower fish biomasses at the cage, farm and regional levels. Availability of systems to oxygenate the water when necessary will be required for animal welfare reasons and to prevent weakened immune responsiveness caused by hypoxic stress. Finally, the implementation of farms, or research units, with water recirculation may also be useful to prevent and control this disease in these particular facilities. In this regard, seawater disinfection with ultraviolet radiation has recently been demonstrated to be effective in inactivating *P. salmonis* (Jones *et al.*, 2020).

14.7 Conclusions

Aquatic diseases are closely related to the environmental conditions and, therefore, physicochemical changes in the aquatic habitats can have profound effects on the health status of wild and cultured fish populations. Modification of the water temperature has a significant influence on the quality of the immune response, particularly in poikilothermic animals such as teleost fishes, and has crucial consequences on the efficiency of the innate and specific defence mechanisms against infectious agents such as *P. salmonis*. The growth rate of *P. salmonis* inside its host cells is probably modified *in vivo* with the temperature elevation, as it has been evidenced in cell cultures under *in vitro* conditions, and this phenomenon may also have implications in the clinical presentation and outcome of piscirickettsiosis. The expected effect of higher temperatures in the sea surface on the presentation of this fish disease is variable and its outcome will depend on a complex interaction of several biotic and abiotic factors. Some of these are the host fish species, the specific families or strains within a fish species, the health status of host fish, the *P. salmonis* strain(s) features, the occurrence of co-infecting agents, the aquaculture system, the presence of stressors, the season of the year and the geographical location. The elevation of the water temperature is expected to cause an augmentation in the incidence, prevalence and severity of piscirickettsiosis in cultured salmonids in different areas of the world, but an opposite effect could occur with European sea bass farmed in the Mediterranean Sea. Concerning the distribution of the pathogen and the disease from a global perspective, a poleward displacement is foreseen. It is most likely that the predicted environmental disturbances associated with climate change will have important effects on piscirickettsiosis presentation in different ways, but several gaps in the knowledge about the pathogen and the disease biology make it more uncertain to predict the outcome of these phenomena. Some of the issues that require to be investigated are the identification of the reservoirs of this pathogen and their role in the ecology of piscirickettsiosis, the importance of this disease in wild fish populations, the shedding rates of the bacterium from infected fish, and the generation of further knowledge about the particularities of the *P. salmonis* strains from distant areas of the world infecting salmonids and other fish species.

References

Aguayo, J., Miquel, A., Aranki, N., Jamett, A., Valenzuela, P.D.T. and Burzio, L.O. (2002) Detection of *Piscirickettsia salmonis* in fish tissues by an enzyme-linked immunosorbent assay using specific monoclonal antibodies. *Diseases of Aquatic Organisms* 49, 33–38.

Alday-Sanz, V., Rodger, H., Turnbull, T., Adams, A. and Richards, R.H. (1994) An immunohistochemical test for rickettsial disease. *Journal of Fish Diseases* 17, 189–191.

Almendras, F.E., Fuentealba, I., Jones, S.R.M., Markham, F. and Spangler, E. (1997) Experimental infection and horizontal transmission of *Piscirickettsia salmonis* in freshwater-raised Atlantic salmon, *Salmo salar* L. *Journal of Fish Diseases* 20, 409–418.

Arkush, K.D., Edes, H.L., McBride, A.M., Adkison, M.A. and Hedrick, R.P. (2006) Persistence of *Piscirickettsia salmonis* and detection of serum antibodies to the bacterium in white seabass *Atractoscion nobilis* following experimental exposure. *Diseases of Aquatic Organisms* 73, 131–139.

Arriagada, G., Hamilton-West, C., Nekoueid, O., Foerster, C., Müller, A. *et al.* (2019) *Caligus rogercresseyi* infestation is associated with *Piscirickettsia salmonis*-attributed mortalities in farmed salmonids in Chile. *Preventive Veterinary Medicine* 171, 104771.

Bartholomew, J.L., Arkush, K.D. and Soto, E. (2017) *Piscirickettsia salmonis*. In: Woo, P.T.K. and Cipriano, R.C. (eds) *Fish Viruses and Bacteria: Pathobiology and Protection*. CAB International, Wallingford, UK, pp. 272–285.

Berger, E.K. (2014) Piscirickettsia salmonis; characterization and infection in the zebra fish model. MSc thesis, University of Oslo, Oslo.

Birkbeck, T.H., Griffen, A.A., Reid, H.I., Laidler, L.A. and Wadsworth, S. (2004) Growth of *Piscirickettsia salmonis* to high titers in insect tissue culture cells. *Infection and Immunity* 72, 3693–3694.

Branson, E.J. and Nieto Diaz-Munoz, D. (1991) Description of new disease condition occurring in farmed coho salmon, *Oncorhynchus kisutch* (Walbaum), in South America. *Journal of Fish Diseases* 14, 147–156.

Bravo, S. (1994) Piscirickettsiosis in freshwater. *Bulletin of the European Association of Fish Pathologists* 14, 137–138.

Brocklebank, J.R., Evelyn, T.P., Speare, D.J. and Armstrong, R.D. (1993) Rickettsial septicemia in farmed Atlantic and chinook salmon in British Columbia: clinical presentation and experimental transmission. *Canadian Veterinary Journal* 34, 745–748.

Chen, M.F., Yun, S., Marty, G.D., McDowell, T.S., House, M.L. *et al.* (2000) A *Piscirickettsia salmonis*-like bacterium associated with mortality of white seabass *Atractoscion nobilis*. *Diseases of Aquatic Organisms* 43, 117–126.

Contreras-Lynch, S., Olmos, P., Vargas, A., Figueroa, J., González-Stegmaier, R. *et al.* (2015) Identification and genetic characterization of *Piscirickettsia salmonis* in native fish from southern Chile. *Diseases of Aquatic Organisms* 115, 233–244.

Corbeil, S., Hyatt, A.D. and Crane, M.S.J. (2005) Characterisation of an emerging rickettsia-like organism in Tasmanian farmed Atlantic salmon *Salmo salar*. *Diseases of Aquatic Organisms* 64, 37–44.

Crozier, L.G., Hendry, A.P., Lawson, P.W., Quinn, T.P., Mantua, N.J. *et al.* (2008) Potential responses to climate change in organisms with complex life histories: evolution and plasticity in Pacific salmon. *Evolutionary Applications* 1, 252–270.

Cvitanich, J.D., Garate, N. and Smith, C.E. (1991) The isolation of a rickettsia-like organism causing disease and mortality in Chilean salmonids and its confirmation by Koch's postulate. *Journal of Fish Diseases* 14, 121–145.

Dettleff, P., Bravo, C., Patel, A. and Martinez, V. (2015) Patterns of *Piscirickettsia salmonis* load in susceptible and resistant families of *Salmo salar*. *Fish and Shellfish Immunology* 45, 67–71.

Estévez, R.A., Mardones, F.O., Álamos, F., Arriagada, G., Carey, J. *et al.* (2019) Eliciting expert judgements to estimate risk and protective factors for piscirickettsiosis in Chilean salmon farming. *Aquaculture* 507, 402–410.

Evelyn, T.P.T., Kent, M.L., Poppe, T.T. and Bustos, P. (1998) Salmonid rickettsial septicemia. In: Kent, M.L. and Poppe, T.T. (eds) *Diseases of Seawater Netpen-Reared Salmonid Fishes*. Department of Fisheries and Oceans, Nanaimo, Canada, pp. 31–33.

Fryer, J.L. and Hedrick, R.P. (2003) *Piscirickettsia salmonis*: a Gram-negative intracellular bacterial pathogen of fish. *Journal of Fish Diseases* 26, 251–262.

Fryer, J.L. and Lannan, C.N. (2005) Family II. Piscirickettsiaceae fam. nov. In: Garrity, G.M. (ed.) *Bergey's Manual of Systematic Bacteriology*. Springer, New York, pp. 180–184.

Fryer, J.L., Lannan, C.N., Garcés, L.H., Larenas, J.J. and Smith, P.A. (1990) Isolation of a rickettsiales-like organism from diseased coho salmon (*Oncorhynchus kisutch*) in Chile. *Fish Pathology* 25, 107–114.

Fryer, J.L., Lannan, C.N., Giovanonni, S.J. and Wood, N.D. (1992) *Piscirickettsia salmonis* gen. nov., sp. nov., the causative agent of an epizootic disease in salmonid fishes. *International Journal of Systematic Bacteriology* 42, 120–126.

Gaete-Carrasco, A., Rosenfeld, C. and Gallardo, A. (2019) Análisis epidemiológico del programa de vigilancia activa de *Piscirickettsia salmonis* del Servicio Nacional de Pesca y Acuicultura de Chile. *Revue Scientifique et Technique (International Office of Epizootics)* 38(3). Available at: https://www.oie.int/fileadmin/Home/eng/Publications_%26_Documentation/docs/pdf/revue_plurithematique/2019/23042019-00145-ES_Gaete-

Carrasco-Rosenfeld_ESP.pdf (accessed 26 March 2020).

Garcés, L.H., Larenas, J.J., Smith, P.A., Sandino, S., Lannan, C.N. and Fryer, J.L. (1991) Infectivity of a rickettsia isolated from coho salmon *Oncorhynchus kisutch*. *Diseases of Aquatic Organisms* 11, 93–97.

Garcés, L.H., Correal, P., Larenas, J.J., Contreras, J.R., Oyanedel, S. *et al.* (1994) Finding of *Piscirickettsia salmonis* in *Ceratothoa gaudichaudii*. In: Hedrick, R.P. and Winton, J.R. (eds) *International Symposium on Aquatic Animal Health. Program and Abstracts*. Fish Health Section of the American Fisheries Society, Seattle, Washington, p. 109.

Gias, E., Brosnahan, C.L., Orr, D., Binney, B., Ha, H.J. *et al.* (2018) *In vivo* growth and genomic characterization of rickettsia-like organisms isolated from farmed chinook salmon (*Oncorhynchus tshawytscha*) in New Zealand. *Journal of Fish Diseases* 41, 1235–1245.

Grant, A.N., Brown, A.G., Cox, D.I., Birkbeck, T.H. and Griffen, A.A. (1996) Rickettsia-like organism in farmed salmon. *Veterinary Record* 138, 423.

ICES (International Council for the Exploration of the Sea) (2018) Report of the Working Group on Pathology and Diseases of Marine Organisms (WGPDMO). *ICES CM 2018/ASG:01*. ICES, Riga.

Irgang, R., Poblete-Morales, M., Gomez-Uchida, D. and Avendaño-Herrera, R. (2019) First description of health status and associated bacterial diversity from non-native Chinook salmon (*Oncorhynchus tshawytscha*, Walbaum 1792) naturalised in a large river system from southern Chile. *Bulletin of the European Association of Fish Pathologists* 39, 60–69.

Jones, S.R.M. (2019) Characterization of *Piscirickettsia salmonis* and salmonid rickettsial septicaemia to inform pathogen transfer risk assessments in British Columbia. *Research Document No. 2019/020*. Fisheries and Oceans Canada, Canadian Science Advisory Secretariat, Ottawa.

Jones, S.R.M., Markham, R.J., Groman, D.B. and Cusack, R.R. (1998) Virulence and antigenic characteristics of a cultured rickettsia-like organism isolated from Atlantic salmon *Salmo salar* in eastern Canada. *Diseases of Aquatic Organisms* 33, 25–31.

Jones, S.R.M., Long, A., McWilliams, C., Polinski, M. and Garver, K. (2020) Factors associated with severity of naturally occurring piscirickettsiosis in netpen- and tank-reared juvenile Atlantic salmon at a research aquarium in western Canada. *Journal of Fish Diseases* 43, 49–55.

Karatas, S., Mikalsen, J., Steinum, T.M., Taksdal, T., Bordevik, M. and Colquhoun, D.J. (2008) Real time PCR detection of *Piscirickettsia salmonis* from formalin-fixed paraffin-embedded tissues. *Journal of Fish Diseases* 31, 747–753.

Kovach, R.P., Ellison, S.C., Pyare, S. and Tallmon, D.A. (2015) Temporal patterns in adult salmon migration timing across southeast Alaska. *Global Change Biology* 21, 1821–1833.

Lannan, C.N. and Fryer, J.L. (1991) Recommended methods for inspection of fish for salmonid rickettsia. *Bulletin of the European Association of Fish Pathologists* 11, 135–136.

Lannan, C.N. and Fryer, J.L. (1993) *Piscirickettsia salmonis*, a major pathogen of salmonid fish in Chile. *Fisheries Research* 17, 115–121.

Lannan, C.N. and Fryer, J.L. (1994) Extracellular survival of *Piscirickettsia salmonis*. *Journal of Fish Diseases* 17, 545–548.

Lannan, C.N., Ewing, S.A. and Fryer, J.L. (1991) A fluorescent antibody test for detection of the rickettsia causing disease in Chilean salmonids. *Journal of Aquatic Animal Health* 3, 229–234.

Larenas, J.J. (1999) Evaluación experimental clínico patológica del efecto de la densidad poblacional, temperatura e infección concomitante con *Renibacterium salmoninarum* sobre la presentación de piscirickettsiosis. MSc thesis. Universidad de Chile, Santiago.

Larenas, J.J., Contreras, J., Oyanedel, S., Morales, M.A. and Smith, P. (1997) Efecto de la densidad poblacional y temperatura en truchas arco iris (*Oncorhynchus mykiss*) inoculadas con *Piscirickettsia salmonis*. *Archivos de Medicina Veterinaria* 29, 113–119.

Larenas, J., Gatica, C., Galleguillos, M., Adarmes, H. and Smith, P. (2012) Comparación de la cinética de la infección de ovas de trucha arcoíris (*Oncorhynchus mykiss*) con dos cepas de *Piscirickettsia salmonis* detectada mediante dot-blot. *Avances en Ciencias Veterinarias* 2, 1–9.

Leal, M.A. (2017) Caracterización epidemiológica de pisciricekttsiosis en el año 2013 en Chile. MSc thesis, Universidad de Chile, Santiago.

Lhorente, J.P., Gallardo, J.A., Villanueva, B., Carabano, M.J. and Neira, R. (2014) Disease resistance in Atlantic salmon (*Salmo salar*): coinfection of the intracellular bacterial pathogen *Piscirickettsia salmonis* and the sea louse *Caligus rogercresseyi*. *PLoS ONE* 9, e95397.

McCarthy, Ú., Steiropoulos, N.A., Thompson, K.D., Adams, A., Ellis, A.E. and Ferguson, H.W. (2005) Confirmation of *Piscirickettsia salmonis* as a pathogen in European sea bass *Dicentrarchus labrax* and phylogenetic comparison with salmonid strains. *Diseases of Aquatic Organisms* 64, 107–119.

Maisey, K., Montero, R. and Christodoulides, M. (2017) Vaccines for piscirickettsiosis (salmonid rickettsial septicaemia, SRS): the Chile perspective. *Expert Review of Vaccines* 16, 215–228.

Makrinos, L. and Bowden, T.J. (2017) Growth characteristics of the intracellular pathogen, *Piscirickettsia salmonis*, in tissue culture and cell-free media. *Journal of Fish Diseases* 40, 1115–1127.

Maquera, O.Y. (2017) Detección mediante inmunohistoquímica de Piscirickettsia salmonis en cortes de tejido de Caligus rogercresseyi. MSc thesis, Universidad de Chile, Santiago.

Marcos-López, M., Ruane, N.M., Scholz, F., Bolton-Warberg, M., Mitchell, S.O. *et al.* (2017) *Piscirickettsia salmonis* infection in cultured lumpfish (*Cyclopterus lumpus* L.). *Journal of Fish Diseases* 40, 1625–1634.

Marshall, S., Heath, S., Henríquez, V. and Orrego, C. (1998) Minimally invasive detection of *Piscirickettsia salmonis* in cultivated salmonids via the PCR. *Applied and Environmental Microbiology* 64, 3066–3069.

Mauel, M.J. and Fryer, J.L. (2001) Amplification of a *Piscirickettsia salmonis*-like 16S rDNA product from bacterioplankton DNA collected from the coastal waters of Oregon, USA. *Journal of Aquatic Animal Health* 13, 280–284.

Mauel, M.J., Giovannoni, S.J. and Fryer, J.L. (1996) Development of polymerase chain reaction assays for detection, identification, and differentiation of *Piscirickettsia salmonis*. *Diseases of Aquatic Organisms* 26, 189–195.

Mauel, M.J., Ware, C. and Smith, P.A. (2008) Culture of *Piscirickettsia salmonis* on enriched blood agar. *Journal of Veterinary Diagnostic Investigation* 20, 213–214.

Meza, K., Inami, M., Dalum, A.S., Bjelland, A.M., Sørum, H. and Løvoll, M. (2019) Development of piscirickettsiosis in Atlantic salmon (*Salmo salar* L.) smolts after intraperitoneal and cohabitant challenge using an EM90-like isolate: a comparative study. *Journal of Fish Diseases* 42, 1001–1011.

Mikalsen, J., Skjaervik, O., Wiik-Nielsen, J., Wasmuth, M.A. and Colquhoun, D.J. (2008) Agar culture of *Piscirickettsia salmonis*, a serious pathogen of farmed salmonid and marine fish. *FEMS Microbiology Letters* 278, 43–47.

Miranda, C.D., Godoy, F.A. and Matthew, R.L. (2018) Current status of the use of antibiotics and the antimicrobial resistance in the Chilean salmon farms. *Frontiers in Microbiology* 9, 1284.

Nekouei, O., Vanderstichel, R., Ming, T., Kaukinen, K.H., Thakur, K. *et al.* (2018) Detection assessment of the distribution of infectious agents in juvenile Fraser River sockeye salmon, Canada, in 2012 and 2013. *Frontiers in Microbiology* 9, 3221.

OIE (World Organization for Animal Health) (2003) Piscirickettsiosis *(Piscirickettsia salmonis). In: Manual of Diagnostic Tests for Aquatic Animals*, 4th edn. OIE, Paris, pp. 193–199.

Olivares, J. and Marshall, S.H. (2010) Determination of minimal concentration of *Piscirickettsia salmonis* in water columns to establish a fallowing period in salmon farms. *Journal of Fish Diseases* 33, 261–266.

Olsen, A.B., Melby, H.P., Speilberg, L., Evensen, O. and Hastein, T. (1997) *Piscirickettsia salmonis* infection in Atlantic salmon *Salmo salar* in Norway – epidemiological, pathological and microbiological findings. *Diseases of Aquatic Organisms* 31, 35–48.

Palmer, R., Ruttledge, M., Callanan, K. and Drinan, E. (1996) A piscirickettsiosis-like disease in farmed Atlantic salmon in Ireland – isolation of the agent. *Bulletin of the European Association of Fish Pathologists* 17, 68–72.

Pastor, F., Valiente, J.A. and Palau, J. (2018) Sea surface temperature in the Mediterranean: trends and spatial patterns (1982–2016). *Pure and Applied Geophysics* 175, 4017–4029.

Pérez, B.A., Alert, A.A., Contreras, J.R. and Smith, P.A. (1998) Detection of *Piscirickettsia salmonis* in upstream-migrating coho salmon, *Oncorhynchus kisutch*, in Chile. *Bulletin of the European Association of Fish Pathologists* 18, 189–191.

Price, D., Stryhn, H., Sanchez, J., Ibarra, R., Tello, A. and St-Hilaire, S. (2016) Retrospective analysis of antibiotic treatments against piscirickettsiosis in farmed Atlantic salmon *Salmo salar* in Chile. *Diseases of Aquatic Organisms* 118, 227–235.

Rees, E.E., Ibarra, R., Medina, M., Sanchez, J., Jakov, E. *et al.* (2014) Transmission of *Piscirickettsia salmonis* among salt water salmonid farms in Chile. *Aquaculture* 428/429, 189–194.

Rozas-Serri, M., Ildefonso, R., Peña, A., Enríquez, R., Barrientos, S. and Maldonado, L. (2017) Comparative pathogenesis of piscirickettsiosis in Atlantic salmon (*Salmo salar* L.) post-smolt experimentally challenged with LF-89-like and EM-90-like *Piscirickettsia salmonis* isolates. *Journal of Fish Diseases* 40, 1451–1472.

Saavedra, J., Grandón, M., Villalobos-González, J., Bohle, H., Bustos, P. and Mancilla, M. (2018) Isolation, functional characterization and transmissibility of p3PS10, a multidrug resistance plasmid of the fish pathogen *Piscirickettsia salmonis*. *Frontiers in Microbiology* 9, 923.

Sernapesca (Servicio Nacional de Pesca y Acuicultura) (2019) *Informe Sanitario de Salmonicultura en Centros Marinos año 2018*. Departamento de Salud Animal, Subdirección de Acuicultura, Sernapesca, Valparaíso, Chile. Available at: http://www.sernapesca.cl/sites/default/files/informe_sanitario_salmonicultura_en_centros_marinos_2018_final.pdf (accessed 19 March 2020).

Smith, P.A., Vecchiola, I.M., Oyanedel, S., Garcés, L.H., Larenas, J. and Contreras, J. (1996) Antimicrobial sensitivity of four isolates of *Piscirickettsia salmonis*. *Bulletin of the European Association of Fish Pathologists* 16, 164–168.

Smith, P.A., Manneschi, G., Guajardo, A., Rojas, M.E., Martínez, V. *et al.* (2009) Susceptibility comparison among *Salmo salar* families experimentally challenged with *Piscirickettsia salmonis*. In: *Proceedings of the 14th EAFP International Conference on Diseases of Fish and Shellfish.* European Association of Fish Pathologists, Prague, p. 163.

Smith, P.A., Contreras, J.R., Rojas, M.E., Guajardo, A., Díaz, S. and Carbonero, A. (2015a) Infectivity of *Piscirickettsia salmonis* in immersion-bath exposed rainbow trout *Oncorhynchus mykiss* (Walbaum) fry. *Journal of Fish Diseases* 38, 765–770.

Smith, P.A., Díaz, F.E., Rojas, M.E., Díaz, S., Galleguillos, M. and Carbonero, A. (2015b) Effect of *Piscirickettsia salmonis* inoculation on the ASK continuous cell line. *Journal of Fish Diseases* 38, 321–324.

Smith, P.A., Elliott, D.G., Bruno, D.W. and Smith, S.A. (2019) Skin and fin diseases. In: Smith, O. (ed.) *Fish Diseases and Medicine*. CRC Press, Boca Raton, Florida, pp. 97–133.

Tandberg, J., Oliver, C., Lagos, L., Gaarder, M., Yáñez, A.J. *et al.* (2017) Membrane vesicles from *Piscirickettsia salmonis* induce protective immunity and reduce development of salmonid rickettsial septicemia in an adult zebrafish model. *Fish and Shellfish Immunology* 67, 189–198.

Thakur, K.K., Vanderstichel, R., Kaukinen, K., Nekouei, O., Laurin, E. and Miller, K.M. (2019) Infectious agent detections in archived Sockeye salmon (*Oncorhynchus nerka*) samples from British Columbia, Canada (1985–94). *Journal of Fish Diseases* 42, 533–547.

Venegas, C.A., Contreras, J.R., Larenas, J.J. and Smith, P.A. (2004) DNA hybridization assays for the detection of *Piscirickettsia salmonis* in salmonid fish. *Journal of Fish Diseases* 27, 431–433.

Yañez, A.J., Silva, H., Valenzuela, K., Pontigo, J.P., Godoy, M. *et al.* (2013) Two novel blood-free solid media for the culture of the salmonid pathogen *Piscirickettsia salmonis*. *Journal of Fish Diseases* 36, 587–591.

Yáñez, J.M., Bangera, R., Lhorente, J.P., Oyarzún, M. and Neira, R. (2013) Quantitative genetic variation of resistance against *Piscirickettsia salmonis* in Atlantic salmon (*Salmo salar*). *Aquaculture* 414–415, 155–159.

Zrnčić, S., Vendramin, N., Boutrup, T.S., Boye, M., Bruun, M.S. *et al.* (2015) Piscirickettsia salmonis infection in European sea bass – an emerging disease in Mediterranean mariculture. In: *Proceedings of the 17th EAFP International Conference on Diseases of Fish and Shellfish.* European Association of Fish Pathologists, Las Palmas de Gran Canaria, Spain, p. 153.

15 Epizootic Ulcerative Syndrome (*Aphanomyces invadans*)

Dibyendu Kamilya[1]* and Kollanoor Riji John[2]

[1]*Central Agricultural University, Lembucherra, India;* [2]*Kerala University of Fisheries and Ocean Studies, Kochi, India*

15.1 Introduction

Epizootic ulcerative syndrome (EUS) is a severe disease of epidemic dimension affecting a large number of wild and farmed finfish. The disease was first described in 1971 during an outbreak in farmed freshwater ayu (*Plecoglossus altivelis*) in Japan (Egusa and Masuda, 1971). Since then, the disease has spread progressively in Australia, the Asia-Pacific region, North America and, most recently, Africa. The outbreak of EUS has been reported in several countries from four continents (Kamilya and Baruah, 2014). However, no confirmed outbreak of the disease has been reported from Europe and South America.

A number of aetiologies had been proposed to be the primary agent for the outbreak of EUS. However, an oomycete, *Aphanomyces invadans*, has been consistently found to be linked with the disease. The oomycete was originally described as *Aphanomyces invaderis*, an invasive fungus (Willoughby *et al.*, 1995). The causative organism was also referred to as *Aphanomyces piscicida* in Japan causing mycotic granulomatosis (Hatai, 1980), *Aphanomyces* sp. in Australia causing red spot disease (Lilley and Roberts, 1997) and EUS-related *Aphanomyces* species (ERA) in the Philippines (Lumanlan-Mayo *et al.*, 1997). All these pathogenic isolates and several other isolates from different geographic locations were compared using conventional, biochemical and molecular techniques and confirmed as the same species, and the name *Aphanomyces invadans* was adopted. Baldock *et al.* (2005) proposed to change the name of the disease from EUS to epizootic granulomatous aphanomycosis (EGA). However, most scientists still use EUS when describing the ulcerative disease of fish caused by *A. invadans*.

This disease has been described in more than 125 species of fish worldwide (Kamilya and Baruah, 2014). Recent studies report susceptibility of new fish species (*Ctenopoma multispine*, *Pollimyrus isidori* and *Enteromius trimaculatus*) from Africa showing lesions presumptive to EUS (Huchzermeyer *et al.*, 2018). Certain genera such as *Channa*, *Puntius* and some other food species are particularly susceptible to EUS (Roberts *et al.*, 1994). However, the common carp (*Cyprinus carpio*), Nile tilapia (*Oreochromis niloticus*), milkfish (*Chanos chanos*), Japanese eel (*Anguilla japonica*) and European eel (*Anguilla anguilla*) are considered to be naturally resistant (Hatai, 1994; Wada *et al.*, 1996; Lilley *et al.*, 1998; Oidtmann *et al.*, 2008). In general, species affected during natural outbreaks are predominantly either bottom-dwellers or possess air-breathing organs (Llobrera and Gacutan, 1987; Roberts *et al.*, 1994). Nevertheless, other species occupying different ecological niches or without having air-breathing organs may also be affected. Whereas for some species like snakeheads (*Channa striata*), the size or age does not influence the susceptibility towards EUS (Cruz-Lacierda and Shariff, 1995), species such as Indian major carps are more susceptible at the younger stage (Roberts *et al.*, 1989; Pradhan *et al.*, 2008a).

EUS has been responsible for causing huge economic losses worldwide. Besides fish resources, the livelihood of fish farmers has been adversely impacted by outbreaks of the disease in enzootic areas, particularly in the Asia-Pacific region and Africa. Although considerable scientific advancements have been made in understanding EUS in fish, the likely impact of impending climate change on EUS and the associated economic and social

*dibyendu.kamilya@gmail.com

ramifications remain unclear and require prioritized interventions.

15.2 Diagnosis of the Pathogen/Disease

15.2.1 Morphological features of the pathogen and life cycle

A. invadans has an aseptate fungal-like mycelial structure and has been assigned to the genus *Aphanomyces* within the family *Saprolegniaceae* on the basis of its pattern of asexual spore morphogenesis (Lilley *et al.*, 1998). The life cycle (Fig. 15.1) begins with the formation of zoosporangia from the mycelium and 30–50 primary zoospores are discharged from lateral evacuation tubes or from the opening of the terminal sporangia. The number of evacuation tubes per sporangium varies from one to four. After the primary zoospores are discharged, they encyst to form achlyoid clusters at the opening

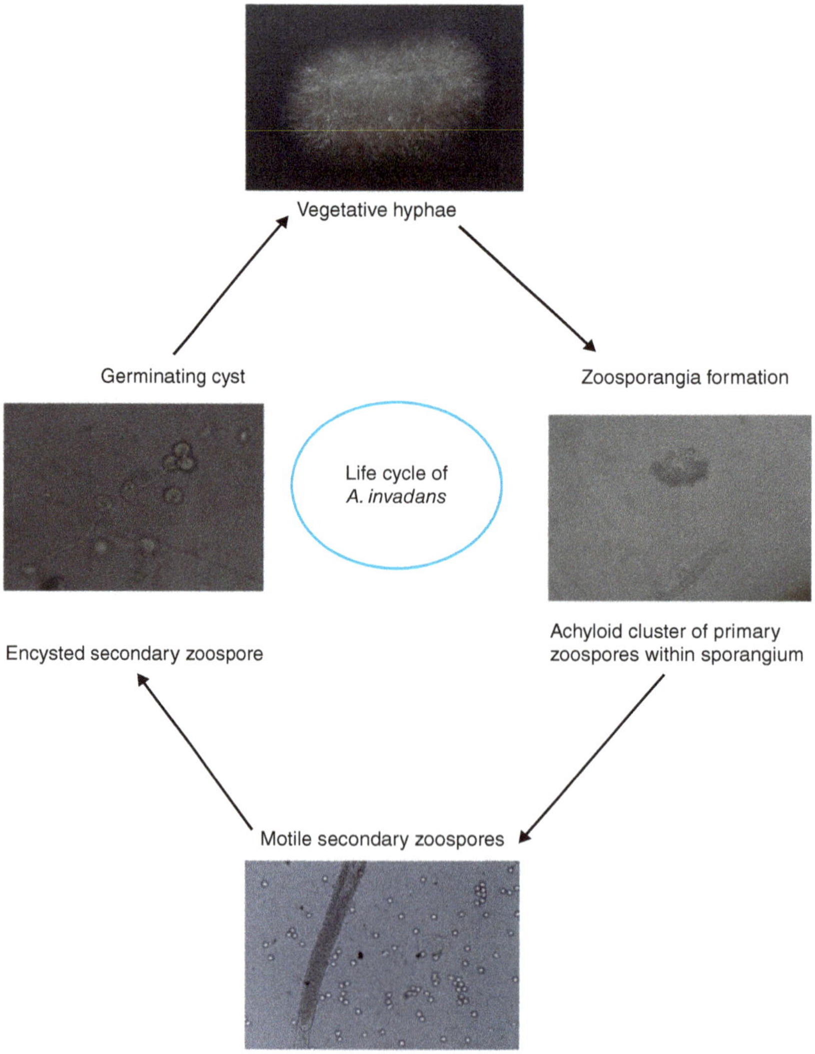

Fig. 15.1. Asexual life cycle of *Aphanomyces invadans*.

of the sporangium. The primary zoospore quickly transforms into the free-swimming, subspherical and biflagellate secondary form about 6 µm in diameter (Willoughby *et al.*, 1995; Lilley *et al.*, 1998). These motile secondary zoospores attach to the fish skin and germinate into vegetative aseptate hyphae, which then invade the fish skin and muscular tissue and may reach the internal organs (Lilley *et al.*, 1998). The secondary zoospore remains motile for a period that depends on the environmental conditions and the location of the fish host or substratum (Lilley *et al.*, 1998). Typically, encysted (about 6.5 µm in diameter) secondary zoospores germinate by forming germ tubes, which eventually develop into mycelium. However, they are capable of limited polyplanetism where new zoospore generation can be formed through repeated zoospore emergence, encystment and re-emergence (Willoughby *et al.*, 1995; Lilley *et al.*, 1998).

15.2.2 Clinical signs

The behavioural signs of EUS-infected fish are not specific for the disease. The infected fish usually lose appetite, become lethargic, float listlessly or swim with the head projecting out of the water. The penetration of hyphae into the skin or underlying tissue results in skin lesions which may range from tiny red spots to severe dermal ulcers. The lesions develop due to a variety of conditions ranging from simple mechanical damage to infections by other pathogenic microorganisms. Therefore, clinical signs alone do not provide a valid diagnostic feature for EUS (Oidtmann, 2012). Although the clinical signs are not considered as pathognomonic for the disease, their development and the progression of ulceration over the body surface have been used for presumptive diagnosis of the disease.

15.2.3 External/internal macroscopic and microscopic lesions

In general, lesions developed on the EUS-affected fish can be grouped into three stages on the basis of their sequential development and gross appearance (Callinan *et al.*, 1989; Viswanath *et al.*, 1997a; Lilley *et al.*, 1998).

During the initial stage of the disease, infected fish have tiny red spots on various parts of the body with no prominent ulcerative lesions. The skin around the red spots seems normal with no discoloration. Epithelial necrosis, haemorrhaging of the underlying dermis and infiltration of inflammatory cells usually occur in the early lesion areas. During the moderate stage, the haemorrhagic spots turn into dermal ulcerative lesions (1–4 cm in diameter) with associated loss of epidermis and scales (Fig. 15.2). The ulcerative area may become circular, oedematous with grey to red discoloration. Significant degenerative and necrotic changes occur in the muscle tissue and skin. The advanced-stage lesions are characterized by the presence of large haemorrhagic and necrotic open ulcers on the body surface extending into the skeletal musculature (Fig. 15.3a and b). The dermal ulcers can be circular or oval shaped with sharply defined margins. In severe cases, there can be erosion of caudal peduncle, head tissue erosion, or even exposure of bones and viscera. The underlying muscle fibres are subjected to massive necrotizing granulomatous mycosis. As the lesion advances, the oomycete hyphae may become progressively enveloped by thick sheaths of host epithelioid cells, forming granuloma. Mycotic granuloma accompanied with degenerative and necrotic changes may also take place in several internal organs.

15.2.4 Conventional diagnosis

EUS in susceptible fish species is histopathologically characterized by epithelioid granuloma in different

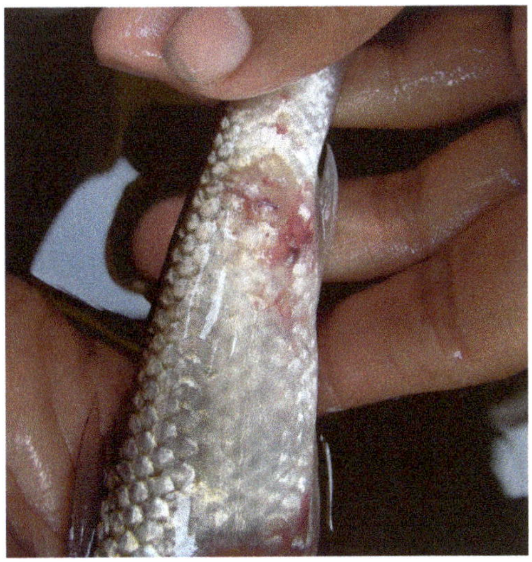

Fig. 15.2. Moderate-stage macroscopic lesion in EUS-infected bata (*Labeo bata*).

organs including skeletal muscle, liver, kidney and other visceral organs (Figs 15.4 and 15.5) (Viswanath *et al.*, 1997b, 1998; Lilley *et al.*, 1998). However, the granulomatous response is not always associated with the infection by *A. invadans*. In several fish species (e.g. bluegill (*Lepomis macrochirus*), largemouth bass (*Micropterus salmoides*), American shad (*Alosa sapidissim*), channel catfish (*Ictalurus punctatus*), black bullhead (*Ameiurus melas*) and European catfish (*Silurus glanis*)), *A. invadans* was isolated from the infected tissue sample, and even the oomycete hyphae were present in the histological sections, but granulomas were not observed (Hawke *et al.*, 2003; Sosa *et al.*, 2007; Oidtmann *et al.*, 2008). Moreover, highly susceptible species may die before the development of a robust immune response due to the acuteness and severity of infection (Chinabut and Roberts, 1999). Mycotic granulomas may not be observed in histological sections if tissue samples are taken far from the skin lesion area or if there is a high degree of necrosis and degeneration of the lesion (Oidtmann, 2012). Additionally, granulomas may also not be present if tissue samples are collected before their formation. For example, fingerlings of catla (*Catla catla*) produced a well-developed epithelioid cell granuloma after 4 days in an experimental challenge with *A. invadans* spores (Pradhan *et al.*,

Fig. 15.3. Advanced-stage macroscopic lesion in EUS-infected (A) striped snakehead (*Channa striata*) and (B) mrigal (*Cirrhinus mrigala*).

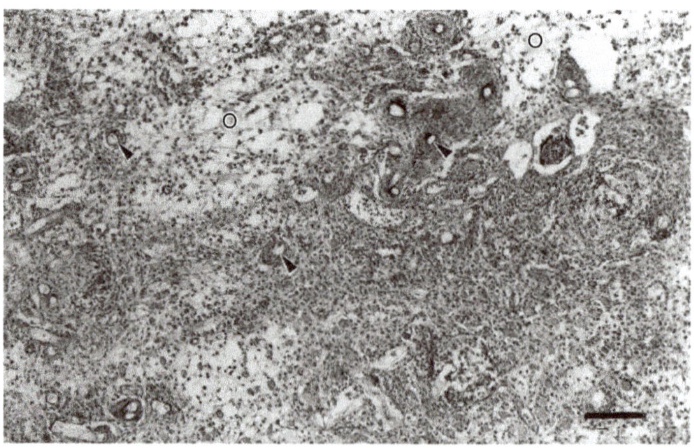

Fig. 15.4. Photomicrograph from three-spot gourami (*Trichogaster trichopterus*) at day 6 post-zoospore inoculation showing extensive cellular infiltration and oedema (O) in the muscle area; and fungal hyphae (arrowheads) enclosed within three or four layers of epithelioid cells. Periodic acid–Schiff stain; scale bar = 100 µm. (Reprinted from Catap, E.S. and Munday, B.L. (2002) Development of a method for reproducing epizootic ulcerative syndrome using controlled doses of *Aphanomyces invadans* in species with different salinity requirements. *Aquaculture* 209, 35–47, with permission from Elsevier.)

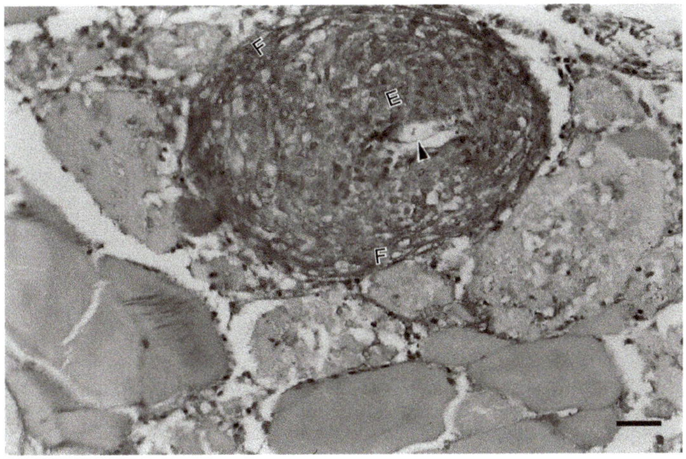

Fig. 15.5. Higher magnification of a mycotic granuloma in sand whiting (*Sillago ciliata*) at day 8 post-zoospore inoculation showing the fungal hypha (arrowhead) in the centre enveloped within layers of epithelioid cells (E). Note the fibrous nature of the peripheral layers (F). Macrophages and lymphocytes infiltrate the surrounding area with degenerating muscle fibres. Periodic acid–Schiff stain; scale bar = 50 µm. (Reprinted from Catap, E.S. and Munday, B.L. (2002) Development of a method for reproducing epizootic ulcerative syndrome using controlled doses of *Aphanomyces invadans* in species with different salinity requirements. *Aquaculture* 209, 35–47, with permission from Elsevier.)

2008b). Thus, the absence of epitheloid granulomas should not be considered an indication that the inspected condition is not EUS (Oidtmann, 2012).

Diagnosis of *A. invadans* can also be carried out by isolation of the oomycete in pure culture, followed by morphological and macroscopic studies of hyphae and zoosporangia, patterns of zoospore release and growth characteristics at specific temperatures (Oidtmann, 2012). However, identification of this oomycete to the species level is difficult as it requires characterization of sexual reproductive stages which are absent in *A. invadans*. Therefore, identification based on morphological and macroscopic criteria is possible only up to the genus level (Lilley *et al.*, 1998).

15.2.5 Diagnosis using molecular/other techniques

A number of methods based on polymerase chain reaction (PCR) have been developed which are highly sensitive and specific for early detection of *A. invadans*. Sequencing the internal transcribed spacer (ITS) region of the oomycete is a specific and sensitive technique for molecular identification of *A. invadans*. Specific PCR assays using primers targeting the ITS or nearby located DNA regions of *A. invadans* have been reported by several researchers for rapidly identifying and screening *A. invadans* (Blazer *et al.*, 2002; Lilley *et al.*, 2003; Phadee *et al.*, 2004; Vandersea *et al.*, 2006; Sosa *et al.*, 2007; Oidtmann *et al.*, 2008; Boys *et al.*, 2012; Huchzermeyer *et al.*, 2018). Recently, the genome of *A. invadans* has been sequenced by Aphanomyces WGS Initiative of the Broad Institute (broadinstitute.org). Moreover, the assembly and annotation of the mitochondrial genome of *A. invadans* have also been completed from the whole-genome shotgun sequence reads (Makkonen *et al.*, 2016). These can serve as a valuable genetic resource for further biodiversity, phylogenetic and other studies.

Another species-specific molecular method is fluorescent peptide nucleic acid *in-situ* hybridization (FISH), which was successfully tested in Atlantic menhaden (*Brevoortia tyrannus*) with ulcerative mycotic lesions. The assay provided visual confirmation of *A. invadans* hyphae in the ulcerated lesions (Vandersea *et al.*, 2006). Other methods used for identification of the pathogen include electrophoresis and Western blotting (Lilley *et al.*, 1997), pyrolysis mass spectrometry (Lilley *et al.*, 2001) and monoclonal antibody-based detection (Miles *et al.*, 2003; Naik *et al.*, 2008).

15.3 Expected/Potential Spread of the Pathogen

EUS has spread across Asia, particularly South-East and South Asia, and has also been recorded from Australia, North America and Africa (Kamilya and Baruah, 2014; Iberahim *et al.*, 2018). Since the first report of the disease in the Zambezi and Chobe river system of Africa in 2006 (Andrew *et al.*, 2008), it has spread across several new locations of Southern Africa with the latest occurrence in the Bangweulu Wetlands, Zambia (Huchzermeyer *et al.*, 2018). Owing to the epizootic nature and extensive susceptible host range, the disease has the potential to spread into new geographical locations. There are limited possibilities for introduction of the EUS pathogen into an importing country via trading of fish products, especially if the product is temperature-treated (e.g. by freezing or cooking) (Oidtmann, 2012). The most probable route by which the disease can spread to unaffected countries is through international trade of live fish. The likelihood of introduction is substantially high if the live fish are imported from a region or country where EUS is enzootic. In fact, EUS infections have been recorded from imported fish (Balasuriya, 1994; Hatai *et al.*, 1994; Hanjavanit *et al.*, 1997). Apart from the infected fish itself, the presence of infective zoospores in transportation water can also serve as the potential source of infection. The recent outbreaks in Southern African countries may be attributed to extensive movement of live fish within and between neighbouring countries, as these countries have been experiencing rapid growth in aquaculture practices in recent times (Huchzermeyer *et al.*, 2018). Spread of the pathogen is also possible through natural migration of fish, particularly in regions of interconnected aquatic ecosystems. Such interconnected aquatic ecoregions comprising river catchment areas, flood plains, inland delta, lakes and swamps are present in Southern Africa, which has been experiencing EUS outbreaks in recent times.

The likely impact of climate change has been investigated in humans and livestock animals, including marine organisms, indicating its influence on disease emergence (Marcos-Lopez *et al.*, 2010). However, climate change's impact on aquatic animals, particularly fish of freshwater origin, is not well studied. The global warming may have some impact on the potential spread of various pathogens including *A. invadans*, although there is no emphatic evidence. The natural outbreak of EUS has been observed in both low (e.g. 10–15°C; Hawke *et al.*, 2003) and high temperatures (e.g. 33°C; Bondad-Reantaso *et al.*, 1992), and low temperature is often attributed as one of the important predisposing factors for some, but not all, EUS outbreaks (Lilley *et al.*, 1998). This may be due to the compromised protective response of the infected fish against the invading oomycete at low temperature (Chinabut *et al.*, 1995). However, 'low temperature' for a tropical climate (the majority of EUS-susceptible fish are tropical species) can be as high as 18–22°C (Bondad-Reantaso *et al.*, 2018). Moreover, outbreaks of EUS have also been observed during warmer months (e.g. 27–33°C) in the tropics, implying no consistent relationship between low temperatures and EUS outbreaks (Lilley *et al.*, 1998). In artificial infection studies, the disease has been induced in roach (*Rutilus rutilus*) (Khan *et al.*, 1998) and ayu (*P. altivelis*) (Wada *et al.*, 1996) at temperatures as low as 11–16°C and 15–16°C, respectively. The highest temperature at which EUS was experimentally induced in fish (snakehead) was 31°C (Chinabut *et al.*, 1995). In a study, the effect of temperature (between 6 and 42°C) on the growth of several *A. invadans* isolates was tested *in vitro*. All the isolates grew at 6°C but failed to survive at 37°C. The optimal growth was observed at 26–30°C, and the highest temperature at which good growth was still observed was 34°C (Lilley and Roberts, 1997). Therefore, it is not easy to predict the impact of global warming on the severity or spread of the EUS pathogen as the infection and subsequent expression of the disease can happen within a wide temperature range. However, climate change-mediated weather extremes (e.g. heavy rainfall and flooding, droughts, storms, etc.) can result in environmental deterioration, which can exert stress on fish and render them more susceptible to infection by the oomycete. Moreover, extensive flooding can facilitate transmission and potential spread of the pathogen to newer locations. Large river systems with interconnecting swamps, inland deltas and other aquatic ecoregions (e.g. Zambezi and Congo river system in Africa) are particularly vulnerable to spread of the pathogen by such extreme weather events.

The consequences of climate change are not limited to temperature but may also result in increased acidification, particularly in headwater streams (Marcogliese, 2008). Slightly acidic water (pH 6.0–7.0) is often considered one of the predisposing factors for infection by *A. invadans* (Callinan *et al.*, 1996; Lilley *et al.*, 1998; Choongo *et al.*, 2009). Low pH may initiate skin damage which subsequently may provide a portal for *A. invadans* infection (Lilley *et al.*, 1998). Therefore, climate change-mediated acidification may also exacerbate infection by *A. invadans*.

EUS-free regions may experience the emergence of the disease if the temperature thresholds for infection and subsequent disease expression are breached. Higher temperature can result in greater virulence of the pathogen, which may be directly or indirectly attributed to compromised resistance of the host due to temperature-mediated stress, enhanced virulence factors or elevated transmission potential of the pathogen (Harvell *et al.*, 2002; Marcogliese, 2008). The introduction of pathogens or their infective spores with greater virulence into new geographic locations may, thus, facilitate EUS emergence in new hosts. Moreover, with the increase in water temperatures, especially in the temperate regions, the demand to culture new fish species would increase. Higher water temperatures could enhance the chances of their survival, as well as the survival of other exotic species and their associated pathogens. This situation may also favour the emergence of the disease either through the introduction of new pathogens to new hosts or through host-switching (Brooker *et al.*, 2007; Marcos-Lopez *et al.*, 2010).

15.4 Expected/Potential Spread of the Reservoir Host

EUS may not occur in an area for a substantial period (months or even years) between outbreaks (Chinabut, 1998). It is unclear how *A. invadans* survives during the periods of no outbreaks and outside the fish host. It is also not known whether fish surviving an outbreak can act as the reservoir for subsequent infection (Oidtmann, 2012). The presence of any reservoir host of *A. invadans* in the aquatic environment is very difficult to establish and remains unresolved. Therefore, it is notoriously difficult to draw any reasonable correlation between the potential spread of the reservoir host in the new geographical area and the emergence of EUS.

15.5 Population Dynamics of Fish

It is well known that environmental stress can lead to suppression of the immune system, rendering fish more susceptible to infection by pathogenic microorganisms. Climate change-associated environmental stress, particularly fluctuation in temperature and salinity, low pH or low dissolved oxygen concentration, can significantly affect the functioning of the fish immune system. The likelihood of enhanced virulence potential of the pathogen and reduced ability of the host to mount a robust protective response against the infecting pathogen may increase the incidence of disease outbreaks, leading to increased morbidity and mortality of fish and, hence, to greater production loss. Water quality variation and habitat alteration due to shift in the hydrodynamic characteristics of a particular aquatic ecosystem can disturb the trophic interaction with the consequent alteration in dynamics of the host population. Poor physicochemical properties of water, particularly low dissolved oxygen concentration (due to higher temperature), can force farmers to lower population density, resulting in less production (Marcos-Lopez *et al.*, 2010). Climate change-associated extreme weather events are of particular concern as they can significantly alter the population dynamics of the host as a result of the change in dynamics of the host–pathogen–environment interactions. The occurrence of droughts may lead to the congregation of fish in the deeper channel of a water body, resulting in an enhanced rate of disease transmission. Many other environmental perturbations including eutrophication, increased stratification, acidification, etc. may also affect the diversity and abundance of fish in different aquatic ecosystems (Marcogliese, 2008).

15.6 Increase in Pathogen Mortality

In general, it is expected that higher temperatures will lead to more generation of pathogens annually with increased transmission rates, resulting in increased disease outbreaks (Marcogliese, 2008). However, this analogy may not be fully applicable to pathogens of poikilotherms, or those with complex life cycles, and such pathogens may be affected disproportionately by higher temperature (Harvell, 2002). As a general prediction, an increase in diversity and abundance of pathogens is expected in a climate change scenario. Global warming, on the contrary, may also lead to increased pathogen mortality in certain circumstances. Occurrence and abundance of pathogens with distinct thermal preferences may decline if the upper thermal tolerance limit is breached. For example, it is speculated that the occurrence of the chytrid fungus *Batrachochytrium dendrobatidis*, responsible for mass mortalities and population declines of amphibian species, may decline with global warming owing to its distinct thermal preference

(Harvell, 2002; Kriger and Hero, 2007). Unfortunately, there are not many demonstrations of the effect of climate change on the increased mortality of fish pathogens. The maximum temperature at which an EUS outbreak has been observed in field conditions is 33°C in an irrigated rice–fish pond in the Philippines (Bondad-Reantaso *et al.*, 1992) and at 31.6°C in fish ponds in India (Pradhan *et al.*, 2014). However, further observations in the Philippines indicate that outbreaks of EUS generally do not occur in the rice fields if the daily temperature remains above 30°C (Lumanlan-Mayo *et al.*, 1997). Other published data suggest very poor growth of *A. invadans* hyphae at temperatures >31°C and no growth at 37°C (Fraser *et al.*, 1992; Roberts *et al.*, 1993). Consequently, >31°C can be considered as the temperature threshold beyond which the oomycete is substantially inactive. Given that the water temperature can be as high as 36°C in certain regions of Asia, it is unclear how *A. invadans* survives in the aquatic environment during periods of no outbreaks. One of the possible explanations is the formation of a resistance stage in the environment or within some unknown carrier (Chinabut *et al.*, 1995). Moreover, the highest temperature at which *A. invadans* can grow *in vitro* is 34°C (Lilley and Roberts, 1997). These indicate that the permissive upper thermal limit may be broader than what has been reported under field conditions (Oidtmann, 2012). In addition, the pathogen itself can evolve into a more temperature-tolerant strain. Therefore, it is very difficult to predict the impact of higher temperature on the mortality of *A. invadans*. Nevertheless, higher temperatures may lead to increased pathogen mortality, particularly in certain regions of tropical Asia. Conversely, global warming may pose an opposite scenario in EUS-free temperate regions where higher temperatures may be responsible for increased survival of the pathogen and hence the enhanced risk of outbreaks if introduced by any means.

15.7 Control and Prevention

A wide range of options have been described and tested to control EUS outbreaks. Given that EUS causes devastating outbreaks in wild fish populations, control of outbreaks in natural water bodies is practically impossible. However, in closed aquaculture systems, risks of EUS outbreaks and fish mortality can be minimized by adopting common fish health management measures. These include removal of infected fish, drying and liming of ponds, disinfection of contaminated equipment, prophylactic treatment of fish (e.g. with a 1% NaCl bath) before introduction to ponds, avoiding introduction of wild fish, use of water from a EUS-free source, and maintaining proper stocking density and water quality (Lilley *et al.*, 1998). In regions where EUS is not enzootic but potentially susceptible fish species are farmed, preventing the introduction of the pathogen should be the first priority. This can be achieved by implementing strict quarantine and health certification guidelines in the context of import of live fish from a country or region where EUS is enzootic. Moreover, undertaking a comprehensive import risk analysis framework can be very useful for EUS-free countries or regions to prevent the likelihood of disease outbreaks. For instance, an assessment of the risk of entry, release, spread and possible impact of EUS in Europe was conducted by the European Food Safety Authority (EFSA) Panel on Animal Health and Welfare (AHAW) (EFSA Panel on Animal Health and Welfare, 2011). The AHAW assessment recommended adoption of scientifically based surveys for aquaculture, live imports and wild fish; implementing a surveillance system related to health certification by the exporting countries; undertaking differential diagnosis of *A. invadans* infection; exploring scientific evidence regarding a wider temperature range for disease expression; and adopting adequate biosecurity measures to minimize the risk of releasing *A. invadans* from closed ornamental facilities.

The treatment options to prevent the spread of EUS and minimize mortality of fish are limited. In one study, as many as 49 compounds comprising different chemicals, commercial fungicides, natural antimicrobial products and surfactants were tested *in vitro* for assessing their fungicidal activity against *A. invadans* (Campbell *et al.*, 2001). Among the tested compounds, malachite green (considered unsuitable for use in aquaculture due to potential health hazard) was found to be most effective against the oomycete. Other compounds which showed some promising anti-*A. invadans* activity include formalin, hydrogen peroxide and chitosan (Lilley and Inglis, 1997; Campbell *et al.*, 2001; Mari *et al.*, 2014). Attempts at using dietary immunostimulants against *A. invadans* have shown variable outcomes to control EUS in tank trials (Miles *et al.*, 2001; Mari *et al.*, 2014).

A protective vaccine against EUS is yet to be developed, and only limited vaccination trials have been conducted against the oomycete. Snakehead and rainbow trout (*Oncorhynchus mykiss*) were immunized with different oomycete preparations (microwaved spores, oomycete extract and extracellular products) and specific antibody responses were observed against *A. invadans* (Thompson *et al.*, 1997, 1999). In another trial, antigenic preparations from *A. invadans* were used to vaccinate catla (*C. catla*) (Saikia and Kamilya, 2012). The vaccinated fish showed significant antibody response, but the protection against *A. invadans* challenge was not significant.

The published data show very limited progress in developing effective control strategies for EUS. With the advent of molecular techniques, methods such as gene silencing or genome editing should be extensively researched to develop effective control strategies. Recently, the clustered regularly interspaced short palindromic repeats (CRISPR)/CRISPR-associated protein 9 (Cas9) system was successfully tested in *A. invadans* to edit the serine protease gene, responsible for its pathogenicity (Majeed *et al.*, 2018).

15.8 Conclusions and Suggestions for Future Studies

During the last five decades, EUS has spread in several geographical locations of the world excluding Europe and South American territories. There has been a continuous increase in the number of susceptible species being reported. Given the epizootic nature of the disease, it is expected that the number is likely to increase further. This can pose a potential risk to the countries which are currently free from EUS. Development of a suitable import risk framework and putting it in place are necessary to avoid the risk of introduction of the disease. Execution of risk mitigation approaches and implementation of surveillance and biosecurity measures, based on the perceived risks, will facilitate a country to tackle the introduction, spread or emergence of EUS. The likelihood of emergence and spread of the disease may further be aggravated in a climate change scenario. Climate change and associated extreme weather events are predicted to influence the geographic distribution, prevalence, abundance and virulence of *A. invadans*. EUS-free temperate regions are likely to become more vulnerable to the emergence of EUS. The problems associated with EUS in the tropics may also be exacerbated due to environment deterioration and inadequacy in implementing stringent regulatory and mitigation measures. Therefore, a comprehensive understanding of the likely impact of climate change on EUS emergence is needed. Future policy and research interventions should focus on the likely interactions between the altered environment, host and the pathogen; and identification of potential risks and development of climate-resilient mitigation strategies. In addition, predictive mathematical modelling can be a useful tool to delineate the impact of complex biotic (host and pathogen) and abiotic (environment) interactions on disease emergence in a climate change perspective.

Even though significant progress has been made in understanding the biology of *A. invadans* and in developing effective diagnostics, effective control strategies are still limited. Future research efforts should be focused on the development of novel control strategies using biotechnological and molecular techniques such as gene silencing (RNAi) or genome editing (CRISPR/Cas9), the production of temperature-tolerant or EUS-resistant fish through selective breeding or genetic engineering programmes and the development of novel vaccines or chemotherapeutic agents.

References

Andrew, T.G., Huchzermeyer, K.D.A., Mbeha, B.C. and Nengu, S.M. (2008) Epizootic ulcerative syndrome affecting fish in the Zambezi river system in Southern Africa. *Veterinary Record* 163, 629–631.

Balasuriya, L.K.S.W. (1994) epizootic ulcerative syndrome in fish in Sri Lanka, country status report. In: Roberts, R.J., Campbell, B. and MacRae, I.H. (eds) *Proceedings of the ODA Regional Seminar on Epizootic Ulcerative Syndrome*. Aquatic Animal Health Research Institute, Bangkok, pp. 39–47.

Baldock, F.C., Blazer, V., Callinan, R., Hatai, K., Karunasagar, I. *et al.* (2005) Outcomes of a short expert consultation on epizootic ulcerative syndrome (EUS): re-examination of casual factors, case definition and nomenclature. In: Walker, P. and Bondad-Reantaso, M.G. (eds) *Diseases in Asian Aquaculture V*. Fish Health Section, Asian Fisheries Society, Manila, pp. 555–585.

Blazer, V.S., Lilley, J.H., Schill, W.B., Kiryu, Y., Densmore, C.L. *et al.* (2002) *Aphanomyces invadans* in Atlantic menhaden along the east coast of the United States. *Journal of Aquatic Animal Health* 14, 1–10.

Bondad-Reantaso, M.G., Lumanlan, S.C., Natividad, J.M. and Phillips, M.J. (1992) Environmental monitoring of the epizootic ulcerative syndrome (EUS) in fish from Munoz, Nueva Ecija in the Philippines. In: Shariff, M., Subasinghe, R.P. and Arthur, J.R. (eds) *Diseases in Asian Aquaculture 1*. Fish Health Section, Asian Fisheries Society, Manila, pp. 475–490.

Bondad-Reantaso, M.G., Garrido-Gamarro, E. and McGladdery, S.E. (2018) Climate change-driven hazards on food safety and aquatic animal health. In: Barange, M., Bahri, T., Beveridge, M.C.M., Cochrane, K.L., Funge-Smith, S. *et al.* (eds) Impacts of climate change on fisheries and aquaculture: synthesis of current knowledge, adaptation and mitigation options. *FAO Fisheries and Aquaculture Technical Paper No. 627*. Food and Agriculture Organization of the United Nations, Rome, pp. 517–533.

Boys, C.A., Rowland, S.J., Gabor, M., Gabor, L., Marsh, I.B. *et al.* (2012) Emergence of epizootic ulcerative syndrome in native fish of the Murray–Darling river system, Australia: hosts, distribution and possible vectors. *PLoS ONE* 7, e35568.

Brooker, R.W., Travis, J.M.J., Clark, E.J. and Dytham, C. (2007) Modelling species' range shifts in a changing climate: the impacts of biotic interactions, dispersal distance and the rate of climate change. *Journal of Theoretical Biology* 245, 59–65.

Callinan, R.B., Fraser, G.C. and Virgona, J.L. (1989) Pathology of red spot disease in sea mullet, *Mugil cephalus* L., from eastern Australia. *Journal of Fish Diseases* 12, 467–479.

Callinan, R.B., Sammut, J. and Fraser, G.C. (1996) Epizootic ulcerative syndrome (red spot disease) in estuarine fish – confirmation that exposure to acid sulfate soil runoff and invasive aquatic fungus, *Aphanomyces* sp., are causative factors. In: Smith, R.J. and Smith, H.J. (eds) *Proceedings of the Second National Conference on Acid Sulfate Soils, Coffs Harbour, Australia, 5–6 September 1996.* Roberts J Smith and Associates, Alstonville, Australia, pp. 146–151.

Campbell, R.E., Lilley, J.H., Taukhid, Panyawachira, V. and Kanchanakhan, S. (2001) In vitro screening of novel treatments for *Aphanomyces invadans. Aquaculture Research* 32, 223–233.

Chinabut, S. (1998) Epizootic ulcerative syndrome: information up to 1997. *Fish Pathology* 33, 321–326.

Chinabut, S. and Roberts, R.J. (1999) *Pathology and Histopathology of Epizootic Ulcerative Syndrome (EUS).* Aquatic Animal Health Research Institute, Bangkok.

Chinabut, S., Roberts, R.J., Willoughby, L.G. and Pearson, M.D. (1995) Histopathology of snakehead, *Channa striatus* (Bloch), experimentally infected with the specific *Aphanomyces* fungus associated with epizootic ulcerative syndrome (EUS) at different temperatures. *Journal of Fish Diseases* 18, 41–47.

Choongo, K., Hang'ombe, B., Samui, K.L., Syachaba, M., Phiri, H. *et al.* (2009) Environmental and climatic factors associated with epizootic ulcerative syndrome (EUS) in fish from the Zambezi floodplains, Zambia. *Bulletin of Environmental Contamination and Toxicology* 83, 474–478.

Cruz-Lacierda, E.R. and Shariff, M. (1995) Experimental transmission of epizootic ulcerative syndrome (EUS) in snakehead, *Ophicephalus striatus*. In: Shariff, M., Arthur, J.R. and Subasinghe, R.P. (eds) *Diseases in Asian Aquaculture II*. Fish Health Section, Asian Fisheries Society, Manila, pp. 327–336.

EFSA (European Food Safety Authority) Panel on Animal Health and Welfare (2011) Scientific opinion on epizootic ulcerative syndrome. *EFSA Journal* 9, 2387.

Egusa, S. and Masuda, N. (1971) A new fungal disease of *Plecoglossus altivelis*. *Fish Pathology* 6, 41–46.

Fraser, G.C., Callinan, R.B. and Calder, L.M. (1992) Aphanomyces species associated with red spot disease: an ulcerative disease of estuarine fish from eastern Australia. *Journal of Fish Diseases* 15, 173–181.

Hanjavanit, C., Suda, H. and Hatai, K. (1997) Mycotic granulomatosis found in two species of ornamental fishes imported from Singapore. *Mycoscience* 38, 433–436.

Harvell, C.D., Mitchell, C.E., Ward, J.R., Altizer, S., Dobson, A.P. *et al.* (2002) Climate warming and disease risks for terrestrial and marine biota. *Science* 296, 2158–2162.

Hatai, K. (1980) *Studies on pathogenic agents of saprolegniasis in freshwater fishes. Special Report No. 8.* Nagasaki Prefectural Institute of Fisheries, Nagasaki, Japan. (in Japanese)

Hatai, K. (1994) Mycotic granulomatosis in ayu (*Plecoglossus altivelis*) due to *Aphanomyces piscicida*. In: Roberts, R.J., Campbell, B. and MacRae, I.H. (eds) *Proceedings of the ODA Regional Seminar on Epizootic Ulcerative Syndrome*. Aquatic Animal Health Research Institute, Bangkok, pp. 101–108.

Hatai, K., Nakamura, K., Rha, S.A., Yuasa, K. and Wada, S. (1994) Aphanomyces infection in dwarf gourami (*Colisa lalia*). *Fish Pathology* 29, 95–99.

Hawke, J.P., Grooters, A.M. and Camus, A.C. (2003) Ulcerative mycosis caused by *Aphanomyces invadans* in channel catfish, black bullhead, and bluegill from southeastern Louisiana. *Journal of Aquatic Animal Health* 15, 120–127.

Huchzermeyer, C.F., Huchzermeyer, K.D.A., Christison, K.W., Macey, B.M., Colly, P.A. *et al.* (2018) First record of epizootic ulcerative syndrome from the Upper Congo catchment: an outbreak in the Bangweulu swamps, Zambia. *Journal of Fish Diseases* 41, 87–94.

Iberahim, N.A., Trusch, F. and van West, P. (2018) *Aphanomyces invadans*, the causal agent of epizootic ulcerative syndrome, is a global threat to wild and farmed fish. *Fungal Biology Reviews* 32, 118–130.

Kamilya, D. and Baruah, A. (2014) Epizootic ulcerative syndrome (EUS) in fish: history and current status of understanding. *Reviews in Fish Biology and Fisheries* 24, 369–380.

Khan, M.H., Marshall, L., Thompson, K.D. and Lilley, J.H. (1998) Susceptibility of five fish species (Nile tilapia, rosy barb, rainbow trout, stickleback and roach) to intramuscular injection with the oomycete fish pathogen, *Aphanomyces invadans*. *Bulletin of the European Association of Fish Pathologists* 18, 192–197.

Kriger, K.M. and Hero, J.M. (2007) Large-scale seasonal variation in the prevalence and severity of chytridiomycosis. *Journal of Zoology* 271, 352–359.

Lilley, J.H. and Inglis, V. (1997) Comparative effects of various antibiotics, fungicides and disinfectants on *Aphanomyces invaderis* and other saprolegniaceous fungi. *Aquaculture Research* 28, 461–469.

Lilley, J.H. and Roberts, R.J. (1997) Pathogenicity and culture studies comparing the *Aphanomyces* involved in epizootic ulcerative syndrome (EUS) with other similar fungi. *Journal of Fish Diseases* 20, 135–144.

Lilley, J.H., Thompson, K.D. and Adams, A. (1997) Characterization of *Aphanomyces invadans* by electrophoretic and Western blot analysis. *Diseases of Aquatic Organisms* 30, 187–197.

Lilley, J.H., Callinan, R.B., Chinabut, S., Kanchanakhan, S., MacRae, I.H. *et al.* (1998) *Epizootic Ulcerative Syndrome (EUS) Technical Handbook*. Aquatic Animal Health Research Institute, Bangkok.

Lilley, J.H., Beakes, G.W. and Hetherington, C.S. (2001) Characterization of *Aphanomyces invadans* isolates using pyrolysis mass spectrometry (PyMS). *Mycoses* 44, 383–389.

Lilley, J.H., Hart, D., Panyawachira, V., Kanchanakhan, S., Chinabut, S. *et al.* (2003) Molecular characterization of the fish pathogenic fungus *Aphanomyces invadans*. *Journal of Fish Diseases* 26, 263–275.

Llobrera, A.T. and Gacutan, R.Q. (1987) *Aeromonas hydrophila* associated with ulcerative disease epizootic in Laguna de Bay, Philippines. *Aquaculture* 67, 273–278.

Lumanlan-Mayo, S.C., Callinan, R.B., Paclibare, J.O., Catap, E.S. and Fraser, G.C. (1997) Epizootic ulcerative syndrome (EUS) in rice–fish culture systems: an overview of field experiments 1993–1995. In: Flegel, T.W. and MacRae, I.H. (eds) *Diseases in Asian Aquaculture III*. Fish Health Section, Asian Fisheries Society, Manila, pp. 129–138.

Majeed, M., Soliman, H., Kumar, G., El-Matbouli, M. and Saleh, M. (2018) Editing the genome of *Aphanomyces invadans* using CRISPR/Cas9. *Parasites and Vectors* 11, 554.

Makkonen, J., Vesterbacka, A., Martin, F., Jussila, J., Dieguez-Uribeondo, J. *et al.* (2016) Mitochondrial genomes and comparative genomics of *Aphanomyces astaci* and *Aphanomyces invadans*. *Scientific Reports* 6, 36089.

Marcogliese, D.J. (2008) The impact of climate change on the parasites and infectious diseases of aquatic animals. *Revue Scientifique et Technique (International Office of Epizootics)* 27, 467–484.

Marcos-Lopez, M., Gale, P., Oidtmann, B.C. and Peeler, E.J. (2010) Assessing the impact of climate change on disease emergence in freshwater fish in the United Kingdom. *Transboundary and Emerging Diseases* 57, 293–304.

Mari, L.S.S., Jagruthi, C., Anbazahan, S.M., Yogeshwari, G., Thirumurugan, R. *et al.* (2014) Protective effect of chitin and chitosan enriched diets on immunity and disease resistance in *Cirrhina mrigala* against *Aphanomyces invadans*. *Fish and Shellfish Immunology* 39, 378–385.

Miles, D.J.C., Polchana, J., Lilley, J.H., Kanchanakhan, S., Thompson, K.D. *et al.* (2001) Immunostimulation of striped snakehead *Channa striata* against epizootic ulcerative syndrome. *Aquaculture* 195, 1–15.

Miles, D.J.C., Thompson, K.D., Lilley, J.H. and Adams, A. (2003) Immunofluorescence of the epizootic ulcerative syndrome pathogen, *Aphanomyces invadans*, using a monoclonal antibody. *Diseases of Aquatic Organisms* 55, 77–84.

Naik, M.G., Rajesh, K.M., Sahoo, A.K. and Shankar, K.M. (2008) Monoclonal antibody-based detection of *Aphanomyces invadans* for surveillance and prediction of epizootic ulcerative syndrome (EUS) outbreak in fish. In: Bondad-Reantaso, M.G., Mohan, C.V., Crumlish, M. and Subasinghe, R.P. (eds) *Diseases in Asian Aquaculture VI*. Fish Health Section, Asian Fisheries Society, Manila, pp. 157–168.

Oidtmann, B. (2012) Review of biological factors relevant to import risk assessments for epizootic ulcerative syndrome (*Aphanomyces invadans*). *Transboundary and Emerging Diseases* 59, 26–39.

Oidtmann, B., Steinbauer, P., Geiger, S. and Hoffmann, R.W. (2008) Experimental infection and detection of *Aphanomyces invadans* in European catfish, rainbow trout and European eel. *Diseases of Aquatic Organisms* 82, 195–207.

Phadee, P., Kurata, O., Hatai, K., Hirono, I. and Aoki, T. (2004) Detection and identification of fish-pathogenic *Aphanomyces piscicida* using polymerase chain reaction (PCR) with species-specific primers. *Journal of Aquatic Animal Health* 16, 220–230.

Pradhan, P.K., Mohan, C.V., Shankar, K.M. and Kumar, B.M. (2008a) Susceptibility of fingerlings of Indian major carps to *Aphanomyces invadans*. *Asian Fisheries Science* 21, 369–375.

Pradhan, P.K., Mohan, C.V., Shankar, K.M. and Kumar, B.M. (2008b) Inflammatory response in Indian major carp, *Catla catla* and barb, *Puntius cauveriensis*, to *Aphanomyces invadans* and Freund's complete adjuvant. *Asian Fisheries Science* 21, 385–394.

Pradhan, P.K., Rathore, G., Sood, N., Swaminathan, T.R., Yadav, M.K. *et al.* (2014) Emergence of epizootic ulcerative syndrome: large-scale mortalities of cultured and wild fish species in Uttar Pradesh, India. *Current Science* 106, 1711–1718.

Roberts, R.J., Wootten, R., MacRae, I., Millar, S. and Struthers, W. (1989) *Ulcerative disease survey,*

Bangladesh. Final Report to the Government of Bangladesh and the Overseas Development Administration. Institute of Aquaculture, Stirling University, Stirling, UK.

Roberts, R.J., Willoughby, L.G. and Chinabut, S. (1993) Mycotic aspects of epizootic ulcerative syndrome (EUS) of Asian fishes. *Journal of Fish Diseases* 16, 169–183.

Roberts, R.J., Frerichs, G.N., Tonguthai, K. and Chinabut, S. (1994) Epizootic ulcerative syndrome of farmed and wild fishes. In: Muir, J.F. and Roberts, R.J. (eds) *Recent Advances in Aquaculture*. Blackwell Science, Oxford, pp. 207–239.

Saikia, D. and Kamilya, D. (2012) Immune responses and protection in catla, *Catla catla* vaccinated against epizootic ulcerative syndrome. *Fish and Shellfish Immunology* 32, 353–359.

Sosa, E.R., Landsberg, J.H., Stephenson, C.M., Forstchen, A.B., Vandersea, M.W. *et al*. (2007) *Aphanomyces invadans* and ulcerative mycosis in estuarine and freshwater fish in Florida. *Journal of Aquatic Animal Health* 19, 14–26.

Thompson, K.D., Lilley, J.H., Chinabut, S. and Adams, A. (1997) The antibody response of snakehead, *Channa striata* Bloch, to *Aphanomyces invadans*. *Fish and Shellfish Immunology* 7, 349–353.

Thompson, K.D., Lilley, J.H., Chen, S.C., Adams, A. and Richards, R.H. (1999) The immune response of rainbow trout (*Oncorhynchus mykiss*) against *Aphanomyces invadans*. *Fish and Shellfish Immunology* 9, 195–210.

Vandersea, M.W., Litaker, R., Yonnish, B., Sosa, E., Landsberg, J.H. *et al*. (2006) Molecular assays for detecting *Aphanomyces invadans* in ulcerative mycotic fish lesions. *Applied and Environmental Microbiology* 72, 1551–1557.

Viswanath, T.S., Mohan, C.V. and Shankar, K.M. (1997) Clinical and histopathological characterization of different types of lesions associated with epizootic ulcerative syndrome (EUS). *Journal of Aquaculture in the Tropics* 12, 35–42.

Vishwanath, T.S., Mohan, C.V. and Shankar, K.M. (1997) Mycotic granulomatosis and seasonality are the consistent features of epizootic ulcerative syndrome of fresh and brackishwater fishes of Karnataka, *India. Asian Fisheries Science* 10, 155–160.

Vishwanath, T.S., Mohan, C.V. and Shankar, K.M. (1998) Epizootic ulcerative syndrome (EUS), associated with a fungal pathogen, in Indian fishes: histopathology – 'a cause for invasiveness'. *Aquaculture* 165, 1–9.

Wada, S., Rha, S., Kondoh, T., Suda, H., Hatai, K. *et al*. (1996) Histopathological comparison between ayu and carp artificially infected with *Aphanomyces piscicida*. *Fish Pathology* 31, 71–80.

Willoughby, L.G., Roberts, R.J. and Chinabut, S. (1995) *Aphanomyces invaderis* sp. nov., the fungal pathogen of freshwater tropical fishes affected by epizootic ulcerative syndrome (EUS). *Journal of Fish Diseases* 18, 273–275.

16 Amoebiosis (*Neoparamoeba perurans*)

Jadwiga Sokołowska[1] and Barbara F. Nowak[2]*

[1]*University of Stirling, Stirling, UK;* [2]*University of Tasmania, Launceston, Tasmania, Australia*

16.1 Introduction

16.1.1 Description of the disease

Amoebioses include a range of diseases caused by amoebae. Amoebae belong to a heterogeneous group of protozoans common in water and soil. While amoebae have been isolated from various organs of fish with no clinical signs, they can be pathogenic and cause amoebioses under particular conditions (Lom and Dyková, 1992). Those amoebae are either endocommensals or free-living before they become parasitic. Amoebic infections can be difficult to diagnose and, as a result, amoebic diseases have been most likely under-reported. However, some amoebioses can cause significant losses in the aquaculture industry (Nowak *et al.*, 2014).

Amoebic gill disease (AGD) is the most important amoebic disease in fish and is the focus of this chapter. AGD is currently one of the serious parasitic threats to the marine salmon industry worldwide (Shinn *et al.*, 2015). It is caused by an opportunistic pathogen, the marine amoeba *Neoparamoeba perurans* (see Young *et al.*, 2007; Crosbie *et al.*, 2012a). The main risk factors for AGD outbreaks are elevated water temperature and high salinity (Clark and Nowak, 1999). The disease is manifested as white, mucoid gill lesions (Zilberg and Munday, 2000; Taylor *et al.*, 2009). Microscopically, small areas of hyperplastic epithelial tissue expand with progression of the disease, increasing the coverage of the gill surface (Zilberg and Munday, 2000; Adams and Nowak, 2003). Histopathological observations revealed a series of consecutive signs of AGD development (Adams and Nowak, 2001). After initial attachment of the amoebae, focal hyperplasia of the epithelium results in thickening of the lamellae, followed by lamellar fusion and further extension of epithelial hyperplasia and oedema (Fig. 16.1; Adams and Nowak, 2003). Interlamellar cysts could be observed within AGD lesions, often containing necrotic amoebae (Fig. 16.1; Adams and Nowak, 2001).

AGD affects fish cultured in the marine environment, causing financial losses (e.g. in Scotland in 2011, 10–20% mortality due to AGD caused losses of up to $US81 million) to the aquaculture industry (Kube *et al.*, 2012; Shinn *et al.*, 2015). AGD management on salmon farms requires regular gill scoring and fresh water or hydrogen peroxide bath to minimize mortalities. As neither of those treatments protects from reinfection, baths are repeated several times in one season, contributing to increased production cost (Clark and Nowak, 1999; Adams *et al.*, 2012). Although the treatments reduce mortalities significantly, the requirement for specialized infrastructure and an extensive workforce to monitor AGD and perform the treatments greatly increases production costs (Kube *et al.*, 2012; APEM, 2014).

AGD can contribute to complex gill disease (CGD), which is a multifactorial gill condition in Atlantic salmon, *Salmo salar*, farmed in Norway, Scotland and Ireland. CGD involves a variety of parasitic (e.g. *Desmozoon lepeophtherii*), bacterial (e.g. *Candidatus* Branchiomonas cysticola) and viral (e.g. salmon gill poxvirus (SGPV)) pathogens (Herrero *et al.*, 2018). Recent molecular analysis of gill samples collected during the first recorded AGD outbreak in Norway confirmed the presence of SGPV in all of the samples, suggesting that SGPV infection preceded AGD and, potentially, facilitated the secondary infection by compromising the immunity of the host (Gjessing *et al.*, 2017). However, this may depend on geographical area and environmental conditions. For example, a study in Ireland showed that SGVP detection was

*b.nowak@utas.edu.au

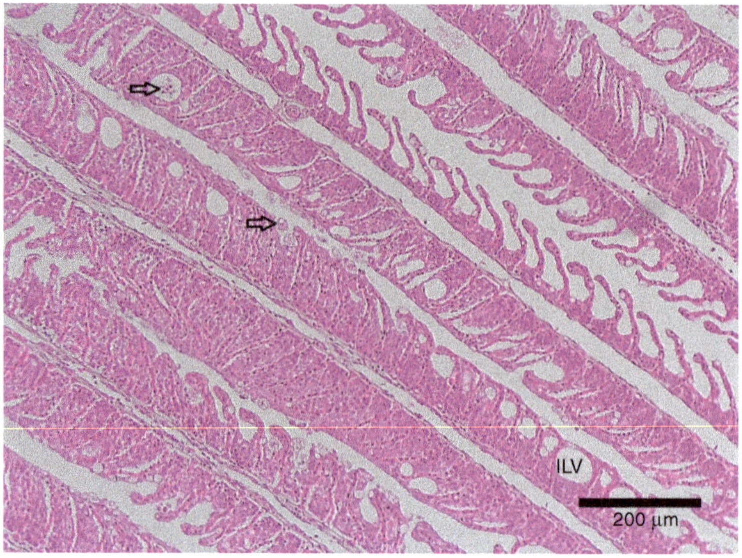

Fig. 16.1. Amoebic gill disease lesions in Atlantic salmon 10 days after experimental infection with 1200 *Neoparamoeba perurans*/l. The amoebae were harvested from the gills of AGD-affected salmon; arrows, examples of amoebae; ILV, interlamellar vesicles. Sections were cut at 4 µm and stained with haematoxylin and eosin. Samples from the same fish were positive for *N. perurans* based on species-specific qPCR.

only sporadic and not preceding AGD (Downes *et al.*, 2018). Atlantic salmon gills were colonized by the pathogens 12–16 weeks after transfer to sea and while *D. lepeophtherii* and *Candidatus* B. cysticola were most common, *N. perurans* was the main pathogen consistently associated with higher gill pathology score (Downes *et al.*, 2018). Furthermore, AGD can be induced in Atlantic salmon via experimental exposure to *N. perurans* (see Crosbie *et al.*, 2012a).

16.1.2 Species and geographic distribution of AGD

Since the first AGD outbreak in Tasmania, Australia (Munday, 1986) and the USA (Kent *et al.*, 1988), the disease has been reported from many hosts and numerous countries in Europe, North and South America, Asia and Africa (Table 16.1). In several farmed species and geographical locations, AGD is a recurring problem (Rodger, 2014; Oldham *et al.*, 2016).

Although *N. perurans* is not host-specific, most of AGD outbreaks affect salmonids farmed in sea cages (Table 16.1). There have been no reports of AGD outbreaks in wild fish although the parasite was detected occasionally in histological sections or its DNA was found in gills of wild fish (Adams *et al.*, 2008; Stagg *et al.*, 2015; Hellebø *et al.*, 2017; Steigen *et al.*, 2018). Therefore, farmed fish are at a higher risk of contracting the infection. Several studies proved that a high density of fish supported faster spread of AGD (Douglas-Helders *et al.*, 2004; Crosbie *et al.*, 2010a). This might explain why outbreaks occur only in farmed fish rather than in wild fish populations. Moreover, Nowak (2007) pointed out similarities between marine cage farming and serial passage experiments with low genetic diversity of the host, frequent introduction of naïve fish and increased host growth rates. All of these promote an increase of virulence in evolving parasites. This is of particular relevance, considering that *N. perurans* is a free-living amoeba which can colonize fish gills and become parasitic.

16.1.3 Causative agent of AGD

Despite many years of research on AGD, the causative agent of AGD was identified just over a decade ago (Young *et al.*, 2007). Until then, two different

Table 16.1. Hosts, geographic distribution and temperature at first reported outbreak of amoebic gill disease.

Species	Country	Temperature at first outbreak (°C)	Impact (% mortality)	Reference
Atlantic salmon (*Salmo salar*)	Australia, Tasmania	12–20	Up to 50	Munday (1986); Young *et al.* (2007)
	Canada, British Columbia	n/a	Minor	ICES (2015)
	Chile	9–12	Up to 53.8	Bustos *et al.* (2011)
	Faroe Islands	n/a	n/a	Oldham *et al.* (2016)
	France	n/a	Minor	Findlay *et al.* (1995); Rodger and McArdle (1996)
	Ireland	12–17	Up to 10	Rodger and McArdle (1996); Young *et al.* (2008a)
	Norway	7–14	Up to 82	Steinum *et al.* (2008)
	Scotland	7.5–13.5	Up to 70	Young *et al.* (2008a)
	South Africa	15	~5 per annum	Mouton *et al.* (2014)
	Spain	n/a	Major	Rodger and McArdle (1996)
	USA, Washington	9–12	Up to 21	Douglas-Helders *et al.* (2001); Young *et al.* (2008a)
Coho salmon (*Oncorhynchus kisutch*)	USA, Washington	9–12	~25	Kent *et al.* (1988)
	Korea	15–17	~70	Kim *et al.* (2016)
Rainbow trout (*Oncorhynchus mykiss*)	Australia, Tasmania	12–20	Up to 50	Munday *et al.* (1990)
Chinook salmon (*Oncorhynchus tshawytscha*)	New Zealand	n/a	Minor	Young *et al.* (2008a)
Brown trout (*Salmo trutta*)	France	n/a	n/a	Munday *et al.* (2001)
Turbot (*Scophthalmus maximus*)	South Africa	14–16.7	~5 per annum	Mouton *et al.* (2014)
	Spain	n/a	Up to 20	Dyková *et al.* (1995, 1998)
Sea bass (*Dicentrarchus labrax*)	Mediterranean	n/a	n/a	Dyková *et al.* (2000)
Sharpsnout sea bream (*Diplodus puntazzo*)	Europe	n/a	n/a	Dyková and Novoa (2001)
Ayu (*Plecoglossus altivelis*)	Japan	14	49.4	Crosbie *et al.* (2010b)
Olive flounder (*Paralichthys olivaceus*)	Korea	n/a	n/a	Kim *et al.* (2005)
Black sea bream (*Acanthopagrus schlegelli*)	Korea	13–14	60	Kim *et al.* (2017)
Rock bream (*Oplegnathus fasciatus*)	Korea	13–14	10	Kim *et al.* (2017)
Grey mullet (*Mugil cephalus*)	Korea	11–17	6.7	Kim *et al.* (2017)
Corkwing wrasse (*Symphodus melops*)	Norway	n/a	n/a	VKM (2014)
Halibut (*Hippoglossus hippoglossus*)	Scotland	12	<0.01 per week	Rodger (2019)

n/a, not available.

species – *Neoparamoeba pemaquidensis* and *Neoparamoeba branchiphila* – were considered as a possible cause of the infection. Although the lack of knowledge of the causative agent caused some limitations, many of the early studies gave a valuable insight into the pathology of AGD and the immune response of the host during the course of the disease (Zilberg and Munday, 2001; Adams and Nowak, 2004; Adams *et al.*, 2004; Gross *et al.*, 2005).

The analysis of 18S rRNA and 28S rRNA gene sequences, obtained from non-cultured amoebae isolated from the gills of an AGD-affected fish, facilitated the identification of a new species from the genus *Neoparamoeba* which was named *N. perurans* (Young *et al.*, 2007). It was soon discovered that out of the three species investigated in connection with the disease, only *N. perurans* was detected in gill samples originating from four host species affected with AGD in six different countries (Young *et al.*, 2008a). The relationship between AGD and *N. perurans* was ultimately confirmed when an experimental infection using *in vitro* cultured *N. perurans* was successfully established and Koch's postulates were fulfilled (Crosbie *et al.*, 2012a).

The main morphological characteristic of all paramoebae is the presence of one or more parasomes inside the cell (Fig. 16.2; Page, 1987). The parasome is an endosymbiont – a *Perkinsela*-like kinetoplastid (Dyková *et al.*, 2003). Phylogenetic studies of *N. perurans* and other *Neoparamoeba* spp. together with their endosymbionts suggested co-evolution of amoeba and endosymbiont (Young *et al.*, 2014; Sibbald *et al.*, 2017). This was supported by metabolic and cellular interdependence between the amoeba and the parasome revealed by analysis of the genome sequences of *N. pemaquidensis* and its endosymbiont (Tanifuji *et al.*, 2017). *Neoparamoeba* do not have microscales covering the cell membrane which are present in species from the genus *Paramoeba* (Page, 1987). It has been suggested that *Neoparamoeba* and *Paramoeba* are the same genus and thus *N. perurans* is a synonym of *Paramoeba perurans* (see Feehan *et al.*, 2013); however, there is insufficient evidence due to the lack of molecular data from species of paramoebae without scales (Kudryavtsev *et al.* 2011; Young *et al.*, 2014).

Under *in vitro* conditions *N. perurans* can assume several morphologies, from rounded cells to trophozoites presenting a varying number of pseudopodia of different lengths. Certain ultrastructural features can be distinguished within the parasite: a nucleus with closely associated parasome and plasma containing endocytotoic vesicles (Wiik-Nielsen *et al.*, 2016). The mechanism of primary attachment of *N. perurans* to a gill remains unclear.

Other species of amoebae can be isolated or detected from gills of Atlantic salmon (English *et al.*, 2019), but so far only *N. perurans* can induce AGD in laboratory experiments (Morrison *et al.*, 2005; Vincent *et al.*, 2007; Crosbie *et al.*, 2012a; Nowak and Archibald, 2018). However, it is possible that the

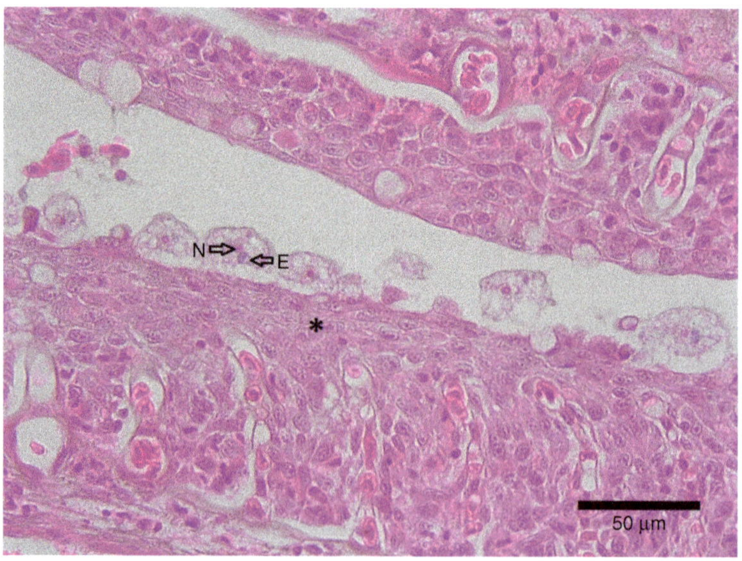

Fig. 16.2. Amoebic gill disease lesion in Atlantic salmon 10 days after experimental infection with 1200 *Neoparamoeba perurans*/l. The amoebae were harvested from the gills of AGD-affected salmon showing amoebae; N, nucleus; E, endosymbiont; asterisk, example of epithelial hyperplasia. Sections were cut at 4 μm and stained with haematoxylin and eosin. Samples from the same fish were positive for *N. perurans* based on species-specific qPCR.

presence of other amoebae may contribute to dysbiosis and gill diseases (Nowak and Archibald, 2018; English *et al.*, 2019).

16.1.4 Risk factors

A number of risk factors have been associated with AGD outbreaks (Table 16.2). Two critical environmental variables are water temperature and salinity. Although a significant correlation between AGD outbreaks and these factors was established (Clark and Nowak, 1999), further analysis of reports from different geographic locations revealed that water temperatures recorded during outbreaks varied considerably. It was therefore suggested that not a particular temperature threshold but rather an unusually high water temperature affects fish, making them more susceptible to the infection due to thermal immunosuppression, rather than predispose *N. perurans* to become a pathogen (Oldham *et al.*, 2016). Atlantic salmon exposed to increased water temperature (19°C) for 8 weeks showed reduced complement activity when compared with control salmon kept at 15°C (Jokinen *et al.*, 2011). At 21 days after experimental infection with *N. perurans*, Atlantic salmon exposed to 15°C showed a higher median gill score (score = 3) than fish kept at 10°C (score = 2), which was reflected by more severe histological lesions and higher amoebae load at the higher temperature (Benedicenti *et al.*, 2019b).

AGD outbreaks often occurred when water salinity increased (Munday *et al.*, 1990). This was confirmed by several reports of unusually low rainfalls prior to the first disease outbreaks in different geographical locations, e.g. Ireland and Chile (Oldham *et al.*, 2016). Although Clark and Nowak (1999) recorded salinity as low as 7.2 ppt during an AGD outbreak, in that case the infection was likely established prior to the drop in salinity.

Similar to other infectious diseases, husbandry practices play an important role in mitigating the impact of AGD. There was a negative correlation between the number of net changes and prevalence of the disease (Clark and Nowak, 1999). While biofouling was suggested to be a risk factor for AGD and a reservoir of the pathogen (Tan *et al.*, 2002), a laboratory exposure to *Ectopleura larynx* hydroids did not affect infection rates or AGD progression rates in Atlantic salmon smolt (Bloecher *et al.*, 2018). Additionally, stocking density can have an impact on the survival of AGD-affected fish. Tanks with low biomass (1.7 kg/m^3) had a significantly higher survival rate than those with high biomass (5.0 kg/m^3) (Crosbie *et al.*, 2010a).

Table 16.2. Risk factors related to amoebic gill disease.

Factor	Risk	Reference
Environmental		
Increased water temperature	Host thermal tolerance compromised, disease outbreak	Munday (1986); Findlay *et al.* (1995); Rodger and McArdle (1996); Steinum *et al.* (2008); Mouton *et al.* (2014); Rodger (2014); Oldham *et al.* (2016)
Increased salinity	Disease outbreak	Munday *et al.* (1990); Clark and Nowak (1999); Oldham *et al.* (2016)
Low rainfall	Increased salinity, disease outbreak	Clark and Nowak (1999)
Low level of dissolved oxygen	Disease outbreak	Clark and Nowak (1999)
Management and husbandry practices		
Biofouling	Pathogen reservoir, reduction in dissolved oxygen level	Clark and Nowak (1999); Rodger (2014)
Infrequent net changes	Reduced water flow, increased biofouling	Clark and Nowak (1999)
Overcrowded cages	Rapid spread of the disease	Douglas-Helders *et al.* (2004); Nowak (2007); Crosbie *et al.* (2010a)
Cleaner fish	Cross-infection	Karlsbakk *et al.* (2013); Haugland *et al.* (2017)
Other gill conditions	Compromised immunity of the host	Gjessing *et al.* (2017)

Cleaner fish (including lumpfish and ballan wrasse) used by the salmon industry to control sea lice are also potential reservoirs of *N. perurans*. In 2013, the presence of the parasite was reported in cultured ballan wrasse, *Labrus bergylta* (see Karlsbakk *et al.*, 2013). More recently, AGD was experimentally induced in lumpfish, *Cyclopterus lumpus* (Haugland *et al.*, 2017). Although lumpfish showed lower susceptibility to AGD than salmon, Atlantic salmon were infected by experimentally infected lumpfish in a cohabitation challenge. This means that cleaner fish can be a potential source of cross-infection and act as a reservoir of the pathogen.

16.2 Diagnosis of the Pathogen/Disease

16.2.1 Morphological features of pathogen and clinical signs of AGD

The most characteristic feature of all paramoebae is the presence of parasomes, which are visible in wet preparations, stained smears (Giemsa) and histological gill sections under light microscopy and ultrathin gill sections under transmission electron microscopy (Munday, 1986; Kent *et al.*, 1988; Zilberg *et al.*, 1999; Wiik-Nielsen *et al.*, 2016). As in other amoebae, *N. perurans* does not have well-defined morphology; therefore, use of molecular methods is required for species identification. Nevertheless, morphological methods are valuable tools for diagnosing disease and investigating host–pathogen interactions and pathogen biology (Adams and Nowak, 2003; Wiik-Nielsen, *et al.* 2016).

Clinically, AGD is recognized by the presence of gross gill lesions and gill assessment is the main tool used by salmon farmers for disease management (Taylor *et al.*, 2009; Rodger, 2014). Behavioural changes observed in diseased fish include lethargy and respiratory distress exhibited by increased opercular movement (Munday, 1986). Despite that, there is no evidence that AGD causes respiratory failure of the host (Leef *et al.*, 2005), however in those experiments the lesions were not severe (Nowak, 2012). It was suggested that affected fish compensated for the reduced gill surface with increased blood flow and gill perfusion (Powell, 2006). Importantly, none of those signs are specific to AGD (Adams *et al.*, 2004; Gjessing *et al.*, 2015). Consequently, the use of laboratory techniques is necessary to confirm a diagnosis.

16.2.2 Available biochemical/immunological/molecular techniques to confirm the clinical diagnosis

Before the development of molecular techniques specifically targeting *N. perurans* or *Neoparamoeba* spp., histology was a basic confirmatory technique that facilitated AGD diagnosis (Munday, 1986). This method allows for verification of the presence of the parasite in a gill section and visualization of the parasomes within the parasite (Adams and Nowak, 2003; Adams *et al.*, 2004). However, to identify the amoeba as *N. perurans*, molecular techniques need to be applied.

Identification of *N. perurans* as a primary cause of AGD facilitated development of species-specific molecular tests. Several assays detecting the 18S rRNA gene of this pathogen in different types of samples have been published (Table 16.3). The first polymerase chain reaction (PCR) assay, which was established by Young *et al.* (2008b), is still routinely used for confirmation of new AGD cases (Mouton *et al.*, 2014; Kim *et al.*, 2016).

Assays developed for gill swabs are especially useful, as this sampling technique allows diagnosis of infection without sacrificing the fish (Downes *et al.*, 2017). Other proposed assays were developed to detect the pathogen in gill samples and histological sections (Fringuelli *et al.*, 2012; Downes *et al.*, 2015). Such techniques are more relevant for research than for routine diagnostic screening, as they require killing the host. Two assays were designed to detect *N. perurans* in the water column for monitoring of pathogen distribution in the environment (Bridle *et al.*, 2010; Wright *et al.*, 2015). The results from real-time quantitative PCR (qPCR) are reported either as Ct values (Fringueli *et al.*, 2012; Downes *et al.*, 2015, 2018) or copy numbers (Bridle *et al.*, 2015; Wright *et al.*, 2015, 2017).

16.2.3 External/internal macroscopic and microscopic lesions

Gill lesions are the characteristic clinical sign of AGD. They appear as white, mucoid spots that grow in size and number as infection develops (Zilberg and Munday, 2000; Adams and Nowak, 2001; Taylor *et al.*, 2009). Gill scoring systems have been developed based on their presence, number and size (Taylor *et al.*, 2009). This allows salmon farmers to assess infection prevalence and thus gill scoring is a vital tool used in AGD management. As gill scoring

Table 16.3. Molecular techniques developed for the detection of *Neoparamoeba perurans*.

Technique	Primers	Sample type	Reference
PCR	F: 5′-ATC TTG ACY GGT TCT TTC GRG A-3′ R: 5′-ATA GGT CTG CTT ATC ACT YAT TCT-3′	Gills and *in vitro* cultures	Young *et al.* (2008b)
	Np18sF1: 5′-CTT ACT AGA CTT TCA CTA TTA CAC-3′ Np18sR2: 5′-TCT AAG CAG AAC GAA CTT TC-3′	Gills	Rozas *et al.* (2011)
Real-time PCR			
SYBR Green®-based method	QNperF3: 5′-GTT TAC ATA TTA TGA CCC ACT-3′	Seawater and gill swabs	Bridle *et al.* (2010)
CFX Connect™ Detection System	QNperR3: 5′-TAA ACC CAA TAG GTC TGC-3′	Seawater, gill swabs, gills	Bridle *et al.* (2015); Wright *et al.* (2015)
TaqMan-based method	NP1: 5′-AAA AGA CCA TGC GAT TCG TAA AGT-3′ NP2: 5′-CAT TCT TTT CGG AGA GTG GAA ATT-3′	Gills	Downes *et al.* (2015)
RT-qPCR	Pspp-F: TTG TCA GAG GTG AAA TTC TTG GAT T Pspp-probe: ATG AAA GAC GAA CTT CTG Pspp-R: TGA AAA CAT CTT TGG CAA ATG C	Gills	Steigen *et al.* (2018)
Duplex real-time PCR			
TaqMan-based method	Peru for: GTT CTT TCG GGA GCT GGG AG Peru rev: GAA CTA TCG CCG GCA CAA AAG	Formalin-fixed, paraffin-embedded gills	Fringuelli *et al.* (2012)
In situ hybridization	TGA CYG GTT CTT TCG RGA GCT G	Formalin-fixed, paraffin-embedded gill samples	Young *et al.* (2007)

systems underestimate the presence of amoebae on gills before the development of gross lesions (Clark and Nowak, 1999; Adams and Nowak, 2003), gross examination of the fish gills to score them (Taylor *et al.*, 2009) has to be carried out on a regular basis to spot signs of infection at an early stage.

The distribution of lesions and amoebae on a gill is influenced by the water flow (Adams and Nowak, 2001; Wiik-Nielsen *et al.*, 2016). According to these studies, gill regions of restricted water flow, e.g. between hemibranchs, are more likely to be colonized by amoebae and to develop lesions. Microscopic changes can be detected several days before gross signs of AGD appear. These are characterized by epithelial hyperplasia and lamellar fusion, with *N. perurans* trophozoites associated with the hyperplastic tissue and necrotic amoebae enclosed within interlamellar cysts (Adams *et al.*, 2004; Wiik-Nielsen *et al.*, 2016).

16.3 Expected/Potential Spread of the Pathogen

The geographical distribution of *N. perurans* may be affected by climate change because of higher water temperature and salinity. It is possible that this free-living amoeba will spread into areas which are currently too cold and will become warmer as a result of climate change. Similarly, increasing temperature (rapid short-term warming to >15°C) and increased frequency of storms along the Atlantic coast of Nova Scotia, Canada resulted in increased mortalities in sea urchin (*Strongylocentrotus droebachiensis*) caused by the closely related *Neoparamoeba invadens* (see Feehan *et al.*, 2016). However, it is also possible that the current areas where *N. perurans* is abundant may become too warm for this species. Presently *N. perurans* infects cultured fish, so any changes to where the fish are farmed may also affect the spread of the amoeba.

Flooding, which is predicted to result from climate changes in some areas, will reduce salinity. Presence of *N. perurans* will be minimized in these areas. There is no published information on the effects of other climate change-related environmental changes, for example pH reduction, on *N. perurans*.

16.4 Expected/Potential Spread of Reservoir Host(s) to New Geographical Areas

While *N. perurans*-positive wild fish have been reported, they are unlikely to be a significant reservoir due to very low prevalence of infection (Adams *et al*., 2008; Stagg *et al*., 2015; Hellebø *et al*., 2017). As *N. perurans* is free-living it can spread without relying on a host moving to new geographical areas.

16.5 Population Dynamics of Fish

As the disease affects mostly fish farmed in sea pens, to ensure the economic viability of fish farming, the industry will need to respond to environmental changes. Either new technology will have to be used to control water temperature and oxygen, or the farming will have to move to cooler areas as the current farming areas become too warm to continue culturing the same fish species. At the same time, hypoxia can become a confounding factor as it is already recorded at aquaculture farms at an increasing rate (Oldham *et al*., 2018). The main species affected by AGD, Atlantic salmon, can be significantly affected by low dissolved oxygen during farming in sea pens (Oldham *et al*., 2017, 2019), particularly if the fish are suffering from AGD. Maximum rate of oxygen uptake was reduced in Atlantic salmon experimentally infected with *N. perurans* (see Hvas *et al*., 2017). Furthermore, increased temperature can affect Atlantic salmon reproduction (Pankhurst *et al*., 2011; Anderson *et al*., 2017) or increase morbidity/mortality (Battaglene *et al*., 2008). However, as the viability of the aquaculture industry relies on the performance of the species, the industry will have to adapt to climate change.

16.6 Increase in Pathogen Mortality

16.6.1 Environmental changes

Environmental conditions affected by climate change, in particular increasing temperature, are likely to increase the impact of AGD on the salmon industry. In most geographical locations the first outbreak of AGD was reported when seawater temperature was above the recorded average (Table 16.1; Oldham *et al*., 2016). Another species from the same genus, *N. invadens*, causes epizootics and mortalities of sea urchin which are associated with an increased temperature (Jellet and Scheibling, 1988) and have become more common in the last 30 years (Feehan *et al*., 2012). However, all species of *Neoparamoeba* are recorded from temperate or cold climates (Nowak and Archibald, 2018), so the species may have to adapt to living at higher temperatures.

Cultured *N. perurans* showed optimal growth over 15 days at 15°C and 35 ppt salinity (Collins *et al*., 2019). However, the maximum temperature tested in this experiment was 18°C and upper temperature limits were not determined. In another *in vitro* experiment, growth in 21 days and cultured suspended *N. perurans* numbers were higher at 10°C than at 15°C after 3 months' acclimatization to those temperatures (Benedicenti *et al*., 2019a). There was a significant difference between the two clones tested and between the clones and a polyclonal culture (Benedicenti *et al*., 2019a), suggesting high variability between individual amoebae in their ability to adapt to different temperatures. Both experiments used Scottish isolates. It is possible that the optimum temperature is strain- or isolate-specific and that the origin of isolates and culture conditions affects optimum temperature range. An attempt to test the upper thermal limit of Tasmanian isolates showed that *N. perurans* cultured on marine yeast broth (MYB) plates could not survive for 24 h at 25°C while amoebae at 20°C were alive when the experiment was terminated at 24 h (Crosbie *et al*., 2012b).

In addition to affecting the amoeba directly, environmental changes may affect the host gill microbiome and as a result make the microenvironment more favourable or more hostile to the amoeba. In *N. perurans* cultures, bacterial community composition and abundance were different at 10 and 15°C (Benedicenti *et al*., 2019a). There was some variability in the microbiome associated with different isolates of *N. perurans* (see Benedicenti *et al*., 2019a).

The extent to which *N. perurans* can survive environmental changes and adapt to them is not fully understood. *N. perurans* can form pseudocysts which together with contractile vacuoles allow some trophozoites to survive exposure to fresh water (Lima *et al*., 2016, 2017). While it was only a small percentage of cells that fully recovered after 2 h exposure to fresh water and 5 h after seawater recovery (Lima *et al.*, 2017), this indicates that at least some individuals of this species have the potential to survive adverse conditions and as a result the species could adapt to environmental changes.

16.6.2 Use of more effective chemotherapeutic agents and/or vaccines

As *N. perurans* is a free-living amoeba affecting fish farmed in sea pens it is unlikely that more effective chemotherapeutic agents and/or vaccines will eradicate the disease. However, it is possible that development of effective treatments or vaccines will reduce the impact of AGD on the salmon industry. While there have been no publications on new chemotherapeutic agents or experimental vaccines for 8 years (Web of Science, 15 April 2019), there is no doubt that testing and evaluating treatments under *in vitro* conditions have continued in this area.

16.7 Control and/or Prevention

16.7.1 Selective breeding and other genetic manipulations

Atlantic salmon breeding programmes date back to the beginnings of commercial production of salmon. Breeding for AGD resistance is a crucial part of the Tasmanian salmon breeding programme which started in 2004 (Elliott and Kube, 2009; Kube *et al.*, 2012). This programme uses gill score (at first infection after transfer to seawater and at second to fifth infections after transfer to seawater) as the selection trait (Elliott and Kube, 2009). Selective breeding based on genomic selection has been successfully used by the Atlantic salmon industry in Tasmania to improve the resistance of Atlantic salmon to AGD (Elliott and Kube, 2009; Kube *et al.*, 2012). Salmonid heterosis for resistance to AGD was suggested as an option for the salmon industry, with the brown trout ♀ × Atlantic salmon ♂ (TS) population exhibiting the highest levels of resistance (Maynard *et al.*, 2016).

Resistance to AGD using gill damage and amoebic load has been included as a trait for genomic selection in the Atlantic salmon breeding programmes in other countries. Both gill scores and amoebic load were considered good indicators of resistance to AGD in Atlantic salmon farmed in Norway. The resistance to AGD maybe polygenic; however, two regions on chromosome 18 were associated with both AGD resistance traits (Robledo *et al.*, 2018). Nevertheless, the heritability was lower than the 80–100% for infectious pancreatic necrosis resistance, which is the only commercially successful quantitative trait locus (QTL) of economic importance used in Atlantic salmon aquaculture so far (Houston and Macqueen, 2019). Genes important in target recognition, cell adhesion and inhibition of inflammatory effects of the cytokine interleukin-18 were linked with AGD resistance in Atlantic salmon in Norway (Boison *et al.*, 2019). Use of genomic information instead of pedigree information significantly increased prediction accuracy by 18% and reduced bias for estimated breeding values (Robledo *et al.*, 2018; Boison *et al.*, 2019).

Increased temperature will be another challenge of climate change which can be addressed by selective breeding. Thermal tolerance has been reported from salmonids farmed in higher temperatures. For example, rainbow trout (*Oncorhynchus mykiss*) farmed in Australia (Molony *et al.*, 2004), Japan and Patagonia showed higher thermal preference and higher thermal tolerance than fish reared in lower temperatures (Molony *et al.*, 2004; Ineno *et al.*, 2005, 2018; Crozier and Hutchings, 2014; Oku *et al.*, 2014; Crichigno *et al.*, 2018). The QTL for thermal tolerance was identified in rainbow trout (Perry *et al.*, 2005).

Gene editing will contribute to programming of disease resistance in farmed fish in the future (Diwan *et al.*, 2017; Elaswad and Dunham, 2018; Houston and Macqueen, 2019). While gene editing has been successfully used in research, there are some regulatory and consumer acceptance challenges (Houston and Macqueen, 2019). However, as a genetically modified salmon strain was approved for human consumption in the USA and Canada (Waltz, 2017), there is a potential for genome-edited fish to be available on the market in the future. Obviously, gene editing to increase salmon resistance to AGD would be only possible when the genes responsible for the resistance are identified. So far it appears that AGD resistance is polygenic, limiting the potential for the application of this approach.

16.7.2 Development of vaccines and use of novel chemotherapeutants

Several studies investigating potential vaccines against AGD were carried out before the identification of *N. perurans* as the causative agent of the disease. Many of these used antigens obtained from wild-type amoebae, isolated from the gills of AGD-positive fish (Table 16.4). Despite limitations caused by the initial pathogen misidentification, this research provided valuable information and foundations for further vaccine development attempts.

Over the years variable immunization strategies were tested along with different antigen preparations (Table 16.4). Regardless of vaccine application and composition, none of those tested before identification of *N. perurans* resulted in fish resistance to AGD. Recognition of the causative agent opened the door for the use of more advanced technologies and design of specific targets. Despite those improvements there have been no reports of a successful experimental vaccine.

16.7.3 Novel culture technologies

Snorkel cages

Novel ways to manage AGD include surface modifications; for example, access and environmental change (water salinity) at the water surface can be achieved in snorkel cages. Snorkel cages were developed for the management of sea lice infections and include controlled and limited surface access, which can vary environmental conditions (e.g. salinity) (Wright *et al.*, 2017). A reduced proportion of salmon with an elevated AGD gross gill score was observed in seawater and freshwater snorkel cages. There was a decrease by 65% in salmon positive for *N. perurans* in the freshwater snorkel cage (Wright *et al.*, 2018). While the salmon in snorkel cages had reduced AGD lesions, the cages would require particular environmental conditions and the effect on AGD might be related to other factors (Wright *et al.*, 2017).

Alternative culture systems

Future Atlantic salmon farming may be carried out in recirculating aquaculture systems (RAS) on land. Several commercial facilities have already been established to produce market-sized Atlantic salmon in RAS in various countries, including Japan, USA, Russia and Poland. Alternatively, larger Atlantic salmon could be grown in RAS so that transfer to sea pens would be delayed and shorter time in the sea would reduce the risks of AGD. While most of the current Atlantic salmon production in RAS occurs in fresh water (Davidson *et al.*, 2016), application of seawater RAS is being considered. The use of RAS allows water treatment and exclusion of pathogens, such as *N. perurans*, from marine farming if seawater RAS is used. Furthermore, the use of RAS for salmon farming would preclude climate change effects (e.g. water temperature, pH changes, more frequent incidents of storms). However, care would have to be taken to ensure full biosecurity so that the biofilters do not become reservoirs of pathogens, including the amoeba.

Offshore farming

While *N. perurans* is widely spread in the coastal environment, moving salmon farming further offshore could potentially reduce parasitic infections including the impact of AGD. For example, ranching southern bluefin tuna (*Thunnus maccoyii*) offshore eliminated parasitic infections, especially blood fluke (*Cardicola* spp.) and sea lice (*Caligus chiastos*). This is most likely due to the greater distance between the bottom of the pen and the ocean floor, increased water flow due to currents and fewer encounters with wild fish (Kirchhoff *et al.*, 2011). Offshore farming of Atlantic salmon is under development in a number of countries, e.g. Norway (Kristiansen *et al.*, 2017). While environmental conditions can be more challenging for the infrastructure and husbandry, the risk of disease outbreaks is expected to be lower than in nearshore locations.

16.8 Conclusions with Suggestions for Future Studies

Infectious disease is a result of an interaction between host, pathogen and environment. Changing environmental conditions will not only affect the pathogen but also the host. In the case of AGD, it is likely that temperature changes by a few degrees will have more adverse effects on the host and will outplay the effects on the pathogen. More research on the effects of changing environment on CGD will reveal if AGD will be the dominant component of that condition. Climate change may also increase the frequency of microbial dysbiosis, however the involvement of dysbiosis in AGD is not understood. All climate change effects on AGD could be significantly reduced or eliminated if RAS replaced cage farming, but currently this is unlikely to be economically viable. So far, selective breeding has shown the highest promise for AGD control. This approach allows the development of thermal resistance in parallel to disease resistance. Future solutions will be based on new knowledge, interdisciplinary collaborations and a big picture approach.

(continued)

Table 16.4. Experimental vaccines against amoebic gill disease.

Vaccine	Adjuvant	Application	Booster	Challenge	Result	Reference
Wild-type amoebae antigens (164,900 cell equivalents/l)	NA	Bath (6 h, 17°C), fw	NA	Challenge 27 days later with 2867 amoebae/l	ND	Morrison and Nowak (2005)
NP251002 antigens (643,889 cell equivalents/l)	NA	Bath (6 h 17°C), fw	NA	Challenge 27 days later with 2867 amoebae/l	ND	Morrison and Nowak (2005)
2000 gill-isolated amoebae (live)/100 g salmon	NA	Anal intubation	Booster 3 weeks post-vaccination	Challenge 2 weeks post-booster by cohabitation	ND	Zilberg and Munday (2001)
Sonicate of 2000 gill-isolated amoebae/100 g salmon		Anal intubation	Booster 3 weeks post-vaccination	Challenge 2 weeks post-booster by cohabitation	ND	Zilberg and Munday (2001)
2000 gill-isolated amoebae (live)/100 g salmon	NA	IP injection	Booster 3 weeks post-vaccination	Challenge 2 weeks post-booster by cohabitation	ND	Zilberg and Munday (2001)
Sonicate of 2000 gill-isolated amoebae/100 g salmon	Mon.	IP injection	Booster 3 weeks post-vaccination	Challenge 2 weeks post-booster by cohabitation	ND	Zilberg and Munday (2001)
Sonicate of 20,000 gill-isolated amoebae/250 g salmon	Mon.	IP injection	Booster 3 weeks post-vaccination	Challenge 2 weeks post-booster by cohabitation	ND	Zilberg and Munday (2001)
Sonicate of 20,000 cultured *Paramoeba* spp./250 g salmon	Mon.	IP injection	Booster 3 weeks post-vaccination	Challenge 2 weeks post-booster by cohabitation	ND	Zilberg and Munday (2001)
Six DNA antigens (6 μg DNA/fish)	None	Injection	Booster 14 days post-vaccination	Challenge 42 days post-booster (500 amoebae/l)	40% increase in protection (but variable between tanks and trials)	Cook *et al.* (2008)
Recombinant protein rr22C03	FCA	IP injection (250 μg dose of r22C03 in 200 μl total volume) followed by booster 5 weeks later (either IP or dip)	IP booster FIA 250 μg dose of r22C03 in 200 μl total volume Dip for 1 min in a solution containing 50 mg/l of the recombinant protein dissolved in PBS	500 amoebae/l Challenge 16 weeks after vaccination aborted after 1 week, fw bath and antibiotics, second challenge 5 weeks after first	ND Increased serum antibody, dip booster antibody in skin mucus Fish co-infected by *Yersinia ruckeri*	Valdenegro-Vega *et al.* (2015)

FCA, Freund's complete adjuvant; FIA, Freund's incomplete adjuvant; fw, fresh water; IP, intraperitoneal; Mon., Montanide; NA, not used; ND, no difference between vaccinated and control fish; PBS, phosphate-buffered saline.

Acknowledgements

We would like to acknowledge Mai Dang for her assistance with preparing the histology slides and taking the photographs. The first author has received funding from the European Union's Horizon 2020 research and innovation programme under Grant Agreement no. 634429 (ParaFishControl). This output reflects only the author's view and the European Union cannot be held responsible for any use that may be made of the information contained herein. The second author would like to thank Professor Gretta Pecl, Centre for Marine Socioecology, Institute of Marine and Antarctic Studies, University of Tasmania and Future Sea project, for inspiration and providing a bigger picture of the effects of climate change on aquaculture and fisheries. This research was supported partially by the Australian Government through the Australian Research Council's Linkage Projects funding scheme (project LP160101762). The views expressed herein are those of the authors and are not necessarily those of the Australian Government or Australian Research Council.

References

Adams, M.B. and Nowak, B.F. (2001) Distribution and structure of lesions in the gills of Atlantic salmon, *Salmo salar* L., affected with amoebic gill disease. *Journal of Fish Diseases* 24(9), 535–542.

Adams, M.B. and Nowak, B.F. (2003) Amoebic gill disease: sequential pathology in cultured Atlantic salmon, *Salmo salar* L. *Journal of Fish Diseases* 26(10), 601–614.

Adams, M.B. and Nowak, B.F. (2004) Sequential pathology after initial freshwater bath treatment for amoebic gill disease in cultured Atlantic salmon, *Salmo salar* L. *Journal of Fish Diseases* 27(3), 163–173.

Adams, M.B., Ellard, K. and Nowak, B.F. (2004) Gross pathology and its relationship with histopathology of amoebic gill disease (AGD) in farmed Atlantic salmon, *Salmo salar* L. *Journal of Fish Diseases* 27(3), 151–161.

Adams, M.B., Villavedra M. and Nowak, B.F. (2008) An opportunistic detection of amoebic gill disease (AGD) in blue warehou (*Seriolella brama* Günther) collected from an Atlantic salmon (*Salmo salar* L.) production cage in south eastern Tasmania. *Journal of Fish Diseases* 31(9), 713–717.

Adams, M.B., Crosbie, P.B.B. and Nowak, B.F. (2012) Preliminary success using hydrogen peroxide to treat Atlantic salmon, *Salmo salar* L., affected with experimentally induced amoebic gill disease (AGD). *Journal of Fish Diseases* 35(11), 839–848.

Anderson, K., Pankhurst, N., King, H. and Elizur, A. (2017) Effect of thermal challenge on the expression of genes involved in ovarian steroidogenesis in Tasmanian Atlantic salmon (*Salmo salar*). *Aquaculture* 479, 474–478.

APEM (2014) Availability and use of freshwater resources for the treatment of amoebic gill disease at Scottish salmon farms. *APEM Report No. 413028 to the Scottish Aquaculture Research Forum*. APEM Ltd, Begbroke and Edinburgh, UK.

Battaglene, S., Carter, C., Hobday, A.J., Lyne, V. and Nowak, B. (2008) Scoping study into adaptation of the Tasmanian salmonid aquaculture industry to potential impacts of climate change. *National Agriculture and Climate Change Action Plan: Implementation Programme Report*. Tasmanian Aquaculture and Fisheries Institute, Hobart. Australia.

Benedicenti, O., Secombes, C.J. and Collins, C. (2019a) Effects of temperature on *Paramoeba perurans* growth in culture and the associated microbial community. *Parasitology* 146, 533–542.

Benedicenti, O., Pottinger, T.G., Collins, C. and Secombes, C.J. (2019b) Effects of temperature on amoebic gill disease development: does it play a role? *Journal of Fish Diseases* 42(9), 1241–1258.

Bloecher, N., Powell, M., Hytterød, S., Gjessling, M., Wiik-Nielsen, J. *et al.* (2018) Effects of cnidarian biofouling on salmon gill health and development of amoebic gill disease. *PLoS ONE* 13(7), e0199842.

Boison, S.A., Gjerde, B., Hillestad, B., Makvandi-Nejad, S. and Moghadam, H.K. (2019) Genomic and transcriptomic analysis of amoebic gill disease resistance in Atlantic salmon (*Salmo salar* L.). *Frontiers in Genetics* 10, 68.

Bridle, A.R., Crosbie, P.B.B., Cadoret, K. and Nowak, B.F. (2010) Rapid detection and quantification of *Neoparamoeba perurans* in the marine environment. *Aquaculture* 309(1–4), 56–61.

Bridle, A.R., Davenport, D.L., Crosbie, P.B.B., Polinski, M. and Nowak, B.F. (2015) *Neoparamoeba perurans* loses virulence during clonal culture. *International Journal for Parasitology* 45(9–10), 575–578.

Bustos, P.A., Young, N.D., Rozas, M.A., Bohle, H.M., Ildefonso, R.S. *et al.* (2011) Amoebic gill disease (AGD) in Atlantic salmon (*Salmo salar*) farmed in Chile. *Aquaculture* 310(3–4), 281–288.

Clark, A. and Nowak, B.F. (1999) Field investigations of amoebic gill disease in Atlantic salmon, *Salmo salar* L., in Tasmania. *Journal of Fish Diseases* 22(6), 433–443.

Collins, C., Hall, M., Fordyce, M.J. and White, P. (2019) Survival and growth *in vitro* of *Paramoeba perurans* populations cultured under different salinities and temperatures. *Protist* 170(2), 153–167.

Cook, M., Elliott, N., Campbell, G., Patil, J., Holmes, B. *et al.* (2008) Amoebic gill disease (AGD) vaccine development phase II – molecular basis of host–pathogen interactions in amoebic gill disease. *Aquafin CRC Project 3.4.4(2) (FRDC Project 2004/217)*. CSIRO Marine and Atmospheric Research, Hobart, Australia.

Crichigno, S.A., Becker, L.A., Orellana, M., Larraze, R., Mirenna, G. *et al.* (2018) Rainbow trout adaptation to a warmer Patagonia and its potential to increase temperature tolerance in cultured stocks. *Aquaculture Reports* 9, 82–88.

Crosbie, P.B.B., Bridle, A.R., Leef, M.J. and Nowak, B.F. (2010a) Effects of different batches of *Neoparamoeba perurans* and fish stocking densities on the severity of amoebic gill disease in experimental infection of Atlantic salmon, *Salmo salar* L. *Aquaculture Research* 41, e505–e516.

Crosbie, P.B.B., Ogawa, K., Nakano, D. and Nowak, B.F. (2010b) Amoebic gill disease in hatchery-reared ayu, *Plecoglossus altivelis* (Temminck & Schlegel), in Japan is caused by *Neoparamoeba perurans*. *Journal of Fish Diseases* 33(5), 455–458.

Crosbie, P.B.B., Bridle, A.R., Cadoret, K. and Nowak, B.F. (2012a) *In vitro* cultured *Neoparamoeba perurans* causes amoebic gill disease in Atlantic salmon and fulfils Koch's postulates. *International Journal for Parasitology* 42(5), 511–515.

Crosbie, P.B.B., Bridle, A.R. and Nowak, B.F. (2012b) FRDC Salmon Subprogram – Extension funding application – AGD Vaccine Phase III. *FRDC Project Number 2008/218 Final Report*. University of Tasmania, Hobart, Australia.

Crozier, L.G. and Hutchings, J.A. (2014) Plastic and evolutionary responses to climate change in fish. *Evolutionary Applications* 7, 68–87.

Davidson, J., May, T., Good, C., Waldrop, T., Kenney, B. *et al.* (2016) Production of market-size North American strain Atlantic salmon *Salmo salar* in a land-based recirculation aquaculture system using freshwater. *Agricultural Engineering* 74, 1–16.

Diwan, A.D., Ninawe, A.S. and Harke, S.N. (2017) Gene editing (CRISPR-Cas) technology and fisheries sector. *Canadian Journal of Biotechnology* 1, 65–72.

Douglas-Helders, M., Saksida, S., Raverty, S. and Nowak, B.F. (2001) Temperature as a risk factor for outbreaks of amoebic gill disease in farmed Atlantic salmon (*Salmo salar*). *Bulletin of the European Association of Fish Pathologists* 21(3), 114–116.

Douglas-Helders, G.M., Weir, I.J., O'Brien, D.P., Carson, J. and Nowak, B.F. (2004) Effects of husbandry on prevalence of amoebic gill disease and performance of reared Atlantic salmon (*Salmo salar* L.). *Aquaculture* 241(1–4), 21–30.

Downes, J.K., Henshilwood, K., Collins, E.M., Ryan, A., O'Connor, I. *et al.* (2015) A longitudinal study of amoebic gill disease on a marine Atlantic salmon farm utilising a real-time PCR assay for the detection of *Neoparamoeba perurans*. *Aquaculture Environment Interactions* 7, 239–251.

Downes, J.K., Rigby, M.L., Taylor, R.S., Maynard, B.T., MacCarthy, E. *et al.* (2017) Evaluation of non-destructive molecular diagnostics for the detection of *Neoparamoeba perurans*. *Frontiers in Marine Science* 4, 61.

Downes, J.K., Yatabe, T., Marcos-Lopez, M., Rodger, H.D., MacCarthy, E. *et al.* (2018) Investigation of co-infections with pathogens associated with gill disease in Atlantic salmon during an amoebic gill disease outbreak. *Journal of Fish Diseases* 41(8), 1217–1227.

Dyková, I. and Novoa, B. (2001) Comments on diagnosis of amoebic gill disease (AGD) in turbot, *Scophthalmus maximus*. *Bulletin of the European Association of Fish Pathologists* 21(1), 40-44.

Dyková, I., Figueras, A. and Novoa, B. (1995) Amoebic gill infection of turbot, *Scophthalmus maximus*. *Folia Parasitologica* 42(2), 91–96.

Dyková, I., Figueras, A., Novoa, B. and Casal, J.F. (1998) *Paramoeba* sp., an agent of amoebic gill disease of turbot *Scophthalmus maximus*. *Diseases of Aquatic Organisms* 33(2), 137–141.

Dyková, I., Figueras, A. and Peric, Z. (2000) *Neoparamoeba* Page, 1987: light and electron microscopic observations on six strains of different origin. *Diseases of Aquatic Organisms* 43(3), 217–223.

Dyková, I., Fiala, I., Lom, J. and Lukeš, J. (2003) *Perkinsella amoebae*-like endosymbionts of *Neoparamoeba* spp. relatives of kinetoplastid Ichthyobodo. *European Journal of Protistology* 39, 37–52.

Elaswad, A. and Dunham, R. (2018) Disease reduction in aquaculture with genetic and genomic technology: current and future approaches. *Reviews in Aquaculture* 10, 876–898.

Elliott, N.G. and Kube, P.D. (2009) Development and early results of the Tasmanian Atlantic salmon breeding program. *Proceedings of the Association for the Advancement in Animal Breeding and Genetics* 18, 362–365.

English, C.J., Tyml, T., Botwright, N., Barnes, A.C., Wynne, J.W. *et al.* (2019) A diversity of amoebae colonise the gills of farmed Atlantic salmon (*Salmo salar*) with amoebic gill disease (AGD). *European Journal of Protistology* 67, 27–45.

Feehan, C., Scheibling, R.E. and Lauzon-Guay, J.S. (2012) An outbreak of sea urchin disease associated with a recent hurricane: support for the 'killer storm hypothesis' on a local scale. *Journal of Experimental Marine Biology and Ecology* 413, 159–168.

Feehan, C.J., Johnson-Mackinnon, J., Scheibling, R.E., Lauzon-Guay, J.S. and Simpson, A.G.B. (2013) Validating the identity of *Paramoeba invadens*, the causative agent of recurrent mass mortality of sea urchins in Nova Scotia, Canada. *Diseases of Aquatic Organisms* 103(3), 209–227.

Feehan, C.J., Scheibling, R.E., Brown, M.S. and Thompson, K.R. (2016) Marine epizootics linked to storms: mechanisms of pathogen introduction and persistence inferred from coupled physical and biological time-series. *Limnology and Oceanography* 61, 316–329.

Findlay, V.L., Helders, M., Munday, B.L. and Gurney, R. (1995) Demonstration of resistance to reinfection with

Paramoeba sp. by Atlantic salmon, *Salmo salar* L. *Journal of Fish Diseases* 18(6), 639–642.

Fringuelli, E., Gordon, A.W., Rodger, H., Welsh, M.D. and Graham, D.A. (2012) Detection of *Neoparamoeba perurans* by duplex quantitative Taqman real-time PCR in formalin-fixed, paraffin-embedded Atlantic salmonid gill tissues. *Journal of Fish Diseases* 35(10), 711–724.

Gjessing, M.C., Yutin, N., Tengs, T., Senkevich, T., Koonin, E. *et al.* (2015) Salmon gill poxvirus, the deepest representative of the *Chordopoxvirinae*. *Journal of Virology* 89(18), 9348–9367.

Gjessing, M.C., Thoen, E., Tengs, T., Skotheim, S.A. and Dale, O.B. (2017) Salmon gill poxvirus, a recently characterized infectious agent of multifactorial gill disease in freshwater- and seawater-reared Atlantic salmon. *Journal of Fish Diseases* 40(10), 1253–1265.

Gross, K.A., Powell, M.D., Butler, R., Morrison, R.N. and Nowak, B.F. (2005) Changes in the innate immune response of Atlantic salmon, *Salmo salar* L., exposed to experimental infection with *Neoparamoeba* sp. *Journal of Fish Diseases* 28(5), 293–299.

Haugland, G.T., Olsen, A.B., Rønneseth, A. and Andersen, L. (2017) Lumpfish (*Cyclopterus lumpus* L.) develop amoebic gill disease (AGD) after experimental challenge with *Paramoeba perurans* and can transfer amoebae to Atlantic salmon (*Salmo salar* L.). *Aquaculture* 478, 48–55.

Hellebø, A., Stene, A. and Aspehaug, V. (2017) PCR survey for *Paramoeba perurans* in fauna, environmental samples and fish associated with marine farming sites for Atlantic salmon (*Salmo salar* L.). *Journal of Fish Diseases* 40(5), 661–670.

Herrero, A., Thompson, K.D., Ashby, A., Rodger, H.D. and Dagleish, M.P. (2018) Complex gill disease: an emerging syndrome in farmed Atlantic salmon (*Salmo salar* L.). *Journal of Comparative Pathology* 163, 23–28.

Houston, R.D. and Macqueen, D.J. (2019) Atlantic salmon (*Salmo salar* L.) genetics in the 21st century: taking leaps forward in aquaculture and biological understanding. *Animal Genetics* 50, 3–14.

Hvas, M., Karlsbakk, E., Maehle, S., Wright, D.W. and Oppedal, F. (2017) The gill parasite *Paramoeba perurans* compromises aerobic scope, swimming capacity and ion balance in Atlantic salmon. *Conservation Physiology* 5(1), cox066.

ICES (International Council for the Exploration of the Sea) (2015) Report of the Working Group on Pathology and Diseases of Marine Organisms (WGPDMO). *ICES CM 2015/SSGEPI:01*. ICES, Helsinki.

Ineno, T., Tsuchida, S., Kanda, M. and Watabe, S. (2005) Thermal tolerance of a rainbow trout *Oncorhynchus mykiss* strain selected by high-temperature breeding. *Fisheries Science* 71(4), 767–775.

Ineno, T., Tamaki, K., Yamada, K., Kodama, R., Tsuchida, S. *et al.* (2018) Thermal tolerance of a thermally selected strain of rainbow trout *Oncorhynchus mykiss* and the pedigrees of its F1 and F2 generations indicated by their critical thermal maxima. *Fisheries Science* 84, 671–679.

Jellet, J.F. and Scheibling, R.E. (1988) Effect of temperature and prey availability on growth of *Paramoeba invadens* in monoxenic culture. *Applied and Environmental Microbiology* 54, 1848–1854.

Jokinen, I.E., Salo, H.M., Markkula, E., Rikalainen, K., Arts, M.T. and Browman, H.I. (2011) Additive effects of enhanced ambient ultraviolet B radiation and increased temperature on immune function, growth and physiological condition of juvenile (parr) Atlantic salmon, *Salmo salar*. *Fish and Shellfish Immunology* 30, 102–108.

Karlsbakk, E., Olsen, A.B., Einen, A.-C.B., Mo, T.A., Fiksdal, I.U. *et al.* (2013) Amoebic gill disease due to *Paramoeba perurans* in ballan wrasse (*Labrus bergylta*). *Aquaculture* 412, 41–44.

Kent, M.L., Sawyer, T.K. and Hedrick, R.P. (1988) *Paramoeba pemaquidensis* (Sarcomastigophora: Paramoebidae) infestation of the gills of coho salmon *Oncorhynchus kisutch* reared in sea water. *Diseases of Aquatic Organisms* 5(3), 163–169.

Kim, H.J., Cho, J.B., Lee, M.K., Huh, M.D. and Kim, K.H. (2005) *Neoparamoeba* sp. infection on gills of olive flounder, *Paralichthys olivaceus in Korea. Journal of Fish Pathology* 18(2), 125–131.

Kim, W.S., Kong, K.H., Kim, J.O. and Oh, M.J. (2016) Amoebic gill infection in coho salmon *Oncorhynchus kisutch* farmed in Korea. *Diseases of Aquatic Organisms* 121(1), 75–78.

Kim, W.S., Kong, K.H., Kim, J.O., Jung, S.J., Kim, J.H. and Oh, M.J. (2017) Amoebic gill disease outbreak in marine fish cultured in Korea. *Journal of Veterinary Diagnostic Investigation* 29(3), 357–361.

Kirchhoff, N.T., Rough, K. and Nowak, B.F. (2011) Moving cages further offshore: effects on southern bluefin tuna, *T. maccoyii*, parasites, health and performance. *PLoS ONE* 6(8), e23705.

Kristiansen, D., Aksnes, V., Su, B., Lader, P. and Bjelland, H.V. (2017) Environmental description in the design of fish farms at exposed locations. In: *Proceedings of the ASME 2017 36th Conference on Ocean, Offshore and Arctic Engineering*. Vol. 6. *Ocean Space Utilization*. Trondheim, Norway, 25–30 June 2017. American Society of Mechanical Engineers, New York, paper OMAE2017-61531, V006T05A003.

Kube, P.D., Taylor, R.S. and Elliott, N.G. (2012) Genetic variation in parasite resistance of Atlantic salmon to amoebic gill disease over multiple infections. *Aquaculture* 364–365, 165–172.

Kudryavtsev, A., Pawlowski, J. and Hausmann, K. (2011) Description of *Paramoeba atlantica* n. sp. (Amoebozoa, Dactylopodida) – a marine amoeba from the eastern Atlantic, with emendation of the dactylopodid families. *Acta Protozoologica* 50, 239–253.

Leef, M.L., Harris, J.O. and Powell, M.D. (2005) Respiratory pathogenesis of amoebic gill disease (AGD) in experimentally infected Atlantic salmon

Salmo salar. *Diseases of Aquatic Organisms* 66(3), 205–213.

Lima, P.C., Taylor, R.S. and Cook, M. (2016) Involvement of contractile vacuoles in the osmoregulation process of the marine parasitic amoeba *Neoparamoeba perurans*. *Journal of Fish Diseases* 39(5), 629–633.

Lima, P.C., Taylor, R.S. and Cook, M. (2017) Pseudocyst formation in the marine parasitic amoeba *Neoparamoeba perurans*: a short-term survival strategy to abrupt salinity variation. *Journal of Fish Diseases* 40(8), 1109–1113.

Lom, J. and Dyková, I. (1992) *Protozoan Parasites of Fishes*. Elsevier, Amsterdam.

Maynard, B.T., Taylor, R.S., Kube, P.D., Cook, M.T. and Elliott, N.G. (2016) Salmonid heterosis for resistance to amoebic gill disease (AGD). *Aquaculture* 451, 106–112.

Molony, B.W., Church, A.R. and Maguire, G.B. (2004) A comparison of the heat tolerance and growth of a selected and non-selected line of rainbow trout *Oncorhynchus mykiss*, in Western Australia. *Aquaculture* 241, 655–665.

Morrison, R.N. and Nowak, B.F. (2005) Bath treatment of Atlantic salmon (*Salmo salar*) with amoebae antigens fails to affect survival to subsequent amoebic gill disease (AGD) challenge. *Bulletin of the European Association of Fish Pathologists* 25(4), 155–160.

Morrison, R.N., Crosbie, P., Adams, M.B., Cook, M.T. and Nowak, B.F. (2005) Cultured gill derived *Neoparamoeba pemaquidensis* fail to elicit AGD in Atlantic salmon (*Salmo salar*). *Diseases of Aquatic Organisms* 66(2), 135–144.

Mouton, A., Crosbie, P., Cadoret, K. and Nowak, B. (2014) First record of amoebic gill disease caused by *Neoparamoeba perurans* in South Africa. *Journal of Fish Diseases* 37(4), 407–409.

Munday, B.L. (1986) Diseases of salmonids. In: Humphrey, J.D. and Langdon, J.S. (eds) *Proceedings of the Workshop on Diseases of Australian Fish and Shellfish*. Department of Agriculture and Rural Affairs, Benalla, Australia, pp. 127–141.

Munday, B.L., Foster, C.K., Roubal, F.R. and Lester, R.G.J. (1990) Paramoebic gill infection and associated pathology of Atlantic salmon, *Salmo salar*, and rainbow trout, *Salmo gairdneri*, in Tasmania. In: Perkins, F.O. and Cheng, T.C. (eds) *Pathology in Marine Science*. Academic Press, San Diego, California, pp. 215–222.

Munday, B.L., Zilberg, D. and Findlay, V. (2001) Gill disease of marine fish caused by infection with *Neoparamoeba pemaquidensis*. *Journal of Fish Diseases* 24(9), 497–507.

Nowak, B.F. (2007) Parasitic diseases in marine cage culture – an example of experimental evolution of parasites? *International Journal for Parasitology* 37(6), 581–588.

Nowak, B.F. (2012) *Neoparamoeba perurans*. In: Woo, P.T.K. and Buchmann, K. (eds) *Fish Parasites: Pathobiology and Protection*. CAB International, Wallingford, UK, pp. 1–18.

Nowak, B.F. and Archibald, J.M. (2018) Opportunistic but lethal: the mystery of Paramoebae. *Trends in Parasitology* 34(5), 404–419.

Nowak, B., Valdenegro-Vega, V., Crosbie, P. and Bridle, A. (2014) Immunity to amoeba. *Developmental and Comparative Immunology* 43(2), 257–267.

Oku, H., Tokuda, M., Matsunari, H., Furuita, H., Murashita, K. and Yamamoto, T. (2014) Characterization of differentially expressed genes in liver in response to the rearing temperature of rainbow trout *Oncorhynchus mykiss* and their heritable differences. *Fish Physiology and Biochemistry* 40, 1757–1769.

Oldham, T., Rodger, H. and Nowak, B.F. (2016) Incidence and distribution of amoebic gill disease (AGD) – an epidemiological review. *Aquaculture* 457, 35–42.

Oldham, T., Dempster, T., Fosse, J.O. and Oppedal, F. (2017) Oxygen gradient affect behavior of caged Atlantic salmon *Salmo salar*. *Aquaculture Environment Interactions* 9, 145–153.

Oldham, T., Oppedal, F. and Dempster, T. (2018) Cage size affects dissolved oxygen distribution in salmon aquaculture. *Aquaculture Environment Interactions* 10, 149–156.

Oldham, T., Nowak, B., Hvas, M. and Oppedal, F. (2019) Metabolic and functional impacts of hypoxia vary with size in Atlantic salmon. *Comparative Biochemistry and Physiology Part A: Molecular and Integrative Physiology* 231, 30–38.

Page, F.C. (1987) The classification of 'naked' amoebae (phylum Rhizopoda). *Archiv für Protistenkunde* 133, 199–217.

Pankhurst, N.W., King, H.R., Anderson, K., Elizur, A., Pankhurst, P.M. and Ruff, N. (2011) Thermal impairment of reproduction is differentially expressed in maiden and repeat spawning Atlantic salmon. *Aquaculture* 316, 77–87.

Perry, G.M., Martyniuk, C.M., Ferguson, M.M. and Danzmann, R.G. (2005) Genetic parameters for upper thermal tolerance and growth-related traits in rainbow trout (*Oncorhynchus mykiss*). *Aquaculture* 250, 120–128.

Powell, M.D. (2006) Physiological and clinical pathology. In: Ferguson, H.W. (ed.) *Systemic Pathology of Fish*, 2nd edn. Scotian Press, London, pp. 336–355.

Robledo, D., Matika, O., Hamilton, A. and Houston, R.D. (2018) Genome-wide association and genomic selection for resistance to amoebic gill disease in Atlantic salmon. *G3: Genes, Genomes, Genetics* 8, 1195–1203.

Rodger, H.D. (2014) Amoebic gill disease (AGD) in farmed salmon (*Salmo salar*) in Europe. *Fish Veterinary Journal* (14), 16–27.

Rodger, H.D. (2019) Amoebic gill disease in farmed halibut (*Hippoglossus hippoglossus*) in the United Kingdom. *Veterinary Record Case Reports* 7, e000797.

Rodger, H.D. and McArdle, J.F. (1996) An outbreak of amoebic gill disease in Ireland. *Veterinary Record* 139, 348–349.

Rozas, M., Bohle, H., Ildefonso, R. and Bustos, P. (2011) Development of PCR assay for detection of *Neoparamoeba perurans* and comparison of histological diagnosis. *Bulletin of the European Association of Fish Pathologists* 31(6), 211–218.

Shinn, A.P., Pratoomyot, J., Bron, J.E., Paladini, G., Brooker, E.E. and Brooker, A.J. (2015) Economic costs of protistan and metazoan parasites to global mariculture. *Parasitology* 142, 196–270.

Sibbald, S.J., Cenci, U., Colp, M., Eglit, Y., O'Kelly, C.J. and Archibald, J.M. (2017) Diversity and evolution of *Paramoeba* spp. and their kinetoplastid endosymbionts. *Eukaryotic Microbiology* 64(5), 598–607.

Stagg, H.E.B., Hall, M., Wallace, I.S., Pert, C.C., Garcia Perez, S. and Collins, C. (2015) Detection of *Paramoeba perurans* in Scottish marine wild fish populations. *Bulletin of the European Association of Fish Pathologists* 35(6), 217–226.

Steigen, A., Nylund, A., Plarre, H., Watanabe, K., Karlsbakk, E. and Brevik, Ø. (2018) Presence of selected pathogens on the gills of five wrasse species in western Norway. *Diseases of Aquatic Organisms* 128(1), 21–35.

Steinum, T., Kvellestad, A., Ronneberg, L.B., Nilsen, H., Asheim, A. *et al.* (2008) First cases of amoebic gill disease (AGD) in Norwegian seawater farmed Atlantic salmon, *Salmo salar* L., and phylogeny of the causative amoeba using 18S cDNA sequences. *Journal of Fish Diseases* 31(3), 205–214.

Tan, C.K.F., Nowak, B.F. and Hodson, S.L. (2002) Biofouling as a reservoir of *Neoparamoeba pemaquidensis* (Page, 1970), the causative agent of amoebic gill disease in Atlantic salmon. *Aquaculture* 210, 49–58.

Tanifuji, G., Cenci, U., Moog, D., Dean, S., Nakayama, T. *et al.* (2017) Genome sequencing reveals metabolic and cellular interdependence in an amoeba–kinetoplastid symbiosis. *Scientific Reports* 7, 11688.

Taylor, R.S., Muller, W.J., Cook, M.T., Kube, P.D. and Elliott, N.G. (2009) Gill observations in Atlantic salmon (*Salmo salar*, L.) during repeated amoebic gill disease (AGD) field exposure and survival challenge. *Aquaculture* 290(1–2), 1–8.

Valdenegro-Vega, V.A., Cook, M., Crosbie, P., Bridle, A.R. and Nowak, B.F. (2015) Vaccination with recombinant protein (r22C03), a putative attachment factor of *Neoparamoeba perurans*, against AGD in Atlantic salmon (*Salmo salar*) and implications of a co-infection with *Yersinia ruckeri*. *Fish and Shellfish Immunology* 44(2), 592–602.

Vincent, B.N., Adams, M.B., Crosbie, P.B.B., Nowak, B.F. and Morrison, R.N. (2007) Atlantic salmon (*Salmo salar* L.) exposed to cultured gill-derived *Neoparamoeba branchiphila* fail to develop amoebic gill disease (AGD). *Bulletin of the European Association of Fish Pathologists* 27(3), 112–115.

VKM (Norwegian Scientific Committee for Food Safety) (2014) Risk assessment of amoebic gill disease. *VKM Report No. 2014:11*. VKM Panel on Animal Health and Welfare, Oslo.

Waltz, E. (2017) First genetically engineered salmon sold in Canada. *Nature* 548, 148.

Wiik-Nielsen, J., Mo, T.A., Kolstad, H., Mohammad, S.N., Hytterød, S. and Powell, M.D. (2016) Morphological diversity of *Paramoeba perurans* trophozoites and their interaction with Atlantic salmon, *Salmo salar* L., gills. *Journal of Fish Diseases* 39(9), 1113–1123.

Wright, D.W., Nowak, B.F., Oppedal, F., Bridle, A.R. and Dempster, T. (2015) Depth distribution of the amoebic gill disease agent, *Neoparamoeba perurans*, in salmon sea-cages. *Aquaculture Environment Interactions* 7, 67–74.

Wright, D., Nowak, B., Oppedal, F., Bridle, A. and Dempster, T. (2017) Free-living *Neoparamoeba perurans* depth distribution is mostly uniform in salmon cages, but reshaped by stratification and potentially extreme fish crowding. *Aquaculture Environment Interactions* 9, 269–279.

Wright, D.W., Geitung, L., Karsbakk, E., Stien, L.H., Dempster, T. *et al.* (2018) Surface environment modification in Atlantic salmon sea-cages: effects on amoebic gill disease, salmon lice growth and welfare. *Aquaculture Environment Interactions* 10, 255–265.

Young, N.D., Crosbie, P.B.B., Adams, M.B., Nowak, B.F. and Morrison, R.N. (2007) *Neoparamoeba perurans* n. sp., an agent of amoebic gill disease of Atlantic salmon (*Salmo salar*). *International Journal for Parasitology* 37(13), 1469–1481.

Young, N.D., Dyková, I., Snekvik, K., Nowak, B.F. and Morrison, R.N. (2008a) *Neoparamoeba perurans* is a cosmopolitan aetiological agent of amoebic gill disease. *Diseases of Aquatic Organisms* 78(3), 217–223.

Young, N.D., Dyková, I., Nowak, B.F. and Morrison, R.N. (2008b) Development of a diagnostic PCR to detect *Neoparamoeba perurans*, agent of amoebic gill disease. *Journal of Fish Diseases* 31(4), 285–295.

Young, N.D., Dyková, I., Crosbie, P.B.B., Wolf, M., Morrison, R.N. *et al.* (2014) Support for the coevolution of *Neoparamoeba* and their endosymbionts, *Perkinsela amoebae*-like organisms. *European Journal of Parasitology* 50(5), 509–523.

Zilberg, D. and Munday, B.L. (2000) Pathology of experimental amoebic gill disease in Atlantic salmon, *Salmo salar* L., and the effect of pre-maintenance of fish in sea water on the infection. *Journal of Fish Diseases* 23(6), 401–407.

Zilberg, D. and Munday, B.L. (2001) Responses of Atlantic salmon, *Salmo salar* L., to *Paramoeba* antigens administered by a variety of routes. *Journal of Fish Diseases* 24(3), 181–183.

Zilberg, D., Nowak, B., Carson, J. and Wagner, T. (1999) Simple gill smear staining for diagnosis of amoebic gill disease. *Bulletin of the European Association of Fish Pathologists* 19(5), 186–189.

17 Scuticociliatosis

Jesús Lamas* and José Manuel Leiro

University of Santiago de Compostela, Santiago de Compostela, Spain

17.1 Introduction

Rising levels of atmospheric carbon dioxide and other greenhouse gases, as a consequence of industrial and agricultural activities, cause climate change and acidification of aquatic ecosystems (Rhein *et al.*, 2013). Carbon dioxide dissolves in seawater, making the water more acidic (Caldeira and Wickett, 2003). Acidification and higher seawater temperatures will both have profound effects on marine ecosystems (Pörtner, 2008). Warming will also lower the amount of dissolved oxygen, as warmer waters hold less oxygen (Rhein *et al.*, 2013). Climate change will have important effects on parasitisms and parasitoses in aquatic ecosystems and those pathologies will increase significantly in aquaculture systems (Marcogliese, 2008; De Silva and Soto, 2009). Scuticociliatosis is caused by parasitic ciliates on marine fish and it will probably expand its geographical range due to ocean warming. The disease may be more severe because of thermal stress on the host, especially in cultured fish species that develop optimally at low temperatures. The present chapter describes the main characteristics of fish scuticociliatosis and the effects of temperature, pH and other factors on scuticociliates, and considers how climate change may affect the importance of this disease in the future.

Scuticociliatosis is a disease that affects marine animals, including fish, crustacean and mollusc species. The disease is caused by parasitic ciliates (subclass Scuticociliatia Small, 1967), which are characterized by a scutica, a transient kinetosomal structure that appears during stomatogenesis (Lynn, 2008). Scuticociliates are common in marine habitats worldwide, although they can also be found in fresh water and even in terrestrial habitats (Corliss, 1979; Beaver and Crisman, 1989; Kisand and Zingel, 2000). They are particularly abundant in eutrophic coastal waters of different salinity (Urrutxurtu *et al.*, 2003; Fenchel, 2013), as free-living organisms or as endosymbionts of several hosts (Corliss, 1979; Lynn, 2008). Although scuticociliates feed on bacteria, microalgae or other protozoa, they are facultative parasites and under certain conditions they can infect invertebrate and vertebrate species, including fish, causing scuticociliatosis. Outbreaks of scuticociliatosis have been described in several fish species worldwide, e.g. in Asia, Australia, Europe and America (Jung and Woo, 2012; Piazzon *et al.*, 2013). The scuticociliate *Cyclidium glaucoma* was first described by the Danish naturalist Otto Müller in 1773, and since then more than 300 species of this subclass of ciliates have been described (Grolière, 1980; Whang *et al.*, 2013). However, the systematic relationships between the different species remain unclear. A recent phylogenetic analysis of nuclear ribosomal DNA and mitochondrial DNA gene sequences has divided the subclass Scuticociliatia into three orders, namely Philasterida, Pleuronematida and Loxocephalida, although the status of the latter is uncertain (Zhang *et al.*, 2019). In a few studies, the species causing scuticociliatosis were not identified (Yoshinaga and Nakazoe, 1993; Dyková and Figueras, 1994; Lee *et al.*, 1994; Umehara *et al.*, 2003; Ramos *et al.*, 2007). The ciliates responsible for causing scuticociliatosis mainly belong to the genera *Miamiensis*, *Philasterides*, *Pseudocohnilembus* and *Uronema* (Table 17.1). Most outbreaks have been reported to be caused by *Miamiensis avidus* and *Philasterides dicentrarchi*, which have been considered synonymous (Song and Wilbert, 2000; Paramá *et al.*, 2006; Jung *et al.*, 2007). However, this synonymy has recently been rejected (de Felipe *et al.*, 2017) and it has been suggested that they may be cryptic species (Zhang *et al.*, 2019). Other

*jesus.lamas@usc.es

ciliate species responsible for fish scuticociliatosis include *Uronema marinum*, *Uronema nigricans*, *Pseudocohnilembus persalinus*, *Pseudocohnilembus longisetus* and *Pseudocohnilembus hargasi*, with the first being responsible for causing scuticociliatosis in several fish species (Table 17.1).

Scuticociliatosis has been reported in at least 30 fish species; however, many outbreaks have occurred in farmed flatfish, causing high mortalities in: olive flounder (*Paralichthys olivaceus*) (Zhou *et al.*, 2000; Jee *et al.*, 2001; Kim *et al.*, 2004a; Jung *et al.*, 2005, 2007; Song *et al.*, 2009a; Harikrishnan *et al.*, 2010a, 2012a; Moustafa *et al.*, 2010a); turbot (*Scophthalmus maximus*) (Dyková and Figueras, 1994; Sterud *et al.*, 2000; Iglesias *et al.*, 2001; Alvarez-Pellitero *et al.*, 2004; Wang *et al.*, 2005; Ramos *et al.*, 2007; Budiño *et al.*, 2011b; Kayis *et al.*, 2011; Du *et al.*, 2019); fine flounder (*Paralichthys adspersus*) (Medina *et al.*, 2016; de Felipe *et al.*, 2017); and spotted knifejaw-eye flounder (*Pleuronichthys cornutus*) (Jung *et al.*, 2011a). Several studies have demonstrated the importance, in terms of prevalence and mortality rates, of scuticociliatosis in farmed flatfish (Buchmann, 2015). An 11-year study (January 1991 to December 2001) of olive flounder farms on Jeju Island (South Korea) showed that scuticociliatosis accounted for 10% of the fish mortalities between 1991 and 1996, but the figure increased to 34.5% in 2001 (Jin *et al.*, 2003). Similar findings were obtained in another study carried out on the same island between 1995 and 2004 (Jin *et al.*, 2007). More recently, Kang *et al.* (2015) reported that mortality rates due to scuticociliatosis in olive flounder farms on Jeju Island, between 2007 and 2014, varied between 23.8 and 36.4%. Studies of inland flounder farms in South Korea also revealed a high prevalence of scuticociliatosis. After monitoring the fish mortalities at 70 farms cultivating olive flounder, rockfish, rainbow trout, Japanese eel, white shrimp and abalone in South Korea, Kim *et al.* (2012) found that the cumulative mortality from May to November 2011 was 27.9%, and that 56.64% of the mortalities in farmed olive flounder populations were due to scuticociliatosis. A later survey (May to October 2012) of 565 inland olive flounder farms in South Korea found that the mortality due to infectious diseases was 22.64%, and that the highest mortality was caused by scuticociliatosis (Jee *et al.*, 2014). Moustafa *et al.* (2010a) reported that mortality reached 70–80% in an outbreak of scuticociliatosis in Japanese flounder in July 2005 in Japan. Regarding turbot (*S. maximus*) farms, mortalities of 100% in individual fry units and 30% in the most heavily infected ongrower units were observed in southern Norway (Sterud *et al.*, 2000). Two outbreaks of scuticociliatosis (summer 1999 and spring 2000) were reported in a turbot on-growing facility in northwest Spain, with mortality rates reaching up to 100 to 150 fish per day, and 100% mortality in some of the tanks (Iglesias *et al.*, 2001). Wang *et al.* (2005) conducted an intensive study of turbot farms along the coast of Shandong Province (China) between 2003 and 2004 and reported several outbreaks of scuticociliatosis associated with high mortality rates, sometimes reaching 90%. Several outbreaks of scuticociliatosis occurred in the north of Portugal between March 2004 and February 2005, causing mortality rates of 3–6% in May, July and August (Ramos *et al.*, 2007). More recently, Du *et al.* (2019) reported mass mortality of cultured juvenile turbots in Laizhou City (China).

The causes of the high prevalence of scuticociliatosis in flatfish are not known. Other species such as sea bass are much more resistant than turbot to the disease (Santos *et al.*, 2010). A few authors have suggested that transmission of the pathogen may be facilitated by the tendency of flatfish to aggregate at the bottom of tanks, which increases the proximity of the fish to each other and skin-to-skin contact (Jung *et al.*, 2007). In addition, we have found that *P. dicentrarchi* can feed on fish food suspended in seawater (J. Lamas and J.M. Leiro, 2019, unpublished results). Surplus food is likely to accumulate on the bottom of tanks and come in contact with the fish, thus generating a risk factor for scuticociliatosis infection. Episodes of scuticociliatosis also occur in fish held in aquaria, including several teleost species (e.g. seahorses and sea dragons) and elasmobranchs (e.g. sharks). In a retrospective study (1994 to 2012) carried out to determine the range and occurrence of diseases in aquarium-held sea dragons in Japan, scuticociliatosis was found to account for 34% of the cases (Bonar *et al.*, 2013). Information about the occurrence of scuticociliatosis in wild fish populations is scarce. Mass mortality of leopard sharks (*Triakis semifasciata*) has recently been reported and attributed to *M. avidus* infections, and it was suggested that the episode of scuticociliatosis in this fish species was not an isolated event (Retallack *et al.*, 2019).

Table 17.1. Scuticociliatosis outbreaks reported in fish.

Scuticociliate species	Host	Geographical location	Reference
Philasterides dicentrarchi	Sea bass (*Dicentrarchus labrax*)	Mediterranean Sea (France)	Dragesco *et al.* (1995)
	Turbot (*Scophthalmus maximus*)	Atlantic Ocean (Spain)	Iglesias *et al.* (2001)
	Turbot (*S. maximus*)	Black Sea (Turkey)	Kayis *et al.* (2011)
	Olive flounder (*Paralichthys olivaceus*)	Pacific Ocean (South Korea)	Kim *et al.* (2004a); Jin *et al.* (2006)
	Fine flounder (*Paralichthys adspersus*)	Pacific Ocean (Peru)	de Felipe *et al.* (2017)
	Sea dragon (*Phycodurus eques*)	Pacific Ocean (Japan)	Umehara *et al.* (2003)
	Sea dragon (*Phyllopteryx taeniolatus*)	Aquarium (Switzerland)	Rossteuscher *et al.* (2008)
	Australian potbellied seahorse (*Hippocampus abdominalis*)	Pacific Ocean (Canada)	Di Cicco *et al.* (2013)
	Spotted seahorse (*Hippocampus kuda*)	Pacific Ocean (South Korea)	Shin *et al.* (2011)
	Zebra shark (*Stegostoma fasciatum*)	European aquarium	Stidworthy *et al.* (2014); Su *et al.* (2017)
	Port Jackson shark (*Heterodontus portusjacksoni*)	North American aquarium	
	Japanese horn shark (*Heterodontus japonicus*)	Pacific Ocean (Taiwan)	
Miamiensis avidus	Olive flounder (*P. olivaceus*)	Pacific Ocean (South Korea and Japan)	Jee *et al.* (2001); Jung *et al.* (2005, 2007); Song *et al.* (2009a); Moustafa *et al.* (2010a)
	Fine flounder (*P. adspersus*)	Pacific Ocean (Peru)	Medina *et al.* (2016)
	Spotted knifejaw-eye flounder (*Pleuronichthys cornutus*)	Pacific Ocean (South Korea and Japan)	Jung *et al.* (2011a)
	Spotted knifejaw (*Oplegnathus fasciatus*)		
	New Zealand groper (*Polyprion oxygeneios*)	Pacific Ocean (New Zealand)	Salinas *et al.* (2012)
	Kingfish (*Seriola lalandi*)	Pacific Ocean (New Zealand)	Smith *et al.* (2009)
	Lined seahorse (*Hippocampus erectus*)	Atlantic Ocean (USA)	Thompson and Moewus (1964)
	Leopard shark (*Triakis semifasciata*)	Pacific Ocean (USA)	Retallack *et al.* (2019)
	Dentex (*Dentex dentex*)	Mediterranean Sea (Turkey)	Turgay *et al.* (2015)
	Zebra shark (*S. fasciatum*)	Pacific Ocean (Taiwan)	Li *et al.* (2017)
P. dicentrarchi/ M. avidus	Turbot (*S. maximus*)	Atlantic Ocean (Spain, France, Portugal)	Alvarez-Pellitero *et al.* (2004)
Pseudocohnilembus persalinus	Olive flounder (*P. olivaceus*)	Pacific Ocean (South Korea)	Kim *et al.* (2004b); Song *et al.* (2009a)
Pseudocohnilembus hargasi	Olive flounder (*P. olivaceus*)	Pacific Ocean (South Korea)	Song *et al.* (2009a)
Pseudocohnilembus longisetus	Black rockfish (*Sebastes schlegelii*)	Pacific Ocean (South Korea)	Whang *et al.* (2011)

(*Continued*)

Table 17.1. Continued.

Scuticociliate species	Host	Geographical location	Reference
Uronema marinum	Olive flounder (*P. olivaceus*)	Pacific Ocean (South Korea)	Song *et al.* (2009a)
	Pomfret (*Pampus argenteus*)	Indian Ocean (Kuwait)	Azad *et al.* (2007)
	California sheepshead wrasse (*Pimelometopon pulchrun*)	Atlantic Ocean (USA)	Cheung *et al.* (1980)
	Cunner (*Tautogolabrus adspersus*)		
	Lined seahorse (*H. erectus*)		
	Spotted seahorse (*H. kuda*)		
	Garibaldi (*Hypsypops rubicunda*)		
	Teardrop butterfly (*Chaetodon unimaculatus*)		
	Diagonal butterfly (*Chaetodon auriga*)		
	Cooper-band butterfly (*Chelmon rostratus*)		
	Royal coachman (*Heniochus acuminatus*)		
	Turbot (*S. maximus*)	Pacific Ocean (China)	Du *et al.* (2019)
Uronema nigricans	Southern bluefin tuna (*Thunnus maccoyii*)	Pacific Ocean (Australia)	Munday *et al.* (1997, 2003); Deveney *et al.* (2005)
Uronema sp.	Turbot (*S. maximus*)	Atlantic Ocean (Norway)	Sterud *et al.* (2000)
	Sand whiting (*Sillago ciliata*)	Pacific Ocean (Australia)	Gill and Callinan (1997)
	Vanderbilt's chromis (*Chromis vanderbilti*)	Atlantic Ocean (Brazil)	Cardoso *et al.* (2017)
	Blue-green damselfish (*Chromis viridis*)		
	Sea goldie (*Pseudanthias squamipinnis*)		
Porpostoma notata	Seahorse (*Hippocampus hippocampus*)	Atlantic Ocean (Spain)	Ofelio *et al.* (2014)
Mesanophrys carcini	Turbot (*S. maximus*)	Pacific Ocean (China)	Wang *et al.* (2005)
Paralembus digitiformis	Olive flounder (*P. olivaceus*)	Pacific Ocean (China)	Zhou *et al.* (2000)

17.2 Diagnosis of Scuticociliatosis

17.2.1 Microscopic methods

Scuticociliates were first defined by Small (1967) on the basis of similarities in the morphogenesis and stomatogenesis (Lynn and Strüder-Kypke, 2005; Gao *et al.*, 2013; Foissner, 2014; Liu *et al.*, 2017). Scuticociliates have a small to medium body size, with uniform body ciliature, sometimes sparse, and one to several distinct caudal cilia are typically present (Corliss, 1979). They are characterized by having dikinetids in the buccal apparatus (paraoral dikinetids) that are located in several (usually three) membranelles or polykineties. A characteristic of all members is the presence of a scutica, a hook-shaped structure formed during stomatogenesis that contains kinetosomes (Fig. 17.1A to D), which constitutes the biggest synapomorphy of the group (Corliss, 1979; Lynn, 2008; Pan *et al.*, 2016). However, due to the high degree of morphological similarity between scuticociliate species, their identification is difficult, often leading to misdiagnosis (Pan *et al.*, 2016; de Felipe *et al.*, 2017; Zhang *et al.*, 2019). Scuticociliatosis has traditionally been

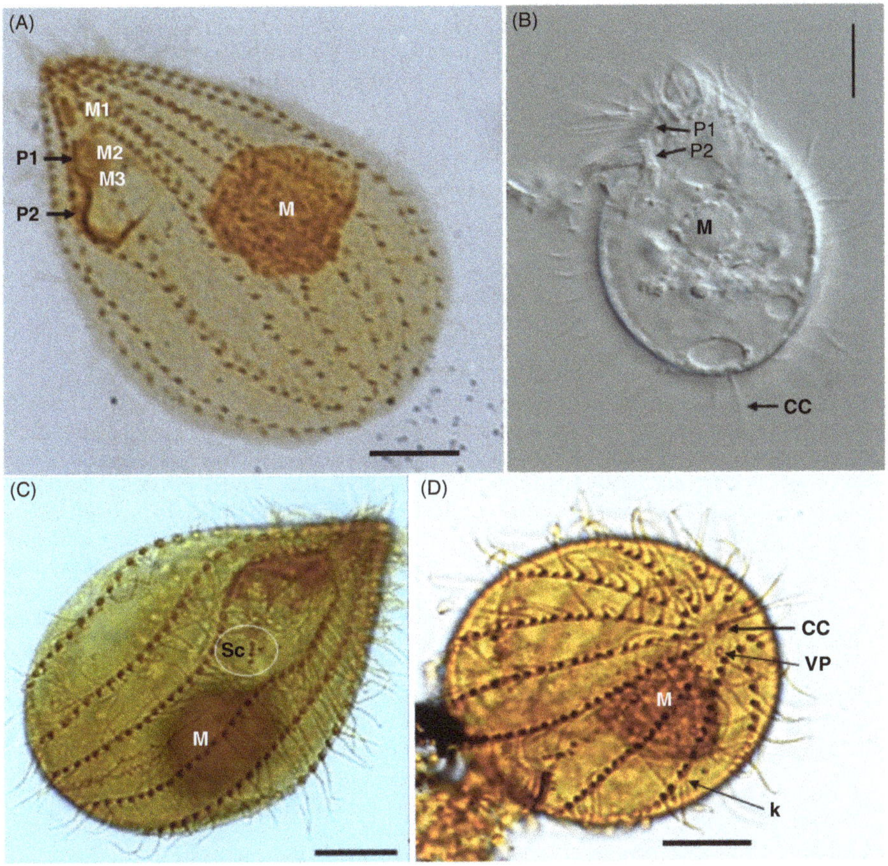

Fig. 17.1. (A, C, D) Silver carbonate-impregnated scuticociliates (*Philasterides dicentrarchi*). (A) Detail of the oral infraciliature showing the three oral polykinetids (M1, M2 and M3) and two paraoral membranes (P1 and P2). (B) Differential interference contrast (DIC) microscopy of a ciliate showing the two paraoral membranes (P1 and P2), the macronucleus (M) and the caudal cilium (CC). (C) Trophont showing the scutica (Sc) at the base of the cytostome. (D) Posterior end where 12 kineties (k) can be counted, and the caudal cilium and the pore (VP) of the contractile vacuole are shown. Scale bars = 10 µm.

diagnosed on the basis of microscopic observations of living cells by using Nomarski or phase-contrast microscopy, or after staining the ciliates with silver impregnation methods to reveal the infraciliature and silverline system (Dragesco *et al.*, 1995; Munday *et al.*, 1997; Song, 2000; Iglesias *et al.*, 2001; Jee *et al.*, 2001; Kim *et al.*, 2004b; Jung *et al.*, 2007). The most common silver impregnation methods used are the Chatton–Lwoff silver nitrate method (Foissner, 1991), the silver carbonate method (Ma *et al.*, 2003), the ammoniacal silver carbonate method (Fernández-Galiano, 1994; Budiño *et al.*, 2011a) and the Wilbert's protargol method (Wilbert, 1975). Other stains have also proved useful for identifying scuticociliates: the Feulgen stain and 4′,6-diamidine-2-phenylindone (DAPI), which stain the nuclear apparatus (de Felipe *et al.*, 2017); and the supravital stain methyl green-pyronin, used to reveal the nuclear morphology and to stain mucocysts in many ciliates (Foissner, 1979, 2014; Abraham *et al.*, 2019). The most important morphological characteristics used to identify scuticociliates include the shape and dimensions of the body (body length, body width and body length/body width ratio), oral ciliature (length of buccal field, length of buccal field/body length ratio, length and number of paraoral membranes, and length of oral polykinetids), somatic ciliature (number of somatic kineties (SK), number of basal bodies on SK and in the scutica, and position of contractile vacuole pore) and the nuclear size (length and width of macronucleus and micronucleus) (Song, 2000; Kim *et al.*, 2004b; Budiño *et al.*, 2011a; Fan *et al.*, 2011). Scanning and transmission electron microscopy techniques have also been used to analyse three-dimensional characteristics of the ciliate, the oral apparatus or the ultrastructure of the trichocysts and other extrusomes, as well as the subpellicular fibre system of several species of scuticociliates (Kaneshiro and Holz, 1976; Peck, 1977; Foissner, 1991; Alvarez-Pellitero *et al.*, 2004; Paramá *et al.*, 2006; Katsaros *et al.*, 2011; Chen *et al.*, 2014).

17.2.2 Molecular methods

Diagnosing scuticociliatosis on the basis of morphological characteristics can be a difficult task due to the small size of the ciliates and the similar infraciliature and silverline patterns in closely related species (Pan *et al.*, 2016). Thus, although the morphological analysis of ciliates remains useful, interpretation of the results is subjective and may lead to misidentification (Whang *et al.*, 2013; Abraham *et al.*, 2019). In addition, identification of the species responsible for causing scuticociliatosis may be further complicated by the high degree of morphological variation in the scuticociliates cultured *in vitro* (Jung *et al.*, 2005; Miao *et al.*, 2008; Budiño *et al.*, 2011a; Salinas *et al.*, 2012; Gao *et al.*, 2013; de Felipe *et al.*, 2017). Scuticociliates are among the most problematic ciliate taxa regarding their systematic relationships (Gao *et al.*, 2010) and their identification may also be complicated by the existence of cryptic species (Zhang *et al.*, 2019). At present, two major approaches are being used to identify ciliates: morphological methods and (more recently) DNA-based methods (Abraham *et al.*, 2019). Nuclear and mitochondrial gene sequences are being used to identify species and to establish the phylogenetic relationship between scuticociliates (Whang *et al.*, 2013; Zhang *et al.*, 2019). In the case of nuclear genes, the sequences most commonly used are those of the small-subunit ribosomal RNA (SSU-rRNA) gene (Kim *et al.*, 2004b; Jung *et al.*, 2005, 2011a,b) and, to a lesser extent, the sequences corresponding to the ITS1-5.8S-ITS2 region (Ma *et al.*, 2018). In order to improve the identification and resolution of the phylogenetic relationships, a combination of three nucleotide sequences (the ITS1-5.8S-ITS2 region, the large-subunit ribosomal RNA (LSU-rRNA) and SSU-rRNA) (Gao *et al.*, 2012a,b, 2013, 2014) or the α- and β-tubulin gene sequences (de Felipe *et al.*, 2017) have been used. In relation to the mitochondrial genes, the mitochondrial cytochrome c oxidase subunit 1 (Cox1) has been used for phylogenetic analysis (Zhang *et al.*, 2019), as well as to analyse intraspecific differences between isolates (Budiño *et al.*, 2011a; Jung *et al.*, 2011b; Whang *et al.*, 2013; Garza *et al.*, 2017).

In order to diagnose scuticociliatosis by these methods, DNA is extracted from fresh fish organs or from paraffin-embedded tissue, and a fragment of a specific gene is amplified using polymerase chain reaction (PCR) (with specific or universal primers for Ciliophora) and then cloned and sequenced. Finally, the sequences are compared with those available in the GenBank nucleotide sequence database for similarity, after a BLAST search to identify the closest matching species (Rossteuscher *et al.*, 2008; Smith *et al.*, 2009; Shin *et al.*, 2011; Ofelio *et al.*, 2014; Stidworthy *et al.*, 2014; Li *et al.*, 2017). In some cases, a primer set is used for specific detection of gene sequences of particular scuticociliate species, and diagnosis is carried out by PCR, multiplex PCR or even quantitative PCR (Kim *et al.*, 2004a; Tange *et al.*, 2010; Power *et al.*, 2019). Other methods of ciliate identification include random amplified polymorphic DNA-fingerprinting (RAPD) (Smith *et al.*, 2009; Liu *et al.*, 2017), PCR/restriction fragment length polymorphisms (RFLP) (Lynn and Stüder-Kypke, 2005) and even metagenomic next-generation sequencing (Retallack *et al.*, 2019). *In situ* hybridization of an SSU-rRNA-targeted oligonucleotide probe labelled with a fluorochrome has also been used to identify scuticociliates (Zhan *et al.*, 2014). Immunological techniques such as Western blot and enzyme-linked immunosorbent assays (ELISAs) have mainly been used to identify serotypes of scuticociliate isolates/strains (Iglesias *et al.*, 2003a; Palenzuela *et al.*, 2009; Song *et al.*, 2009b; Budiño *et al.*, 2012a; de Felipe *et al.*, 2017). Identification of scuticociliates at serotype level is important for the development of vaccines (Piazzon *et al.*, 2008; Budiño *et al.*, 2012a). In this respect, and although it would be costly and time-consuming, a combination of morphological, biological, molecular (by multigene analysis) and serological techniques is required for the complete identification and characterization of scuticociliate parasites of fish (de Felipe *et al.*, 2017).

17.3 Pathology of Scuticociliatosis

Pathogenic scuticociliate species are highly histophagous and cause severe damage in infected fish. Clinical signs can include anorexia, lethargy, irregular respiration and death (Di Cicco *et al.*, 2013). A few studies have also reported swimming disturbances, such as spinning and other movements (Sterud *et al.*, 2000; Iglesias *et al.*, 2001; Moustafa and Mousa-Balabel, 2011). External macroscopic observations of infected fish usually include skin lesions (Fig. 17.2A), which can include discoloration with pale epithelial erosions and ulcers of variable size and shape, some of them haemorrhagic (Zhou *et al.*, 2000; Iglesias *et al.*, 2001; Azad *et al.*, 2007; Ramos *et al.*, 2007; Rossteuscher *et al.*, 2008; Moustafa *et al.*, 2010a; Kubiski *et al.*, 2011; Turgay *et al.*, 2015; Cardoso *et al.*, 2017; de Felipe *et al.*, 2017). These lesions can also affect the fins and ocular or cranial regions (Turgay *et al.*, 2015). In some fish, haemorrhagic skin ulcers extend into the underlying muscle (Iglesias *et al.*, 2001; Jung *et al.*, 2007; Ramos *et al.*, 2007). Other clinical signs include pale gills, gill erosions, exophthalmia and distended abdominal cavity (Fig. 17.2B; Dragesco *et al.*, 1995; Sterud *et al.*, 2000; Iglesias *et al.*, 2001; Azad *et al.*, 2007; Jung *et al.*, 2007; Ramos *et al.*, 2007; Stidworthy *et al.*, 2014; Turgay *et al.*, 2015). However, fish can also die without showing any clinical or gross pathological changes (Sterud *et al.*, 2000; Iglesias *et al.*, 2001). Ciliates can be found in almost any organ of infected fish. The parasite enters the fish through lesions in the gills or skin, or through the periorbital area or the nasal region (Paramá *et al.*, 2003; Jin *et al.*, 2009; Moustafa *et al.*, 2010b). During the second phase of infection, the parasites generate a systemic infection that affects different organs (brain, liver, digestive tract, kidney, spleen and heart) (Fig. 17.3A to D) and generates ascites (Dyková and Figueras, 1994; Dragesco *et al.*, 1995; Munday *et al.*, 1997; Iglesias *et al.*, 2001; Jee *et al.*, 2001; Padrós *et al.*, 2001; Deveney *et al.*, 2005; Ramos *et al.*, 2007;

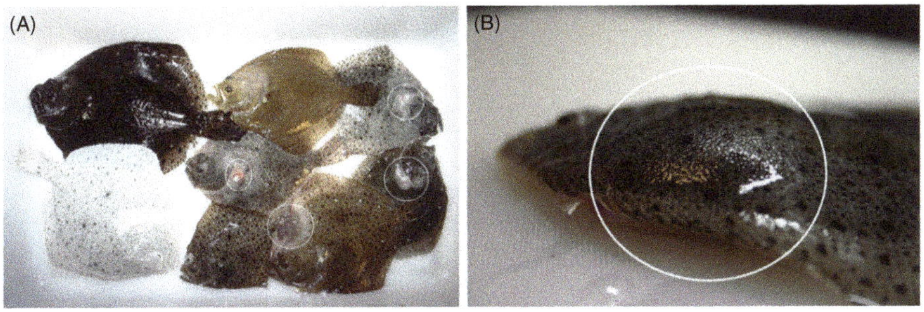

Fig. 17.2. (A) Naturally infected turbot showing characteristic symptoms of scuticociliatosis: discoloration and skin lesions localized in the vicinity of the operculum (circles). (B) Turbot infected experimentally showing prominent swollen abdomen due to the accumulation of ascites (circle).

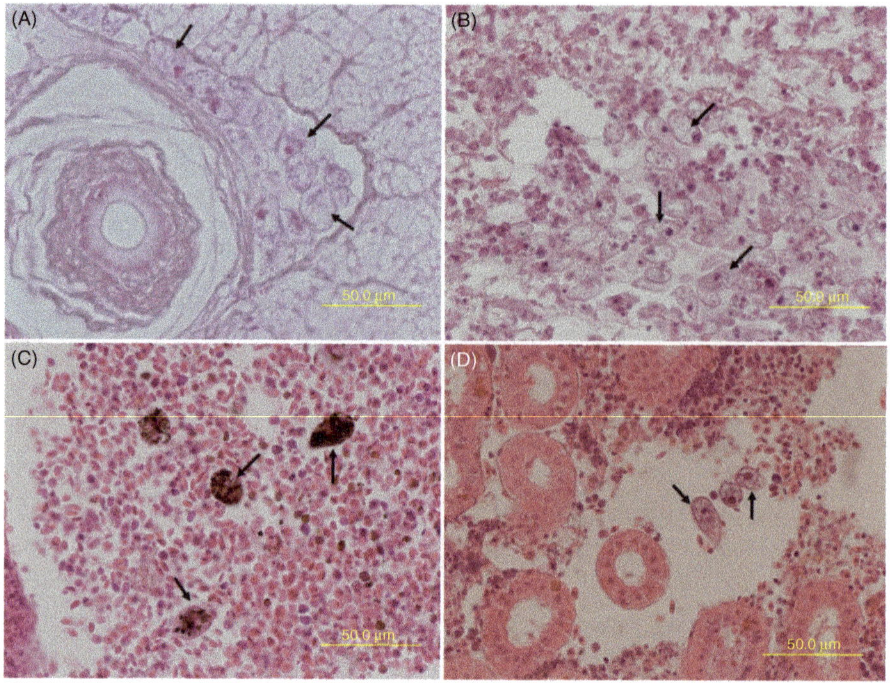

Fig. 17.3. Histological sections from turbot infected with the scuticociliate *Philasterides dicentrarchi*, showing the presence of ciliates (arrows) in the liver (A), pericardium (B), blood vessel (C) and kidney (D). Haematoxylin and eosin stain.

Rossteuscher *et al.*, 2008; Jin *et al.*, 2009; Moustafa *et al.*, 2010a; Harikrishnan *et al.*, 2012a).

Scuticociliates proliferate in the epidermis, dermis and the subjacent muscle, leading to haemorrhaging, epidermis detachment and necrosis, as well as dermal and muscle necrosis, sometimes accompanied by leucocyte infiltration (Lee *et al.*, 1994; Jee *et al.*, 2001; Padrós *et al.*, 2001; Umehara *et al.*, 2003; Azad *et al.*, 2007; Jung *et al.*, 2007; Ramos *et al.*, 2007; Rossteuscher *et al.*, 2008; Moustafa *et al.*, 2010a,b). In the gills, ciliates can occur as free-living forms on the exterior epithelial cells, without causing apparent damage (Padrós *et al.*, 2001), or within the organs (Dyková and Figueras, 1994; Azad *et al.*, 2007; Jung *et al.*, 2007; Ramos *et al.*, 2007). Scuticociliates have been described in the subepithelial connective tissue of the digestive tract, in the gut submucosa and in the gastric and intestinal serosa (Padrós *et al.*, 2001; Jung *et al.*, 2007). They produce necrosis of the epithelium, loss of the mucosa and degeneration of the gastric musculature (Azad *et al.*, 2007). In other cases, no obvious pathology was detected in the epithelium or muscular layers of the intestine, despite fish being heavily infected with ciliates (Jung *et al.*, 2007). In relation to the inflammatory response, mononuclear cell infiltration has also been observed in the lamina propria of the gastrointestinal tract (Moustafa *et al.*, 2010a). In the liver, abundant ciliates have been observed in the hepatic parenchyma and in outermost areas of connective tissue, around the bile ducts and veins (Sterud *et al.*, 2000; Padrós *et al.*, 2001; Harikrishnan *et al.*, 2010a, 2012a; Stidworthy *et al.*, 2014). Ciliates have also been observed in the anterior kidney (Padrós *et al.*, 2001; Moustafa *et al.*, 2010a) and inside the renal tubules, causing severe necrosis of the tubular epithelium (Azad *et al.*, 2007). Parasites have often been detected in the nervous system of infected fish, causing haemorrhaging and liquefactive necrosis (Lee *et al.*, 1994; Munday *et al.*, 1997; Sterud *et al.*, 2000; Iglesias *et al.*, 2001; Padrós *et al.*,

2001; Azad *et al.*, 2007; Jung *et al.*, 2007; Rossteuscher *et al.*, 2008; Moustafa *et al.*, 2010b; Harikrishnan *et al.*, 2012a), as well as meningoencephalitis (Stidworthy *et al.*, 2014; Li *et al.*, 2017). Because scuticociliates are common in the brain of infected fish, direct neural invasion through the nasal cavity/ampullary system has been suggested (Moustafa *et al.*, 2010b; Li *et al.*, 2017). The parasites have also been observed in the heart (Puig *et al.*, 2007), appearing in the epicardium, endocardium and inside the ventricle (Sterud *et al.*, 2000; Padrós *et al.*, 2001). Scuticociliates have also been found in the pancreas and gonads, mainly occupying zones rich in connective tissue (Dyková and Figueras, 1994; Padrós *et al.*, 2001).

17.4 Propagation and Transmission

Scuticociliates are normally free-living organisms that can be commensals or pathogens of fish, crustaceans and molluscs (Lynn and Strüder-Kypke, 2005; Harikrishnan *et al.*, 2010b; Pan, 2016; Tao *et al.*, 2016; Liu *et al.*, 2017). The ciliate species that cause scuticociliatosis in marine invertebrates include species within the genera *Metanophrys* (Sahoo *et al.*, 2018), *Mesanophrys* (Small *et al.*, 2005), *Paramesanophrys* (Lynn, 2008), *Anophryoides* (Cawthorn *et al.*, 1996; Athanasopoulou *et al.*, 2004) and *Miamiensis* (Tao *et al.*, 2016), among others. Several scuticociliate species (Table 17.1), including *P. dicentrarchi* (Dragesco *et al.*, 1995), *M. avidus* (Thompson and Moewus, 1964), *P. persalinus* (Kim *et al.*, 2004b), *P. hargasi* (Song *et al.*, 2009a), *P. longisetus* (Whang *et al.*, 2011), *U. marinum* (Jee *et al.*, 2001), *U. nigricans* (Deveney *et al.*, 2005), *Uronema* sp. (Sterud *et al.*, 2000), *Porpostoma natata* (Ofelio *et al.*, 2014), *Mesanophrys carcini* (Smith *et al.*, 2009) and *Paralembus digitiformis* (Gong and Li, 2007), can cause scuticociliatosis in fish. A few species such as *M. avidus* can cause mortalities in both invertebrates and fish (Tao *et al.*, 2016). It is possible that episodes of scuticociliatosis in wild fish, crustaceans and molluscs in the surroundings of fish farms could facilitate transmission of the parasites to farmed fish. However, the prevalence of scuticociliatosis in wild marine organisms has not been documented, and transmission of pathogenic scuticociliates between farmed and wild marine fish or invertebrates surrounding fish farms has not been demonstrated. We have sampled the main feeder channel and fish tanks in two turbot farms, with the aim of detecting *P. dicentrarchi*. The parasites are always present, and they are much more abundant inside the tanks than at the entrance, indicating that they proliferate inside the tanks (J. Lamas and J.M. Leiro, 2019, unpublished results). Scuticociliates are usually scavengers that feed on bacteria and dead tissue and the best method of preventing infection is to improve tank hygiene (Yokohama *et al.*, 2015). Cleaning reduces the number of parasites in the tanks and also decreases the risk of infection; however, it may not be sufficient to prevent it. Scuticociliates probably move continuously between the fish farm and the environment. Inland fish farms release organic matter that can accumulate within a radius of 1 km around the farms (Wu, 1995), generating anoxic conditions and modifying benthic communities by e.g. increasing the numbers of bacteria (Bongiorni *et al.*, 2005). Scuticociliates are microaerophilic, benthic organisms that live close to the sediment–water interface, and they are highly adapted to low levels of dissolved oxygen (Fenchel, 1968, 1990; Wilbert, 1995; Shimeta and Sisson, 1999; Mallo *et al.*, 2013; Weisse, 2017). However, some scuticociliates such as *P. dicentrarchi* can adapt to a wide range of oxygen levels, which may allow them to be endoparasites (Mallo *et al.*, 2013, 2014). The increase in the numbers of scuticociliates in areas surrounding fish farms probably increases the number of ciliates that reach the tanks, where nutrients are also available. In this respect, and because free-living scuticociliates can feed on bacteria or microalgae (Yokohama *et al.*, 2015), the presence of organic matter or bacteria enables the numbers of scuticociliates to increase. The abundance of bacteria and protists is generally expected to increase with global warming, but significant shifts in community composition may also occur (Smale *et al.*, 2017). *Uronema* is highly dependent on the density of bacteria in the medium (Crosbie and Munday, 1999; Zhang *et al.*, 2001; Al-Marzouk and Azad, 2007). Moreover, silver pomfret mortalities caused by *Uronema* sp. have been found to be related to increased seawater temperature and an increased total bacterial load in the water in rearing tanks (Al-Marzouk and Azad, 2007).

17.4.1 Effect of increased temperature on ciliate multiplication

Temperature has a direct effect on scuticociliate multiplication. Thus, an increase in temperature caused higher growth rates in the scuticociliates

Orchitophrya cf. *stellarum* (Bates *et al.*, 2010), *Paranophrys magna* (Zhang and Song, 2000), *P. dicentrarchi* (Iglesias *et al.*, 2003b) and *Uronema* sp. (Al-Marzouk and Azad, 2007). In addition, ocean warming will exacerbate biofouling (Khosravi *et al.*, 2019), which has been shown to act as a reservoir of some scuticociliates (Power *et al.*, 2019). In the light of these observations, an increase in scuticociliate populations in the environment is expected to occur as global warming proceeds. However, the problem may be rather more complex, because of the high diversity of marine ciliates in seawater (Song *et al.*, 2009; Liu *et al.*, 2017) which may compete with each other. Environmental factors such as temperature strongly impact ciliate communities, determining species distribution (Finlay *et al.*, 1997). In addition, temperature also affects the multiplication of dinoflagellates (Jeong *et al.*, 2018), some of which are toxic to scuticociliates (Kim *et al.*, 2017). For all these reasons, further studies are needed to evaluate the impact of rising temperature on scuticociliate populations in marine ecosystems. Another important aspect is what occurs on fish farms, mainly inland farms. In a two-year-long study, we determined the density of *P. dicentrarchi*, the ratio *P. dicentrarchi*/other ciliates and the density of bacteria in seawater samples from several tanks on two turbot farms, and observed a good correlation between the water temperature and the density of *P. dicentrarchi*. The density of both *P. dicentrarchi* and bacteria and the ratio of *P. dicentrarchi*/other protozoa increased with temperature, indicating that scuticociliates may have some advantage inside the fish tanks, relative to other ciliate species (J. Lamas and J.M. Leiro, 2019, unpublished results).

17.4.2 Effect of temperature on ciliate physiology and virulence

Temperature has an important effect on parasite virulence (Lõhmus and Björklund, 2015; Sheath *et al.*, 2016). There is limited information on the mechanisms that are involved in virulence. However, proteases have been described as virulence factors in scuticociliates (Paramá *et al.*, 2004a, 2007a,b; Piazzon *et al.*, 2011b; Narasaki *et al.*, 2018). Higher proteolytic activity has been observed in ciliates from infected fish than in ciliates cultured *in vitro* (Al-Marzouk and Azad, 2007). Similarly, the expressions of genes encoding cysteine peptidases and leishmanolysin-like peptidases were upregulated in the cell-fed ciliates and in ciliates from an infection relative to the starved and cultured ciliates, respectively (Seo *et al.*, 2013; Mallo *et al.*, 2017), which suggests that active ciliates produce more proteases than resting ones. The proteolytic activity in short-term scuticociliate cultures is higher than in long-term cultures, which are usually less virulent (Kwon *et al.*, 2003). Piazzon *et al.* (2011b) showed that *P. dicentrarchi* proteases may favour infection because they can destroy fish complement and antibodies. As enzyme activity increases with temperature, an increase in water temperature will probably increase protease activity during infection. In summary, temperature will probably affect the three factors most likely to influence infection success: increased numbers of ciliates in seawater, decreased fish immune response (after long-term exposure to high temperatures) and increased virulence of scuticociliates.

17.4.3 Effect of temperature on scuticociliatosis

Disease prevalence

Temperature affects the prevalence and fish mortality due to scuticociliatosis. A survey of scuticociliatosis in farmed olive flounder on Jeju Island, carried out between 1995 and 2004, showed that the prevalence was highest between May and September, when water temperature was high (Jin *et al.*, 2007). Scuticociliatosis caused by *M. avidus* also occurred in July 2005 in the same fish species cultured in Japan (Moustafa *et al.*, 2010a). An experimental study showed that cumulative mortality caused by *M. avidus* was low at 10°C and increased at higher water temperatures (Bae *et al.*, 2009). *M. avidus* has also been shown to cause mortalities in *Dentex dentex* at 18°C (Turgay *et al.*, 2015). Scuticociliatosis outbreaks by *P. dicentrarchi* occurred at 18–20°C in turbot but some of them were preceded by periods of high water temperatures (>20°C) (Iglesias *et al.*, 2001; Alvarez-Pellitero *et al.*, 2004). Another study showed that turbot mortality due to *P. dicentrarchi* infections was particularly high during the summer months (Ramos *et al.*, 2007). Infections also occasionally occurred at lower temperatures. An outbreak caused by *P. dicentrarchi* caused mortality in sea dragons maintained in aquaria at temperatures of between 14 and 18.5°C (Rossteuscher *et al.*, 2008). Outbreaks of scuticociliatosis due to *P. dicentrarchi* were also reported in fine flounder (*P. adspersus*) farmed in Peru at water temperatures higher than 20°C, which lasted for several months and were associated with the El Niño climate cycle (de Felipe *et al.*, 2017). Temperature is also a very important factor in fish infections caused by other scuticociliates. Thus, *Uronema* sp. generated mortalities in turbot at 17°C (Sterud *et al.*, 2000) and in silver pomfret (*Pampus argenteus*) at 20–22°C (Azad *et al.*, 2007). Mortalities associated with an unidentified scuticociliate have also been reported in squarespot anthias (*Pseudanthias pluerotaenia*) at 24°C (Kubiski *et al.*, 2011).

Occurrence of infection

Establishing how the disease starts and how fish-to-fish transmission of parasites occurs is important for understanding the occurrence of scuticociliatosis on fish farms. In experimental infections, external lesions may favour entry of parasites into the fish (Paramá *et al.*, 2003; Lim *et al.*, 2005; Al-Marzouk and Azad, 2007), although other studies have shown that fish apparently lacking lesions on the skin and gills also became infected (Jung *et al.*, 2007; Jin *et al.*, 2009). Nevertheless, the health of skin and gills is important for entry of the parasite. Blood is a potent chemoattractant for scuticociliates, and the existence of microhaemorrhages may help the trophonts to locate lesions and penetrate the fish (Paramá *et al.*, 2004a). In addition, the active forms of scuticociliates display a higher capacity to produce proteases than the free-living forms (Xiong *et al.*, 2015), which may facilitate the entry and dissemination of parasites into the host via digestion of tissues (Paramá *et al.*, 2004b). Scuticociliatosis has been described in a large number of fish species (Table 17.1) and ocean warming may have different effects at physiological and immunological levels. An increase in temperature caused moderate damage to gill tissue in tropical fish species such as *Lophiosilurus alexandri* (Takata *et al.*, 2018). In 2015/16, El Niño generated a 2.5°C increase in seawater temperature in the vicinity of the Galapagos Islands; the increase lasted 6 months and coincided with the appearance of ulcerative skin disease affecting 18 teleost species (e.g. *Holacanthus passer*, *Microspathodon dorsalis*, *Chromis atrilobata*, *Epinephelus labriformis*, *Muraena lentiginosa*) from 13 different families of Galapagos marine fishes (Lamb *et al.*, 2018). This type of episode is likely to increase in the future (Wang *et al.*, 2017), and alterations to gills or skin health are likely to affect the development of scuticociliatosis.

Host immune responses

Temperature has an important impact on fish physiology, including the immune system (Bly and Clem, 1992; Makrinos and Bowden, 2016). In general, an increase in temperature, within physiological limits, improves the innate and adaptive immune responses (Rijkers *et al.*, 1980; Secombes *et al.*, 1991; Martínez *et al.*, 2018). However, variations in water temperature outside the physiological range can modify the immune response, generating immunosuppression and making fish more susceptible to infections by pathogens (Woo *et al.*, 1983; Bly and Clem, 1991, 1992; Houston *et al.*, 1996; Ndong *et al.*, 2007; Verma *et al.*, 2007). Nevertheless, the relationship between temperature and fish immune responses is complex (Makrinos and Bowden, 2016) and determining how temperature affects the host–pathogen interaction may help to clarify the effects on a particular disease. The complement system and the coagulation system are key components in fish defence against *P. dicentrarchi* (Leiro *et al.*, 2008; Piazzon *et al.*, 2011a, 2013; Blanco-Abad *et al.*, 2018). Summer conditions in farmed Russian sturgeon were found to induce a significant decrease in the alternative complement pathway (Castellano *et al.*, 2017). As mentioned above, most of these outbreaks of scuticociliatosis in farmed flatfish occurred in the summer months. However, the water temperature was only occasionally reported, and it is not possible to establish whether it was outside the physiological ranges for the species affected. Flatfish species, such as the olive flounder (*P. olivaceus*) and the fine flounder (*P. adspersus*), which often suffer outbreaks of scuticociliatosis, can grow at temperatures between 10 and 30°C, although the optimum is around 20–25°C (Iwata *et al.*, 1994; Dou *et al.*, 2005; Silva and Oliva, 2010; Cho *et al.*, 2012), i.e. a wide range. In this respect, *P. adspersus* was affected by an outbreak of scuticociliatosis after being held for several months at temperatures >20°C, although the highest temperature reached was not reported (de Felipe *et al.*, 2017). Similarly, turbot (*S. maximus*), which is also highly susceptible to scuticociliatosis, can grow at temperatures between 6 and 22°C, with the optimal temperature ranging between 12 and 18°C (Burel *et al.*, 1996; Martínez-Tapia and

Fernández-Pato, 1991; Kimsland *et al.*, 2001; Sahin, 2001; Aksungur *et al.*, 2007). Outbreaks in farmed turbot have been reported to occur at temperatures higher than 20°C (Iglesias *et al.*, 2001) but also at 18–20°C (Alvarez-Pellitero *et al.*, 2004), suggesting that, at least in the last case, fish were not immunocompromised as a result of temperature. However, because the scuticociliatosis outbreaks occurred mainly in summer months, some of them were probably facilitated by thermal stress effects, alone or in combination with other factors such as stock density. In this respect, climate change would probably make some fish species even more susceptible to scuticociliatosis, and currently resistant fish may become susceptible.

17.4.4 Effects of other environmental factors (salinity, pH) on scuticociliatosis

Other environmental factors such as salinity also seem to affect scuticociliate infection. Hyposaline conditions can favour the development of scuticociliatosis caused by *M. avidus* in olive flounder (Takagishi *et al.*, 2009); salinities between 22 and 28 ‰ favour proliferation of this scuticociliate. Despite their preference for a range of salinities, the pathogenic scuticociliates seems to be euryhaline, as observed in *M. avidus* (Kaneshiro *et al.*, 1969), *P. dicentrarchi* (Iglesias *et al.*, 2003b) and *U. marinum*, which can proliferate at a wide range of salinities (10 to 35 ‰) (Jee *et al.*, 2001).

Ocean acidification is expected to increase as a consequence of increasing atmospheric carbon dioxide levels (Rhein *et al.*, 2013). The average pH of ocean surface waters has already fallen by about 0.1 units, from about 8.2 to 8.1 (total scale), since the beginning of the industrial revolution (Orr *et al.*, 2005). Scuticociliates appear to tolerate a wide range of pH. In the case of *U. marinum*, growth was observed over a pH range of 6.6 to 8.3, with the optimum close to 7.3 (Hanna and Lilly, 1970). Similarly, *P. dicentrarchi* can survive and proliferate at pH 6.2 to 8.2, although the greatest proliferation was at pH 7.2 (Iglesias *et al.*, 2003b). *M. avidus* has also been reported to survive at pH 5 to 10 (Jung *et al.*, 2007). Due to wide pH tolerance, the impact of ocean acidification on scuticociliates may be rather limited or might even favour ciliate proliferation, as optimum growth seems to occur at around pH 7.2–7.3. Nevertheless, it is also possible that pH could affect the fish immune system and resistance to scuticociliates. However, not much is known about the effects water acidification has on the fish immune system (Makrinos and Bowden, 2016).

17.5 Control and Prevention of Scuticociliatosis

17.5.1 Chemicals, antibiotics and other compounds

One measure used to prevent or decrease fish mortality due to scuticociliatosis on fish farms is to kill the scuticociliates present in seawater. Numerous compounds, including antibiotics, antifungals and antiparasitics, have been tested *in vitro*, and many have been shown to be toxic to scuticociliates (Table 17.2).

Formalin and hydrogen peroxide are effective against scuticociliates and could be used to decrease the levels of the parasites in seawater. A survey carried out on South Korea olive flounder farms between May and October 2012 found that formalin was commonly used to treat external parasites (Jee *et al.*, 2014). Formalin baths can effectively reduce scuticociliate populations in fish tanks, preventing the spread of ciliates to uninfected fish. In this respect, bath treatment of olive flounder with 100 ppm formalin did not eliminate the internal parasites but was effective for preventing the spread of scuticociliatosis (Jee and Jo, 2002). Formalin treatment (100 ppm for 45 min) did not prevent mortalities due to scuticociliatosis in infected silver pomfret (*P. argenteus*), but it was effective for fish that had no clinical symptoms and were reared in separate tanks (Azad *et al.*, 2007). Olive flounder farms manage scuticociliatosis by treating infected fish with antibiotics (oxytetracycline, gentamycin, tetracycline, amoxycillin and cefazolin), alone or in combination with formalin, Jenoclean or sodium chloride, at concentrations of 350–150 ppm (Jin *et al.*, 2010). However, formaldehyde is considered a carcinogenic, mutagenic and reproductive toxicant and its use is restricted in many countries (Hunt and Dale, 2018). Hydrogen peroxide could be used as an alternative to formalin. It is currently used as a biocide in veterinary medicine and also to treat water, and it can kill *U. marinum* when applied at 100 ppm for 90 min (Jee *et al.*, 2002) or *P. dicentrarchi* when applied at 300 ppm for 60 min (Harikrishnan *et al.*, 2010c) or at 25–80 ppm, depending on the strain, for 24 h (Budiño *et al.*, 2012b).

Table 17.2. Compounds with anti-scuticociliate activity.

Species	Chemotherapeutic agent	Host	*In vitro* treatment	*In vivo* treatment	Reference
Uronema nigricans	Formalin	–	100, 200 ppm for 120 and 60 min, respectively	–	Crosbie and Munday, (1999)
	Hydrogen peroxide + malachite green		250 ppm + 2 ppm for 60 min		
Uronema marinum	Formalin	–	50 ppm for 90 min	–	Jee *et al.* (2002)
	Hydrogen peroxide		100 ppm for 90 min		
	Copper sulfate		100 ppm for 90 min		
Uronema sp.	Formalin + neomycin sulfate	Vanderbilt's chromis (*Chromis vanderbilti*) Blue-green damselfish (*Chromis viridis*) Sea goldie (*Pseudanthias squamipinnis*)	–	Three baths at 46.25 ppm for 60 min + 60 ppm for 5 days	Cardoso *et al.* (2017)
Philasterides dicentrarchi	Formalin	–	62 ppm for 120 min	–	Iglesias *et al.* (2002)
		–	300 ppm for 30 min, 200 ppm for 60 min, 100 ppm for 90 min	–	Jin *et al.* (2010)
		–	400 ppm	–	Harikrishnan *et al.* (2010c)
	Hydrogen peroxide	–	200 ppm	–	
	Jenoclean	–	100 ppm	–	
	Niclosamide	–	0.8 ppm for 120 min	–	Iglesias *et al.* (2002)
	Oxyclozanide	–	0.8 ppm for 24 h	–	
	Bithionol sulfoxide	–	25 ppm for 24 h	–	
	Toltrazuril	–	50 ppm for 24 h	–	
	N-(2′-hydroxy-5′-chloro-benzoyl)2-chloro-4-nitroaniline	–	6 ppm for 24 h	–	
	Furaltadone	–	25 ppm for 24 h	–	
	Carnidazole	–	100 ppm for 24 h	–	
	Pyrimethamine	–	100 ppm for 24 h	–	
	Quinacrine sulfate	–	100 ppm for 24 h	–	
	Pyridothienotriazine	–	0.8 ppm for 24 h	–	Quintela *et al.* (2003); Paramá *et al.* (2004c)
	Metronidazole	Seahorse (*Hippocampus abdominalis*)	–	50 ppm (bath) for 10 days	Di Cicco *et al.* (2013)
	Indomethacin	–	32 ppm for 72 h[a]	–	Paramá *et al.* (2007c)
	Pamidronate	–	220 ppm for 24 h[a]	–	Mallo *et al.* (2015)

(Continued)

Table 17.2. Continued.

Species	Chemotherapeutic agent	Host	*In vitro* treatment	*In vivo* treatment	Reference
	Chloroquine	–	20 ppm for 48 h[a]	–	Mallo *et al.* (2016a,b,c)
	Artemisinin	–	45 ppm for 72 h[a]	–	
	Propyl gallate	–	5 ppm for 72 h[a]	–	Mallo *et al.* (2014)
	Resveratrol	–	12 ppm for 24 h in seawater[a]	–	Morais *et al.* (2009)
		–	17 ppm for 7 days in L-15 medium[a]	–	Leiro *et al.* (2004)
	Mangiferin	–	200 ppm for 7 days in L-15 medium[a]	–	
	(–)-Epigallocatechin-3-gallate	–	100 ppm for 7 days in L-15 medium[a]	–	
	Salicyl hydroxamic acid	–	80 ppm for 72 h[a]	–	Mallo *et al.* (2013)
Miamiensis avidus	Doxycicline	Olive flounder (*Paralichthys olivaceus*)	–	2% in diet for 6 days	Kang *et al.* (2013)
		Olive flounder (*P. olivaceus*)	200 ppm for 30 min	–	
	Albendazole	Olive flounder (*P. olivaceus*)	–	2% in diet for 6 days	
		Olive flounder (*P. olivaceus*)	–	200 ppm for 30 min	
	Benzalkonium chloride + bronopol	Olive flounder (P. *olivaceus*)	–	50 ppm + 500 ppm (bath) for 14 and 27 h, respectively	Park *et al.* (2014)

[a]The half maximal inhibitory concentration (IC50).

Bath treatments with methylene blue or nitrofurazone are useful for treating skin lesions caused by *U. marinum* (Cheung *et al.*, 1980; Bassleer, 1983). Park *et al.* (2014) used a combination of benzalkonium chloride (to eliminate fish mucus) and bronopol (to kill the ciliates) to treat infected olive flounder. Treating fish with benzalkonium chloride (50 ppm) and then with bronopol (500 ppm) reduced the mortality in fish naturally exposed to scuticociliates. Dragesco *et al.* (1995) used dimetridazole to treat sea bass (*Dicentrarchus labrax*) infected with *P. dicentrarchi*; however, incomplete information about how the treatments were used is reported. As scuticociliates are endoparasites that cause systemic infections, successful treatment of infected fish is more difficult and no commercial treatments for controlling this disease in fish are available.

Compounds of natural origin

Several natural compounds are toxic to scuticociliates and may represent a useful alternative method of preventing the appearance of this disease. The polyphenols resveratrol, mangiferin and (−)-epigallocatechin-3-gallate have anti-*P. dicentrarchi* activity, which is particularly high in resveratrol (LD_{100}, 60–70 ppm for 24 h) (Leiro *et al.*, 2004; Budiño *et al.*, 2012b). Resveratrol alters the detoxification mechanisms of reactive oxygen species and damages scuticociliate mitochondria (Lamas *et al.*, 2009; Morais *et al.*, 2009, 2013; Mallo *et al.*, 2013), being a good candidate to be used to decrease ciliate levels in seawater. The polyphenols propyl gallate and curcumin have also displayed anti-*P. dicentrarchi* activity *in vitro*, both having a clear cytotoxic effect at a concentration of 100 μM (Mallo *et al.*, 2013, 2014, 2017). Although these natural compounds have anti-scuticociliate activity *in vitro*, the protocols of administration, to use them to decrease parasite concentration in the tanks or even in diets, have yet to be determined.

Compounds tested in diets or by injection

Oral administration or injection of some compounds seems to reduce mortality rates in infected fish. Clioquinol administered intramuscularly (40 mg/kg) significantly reduced mortality in fish that were experimentally and naturally infected with *M. avidus* (Lee *et al.*, 2017). Other studies have shown that it is possible to reduce fish mortality by administering treatments in the diet. Relative to controls, survival rates were higher in olive flounder exposed to *M. avidus* and then fed a diet containing doxycycline or bathed in fresh water containing doxycycline. However, treatment with albendazole was less effective (Kang *et al.*, 2013). Another interesting approach is to use recombinant proteins with anti-scuticociliate activity. A recent study has shown that the turbot antimicrobial peptide NK-lysin displays anti-*P. dicentrarchi* activity (Lama *et al.*, 2018).

17.5.2 Alternative methods to control scuticociliatosis

Bathing fish in fresh water may be a good method of controlling scuticociliatosis (Cardoso *et al.*, 2017). Adding dinoflagellates to seawater eliminates scuticociliates from tanks. After testing the toxicity of several dinoflagellate species, Kim *et al.* (2017) estimated that the use of 1 m^3 of the stock culture containing 20,000 *Alexandrium andersonii*/ml eliminated all *M. avidus* in 7 m^3 of seawater and all *Miamiensis* spp. in 19 m^3 of seawater within 48 h, without affecting *Artemia salina* nauplii or juvenile olive flounder. Although these methods must be standardized, they may be good ways of decreasing ciliate levels without adding chemicals to the water. Microspheres that can be phagocytosed by the ciliates can also be used as treatments. Paramá *et al.* (2005) reported that chitosan microspheres cross-linked with glutaraldehyde and containing β-cyclodextrin were toxic to *P. dicentrarchi* and could be used to control scuticociliatosis, as an alternative to other treatments.

17.5.3 Control of scuticociliatosis by management methods

Good farm management is essential to prevent or minimize the risk of scuticociliatosis infections. Basic measures such as ensuring good water quality, cleaning tanks or ponds regularly and maintaining low stocking densities (avoiding overcrowding in tanks) can help to prevent scuticociliatosis (Qin *et al.*, 2007; Datta, 2012; Cardoso *et al.*, 2017; Yanagisawa *et al.*, 2018). These measures are particularly important when the water temperature is higher than 18°C (Turgay *et al.*, 2015). Scuticociliates are free-living organisms that can feed on bacteria or suspended organic matter, so having clean water in tanks/ponds helps minimize their concentrations. In addition, as scuticociliates may enter the fish through lesions (Paramá *et al.*, 2003), overcrowding may increase stress and the appearance of microlesions on fish skin, thus increasing the risk of scuticociliatosis.

17.5.4 Prevention of scuticociliatosis by genetic selection of the hosts

The best method of preventing diseases is to use resistant fish. Genetic selection based on challenge tests, as well as searches for quantitative trait loci (QTLs) and the use of marker-assisted or genomic selection are promising methods for reducing mortality in aquaculture systems (Gjedrem and Rye, 2016). Attempts have been made to identify fish that display resistance to scuticociliatosis. Thus, Rodríguez-Ramilo *et al.* (2013) have searched for QTL that affect resistance to *P. dicentrarchi* in several turbot families. They found several genomic regions associated with turbot resistance to *P. dicentrarchi*. These genomic regions are generally similar to those found for *Aeromonas salmonicida*, suggesting their involvement in unspecific resistance to both bacteria and parasites (Rodríguez-Ramilo *et al.*, 2011). In addition, some of those regions are also involved in resistance to viruses (Rodríguez-Ramilo *et al.*, 2014). A recent study has identified a candidate QTL region in the turbot genome which appears to be involved in resilience to *P. dicentrarchi* and explains 33% of the additive genetic variance (Saura *et al.*, 2019). The region contains 33 genes related to immune response and defence mechanisms, some of which have previously been found to be activated in turbot infections caused by *P. dicentrarchi*. Consequently, genomic selection in turbot is being evaluated with the aim of increasing resistance to *P. dicentrarchi* (Martínez *et al.*, 2016).

17.5.5 Use of immunostimulants to prevent scuticociliatosis

Immunostimulants are usually administered in the diet and can modify the activity of immune factors, having a dose-dependent effect (Vallejos-Vidal *et al.*, 2016).

Animal, plants and fungal components

Numerous studies, mainly carried out in South Korea, have shown that the administration of some herbal or fungal extracts (e.g. *Punica granatum*, *Chrysanthemum cinerariaefolium*, *Zanthoxylum schinifolium*, *Hericium erinaceus*) in the diet can increase resistance to scuticociliatosis. Thus, the enriched diets stimulated several immune activities and increased survival rates in olive flounder infected with *P. dicentrarchi* (Harikrishnan *et al.*, 2010d, 2011a, 2012b) or with *U. marinum* (Harikrishnan *et al.*, 2010e, 2011b). Dietary administration of chitin and chitosan to kelp grouper (*Epinephelus bruneus*) produced several haematological changes, enhanced the activity of several immune parameters and also increased resistance to *P. dicentrarchi* (Harikrishnan *et al.*, 2012c).

Nucleotides, vitamins

Intraperitoneal administration of unmethylated cytosine-phosphate-guanine dinucleotides increased the serum scuticocidal activity of olive flounder against *P. dicentrarchi* (Lee and Kim, 2009) and had a therapeutic effect against *M. avidus* in the same fish species (Kang *et al.*, 2014). Vitamin C, which had a positive effect on fish protection against fish pathogens, did not protect groper (*Polyprion oxygeneios*) against *M. avidus* (Salinas *et al.*, 2012).

Probiotics

A few probiotics also seem to increase fish resistance to scuticociliatosis. Olive flounder fed diets containing *Lactobacillus plantarum*, *Lactobacillus acidophilus* or *Saccharomyces cerevisiae* showed increased survival rates in *U. marinum* experimental infections (Harikrishnan *et al.*, 2011b). The use of dietary immunostimulants is a good approach to increase fish resistance to scuticociliatosis, with several substances appearing to have a positive effect; however, we do not know if dietary immunostimulants are being used routinely as a prophylactic method on fish farms.

17.5.6 Vaccines against scuticociliatosis

Vaccination can greatly reduce fish mortalities associated with pathogens and is a key factor for the success of aquaculture. Vaccines are even more important to prevent diseases for which treatments are not available. Moreover, the use of vaccines can help to minimize the need to control diseases by using toxic compounds that may be harmful to the environment. In previous sections, we described several methods that can be used to minimize the risk of scuticociliatosis on fish farms and also some treatments used to kill scuticociliates. Many compounds have been found to be toxic to scuticociliates *in vitro*, and can be used to decrease the levels of scuticociliates in tanks or on the surface of fish; however, there are no treatments available to control the disease in infected fish, which usually suffer systemic infection.

Antigen production

One advantage associated with the development of a vaccine to prevent scuticociliatosis (rather than other diseases) is that it is possible to culture the ciliates and thus generate large amounts of antigen in a relatively short period of time (Pinheiro and Bols, 2013). Pathogenic scuticociliates can be cultured in seawater containing different nutrients. In some cases, culture medium containing bacteria has been used (Parker, 1976; Crosbie and Munday, 1999). However, ciliates used in vaccines must be cultured in an axenic medium. *U. nigricans*, *Parauronema virginianum*, *M. avidus* and *Miamiensis* spp. have been cultured in seawater containing cerophyl extract, proteose peptone, trypticase, yeast nucleic acid, biotin, calcium pantothenate folic acid, nicotinamide, pyridoxal HCl, riboflavin, thiamine HCl and D,L-thioctic acid (Soldo and Merlin, 1972). A ciliate identified as causing scuticociliatosis in *P. olivaceus* was cultured in peptone, yeast extract, Millport saline and distilled water (Yoshinaga and Nakazoe, 1993). The culture method can be simplified by using commercially available cell culture media containing some additional sources of proteins, lipids, nucleotides and carbohydrates. Antibiotics and antifungal agents can be added to the culture medium to prevent bacterial and fungal contamination. *P. dicentrarchi* has been cultured in L-15 Leibovitz medium supplemented with fetal bovine serum (FBS), glucose, nucleotides and phospholipids (Iglesias *et al.*, 2003b), and a similar culture medium has been used to grow *U. marinum* (Anderson *et al.*, 2009). In other cases, ciliates were grown in minimum essential medium (MEM) supplemented with yeast extract and FBS (Harikrishnan *et al.*, 2010e). *M. avidus* has been cultured in Eagle minimum essential medium (EMEM) containing the salmon cell line CHSE-214 and FBS (Lee and Kim, 2008a,b, 2009; Jung *et al.*, 2011b). This medium has been used to grow ciliates of several species, including *M. avidus*, *P. persalinus*, *P. hargasi* and *U. marinum* (Song *et al.*, 2009a), showing that the majority of pathogenic scuticociliates can probably grow in similar media. Scuticociliates obtained from turbot have also been cultured in 1651 MA medium (American Type Culture Collection, ATCC) containing FBS (Alvarez-Pellitero *et al.*, 2004). The culture medium may contain animal tissue homogenates to improve growth. *U. marinum* has been cultured in MEM containing host brain homogenate (Kwon *et al.*, 2003). Similarly, *P. dicentrarchi* has been cultured in L-15 medium containing autoclaved host tissue homogenates (Castro *et al.*, 2007); in this case, the richest medium induced the highest ciliate proliferation. However, it is probably not advisable to maintain ciliates in a very rich medium if they are not being cultured for inclusion in vaccines, as otherwise they must be subcultured continuously. The composition of the culture medium can be modified to increase ciliate proliferation and thus yield large quantities of antigen. To produce a vaccine against *P. dicentrarchi*, ciliates were cultured in L-15 medium containing homogenized fish tissues (1:200) for 5 days and then for another 2 days in L-15 without fish tissue (Sanmartín *et al.*, 2008).

One important aspect of the development of vaccines against scuticociliatosis is how the culture process affects the levels of protective antigens and the optimum conditions under which the scuticociliates are included in the vaccine. The virulence of scuticociliates has been shown to decrease when assayed after many passages in axenic culture (Alvarez-Pellitero *et al.*, 2004), and subculture may also affect the amount/type of antigens expressed by the parasite. Paramá *et al.* (2003) suggested that it is important to use virulent ciliates in vaccines against scuticociliatosis, by passage of the parasites through fish before cultivation for vaccine preparation. This method has been used in the preparation of several vaccines against *P. dicentrarchi* (Lamas *et al.*, 2008; Sanmartín *et al.*, 2008). It is possible to maintain scuticociliate virulence by cryopreservation and protocols have been developed to maintain scuticociliates in liquid nitrogen without loss of virulence, thus enabling the ciliates to be stored indefinitely and then defrosted immediately before vaccine preparation (Folgueira *et al.*, 2018; Liu *et al.*, 2019).

Vaccines containing formalin-killed ciliates as antigen

Several vaccine formulations against scuticociliatosis have been shown to induce a good immune response and protection in vaccinated fish. In turbot vaccinated against scuticociliatosis and challenged 1 month after the booster injection, the mortality rate was 21.4% in control fish, 16.6% in the group injected with the adjuvant, and zero in fish injected with antigen (formalin-killed ciliates) or antigen + adjuvant (Sitjà-Bobadilla *et al.*, 2008). After carrying out several trials in turbot to test vaccines containing several types of antigens and a different adjuvant (GERBU 734), Palenzuela *et al.* (2009) obtained contradictory results, generally with a lack of correlation between serum antibody levels and protection. Formalin-killed ciliates have also been used in other vaccines against scuticociliatosis with very good results. A vaccine containing 10^5 ciliates (*P. dicentrarchi*)/ml and the oil-based adjuvant Montanide ISA 763 A, administered intraperitoneally in two doses (given 1 month apart), induced high antibody levels and protection in turbot. After experimental challenge, the cumulative mortality of fish injected with the vaccine was 6%, while the mortality in the group injected with phosphate-buffered saline was 30%. The vaccine induced protection for at least 5 months after the second dose (Sanmartín *et al.*, 2008). After optimization of the vaccine formulation, by modifying the amount of adjuvant, antigen and formalin, the formulation containing 10^6 trophozoites/ml, 0.2% formalin and 50% adjuvant generated the highest levels of specific antibodies in serum and 100% protection in the vaccinated fish (Lamas *et al.*, 2008).

Subunit vaccines

Other alternatives to vaccines containing formalin-killed ciliates have been tested against scuticociliatosis. A vaccine containing *U. marinum* antigens encapsulated in poly(D,L-lactide-*co*-glycolic acid) microspheres increased resistance to the parasite in *E. bruneus*. After challenge, a decrease in mortality was observed in vaccinated fish (20%), relative to fish injected with microspheres (40%) or with antigen alone (30%) (Harikrishnan *et al.*, 2012d). Another vaccine made of microspheres composed of two biodegradable polymers (chitosan and Gantrez) and containing *P. dicentrarchi* surface antigens, which were either encapsulated or attached to the surface, induced high IgM (immunoglobulin M) levels in the serum of vaccinated turbot, similar to those induced by Freund's complete adjuvant, and a good correlation was found between antibody and survival levels after a challenge with the parasite homologous strain (León-Rodríguez *et al.*, 2012). Due to the toxicity of some oil-based adjuvants in fish vaccines (Noia *et al.*, 2014), it has been suggested that this vaccine may a better alternative to oil-based adjuvants for the immunoprophylaxis of scuticociliatosis in turbot (León-Rodríguez *et al.*, 2012). In addition, this vaccine formulation stimulated the turbot innate immune response (León-Rodríguez *et al.*, 2013; Fontenla *et al.*, 2016).

Strain selection and protective antigens

Another important factor in the development of vaccines against scuticociliates is the selection of the strain(s) used as antigen. A vaccine generated against *P. dicentrarchi* induced good protection against the homologous strain but not against the heterologous ones (Piazzon *et al.*, 2008; Budiño *et al.*, 2012a), making it difficult to produce a universal vaccine against scuticociliatosis caused by this parasite. These results were confirmed using immobilization assays with turbot and mice antisera, which agglutinated only homologous *P. dicentrarchi* strains (Piazzon *et al.*, 2008). Cross-protection between heterologous *M. avidus* strains was not observed in Japanese flounder (Yanagisawa *et al.*, 2018). Budiño *et al.* (2011a) observed important genetic variability among seven *P. dicentrarchi* isolates obtained from infected turbot. Western blot analysis with antisera from immunized mice, which yielded a higher discriminatory capacity than turbot serum, revealed differences in the seven isolates in the 50–64 kDa region, which could be separated into three groups on the basis of the band pattern. However, after cross-protection experiments in turbot, the isolates were divided into two main groups (Budiño *et al.*, 2012a). Agglutination methods enabled division of six *M. avidus* isolates, obtained from infected Japanese flounder, into three serotypes (Song *et al.*, 2009b). After genetic and serological analysis, 21 *M. avidus* isolates obtained from different fish species affected by scuticociliatosis were grouped into five genotypes, after comparison of the mitochondrial cytochrome c oxidase subunit 1 gene sequences. Interestingly, the genotype groups matched the serotype groups, but no associations between those groups and the geographical origin, host species or pathogenicity were found (Jung *et al.*, 2011b). In this respect, Budiño *et al.* (2011b) reported the coexistence of several serotypes of *P. dicentrarchi* that infected turbot on the same fish farm. These findings clearly indicate that a vaccine containing antigen from only one strain may not be effective for preventing scuticociliatosis on a particular farm. In order to generate a universal vaccine, the number of serotypes involved in scuticociliatosis must be determined, in order to include the ciliates or their antigens in the same vaccine. It may also be possible to identify protective antigens for inclusion of antigen mixtures in recombinant vaccines. Information about which antigens provide protection against scuticociliatosis and where they are located is also required. It has been suggested that β-tubulin could be used as antigen, as the serum of mice injected with *P. persalinus* recombinant β-tubulin displayed higher killing activity than control serum (Kim *et al.*, 2006). However, it is more generally considered that protective antigens are located on the cell membrane of scuticociliates (surrounding cilia or in the interciliary regions). Iglesias *et al.* (2003a) reported that serum from turbot that had survived infection by *P. dicentrarchi* contained antibodies that recognized ciliary antigens and induced ciliate agglutination. Interestingly, ciliates became agglutinated/immobilized after incubation with heat-inactivated immune serum but, after a few hours, they escaped from the masses of agglutinated ciliates, indicating that *P. dicentrarchi* is capable of shedding surface antigens (Iglesias *et al.*, 2003a). Similar results were obtained with heat-inactivated antiserum from immunized olive flounder; however, the immune serum agglutinated cultured ciliates but not ciliates obtained from an infected fish. This suggests that expression of immobilization antigens may vary depending on the infection state and that surface constitutive antigens should be used in vaccines (Lee and Kim, 2008a). Although the antigens that protect against scuticociliates seem to be located on the plasma membrane, the molecules responsible have not been identified. In this respect, identifying the protective antigens may be the first step towards generating a universal vaccine against pathogenic scuticociliates.

17.6 Conclusions and Future Directions

Scuticociliatosis is an emerging infectious disease that affects a wide range of fish (especially captive fish), generating high mortality rates and economic losses in aquaculture. The causative agents are cosmopolitan, highly diverse, free-living ciliates that occupy a wide range of ecosystems and are particularly abundant in benthic and coastal eutrophic habitats. A few species infect fish and other marine animals to become facultative parasites. They proliferate inside the host and cause systemic infection and death in a short period of time. However, very little is known about the mechanisms of adaptation to parasitism in pathogenic scuticociliates. Proteases may be important in relation to pathogenicity, but the ciliates must have mechanisms to avoid the fish immune system. Further research on host–parasite interactions is required to better understand the disease and its mechanism(s).

The increase in greenhouse gas emissions, as a consequence of industrial and agricultural activities, causes acidification and warming of the ocean surface. Warming will also decrease dissolved oxygen in surface waters. Higher water temperature has an important influence on ciliate activity and increases scuticociliatosis outbreaks. Within their physiological range, scuticociliates increase proliferation with higher temperatures. In a global warming scenario, an increase in seawater temperature would probably affect many physiological parameters in scuticociliates, making them more virulent, as well as affecting the host immune responses (as a consequence of thermal stress). Under these conditions, some susceptible fish species may become more susceptible to scuticociliatosis, and currently resistant fish may become susceptible. Thus, unless better methods of prevention and control are developed, it seems reasonable to expect increases in the frequency of outbreaks, higher fish mortality rates and greater economic losses due to scuticociliatosis on fish farms. Scuticociliates can adapt to changes in water pH and oxygen concentrations. However, it is not clear how changes in these parameters would affect the virulence of the parasites. In addition, more data are also needed about how variations in temperature, pH and oxygen levels would affect ciliate populations (particularly scuticociliates) in seawater.

Many chemicals are toxic to scuticociliates under *in vitro* conditions. However, the toxicity of most of these compounds to fish or other organisms has not yet been evaluated. In addition, protocols for administration of the treatments (including the concentration of the compounds and duration of the treatments) have not been developed. As toxic chemicals should not be released into the aquatic environment, less aggressive compounds must be found for treating scuticociliatosis. Research to discover new, natural compounds that are toxic to scuticociliates, but not to other organisms, and to establish protocols for application on fish farms are clearly required. Finally, compounds that can be included in the diet to treat scuticociliatosis in infected fish are needed.

Vaccines are available against scuticociliatosis; however, the protection is only against homologous serotypes and not against heterologous serotypes in the same ciliate species. Several serotypes in some species are known, but their number distribution is not clear worldwide. Identifying the protective antigens(s) is also important to enable development of recombinant vaccines. The use of formalin-killed ciliates is problematic as ciliates must express those antigens when they are processed for use in vaccines. This problem could be solved by the use of recombinant vaccines which would include antigens from several serotypes to produce a 'universal vaccine'. Finally, genetic selection programmes aimed at identifying more resistant fish families would also be useful.

Acknowledgements

Some of our research reported in this review has been financially supported by grant AGL2017-83577-R awarded by the Ministerio de Economía y Competitividad (Spain) and Fondo Europeo de Desarrollo Regional (FEDER) (European Union), by grant ED431C2017/31 from the Xunta de Galicia (Spain), and by the PARAFISHCONTROL project, which received funding from the European Union's Horizon 2020 research and innovation programme under Grant Agreement no. 634429.

References

Abraham, J.S., Sripoorna, S., Maurya, S., Makhija, S., Gupta, R. and Toteja, R. (2019) Techniques and tools for species identification in ciliates, a review. *International Journal of Systematic and Evolutionary Microbiology* 69, 877–894.

Aksungur, N., Aksungur, M., Akbulut, B. and Kutlu, I. (2007) Effects of stocking density on growth performance, survival and food conversion ratio of turbot (*Psetta maxima*) in the net cages on the southeastern coast of the Black Sea. *Turkish Journal of Fisheries and Aquatic Sciences* 7, 147–152.

Al-Marzouk, A. and Azad, I.S. (2007) Growth kinetics, protease activity and histophagous capability of *Uronema* sp. infesting cultured silver pomfret *Pampus argenteus* in Kuwait. *Diseases of Aquatic Organisms* 76, 49–56.

Alvarez-Pellitero, P., Palenzuela, O., Padrós, F., Sitjà-Bobadilla, A., Riaza, A. *et al.* (2004) Histophagous scuticociliatids (Ciliophora) parasitizing turbot *Scophthalmus maximus*, morphology, *in vitro* culture and virulence. *Folia Parasitologica* 51, 177–187.

Anderson, S.A., Hulston, D.A., McVeagh, S.M., Webb, V.L. and Smith, P.J. (2009) *In vitro* culture and cryopreservation of *Uronema marinum* isolated from farmed New Zealand groper (*Polyprion oxygeneios*). *Journal of Microbiological Methods* 79, 62–66.

Athanasopoulou, F., Speare, D., Cawthorn, R.J., MacMillan, R. and Despres, B. (2004) Pathology of *Anophryoides haemophila* (Scuticociliatida, Orchitophryidae), parasite of American lobster *Homarus americanus* kept under experimental conditions. *Aquaculture* 236, 103–117.

Azad, I.S., Al-Marzouk, A., James, C.M., Almatar, S. and Al-Gharabally, H. (2007) Scuticociliatosis-associated mortalities and histopathology of natural infection in cultured silver pomfret (*Pampus argentus* Euphrasen) in Kuwait. *Aquaculture* 262, 202–210.

Bae, M.J., Im, E.H., Kim, H.Y. and Jung, S.J. (2009) The effect of temperature to scuticociliatida *Miamiensis avidus* proliferation, and to mortality of infected olive flounder *Paralichthys olivaceus*. *Journal of Fish Pathology* 22, 97–105.

Bassleer, G. (1983) *Uronema marinum*. A new and common parasite on tropical salt-water fishes. *Freshwater and Marine Aquarium* 6, 78–81.

Bates, A.E., Stickle, W.B. and Harley, C.D.G. (2010) Impact of temperature on an emerging parasitic association between a sperm-feeding scuticociliate and Northeast Pacific sea stars. *Journal of Experimental Marine Biology and Ecology* 384, 44–50.

Beaver, J.R. and Crisman, T.L. (1989) The role of ciliated protozoa in pelagic freshwater ecosystems. *Microbial Ecology* 17, 111–136.

Blanco-Abad, V., Noia, M., Valle, A., Fontenla, F., Folgueira, I. *et al.* (2018) The coagulation system helps control infection caused by the ciliate parasite *Philasterides dicentrarchi* in the turbot *Scophthalmus maximus* (L.). *Developmental and Comparative Immunology* 87, 147–156.

Bly, J.E. and Clem, L.W. (1991) Temperature-mediated processes in teleost immunity, *in vitro* immunosuppression induced by *in vivo* low temperature in channel catfish. *Veterinary Immunology and Immunopathology* 28, 365–377.

Bly, J.E. and Clem, L.W. (1992) Temperature and teleost immune functions. *Fish and Shellfish Immunology* 2, 159–171.

Bonar, C.J., Garner, M.M., Weber, E.S., Keller, C.J., Murray, M. *et al.* (2013) Pathologic findings in weedy (*Phyllopteryx taeniolatus*) and leafy (*Phycodurus eques*) seadragons. *Veterinary Pathology* 50, 368–376.

Bongiorni, L., Mirto, S., Pusceddu, A. and Danovaro, R. (2005) Response of benthic protozoa and thraustochytrid protists to fish farm impact in seagrass (*Posidonia oceanica*) and soft-bottom sediments. *Microbial Ecology* 50, 268–276.

Buchmann, K. (2015) Impact and control of protozoan parasites in maricultured fishes. *Parasitology* 142, 168–177.

Budiño, B., Lamas, J., Pata, M.P., Arranz, J.A., Sanmartín, M. and Leiro, J. (2011a) Intraspecific variability in several isolates of *Philasterides dicentrarchi* (syn. *Miamiensis avidus*), a scuticociliate parasite of farmed turbot. *Veterinary Parasitology* 175, 260–272.

Budiño, B., Lamas, J., González, A., Pata, M.P., Devesa, S. *et al.* (2011b) Coexistence of several *Philasterides dicentrarchi* strains on a turbot fish farm. *Aquaculture* 322–323, 23–32.

Budiño, B., Leiro, J., Cabaleiro, S. and Lamas, J. (2012a) Characterization of *Philasterides dicentrarchi* isolates that are pathogenic to turbot. Serology and cross-protective immunity. *Aquaculture* 364–365, 130–136.

Budiño, B., Pata, M.P., Leiro, J. and Lamas, J. (2012b) Differences in the *in vitro* susceptibility to resveratrol and other chemical compounds among several *Philasterides dicentrarchi* isolates from turbot. *Parasitology Research* 110, 1573–1578.

Burel, C., Person-Le Ruyet, P., Gaumet, F., Le Roux, A., Severe, A. and Boeuf, G. (1996) Effects of temperature on growth and metabolism in juvenile turbot. *Journal of Fish Biology* 49, 678–692.

Caldeira, K. and Wickett, M.E. (2003) Oceanography: anthropogenic carbon and ocean pH. *Nature* 425, 365.

Cardoso, P.H.M., de Carvalho-Balian, S., Reiko-Matushima, E., Benites de Pádua, S. and Laterça-Martins, M. (2017) First report of scuticociliatosis caused by *Uronema* sp. in ornamental reef fish imported into Brazil. *Brazilian Journal of Veterinary Parasitology* 26, 491–495.

Castellano, M., Silva-Álvarez, V., Fernández López, E., Mauris, V., Conijeski, D. *et al.* (2017) Russian sturgeon cultured in a subtropical climate shows weaken innate defences and a chronic stress response. *Fish and Shellfish Immunology* 68, 443–451.

Castro, R., Paramá, A., Barja, J.L., Leiro, J., Sanmartin, M.L. and Lamas, J (2007) Culture of the histophagous ciliate *Philasterides dicentrarchi* (Ciliophora, Scuticociliatia) in fish tissue. *Journal of Fish Diseases* 30, 239–242.

Cawthorn, R.J., Lynn, D., Despres, B., MacMillan, R., Maloney, R. *et al.* (1996) Description of *Anophryoides haemophila* n. sp. (Scuticociliatida, Orchitophryidae), a pathogen of American lobsters *Homarus americanus*. *Diseases of Aquatic Organisms* 24, 143–148.

Chen, Y., Zhao, Y., Pan, X., Ding, W., Al-Rasheid, K.A. and Quiu, Z. (2014) Morphology of new *Frontonia* ciliate, *F. paramagna* spec. nov. (Ciliophora, Peniculida) from Harbin, Northeast China. *Zootaxa* 3827, 375–386.

Cheung, P.J., Nigrelli, R.F. and Ruggieri, G.D. (1980) Studies on the morphology of *Uronema marinum* Dujardin (Ciliatea, Uronematidae) with a description of the histopathology of the infection in marine fishes. *Journal of Fish Diseases* 3, 295–303.

Cho, S.H., Kim, K.T., Choi, I.C., Jeon, G.H. and Kim, D.S. (2012) Compensatory growth of grower olive flounder (*Paralichthys olivaceus*) with different feeding regime at suboptimal temperature. *Asian-Australian Journal of Animal Science* 25, 272–277.

Corliss, J.O. (1979) *The Ciliated Protozoa, Characterization, Classification, and Guide to the Literature*, 2nd edn. Pergamon Press, Oxford.

Crosbie, P.B. and Munday, B.L. (1999) Environmental factors and chemical agents affecting the growth of the pathogenic marine ciliate *Uronema nigricans*. *Diseases of Aquatic Organisms* 36, 213–219.

Datta, S. (2012) Management of water quality in intensive aquaculture. *Research in Environment and Life Sciences* 5, 1–16.

de Felipe, A.P., Lamas, J., Sueiro, R.A., Folgueira, I. and Leiro, J.M. (2017) New data on flatfish scuticociliatosis reveal that *Miamiensis avidus* and *Philasterides dicentrarchi* are different species. *Parasitology* 29, 1–18.

De Silva, S.S. and Soto, D. (2009) Climate change and aquaculture: potential impacts, adaptation and mitigation. In: Cochrane, K., De Young, C., Soto, D. and Bahri, T. (eds) Climate change implications for fisheries and aquaculture. Overview of current scientific knowledge. *FAO Fisheries and Aquaculture Technical Paper No. 530*. Food and Agriculture Organization of the United Nations, Rome, pp. 151–212.

Deveney, M.R., Bayly, T.J., Johnston, C.J. and Nowak, B.F. (2005) A parasite survey of farmed Southern bluefin tuna, *Thunnus maccoyii* (Castelnau). *Journal of Fish Diseases* 28, 279–284.

Di Cicco, E., Paradis, E., Stephen, C., Turba, M.E. and Rossi, G. (2013) Scuticociliatid ciliate outbreak in Australian potbellied seahorse, *Hippocampus abdominalis* (Lesson, 1827): clinical signs, histopathologic findings, and treatment with metronidazole. *Journal of Zoo and Wildlife Medicine* 44, 435–440.

Dou, S.Z., Masudaş, R., Tanaka, M. and Tsukamoto, K. (2005) Effects of temperature and delayed initial feeding on the survival and growth of Japanese flounder larvae. *Journal of Fish Biology* 66, 362–377.

Dragesco, A., Dragesco, J., Coste, F., Gasc, C., Romenstad, B. *et al.* (1995) *Philasterides dicentrarchi*, n. sp. (Ciliophora, Scuticociliatida), a histophagous opportunistic parasite of *Dicentrarchus labrax* (Linnaeus, 1758), a reared marine fish. *European Journal of Protistology* 31, 327–340.

Du, G., Qu, L., Shang, K., Sun, C., Wang, C. and Gao, P. (2019) Ciliate *Uronema marinum* is the causative agent of scuticociliatosis in farm raised turbot *Scophthalmus maximus*. *Journal of Oceanology and Limnology* 37, 1726–1735.

Dyková, I. and Figueras, A. (1994) Histopathological changes in turbot *Scophthalmus maximus* due to a histophagous ciliate. *Diseases of Aquatic Organisms* 18, 5–9.

Fan, X., Hu, X., Al-Farraj, S.A., Clamp, J.C. and Song, W. (2011) Morphological description of three marine ciliates (Ciliophora, Scuticociliatia), with establishment of a new genus and two new species. *European Journal of Protistology* 47, 186–196.

Fenchel, T. (1968) The ecology of marine microbenthos II. The food of marine benthic ciliates. *Ophelia* 5, 73–121.

Fenchel, T. (1990) Adaptive significance of polymorphic life cycles in Protozoa: responses to starvation and refeeding in two species of marine ciliates. *Journal of Experimental Marine Biology and Ecology* 136, 159–177.

Fenchel, T. (2013) *Ecology of Protozoa: The Biology of Free-living Phagotropic Protists. Brock/Springer Series in Contemporary Bioscience*. Springer, Berlin.

Fernández-Galiano, D. (1994) The ammoniacal silver carbonate method as a general procedure in the study of protozoa from sewage (and other) waters. *Water Research* 28, 495–496.

Finlay, B.J., Maberly, S.C. and Cooper, I.J. (1997) Microbial diversity and ecosystem function. *Oikos* 80, 209–213.

Foissner, W. (1979) Methylgrün-Pyronin: Seine Eignung zur supravitalen Ubersichtsfärbung von Protozoen, *besonders ihrer Protrichocysten. Mikroskopie* 35, 108–115.

Foissner, W. (1991) Basic light and scanning electron microscopic methods for taxonomic studies of ciliated protozoa. *European Journal of Protistology* 27, 313–330.

Foissner, W. (2014) An update of 'basic light and scanning electron microscopic methods for taxonomic studies of ciliated protozoa'. *International Journal of Systematic and Evolutionary Microbiology* 64, 271–292.

Folgueira, I., de Felipe, A.P., Sueiro, R.A., Lamas, J. and Leiro, J. (2018) Protocol for cryopreservation of the turbot parasite *Philasterides dicentrarchi* (Ciliophora, Scuticociliatia). *Cryobiology* 80, 77–83.

Fontenla, F., Blanco-Abad, V., Pardo, B.G., Folgueira, I., Noia, M. *et al.* (2016) Vaccine-induced modulation of gene expression in turbot peritoneal cells. A microarray approach. *Molecular Immunology* 75, 188–199.

Gao, F., Fan, X., Yi, Z., Strüder-Kypke, M. and Song, W. (2010) Phylogenetic consideration of two scuticociliate genera, *Philasterides* and *Boveria* (Protozoa, Ciliophora) based on 18S rRNA gene sequences. *Parasitology International* 59, 549–555.

Gao, F., Strüder-Kypke, M., Yi, Z., Miao, M., Al-Farraj, S.A. and Song, W. (2012a) Phylogenetic analysis and taxonomic distinction of six genera of pathogenic scuticociliates (Protozoa, Ciliophora) inferred from small-subunit rRNA gene sequences. *International Journal of Systematic and Evolutionary Microbiology* 62, 246–256.

Gao, F., Katz, L.A. and Song, W. (2012b) Insights into the phylogenetic and taxonomy of philasterid ciliates (Protozoa, Ciliophora, Scuticociliatia) based on analyses of multiple molecular markers. *Molecular Phylogenetics and Evolution* 64, 308–317.

Gao, F., Katz, L.A. and Song, W. (2013) Multigene-based analyses on evolutionary phylogeny of two controversial ciliate orders: Pleuronematida and Loxocephalida (Protista, Ciliophora, Oligohymenophorea). *Molecular Phylogenetics and Evolution* 68, 55–63.

Gao, F., Gao, S., Wang, P., Katz, L.A. and Song, W. (2014) Phylogenetic analyses of cyclidiids (Protista, Ciliophora, Scuticociliatia) based on multiple genes suggest their close relationship with thigmotrichids. *Molecular Phylogenetics and Evolution* 75, 219–226.

Garza, J.B., Bott, N.J., Hammond, M.D., Shepherd, N. and Nowak, B.F. (2017) Molecular characterisation of *Miamiensis avidus* (Ciliophora: Scuticociliata) from ranched Southern bluefin tuna, *Thunnus maccoyii* of Port Lincoln, South Australia. *Aquaculture* 469, 44–49.

Gill, P.A. and Callinan, R.B. (1997) Ulcerative dermatitis associated with *Uronema* sp. infection of farmed sand whiting *Sillago ciliate*. *Australian Veterinary Journal* 75, 357.

Gjedrem, T. and Rye, M. (2016) Selection response in fish and shellfish: a review. *Reviews in Aquaculture* 10, 168–179.

Gong, C.-G. and Li, F.-C. (2007) Chemical treatment on protozoon *Paralembus digitiformis* in farmed Japanese flounder. *Fisheries Science* 2007-09.

Grolière, C.A. (1980) Morphologie et stomatogenèse chez deux ciliés Scuticociliatida des genres *Philasterides* Kahl, 1926 et *Cyclidium* O. F. Müller, 1786. *Acta Protozoologica* 19, 195–206.

Hanna, B.A. and Lilly, D.M. (1970) Axenic culture of *Uronema marinum*. *American Zoologist* 10, 539–540.

Harikrishnan, R., Jin, C.N., Kim, M.C., Kim, J.H., Balasundarann, C. and Heo, M.S. (2010a) Histopathology and mortality in olive flounder infected by scuticociliates caused by *Philasterides dicentrarchi. Israeli Journal of Aquaculture – Bamidgeh* 62, 202–211.

Harikrishnan, R., Balasundaram, C. and Heo, M.S. (2010b) Scuticociliatosis and its recent prophylactic measures in aquaculture with special reference to South Korea: Taxonomy, diversity and diagnosis of scuticociliatosis: Part I Control strategies of scuticociliatosis: Part II. *Fish and Shellfish Immunology* 29, 15–31.

Harikrishnan, R., Jin, C.N., Kim, M.C., Kim, J.S., Balasundaram, C. and Heo, M.S. (2010c) Effectiveness and immunomodulation of chemotherapeutants against scuticociliate *Philasterides dicentrarchi* in olive flounder. *Experimental Parasitology* 124, 306–314.

Harikrishnan, R., Balasundaran, C., Kim, M.C., Kim, J.S., Han, Y.J. and Heo, M.S. (2010d) Effect of a mixed herb-enriched diet on the innate immune response and disease resistance of *Paralichthys olivaceus* against *Philasterides dicentrarchi* infection. *Journal of Aquatic Animal Health* 22, 235–243.

Harikrishnan R., Heo, J., Balasundaram, C., Kim, M.C., Kim, J.S. *et al.* (2010e) Effect of traditional Korean medicinal (TKM) triherbal extract on the innate immune system and disease resistance in *Paralichthys olivaceus* against *Uronema marinum*. *Veterinary Parasitology* 170, 1–7.

Harikrishnan, R., Kim, J.S., Kim, M.C. and Heo, M.S. (2011a) *Hericium erinaceum* enriched diets enhance the immune response in *Paralichthys olivaceus* and protect from *Philasterides dicentrarchi* infection. *Aquaculture* 318, 48–53.

Harikrishnan, R., Kim, M.C., Kim, J.S., Balasundaram, C. and Heo, M.S. (2011b) Immunomodulatory effect of probiotics enriched diets on *Uronema marinum* infected olive flounder. *Fish and Shellfish Immunology* 30, 964–971.

Harikrishnan, R., Jin, C.N., Kim, J.S., Balasundaram, C. and Heo, M.S. (2012a) *Philasterides dicentrarchi*, a histophagous ciliate causing scuticociliatosis in olive flounder, *Philasterides dicentrarchi* – histopathology investigations. *Experimental Parasitology* 130, 239–245.

Harikrishnan, R., Kim, J.S., Kim, M.C., Balasundaram, C. and Heo, M.S. (2012b) Pomegranate enriched diet enhances the hematology, innate immune response, and disease resistance in olive flounder against *Philasterides dicentrarchi*. *Veterinary Parasitology* 187, 147–156.

Harikrishnan, R., Kim, J.S., Balasundaram, C. and Heo, M.S. (2012c) Dietary supplementation with chitin and chitosan on haematology and innate immune response in *Epinephelus bruneus* against *Philasterides dicentrarchi*. *Experimental Parasitology* 131, 116–124.

Harikrishnan, R., Balasundaram, C., Heo, M.S. and Sa, M. (2012d) Poly D,L-lactide-co-glycolic acid (PLGA)-encapsulated vaccine on immune system in *Epinephelus bruneus* against *Uronema marinum*. *Experimental Parasitology* 131, 325–332.

Houston, A.H., Dobric, N. and Kahurananga, R. (1996) The nature of hematological response in fish: studies on rainbow trout *Oncorhynchus mykiss* exposed to stimulated winter, spring and summer conditions. *Fish Physiology and Biochemistry* 15, 339–347.

Hunt, A. and Dale, N. (2018) Economic valuation in formaldehyde regulation. *OECD Environment Working Papers No. 134*. OECD Publishing, Paris.

Iglesias, R., Paramá, A., Álvarez, M.F., Leiro, J., Fernández, J. and Sanmartín, M.L. (2001) *Philasterides dicentrarchi* (Ciliophora, Scuticociliatida) as the causative agent of scuticociliatosis in farmed turbot *Scophthalmus maximus* in Galicia (NW Spain). *Diseases of Aquatic Organisms* 46, 47–55.

Iglesias, R., Paramá, A., Álvarez, M.F., Leiro, J. and Sanmartín, M.L. (2002) Antiprotozoals effective *in vitro* against the scuticociliate fish pathogen *Philasterides dicentrarchi*. *Diseases of Aquatic Organisms* 49, 191–197.

Iglesias, R., Paramá, A., Álvarez, M.F., Leiro, J., Ubeira, F.M. and Sanmartín, M.L. (2003a) *Philasterides dicentrarchi* (Ciliophora: Scuticociliatida) expresses surface immobilization antigens that probably induce protective immune responses in turbot. *Parasitology* 126, 125–134.

Iglesias, R., Paramá, A., Álvarez, M.F., Leiro, J., Aja, C. and Sanmartín, M.L. (2003b) *In vitro* growth requirements for the fish pathogen *Philasterides dicentrarchi* (Ciliophora, Scuticociliatida). *Veterinary Parasitology* 111, 19–30.

Iwata, N., Kikuchi, K., Honda, H., Kiyono, M. and Kurokura, H. (1994) Effects of temperature on the growth of Japanese flounder, *Paralichthys olivaceus*. *Fisheries Sciences* 60, 527–531.

Jee, B.Y. and Jo, M.R. (2002) Trials for the control of scuticociliatosis in the cultured olive flounder (*Paralichthys olivaceus*) by bath treatment. *Journal of Fish Pathology* 15, 93–97.

Jee, B.Y., Kim, Y.C. and Park, M.S. (2001) Morphology and biology of parasite responsible for scuticociliatosis of cultured olive flounder *Paralichthys olivaceus*. *Diseases of Aquatic Organisms* 47, 49–55.

Jee, B.Y., Jo, M.R., Kim, J., Woo, P. and Mi, S. (2002) *In vitro* efficacy of formalin, hydrogen peroxide and copper sulfate on the scuticocilliate *Uronema marinum* at low salinity. *Journal of Fish Pathology* 15, 111–115.

Jee, B.Y., Shin, K.W., Lee, D.W. and Lee, M.K. (2014) Monitoring of the mortalities and medications in the inland farms of olive flounder, *Paralichthys olivaceus*, in South Korea. *Journal of Fish Pathology* 27, 77–83.

Jeong, H.J., Lee, K.H., Yoo, Y.D., Kang, N.S., Song, J.Y. *et al.* (2018) Effects of light intensity, temperature, and salinity on the growth and ingestion rates of the red tide mixotrophic dinoflagellate *Paragymnodinium shiwhaense*. *Harmful Algae* 80, 46–54.

Jin, C.N., Lee, C.H., Oh, S.P., Jung, Y.U., Song, C.B. *et al.* (2003) Scuticociliatosis in flounder farms of Jeju Island. *Journal of Fish Pathology* 16, 135–138.

Jin, C.N., Kang, H.S., Lee, C.H., King, S.K., Lee, Y.D. *et al.* (2006) The pathogenicity of scuticociliate *Philasterides dicentrarchi* isolated from cultured olive flounder *Paralichthys olivaceus*. *Journal of Fish Pathology* 19, 87–97.

Jin, C.N., Kang, H.S., Lee, C.H., Lee, Y.D, Lee, J.H. and Heo, M.S. (2007) Biological characteristics of scuticociliate, *Philasterides dicentrarchi* isolated from cultured olive flounder, *Paralichthys olivaceus*. *Journal of Aquaculture* 20, 106-133.

Jin, C.N., Harikrishnan, R., Moon, Y.G., Kim, M.C., Kim, J.S. *et al.* (2009) Histopathological changes of Korea cultured olive flounder *Paralichthys olivaceus* due to scuticociliatosis caused by histophagous scuticociliate, *Philasterides dicentrarchi*. *Veterinary Parasitology* 161, 292–301.

Jin, C.N., Harikrishnan, R., Moon, Y.G., Kim, M.C., Kim, J.S. *et al.* (2010) Effectiveness of chemotherapeutants against scuticociliate *Philasterides dicentrarchi*, a parasite of olive flounder. *Veterinary Parasitology* 168, 19–24.

Jung, S.J. and Woo, P.T.K. (2012) *Miamiensis avidus* and related species. In: Woo, P.T.K. and Buchmann, K. (eds) *Fish Parasites: Pathobiology and Protection*. CAB International, Wallingford, UK, pp. 73–91.

Jung, S.J., Kitamura, S.I., Song, J.Y., Jeong, I.Y. and Oh, M.J. (2005) Complete small subunit rRNA gene sequence of the scuticociliate *Miamiensis avidus* pathogenic to olive flounder *Paralichthys olivaceus*. *Diseases of Aquatic Organisms* 64, 159–162.

Jung, S.J., Kitamura, S.I., Song, J.Y. and Oh, M.J. (2007) *Miamiensis avidus* (Ciliophora: Scuticociliatida) causes systemic infection of flounder *Paralichthys olivaceus* and is a senior synonym of *Philasterides dicentrarchi*. *Diseases of Aquatic Organisms* 73, 227–234.

Jung, S.J., Bae, M.J., Oh, M.J. and Lee, J. (2011a) Sequence conservation in the internal transcribed spacers and 5.8S ribosomal RNA of parasitic scuticociliates *Miamiensis avidus* (Ciliophora, Scuticociliatia). *Parasitology International* 60, 216–219.

Jung, S.J., Im, E.Y., Strüder-Kypke, M.C., Kitamura, S.I. and Woo, P.T.K. (2011b) Small subunit ribosomal RNA and mitochondrial cytochrome c oxidase subunit 1 gene sequences of 21 strains of the parasitic scuticociliate *Miamiensis avidus* (Ciliophora, Scuticociliatia). *Parasitology Research* 108, 1153–1161.

Kaneshiro, E.S. and Holz, G.G. Jr (1976) Observations on the ultrastructure of *Uronema* spp., marine scuticociliates. *Journal of Protozoology* 23, 503–517.

Kaneshiro, E.S., Dunham, P.B. and Holz, G.G. (1969) Osmoregulation in a marine ciliate, *Miamiensis avidus*. I. Regulation of inorganic ions and water. *The Biological Bulletin* 136, 63–75.

Kang, B.J., Jang, Y.H., Jhon, B.K., Park, B.H. and Jin, C.N. (2015) Monitoring of scuticociliatosis of olive flounder (*Paralichthys olivaceus*) farm in Jeju, Korea from 2007 to 2014. *Journal of Fish Pathology* 28, 165–169.

Kang, Y.J., Kim, D.S. and Kim, K.H. (2013) Evaluation of treatment efficacy of doxycycline and albendazole against scuticociliatosis in olive flounder (*Paralichthys olivaceus*). *Aquaculture* 416–417, 192–195.

Kang, Y.J., Choi, S.H. and Kim, K.H. (2014) Preventive and therapeutic effects of auxotrophic *Edwardsiella tarda* mutant harboring CpG 1668 motif-enriched plasmids against scuticociliatosis in olive flounder (*Paralichthys olivaceus*). *Experimental Parasitology* 144, 34–38.

Katsaros, C.I., Varvarigos, V., Gachon, C.M.M., Brand, J., Motomura, T. *et al.* (2011) Comparative immunofluorescence and ultrastructural analysis of microtubule organization in *Uronema* sp., *Klebsormidium flaccidum*, *K. subtilissimum*, *Stichococcus bacillaris* and *S. chloranthus* (Chorophita). *Protist* 162, 315–331.

Kayis, S., Yandi, I., Altinok, I. and Capkin, E. (2011) Treatment by vinegar of *Philasterides dicentrarchi* (Ciliophora: Scuticociliatida) infestation in cultured juvenile turbot (*Psetta maxima*). *Israeli Journal of Aquaculture – Bamidgeh* 63, 627.

Khosravi, M., Nasrolahi, A., Shokri, M.R., Dobretsov, S. and Pansch, D.C. (2019) Impact of warming on biofouling communities in the northern Persian Gulf. *Journal of Thermal Biology* 85, 102403.

Kim, J.H., Jeong, H.J., Lim, A.S., Kwon, J.E., Lee, K.H. *et al.* (2017) Removal of two pathogenic scuticociliates *Miamiensis avidus* and *Miamiensis* sp. using cells or culture filtrates of the dinoflagellate *Alexandrium andersonii*. *Harmful Algae* 63, 133–145.

Kim, J.W., Lee, H.N., Jee, B.Y., Woo, S.H., Kim, Y.J. and Lee, M.K. (2012) Monitoring of the mortalities in the aquaculture farms of South Korea. *Journal of Fish Pathology* 25, 271–277.

Kim, S.M., Cho, J.B., Kim, S.K., Nam, Y.K. and Kim, K.H. (2004a) Occurrence of scuticociliatosis in olive flounder *Paralichthys olivaceus* by *Philasterides dicentrarchi* (Ciliophora: Scuticociliatida). *Diseases of Aquatic Organisms* 62, 233–238.

Kim, S.M., Cho, J.B., Lee, E.H., Kwon, S.R., Kim, S.K. *et al.* (2004b) *Pseudocohnilembus persalinus* (Ciliophora: Scuticociliatida) is an additional species causing scuticociliatosis in olive flounder *Paralichthys olivaceus*. *Diseases of Aquatic Organisms* 62, 239–244.

Kim, S.M., Lee, E.H., Kwon, S.R., Lee, S.J., Kim, S.K. *et al.* (2006) Preliminary analysis of recombinant β-tubulin of *Pseudocohnilembus persalinus* (Ciliophora: Scuticociliatida) as a vaccine antigen candidate against scuticociliatosis. *Aquaculture* 260, 21–26.

Kimsland, A., Foss, A., Gnnarsson, S., Bemtssen, M.H.G., Fitz, R. *et al.* (2001) The interaction of temperature and salinity on growth and food conversion in juvenile turbot (*Scophthalmus maximus*). *Aquaculture* 198, 353–367.

Kisand, V. and Zingel, P. (2000) Dominance of ciliate grazing on bacteria during spring in a shallow eutrophic lake. *Aquatic Microbial Ecology* 22, 135–142.

Kubiski, S.V., Howerth, E.W., Clauss, T.M., Berliner, A.L. and Camus, A.C. (2011) Pathology in practice. Scuticociliatosis. *Journal of the American Veterinary Medical Association* 238, 301–303.

Kwon, S.R., Kim, C.S. and Kim, K.H. (2003) Differences between short- and long-term cultures of *Uronema marinum* (Ciliophora: Scuticociliatida) in chemiluminescence inhibitory activity, antioxidative enzyme and protease activity. *Aquaculture* 221, 107–114.

Lama, R., Pereiro, P., Costa, M.M., Encinar, J.A., Medina-Gali, R.M. *et al.* (2018) Turbot (*Scophthalmus maximus*) Nk-lysin induces protection against the pathogenic parasite *Philasterides dicentrarchi* via membrane disruption. *Fish and Shellfish Immunology* 82, 190–199.

Lamas, J., Sanmartín, M.L., Paramá, A.I., Castro R., Cabaleiro, S. *et al.* (2008) Optimization of an inactivated vaccine against a scuticociliate parasite of turbot: effect of antigen, formalin and adjuvant concentration on antibody response and protection against the pathogen. *Aquaculture* 278, 22–26.

Lamas, J., Morais, P., Arranz, J.A., Sanmartín, M.L., Orallo, F. and Leiro, J. (2009) Resveratrol promotes an inhibitory effect on the turbot scuticociliate parasite *Philasterides dicentrarchi* by mechanisms related to cellular detoxification. *Veterinary Parasitology* 161, 307–315.

Lamb, R.W., Smith, F., Aued, A.W., Salinas-de-León, P., Suárez, J. *et al.* (2018) El Niño drives a widespread ulcerative skin disease outbreak in Galapagos marine fishes. *Scientific Reports* 8, 16602.

Lee, E.H. and Kim, K.H. (2008a) Can the surface immobilization antigens of *Philasterides dicentrarchi* (Ciliophora: Scuticociliatida) be used as target antigens to develop vaccines in cultured fish? *Fish and Shellfish Immunology* 24, 142–146.

Lee, E.H. and Kim, K.H. (2008b) Immobilization antigen-independent protection of olive flounder (*Paralichthys olivaceus*) against *Philasterides dicentrarchi* (Ciliophora: Scuticociliatia) infection. *Aquaculture* 279, 211–213.

Lee, E.H. and Kim, K.H. (2009) CpG-ODN increases resistance of olive flounder (*Paralichthys olivaceus*) against *Philasterides dicentrarchi* (Ciliophora: Scuticociliatia) infection. *Fish and Shellfish Immunology* 26, 29–32.

Lee, J.-H., Park, J.-J., Choi, J.-H., Shin, D.-H. and Park, K.H. (2017) Anti-scuticociliate effects of aquatic hydrogen peroxide preparation in olive flounder *Paralichthys olivaceus*. *Journal of Fish Pathology* 30, 107-114.

Lee, N.-S., Park, J.-H., Han, K.-S. and Huh, M.D. (1994) Histopathological changes in fingerlings of Japanese flounder, *Paralichthys olivaceus*, with severe scuticociliatosis. *Journal of Fish Pathology* 7, 151–160.

Leiro, J., Arranz, J.A., Paramá, A., Álvarez, M.F. and Sanmartín, M.L. (2004) *In vitro* effects of the polyphenols resveratrol, mangiferin and (–)-epigallocatechin-3-gallate on the scuticociliate fish pathogen *Philasterides dicentrarchi*. *Diseases of Aquatic Organisms* 59, 171–174.

Leiro, J., Piazzón, M.C., Budiño, B., Sanmartín, M.L. and Lamas, J. (2008) Complement-mediated killing of *Philasterides dicentrarchi* (Ciliophora) by turbot serum: relative importance of alternative and classical pathways. *Parasite Immunology* 30, 535–543.

León-Rodríguez, L., Luzardo-Álvarez, A., Blanco-Méndez, J., Lamas, J. and Leiro, J. (2012) A vaccine based on biodegradable microspheres induces protective immunity against scuticociliatosis without producing side effects in turbot. *Fish and Shellfish Immunology* 33, 21–27.

León-Rodríguez, L., Luzardo-Álvarez, A., Blanco-Méndez, J., Lamas, J. and Leiro, J. (2013) Biodegradable microparticles covalently linked to surface antigens of the scuticociliate parasite *P. dicentrarchi* promote innate immune responses *in vitro*. *Fish and Shellfish Immunology* 34, 236–243.

Li, W.T., Lo, C., Su, C.Y., Kuo, H., Lin, S.J. *et al.* (2017) Locally extensive meningoencephalitis caused by *Miamiensis avidus* (syn. *Philasterides dicentrarchi*) in a zebra shark. *Diseases of Aquatic Organisms* 126, 167–172.

Lim, S.U., Seo, J.S., Kim, M.S., Ahn, S.J., Jeong, H.D. *et al.* (2005) Molecular cloning and characterization of cathepsin B from a scuticociliate, *Uronema marinum*. *Comparative Biochemistry and Physiology Part B: Biochemistry and Molecular Biology* 142, 283–292.

Liu, M., Li, L., Qu, Z., Luo, X., Al-Farraj, S.A. *et al.* (2017a) Morphological redescription and SSU rDNA-based phylogeny of two freshwater ciliates, *Uronema nigricans* and *Lembadion lucens* (Ciliophora, Oligohymenophorea), with discussion on the taxonomic status of *Uronemita sinensis*. *Acta Protozoologica* 56, 17–37.

Liu, W., Jiang, J., Xu, Y., Pan, X., Qu, Z. *et al.* (2017b) Diversity of free-living marine ciliates (Alveolata, Ciliophora): faunal studies in coastal waters of China during the years 2011–2016. *European Journal of Protistology* 61, 424-438.

Liu, Y., Nan, B., Duan, L., Cheng, T., Bourland, W.A. *et al.* (2019) Simple and rapid cryopreservation technique for ciliates: a long-term storage procedure used for marine scuticociliates. *Journal of Eukaryotic Microbiology* 66, 836–848.

Lõhmus, M. and Björklund, M. (2015) Climate change: what will it do to fish–parasite interactions? *Biological Journal of the Linnean Society* 116, 397–411.

Lynn, D. (2008) *The Ciliated Protozoa: Characterization, Classification, and Guide to the Literature*, 3rd edn. Springer, Heidelberg, Germany.

Lynn, D.H. and Strüder-Kypke, M. (2005) Scuticociliate endosymbionts of echinoids (phylum Echinodermata): phylogenetic relationships among species in the genera *Entodiscus*, *Plagiopyliella*, *Thyrophylax*, and *Entorhipidium* (phylum Ciliophora). *Journal of Parasitology* 91, 1190–1199.

Lynn, D.H. and Strüder-Kypke, M.C. (2006) Species of *Tetrahymena* identical by small subunit rRNA gene sequences are discriminated by mitochondrial cytochrome c oxidase I gene sequences. *Journal of Eukaryotic Microbiology* 53, 385–387.

Ma, H., Choi, J.K. and Song, W. (2003) An improved silver carbonate impregnation for marine ciliated protozoa. *Acta Protozoologica* 42, 161–164.

Ma, M., Lu, B., Fan, X., Shi, X. and Chen, X. (2018) Taxonomic clarification of a well-known pathogenic scuticociliate, *Miamiensis avidus* Thompson & Moewus, 1964 (Ciliophora, Scuticociliatia). *Journal of Ocean University of China* 17, 1231–1242.

Makrinos, D.L. and Bowden, T.J. (2016) Natural environmental impacts on teleost immune function. *Fish and Shellfish Immunology* 53, 50–57.

Mallo, N., Lamas, J. and Leiro, J.M. (2013) Evidence of an alternative oxidase pathway for mitochondrial respiration in the scuticociliate *Philasterides dicentrarchi*. *Protist* 164, 824–836.

Mallo, N., Lamas, J. and Leiro, J.M. (2014) Alternative oxidase inhibitors as antiparasitic agents against scuticociliatosis. *Parasitology* 141, 1311–1321.

Mallo, N., Lamas, J., Piazzon, C. and Leiro, J.M. (2015) Presence of a plant-like proton-translocating pyrophosphatase in a scuticociliate parasite and its role as a possible drug target. *Parasitology* 142, 449–462.

Mallo, N., Lamas, J., de Felipe, A.P., de Castro, M.E., Sueiro, R.A. and Leiro, J.M. (2016a) Presence of an isoform of H^+-pyrophosphatase located in the alveolar sacs of a scuticociliate parasite of turbot: physiological consequences. *Parasitology* 143, 576–587.

Mallo, N., Lamas, J., de Felipe, A.P., Sueiro, R.A., Fontenla, F. and Leiro, J.M. (2016b) Enzymes involved in pyrophosphate and calcium metabolism as targets for anti-scuticociliate chemotherapy. *Journal Eukaryotic Microbiology* 63, 505–515.

Mallo, N., Lamas, J., de Felipe, A.P., Sueiro, R.A., Fontenla, F. and Leiro, J.M. (2016c) Role of H^+-pyrophosphatase activity in the regulation of intracellular pH in a scuticociliate parasite of turbot: physiological effects. *Experimental Parasitology* 169, 59–68.

Mallo, N., de Felipe, A.P., Folgueira, I., Sueiro, R.A., Lamas, J. and Leiro, J.M. (2017) Combined antiparasitic and anti-inflammatory effects of the natural polyphenol curcumin on turbot scuticociliatosis. *Journal of Fish Diseases* 40, 205–217.

Marcogliese, D.J. (2008) The impact of climate change on the parasites and infectious diseases of aquatic animals. *Revue Scientifique et Technique (International Office of Epizootics)* 27, 467–484.

Martínez, D., Vargas-Lagos, C., Oyarzún, R., Loncoman, C.A., Pontigo, J.P. *et al.* (2018) Temperature modulates the immunological response of the sub-Antarctic notothenioid fish *Eleginops maclovinus* injected with *Piscirickettsia salmonis*. *Fish and Shellfish Immunology* 82, 492-503.

Martínez, P., Robledo, D., Rodríguez-Ramilo, D.S., Hermida, M., Taboada, X. *et al.* (2016) Turbot (*Scophthalmus maximus*) genomic resources: application for boosting aquaculture production. In: MacKenzie, S.A. and Jentof, S. (eds) *Genomics in Aquaculture*. Academic Press, Amsterdam, pp. 131–163.

Martínez-Tapia, C. and Fernández-Pato, C.A. (1991) Influence of stock density on turbot (*Scophthalmus maximus* L.) growth. *ICES Mariculture Committee* 20, 1–7.

Medina, M., Sotil, G., Flores, V. and Fernández, C. (2016) Occurrence of scuticociliates in the flounder *Paralichthys adspersus* caused by *Miamiensis avidus*, in Peru. *Revista Peruana de Biología* 23, 261–270.

Miao, M., Warren, A., Song, W., Wang, S., Shang, H. and Chen, Z. (2008) Analysis of internal transcribed spacer 2 (ITS2) region of scuticociliates and related taxa (Ciliophora, Oligohymenophorea) to infer their evolution and phylogeny. *Protist* 159, 519–533.

Morais, P., Lamas, J., Sanmartín, M.L., Orallo, F. and Leiro, J. (2009) Resveratrol induces mitochondrial alterations, autophagy and a cryptobiosis-like state in scuticociliates. *Protist* 160, 552–564.

Morais, P., Piazzon, C., Lamas, J., Mallo, N. and Leiro, J.M. (2013) Effect of resveratrol on oxygen consumption by *Philasterides dicentrarchi*, a scuticociliate parasite of turbot. *Protist* 164, 206–217.

Moustafa, E.M.M. and Mousa-Balabel, T.M. (2011) Behavioural and pathological changes in flat fish (*Paralichthys olivaceus*) infected with *Miamiensis avidus*. *Kafr El-Sheikh Veterinary Medical Journal* 9, 335–367.

Moustafa, E.M., Naota, M., Morita, T., Tange, N. and Shimada, A. (2010a) Pathological study on the scuticociliatosis affecting farmed Japanese flounder (*Paralichthys olivaceus*) in Japan. *Journal of Veterinary Medical Science* 72, 1359–1362.

Moustafa, E.M., Tange, N., Shimada, A. and Morita, T. (2010b) Experimental scuticociliatosis in Japanese flounder (*Paralichthys olivaceus*) infected with *Miamiensis avidus*: pathological study on the possible neural routes of invasion and dissemination of the scuticociliate inside the fish body. *Journal of Veterinary Medical Science* 72, 1557–1563.

Munday, B.L., O'Donoghue, P.J., Watts, M., Rough, K. and Hawkesford, T. (1997) Fatal encephalitis due to the scuticociliate *Uronema nigricans* in sea-caged, southern bluefin tuna *Thunnus maccoyii*. *Diseases of Aquatic Organisms* 30, 17–25.

Munday, B.L., Sawada, Y., Cribb, T. and Hayward, C.J. (2003) Diseases of tunas, *Thunnus* spp. *Journal of Fish Diseases* 26, 187–206.

Narasaki, Y., Obayashi, Y., Ito, S. and Kitamura, S.-I. (2018) Extracellular proteinases of *Miamiensis avidus* causing scuticociliatosis are potential virulence factors. *Fish Pathology* 53, 1–9.

Ndong, D., Chen, Y.Y., Lin, Y.H., Vaseeharan, B. and Chen, J.C. (2007) The immune response of tilapia *Oreochromis mossambicus* and its susceptibility to *Streptococcus iniae* under stress in low and high temperatures. *Fish and Shellfish Immunology* 22, 686–694.

Noia, M., Domínguez, B., Leiro, J., Blanco-Méndez, J., Luzardo-Álvarez, A. and Lamas, J. (2014) Inflammatory responses and side effects generated by several adjuvant-containing vaccines in turbot. *Fish and Shellfish Immunology* 38, 244–254.

Ofelio, C., Blanco, A., Roura, Á., Pintado, J., Pascual, S. and Planas, M. (2014) Isolation and molecular identification of the scuticociliate *Porpostoma notata* Moebius, 1888 from moribund reared *Hippocampus hippocampus* (L.) seahorses, by amplification of the SSU rRNA gene sequences. *Journal of Fish Diseases* 37, 1061–1065.

Orr, J.C., Fabry, V.J., Aumont, O., Bopp, L., Doney, S.C. *et al.* (2005) Anthropogenic ocean acidification over the twenty-first century and its impact on calcifying organisms. *Nature* 437, 681–686.

Padrós, F., Zarza, C. and Crespo, S. (2001) Infecciones por ciliados histiófagos en acuicultura marina: aspectos histopatológicos. *Monografías del Instituto Canario de Ciencias Marinas* 4, 500–512.

Palenzuela, O., Sitjà-Bobadilla, A., Riaza, A., Silva, R., Arán, J. and Álvarez-Pellitero, P. (2009) Antibody responses of turbot *Psetta maxima* against various antigen formulations of scuticociliates Ciliophora. *Diseases of Aquatic Organisms* 86, 123–134.

Pan, X. (2016) Seven scuticociliates (Protozoa, Ciliophora) from Alabama, USA, with descriptions of two parasitic species isolated from a freshwater mussel *Potamilus purpuratus*. *European Journal of Taxonomy* 249, 1–19.

Pan, X., Fan, X., Al-Farraj, S.A., Gao, S. and Chen, Y. (2016) Taxonomy and morphology of four '*ophrys*-related' scuticociliates (Protista, Ciliophora, Scuticociliatia), with the description of a new genus, *Paramesanophrys* gen. nov. *European Journal of Taxonomy* 191, 1–18.

Paramá, A., Iglesias, R., Álvarez, M.F., Leiro, J., Aja, C. and Sanmartín, M.L. (2003) *Philasterides dicentrarchi* (Ciliophora, Scuticociliatida): experimental infection and posible routes of entry in farmed turbot (*Scophthalmus maximus*). *Aquaculture* 217, 73–80.

Paramá, A., Iglesias, R., Álvarez, M.F., Sanmartín, M.L. and Leiro, J. (2004a) Chemotactic responses of the fish-parasitic scuticociliate *Philasterides dicentrarchi* to blood and blood components of the turbot *Scophthalmus maximus*, evaluated using a new microplate multiassay. *Journal of Microbiological Methods* 58, 361–366.

Paramá, A., Iglesias, R., Álvarez, M.F., Leiro, J., Ubeira, F.M. and Sanmartín, M.L. (2004b) Cysteine proteinase activities in the fish pathogen *Philasterides dicentrarchi* (Ciliophora: Scuticociliatida). *Parasitology* 128, 541–548.

Paramá, A., Iglesias, R., Álvarez, F., Leiro, J.M., Quintela, J.M. *et al.* (2004c) *In vitro* efficacy of new antiprotozoals against *Philasterides dicentrarchi* (Ciliophora, Scuticociliatida). *Diseases of Aquatic Organisms* 62, 97–102.

Paramá, A., Luzardo, A., Blanco-Méndez, J., Sanmartín, M.L. and Leiro, J. (2005) *In vitro* efficacy of glutaraldehyde-crosslinked chitosan microspheres against the fish-pathogenic ciliate *Philasterides dicentrarchi*. *Diseases of Aquatic Organisms* 64, 151–158.

Paramá, A., Arranz, J.A., Álvarez, M.F., Sanmartín, M.L. and Leiro, J. (2006) Ultrastructure and phylogeny of *Philasterides dicentrarchi* (Ciliophora, Scuticociliatia) from farmed turbot in NW Spain. *Parasitology* 132, 555–564.

Paramá, A., Castro, R., Lamas, J., Sanmartín, M.L., Santamarina, M.T. and Leiro, J. (2007a) Scuticociliate proteinases may modulate turbot immune response by inducing apoptosis in pronephric leucocytes. *International Journal for Parasitology* 37, 87–95.

Paramá, A., Castro, R., Arranz, J.A., Sanmartín, M.L., Lamas, J. and Leiro, J. (2007b) Scuticociliate cysteine proteinases modulate turbot leucocyte functions. *Fish and Shellfish Immunology* 23, 945–956.

Paramá, A., Piazzon, M.C., Lamas, J., Sanmartín, M.L. and Leiro, J. (2007c) *In vitro* activity of the nonsteroidal anti-inflammatory drug indomethacin on a scuticociliate parasite of farmed turbot. *Veterinary Parasitology* 148, 318–324.

Park, S.B., Jang, H.B., Fagutao, F.F., Kim, Y.K., Nho, S.W. *et al.* (2014) Combination treatment against scuticociliatosis by reducing the inhibitor effect of mucus in olive flounder, *Paralichthys olivaceus*. *Fish and Shellfish Immunology* 38, 282–286.

Parker, J.G. (1976) Cultural characteristics of the marine ciliated protozoan, *Uronema marinum* Dujardin. *Journal of Experimental Marine Biology and Ecology* 24, 213–226.

Peck, R.K. (1977) Cortical ultrastructure of the scuticociliates *Dexiotricha media* and *Dexiotricha colpidiopsis* (Hymenostomata). *Journal of Protozoology* 24, 122–134.

Piazzon, C., Lamas, J., Castro, R., Budiño, B., Cabaleiro, S. *et al.* (2008) Antigenic and cross-protection studies on two turbot scuticociliate isolates. *Fish and Shellfish Immunology* 25, 417–424.

Piazzon, M.C., Wiegertjes, G.F., Leiro, J. and Lamas, J. (2011a) Turbot resistance to *Philasterides dicentrarchi* is more dependent on humoral than on cellular immune responses. *Fish and Shellfish Immunology* 30, 1339–1347.

Piazzon, C., Lamas, J. and Leiro, J.M. (2011b) Role of scuticociliate proteinases in infection success in turbot, *Psetta maxima* (L.). *Parasite Immunology* 33, 535–344.

Piazzon, M.C., Leiro, J. and Lamas, J. (2013) Fish immunity to scuticociliate parasites. *Developmental and Comparative Immunology* 41, 248–256.

Pinheiro, M.D.O. and Bols, N.C. (2013) Use of cell cultures to study the interactions of ciliates with fish. *Springer Science Reviews* 1, 95–113.

Pörtner, H.-O. (2008) Ecosystem effects of ocean acidification in times of ocean warming: a physiologist's view. *Marine Ecology Progress Series* 373, 203–217.

Power, C., Balli-Garza, J., Evans, D., Nowak, B.F., Bridle, A.R. and Bott, N.J. (2019) Detection of *Miamiensis avidus* (Ciliophora: Scuticociliatia) and *Cardicola* spp. (Trematoda: Aporocotylidae) DNA in biofouling from southern bluefin tuna, *Thunnus maccoyii* pontoons off Port Lincoln, South Australia. *Aquaculture* 502, 128–133.

Puig, L., Traveset, R., Palenzuela, O. and Padrós, F. (2007) Histopathology of experimental scuticociliatosis in turbot *Scophthalmus maximus*. *Diseases of Aquatic Organisms* 76, 131–140.

Qin, L., Wang, Y.-G., Zhang, L.-J. and Dai, J.-X. (2007) Histopathology of turbot associated with *Mesanophrys carcini* parasite. *Acta Hydrobiologica Sinica* 5, 618–628.

Quintela, J.M., Peinador, C., González, L., Iglesias, R., Paramá, A. *et al.* (2003) Piperazine *N*-substituted naphthyridines, pyridothienopyrimidines and pyridothienotriazines: new antiprotozoals active against *Philasterides dicentrarchi*. *European Journal of Medical Chemistry* 38, 265–275.

Ramos, M.F., Costa, A.R., Barandela, T., Saraiva, A. and Rodrigues, P.N. (2007) Scuticociliate infection and pathology in cultured turbot *Scophthalmus maximus* from the north of Portugal. *Diseases of Aquatic Organisms* 74, 249–253.

Retallack, H., Okihiro, M.S., Britton, E., Sommeran, S.V. and De Risi, J.L. (2019) Metagenomic next-generation sequencing reveals *Miamiensis avidus* (Ciliophora: Scuticociliatida) in the 2017 epizootic of leopard sharks (*Triakis semifasciata*) in San Francisco Bay, California, USA. *Journal of Wildlife Diseases* 55, 375–386.

Rhein, M., Rintoul, S.R., Aoki, S., Campos, E., Chambers, D. *et al.* (2013) Observations: ocean. In: Stocker, T.F., Qin, D., Plattner, G.-K., Tignor, M., Allen, S.K. *et al.* (eds.) *Climate Change 2013: The Physical Science Basis. Contribution of Working Group I to the Fifth Assessment Report of the Intergovernmental Panel on Climate Change*. Cambridge University Press, Cambridge, UK, pp. 255–315.

Rijkers, G.T., Frederix-Walters, E.M.H. and van Muiswinkel, W.B. (1980) The immune system of cyprinid fish. Kinetics and temperature dependence of antibody producing cells in carp (*Cyprinus carpio*). *Immunology* 41, 91–97.

Rodríguez-Ramilo, S.T., Toro, M.A., Bouza, C., Hermida, M., Pardo, B.G. *et al.* (2011) QTL detection for *Aeromonas salmonicida* resistance related traits in turbot (*Scophthalmus maximus*). *BMC Genomics* 12, 541.

Rodríguez-Ramilo, S.T., Fernández, J., Toro, M.A., Bouza, C., Hermida, M. *et al.* (2013) Uncovering QTL for resistance and survival time to *Philasterides dicentrarchi* in turbot (*Scophthalmus maximus*). *Animal Genetics* 44, 149–157.

Rodríguez-Ramilo, S.T., De La Herrán, R., Ruiz-Rejón, C., Hermida, M., Fernández, C. *et al.* (2014) Identification of quantitative trait loci associated with resistance to viral haemorrhagic septicaemia (VHS) in turbot (*Scophthalmus maximus*): a comparison between bacterium, parasite and virus diseases. *Marine Biotechnology (NY)* 16, 265–276.

Rossteuscher, S., Wenker, C., Jermann, T., Wahli, T., Oldenberg, E. and Schmidt-Posthaus, H. (2008) Severe scuticociliate (*Philasterides dicentrarchi*) infection in a population of sea dragons (*Phycodurus eques* and *Phyllopteryx taeniolatus*). *Veterinary Pathology* 45, 546–550.

Sahin, T. (2001) Effect of water temperature on growth of hatchery reared Black Sea turbot, *Scophthalmus maximus* (Linnaeus, 1758). *Turkish Journal of Zoology* 25, 183–186.

Sahoo, P.K., Pattanayak, S., Paul, A., Sahoo, M.K., Rajesh Kumar, P. *et al.* (2018) First record of *Metanophrys sinensis* (Protozoa: Ciliophora: Scuticociliatida) from India causing large scale mortality in a new host *Macrobrachium rosenbergii* larvae. *Journal of Fish Diseases* 41, 1303–1307.

Salinas, I., Anderson, S.A., Wright, J. and Webb, V.L. (2012) *In vivo* innate immune responses of groper (*Polyprion oxygeneios*) against *Miamiensis avidus* infection and lack of protection following dietary vitamin C administration. *Fish and Shellfish Immunology* 32, 8–15.

Sanmartín, M.L., Paramá, A., Castro, R., Cabaleiro, S., Leiro, J. *et al.* (2008) Vaccination of turbot, *Psetta maxima* (L.), against the protozoan parasite *Philasterides dicentrarchi*: effects on antibody production and protection. *Journal of Fish Diseases* 31, 135–140.

Santos, M.J., Cavaleiro, F., Campos, P., Sousa, A., Teixeira, F. and Martins, M. (2010) Impact of amoeba and scuticociliatidia infections on the aquaculture European sea bass (*Dicentrarchus labrax* L.) in Portugal. *Veterinary Parasitology* 171, 15–21.

Saura, M., Carabaño, M.J., Fernández, A., Cabaleiro, S., Doeschl-Wilson, A.B. *et al.* (2019) Disentangling genetic variation for resistance and endurance to scuticociliatosis in turbot using pedigree and genomic information. *Frontiers in Genetics* 10, 539.

Secombes, C.J., White, A., Fletcher, T.C. and Houlihan, D.F. (1991) The development of an ELISPOT assay to quantify total and specific antibody-secreting cells in dab *Limanda limanda* (L). *Fish and Shellfish Immunology* 1, 87–97.

Seo, J.S., Jeon, E.J., Jung, S.H., Park, M.A., Kim, J.W. *et al.* (2013) Molecular cloning and expression analysis of peptidase genes in the fish-pathogenic scuticociliate *Miamiensis avidus*. *BMC Veterinary Research* 9, 10.

Sheath, D.J., Andreou, D. and Britton, J.R. (2016) Interactions of warming and exposure affect susceptibility to parasite infection in a temperate fish species. *Parasitology* 143, 1340–1346.

Shimeta, J. and Sisson, J.D. (1999) Taxon-specific tidal resuspension of protist into the subtidal benthic boundary layer of a coastal embayment. *Marine Ecology Progress Series* 177, 51–62.

Shin, P.S., Han, J.E., Gómez, D.K., Kim, J.H., Choresca, C.H. Jr *et al.* (2011) Identification of scuticociliate *Philasterides dicentrarchi* from Indo-Pacific seahorses *Hippocampus kuda*. *African Journal of Microbiology Research* 5, 738–741.

Silva, A. and Oliva, M. (2010) Review of the biology and cultivation of Chilean flounder (*Paralichthys adspersus*). *Latin American Journal of Aquatic Research* 38, 377–386.

Sitjà-Bobadilla, A., Palenzuela, O. and Alvarez-Pellitero, P. (2008) Immune response of turbot, *Psetta maxima* (L.) (Pisces: Teleostei), to formalin-killed scuticociliates (Ciliophora) and adjuvanted formulations. *Fish and Shellfish Immunology* 24, 1–10.

Smale, D.A., Taylor, J.D., Coombs, S.H., Moore, G. and Cunliffe, M. (2017) Community responses to seawater warming are conserved across diverse biological groupings and taxonomic resolutions. *Proceedings of the Royal Society B: Biological Sciences* 284, 20170534.

Small, E.B. (1967) The Scuticoliatida, a new Order of the Class Ciliatea (Phylum Protozoa, Subphylum Ciliophora). *Transactions of the American Microscopical Society* 86, 354–370.

Small, H.J., Neil, D.M., Taylor, A.C., Bateman, K. and Coombs, G.H. (2005) A parasitic scuticociliate infection in the Norway lobster (*Nephrops norvegicus*). *Journal of Invertebrate Pathology* 90, 108–117.

Smith, P.J., McVeagh, S.M., Hulston, D., Anderson, S.A. and Gublin, Y. (2009) DNA identification of ciliates associated with disease outbreaks in a New Zealand marine fish hatchery. *Diseases of Aquatic Organisms* 86, 163–167.

Soldo, A.T. and Merlin, E.J. (1972) The cultivation of symbiote-free marine ciliates in axenic medium. *Journal of Protozoology* 19, 519– 524.

Song, J.Y., Kitamura, S., Oh, M.J., Kang, H.S., Lee, J.H. *et al.* (2009a) Pathogenicity of *Miamiensis avidus* (syn. *Philasterides dicentrarchi*), *Pseudocohnilembus persalinus*, *Pseudocohnilembus hargisi* and *Uronema marinum* (Ciliophora, Scuticociliatida). *Diseases of Aquatic Organisms* 83, 133–143.

Song, J.Y., Sasaki, K., Okada, T., Sakashita, M., Kawakami, H. *et al.* (2009b) Antigenic differences of the scuticociliate *Miamiensis avidus* from Japan. *Journal of Fish Diseases* 32, 1027–1034.

Song, W. (2000) Morphological and taxonomical studies on some marine scuticociliates from China Sea, with description of two new species, *Philasterides armatalis* sp. n. and *Cyclidium varibonneti* sp. n. (Protozoa: Ciliophora: Scuticociliatida). *Acta Protozoologica* 39, 295–322.

Song, W. and Wilbert, N. (2000) Redefinition and redescription of some marine scuticociliates from China, with report of a new species, *Metanophrys sinensis* nov. spec. (Ciliophora, Scuticociliatida). *Zoologischer Anzeiger* 239, 45–74.

Song, W., Warren, A. and Hu, X. (2009) *Free-living Ciliates in the Bohai and Yellow Seas, China*. Science Press, Beijing.

Sterud, E., Hansen, M.K. and Mo, T.A. (2000) Systemic infection with *Uronema*-like ciliates in farmed turbot *Scophthalmus maximus*. *Journal of Fish Diseases* 23, 33–37.

Stidworthy, M.F., Garner. M.M., Bradway, D.S., Westfall, B.D., Joseph, B. *et al.* (2014) Systemic scuticociliatosis (*Philasterides dicentrarchi*) in sharks. *Veterinary Pathology* 51, 628–632.

Su, C.-Y., Lo, C. and Li, W.-T. (2017) Scuticociliatosis (*Philasterides dicentrarchi*) in central nervous system of zebra shark (*Stegostoma fasciatum*). Presented at: *IAAAM 2017, 48th Annual Meeting & Conference of the International Association for Aquatic Animal Medicine, Cancún, Mexico, 20 – 24 May 2017.*

Takagishi, N., Yoshinaga, T. and Ogawa, K. (2009) Effect of hyposalinity on the infection and pathogenicity of *Miamiensis avidus* causing scuticociliatosis in olive flounder *Paralichthys olivaceus*. *Diseases of Aquatic Organisms* 86, 175–179.

Takata, R., Nakayama, C.L., de Souza, E., Silva, W., Bazzoli, N. and Luz, R.K. (2018) The effect of water temperature on muscle cellularity and gill tissue of larval and juvenile *Lophiosilurus alexandri*, a neotropical freshwater fish. *Journal of Thermal Biology* 76, 80–88.

Tange, N., Song, J.-Y. and Kitamura, S.-I. (2010) Detection and identification of *Miamiensis avidus* causing scuticociliatosis by PCR. *Fish Pathology* 45, 130–132.

Tao, Z., Liu, L., Chen, X., Zhou, S. and Wang, G. (2016) First isolation of *Miamiensis avidus* (Ciliophora: Scuticociliatida) associated with skin ulcers from reared pharaoh cuttlefish *Sepia pharaonis*. *Diseases of Aquatic Organisms* 122, 67–71.

Thompson, J.C. and Moewus, L. (1964) *Miamiensis avidus* n.g., n. sp., a marine facultative parasite in the ciliate order Hymenostomatida. *Journal of Protozoology* 11, 378–381.

Turgay, E., Steinum, T.M., Güll, A.E. and Karatas, S. (2015) An outbreak of scuticociliatosis in cultured common dentes (*Dentex dentex*) in Turkey. *Bulletin of the European Association of Fish Pathologists* 35, 104–111.

Umehara, A., Kosuga, Y. and Hirose, H. (2003) Scuticociliata infection in the weedy sea dragon *Phyllopteryx taeniolatus*. *Parasitology International* 52, 165–168.

Urrutxurtu, I., Orive, E. and Sota, A.D.L. (2003) Seasonal dynamics of ciliated protozoa and their potential food in an eutrophic estuary (Bay of Biscay). *Estuarine, Coastal and Shelf Science* 57, 1169–1182.

Vallejos-Vidal, E., Reyes-López, F., Teles, M. and MacKenzie, S. (2016) The response of fish to immunostimulant diets. *Fish and Shellfish Immunology* 56, 34–69.

Verma, A.K., Pal, A.K., Manush, S.M., Das, T., Dalvi, R.S. *et al.* (2007) Persistent sub-lethal chlorine exposure augments temperature induced immunosuppression in *Cyprinus carpio* advanced fingerlings. *Fish and Shellfish Immunology* 22, 547–555.

Wang, G., Cai, W.B., Wu, L., Santoso, A., Lin, X., Chen, Z. and McPhaden, M.J. (2017) Continued increase of extreme El Niño frequency long after 1.5°C warming stabilization. *Nature Climate Change* 7, 568–572.

Wang, Y., Chen, J. and Qin, L. (2005) *Mesanophrys carcini* causing severe scuticociliatosis in farmed turbot (*Scophthalmus maximus*) in China. *Journal of Fishery Sciences of China* 12, 594–601.

Weisse, T. (2017) Functional diversity of aquatic ciliates. *European Journal of Protistology* 61, 331–358.

Whang, I., Kang, H.S. and Lee, J. (2011) Morphological and molecular characterization of *Pseudocohnilembus longisetus* Thompson, 1965 from farmed black rockfish *Sebastes schlegelii* in Korea. *Veterinary Parasitology* 179, 227–233.

Whang, I., Kang, H.S. and Lee, J. (2013) Identification of scuticociliates (*Pseudocohnilembus persalinus*, *P. longisetus*, *Uronema marinum* and *Miamiensis avidus*) based on the *cox1* sequence. *Parasitology International* 62, 7–13.

Wilbert, N. (1975) Eine verbesserte Technik der Protargolim prägnation für *Ciliaten. Mikrokosmos* 6, 171–179.

Wilbert, N. (1995) Benthic ciliates of salt lakes. *Acta Protozoologica* 34, 271–288.

Woo, P.T.K., Wehnert, S.D. and Rodgers, D. (1983) The susceptibility of fishes to haemoflagellates at different ambient temperatures. *Parasitology* 87, 385–392.

Wu, R.S.S. (1995) The environmental impact of marine fish culture: towards a sustainable future. *Marine Pollution Bulletin* 31, 159–166.

Xiong, J., Wang, G., Cheng, J., Tian, M., Pan, X. *et al.* (2015) Genome of the facultative scuticociliatosis pathogen *Pseudocohnilembus persalinus* provides insight into its virulence through horizontal gene transfer. *Scientific Reports* 5, 15470.

Yanagisawa, M., Kaneko, A., Kino, S., Osawa, A., Yoshizawa, M. *et al.* (2018) Serotype surveillance of *Miamiensis avidus* causing scuticociliatosis in Japanese aquariums. *Presented at: 10th International Aquarium Congress, Fukushima, Japan, 5–10 November 2018.*

Yokohama, H., Itoh, N. and Ogawa, K. (2015) Fish and shellfish diseases caused by marine protists. In: Ohtsuka, S., Suzaki, T., Horiguchi, T., Suzuki, N. and Not, F. (eds) *Marine Protists: Diversity and Dynamics*. Springer, Tokyo, pp. 533–549.

Yoshinaga, T. and Nakazoe, J. (1993) Isolation and *in vitro* cultivation of an unidentified ciliate causing scuticociliatosis in Japanese flounder (*Paralichthys olivaceus*). *Gyobyo Kenkyu* 28, 131–134.

Zhan, Z., Stoeck, T., Dunthorn, M. and Xu, K. (2014) Identification of the pathogenic ciliate *Pseudocohnilembus persalinus* (Oligohymenophorea: Scuticociliatia) by fluorescence *in situ* hybridization. *European Journal of Protistology* 50, 16–24.

Zhang, S.L. and Song, W.B. (2000) Experimental ecology studies on the marine ciliate *Paranophrys magna* II. The effects of temperature and salinity on the population growth. *Chinese Journal of Applied and Environmental Biology* 6, 227–231.

Zhang, S.-L., Ma, H.-G., Xu, H.-L. and Song, W.-B. (2001) On experimental ecology of the marine ciliate *Paranophrys magna* IV: effects of competition on population growth. *Acta Ecologica Sinica* 21, 2039–2044.

Zhang, T., Fan, X., Gao, F., Al-Farraj, S.A., El-Serehy, H.A. and Song, W. (2019) Further analyses on the phylogeny of the subclass Scuticociliatia (Protozoa, Ciliophora) based on both nuclear and mitochondrial data. *Molecular Phylogenetics and Evolution* 139, 106565.

Zhou, L., Zhan, W. and Song, W. (2000) Observations on the histopathology of the external ulcerative disease of Japanese flounder, *Paralichthys olivaceus*. *Journal of Ocean University of Qingdao* 30, 593–597.

18 Ichthyophthiriosis (*Ichthyophthirius multifiliis*)

Louise von Gersdorff Jørgensen* and Kurt Buchmann

University of Copenhagen, Frederiksberg, Denmark

18.1 Introduction

Ichthyophthirius multifiliis is a protozoan ciliate that infects freshwater fishes worldwide. It was named by Fouquet in 1876, who was impressed with its rapid multiplication on brown trout (*Salmo trutta*) in ponds when the water temperature increased in the spring. Accordingly, he applied the Greek words for fish ('ichthyos'), louse ('phthirius'), many ('multi') and sons or children ('filiis') to indicate that the parasite appeared as a louse (which it is not) in the epidermis of fish and its rapid multiplication to produce numerous cells.

18.1.1 History and taxonomy

Despite its relatively recent scientific description, *I. multifiliis* was probably recognized in ancient China (Dashu and Lien-siang, 1960) based on clinical signs (epidermal white spots) on pond fishes. The parasite has spread worldwide probably due to anthropogenic translocation of infected hosts (Matthews, 2005) and it infects a wide variety of freshwater fish in different climatic zones. The disease, ichthyophthiriosis, is a well-known problem associated with significant economic losses in fish production especially in food and ornamental fishes (Dickerson and Dawe, 1995; Jørgensen, 2017). *I. multifiliis* has cilia on all life cycle stages (Figs 18.1 and 18.2). It belongs to the phylum Ciliophora, class Oligohymenophorea (Matthews, 2005), order Hymenostomatida due to the organelle of Lieberkühn (Lynn *et al.*, 1991) and the family Ichthyophthiridae. The parasite is related to *Tetrahymena* and *Paramecium* and has a marine counterpart *Cryptocaryon irritans*, family Cryptocaryonidae (Wright and Colorni, 2002). Five serotypes of *I. multifiliis* have been defined based on *in vitro* immobilization analyses using specific antisera that target surface epitopes, termed I-antigens (Dickerson *et al.*, 1993; Dickerson and Clark, 1998).

18.1.2 Morphology and life cycle

All life cycle stages of the parasite have cilia (Figs 18.1 and 18.2). The infective stage, the theront, is a free-swimming, short-lived, cylindrically shaped organism measuring 30–50 μm in length (MacLennon, 1942; Matthews, 2005), although theronts produced at a low temperature (5°C) may reach a diameter of 57 μm (Aihua and Buchmann, 2001).

The anterior end of the theront is equipped with a perforatorium, which is the first point of parasite–host contact and has been suggested to assist penetration of the host surface (Roque *et al.*, 1967; Canella and Rocchi-Canella, 1976; Matthews, 1994). The posterior end holds a caudal cilium (Fig. 18.2B) which has been proposed to function as a rudder (Geisslinger, 1987).

The theront is non-feeding and depending on temperature it has energy for up to 24 h (Matthews, 2005). It will penetrate the fish epidermis, transform into a trophont and develop a cytostome which assists feeding on host cells and debris (Maclennon, 1935; Ewing *et al.*, 1985; Dickerson and Dawe, 1995; Jørgensen, 2016b). The ciliated trophont is covered by one to several host cell layers (Fig. 18.3) and is therefore regarded as a true endoparasite although it is visible externally (Fig. 18.4).

*lvgj@sund.ku.dk

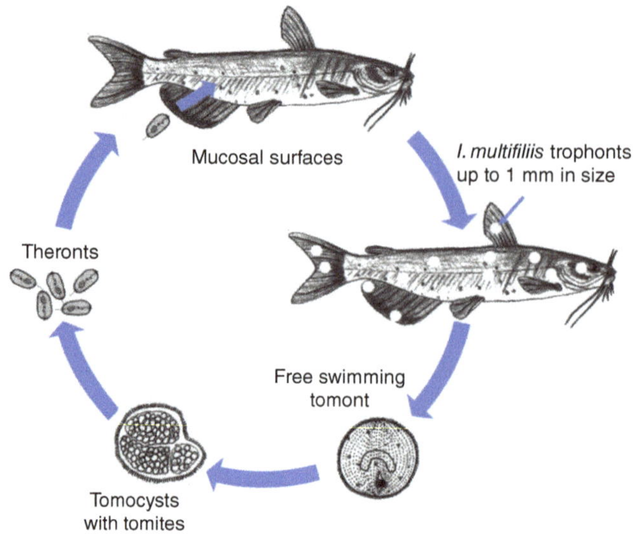

Fig. 18.1. The life cycle of *Ichthyophthirius multifiliis*.

The trophont in the epidermis is rounded to elongated but flexible, located in an interstitial space, and its volume may increase up to 37,000 times from the early post-penetration trophont stage (around 30 µm) to the late trophont stage (0.5 up to 1 mm) ready to leave the fish (Wagner, 1960; Dickerson and Dawe, 1995; Jørgensen *et al.*, 2018). It contains mucocysts, food vacuoles, contractile vacuoles and one macronucleus, as well as at least one micronucleus (Matthews, 2005). It continuously rotates in its interstitial space. The trophont transforms into a tomont when it ceases to feed and disengages from the epithelium. During this process huge mortalities may occur in fish populations. After the tomont leaves the fish, it searches for appropriate substrates for attachment and subsequently encysts by secretion of a gelatin capsule. Asexual reproduction (binary fissions) in the cyst leads to production of 50 to 1000 tomites (Dickerson and Dawe, 1995; Matthews, 2005), which eventually escape the cyst as free-swimming theronts. The life cycle is completed when theronts

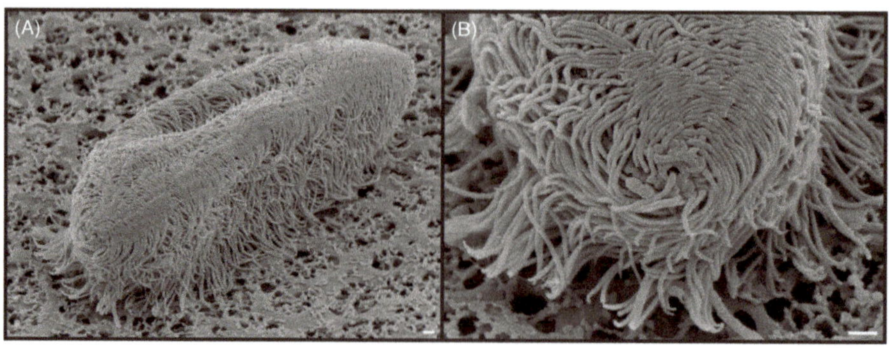

Fig. 18.2. (A) *Ichthyophthirius multifiliis* theront covered with cilia on the surface. This elongated infective stage of the parasite is approximately 30 µm long. (B) The caudal cilium at the posterior end of the theront. Scale bars = 1 µm. (Image courtesy of Dr Ole Sten Møller.)

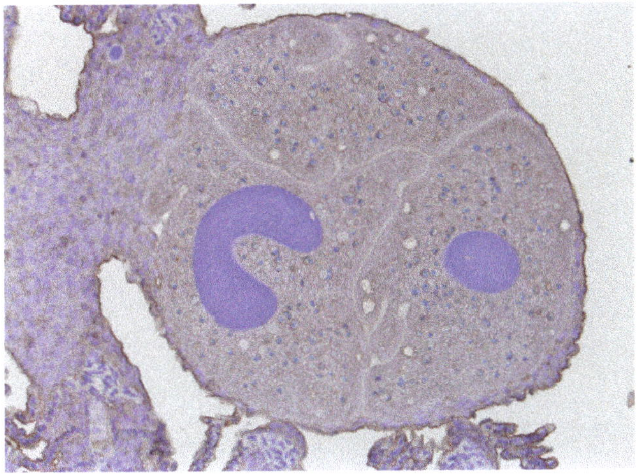

Fig. 18.3. Histological section of one or several *Ichthyophthirius multifiliis* trophonts located under one to two layers of gill epithelial cells. A horseshoe-shaped macronucleus is visible.

Fig. 18.4. (A) Rainbow trout (*Oncorhynchus mykiss*) and (B) zebrafish (*Danio rerio*) infected with *Ichthyophthirius multifiliis*. Every white spot represents one parasite and (B) is a closeup of the caudal fin.

locate and penetrate a fish surface. Temperature is a central parameter in the persistence of infection in fish populations and has a major influence on the length of the life cycle.

18.2 Hosts

I. multifiliis parasitizes a wide range of freshwater fish species and is known to cause epizootics in wild and cultured fish including ornamental fish.

18.2.1 Warm-water fish (20–30°C)

I. multifiliis causes problems in a great variety of ornamental as well as fish productions in warm freshwater. The life cycle of the parasite takes only a few days in the temperature range 25–30°C, leading to high infection levels shortly following introduction of the parasite in a fish tank system. Highly susceptible ornamental species include clown loach (*Chromobotia macracanthus*), tetra (*Paracheirodon* sp.), guppies and mollies (*Poecilia* sp.), catfish (*Ancistrus* sp.), barbs (*Barbodes* sp.) and cichlids, whereas zebrafish are relatively resistant and rarely succumb to the infection (Fig. 18.4). Susceptible species of warm-water fish used in production include channel catfish (*Ictalurus punctatus*) (Clark *et al.*, 1987), carp (*Cyprinus carpio*) (Gonzalez *et al.*, 2007b), grass carp (*Ctenopharyngodon idella*) (Lin *et al.*, 2016), tilapia (family Cichlidae) (Xu *et al.*, 2008) and eel (*Anguilla* spp.) (Aguilar *et al.*, 2005).

18.2.2 Cold-water fish (5–20°C)

Rainbow trout is an important cold-water fish, which is highly susceptible to *I. multifiliis* (Buchmann *et al.*, 2001). The developmental rate of all stages in the life cycle of the parasite is decreased at low temperatures compared with warm waters, whereby fish farmers may initiate control operations before infection becomes severe. Other susceptible cold-water fish species include Atlantic salmon (*Salmo salar*) (Valtonen and Keranen, 1981) as well as perch (Grignard *et al.*, 1996), pikeperch (Németh *et al.*, 2013) and European catfish (Roohi *et al.*, 2014).

18.3 Diagnosis and Pathology

Macroscopically, visible clinical signs of ichthyophthiriosis are white spots appearing on the skin and fins of the fish. Light microscopy of skin scrapings will confirm the diagnosis if the sampled organisms are ciliated and contain a horseshoe-shaped nucleus. Molecular diagnostics are also available (Jørgensen *et al.*, 2018) based on DNA extraction, polymerase chain reaction (PCR) and sequencing of a surface protein of the parasite (I-antigen). Behavioural changes appear during heavy infections, leading the fish to flash and scrape their surfaces against any objects and substrates. Developing trophonts elicit extensive hyperplasia of epithelial and mucous cells (Hines and Spira, 1974c; Ventura and Paperna, 1985; Jørgensen, 2016a). Epidermal surfaces become emaciated due to penetration of theronts and exit of trophonts leaving microscopical lesions, which probably challenge osmoregulation and may allow opportunistic microorganisms (bacteria and fungi) to invade the fish (Hines and Spira, 1974c). Severe clinical signs include lethargy, anorexia and hyperventilation, which indicate decreased respiratory surface areas of gills (Matthews, 2005). At fish farms, infected fish congregate near water inlets, seeking water currents with higher oxygen (O_2) levels. Trophonts in skin, fins and gills evoke proliferation of epithelial cells (Ventura and Paperna, 1985) and leucocyte infiltration dominated by neutrophils, eosinophils, basophils, B cells, T cells and macrophages in carp (*C. carpio*), zebrafish (*Danio rerio*) and rainbow trout (*Oncorhynchus mykiss*) (Hines and Spira, 1974b; Jørgensen *et al.*, 2011; Olsen *et al.*, 2011; Jørgensen, 2016b). Fish may die from invasions by a high number of theronts, although the usual cause of death is the escape of the fully developed trophonts at the end of the epidermal feeding stage (Ewing and Kocan, 1992).

18.4 Climate Change

18.4.1 Expected/potential spread of the pathogen

Climate change is ongoing and includes rising temperatures, an increase of atmospheric carbon dioxide (CO_2), and increased frequency and intensity of droughts and severe flow events (Kundzewicz *et al.*, 2007; Woodward *et al.*, 2010). Freshwater systems are especially vulnerable to climate change because of their relative isolation, physical fragmentation and human influence within a largely terrestrial environment (Woodward *et al.*, 2010). The development of all life cycle stages of *I. multifiliis* is highly temperature dependent in the temperature range between 5 and 30°C (Wagner, 1960; Aihua and Buchmann, 2001). Completion of the life cycle is reached within a few days (Dickerson and Dawe, 1995) at temperatures between 25 and 28°C, whereas it takes approximately 12 days in rainbow trout at 15°C (Aihua and Buchmann, 2001) and completion may take several months at 5–10°C (Matthews, 2005). In climatic regions such as Europe where the temperature potentially will increase by 1–5°C by the end of the 21st century due to climate change (Dittmar *et al.*, 2014; King and Karoly, 2017), the infection pressure may increase correspondingly and subject new hosts to infection. Fitness of the parasite decreases at temperatures exceeding 30°C (Camacho, 2010) and it could be suggested that freshwaters, which presently have reached or exceeded this temperature would limit infection pressure as warming continues. However, the selective pressure on the parasite would favour strains adapted for warmer climates in areas where the temperature rises significantly. In climate zones where the temperature rises from a baseline of 5–10°C (where the parasite normally would be absent or have a very low prevalence), an increased spread of the parasite is expected. The effects of warming are expected to be especially noticeable in colder high-altitude systems (Woodward *et al.*, 2010) and in such systems there is a possibility that ichthyophthiriosis will spread and become more severe. Thus, naïve fish species would become exposed and very likely infected because of the wide host preference of the parasite.

In warm climate zones (20–30°C) infection pressure may increase following warming until 30°C but exposure of new host species is not expected. The life cycle stages are readily affected and adapt to changing temperatures. Lower temperatures induce production of fewer but larger sized theronts and higher temperatures lead to smaller theronts. Thus, at 5°C an average theront size may be 57.42 µm × 28.6 µm whereas at 30°C the average size is 28.64 µm × 20 µm (Aihua and Buchmann, 2001).

With a changing climate comes new environmental microbiota (Woodward *et al.*, 2010). Some bacterial products (e.g. surfactant of *Pseudomonas*, strain H6) are lethal to parasites (Al-Jubury *et al.*, 2018), thus the microbiota may influence the abundance of *I. multifiliis*. Floods and increased precipitation may lower salinities in certain areas such as fjords with brackish water and if the salinity becomes less than 5 ppt, *I. multifiliis* may invade these habitats and infect susceptible host species.

18.4.2 Expected/potential spread of hosts

Fish respond to climate change and fish species' migrations and/or translocations may be expected (Laidre *et al.*, 2008). Some species will adapt to the new environment whereas others will have to migrate or will locally become extinct (Moller *et al.*, 2008). It is believed that rising water temperatures will lead to shifts in distributions of freshwater species and exacerbate existing problems in water quality, especially in those systems experiencing high anthropogenic loading of nutrients (Bahri *et al.*, 2018). Fish with specific temperature preferences will likely invade new habitats with altered but suitable temperature conditions. *I. multifiliis* will thrive on almost any freshwater host in climate zones within a range of 5–30°C, so it is expected that migrating fish will spread *I. multifiliis* to new areas. Inland fisheries are heterogeneous with large regional differences reflecting geographical location and the environmental situation, and are sensitive to climate change (Harrod *et al.*, 2018). Aquaculturists may experience problems with a changing environment and outbreaks of ichthyophthiriosis and may have to translocate their production facilities to new geographical areas.

18.4.3 Population dynamics of hosts

Climate change (e.g. temperature, pH, dissolved CO_2) induces both long- and short-term changes in freshwater habitats. Hence the physiology of fish including reproduction, growth and immunity will change, whether wild or cultured. Alterations in water temperature and physical conditions will determine the major part of fish responses to climate change (Harrod *et al.*, 2018). These changes may negatively impact natural fish populations, fish farm economy and biodiversity through elevated stress levels and disease pressures. Examples of short-term impacts on cultured fish include destruction of infrastructure leading to production losses due to extreme weather events, diseases, toxic algae and parasites. Long-term examples include limited access to freshwater, limited access to feeds from marine and terrestrial sources, a lower productivity because of worse farming conditions and eutrophication (Dabbadie *et al.*, 2018). However, fish growth is temperature dependent and will accelerate (within limits) with increasing temperatures (Barrow *et al.*, 2018), whereby fish farmers may be able to harvest fish after a shorter growth period and thereby increase production. Acidification of freshwater, which can be caused by climate change, reduces growth, yolk-to-tissue conversion and maximal O_2 uptake capacity of pink salmon yolk sac larvae (Ou *et al.*, 2015). Since *I. multifiliis* is a generalist, it is not possible to describe a general trend in the dynamics of its host populations during climate change. However, the immune response towards the parasite has been described in several host species and the following section focuses on the immune system in relation to temperature changes.

The activity and efficiency of the immune system of ectothermic animals such as fish are highly dependent on temperature (Ellis, 2001). Fish may actively seek suitable temperatures or actively increase metabolic activity to elevate body temperature. Cold-blooded animals actively seek warmer external temperatures to increase their internal temperature to induce a more efficient protective response against pathogens (Boltaña *et al.*, 2013). Behavioural fever may be regarded as a defensive reaction of the innate immune system (Bicego *et al.*, 2007) aiming at increasing the resistance to or recovery from disease (Boltana *et al.*, 2018). However, a chronic exposure to elevated temperatures may induce stress, exhaust the immune system and eventually lead to decreased immunity (Boltana *et al.*, 2018). With climate change some regions (e.g. in Europe) may experience higher temperatures (King and Karoly, 2017) whereas others (e.g. eastern USA) may become colder (Cohen *et al.*, 2018). In the latter case, some fish hosts may suffer from lowered immune responses. In channel catfish a lower temperature (15°C compared with 20, 25 and 30°C) hampered the production of antibodies following immunization against *I. multifiliis*, leading to higher mortality rates (Martins *et al.*, 2011). Other host species adapted to lower temperatures such as the three-spined stickleback (*Gasterosteus aculeatus*) are negatively influenced by high temperatures and show improved reactions at relatively low temperatures (Dittmar *et al.*, 2014).

Overall, increasing temperature and atmospheric CO_2 affect freshwater habitats and, combined with anthropogenic impact, may stress the fish, which increases susceptibility to *I. multifiliis* (Cherry, 2003).

18.4.4 Host immune responses

Innate responses

Following penetration of the host surface, the parasite starts feeding in the epidermis or under the gill epithelium. The fish reacts by initiating an inflammatory response involving a series of cytokines, including interleukin (IL)-1β, tumour necrosis factor alpha (TNF-α) and IL-8 (Hines and Spira, 1973; Sigh *et al.*, 2004; Gonzalez *et al.*, 2007c; Xu *et al.*, 2017). In common carp and rainbow trout a series of genes encoding complement factors are upregulated a few hours after infection, indicating that complement pathways are involved in the innate response (Gonzalez *et al.*, 2007a; Jørgensen *et al.*, 2008). The CXC receptor 1 (CXCR1) associated with leucocyte and especially neutrophil recruitment is upregulated from 3 h in carp. A similar upregulation of IL-1β and CXCR1 was found following mechanical tissue damage mimicking an *I. multifiliis* infection, confirming the role as a non-specific reaction (Gonzalez *et al.*, 2007c). Acute-phase proteins such as serum amyloid A (SAA), hepcidin and pre-cerebellin are also upregulated during the primary response towards the parasite (Gonzalez *et al.*, 2007b; Jørgensen *et al.*, 2008) and play an unknown role in the acute innate reaction. The cellular responses in affected fish surfaces have been described for a variety of fish species with the main focus on channel catfish, carp and rainbow trout. Hines and Spira (1973) were the first to describe the recruitment of neutrophils at infected locations in the skin of carp, but since then various authors have confirmed that neutrophils are attracted to the affected location (Ventura and Paperna, 1985; Cross, 1994; Jørgensen, 2016b). Histologically, leucocytes engaged in the response against *I. multifiliis* are often attracted to the interstitial spaces occupied by the parasite and not in direct contact with the parasite. Cross (1994) suggested that neutrophils are degraded in the necrotic layer surrounding the parasite. A recent study, however, has shown that the parasites ingest whole functional neutrophils and this may be the reason why there are few or no immune cells in close contact with the parasite in the skin or on the fins (Jørgensen, 2016b). In addition, neutrophils produce extracellular traps of DNA and histone proteins, which may present a hostile microenvironment for the parasite (Papayannopoulos, 2018). Other cells such as lymphocytes, eosinophils and basophils arrive a bit later and in lower numbers (Hines and Spira, 1973; Ventura and Paperna, 1985; Cross, 1994). Macrophage-like cells expressing major histocompatibility complex II ($MHCII^+$) surrounded the parasite in the gills of rainbow trout 4 days after infection (Olsen *et al.*, 2011). Following infection or vaccination with live parasites, transcription of immunoglobulin (Ig) genes in channel catfish increased already 4 h after immunization (Xu *et al.*, 2017). IgM and IgD transcripts increase and show a continued upregulation during the following days in both rainbow trout and channel catfish (Sigh *et al.*, 2004; Xu *et al.*, 2017).

In hosts such as rainbow trout, immune gene expression is positively correlated to temperature (Raida and Buchmann, 2007), and the innate mechanisms may accelerate in the fish with global warming. The parasite will, however, also display increased multiplication with elevated temperatures and it is expected that a new balance between fish and parasite will be established due to climate change.

Adaptive responses

Fish hosts are able to acquire immunity against ichthyophthiriosis (Buschkiel, 1910; Bauer, 1953; Buchmann *et al.*, 2001) and with rising temperatures, protective immunity may develop faster in certain hosts (e.g. rainbow trout). Antibodies play a major role in the protective response of fish against *I. multifiliis* but other factors may also contribute to the immune protection (Hines and Spira,

1974a; Clark *et al.*, 1988). In channel catfish, rainbow trout and carp, IgM levels in the blood are elevated following immunization with the parasite and sera from immune fish immobilize the parasites under *in vitro* conditions (Clark *et al.*, 1987; Jørgensen *et al.*, 2011, 2017). In rainbow trout the mucosal immunoglobulin IgT plays a significant role in protection (Xu *et al.*, 2013, 2016) and both IgM and IgT (Fig. 18.5) bind directly to the surface of the parasite (Jørgensen *et al.*, 2011). The major part of IgM antibodies in immunized fish targets a surface glycosylphosphatidylinositol (GPI)-anchored protein Iag52b; immunizations with this protein purified from the parasite together with Freund's complete adjuvant induce a high level of protection but only towards the homologous serotype (Clark *et al.*, 1996; Wang *et al.*, 2002). Intraperitoneal injection of monoclonal IgG antibodies targeting Iag52b will confer some protection by inducing penetrating theronts and early trophonts to leave host fish (Clark *et al.*, 1996), supporting the immunization potential of Iag52b. The protein is complex, however, and classical expression systems such as *Escherichia coli* and insect cells have proven unsuitable to recombinantly express the protein, thus hampering vaccine development (Jørgensen *et al.*, 2012).

The cellular adaptive immune response to *I. multifiliis* is induced shortly after infection. T cells, MHCII$^+$ cells, IgM and IgT lymphocytes have been demonstrated around the developing trophont (Olsen *et al.*, 2011). It is not known how the various humoral and cellular immune factors account for the immunity in primed fish, but the responses create a microhabitat, which forces early trophonts to prematurely exit the invaded site. It cannot be excluded that a combination of various innate and adaptive response mechanisms (humoral and cellular) may be involved in the protective response. In carp the response includes an accumulation of eosinophilic granular cells, basophils and neutrophils near infection sites (Cross and Matthews, 1993) and it has been demonstrated that upregulation of IgM and IgZ in immunized zebrafish occurs concomitantly with a significantly increased occurrence of neutrophils (Jørgensen *et al.*, 2018).

Global warming can, due to the temperature-dependent immune reactivity in fish, lead to increased reactivity in some fishes and this may induce increased resistance towards infection. Temperature is believed to be a strong modifier of host–parasite interactions (Ellis, 2001) but it remains undetermined to what extent various fish species in different habitats may suffer or gain from temperature changes. In a production facility, however, the accelerated development of immunity may be beneficial provided the parasite intensity is kept unchanged.

18.4.5 Increase in pathogen mortality with changing temperatures

Abiotic factors

The majority of physical, chemical and biological processes in aquatic systems are driven by temperature, and both abiotic and biotic conditions are influenced by temperature (Harrod *et al.*, 2018). Warming and acidification of the water reduce the solubility of dissolved O_2, and different host and

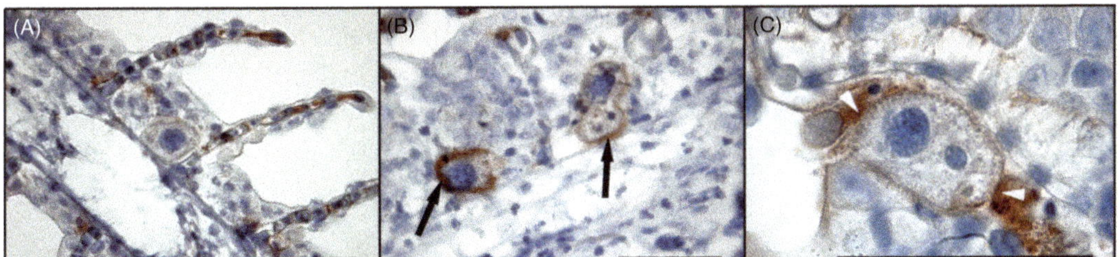

Fig. 18.5. Gills from one naïve and two immunized rainbow trout (*Oncorhynchus mykiss*) 2 h after a challenge with *Ichthyophthirius multifiliis*. Immunohistochemical staining was conducted detecting IgM and IgT near parasites. (A) Naïve gill tissue stained against IgM; (B) gill tissue from immune fish stained against IgM; and (C) gill tissue from immune fish stained against IgT. Scale bars = 50 μm; black arrows and white arrowheads indicate IgM and IgT, respectively associated to the surface of the parasite.

parasite species and taxonomic groups tolerate varying O_2 levels (Bahri *et al.*, 2018). For example, *I. multifiliis* is highly sensitive to O_2 deficit and will not continue development from the trophont stage into the reproducing tomont stage under suboptimal O_2 conditions. It has been estimated that the parasite requires an O_2 level of at least 0.6–0.8 mg/l at 15°C (Wagner, 1960). Climate change involves temperature increases and induces an ongoing deoxygenation of oceans and freshwater bodies (Breitburg *et al.*, 2018), and O_2 depletion may reduce the infection level at specific sites. Most fish host species such as channel catfish, will, however, also be negatively affected by decreasing O_2 levels (Steeby *et al.*, 2004) and are expected to migrate away from hypoxic water bodies.

The pH of water is generally considered critical for the development and persistence of aquatic organisms and will correspondingly affect aquatic parasites. Water absorbs atmospheric CO_2 which makes it more acidic. The concentration of atmospheric CO_2 has increased from 278 ppm in the middle of the 18th century to the current level of around 400 ppm (Ciais *et al.*, 2013). In the last 35 years the increase of atmospheric CO_2 correlated well with acidification (pH changes from 0.3 to 4) of freshwater reservoirs in Germany (Weiss *et al.*, 2018). Freshwater systems have relatively low buffering capacity and are sensitive to climate changes (Harrod *et al.*, 2018) and freshwater acidification may be faster than ocean acidification. However, acidification may be slower in some freshwater habitats with a better buffering capacity (e.g. from limestone). Stages of *I. multifiliis* can survive within a pH range of 5.5–10.1 (Wagner, 1960), which indicates that elevation of atmospheric CO_2 may reduce infection levels. Fish species are also sensitive to acidification (Bahri *et al.*, 2018). Water hardness influences the effect of pH; silver catfish (*Rhamdia quelen*) had higher infection intensities at around pH 7 and this depended on water hardness (Garcia *et al.*, 2011).

I. multifiliis does not tolerate salinities higher than 5 ppt. At 7.5 ppt, trophonts may divide but no theronts escape the tomocysts. At 5 ppt, 33 % of the trophonts released from an infected fish resulted in successful tomocyst development but lower numbers of theronts were produced. Adverse effects on parasite reproduction were demonstrated at 3 ppt (Aihua and Buchmann, 2001). Droughts caused by climate change and anthropogenic pressure lead to salinization of freshwater bodies (Canedo-Arguelles *et al.*, 2019), whereas floods lead to a periodical decrease in salinity.

Climate change also affects primary production (e.g. algae), which is an integrator of light, temperature and nutrient changes (Bahri *et al.*, 2018). Hence, *I. multifiliis* may also be affected indirectly as the parasite responds to light (Dickerson, 2006) and temperature (Aihua and Buchmann, 2001).

18.5 Control and/or Prevention

18.5.1 Selective breeding

The susceptibility to infection varies considerably between individuals in a fish population. This indicates potential for selection of fish with a better innate resistance to produce fish with higher disease resistance. This approach will be even more relevant in the future where the ambient temperature will rise due to global warming. Classical breeding programmes for salmonids suffer from a long generation time. Following selection of individual fish with high natural resistance to a certain pathogen, it will take 3–4 years before the fish reach sexual maturity and can be used as breeders for the next generation. Novel DNA technology allows breeders to accelerate this process. By use of a single nucleotide polymorphism (SNP) technology (LaFramboise, 2009) developed specifically for certain fish species, such as rainbow trout, markers associated with resistance for specific diseases can be identified using DNA-typing of surviving fish. Consequently, it will be possible to pinpoint parental fish carrying genes for *I. multifiliis* resistance (Gonzalez-Pena *et al.*, 2016). Marker-assisted selection (MAS) can then be applied in focused breeding programmes, which will accelerate successful breeding and decrease the delivery time of novel products to the market to a few years. This accelerated selective breeding system will be highly relevant in a changing environment as it can be adapted to select temperature-tolerant fish with resistance to a particular pathogen.

18.5.2 New vaccines

Vaccination of fish confers protection against a range of specific bacterial and viral diseases. At present no effective vaccines against parasitic diseases in fish are commercially available but laboratory studies have indicated the potential for this beneficial approach. Fish that survive infections with *I. multifiliis* develop protective immunity (Clark *et al.*, 1987; Cross and Matthews, 1992; Jørgensen *et al.*, 2008), which indicates that the fundamental immunological mechanisms are present for the development of vaccines. The simplest vaccines contain dead parasites or parasitic material and these vaccines induce a moderate to high protection (Alishahi and Buchmann, 2006; Jørgensen, 2017). Injection of live *I. multifiliis* theronts into the body cavity induces a highly protective response in rainbow trout (Alishahi and Buchmann, 2006) but production of live parasites relies on live fish, making production impractical, laborious and expensive. Advanced experimental vaccines include DNA and subunit vaccines, but these vaccines induced no or merely moderate protection (Jørgensen *et al.*, 2012, 2017). The most efficient current vaccine, besides using live parasites, contains purified Iag52b and Freund's complete adjuvant (Wang *et al.*, 2002). Immunoprophylactic approaches using specific vaccines against *I. multifiliis* is a future path to control ichthyophthiriosis also during climate change.

18.5.3 New chemicals and drugs

Historically, ichthyophthiriosis has been controlled by addition of various chemicals (which affect one or more stage of the parasite) to the pond water. Several of these efficient but hazardous substances, such as malachite green and methylene blue, are not approved for use in most countries (Alderman, 1985; Tieman and Goodwin, 2001). Formaldehyde, in spite of its carcinogenicity, may still be used as it is lethal to the free-swimming stages of the parasite and is tolerated by several fish species such as rainbow trout (Heinecke and Buchmann, 2009; Forwood *et al.*, 2014). Copper sulfate is efficient against the parasite (Schlenk *et al.*, 1998) but it has a negative environmental impact, especially on plants, algae and invertebrates. Environmentally friendly substances such as peracetic acid and sodium percarbonate are effective mainly due to their release of hydrogen peroxide when added to water (Heinecke and Buchmann, 2009; Pedersen and Henriksen, 2017). Correspondingly, free hydrogen peroxide may be applied as it kills theronts within seconds (Rach *et al.*, 2000). In recent years a series of laboratory investigations has suggested using plant extracts like garlic (Buchmann *et al.*, 2003; Lin *et al.*, 2016) and bacterial products (Al-Jubury *et al.*, 2018) as highly efficacious treatment methods. These are not licensed but there is a potential to use various environmentally friendly compounds for control of ichthyophthiriosis. The efficacy of different compounds used in fish farms (at present and in the future) is highly dependent on physical and chemical conditions such as temperature, pH and salinity. With a changing climate, efficacy of compounds for treatment will also change.

18.5.4 Disruption of pathogen transmission

The sizes of different developmental stages of the parasite are well known; consequently, the use of a water filtration system is one method to control ichthyophthiriosis. Management procedures using mechanical filters with relevant mesh sizes are currently applied in fish farms to trap free-living stages of *I. multifiliis*. Tomonts with diameters of more than 100 μm are easily removed using 80 μm filters; for each tomont removed, the infection pressure 24 h later will be lowered by up to 1000 theronts (Heinecke and Buchmann, 2009). Clogging of these filters by other suspended materials in the water column may present a significant problem. The sizes of the life cycle stages are temperature dependent (Wagner, 1960; Aihua and Buchmann, 2001) and the temperature shifts resulting from climate change may require further adjustment of the size of filters used.

18.6 Conclusions with Suggestions for Future Studies

Global warming will change the physical and chemical conditions of water bodies and these will affect fish hosts and pathogens. It is expected that many local artificial or natural ecosystems will face new challenges, especially in situations with parasites and hosts adapting to ongoing changes to the environment. The physiology including the immune response of the fish host will be markedly influenced by the climate. The life cycle stages of *I. multifiliis* will also be affected. The severity of infections with *I. multifiliis* may depend on the ability of the parasite and the hosts to adapt to changes in temperature, CO_2, O_2 and pH levels in the aquatic system. Since *I. multifiliis* includes many different strains, a possible scenario is that strains with a preference for a specific temperature, CO_2 or O_2 concentration or pH level will be favoured, while the hosts will adapt more slowly to environmental changes. When shifting parameters such as temperature, CO_2 and pH are analysed with regard to their influence on hosts and parasites, a synergistic effect of the parameters is difficult to predict. The overall effect may become more pronounced than estimated from individual parameters combined, in response to climate change. Small changes can lead to unforeseen scenarios, which make predictions very difficult. Therefore, there is a need for further integrated studies on infected fish species with a focus on the combined effects of several environmental changes in a natural or fish production-like environment.

Acknowledgements

This project was supported by the European Union's Horizon 2020 research and innovation programme (Grant Agreement no. 634429; ParaFishControl). The views expressed in this review are those of the authors and not the European Union. Thanks go to Ole Sten Møller for kindly providing the scanning electron micrograph images of *Ichthyophthirius multifiliis*.

References

Aguilar, A., Alvarez, M.F., Leiro, J.M. and Sanmartin, M.L. (2005) Parasite populations of the European eel (*Anguilla anguilla* L.) in the Rivers Ulla and Tea (Galicia, northwest Spain). *Aquaculture* 249(1–4), 85–94. https://doi.org/10.1016/j.aquaculture.2005.04.052

Aihua, L. and Buchmann, K. (2001) Temperature- and salinity-dependent development of a Nordic strain of *Ichthyophthirius multifiliis* from rainbow trout. *Journal of Applied Ichthyology* 17(6), 273–276. https://doi.org/10.1046/j.1439-0426.2001.00279.x

Alderman, D. (1985) Malachite green: a review. *Journal of Fish Diseases* 8(3), 289–298. https://doi.org/10.1111/j.1365-2761.1985.tb00945.x

Alishahi, M. and Buchmann, K. (2006) Temperature-dependent protection against *Ichthyophthirius multifiliis* following immunisation of rainbow trout using live theronts. *Diseases of Aquatic Organisms* 72(3), 269–273. https://doi.org/10.3354/dao072269

Al-Jubury, A., Lu, C., Kania, P., Jorgensen, L., Liu, Y. *et al.* (2018) Impact of *Pseudomonas* H6 surfactant on all external life cycle stages of the fish parasitic ciliate *Ichthyophthirius multifiliis*. *Journal of Fish Diseases* 41(7), 1147–1152. https://doi.org/10.1111/jfd.12810

Bahri, T., Barange, M. and Moustahfid, H. (2018) Climate change and aquatic systems. In: Barange, M., Bahri, T., Beveridge, M.C.M., Cochrane, K.L., Funge-Smith, S. and Poulain, F. (eds) Impacts of climate change on fisheries and aquaculture. Synthesis of current knowledge, adaptation and mitigation options. *FAO Fisheries and Aquaculture Technical Paper No. 627*. Food and Agriculture Organization of the United Nations, Rome, pp. 1–18.

Barrow, J., Ford, J., Day, R. and Morrongiello, J. (2018) Environmental drivers of growth and predicted effects of climate change on a commercially important fish, *Platycephalus laevigatus*. *Marine Ecology Progress Series* 598, 201–212. https://doi.org/10.3354/meps12234

Bauer, O.N. (1953) Immunity of fish occurring in infections with *Ichthyophthirius multifiliis* Fouquet, 1876. *Doklady Novaia Erviia* 93, 377–379.

Bicego, K., Barros, R. and Branco, L. (2007) Physiology of temperature regulation: comparative aspects. *Comparative Biochemistry and Physiology Part A: Molecular and Integrative Physiology* 147(3), 616–639. https://doi.org/10.1016/j.cbpa.2006.06.032

Boltaña, S., Rey, S., Roher, N., Vargas, R., Huerta, M. *et al.* (2013) Behavioural fever is a synergic signal amplifying the innate immune response. *Proceedings of the Royal Society B: Biological Sciences* 280(1766), 20131381. https://doi.org/10.1098/rspb.2013.1381

Boltana, S., Aguilar, A., Sanhueza, N., Donoso, A., Mercado, L. *et al.* (2018) Behavioral fever drives epigenetic modulation of the immune response in fish. *Frontiers in Immunology* 9, 1241. https://doi.org/10.3389/fimmu.2018.01241

Breitburg, D., Levin, L., Oschlies, A., Gregoire, M., Chavez, F. *et al.* (2018) Declining oxygen in the global ocean and coastal waters. *Science* 359(6371), eaam7240. https://doi.org/10.1126/science.aam7240

Buchmann, K., Sigh, J., Nielsen, C.V. and Dalgaard, M. (2001) Host responses against the fish parasitizing ciliate *Ichthyophthirius multifiliis*. *Veterinary Parasitology* 100(1–2), 105–116. https://doi.org/10.1016/S0304-4017(01)00487-3

Buchmann, K., Jensen, P. and Kruse, K. (2003) Effects of sodium percarbonate and garlic extract on *Ichthyophthirius multifiliis* theronts and tomocysts: *in vitro* experiments. *North American Journal of Aquaculture* 65(1), 21–24. https://doi.org/10.1577/1548-8454(2003)065<0021:EOSPAG>2.0.CO;2

Buschkiel, A.L. (1910) Beiträge zur Kenntnis des *Ichthyophthirius multifiliis* Fouquet. *Archiv für Protistenkunde* 21, 61–102.

Camacho, S.M.P. (2010) Developing strategies for the control of *Ichthyophthirius multifiliis* Fouquet, 1876 (Ciliophora). PhD thesis, University of Stirling, Stirling, UK. Available at: http://hdl.handle.net/1893/3032 (accessed 24 March 2020).

Canedo-Arguelles, M., Kefford, B. and Schafer, R. (2019) Salt in freshwaters: causes, effects and prospects – introduction to the theme issue. *Philosophical*

Transactions of the Royal Society B: Biological Sciences 374(1764), 20180002. https://doi.org/10.1098/rstb.2018.0002

Canella, M.F. and Rocchi-Canella, I. (1976) Contributions a la connaissance des Cilies. VII Biologie des Ophryoglenina (Cilies Hymentostoes Histophages). *Annals of the University of Ferrara (Italy), Nuova Serie, sezione III*, 1–510.

Cherry, B. (2003) Laboratory infection of zebrafish *(Danio rerio)* and channel catfish *(Ictalurus punctatus)* with the protozoan parasite *Ichthyophthirius multifiliis*: a model for parasite persistence. PhD thesis, University of Pennsylvania, Pittsburgh, Pennsylvania. Available at: https://repository.upenn.edu/dissertations/AAI3109164 (accessed 24 March 2020).

Ciais, P., Sabine, C., Bala, G., Bopp, L., Brovkin, V. *et al.* (2013) Carbon and other biogeochemical cycles. In: Stocker, T.F., Qin, D., Plattner, G.K., Tignor, M.M.B., Allen, S.K. *et al.* (eds) *Climate Change 2013: The Physical Science Basis. Contribution of Working Group I to the Fifth Assessment Report of the Intergovernmental Panel on Climate Change*. Cambridge University Press, Cambridge and New York, pp. 465–570.

Clark, T., Dickerson, H., Gratzek, J. and Findly, R. (1987) *In vitro* response of *Ichthyophthirius multifiliis* to sera from immune channel catfish. *Journal of Fish Biology* 31(sA), 203–208. https://doi.org/10.1111/j.1095-8649.1987.tb05314.x

Clark, T., Dickerson, H. and Findly, R. (1988) Immune response of channel catfish to ciliary antigens of *Ichthyophthirius multifiliis*. *Developmental and Comparative Immunology* 12(3), 581–594. https://doi.org/10.1016/0145-305X(88)90074-2

Clark, T., Lin, T. and Dickerson, H. (1996) Surface antigen cross-linking triggers forced exit of a protozoan parasite from its host. *Proceedings of the National Academy of Sciences USA* 93(13), 6825–6829. https://doi.org/10.1073/pnas.93.13.6825

Cohen, J., Pfeiffer, K. and Francis, J.A. (2018) Warm Arctic episodes linked with increased frequency of extreme winter weather in the United States. *Nature Communications* 9, 869. https://doi.org/10.1038/s41467-018-02992-9

Cross, M.L. (1994) Localized cellular responses to *Ichthyophthirius multifiliis*: protection or pathogenesis? *Parasitology Today* 10(9), 364–368. https://doi.org/10.1016/0169-4758(94)90253-4

Cross, M.L. and Matthews, R.A. (1992) Ichthyophthiriasis in carp, *Cyprinus carpio* L.: fate of parasites in immunized fish. *Journal of Fish Diseases* 15(6), 497–505. https://doi.org/10.1111/j.1365-2761.1992.tb00681.x

Cross, M.L. and Matthews, R.A. (1993) Localized leucocyte response to *Ichthyophthirius multifiliis* establishment in immune carp *Cyprinus carpio* L. *Veterinary Immunology and Immunopathology* 38(3–4), 341–358. https://doi.org/10.1016/0165-2427(93)90092-I

Dabbadie, L., Aguilar-Manjarrez, J., Beveridge, M.C.M., Bueno, P.B., Ross, L.G. *et al.* (2018) Effects of climate change on aquaculture: drivers, impacts and policies. In: Barange, M., Bahri, T., Beveridge, M.C.M., Cochrane, K.L., Funge-Smith, S. and Poulain, F. (eds) Impacts of climate change on fisheries and aquaculture. Synthesis of current knowledge, adaptation and mitigation options. *FAO Fisheries and Aquaculture Technical Paper No. 627*. Food and Agriculture Organization of the United Nations, Rome, pp. 449–464.

Dashu, N. and Lien-siang, L. (1960) Studies on the morphology and life cycle of *Ichthyophthirius multifiliis* and its control with a description of a new species. *Acta Hydrobiologica* 2, 197–225.

Dickerson, H.W. (2006) *Ichthyophthirius multifiliis* and *Cryptocaryon irritans* (phylum Ciliophora). In: Woo, P.T.K. (ed.) *Fish Diseases and Disorders*. Vol. 1. *Protozoan and Metazoan Infections*. CAB International, Wallingford, UK, pp. 116–153.

Dickerson, H. and Clark, T. (1998) *Ichthyophthirius multifiliis*: a model of cutaneous infection and immunity in fishes. *Immunological Reviews* 166(1), 377–84. https://doi.org/10.1111/j.1600-065X.1998.tb01277.x

Dickerson, H.W. and Dawe, D.L. (1995) *Ichthyophthirius multifiliis* and *Cryptocaryon irritans* (phylum Ciliophora). In: Woo, P.T.K. (ed.) *Fish Diseases and Disorders*. Vol. 1. *Protozoan and Metazoan Infections*. CAB International, Wallingford, UK, pp. 181–228.

Dickerson, H., Clark, T. and Leff, A. (1993) Serotypic variation among isolates of *Ichthyophthirius multifiliis* based on immobilization. *Journal of Eukaryotic Microbiology* 40(6), 816–820. https://doi.org/10.1111/j.1550-7408.1993.tb04480.x

Dittmar, J., Janssen, H., Kuske, A., Kurtz, J. and Scharsack, J. (2014) Heat and immunity: an experimental heat wave alters immune functions in three-spined sticklebacks (*Gasterosteus aculeatus*). *Journal of Animal Ecology* 83(4), 744–757. https://doi.org/10.1111/1365-2656.12175

Ellis, A. (2001) Innate host defense mechanisms of fish against viruses and bacteria. *Developmental and Comparative Immunology* 25(8–9), 827–839. https://doi.org/10.1016/S0145-305X(01)00038-6

Ewing, M. and Kocan, K. (1992) Invasion and development strategies of *Ichthyophthirius multifiliis*, a parasitic ciliate of fish. *Parasitology Today* 8(6), 204–208. https://doi.org/10.1016/0169-4758(92)90265-4

Ewing, M., Kocan, K. and Ewing, S. (1985) *Ichthyophthirius multifiliis* (Ciliophora) invasion of gill epithelium. *Journal of Protozoology* 32(2), 305–310. https://doi.org/10.1111/j.1550-7408.1985.tb03055.x

Forwood, J.M., Harris, J.O., Landos, M. and Deveney, M.R. (2014) Minimum effective concentrations of formalin and sodium percarbonate on the free-living stages of an Australian isolate of *Ichthyophthirius multifiliis*. *Parasitology Research* 113(9), 3251–3258. https://doi.org/10.1007/s00436-014-3987-5

Garcia, L.D., Becker, A.G., Cunha, M.A., Baldisserotto, B., Copatti, C.E. *et al.* (2011) Effects of water pH and hardness on infection of silver catfish, *Rhamdia quelen*, fingerlings by *Ichthyophthirius multifiliis*. *Journal of the World Aquaculture Society* 42(3), 399–405. https://doi.org/10.1111/j.1749-7345.2011.00479.x

Geisslinger, M. (1987) Observations on the caudal cilium of the tomite of *Ichthyophthirius multifiliis* Fouquet, 1876. *Journal of Protozoology* 34(2), 180–182. https://doi.org/10.1111/j.1550-7408.1987.tb03157.x

Gonzalez, S.F., Buchmann, K. and Nielsen, M.E. (2007a) Complement expression in common carp (*Cyprinus carpio* L.) during infection with *Ichthyophthirius multifiliis*. *Developmental and Comparative Immunology* 31(6), 576–586. https://doi.org/10.1016/j.dci.2006.08.010

Gonzalez, S.F., Buchmann, K. and Nielsen, M.E. (2007b) *Ichthyophthirius multifiliis* infection induces massive up-regulation of serum amyloid A in carp (*Cyprinus carpio*). *Veterinary Immunology and Immunopathology* 115(1–2), 172–178. https://doi.org/10.1016/j.vetimm.2006.09.007

Gonzalez, S.F., Huising, M.O., Stakauskas, R., Forlenza, M., Lidy Verburg-van Kemenade, B.M. *et al.* (2007c) Real-time gene expression analysis in carp (*Cyprinus carpio* L.) skin: inflammatory responses to injury mimicking infection with ectoparasites. *Developmental and Comparative Immunology* 31(3), 244–254. https://doi.org/10.1016/j.dci.2006.06.010

Gonzalez-Pena, D., Gao, G., Baranski, M., Moen, T., Cleveland, B.M. *et al.* (2016) Genome-wide association study for identifying loci that affect fillet yield, carcass, and body weight traits in rainbow trout (*Oncorhynchus mykiss*). *Frontiers in Genetics* 7, 203. https://doi.org/10.3389/fgene.2016.00203

Grignard, J.C., Melard, C. and Kestemont, P. (1996) A preliminary study of parasites and diseases of perch in an intensive culture system. *Journal of Applied Ichthyology* 12(3–4), 195–199. https://doi.org/10.1111/j.1439-0426.1996.tb00089.x

Harrod, C., Ramírez, A., Valbo-Jørgensen, J. and Funge-Smith, S. (2018) How climate change impacts inland fisheries. In: Barange, M., Bahri, T., Beveridge, M.C.M., Cochrane, K.L., Funge-Smith, S. and Poulain, F. (eds) Impacts of climate change on fisheries and aquaculture. Synthesis of current knowledge, adaptation and mitigation options. *FAO Fisheries and Aquaculture Technical Paper No. 627*. Food and Agriculture Organization of the United Nations, Rome, pp. 375–392.

Heinecke, R. and Buchmann, K. (2009) Control of *Ichthyophthirius multifiliis* using a combination of water filtration and sodium percarbonate: dose–response studies. *Aquaculture* 288(1–2), 32–35. https://doi.org/10.1016/j.aquaculture.2008.11.017

Hines, R.S. and Spira, D. (1973) Ichthyophthiriasis in mirror carp. II. Leucocyte response. *Journal of Fish Biology* 5(4), 527–534. https://doi.org/10.1111/j.1095-8649.1973.tb04484.x

Hines, R.S. and Spira, D. (1974a) Ichthyophthiriasis in mirror carp *Cyprinus carpio* (L). V. Acquired immunity. *Journal of Fish Biology* 6(4), 373–378. https://doi.org/10.1111/j.1095-8649.1974.tb04554.x

Hines, R.S. and Spira, D.T. (1974b) Ichthyophthiriasis in the mirror carp *Cyprinus carpio* (L.) IV. Physiological dysfunction. *Journal of Fish Biology* 6, 365–371. https://doi.org/10.1111/j.1095-8649.1974.tb04553.x

Hines, R.S. and Spira, D. (1974c) Ichthyophthiriasis in mirror carp *Cyprinus carpio* (L). III. Pathology. *Journal of Fish Biology* 6(2), 189–196. https://doi.org/10.1111/j.1095-8649.1974.tb04536.x

Jørgensen, L.v.G. (2016a) Infection and immunity against *Ichthyophthirius multifiliis* in zebrafish (*Danio rerio*). *Fish and Shellfish Immunology* 57, 335–339. https://doi.org/10.1016/j.fsi.2016.08.042

Jørgensen, L.v.G. (2016b) The dynamics of neutrophils in zebrafish (*Danio rerio*) during infection with the parasite *Ichthyophthirius multifiliis*. *Fish and Shellfish Immunology* 55, 159–164. https://doi.org/10.1016/j.fsi.2016.05.026

Jørgensen, L.v.G. (2017) The fish parasite *Ichthyophthirius multifiliis* – host immunology, vaccines and novel treatments. *Fish and Shellfish Immunology* 67, 586–595. https://doi.org/10.1016/j.fsi.2017.06.044

Jørgensen, L.v.G., Nemli, E., Heinecke, R.D., Raida, M.K. and Buchmann, K. (2008) Immune-relevant genes expressed in rainbow trout following immunisation with a live vaccine against *Ichthyophthirius multifiliis*. *Diseases of Aquatic Organisms* 80(3), 189–197. https://doi.org/10.3354/dao01935

Jørgensen, L.v.G., Heinecke, R., Skjodt, K., Rasmussen, K. and Buchmann, K. (2011) Experimental evidence for direct *in situ* binding of IgM and IgT to early trophonts of *Ichthyophthirius multifiliis* (Fouquet) in the gills of rainbow trout, *Oncorhynchus mykiss* (Walbaum). *Journal of Fish Diseases* 34(10), 749–755. https://doi.org/10.1111/j.1365-2761.2011.01291.x

Jørgensen, L.v.G., Sigh, J., Kania, P., Holten-Andersen, L., Buchmann, K. *et al.* (2012) Approaches towards DNA vaccination against a skin ciliate parasite in fish. *PLoS ONE* 7(11), e48129. https://doi.org/10.1371/journal.pone.0048129

Jørgensen, L.v.G., Kania, P.W., Rasmussen, K.J., Mattsson, A.H., Schmidt, J.G. *et al.* (2017) Rainbow trout (*Oncorhynchus mykiss*) immune response towards a recombinant vaccine targeting the parasitic ciliate *Ichthyophthirius multifiliis*. *Journal of Fish Diseases* 40(12), 1815–1821. https://doi.org/10.1111/jfd.12653

Jørgensen, L.v.G., Korbut, R., Jeberg, S., Kania, P.W. and Buchmann, K. (2018) Association between adaptive immunity and neutrophil dynamics in zebrafish (*Danio rerio*) infected by a parasitic ciliate. *PLoS ONE* 13(9), e0203297. https://doi.org/10.1371/journal.pone.0203297

King, A. and Karoly, D. (2017) Climate extremes in Europe at 1.5 and 2 degrees of global warming. *Environmental Research Letters* 12(11), 114031. https://doi.org/10.1088/1748-9326/aa8e2c

Kundzewicz, Z.W., Mata, L.J., Arnell, N.W., Döll, P., Kabat, P. *et al.* (2007) Freshwater resources and their management. In: Becker, A. and Bruce, J. (eds) *Climate Change 2007: Impacts, Adaptation and Vulnerability. Contribution of Working Group II to the Fourth Assessment Report of the Intergovernmental Panel on Climate Change*. Cambridge University Press, Cambridge, UK, pp. 173–210.

LaFramboise, T. (2009) Single nucleotide polymorphism arrays: a decade of biological, computational and technological advances. *Nucleic Acids Research* 37(13), 4181–4193. https://doi.org/10.1093/nar/gkp552

Laidre, K., Stirling, I., Lowry, L., Wiig, O., Heide-Jorgensen, M. *et al.* (2008) Quantifying the sensitivity of arctic marine mammals to climate-induced habitat change. *Ecological Applications* 18(sp2), S97–S125. https://doi.org/10.1890/06-0546.1

Lin, D., Hua, Y., Zhang, Q., Xu, D., Fu, Y. *et al.* (2016) Evaluation of medicated feeds with antiparasitical and immune-enhanced Chinese herbal medicines against *Ichthyophthirius multifiliis* in grass carp (*Ctenopharyngodon idellus*). *Parasitology Research* 115(6), 2473–2483. https://doi.org/10.1007/s00436-016-5000-y

Lynn, D., Frombach, S., Ewing, M. and Kocan, K. (1991) The organelle of Lieberkühn as a synapomorphy for the Ophryoglenina (Ciliophora, Hymenostomatida). *Transactions of the American Microscopical Society* 110(1), 1–11. https://doi.org/10.2307/3226734

MacLennon, R.F. (1935) Observations on the life cycle of *Ichthyophthirius*, a ciliate parasitic on fish. *North-Western Scientist* 9, 12–14.

MacLennon, R.F. (1942) Growth in the ciliate *Ichthyophthirius*. II. Volume. *Journal of Experimental Zoology* 91(1), 1–13.

Martins, M., Xu, D., Shoemaker, C. and Klesius, P. (2011) Temperature effects on immune response and hematological parameters of channel catfish *Ictalurus punctatus* vaccinated with live theronts of *Ichthyophthirius multifiliis*. *Fish and Shellfish Immunology* 31(6), 774–780. https://doi.org/10.1016/j.fsi.2011.07.015

Matthews, R.A. (1994) *Ichthyophthirius multifiliis* Fouquet, 1876: infection and proactive response within the fish host. In: Pike, A.W. and Lewis, J.W. (eds) *Parasitic Diseases of Fish*. Samara Publishing, Tresaith, UK, pp. 17–42.

Matthews, R.A. (2005) *Ichthyophthirius multifiliis* Fouquet and ichthyophthiriosis in freshwater teleosts. *Advances in Parasitology* 59, 159–241. https://doi.org/10.1016/S0065-308X(05)59003-1

Moller, A., Rubolini, D. and Lehikoinen, E. (2008) Populations of migratory bird species that did not show a phenological response to climate change are declining. *Proceedings of the National Academy of Sciences USA* 105(42), 16195–16200. https://doi.org/10.1073/pnas.0803825105

Németh, S., Horváth, Z., Felföldi, Z., Beliczky, G. and Demeter, K. (2013) The use of permitted ectoparasite disinfection methods on young pike-perch (*Sander lucioperca*) after transition from over-wintering lake to RAS. *Aquaculture, Aquarium, Conservation and Legislation* 6(1), 1.

Olsen, M.M., Kania, P.W., Heinecke, R.D., Skjoedt, K., Rasmussen, K.J. *et al.* (2011) Cellular and humoral factors involved in the response of rainbow trout gills to *Ichthyophthirius multifiliis* infections: molecular and immunohistochemical studies. *Fish and Shellfish Immunology* 30(3), 859–869. https://doi.org/10.1016/j.fsi.2011.01.010

Ou, M., Hamilton, T.J., Eom, J., Lyall, E.M., Gallup, J. *et al.* (2015) Responses of pink salmon to CO_2-induced aquatic acidification. *Nature Climate Change* 5, 950–955. https://doi.org/10.1038/nclimate2694

Papayannopoulos, V. (2018) Neutrophil extracellular traps in immunity and disease. *Nature Reviews Immunology* 18(2), 134–147. https://doi.org/10.1038/nri.2017.105

Pedersen, L. and Henriksen, N.H. (2017) Semi-continuously addition of peracetic acid to a flow-through fish farm. *Journal of Cleaner Production* 142(Pt4), 2606–2608. https://doi.org/10.1016/j.jclepro.2016.11.015

Rach, J., Gaikowski, M. and Ramsay, R. (2000) Efficacy of hydrogen peroxide to control parasitic infestations on hatchery-reared fish. *Journal of Aquatic Animal Health* 12(4), 267–273. https://doi.org/10.1577/1548-8667(2000)012<0267:EOHPTC>2.0.CO;2

Raida, M. and Buchmann, K. (2007) Temperature-dependent expression of immune-relevant genes in rainbow trout following *Yersinia ruckeri* vaccination. *Diseases of Aquatic Organisms* 77(1), 41–52. https://doi.org/10.3354/dao01808

Roohi, J.D., Sattari, M., Asgharnia, M. and Rufchaei, R. (2014) Occurrence and intensity of parasites in European catfish, *Silurus glanis* L., 1758 from the Anzali wetland, southwest of the Caspian Sea, Iran. *Croatian Journal of Fisheries* 72(1), 25–31. https://doi.org/10.14798/72.1.710

Roque, M., De Puytorac, P. and Lom, J. (1967) L'architecture buccale et la stomatogenese d'*Ichthyophthirius multifiliis* Fouquet, *1876. Protistologica* 3, 79–89.

Schlenk, D., Gollon, J. and Griffin, B. (1998) Efficacy of copper sulfate for the treatment of ichthyophthiriasis in channel catfish. *Journal of Aquatic Animal Health* 10(4), 390–396. https://doi.org/10.1577/1548-8667(1998)010<0390:EOCSFT>2.0.CO;2

Sigh, J., Lindenstrøm, T. and Buchmann, K. (2004) The parasitic ciliate *Ichthyophthirius multifiliis* induces expression of immune relevant genes in rainbow trout, *Oncorhynchus mykiss* (Walbaum). *Journal of Fish Diseases* 27(7), 409–417. https://doi.org/10.1111/j.1365-2761.2004.00558.x

Steeby, J., Hargreaves, J., Tucker, C. and Cathcart, T. (2004) Modeling industry-wide sediment oxygen demand and estimation of the contribution of sediment to total respiration in commercial channel catfish ponds. *Aquacultural Engineering* 31(3–4), 247–262. https://doi.org/10.1016/j.aquaeng.2004.05.006

Tieman, D. and Goodwin, A. (2001) Treatments for ich infestations in channel catfish evaluated under static and flow-through water conditions. *North American Journal of Aquaculture* 63(4), 293–299. https://doi.org/10.1577/1548-8454(2001)063<0293:TFIIC>2.0.CO;2

Valtonen, E.T. and Keranen, A.L. (1981) Ichthyophthiriasis of Atlantic salmon, *Salmo salar* L, at the Montta Hatchery in northern Finland in 1978–1979. *Journal of Fish Diseases* 4(5), 405–411. https://doi.org/10.1111/j.1365-2761.1981.tb01150.x

Ventura, M. and Paperna, I. (1985) Histopathology of *Ichthyophthirius multifiliis* infections in fishes. *Journal of Fish Biology* 27(2), 185–203. https://doi.org/10.1111/j.1095-8649.1985.tb04020.x

Wagner, G. (1960) Der Entwicklungszyklus von *Ichthyophthirius multifiliis* Bouquet und der Einfluss physikalischer und chemischer Aussenfaktoren. *Zeitschrift für Fischerei und deren Hilfswissenschaften* 9, 425–443.

Wang, X., Clark, T., Noe, J. and Dickerson, H. (2002) Immunisation of channel catfish, *Ictalurus punctatus*, with *Ichthyophthirius multifiliis* immobilisation antigens elicits serotype-specific protection. *Fish and Shellfish Immunology* 13(5), 337–350. https://doi.org/10.1006/fsim.2001.0410

Weiss, L., Potter, L., Steiger, A., Kruppert, S., Frost, U. *et al.* (2018) Rising pCO_2 in freshwater ecosystems has the potential to negatively affect predator-induced defenses in *Daphnia*. *Current Biology* 28(2), 327–332. https://doi.org/10.1016/j.cub.2017.12.022

Woodward, G., Perkins, D. and Brown, L. (2010) Climate change and freshwater ecosystems: impacts across multiple levels of organization. *Philosophical Transactions of the Royal Society B: Biological Sciences* 365(1549), 2093–2106. https://doi.org/10.1098/rstb.2010.0055

Wright, A. and Colorni, A. (2002) Taxonomic re-assignment of *Cryptocaryon irritans*, a marine fish parasite. *European Journal of Protistology* 37(4), 375–378. https://doi.org/10.1078/0932-4739-00858

Xu, D., Klesius, P. and Shoemaker, C. (2008) Protective immunity of Nile tilapia against *Ichthyophthirius multifiliis* post-immunization with live theronts and sonicated trophonts. *Fish and Shellfish Immunology* 25(1–2), 124–127. https://doi.org/10.1016/j.fsi.2008.03.012

Xu, D., Moreira, G., Shoemaker, C., Zhang, D. and Beck, B. (2017) Expression of immune genes in systemic and mucosal immune tissues of channel catfish vaccinated with live theronts of *Ichthyophthirius multifiliis*. *Fish and Shellfish Immunology* 66, 540–547. https://doi.org/10.1016/j.fsi.2017.05.051

Xu, Z., Parra, D., Gomez, D., Salinas, I., Zhang, Y.A. *et al.* (2013) Teleost skin, an ancient mucosal surface that elicits gut-like immune responses. *Proceedings of the National Academy of Sciences USA* 110(32), 13097–13102. https://doi.org/10.1073/pnas.1304319110

Xu, Z., Takizawa, F., Parra, D., Gómez, D., Jørgensen, L.v.G. *et al.* (2016) Mucosal immunoglobulins at respiratory surfaces mark an ancient association that predates the emergence of tetrapods. *Nature Communications* 7, 10728. https://doi.org/10.1038/ncomms10728

19 Microsporidiosis (*Loma salmonae*)

DAVID J. SPEARE*

University of Prince Edward Island, Charlottetown, Prince Edward Island, Canada

19.1 Brief Introduction

Loma salmonae, a microsporidian, characteristically completes its life cycle through the formation of a host–parasite structure referred to as a xenoma. It is the causative agent of microsporidial gill disease of salmon (MGDS) in Pacific salmon on the west coast of Canada. MGDS is a complicating factor for the farming of Pacific salmon; *L. salmonae* is also reported in rainbow trout *Oncorhynchus mykiss* (farmed and wild) and wild stocks of most species of Pacific salmon, but in these latter instances the degree of infection is often mild and the pathophysiological significance is unknown. Found within marine and freshwater environments, this disease agent is capable of transmitting within both and is thus somewhat distinguished from other key pathogens affecting teleost aquaculture.

The first description of *L. salmonae* in farmed salmon involved juvenile coho salmon (*Oncorhynchus kisutch*) recently stocked in salt water. Although high levels of *L. salmonae* presented in the gills of these fish, the salmon also suffered from severe gill disease arising from impaction of silica-rich setae-bearing frustules of a non-colonial diatom (Speare *et al.*, 1989) between gill lamellae. The overall effect on the gill proved highly lethal, and recovery of the remaining stock was delayed due to the long persistence of microsporidian spores and embedded diatom frustules within gill tissue. A full examination of fish, for all pathogens, and an awareness of limiting environmental conditions are therefore always recommended. The detection of a small number of xenomas within gill tissue is unlikely to have an effect on fish, although the finding remains noteworthy as a potential predictor of future problems.

Presently, although *L. salmonae* has been reported from several parts of the world, it is considered a significant disease agent only in British Columbia (Canada) within Pacific salmon. *L. salmonae* also infects rainbow trout but rarely causes disease (e.g. see Gandhi *et al.*, 1995). MGDS has not yet been reported from other parts of the world where Pacific salmon are cultivated commercially, although it should remain a concern wherever chinook or coho salmon are farmed, particularly if distribution patterns of wild salmon (reservoirs) become influenced by climate change.

Microsporidians are highly vexing disease-causing agents in fish and insects and in recent years have become newsworthy as opportunistic pathogens in humans. However, research into the respective and linked features of the host–pathogen–environment triad is largely unknown. Pebrine disease of the silkworm (*Bombyx mori* L.), caused by the microsporidian *Nosema bombycis* is a well-recognized current and historical problem of the silk industry. Despite its devastating role in sericulture of silkworm in France and Italy, environmental conditions promoting or limiting this disease have only recently been the subject of evidence-based translational study (Rahmathulla *et al.*, 2012). The question of where microsporidians fit in the tree of life is also not without controversy. Recent studies indicate their close connections to fungi; despite this, most efforts towards treatment and control doggedly are borrowed from methods used for protozoan parasites. In general, little substantive progress has been made (Bacchi *et al.*, 2002; Keeling and Fast, 2002). Aspects of their life cycle, bioenergetics, transmission and persistence within different environments remain to be studied. Their relationship to host metabolism is particularly interesting, as it is for other obligate intracellular pathogens, and this may ultimately hold the key concerning discoveries of effective drugs (Lovy *et al.*, 2006).

Applied progress on the treatment and control of microsporidian-induced diseases of animals and

*speare@upei.ca

man is hampered by difficulties establishing these agents in cell lines (Williams, 2009); inconsistent *in vivo* models also limit the value of making comparisons between trials. Focused work on the host–pathogen–environment interactions associated with MGDS developed (Shaw *et al.*, 1998; Speare *et al.*, 1998a) due to persistent economic effects of MGDS within aquaculture-reared chinook salmon (*Oncorhynchus tshawytscha*) along coastal British Columbia, Canada. An important result of these efforts – the establishment of reproducible *in vivo* infection models – provides data through which we can consider various ways that climate change may alter the occurrence and effects of *L. salmonae* on stocks of wild and farmed salmon.

L. salmonae has been described from many regions of the world in both fresh- and salt-water environments (see Becker and Speare, 2007). Transmission is direct from fish to fish and occurs under a variety of environmental conditions (Shaw *et al.*, 1998; Ramsay *et al.*, 2002). The ease of direct experimental transmission contrasts sharply with other microsporidial diseases of fish (Fujiyama *et al.*, 2002). The rainbow trout is the best-studied host model for MGDS, and a collection of studies indicate that naïve cohorts become infected by cohabitants only when spores are released into the water column from spore-filled xenomas. Naïve fish ingest these spores and thereby become infected (Becker *et al.*, 2005b). Within a heavily stocked aquaculture setting, this is assumed to be the method by which amplification of infection arises although the initial source of the spores is rarely known. A reasonable hypothesis centres on the likelihood that wild salmon, in proximity to aquaculture pens, are the initial source of infection. *L. salmonae* is frequently detected within wild Pacific salmon (Kent *et al.*, 1998), both from surveys of salmon while at sea and also in salmon sampled during pre-spawning migrations in rivers. Since it is in most species of Pacific salmon it is reasonable to consider these species as reservoirs of the parasite; generally few xenomas are in wild salmon when at sea, but xenomas are at much higher levels such that the parasite contributes to pre-spawning mortalities as salmon return to spawning beds. It remains to be determined whether wild stocks of Pacific salmon initially encounter *L. salmonae* during early parts of their life cycle in fresh water, or within the marine environment, or upon subsequent return to fresh water. Does the microsporidian remain as an undetectable latent infection for a protracted period only to emerge when returning salmon undergo migration-related stress, or salmon pick up a heavy load of spores as they return to fresh water? Learning more about the distribution and persistence of *L. salmonae* within natural habitats will help us to predict whether Pacific salmon migration patterns and behaviour, as affected by climate change events, will lead to an enhanced role of *L. salmonae* as a disease issue within wild stocks of fish. Based on the general principle that changes in climate will drive changes in the distribution, seasonality and severity of infectious diseases (Hughes, 2000), specific studies on the pathogen development rate and microevolution (de la Rocque *et al.*, 2008) should become a priority. The effects of temperature on the life cycle of *L. salmonae* have been partially studied, and these results provide insights into how climate change might enhance the role of this pathogen in farmed and wild stocks of salmon. However, caution is needed to avoid arriving at over-simplistic conclusions brought about by considering only one environmental variable (Rogers and Randolph, 2006; Rohr *et al.*, 2011).

19.2 Diagnosis of the Pathogen and Disease

Within farmed chinook salmon populations, *L. salmonae* infections cause severe disease referred to MGDS, characterized by very high mortality rates. Although the descriptive epidemiology of MGDS is poorly characterized, a commonly reported disease history includes clinical signs arising concurrently with rising water temperatures and often starting within July and August of a salmon population's second summer within sea pens (Becker and Speare, 2007). Morbidity and mortality persist until late autumn. This timing enhances the economic significance of MGDS because fish are nearing market weight and at a time when cost inputs are maximal. Mortalities of up to 80% of stock within an affected sea pen have been documented (Constantine, 1999). Salmon within net pens are often affected simultaneously by several other diseases. Therefore, it is sometimes challenging to relate morbidity and mortality to a single agent or to determine the extent to which a particular agent is additive to the presenting conditions (Speare *et al.*, 1989).

MGDS can be readily diagnosed in farmed fish through the combination of clinical signs (diseased fish are near the surface and edge of net pens with laboured and rapid opercular movements) and full examination of gill and other organ tissues. *L. salmonae* forms characteristic xenomas within the gill tissue that are just visible to the naked eye, appearing as a miliary distribution of white dots the size of grains of salt against the backdrop of red gill tissue (Fig. 19.1). As such, it is easily misdiagnosed, at the level of gross observation, as *Ichthyophthirius multifiliis*. In specimens that undergo autolytic changes, xenomas will not be visible with the naked eye since the background gill tissue will have turned pale and the lack of contrast between the pale xenomas and the pale background obscures the finding. An unstained whole-mount of gill tissue, when examined with light microscopy, is sufficient to help detect xenomas. They appear as dark granular masses both within lamellae and filaments; although the xenomas are usually single, they can be found in clusters. Slight pressure on the coverslip is sometimes useful to rupture xenomas, which permits observations of released spores. Histopathology has proven valuable in detecting xenomas and interpreting the degree of gill damage related to infection. An unusual, yet characteristic, finding is that there is often limited host reaction surrounding intact xenomas, whereas there can be dramatic inflammatory responses following the rupture of xenomas and release of spores (Kent *et al.*, 1995). Whereas xenomas typically are restricted to the gills of rainbow trout, xenomas are also found in internal organs (e.g. in the heart in chinook salmon) (Kent *et al.*, 1995, 1998; Ramsay *et al.*, 2002). Therefore, it is critical to collect samples from several organ sites to make a complete assessment of the pathology.

In contrast to the relative ease of detecting *L. salmonae* and diagnosing MGDS, many other microsporidians and the diseases they cause present challenges for diagnosticians particularly in those instances where, rather than spores forming within readily detectable xenomas, small numbers of spores develop within individual host cells. Hence, in human medicine, they went virtually unnoticed until very recently. Detection of the characteristically small (2–4 μm) bean-shaped spores can be enhanced during screening procedures using histochemical stains or with polarized light (Tiner, 1988). Transmission electron microscopy is a valued approach to gain further details of meront development and transition into spore-forming structures (Fig. 19.1). Spores provide excellent differentiating information such as overall dimension and characteristics of the

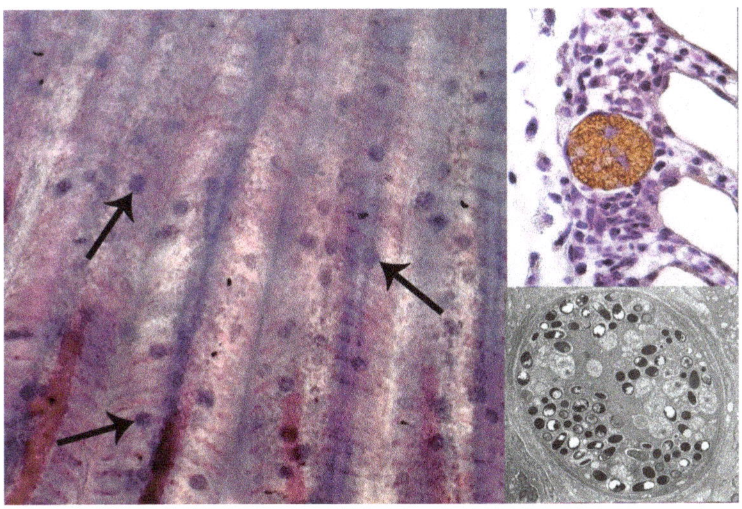

Fig. 19.1. Arrows point to circular xenomas randomly distributed along the length of the gill filament; these are easily detected within an unstained squash preparation of a dissected gill arch. The intensity of infection can be quantified by counting the number of xenomas per gill. Upper right image shows the use of immunolabelling of *Loma* spores within a single xenoma within a histological 6 μm section of gill. Lower right image of a single xenoma, within a host pillar cell, demonstrates the advantages of using transmission electron microscopy (TEM) for assessing the size and shape of mature spores and features of pre-spore development. (TEM image courtesy of Dr Jan Lovy, New Jersey Division of Fish and Wildlife.)

polar filaments. Recently developed immunological approaches and molecular techniques can be used as diagnostic screening tools and also as approaches to better understand the genetic relationships among microsporidians (Speare *et al.*, 1998b; Sanchez *et al.*, 1999; Brown *et al.*, 2010).

19.3 *Loma salmonae* Life Cycle and Influence of Water Temperature

Although microsporidians are obligate intracellular organisms with no known ability for motility, a curious question surrounding their biology is the manner through which they move from the gastrointestinal tract (GIT) to their final host cell. Many studies support the theory that microsporidians generally infect their hosts following oral consumption. The environment of the GIT triggers polar tube eversion by the spore; however, spores are also infective when injected into the peritoneum of fish where presumably they are picked up by macrophages and intracellular mechanisms (such as phagolysis) trigger polar tube eversion.

For intestine-inhabiting microsporidians, localization within the GIT would require no unique transport mechanism following initial infection. However, many microsporidians locate in distant parts of the body, with distinct pathognomic cellular and tissue tropism which often aids diagnosis. With *L. salmonae*, the location of xenomas varies somewhat between different salmonid species. In the rainbow trout, there is a high level of cellular and tissue tropism such that xenomas are found almost exclusively within pillar cells of the gill. In other Pacific salmon, xenomas are found in pillar cells in high numbers too, but also within endothelial cells of the afferent and efferent filament vasculature including the central venous sinusoid of the gill (Fig. 19.2), and also (presumably in endothelial cells) within the heart and spleen in large numbers (Ramsay *et al.*, 2002).

Using a combination of molecular techniques, immunostaining and electron microscopy it has been shown that *L. salmonae* initially, after oral ingestion as spores, enters the lamina propria of the gut mucosa (Sanchez *et al.*, 2001d). Days later it is found exclusively in the heart where it presumably completes stages of merogony. One week later it is found exclusively in the gills, where its life cycle completes with the formation of xenomas. Immature

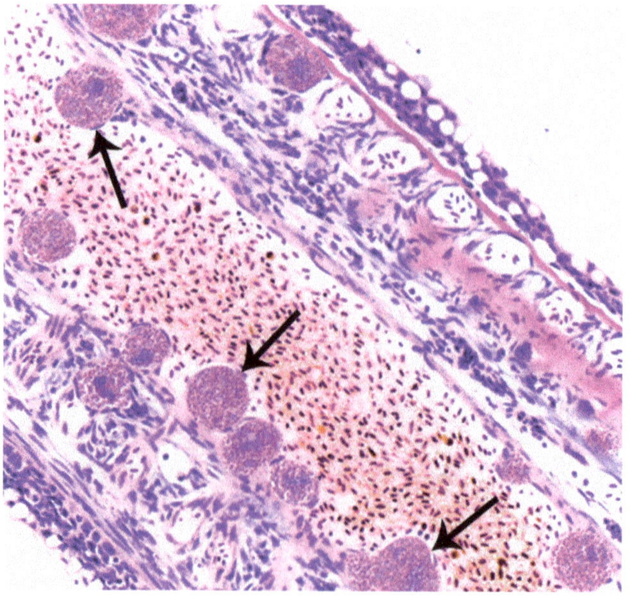

Fig. 19.2. Xenoma development (arrows point to three examples) within endothelial cells of the gill's central venous sinus. Histological section from an infected chinook salmon. (Image courtesy of Dr Jan Lovy, New Jersey Division of Fish and Wildlife.)

stages transit from the heart to the gill within macrophages. Moreover, it appears that macrophages cooperate with pillar cells (which are also known to be phagocytic) to allow the parasite to transfer from one cell to the next through a process of cellular internalization. An infected macrophage docks alongside, and then is taken up by, a pillar cell (Sanchez *et al.*, 1999, 2000; Rodriguez-Tovar *et al.*, 2002). An unusual finding is the occasional presence of developing xenomas, or fully formed spores, within gill epithelial cells or gill mucous cells (Fig. 19.3). Although this may simply represent a happenstance or accidental association with a cell type usually not permissive to final stages of *L. salmonae* development, an alternative possibility is that this unexpected cellular tropism reflects different strains or subpopulations of *L. salmonae* capable of engineering unique mechanisms defining cellular tropism. This finding raises the possibility of other unexpected future findings such as a broadening of the host range for *L. salmonae*, as well as differences in xenoma development rates. The former may arise from a drift in host cell tropism while the latter may reflect differences in the metabolic potential, or normal lifespan, of the various host cells supporting xenoma development.

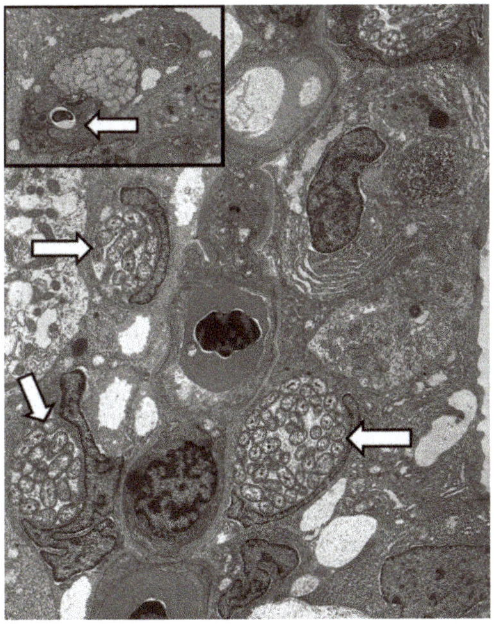

Fig. 19.3. Aberrant host cell location selection by *Loma salmonae*. Arrows (main image) show stages of *L. salmonae* developing within lamellar epithelial cells. Inset image shows several fully formed *Loma* spores within the cytoplasm of a gill mucous cell. (Images courtesy of Dr Jan Lovy, New Jersey Division of Fish and Wildlife.)

19.3.1 Influence of water temperature on the life-cycle kinetics of *Loma salmonae*

Life-cycle kinetics of *L. salmonae* depend greatly on water temperature (Beaman and Speare, 1999; Speare *et al.*, 1999b) and the overall picture emerging from laboratory studies provides a reasonable explanation for the seasonal incidence of MGDS (Fig. 19.4). Water temperature is generally considered to be the key driver of many infectious diseases of colder-water fish species (Karvonen *et al.*, 2010). As shown for the incidence of the microsporidian *Glugea stephani* within winter flounder (*Pseudopleuronectes americanus*), water temperature was the only environmental factor statistically correlated with *G. stephani* infections in the winter flounder population of the New York–New Jersey lower bay complex (Cali and Takvorian, 1991). In contrast to Cali and Takvorian's study (1991) in which higher water temperatures favoured the incidence of *G. stephani*, an opposite finding has been described by Takahashi and Ogawa (1997) when examining experimental infections of ayu (*Plecoglossus altivelis*) with *Glugea plecoglossi*. In the latter study, elevating the water temperature at specific periods post-exposure was highly protective to ayu, such that it could be considered a treatment-prevention approach. This marked difference between the two studies may serve to highlight the challenge in drawing comparative conclusions, when colder-water host–parasite interactions are compared with those typically unfolding at higher water temperatures. As such, it appears entirely possible that water temperature increment due to climate change could effect opposite results even for what might appear to be relatively similar host–pathogen interactions. An earlier study (Olson, 1981) examining the restrictive effect of colder water temperatures on the development of *G. stephani* in the English sole (*Paraphrys vetulus*) concluded that fish infected within warmer estuaries become protected from disease development when they migrate to the ocean and colder water temperatures, although the microsporidian likely merely remains dormant. This host–pathogen balance could be

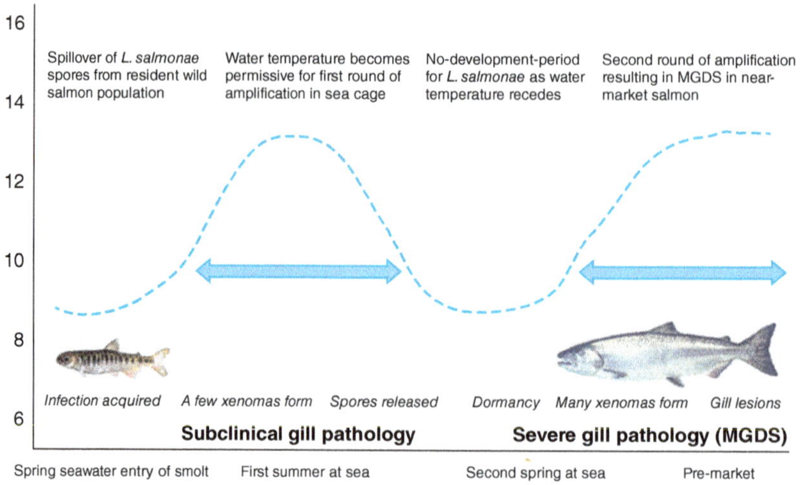

Fig. 19.4. Parasite–temperature interaction leading to microsporidial gill disease of salmon (MGDS) in year 2 of the chinook salmon production cycle. The *y*-axis denotes the daily mean seawater temperature at farm site; the *x*-axis follows the salmon production cycle from seawater entry through to 5 kg market-sized fish. Hatchery-derived salmon, free of *Loma salmonae*, enter net pens as naïve smolt and acquire initial low-level *L. salmonae* from nearby infected wild salmon as a 'spillover' event. Within-cage amplification of *L. salmonae* unfolds during the first summer in net pens as water temperatures move into the permissive range for *L. salmonae* xenoma development (double-sided arrow indicates the threshold water temperature of 11°C). Receding water temperature interrupts within-cage amplification until the second summer season as the salmon enter a pre-market size range.

sensitive to climate change events, altering fish residence time in different environments.

The preferred temperature range for *L. salmonae* lies between 13 and 17°C (Beaman and Speare, 1999; Becker *et al.*, 2003; Becker and Speare, 2004b). Within this temperature range, the maximum number of xenomas form relative to the exposure dose of spores. A reduction of temperature by 2°C (to 11°C), or an increase of temperature by 2°C (to 19°C), results in a sharp reduction in the number of xenomas forming and the rate of development becomes unpredictable. An additional temperature change of as little as 1°C at either end of the range (i.e. at 10°C or below, or at 20°C or above) has a dramatic effect (Beaman and Speare, 1999) such that xenomas do not form. Instead, parasite development is arrested within cardiac subendothelial macrophages and does not proceed to the gill (Sanchez *et al.*, 2000). However, if the water temperature is brought back within the permissive range, further development is reinitiated and delayed xenoma formation is noted (Speare *et al.*, 1999b). The role of water temperature and water temperature changes has also been examined for the interaction of another microsporidian, *Enterocytozoon salmonis*, with the chinook salmon, and results are quite similar to those noted for *L. salmonae* in rainbow trout. The highest post-exposure mortality rates were in fish kept at water temperatures between 15 and 18°C, whereas lower temperatures delayed mortality rates and reduced them overall. Also, moving fish from non-permissive colder water temperatures to higher water temperatures led to fish mortality rates similar to those when held continuously at the higher water temperatures (Antonio and Hedrick, 1995). Fujiyama *et al.* (2002) described a linkage between an unusual increase in water temperature and massive infections with *Kabatana takedai* affecting masu salmon (*Oncorhynchus masou*) within a hatchery system, and subsequent work (Zenke *et al.*, 2005) revealed a picture for the host–parasite–temperature interactions mostly similar to that described for *L. salmonae*. Temperature manipulation was suggested as a partially effective means to protect fish from *K. takedai*.

Through the permissive water temperature range of *L. salmonae*, xenoma development rate has a polynomial relationship to temperature which theoretically could be used to predict when MGDS might first arise on a farm. However, the practicality of this is somewhat limited as it requires a consistent water temperature (Beaman and Speare, 1999). A more practical model is based on degree-days or thermal units, similar to what is used to predict hatching of insects, corrected with a 'no development temperature' of 7°C; development of xenomas can be predicted based on the accumulated thermal units and treatment strategies initiated at key control points (Speare *et al.*, 1999a,b). Relatedly, Rahmathulla *et al.* (2012) have proposed using temperature and humidity data to develop forecast models to support 'red alert' notices for the onset of Pebrine disease for silkworm producers.

Spores represent a life cycle stage that is inherently stable under diverse environmental conditions. Early work by Shaw *et al.* (2000a) demonstrated that *L. salmonae* spores retained viability for up to 100 days when kept at 4°C. Recent work by McConnachie *et al.* (2013) further demonstrated that *L. salmonae* spores, taken up by bivalves in the vicinity of a fish farm, could retain infectivity (and remain in mussels) for several days. Spore viability remained high when expelled from mussels within pseudofaeces. Within fish tissues, spores can also remain stable and infective for an extended period (Ramsay *et al.*, 2002, 2003). Few other studies have examined the duration of viability of microsporidian spores at different temperatures although this information is obviously core to our understanding of how microsporidian spores will persist in the environment and how this may become altered through climate change.

Rohr *et al.* (2011) describe an unusual gap in our theoretical approach to understanding the effect climate change has on host–parasite interactions. The work highlights the need to better understand the metabolic theory of ecology and the role it may have when looking at interactions between organisms that differ markedly in size. Furthermore, this may have particular value for poikilothermic animals and their parasites (Brown, 2004; Paull and Johnson, 2011). Parasite adaption to temperature change and increased diurnal fluctuation may be more rapid than in their hosts, thus providing them with advantages in the host–parasite interaction. Parasite generation time may be reduced. These issues should be considered when species expansion of *L. salmonae* is of concern, particularly for hosts where *L. salmonae* is shown to be infective although the life cycle is not completed. An example of this is the Atlantic salmon (*Salmo salar*).

19.4 Host Range and Current Geographic Range, and Anticipated Effects of Temperature Change

19.4.1 Experimentally documented host range and transmission studies

Several studies have documented the host range for *L. salmonae* since the initial findings of Putz, 1965, in freshwater hatchery-raised rainbow trout. Morrison and Sprague (1983) differentiated developmental characteristics and spore morphology of *L. salmonae* obtained from rainbow trout reared in a freshwater hatchery from California, from *Loma fontinalis* arising in freshwater-reared brook trout, and further separated both of these from *Loma morhua* and *Loma branchialis* – the latter being marine gill pathogens of cod and haddock. As detailed by Shaw *et al.* (2000a,b,c), the host range for *L. salmonae* is not limited to rainbow trout, and also involves chinook salmon, coho salmon, sockeye and kokanee salmon, with additional reports of its presence in brown trout and brook trout. To define this more specifically, Shaw *et al.* (2000a,b,c) detail experimental exposure studies showing that a range of non-salmonid species could not be infected. In summary, all six species of Pacific salmon in the eastern Pacific are susceptible to *L. salmonae*. Furthermore, susceptibility was not limited to the genus *Oncorhynchus*. Shaw *et al.* (2000a,b,c) review and report on the presence and effects of *L. salmonae* on Pacific salmon returning to spawning sites and also in smolts migrating out to sea. From this a theory emerges that wild Pacific salmon initially become infected shortly after hatch from spores released from breeding salmon; subclinical infections are perhaps maintained through their life in seawater. The infection reactivates to cause clinical disease during return migration as fish undergo sexual maturation and encounter warmer waters during upstream migration.

The possibility that Pacific salmon may carry two variants of *L. salmonae* is based on studies by Sanchez *et al.* (2001c). When gill material from salmon farmed in British Columbia was given to rainbow trout and brook trout, a high level of infection developed in rainbow trout and a very low level of infection with low incidence developed in brook trout. However, when experimentally infected brook trout were used as sources of xenoma-derived spores, the infectious material was unable to infect rainbow trout, but, in contrast, induced high levels of infection in brook trout and low levels of infection in chinook salmon. From this, it was concluded that farmed salmon carried a dominant strain typically infective to salmonids in the genus *Oncorhynchus* and a subdominant strain highly infectious to salmonids in the genus *Salvelinus* and only minimally infective to *Oncorhynchus*. Although subsequent work by Brown *et al.* (2010) raised the possibility that the isolated SV strain may have been endemic within the experimental brook trout used, this suggestion was not borne out by subsequent thorough assessment of fish stocks and control fish. The discovery of the SV strain highlights the issue that further strain variation is likely to develop. Through this, a broader host range may be documented in the future. This possibility is further supported by observations that non-permissive hosts such as the Atlantic salmon do become 'infected' after ingestion of live spores, but the parasite development stalls as it encounters a merogony phase within the heart (Sanchez *et al.*, 2001a). To consider this host range a bit further, particularly as it relates to Atlantic salmon, it has only been in the last couple of decades that Atlantic salmon have been raised in large numbers within the Pacific Ocean. Now, they are the critical aquaculture species in the Pacific North-West, and the Patagonian. During biotic expansion phases of parasites, hosts come into contact with parasites that they have not previously contacted, and this sets the stage for host switching happening rapidly without evolutionary pressures (Brooks and Hoberg, 2007). It is curious that *L. salmonae*, within Atlantic salmon, can complete portions of its life cycle, reaching the heart, but not proceeding to the gill. It remains speculative whether this indicates a degree of innate resistance or, instead, a misdirection of the stages of *L. salmonae* which generally would colonize gill pillar cells and endothelial cells. If the latter, one might expect the emergence of a strain capable of completing its life cycle in Atlantic salmon.

Within the range of *L. salmonae*-susceptible species of salmonids, there is a considerable difference in the degree of susceptibility and also the distribution and persistence of the pathogen within the host. Ramsay *et al.* (2002) noted that after a standard challenge with infective spores, Pacific salmon (chinook and coho) developed much more severe gill infections compared with rainbow trout. Although there was some additional difference in the time from infection until xenoma formation, a much more dramatic difference was noted in the range of organs infected and also the persistent infectivity of tissues from these organs. Nevertheless, in Pacific salmon, there was no evidence of autoinfection, and indeed a robust protective response after the infection has repeatedly been demonstrated (Speare *et al.*, 2007). In contrast, *Loma* infections of brook trout can lead to repeat infections (Speare and Daley, 2003) and the process of autoinfection has been clearly demonstrated for cod infected with *L. morhua* (Rodriguez-Tovar *et al.*, 2003a).

Various infection models have been developed to study *L. salmonae*'s interaction with salmonids. *Per os* exposure is a standard approach, and a single exposure to a high spore dose leads to high levels of xenomas on the gill in a shorter period compared with cohabitation infection models (Ramsay *et al.*, 2001). Alongside this, the rate at which xenomas clear from the gills is also more protracted in cohabitation models compared with a single high-dose oral exposure to spores. Transmission from tanks housing *Loma*-infected salmonids to naïve salmonids (Becker and Speare, 2004b) adds a further delay, even compared with cohabitation models, for xenoma development. In total, these studies show the ease by which *L. salmonae* can transmit between cohorts housed within the same tank and also merely through the release of spores within the effluent flow from infected populations of fish. Furthermore, as shown by Becker *et al.* (2005a,b), even a brief period of exposure to infective spores is sufficient for transmission, as demonstrated by infections developing in naïve trout after spending only 1 h with lightly infected cohorts.

19.4.2 Documented geographic range

L. salmonae has frequently been reported from the Pacific coast of North America, particularly within marine and freshwater regions of the Pacific North-West. These regions are considered endemic for the parasite, and this is in line with the geographic range of most of the susceptible species of Pacific salmon. Although *L. salmonae* has not been reported in Pacific salmon from other regions (e.g. Russia, Japan), it may also be found in those areas. So too, the expansion of chinook salmon populations into Patagonia has not yet led to reports of *L. salmonae* in either naturalizing populations or in farmed Pacific salmon in this region. Sporadic reports of *L. salmonae* indicate it is in England (farmed and wild rainbow trout and wild brown trout; Poynton, 1986), Scotland (rainbow trout;

Bruno *et al.*, 1995) and the eastern USA (rainbow trout; Markey *et al.*, 1994). It is likely that *L. salmonae* in wild populations of salmonids is, as yet, under-reported and the geographical range may, therefore, appear oddly restricted. However, it is likely that morbidity-inducing outbreaks of *L. salmonae* infections in captive, and other highly studied, fish populations would be reported in the scientific literature. In the absence of such reporting, *L. salmonae* as a cause of fish mortality appears to be a unique problem within Pacific salmon farming in the Pacific North-West, and of surveillance concern within wild populations of salmon during their spawning runs back into freshwater habitats as it may contribute to our understanding of pre-spawning mortalities.

Although the species range of *L. salmonae* appears governed by innate species factors, rather than environmental factors, situations of increasing water temperature are very likely to influence the geographical range of *L. salmonae*. Changes to salmon migration patterns, and also relocation and geographical expansion of Pacific salmon farming operations based on water temperature, economics and industry expansion, should be considered. With specific reference to *L. salmonae* as a factor in pre-spawning mortality rates of Pacific salmon during their return to fresh water, any increments in water temperature would allow for more rapid development of *L. salmonae* to the xenoma stage and subsequent rupture would initiate gill pathology. The latter may restrict the physiological capacity and therefore interfere with upstream migration.

As noted previously (Section 19.4.1), farmed salmon in British Columbia carry two variants of *L. salmonae*. The type strain is highly infectious to salmon in the genus *Oncorhynchus*, whereas the atypical strain has a much stronger affinity for brook trout and – although as yet untested – may favour other members of the genus *Salvelinus*. Although the temperature preferences of the typical strain have been thoroughly studied, this is not the case for the variant strain. In a parallel situation in which the biotic potentials of the microsporidians *Nosema apis* and *Nosema ceranae* have been examined relative to their effects on the honeybee (*Apis mellifera*), it has been shown that higher temperatures favoured *N. ceranae* over *N. apis*. At higher temperature, *N. ceranae* exhibited augmented exponential growth and therefore would have advantages in transmission. Both of these microsporidians are known pathogens of the honeybee and are causative agents of the disease known as nosemosis. *N. ceranae* appears to be the more pathogenic and may contribute to colony collapse (Paxton *et al.*, 2007). The enhanced adaption of *N. ceranae* compared with *N. apis* allows it to complete its endogenous cycle with higher biotic index and this helps explain its persistence in honeybees throughout the year as compared with *N. apis*, which is seasonal (Martin-Hernandez *et al.*, 2009). With this in mind, further studies of the comparative temperature effect differences between the typical and atypical variants of *L. salmonae* should be conducted to provide insights into how temperature increases (e.g. due to climate change) lead to differential selection on these two variants, and to what extent this may also extend the seasonal pattern of disease.

19.5 Pathobiology of *Loma salmonae* and Anticipated Effects of Water Temperature

19.5.1 Cellular and tissue pathology

L. salmonae shows a degree of diversity during its life cycle and also for its preferred location for xenoma development. The target cell for *L. salmonae* in rainbow trout is the pillar cell (Fig. 19.1). These are unique contractile cells forming vascular channels (pillar channels) within the lamellae (Rodriguez-Tovar *et al.*, 2002). However, in other susceptible salmonids, the target cell, in addition to pillar cells, can be endothelial cells within the afferent or efferent vasculature of the gill filament (Fig. 19.2) and also endothelial cells elsewhere in the body, especially the heart. However, as shown by Rodriguez-Tovar *et al.* (2002), these final destination cells are reached by *L. salmonae* through transit within leucocytes.

Xenomas are unique host–parasite structures that remain poorly understood (Williams, 2009). Development of xenomas takes several weeks before infective spores form, and through some mechanism – perhaps a delay in apoptosis – the lifespan of the infected cell is protracted to allow for parasite development through to mature spores. As proposed by Lovy *et al.* (2006), microsporidia may exploit host cellular physiology during the transition from meronts to sporonts by explicitly exploiting the process of cellular autophagy. Meronts of *L. salmonae* become enclosed by host endoplasmic reticulum (ER) as would occur during autophagy; however, rather than leading to degradation of the parasite, *L. salmonae* instead uses host ER membranes for development of outer parasite membranes and the limiting membrane of the parasitophorous vacuole. In addition to providing structural features for the parasite, this may also provide proteins, perhaps degraded to various degrees, not generally available within the cell's cytoplasm.

As with many other intracellular pathogens, host recognition may be minimal as evidenced by little or no inflammatory response. The xenomas of *L. salmonae* initiate very little inflammatory response through the lengthy development cycle until late in their maturity, when, as noted by Rodriguez-Tovar *et al.* (2003b), inflammatory cells begin to encircle and then invade the xenoma. This event may have a pivotal role in xenoma rupture. As noted by Ramsay *et al.* (2002), there may be species differences in this rate of recognition, in that rainbow trout xenomas begin to degrade by as early as week 4 following spore exposure (although maximally around week 8), whereas this may not occur until week 12 in chinook salmon, even though xenomas form at the same rate in both species. Although a leucocyte reaction is delayed in response to xenomas in all host species, destruction of the xenoma wall follows upon leucocyte invasion of the xenoma. Release of spores into gill tissue provokes an acute neutrophil-rich suppurative response that transitions to a chronic macrophage-rich granulomatous response (Lovy *et al.*, 2007b).

Xenoma localization within pillar cells suggests parasite selection of a host cell providing suitable egress to the water for released spores. However, as noted by Lovy *et al.* (2007b), many of the spores released from xenomas are retained within gill tissue (Fig. 19.5) and spore retention is even more pronounced in chinook salmon where the majority of the xenomas form more deeply within gill interstitium. Retained spores initiate marked inflammatory responses and spores can be found within leucocytes where the spores appear resistant to destruction within the phagosome. Subsequently, blisters form on the gill, and it appears that leucocytes containing infective spores are released into the water.

Although gill damage does not begin until xenomas mature and rupture, gill pathology – once initiated – persists for several weeks following xenoma rupture (Fig. 19.5). Gill lesions mostly centre on gill vasculature, and this can be dramatic in chinook salmon with massive infections. Vasculitis and perivasculitis are accompanied by thrombosis and subsequent neovascularization (Lovy *et al.*, 2007b). Lesions are far less dramatic in rainbow trout. Healing of branchial tissue is remarkable, and in contrast to the manner in which airways of higher animals would heal from similar damage, there is a noticeable lack of fibrosis and a remarkable ability to regenerate destroyed lamellae such that the appearance of the gill can return to a normal state.

19.5.2 Clinical pathology

MGDS is an inflammatory disease of the gill. The leucocrit of affected fish informs us of a progressive

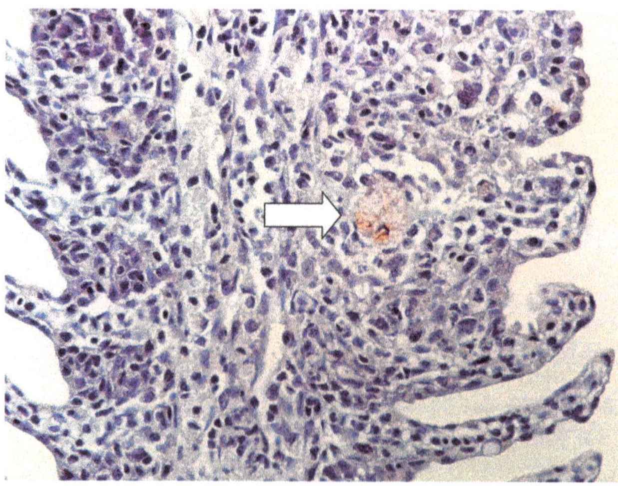

Fig. 19.5. Histological section showing fusion of several adjacent gill lamellae and presence of inflammatory cells surrounding the remnant of a ruptured xenoma (arrow). The use of immunolabelling helps identify the presence of *Loma* spores remaining in gill tissue following xenoma rupture.

decline in leucocyte numbers as infection proceeds. This decline suggests that consumption of circulating inflammatory cells directed towards gill inflammation exceeds upregulation and replacement (Powell *et al.*, 2006). Despite the degree of gill damage, affected fish retain the ability to maintain normal plasma electrolyte levels, but it is unknown to what degree they retain a capacity to respond quickly and efficiently to changes in water chemistry (Powell *et al.*, 2006). Whereas rainbow trout with MGDS increase their routine metabolic rate, the opposite is true for brook trout (Powell *et al.*, 2006). Important studies remain to be completed in this area for chinook salmon and other migratory salmon species as this could have a considerable impact on their ability to survive lengthy migrations and pre-spawning stressors, since it has been shown that maximum post-exercise metabolic rate in rainbow trout with MGDS exceeds that of uninfected controls. In support of this concern, it has been shown that elevated water temperature, a factor which induces more severe host reactions in rainbow trout affected with the pathogen *Ichthyophonus*, caused infected fish to have reduced swimming stamina compared with uninfected controls. The authors point to the implications this could have on spawning migration success of salmonids if they face rising water temperatures (Kocan *et al.*, 2009).

19.5.3 Growth

Although the effects of MGDS on growth indices of chinook salmon in a farmed setting have not been carried out, several insights can be gathered from the rainbow trout model of MGDS (Fig. 19.6). Of interest, MGDS appears to have no effect on growth until the period when xenomas begin to rupture. From this point, growth rate deflects downwards and remains persistently reduced for another 6 weeks (Speare *et al.*, 1998b). Growth deceleration is attributed to initial declines in fish appetite, which was reduced by 40% compared with control fish, alongside a delayed impairment of feed conversion rate (FCR). The latter was inhibited by up to 95% compared with control fish (Ramsay *et al.*, 2004). The dramatic effect on FCR may be partly

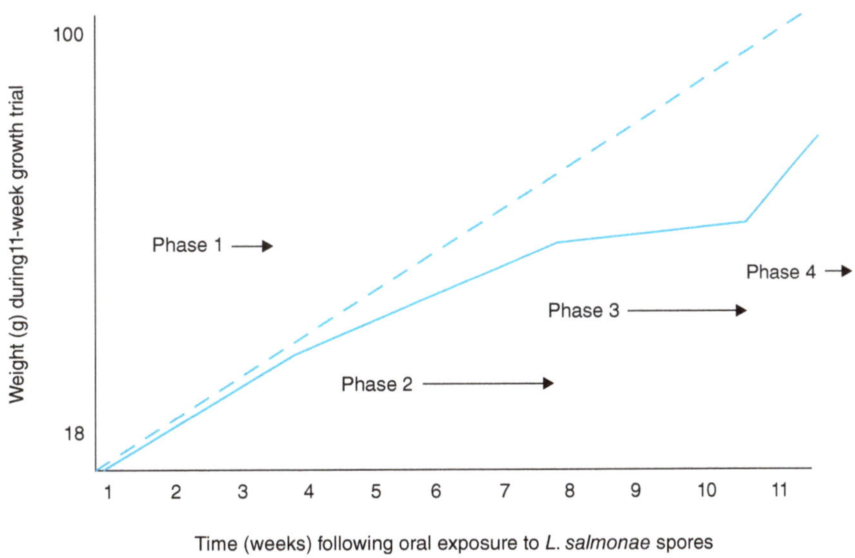

Fig. 19.6. Temporal growth disturbance pattern in juvenile rainbow trout experimentally infected with *Loma salmonae* (data derived from Ramsay *et al.*, 2004). Solid line depicts infected trout, dotted line depicts controls. During the first 4 weeks of *Loma* development (phase 1) trout show no signs of growth inhibition. However, by the 4th week, when a low percentage of xenomas rupture (phase 2), the growth rate of infected fish shows first evidence of decline. During phase 3, the majority of xenomas rupture, cellular inflammation in the gill is maximal and fish growth rate is markedly affected due to significant reduction in fish appetite. In phase 4, gill inflammation subsides, and the gill begins its repair and regeneration. Fish enter into compensatory growth mode marked by increased appetite and very rapid growth rate.

owing to energy requirements for redevelopment and remodelling of gill tissue. Although it remains to be studied, the more severe level of infection in chinook salmon with MGDS is likely to provide a more dramatic effect on growth rates compared with what has been observed in rainbow trout. So too, the protracted persistence of spores in the gills of chinook salmon following xenoma rupture would suggest that declines in growth rate would return to normal much more slowly.

19.5.4 Protective responses

An unexpected finding is the development of robust host protective responses following *L. salmonae* infection. This has been demonstrated in rainbow trout and chinook salmon following experimental infection, but the opposite has been shown for brook trout (Speare and Daley, 2003). Protective immune responses appear to be cell-mediated rather than humoral; these develop by week 4 following initial infection and protection remains robust even when fish are immunosuppressed by cortisol administration (Rodriguez-Tovar *et al.*, 2006a,b). Notably, cortisol administered to fish before any exposure to *L. salmonae* makes them extremely prone to super-infections (Lovy *et al.*, 2008). Protection appears to be elicited by host response to early stages of *L. salmonae*. Protection is generated even when fish are exposed to *L. salmonae* at temperatures below the permissive range of xenomas and also when fish receive only a minimal initial challenge dose as might occur during natural cohabitation transmission (Speare *et al.*, 1998a; Beaman *et al.*, 1999). In naïve fish, a marked degree of protection can also be provided through the oral administration of immunostimulants such as β-glucan (Guselle *et al.*, 2006), although this is effective only when fish receive β-glucan prior to exposure to *L. salmonae*. In total, these studies show that a functioning immune response is a significant factor in protecting salmonids from reinfections.

19.6 Control and Prevention: Modifications Imposed by Expected Temperature Change

There are no approved licensed medications available to use against microsporidians in aquaculture; therapeutic agents, in general, are not particularly effective against microsporidians even in the situation where human patients are treated aggressively against opportunistic microsporidians. Biosecurity, therefore, would appear to be the most effective approach to limit the effects of MGDS within a farmed population of fish. However, this could only be feasible within land-based cultivation systems where water supplies are either pathogen-free or could be rendered pathogen-free. Well-water is likely to be free of *L. salmonae* spores, whereas river and lake water may be contaminated depending on the status of any resident salmonid populations. Pump-ashore operations using seawater are best viewed as vulnerable given the likelihood of spores being introduced from the ocean to the farm. Depending on water flow rates, the possibility of using ultraviolet (UV) treatment of incoming water could be considered; it has been shown that even high concentrations of spores can be rendered non-infectious when treated with UV radiation (Becker and Speare, 2004a). Land-based farming operations typically use recirculation systems to limit the amount of water being pumped from natural sources. Although this is likely to result in the build-up of spores within a system afflicted by MGDS, the coupling of UV treatment into such a system may be effective at reducing horizontal transmission among tanks of fish linked to the same recirculation system.

With the current use of marine net pens in open oceans, biosecurity is impossible. *L. salmonae* has been shown to be readily transmitted from fish to fish in both fresh- and salt-water environments even when contacts times are brief (Shaw *et al.*, 1998; Becker *et al.*, 2003, 2005a,b). The degree to which MGDS relates to transmission from wild fish to farmed fish (spillover) is unknown, but this is assumed to be the initial source of infection to farmed stocks especially when *L. salmonae* is not detected in hatchery stock before transport to marine sites (Fig. 19.4). The observed pattern of MGDS mortalities in farmed chinook salmon suggests that following initial exposure of fish during the first summer, there is an amplification of *L. salmonae* within the farm setting thereby leading to an 'outbreak' of MGDS during the second summer. To estimate the pattern of spore release from a population of affected fish, one might consider the farm size, stocking density, numbers of xenomas per fish, the numbers of spores per xenoma, the duration through which xenomas rupture, local water current flow rates and the depth of water below net pens. Such work has yet to be carried out, but this remains a vital area of interest relative to year-over-year management of MGDS. Very little is known about the behaviour of spores within water. Although they have been shown to remain infective over 3 months at low water temperatures (Shaw *et al.*, 2000a), it is not known how long they survive at warmer water temperatures or in conditions where water temperature fluctuates. Nor is it known how readily the spores disperse or settle. If they settle, is their survival enhanced or reduced by the nature of the substrate that they find themselves in? These are critical questions when we consider the use of site fallowing as a method to break infection cycles. If spore survival is limited, a fallowing period of one growing season may be enough time for the spores to lose infectivity. If survival is extensive, fallowing may be useless. As yet it is unknown whether long-term survival of *Loma* spores could be affected by changes in ocean pH as might arise during climate change as elevated levels of atmospheric carbon dioxide dissolve in water. Spore survival, in general, has not been thoroughly studied and remains an open area with respect to our understanding of biosecurity.

Year-class separation is a standard approach to limiting fish diseases in situations where infective agents remain persistent within previously infected fish. Studies have also pointed to this being an essential approach for MGDS control because infected fish can retain a low number of xenomas for a prolonged period – long after the majority of xenomas have ruptured (Ramsay *et al.*, 2001, 2002). Furthermore, spores which have been released from xenomas, but have been retained in host tissues, can remain infective for more than 20 weeks. In situations where integrated multi-trophic aquaculture (IMTA) is being practised, an additional challenge would involve how best to deal with commercial bivalves cultivated near infected net pens. Recent work (McConnachie *et al.*, 2013) has shown that bivalves are remarkably efficient at clearing waters of *Loma* spores – but unfortunately, a large percentage of these imbibed spores can remain infective. This also points to challenges with fallowing non-IMTA sites. Natural populations of shellfish living alongside fish farms should be considered a source of future contamination.

Although not yet implemented, some progress has been made identifying potential medications against *L. salmonae*. The overall strategic goal of therapy would be to apply it early to a population of fish such that uninfected fish would be protected from horizontal transmission from infected cohorts. It remains unlikely that any therapeutic agent would have a benefit when fish are maximally affected by xenomas, and even the application of an anti-inflammatory agent (cycloxygenase inhibitor) during this period has been shown ineffective in protecting the gill from damage (Lovy *et al.*, 2007a). When applied early in infection, fumagillin, albendazole and cationic ionophores (Kent and Dawe, 1994; Speare *et al.*, 1999a, 2000) have been shown to markedly reduce the numbers of xenomas forming on the gill after a standard exposure to spores. Cationic ionophores, which may block the formation of microsporidial polar tubes, reduced xenoma formation by 93% compared with controls while at the same time having no effect on fish growth. In contrast, other drugs tested had either no effect or the effect of causing xenoma development to be somewhat later than usual (Speare *et al.*, 1998c).

Targeted manipulation of the host immune system through the use of immunostimulants, selective breeding or vaccination holds great promise in the control of MGDS. In-feed administration of β-glucans has been shown to be highly effective as long as treatment is initiated several weeks before fish are anticipated to be exposed to spores (Guselle *et al.*, 2006, 2010). Use of β-glucans could, therefore, be coupled with disease screening initiatives to determine at what time of the year MGDS is first beginning to develop within farmed populations; alternatively, based on water temperature history, a target date for treatment initiation could be developed as a standard practice. Strain selection is a viable consideration and, given the role of cell-mediated immunity in the control of *L. salmonae*, we may see improvements coincident with genetic selection against bacterial kidney disease. In both diseases, macrophage function is key to disease evolution.

Vaccination also holds tremendous promise given the high degree of cell-mediated immune response protection that develops within 3 weeks following exposure of fish to low doses of either live (virulent or avirulent strains of *Loma*) or dead spores (Speare *et al.*, 1998c; Kent *et al.*, 1999; Sanchez *et al.*, 2001b,d; Rodriguez-Tovar *et al.*, 2006a,b). Within immunized fish, *L. salmonae* development is blocked within subendocardial macrophages and does not transfer subsequently to the gill (Sanchez *et al.*, 2001b,d). Furthermore, spores from *Glugea* spp. can also be used to develop a cross-protective vaccine against *L. salmonae* in salmonids (Harkness *et al.*, 2013). The latter results, which are surprising, may point to the effectiveness of any augmentation of the cell-mediated immune system and perhaps mirror the effectiveness of treatments such as β-glucan.

Where practical, environmental modulation may also prove useful in controlling MGDS. Given the regulatory effects of water temperature, a non-permissive temperature could be used to block the development of *L. salmonae*. Water temperature reduction would need to be persistent. Reducing water temperature once MGDS emerges is likely to have little to no effect since it has been shown that only the very early stages of *L. salmonae* can be blocked by low water temperatures (Speare *et al.*, 1999b). However, water temperature reduction has been shown effective for *K. takedai* control in masu salmon (*O. masou*) and therefore further studies are warranted to determine where this tactic may be useful. The need for prolonged temperature reduction can be deduced from studies of the interactions of *Enterocytozoon salmonis* and chinook salmon at various consistent and changing water temperatures (Antonio and Hedrick, 1995).

19.7 Conclusions with Suggestions for Future Studies

The salmon farming industry in British Columbia has changed considerably since its early beginnings in 1970 when the industry was dominated by the culture of regional salmonid species – chinook and coho salmon; since 1990 there has been a dramatic shift towards the non-indigenous Atlantic salmon. Despite this, chinook salmon production remains significant and perhaps may undergo a resurgence as public pressure continues to mount against the culture of Atlantic salmon – a species that is not native to the Pacific North-West. Additionally, hatchery production of many species of Pacific salmon continues at a very high rate in the Pacific North-West as part of the Salmon Enhancement programmes through which juvenile hatchery-reared Pacific salmon, derived from returning wild broodstocks, are released to support capture fisheries.

The current production models for chinook salmon and Atlantic salmon are overall quite similar. Hatchery rearing in freshwater land-based systems, at high stocking densities, is used to bring the juveniles to a 'smolt' stage replete with metaplastic changes to the gills favouring chloride cells. These cells provide salmon the capacity to deal with elevated sodium chloride levels. Smolt are transferred to net pen enclosures in marine settings. Here, they will spend up to 24 months until they reach market size, and it is during this relatively long production period when they are sharing the environment with wild salmon carrying *L. salmonae*. In many instances, young wild salmon pass through the mesh of net pens and cohabitate with farmed fish.

Production of salmon within marine net pens has proven to be a relatively low-cost method of farming despite the various issues arising such as predators, damage to net pens due to storms, disease transfer from wild to farmed stocks and mortalities stemming from harmful algal blooms. With the maturation of the industry and identification of year-over-year environmental hazards, aquaculture site selection has already changed in British Columbia in response to algal blooms. Severe algal blooms also hazard the Chilean industry. With expectations of water temperature increments, it is reasonable to expect salmon aquaculture production ranges to expand poleward. It is challenging to adequately anticipate whether there will be a phenological match or mismatch, as per the arguments of Lafferty (2009) and Rohr *et al.* (2011), between farming production sites, migration patterns and periods of residence of wild salmon, and the spillover and spillback potential for *L. salmonae*.

There has been an increasing concern about the potential adverse effects of net pen salmon fish farming on the environment. The concerns are varied and range from pollution (e.g. released feeds, fish faeces, antibiotics and antiparasitic agents) to more complex issues such as the potential for net pen salmon to contribute to the load of pathogens that would be encountered especially by juvenile wild salmon as they emerge from freshwater systems to enter the sea. So, too, accidental release of salmon (especially Atlantic salmon in regions where they are not indigenous) is a significant topic of concern in its own right. When these issues are addressed collectively, and in the face of threatened stocks of wild salmon due to cumulative climate change phenomena, it is reasonable that over the next two decades there will be a movement towards the use of 'pump-ashore' facilities. Here, salmon are no longer raised in marine net pens, but rather in large land-based tanks supplied by salt water pumped from the sea. The economics of this have yet to be determined, but generally require the use of recirculating biological filtration systems to help reduce the expense of water pumping. The specifics of an aquaculture setting dictate the specific diseases/pathogens that are likely to flourish. Microsporidians, due to their resistant spores, are expected to become very significant within recirculating systems, similar to problems associated with the environmentally resistant mycobacteria in similar systems.

The scarcity of long-term empirical data dramatically limits our understanding of the relationships between climate change and aquatic disease dynamics (Lafferty, 2009; Karvonen *et al.*, 2010). Aquaculture facilities may become an ideal model through which to understand evolving host–pathogen interactions because they are a highly monitored environment. Water characteristics are recorded, and data are stored. Disease detection and record keeping have become the norm. Population densities are high, and the population remains 'in place' and thus easily monitored during substantial periods of the life cycle. Many farms – such as net pens – are directly connected to natural systems, and multiple net pens and farm sites provide replication levels needed to draw evidence-based conclusions. However, drawing conclusions as to the increased incidence of particular diseases should also be tempered by the biasing effects of increased surveillance. Furthermore, reporting of disease in what appears to be new geographical areas is more likely to be published than the findings of absence of disease in new or endemic areas. Given the presence of aquaculture sites, monitoring equipment and *in situ* staff, it may be prudent for government (and other) research agencies to consider the value arising in partnering with these commercial operations for mutual benefit. Large data sets collected by farming operations over lengthy time spans could be valuable for data mining approaches. These data sets would conceivably include items of oceanographic interest alongside disease events arising within well-monitored fish populations entrained to the region through captive rearing during their production cycle. Furthermore, these same fish could be followed through their reproduction cycle and intergenerational effects deduced from performance of progeny.

References

Antonio, D.B. and Hedrick, R.P. (1995) Effect of water temperatures on infections with the microsporidian *Enterocytozoon salmonis* in chinook salmon. *Diseases of Aquatic Organisms* 22, 233–236.

Bacchi, C.J., Weiss, L.M., Lane, S. and Wittner, M. (2002) Novel synthetic polyamines are effective in the treatment of experimental microsporidiosis, an opportunistic AIDS-associated infection. *Antimicrobial Agents and Chemotherapy* 46, 55–61.

Beaman, H.J. and Speare, D.J. (1999) Regulatory effects of water temperature on *Loma salmonae* (Microspora) development in rainbow trout. *Journal of Aquatic Animal Health* 11, 237–245.

Beaman, H.J., Speare, D.J., Brimacombe, M. and Daley, J. (1999) Evaluating protection against *Loma salmonae* generated from primary exposure of rainbow trout, *Oncorhynchus mykiss* (Walbaum) outside of the xenoma-expression temperature boundaries. *Journal of Fish Diseases* 22, 445–450.

Becker, J.A. and Speare, D.J. (2004a) Ultraviolet light control of horizontal transmission of *Loma salmonae*. *Journal of Fish Diseases* 27, 177–180.

Becker, J.A. and Speare, D.J. (2004b) Impact of water temperature shift on xenoma clearance and recovery time during a *Loma salmonae* (Microsporidia) infection in rainbow trout *Oncorhynchus mykiss*. *Diseases of Aquatic Organisms* 58, 185–191.

Becker, J.A. and Speare, D.J. (2007) Transmission of the microsporidian gill parasite, *Loma salmonae*. *Animal Health Research Reviews* 8, 59–68.

Becker, J.A., Speare, D.J. and Dohoo, I.R. (2003) Effect of water temperature and flow rate on the transmission of microsporidial gill disease caused by *Loma salmonae* in rainbow trout *Oncorhynchus mykiss*. *Fish Pathology* 38, 105–112.

Becker, J.A., Speare, D.J. and Dohoo, I.R. (2005a) Influence of feeding ratio and size on susceptibility to microsporidial gill disease caused by *Loma salmonae* in rainbow trout, *Oncorhynchus mykiss* (Walbaum). *Journal of Fish Diseases* 28, 173–180.

Becker, J.A., Speare, D.J. and Dohoo, I.R. (2005b) Effects of the number of infected fish and acute exposure period on the horizontal transmission of *Loma salmonae* (Microsporidia) in rainbow trout, *Oncorhynchus mykiss*. *Aquaculture* 244, 1–9.

Brooks, D.R. and Hoberg, E.P. (2007) How will climate change affect parasite–host assemblages? *Trends in Parasitology* 23, 571–573.

Brown, A.M.V., Kent, M.L. and Adamson, M.L. (2010) Low genetic variation in the salmon and trout parasite *Loma salmonae* (Microsporidia) supports marine transmission and clarifies species boundaries. *Diseases of Aquatic Organisms* 91, 35–46.

Brown, J.H. (2004) Toward a metabolic theory of ecology. *Ecology* 85, 1771–1789.

Bruno, D.W., Collins, R.O. and Morrison, C.M. (1995) The occurrence of *Loma salmonae* (Protozoa: Microspora) in farmed rainbow trout, *Oncorhynchus mykiss* Walbaum, in Scotland. *Aquaculture* 133, 341–344.

Cali, A. and Takvorian, P.M. (1991) The incidence of *Glugea stephani* (Protozoa: Microsporidia) in winter flounder, *Pseudopleuronectes americanus*, from the New York–New Jersey lower bay complex and factors influencing it. *Canadian Journal of Zoology* 69, 317–321.

Constantine, J. (1999) *Estimating the Cost of Loma salmonae to B.C. Aquaculture*. Animal Health Branch, Ministry of Agriculture and Food, Abbotsford, British Columbia, Canada.

de la Rocque, S., Rioux, J.A. and Slingenbergh, J. (2008) Climate change: effects on animal disease systems and implications for surveillance and control. *Revue Scientifique et Technique (Office International Epizootics)* 27, 339–354.

Fujiyama, I., Urawa, S., Yokoyama, H. and Ogawa, K. (2002) Investigation of the transmission stage of the microsporidian *Kabatana takedai* in salmonids. *Bulletin of the National Salmon Resources Center* 5, 1–6.

Gandhi, S., Locatelli, L. and Feist, S.W. (1995) Occurrence of *Loma* sp. (Microsporidia) in farmed rainbow trout (*Oncorhynchus mykiss*) at a farm site in south west England. *Bulletin of the European Association of Fish Pathologists* 15, 58–60.

Guselle, N.J., Markham, R.J.F. and Speare, D.J. (2006) Intraperitoneal administration of β-1,3/1,6-glucan to rainbow trout, *Oncorhynchus mykiss* (Walbaum), protects against *Loma salmonae*. *Journal of Fish Diseases* 29, 375–381.

Guselle, N.J., Speare, D.J. and Markham, R.J.F. (2010) Efficacy of intraperitoneally and orally administered ProVale, a yeast β-(1,3)/(1,6)-D-glucan product, in inhibiting xenoma formation by the microsporidian *Loma salmonae* on rainbow trout gills. *North American Journal of Aquaculture* 72, 65–72.

Harkness, J.E., Guselle, N.J. and Speare, D.J. (2013) Demonstrated efficacy of a pilot heterologous whole-spore vaccine against microsporidial gill disease in rainbow trout. *Clinical and Vaccine Immunology* 20, 1483–1484.

Hughes, L. (2000) Biological consequences of global warming: is the signal already there? *Trends in Ecology and Evolution* 15, 56–61.

Karvonen, A., Rintamaki, P., Jokela, J. and Valtonen E.T. (2010) Increasing water temperature and disease risks in aquatic systems: climate change increases the risk of some, but not all, diseases. *International Journal for Parasitology* 40, 1483–1488.

Keeling, P.J. and Fast, N.M. (2002) Microsporidia: biology and evolution of highly reduced intracellular parasites. *Annual Reviews in Microbiology* 56, 93–116.

Kent, M.L. and Dawe, S.C. (1994) Efficacy of Fumagillin DCH against experimentally induced *Loma salmonae* (Microsporea) infections in chinook salmon *Oncorhynchus tshawytscha*. *Diseases of Aquatic Organisms* 20, 231–233.

Kent, M.L., Dawe, S.C. and Speare, D.J. (1995) Transmission of *Loma salmonae* (Microsporea) to chinook salmon in seawater. *Canadian Veterinary Journal* 36, 98–101.

Kent, M.L., Traxler, G.S., Kieser, D., Richard, J., Dawe S.C. *et al.* (1998) Survey of salmonid pathogens in ocean-caught fishes in British Columbia, Canada. *Journal of Aquatic Animal Health* 10, 211–219.

Kent, M.L., Dawe, S.C. and Speare, D.J. (1999) Resistance to reinfection in chinook salmon *Oncorhynchus tshawytscha* to *Loma salmonae* (Microsporidia). *Diseases of Aquatic Organisms* 37, 205–208.

Kocan, R., Hershberger, P., Sanders, G. and Winton, J. (2009) Effects of temperature on disease progression and swimming stamina in *Ichthyophonus*-infected rainbow trout, *Oncorhynchus mykiss* (Walbaum). *Journal of Fish Diseases* 32, 835–843.

Lafferty, K.D. (2009) The ecology of climate change and infectious diseases. *Ecology* 90, 888–900.

Lovy, J., Wright, G.M., Wadowska, D.W. and Speare, D.J. (2006) Ultrastructural morphology suggesting a new hypothesis for development of microsporidians seen in *Loma salmonae* infecting the gills of trout. *Journal of Fish Biology* 68, 450–457.

Lovy, J., Wright, G.M. and Speare, D.J. (2007a) Pathological effects caused by chronic treatment of rainbow trout with indomethacin. *Journal of Aquatic Animal Health* 19, 94–98.

Lovy, J., Wright, G.M. and Speare, D.J. (2007b) Ultrastructural examination of the host inflammatory response within gills of netpen reared chinook salmon (*Oncorhynchus tshawytscha*) with microsporidial gill disease. *Fish and Shellfish Immunology* 22, 131–149.

Lovy, J., Speare, D.J., Stryhn, H. and Wright, G.M. (2008) Effects of dexamethasone on host innate and adaptive immune responses and parasite development in rainbow trout *Oncorhynchus mykiss* infected with *Loma salmonae*. *Fish and Shellfish Immunology* 24, 649–658.

McConnachie, S.H., Guselle, N.J. and Speare, D.J. (2013) Retention of viable microsporidial (*Loma salmonae*) spores within the blue mussel (*Mytilus edulis*): use of an experimental laboratory model probing pathogen transfer within a multi-trophic aquaculture setting. *Aquaculture* 376–379, 1–5.

Markey, P.T., Blazer, V.S., Ewing, M.S. and Kocan, K.M. (1994) *Loma* sp. in salmonids from the eastern United States: associated lesions in rainbow trout. *Journal of Aquatic Animal Health* 6, 318–328.

Martin-Hernandez, R., Meana, A., Garcia-Palencia, P., Marin, P., Botias, C. *et al.* (2009) Effects of temperature on the biotic potential of honeybee microsporidia. *Applied and Environmental Microbiology* 75, 2554–2557.

Morrison, C.M. and Sprague, V. (1983) *Loma salmonae* (Putz, Hoffman and Dunbar, 1965) in the rainbow trout, *Salmo gairdneri* Richardson, and *L. fontinalis* sp. nov. (Microsporida) in the brook trout, *Salvelinus fontinalis* (Mitchill). *Journal of Fish Diseases* 6, 345–353.

Olson, R.E. (1981) Effect of low temperature on the development of the microsporidian *Glugea stephani* in English sole (*Parophrys vetulus*). *Journal of Wildlife Diseases* 17, 559–562.

Paull, S.H. and Johnson, P.T.J. (2011) High temperature enhances host pathology in a snail–trematode system: possible consequences of climate change for the emergence of disease. *Freshwater Biology* 56, 767–778.

Paxton, R.J., Klee, S., Korpela, S. and Fries, L. (2007) *Nosema ceranae* has infected *Apis mellifera* in Europe since at least 1998 and may be more virulent than *Nosema apis*. *Apidologie* 38, 558–565.

Powell, M.D., Speare, D.J. and Becker, J.A. (2006) Whole body net ion fluxes, plasma electrolyte concentrations and haematology during a *Loma salmonae* infection in juvenile rainbow trout, *Oncorhynchus mykiss* (Walbaum). *Journal of Fish Diseases* 29, 727–735.

Poynton, S.L. (1986) Distribution of the flagellate *Hexamita salmonis* Moore, 1922 and the microsporidian *Loma salmonae* Putz, Hoffman and Dunbar, 1965 in brown trout, *Salmo trutta* L. and rainbow trout, *Salmo gairdneri* Richardson, in the River Itchen (UK) and three of its fish farms. *Journal of Fish Biology* 29, 417–429.

Rahmathulla, V.K., Kishor Kumar, C.M., Angadi, B.S. and Sivaprasad V. (2012) Influence of weather factors on incidence and intensity of microsporidiosis in silkworm (*Bombyx mori* L.) *Journal of Entomology* 9, 266–273.

Ramsay, J.M., Speare, D.J., Sanchez, J.G. and Daley, J. (2001) The transmission potential of *Loma salmonae* (Microspora) in the rainbow trout, *Oncorhynchus mykiss* (Walbaum), is dependent upon the method and timing of exposure. *Journal of Fish Diseases* 24, 453–460.

Ramsay, J.M., Speare, D.J., Dawe, S.C. and Kent, M.L. (2002) Xenoma formation during microsporidial gill disease of salmonids caused by *Loma salmonae* is affected by host species (*Oncorhynchus tshawytscha*, *O. kisutch*, *O. mykiss*) but not by salinity. *Diseases of Aquatic Organisms* 48, 125–131.

Ramsay, J.M., Speare, D.J., Becker, J.A. and Daley, J. (2003) *Loma salmonae*-associated xenoma onset and clearance in rainbow trout, *Oncorhynchus mykiss* (Walbaum): comparisons of *per os* and cohabitation exposure using survival analysis. *Aquaculture Research* 34, 1329–1335.

Ramsay, J.M., Speare, D.J. and Daley, J. (2004) Timing of changes in growth rate, feed intake and feed conversion in rainbow trout, *Oncorhynchus mykiss* (Walbaum), experimentally infected with *Loma salmonae* (Microspora). *Journal of Fish Diseases* 27, 425–429.

Rogers, D.J. and Randolph, S.E. (2006) Climate change and vector-borne diseases. *Advances in Parasitology* 62, 345–381.

Rodriguez-Tovar, L.E., Wright, G.M., Wadowska, D.W., Speare, D.J. and Markham, R.J.F. (2002) Ultrastructural study of the early development and localization of *Loma salmonae* in the gills of experimentally infected rainbow trout. *Journal of Parasitology* 88, 244–254.

Rodriguez-Tovar, L.E., Wadowska, D.W., Wright, G.M., Groman, D.B., Speare, D.J. and Whelan, D.S. (2003a) Ultrastructural evidence of autoinfection in the gills of Atlantic cod *Gadus morhua* infected with *Loma* sp. (phylum Microsporidia). *Diseases of Aquatic Organisms* 57, 227–230.

Rodriguez-Tovar, L.E., Wright, G.M., Wadowska, D.W., Speare, D.J. and Markham, R.J.F. (2003b) Ultrastructural study of the late stages of *Loma salmonae* development in the gills of experimentally infected rainbow trout. *Journal of Parasitology* 89, 464–474.

Rodriguez-Tovar, L.E., Becker, J.A., Markham, R.J.F. and Speare, D.J. (2006a) Induction time for resistance to microsporidial gill disease caused by *Loma salmonae* following vaccination of rainbow trout (*Oncorhynchus mykiss*) with a spore-based vaccine. *Fish and Shellfish Immunology* 21, 170–175.

Rodriguez-Tovar, L.E., Markham, R.J.F., Speare, D.J. and Sheppard, J. (2006b) Cellular immunity in salmonids infected with the microsporidian parasite *Loma salmonae* or exposed to non-viable spores. *Veterinary Immunology and Immunopathology* 114, 72–83.

Rohr, J.R., Dobson, A.P., Johnson, P.T.J., Kilpatrick, A.M., Paull, S.H. *et al.* (2011) Frontiers in climate change–disease research. *Trends in Ecology and Evolution* 25, 272–277.

Sanchez, J.G., Speare, D.J. and Markham, R.J.F. (1999) Nonisotopic detection of *Loma salmonae* in rainbow trout (*Oncorhynchus mykiss*) gills by in situ hybridization. *Veterinary Pathology* 36, 610–612.

Sanchez, J.G., Speare, D.J. and Markham, R.J.F. (2000) Normal and aberrant tissue distribution of *Loma salmonae* (Microspora) within rainbow trout (*Oncorhynchus mykiss*) following experimental infection at water temperatures within and outside of the xenoma-expression temperature boundaries. *Journal of Fish Diseases* 23, 235–242.

Sanchez, J.G., Speare, D.J. and Markham, R.J.F. (2001a) Altered tissue distribution of *Loma salmonae*: effects of natural and acquired resistance. *Journal of Fish Diseases* 24, 33–40.

Sanchez, J.G., Speare, D.J., Markham, R.J.F. and Jones, S.R.M. (2001b) Experimental vaccination of rainbow trout against *Loma salmonae* using a live low-virulence variant of *L. salmonae*. *Journal of Fish Biology* 59, 427–441.

Sanchez, J.G., Speare, D.J., Markham, R.J.F. and Jones S.R.M. (2001c) Isolation of a *Loma salmonae* variant: biological characteristics and host range. *Journal of Fish Biology* 59, 427–441.

Sanchez, J.G., Speare, D.J., Markham, R.J.F., Wright, G.M. and Kibenge, F.S.B. (2001d) Localization of the initial developmental stages of *Loma salmonae* in rainbow trout (*Oncorhynchus mykiss*). *Veterinary Pathology* 38, 540–546.

Shaw, R.W., Kent, M.L. and Adamson, M.L. (1998) Modes of transmission of *Loma salmonae* (Microsporidia). *Diseases of Aquatic Organisms* 33, 151–156.

Shaw, R.W., Kent, M.L. and Adamson, M.L. (2000a) Viability of *Loma salmonae* (Microsporidia) under laboratory conditions. *Parasitology Research* 86, 978–981.

Shaw, R.W., Kent, M.L. and Adamson, M.L. (2000b) Innate susceptibility differences in chinook salmon *Oncorhynchus tshawytscha* to *Loma salmonae* (Microsporidia). *Diseases of Aquatic Organisms* 43, 49–53.

Shaw, R.W., Kent, M.L., Brown A.M.V., Whipps, C.M. and Adamson, M.L. (2000c) Experimental and natural host specificity of *Loma salmonae* (Microsporidia). *Diseases of Aquatic Organisms* 40, 131–136.

Speare, D.J. and Daley, J. (2003) Failure of vaccination in brook trout *Salvelinus fontinalis* against *Loma salmonae* (Microspora). *Fish Pathology* 38, 27–28.

Speare, D.J., Brackett, J. and Ferguson, H.W. (1989) Sequential pathology of the gills of coho salmon with a combined diatom and microsporidian gill infection. *Canadian Veterinary Journal* 30, 571–576.

Speare, D.J., Arsenault, G.J. and Buote, M.A. (1998a) Evaluation of rainbow trout as a model species for studying the pathogenesis of the branchial microsporidian *Loma salmonae*. *Contemporary Topics in Laboratory Animal Science* 37, 55–58.

Speare, D.J., Daley, J., Markham, R.J.F., Sheppard, J., Beaman, H.J. and Sanchez, G.J. (1998b) *Loma salmonae*-associated growth rate suppression in rainbow trout (*Oncorhynchus mykiss*) occurs during early-onset xenoma dissolution as determined by *in situ* hybridization and immunohistochemistry. *Journal of Fish Diseases* 21, 345–354.

Speare, D.J., Ritter, G. and Schmidt, H. (1998c) Quinine hydrochloride treatment delays xenoma formation and dissolution in rainbow trout challenged with *Loma salmonae*. *Journal of Comparative Pathology* 119, 459–465.

Speare, D.J., Athanassopoulou, F., Daley, J. and Sanchez, J.G. (1999a) A preliminary investigation of alternatives to fumagillin for the treatment of *Loma salmonae* infection in rainbow trout. *Journal of Comparative Pathology* 121, 241–248.

Speare, D.J., Beaman, H.J. and Daley, J. (1999b) Effect of water temperature manipulation on a thermal unit predictive model for *Loma salmonae*. *Journal of Fish Diseases* 22, 277–283.

Speare, D.J., Daley, J., Dick. P., Novilla, M. and Poe, S. (2000) Ionophore-mediated inhibition of xenoma-expression in trout challenged with *Loma salmonae* (Microspora). *Journal of Fish Diseases* 23, 231–233.

Speare, D.J., Markham, R.J.F. and Guselle, N.J. (2007) Development of an effective whole-spore vaccine using a low-virulence strain of *Loma salmonae* to protect against microsporidial gill disease in rainbow trout (*Oncorhynchus mykiss*). *Clinical and Vaccine Immunology* 14, 12–18.

Takahashi, S. and Ogawa, K. (1997) Efficacy of elevated water temperature of ayu infected with the microsporidian *Glugea plecoglossi*. *Fish Pathology* 32, 193–198.

Tiner, J.D. (1988) Birefringent spores differentiate *Encephalitozoon* and other microsporidia from coccidian. *Veterinary Pathology* 25, 227–230.

Williams, B.A.P. (2009) Unique physiology of host–parasite interactions in microsporidia infections. *Cellular Microbiology* 11, 1551–1560.

Zenke, K., Urawa, S., Fujiyama, I., Yokoyama, H. and Ogawa, K. (2005) Effects of water temperature on infection of the microsporidian *Kabatana takedai* in salmonids. *Fish Pathology* 40, 119–123.

20 Myxoboliosis (*Myxobolus cerebralis*)

Julie D. Alexander* and Jerri L. Bartholomew

Oregon State University, Corvallis, Oregon, USA

20.1 Introduction

This chapter provides an overview of salmonid whirling disease (SWD) and discusses disease aetiology in the context of climate change. SWD is caused by the myxozoan parasite *Myxobolus cerebralis*. Predicting the effects of climate change on SWD is complicated by the involvement of two obligate host species and two separate parasite spore stages, each of which has its own environmental optima and tolerances. The chapter highlights this complexity by discussing the potential effects of climate-driven changes due to temperature and precipitation on *M. cerebralis* spore stages and both hosts. We present a degree-day model to describe parasite dynamics under different climate scenarios and discuss potential future disease dynamics. Throughout the chapter, we interpret predictions based on available data and discuss implications for *M. cerebralis* and SWD risk.

20.2 Description of Disease, Current Distribution, Fish Host Species

20.2.1 Description of disease

M. cerebralis is a metazoan endoparasite thought to have evolved as a non-pathogenic parasite of *Salmo trutta* (brown trout) in central Europe and northern Asia (Bogdanova, 1969; Hoffman, 1970). The first description of SWD was in Germany following an outbreak in allopatric (non-native) rainbow (*Oncorhynchus mykiss*) and brook (*Salvelinus fontinalis*) trout imported for pond culture (Höfer, 1903 in Bartholomew and Reno, 2002); however, the parasite's heteroxenous life cycle remained undescribed until nearly a century later (Markiw and Wolf, 1983; Wolf and Markiw, 1984). Parasite myxospores produced in cartilage tissues of infected salmonids are released following the death and decay of the fish host and go on to infect the obligate invertebrate host *Tubifex tubifex* (Müller, 1774) (Wolf *et al.*, 1986) (Fig. 20.1). Parasite stages proliferate and develop into actinospores (triactinomyxons (TAMs)), which can go on to infect salmonids after they are released from the invertebrate hosts. Severely infected salmonids exhibit a range of clinical signs (Fig. 20.2) including the characteristic whirling behaviour for which the disease is named ('Drehkrankheit' which translates to 'whirling disease').

20.2.2 Current distribution

M. cerebralis is enzootic in cultured and free-ranging salmonids in areas where salmonid aquaculture occurs (Africa: Lebanon, Morocco, South Africa; Eurasia: Russia, Georgia; Europe: Austria, Belgium, Bulgaria, Czech Republic, Denmark, England, France, Hungary, Ireland, Italy, Liechtenstein, Luxemburg, Netherlands, Norway, Poland, Scotland, Spain, Sweden, Yugoslavia; North America: Canada, USA; and Oceania: New Zealand), with the exception of South America (Uspenskaya, 1955, 1957; Bogdanova, 1960, 1964, 1968; Christensen, 1972; Halliday, 1974; Hoffman, 1990; Margolis *et al.*, 1996; Canadian Food Inspection Agency, 2016).

The parasite has achieved this cosmopolitan distribution through anthropogenic activities, an abundance of susceptible naïve fish hosts and a widely distributed invertebrate host species. In North America, the parasite has been introduced multiple times (Bartholomew and Reno, 2002) and has become established in some populations following detection, sometimes with population-level losses. SWD outbreaks in Colorado and Montana's

*alexanju@oregonstate.edu

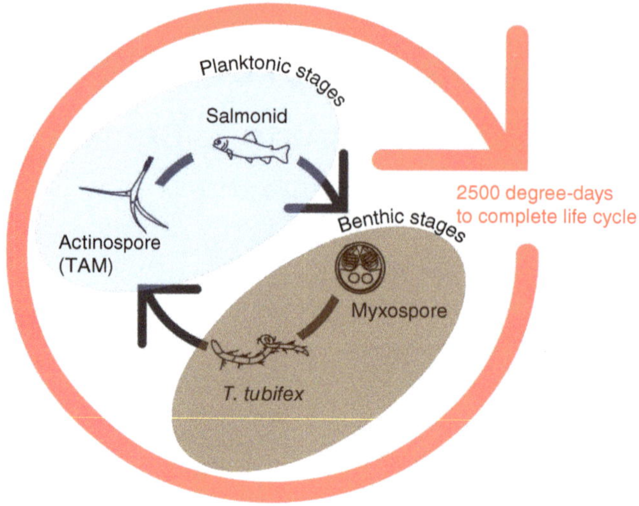

Fig. 20.1. *Myxobolus cerebralis* has two obligate hosts, *Tubifex tubifex* and a salmonid fish, and two microscopic waterborne spore stages, myxospores (8–10 μm diameter) and actinospores (triactinomyxons (TAMs); 150–180 μm tip to tip). Myxospores infect *Tubifex tubifex* and TAMs infect salmonids. The life cycle requires approximately 2500 degree-days for completion, thus shifts in phenology are expected in response to rising water temperatures.

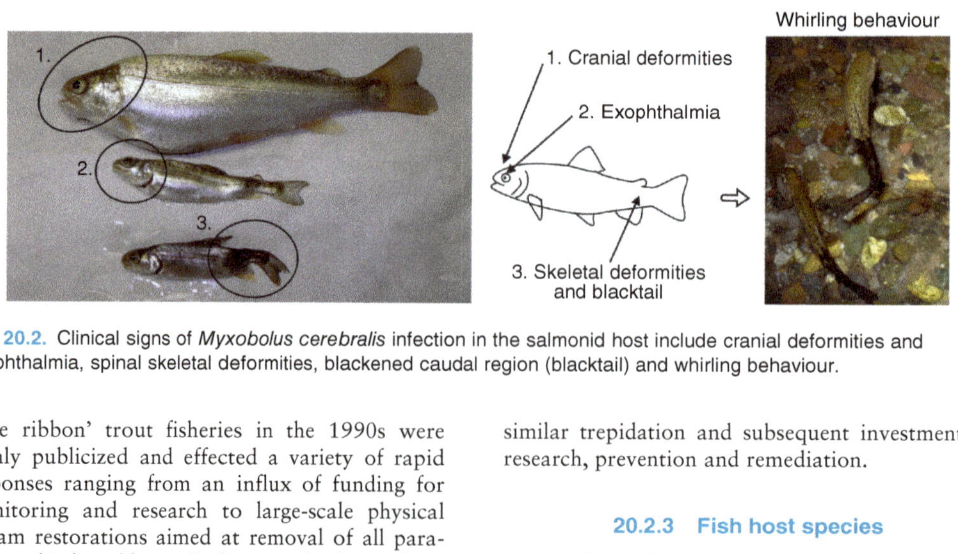

Fig. 20.2. Clinical signs of *Myxobolus cerebralis* infection in the salmonid host include cranial deformities and exophthalmia, spinal skeletal deformities, blackened caudal region (blacktail) and whirling behaviour.

'blue ribbon' trout fisheries in the 1990s were highly publicized and effected a variety of rapid responses ranging from an influx of funding for monitoring and research to large-scale physical stream restorations aimed at removal of all parasites and infected hosts. Unfortunately, these efforts were ineffective in halting parasite expansion and *M. cerebralis* was detected in the iconic Yellowstone cutthroat trout in the late 1990s. Recent emergence of the pathogen in Alberta, Canada (Canadian Food Inspection Agency, 2016) has been met with similar trepidation and subsequent investment in research, prevention and remediation.

20.2.3 Fish host species

Most salmonids are susceptible to *M. cerebralis*, but widespread stocking of salmonids worldwide has confounded our ability to delineate salmonid hosts into susceptibility categories based on allopatry or sympatry. In resistant fish, the parasite can infect (invade), but no clinical disease or myxospores develop, regardless of parasite dose. In susceptible fish, myxospores develop but infection severity ranges from subclinical to clinical. Susceptibility to infection and development of disease vary among genus, species, strains and individual salmonids within a population. For example, *Oncorhynchus* spp. resident trout are highly susceptible, but the anadromous salmonids native to the Pacific Coast of North America exhibit a wide range of susceptibilities: anadromous steelhead trout (*O. mykiss*) are highly susceptible, coho (*Oncorhynchus kisutch*) salmon are less susceptible than steelhead, sockeye (*Oncorhynchus nerka*) or Chinook salmon (*Oncorhynchus tshawytscha*) (Hedrick *et al.*, 2001a). The *Salvelinus* spp. range from highly susceptible (brook trout) to completely resistant (lake trout) (O'Grodnick, 1979; Hedrick *et al.*, 1999b, 2001b) (Table 20.1). Among the susceptible salmonids, development and severity of clinical signs depend on the age, size and exposure conditions (dose and environmental conditions, see below) (Hoffman and Putz, 1969; O'Grodnick, 1979; Buchanan and Sanders, 1983; Ibarra *et al.*, 1991; Hedrick *et al.*, 1999a,b, 2001b, 2003; Downing *et al.*, 2002; Sollid *et al.*, 2002; Vincent, 2002; Bartholomew *et al.*, 2003; Koel *et al.*, 2006; Murcia *et al.*, 2006; DuBey *et al.*, 2007). Clinical signs appear approximately 3–8 weeks after exposure to TAMs (MacConnell and Vincent, 2002). Whirling disease severity results from the combination of the salmonid host's immunological response to parasite damage and the actual physical damage resulting from parasite proliferation (see Sections 20.3.2 and 20.3.4).

Population-level losses following detection of *M. cerebralis* have not been reported from systems with anadromous fish. Variation in susceptibility aside, anadromous salmonids could be less likely to suffer population-level effects of disease than their non-anadromous counterparts if smoltification and outmigration result in a shorter exposure window. In this context, anadromous salmonids could act as parasite sinks by removing parasites and decreasing system-specific whirling disease risk. However, population-level effects of whirling disease may occur but remain undetected because of the challenges associated with monitoring disease in anadromous populations.

In general, climate change-related effects on whirling disease will largely be driven by thermal

Table 20.1. Susceptibility to whirling disease assessed by natural or laboratory exposure to *Myxobolus cerebralis* at vulnerable life stages.

Susceptibility[a]	Genus	Species	Common name
S_1	*Oncorhynchus*	*mykiss*	Rainbow, steelhead trout
		clarki spp.	Cutthroat trout: Yellowstone, westslope, Colorado River, Rio Grande, greenback
		nerka	Sockeye salmon
		tshawytscha	Chinook salmon
	Salvelinus	*fontinalis*	Brook trout
	Salmo	*salar*	Atlantic salmon
	Hucho	*hucho*	Danube salmon
	Prosopium	*willamsoni*	Mountain whitefish
	Thymallus	*thymallus*	European grayling
S_0	*Oncorhynchus*	*keta*	Chum salmon
		gorbuscha	Pink salmon
		masu	Cherry salmon
		kisutch	Coho salmon
	Salvelinus	*malma*	Dolly Varden trout
		confluentus	Bull trout
		trutta	Brown trout
R	*Salvelinus*	*namayacush*	Lake trout
	Thymallus	*arcticus*	Arctic grayling

[a]S_1, moderate to high susceptibility and parasite amplification; clinical disease signs occur following low to moderate parasite doses and myxospore formation occurs, often several orders of magnitude greater than exposure dose (= parasite amplification). S_0, low susceptibility and parasite amplification; clinical disease signs and myxospore formation possible at high parasite doses (parasite production but not necessarily amplification). R, resistant; no clinical disease signs or myxospores formed, parasite sinks/biological filters.

and hydrological changes (see Section 20.4) to fish habitat. However, some host populations and species will be better equipped to deal with the challenges of changing environments and whirling disease risk as a function of their specific climate vulnerability attributes including sensitivity, geographical location and projected future climate conditions, and adaptive capacity (Crozier *et al.*, 2008, 2019; Lynch *et al.*, 2016).

20.3 Diagnosis of the Pathogen

20.3.1 Morphological features

Myxosporean parasites are commonly identified based on morphological and morphometric characteristics of the myxospore in the vertebrate host because the spores are persistent and conspicuous. Like the majority of myxosporeans, *M. cerebralis* has two morphologically distinct spores: the actinospore (or TAM) and the myxospore (Fig. 20.1). The myxospore is oval shaped and 8.7 μm long, 8.2 μm wide and 6.3 μm thick. Two hard valve shells encapsulate a binucleate sporoplasm and two polar capsules, each of which contains an extrudable polar filament (Lom and Hoffman, 1971). In contrast to the smaller myxospore, the actinospore is morphologically similar to a triradially symmetric grappling hook. Three valve cells form an axis (~150 μm) with three caudal processes (~195 μm each) and within the apex of the axis are polar capsules, each containing an extrudable polar filament and a multinucleate sporoplasm (El-Matbouli and Hoffmann, 1998).

20.3.2 Clinical signs including behavioural changes

A range of clinical signs and behavioural changes manifest in salmonid hosts when infections are severe. Clinical signs can include skeletal deformities, blacktail, stunted growth and mortality (Fig. 20.2). Skeletal deformities include shortened opercula, skull anomalies (reduced nose, reduced or misaligned jaws, indented skull) and crooked spine resulting from disrupted osteogenesis following cartilage damage and inflammation (MacConnell and Vincent, 2002). Blacktail, literally a blackened caudal region, is due to inflammation-related pressure on root ganglia that control skin melanocytes in response to parasite development in posterior spinal cartilage (Halliday, 1976; El-Matbouli *et al.*, 1995). Fish growth may be retarded during parasite proliferation but typically resumes post-myxospore formation except in severely infected fish in which too much cartilage has been consumed (Hedrick *et al.*, 2001b; MacConnell and Vincent, 2002).

Behavioural changes include tail-chasing or circular swimming ('whirling') followed by episodes of anterior body contractions. The behaviours are attributed to the host's inflammatory response to parasite growth and development compressing regions of the brainstem and spinal cord (Rose *et al.*, 2000). Mortality can result directly from the inflammatory response to parasite-related physical damage, as well as indirectly from skeletal deformities that alter behaviour or impede or prevent foraging and swimming efficacy (Hedrick *et al.*, 1998; Steinbach *et al.*, 2009).

20.3.3 Available techniques to confirm diagnosis

A variety of methods and tools exist for detection of *M. cerebralis* and diagnosis of SWD. The suitability and efficacy of each depends upon the purpose (e.g. diagnosis of disease or detection) and the sample type (environmental: soil or water; tissue: fish or worm). Diagnostic methods for fish host tissues depend on the purpose; guidelines for inspection purposes exist for North America (MacConnell and Bartholomew, 2012), with presumptive morphological identification confirmed using histological or molecular methods. However, molecular assays are increasingly used to document parasite distribution and infection in both hosts. Methods for detecting parasite stages from water (neutrally buoyant TAMs) or substrate (negatively buoyant myxospores) samples are not as well developed or standardized as those developed for use with host tissues, and reliance on morphological characteristics of spore stages is highly prone to error due to the number of closely related and morphologically similar *Myxobolus* spp., thus molecular confirmation is necessary.

The majority of molecular assays in use are underpinned by variable regions of the 18S rDNA gene commonly used in phylogenetic studies, but assays have also been developed for the internal transcribed spacer region ITS-1 and the heat shock protein Hsp70 (Andree *et al.*, 1997, 1998, 1999, 2002; Cavender *et al.*, 2004; Kelley *et al.*, 2004). In addition, loop-mediated isothermal amplification

(LAMP) and *in situ* hybridization (ISH) assays are available (El-Matbouli *et al.*, 1995; Antonio *et al.*, 1998). The surging popularity of environmental DNA (eDNA) as a detection tool may also warrant consideration. In general, eDNA may be most applicable as an early warning approach if stream sediments or water are being collected (Richey *et al.*, 2018) but could also have non-lethal diagnostic sampling applications with further development (e.g. detection of parasite eDNA in non-lethally collected blood in contrast to samples collected in fish host abdominal cavity) (Berger and Aubin-Horth, 2018). However, it is important to consider the ecological context of the eDNA target: waterborne spores, DNA excreted by salmonid hosts or sediment-dwelling invertebrates.

For inspection and certification purposes, diagnosis of SWD requires identification of *M. cerebralis* myxospores (mature spores) in cartilage (MacConnell and Bartholomew, 2012). The identification of myxospores requires samples be collected on an appropriate temporal scale to ensure a sufficient number of degree-days for the parasite to reach the cartilage tissue, replicate and form spores. There are clear guidelines for diagnostics and inspections for fish movement (e.g. American Fisheries Society Fish Health Section for US samples, Canadian Food Inspection Agency for Canadian samples) that direct the use of the pepsin–trypsin dextrose digest centrifugation method (PTD) and microscopic (morphological) identification of myxospores. However, as stated earlier, myxospore morphological characteristics are difficult to differentiate from closely related *Myxobolus* spp. Over 700 *Myxobolus* spp. have been described, six of which exhibit a tropism for salmonid cranial tissue (Markiw, 1992; Lom and Dyková, 2006; Hogge *et al.*, 2008). Thus, molecular confirmation – polymerase chain reaction (PCR), quantitative PCR (qPCR), ISH – should be combined with morphological identification when documenting *M. cerebralis* from new locations or host species.

20.3.4 External/internal macroscopic and microscopic lesions

M. cerebralis causes internal microscopic myxospore-filled lesions in the fish host as it matures. It exhibits a tropism for cartilage, but lesions can also form in peripheral nerves and epineurium as the parasite migrates to the cartilage (Baldwin *et al.*, 2000). External cysts containing myxospores are uncommon putatively on opercula and fins and confirmed using molecular techniques (Taylor and Haber, 1974).

Myxospore formation requires approximately 1000–1500 degree-days (approximately 3 months at ~12°C) from the date of infection. During this period, the parasite's proliferative stages consume cartilage, resulting in necrotic lesions. The early developmental stages do not appear to elicit an immune response (about 3 weeks) as they replicate and migrate to cartilage tissue via the nervous system (MacConnell and Vincent, 2002). However, once in the cartilage, trophozoites digest chondrocytes which elicits an intense inflammatory response in highly susceptible hosts (El-Matbouli and Hoffmann, 1998; Hedrick and El-Matbouli, 2002; MacConnell and Vincent, 2002). As the infection progresses and parasite replication amplifies, surrounding tissues are invaded and granulomatous inflammation occurs. Large granulomatous lesions typically have necrotic centres full of myxospores. When infection is severe (e.g. highly susceptible rainbow trout), granulomatous inflammation can be so extensive that cartilage framework is destroyed, resulting in gross skeletal deformities; Fig. 20.2). In hosts that survive the infection, myxospores become encased in bone as ossification of the skeleton progresses and will not be released until after host death, following decomposition of skeletal tissue.

Lesions are most common in juvenile fish because development and severity of lesions decrease as the skeleton ossifies during fish growth (Hoffman, 1974). Tissue tropism is largely restricted to cartilage, but lesion locations vary among fish host species. For example, lesions are most abundant in cranial and brainstem cartilage in rainbow trout but can be found throughout the body in other cartilage tissue (MacConnell and Vincent, 2002). In contrast, lesions are rarely observed in brown trout (less susceptible) cranial cartilage but are common in gill-arch and fin-ray cartilage (Hedrick *et al.*, 1999a; Baldwin *et al.*, 2000). In general, lesions in less susceptible fish hosts tend to be smaller and contain fewer myxospores than lesions in more susceptible hosts (Baldwin *et al.*, 2000).

The immune response to *M. cerebralis* varies widely among salmonids. Some fish mount effective responses early, eliminating infections through recognition and destruction of *M. cerebralis* proliferative stages (e.g. brown trout). In contrast, highly susceptible hosts (e.g. rainbow trout) typically mount extensive ineffective inflammatory responses that contribute to disease pathology and host death (Baldwin *et al.*, 2000). Once formed, mature myxospores do not appear to elicit a cellular immune response (Halliday, 1974). Hosts appear to develop acquired immunity only following development of lesions (MacConnell and Vincent, 2002).

20.4 Climate Change and Whirling Disease

Climate projections forecast important hydrological changes for freshwater salmonid habitat, and these will have both direct and indirect effects on *M. cerebralis* and whirling disease dynamics (Fig. 20.3). In the intermountain west and western North America in particular, climate forecast models indicate winters will become warmer and precipitation will shift from snow to rain (Mote, 2003; Mote *et al.*, 2003). Warming has already been linked with earlier snowmelt runoff for many rivers in the Pacific North-West (Stewart *et al.*, 2005). Continued warming is expected to drive ongoing shifts towards earlier snowmelt in streams that snowpack reserves historically provided water inputs into the spring and summer seasons. This will result in greater hydrological extremes including higher overall flooding intensities in winter and lower, warmer base flows in summer and autumn. These thermal and hydrological changes will influence *M. cerebralis* and the aetiology of whirling disease. The remainder of this section focuses on direct and indirect effects of (i) higher-magnitude floods, (ii) earlier runoff, lower and warmer spring flows and (iii) extended periods of lower and warmer base flow on the distribution, phenology, transmission and persistence of *M. cerebralis* and whirling disease (Fig. 20.4).

20.4.1 Changes in distribution

M. cerebralis has already achieved a cosmopolitan distribution. The dramatic outbreaks of SWD in native and wild trout populations in the intermountain west region of North America provide clear evidence of the parasite's ability to invade and proliferate to detectable levels in new fish populations over short temporal periods when conditions are suitable. Consequently, data on *M. cerebralis* exist from a variety of systems spread across broad environmental gradients, which we draw upon to describe potential spread and interpret how the parasite and disease severity might respond to climate-related environmental changes. In general, geographic range changes in *M. cerebralis* distribution are expected to follow changes in salmonid host range, and these will be underpinned by changes in thermal

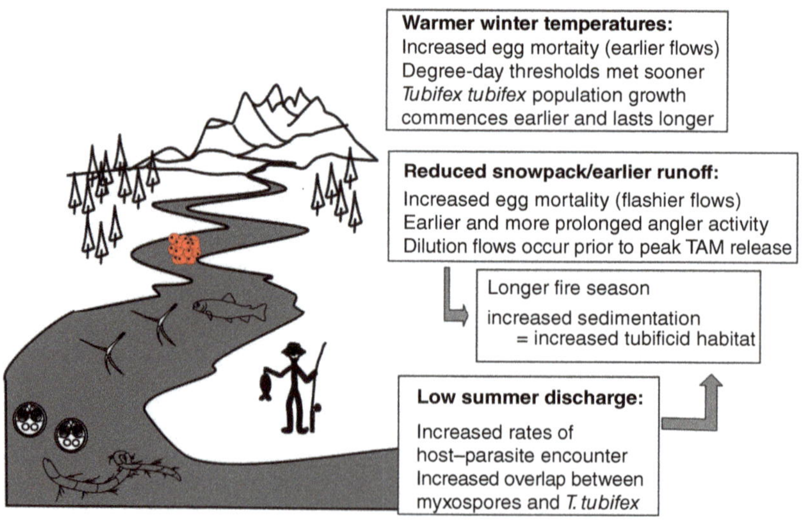

Fig. 20.3. Climate projections, including (i) warmer winter water temperatures, (ii) shifts in the timing and composition of hydrograph/spring runoff and (iii) low summer flows, will affect disease risk through effects on salmonid hosts (egg survival), rates of parasite encounter (reduced dilution of triactinomyxons (TAMs) or increased myxospore dispersal) and oligochaete hosts (increased reproductive period).

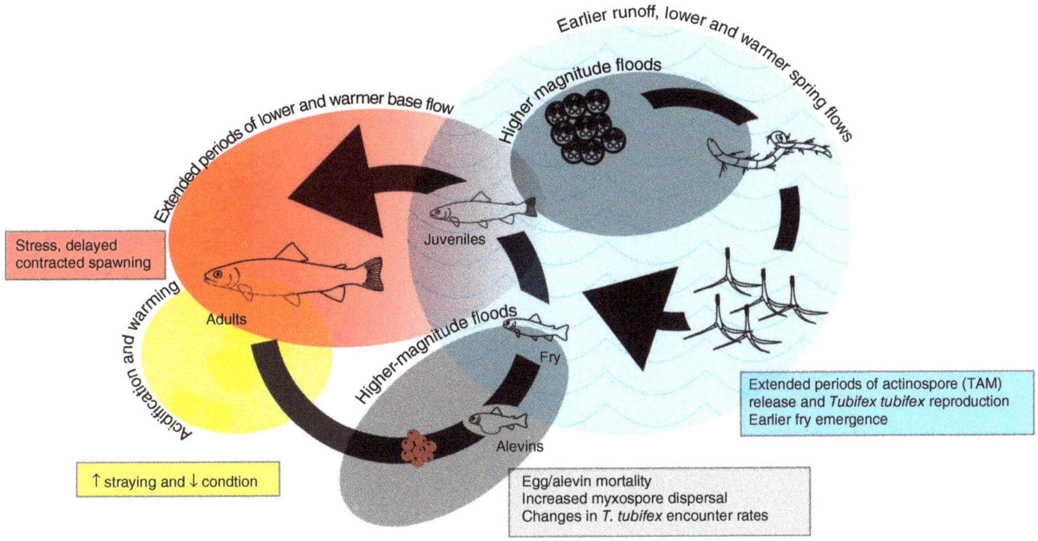

Fig. 20.4. Projected climate-related changes, including (i) higher-magnitude flooding in winter (grey), (ii) earlier runoff and warmer spring flows (blue), (iii) extended periods of lower and warmer base flows in summer and autumn (red) and (iv) oceanic changes (for anadromous salmonid life stages, yellow), will influence all phases of the *Myxobolus cerebralis* life cycle. TAM, triactinomyxon.

and hydrological regimes. Changes in salmonid host distribution are covered in this section.

Spread to new geographical areas

M. cerebralis is already present in many regions of the USA (Bartholomew and Reno, 2002) and its geographic range is expected to expand as fish are stocked into new drainages by humans, or as fish migrate to previously uninhabited or inaccessible regions seeking thermal refuge as climate change pressures intensify. The geographic range of *M. cerebralis* may also contract if climate changes cause current areas to become non-permissive. For example, *M. cerebralis* could be extirpated from southern or lower-elevation ranges if future thermal regimes exceed thermal tolerance ranges of the salmonid hosts or waterborne parasite stages. Although the net change in distribution is difficult to predict, *M. cerebralis* exhibits the capacity to expand further north or at higher elevations following altered thermal and hydrological regimes or fish distribution changes. For example, *M. cerebralis* was detected in hatchery rainbow trout in Alaska in 2007 (Arsan *et al.*, 2007). At that time, the closest enzootic area was the Snake River basin in northeastern Oregon (which is >4000 km south of Alaska). The Alaskan facility operated on heated water from a nearby power plant and the parasite was not detected following closure of the hatchery, which suggests the parasite was unable to establish in absence of the altered thermal regime. However, the recent detection of clinically infected trout in Alberta, Canada (Canadian Food Inspection Agency, 2016) and subsequent detections of *M. cerebralis* in at least four (North Saskatchewan, Red Deer, Oldman, and Bow – including Johnson Lake in Banff National Park) watersheds demonstrate range expansion, perhaps related to the ongoing warm–dry period in the region.

ANTHROPOGENIC DISPERSAL PATHWAYS. Potential routes for introduction and dissemination of *M. cerebralis* include anthropogenic and natural movement of infected fish, and anthropogenic movement of parasite spores or infected invertebrates (Fig. 20.5) (Bruneau, 2001; Bartholomew *et al.*, 2005; Arsan and Bartholomew, 2008, 2009).

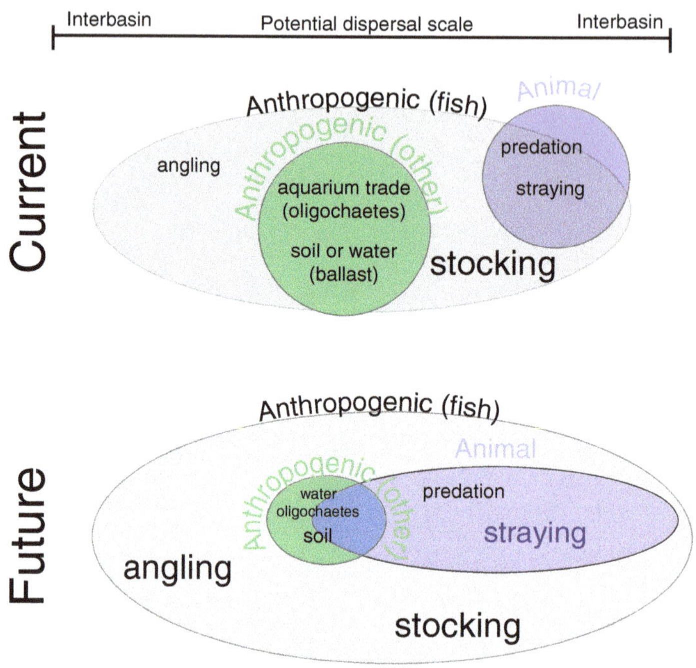

Fig. 20.5. Routes for the introduction and dissemination of *Myxobolus cerebralis*, including anthropogenic and natural movement of infected fish, and anthropogenic movement of parasite spores or infected invertebrates, may shift under future climate contexts. The relative importance of each route is illustrated by text and shape size under each climate context.

In general, human movement of infected fish hosts presents the greatest risk factor for introduction and parasite establishment over broad geographic scales. Introductions of live subclinically infected fish constitute such a high risk for parasite establishment because fish are typically moved in large numbers, providing a greater infective source, often multiple times, and into locations where other (at risk) salmonid species are typically present (Hoffman, 1990; Bartholomew and Reno, 2002). The movement of fish carcasses for stream enrichment (Wipfli *et al.*, 1998, 2010) also represents a potential dispersal pathway if carcasses are not treated (e.g. frozen) or screened in advance. Despite the availability and use of more sensitive diagnostic methods and changes in fish stocking practices, stocking of subclinically infected salmonids (legal and illegal) will almost certainly remain an important dispersal pathway in the future in response to sport-fishing demands in areas where native stocks have declined.

In addition, recreational angling represents an important pathway for parasite introduction. As regulations and the sensitivity and specificity of assays to detect infected fish increase, the risk of introductions from (legal) stocking should decrease, thus the relative risk of introduction from human recreational activities may increase. In particular, dispersal of myxospores with angling equipment (Gates *et al.*, 2008) has the potential to move the parasite into areas that may become more permissive to parasite establishment as temperatures warm and declines in snowpack open new areas to recreation, or open less accessible areas earlier and for longer periods (Fig. 20.5). Consequently, the potential for broad geographic dispersal exists if portable gear (e.g. wading boots, bait containers) is not disinfected and dried. The sale (and movement) of freshwater oligochaetes (mixed species) as a food source for aquarium fish represents a potential pathway for broad dissemination of myxozoan parasites (Hallett *et al.*, 2006) and may increase as the human population grows and expands at higher elevations following changes in the climate. Similarly, movements of water or sediments containing parasite spores may increase in importance under future climate conditions, as discussed above (e.g. recreational angling, human population expansion) (Steinbach *et al.*, 2009).

ANIMAL DISPERSAL PATHWAYS: POTENTIAL CHANGES IN PARASITE DISSEMINATION BY ANIMALS. Dissemination through fish movement (behaviour) represents a lesser risk for *M. cerebralis* introduction and dispersal than anthropogenic movement of infected fish hosts. However, changes in fish movement may become more important as salmonid life histories and behaviours change with future climate conditions. Some of the most dramatic fish population responses linked to climate change include changes in species' spatial distributions and phenology, or the timing of key behaviours, e.g. migration and spawning (Parmesan and Yohe, 2003; Crozier and Hutchings, 2014). In general, milder winters, earlier spring warming and warmer summers are associated with earlier fry emergence, migration and spawning (Visser and Both, 2005; Kovach *et al.*, 2013a,b, 2015). Changes in phenology including earlier fry emergence, delayed reproduction (longer periods of base flows will delay reproduction) and altered migration behaviour may result in shorter, more episodic periods of host–parasite overlap, potentially overlapping with high parasite densities (Pauls *et al.*, 2013; Kovach *et al.*, 2016; Lynch *et al.*, 2016). This in turn may manifest as increased stochasticity; avoidance of disease in some water years and near total (fry) losses in others.

The introduction of *M. cerebralis* into new areas may increase due to climate-related changes. As thermal refugia become smaller and less abundant, increased overlap among salmonids using these habitats may present opportunities for *M. cerebralis* emergence into new fish host populations. For example, rapid upstream migration of non-native rainbow trout into native westslope cutthroat trout habitat has occurred in response to increased stream temperatures in the Flathead River, Montana (Hitt *et al.*, 2003) and, in turn, *M. cerebralis* has extended further upstream (McGinnis and Kerans, 2013).

Anadromous species hatch in fresh water, migrate to the ocean to forage and grow, and return to fresh water to spawn, sometimes traveling >1500 km (Melnychuk *et al.*, 2010; Keefer *et al.*, 2019). Introductions may occur as thermally stressed infected fish die in cold-water refugia during periods of extreme low or high flows along their migration routes (anadromous species), or following movements into new areas in response to changes in fish barrier distribution (e.g. dam removal, freshwater-resident and anadromous species), or if uninfected fish seek refuge in *M. cerebralis*-positive cold-water refugia.

Straying behaviour of anadromous salmonids also represents a potential introduction risk that may change in relative importance, as straying increases in response to changes in migration timing following altered temperature or flow cues or ocean acidification (Crozier *et al.*, 2019). Although the probability of *M. cerebralis* introduction from straying or refugia-seeking salmonids is likely low between watersheds, the introduction and establishment of *M. cerebralis* in tributaries of the lower Columbia River basin (USA) was attributed to straying adult salmon and steelhead trout within the basin (Engelking, 2002; Zielinski *et al.*, 2010).

The dispersal of *M. cerebralis*-infected fish tissue by other animals represents another relatively low risk that could increase following environmental change. For example, the ingestion of *M. cerebralis*-infected fish tissue by avian piscivores may be of relevance along migratory pathways or as animals increasingly utilize higher-elevation habitats (Arsan and Bartholomew, 2009; Koel *et al.*, 2010).

ANIMAL DISPERSAL: RISK OF DISSEMINATION VIA THE INVERTEBRATE HOST. Relative to the movement of infected fish or infected fish tissues, dissemination of *M. cerebralis* through the movement of the invertebrate intermediate host represents a low introduction and dispersal risk. With the exception of oligochaete movements driven by aquarium trade sales (Lowers and Bartholomew, 2003; Hallett *et al.*, 2005, 2006), the invertebrate host dispersal is limited. However, future invertebrate host responses to changing environmental conditions could also allow *M. cerebralis* (and other myxozoans) to expand spatially (e.g. further upstream or to previously marginal locations) and temporally (e.g. longer periods of host persistence and in turn parasite release) (see Section 20.4.2). For example, increased range and/or range shifts to more northern latitudes or higher elevations for the myxozoan, *Tetracapsuloides bryosalmonae*, in response to the combination of warmer temperatures and changes in invertebrate host distribution and densities are expected (Okamura and Feist, 2011; Okamura *et al.*, 2011).

20.4.2 Phenology

Increased water temperatures may drive changes in fish and invertebrate host phenology resulting in changes in disease risk through altered temporal overlap. Changes in timing of emergence and migration in juvenile and adult salmonids have been linked to altered thermal regimes (Crozier *et al.*, 2008; Kovach *et al.*, 2013a,b, 2015; Crozier and Hutchings, 2014). The timing of TAM release drives disease severity in juvenile fish stages because fry life stages (up to 9 weeks post-hatch) are most susceptible to infection (Ryce *et al.*, 2004, 2005). Larger (older) salmonids can still become infected and produce myxospores but have much higher dose thresholds than fry. Thus, if temperature changes extend the seasonality of TAM release from the invertebrate host, earlier emergence of fry could result in increased risk of infection.

If earlier fry emergence and emigration reduce the overall risk of TAM encounter, anadromous salmonids may be at lower risk of whirling disease than freshwater-resident salmonids. In contrast to potentially driving emergence into previously *M. cerebralis*-negative fish populations, water temperature-related changes in the timing of peak TAM release from the invertebrate host could also buffer those populations. For example, earlier emigration has also been observed (sockeye salmon: Crozier *et al.*, 2011; pink salmon (*Oncorhynchus keta*): Kovach *et al.*, 2012) in response to increased temperatures. Limited overlap between host and parasite stages may partially explain differences in disease impacts among wild salmonids populations. Although *M. cerebralis* has been introduced and even become established in river drainages in eastern and western North America and the former eastern USSR, salmonid population-level losses have not been documented, suggesting these populations may not have suffered declines similar to those observed in populations in the US intermountain west. However, disease risk may be buffered or intensified as a function of salmonid responses to shifting thermal regimes (e.g. earlier outmigration, increased straying or use of *M. cerebralis*-positive cold-water refugia along their migration routes).

Changes in temperature will also alter invertebrate host phenology and affect disease risk through changes in host–parasite overlap. In general, increased temperature is associated with the earlier occurrence of emergence and reproduction in aquatic species (Parmesan and Yohe, 2003). Tubificids mature earlier and more rapidly, are fecund longer and produce more progeny at higher temperatures (Bonacina *et al.*, 1989a,b). Thus, under warmer conditions *T. tubifex* hosts may be more abundant and available to encounter and ingest myxospores. Consequently, warming temperatures may extend the temporal window of TAM release from the invertebrate host. *M. cerebralis*-infected *T. tubifex* can release TAMs at temperatures between 5 and 20°C (El-Matbouli and Hoffmann, 1989; El-Matbouli *et al.*, 1995, 1999; Kerans *et al.*, 2005) and peak release has been described in the spring warming and autumn cooling (Gilbert and Granath, 2003), during winter in high-altitude basins (Thompson and Nehring, 2000) and during late autumn–winter in Montana spring creeks (Neudecker *et al.*, 2012). These suggest that TAM release is driven by the accumulation of degree-days, beginning with myxospore infection of the oligochaete host, rather than a thermal threshold.

Water temperature-driven changes in parasite propagation in hosts: using degree-days as a conceptual model for parasite production under different thermal regimes

Changing water temperatures will have both positive and negative effects on all phases of the *M. cerebralis* development in both hosts. For example, increased water temperatures are positively correlated with parasite proliferation in both hosts (El-Matbouli *et al.*, 1999; Baldwin *et al.*, 2000; Kerans *et al.*, 2005) and reduced viability of both spore stages (El-Matbouli and Hoffmann, 1991a; El-Matbouli *et al.*, 1999; Hedrick *et al.*, 2008; Kallert and El-Matbouli, 2008). One approach that may be useful for capturing the net effects of increased water temperatures is conceptualizing the accumulation of thermal units or degree-days. We present a conceptual model for *M. cerebralis* here to illustrate the potential effects of altered thermal regimes on the life cycle.

In the salmon host, myxospore development was reported as 52 days at 17°C (884 degree-days) and 90 days at 12°C (~1180 degree-days) in rainbow trout (Halliday, 1973; Murcia *et al.*, 2006); approximately 1000 degree-days. In the tubificid host, release of actinospores occurred 168 days (1344 degree-days) and 106–108 days (1590–1620 degree-days) post-exposure at 8 and 15°C, respectively, in laboratory challenges (Kerans *et al.*, 2005); approximately 1500 degree-days. Thus, one complete generation of *M. cerebralis* requires a minimum of about 2500 degree-days, if transmission to the next host was instantaneous (Fig. 20.1). Consequently, an increase of 2–5°C could narrow the development time window by 1 month (+2°C) to 2 months (+5°C), resulting in the potential for another life cycle completion and shortening the development window by 12–25 days within the fish host and 18–37 days in the oligochaete host.

Increases in winter water temperatures may have equal, if not more profound, impacts on the *M. cerebralis* life cycle than increases in summer temperatures. Climate projections indicate winter water temperature changes will be extreme, with increases of up to 5°C expected in the Pacific North-West in the near future (Mote *et al.*, 2003), which is important because myxospore survival in stream sediments may be increased at elevated winter temperatures (freezing reduces spore viability) (Hedrick *et al.*, 2008; and see Sections 20.4.3 and 20.4.4 below). In addition, although simplified above, instantaneous transmission particularly at the myxospore stage is unlikely, if not impossible, because breakdown of cartilage tissue and release of myxospores likely occur gradually in winter months and the rate of decomposition, and thus release, may be strongly influenced by warming thermal regimes. The development timeline and magnitude of release for TAM stages may also respond to changes in thermal regimes depending upon the invertebrate hosts' thermal optima, because *T. tubifex* strains vary in their ability to amplify *M. cerebralis* and exhibit variability in TAM production amplitude at different water temperatures (Kerans *et al.*, 2005). Although the mechanism is unclear, these differences may reflect differences among hosts' thermal optima and have important disease risk implications for cold-adapted strains. For example, although the total degree-days to onset of TAM release was similar (1320–1456 degree-days) between the strains when held at 8 and 15°C, Mount Whitney *T. tubifex* released comparable numbers of TAMs, but Madison River *T. tubifex* released >5 times more TAMs at the warmer temperature than when held at the cooler temperature (calibrated for equivalent degree-days).

20.4.3 Effects of precipitation and water quality on transmission

The effects of climate-driven changes in precipitation (considered as stream discharge) on phases of the myxozoan life cycle may be just as important as those of temperature. The role of precipitation is less clear in the context of future climate scenarios but will almost certainly play an important role in parasite transmission. The predicted shifts in precipitation from snowpack runoff to rain will affect freshwater habitat stability through warming thermal regimes as well as significant changes in the timing and magnitude of discharge. Decreased-magnitude discharges may increase habitat (e.g. fine sediment) available for invertebrate hosts (Marcogliese, 2001, 2008). The concomitant lower water levels may also cause vertebrate hosts to aggregate in greater densities (see Section 20.4.1 above for discussion of thermal refugia use and disease risk). An increased overlap between high densities of hosts (vertebrate and invertebrate) and parasites can lead to greater infection prevalence and disease severity (DuBey and Caldwell, 2004; Nehring *et al.*, 2013; Searle *et al.*, 2013). Water velocity is correlated with discharge, and several studies have examined its effects on interactions between myxozoans and their hosts. When transmission and infection dynamics of *M. cerebralis* were examined at low and high velocity in a laboratory experiment, both prevalence of infection in *T. tubifex* (invertebrate host) and actinospore densities in water were higher in the low-velocity treatment (Hallett and Bartholomew, 2008). Prevalence and severity of infection in the fish hosts were also higher in the low-velocity treatment. This inverse relationship was also described for transmission of *Ceratomyxa shasta* to its fish host (Ray and Bartholomew, 2013) and to its invertebrate host, *Manayunkia speciosa* (Bjork and Bartholomew, 2009). Although the mechanism(s) have not been identified, spore attachment to fish hosts (e.g. Ray and Bartholomew, 2013) and invertebrate host density and ingestion of infectious spores are likely reduced at higher velocities, thus transmission efficiency may be reduced if peak periods of TAM or myxospore availability overlap with high-magnitude discharge events.

Climate-related changes in water quality such as increased eutrophication also have the potential to alter disease dynamics. Nutrient loading in particular may increase disease risk by increasing invertebrate host abundance because the amount of organic matter and sediments in streams is positively correlated with high densities of *T. tubifex* (Allen and Bergersen, 2002; Kaeser *et al.*, 2006; McGinnis and Kerans, 2013). Higher nutrient load (measured as conductivity) has also been correlated with higher prevalence and intensity of infection in salmonid hosts in the Lostine River (Sandell *et al.*, 2001). The mechanism was not described but could be related to higher abundances of the invertebrate hosts or enhanced TAM transmission. Eutrophication has also been linked with outbreaks of salmonid proliferative kidney disease (PKD) caused by the myxozoan parasite *T. bryosalmonae*, likely through increased invertebrate host (filter-feeding organisms) densities (Okamura *et al.*, 2011). Following diversion of effluent from sewage treatment, the prevalence of *T. bryosalmonae* (cause of PKD) infection was reduced in hatchery and wild fish sampled downstream (El-Matbouli and Hoffmann, 2002). Thus, increased nutrient input or reduced water quality may exacerbate disease risk indirectly as a function of increased population growth or carrying capacity of invertebrate hosts.

20.4.4 Persistence

Predicting the net effects of climate change on parasite persistence is complicated by the parasite's multiple 'habitats', which include the aquatic environment and host tissue(s). In general, parasite spore persistence in the environment is likely to be negatively impacted by climate changes but these effects may be buffered by increases in the overall magnitude or timing of production in hosts. However, because the myxospore and actinospore stages exhibit differences in thermal and desiccation tolerances in addition to dispersal strategies, the effects of specific changes on parasite persistence will differ for each stage.

Predicting how host persistence will change in the salmonid host is complicated because climate change effects can be dual and opposing. Salmonid host persistence is likely to be bolstered in populations that have retained genetic variation and less so in populations subjected to major population losses attributed to *M. cerebralis* in addition to other stressors. Where cold water temperatures currently limit habitat suitability and distributions of some suitable fish host species (e.g. higher latitudes and elevations) (Nakano *et al.*, 1996; Coleman and Fausch, 2007), a warming climate may gradually increase the quality and extent of suitable habitat, facilitating fish host movement into these areas. Along with these distributional changes, *M. cerebralis* may establish in host populations where it is currently unable to complete its life cycle.

In contrast to their less-tolerant salmonid host counterparts, *T. tubifex* strains are highly tolerant of degraded habitat conditions and may benefit from climate-related changes in the form of competitive release (changes in invertebrate community) or improved habitat (organic enrichment). Because of this host's broad environmental tolerances and plastic reproductive strategies (see below), *M. cerebralis* persistence may increase with climate change. Moreover, where environmental conditions currently limit distribution (e.g. as above, higher latitudes and elevations), climate-related changes in human movements or pressure (e.g. private aquaculture or recreational activities), suitable tubificid hosts may be disseminated more broadly, which in turn would increase the probability of *M. cerebralis* persistence.

In salmonid hosts, *M. cerebralis* exhibits a thermal tolerance of approximately 6–15°C (MacConnell and Vincent, 2002). Salmonids tolerate water temperatures well above this range so although this may indicate that parasite proliferation in the vertebrate could be negatively impacted by warming stream conditions, reduced parasite proliferation above this temperature range has not been demonstrated experimentally.

M. cerebralis appears to exhibit broader temperature tolerance in the invertebrate host. The development and release of *M. cerebralis* TAMs by *T. tubifex* increases with temperature up to ~20°C (El-Matbouli *et al.*, 1999; Blazer *et al.*, 2003; Kerans *et al.*, 2005), suggesting *M. cerebralis* may have an upper tolerance of ~20°C in its invertebrate host. Although the biological reason for the differences in tolerance is not clear, the ability to persist across a broad temperature range may help buffer the parasite from the adverse effects of temperature. However, temperature regime changes that exceed *M. cerebralis*' upper thermal tolerance may result in exclusion of the parasite from some drainages. Water temperatures >20°C halted TAM release by infected *T. tubifex* (El-Matbouli *et al.*, 1999).

Warmer water temperatures during peak periods of parasite transmission are expected to increase parasite spore mortality but increase transmission. The longevity of myxozoan stages in the environment is inversely correlated with water temperature (Yokoyama *et al.*, 1997; El-Matbouli *et al.*, 1999; Kallert and El-Matbouli, 2008). Actinospores have a shorter persistence period than myxospores, which are comparatively robust due to the hardened valves that surround the sporoplasm (Hedrick *et al.*, 2008). However, longevity of both spore stages decreases with increasing temperature: *M. cerebralis* actinospores remain viable for ~15 days at 15°C but only for 1 day at 23°C; myxospores appear to remain viable for >60 days at <10°C and for 7 days at 22°C (El-Matbouli and Hoffmann, 1991a; Hedrick *et al.*, 2008).

Changes in host mortality due to increased fluctuations in dissolved gases

Decreases in dissolved oxygen levels as a result of climate change are likely to have limiting effects on the salmonid host, as biological oxygen demand increases with temperature. Low dissolved oxygen is commonly cited as a factor limiting production in fish culture systems and has been linked to decreased growth and immunosuppression in fish (Kindschi and Koby, 1994; Niklasson *et al.*, 2011; Abdel-Tawwab *et al.*, 2019). Thus, fish exposed to hypoxic conditions may be more vulnerable to disease. To date, there has been little research on the effects of dissolved gases on *M. cerebralis* infection with the exception of a study that examined oxygen supersaturation as a stressor, finding it had a fairly weak effect on fish host mortality (Schisler *et al.*, 2000). In contrast, the oligochaete host exhibits a broad tolerance for 'stressful' environmental conditions (Brinkhurst, 1980, 1996) and thus may gain a competitive advantage and/or additional habitat, and in turn contribute to increased disease risk through higher abundance.

Increase in pathogen mortality due to environmental change

The environmental spore stages of *M. cerebralis* are vulnerable to fluctuations in temperature and river discharge (changes in flow). In addition, there will be indirect effects on spores' viability and their ability to transmit to their respective hosts.

Myxospores can overwinter within fish carcasses although freezing may reduce viability (El-Matbouli and Hoffmann, 1991a; Hedrick *et al.*, 2008). Cryptic stages of malacosporeans can overwinter in bryozoan statoblasts (Abd-Elfattah *et al.*, 2014) and myxosporeans likely similarly persist within their invertebrate hosts over winter (or other periods of suboptimal temperatures). *T. tubifex* can encyst to avoid desiccation or starvation (Anlauf, 1994), but whether *M. cerebralis* can remain viable within the encysted worm is unknown.

As the causative agent of infection for fish hosts, the actinospore stage directly drives infection risk. In addition, it is the more vulnerable to disruption/deactivation of the two spore stages. Viability staining indicates actinospores are rendered non-viable through: (i) freezing (<1 h); (ii) drying (1 h); (iii) temperatures >75°C (5 min minimum); or (iv) treatment with chlorine (130 ppm) or hydrogen peroxide (>10% v/v solution) (Wagner, 2002; Wagner *et al.*, 2003).

There have been no direct studies on how changes in water chemistry, dissolved gases or nutrients affect the survival of environmental spore stages of *M. cerebralis*. Many of these variables are expected to have a greater effect on the hosts than the parasite. One study noted that prevalence of infection in sentinel fish increased in streams with high specific conductivity (conductivity is linearly related to osmolarity and levels are primarily driven by geology and nutrient inputs, e.g. wastewater runoff) (Sandell *et al.*, 2001), and it was suggested that conductivity may increase transmission efficiency by increasing TAM recognition of host tissues. Although it is likely that warming water temperatures will result in increased conductivity and decreased oxygen levels, it is unclear how these changes will affect *M. cerebralis* transmission.

20.5 Control, Prevention and Mitigation

Prevention measures are paramount for natural systems because once *M. cerebralis* is established, current options for eradication are impractical or ineffective. Although a variety of viable control strategies have been tested experimentally, none have been implemented successfully on a large scale (Wagner, 2002), including stocking resistant strains of *T. tubifex* (Thompson, 2011). Consequently, the best control strategies for natural systems include preventing introduction and spread of *M. cerebralis* by employing appropriate techniques to detect and diagnose the pathogen in subclinically infected animals prior to stocking/hatchery release and investing in public education to reduce inadvertent dissemination through recreational activities. Unfortunately, once *M. cerebralis* is established in natural systems, few options exist for eradication, and the focus instead shifts to reducing disease risk and severity or mitigating the effects of disease.

Although mitigation tools are not well developed for *M. cerebralis*, approaches that have been useful for control of other aquatic pathogens may be suitable. For example, habitat (flow) manipulation has been effective for reducing the salmonid risk of enteronecrosis through reduced invertebrate host distribution and dilution of spores (Alexander *et al.*, 2016; Bartholomew *et al.*, 2017). In addition, application of molluscicides has been effective for reducing schistosomiasis by targeting of invertebrate hosts (Patz *et al.*, 2003; Sokolow *et al.*, 2018). Management by stocking refractory hosts (whether species or life stages) has been used with mixed success. For example, in Colorado, SWD risk was not reduced following stocking of resistant tubificids (Thompson, 2011) but stocking resistant fish crosses (and likely, age classes) appears more promising. Thus, risk assessments, particularly if used in tandem with climate prediction or species distribution models, likely represent the best tools for prevention through identification of locations at high risk of parasite introduction, establishment and disease.

Fish hatcheries and aquaculture facilities offer comparatively abundant opportunities for prevention and control. Measures that disrupt the parasite's life cycle by eliminating parasite spores or the invertebrate host are straightforward and effective. Effective measures include: (i) elimination of invertebrate host habitat (paving raceways, regular removal of organic material) or disinfection of rearing ponds (calcium cyanamide, calcium cyanide or chlorine) to kill the invertebrate host and parasite spore stages; and (ii) use of pathogen-free water (i.e. converting from surface to groundwater supply, or treatment of incoming water with ultraviolet (UV) light, ozonation, chlorination or filtration) (see Wagner, 2002; Steinbach *et al.*, 2009; Hedrick *et al.*, 2012).

20.5.1 Use of prophylaxes or chemicals

No drug or therapeutant treatment exists for *M. cerebralis* in fish or invertebrate hosts. Prophylactic and treatment drugs (e.g. furazolidone medicated pellets; Taylor *et al.*, 1973) have been tested in fish but none prevented or eliminated infection (Wagner, 2002). Previous research suggested the antibiotic fumagillin, which is effective for slowing *C. shasta*-induced mortality (Ibarra *et al.*, 1990), was effective for prevention of whirling disease (El-Matbouli and Hoffmann, 1991b). However, subsequent studies refuted its efficacy (Staton *et al.*, 2002) and drug development has not since been a high priority given the logistical challenges associated with their application to wild fish (Wagner, 2002; Steinbach *et al.*, 2009). Although vaccines are not currently available (the parasite cannot be cultured outside hosts and protective antigens have not been identified), UV radiation appears to function as a putative vaccine, at least in rainbow trout, by reducing the severity of whirling disease (Hedrick *et al.*, 2012). The mechanism(s) are not thoroughly understood but may include compromised parasite development post-invasion or stimulation of the host immune system (Sarker *et al.*, 2015), and certainly warrants additional research. In addition, recent work with transcriptional-based approaches using qPCR and expression-based studies using proteomics in other aquatic parasite systems suggest there may be multiple mechanisms to target throughout the infection process. For example, gene expression profiling has been applied to the ciliate parasite *Ichthyophthirius multifiliis* to identify genes that are differentially regulated during the different life stages of the parasite (Abernathy *et al.*, 2011; Cassidy-Hanley *et al.*, 2011). In another approach, transcripts obtained from qPCR-based methods have been used in tandem with protein-based studies to follow *T. bryosalmonae* protein and gene expression during host–parasite interactions (Kumar *et al.*, 2015a,b,c) and vaccine development is underway as a result. In addition, emerging small nucleotide-based approaches including small interfering RNA (siRNA) and clustered regularly interspaced short palindromic repeats (CRISPR)/ CRISPR-associated protein (Cas) may provide an alternative for some aquatic parasites, including *M. cerebralis* (Sarker and El-Matbouli, 2015). For example, incubating *T. tubifex* in a solution containing double-stranded RNA targeting the serine protease of *M. cerebralis* prevented infection in the subsequent (rainbow trout) host (Sarker *et al.*, 2017).

20.5.2 Stocking of resistant fish

One of the least costly but most effective management strategies is simply to stock larger fish (Steinbach *et al.*, 2009). Although larger fish can still become infected with *M. cerebralis*, they are less susceptible and produce fewer myxospores (Steinbach *et al.*, 2009). Another approach is to stock species or strains of salmonids that are resistant to whirling disease, or whose life histories reduce or limit the overlap of susceptible life stages (fry) with seasonal peaks of waterborne actinospores (Neudecker *et al.*, 2012). However, this approach may be impractical because of future shifts in thermal or hydrological regimes.

20.5.3 Selective breeding programmes

The development of resistance in a previously susceptible strain of rainbow trout (Höfer strain; Hedrick *et al.*, 2003) has offered opportunities both for research (Baerwald *et al.*, 2008, 2011) and development of selective breeding programmes for stocking. Two strategies employed in the US intermountain west involve crossing: (i) wild fish populations with high genetic diversity (Miller and Vincent, 2008; Steinbach *et al.*, 2009); and (ii) vulnerable native populations with *M. cerebralis*-resistant fish stocks, such as the domesticated German Höfer strain (Schisler *et al.*, 2006) or the Harrison Lake and deSmet strains – Montana strains that exhibit resistance to *M. cerebralis* (Vincent, 2002; Wagner *et al.*, 2006). The aims are to produce progeny with resistance to whirling disease while retaining genetic traits important for survival in the wild, but the approach differs. For example, several strains including the European hatchery-derived Höfer and Harrison Lake (Montana) strains, and crosses, are maintained as broodstock in the Colorado state hatchery system and are produced and stocked for recreational purposes statewide (Schisler *et al.*, 2006). Although these strains can be infected with *M. cerebralis*, they survive and reproduce and parasite loads are usually low (Hedrick *et al.*, 2003; Schisler *et al.*, 2006; Fetherman *et al.*, 2012), meaning they could possibly function as parasite sinks, or at least not amplify the parasite/increase disease risk for coexisting salmonids. The incorporation of locally adapted strains could be expanded to include more temperature-tolerant strains as a bet-hedging strategy for warming climate scenarios.

20.6 Conclusions with Suggestions for Future Studies

Whirling disease presents a serious threat to the future of wild salmonid populations and we have significant gaps in our understanding of how climate change will alter interactions between the hosts and the parasite under future conditions. The magnitude and directions of the net effects will require more environmental data on transmission and persistence and experimental data to fuel risk assessments. Watersheds proximal to the margins of the parasite's current distribution range will be at the greatest risk in the short term – and may present the best opportunities for understanding linkages. Development of quantitative methods for measuring parasites and hosts (as appropriate, see Section 20.3.3) using eDNA could result in tools that would improve predictions of which watersheds are most vulnerable to the emergence of whirling disease.

20.6.1 What we can apply from management of other aquatic pathogens?

One option for managing parasites with complex life cycles is to target an intermediate host, rather than the parasite. Decreasing the density of one obligate host can decrease or disrupt parasite replication, reducing disease risk for subsequent hosts involved in the life cycle. Methods for decreasing host abundance include chemical application (e.g. molluscicides for schistosomiasis) (McCullough *et al.*, 1980; Sokolow *et al.*, 2018), physical separation of hosts and vectors (e.g. mosquito nets for malaria) (McCullough *et al.*, 1980; Alonso *et al.*, 1993) and habitat manipulation (e.g. replacement of earthen ponds with concrete raceways in hatcheries for whirling disease in fish) (Hoffman, 1990).

Manipulating host habitat in natural systems is more complicated than in aquaculture facilities but may be applied with success when the ecologies of the hosts and parasite are reasonably well understood. Manipulating river flow dynamics as an approach to manage risk of infection by the myxozoan parasite *C. shasta* in salmonids has recently been applied in the Klamath River, California (True *et al.*, 2013), where population declines have been attributed to the parasite (Fujiwara *et al.*, 2011). Phases of the *C. shasta* life cycle susceptible to alteration or interruption via flow management include transmission to the fish or invertebrate hosts. To date, approaches have included dilution flows in spring aimed at reducing water temperature and diluting actinospores to reduce disease risk and severity in the fish hosts, and surface flushing flows in late winter/early spring to reduce densities of infected invertebrate hosts by reducing either (i) myxospore transmission to invertebrate hosts or (ii) invertebrate host populations through disturbance. Management of other salmonid parasites with complex life cycles also targets the invertebrate hosts (Carraro *et al.*, 2017, 2018).

20.6.2 Data gaps and outstanding questions

- Will changes in precipitation be more influential than temperature changes at host or parasite range limits?
- How will changes in our approaches to fisheries management (e.g. as management aims shift from stocking for sport fishing to native restoration) affect disease monitoring and management?
- Will the effects of anthropogenic factors (e.g. dams, pollution, population growth) override the predicted effects of climate change on disease?
- How will snowpack changes influence water availability and how will water quantity affect salmonid disease risk?
- Will climate-related variability in temperatures or flow consistency cause changes in parasite virulence?

20.6.3 Filling the knowledge gaps

Future studies that would begin to fill knowledge gaps should:

- Aim to assess the interactions among climate drivers, ecosystem changes, water quality and whirling disease, with a focus on individual-, population- and community-level impacts on salmonid hosts (Fig. 20.6).
- Develop conceptual models using empirical data to predict the effects of climate-related parameters on host and parasite life cycle stages.
- Determine the hydrological (discharge, flow residence time, mixing) thresholds for predicting epidemics/outbreaks.
- Determine how water-use management and infrastructure (e.g. diversion of water for irrigation and flood prevention/mitigation) affect whirling disease risk to facilitate decision making and mitigate against future problems.
- Examine how changing ocean conditions will affect returning anadromous salmon and subsequent risk of disease.
- Examine relationships between climate change and host population dynamics, evolution and interspecific interactions, particularly in freshwater systems in North America. Much of climate change-related research has focused on species phenology and distributions, thus important knowledge gaps exist how such changes will manifest over longer time scales.
- Validate model predictions with empirical data, which requires funding monitoring studies.
- Increase the breadth of taxonomic focus; rather than simply focusing on host responses, what

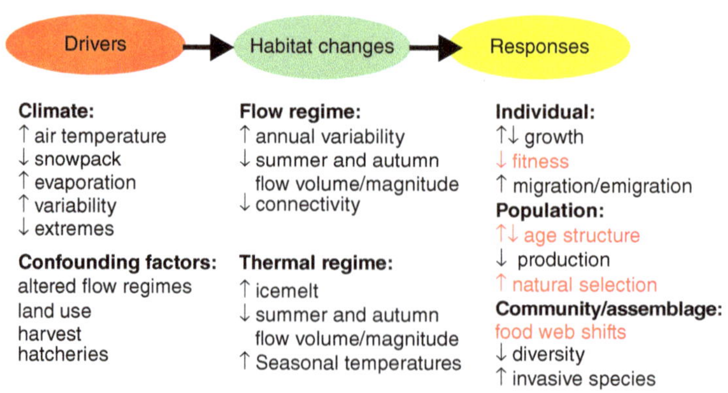

Fig. 20.6. Drivers, resultant changes impacting salmonid and oligochaete host habitats, and potential responses. Red illustrates potential host responses to climate change that are expected to be exacerbated by *Myxobolus cerebralis* and should be prioritized for future studies.

can we learn from examining the population dynamics and drivers of parasites or pathogens in the context of climate change?

- Study populations that coexist with parasites or have rebounded to better understand factors contributing to resilience to test responses of salmonids to climate change (Fig 20.6).

References

Abd-Elfattah, A., Fontes, I., Kumar, G., Soliman, H., Hartikainen, H. *et al.* (2014) Vertical transmission of *Tetracapsuloides bryosalmonae* (Myxozoa), the causative agent of salmonid proliferative kidney disease. *Parasitology* 141, 482–490.

Abdel-Tawwab, M., Monier, M.N., Hoseinifar, S.H. and Faggio, C. (2019) Fish response to hypoxia stress: growth, physiological, and immunological biomarkers. *Fish Physiology and Biochemistry* 45, 997–1013.

Abernathy, J., Xu, D.H., Peatman, E., Kucuktas, H., Klesius, P. and Liu, Z.J. (2011) Gene expression profiling of a fish parasite *Ichthyophthirius multifiliis*: insights into development and senescence-associated avirulence. *Comparative Biochemistry and Physiology Part D: Genomics and Proteomics* 6, 382–392.

Alexander, J.D., Bartholomew, J.L., Wright, K.A., Som, N.A. and Hetrick, N.J. (2016) Integrating models to predict distribution of the invertebrate host of myxosporean parasites. *Freshwater Science* 35, 1263–1275.

Allen, M.B. and Bergersen, E. (2002) Factors influencing the distribution of *Myxobolus cerebralis*, the causative agent of whirling disease in the Cache la Poudre River, Colorado. *Diseases of Aquatic Organisms* 49, 51–60.

Alonso, P.L., Lindsay, S.W., Schellenberg, J.R.M.A., Konteh, M., Keita, K. *et al.* (1993) A malaria control trial using insecticide-treated bed nets and targeted chemoprophylaxis in a rural area of The Gambia, West Africa. 5. Design and implementation of the trial. *Transactions of the Royal Society of Tropical Medicine and Hygiene* 87, 31–36.

Andree, K.B., Gresoviac, S.J. and Hedrick, R.P. (1997) Small subunit ribosomal RNA sequences unite alternate actinosporean and myxosporean states of *Myxobolus cerebralis* the causative agent of whirling disease in salmonid fish. *Journal of Eukaryotic Microbiology* 44, 208–215.

Andree, K.B., MacConnell, E. and Hedrick, R.P. (1998) A nested polymerase chain reaction for the detection of genomic DNA of *Myxobolus cerebalis* in rainbow trout *Oncorhynchus mykiss*. *Diseases of Aquatic Organisms* 34, 145–154.

Andree, K.B., El-Matbouli, M., Hoffman, R.W. and Hedrick, R.P. (1999) Comparison of 18S and ITS-1 rDNA sequences of selected geographic isolates of *Myxobolus cerebralis*. *International Journal for Parasitology* 29, 771–775.

Andree, K.B., Hedrick, R.P. and MacConnell, E. (2002) A review of the approaches to detect *Myxobolus cerebralis*, the cause of salmonid whirling disease. In: Bartholomew, J. and Wilson, C. (eds) *Whirling Disease: Reviews and Current Topics. American Fisheries Society Symposium 29*. American Fisheries Society, Bethesda, Maryland, pp. 197–212.

Anlauf, A. (1994) Some characteristics of genetic variants of *Tubifex tubifex* (Müller, 1774) (Oligochaeta: Tubificidae) in laboratory cultures. *Hydrobiologia* 278, 1–6.

Antonio, D.B., Andree, K.B., McDowell, T.S. and Hedrick, R.P. (1998) Detection of *Myxobolus cerebralis* in rainbow trout and oligochaete tissues by using a non-radioactive *in situ* hybridization (ISH) protocol. *Journal of Aquatic Animal Health* 10, 338–347.

Arsan, E.L. and Bartholomew, J.L. (2008) Potential for dissemination of the non-native salmonid parasite *Myxobolus cerebralis* in Alaska. *Journal of Aquatic Animal Health* 20, 136–149.

Arsan, E.L. and Bartholomew, J.L. (2009) Potential dispersal of the non-native parasite *Myxobolus cerebralis* in the Willamette River Basin, Oregon: a qualitative analysis of risk. *Reviews in Fisheries Science* 17, 360–372.

Arsan, E.L., Hallett, S.L. and Bartholomew, J.L. (2007) *Tubifex tubifex* from Alaska: distribution and susceptibility to *Myxobolus cerebralis*. *Journal of Parasitology* 93, 1332–1342.

Baerwald, M.R., Welsh, A.B., Hedrick, R.P. and May, B. (2008) Discovery of genes implicated in whirling disease infection and resistance in rainbow trout using genome-wide expression profiling. *BMC Genomics* 9, 37.

Baerwald, M.R., Petersen, J.P., Hedrick, R.P., Schisler, G.J. and May, B. (2011) A major effect quantitative trait locus for whirling disease resistance identified in rainbow trout (*Oncorhynchus mykiss*). *Heredity* 106, 920–926.

Baldwin, T.J., Vincent, E.R., Silflow, R.M. and Stanek, D. (2000) *Myxobolus cerebralis* infection in rainbow trout (*Oncorhynchus mykiss*) and brown trout (*Salmo trutta*) exposed under natural stream conditions. *Journal of Veterinary Diagnostic Investigation* 12, 312–321.

Bartholomew, J.L. and Reno, P.W. (2002) The history and dissemination of whirling disease. In: Bartholomew, J.L. and Wilson, J.C. (eds) *Whirling Disease: Reviews and Current Topics. American Fisheries Society Symposium* 29. American Fisheries Society, Bethesda, Maryland, pp. 3–24.

Bartholomew, J.L., Lorz, H.V., Sollid, S.A. and Stevens, D.G. (2003) Susceptibility of juvenile and yearling bull trout to *Myxobolus cerebralis*, and effects of sustained parasite challenges. *Journal of Aquatic Animal Health* 15, 248–255.

Bartholomew, J.L., Kerans, B.L., Hedrick, R.P., MacDiarmid, S.C. and Winton, J.R. (2005) A risk assessment based approach for the management of whirling disease. *Reviews in Fisheries Science* 13, 205–230.

Bartholomew, J., Hallett, S.L., Holt, R., Alexander, J.D., Atkinson, S.D. *et al.* (2017) Klamath River Fish Health Studies: Salmon Disease Monitoring and Research. Oregon State University, BOR/USGS Interagency Agreement #R15PG00065. FY2017 April 01, 2017–March 31, 2018. Annual Report. Available at: https://microbiology.science.oregonstate.edu/sites/microbiology.science.oregonstate.edu/files/bartholomew/KlamathRiverFishHealth_FY2017_annualreport_May31.pdf (accessed 22 November 2019).

Berger, C.S. and Aubin-Horth, N. (2018) An eDNA-qPCR assay to detect the presence of the parasite *Schistocephalus solidus* inside its threespine stickleback host. *Journal of Experimental Biology* 221, jeb178137.

Bjork, S.J. and Bartholomew, J.L. (2009) The effects of water velocity on the *Ceratomyxa shasta* infectious cycle. *Journal of Fish Diseases* 32, 131–142.

Blazer, V.S., Waldrop, T.B., Schill, W.B., Densmore, C.L. and Smith, D. (2003) Effects of water temperature and substrate type on spore production and release in eastern *Tubifex tubifex* worms infected with *Myxobolus cerebralis*. *Journal of Parasitology* 89, 21–26.

Bogdanova, E.A. (1960) Natural habitat of the myxosporidian (*Myxosoma cerebralis*, whirling disease) at Sakhalin (SE Russia). *Doklady Akademii Nauk SSSR* 134, 1501–1503.

Bogdanova, E.A. (1964) *Parasites and diseases of fingerling European chum, and humpback salmon in fish rearing on the Kola Peninsula. Session of the Scientific Council on the problems 'Theoretical basis of rational use of the resources of the White Sea and the inland reservoirs of Karelia'*, Petrozavodsk, USSR.

Bogdanova, E.A. (1968) Modern data on the distribution of *Myxosoma cerebralis* (Protozoa, Cnidosporidia) as an agent of whirling disease of salmonids. *Bulletin de Office International des Épizooties* 69, 1499–1506.

Bogdanova, E.A. (1969) New data on the distribution of *Myxosoma cerebralis* and peculiarities of its ecology depending on biotic and abiotic factors. *Progress in Protozoology* 2, 226.

Bonacina, C., Bonomi, G. and Monti, C. (1989a) Density-dependent processes in cohorts of *Tubifex tubifex*, with special emphasis on the control of fecundity. *Hydrobiologia* 180, 135–141.

Bonacina, C., Bonomi, G. and Monti, C. (1989b) Population analysis in mass cultures of *Tubifex tubifex*. *Hydrobiologia* 180, 127–134.

Brinkhurst, R.O. (1980) The production biology of the Tubificidae (Oligochaeta). In: Brinkhurst, R.O. and Cook, D.G. (eds) *Aquatic Oligochaete Biology*, 1st ed. New York, Plenum Press, pp. 205–209.

Brinkhurst, R.O. (1996) On the role of tubificid oligochaetes in relation to fish disease with special reference to the Myxozoa. *Annual Review of Fish Diseases* 6, 29–40.

Bruneau, N.A. (2001) A quantitative risk assessment for the introduction of *Myxobolus cerebralis* to Alberta, Canada, through the importation of live farmed salmonids. In: Rogers, C.J. (ed.) *Risk Analysis in Aquatic Animal Health*. Office International des Épizooties, Paris, pp. 41–50.

Buchanan, D.V. and Sanders, J.E. (1983) Relative susceptibility of four strains of summer steelhead to infection by *Ceratomyxa shasta*. *Transactions of the American Fisheries Society* 14, 541–543.

Canadian Food Inspection Agency (2016) Confirmed detections of whirling disease – Alberta 2016. Available at: https://www.inspection.gc.ca/animal-health/aquatic-animals/diseases/reportable-diseases/whirling-disease/alberta/eng/1520358140712/1520358141169 (accessed 15 January 2019).

Carraro, L., Bertuzzo, E., Mari, L., Fontes, I., Hartikainen, H. *et al.* (2017) Integrated field, laboratory, and theoretical study of PKD spread in a Swiss prealpine river. *Proceedings of the National Academy of Sciences USA* 114, 11992–11997.

Carraro, L., Mari, L., Gatto, M., Rinaldo, A. and Bertuzzo, E. (2018) Spread of proliferative kidney disease in fish along stream networks: a spatial metacommunity framework. *Freshwater Biology* 63, 114–127.

Cassidy-Hanley, D.M., Cordonnier-Pratt, M.M., Pratt, L.H., Devine, C., Hossain, M.M. *et al.* (2011) Transcriptional profiling of stage specific gene expression in the parasitic ciliate *Ichthyophthirius multifiliis*. *Molecular and Biochemical Parasitology* 178, 29–39.

Cavender, W.P., Wood, J.S., Powell, M.S., Overturf, K.E. and Cain, K.D. (2004) Real-time quantitative polymerase chain reaction (qPCR) to identify *Myxobolus cerebralis* in rainbow trout *Oncorhynchus mykiss*. *Diseases of Aquatic Organisms* 60, 205–213.

Christensen, N.O. (1972) Some diseases of trout in Denmark. *Symposia of the Zoological Society of London* 30, 83–88.

Coleman, M.A. and Fausch, K.D. (2007) Cold summer temperature limits recruitment of age-0 cutthroat trout in high-elevation Colorado streams. *Transactions of the American Fisheries Society* 136, 1231–1244.

Crozier, L.G. and Hutchings, J.A. (2014) Plastic and evolutionary responses to climate change in fish. *Evolutionary Applications* 7, 68–87.

Crozier, L.G., Hendry, A.P., Lawson, P.W., Quinn, T.P., Mantua, N.J. *et al.* (2008) Potential responses to climate change in organisms with complex life histories:

evolution and plasticity in Pacific salmon. *Evolutionary Applications* 1, 252–270.

Crozier, L.G., Scheuerell, M.D. and Zabel, R.W. (2011) Using time series analysis to characterize evolutionary and plastic responses to environmental change: a case study of a shift toward earlier migration date in sockeye salmon. *American Naturalist* 178, 755–773.

Crozier, L.G., McClure, M.M., Beechie, T., Bograd, S., Boughton, D. *et al.* (2019) Climate vulnerability assessment for Pacific salmon and steelhead in the California Current Large Marine Ecosystem. *PLoS ONE* 14, e0217711.

Downing, D.C., McMahon, T.E., Kerans, B.L. and Vincent, E.R. (2002) Relation of spawning and rearing life history of rainbow trout and susceptibility to *Myxobolus cerebralis* infection in the Madison River, Montana. *Journal of Aquatic Animal Health* 14, 191–203.

DuBey, R. and Caldwell, C. (2004) Distribution of *Tubifex tubifex* lineages and *Myxobolus cerebralis* infection in the tailwater of the San Juan River, New Mexico. *Journal of Aquatic Animal Health* 16, 179–185.

DuBey, R.J., Caldwell, C.A. and Gould, W.R. (2007) Relative susceptibility and effects of on performance of Rio Grande cutthroat trout and rainbow trout challenged with *Myxobolus cerebralis*. *Transactions of the American Fisheries Society* 136, 1406–1414.

El-Matbouli, M. and Hoffmann, R. (1989) Experimental transmission of two *Myxobolus* spp. developing bisporogeny via tubificid worms. *Parasitology Research* 75, 461–464.

El-Matbouli, M. and Hoffmann, R.W. (1991a) Effects of freezing, aging, and passage through the alimentary canal of predatory animals on the viability of *Myxobolus cerebralis* spores. *Journal of Aquatic Animal Health* 3, 260–262.

El-Matbouli, M. and Hoffmann, R.W. (1991b) Prevention of experimentally induced whirling disease in rainbow trout *Oncorhynchus mykiss* by fumagillin. *Diseases of Aquatic Organisms* 10, 109–113.

El-Matbouli, M. and Hoffmann, R.W. (1998) Light and electron microscopic studies on the chronological development of *Myxobolus cerebralis* to the actinosporean stage in *Tubifex tubifex*. *International Journal for Parasitology* 28, 195-217.

El-Matbouli, M. and Hoffmann, R.W. (2002) Influence of water quality on the outbreak of proliferative kidney disease – field studies and exposure experiments. *Journal of Fish Diseases* 25, 459–467.

El-Matbouli, M., Hoffmann, R.W. and Mandok, C. (1995) Light and electron microscopic observations on the route of the triactinomyxon-sporoplasm of *Myxobolus cerebralis* from epidermis into rainbow trout cartilage. *Journal of Fish Biology* 46, 919–935.

El-Matbouli, M., McDowell, T.S., Antonio, D.B., Andree, K.B. and Hedrick, R.P. (1999) Effect of water temperature on the development, release and survival of the triactinomyxon stage of *Myxobolus cerebralis* in its oligochaete host. *International Journal for Parasitology* 29, 627–641.

Engelking, H.M. (2002) Potential for introduction to *Myxobolus cerebralis* into the Deschutes River watershed in central Oregon from adult anadromous salmonids. In: Bartholomew, J.L. and Wilson, J.C. (eds) *Whirling Disease: Reviews and Current Topics. American Fisheries Society Symposium* 29. American Fisheries Society, Bethesda, Maryland, pp. 25–32.

Fetherman, E.R., Winkelman, D.L., Schisler, G.J. and Antolin, M.F. (2012) Genetic basis of differences in myxospore count between whirling disease-resistant and -susceptible strains of rainbow trout. *Diseases of Aquatic Organisms* 102, 97–106.

Fujiwara, M., Mohr, M.S., Greenberg, A., Foott, J.S. and Bartholomew, J. (2011) Effects of ceratomyxosis on population dynamics of Klamath fall-run chinook salmon. *Transactions of the American Fisheries Society* 140, 1380–1391.

Gates, K.K., Guy, C.S., Zale, A.V. and Horton, T.B. (2008) Adherence of *Myxobolus cerebralis* myxospores to waders: implications for disease dissemination. *North American Journal of Fisheries Management* 28, 1453–1458.

Gilbert, M.A. and Granath, W.O. Jr (2003) Whirling disease of salmonid fish: life cycle biology, and disease. *Journal of Parasitology* 89, 658–667.

Hallett, S.L. and Bartholomew, J.L. (2008) Effects of water flow on the infection dynamics of *Myxobolus cerebralis*. *Parasitology* 135, 371–684.

Hallett, S.L., Atkinson, S.D., Erséus, C. and El-Matbouli, M. (2005) Dissemination of triactinomyxons (Myxozoa) via oligochaetes used as live food for aquarium fishes. *Diseases of Aquatic Organisms* 65, 137–152.

Hallett, S.L., Atkinson, S.D., Erséus, C. and El-Matbouli, M. (2006) Myxozoan parasites disseminated via oligochaete worms as live food for aquarium fishes: descriptions of aurantiactinomyxon and raabeia actinospore types. *Diseases of Aquatic Organisms* 69, 213–225.

Halliday, M.M. (1973) Studies on *Myxosoma cerebralis*, a parasite of salmonids. II. Development and pathology of *Myxosoma cerebralis* in experimentally infected rainbow trout (*Salmo gairdneri*) fry reared at different water temperatures *Nordisk Veterinaermedicin* 25, 349–358.

Halliday, M.M. (1974) Studies on *Myxosoma cerebralis*, a parasite of salmonids III. Some studies on the epidemiology of *Myxosoma cerebralis* in Denmark, Scotland and Ireland. *Nordisk Veterinaermedicin* 26, 165–172.

Halliday, M.M. (1976) The biology of *Myxosoma cerebralis*: the causative organism of whirling disease of salmonids. *Journal of Fish Biology* 9, 399–357.

Hedrick, R.P. and El-Matbouli, M. (2002) Recent advances with taxonomy, life cycle, and development of *Myxobolus cerebralis* in the fish and oligochaete hosts. In: Bartholomew, J.L. and Wilson, J.C. (eds) *Whirling Disease: Reviews and Current Topics. American Fisheries Society Symposium* 29. American Fisheries Society, Bethesda, Maryland, pp. 45–54.

Hedrick, R.P., El-Matbouli, M., Adkinson, M.A. and MacConnell, E. (1998) Whirling disease: re-emergence among wild trout. *Immunological Reviews* 166, 365–376.

Hedrick, R.P., McDowell, T.S., Gay, M., Marty, G.D., Gerorgiadis, M.P. *et al.* (1999a) Comparative susceptibility of rainbow trout *Oncorhynchus mykiss* and brown trout *Salmo trutta* to *Myxobolus cerebralis*, the cause of salmonid whirling disease. *Diseases of Aquatic Organisms* 37, 173–183.

Hedrick, R.P., McDowell, T.S., Mukkatira, K., Gerorgiadis, M.P. and MacConnell, E. (1999b) Susceptibility of selected inland salmonids to experimentally induce infections with *Myxobolus cerebralis*, the causative agent of whirling disease. *Journal of Aquatic Animal Health* 11, 330–339.

Hedrick, R.P., McDowell, T.S., Mukkatira, K., Georgiadis, M.P. and MacConnell, E. (2001a) Salmonids resistant to *Ceratomyxa shasta* are susceptible to experimentally induced infections with *Myxobolus cerebralis*. *Journal of Aquatic Animal Health* 13, 35–42.

Hedrick, R.P., McDowell, T.S., Mukkatira, K., Georgiadis, M.P. and MacConnell, E. (2001b) Susceptibility of three species of anadromous salmonids to experimentally induced infections with *Myxobolus cerebralis*, the causative agent of whirling disease. *Journal of Aquatic Animal Health* 13, 43–50.

Hedrick, R.P., McDowell, T.S., Marty, G.D., Fosgate, G.T., Mukkatira, K. *et al.* (2003) Susceptibility of two strains of rainbow trout (one with suspected resistance to whirling disease) to *Myxobolus cerebralis* infection. *Diseases of Aquatic Organisms* 55, 37–44.

Hedrick, R.P., McDowell, T.S., Mukkatira, K., MacConnell, E. and Petri, B. (2008) Effects of freezing, drying, ultraviolet irradiation, chlorine, and quaternary ammonium treatments on the infectivity of myxospores of *Myxobolus cerebralis* for *Tubifex tubifex*. *Journal of Aquatic Animal Health* 20, 116–125.

Hedrick, R.P., McDowell, T.S., Adkison, M.A., Myklebust, K.A., Mardones, F.O. *et al.* (2012) Invasion and initial replication of ultraviolet irradiated waterborne infective stages of *Myxobolus cerebralis* results in immunity to whirling disease in rainbow trout. *International Journal for Parasitology* 42, 657–666.

Hitt, N.P., Frissell, C.A., Muhlfeld, C.C. and Allendorf, F.W. (2003) Spread of hybridization between native westslope cutthroat trout, *Oncorhynchus clarki lewisi*, and nonnative rainbow trout, *Oncorhynchus mykiss*. *Canadian Journal of Fisheries and Aquatic Sciences* 60, 1440–1451.

Höfer, B. (1903) Ueber die Drehkrankheit der Regenborgenforelle. *Allgemeine Fischerei Zeitschrift* 28, 7–8.

Hoffman, G.L. (1970) Intercontinental and transcontinental dissemination and transfaunation of fish parasites with emphasis on whirling disease (*Myxosoma cerebralis*) and its effect on fish. In: Snieszko, S.F. (ed.) *Symposium of Diseases of Fishes and Shellfishes*. American Fisheries Society, Bethesda, Maryland, pp. 69–81.

Hoffman, G.L. (1974) Disinfection of contaminated water by ultraviolet irradiation, with emphasis on whirling disease (*Myxosoma cerebralis*) and its effect on fish. *Transactions of the American Fisheries Society* 103, 541–550.

Hoffman, G.L. (1990) *Myxobolus cerebralis*, a worldwide cause of salmonid whirling disease. *Journal of Aquatic Animal Health* 2, 30–37.

Hoffman, G.L. and Putz, R.E. (1969) Host susceptibility and the effects of aging, freezing, heat and chemicals on spores of *Myxosoma cerebralis*. *The Progressive Fish-Culturist* 31, 35–37.

Hogge, C. I., Campbell, M.R. and Johnson, K.A. (2008) A new species of myxozoan (Myxosporea) from the brain and spinal cord of rainbow trout (*Oncorhynchus mykiss*) from Idaho. *Journal of Parasitology* 94, 218–222.

Ibarra, A.M., Gall, G.A.E. and Hedrick, R.P. (1990) Trials with fumagillin DHC and malachite green to control ceratomyxosis in rainbow trout (*Oncorhynchus mykiss*). *Fish Pathology* 25, 217–223.

Ibarra, A.M., Gall, G.A.E. and Hedrick, R.P. (1991) Susceptibility of two strains of rainbow trout *Oncorhynchus mykiss* to experimentally induced infections with the myxosporean *Ceratomyxa shasta*. *Diseases of Aquatic Organisms* 10, 191–194.

Kaeser, A.J., Rasmussen, C. and Sharpe, W.E. (2006) An examination of environmental factors associated with *Myxobolus cerebralis* infection of wild trout in Pennsylvania. *Journal of Aquatic Animal Health* 18, 90–100.

Kallert, D.M. and El-Matbouli, M. (2008) Differences in viability and reactivity of actinospores of three myxozoan species upon ageing. *Folia Parasitologica* 55, 105–110.

Keefer, M.L., Clabough, T.S., Jepson, M.A., Bowerman, T. and Caudill, C.C. (2019) Temperature and depth profiles of Chinook salmon and the energetic costs of their long-distance homing migrations. *Journal of Thermal Biology* 79, 155–165.

Kelley, G.O., Zagmutt-Vergara, F.J., Leutenegger, C.M., Myklebust, K.A., Adkinson, M.A. *et al.* (2004) Evaluation of five diagnostic methods of the detection and quantification of *Myxobolus cerebralis*. *Journal of Veterinary Diagnostic Investigation* 16, 202–211.

Kerans, B.L., Stevens, R.I. and Lemmon, J.C. (2005) Water temperature affects a host–parasite interaction:

Tubifex tubifex and *Myxobolus cerebralis*. *Journal of Aquatic Animal Health* 17, 216–221.

Kindschi, G.A. and Koby, R.F. (1994) Performance and oxygen consumption of Snake River cutthroat trout reared at four densities with supplemental oxygen. *The Progressive Fish-Culturist* 56, 13–18.

Koel, T.M., Mahony, D.L., Kinnan, K.L., Rasmussen, C., Hudson, C.J. *et al.* (2006) *Myxobolus cerebralis* in native cutthroat trout of the Yellowstone Lake ecosystem. *Journal of Aquatic Animal Health* 18, 157–175.

Koel, T.M., Kerans, B.L., Barras, S.C., Hanson, K.C. and Wood, J.S. (2010) Avian piscivores as vectors for *Myxobolus cerebralis* in the greater Yellowstone ecosystem. *Transactions of the American Fisheries Society* 139, 976–988.

Kovach, R.P., Gharrett, A.J. and Tallmon, D.A. (2012) Genetic change for earlier migration timing in a pink salmon population. *Proceedings of the Royal Society B: Biological Sciences* 279, 3870–3878.

Kovach, R.P., Gharrett, A.J. and Tallmon, D.A. (2013a) Temporal patterns of genetic variation in a salmon population undergoing rapid change in migration timing. *Evolutionary Applications* 6, 795–807.

Kovach, R.P., Joyce, J.E., Echave, J.D., Lindberg, M.S. and Tallmon, D.A. (2013b) Earlier migration timing, decreasing phenotypic variation, and biocomplexity in multiple salmonid species. *PLoS ONE* 8, e53807.

Kovach, R.P., Ellison, S.C., Pyare, S. and Tallmon, D.A. (2015) Temporal patterns in adult salmon migration timing across southeast Alaska. *Global Change Biology* 21, 1821–1833.

Kovach, R.P., Muhlfeld, C.C., Al-Chokhachy, R., Dunham, J.B., Letcher, B.H. *et al.* (2016) Impacts of climatic variation on trout: a global synthesis and path forward. *Reviews in Fish Biology and Fisheries* 26, 135–151.

Kumar, G., Abd-Elfattah, A. and El-Matbouli, M. (2015a) Identification of differentially expressed genes of brown trout (*Salmo trutta*) and rainbow trout (*Oncorhynchus mykiss*) in response to *Tetracapsuloides bryosalmonae* (Myxozoa). *Parasitology Research* 114, 929–939.

Kumar, G., Gotesman, M. and El-Matbouli, M. (2015b) Interaction of *Tetracapsuloides bryosalmonae*, the causative agent of proliferative kidney disease, with host proteins in the kidney of *Salmo trutta*. *Parasitology Research* 114, 1721–1727.

Kumar, G., Sarker, S., Menanteau-Ledouble, S. and El-Matbouli, M. (2015c) *Tetracapsuloides bryosalmonae* infection affects the expression of genes involved in cellular signal transduction and iron metabolism in the kidney of the brown trout *Salmo trutta*. *Parasitology Research* 114, 2301–2308.

Lom, J. and Dyková, I. (2006) Myxozoan genera: definition and notes on taxonomy, life-cycle terminology and pathogenic species. *Folia Parasitologica* 53, 1–36.

Lom, J. and Hoffman, G.L. (1971) Morphology of the spores of *Myxobolus cerebralis* (Höfer, 1903) and *M. cartilaginis* (Hoffman, Putz, and Dunbar, 1965). *Journal of Parasitology* 56, 1302–1308.

Lowers, J.M. and Bartholomew, J.L. (2003) Detection of myxozoan parasites in oligochaetes imported as food for ornamental fish. *Journal of Parasitology* 89, 84–91.

Lynch, A.J., Myers, B.J.E., Chu, C., Eby, L.A., Falke, J.A. *et al.* (2016) Climate change effects on North American inland fish populations and assemblages. *Fisheries* 41, 346–361.

MacConnell, E. and Vincent, E.R. (2002) The effects of *Myxobolus cerebralis* on the salmonid host. In: Bartholomew, J.L. and Wilson, J.C. (eds) *Whirling Disease: Reviews and Current Topics. American Fisheries Society Symposium* 29. American Fisheries Society, Bethesda, Maryland, pp. 95–108.

MacConnell, E. and Bartholomew, J.L. (2012) Whirling disease of salmonids. In: *AFS-FHS Blue Book Section 1: Diagnostic Procedures for Finfish and Shellfish Pathogens*. American Fisheries Society Fish Health Section, Bethesda, Maryland.

McCullough, F.S., Gayral, P., Duncan, J. and Christie, J.D. (1980) Molluscicides in schistosomiasis control. *Bulletin of the World Health Organization* 58, 681–689.

McGinnis, S. and Kerans, B.L. (2013) Land use and host community characteristics as predictors of disease risk. *Landscape Ecology* 28, 29–44.

Marcogliese, D.J. (2001) Implications for climate change for parasitism of animals in the aquatic environment. *Canadian Journal of Zoology* 79, 1331–1352.

Marcogliese, D. (2008) The impact of climate change on the parasites and infectious diseases of aquatic animals. *Revue Scientifique et Technique (International Office of Epizootics)* 27, 467–484.

Margolis, M.L., Kent, M.L. and Bustos, P. (1996) Diseases of salmonids resembling myxosporean whirling disease, and the absence of *Myxosoma cerebralis*, in South America. *Diseases of Aquatic Organisms* 25, 33–37.

Markiw, M.E. and Wolf, K. (1983) *Myxosoma cerebralis* (Myxozoa: Myxosporea) etiologic agent of salmonid whirling disease requires tubificid worm (Annelida: Oligochaeta) in its life cycle. *Journal of Protozoology* 30, 561–564.

Markiw, M.E. (1992) *Salmonid whirling disease. Leaflet No. 17*. US Fish and Wildlife Service, Washington, DC.

Melnychuk, M.C., Welch, D.W. and Walters, C.J. (2010) Spatio-temporal migration patterns of Pacific salmon smolts in rivers and coastal marine waters. *PLoS ONE* 5, e12916.

Miller, M.P. and Vincent, E.R. (2008) Rapid natural selection for resistance to an introduced parasite of rainbow trout. *Evolutionary Applications* 1, 336–341.

Mote, P.W. (2003) Trends in temperature and precipitation in the Pacific Northwest. *Northwest Science* 77, 271–282.

Mote, P.W., Parson, E., Hamlet, A.F., Keeton, W.S., Lettenmaier, D. *et al.* (2003) Preparing for climatic change: the water, salmon, and forests of the Pacific Northwest. *Climatic Change* 61, 45–88.

Murcia, S., Kerans, B.L., MacConnell, E. and Koel, T.M. (2006) *Myxobolus cerebralis* infection patterns in Yellowstone cutthroat trout after natural exposure. *Diseases of Aquatic Organisms* 71, 191–199.

Nakano, S., Kitano, F. and Maekawa, K. (1996) Potential fragmentation and loss of thermal habitats for charrs in the Japanese archipelago due to climatic warming. *Freshwater Biology* 36, 711–722.

Nehring, R.B., Hancock, B., Catanese, M., Stinson, M.E.T., Winkelman, D. *et al.* (2013) Reduced *Myxobolus cerebralis* actinospore production in a Colorado reservoir may be linked to changes in *Tubifex tubifex* population structure. *Journal of Aquatic Animal Health* 25, 205–220.

Neudecker, R.A., McMahon, T.E. and Vincent, E.R. (2012) Spatial and temporal variation of whirling disease risk in Montana spring creeks and rivers. *Journal of Aquatic Animal Health* 24, 201–212.

Niklasson, L., Sundh, H., Fridell, F., Taranger, G.L. and Sundell, K. (2011) Disturbance of the intestinal mucosal immune system of farmed Atlantic salmon (*Salmo salar*), in response to long-term hypoxic conditions. *Fish and Shellfish Immunology* 31, 1072–1080.

O'Grodnick, J.J. (1979) Susceptibility of various salmonids to whirling disease (*Myxosoma cerebralis*). *Transactions of the American Fisheries Society* 108, 187–190.

Okamura, B. and Feist, S.W. (2011) Emerging diseases in freshwater systems. *Freshwater Biology* 56, 627–637.

Okamura, B., Hartikainen, H., Schmidt-Posthaus, H. and Wahli, T. (2011) Life cycle complexity, environmental change and the emerging status of salmonid proliferative kidney disease. *Freshwater Biology* 56, 735–753.

Parmesan, C. and Yohe, G. (2003) A globally coherent fingerprint of climate change impacts across natural systems. *Nature* 421, 37–42.

Patz, J.A., Githeco, A.K., McCarty, J.P., Hussein, S., Confalonieri, U. *et al.* (2003) Climate change and infectious diseases. In: McMichael, A., Campbell-Lendrum, D., Corvalán, C., Ebi, K.L., Githeko, A. *et al.* (eds) *Climate Change and Human Health: Risks and Responses*. World Health Organization, Geneva, Switzerland, pp. 103–132.

Pauls, S.U., Nowak, C., Balint, M. and Pfenninger, M. (2013) The impact of global climate change on genetic diversity within populations and species. *Molecular Ecology* 22, 925–946.

Ray, R.A. and Bartholomew, J.L. (2013) Estimation of transmission dynamics of the *Ceratomyxa shasta* actinospore to the salmonid host. *Parasitology* 140, 907–916.

Richey, C.A., Kenelty, K.V., Hopkins, K.V., Stevens, B.N., Martinez-Lopez, B. *et al.* (2018) Distribution and prevalence of *Myxobolus cerebralis* in postfire areas of Plumas National Forest: utility of environmental DNA sampling. *Journal of Aquatic Animal Health* 30, 130–143.

Rose, J.D., Marrs, G.S., Lewis, C. and Schisler, G. (2000) Whirling disease behavior and its relation to pathology of brain stem and spinal cord in rainbow trout. *Journal of Aquatic Animal Health* 12, 107–118.

Ryce, E.K.N., Zale, A.V. and MacConnell, E. (2004) Effects of fish age and parasite dose on the development of whirling disease in rainbow trout. *Diseases of Aquatic Organisms* 59, 225–233.

Ryce, E.K.N., Zale, A.V., MacConnell, E. and Nelson, M. (2005) Effects of fish age versus size on the development of whirling disease in rainbow trout. *Diseases of Aquatic Organisms* 63, 69–76.

Sandell, T.A., Lorz, H.L., Stevens, D.G. and Bartholomew, J.L. (2001) Dynamics of *Myxobolus cerebralis* in the Lostine River, Oregon: implications for resident and anadromous salmonids. *Journal of Aquatic Animal Health* 13, 142–150.

Sarker, S. and El-Matbouli, M. (2015) Can RNAi target salmonid whirling disease *in vivo*? *Nucleic Acid Therapeutics* 25, 285–286.

Sarker, S., Kallert, D.M., Hedrick, R.P. and El-Matbouli, M. (2015) Whirling disease revisited: pathogenesis, parasite biology and disease intervention. *Diseases of Aquatic Organisms* 114, 155–175.

Sarker, S., Menanteau-Ledouble, S., Kotob, M.H. and El-Matbouli, M. (2017) A RNAi-based therapeutic proof of concept targets salmonid whirling disease *in vivo*. *PLoS ONE* 12, e0178687.

Schisler, G.J., Bergersen, E.P. and Walker, P.G. (2000) Effects of multiple stressors on morbidity and mortality of the fingerling rainbow trout infected with *Myxobolus cerebralis*. *Transactions of the American Fisheries Society* 129, 859–865.

Schisler, G.J., Myklebust, K.A. and Hedrick, R.P. (2006) Inheritance of *Myxobolus cerebralis* resistance among F1-generation crosses of whirling disease resistant and susceptible rainbow trout strains. *Journal of Aquatic Animal Health* 18, 109-115.

Searle, C.L., Xie, G.Y. and Blaustein, A.R. (2013) Development and infectious disease in hosts with complex life cycles. *PLoS ONE* 8, e60920.

Sokolow, S.H., Wood, C.L., Jones, I.J., Lafferty, K.D., Kuris, A.M. *et al.* (2018) To reduce the global burden of human schistosomiasis, use 'old fashioned' snail control. *Trends in Parasitology* 34, 23–40.

Sollid, S.A., Lorz, H.V., Stevens, D.G. and Bartholomew, J.L. (2002) Relative susceptibility of selected

Deschutes River, Oregon, salmonid species to experimentally induced infection by *Myxobolus cerebralis*. In: Bartholomew, J.L. and Wilson, J.C. (eds) *Whirling Disease: Reviews and Current Topics. American Fisheries Society Symposium* 29. American Fisheries Society, Bethesda, Maryland, pp. 117–124.

Staton, L., Erdahl, D. and El-Matbouli, M. (2002) Efficiency of fumagillin and TNP-470 to prevent experimentally induced whirling disease in rainbow trout *Oncorhynchus mykiss*. In: Bartholomew, J.L. and Wilson, J.C. (eds) *Whirling Disease: Reviews and Current Topics. American Fisheries Society Symposium* 29. American Fisheries Society, Bethesda, Maryland, pp. 239–247.

Steinbach, E.L.C., Stromberg, K.E., Ryce, E.K.N. and Bartholomew, J.L. (2009) *Whirling Disease in the United States. A Summary of Progress in Research and Management 2009*. Montana Water Center, Montana State University, Bozeman, Montana.

Stewart, I.T., Cayan, D.R. and Dettinger, M.D. (2005) Changes toward earlier streamflow timing across western North America. *Journal of Climate* 18, 1136–1155.

Taylor, R.E.L. and Haber, M.H. (1974) Opercular cyst formation in trout infected with *Myxobolus cerebralis*. *Journal of Wildlife Diseases* 10, 347–351.

Taylor, R.E.L., Coli, S.J. and Junell, D.R. (1973) Attempts to control whirling disease by continuous drug feeding. *Journal of Wildlife Diseases* 9, 302–305.

Thompson, K.G. (2011) Evaluation of small-scale habitat manipulation to reduce the impact of the whirling disease parasite in streams. *Aquatic Ecosystem Health and Management* 14, 305–317.

Thompson, K.G. and Nehring, R.B. (2000) A simple technique used to filter and quantify the actinospore of *Myxobolus cerebralis* and determine its seasonal abundance in the Colorado River. *Journal of Aquatic Animal Health* 12, 316–323.

True, K., Bolick, A. and Foott, J. (2013) *Myxosporean Parasite (Ceratomyxa shasta and Parvicapsula minibicornis) Annual Prevalence of Infection in Klamath River Basin Juvenile Chinook Salmon, Apr–Aug 2012*. US Fish and Wildlife Service, California–Nevada Fish Health Center, Anderson, California.

Uspenskaya, A.V. (1955) Biology, distribution and economic importance of *Myxosoma cerebralis*, the causative agent of whirling disease of trout. *Lectures of the Academy of Science USSR* 105, 1132–1135.

Uspenskaya, A.V. (1957) The ecology and spreading of the pathogen of trout whirling disease –*Myxosoma cerebralis* (Hofer, 1903, Plehn, 1905) in the fish ponds of the Soviet Union. In: Petrushevski, G.K. (ed.) *Parasites and Diseases of Fish*. All Union Institute of Freshwater Fisheries, Leningrad, USSR, pp. 47–55.

Vincent, E.R. (2002) Relative susceptibility of various salmonids to whirling disease with emphasis on rainbow and cutthroat trout. In: Bartholomew, J.L. and Wilson, J.C. (eds) *Whirling Disease: Reviews and Current Topics. American Fisheries Society Symposium* 29. American Fisheries Society, Bethesda, Maryland, pp. 109–116.

Visser, M.E. and Both, C. (2005) Shifts in phenology due to global climate change: the need for a yardstick. *Proceedings of the Royal Society B: Biological Sciences* 272, 2561–2569.

Wagner, E.J. (2002) Whirling disease prevention, control, and management: a review. In: Bartholomew, J.L. and Wilson, J.C. (eds) *Whirling Disease: Reviews and Current Topics. American Fisheries Society Symposium* 29. American Fisheries Society, Bethesda, Maryland, pp. 217–226.

Wagner, E.J., Smith, M., Arndt, R. and Roberts, D.W. (2003) Physical and chemical effects on viability of the *Myxobolus cerebralis* triactinomyxon. *Diseases of Aquatic Organisms* 53, 133–142.

Wagner, E.J., Wilson, C., Arndt, R., Goddard, P., Miller, M. *et al.* (2006) Evaluation of disease resistance of the Fish Lake-DeSmet, Wounded Man, and Harrison Lake strains of rainbow trout exposed to *Myxobolus cerebralis*. *Journal of Aquatic Animal Health* 18, 128–135.

Wipfli, M.S., Hudson, S.J. and Caouette, J. (1998) Influence of salmon carcasses on stream productivity: response of biofilm and benthic macroinvertebrates in southeastern Alaska, USA. *Canadian Journal of Fisheries and Aquatic Sciences* 55, 1503–1511.

Wipfli, M.S., Hudson, J.P., Caouette, J.P., Mitchell, N.L., Lessard, J.L. *et al.* (2010) Salmon carcasses increase stream productivity more than inorganic fertilizer pellets: a test on multiple trophic levels in streamside experimental channels. *Transactions of the American Fisheries Society* 139, 824–839.

Wolf, K. and Markiw, M.E. (1984) Biology contravenes taxonomy in the Myxozoa: new discoveries show alternation of the invertebrate and vertebrate hosts. *Science* 225, 1449–1452.

Wolf, K., Markiw, M.E. and Hiltunen, J.K. (1986) Salmonid whirling disease: *Tubifex tubifex* (Müller) identified as the essential oligochaete in the protozoan life cycle. *Journal of Fish Diseases* 9, 83–85.

Yokoyama, H., Danjo, T., Ogawa, K. and Wakabayashi, H. (1997) A vital staining technique with fluorescein diacetate (FDA) and propidium iodide (PI) for the determination of viability of myxosporean and actinosporean spores. *Journal of Fish Diseases* 20, 281–286.

Zielinski, C.M., Lorz, H.V. and Bartholomew, J.L. (2010) Detection of *Myxobolus cerebralis* in the lower Deschutes River basin, Oregon, USA. *North American Journal of Fisheries Management* 30, 1032–1040.

21 Gyrodactylosis (*Gyrodactylus salaris*)

Tor Atle Mo*

Norwegian Institute for Nature Research, Oslo, Norway

21.1 Brief Introduction

In 1952, the Swedish Fish Inspector B. Svenonius observed *Gyrodactylus* specimens on fins and skin of salmon parr of a Baltic strain of Atlantic salmon (*Salmo salar* L.) in a fish farm located in Hölle close to the River Indalsälven (Malmberg, 1957). A few *Gyrodactylus* specimens were sent to the scientist Göran Malmberg and his studies resulted in the description of *Gyrodactylus salaris* Malmberg, 1957. No disease nor mortality of infected Baltic salmon parr was reported.

In August 1975, *G. salaris* was found on a few parr of Atlantic salmon in the River Lakselva in Misvær, northern Norway (Johnsen, 1978). One year later, all the parr in the river were infested and with heavy mortality. Almost all salmon parr had disappeared from the river in August 1977 (Johnsen, 1978). During the late 1970s and early 1980s, massive mortalities caused by *G. salaris* were observed in several Norwegian rivers; it was concluded that *G. salaris* was not native in the affected rivers but had recently been introduced with imports of live fish from Sweden (Johnsen and Jensen, 1986). During the next 20 years, *G. salaris* spread to many fish farms and rivers and was one of the biggest disasters that hit wild Atlantic salmon in Norway. These observations led to numerous studies and management measures of a parasite which otherwise probably would had been given little scientific attention (Kuusela *et al.*, 2005).

Based on genetic studies, *G. salaris* was synonymized with *Gyrodactylus thymalli* Žitňan, 1960 (Fromm *et al.*, 2014), even when this conspecificity was not supported in host susceptibility experiments (Soleng and Bakke, 2001; Sterud *et al.*, 2002). Consequently, *G. salaris* is now a widespread parasite with numerous strains and genotypes occurring on many salmonid species in Europe (see below). Synonymizing *G. salaris* and *G. thymalli* has some management implications as formerly *G. salaris*-free countries, such as Great Britain, are now within the natural range of *G. salaris* (Fromm *et al.*, 2014). However, the grayling-specific strains have been found only on European grayling (*Thymallus thymallus* (L.)) (Hansen *et al.*, 2003, 2016; Kuusela *et al.*, 2005; Anttila *et al.*, 2008; Pettersen *et al.*, 2015; Mieszkowska *et al.*, 2018) and in laboratory experiments, the grayling-specific strains are non-pathogenic to Atlantic salmon (Sterud *et al.*, 2002) and likely to other salmonids. The national managements of infectious diseases focus only on strains of causative agents that are pathogenic to wildlife or domestic hosts, and thus the *G. salaris*/*G. thymalli* (grayling-specific) strains are of no concern for fish health management. In addition, several *G. salaris* strains have been isolated and validated as non-pathogenic which will add to the management problem complex. This chapter focuses on strains of *G. salaris* found on farmed and wild salmonids other than thymallids, and mostly strains of *G. salaris* that cause gyrodactylosis in Atlantic salmon parr. In the following discussion, the use of haplotypes (A–F) for the strains of *G. salaris* is in accordance with Hansen *et al.* (2003) unless other references are given.

21.1.1 Geographical distribution of non-grayling strains of *Gyrodactylus salaris*

The enzootic area for the non-grayling strains of *G. salaris* (Fig. 21.1) is likely lakes and rivers draining into the Baltic Sea (Malmberg and Malmberg,

*tor.mo@nina.no

 Climate Change and Infectious Fish Diseases (P.T.K. Woo, J.-A. Leong and K. Buchmann)

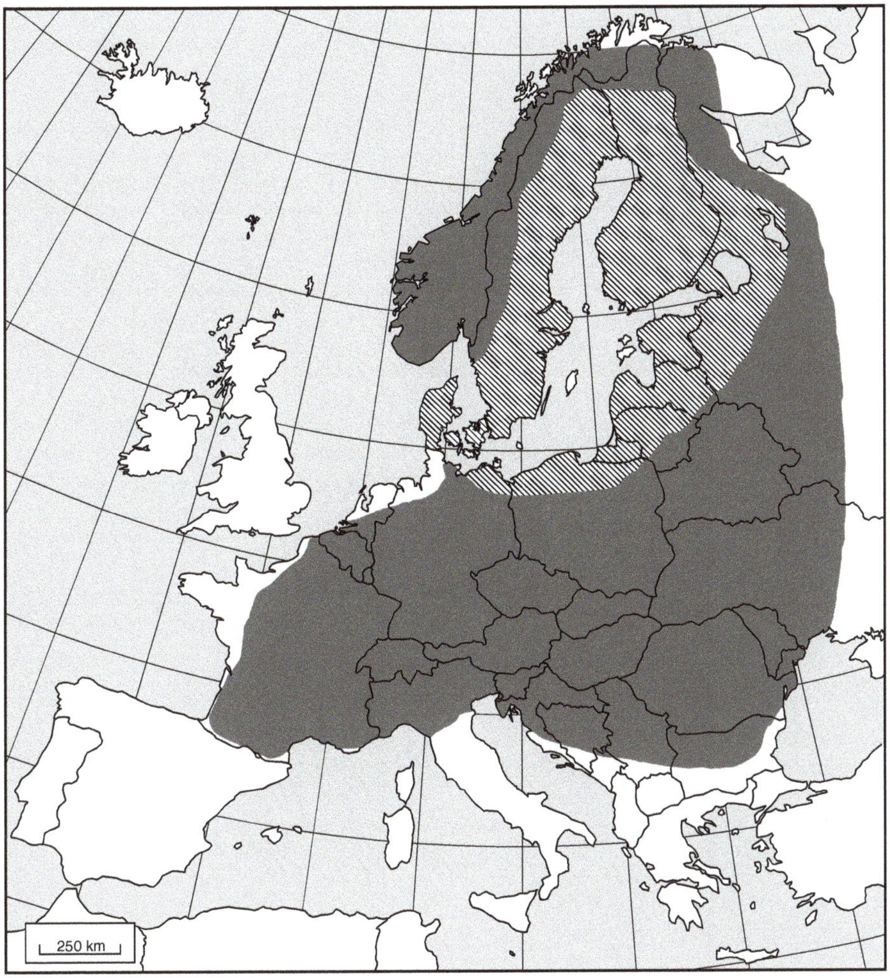

Fig. 21.1. Estimated geographical distribution of non-grayling strains of *Gyrodactylus salaris*. The enzootic area is hatched, while the parasite is considered introduced to the grey area.

1993; Kudersky *et al.*, 2003; Meinilä *et al.*, 2004; Kuusela *et al.*, 2007, 2009). This assumption is based on laboratory experiments showing that Baltic salmon stocks in general have an innate and acquired resistance against *G. salaris* (Bakke *et al.*, 1990a, 2002; Dalgaard *et al.*, 2003), although the Baltic salmon in the River Indalsälven is an exception (Bakke *et al.*, 2004). Furthermore, clinical outbreaks of gyrodactylosis have not been reported in wild or farmed salmon parr in the Baltic area (Rintamäki, 1989; Rintamäki-Kinnunen and Valtonen, 1996; Anttila *et al.*, 2008), although high intensities of *G. salaris* may occur in farmed Baltic salmon (Rintamäki, 1989; Ozerov *et al.*, 2010). The presence of non-grayling strains on farmed and wild Baltic salmon is confirmed in Sweden (Malmberg, 1957; Malmberg and Malmberg, 1993), Finland (Rintamäki, 1989; Rintamäki-Kinnunen and Valtonen, 1996; Anttila *et al.*, 2008; Kuusela *et al.*, 2009), Russia (Ergens, 1983; Kudersky *et al.*, 2003; Ieshko *et al.*, 2016), Estonia (Ozerov *et al.*, 2010) and Latvia (Hansen *et al.*, 2003).

The enzootic area for *G. salaris* on Atlantic salmon likely also include rivers draining into the Kattegat area (Malmberg *et al.*, 1995; Degerman *et al.*, 2012). Kattegat is located north of the outlet of the Baltic Sea, and the salmon in the rivers draining into Kattegat belong to the east Atlantic salmon which is genetically different from Baltic salmon (Ståhl, 1987; Bourret *et al.*, 2013). Several haplotypes of *G. salaris* have been found in Swedish rivers draining into Kattegat (Hansen *et al.*, 2003). Likely, *G. salaris* was first spread to some of these rivers with migrating salmonids through brackish water from the Baltic Sea drainage as *G. salaris* is not present in any of the rivers north of Kattegat apart from a recent spread of another *G. salaris* haplotype to four rivers in the Oslo Fjord (see below). However, the haplotypes that occur in the River Göta älv likely have spread downstream from Lake Vänern. This lake houses landlocked Baltic salmon that is infested with the same haplotypes as in Göta älv (Olstad *et al.*, 2013). Alternatively, *G. salaris* could have spread with migration of Baltic salmon from the freshwater Ancylus Lake through the opening in mid-Sweden to the west coast. However, if this is so, the haplotype of *G. salaris* in Vänern should occur in several of the Swedish rivers. Even if the Atlantic salmon in the rivers draining into Kattegat are genetically close to the susceptible strains of Atlantic salmon in Norwegian and Scottish rivers (Bakke *et al.*, 1990a; Bakke and Mackenzie, 1993), no epizootics due to *G. salaris* have been reported in the Swedish rivers draining into Kattegat (Malmberg *et al.*, 1995), but hundreds of *G. salaris* specimens may be found on salmon parr in some of the rivers (Malmberg, 1993; Malmberg and Malmberg, 1993). In Denmark, several non-pathogenic strains of *G. salaris* have been found on wild and farmed Atlantic salmon and rainbow trout (*Oncorhynchus mykiss* (Walbaum)) in watercourses draining into Kattegat and the North Sea (Lindenstrøm *et al.*, 2003; Jørgensen *et al.*, 2007, 2008) which could also be natural occurrences of *G. salaris*.

G. salaris has been found on farmed salmonids, mostly rainbow trout, in many European countries including (in alphabetical order) Bosnia-Herzegovina (Žitňan and Čankovič, 1970), Denmark (Buchmann *et al.*, 1995; Buchmann and Bresciani, 1997), Finland (Malmberg, 1993), France (Johnston *et al.*, 1996), Germany (Lux, 1990; Cunningham *et al.*, 2003), Italy (Paladini *et al.*, 2009b), Macedonia (Ziętara *et al.*, 2010), Norway (Mo, 1991c, 1994), Poland (Rokicka *et al.*, 2007; Ziętara *et al.*, 2010), Portugal (Johnston *et al.*, 1996), Spain (Santamarina *et al.*, 1991; Tojo *et al.*, 1992), Sweden (Malmberg, 1993; Cunningham *et al.*, 2003) and Romania (Hansen *et al.*, 2016). According to Lucký (1963) and Řehulka (1973), *G. salaris* has been found in several localities in the Czech Republic, but the presented drawings and measurements of the anchors, marginal hooks and ventral membrane are not in accordance with those in *G. salaris* (Mo, 1991a,b,c). Several different genetic strains of *G. salaris* have been found on farmed salmonids. However, the origins of these strains may be difficult to trace because rainbow trout has been moved extensively between farms over large areas in Europe since its introduction from North America in the second half of the 19th century (MacCrimmon, 1971). In Great Britain and Ireland, Atlantic salmon in rivers and farms have been thoroughly examined for *G. salaris* and non-grayling strains of *G. salaris* were not found (Peeler and Thrush, 2004; Peeler and Oidtmann, 2008).

21.1.2 Salmonids susceptible to *Gyrodactylus salaris*

Laboratory experiments have shown that pathogenic strains of *G. salaris* can attach to and reproduce on many species of *Salmo*, *Salvelinus*, *Oncorhynchus* and *Thymallus* (e.g. Bakke, 1991; Bakke *et al.*, 2002, 2007; Paladini *et al.*, 2014). Except for east Atlantic salmon, and possibly for the west Atlantic salmon, other salmonid species usually show no clinical signs and most of them are reservoir hosts or carriers of the parasite (Bakke *et al.*, 2002). It is likely that *G. salaris* can infest more salmonid species than those studied so far. Apart from Atlantic salmon, rainbow trout seem to the most suitable host. This has been shown in laboratory experiments (e.g. Bakke *et al.*, 1991a) and in fish farms (e.g. Mo, 1991c). Interestingly, brown trout (*Salmo trutta* L.) is usually not infested with *G. salaris* in Norwegian rivers (Johnsen and Jensen, 2003), except during the first year of the epizootics when the infestation pressure is very high and brown trout frequently feed on dead and moribund Atlantic salmon parr (T.A. Mo, 1980s and 1990s, unpublished results). The resistance of brown trout to *G. salaris* has been confirmed under laboratory conditions (Jansen and Bakke, 1995), but some specimens may survive on brown trout for more than 100 days (Paladini *et al.*, 2014). Based on the close relationship between Atlantic salmon and brown trout, it is expected that *G. salaris* occurs more frequently on brown trout than on other salmonid genera. The reason for this comparatively very low susceptibility in brown trout is unknown and further studies are suggested.

Several other salmonids from different genera have been tested in laboratories as reservoir hosts for pathogenic strains of *G. salaris*. In most studies, the infection follows a similar pattern: an initial increase in the number of parasites, followed by a decrease until disappearance. European grayling has a similar susceptibility as brown trout. Most parasites disappear after 6–7 weeks (Soleng and Bakke, 2001; Sterud *et al.* 2002), but some parasites may survive on individual European grayling for more than 100 days (Paladini *et al.*, 2014). Anadromous Arctic charr (*Salvelinus alpinus* (L.)) may remain infected for up to 280 days (Bakke *et al.*, 1996) and this is in accordance with the observations done in rivers (Winger *et al.*, 2008; Robertsen *et al.*, 2008). In lake trout (*Salvelinus namaycush* (Walbaum)) and brook trout (*Salvelinus fontinalis* (Mitchill)), *G. salaris* mostly disappeared after 4 and 10 weeks, respectively (Bakke *et al.*, 1992a,b). Rainbow trout is the only species in the genus *Oncorhynchus* that has been studied. The susceptibility of pink salmon (*Oncorhynchus gorbuscha* (Walbaum)) is of particular interest as this is an invasive species in the North-East Atlantic, Barents Sea and White Sea. It could contribute to a further spread of *G. salaris* between Atlantic salmon rivers (see Section 21.3.2).

Only a few strains of *G. salaris* have been included in the laboratory experiments and thus it is not known if each salmonid species is equally susceptible to all the different *G. salaris* strains. In addition, only one or few stocks of a fish species have been studied. The susceptibility to *G. salaris* may vary between the different fish stocks as Bakke *et al.* (1996) found different susceptibility in anadromous and resident populations of Arctic charr.

Several strains of *G. salaris* have been found on wild fish other than Atlantic salmon. In northern Norway, *G. salaris* frequently occurred (the parasite is now eradicated) on anadromous Arctic charr living sympatric with infested Atlantic salmon (Winger *et al.*, 2008). In southern Norway, *G. salaris* occurs on resident Arctic charr in several lakes (Robertsen *et al.*, 2008). However, this strain is non-pathogenic to Atlantic salmon (Olstad *et al.*, 2007; Ramirez *et al.*, 2014) and may be the same strain of *G. salaris* as in Denmark (Jørgensen *et al.*, 2007) and Poland (Rokicka *et al.*, 2007).

21.1.3 Non-salmonids susceptible to *Gyrodactylus salaris*

G. salaris has been experimentally transmitted to Eurasian minnow (*Phoxinus phoxinus* (L.)) (Bakke and Sharp, 1990), roach (*Rutilus rutilus* (L.)) (Bakke *et al.*, 1990b), European perch (*Perca fluviatilis* L.) (Bakke *et al.*, 1990b), European brook lamprey (*Lampetra planeri* (Bloch)) (Bakke *et al.*, 1990b), European eel (*Anguilla anguilla* (L.)) (Bakke *et al.*, 1991b), three-spined stickleback (*Gasterosteus aculeatus* L.) (Soleng and Bakke, 1998), nine-spined stickleback (*Pungitius pungitius* (L.)) (Soleng and Bakke, 1998) and flounder (*Platichthys flesus* (L)) (Soleng and Bakke, 1998). In all these non-salmonids, *G. salaris* survives a few days and the survival times are similar to detached specimens (Olstad *et al.*, 2006). All non-salmonids seem to be innately resistant to *G. salaris*. In Norwegian rivers with ongoing epizootics, *G. salaris* has occasionally been found on flounders (Mo, 1987). As flounders live close to the river bottom, they are likely infected by detached parasites that drift to the riverbed.

21.1.4 The Norwegian story and the current situation

In the early 1970s, a pathogenic strain of *G. salaris* (haplotype A) was unintentionally introduced to Norway with imports of live infested Baltic salmon from Sweden to a Norwegian hatchery (Johnsen and Jensen, 1986; Mo, 2004). Within the next 10 years, *G. salaris* was introduced several times with imports of live Baltic salmon and rainbow trout (Mo, 1994; Johnsen *et al.*, 1999; Hansen *et al.*, 2003). Before it was detected, *G. salaris* had spread with stocks from infested Atlantic salmon to 14 rivers in different Norwegian regions (Johnsen and Jensen, 1986; Mo, 1994; Johnsen *et al.*, 1999). From these primary infected rivers, *G. salaris* spread to 29 more rivers, mostly with migrating salmonids (Soleng *et al.*, 1998; Johnsen *et al.*, 1999; Jansen *et al.*, 2007). Although *G. salaris* is a freshwater parasite, it can tolerate low water salinities for a period (Soleng and Bakke, 1997) and thus it can be spread with salmonids that migrate between rivers within fjords which periodically have low salinities in the surface layers, especially during spring floods. In 1975, another pathogenic strain of *G. salaris* (haplotype B) had spread to a river in northern Norway because of illegal stocking of Baltic salmon (Johnsen *et al.*, 1999) and from this river the parasite spread to two more rivers. Finally, a third pathogenic strain of *G. salaris* (haplotype F) had spread to a river in southern Norway with escaped, infested farmed fish (Mo, 1991c) and this later spread to three more rivers. In one river, located in western Norway, *G. salaris* (haplotype F) likely came with stocked, infested rainbow trout (T.A. Mo, 2000, unpublished results). At present *G. salaris* has been detected on Atlantic salmon parr in 51 Norwegian rivers; 43 rivers with haplotype A, three rivers with haplotype B and five rivers with haplotype F.

The parasite has caused severe epizootics and mortality among Atlantic salmon parr in Norwegian rivers. The parr may succumb due to osmoregulatory disturbances (Pettersen *et al.*, 2013) and secondary fungal (*Saprolegnia* sp.) infections (Johnsen, 1978; Johnsen and Jensen, 1988), mostly in their first (0+) and second (1+) year. However, many diseased parr are eaten by older salmon parr and brown trout before death occurs (T.A. Mo, 1980s and 1990s, unpublished results). As a result, few salmon smolts leave the rivers and thus few adult salmon return to spawn. Within a few years of the introduction of *G. salaris*, salmon populations have declined drastically and the mean mortality among salmon parr is estimated to be 86% (Johnsen *et al.*, 1999). Likely, the mortality can be explained by the lack of effective responses in Norwegian salmon against *G. salaris* (Bakke *et al.*, 1990a, 2004), but the survival of salmon parr has been unusually high (30–40 %) in a few Norwegian rivers (Johnsen *et al.*, 1999). In laboratory experiments, salmon parr from these rivers have a similar mortality to salmon parr from rivers with high mortality (T.A. Mo, 1980s and 1990s, unpublished results). The reason for the better survival in some rivers is unknown but variable water chemistry is probably involved. Laboratory experiments have shown that *G. salaris* is more sensitive to various metal ions than salmon parr (Poléo *et al.*, 2004a), especially aluminium (Soleng *et al.*, 1999, 2005). Likely, the water chemistry or other factors in the environment are of importance for the outcome of the Atlantic salmon–*G. salaris* association.

G. salaris is one of the major threats to Atlantic salmon in Norway (Forseth *et al.*, 2017) and the Norwegian Environment Agency has decided to eradicate pathogen strains from all rivers where it is possible, both to reduce the risk for further spread and to re-establish the salmon populations in the affected rivers (see Section 21.5.2).

The previously mentioned Norwegian hatchery that introduced *G. salaris*-infested Baltic salmon sold salmon parr to other hatcheries and these also became infested. Because of the epizootics in many rivers, surveillance for *G. salaris* in Norwegian hatcheries began in the early 1980s. In 1983, *G. salaris* became a notifiable pathogen in Norway and authorities could demand eradication of the parasite. As eradication of *G. salaris* in fish farms with chemicals is challenging, the practical measure was to slaughter all fish in an infected farm or hatchery. The transports of live salmon between freshwater hatcheries and farms were also highly regulated to prevent the parasite from being dispersed by fish transports. Until 1990, *G. salaris* was detected in 11 salmon hatcheries and 26 inland rainbow trout farms, and because of a fear for spread to local Atlantic salmon populations, the parasite was eradicated from all fish farms (Mo, 1994). Later, *G. salaris* has been detected in only two salmon hatcheries in 2002 and the parasite was quickly eradicated in both hatcheries. The detections of *G. salaris* in Norwegian fish farms were done before the strains could be identified by genetic methods (Hansen *et al.*, 2003) and the haplotypes that occurred in those fish farms are unknown. However, based on information about fish transports and their origin, haplotype A most likely infested salmon in the hatcheries while haplotype F likely occurred on farmed rainbow trout. All *G. salaris*-infested farms have been declared parasite-free after extensive examinations and today there is no known occurrence of *G. salaris* in Norwegian fish farms.

Due to the disastrous outbreaks *G. salaris* causes in Norwegian rivers, several monitoring programmes are carried out to map the occurrence of *G. salaris* and prevent further spread of the parasite. A risk-based selection of rivers (approximately 70 out of 430 salmon rivers) is examined every year while all freshwater farms and hatcheries producing Atlantic salmon or rainbow trout are examined every other year (Hytterød *et al.*, 2018).

21.2 Diagnosis of the Pathogen/Disease

Clinical gyrodactylosis caused by *G. salaris* has been detected on freshwater pre-smolt stages of farmed or wild Atlantic salmon (Johnsen, 1978; Johnsen and Jensen, 1991, 1992; Mo, 1992, 1994; Appleby and Mo, 1997). Large numbers of *G. salaris* have occasionally been seen on farmed Baltic salmon but the fish generally are not considered diseased (Rintamäki, 1989; Ozerov *et al.*, 2010).

G. salaris can also become numerous on wild adult Atlantic salmon within a few weeks after returning to rivers (Mo, 1994) and on wild Arctic charr (Winger *et al.*, 2008), but these fishes do not have gyrodactylosis. In some European areas (e.g. Italy) gyrodactylosis represents a common and economically significant disease of rainbow trout (Paladini *et al.*, 2009a), while in other areas (e.g. Sweden) rainbow trout rarely suffer from the disease (G. Malmberg, 1983, unpublished results). It is not known if this is due to variable susceptibility of the rainbow trout to strains of *G. salaris*, or because of variable response between rainbow trout strains, or variable environmental conditions between the farms. On other farmed salmonids, the numbers of *G. salaris* are usually low (1–50 specimens per fish).

G. salaris (Fig. 21.2) mostly infests the fins, the skin and the head including eyes and the nostrils of fish (Jensen and Johnsen, 1992; Mo, 1992). The gills are rarely infested, except when the fish has hundreds or even thousands of *G. salaris*, and this is usually in late summer and autumn (Mo, 1992; Appleby and Mo, 1997). Atlantic salmon parr suffering from gyrodactylosis commonly rub their sides against the bottom of the tank or objects in the river to get rid of parasites. This is called flashing and is typical for fish with parasitic skin infestations. If not treated, the fish may become lethargic and even anorectic before they become moribund and lie on the bottom. This process may take several weeks, depending on the water temperature. Typical macroscopic changes include greyish appearance of the skin caused by an increased mucus production in addition to hundreds and even thousands of parasites. At the outer edge of the fins, the skin between the rays may be eroded. The parasite uses an opisthaptor, with 16 small marginal hooks and two large central hooks, to attach to the host (Fig. 21.3). The tips of the two large central hooks can be withdrawn, but when the parasite needs a firmer attachment, the tips of the large hooks are protruded through small holes and burrowed into the host epidermis (Fig. 21.4). The damage to the skin due to grazing by the parasite and the holes made by the hooks allow inflow of fresh water into the body of the fish, which may suffer from osmoregulatory

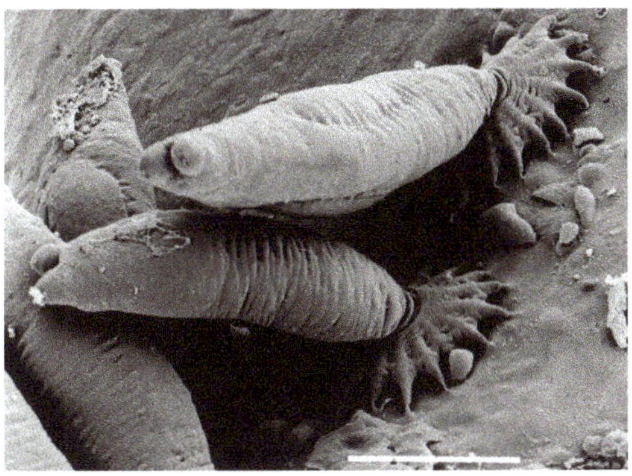

Fig. 21.2. *Gyrodactylus salaris* attached to the skin of an Atlantic salmon parr in a scanning electron microscope. Scale bar = 100 µm.

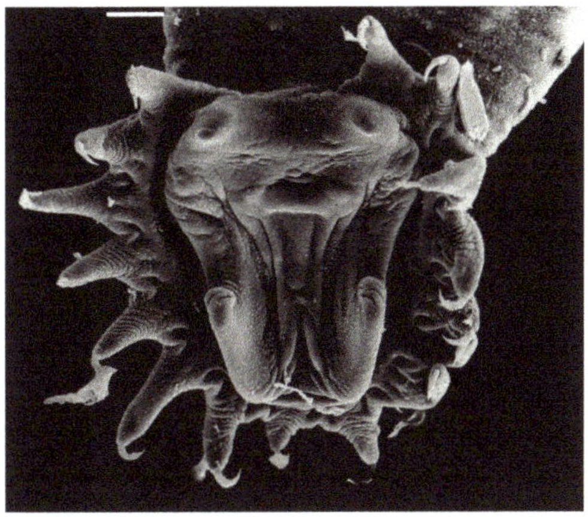

Fig. 21.3. Ventral side of the attachment organ, the opisthaptor, of *Gyrodactylus salaris* in a scanning electron microscope. Scale bar = 10 µm.

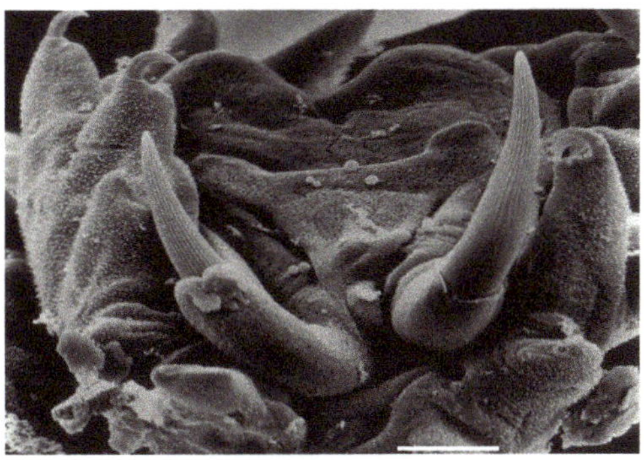

Fig. 21.4. Ventral side of the attachment organ of *Gyrodactylus derjavinoides*, with points of the two large hooks released, in a scanning electron microscope. Scale bar = 10 µm.

failure and eventually die, especially when the number of parasites is critically high (Pettersen *et al.*, 2013).

For identification, specimens of parasite are removed using a fine pair of forceps from either a dead or anesthetized fish. These specimens are identified to species either based on the shape and size of the hooks and bars in the attachment organ or using molecular techniques (OIE, 2019b).

Each *Gyrodactylus* specimen can be prepared for both morphologic/morphometric and molecular identification. A (new) scalpel is used to cut the specimen in two halves. This is most easily done when the specimen is preserved in ethanol. The posterior part including the opisthaptor (Fig. 21.5) is used for morphological/morphometric identification and the anterior part for molecular identification. The opisthaptor can be fixed and mounted on a slide by adding ammonium picrate-glycerine (Malmberg's fixative) to the edge of the cover glass (Malmberg, 1957). Alternatively, the more commonly used technique is to add a drop of Proteinase K to the edge of the cover glass. This will digest the soft tissue and release the hooks. They will now more visible and better oriented for species identification. The anterior part of the parasite can be processed for molecular identification (see below).

For storage of *Gyrodactylus* specimens and e.g. deposition to a museum, the preparation in Malmberg's fixative is sealed by adding nail polish to the edge of the cover glass or the fixed specimen will be remounted in glycerine-gelatin or in Canada balsam (Malmberg, 1970).

The morphological and morphometric variations have been thoroughly studied (Mo, 1991a,b,c; Mo *et al.*, 2001, 2004, 2010; Olstad *et al.*, 2009). In general, the hooks and the ventral bar in the opisthaptor increase in size at decreasing water temperatures and vice versa (Mo, 1991a,b,c). Thus, it is expected the size of the opisthaptor will be affected due to climate change. However, independent of size, the morphological characters can only identify *G. salaris* to species and cannot be used for discrimination between different strains. Thus, molecular methods have mostly taken over in the identification of *G. salaris* and the different strains. However, in many cases the morphological methods can quickly be used to identify specimens to species and avoid the costly and laborious molecular processes.

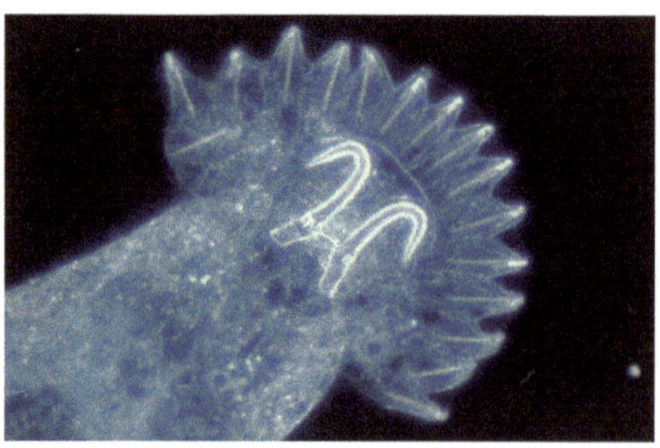

Fig. 21.5. The opisthaptor in *Gyrodactylus salaris* seen in a light microscope.

A part of the internal transcribed spacer (ITS) region of ribosomal DNA is used for molecular identification of a *Gyrodactylus* species (Cunningham, 1997), while regions of the mitochondrial DNA are used to separate strains of *G. salaris* (Meinilä *et al.*, 2002, 2004; Hansen *et al.*, 2003). *G. salaris* is considered a severe pathogen by the World Organization for Animal Health (OIE) (OIE, 2019a) and an OIE reference laboratory is appointed to be responsible for a manual concerning this parasite. In a suspected case, it is recommended that the identification of *G. salaris* follows the procedures in the OIE Manual of Diagnostic Tests for Aquatic Animals (OIE, 2019b), which can be accessed online at www.oie.int (accessed 1 April 2020). Upon request, the OIE reference laboratory will provide advice and technical support and send preserved parasite specimens for morphological comparisons or DNA extracts for positive control in genetic analysis. The OIE member countries have to report positive observations of *G. salaris* on wild and farmed fish, and their diagnostic laboratories may send specimens to the OIE reference laboratory for verification of their diagnosis.

The surveillances of wild and farmed fish are based on traditional examination of whole fish or excised fins under a stereo microscope (Mo *et al.*, 2004). Recently, environmental DNA (eDNA) methods have been developed to detect the presence of *G. salaris* in water samples (Rusch *et al.*, 2018; Fossøy *et al.*, 2020). These methods, which are non-lethal to fish, are less costly and at least as sensitive as the traditional examination methods, and thus an eDNA method will probably soon replace the traditional methods.

There is no established correlation between pathogenicity and the genetic strains (e.g. haplotypes) of *G. salaris*. For example, the haplotype F which has been named the 'rainbow trout *G. salaris*' consists of both pathogenic (Hansen *et al.*, 2003) and non-pathogenic variants (Olstad *et al.*, 2007; Ramirez *et al.*, 2014). Several haplotypes are found on rainbow trout, and thus the 'rainbow trout *G. salaris*' consists of a complex of haplotypes. The pathogenicity of *G. salaris* strains found on farmed salmonids in Europe is mostly unknown. Probably, most of them are non-pathogenic to local wild salmonids but some may cause severe epizootics in farmed and wild fish if translocated to new areas.

21.3 Expected/Potential Spread of the Pathogen

G. salaris has a direct life cycle with only one host. As in all gyrodactylids, *G. salaris* gives birth and both the mother and newborn, which is almost of the same size as its mother (Fig. 21.6), occur on the same host. Transmission to a new host may happen by a 'jump over' when two fish have physical contact or are close to each other; for example, when living in shoals or when fish aggregate during the spawning season. Transmission when fish are nipping at an infested dead host may also occur (Olstad *et al.*, 2006). Alternatively, gyrodactylids can detach from the host and accidentally find a new host in the water column, or the parasite can sink to the bottom and actively search for a host living near or in contact with the substrate. This transmission may be more common for solitary fish or fish living close to the bottom. A searching activity at the bottom of a Petri dish and attachment to cut-off fins has been observed for *G. salaris* (T.A. Mo, 1985, unpublished results). As salmon parr live close to and occasionally in contact with the river bottom, this may explain why *G. salaris* is frequently observed on pectoral, anal and pelvic fins of this host species (Jensen and Johnsen, 1992; Mo, 1992).

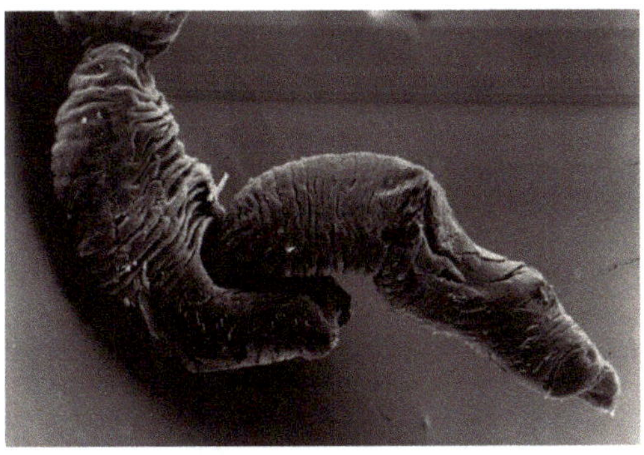

Fig. 21.6. The birth of *Gyrodactylus salaris* in a scanning electron microscope. Note that the newborn is almost the same size as its mother.

The reproduction and survival of *G. salaris* are largely dependent on the water temperature. The parasite gives birth at temperatures between 2.5 and 19°C (Jansen and Bakke, 1991) and its temperature optimum is likely somewhere around 10°C. A rise in the water temperature closer to the temperature optimum (e.g. due to climate change) may result in more adaptable parasites which can survive in new habitats and new geographical areas or even on new hosts.

21.3.1 Potential spread of *Gyrodactylus salaris* to new geographical areas

A natural spread of *G. salaris* to new geographical areas is dependent on migration of the fish host. *G. salaris* is adapted to fresh water but survives for some time in brackish water dependent on salinity and water temperature (Soleng and Bakke, 1997). The salinity tolerance of *G. salaris* has been studied in detail only for one strain (haplotype F). However, as all three pathogenic strains of *G. salaris* in Norway have spread with migrating salmonids between rivers in fjords, the salinity tolerance may be similar for most, if not all, non-grayling *G. salaris* strains. Potentially, *G. salaris* has a natural occurrence in all salmon rivers draining into the Baltic Sea, but studies confirming the presence or absence of the parasite in many of the Baltic salmon rivers seem to be lacking. Outside the enzootic area, a further natural spread of *G. salaris* depends on host migration in coastal areas and, so far, high salinities in these areas has limited this spread. However, climate change may result in heavier and more prolonged rainfalls in north-east Europe (Mäkinen *et al.*, 2018). This results in increased river floods (Thober *et al.*, 2018) and decreased salinity in the coastal areas. Thus, *G. salaris* can survive for longer periods and longer distances on migrating Atlantic salmon parr and smolt than previously. Possibly, this explains the recent northward spread of *G. salaris* to more rivers on the Swedish Kattegat coast and it is feared that the parasite will spread further north towards larger Atlantic salmon rivers in Norway. A similar spread can be expected from the River Keret to more rivers draining into the White Sea and from the Tuloma and Kola rivers to more rivers draining into the Barents Sea. Furthermore, increased yearly mean water temperatures and shorter ice-covered periods may result in the establishment of Atlantic salmon in northern rivers that are currently too cold, such as in rivers east of the White Sea. Potentially, *G. salaris* may also establish together with the salmon.

Climate warming in northern areas including the Arctic is expected to facilitate more human activities at progressively higher latitudes (Ricciardi *et al.*, 2017) and there will be more human-mediated movements of non-indigenous fish to northern areas and a further natural spread of introduced species (Chan *et al.*, 2019). The spread of *G. salaris* to new geographic areas and localities, including fish farms, lakes and rivers, has mostly been due to human translocations of infested salmonids. In some rivers, the introduction of pathogenic strains of *G. salaris* has resulted in epizootics and significant host mortality, while in other localities introduced non-pathogenic strains of *G. salaris* have no known negative effect on the host population. In Russia, translocation of *G. salaris*-infested fish caused an epizootic outbreak in the Atlantic salmon population in the River Keret which drains into the White Sea (Kuusela *et al.*, 2005). Because of increased water floods in the rivers and the resulting reduced salinity in the coastal areas, it is feared that *G. salaris* will be spread to more rivers that drain into the White Sea in the future. However, if the epizootics continue to kill most of the parr in the River Keret, a low number of smolt leave the river and this slows down the potential for further spread of the parasite to more rivers in the area. On the other hand, stocking of hatchery-reared salmon parr to compensate for the mortality will increase the risk for further spread. In the White Sea drainages, *G. salaris* has also been spread to the landlocked salmon in the Russian River Pista, most likely from a fish farm upstream in Finland (Artamonova *et al.*, 2011). However, the first observations indicate that the salmon stock is parasite tolerant (Kuusela *et al.*, 2005) and the potential for further spread is small. Recently, *G. salaris* was found on Atlantic salmon parr in the Russian rivers Tuloma and Kola draining into the Barents Sea near the city of Murmansk (H. Hansen, Oslo, 2019, personal communication). Likely, translocation of infested rainbow trout explains the spread. So far, it is not known whether the introduction caused an epizootic or not. Again, increased river floods because of climate change may increase the potential for further spread to more rivers draining into the Barents Sea. In these rivers in the northernmost part of the Atlantic salmon distribution, an increase in the yearly mean water temperature may increase the severity of the epizootics.

In addition to translocation of infested fish, *G. salaris* can be spread with equipment, such as fishing nets, waders, canoes etc., if they are not completely dried before they are used again.

21.3.2 Potential spread of *Gyrodactylus salaris* to new host species

G. salaris is also able to switch to and establish in new hosts; for example, migrating adult salmon in northern Norway spread *G. salaris* (haplotype A) to Arctic charr populations in three lakes. After salmon ladders were closed, *G. salaris* was maintained on resident charr in the lakes for more than 20 years until the parasite was simultaneously eradicated from all three lakes and the river downstream. Further north in Norway, another *G. salaris* strain (haplotype B) was frequently found on anadromous Arctic charr living sympatric with Atlantic salmon (Winger *et al.*, 2008) in three rivers until the parasite was simultaneously eradicated. In southern Norway, a variant of haplotype F of *G. salaris* is found on Arctic charr in several lakes draining into the Atlantic salmon river Numedalslågen. Probably, infested rainbow trout were stocked into the lakes and *G. salaris* switched to resident Arctic charr. Rainbow trout could not establish self-sustaining populations and died out, and *G. salaris* has since survived in low numbers on the resident Arctic charr populations (Robertsen *et al.*, 2008). Based on the mitochondrial cytochrome oxidase 1 gene, this *G. salaris* strain is identical to haplotype F but differs by a single base change in the ribosomal segment (Ramirez *et al.*, 2014), i.e. in the ITS1, 5.8S rDNA and ITS2, commonly used for species identification in the genus *Gyrodactylus* (Cunningham, 1997). Laboratory experiments have shown that this variant of haplotype F is non-pathogenic on Atlantic salmon, but it can survive on salmon parr in low numbers for many weeks (Olstad *et al.*, 2007; Ramirez *et al.*, 2014). This strain has not been observed on hundreds of examined Atlantic salmon parr in the River Numedalslågen downstream from the lakes. The same non-pathogenic variant of haplotype F has been found in farmed rainbow trout in Denmark (Lindenstrøm *et al.*, 2003). In general, Arctic charr can be a reservoir host for pathogenic and non-pathogenic strains of *G. salaris* although some Arctic charr populations are resistant (Bakke *et al.*, 1996).

The introduction of North American salmonids to Europe has increased the number of potential reservoir hosts for *G. salaris*. In addition to the numerous freshwater rainbow trout farms throughout Europe, rainbow trout has established self-reproducing populations in several European countries (Stanković *et al.*, 2015). Many of these wild rainbow trout populations need to be examined. *G. salaris* may occur in some of them, which potentially can contribute to further spread of the parasite with fish migration and establishment in even more new geographical areas. Another potential risk is associated with the introduction of pink salmon from the Pacific Ocean to rivers draining into the White Sea and Barents Sea during the second half of the 20th century (Niemelä *et al.*, 2016). In 2017, pink salmon spawned in more than 200 rivers along the Norwegian coast and this may indicate that pink salmon is in the process of increasing its geographical range in rivers draining into the Atlantic Ocean (Mo *et al.*, 2018; Sandlund *et al.*, 2018). As *G. salaris* survives and reproduces on rainbow trout, the parasite may also use the closely related pink salmon as a reservoir host. If this is so, pink salmon may contribute to the spread of *G. salaris* as they migrate in estuaries and coastal areas during their first weeks after leaving their natal river (Sandlund *et al.*, 2018).

21.4 Potential Effects of Climate Change on *Gyrodactylus salaris*

The development and reproduction of *G. salaris*, and thus the population dynamics, are largely dependent on water temperature as reproduction is positively correlated to temperatures between 2.5 and 19°C (Jansen and Bakke, 1991). This results in significant seasonal variations in populations sizes (parasite abundances) on Atlantic salmon parr in Norwegian rivers (Mo, 1992; Jansen and Bakke, 1993; Appleby and Mo, 1997). Parasite abundances are lowest in the late winter and increase during the spring and summer. Because of the very high abundances, mortality of Atlantic salmon parr occurs in the late autumn and affects mainly parr in their first (0+) and second year (1+). Comparable studies have not been done in the natural range of *G. salaris* but similar seasonal variations can be expected although at much lower abundances.

It is not known how an increase in the water temperature due to climate change will affect *G. salaris* and outbreaks of clinical gyrodactylosis. The reproduction of the parasite will increase but this can be offset by a better host response against the parasite. However, a rise in the water temperature may result in a longer period of the year with temperatures above 3–4°C, with a net increase in the number of parasites. Thus, an increased negative effect on the host populations can be expected. This may apply to both farmed and wild fish. Therefore, outbreaks of clinical gyrodactylosis in fish farms may become more severe and frequent, resulting in an increased need for chemotherapeutic treatments.

A lowered pH as a consequence of increased precipitation and more melting of frozen structures (e.g. glaciers, the Poles), especially in interaction with metals, might affect Atlantic salmon parr negatively (Poléo, 1995); consequently, salmon parr could be more vulnerable to *G. salaris*. However, a lowered pH might also affect *G. salaris* negatively, especially if the content of metals (e.g. aluminium, zinc) increases in the water because of more leaching from the soil (Soleng *et al.*, 1999, 2005; Poléo *et al.*, 2004a). Levels of aluminium ions have increased in some rivers, mainly after heavy rainfalls. In addition, field observations indicate that the toxicity of aluminium increases with increasing water temperature (Poléo *et al.*, 2004b). Thus, the outcome of a lowered pH could be increased *G. salaris*-induced host mortality, but it could also lead to reduced parasite survival and reproduction.

21.5 Control and Prevention of *Gyrodactylus salaris*

21.5.1 Chemotherapeutic treatments of fish in captivity

Historically, treatments with salt water (20–25% salinity for 30 min) and formalin (1:4000–6000 for 30 min) have been used against ectoparasites including *Gyrodactylus* spp. Such treatments normally do not eradicate the parasite from a farm and multiple and new treatments must be carried out on a regular basis. Numerous compounds have been studied to kill *Gyrodactylus* spp. in farmed and ornamental fish; Schelkle *et al.* (2009) had a comprehensive presentation of the various compounds. More compounds such as hydrogen peroxide (Pietrak and Backman, 2018), the anthelminthic plant extract arctigenin (Tu *et al.*, 2018) and the insecticide Timor C (Zorin *et al.*, 2019) have been studied. Although most compounds tested against *Gyrodactylus* spp. are effective, 100% efficacy has not been achieved without toxicity to hosts. Leaving just one (hermaphrodite, viviparous) worm can be enough to initiate a new disease outbreak (Schelkle *et al.*, 2009).

Because many *Gyrodactylus* species cause gyrodactylosis in several farmed and ornamental fish species, new chemotherapeutic agents can be expected to be tested and developed in the future. Some of these may also be effective against *G. salaris* if salmonids can tolerate the treatment.

21.5.2 Eradication of *Gyrodactylus salaris* on wild fish

The piscicide rotenone has mainly been used for the eradication of *G. salaris* from Norwegian rivers (Johnsen *et al.*, 1989; Sandodden *et al.*, 2004, 2018). In 42 rivers, rotenone has been used to kill all the fish and thus the parasite, which has no intermediate host or resting stages (Sandodden *et al.*, 2018). After a treatment, each river is stocked with offspring of the original Atlantic salmon population which has been kept in a *G. salaris*-free hatchery during the treatment period. Of the 42 rotenone-treated rivers, 37 rivers are declared free from *G. salaris* and five are in a 5-year surveillance period before they can be declared free. In addition, one river has been declared free from *G. salaris* after the use of aluminium sulfate and sulfuric acid which kills *G. salaris* but not its host (Hindar *et al.*, 2014). Thus, currently (2020), pathogenic strains of *G. salaris* are present in only eight Norwegian rivers. All eight rivers are in a process for the eradication of the parasite. Because two of the rivers are relatively large and complicated for a full chemical eradication, the processes will take longer than usual in other rivers. The first goal is to reduce the spread of *G. salaris* in the watercourses. In one river, a very large salmonid migration barrier has been built to prevent spawning of anadromous salmonids above the barrier (Fig. 21.7). After some years (i.e. the maximum smolt age in each river) all Atlantic salmon parr above the barrier will disappear and thus also *G. salaris* which is not able to survive on the resident brown trout, the only other salmonid in the rivers. Then, *G. salaris* can be chemically eradicated in a restricted lower part of the watercourse.

As rotenone also kills numerous invertebrates, a more environmentally friendly alternative has been in demand – preferably an alternative that kills only *G. salaris* and not fish and other invertebrates. This inspired scientists to develop the aluminium sulfate (AlS) method (Soleng *et al.*, 1999, 2005; Poléo *et al.*, 2004b) which kills *G. salaris* but not Atlantic salmon or other fish in the river, and only temporarily affects invertebrates. In 2011 and 2012, the River Lærdalselva, located in western Norway, was treated in two 14-day periods each year with aluminium (25–30 μg Al/l). To achieve the desired effect of aluminium, the pH must be reduced to 5.7–5.9 and this was achieved by adding sulfuric acid (Hindar *et al.*, 2014). Five years later, in 2017, the River Lærdalselva was declared free from *G. salaris*. In large rivers (>100 m^3/s) the AlS method may not be applicable because of the large volumes of sulfuric acid needed. Thus, another environmentally friendly method based on chlorine (Hagen *et al.*, 2014) is under development. Low concentrations of sodium hypochlorite (<30 μg Cl/l), much lower than those commonly added to drinking water (>200 μg Cl/l), are toxic to *G. salaris* while this has no measurable effects on Atlantic salmon parr (Hagen *et al.*, 2014). If the further development is successful, the 'chlorine method' may be used for future eradication of *G. salaris* in the remaining infested rivers in Norway.

The main reason for eradicating *G. salaris* from Norwegian rivers is to preserve the iconic and vulnerable Atlantic salmon. However, in some rivers the Atlantic salmon parr are hosts for the parasitic glochidia stages of the vulnerable freshwater pearl mussel (*Margaritifera margaritifera* L.). Because of the dramatic decline in number of hosts, the reproduction and survival of the freshwater pearl mussel are threatened. A similar decline in salmon hosts for glochidia larvae has been observed in the River Keret, north-west Russia (Kudersky *et al.*, 2003; Makhrov *et al.*, 2011). Thus, the eradication of *G. salaris* will therefore have a double effect in some rivers by saving two vulnerable species for which at least Norway has a special conservation responsibility.

21.5.3 Preventive measures against *Gyrodactylus salaris*

Although there is resistance (innate and acquired) against *G. salaris* (Bakke *et al.*, 1990a, 2002; Dalgaard *et al.*, 2003), clinical outbreaks of gyrodactylosis in farmed salmonids are not sufficiently serious to cover the costs of selective breeding or the development of a vaccine, and the outcome of a breeding programme or vaccine development may be uncertain. In most cases, clinical gyrodactylosis in farmed fish can be controlled by chemotherapeutic treatments (see Section 21.5.1).

Fig. 21.7. Fish migration barrier in the River Driva, western Norway, to eliminate potential hosts for *Gyrodactylus salaris* upstream of the barrier. (Photograph by Sigurd Hytterød.)

Selective breeding of wild Atlantic salmon and stocking more *G. salaris*-tolerant fish in the affected rivers has been discussed (Salte *et al.*, 2010; Zueva *et al.*, 2018). This could potentially be successful, but the process would be costly and time-consuming. In addition, the Atlantic salmon populations are managed separately in each river in accordance with national and international recommendations, such as those of the North Atlantic Salmon Conservation Organization (NASCO). Selective breeding would have to be done in each *G. salaris*-affected river and would conflict with the recommended conservation of intact population gene pools. Furthermore, rivers with a future resistant salmon population would still be infested and *G. salaris* and the parasite would likely spread to more rivers to cause new epizootics.

21.5.4 Disinfection

Sport fishing, paddling and many more human activities can spread organisms infectious to fish, including *G. salaris*, if the equipment used in infected waters is not sufficiently disinfected. In the case of *G. salaris*, freezing or complete drying of the equipment efficiently kills all the parasites which have no eggs or other tolerant stages. Detached *G. salaris* could survive for many hours or even days (Olstad *et al.*, 2006), and thus wet equipment should be disinfected to avoid spreading *G. salaris* to parasite-free farms, lakes or rivers. A commonly used disinfectant, Virkon S®, effectively kills *G. salaris* in less than 30 s in a 1% solution (Koski *et al.*, 2016). However, many anglers fear that a disinfectant may harm their valuable equipment. A preferred alternative method is hot water (45–50°C), which also kills *G. salaris* within a few seconds (Koski *et al.*, 2016). In addition, hot water is an environmentally friendly alternative and, in most cases, much cheaper than a chemical disinfectant.

21.6 Conclusions and Suggestions for Future Studies

In Norway, the authorities have decided to eradicate *G. salaris* in accordance with the Convention on Biological Diversity (www.cbd.int (accessed 1 April 2020)) and this parasite is now eradicated from all (39) fish farms and currently (2020) from 43 of the 51 infected rivers. The remaining eight infested rivers are in the process of eradication of the parasite. In other geographical regions there are no similar eradication plans. Thus, there is a potential for further spread of the parasite. Hence there is a need for active surveillance for early detection. Preferably, *G. salaris*, and not only clinical gyrodactylosis, should be notifiable in relevant countries to avoid a spread of the parasite with fish transports especially in the years ahead.

The OIE has included *G. salaris* in the list of notifiable aquatic animal diseases (OIE, 2019a) and OIE member countries are required to report detections of *G. salaris* in farmed or wild fish. However, as several strains of *G. salaris* are non-pathogenic, discrimination between pathogenic and non-pathogenic strains is needed. This is especially important if the synonymizing of *G. salaris* and *G. thymalli* is accepted. The OIE has not accepted the synonymizing, partly because the International Code of Zoological Nomenclature (www.iczn.org (accessed 1 April 2020)) has not been followed and partly for practical reasons. The synonymizing will result in a comprehensive revision of both the OIE code (OIE, 2019a) and the OIE manual (OIE, 2019b). A separation between genetic clades and haplotypes is not a useful tool to separate pathogenic from non-pathogenic strains because at least one haplotype (F) includes both pathogenic and non-pathogenic strains. Thus, studies to determine the connection between pathogenicity and genetics are needed. At the same time, it is necessary to study the pathogenicity of the many *G. salaris* strains in Europe as non-pathogenic strains are of no or limited concern to the national fish health managements and it is not necessary to report their occurrence to the OIE.

Studies including the ecological effects of *G. salaris* are lacking. So far, only the direct effects on Atlantic salmon parr and the salmon population have been studied. The Atlantic salmon is an important predator in many northern European rivers and a significant reduction of this fish species may have several consequences for other fish, prey species, and for the freshwater pearl mussel which is dependent on salmon parr to complete its life cycle. Thus, the introduction of *G. salaris* may affect the entire ecosystem in a river. This should be studied and can also give us an increased understanding of the significance of parasites on the animal biodiversity in aquatic ecosystems.

It should also be studied because relevant climate factors, such as water temperature, pH, increased precipitation and decreased salinity in coastal areas, will affect the *G. salaris*–Atlantic salmon association. Potentially *G. salaris* may adapt to climate changes much faster than Atlantic salmon; for example, resulting in more virulent parasite strains that can threaten salmon stocks which are tolerant today or resulting in more salinity-tolerant parasite strains with increased potential for further spread to new rivers.

References

Anttila, P., Romakkaniemi, A., Kuusela, J. and Koski, P. (2008) Epidemiology of *Gyrodactylus salaris* (Monogenea) in the River Tornionjoki, a Baltic wild salmon river. *Journal of Fish Diseases* 31, 373–382.

Appleby, C. and Mo, T.A. (1997) Population dynamics of *Gyrodactylus salaris* (Monogenea) infecting Atlantic salmon, *Salmo salar*, parr in the River Batnfjordselva, Norway. *Journal of Parasitology* 83, 23–30.

Artamonova, V.S., Makhrov, A.A., Shulman, B.S., Khaimina, O.V., Yurtseva, A.O. *et al.* (2011) Response of the Atlantic salmon (*Salmo salar* L.) population of the Keret River to the invasion of parasite *Gyrodactylus salaris* Malmberg. *Russian Journal of Biological Invasions* 2, 73–80.

Bakke, T.A. (1991) A review of the inter- and intraspecific variability in salmonid hosts to laboratory infections with *Gyrodactylus salaris* Malmberg. *Aquaculture* 98, 303–310.

Bakke, T.A. and Mackenzie, K. (1993) Comparative susceptibility of native Scottish and Norwegian stocks of Atlantic salmon, *Salmo salar* L., to *Gyrodactylus salaris* Malmberg: laboratory experiments. *Fisheries Research* 17, 69–85.

Bakke, T.A. and Sharp, L.A. (1990) Susceptibility and resistance of minnows, *Phoxinus phoxinus* (L.) to *Gyrodactylus salaris* Malmberg, 1957 (Monogenea) under laboratory conditions. *Fauna Norwegica, Serie A* 11, 51–55.

Bakke, T.A., Jansen, P.A. and Hansen, L.P. (1990a) Differences in the host resistance of Atlantic salmon, *Salmo salar* L., stocks to the monogenean *Gyrodactylus salaris* Malmberg 1957. *Journal of Fish Biology* 37, 577–587.

Bakke, T.A., Jansen, P.A. and Brabrand, Å. (1990b) Susceptibility and resistance of brook lamprey, *Lampetra planeri* (Bloch), roach, *Rutilus rutilus* (L.) and perch, *Perca fluviatilis* L. to *Gyrodactylus salaris* Malmberg (Monogenea). *Fauna Norwegica, Serie A* 11, 23–26.

Bakke, T.A., Jansen, P.A. and Kennedy, C.R. (1991a) The host specificity of *Gyrodactylus salaris* Malmberg (Platyhelminthes, Monogenea): susceptibility of *Oncorhynchus mykiss* (Walbaum) under experimental conditions. *Journal of Fish Biology* 39, 45–57.

Bakke, T.A., Jansen, P.A. and Hansen, L.P. (1991b) Experimental transmission of *Gyrodactylus salaris* Malmberg, 1957 (Platyhelminthes, Monogenea) from the Atlantic salmon (*Salmo salar*) to the European eel (*Anguilla anguilla*). *Canadian Journal of Zoology* 69, 733–737.

Bakke, T.A., Harris, P.D. and Jansen, P.A. (1992a) The susceptibility of *Salvelinus fontinalis* (Mitchill) to *Gyrodactylus salaris* Malmberg (Platyhelminthes; Monogenea) under experimental conditions. *Journal of Fish Biology* 41, 499–507.

Bakke, T.A., Jansen, P.A. and Grande, M. (1992b) The susceptibility of *Salvelinus namaycush* (Walbaum) to *Gyrodactylus salaris* Malmberg (Platyhelminthes; Monogenea) under experimental conditions. *Fauna Norwegica, Serie A* 13, 1–7.

Bakke, T.A., Jansen, P.A. and Harris, P.D. (1996) Differences in susceptibility of anadromous and resident stocks of Arctic charr to infections of *Gyrodactylus salaris* under experimental conditions. *Journal of Fish Biology* 48, 341–351.

Bakke, T.A., Harris, P.D. and Cable, J. (2002) Host specificity dynamics: observations on gyrodactylid monogeneans. *International Journal for Parasitology* 32, 281–308.

Bakke, T.A., Harris, P.D., Hansen, H., Cable, J. and Hansen, L.P. (2004) Susceptibility of Baltic and east Atlantic salmon *Salmo salar* stocks to *Gyrodactylus salaris* (Monogenea). *Diseases of Aquatic Organisms* 58, 171–177.

Bakke, T.A., Cable, J. and Harris, P.D. (2007) The biology of gyrodactylid monogeneans: the 'Russian-doll killers'. *Advances in Parasitology* 64, 161–376.

Bourret, V., Kent, M.P., Primmer, C.R., Vasemagi, A., Karlsson, S. *et al.* (2013) SNP-array reveals genome-wide patterns of geographical and potential adaptive divergence across the natural range of Atlantic salmon (*Salmo salar*). *Molecular Ecology* 22, 532–551.

Buchmann, K. and Bresciani, J. (1997) Parasitic infections in pond-reared rainbow trout *Oncorhynchus mykiss* in Denmark. *Diseases of Aquatic Organisms* 28, 125–138.

Buchmann, K., Uldal, A. and Lyholt, H.C.K. (1995) Parasite infections in Danish trout farms. *Acta Veterinaria Scandinavica* 36, 283–298.

Chan, F.T., Stanislawczyk, K., Sneekes, A.C., Dvoretsky, A., Gollasch, S. *et al.* (2019) Climate change opens new frontiers for marine species in the Arctic: current trends and future invasion risks. *Global Change Biology* 25, 25–38.

Cunningham, C.O. (1997) Species variation within the internal transcribed spacer (ITS) region of *Gyrodactylus* (Monogenea: Gyrodactylidae) ribosomal RNA genes. *Journal of Parasitology* 83, 215–219.

Cunningham, C.O., Collins, C.M., Malmberg, G. and Mo, T.A. (2003) Analysis of ribosomal RNA intergenic spacer (IGS) sequences in species and populations of *Gyrodactylus* (Platyhelminthes: Monogenea) from salmonid fish in northern Europe. *Diseases of Aquatic Organisms* 57, 237–246.

Dalgaard, M.B., Nielsen, C.V. and Buchmann, K. (2003) Comparative susceptibility of two races of *Salmo salar* (Baltic Lule river and Atlantic Conon river strains) to infection with *Gyrodactylus salaris*. *Diseases of Aquatic Organisms* 53, 173–176.

Degerman, E., Petersson, E., Jacobsen, P.-E., Karlsson, L., Lettevall, E. and Nordwall, F. (2012) Laxparasiten *Gyrodactylus salaris* i västkustens laxåar. *Aqua Reports* 8, 1–68. (in Swedish)

Ergens, R. (1983) *Gyrodactylus* from Eurasian freshwater Salmonidae and Thymallidae. *Folia Parasitologica* 30, 15–26.

Forseth, T., Barlaup, B.T., Finstad, B., Fiske, P., Gjøsæter, H. *et al.* (2017) The major threats to Atlantic salmon in Norway. *ICES Journal of Marine Science* 74, 1496–1513.

Fossøy, F., Brandsegg, H., Sivertsgård, R., Pettersen, O., Sandercock, B.K. *et al.* (2020) Monitoring presence and abundance of two gyrodactylid ectoparasites and their salmonid hosts using environmental DNA. *Environmental DNA* 2, 53–62.

Fromm, B., Burow, S., Hahn, C. and Bachmann, L. (2014) MicroRNA loci support conspecificity of *Gyrodactylus salaris* and *Gyrodactylus thymalli* (Platyhelminthes: Monogenea). *International Journal for Parasitology* 44, 787–793.

Hagen, A.G., Hytterød, S. and Olstad, K. (2014) Low concentrations of sodium hypochlorite affect population dynamics in *Gyrodactylus salaris* (Malmberg, 1957): practical guidelines for the treatment of the Atlantic salmon, *Salmo salar* L. parasite. *Journal of Fish Diseases* 37, 1003–1011.

Hansen, H., Bachmann, L. and Bakke, T.A. (2003) Mitochondrial DNA variation of *Gyrodactylus* spp. (Monogenea, Gyrodactylidae) populations infecting Atlantic salmon, grayling, and rainbow trout in Norway and Sweden. *International Journal for Parasitology* 33, 1471–1478.

Hansen, H., Cojocaru, C.-D. and Mo, T.A. (2016) Infections with *Gyrodactylus* spp. (Monogenea) in Romanian fish farms: *Gyrodactylus salaris* Malmberg, 1957 extends its range. *Parasites and Vectors* 9, 444.

Hindar, A., Garmo, Ø., Hagen, A.G., Hytterød, S., Høgberget, R. *et al.* (2014) Tiltak med AlS for utryddelse av lakseparasitten *Gyrodactylus salaris* i Lærdalselva i 2011 og 2012. *NIVA Rapport Nr. 6701-2014*. Norsk institutt for vannforskning, Oslo. (in Norwegian)

Hytterød, S., Darrud, M., Johansen, K., Larsen, S., Mohammad, S.N. and Hansen, H. (2018) The surveillance programme for *Gyrodactylus salaris* in Atlantic salmon and rainbow trout in Norway 2017. *Surveillance Programmes in Norway Report from the Norwegian Veterinary Institute*. Norwegian Veterinary Institute, Oslo.

Ieshko, E., Barskaya, Y., Parshukov, A., Lumme, J. and Khlunov, O. (2016) Occurrence and morphogenetic characteristics of *Gyrodactylus* (Monogenea: Gyrodactylidae) from a rainbow trout farm (Lake Ladoga, Russia). *Acta Parasitologica* 61, 151–157.

Jansen, P.A. and Bakke, T.A. (1991) Temperature-dependent reproduction and survival of *Gyrodactylus salaris* Malmberg, 1957 (Platyhelminthes: Monogenea) on Atlantic salmon (*Salmo salar* L.). *Parasitology* 102, 105–112.

Jansen, P.A. and Bakke, T.A. (1993) Regulatory processes in the monogenean *Gyrodactylus salaris* Malmberg–Atlantic salmon (*Salmo salar* L.) association. I. Field studies in southeast Norway. *Fisheries Research* 17, 87–101.

Jansen, P.A. and Bakke, T.A. (1995) Susceptibility of brown trout to *Gyrodactylus salaris* (Monogenea) under experimental conditions. *Journal of Fish Biology* 46, 415–422.

Jansen, P.A., Matthews, L. and Toft, N. (2007) Geographic risk factors for inter-river dispersal of *Gyrodactylus salaris* in fjord systems in Norway. *Diseases of Aquatic Organisms* 74, 139–149.

Jensen, A.J. and Johnsen, B.O. (1992) Site specificity of *Gyrodactylus salaris* Malmberg, 1957 (Monogenea) on Atlantic salmon (*Salmo salar* L) in the River Lakselva, Northern Norway. *Canadian Journal of Zoology* 70, 264–267.

Johnsen, B.O. (1978) The effect of an attack by the parasite *Gyrodactylus salaris* on the population of salmon parr in the river Lakselva, Misvær in Northern Norway. *Astarte* 11, 7–9.

Johnsen, B.O. and Jensen, A.J. (1986) Infestations of Atlantic salmon, *Salmo salar*, by *Gyrodactylus salaris* in Norwegian rivers. *Journal of Fish Biology* 29, 233–241.

Johnsen, B.O. and Jensen, A.J. (1988) Introduction and establishment of *Gyrodactylus salaris* Malmberg, 1957, on Atlantic salmon, *Salmo salar* L., fry and parr in the River Vefsna, northern Norway. *Journal of Fish Diseases* 11, 35–45.

Johnsen, B.O. and Jensen, A.J. (1991) The *Gyrodactylus* story in Norway. *Aquaculture* 98, 289–302.

Johnsen, B.O. and Jensen, A.J. (1992) Infection of Atlantic salmon, *Salmo salar* L., by *Gyrodactylus salaris*, Malmberg 1957, in the River Lakselva, Misvær in northern Norway. *Journal of Fish Biology* 40, 433–444.

Johnsen, B.O. and Jensen, A.J. (2003) *Gyrodactylus salaris* in Norwegian rivers. In: Veselov, A.J., Ieshko, E.P., Nemova, N.N., Sterligova, O.P. and Shustov, Y.A. (eds) *Atlantic Salmon: Biology, Conservation and Restoration*. Russian Academy of Sciences, Karelian Research Center, Institute of Biology, Petrozavodsk, Russia, pp. 38–44.

Johnsen, B.O., Jensen, A.J. and Sivertsen, B. (1989) Extermination of *Gyrodactylus salaris*-infected Atlantic salmon *Salmo salar* by rotenone treatment in the river Vikja, western Norway. *Fauna Norwegica, Serie A* 10, 39–43.

Johnsen, B.O., Møkkelgjerd, P.I. and Jensen, A.J. (1999) Parasitten *Gyrodactylus salaris* på laks i norske vassdrag, statusrapport ved inngangen til år 2000. *NINA Oppdragsmelding Nr. 617*. Norsk institutt for naturforskning, Trondheim, Norway. (in Norwegian)

Johnston, C., Mackenzie, K., Cunningham, C.O., Eiras, J.C. and Bruno, D.W. (1996) Occurrence of *Gyrodactylus salaris* Malmberg, 1957, in Portugal. *Bulletin of the European Association of Fish Pathologists* 16, 89–91.

Jørgensen, L.V., Heinecke, R.D., Kania, P. and Buchmann, K. (2008) Occurrence of gyrodactylids on wild Atlantic salmon, *Salmo salar* L., in Danish rivers. *Journal of Fish Diseases* 31, 127–134.

Jørgensen, T.R., Larsen, T.B., Jørgensen, L.G., Bresciani, J., Kania, P.W. and Buchmann, K. (2007) Characterisation of a low pathogenic form of *Gyrodactylus salaris* from rainbow trout. *Diseases of Aquatic Organisms* 73, 235–244.

Koski, P., Anttila, P. and Kuusela, J. (2016) Killing of *Gyrodactylus salaris* by heat and chemical disinfection. *Acta Veterinaria Scandinavica* 58, 1–6.

Kudersky, L.A., Ieshko, E. and Schulman, B. (2003) Distribution range formation history of the monogenean *Gyrodactylus salaris* Malmberg, 1957 – a parasite of juvenile Atlantic salmon *Salmo salar* Linnaeus, 1758. In: Veselov, A.J., Ieshko, E.P., Nemova, N.N., Sterligova, O.P. and Shustov, Y.A. (eds) *Atlantic Salmon: Biology, Conservation and Restoration*. Russian Academy of Sciences, Karelian Research Center, Institute of Biology, Petrozavodsk, Russia, pp. 77–83.

Kuusela, J., Holopainen, R., Meinilä, M., Veselov, A., Shurov, I. *et al.* (2005) Potentially dangerous *Gyrodactylus salaris* in Russian Karelia: harmless and harmful combinations of host species and parasite strains. In: *Salmonid Fishes of Eastern Fennoscandia*, pp. 47–55. Available at: http://resources.krc.karelia.ru/krc/doc/publ/Salmonid_fishes_47-55.pdf (accessed 3 April 2020).

Kuusela, J., Zietara, M.S. and Lumme, J. (2007) Hybrid origin of Baltic salmon-specific parasite *Gyrodactylus salaris*: a model for speciation by host switch for hemiclonal organisms. *Molecular Ecology* 16, 5234–5245.

Kuusela, J., Holopainen, R., Meinila, M., Anttila, P., Koski, P. *et al.* (2009) Clonal structure of salmon parasite *Gyrodactylus salaris* on a coevolutionary gradient on Fennoscandian salmon (*Salmo salar*). *Annales Zoologici Fennici* 46, 21–33.

Lindenstrøm, T., Collins, C.M., Bresciani, J., Cunningham, C.O. and Buchmann, K. (2003) Characterization of a *Gyrodactylus salaris* variant: infection biology, morphology and molecular genetics. *Parasitology* 127, 165–177.

Lucký, Z. (1963) Occurence of *Gyrodactylus salaris* Malmberg, 1956 on the skin of *Salmo gairdnerii*. *Acta Universitatis Agriculturae, Brno* 11, 127–130. (in Czech)

Lux, E. (1990) *Gyrodactylus salaris* – Parazitierung von Salmoniden, ein diagnostisch-taxonomischen problem. Deutsche Veterinärmedizinische Gesellschaft eV, Giessen, Germany, pp. 86–98. (in German)

MacCrimmon, H.R. (1971) World distribution of rainbow trout (*Salmo gairdneri*). *Journal of the Fisheries Research Board of Canada* 28, 663–704.

Makhrov, A.A., Ieshko, E.P., Shchurov, I.L. and Shirokov, V.A. (2011) Distribution of the freshwater pearl mussel (*Margaritifera margaritifera*) in the Republic of Karelia (North-West Russia). *Toxicological and Environmental Chemistry* 93, 1731–1747.

Mäkinen, H., Kaseva, J., Trnka, M., Balek, J., Kersebaum, K.C. *et al.* (2018) Sensitivity of European wheat to extreme weather. *Field Crops Research* 222, 209–217.

Malmberg, G. (1957) Om förekomsten av *Gyrodactylus* på svenska fiskar. *Skrifter utgivan av Södra Sveriges Fiskeriförening, Årsskrift 1956*, 19–76. (in Swedish)

Malmberg, G. (1970) The excretory systems and the marginal hooks as a basis for the systematics of *Gyrodactylus* (Trematoda, Monogenea). *Arkiv för Zoolgi* 23, 1–235.

Malmberg, G. (1993) Gyrodactylidae and gyrodactylosis of Salmonidae. *Bulletin Français de la Peche et de la Pisciculture* 328, 5–46.

Malmberg, G. and Malmberg, M. (1993) Species of *Gyrodactylus* (Platyhelminthes, Monogenea) on salmonids in Sweden. *Fisheries Research* 17, 59–68.

Malmberg, G., Malmberg, M. and Carlstrand, H. (1995) *Gyrodactylus salaris* on salmon on the Swedish west coast. *Bulletin of the Scandinavian Society for Parasitology* 5, 45–45.

Meinilä, M., Kuusela, J., Zietara, M. and Lumme, J. (2002) Primers for amplifying ~820 bp highly polymorphic mitochondrial CO1 gene for *Gyrodactylus salaris*. *Hereditas* 137, 72–74.

Meinilä, M., Kuusela, J., Zietara, M.S. and Lumme, J. (2004) Initial steps of speciation by geographic isolation and host switch in salmonid pathogen *Gyrodactylus salaris* (Monogenea: Gyrodactylidae). *International Journal for Parasitology* 34, 515–526.

Mieszkowska, A., Górniak, M., Jurczak-Kurek, A. and Ziętara, M.S. (2018) Revision of *Gyrodactylus salaris* phylogeny inspired by new evidence for Eemian crossing between lineages living on grayling in Baltic and White sea basins. *PeerJ* 6, e5167.

Mo, T.A. (1987) Taxonomiske og biologiske undersakelser. Virksomheten i 1986 og forslag til virksornheten i 1987. *Gyrodactylus*-undersakelsene ved Zoologisk Museum, *Universitetet i Oslo. Rapport Nr.* 2, 1–69. (in Norwegian)

Mo, T.A. (1991a) Seasonal variations of opisthaptoral hard parts of *Gyrodactylus salaris* Malmberg, 1957 (Monogenea: Gyrodactylidae) on parr of Atlantic salmon *Salmo salar* L. in the River Batnfjordselva, Norway. *Systematic Parasitology* 19, 231–240.

Mo, T.A. (1991b) Variations of opisthaptoral hard parts of *Gyrodactylus salaris* Malmberg, 1957 (Monogenea: Gyrodactylidae) on parr of Atlantic salmon *Salmo salar* L. in laboratory experiments. *Systematic Parasitology* 20, 11–19.

Mo, T.A. (1991c) Variations of opisthaptoral hard parts of *Gyrodactylus salaris* Malmberg, 1957 (Monogenea: Gyrodactylidae) on rainbow trout *Oncorhynchus mykiss* (Walbaum, 1792) in a fish farm, with comments on the spreading of the parasite in south-eastern Norway. *Systematic Parasitology* 20, 1–9.

Mo, T.A. (1992) Seasonal variations in prevalence and infestation intensity of *Gyrodactylus salaris* Malmberg, 1957 (Monogenea: Gyrodactylidae) on Atlantic salmon parr, *Salmo salar* L., in the River Batnfjordselva, Norway. *Journal of Fish Biology* 41, 697–707.

Mo, T.A. (1994) Status of *Gyrodactylus salaris* problems and research in Norway. In: Pike, A.W. and Lewis, J.W. (eds) *Parasitic Diseases of Fish*. Samara Publishing Ltd, Tresaith, UK, pp. 43–56.

Mo, T.A. (2004) Innførsel av *Gyrodactylus salaris* til Norge og egenskaper hos parasitten av betydning for valg av strategi og tiltak mot den. *Norsk Veterinærtidsskrift* 116, 164–167. (in Norwegian)

Mo, T.A., Hellesnes, I. and Norheim, K. (2004) Overvåkings- og kontrollprogram for *Gyrodactylus salaris* på laks og regnbueørret i Norge. *Norsk Veterinærtidsskrift* 116, 157–163. (in Norwegian)

Mo, T.A., Thorstad, E.B., Sandlund, O.T., Berntsen, H.H., Fiske, P. and Uglem, I. (2018) The pink salmon invasion: a Norwegian perspective. *Journal of Fish Biology* 93, 5–7.

Niemelä, E., Johansen, N., Zubchenko, A.V., Dempson, J.B., Veselov, A. *et al.* (2016) *Pink salmon in the Barents region. With special attention to the status in the transboundary rivers Tana and Neiden, rivers in North West Russia and in East Canada. Report 3–2016.* Office of the Finnmark County Governor, Department of Environmental Affairs, Vadsø, Norway.

OIE (World Organization for Animal Health) (2019a) Chapter 10.3: Infection with *Gyrodactylus salaris*. In: *Aquatic Animal Health Code. Available at*: https://www.oie.int/en/standard-setting/aquatic-code/ (accessed 1 April 2020).

OIE (World Organization for Animal Health) (2019b) Chapter 2.3.3: Infection with *Gyrodactylus salaris*. In: Manual of Diagnostic Test for Aquatic Animals. Available at: https://www.oie.int/en/standard-setting/aquatic-manual/ (accessed 1 April 2020).

Olstad, K., Cable, J., Robertsen, G. and Bakke, T.A. (2006) Unpredicted transmission strategy of *Gyrodactylus salaris* (Monogenea: Gyrodactylidae): survival and infectivity of parasites on dead hosts. *Parasitology* 133, 33–41.

Olstad, K., Robertsen, G., Bachmann, L. and Bakke, T.A. (2007) Variation in host preference within *Gyrodactylus salaris* (Monogenea): an experimental approach. *Parasitology* 134, 589–597.

Olstad, K., Bachmann, L. and Bakke, T.A. (2009) Phenotypic plasticity of taxonomic and diagnostic structures in gyrodactylosis-causing flatworms (Monogenea, Platyhelminthes). *Parasitology* 136, 1305–1315.

Olstad, K., Hytterød, S. and Hansen, H. (2013) Risiko for spredning av *Gyrodactylus salaris* fra Vänern og Klarälven til norske vassdrag ved reetablering av laks i Trysil- /Femundselva. *NINA Rapport Nr. 991*. Norsk institutt for naturforskning, Trondheim, Norway. (in Norwegian)

Ozerov, M.Y., Lumme, J., Päkk, P., Rintamäki, P., Zietara, M.S. *et al.* (2010) High *Gyrodactylus salaris* infection rate in triploid Atlantic salmon *Salmo salar*. *Diseases of Aquatic Organisms* 91, 129–136.

Paladini, G., Cable, J., Fioravanti, M.L., Faria, P.J., Di Cave, D. and Shinn, A.P. (2009a) *Gyrodactylus orecchiae* sp. n. (Monogenea: Gyrodactylidae) from farmed populations of gilthead seabream (*Sparus aurata*) in the Adriatic Sea. *Folia Parasitologica (Ceske Budejovice)* 56, 21–28.

Paladini, G., Gustinelli, A., Fioravanti, M.L., Hansen, H. and Shinn, A.P. (2009b) The first report of *Gyrodactylus salaris* Malmberg, 1957 (Platyhelminthes, Monogenea) on Italian cultured stocks of rainbow trout (*Oncorhynchus mykiss* Walbaum). *Veterinary Parasitology* 165, 290–297.

Paladini, G., Hansen, H., Williams, C.F., Taylor, N.G.H., Rubio-Mejía, O.L. *et al.* (2014) Reservoir hosts for *Gyrodactylus salaris* may play a more significant role in epidemics than previously thought. *Parasites and Vectors* 7, 576.

Peeler, E.J. and Oidtmann, B.C. (2008) Demonstrating freedom from *Gyrodactylus salaris* (Monogenea: Gyrodactylidae) in farmed rainbow trout *Oncorhynchus mykiss*. *Diseases of Aquatic Organisms* 79, 47–56.

Peeler, E.J. and Thrush, M.A. (2004) Qualitative analysis of the risk of introducing *Gyrodactylus salaris* into the United Kingdom. *Diseases of Aquatic Organisms* 62, 103–113.

Pettersen, R.A., Hytterød, S., Vøllestad, L.A. and Mo, T.A. (2013) Osmoregulatory disturbances in Atlantic salmon, *Salmo salar* L., caused by the monogenean *Gyrodactylus salaris*. *Journal of Fish Diseases* 36, 67–70.

Pettersen, R.A., Mo, T.A., Hansen, H. and Vollestad, L.A. (2015) Genetic population structure of *Gyrodactylus thymalli* (Monogenea) in a large Norwegian river system. *Parasitology* 142, 1693–1702.

Pietrak, M. and Backman, S. (2018) Treatment of lumpfish (*Cyclopterus lumpus* L.) infected with *Gyrodactylus* cyclopteri (Scyborskaya 1948). *Journal of Fish Diseases* 41, 721–723.

Poléo, A.B.S. (1995) Aluminium polymerization – a mechanism of acute toxicity of aqueous aluminium to fish. *Aquatic Toxicology* 31, 347–356.

Poléo, A.B.S., Schjolden, J., Hansen, H., Bakke, T.A., Mo, T.A. *et al.* (2004a) The effect of various metals on *Gyrodactylus salaris* (Platyhelminthes, Monogenea) infections in Atlantic salmon (*Salmo salar*). *Parasitology* 128, 169–177.

Poléo, A.B.S., Lydersen, E. and Mo, T.A. (2004b) Aluminium mot lakseparasitten *Gyrodactylus salaris*. *Norsk Veterinærtidsskrift* 116, 176–180. (in Norwegian)

Ramirez, R., Bakke, T.A. and Harris, P.D. (2014) Same barcode, different biology: differential patterns of infectivity, specificity and pathogenicity in two almost identical parasite strains. *International Journal for Parasitology* 44, 543–549.

Řehulka, J. (1973) Remarks on the occurrence of *Gyrodactylus salaris* Malmberg, 1957 *sensu* Ergens, 1961 (Monogenoidea; Gyrodactylidae). *Věstník Československé Společnosti Zoologické* 37, 293–295.

Ricciardi, A., Blackburn, T.M., Carlton, J.T., Dick, J.T.A., Hulme, P.E. *et al.* (2017) Invasion science: a horizon scan of emerging challenges and opportunities. *Trends in Ecology and Evolution* 32, 464–474.

Rintamäki, P. (1989) *Gyrodactylus salaris* at a fish farm in Northern Finland. In: *Proceedings of Soviet–Finnish Symposium, Petrozavodsk, January* 1988, pp. 123–130.

Rintamäki-Kinnunen, P. and Valtonen, E.T. (1996) Finnish salmon resistant to *Gyrodactylus salaris*: a long-term study at fish farms. *International Journal for Parasitology* 26, 723–732.

Robertsen, G., Olstad, K., Plaisance, L., Bachmann, L. and Bakke, T.A. (2008) *Gyrodactylus salaris* (Monogenea, Gyrodactylidae) infections on resident Arctic charr (*Salvelinus alpinus*) in southern Norway. *Environmental Biology of Fishes* 83, 99–105.

Rokicka, M., Lumme, J. and Ziętara, M.S. (2007) Identification of *Gyrodactylus* ectoparasites in Polish salmonid farms by PCR-RFLP of the nuclear ITS segment of ribosomal DNA (Monogenea, Gyrodactylidae). *Acta Parasitologica* 52, 185–195.

Rusch, J.C., Hansen, H., Strand, D.A., Markussen, T., Hytterod, S. and Vralstad, T. (2018) Catching the fish with the worm: a case study on eDNA detection of the monogenean parasite *Gyrodactylus salaris* and two of its hosts, Atlantic salmon (*Salmo salar*) and rainbow trout (*Oncorhynchus mykiss*). *Parasites and Vectors* 11, 333.

Salte, R., Bentsen, H.B., Moen, T., Tripathy, S., Bakke, T.A. *et al.* (2010) Prospects for a genetic management strategy to control *Gyrodactylus salaris* infection in wild Atlantic salmon (*Salmo salar*) stocks. *Canadian Journal of Fisheries and Aquatic Sciences* 67, 121–129.

Sandlund, O.T., Berntsen, H.H., Fiske, P., Kuusela, J., Muladal, R. *et al.* (2018) Pink salmon in Norway: the reluctant invader. *Biological Invasions* 21, 1033–1054.

Sandodden, R., Mo, T.A. and Skår, K. (2004) Bekjempelse av *Gyrodactylus salaris* med bruk av kombinerte løsninger. *Norsk Veterinærtidsskrift* 116, 182–185. (in Norwegian)

Sandodden, R., Brazier, M., Sandvik, M., Moen, A., Wist, A.N. and Adolfsen, P. (2018) Eradication of *Gyrodactylus salaris* infested Atlantic salmon (*Salmo salar*) in the Rauma River, Norway, using rotenone. *Management of Biological Invasions* 9, 67–77.

Santamarina, M.T., Tojo, J., Ubeira, F.M., Quinteiro, P. and Sanmartin, M.L. (1991) Anthelmintic treatment against *Gyrodactylus* sp. infecting rainbow trout *Oncorhynchus mykiss*. *Diseases of Aquatic Organisms* 10, 39–43.

Schelkle, B., Shinn, A.P., Peeler, E. and Cable, J. (2009) Treatment of gyrodactylid infections in fish. *Diseases of Aquatic Organisms* 86, 65–75.

Shinn, A.P., Gibson, D.I. and Sommerville, C. (2001) Morphometric discrimination of *Gyrodactylus salaris* Malmberg (Monogenea) from species of *Gyrodactylus* parasitising British salmonids using novel parameters. *Journal of Fish Diseases* 24, 83–97.

Shinn, A.P., Hansen, H., Olstad, K., Bachmann, L. and Bakke, T.A. (2004) The use of morphometric characters to discriminate specimens of laboratory-reared and wild populations of *Gyrodactylus salaris* and *G. thymalli* (Monogenea). *Folia Parasitologica* 51, 239–252.

Shinn, A.P., Collins, C., Garcia-Vasquez, A., Snow, M., Matejusova, I. *et al.* (2010) Multi-centre testing and validation of current protocols for the identification of *Gyrodactylus salaris* (Monogenea). *International Journal for Parasitology* 40, 1455–1467.

Soleng, A. and Bakke, T.A. (1997) Salinity tolerance of *Gyrodactylus salaris* (Platyhelminthes, Monogenea): laboratory studies. *Canadian Journal of Fisheries and Aquatic Sciences* 54, 1837–1845.

Soleng, A. and Bakke, T.A. (1998) The susceptibility of three-spined stickleback (*Gasterosteus aculeatus*), nine-spined stickleback (*Pungitius pungitius*) and flounder (*Platichthys flesus*) to experimental infections with the monogenean *Gyrodactylus salaris*. *Folia Parasitologica* 45, 270–274.

Soleng, A. and Bakke, T.A. (2001) The susceptibility of grayling (*Thymallus thymallus*) to experimental infections with the monogenean *Gyrodactylus salaris*. *International Journal for Parasitology* 31, 793–797.

Soleng, A., Bakke, T.A. and Hansen, L.P. (1998) Potential for dispersal of *Gyrodactylus salaris* (Platyhelminthes, Monogenea) by sea-running stages of the Atlantic salmon (*Salmo salar*): field and laboratory studies. *Canadian Journal of Fisheries and Aquatic Sciences* 55, 507–514.

Soleng, A., Poléo, A.B.S., Alstad, N.E.W. and Bakke, T.A. (1999) Aqueous aluminium eliminates *Gyrodactylus salaris* (Platyhelminthes, Monogenea) infections in Atlantic salmon. *Parasitology* 119, 19–25.

Soleng, A., Poléo, A.B.S. and Bakke, T.A. (2005) Toxicity of aqueous aluminium to the ectoparasitic monogenean *Gyrodactylus salaris*. *Aquaculture* 250, 616–620.

Ståhl, G. (1987) Genetic population structure of Atlantic salmon. In: Rymann, N. and Utter, F. (eds) *Population Genetics and Fisheries Management*. University of Washington Press, Seattle, Washington, pp. 212–240.

Stanković, D., Crivelli, A.J. and Snoj, A. (2015) Rainbow trout in Europe: introduction, naturalization, and impacts. *Reviews in Fisheries Science and Aquaculture* 23, 39–71.

Sterud, E., Mo, T.A., Collins, C.M. and Cunningham, C.O. (2002) The use of host specificity, pathogenicity, and molecular markers to differentiate between *Gyrodactylus salaris* Malmberg, 1957 and *G. thymalli* Zitnan, 1960 (Monogenea: Gyrodactylidae). *Parasitology* 124, 203–213.

Thober, S., Kumar, R., Wanders, N., Marx, A., Pan, M. *et al.* (2018) Multi-model ensemble projections of European river floods and high flows at 1.5, 2, and 3 degrees global warming. *Environmental Research Letters* 13, 014003.

Tojo, L., Santamarina, M.T., Ubeira, F.M., Estevez, J. and Sanmartin, M.L. (1992) Anthelmintic activity of benzimidazoles against *Gyrodactylus* sp. infecting rainbow trout *Oncorhynchus mykiss*. *Diseases of Aquatic Organisms* 12, 185–189.

Tu, X., Huang, A., Hu, Y., Ling, F. and Wang, G. (2018) Arctigenin: an emerging candidate against infections of *Gyrodactylus*. *Aquaculture* 495, 983–991.

Winger, A.C., Primicerio, R., Kristoffersen, R., Siikavuopio, S.I. and Knudsen, R. (2008) *Gyrodactylus salaris* infecting allopatric Arctic charr *Salvelinus alpinus* fry: an experimental study of host survival. *Journal of Fish Biology* 73, 2198–2209.

Ziętara, M.S., Rokicka, M., Stojanovski, S. and Lumme, J. (2010) Introgression of distant mitochondria into the genome of *Gyrodactylus salaris*: nuclear and mitochondrial markers are necessary to identify parasite strains. *Acta Parasitologica* 55, 20–28.

Žitňan, R. and Čankovič, M. (1970) Comparison of the epizootological importance of the parasites of *Salmo gairdneri irideus* in the two coast areas of Bosnia and Herzegovina. *Helminthologia* 11, 161–166.

Zorin, B., Gibson-Kueh, S. and Zilberg, D. (2019) A novel treatment against the monogenean parasite, *Gyrodactylus turnbulii*, infecting guppies (*Poecilia reticulata*), using a plant-based commercial insecticide Timor C. *Aquaculture* 501, 313–318.

Zueva, K.J., Lumme, J., Veselov, A.E., Kent, M.P. and Primmer, C.R. (2018) Genomic signatures of parasite-driven natural selection in north European Atlantic salmon (*Salmo salar*). *Marine Genomics* 39, 26–38.

22 Eubothriosis

Ken MacKenzie*

The University of Aberdeen, Aberdeen, UK

22.1 Introduction

'Eubothriosis' refers to the pathological effects caused by adult bothriocephalidean cestodes of the genus *Eubothrium* Nybelin, 1922 on their fish hosts. The term is most frequently used with reference to *Eubothrium crassum* (Bloch, 1799), which is a parasite of commercially important salmonid fish.

E. crassum is very similar morphologically to its sympatric congener *Eubothrium salvelini* (Schrank, 1790), but they have different host ranges: *E. crassum* is mainly a parasite of salmonid fishes of the genus *Salmo*, whereas *E. salvelini* mainly infects members of the genera *Salvelinus* and *Oncorhynchus* (Kuchta *et al.*, 2008). Both parasites are largely host specific but mature forms of *E. crassum* have been found infecting a wide spectrum of other salmonid species, including members of the genera *Coregonus*, *Hucho*, *Salvelinus* and *Thymallus*, although Scholz *et al.* (2003) considered the European reports from Arctic charr, *Salvelinus alpinus*, and grayling, *Thymallus thymallus*, to be misidentifications. The freshwater form of *E. crassum* is distributed from the Arctic coast of Norway to southern France and from Ireland to eastern Eurasia, but there is no reliable evidence of its occurrence in North America (Kennedy, 1978b). Its preferred host is trout, *Salmo trutta*. The Atlantic marine form infects sea trout and salmon, *Salmo salar*, while the Pacific marine form infects Pacific salmon, *Oncorhynchus* spp. (see Akhmerov, 1962, 1963; Kuperman, 1978). It was suggested by Kennedy (1978b) that freshwater forms of *E. crassum* do not survive for long in the sea and marine forms do not survive for long in fresh water. The freshwater race of *E. crassum* does not appear to be capable of becoming gravid in sea trout although it does so in resident brown trout (Kennedy, 1978a).

High rates of prevalence and mean intensity of *E. crassum* have been reported in wild trout and salmon. For salmon caught at sea, 100% prevalence and a mean intensity of 15 (maximum 36 worms) were reported in the White Sea (Shulman and Shulman-Albova, 1953), 53.8% prevalence with a maximum intensity of 81 off the north-west coast of Norway (Kennedy, 1978a), and 36% prevalence and a mean intensity of 8.1 off western Norway (Bristow and Berland, 1991a). Pippy (1969) reported 47% prevalence in wild salmon caught off Greenland. The heaviest infections of 100% prevalence and 172.6 mean intensity were reported from the Baltic salmon stock by Setyawan *et al.* (2019). Prevalences of 71.4 and 100% and intensities of up to 250 and 72 per fish were reported from sea trout in the Baltic Sea by Buchmann (1987) and Unger and Palm (2016), respectively. In fresh water, prevalences of 72.2 to 100% have been reported in trout from Scottish, English and German freshwater systems (Wootten, 1972; Campbell, 1974; Kennedy, 1996; Unger and Palm, 2016).

This chapter reviews the biology, ecology and effects on its hosts of *E. crassum* and attempts to predict how this parasite is likely to be affected by climate change.

22.2 Diagnosis of the Pathogen

22.2.1 Identification

Nine species are currently recognized in the genus *Eubothrium* (see Andersen and Kennedy, 1983; Kuchta *et al.*, 2008), which is unusual among cestodes in that some species occur in the sea and others in fresh water (Scholz *et al.*, 2003). Three ecological forms or 'races' are recognized in *E. crassum*, one freshwater and two marine, with separate marine forms occurring in the North Atlantic and North Pacific Oceans (Kennedy, 1978b).

*k.mackenzie@abdn.ac.uk

Eubothrium spp. are medium- to large-sized worms (Fig. 22.1). The elongate scolex is oval with elongate bothria and an apical disc with a margin which is either entire or with two to several incisions (Fig. 22.2). The strobila has markedly craspedote, trapeziform segments which are much wider than long (Kuchta *et al.*, 2008). Nybelin (1922) used scolex size and shape, together with some internal morphological features (size of cirrus sac, continuity of testes, position of vitellaria), to separate the different species of *Eubothrium*, but Kennedy (1978b) considered many of these characters to be unreliable, particularly for distinguishing between *E. crassum* and *E. salvelini*. Andersen and Kennedy (1983), however, found scolex shape using scanning electron microscopy to be a valuable specific character. They described the marine races of *E. crassum* as massive worms which often lie coiled in a lump in the anterior part of the gut or caecum. When contracted they are up to 400 mm long by up to 6 mm wide (Fig. 22.1). The scolex appears compressed, with an oval and four-lobed apical disc (Andersen, 1979) (Fig. 22.2). Laterally, the scolex is very broad with a clear median groove. Andersen and Kennedy (1983) found only minor morphological differences between the freshwater and marine races of *E. crassum*, but the freshwater race tended to be smaller at up to 150 mm by 4.5 mm when contracted. *E. salvelini* are generally smaller than *E. crassum* at up to 120 mm long and 3.5 mm wide when contracted and have a bilobed apical disc with shallow indentations or incisions above each bothrium.

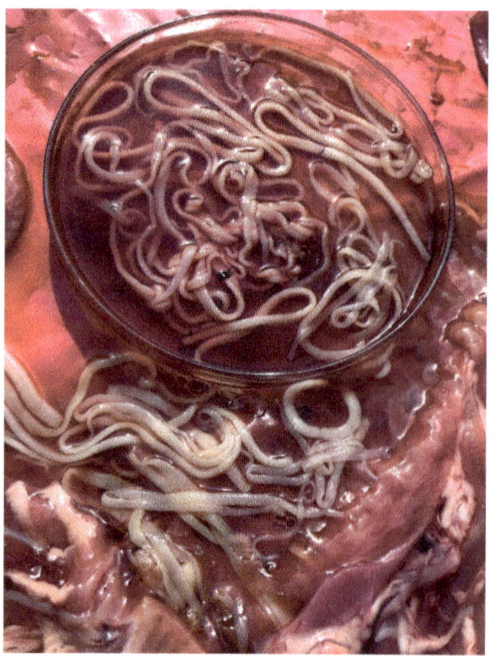

Fig. 22.1. *Eubothrium crassum* from the pyloric caeca of a Baltic salmon. (Photograph by Kurt Buchmann.)

There is some disagreement concerning the genetic relationship between the Atlantic marine and freshwater forms of *E. crassum*. Bristow and Berland (1989), using allozymes, claimed that the two forms differ genetically and are sibling species. However, Kráľová-Hromadová *et al.* (2003) found the internal transcribed spacer ITS1 sequences of marine and freshwater forms to be identical.

22.2.2 Life cycle

The role of copepods in the life cycle of *E. crassum* was first observed by Rosen (1918), who described the freshwater life cycle of *E. crassum* (as *Abothrium infundibuliformis*) with *Cyclops* spp. as first intermediate hosts and perch, *Perca fluviatilis*, as second intermediate host. Vik (1963) showed that sticklebacks, *Gasterosteus aculeatus*, along with perch, were transport rather than intermediate hosts, and that trout, *S. trutta*, could also become infected by consuming infected copepods. A wide range of other fish species have since been reported as hosts of *E. crassum* in Europe (Table 22.1). With the exception of the two species of Pacific salmon *Oncorhynchus*

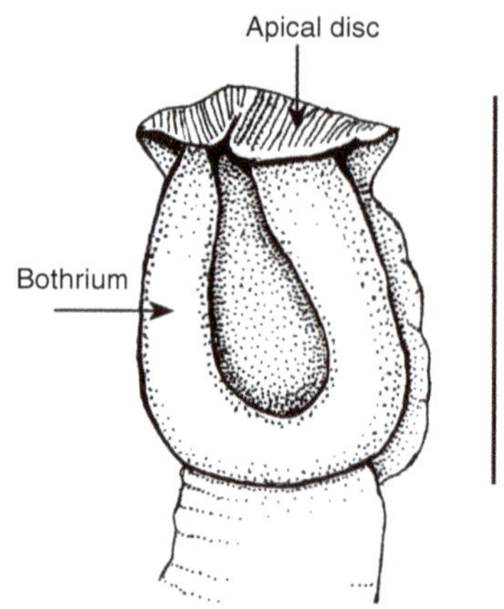

Fig. 22.2. Drawing of the scolex of *Eubothrium crassum*. Scale bar = 1 mm. (Modified from Nybelin, 1922.)

Table 22.1. Reports of fish other than *Salmo* spp. infected with *Eubothrium crassum* in Europe.

Host species	Location	Reference
Perch (*Perca fluviatilis*)	Switzerland	Rosen (1918)
	Ireland	Kane (1966)
	Danube Basin	Kulakovskaya and Koval (1973)
Rainbow trout (*Oncorhynchus mykiss*)	Ireland	Kane (1966)
	England	Wootten (1972)
	Denmark	Buchmann *et al.* (1995); Buchmann and Bresciani (1997)
	Scotland	Dorucu *et al.* (1995); Kráľová-Hromadová *et al.* (2003)
Pink salmon (*Oncorhynchus gorbuscha*)	Russia	Grozdilova (1974)
Stickleback (*Gasterosteus aculeatus*)	Norway	Vik (1963); Kuhn *et al.* (2015)
	Baltic Sea	Rolbiecki *et al.* (2002)
	Scotland	de Roij and MacColl (2015)
Bream (*Abramis brama*)	Ireland	Kane (1966)
Carp (*Cyprinus carpio*)	Ireland	Kane (1966)
Pike (*Esox lucius*)	Ireland	Kane (1966)
Rudd (*Scardinius erythrophthalmus*)	Ireland	Kane (1966)
Ruffe (*Gymnocephalus cernua*)	England	Wootten (1972)
Asp (*Aspius aspius*)	Danube Basin	Kulakovskaya and Koval (1973)
Smelt (*Osmerus eperlanus*)	Baltic Sea	Rokicki (1975); Pilecka-Rapasz *et al.* (2017)
Herring (*Clupea harengus*)	Baltic Sea	Gaevskaya and Shapiro (1981); Petkevičiūtė and Bondarenko (2001)
Round goby (*Neogobius melanostomus*)	Baltic Sea	Rolbiecki (2006)
Lumpsucker (*Cyclopterus lumpus*)	Baltic Sea	Rolbiecki and Rokicki (2008)
Vendace (*Coregonus albula*)	Russia	Anikieva *et al.* (2016)

spp., all the worms from these hosts appear to be juveniles. Kennedy (1978b) considered fish of genera other than *Salmo* reported as hosts of freshwater *E. crassum* to be 'accidental' rather than transport hosts. These 'accidental hosts' may, however, still serve as transport hosts if predated upon by salmonids. All the reports in Table 22.1 except that from pink salmon in the White Sea are from fish caught in fresh- or brackish-water environments and are presumed to be the freshwater form of *E. crassum*. No accidental or transport fish hosts appear to have been reported from a purely marine environment.

The life cycle of the marine form of *E. crassum* was investigated experimentally in Norway by Hodneland and Solberg (1995) and Saksvik *et al.* (2001b). The copepod *Acartia tonsa* was experimentally infected with *E. crassum* (as *Eubothrium* sp.) eggs from farmed Atlantic salmon, and eggs from marine *E. crassum* could infect freshwater copepods and conversely eggs from the freshwater form could infect a marine copepod, suggesting that the two life cycles may not be completely separate. Saksvik *et al.* (2001b) maintained their infected salmon in the laboratory for 11 months and saw no evidence of cestode mortality during that period. This suggests a lifespan of at least 1 year for the marine form of *E. crassum*, thereby supporting the conclusion of Fahy (1980) that the lifespan at sea of the marine form in sea trout is at least 1 year but growth continues for at least 2 years and possibly for longer. However, this lifespan will be achieved only if the host remains at sea and does not migrate into fresh water.

Life cycle studies reveal a degree of confusion regarding the taxonomic status of the different forms of *E. crassum*. The question of what a species is has long been a subject of much debate among biologists, with recent advances in genetics questioning the status of many previously established taxonomic relationships. The taxonomic confusion that exists today is illustrated by the fact that there are currently more than 30 different definitions of a species (Zachos, 2016; Barras, 2019). In essence, a 'species' comes into existence at a certain point in the process of a population of organisms dividing into separate subpopulations or 'lineages' characterized by different behavioural patterns. As this evolutionary process continues, the different lineages become increasingly isolated from one another until at some point, for our convenience, we decide to call them different species. As far as *E. crassum* is concerned, Kennedy (1978b) considered *Eubothrium* to be primarily a marine genus, some species of which have entered fresh waters together with their anadromous hosts and have then either developed into freshwater races or evolved into different freshwater species. If we accept Kennedy's suggestion, the life cycle studies on *E. crassum* suggest that the marine and freshwater races must have diverged from a common marine ancestor fairly recently in geological time and that both share an earlier common marine ancestor with *E. salvelini*.

22.2.3 Macroscopic and microscopic changes in infected hosts

While *Eubothrium* infections do not usually produce serious pathology in the host fish, Mitchell (1993) nevertheless referred to a reduction in the host's condition factor, estimated at a potential loss of growth of between 10 and 20%, reduced levels of circulating red blood cells and reduced haemoglobin levels, but provided no source references for his information. Bristow and Berland (1991b) also reported a loss of condition in farmed salmon infected with *E. crassum*, which were significantly lighter than uninfected fish from the same population. A similar effect was demonstrated by Saksvik *et al.* (2001a) in a controlled experiment which showed significant differences in growth between infected and uninfected salmon, but not until several months post-infection, and haematocrits were significantly lower in the infected salmon. Bosi *et al.* (2005) detected cellular deviations in the caeca of naturally infected rainbow trout: an enhanced mucus production with epithelial cellular desquamation, a leucocytic infiltration of the lamina propria–submucosa and vacuolization of the intestinal epithelial cells. Buchmann and Bresciani (1997), however, reported no pathology associated with very low intensities of *E. crassum* infection in pond-reared rainbow trout in Denmark.

Young sockeye salmon in Canada infected with *E. salvelini* showed reduced growth (Dombroski, 1955; Boyce, 1979), reduced survival and swimming performance (Smith and Margolis, 1970; Boyce, 1979), and reduced ability to adapt to seawater (Boyce and Clarke, 1983). In Europe, increasing intensity of infection with *E. salvelini* in Arctic charr was associated with loss of condition and chronic haemolytic anaemia (Hoffman *et al.*, 1986), and with high mortalities in juvenile char (Gerdeaux *et al.*, 1995).

An experimental study carried out by Poulin *et al.* (1992) on the effects of *E. salvelini* infection on the behaviour of the copepod *Cyclops vernalis* showed that infected copepods started to swim more actively than uninfected controls and that they became more likely to be captured by their fish host, brook trout *Salvelinus fontinalis*, than uninfected copepods.

22.3 Expected/Potential Spread of the Pathogen Due to Climate Change

22.3.1 General comments

Trying to predict the effects of climate change on any organism is a difficult task involving multiple possibilities. The would-be prophet must base his or her predictions on the assumption that current changes will continue into the future, which is by no means certain. While this is a difficult problem when the subject is a free-living organism, it becomes increasingly more complex when it is a parasite, especially one with a complex multi-host life cycle. In such a situation, the effects on all the different hosts in the life cycle must be considered. Climate change will thus affect parasites in two ways: through direct effects on the parasite itself, and through indirect effects on other hosts in its life cycle. The probable effects of climate change on aquatic parasites were reviewed by Marcogliese (2001, 2008) and Lõhmus and Björklund (2015).

22.3.2 Increase in water temperature

Eubothrium is considered holarctic and circumpolar in distribution (Shulman, 1961; Kennedy, 1978b). While the freshwater form of *E. crassum* has been found in Europe as far south as southern France (Joyeux and Baer, 1936), there are no reports of its occurrence in other parts of Europe bordering the Mediterranean. The few records of levels of infection with the Atlantic marine form of *E. crassum* show an increase with increasing latitude and decreasing water temperature (Shulman and Shulman-Albova, 1953; Kennedy, 1978a; Bristow and Berland, 1991a). Although Kennedy (1978a) did not find freshwater *E. crassum* on Spitzbergen because of the absence of its salmonid hosts, he did find *E. salvelini* infecting Arctic charr there. As the Arctic continues to warm, it seems only a matter of time before trout are introduced to Spitzbergen, in which case *E. crassum* is likely to follow. As there is very little land at a higher latitude than Spitzbergen, *E. crassum* will then have reached the limit of its northward expansion. The effects of global warming on its distribution are thus likely to be more pronounced at the southernmost range of its distribution, where either the parasite itself or its salmonid hosts, or both, may find the higher temperatures difficult to tolerate. These effects will be more severe for the freshwater form of *E. crassum* because freshwater fishes are more vulnerable to climate change than marine species due to freshwater habitats being more constrained and more exposed to anthropogenic stressors (Marcogliese, 2001). It can therefore be predicted that the geographical distributions of both the marine and freshwater forms of *E. crassum* will shift northwards.

One of the general predictions regarding the probable effects of increasing temperatures on parasites and diseases is that longer growing seasons and higher temperatures should lead to more generations of parasites annually (Marcogliese, 2008). This may not apply to *E. crassum* given that its geographical distribution suggests an adaptation to low temperatures, plus the evidence of a relatively longer lifespan in its definitive hosts (Fahy, 1980; Saksvik *et al.*, 2001b), although there is the possibility that its lifespan may be reduced at higher temperatures. Increasing temperatures are likely to cause thermal stress in its salmonid hosts, leading to reduced growth, suboptimal behaviours and reduced immunocompetence (Marcogliese, 2008; Dittmar *et al.*, 2014), possibly resulting in increased virulence of the parasite. Increasing temperatures may also reduce the time from infection of the copepod host to infectivity of the procercoid to the definitive host, leading to increased levels of infection. Again, these effects are likely to be felt more in the southern parts of the parasite's distribution.

Salmonid aquaculture is presently limited to locations at latitudes of 40 to 70°N, with most activity concentrated between 45 and 70° (Hermansen and Heen, 2012). Increasing water temperatures may result in farmed salmonids at lower latitudes becoming free of *E. crassum* infection but will also increase the probability of thermal stress on the fish.

Another effect of increasing temperatures in northern regions is a reduction in winter ice cover. Low temperatures may cause fish to reduce or cease feeding (Chubb, 1980; Marcogliese, 2001), leading to reduced parasite recruitment. However, Vik (1963) found that even when trout were fished through the ice in winter, *E. crassum* was still present as larvae, immature and mature worms, suggesting that infection by this parasite is not greatly influenced by ice cover.

22.4 Expected/Potential Spread of Intermediate and/or Reservoir Hosts Due to Climate Change

As atmospheric carbon dioxide continues to increase, more of it is being absorbed by both oceanic and freshwater systems, leading to changes in water chemistry and a continuous reduction in pH, with potentially serious consequences for many aquatic organisms. Increasing acidification will affect *E. crassum* in both marine and fresh waters, and most of the effects will be indirect via the effects on its host species. If current trends continue, it is predicted that many marine organisms, particularly pteropods and crustaceans, will have difficulty maintaining their calcium carbonate exoskeletons (Orr *et al.*, 2005). In freshwater systems, the species diversity of many organisms is significantly decreased by acidification (Marcogliese, 2001). Some benthic macroinvertebrates, such as snails and crustaceans, are particularly acid-sensitive (Muniz, 1991), but their sensitivity varies between species. Studies on copepods are of particular relevance because the planktonic copepod *A. tonsa* was shown to be a suitable intermediate host for this cestode (Saksvik *et al.*, 2001b). Copepods often comprise the majority of the biomass of zooplankton in marine ecosystems, so information on their response to increased acidification is vital to our understanding of how marine ecosystems will respond to this and other climatic changes. Studies conducted to date have shown considerable variations in the responses of different species and different developmental stages to increasing acidification. Earlier studies indicated that copepods may be resistant to acidification, but these tended to be based on short-term exposure on adult females, whereas more recent studies showed that multigenerational exposure had significant negative effects on growth and reproduction (Wang *et al.*, 2018). Different populations of the same species also have different sensitivities and the negative effects are amplified by other stressors such as increased temperature, food deprivation and metal pollution (Mayor *et al.*, 2015; Wang *et al.*, 2018). Studies on *A. tonsa* showed high nauplii mortality whereas other stages were unaffected, but reproduction was negatively impacted (Cripps *et al.*, 2014). Studies on other species of *Acartia* gave the following results.

- Antioxidant capacity and developmental rate of *Acartia bifilosa* increased with increasing acidification but decreased in combination with high temperatures (Vehmaa *et al.*, 2013).
- No significant effects were observed on survival, body size, reproduction or development of *Acartia tsuensis* (Kurihara and Ishimatsu, 2008).
- In *Acartia steurei* and *Acartia erythraea*, hatching rate and nauplii mortality increased and reproduction decreased at higher levels of acidification, but there was little effect on the survival of adult copepods (Kurihara *et al.*, 2004).
- Nauplii mortality of *Acartia* spp. was high, but other stages were not affected. Reproduction was negatively impacted (Vehmaa *et al.*, 2012).

While there have been many studies and reviews on the effects of acidification on free-living invertebrates, there have been few on its effects on parasites and on how it may affect the sensitivity of parasitized hosts. Those few have addressed only digenean parasites and their mollusc and amphipod intermediate hosts (Harland *et al.*, 2015; MacLeod and Poulin, 2015a,b; Guilloteau *et al.*, 2016); there appears to be none that has focused on copepods. This lack of information makes it very difficult to predict if acidification is likely to make infected copepods more susceptible to predation or if increasing mortality of infected copepods will lead to reduced infections in salmonid fish.

Villarino *et al.* (2015) evaluated the impacts of future climate change on community structure, diversity, distribution and phenology of 14 different species of marine copepods in the North Atlantic. Their projections indicated poleward shifts, earlier seasonal peaks and changes in biodiversity spatial patterns, with important range variations between species. Other studies indicated that higher temperatures reduced energy status and decreased copepodite and nauplii abundance, but also that acidification partially counteracted some observed effects of increased temperature while adding to others (Garzke *et al.*, 2016; Pedersen and Hanssen, 2017). Garzke *et al.* (2016) predicted that copepod populations would be more affected by warming than by acidification. However, ocean acidification effects could modify some temperature impacts. Studies on *Acartia* spp., known hosts for *E. crassum* (see Saksvik *et al.*, 2001b), found that higher temperatures resulted in an increase in egg and nauplii production, but had negative effects on egg viability, development and oxidative status, and a decrease in developmental rates in combination with acidification (Vehmaa *et al.*, 2012, 2013). Increasing temperatures may also reduce the time from infection of the copepod host to infectivity of the procercoid to the definitive host, leading to increased levels of infection.

Climate warming has increased expansion of the geographical ranges of fish, with the result that species that have evolved in isolation may be brought into close contact. These host species carry their established parasites with them and expose them to new potential hosts, providing parasites with opportunities to expand their host range. The potential for fish parasites to switch to relatively unrelated hosts due to environmental changes was demonstrated by Konovalov (1995), who found eight parasite species, comprising five protistans, one myxosporean and two monogeneans, that normally infect minnow (*Phoxinus phoxinus*) infecting juvenile pike (*Esox lucius*) in a Siberian river. Minnows had become so rare in this river that it was impossible to obtain a sufficient sample for parasitological study, and it was suggested that those parasites, which are normally specific to minnows, had secondarily infected juvenile pike. Hoberg and Brooks (2015) showed that phylogenetic studies of hosts and parasites demonstrate that there is substantial evidence of host switching in the past in response to climatic and environmental changes, and that shifts on to relatively unrelated hosts are common during phylogenetic diversification of parasite lineages. They further suggested that host shifting is not necessarily led by genetic changes, but that the genetic capabilities for switching are already in the system. The number of species apart from salmonids listed as hosts of *E. crassum* in Table 22.1 is evidence of the potential for this parasite to switch to new hosts. It may require only a small environmental and/or genetic change to turn these 'accidental' hosts into definitive hosts in which *E. crassum* will mature and reproduce.

Among potential new hosts encountered by parasites as a result of the expansion of their hosts' ranges are those that have been introduced, either intentionally by humans or accidentally through climate change. For example, Hemmingsen *et al.* (2005) suggested that the intentional introduction of the red king crab, *Paralithodes camtschaticus*, to the Barents Sea in the 1960s was indirectly responsible for increased transmission of trypanosomes to cod by promoting an increase in the population of the leech vector of the trypanosome. In recent times introduced marine species have become a major environmental and economic problem (Torchin *et al.*, 2002). One example is the introduced round goby in the Baltic Sea, which has been reported as an 'accidental' host of *E. crassum* (see Rolbiecki, 2006). Another from the north-east Atlantic is the introduced pink salmon, *Oncorhynchus gorbuscha*. This native of the North Pacific was introduced to rivers in the Kola Peninsula in north-west Russia in the period 1956–1959 and began to appear in Norwegian rivers from 1960 and in Scottish rivers from 1967 (Berg, 1977; Armstrong *et al.*, 2018; Mo *et al.*, 2018). The numbers of pink salmon spawning in these rivers have greatly increased in recent years. While there were no reports of *E. crassum* from pink salmon in Norway or Scotland, there is a report from the White Sea (Grozdilova, 1974). The freshwater form of *E. crassum* readily infects rainbow trout and the Pacific marine form was reported from pink salmon in eastern Siberia by Akhmerov (1963), so it is probably only a matter of time before *E. crassum* is reported from pink salmon in Norway and Scotland. This invasion was due to a deliberate introduction, but the opening of the sea passage along the north coast of Siberia will inevitably lead to more introductions of North Pacific species into the north-east Atlantic and possibly beyond (Chan *et al.*, 2018). Apart from new fish hosts, these may also include planktonic copepods that may be susceptible to infection by *E. crassum*.

22.5 Population Dynamics of Intermediate Hosts and Fish

Sea levels have been rising worldwide in recent years and are predicted to continue rising as temperatures increase and polar ice caps continue to melt (Romm, 2016). This rise leads to intrusion of seawater into low-lying coastal lakes and larger bodies of water such as the Baltic Sea, increasing their salinity and extending the range of marine organisms. For *E. crassum*, this may lead to a blurring of the distinction between its marine and freshwater forms. We already know that eggs from marine *E. crassum* can infect freshwater copepods and conversely eggs from the freshwater form can infect marine copepods (Saksvik *et al.*, 2001b). Records from the Baltic Sea (Table 22.1) suggest that many species of fish, both freshwater and marine, that are able to tolerate low salinity can become infected with immature forms of what is presumably the freshwater form of *E. crassum*. Whether the evolutionary divergence of the marine and freshwater forms will continue or not under these changing conditions remains to be seen. It can safely be predicted, however, that sea level rise will favour the range extension of the marine forms at the expense of the freshwater form.

In contrast to sea level rise, Marcogliese (2001) referred to the predicted lowering of freshwater levels in north-eastern North America with the changing climate. Water levels are anticipated to fall despite increases in precipitation because evaporation through vegetation will exceed precipitation. Lower water levels generally imply lower flow rates, which would affect the transmission of some parasites by promoting retention of free-swimming infective stages. If similar lowering of water levels was to take place in Eurasia, it would result in increasing retention of *E. crassum* eggs in riverine systems, thereby increasing the chances of successful transmission.

22.6 Increase in Pathogen Mortality

As indicated in Section 22.3.2, the effects of climate change on the distribution of *E. crassum* are likely to be more pronounced at the southernmost range of its distribution, where the parasite may find the higher temperatures difficult to tolerate. Because freshwater habitats are more constrained and more exposed to anthropogenic stressors (Marcogliese, 2001), these effects are likely to be more severe for the freshwater form of *E. crassum*. The combined effects of warming and acidification on its copepod hosts, and consequently on *E. crassum*, are impossible to predict given current knowledge. The effects of these changes on the survival and virulence of the crustacean hosts of copepods have been shown to be antagonistic, with ocean acidification possibly modifying some of the more severe effects of increasing temperature (Garzke *et al.*, 2016).

22.7 Control and/or Prevention

Managing the thermal environment for farmed salmonids in these southern latitudes may be prohibitively expensive, but there is also the potential of mitigating the effects of raised temperatures on salmonid reproduction through endocrine therapies (Pankhurst and King, 2010). Well-managed selective breeding programmes would be required to breed more robust fish which would better withstand increasing temperatures (Sae-Lim *et al.*, 2017). Such breeding programmes should also take account of the susceptibility to parasites and disease of the new strains developed. Bui *et al.* (2019) proposed that natural host behaviour patterns could be harnessed to control parasitic infections, with particular reference to Atlantic salmon and sea lice. The reasoning behind this approach is that because wild salmon have co-evolved with their parasites, so certain behaviour patterns in the fish to avoid infection in the wild should be retained in farmed salmon. To exploit these behavioural patterns, fish farmers must draw on existing knowledge of wild salmon behaviour and also observe the behaviour of farmed salmon. Recognizing the farmed salmon as a species with an evolutionary history and taking advantage of their naturally developed responses to parasites by modifying aquaculture systems accordingly will facilitate management of the health and welfare of farmed fish. This approach combined with selective breeding could signal the future direction of parasite control in salmonid farming.

22.8 Conclusions

It is evident that the different effects of climate change cannot be considered in isolation. Temperature and acidification interact with one another to a large extent and this makes it extremely difficult to predict their net effects on the biology of *E. crassum* in the future. One prediction that can be made with reasonable confidence is that the distribution of *E. crassum* will move northwards and that it is likely to disappear from its current southernmost limits. Although *E. crassum* is at present not regarded as a serious fish pathogen, there is no doubt that it has the potential to become one. This change in distribution has implications for salmonid aquaculture. For example, in southern regions which become free of *E. crassum* the fish may benefit from its absence, but this may be counteracted by possible adverse effects of rising temperatures on fish biology. If climate change results in an increase in the prevalence of *E. crassum* in copepods, measures will have to be taken by aquaculture managers to keep copepod hosts out of the farm system. This would mean efficient filtering of the water entering the system or, in the case of mariculture, moving to onshore facilities where fish would be less vulnerable to infection. One control strategy against *E. crassum* and other parasite pathogens in the future may depend on selective breeding for resistance and the exploitation of natural responses evolved to reduce the risk of infection. Finally, the potential for *E. crassum* to colonize new fish and copepod hosts cannot be ignored. This potential may be enhanced by rising sea levels and the invasion of the North Atlantic by species of North Pacific origin, which may become reservoir hosts for the parasite.

References

Akhmerov, A.Kh. (1962) On the biology of the cestode *Eubothrium crassum* (Bloch, 1779). *Trudy Gelmintologichesko Laboratorii AN SSSR* 12, 5–8. (in Russian)

Akhmerov, A.Kh. (1963) Helminths as biological indicators of local stocks of Amur anadromous salmon (*Oncorhynchus*). *Voprosy Ikhtiologi* 3, 536–555.

Andersen, K.I. (1979) Studies on the scolex morphology of *Eubothrium* spp. with emphasis on characters usable in species discrimination and with brief references on the scolices of *Bothriocephalus* sp. and *Triaenophorus* spp. (Cestoda: Pseudophyllidea). *Zeitshrift für Parasitenkunde* 60, 147–156.

Andersen, K.I. and Kennedy, C.R. (1983) Systematics of the genus *Eubothrium* Nybelin (Cestoda, Pseudophyliidea), with partial redescription of the species. *Zoologica Scripta* 12, 95–105.

Anikieva, L.V., Ieshko, E.P and Rumyantsev, E.A. (2016) Ecological analysis of helminths in vendace and smelt from Lake Onega. *Trudy Karelskogo Nauchnogo Centra Rossiiskoi Akademii Nauk* 4, 37–47.

Armstrong, J.D., Bean, C.W. and Wells, A. (2018) The Scottish invasion of pink salmon in 2017. *Journal of Fish Biology* 93, 8–11.

Barras, C. (2019) The end of species. *New Scientist, 25 January*, 36–39.

Berg, M. (1977) Pink salmon, *Oncorhynchus gorbuscha* (Walbaum) in Norway. *Report of the Institute for Freshwater Research, Drottningholm* 56, 12–17.

Bosi, G., Shinn, A.P., Giari, L., Simoni, E., Pironi, F. and Dezfuli, B.S. (2005) Changes in the neuromodulators of the diffuse endocrine system of farmed rainbow trout, *Oncorhynchus mykiss* (Walbaum), naturally infected with *Eubothrium crassum* (Cestoda). *Journal of Fish Biology* 28, 703–711.

Boyce, N.P. (1979) Effects of *Eubothrium salvelini* (Cestoda: Pseudophyllidea) on the growth and vitality of sockeye salmon, *Oncorhynchus nerka*. *Canadian Journal of Zoology* 57, 597–602.

Boyce, N.P. and Clarke, W.C. (1983) *Eubothrium salvelini* (Cestoda, Pseudophyllidea) impairs seawater adaption of migrant sockeye salmon yearlings (*Oncorhynchus nerka*) from Babine Lake, British Columbia. *Canadian Journal of Fisheries and Aquatic Sciences* 40, 821–824.

Bristow, G.A. and Berland, B. (1989) *Eubothrium crassum* (Cestoda: Pseudophyllidea) – one species or two? Electrophoretic evidence for species status of sea-water and fresh-water forms. *Information* 20, 51.

Bristow, G.A. and Berland, B. (1991a) A report on some metazoan parasites of wild marine salmon (*Salmo salar* L.) from the west coast of Norway with comments on their interactions with farmed salmon. *Aquaculture* 98, 1–3.

Bristow, G.A. and Berland, B. (1991b) The effect of long term, low level *Eubothrium* sp. (Cestoda: Pseudophyllidea) infection on growth in farmed salmon (*Salmo salar* L.). *Aquaculture* 98, 325–330.

Buchmann, K. (1987) Cestodes in migratory trout (*Salmo trutta* L.) from the Baltic Sea. *Bulletin of the European Association of Fish Pathologists* 7, 115–117.

Buchmann, K. and Bresciani, J. (1997) Parasitic infections in pond-reared rainbow trout in Denmark. *Diseases of Aquatic Organisms* 28, 125–138.

Buchmann, K., Uldal, A. and Lyholt, H.C.K. (1995) Parasite infections in Danish trout farms. *Acta Veterinaria Scandinavica* 36, 283–298.

Bui, S., Oppedal, F., Sievers, M. and Dempster, T. (2019) Behaviour in the toolbox to outsmart parasites and improve fish welfare in aquaculture. *Reviews in Aquaculture* 11, 169–186.

Campbell, A.D. (1974) The parasites of fish in Loch Leven. *Proceedings of the Royal Society of Edinburgh, Section B: Biological Sciences* 74, 347–364.

Chan, F.T., Stanislawczyk, K., Sneekes, A.C., Dvoretsky, A., Gollasch, S. *et al.* (2019) Climate change opens new frontiers for marine species in the Arctic: current trends and future invasion risks. *Global Change Biology* 25, 25–38.

Chubb, J.C. (1980) Seasonal occurrence of helminths in freshwater fishes. Part 3. Larval Cestoda and Nematoda. *Advances in Parasitology* 18, 1–120.

Cripps, G., Lindeque, P. and Flynn, K. (2014) Parental exposure to elevated pCO_2 influences the reproductive success of copepods. *Journal of Plankton Research* 36, 1165–1174.

de Roij, J. and MacColl, A.D.C. (2015) Consistent differences in macroparasite community composition among populations of three-spined sticklebacks, *Gasterosteus aculeatus L. Parasitology* 139, 1478–1491.

Dittmar, J., Janssen, H., Kuske, A., Kurtz, J. and Scharsack, J.P. (2014) Heat and immunity: an experimental heat wave alters immune functions in three-spined sticklebacks (*Gasterosteus aculeatus*). *Journal of Animal Ecology* 83, 744–757.

Dombroski, E. (1955) Cestode and nematode infection of sockeye smolts from Babine Lake, British Columbia. *Journal of the Fisheries Research Board of Canada* 12, 93–96.

Dorucu, M., Crompton, D.W.T., Huntingford, F.A. and Walters, DE. (1995) The ecology of endoparasitic helminth infections of brown trout (*Salmo trutta*) and rainbow trout (*Oncorhynchus mykiss*) in Scotland. *Folia Parasitologica* 42, 29–35.

Fahy, E. (1980) *Eubothrium crassum* in migratory trout, *Salmo trutta* L., in the sea. *Journal of Fish Biology* 16, 99–104.

Gaevskaya, A.V. and Shapiro, L.S. (1981) On problems of the Baltic herring (*Clupea harengus membras* L.) location in the Vistula Bay of the Baltic Sea. In: Rikhter, V. (ed.) *Stock State and Principle of the National Fishery in the Atlantic*. Trudy AtlantNIRO, Kaliningrad, USSR, pp. 11–19. (in Russian)

Garzke, J., Hansen, T., Ismar, S.M.H. and Sommer, U. (2016) Combined effects of ocean warming and acidification on copepod abundance, body size and fatty acid content. *PLoS ONE* 11(5), e015952.

Gerdeaux, D., Fillon, M.A. and Van Overmeire, I. (1995) Arctic charr, *Salvelinus alpinus*, of Lake Annecy: yield, growth and parasitism by *Eubothrium salvelini*. *Nordic Journal of Freshwater Research* 71, 245–251.

Grozdilova, T.A. (1974) Parasite fauna of the gorbuscha *Oncorhynchus gorbuscha* of the White Sea. *Parazitologiya* 8, 293–298. (in Russian)

Guilloteau, P., Poulin, R. and MacLeod, C.D. (2016) Impacts of ocean acidification on multiplication and caste organisation of parasitic trematodes in their gastropod host. *Marine Biology* 163, 96.

Harland, H., MacLeod, C.D. and Poulin, R. (2015) Non-linear effects of ocean acidification on the transmission of a marine intertidal parasite. *Marine Ecology Progress Series* 536, 55–64.

Hemmingsen, W., Jansen, P.A. and MacKenzie, K. (2005) Crabs, leeches and trypanosomes: an unholy trinity? *Marine Pollution Bulletin* 50, 336–339.

Hermansen, Ø. and Heen, K. (2012) Norwegian salmonid farming and global warming: socioeconomic impacts. *Aquaculture Economics and Management* 16, 202–221.

Hoberg, E.P. and Brooks, D.R. (2015) Evolution in action: climate change, biodiversity dynamics and emerging infectious disease. *Philosophical Transactions of the Royal Society B: Biological Sciences* 370, 20130553.

Hodneland, K. and Solberg, P. (1995) Infection trials with marine *Eubothrium* in planktonic copepods, with aspects of procercoid development. In: *4th International Symposium of Fish Parasitology, Munich, Germany, 3–7 October 1995. Abstracts of Papers*, p. 51. Institute of Zoology, Fish Biology and Fish Diseases, University of Munich, Munich, Germany.

Hoffman, R., Kennedy, C.R. and Meder, J. (1986) Effects of *Eubothrium salvelini* Shrank, 1790 on Arctic charr, *Salvelinus alpinus* (L.), in an alpine lake. *Journal of Fish Biology* 9, 153–157.

Joyeux, C.H. and Baer, J.G. (1936) Cestodes. *Fauna de France* 30, 1–613.

Kane, M.B. (1966) Parasites of Irish fishes. *Scientific Proceedings of the Royal Dublin Society* 18, 205–220.

Kennedy, C.R. (1978a) Studies on the biology of *Eubothrium salvelini* and *E. crassum* in resident and migratory *Salvelinus alpinus* and *Salmo trutta* and in *S. salar* in North Norway and the islands of Spitzbergen and Jan Mayen. *Journal of Fish Biology* 12, 147–162.

Kennedy, C.R. (1978b) The biology, specificity and habitat of the species of *Eubothrium* (Cestoda: Pseudophylldea), with reference to their use as biological tags: a review. *Journal of Fish Biology* 12, 393–410.

Kennedy, C.R. (1996) Establishment, survival and site selection of the cestode *Eubothrium crassum* in brown trout, *Salmo trutta*. *Parasitology* 112, 347–355.

Konovalov, S.M. (1995) Parasites as indicators of biological processes, with special reference to sockeye salmon (*Oncorhynchus* nerka). *Canadian Journal of Fishes and Aquatic Sciences* 52(Suppl.), 202–212.

Kráľová-Hromadová, I., Scholz, T., Shinn, A.P., Cunningham, C.O., Wootten, R. *et al.* (2003) A molecular study of *Eubothrium rugosum* (Batsch, 1786) (Cestoda: Pseudophyllidea) using ITS rDNA sequences, with notes on the distribution and intraspecific sequence variation of *Eubothrium crassum* (Bloch, 1779). *Parasitology Research* 89, 473–479.

Kuchta, R., Scholz, T. and Bray, R.A. (2008) Revision of the order Bothriocephalidea Kuchta, Scholz, Brabec and Bray, 2008 (Eucestoda) with amended generic diagnoses and keys to families and genera. *Systematic Parasitology* 71, 81–136.

Kuhn, J.A., Kristoffersen, R., Knudsen, R., Jakobsen, J., Marcogliese, D.J. *et al.* (2015) Parasite communities of two three-spined stickleback populations in subarctic Norway – effects of a small spatial-scale host introduction. *Parasitology Research* 114, 1327–1339.

Kulakovskaya, O.P. and Koval, V.P. (1973) *Parasitic Fauna of Fish in the Danube Basin*. Naukova Dumka, Kiev, USSR. (in Russian)

Kuperman, B.I. (1978) The cestodes *Eubothrium salvelini* and *Eubothrium crassum* from salmon of the Kamchatka waters: characteristics of the life cycle and biology. *Biologiya Morya* 1978(4), 53–60. (in Russian)

Kurihara, H. and Ishimatsu, A. (2008) Effects of high CO_2 seawater on the copepod (*Acartia tsuensis*) through all life stages and subsequent generations. *Marine Pollution Bulletin* 56, 1086–1090.

Kurihara, H., Shimode, S. and Shirayama, Y. (2004) Effects of raised CO_2 concentration and early development of two marine copepods (*Acartia steuri* and *Acartia erythraea*). *Marine Pollution Bulletin* 49, 721–727.

Lõhmus, M. and Björklund, M. (2015) Climate change: what will it do to fish–parasite interactions? *Biological Journal of the Linnean Society* 116, 397–411.

MacLeod, C.D. and Poulin, R. (2015a) Differential tolerances to ocean acidification by parasites that share the same host. *International Journal for Parasitology* 45, 485–493.

MacLeod, C.D. and Poulin, R. (2015b) Interactive effects of parasitic infection and ocean acidification on the calcification of a marine gastropod. *Marine Ecology Progress Series* 537, 137–150.

Marcogliese, D.J. (2001) Implications of climate change for parasitism of animals in the aquatic environment. *Canadian Journal of Zoology* 79, 1331–1352.

Marcogliese, D.J. (2008) The impact of climate change on the parasites and infectious diseases of aquatic animals. *Revue Scientifique et Technique (International Office of Epizootics)* 27, 467–484.

Mayor, D.J., Sommer, U., Cook, K.B. and Viant, M.R. (2015) The metabolic response of marine copepods to environmental warming and ocean acidification in the absence of food. *Scientific Reports* 5, 13690.

Mitchell, C.G. (1993) *Eubothrium. Aquaculture Information Series 14*. The Scottish Office Agriculture and Fisheries Department, Aberdeen, UK.

Mo, T.A., Thorstad, E.B., Sandlund, O.T., Berntsen, H.H., Fiske, P. and Uglem, I. (2018) The pink salmon invasion: a Norwegian perspective. *Journal of Fish Biology* 93, 5–7.

Muniz, L.P. (1991) Freshwater acidification: its effects on species and communities of freshwater microbes, plants and animals. *Proceedings of the Royal Society of Edinburgh, Section B: Biological Sciences* 97, 227–254.

Nybelin, O. (1922) Anatomisch-systematischen Studien über Pseudophyllideen. *Göteborgs Kungl. Vetenskaps och Vitterhets Samhälles Handlingar* 26, 1–128.

Orr, J.C., Fabry, V.J., Aumont, O., Bopp, L., Doney, S.C. *et al.* (2005) Anthropogenic ocean acidification over the twenty-first century and its impact on calcifying organisms. *Nature* 437, 681–686.

Pankhurst, N.W. and King, H.R. (2010) Temperature and salmonid reproduction: implications for aquaculture. *Journal of Fish Biology* 76, 69–85.

Pedersen, A.P. and Hanssen, A.E. (2017) Ocean acidification ameliorates harmful effects of warming in primary consumer. *Ecology and Evolution* 8, 396–404.

Petkevičiūtė, R. and Bondarenko, S.K. (2001) Comparative karyological studies on the species of *Eubothrium* Nybelin, 1922 (Cestoda: Pseudophyllidea). *Systematic Parasitology* 50, 127–134.

Pilecka-Rapasz, M., Piasecki, W., Glońko, M., Kesminas, V., Domagatai, J. *et al.* (2017) Parasitological survey of smelt, *Osmerus eperlanus* (Actinopterygii: Osmeridae) from five estuary sites along the southern coast of the Baltic Sea. *International Journal of Oceanography and Hydrobiology* 46, 314–324.

Pippy, J.H.C. (1969) Preliminary report on parasites as biological tags in Atlantic salmon (*Salmo salar*). 1. Investigations 1966–1968. *FRB Technical Report No. 134*. Fisheries Research Board of Canada, Ottawa.

Poulin, R., Curtis, M.A. and Rau, M.E. (1992) Effects of *Eubothrium salvelini* (Cestoda) on the behaviour of *Cyclops vernalis* (Copepoda) and its susceptibility to fish predators. *Parasitology* 105, 265–271.

Rokicki, J. (1975) Helminth fauna of fishes of the Gdansk Bay (Baltic Sea). *Acta Parasitologica Polonica* 23, 37–84.

Rolbiecki, L. (2006) Parasites of the round goby, *Neogobius melanostomus* (Pallas, 1811), an invasive species in the Polish fauna of the Vistula Lagoon ecosystem. *Oeceanologia* 48, 545–561.

Rolbiecki, L. and Rokicki, J. (2008) Helminths of the lumpsucker (*Cyclopterus lumpus*) from the Gulf of Gdansk and Vistula Lagoon (Poland). *Oceanological and Hydrobiological Studies* 37, 53–59.

Rolbiecki, L., Rokicki, J. and Szugaj, K. (2002) Variability of perch, *Perca fluviatilis* L. helminth fauna in the Gulf of Gdansk, Baltic Sea. *Oceanological Studies* 31, 43–50.

Romm, J. (2016) *Climate Change. What Everyone Needs to Know*. Oxford University Press, New York.

Rosen, F. (1918) Recherches sur le développement des cestodes. I. Le cycle évolutif des Bothriocéphales. *Bulletin de la Société neuchetâloise des sciences naturelles* 43, 241–300.

Sae-Lim, P., Kause, A., Mulder, H.A. and Olesen, I. (2017) Breeding and Genetics Symposium: Climate change and selective breeding in aquaculture. *Journal of Animal Science* 95, 1801–1812.

Saksvik, M., Nilsen, F., Nylund, A. and Berland, B. (2001a) Effect of marine *Eubothrium* sp. (Cestoda: Pseudophyllidea) on the growth of Atlantic salmon, *Salmo salar* L. *Journal of Fish Biology* 24, 111–119.

Saksvik, M., Nylund, A., Nilsen, F. and Hodneland, K. (2001b) Experimental infection of Atlantic salmon (*Salmo salar*) with marine *Eubothrium* sp. (Cestoda: Pseudophyllidea): observations on the life cycle, aspects of development and growth of the parasite. *Folia Parasitologica* 48, 118–126.

Scholz, T., Kuchta, R., Shinn, A.P., Šnábel, V. and Hanzelová, V. (2003) Host specificity and geographical distribution of *Eubothrium* in European salmonid fish. *Journal of Helminthology* 77, 255–262.

Setyawan, A.C., Zuo, S., Kania, P.W. and Buchmann, K. (2019) Endoparasitic helminths of Baltic salmon *Salmo salar* L.: ecological implications. *Diseases of Aquatic Organisms* 135, 193–199.

Shulman, S.S. (1961) Zoogeography of parasites of USSR freshwater fishes. In: Dogiel, V.A., Petrushevski, G.K. and Polyanski, Yu.I. (eds.) *Parasitology of Fishes*. Oliver and Boyd, Edinburgh and London, pp. 180–245.

Shulman, S.S. and Shulman-Albova, R.E. (1953) *Parasites of Fishes of the White Sea*. Akademiya Nauk SSSR, Moscow. (in Russian)

Smith, H.D. and Margolis, L. (1970) Some effects of *Eubothrium salvelini* (Schrank, 1790) on sockeye salmon, *Oncorhynchus nerka* (Walbaum), in Babine Lake, British Columbia. *Journal of Parasitology* 56, 321–322.

Torchin, M.E., Lafferty, K.D. and Kuris, A.M. (2002) Parasites and marine invasions. *Parasitology* 124(Suppl.), S137–S151.

Unger, P. and Palm, H.W. (2016). Parasitisation of sea trout (*Salmo trutta* trutta L.) from the spawning ground and German coastal waters off Mecklenburg-Western Pomerania, Baltic Sea. *Parasitology Research* 115, 165–174.

Vehmaa, A., Brutemark, A. and Engström-Øst, J. (2012) Maternal effects may act as an adaptation mechanism for copepods facing pH and temperature changes. *PLoS ONE* 7, e48538.

Vehmaa, A., Hogfors, H., Gorokhova, E., Brutemark, A., Holmborn, T. and Engström-Øst, J. (2013) Projected marine climate change effects on copepod oxidative status and reproduction. *Ecology and Evolution* 3, 4548–4557.

Vik, R. (1963) Studies of the helminth fauna of Norway. IV. Occurrence and distribution of *Eubothrium crassum* (Bloch, 1779) and *E. salvelini* (Schrank, 1790) (Cestoda) in Norway, with notes on their life cycles. *Nytt Magasin for Zoologi* 11, 47–73.

Villarino, E., Chust, G., Licandro, P., Butenschön, M., Ibaibarriaga, L. *et al.* (2015) Modelling the future biogeography of North Atlantic zooplankton communities in response to climate change. *Marine Ecology Progress Series* 531, 121–142.

Wang, M., Jeong, C.-B. Lee, Y.H. and Lee, J.S. (2018) Effects of ocean acidification on copepods. *Aquatic Toxicology* 196, 17–24.

Wootten, R. (1972) Occurrence of *Eubothrium crassum* (Bloch,1979) (Cestoda: Pseudophyllidea) in brown trout *Salmo trutta* L., and rainbow trout *S. gairdneri* Richardson, 1836, from Hanningfield Reservoir, Essex. *Journal of Helminthology XLVI*, 327–339.

Zachos, F.E. (2016) *Species Concepts in Biology: Historical Development, Theoretical Foundations and Practical Relevance*. Springer, Cham, Switzerland.

23 Diplostomiasis (*Diplostomum spathaceum* and Related Species)

Anssi Karvonen[1]* and David J. Marcogliese[2,3]

[1]*University of Jyväskylä, Jyväskylä, Finland;* [2]*Environment and Climate Change Canada, Montreal, Quebec, Canada;* [3]*Fisheries and Oceans Canada, St. Andrews, New Brunswick, Canada*

23.1 Introduction

Diplostomum spp. (Trematoda) are widespread parasites of freshwater and brackish-water fishes (Chappell, 1995) that infect different parts of the fish eye such as the lens, humour and retina (Marcogliese *et al.*, 2001a; Karvonen *et al.*, 2006b; Désilets *et al.*, 2013; Padros *et al.*, 2018). In the ecological literature, species infecting the lens are commonly grouped as a single species, *Diplostomum spathaceum*. However, morphological (Niewiadomska, 1984, 1986; Niewiadomska and Kiseliene, 1994) and particularly molecular studies (Niewiadomska and Laskowski, 2002; Locke *et al.*, 2010a,b, 2015; Rellstab *et al.*, 2011; Blasco-Costa *et al.*, 2014) indicated that *Diplostomum* is a species complex with a number of different species infecting specific parts of fish eyes. However, details of life histories and ecological differences of many of the species are not known. Our present focus is on species infecting the lens and causing diplostomiasis. Most of the published literature is on *D. spathaceum* (in the light of the current knowledge possibly including more than one species) or *Diplostomum pseudospathaceum* (the species was described by Niewiadomska (1984) and verified by Niewiadomska and Laskowski (2002) using molecular techniques). For simplicity and due to the lack of data, we assume here that effects of all parasite species possibly co-infecting a lens are similar and they respond roughly the same way to climate change.

The life cycle of lens-infecting *Diplostomum* spp. includes an avian definitive host, a molluscan first intermediate host and a fish second intermediate host (Chappell *et al.*, 1994; Karvonen, 2012) (Fig. 23.1). Parasites reproduce sexually in the gut of the bird and start producing eggs in 3 days after establishment (Chappell *et al.*, 1994). Eggs are released into the aquatic environment through bird faeces. They hatch into free-swimming miracidia that are non-feeding and short-lived stages, which infect the molluscan intermediate host. This is typically a snail of the genus *Lymnaea*. Within a snail, each miracidium gives rise to a mother sporocyst, which then replicates asexually to multiple daughter sporocysts. Larval cercariae are formed in the sporocyst through asexual reproduction. Thus, cercariae from a single miracidial infection in a snail are genetically identical. However, one snail can be infected with multiple miracidia (Rauch *et al.*, 2005; Louhi *et al.*, 2013) and produce cercariae of different genotypes at the same time. Cercariae are released from an infected snail to the surrounding water in very high numbers (Lyholt and Buchmann, 1996; Karvonen *et al.*, 2004a). Free-swimming cercariae in the water column do not feed after leaving the snail, but rely on glycogen reserves, which last for approximately 24 h (Karvonen *et al.*, 2003). Afterwards, cercariae lose their infectivity and die. If a cercaria encounters a fish, it penetrates gills or skin and migrates as a diplostomulum to the eye lens. Details of the route and mechanisms of migration are unknown (Ratanarat-Brockelman, 1974). The migration typically takes place within 24 h, but this depends on the water temperature (Lyholt and Buchmann, 1996). A diplostomulum that fails to complete the migration exhausts its energy reserves and is eliminated

*anssi.t.karvonen@jyu.fi

 Climate Change and Infectious Fish Diseases (P.T.K. Woo, J.-A. Leong and K. Buchmann)

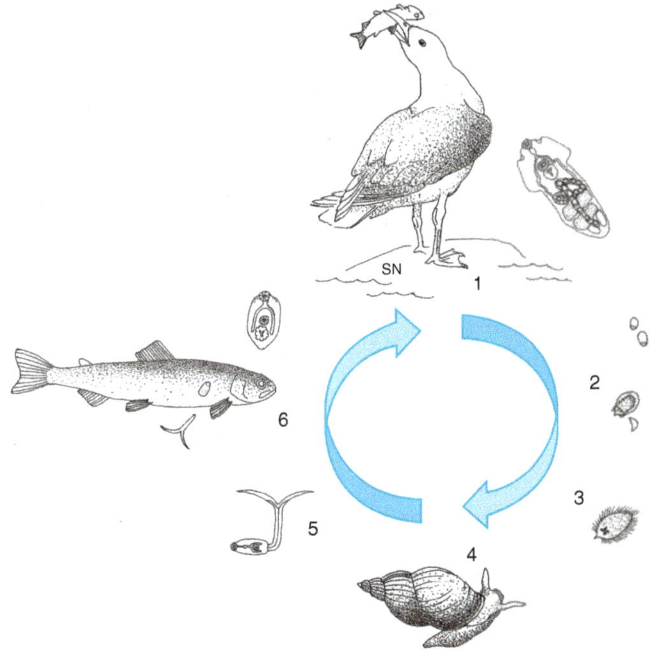

Fig. 23.1. Life cycle of *Diplostomum* spp. includes three hosts. Parasites mature in the intestine of a definitive host (1) and release eggs (2) into the aquatic environment with faeces. Eggs hatch into miracidia (3) that seek the first intermediate host, an aquatic snail. Within the snail (4), parasites reproduce asexually producing high numbers of cercariae (5) that are released to water. Cercariae are short-lived and await a fish host. They penetrate the epithelium of the fish, migrate to the eye lens and develop to metacercariae (6). The life cycle is completed when the definitive host consumes an infected fish. (Drawings courtesy of Sven Nikander. Reproduced with permission from Karvonen, 2012.)

by the fish immune system. Those that reach the lens are at least partly protected from the host immune system, as the eye lens is not directly connected to blood circulation of the fish. In the lens, parasites develop to the final larval stages, metacercariae, within a few weeks and afterwards can probably survive in the lens for years. Consequently, the number of metacercariae in fish tends to increase with time (Marcogliese *et al.*, 2001b). The life cycle is completed when a fish-eating bird consumes an infected fish.

D. spathaceum has been reported from eye lenses of over 100 fish species in Europe (Chappell, 1995), while the global distribution of the *Diplostomum* spp. probably includes many more host species. Thus, the genus is considered one of the most common and abundant parasites of freshwater fishes. Species belonging to Cyprinidae, Percidae, Salmonidae, Coregonidae, Catostomidae and Gasterosteidae, which inhabit littoral waters, are commonly infected (Margolis and Arthur, 1979; McDonald and Margolis, 1995; Valtonen and Gibson, 1997; Seppälä *et al.*, 2011). *Diplostomum* spp. have also been reported from marine fish species in brackish waters including Gadidae and Pleuronectidae (Buchmann, 1986; Koie, 1999). However, the prevalence and abundance of the infection are typically highly variable and can depend, in addition to fish species, on the geographical location, type of water body, habitat within a water body, season, host age and community structure of all the other hosts in the life cycle (Pennycuick, 1971; Sweeting, 1974; Burrough, 1978; Balling and Pfeiffer, 1997; McKeown and Irwin, 1997; Valtonen and Gibson, 1997; Valtonen *et al.*, 1997; Marcogliese and Compagna, 1999; Marcogliese *et al.*, 2001a,b; Karvonen *et al.*, 2004b, 2015). For example, locations within a single lake can have different infection levels (Balling and Pfeiffer, 1997), possibly reflecting factors such as differences in abundance of infected snails. Also, infections in the higher latitudes are seasonal and take place mainly during summer months (McKeown and Irwin, 1997; Marcogliese *et al.*, 2001a; Karvonen *et al.*, 2004b), which results in first infections in eye lenses of young fish and accumulation of infections with fish age.

23.2 Diagnosis

Metacercariae of *Diplostomum* spp. are soft-bodied, flat, bilateral and round or oval-shaped, with a body length of approximately 0.3–0.4 mm when fully developed (note that the morphology and size depend strongly on the age of the metacercaria (see Sweeting, 1974)). Infections are clearly visible from a dissected eye lens under a microscope (Fig. 23.2) and identification is straightforward as all parasite species in eye lenses of a fish belong to this same genus. However, identification at species level is notoriously difficult. Different species are morphologically very similar, especially as larval stages (miracidia, cercaria, metacercaria), and their identification requires particular expertise and experience (Niewiadomska, 1986; Niewiadomska and Kiseliene, 1994). More recently, species identification has been aided using molecular techniques (Niewiadomska and Laskowski, 2002; Moszczynska *et al.*, 2009), while these have resulted in a significant increase in the number of *Diplostomum* species (Locke *et al.*, 2010b; Blasco-Costa *et al.*, 2014).

23.2.1 Parasitic cataracts (diplostomiasis)

The most notable sign of infection in an eye lens is cataract formation due to metacercarial movement and metabolism which damage the structure of the lens. If there are many metacercariae, the damage accumulates and can result in the chronic stage of infection, known as diplostomiasis. A severe condition can be observed visually as the eye lens becomes opaque, grey or whitish. In extreme cases, the lens capsule can rupture, or the lens becomes dislocated, when the fish host loses its eyesight. Fewer cataracts (e.g. small clouds of granules or thread-like formations (Shariff *et al.*, 1980)) and their early stages following development of the metacercariae can be seen reliably only using a microscope, such as an ophthalmoscope (Karvonen *et al.*, 2004c). This type of infection occurs in most of the infected fish species worldwide.

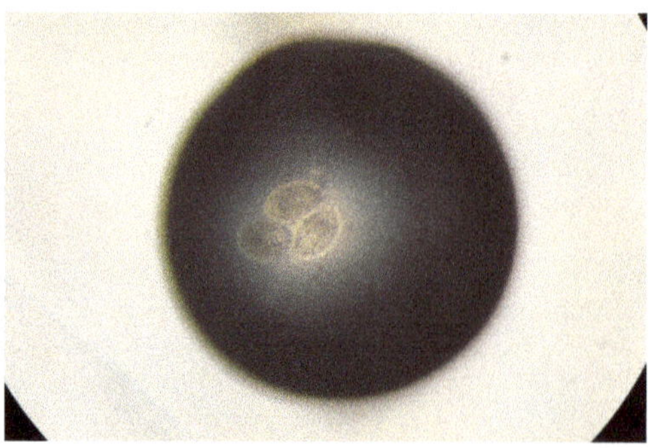

Fig. 23.2. Three metacercariae of *Diplostomum* sp. in an eye lens of Atlantic salmon. (Photograph courtesy of Ines Klemme.)

Cataracts gradually impair the vision of fish and the degree of impairment is linearly related to the number of parasites in the lens (Karvonen *et al.*, 2004c) (Fig. 23.3). In other words, a few parasites rarely cause severe cataracts, except in small fishes, although parasites can remain in the lens for years. However, there are no detailed data on long-term dynamics of cataracts recorded from individual fish. Development of cataracts is also related to the size of the fish and, consequently, size of the eye lens. In a small fish, even a low number of parasites can be sufficient to cause severe pathology (Karvonen and Lindström, 2018). Further, recent evidence suggests that fish may also show differences in their ability to tolerate the deleterious effects of the parasites, i.e. the same number of parasites results in different degrees of cataract formation (Klemme and Karvonen, 2017). Infection can also decrease the lens size directly (Karvonen and Seppälä, 2008a), but the significance of such effects for the visual ability of fish needs further study.

Infections in the eye lens and the subsequent cataract formation can have significant implications for the well-being of fish. Gradual deterioration of eyesight with increasing infection intensity can cause several physiological and behavioural effects in fish. One notable phenotypic sign of infection is darkening of the fish skin as the light intensity entering the eye decreases. This impairs the cryptic coloration of the fish particularly against a light background (Seppälä *et al.*, 2005a), which can lead to increased detection by predators. Eye infection can also decrease the efficiency of fish to detect and harvest prey items (Crowden and Broom, 1980; Owen *et al.*, 1993), which can result in decreased growth (Karvonen and Seppälä, 2008b). Impaired visual abilities have also a range of other effects that relate to social interactions of fish and susceptibility

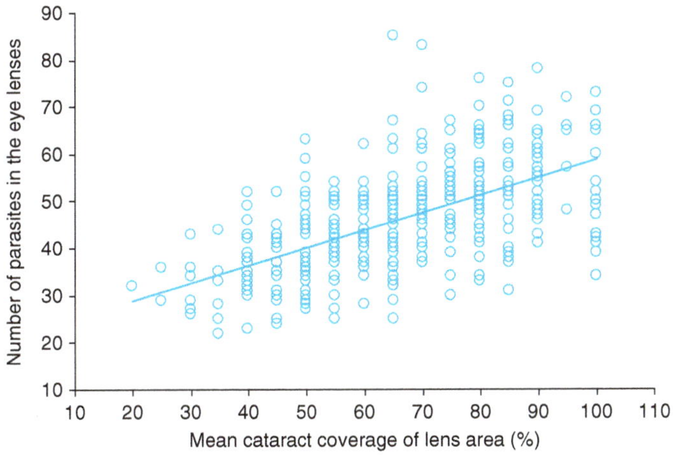

Fig. 23.3. Relationship between cataract coverage and the number of *Diplostomum pseudospathaceum* in the eye lenses of whitefish (*Coregonus lavaretus*). (Data from Karvonen and Seppälä, 2008b. Reproduced with permission from Karvonen, 2012.)

to avian predation, the latter of which is essential for completion of the parasite life cycle. For example, infection reduces group cohesion of shoaling fish (Seppälä *et al.*, 2008), which can render individual fish detectable by predators. Infected fish may also swim closer to the water surface (Crowden and Broom, 1980), although this evidence is not conclusive (Seppälä *et al.*, 2004). Further, infection and cataracts increase catchability of fish in experiments mimicking predation from fish-eating birds plunging into water from the air. These effects also coincide with the metacercariae becoming fully developed (Seppälä *et al.*, 2004, 2005b), which supports the idea that cataracts can enhance parasite transmission to the definitive hosts.

23.2.2 Implications of climate change for the parasite life cycle

Parasitism and disease in aquatic animals may increase with climate change, principally in response to rising temperatures that accelerate parasite development (Marcogliese, 2001, 2008; Lõhmus and Björklund, 2015), although general predictions are difficult to make as effects are species-specific, context-dependent and may vary among hosts (Marcogliese, 2008, 2016; Rohr *et al.*, 2011). The existence of extreme weather events, confounding factors and non-linear thresholds further complicates matters (Marcogliese, 2008, 2016; Rohr *et al.*, 2011; Altizer *et al.*, 2013) and some diseases may actually decrease in occurrence with climate change (Lafferty, 2009; Karvonen *et al.*, 2010b). Nevertheless, temperature is considered the most important abiotic factor that influences parasitic platyhelminths in ectothermic hosts, including fish (Chubb, 1979). In parasites with complex life cycles, such as *Diplostomum* spp., temperature affects all free-living life cycle stages as well as those in ectothermic hosts. Higher temperatures are generally expected to lead to faster growth, development and reproduction, earlier transmission and development in the spring, prolonged transmission in the autumn and more generations per year. However, it may also increase mortality rate among parasites in the aquatic environment (Marcogliese, 2001). Thus, it is important to consider the net effects of temperature within the entire parasite life cycle. As there are no data on relationships between temperature and cataracts (diplostomiasis) per se, and because cataracts are related to parasite numbers (Fig. 23.3), it is relevant to consider temperature-related factors that control the latter.

Our current discussion includes the effects of temperature on potential spread of *Diplostomum* spp., followed by temperature effects on production and transmission of the life cycle stages. Throughout, we explore evidence on temperature effects on *Diplostomum* spp. and other trematode taxa and on general trematode biology, while discussing the net effects of increasing temperature for the parasite life cycle. We also consider ecological evidence from field studies as well as explore the effects of temperature on the physiology and resistance of snails and fish. Besides temperature, we also focus on other environmental changes that are associated with climate change and likely to either increase or decrease parasite population. Finally, we discuss the implications of climate warming for parasite prevention strategies in aquaculture.

23.3 Potential Spread of *Diplostomum* spp. with Increasing Water Temperature

Range shifts of aquatic biota are expected with climate change (Settele *et al.*, 2014). However, given that *Diplostomum* spp. are already widespread in the northern hemisphere (e.g. see Fig. 1 in Locke *et al.*, 2015), large range shifts are unlikely. Lens-infecting *Diplostomum* spp. are generalists infecting a range of fish species (Locke *et al.*, 2010a,b, 2015; Rellstab *et al.*, 2011), so changes in fish species composition as a result of fish host range expansion or contraction should not have large effects on the parasite's distribution. However, the host spectrum in any given habitat could change. For example, increasing temperatures are predicted to have significant negative effects on cold-water stenotherms, such as salmonids and coregonids, contracting their range (Marcogliese, 2001; Chen *et al.*, 2016). These high-latitude cold-water stenotherms may experience an increase in their northern range, but a contraction of their southern boundaries with the expansion northward of temperate fishes (Ficke *et al.*, 2007; Settele *et al.*, 2014). In addition, warm- and cool-water fishes may displace native species as they migrate into higher latitudes in the northern hemisphere and lower latitudes in the southern hemisphere (Ficke *et al.*, 2007). Another potential complicating factor is that snail intermediate host populations may be at risk because freshwater molluscs are predicted to be unable to track high rates of climate change (Settele *et al.*, 2014). Their populations are further compromised by invasive species, habitat modification and contaminants (Settele *et al.*, 2014).

In addition to range shifts of resident fish and invertebrate species, increasing temperature is expected to lead to the invasion of new and potentially susceptible hosts (Marcogliese, 2001; Altizer *et al.*, 2013), including those for *Diplostomum* spp. For example, there have already been numerous introductions of warm-water fish species into the lower Great Lakes, expanding their distributions northward (Marcogliese, 2001). Further warming should facilitate the introduction and expansion of warm-water invaders (Collingsworth *et al.*, 2017). There are at least two possible outcomes. First, invasive species can facilitate parasite transmission by effectively acting as new susceptible hosts and boost the life cycle completion. Second, new hosts can be resistant to infections and act as sinks of infection while diluting infection risk among the native hosts. In Europe, round gobies (*Neogobius melanostomus*) and bighead gobies (*Ponticola kessleri*) were heavily infected with *Diplostomum* spp. in their introduced range in the Danube River, the Rhine River and parts of the south-western Baltic Sea, potentially enhancing transmission of *Diplostomum* spp. to piscivorous birds (Ondračková *et al.*, 2009, 2015; Muhlegger *et al.*, 2010; Francová *et al.*, 2011; Kvach and Winkler, 2011). In contrast, following the introduction of the round goby in the St. Lawrence River, Canada, abundance of *Diplostomum* spp., which at one time was one of the most common fish parasites in that river, declined to extremely low levels within 5 years or less in yellow perch (*Perca flavescens*), golden shiner (*Notemigonus crysoleucas*) and the spottail shiner (*Notropis hudsonius*) (Gendron and Marcogliese, 2017). The authors suggested this was due to gobies acting as incompetent decoy hosts for cercariae and diluting the risk of infection to the native fish. The difference between the capacity of gobies as hosts for *Diplostomum* spp. in Europe and North America may be because invasive gobies in Europe were exposed to widespread European species, while those in North America were exposed to new parasites with which they had no previous experience. This idea is supported by the increase in abundance of *Diplostomum* spp. in round gobies over time (15 years) since their initial invasion into the Great Lakes (Gendron *et al.*, 2012).

23.4 Implications of Increasing Temperature for the Parasite Life Cycle Stages

23.4.1 Effects of elevated temperature on life cycle stages

Life cycle stages of *Diplostomum* spp. outside the endothermic avian host (miracidia, cercariae, and the larval forms residing in ectothermic snails and fish) are potentially influenced by increasing water temperature. However, it is important to note that, in general, physiological tolerance of parasites to temperature not only varies among species, but also among stages of the same species (Chubb, 1979; Marcogliese, 2001). Overall, there are a few experimental studies on effects of temperature on different life cycle stages of *D. spathaceum* and related species (Table 23.1), although much more information is needed to make reliable predictions. For example, swimming velocity of the miracidia hatching from eggs increased at higher temperatures, but the lifespan declined (Harris, 1986). While the latter result is likely due to faster depletion of the finite glycogen reserves, the net effects on transmission are unknown. Considering these effects alone, an increase in temperature should likely promote the encounter between miracidia and potential snail hosts but decrease the infective time period. Further evidence on the snail host has shown that the time to patency decreased and cercarial output increased at higher temperatures (Harris, 1986; Waadu and Chappell, 1991), both of which should promote transmission to the fish host. However, both cercarial activity time and lifespan were reduced at higher temperatures, which should limit transmission (Harris, 1986; Sous, 1992; Lyholt and Buchmann, 1996). Moreover, cercarial penetration and speed of migration to the eyes increased at higher temperatures (Whyte *et al.*, 1988; Lyholt and Buchmann, 1996), but infectivity peaked at the mid-range of the experimental exposure temperatures (Stables and Chappell, 1986b). The latter results suggest that infection success would decrease eventually as temperature increases. To sum up, the contrasting effects of higher temperature would increase parasite reproduction but decrease longevity and infectivity of the transmission stages. Consequently, it is at present difficult to predict the overall effects of temperature on the parasite transmission success (Fig. 23.4).

An early meta-analysis by Poulin (2006) suggested that cercarial emergence could increase 200-fold with a 10°C increase in temperature, prompting the author to suggest that climate change could have a huge influence on parasite populations. However, in a subsequent meta-analysis that accounted for the minimum emergence temperature threshold (the temperature where emergence rates decrease to almost zero) and acclimatization status of infected molluscs, temperature above a particular threshold actually does not appear to affect cercarial development (Morley and Lewis, 2013). In addition, cercarial emergence from molluscan hosts shows a peaked pattern with temperature, at first increasing within low temperature ranges. It was unaffected within the optimum temperature ranges (thermostability), which correspond to the latitudinal range inhabited, but then declined at higher temperatures (Morley and Lewis, 2013). However, there were also geographic strain-specific differences in thermostability within *D. spathaceum* in two lymnaeid species (Morley and Lewis, 2013), which underscores the complexity of making predictions and establishing general rules for *Diplostomum* spp.

Thermostability over a range equivalent to typical summer temperatures for a particular species was also observed for most trematode species in cercarial mortality and glycogen utilization rate over normal temperature ranges encountered (Morley, 2011). A more recent meta-analysis of over 30 trematode species including *D. spathaceum* demonstrated an optimal temperature for both cercarial output and infectivity, while mortality was directly related to temperature (Studer and Poulin, 2014). Specifically, cercarial mortality and glycogen utilization rate increased linearly with temperature in *Diplostomum phoxini* (Morley, 2011). Furthermore, temperature had little influence on miracidial survival and metabolism over normal temperature ranges, suggesting that miracidia are more resistant to temperature changes than cercariae (Morley, 2012). Interestingly, there was little correlation in thermal responses between miracidia and cercariae within geographic strains of the same species (Morley, 2012). Using metabolic measures, Morley and Lewis (2015) showed that in general, trematode miracidia and cercariae show increased infectivity with temperature, maximizing over optimal temperature ranges and then declining at higher temperatures. Infectivity of metacercariae to definitive hosts, in contrast, was highest at low temperatures and declined as temperature increased. The overall conclusion is that temperature is not hugely important for the survival and function of trematode free-living transmission stages. Rather, transmission may depend more on thermal effects of climate change on the target hosts, among other factors (Morley and Lewis, 2015). For example, it is possible that any higher production of infective stages with temperature would be compensated for by their higher mortality, resulting in a roughly stable risk of infection to fish regardless of temperature. Under such conditions, factors such as host age, size and physiological state may affect infectivity more than direct effects of temperature on miracidia or cercariae (Morley and Lewis, 2015).

Table 23.1. Experimental studies on effects of temperature on free-living stages of *Diplostomum spathaceum* and other species, as well as on those stages in gastropods (*Lymnaea* spp.) and rainbow trout (*Oncorhynchus mykiss*). Parasites are *D. spathaceum* unless otherwise indicated.

Trait	Temperature	Comment	Reference
Egg hatch	Delayed at 4°C in *Diplostomum phoxini*	At 4°C, 6-day delay in hatch, but equals rate at 20°C by 14 days	Harris (1986)
Miracidial swimming velocity	Increases with temperature to a maximum at 25°C, then declines in *D. phoxini*	Examined velocity between 5 and 40°C	Harris (1986)
Miracidial lifespan	Maximum at 4°C, then declines with increasing temperature in *D. phoxini*	No survival at 40°C	Harris (1986)
Miracidial infectivity to snails	Declines if exposed at lower temperatures (<14°C) and switched to 20°C	Effect lost if snails all exposed at same temperature (20°C), then switched to lower temperatures	Waadu and Chappell (1991)
Time to patency in snails	Faster at higher temperatures in *D. phoxini*	75 h at 10°C versus 40 h at 20°C	Harris (1986)
	Affected by snail maintenance temperature	Delayed in snails infected at 20°C if held at 14°C, not 20–25°C	Waadu and Chappell (1991)
Cercarial shedding (minimum temperature)	None at <10°C	–	Bauer (1959)
	Occurs at 4–6°C	Stops at 3–5°C	Lyholt and Buchmann (1996)
	None at <9°C	Field-based study	Sous (1992)
Cercarial output	Increases with temperature, peaks at 18°C	–	Bauer (1959)
	Declines at temperatures <10°C	Gradual decline to 5–6°C, then drops rapidly	Lyholt and Buchmann (1996)
	Rate of output increases between 10 and 20°C in *D. phoxini*	–	Harris (1986)
	Rate of output increases between 10 and 27°C	<100/h at 4–14°C; 20–1100/h at 15–20°C; 100–4700/h at 20–27°C	Sous (1992)
Cercarial activity	None at <9–10°C	Move to upper waters at 18–22°C	Bauer (1959)
	Peaks at intermediate temperatures in *D. phoxini*	None at 4°C, maximum at 15°C, then declines	Harris (1986)
Cercarial lifespan	Shorter at higher temperatures (e.g. 72 h at 20°C versus 240 h at 4°C)	Consistent decline between 4 and 25°C	Harris (1986); Sous (1992); Lyholt and Buchmann (1996)
Cercarial penetration	Increases at higher temperatures	*In vitro* system; occurs as low as 4°C	Whyte *et al.* (1988)
	Occurs at 7.5°C	–	Stables and Chappell (1986b)
Cercarial migration to fish eyes	Faster at higher temperature	–	Lyholt and Buchmann (1996)
	Inhibited at <10°C	–	Stables and Chappell (1986b)
Cercarial establishment in eyes	Highest at >18°C	Occurs at 13–16°C	Bauer (1959)
	Maximum in mid-range (17.5°C)	No infections at <10°C if fish maintained at <10°C, but infections obtained at 5°C if fish maintained at 15°C	Stables and Chappell (1986b)
	Greater at high temperature (15°C)	No infections at 5°C	Lyholt and Buchmann (1996)

Fig. 23.4. A dense swarm of cercariae of *Diplostomum pseudospathaceum* released from snail (*Lymnaea stagnalis*). Production and release of cercariae increase significantly from 10 to 20°C. However, cercarial infectivity and lifespan deplete faster at higher temperatures. (Photograph by Anssi Karvonen.)

23.4.2 Ecological evidence from field studies

Comparisons of parasite infections in fish inhabiting areas of elevated temperatures with those under ambient conditions may provide insight into effects of temperature increases at the scale of an entire host–parasite relationship (Marcogliese, 2001, 2008). For example, infection of European perch (*Perca fluviatilis*) by *Diplostomum baeri* occurred earlier in Biotest Lake, a semi-enclosed area in the Baltic Sea heated by nuclear power plant thermal effluent, than at an ambient site, and infections accumulated there to a higher degree at an increased prevalence in 1986/87 (Höglund and Thulin, 1990). Additionally, fish were presumed to show increased mortality in the heated area due to selective predation on heavily infected hosts. In another example of a similar system, the release of cercariae by infected snails (*Helisoma trivolvis*) and recruitment of metacercariae of the eye fluke *Tylodelphys scheuringi* in the mosquitofish (*Gambusia affinis*) were prolonged into the winter months in a thermally altered reservoir compared with ambient areas in South Carolina, USA (Aho *et al.*, 1982). Cercarial release also ceased during the warmest months (e.g. July and August), implying an upper thermal limit to this trait, in agreement with Morley and Lewis (2013). The trematode *Ornithodiplostomum ptychocheilus* released cercariae from infected *Physa* sp. year-round in the same thermally altered reservoir, also with the exception of the warmest months (Camp *et al.*, 1982). However, recruitment by mosquitofish did not differ between the thermally altered and ambient areas, showing that there can be distinct different responses to temperature between phylogenetically related parasites in the same fish host at the same sites.

Systems with natural elevation in water temperature compared with that in the ambient environment may also provide interesting comparisons on the effect of temperature. Karvonen *et al.* (2013) examined *Diplostomum* spp. in three-spined sticklebacks (*Gasterosteus aculeatus*) from two Icelandic lakes that possess natural temperature gradients due to groundwater inflow and geothermic activity. In both lakes, sticklebacks from the warm areas showed a much higher abundance of *D. baeri* than those from cold regions. A second species of *Diplostomum* also had much higher prevalence and abundance in the warm part of one lake compared with the colder part, although it was absent from the second lake (Karvonen *et al.*, 2013).

23.5 Population Dynamics of the Hosts

Since fish growth is temperature-dependent, an extended growing season and reduction in overwintering stress could lead to increases in fish productivity in temperate fishes that are currently limited by suboptimal temperatures for their growth (Ficke *et al.*, 2007). This, however, assumes that temperatures remain within optimal ranges and other conditions are adequate and food is not limiting. Reproduction could be negatively affected in those fishes requiring low overwintering temperatures for spawning, such as salmonids (Ficke *et al.*, 2007). However, the duration of optimal temperatures for growth will likely increase for all thermal guilds of fishes (Collingsworth *et al.*, 2017). Furthermore, recruitment and production of spring and summer spawners can be promoted (Collingsworth *et al.*, 2017). In addition, fish populations could be negatively affected not only by increased temperatures, but also by decreased levels of dissolved oxygen and changes in contaminant concentrations, disease dynamics and hydrography, along with any other associated habitat modifications (Ficke *et al.*, 2007; Collingsworth *et al.*, 2017). Besides fish, similar processes could apply also to other hosts in parasite life cycles, such as snails in the case of *Diplostomum* spp. Without long-term data or epidemiological modelling, however, it is not possible to predict how these changes could affect levels of *Diplostomum* spp. in fishes. Nevertheless, given that most lens-infecting species of *Diplostomum* are generalists, any decrease in the availability of fish intermediate hosts likely will be offset by increases in others.

23.6 Effect of Temperature on Parasite Mortality

Diplostomum spp. metacercariae are generally well protected within the lens of the fish eye, both from the host immune attacks and from the external environment, and there are no experimental data showing metacercarial mortalities in fish directly following environmental perturbations. Thus, effects of the ambient environment on the parasite population are more likely to concern the free-living infective stages, miracidia and cercariae, as well as processes related to host physiology and resistance. As discussed earlier, increasing temperature tends to decrease the longevity of the infective stages as their finite energy reserves are exhausted more rapidly in higher water temperatures (Table 23.1). Similarly, temperature could enhance host immune function to prevent parasite migration in host tissues towards the eye (see below). Whether this results in negative net effects on the parasite population given the probable increase in parasite replication with temperature needs elucidation.

23.6.1 Effects of climate warming on host physiology and immunological resistance

Temperature also controls the physiological functions (e.g. immunity) in the fish hosts, some of which have direct relevance to parasite infections. Early work examining the immune response in fish to *Diplostomum* spp. used the rainbow trout (*Oncorhynchus mykiss*)–*D. spathaceum* system, while more recent studies have explored ecological immunology in three-spined stickleback. In general, immune responses in the eye lenses of fish are considered weak or non-existent as the lens is not directly connected to blood circulation (Sitjá-Bobadilla, 2008). Therefore, the time window for fish to fight off an initial infection is very narrow and consists of the time diplostomules are migrating to the lens, typically within 24 h from exposure (Chappell *et al.*, 1994; Sitjá-Bobadilla, 2008). Given that not all diplostomules reach the eye in an initial exposure, non-specific immune responses of the fish are likely responsible for partly preventing the infection (Whyte *et al.*, 1991). In rainbow trout, these responses include, for example, activity of the alternative pathway of the complement cascade (Whyte *et al.*, 1988, 1989) as well as macrophages (Whyte *et al.*, 1989; Chappell *et al.*, 1994). Fish also display antibody-mediated specific responses to infection with *Diplostomum* spp. that develop within a few weeks from the first exposure and significantly reduce the number of parasites establishing in subsequent exposures (Stables and Chappell, 1986a; Höglund and Thuvander, 1990; Whyte *et al.*, 1990; Karvonen *et al.*, 2005, 2010a; Rellstab *et al.*, 2013). In sticklebacks, *in vitro* experiments have demonstrated that head kidney leucocytes (HKL) exhibit a strong respiratory burst when exposed to antigens of *D. pseudospathaceum* (Franke *et al.*, 2014). However, the HKL respiratory burst activity also drops 1.5 days after exposure, implying that phagocytic cell activation is important for the immune response to *D. pseudospathaceum* (Scharsack and Kalbe, 2014). These authors suggested that the innate immune response, but not the acquired immune response, was activated to defend against *D. pseudospathaceum* in three-spined sticklebacks (Scharsack and Kalbe, 2014). There was also evidence supporting parasite genotype-specific innate immune activity in *G. aculeatus* (Haase *et al.*, 2014), while other studies found no evidence of genotype specificity in the acquired responses (Rellstab *et al.*, 2013; Haase *et al.*, 2016). Further, the immune response in three-spined sticklebacks against *Diplostomum* spp. varies among populations and habitats (Scharsack and Kalbe, 2014; Scharsack *et al.*, 2016). For example, fish sympatric with *D. pseudospathaceum* show a stronger innate response against initial infection than those from uninfected populations (Kalbe and Kurtz, 2006).

Temperature affects basically all physiological functions in ectotherms (Bowden, 2008). In fish, the immune response is stimulated or at least positively correlated with temperature, as shown by lysozyme activity, concentration of circulating immunoglobulin M (IgM), and major histocompatibility complex (MHC) and cytokine gene expression (Tort *et al.*, 2003; Bowden, 2008; Martin *et al.*, 2010; Uribe *et al.*, 2011). Circulating IgM concentration increases in salmonids when acclimatized to 19°C (Uribe *et al.*, 2011). However, it is not known if these processes play a role in defence against *Diplostomum* spp. In contrast, temperature effects on complement activity are inconclusive, effects on haematology vary with cell type and phagocytosis is not greatly affected (Bowden, 2008; Uribe *et al.*, 2011), processes and functions which do play a role in the immune response against *Diplostomum* spp. In most fish species examined, acquired immune activity and immune gene expression are enhanced while innate immune activity is suppressed at the highest temperatures tested (Dittmar *et al.*, 2014). Nevertheless, one might expect resistance to parasites such as *Diplostomum* spp. to increase with climate change (Scharsack *et al.*, 2016). However, higher temperatures also accelerate parasite growth, development and life cycle completion, and it is not clear whether the host or the parasite benefits more under these circumstances (Scharsack *et al.*, 2016). Again, the above aspects well illustrate the complex nature of temperature effects; they potentially elevate transmission and can result in higher parasite numbers (Fig. 23.5), but also interact with different types of temperature effects on the host as well as with many other ecological and evolutionary factors determining host resistance.

Extreme weather events are also predicted to increase with climate change (Marcogliese, 2001). Examination of how host–parasite systems respond to extreme weather such as heatwaves and drought may provide insight into how climate change will affect outbreaks of diseases in ecosystems (Hudson *et al.*, 2006; Poulin and Mouritsen, 2006; Morley and Lewis, 2014). Studies have shown that acute temperature changes experienced during heatwaves can cause immunosuppression in fish (Uribe *et al.*, 2011; Scharsack *et al.*, 2016). Generally, in ectotherms, such changes can suppress various immune functions such as phagocytosis, respiratory burst and antibody production (Martin *et al.*, 2010). Immune function is more efficient if fish are acclimatized to higher or varying temperatures (Martin *et al.*, 2010; Scharsack *et al.*, 2016). Indeed, in simulated heatwave experiments, innate and adaptive immune functions were optimal at 13–17°C compared with 18–24°C in three-spined stickleback (Dittmar *et al.*, 2014). Exposure to a simulated heatwave also caused long-lasting deleterious effects on immune function, but less so if fish were from presumably better-adapted populations (Dittmar *et al.*, 2014). Sticklebacks maintained in artificial enclosures with the lowest parasite load and an intermediate level of MHC class IIb sequence variation survived best, while those with the highest parasite burdens perished during the 2003 European heatwave. This suggests a link between MHC diversity and fitness (Wegner *et al.*, 2008).

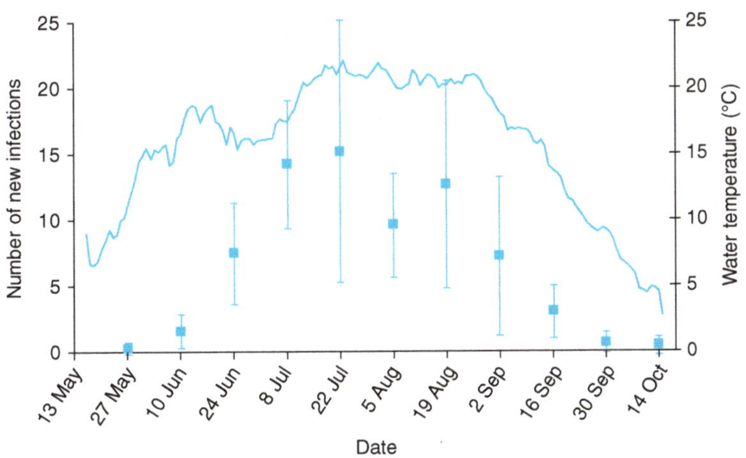

Fig. 23.5. Timing and temperature dependence of transmission of *Diplostomum* spp. to fish in natural conditions. Data show the mean number of new *Diplostomum* spp. infections, with standard deviation represented by vertical bars, in eye lenses of fish caged in an oligotrophic lake during 2-week periods in May–October. The solid line indicates water temperature of the lake. Infections peak naturally in July–August when water temperature exceeds 15°C. Climate warming could potentially enhance parasite reproduction in the snail intermediate hosts and transmission to fish, resulting in a higher number of infections within the current window on transmission. Additionally, higher water temperatures in spring and autumn could prolong the optimal infection period from both ends of the range. (Reproduced with permission from Karvonen *et al.*, 2004b.)

In addition to fish, higher water temperatures can influence resistance of the other poikilothermic intermediate host of *Diplostomum* spp., the snail. Compared with fish, however, there is little information on the immune response of snails to the infection. It has been established that the susceptibility of *Lymnaea stagnalis* to *D. spathaceum* varies with age – young snails being susceptible and older snails more resistant to infection (Chappell *et al*., 1994). Haemocyte profiles also differ between infected and uninfected snails with haemocytes from infected snails displaying reduced phagocytotic capability, and serum showing lower opsonic and agglutinating abilities (Riley and Chappell, 1992). This is consistent with the suggestion that the initial infection decreases immune function in snails, making them more susceptible to accumulate further infections (Louhi *et al*., 2013).

Evidence on the effect of temperature on snail immune function comes mostly from parasite systems other than *Diplostomum* spp. For example, exposure of *L. stagnalis* to simulated heatwaves of 25°C increased infection success of the trematode *Echinoparyphium aconiatum* (Leicht and Seppälä, 2014). Exposure of the snails to 23.5–25°C for more than 1 week also reduced their haemocyte concentration and phenoloxidase-like activity, an oxidative defence against parasites (Leicht *et al*., 2013, 2017; Salo *et al*., 2017). It is likely that similar processes could also influence infections of *Diplostomum* spp. in *L. stagnalis* and in other lymnaeids. Overall, this evidence suggests that increasing temperature could impair the ability of snails to prevent infections, likely resulting in increasing prevalence of *Diplostomum* spp. and other trematode infections. As these infections typically castrate the host, increased likelihood of parasitism would undoubtedly influence also the snail populations, which again would be reflected in the net effects of temperature on the parasite life cycle.

23.6.2 Net effects of increasing temperature on the parasite life cycle

Overall, there are no detailed studies on the net effects of temperature on transmission and pathology of *Diplostomum* spp. However, studies on a similar host–parasite system may shed some light on the question of net effects. *Ribeiroia ondatrae* has a three-host life cycle, infecting snails and birds. One fundamental difference is that amphibian tadpoles are second intermediate hosts, where the parasite causes limb malformations. Nevertheless, the second intermediate host is a freshwater ectothermic vertebrate with pathological consequences. Studies on net effects of temperature on different life history aspects of *R. ondatrae* may provide informative for *Diplostomum* spp. Paull and Johnson (2011) and Paull *et al*. (2012) demonstrated differential effects on different parasite and host life history characteristics (see Marcogliese, 2016). For example, cercarial survival and establishment in the tadpole peaked at low temperatures, while egg development rate, cercarial development rate and cercarial penetration to tadpoles peaked at high temperatures, but metacercarial numbers in the tadpoles were lowest at high temperatures. Growth of snails (*Planorbella trivolvis*), infected or not, and Pacific chorus frog (*Pseudacris regilla*) tadpoles peaked at high temperatures, along with snail fecundity. However, fecundity of infected snails peaked at intermediate temperatures. Their crucial finding was that pathology in the snail in terms of castration and gigantism peaked at high temperatures, but malformations in the tadpoles were maximized at intermediate temperatures and were lowest at high temperatures (Paull and Johnson, 2011; Paull *et al*., 2012). In a year-long mesocosm study, a temperature increase of 3°C induced snails to release cercariae of *R. ondatrae* nine months earlier than at ambient conditions and increased snail mortality fourfold (Paull and Johnson, 2014). However, infections in bullfrog (*Lithobates catesbeianus*) tadpoles peaked two months earlier. In chorus frogs (*Pseudacris triserata*), infections were reduced by half and malformations by two-thirds (Paull and Johnson, 2014). After 1 year, 92% fewer adult snails were releasing cercariae in the thermally altered mesocosm compared with the ambient one (Paull and Johnson, 2014). These results suggest that changes in the impact of parasites on their hosts following global warming depend on the timing and temporal overlap of the temperature-driven changes in the host and parasite populations.

To conclude, the above examples highlight the need to understand the net effects of temperature increases on parasite transmission in general and on *Diplostomum* spp. in particular (Altizer *et al*., 2013; Marcogliese, 2016). It seems clear that elevated temperatures will influence both parasites and hosts, patterns that could show contrasting effects on parasite prevalence and abundance. Untangling these relationships requires rigorous experimental approaches in the laboratory and under field conditions. Due to multiple underlying factors, interpreting the overall effect of climate warming also emphasizes the importance of long-term time-series data on parasite population dynamics. Such data are not available for most systems but would be invaluable as they capture the outcome of the entire process within a host–parasite interaction. Furthermore, the current evidence on temperature effects needs to be interpreted with caution, as it may not have accounted for acclimatization of both hosts and parasites, in addition to infected hosts, or variation in temperature (Morley and Lewis, 2013; Raffel *et al.*, 2013, 2015; Rohr *et al.*, 2013; Altman *et al.*, 2016). This is important, as organisms generally acclimatize their performance after a temperature shift, which could change the interpretation of the temperature effects. Moreover, natural temperatures are rarely constant but variable and even a short-term variation in temperature will change the outcome of a host–parasite interaction (Paaijmans *et al.*, 2010; Raffel *et al.*, 2013). Indeed, variation in temperature is expected to increase with climate warming (Jiménez Cisneros *et al.*, 2014; IPCC, 2018), which emphasizes the importance of incorporating temperature dynamics into studies on disease occurrence. Undoubtedly, short-term temperature variation plays an important role in epidemics of directly transmitted pathogens. However, implications of the temperature variation for macroparasites with complex life cycles, such as *Diplostomum* spp., may be challenging and difficult to predict.

23.7 Other Associated Consequences of Climate Change

Freshwater ecosystems can be expected to undergo numerous changes aside from increased temperature associated with climate change. These include changes in precipitation, salinity, eutrophication, acidification, hydrology and water levels, reduced ice cover, habitat loss, fragmentation, pollution, ultraviolet (UV) radiation and invasive species (Marcogliese, 2001, 2008, 2016), all of which could affect the distribution and abundance of *Diplostomum* spp. (Table 23.2). There is no general unidirectional effect of these environmental changes; they can lead to parasite population increases or declines, emphasizing the importance of confounding factors and context dependency (Rohr *et al.*, 2011; Altizer *et al.*, 2013).

There are some examples of impacts of these factors on *Diplostomum* spp. that may be illuminating. For example, abundance of *Diplostomum* spp. in mudpuppies (*Necturus maculosus*) in the St. Lawrence River was highest in a regulated fluvial lake with stable water levels compared with two other fluvial lakes where levels fluctuated (Marcogliese *et al.*, 2000). Experimental studies also demonstrated that transmission of *D. spathaceum* to rainbow trout was greatly reduced at higher flow rates, with a 10-fold increase in flow rate decreasing infections 30-fold (Stables and Chappell, 1986b). Abundance of a similar parasite, *Posthodiplostomum minimum*, also increased under low flow conditions, but was severely reduced under high flow conditions associated with snowmelt in *Fundulus zebrinus* in the Platte River, Nebraska (Janovy *et al.*, 1997). Further, infection of eye flukes in the snail *Lymnaea peregra* plummeted in a reservoir after it was filled to maximum volume (Moody and Gaten, 1982). Thus, regulation of water bodies and flow rates, coupled with declining water levels, should potentially serve to increase infection levels of *Diplostomum* spp. Indeed, reservoir construction is considered a means of helping to mitigate or adapt to effects of climate change on streams, rivers and wetlands (Jiménez Cisneros *et al.*, 2014; Muller, 2019).

Contaminant concentration may increase under low water conditions and climate change (Johnson *et al.*, 2009; Jiménez Cisneros *et al.*, 2014; Landis *et al.*, 2014; Morley and Lewis, 2014). Lethal and sub-lethal effects of combined exposure of animals to both parasites and contaminants can be greater than the effects of either stressor alone (Marcogliese and Pietrock, 2011). For example, the combined exposure to municipal, agricultural and industrial pollution and infection with *Diplostomum* spp. increased oxidative stress in yellow perch, *P. flavescens* (Marcogliese *et al.*, 2010). Moreover, exposure to increasing temperature concurrently with another stressor may negatively impact an organism's health, leading to population declines in ectotherms (Rohr and Palmer, 2013). Survival of naturally infected snails (*L. stagnalis* and *L. peregra*) was reduced when exposed to cadmium compared with controls (Morley *et al.*, 2003a). Free-living stages of a parasite also are sensitive to environmental contaminants (Morley *et al.*, 2003c; Pietrock and Marcogliese, 2003). Exposure of cercariae of *Diplostomum* spp. to cadmium, chromium, mercury and sediment extracts from the polluted Oder River reduced their lifespan (Pietrock *et al.*, 2001,

Table 23.2. Putative effects of environmental or biological changes associated with climate change. (Derived from Table I in Marcogliese, 2008 on populations of *Diplostomum* spp. in fishes, based on information in Marcogliese, 2001, 2004, 2005, 2008; Marcogliese *et al.*, 2010; Tully *et al.*, 2019.)

Environmental or biological change	General response of *Diplostomum* spp.	Putative cause
Species introductions with change in host range	Population increase or decrease	Introduction of host species should increase generalist and specialist *Diplostomum* spp. of introduced hosts, but decrease specialist *Diplostomum* spp. of native hosts at risk
Loss of habitat due to temperature	Population decline	Applies to specialist species whose hosts lose habitat
Reduced flow rates	Population increase	Retention of free-living infective stages, increased infectivity of fish, promotion of snail habitat
Eutrophication	Population increase	Promotes parasites which use snails as intermediate hosts and birds as definitive hosts
Increased stratification	Population decline	Reduction in snail habitat due to seasonal anoxia in bottom waters
Reduced ice cover	Population increase	Promotes transmission of *Diplostomum* spp. to avian definitive hosts over longer period
Increased acidification in headwater streams	Population decline	Reduced survival of snail intermediate hosts sensitive to acidification
Decreased acidification in lakes	Population increase	Promotes survival of snail intermediate hosts
Increased ultraviolet radiation	Population decline	Mortality of free-living infective stages
Decrease in salinity due to increased precipitation	Population increase	Increase in available habitat due to lower salinity
Rise in sea level	Population decline	Loss of habitat due to intrusion of salt water
Increased concentration of contaminants	Population decline	Combined effects of contaminants and *Diplostomum* spp. infection reduce fish health; transmission to fish reduced through effects on cercariae
Socio-economic adaptation dam construction	Population increase	Replacement of lotic conditions with still or slow-moving waters (see altered hydrology above)
Modifying water withdrawal or delivery	Population decrease	Increased stream flow

2002a,b). Exposure to mixtures of cadmium and zinc, however, increased survival in *D. spathaceum* (Morley *et al.*, 2001, 2002). Notably, cercarial activity of *D. spathaceum* was reduced following exposure to zinc, cadmium and zinc–cadmium mixture at all concentrations tested and cercariae were vulnerable during the period of maximal cercarial infectivity (Morley *et al.*, 2003b). Infectivity of cercariae of both *P. minimum* and *O. ptychocheilus* to fathead minnows (*Pimephales promelas*) was reduced following exposure to cadmium (Pietrock and Goater, 2005). Climate change can also increase the toxicity of chemical contaminants as well as their uptake and an animal's susceptibility (Schiedek *et al.*, 2007; Noyes *et al.*, 2009; Hooper *et al.*, 2013; Stahl *et al.*, 2013). Furthermore, exposure to contaminants may decrease an organism's thermal tolerance to increasing temperature (Noyes *et al.*, 2009), as well interact with other climate-associated stressors (Moe *et al.*, 2013). Thus, any increase in contaminants may decrease *Diplostomum* spp. infections in fish and snail intermediate hosts.

Nutrient pollution is also expected to increase in fresh waters (Ficke *et al.*, 2007; Jiménez Cisneros *et al.*, 2014; Collingsworth *et al.*, 2017), which can have significant implications for parasitism. For example, occurrence of limb malformations and abundance of *R. ondatrae* in anurans in agricultural wetlands were associated with eutrophication through effects on snail species composition and biomass (Johnson and Chase, 2004; Johnson *et al.*, 2007). Eutrophication combined with high temperatures leads to more frequent harmful algal blooms (HABs) of cyanobacteria (Paerl *et al.*, 2011; Moe *et al.*,

2013; Jiménez Cisneros *et al.*, 2014). HABs produce toxins and hypoxic conditions, detrimental to aquatic life (Moe *et al.*, 2013). Interestingly, exposure to low concentrations of the cyanobacterial toxin microcystin-LR increased infection intensities of larval trematodes in leopard frogs, *Rana pipiens* (Milotic *et al.*, 2018). While it did not affect growth or survival of the snail *L. stagnalis*, the intermediate host for *Diplostomum* spp., exposure reduced the fecundity of adult snails (Gérard *et al.*, 2005). Therefore, eutrophication associated with climate change may promote infections of *Diplostomum* spp. in fish; but if allowed to progress, resulting in anoxia and the proliferations of HABs, infections may decrease (see also Budria, 2017). Similar effects of more frequent and widespread hypoxia in the benthos are expected in several water bodies because of longer periods of stratification during summer (Ficke *et al.*, 2007; Collingsworth *et al.*, 2017). Such developments would also negatively affect populations of snail and fish intermediate hosts of *Diplostomum* spp. in deeper waters (Table 23.2).

Exposure to UV radiation in freshwater ecosystems is expected to increase due to enhanced penetration under certain conditions, and it may be most problematic in clear, shallow waters (see Marcogliese, 2001). UV is harmful to invertebrates, including parasites, whose free-living stages such as cercariae are sensitive to environmental stressors (Pietrock and Marcogliese, 2003). While exposure to UV radiation may negatively affect free-living stages of *Diplostomum* spp., it also is immunosuppressive in fish. Exposure of rainbow trout to UV radiation led to increased numbers of *D. spathaceum* compared with controls, presumably because of reduced resistance (Markkula *et al.*, 2007). Exposure to UV is expected to increase in streams with climate change following reduced discharge, lower stream depth and reduced dissolved organic carbon (Clements *et al.*, 2008; Moe *et al.*, 2013). However, conflicting effects on parasites and hosts make any predictions problematic.

Other abiotic parameters that may be affected by climate may also negatively or positively impact the immune response in fish (Uribe *et al.*, 2011). For example, an increase in hypoxia decreased the respiratory burst activity of macrophages and lowered the level of circulating antibodies. In contrast, elevated salinity increased lytic enzyme activity, macrophage respiratory burst activity, HKL phagocytic activity, plasma lysozyme concentration and circulating IgM (Bowden, 2008; Uribe *et al.*, 2011). Effects of pH on immune response, on the other hand, have provided conflicting results (Bowden, 2008; Uribe *et al.*, 2011). Temperature stress combined with contaminants such as nickel and chlorine also causes immunosuppression in fishes, including reduced spleen cellularity, erythrocyte and leucocyte counts, and increased superoxide production (Prophete *et al.*, 2006; Verma *et al.*, 2007).

23.8 Control and Prevention of *Diplostomum* Species

The lens-infecting *Diplostomum* spp. also occur in pond aquaculture as all the necessary hosts of the parasite are commonly present. For example, fish farms typically attract fish-eating birds, the definitive hosts of *Diplostomum* spp., to feed. Earth ponds with vegetation used in rearing aquaculture fish also provide favourable habitats for snail intermediate hosts that become readily infected following parasite output from birds attracted to the ponds. Prevalence of infection in the snails can be high, which results in high infection also in fish (Stables and Chappell, 1986c; Field and Irwin, 1994; Karvonen *et al.*, 2006a). Parasite cercariae can also be brought into a facility with incoming water from upstream water bodies, but this is not considered a significant source of infection in fish compared with transmission occurring within the facility (Field and Irwin, 1994; Karvonen *et al.*, 2006a). High numbers of metacercariae and resulting pathology in the eyes of fish may become a problem if they reduce the desired growth in fish intended for market. Similarly, infected fish for stocking to support natural fish populations may have lower success in the wild, although detailed data on the effects of *Diplostomum* spp. infections in natural fish populations are not available.

It is likely that problems associated with *Diplostomum* spp. in aquaculture are also likely to increase. These may be through increased rate of parasite replication, prolonged period of parasite transmission and metacercarial development, or impaired ability of cold-water species such as salmonids to resist the infection (Hakalahti *et al.*, 2006). Such effects may be manifested as longer and later outbreaks of the disease in the autumn (Fig. 23.5), thus necessitating extra control measures. Control of *Diplostomum* spp. infections, however, can be challenging as there is no effective treatment of the infection in fish and immunizing fish against the infection provides only partial protection against later infections (Höglund and Thuvander, 1990; Karvonen *et al.*, 2005). Studies have shown that immunization alone does not protect fish from the deleterious effects of infection and other means of defence, such as behavioural avoidance of cercariae, may be needed to complement any immune-mediated response (Karvonen *et al.*, 2004b, 2010a). However, such behavioural avoidance is often impossible in the confined space of aquaculture tanks and ponds. In addition, other types of control measures such as treatment or filtering of water are not feasible because of large water volumes and the continuous output of parasite cercariae during summer months. Removal or chemical eradication of snails from the rearing ponds is generally considered the only viable option to control and prevent the infections in fish (Stables and Chappell, 1986c; Field and Irwin, 1994), although this can be system-specific and depends on the magnitude of cercarial input from upstream water bodies. Nevertheless, eradication of snails as the main preventive method should work equally well even with increasing water temperatures, although the positive effect of temperature on reproduction of snails may necessitate more frequent use of the eradication protocols. However, constructing the tanks and ponds in a way that limits establishment of vegetation and snail populations should help in longer-term prevention of infections.

23.9 Conclusions

The current evidence on the relationships between climate warming and infections of *Diplostomum* spp. strongly highlights the difficulty of determining the net effects on the complex parasite life cycle. This is because most, if not all, of the life cycle stages in the aquatic environment likely respond to temperature by increasing parasite replication and infectivity. However, elevated temperature will likely result in increased mortality of the infective stages, possibly accompanied by higher resistance in the fish hosts. Increasing temperature also acts in concert with many other interrelated environmental changes such as alteration in hydrology, increasing eutrophication, pollution and UV radiation, loss of habitats and higher risk of invasive species. All these factors working in concert illustrate the magnitude and scope of environmental effects on *Diplostomum* spp., and on many other host–parasite systems covered in this book. The high number of variables emphasizes the importance of long-term time-series studies, which would adequately provide the influence of all related factors. Pinpointing the importance of individual factors, on the other hand, requires rigorous experimental approaches supported by mathematical models on parasite dynamics with changing temperature. One area of experimental research needed concerns the effect of temperature and the other related factors (Table 23.2) on the severity of *Diplostomum* spp.-induced pathology in fish, which has received relatively little attention. Alongside the effects on the parasite life cycle, it is one of the key factors determining the impact of the parasite on fish populations in nature and in intensive aquaculture units.

Acknowledgements

D.J.M. would like to thank Neil Morley for providing essential information from the Harris (1986) thesis. A.K. was supported by a grant (#310632) from the Academy of Finland.

References

Aho, J.M., Camp, J.W. and Esch, G.W. (1982) Long-term studies on the population biology of *Diplostomulum scheuringi* in a thermally altered reservoir. *Journal of Parasitology* 68, 695–708. https://doi.org/10.2307/3280931

Altizer, S., Ostfeld, R.S., Johnson, P.T.J., Kutz, S. and Harvell, C.D. (2013) Climate change and infectious diseases: from evidence to a predictive framework. *Science* 341, 514–519. https://doi.org/10.1126/science.1239401

Altman, K.A., Paull, S.H., Johnson, P.T.J., Golembieski, M.N., Stephens, J.P. *et al.* (2016) Host and parasite thermal acclimation responses depend on the stage of infection. *Journal of Animal Ecology* 85, 1014–1024. https://doi.org/10.1111/1365-2656.12510

Balling, T.E. and Pfeiffer, W. (1997) Location-dependent infection of fish parasites in Lake Constance. *Journal of Fish Biology* 51, 1025–1032. https://doi.org/10.1111/j.1095-8649.1997.tb01541.x

Bauer, O.N. (1959) Parasites of freshwater fish and the biological basis for their control. *Bulletin of the State Scientific Research Institute of Lake and River Fisheries* 49. (Translated from Russian, Israel Program for Scientific Translations, Jerusalem, 1962).

Blasco-Costa, I., Faltýnková, A., Georgieva, S., Skirnisson, K., Scholz, T. and Kostadinova, A. (2014) Fish pathogens near the Arctic Circle: molecular, morphological and ecological evidence for unexpected diversity of *Diplostomum* (Digenea: Diplostomidae) in Iceland. *International Journal for Parasitology* 44, 703–715. https://doi.org/10.1016/j.ijpara.2014.04.009

Bowden, T.J. (2008) Modulation of the immune system of fish by their environment. *Fish and Shellfish Immunology* 25, 373–383. https://doi.org/10.1016/j.fsi.2008.03.017

Buchmann, K. (1986) Prevalence and intensity of infection of *Cryptocotyle lingua* (Creplin) and *Diplostomum spathaceum* (Rudolphi) – parasitic metacercariae of Baltic cod (*Gadus morhua* L). *Nordisk Veterinaer Medicin* 38, 303–307.

Budria, A. (2017) Beyond troubled waters: the influence of eutrophication on host–parasite interactions. *Functional Ecology* 31, 1348–1358. https://doi.org/10.1111/1365-2435.12880

Burrough, R.J. (1978) Population biology of two species of eyefluke, *Diplostomum spathaceum* and *Tylodelphys clavata*, in roach and rudd. *Journal of Fish Biology* 13, 19–32. https://doi.org/10.1111/j.1095-8649.1978.tb03409.x

Camp, J.W., Aho, J.M. and Esch, G.W. (1982) A long-term study on various aspects of the population biology of *Ornithodiplostomum ptychocheilus* in a South Carolina cooling reservoir. *Journal of Parasitology* 68, 709–718. https://doi.org/10.2307/3280932

Chappell, L.H. (1995) The biology of diplostomatid eyeflukes of fishes. *Journal of Helminthology* 69, 97–102. https://doi.org/10.1017/S0022149X00013961

Chappell, L.H., Hardie, L.J. and Secombes, C.J. (1994) Diplostomiasis: the disease and host–parasite interactions. In: Pike, A.W. and Lewis, J.W. (eds) *Parasitic Diseases of Fish*. Samara Publishing Ltd, Tresaith, UK, pp. 59–86.

Chen, Y.S., Todd, A.S., Murphy, M.H. and Lomnicky, G. (2016) Anticipated water quality changes in response to climate change and potential consequences for inland fishes. *Fisheries* 41, 413–416. https://doi.org/10.1080/03632415.2016.1182509

Chubb, J.C. (1979) Seasonal occurrences of helminths in freshwater fishes. Part II. Trematoda. *Advances in Parasitology* 17, 141–313.

Clements, W.H., Brooks, M.L., Kashian, D.R. and Zuellig, R.E. (2008) Changes in dissolved organic material determine exposure of stream benthic communities to UV-B radiation and heavy metals: implications for climate change. *Global Change Biology* 14, 2201–2214. https://doi.org/10.1111/j.1365-2486.2008.01632.x

Collingsworth, P.D., Bunnell, D.B., Murray, M.W., Kao, Y.C., Feiner, Z.S. *et al.* (2017) Climate change as a long-term stressor for the fisheries of the Laurentian Great Lakes of North America. *Reviews in Fish Biology and Fisheries* 27, 363–391. https://doi.org/10.1007/s11160-017-9480-3

Crowden, A.E. and Broom, D.M. (1980) Effects of the eyefluke, *Diplostomum spathaceum*, on the behavior of dace (*Leuciscus leuciscus*). *Animal Behaviour* 28, 287–294. https://doi.org/10.1016/s0003-3472(80)80031-5

Désilets, H.D., Locke, S.A., McLaughlin, J.D. and Marcogliese, D.J. (2013) Community structure of *Diplostomum* spp. (Digenea: Diplostomidae) in eyes of fish: main determinants and potential interspecific interactions. *International Journal for Parasitology* 43, 929–939. https://doi.org/10.1016/j.ijpara.2013.07.002

Dittmar, J., Janssen, H., Kuske, A., Kurtz, J. and Scharsack, J.P. (2014) Heat and immunity: an experimental heat wave alters immune functions in three-spined sticklebacks (*Gasterosteus aculeatus*). *Journal of Animal Ecology* 83, 744–757. https://doi.org/10.1111/1365-2656.12175

Ficke, A.D., Myrick, C.A. and Hansen, L.J. (2007) Potential impacts of global climate change on freshwater fisheries. *Reviews in Fish Biology and Fisheries* 17, 581–613. https://doi.org/10.1007/s11160-007-9059-5

Field, J.S. and Irwin, S.W.B. (1994) The epidemiology, treatment and control of diplostomiasis on a fish farm in Northern Ireland. In: Pike, A.W. and Lewis, J.W. (eds) *Parasitic Diseases of Fish*. Samara Publishing Ltd, Tresaith, UK, pp. 87–100.

Francová, K., Ondračková, M., Polačik, M. and Jurajda, P. (2011) Parasite fauna of native and non-native populations of *Neogobius melanostomus* (Pallas, 1814) (Gobiidae) in the longitudinal profile of the Danube River. *Journal of Applied Ichthyology* 27, 879–886. https://doi.org/10.1111/j.1439-0426.2010.01582.x

Franke, F., Rahn, A.K., Dittmar, J., Erin, N., Rieger, J.K. *et al.* (2014) *In vitro* leukocyte response of three-spined sticklebacks (*Gasterosteus aculeatus*) to helminth parasite antigens. *Fish and Shellfish Immunology* 36, 130–140. https://doi.org/10.1016/j.fsi.2013.10.019

Gendron, A.D. and Marcogliese, D.J. (2017) Enigmatic decline of a common fish parasite (*Diplostomum* spp.) in the St. Lawrence River: evidence for a dilution effect induced by the invasive round goby. *International Journal for Parasitology: Parasites and Wildlife* 6, 402–411. https://doi.org/10.1016/j.ijppaw.2017.04.002

Gendron, A.D., Marcogliese, D.J. and Thomas, M. (2012) Invasive species are less parasitized than native competitors, but for how long? The case of the round goby in the Great Lakes–St. Lawrence Basin. *Biological Invasions* 14, 367–384. https://doi.org/10.1007/s10530-011-0083-y

Gérard, C., Brient, L. and Le Rouzic, B. (2005) Variation in the response of juvenile and adult gastropods (*Lymnaea stagnalis*) to cyanobacterial toxin (microcystin-LR). *Environmental Toxicology* 20, 592–596. https://doi.org/10.1002/tox.20147

Haase, D., Rieger, J.K., Witten, A., Stoll, M., Bornberg-Bauer, E. *et al.* (2014) Specific gene expression responses to parasite genotypes reveal redundancy of innate immunity in vertebrates. *PLoS ONE* 9, e108001. https://doi.org/10.1371/journal.pone.0108001

Haase, D., Rieger, J.K., Witten, A., Stoll, M., Bornberg-Bauer, E. *et al.* (2016) Immunity comes first: the effect of parasite genotypes on adaptive immunity and immunization in three-spined sticklebacks. *Developmental and Comparative Immunology* 54, 137–144. https://doi.org/10.1016/j.dci.2015.09.008

Hakalahti, T., Karvonen, A. and Valtonen, E.T. (2006) Climate warming and disease risks in temperate regions – *Argulus coregoni* and *Diplostomum spathaceum* as case studies. *Journal of Helminthology* 80, 93–98. https://doi.org/10.1079/joh2006351

Harris, A.L. (1986) Larval trematode infections of the freshwater snail Lymnaea peregra (Muller). MPhil thesis, Queen Mary & Westfield College, University of London, London.

Höglund, J. and Thulin, J. (1990) The epidemiology of the metacercariae of *Diplostomum baeri* and *Diplostomum spathaceum* in perch (*Perca fluviatilis*) from the warm water effluent of a nuclear power station. *Journal of Helminthology* 64, 139–150. https://doi.org/10.1017/S0022149X00012050

Höglund, J. and Thuvander, A. (1990) Indications of non-specific protective mechanisms in rainbow trout *Oncorhynchus mykiss* with diplostomosis. *Diseases of Aquatic Organisms* 8, 91–97.

Hooper, M.J., Ankley, G.T., Cristol, D.A., Maryoung, L.A., Noyes, P.D. and Pinkerton, K.E. (2013) Interactions between chemical and climate stressors: a role for mechanistic toxicology in assessing climate change risks. *Environmental Toxicology and Chemistry* 32, 32–48. https://doi.org/10.1002/etc.2043

Hudson, P.J., Cattadori, M., Boag, B. and Dobson, A.P. (2006) Climate disruption and parasite–host dynamics: patterns and processes associated with warming and the frequency of extreme climatic events. *Journal of Helminthology* 80, 175–182. https://doi.org/10.1079/joh2006357

IPCC (Intergovernmental Panel on Climate Change) (2018) Summary for policymakers. In: Masson-Delmotte, V., Zhai, P., Pörtner, H.-O., Roberts, D., Skea, J. *et al.* (eds) *Global Warming of 1.5°C. An IPCC Special Report on the impacts of global warming of 1.5°C above pre-industrial levels and related global greenhouse gas emission pathways, in the context of strengthening the global response to the threat of climate change, sustainable development, and efforts to eradicate poverty*. World Meteorological Organization, Geneva, Switzerland.

Janovy, J., Snyder, S.D. and Clopton, R.E. (1997) Evolutionary constraints on population structure: the parasites of *Fundulus zebrinus* (Pisces: Cyprinodontidae) in the South Platte River of Nebraska. *Journal of Parasitology* 83, 584–592. https://doi.org/10.2307/3284228

Jiménez Cisneros, B.E., Oki, T., Arnell, N.W., Benito, G., Cogley, J.G. *et al.* (2014) Freshwater resources. In: Field, C.B., Barros, V.R., Dokken, D.J., Mach, K.J., Mastrandrea, M.D. *et al.* (eds) *Climate Change 2014: Impacts, Adaptation, and Vulnerability. Part A: Global and Sectoral Aspects. Contribution of Working Group II to the Fifth Assessment Report of the Intergovernmental Panel on Climate Change*. Cambridge University Press, Cambridge and New York, pp. 229–269.

Johnson, A.C., Acreman, M.C., Dunbar, M.J., Feist, S.W., Giacomello, A.M. *et al.* (2009) The British river of the future: how climate change and human activity might affect two contrasting river ecosystems in England. *Science of the Total Environment* 407, 4787–4798. https://doi.org/10.1016/j.scitotenv.2009.05.018

Johnson, P.T.J. and Chase, J.M. (2004) Parasites in the food web: linking amphibian malformations and aquatic eutrophication. *Ecology Letters* 7, 521–526. https://doi.org/10.1111/j.1461-0248.2004.00610.x

Johnson, P.T.J., Chase, J.M., Dosch, K.L., Hartson, R.B., Gross, J.A. *et al.* (2007) Aquatic eutrophication promotes pathogenic infection in amphibians. *Proceedings of the National Academy of Sciences USA* 104, 15781–15786. https://doi.org/10.1073/pnas.0707763104

Kalbe, M. and Kurtz, J. (2006) Local differences in immunocompetence reflect resistance of sticklebacks against the eye fluke *Diplostomum pseudospathaceum*. *Parasitology* 132, 105-116. https://doi.org/10.1017/s0031182005008681

Karvonen, A. (2012) *Diplostomum spathaceum* and related species. In: Woo, P.T.K. and Buchmann, K. (eds) *Fish Parasites: Pathobiology and Protection*. CAB International, Wallingford, UK, pp. 260–269.

Karvonen, A. and Lindström, K. (2018) Spatiotemporal and gender-specific parasitism in two species of gobiid fish. *Ecology and Evolution* 8, 6114–6123. https://doi.org/10.1002/ece3.4151

Karvonen, A. and Seppälä, O. (2008a) Eye fluke infection and lens size reduction in fish: a quantitative analysis. *Diseases of Aquatic Organisms* 80, 21–26. https://doi.org/10.3354/dao01918

Karvonen, A. and Seppälä, O. (2008b) Effect of eye fluke infection on the growth of whitefish (*Coregonus lavaretus*) – an experimental approach. *Aquaculture* 279, 6–10. https://doi.org/10.1016/j.aquaculture.2008.04.013

Karvonen, A., Paukku, S., Valtonen, E.T. and Hudson, P.J. (2003) Transmission, infectivity and survival of *Diplostomum spathaceum* cercariae. *Parasitology* 127, 217–224. https://doi.org/10.1017/s0031182003003561

Karvonen, A., Kirsi, S., Hudson, P.J. and Valtonen, E.T. (2004a) Patterns of cercarial production from *Diplostomum spathaceum*: terminal investment or bet hedging? *Parasitology* 129, 87–92. https://doi.org/10.1017/s0031182004005281

Karvonen, A., Seppälä, O. and Valtonen, E.T. (2004b) Parasite resistance and avoidance behaviour in preventing eye fluke infections in fish. *Parasitology* 129, 159–164. https://doi.org/10.1017/s0031182004005505

Karvonen, A., Seppälä, O. and Valtonen, E.T. (2004c) Eye fluke-induced cataract formation in fish: quantitative analysis using an ophthalmological microscope. *Parasitology* 129, 473–478. https://doi.org/10.1017/s0031182004006006

Karvonen, A., Paukku, S., Seppälä, O. and Valtonen, E.T. (2005) Resistance against eye flukes: naïve versus previously infected fish. *Parasitology Research* 95, 55–59. https://doi.org/10.1007/s00436-004-1246-x

Karvonen, A., Savolainen, M., Seppälä, O. and Valtonen, E.T. (2006a) Dynamics of *Diplostomum spathaceum* infection in snail hosts at a fish farm. *Parasitology Research* 99, 341–345. https://doi.org/10.1007/s00436-006-0137-8

Karvonen, A., Terho, P., Seppälä, O., Jokela, J. and Valtonen, E.T. (2006b) Ecological divergence of closely related *Diplostomum* (Trematoda) parasites. *Parasitology* 133, 229–235. https://doi.org/10.1017/s0031182006000242

Karvonen, A., Halonen, H. and Seppälä, O. (2010a) Priming of host resistance to protect cultured rainbow trout *Oncorhynchus mykiss* against eye flukes and parasite-induced cataracts. *Journal of Fish Biology* 76, 1508–1515. https://doi.org/10.1111/j.1095-8649.2010.02597.x

Karvonen, A., Rintamäki, P., Jokela, J. and Valtonen, E.T. (2010b) Increasing water temperature and disease risks in aquatic systems: climate change increases the risk of some, but not all, diseases. *International Journal for Parasitology* 40, 1483–1488. https://doi.org/10.1016/j.ijpara.2010.04.015

Karvonen, A., Kristjánsson, B.K., Skúlason, S., Lanki, M., Rellstab, C. and Jokela, J. (2013) Water temperature, not fish morph, determines parasite infections of sympatric Icelandic threespine sticklebacks (*Gasterosteus aculeatus*). *Ecology and Evolution* 3, 1507–1517. https://doi.org/10.1002/ece3.568

Karvonen, A., Lucek, K., Marques, D.A. and Seehausen, O. (2015) Divergent macroparasite infections in parapatric Swiss lake–stream pairs of threespine stickleback (*Gasterosteus aculeatus*). *PLoS ONE* 10, e0130579. https://doi.org/10.1371/journal.pone.0130579

Klemme, I. and Karvonen, A. (2017) Vertebrate defense against parasites: interactions between avoidance, resistance, and tolerance. *Ecology and Evolution* 7, 561–571. https://doi.org/10.1002/ece3.2645

Koie, M. (1999) Metazoan parasites of flounder *Platichthys flesus* (L.) along a transect from the southwestern to the northeastern Baltic Sea. *ICES Journal of Marine Science* 56, 157–163. https://doi.org/10.1006/jmsc.1999.0463

Kvach, Y. and Winkler, H.M. (2011) The colonization of the invasive round goby *Neogobius melanostomus* by parasites in new localities in the southwestern Baltic Sea. *Parasitology Research* 109, 769–780. https://doi.org/10.1007/s00436-011-2321-8

Lafferty, K.D. (2009) The ecology of climate change and infectious diseases. *Ecology* 90, 888–900. https://doi.org/10.1890/08-0079.1

Landis, W.G., Rohr, J.R., Moe, S.J., Balbus, J.M., Clements, W. *et al.* (2014) Global climate change and contaminants, a call to arms not yet heard? *Integrated Environmental Assessment and Management* 10, 483–484. https://doi.org/10.1002/ieam.1568

Leicht, K. and Seppälä, O. (2014) Infection success of *Echinoparyphium aconiatum* (Trematoda) in its snail host under high temperature: role of host resistance. *Parasites and Vectors* 7, 192. https://doi.org/10.1186/1756-3305-7-192

Leicht, K., Jokela, J. and Seppälä, O. (2013) An experimental heat wave changes immune defense and life history traits in a freshwater snail. *Ecology and Evolution* 3, 4861–4871. https://doi.org/10.1002/ece3.874

Leicht, K., Seppälä, K. and Seppälä, O. (2017) Potential for adaptation to climate change: family-level variation in fitness-related traits and their responses to heat waves in a snail population. *BMC Evolutionary Biology* 17, 140. https://doi.org/10.1186/s12862-017-0988-x

Locke, S.A., McLaughlin, J.D., Dayanandan, S. and Marcogliese, D.J. (2010a) Diversity and specificity in *Diplostomum* spp. metacercariae in freshwater fishes revealed by cytochrome c oxidase I and internal transcribed spacer sequences. *International Journal for Parasitology* 40, 333–343. https://doi.org/10.1016/j.ijpara.2009.08.012

Locke, S.A., McLaughlin, J.D. and Marcogliese, D.J. (2010b) DNA barcodes show cryptic diversity and a potential physiological basis for host specificity among Diplostomoidea (Platyhelminthes: Digenea) parasitizing freshwater fishes in the St. Lawrence River, Canada. *Molecular Ecology* 19, 2813–2827. https://doi.org/10.1111/j.1365-294X.2010.04713.x

Locke, S.A., Al-Nasiri, F.S., Caffara, M., Drago, F., Kalbe, M. *et al.* (2015) Diversity, specificity and speciation in larval Diplostomidae (Platyhelminthes: Digenea) in the eyes of freshwater fish, as revealed by DNA barcodes. *International Journal for Parasitology* 45, 841–855. https://doi.org/10.1016/j.ijpara.2015.07.001

Lõhmus, M. and Björklund, M. (2015) Climate change: what will it do to fish–parasite interactions? *Biological Journal of the Linnean Society* 116, 397–411. https://doi.org/10.1111/bij.12584

Louhi, K.-R., Karvonen, A., Rellstab, C., Louhi, R. and Jokela, J. (2013) Prevalence of infection as a predictor of multiple genotype infection frequency in parasites with multiple-host life cycle. *Journal of Animal Ecology* 82, 191–200. https://doi.org/10.1111/j.1365-2656.2012.02028.x

Lyholt, H.C.K. and Buchmann, K. (1996) *Diplostomum spathaceum*: effects of temperature and light on cercarial shedding and infection of rainbow trout. *Diseases of Aquatic Organisms* 25, 169–173. https://doi.org/10.3354/dao025169

McDonald, T.E. and Margolis, L. (1995) Synopsis of the parasites of fishes of Canada: Supplement (1978–1993). *Canadian Special Publication of Fisheries and Aquatic Sciences No. 122*. National Research Council of Canada, Ottawa.

McKeown, C.A. and Irwin, S.W.B. (1997) Accumulation of *Diplostomum* spp. (Digenea: Diplostomatidae)

metacercariae in the eyes of 0+ and 1+ roach (*Rutilus rutilus*). *International Journal for Parasitology* 27, 377–380. https://doi.org/10.1016/s0020-7519(96)00204-4

Marcogliese, D.J. (2001) Implications of climate change for parasitism of animals in the aquatic environment. *Canadian Journal of Zoology* 79, 1331–1352. https://doi.org/10.1139/cjz-79-8-1331

Marcogliese, D.J. (2004) Parasites: small players with crucial roles in the ecological theater. *EcoHealth* 1, 151–164. https://doi.org/10.1007/s10393-004-0028-3

Marcogliese, D.J. (2005) Parasites of the superorganism: are they indicators of ecosystem health? *International Journal for Parasitology* 35, 705–716. https://doi.org/10.1016/j.ijpara.2005.01.015

Marcogliese, D.J. (2008) The impact of climate change on the parasites and infectious diseases of aquatic animals. *Revue Scientifique et Technique (International Office of Epizootics)* 27, 467–484. https://doi.org/10.20506/rst.27.2.1820

Marcogliese, D.J. (2016) The distribution and abundance of parasites in aquatic ecosystems in a changing climate: more than just temperature. *Integrative and Comparative Biology* 56, 611–619. https://doi.org/10.1093/icb/icw036

Marcogliese, D.J. and Compagna, S. (1999) Diplostomatid eye flukes in young-of-the-year and forage fishes in the St. Lawrence River, Quebec. *Journal of Aquatic Animal Health* 11, 275–282. https://doi.org/10.1577/1548-8667(1999)011<0275:DEFIYO>2.0.CO;2

Marcogliese, D.J. and Pietrock, M. (2011) Combined effects of parasites and contaminants on animal health: parasites do matter. *Trends in Parasitology* 27, 123–130. https://doi.org/10.1016/j.pt.2010.11.002

Marcogliese, D.J., Rodrigue, J., Ouellet, M. and Champoux, L. (2000) Natural occurrence of *Diplostomum* sp. (Digenea: Diplostomatidae) in adult mudpuppies and bullfrog tadpoles from the St. Lawrence River, Quebec. *Comparative Parasitology* 67, 26–31.

Marcogliese, D.J., Compagna, S., Bergeron, E. and McLaughlin, J.D. (2001a) Population biology of eyeflukes in fish from a large fluvial ecosystem: the importance of gulls and habitat characteristics. *Canadian Journal of Zoology* 79, 1102–1113. https://doi.org/10.1139/cjz-79-6-1102

Marcogliese, D.J., Dumont, P., Gendron, A.D., Mailhot, Y., Bergeron, E. and McLaughlin, J.D. (2001b) Spatial and temporal variation in abundance of *Diplostomum* spp. in walleye (*Stizostedion vitreum*) and white suckers (*Catostomus commersoni*) from the St. Lawrence River. *Canadian Journal of Zoology* 79, 355–369. https://doi.org/10.1139/z00-209

Marcogliese, D.J., Dautremepuits, C., Gendron, A.D. and Fournier, M. (2010) Interactions between parasites and pollutants in yellow perch (*Perca flavescens*) in the St. Lawrence River, Canada: implications for resistance and tolerance to parasites. *Canadian Journal of Zoology* 88, 247–258. https://doi.org/10.1139/z09-140

Margolis, L. and Arthur, J.R. (1979) Synopsis of the parasites of fishes of Canada. *Bulletin of the Fisheries Research Board of Canada No. 199*. Department of Fisheries and Oceans, Ottawa.

Markkula, S.E., Karvonen, A., Salo, H., Valtonen, E.T. and Jokinen, E.I. (2007) Ultraviolet B irradiation affects resistance of rainbow trout (*Oncorhynchus mykiss*) against bacterium *Yersinia ruckeri* and trematode *Diplostomum spathaceum*. *Photochemistry and Photobiology* 83, 1263–1269. https://doi.org/10.1111/j.1751-1097.2007.00165.x

Martin, L.B., Hopkins, W.A., Mydlarz, L.D. and Rohr, J.R. (2010) The effects of anthropogenic global changes on immune functions and disease resistance. *Annals of the New York Academy of Sciences* 1195, 129–148. https://doi.org/10.1111/j.1749-6632.2010.05454.x

Milotic, M., Milotic, D. and Koprivnikar, J. (2018) Exposure to a cyanobacterial toxin increases larval amphibian susceptibility to parasitism. *Parasitology Research* 117, 513–520. https://doi.org/10.1007/s00436-017-5727-0

Moe, S.J., De Schamphelaere, K., Clements, W.H., Sorensen, M.T., Van den Brink, P.J. and Liess, M. (2013) Combined and interactive effects of global climate change and toxicants on populations and communities. *Environmental Toxicology and Chemistry* 32, 49–61. https://doi.org/10.1002/etc.2045

Moody, J. and Gaten, E. (1982) The population dynamics of eyeflukes *Diplostomum spathaceum* and *Tylodelphys clavata* (Digenea, Diplostomatidae) in rainbow and brown trout in Rutland water – 1974–1978. *Hydrobiologia* 88, 207–209. https://doi.org/10.1007/bf00008315

Morley, N.J. (2011) Thermodynamics of cercarial survival and metabolism in a changing climate. *Parasitology* 138, 1442–1452. https://doi.org/10.1017/s0031182011001272

Morley, N.J. (2012) Thermodynamics of miracidial survival and metabolism. *Parasitology* 139, 1640–1651. https://doi.org/10.1017/s0031182012000960

Morley, N.J. and Lewis, J.W. (2013) Thermodynamics of cercarial development and emergence in trematodes. *Parasitology* 140, 1211–1224. https://doi.org/10.1017/s0031182012001783

Morley, N.J. and Lewis, J.W. (2014) Extreme climatic events and host–pathogen interactions: the impact of the 1976 drought in the UK. *Ecological Complexity* 17, 1–19. https://doi.org/10.1016/j.ecocom.2013.12.001

Morley, N.J. and Lewis, J.W. (2015) Thermodynamics of trematode infectivity. *Parasitology* 142, 585–597. https://doi.org/10.1017/s0031182014001632

Morley, N.J., Crane, M. and Lewis, J.W. (2001) Toxicity of cadmium and zinc to *Diplostomum spathaceum* (Trematoda: Diplostomidae) cercarial survival. *International Journal for Parasitology* 31, 1211–1217. https://doi.org/10.1016/s0020-7519(01)00229-6

Morley, N.J., Crane, M. and Lewis, J.W. (2002) Toxicity of cadmium and zinc mixtures to *Diplostomum spathaceum* (Trematoda: Diplostomidae) cercarial survival. *Archives of Environmental Contamination and Toxicology* 43, 28–33. https://doi.org/10.1007/s00244-002-1244-x

Morley, N.J., Crane, M. and Lewis, J.W. (2003a) Cadmium toxicity and snail–digenean interactions in a population of *Lymnaea* spp. *Journal of Helminthology* 77, 49–55. https://doi.org/10.1079/joh2002148

Morley, N.J., Crane, M. and Lewis, J.W. (2003b) Toxicity of cadmium and zinc to the cercarial activity of *Diplostomum spathaceum* (Trematoda: Diplostomidae). *Folia Parasitologica* 50, 57–60. https://doi.org/10.14411/fp.2003.011

Morley, N.J., Irwin, S.W.B. and Lewis, J.W. (2003c) Pollution toxicity to the transmission of larval digeneans through their molluscan hosts. *Parasitology* 126, S5–S26. https://doi.org/10.1017/s0031182003003755

Moszczynska, A., Locke, S.A., McLaughlin, J.D., Marcogliese, D.J. and Crease, T.J. (2009) Development of primers for the mitochondrial cytochrome c oxidase I gene in digenetic trematodes (Platyhelminthes) illustrates the challenge of barcoding parasitic helminths. *Molecular Ecology Resources* 9, 75–82. https://doi.org/10.1111/j.1755-0998.2009.02634.x

Muhlegger, J.M., Jirsa, F., Konecny, R. and Frank, C. (2010) Parasites of *Apollonia melanostoma* (Pallas 1814) and *Neogobius kessleri* (Guenther 1861) (Osteichthyes, Gobiidae) from the Danube River in Austria. *Journal of Helminthology* 84, 87–92. https://doi.org/10.1017/s0022149x09990095

Muller, M. (2019) Dams have the power to slow climate change. *Nature* 566, 315–317.

Niewiadomska, K. (1984) Present status of *Diplostomum spathaceum* (Rudolphi, 1819) and differentiation of *Diplostomum pseudospathaceum* nom. nov. (Trematoda: Diplostomatidae). *Systematic Parasitology* 6, 81–86. https://doi.org/10.1007/bf02185515

Niewiadomska, K. (1986) Verification of the life-cycles of *Diplostomum spathaceum* (Rudolphi, 1819) and *D. pseudospathaceum* Niewiadomska, 1984 (Trematoda: Diplostomidae). *Systematic Parasitology* 8, 23–31. https://doi.org/10.1007/bf00010306

Niewiadomska, K. and Kiseliene, V. (1994) *Diplostomum cercariae* (Digenea) in snails from Lithuania. II. Survey of species. *Acta Parasitologica* 39, 179–186.

Niewiadomska, K. and Laskowski, Z. (2002) Systematic relationships among six species of *Diplostomum* Nordmann, 1832 (Digenea) based on morphological and molecular data. *Acta Parasitologica* 47, 20–28.

Noyes, P.D., McElwee, M.K., Miller, H.D., Clark, B.W., Van Tiem, L.A. *et al.* (2009) The toxicology of climate change: environmental contaminants in a warming world. *Environment International* 35, 971–986. https://doi.org/10.1016/j.envint.2009.02.006

Ondračková, M., Dávidová, M., Blazek, R., Gelnar, M. and Jurajda, P. (2009) The interaction between an introduced fish host and local parasite fauna: *Neogobius kessleri* in the middle Danube River. *Parasitology Research* 105, 201–208. https://doi.org/10.1007/s00436-009-1384-2

Ondračková, M., Hudcová, I., Dávidová, M., Adámek, Z., Kašný, M. and Jurajda, P. (2015) Non-native gobies facilitate the transmission of *Bucephalus polymorphus* (Trematoda). *Parasites and Vectors* 8, 382. https://doi.org/10.1186/s13071-015-0999-7

Owen, S.F., Barber, I. and Hart, P.J.B. (1993) Low-level infection by eye fluke, *Diplostomum* spp., affects the vision of 3-spined sticklebacks, *Gasterosteus aculeatus*. *Journal of Fish Biology* 42, 803–806. https://doi.org/10.1111/j.1095-8649.1993.tb00387.x

Paaijmans, K.P., Blanford, S., Bell, A.S., Blanford, J.I., Read, A.F. and Thomas, M.B. (2010) Influence of climate on malaria transmission depends on daily temperature variation. *Proceedings of the National Academy of Sciences USA* 107, 15135–15139. https://doi.org/10.1073/pnas.1006422107

Padros, F., Knudsen, R. and Blasco-Costa, I. (2018) Histopathological characterisation of retinal lesions associated to *Diplostomum* species (Platyhelminthes: Trematoda) infection in polymorphic Arctic charr *Salvelinus alpinus*. *International Journal for Parasitology: Parasites and Wildlife* 7, 68–74. https://doi.org/10.1016/j.ijppaw.2018.01.007

Paerl, H.W., Hall, N.S. and Calandrino, E.S. (2011) Controlling harmful cyanobacterial blooms in a world experiencing anthropogenic and climatic-induced change. *Science of the Total Environment* 409, 1739–1745. https://doi.org/10.1016/j.scitotenv.2011.02.001

Paull, S.H. and Johnson, P.T.J. (2011) High temperature enhances host pathology in a snail–trematode system: possible consequences of climate change for the emergence of disease. *Freshwater Biology* 56, 767–778. https://doi.org/10.1111/j.1365-2427.2010.02547.x

Paull, S.H. and Johnson, P.T.J. (2014) Experimental warming drives a seasonal shift in the timing of host–parasite dynamics with consequences for disease risk. *Ecology Letters* 17, 445–453. https://doi.org/10.1111/ele.12244

Paull, S.H., LaFonte, B.E. and Johnson, P.T.J. (2012) Temperature-driven shifts in a host–parasite interaction drive nonlinear changes in disease risk. *Global Change Biology* 18, 3558–3567. https://doi.org/10.1111/gcb.12018

Pennycuick, L. (1971) Differences in the parasite infections in three-spined sticklebacks (*Gasterosteus aculeatus* L.) of different sex, age and size. *Parasitology* 63, 407–418. https://doi.org/10.1017/S0031182000079932

Pietrock, M. and Goater, C.P. (2005) Infectivity of *Ornithodiplostomum ptychocheilus* and *Posthodiplostomum minimum* (Trematoda: Diplostomidae) cercariae following exposure to cadmium. *Journal of Parasitology* 91, 854–856. https://doi.org/10.1645/GE-473R.1

Pietrock, M. and Marcogliese, D.J. (2003) Free-living endohelminth stages: at the mercy of environmental conditions. *Trends in Parasitology* 19, 293–299. https://doi.org/10.1016/s1471-4922(03)00117-x

Pietrock, M., Meinelt, T., Marcogliese, D.J. and Steinberg, C.E.W. (2001) Influence of aqueous sediment extracts from the Oder River (Germany/Poland) on survival of *Diplostomum* sp. (Trematoda: Diplostomidae) cercariae. *Archives of Environmental Contamination and Toxicology* 40, 327–332. https://doi.org/10.1007/s002440010179

Pietrock, M., Marcogliese, D.J. and McLaughlin, J.D. (2002a) Effects of cadmium upon longevity of *Diplostomum* sp. (Trematoda: Diplostomidae) cercariae. *Chemosphere* 47, 29–33. https://doi.org/10.1016/s0045-6535(01)00283-1

Pietrock, M., Marcogliese, D.J., Meinelt, T. and McLaughlin, J.D. (2002b) Effects of mercury and chromium upon longevity of *Diplostomum* sp. (Trematoda: Diplostomidae) cercariae. *Parasitology Research* 88, 225–229. https://doi.org/10.1007/s00436-001-0529-8

Poulin, R. (2006) Global warming and temperature-mediated increases in cercarial emergence in trematode parasites. *Parasitology* 132, 143–151. https://doi.org/10.1017/s0031182005008693

Poulin, R. and Mouritsen, K.N. (2006) Climate change, parasitism and the structure of intertidal ecosystems. *Journal of Helminthology* 80, 183–191. https://doi.org/10.1079/joh2006341

Prophete, C., Carlson, E.A., Li, Y., Duffy, J., Steinetz, B. *et al.* (2006) Effects of elevated temperature and nickel pollution on the immune status of Japanese medaka. *Fish and Shellfish Immunology* 21, 325–334. https://doi.org/10.1016/j.fsi.2005.12.009

Raffel, T.R., Romansic, J.M., Halstead, N.T., McMahon, T.A., Venesky, M.D. and Rohr, J.R. (2013) Disease and thermal acclimation in a more variable and unpredictable climate. *Nature Climate Change* 3, 146–151. https://doi.org/10.1038/nclimate1659

Raffel, T.R., Halstead, N.T., McMahon, T.A., Davis, A.K. and Rohr, J.R. (2015) Temperature variability and moisture synergistically interact to exacerbate an epizootic disease. *Proceedings of the Royal Society B: Biological Sciences* 282, 20142039. https://doi.org/10.1098/rspb.2014.2039

Ratanarat-Brockelman, C. (1974) Migration of *Diplostomum spathaceum* (Trematoda) in the fish intermediate host. *Zeitschrift für Parasitenkunde* 43, 123–134. https://doi.org/10.1007/BF00329170

Rauch, G., Kalbe, M. and Reusch, T.B.H. (2005) How a complex life cycle can improve a parasite's sex life. *Journal of Evolutionary Biology* 18, 1069–1075. https://doi.org/10.1111/j.1420-9101.2005.00895.x

Rellstab, C., Louhi, K.R., Karvonen, A. and Jokela, J. (2011) Analysis of trematode parasite communities in fish eye lenses by pyrosequencing of naturally pooled DNA. *Infection, Genetics and Evolution* 11, 1276–1286. https://doi.org/10.1016/j.meegid.2011.04.018

Rellstab, C., Karvonen, A., Louhi, K.-R. and Jokela, J. (2013) Genotype-specific vs. cross-reactive host immunity against a macroparasite. *PLoS ONE* 8, e78427. https://doi.org/10.1371/journal.pone.0078427

Riley, E.M. and Chappell, L.H. (1992) Effect of infection with *Diplostomum spathaceum* on the internal defense system of *Lymnaea stagnalis*. *Journal of Invertebrate Pathology* 59, 190–196. https://doi.org/10.1016/0022-2011(92)90032-y

Rohr, J.R. and Palmer, B.D. (2013) Climate change, multiple stressors, and the decline of ectotherms. *Conservation Biology* 27, 741–751. https://doi.org/10.1111/cobi.12086

Rohr, J.R., Dobson, A.P., Johnson, P.T.J., Kilpatrick, A.M., Paull, S.H. *et al.* (2011) Frontiers in climate change–disease research. *Trends in Ecology and Evolution* 26, 270–277. https://doi.org/10.1016/j.tree.2011.03.002

Rohr, J.R., Raffel, T.R., Blaustein, A.R., Johnson, P.T.J., Paull, S.H. and Young, S. (2013) Using physiology to understand climate-driven changes in disease and their implications for conservation. *Conservation Physiology* 1, cot022. https://doi.org/10.1093/conphys/cot022

Salo, T., Stamm, C., Burdon, F.J., Räsänen, K. and Seppälä, O. (2017) Resilience to heat waves in the aquatic snail *Lymnaea stagnalis*: additive and interactive effects with micropollutants. *Freshwater Biology* 62, 1831–1846. https://doi.org/10.1111/fwb.12999

Scharsack, J.P. and Kalbe, M. (2014) Differences in susceptibility and immune responses of three-spined sticklebacks (*Gasterosteus aculeatus*) from lake and river ecotypes to sequential infections with the eye fluke *Diplostomum pseudospathaceum*. *Parasites and Vectors* 7, 109. https://doi.org/10.1186/1756-3305-7-109

Scharsack, J.P., Franke, F., Erin, N.I., Kuske, A., Buscher, J. *et al.* (2016) Effects of environmental variation on host–parasite interaction in three-spined sticklebacks (*Gasterosteus aculeatus*). *Zoology* 119, 375–383. https://doi.org/10.1016/j.zool.2016.05.008

Schiedek, D., Sundelin, B., Readman, J.W. and Macdonald, R.W. (2007) Interactions between climate change and contaminants. *Marine Pollution Bulletin* 54, 1845–1856. https://doi.org/10.1016/j.marpolbul.2007.09.020

Seppälä, O., Karvonen, A. and Valtonen, E.T. (2004) Parasite-induced change in host behaviour and susceptibility to predation in an eye fluke–fish interaction. *Animal Behaviour* 68, 257–263. https://doi.org/10.1016/j.anbehav.2003.10.021

Seppälä, O., Karvonen, A. and Valtonen, E.T. (2005a) Impaired crypsis of fish infected with a trophically transmitted parasite. *Animal Behaviour* 70, 895–900. https://doi.org/10.1016/j.anbehav.2005.01.021

Seppälä, O., Karvonen, A. and Valtonen, E.T. (2005b) Manipulation of fish host by eye flukes in relation to cataract formation and parasite infectivity. *Animal*

Behaviour 70, 889–894. https://doi.org/10.1016/j.anbehav.2005.01.020

Seppälä, O., Karvonen, A. and Valtonen, E.T. (2008) Shoaling behaviour of fish under parasitism and predation risk. *Animal Behaviour* 75, 145–150. https://doi.org/10.1016/j.anbehav.2007.04.022

Seppälä, O., Karvonen, A. and Valtonen, E.T. (2011) Eye fluke-induced cataracts in natural fish populations: is there potential for host manipulation? *Parasitology* 138, 209–214. https://doi.org/10.1017/s0031182010001228

Settele, J., Scholes, R., Betts, R., Bunn, S., Leadley, P. *et al.* (2014) Terrestrial and inland water systems. In: Field, C.B., Barros, V.R., Dokken, D.J., Mach, K.J., Mastrandrea, M.D. *et al.* (eds.) *Climate Change 2014: Impacts, Adaptation, and Vulnerability. Part A: Global and Sectoral Aspects. Contribution of Working Group II to the Fifth Assessment Report of the Intergovernmental Panel on Climate Change*. Cambridge University Press, Cambridge and New York, pp. 271–359.

Shariff, M., Richards, R.H. and Sommerville, C. (1980) The histopathology of acute and chronic infections of rainbow trout *Salmo gairdneri* Richardson with eye flukes, *Diplostomum* spp. *Journal of Fish Diseases* 3, 455–465. https://doi.org/10.1111/j.1365-2761.1980.tb00432.x

Sitjá-Bobadilla, A. (2008) Living off a fish: a trade-off between parasites and the immune system. *Fish and Shellfish Immunology* 25, 358–372. https://doi.org/10.1016/j.fsi.2008.03.018

Sous, S.M. (1992) Influence of abiotic factors on emission and survival of *Diplostomum chromatophorum* (Brown, 1931) (Trematoda, Diplostomatidae). *Ecological Parasitology* 1, 154–159.

Stables, J.N. and Chappell, L.H. (1986a) Putative immune response of rainbow trout, *Salmo gairdneri*, to *Diplostomum spathaceum* infections. *Journal of Fish Biology* 29, 115–122. https://doi.org/10.1111/j.1095-8649.1986.tb04931.x

Stables, J.N. and Chappell, L.H. (1986b) *Diplostomum spathaceum* (Rud. 1819) – effects of physical factors on the infection of rainbow trout (*Salmo gairdneri*) by cercariae. *Parasitology* 93, 71–79. https://doi.org/10.1017/s0031182000049830

Stables, J.N. and Chappell, L.H. (1986c) The epidemiology of diplostomiasis in farmed rainbow trout from north-east Scotland. *Parasitology* 92, 699–710. https://doi.org/10.1017/S0031182000065550

Stahl, R.G., Hooper, M.J., Balbus, J.M., Clements, W., Fritz, A. *et al.* (2013) The influence of global climate change on the scientific foundations and applications of Environmental Toxicology and Chemistry: introduction to a SETAC international workshop. *Environmental Toxicology and Chemistry* 32, 13–19. https://doi.org/10.1002/etc.2037

Studer, A. and Poulin, R. (2014) Analysis of trait mean and variability versus temperature in trematode cercariae: is there scope for adaptation to global warming? *International Journal for Parasitology* 44, 403–413. https://doi.org/10.1016/j.ijpara.2014.02.006

Sweeting, R.A. (1974) Investigations into natural and experimental infections of freshwater fish by common eye-fluke *Diplostomum spathaceum* Rud. *Parasitology* 69, 291–300. https://doi.org/10.1017/S0031182000062995

Tort, L., Balasch, J.C. and Mackenzie, S. (2003) Fish immune system. A crossroads between innate and adaptive responses. *Inmunologia* 22, 277–286.

Tully, K., Gedan, K., Epanchin-Niell, R., Strong, A., Bernhardt, E.S. *et al.* (2019) The invisible flood: the chemistry, ecology, and social implications of coastal saltwater intrusion. *BioScience* 69, 368–378. https://doi.org/10.1093/biosci/biz027

Uribe, C., Folch, H., Enriquez, R. and Moran, G. (2011) Innate and adaptive immunity in teleost fish: a review. *Veterinarni Medicina* 56, 486–503.

Waadu, G.D.B. and Chappell, L.H. (1991) Effect of water temperature on the ability of *Diplostomum spathaceum* miracidia to establish in lymnaeid snails. *Journal of Helminthology* 65, 179–185. https://doi.org/10.1017/s0022149x00010671

Valtonen, E.T. and Gibson, D.I. (1997) Aspects of the biology of diplostomid metacercarial (Digenea) populations occurring in fishes in different localities of northern Finland. *Annales Zoologici Fennici* 34, 47–59.

Valtonen, E.T., Holmes, J.C. and Koskivaara, M. (1997) Eutrophication, pollution, and fragmentation: effects on parasite communities in roach (*Rutilus rutilus*) and perch (*Perca fluviatilis*) in four lakes in central Finland. *Canadian Journal of Fisheries and Aquatic Sciences* 54, 572–585. https://doi.org/10.1139/f96-306

Wegner, K.M., Kalbe, M., Milinski, M. and Reusch, T.B.H. (2008) Mortality selection during the 2003 European heat wave in three-spined sticklebacks: effects of parasites and MHC genotype. *BMC Evolutionary Biology* 8, 124. https://doi.org/10.1186/1471-2148-8-124

Verma, A.K., Pal, A.K., Manush, S.M., Das, T., Dalvi, R.S. *et al.* (2007) Persistent sub-lethal chlorine exposure augments temperature induced immunosuppression in *Cyprinus carpio* advanced fingerlings. *Fish and Shellfish Immunology* 22, 547–555. https://doi.org/10.1016/j.fsi.2006.08.001

Whyte, S.K., Chappell, L.H. and Secombes, C.J. (1988) *In vitro* transformation of *Diplostomum spathaceum* (Digenea) cercariae and short-term maintenance of post-penetration larvae *in vitro*. *Journal of Helminthology* 62, 293–302. https://doi.org/10.1017/S0022149X0001169X

Whyte, S.K., Chappell, L.H. and Secombes, C.J. (1989) Cytotoxic reactions of rainbow trout, *Salmo gairdneri* Richardson, macrophages for larvae of the eye fluke *Diplostomum spathaceum* (Digenea).

Journal of Fish Biology 35, 333–345. https://doi.org/10.1111/j.1095-8649.1989.tb02986.x

Whyte, S.K., Chappell, L.H. and Secombes, C.J. (1990) Protection of rainbow trout, *Oncorhynchus mykiss* (Richardson), against *Diplostomum spathaceum* (Digenea): the role of specific antibody and activated macrophages. *Journal of Fish Diseases* 13, 281–291. https://doi.org/10.1111/j.1365-2761.1990.tb00784.x

Whyte, S.K., Secombes, C.J. and Chappell, L.H. (1991) Studies on the infectivity of *Diplostomum spathaceum* in rainbow trout (*Oncorhynchus mykiss*). *Journal of Helminthology* 65, 169–178. https://doi.org/10.1017/s0022149x0001066x

24 Anisakiosis (*Anisakis simplex* s.l.)

Arne Levsen[1*], Paolo Cipriani[1], Miguel Bao[1], Lucilla Giulietti[1], and Simonetta Mattiucci[2]

[1]*Institute of Marine Research, Bergen, Norway;* [2]*Sapienza – University of Rome, Rome, Italy*

24.1 Introduction

Nematodes of the anisakid family (Anisakidae) commonly occur as larvae in many fish species in most marine habitats around the globe. The genus *Anisakis*, giving the root name to the entire family, is probably one of the most extensively studied groups of marine parasites, primarily due to the fact that *Anisakis simplex sensu lato* (s.l.) occurs quite ubiquitously in all oceanographic basins. Also, two *Anisakis* species, i.e. *Anisakis simplex sensu stricto* (s.s.) and *Anisakis pegreffii*, are known to cause accidental infection and hypersensitivity disorders in humans. This zoonotic condition is contracted through accidental ingestion of viable third-stage larvae of the nematode species upon consumption of raw or only lightly processed fishery products. Other potentially zoonotic and relatively common anisakid genera in fish are *Pseudoterranova* and *Contracaecum* (the latter including those species maturing in seals).

The term 'anisakidosis' designates infections with anisakid nematodes in general, while 'anisakiosis' refers exclusively to infections with species belonging to genus *Anisakis*. Due to its frequent and widespread occurrence in many commercially valuable fish species worldwide, except, it seems, strictly Antarctic waters, the present chapter focuses primarily on the two *A. simplex* (s.l.) species. However, climate-related aspects of anisakidosis by extending to species of *Pseudoterranova* or *Contracaecum* are addressed whenever considered relevant.

Based on multilocus molecular assessment and morphological diagnostic features, the genus *Anisakis* presently comprises nine nominal and one still to be described species. The phylogeny of the so far described species of *Anisakis*, as inferred from different molecular markers, shows the existence of four distinct phylogenetic clades, with the first clade containing the members of the *A. simplex* (s.l.) complex, i.e. *A. simplex* (s.s.), *A. pegreffii* and *Anisakis berlandi*. A second clade includes the species *Anisakis ziphidarum* and *Anisakis nascettii*, while a third clade comprises the three sibling species *Anisakis physeteris*, *Anisakis brevispiculata* and *Anisakis paggiae*. Finally, a further distinct clade includes the species *Anisakis typica* and the still undescribed taxon *Anisakis* sp. 1 (Mattiucci *et al.*, 2018).

All *Anisakis* species have complex indirect life cycles involving mostly cetacean whales as definitive hosts, where the mature female worms shed the eggs with the host's faeces. The eggs embryonate and hatch in the water, apparently releasing free-swimming third-stage larvae (Køie *et al.*, 1995) which are eaten by planktonic or semi-planktonic crustaceans such as copepods or krill, thus acting as the first paratenic host. Fish (or squid) become infected by feeding on crustaceans containing third-stage larvae that, after ingestion, bore through the wall of the digestive tract into the visceral cavity where the parasites become encapsulated by host cells (see Smith and Wootten, 1978). Thus, it is widely accepted that fish act as a paratenic host, transporting the larvae from the intermediate host level to a suitable definitive mammalian host, thereby completing the life cycle (Fig. 24.1). While most *Anisakis* species are generalists at the paratenic host level (a few exceptions exist, e.g. *A. ziphidarum*), there seems to be a tendency towards host specificity in most *Anisakis* with regard to their definitive host range, probably as a result of co-evolutionary processes (Mattiucci and Nascetti, 2008; Mattiucci *et al.*, 2018). Based

*arne.levsen@hi.no

 Climate Change and Infectious Fish Diseases (P.T.K. Woo, J.-A. Leong and K. Buchmann)

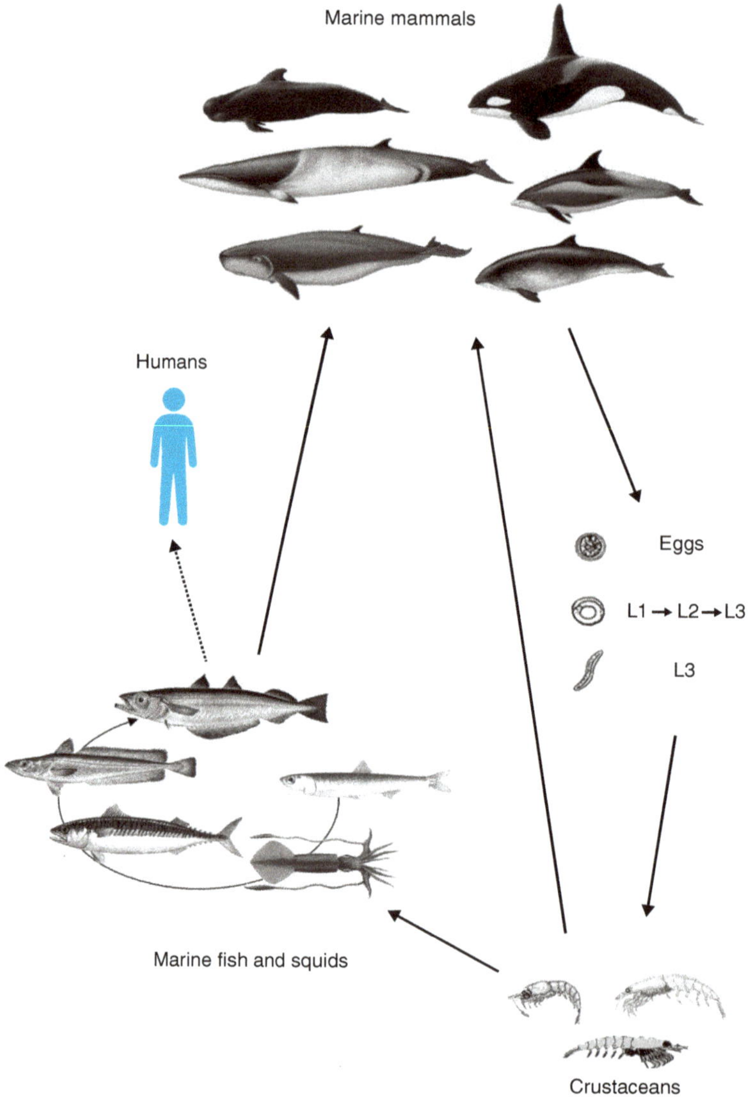

Fig. 24.1. Generalized *Anisakis* spp. life cycle. Development from first-stage larva (L1) to third-stage larva (L3) occurs within the egg.

on co-phylogenetic studies it was suggested that the members of the *A. simplex* (s.l.) complex have likely co-evolved with dolphins of the order Delphinoidea as definitive hosts. Thus, combined with evidence that highest infection levels with the larvae of the *A. simplex* (s.l.) complex are usually found in demersal and pelagic fish and various squid species, it was suggested that these *Anisakis* species have a life cycle mainly associated with the pelagic or benthopelagic realm (Mattiucci and Nascetti, 2008; Mattiucci *et al.*, 2018).

Of epidemiological data on anisakid parasites presently recorded from various fish host species, by far the largest number relate to just two species,

A. simplex (s.s.) and *A. pegreffii*. Both siblings are known to infect a wide range of phylogenetically different fish host species from oceanic waters around the globe. While *A. simplex* (s.s.) has a northerly distribution, apparently restricted to subarctic and temperate waters of the northern hemisphere, its sibling species *A. pegreffii* is found mainly in warmer seas of both hemispheres including the Austral region. The North-East (NE) Atlantic, the waters off the Iberian coast and the western Mediterranean Sea (Alboran Sea), along with southern and central areas of Japanese waters in the North-West Pacific, represent areas of sympatry where both species may co-occur in the same individual paratenic or definitive host (Fig. 24.2). For comprehensive information on the taxonomy and molecular systematics, as well as basic ecological and co-evolutionary host–parasite relationships of *Anisakis* species, see the recent review by Mattiucci *et al.* (2018).

In fish, the larvae of *A. simplex* (s.l.), commonly known as the 'herring worm' or 'whale worm', are typically seen as encapsulated flat and tight coils, measuring approximately 4–5 mm across, on the organs of the visceral cavity. The range of suitable teleost hosts includes economically important species such as herring (*Clupea harengus*), Atlantic mackerel (*Scomber scombrus*), European hake (*Merluccius merluccius*), Atlantic cod (*Gadus morhua*) and (wild) Atlantic salmon (*Salmo salar*). When an infected fish is eaten by another fish, the encapsulated larvae become digested free and, following stomach wall penetration, may establish in this new host, thus repeating the larval fish host cycle. This transfer of larvae up the natural food chain may result in extensive accumulation, especially in larger piscivorous fish species such as European hake and cod which sometimes harbour hundreds or even thousands of encapsulated *A. simplex* (s.l.) larvae (Smith and Wootten, 1978; Levsen *et al.*, 2018a). However, the number of fish host cycles that individual *Anisakis* larvae may undergo without losing infectivity has not yet been elucidated. Although most larvae reside in or on the visceral organs of their fish host, some may migrate into the fish flesh,

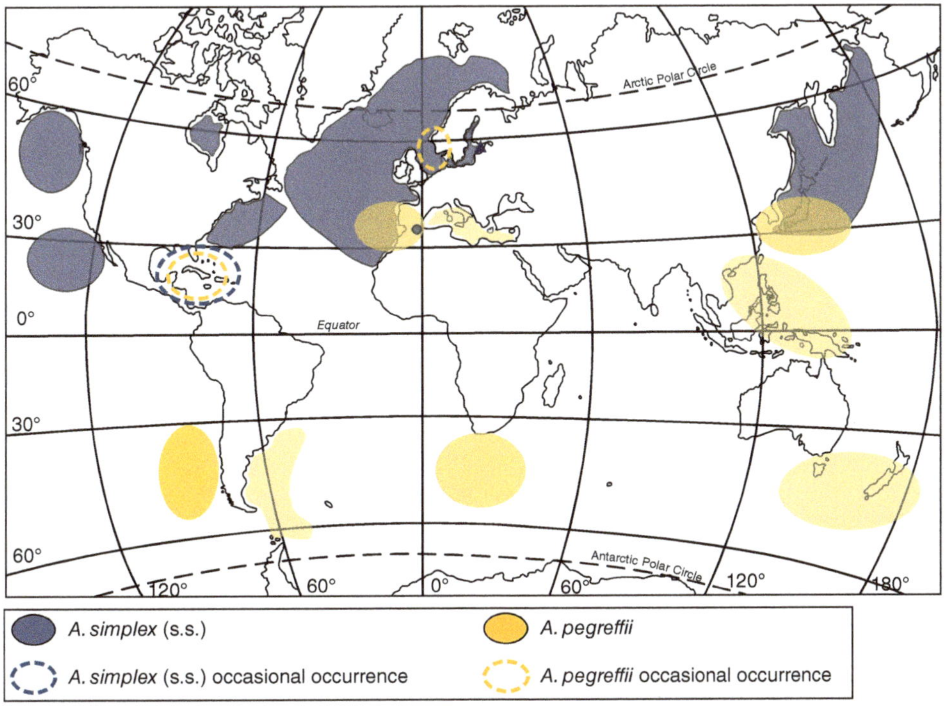

Fig. 24.2. Global distribution of *Anisakis simplex* (s.s.) and *Anisakis pegreffii* based on molecularly confirmed records from various fish host species.

especially the hypaxial part of the body musculature surrounding the visceral organs on both sides, commonly known as the belly flaps. The degree and extent to which individual *Anisakis* larvae exhibit muscle-penetrating behaviour seem to vary with actual *Anisakis* sibling species as well as fish host species. For example, *A. simplex* (s.s.) can frequently be found in the flesh of some of its main fish hosts in the NE Atlantic including European hake, cod and blue whiting (*Micromesistius poutassou*), sometimes at high abundances also in the thicker and central parts of the fillets (Gay *et al*., 2018; Levsen *et al*., 2018a; Pascual *et al*., 2018). Hence, if not frozen or properly heat-treated before consumption, these fish species may pose a particular risk for consumers to acquire anisakiosis.

The other sibling *A. pegreffii* seems to show this flesh-migrating behaviour to a much lesser extent. Suzuki *et al*. (2010) reported 12 times higher mean intensities of *A. simplex* (s.s.) compared with *A. pegreffii* in the fillets of chub mackerel (*Scomber japonicus*) from Japanese waters. Cipriani *et al*. (2015) found six times higher relative proportions of *A. simplex* (s.s.) compared with *A. pegreffii* in fillets of European hake from the Atlantic coast of Spain. Similar results (i.e. about ten times higher relative proportion of *A. simplex* (s.s.) compared with *A. pegreffii*) were recently found in the belly flaps of European hake from the Spanish–Portuguese coast; the proportion of the different genotypes in the viscera was approximately 68% for *A. simplex* (s.s.), 30% for *A. pegreffii* and 1.5% for their hybrids (Pascual *et al*., 2018). Post-mortem migration from the fish viscera into the muscle may occur in both *Anisakis* species, such as observed for *A. pegreffii* in European anchovy (*Engraulis encrasicolus*) from the Mediterranean Sea (Cipriani *et al*., 2016), and simultaneously for *A. simplex* (s.s.) and *A. pegreffii* in mixed infections in chub mackerel. However, the finding of higher penetration rate of *A. simplex* (s.s.) compared with *A. pegreffii* was also documented in chub mackerel (Suzuki *et al*., 2010). Smith and Wootten (1975) reported large scale post-mortem migration of *Anisakis* sp. larvae from the viscera into the flesh of herring in experimental trials, where almost 20% of the overall larval burden was recorded in the flesh after 37 h of cool storage.

Since *Anisakis* species, in particular *A. simplex* (s.s.) and *A. pegreffii*, have primarily received attention as zoonotic parasites, their larvae's ability to cause disease in fish is not equally well elucidated (see Buchmann and Mehrdana, 2016). Although largely aiming to assess the consumer risk imposed by the presence of *Anisakis* larvae in fishery products, recent epidemiological studies of some commercially important fish species from various European fishing grounds have provided a good and comprehensive picture of the current geographic distribution and infection characteristics of these two siblings in the fish host species examined (see Levsen *et al*., 2018a). Thus, the present chapter aims to point out and discuss the possible effects that certain climate-mediated changes, including elevated seawater temperature and changing trophic relationships, may have on the occurrence, distribution and biology of *Anisakis* spp. in various fish host species from different areas and ecosystems. The widely distributed siblings *A. simplex* (s.s.) and *A. pegreffii* may even prove useful biological markers to indicate apparently climate-driven changes in migration patterns of certain fish and/or cetacean host species in some hydrographically and ecologically well studied and defined areas such as the North Sea and Baltic Sea, the Barents Sea and Arctic Sea, as well as the Mediterranean and its tributary seas.

24.2 Diagnosis of the Infection and Effects on Fish Health

24.2.1 Macroscopic appearance of infection

The presence of *Anisakis* spp. in fish is usually easily detected upon visual examination of the visceral organs, mesenteries and peritoneal linings. Depending on various factors such as fish host species, host size and worm abundance, the larvae may occur scattered one by one, or in clusters sometimes containing hundreds of worms, on the organs and mesenteries of the visceral cavity. For a quick diagnosis of anisakiosis in fish, and to clearly distinguish the larvae of the *A. simplex* species complex from other anisakids, fresh or frozen larvae can be studied under a dissection microscope at low magnification. *A. simplex* (s.l.) larvae are readily identified by a comparatively broad and elongate oesophageal ventricle and the presence of a so-called 'mucron' at the very tip of the tail (Fig. 24.3).

A typical feature of *Anisakis* infections in fish is a fibrous connective tissue capsule generated by the host and surrounding each larva, sometimes even forming around clusters of larvae (see Levsen and Berland, 2012). Additionally, extensive melanomacrophage aggregates or melanin deposits have been observed around encapsulated larvae, especially on the liver of flounder (*Platichthys flesus*) (Dezfuli *et al.*, 2007), Atlantic salmon (Murphy *et al.*, 2010) as well as Atlantic cod and blue whiting (Levsen and Berland, 2012) (Fig. 24.4). Aggregations of hundreds of *Anisakis* larvae can sometimes form masses in the visceral cavity of European hake (Fig. 24.5). These clusters are interconnected by degenerated host tissue and infiltrated by leucocytes, including macrophages, showing a strong inflammatory response in the fish (Pascual *et al.*, 2018). Bao *et al.* (2015) observed accumulations of hundreds of *Anisakis* larvae at the posterior end of the terminal blind sac of the stomach of allis shad (*Alosa alosa*) from western Iberian waters. This phenomenon was also observed, although to a lesser extent, in two other clupeids, twaite shad (*Alosa fallax*) from the Atlantic coast of Spain (Bao *et al.*, 2015) and herring from the North Sea and Norwegian Sea (Sluiters, 1974; Tolonen and Karlsbakk, 2003; Bao *et al.*, 2017). This anatomical site, adjacent to the ductus pneumaticus, appeared damaged in some heavily infected allis shad, presumably caused by migrating *Anisakis* larvae. Thus, simultaneous migration of lots of larvae within the visceral cavity may inflict mechanical injury or even cause occlusion of the digestive tract. In general, however, it appears that encapsulation in the body cavity or in association with the mesothelium induces fewer pathological changes in the host compared with situations where the nematode larva invades central organs such as the liver (Klapper *et al.*, 2018).

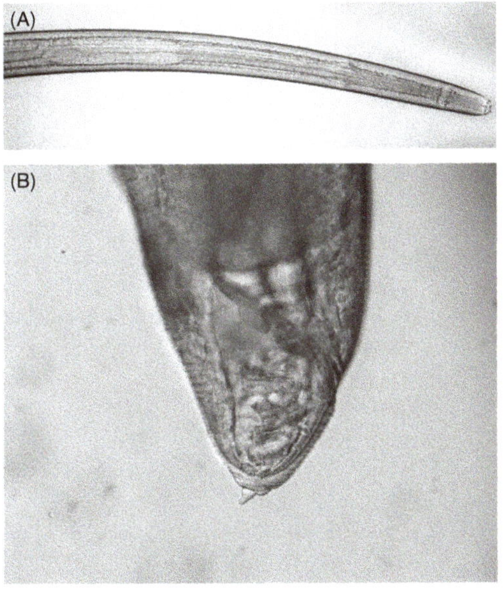

Fig. 24.3. Anterior body (A) and tail (B) of *Anisakis simplex* (s.l.) third-stage larva.

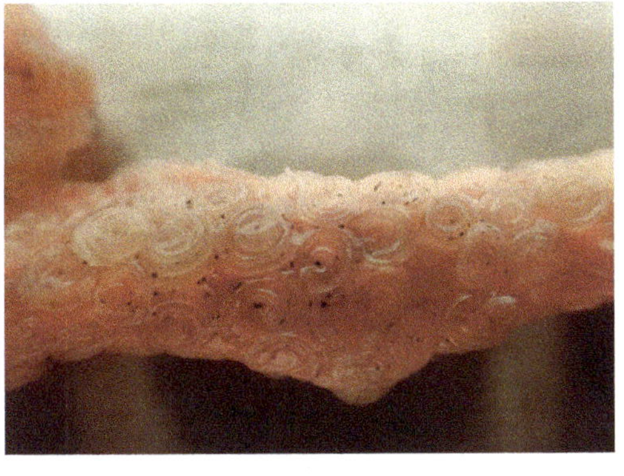

Fig. 24.4. Encapsulated *Anisakis* larvae on the liver of blue whiting (deposits of melanin are visible).

24.2.2 Molecular identification of *Anisakis simplex* (s.l.)

Various nuclear and mitochondrial molecular markers have been developed over the past two to three decades allowing a quick and reliable identification of sibling species within different groups of anisakid nematodes. The identification at species level of individual *Anisakis* larva by allozymes electrophoresis was initially applied to distinguish several species within the genus, and was further used to develop DNA-based approaches for species

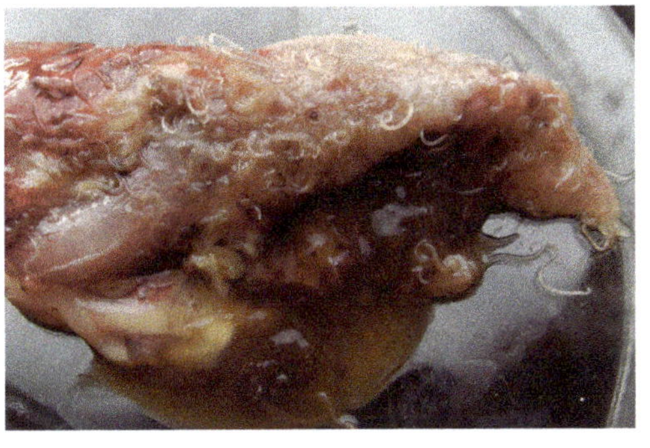

Fig. 24.5. Aggregation of *Anisakis* larvae in the visceral cavity of European hake.

identification, including direct sequencing of nuclear and mitochondrial DNA genes (Valentini *et al.*, 2006; Mattiucci *et al.*, 2014a, 2016, 2018). Thus, a multigene approach including both nuclear and mitochondrial markers permits robust identification of the taxa investigated. At present, the following molecular techniques are available for unambiguous identification of the two sibling species *A. simplex* (s.s.) and *A. pegreffii*: (i) diagnostic allozyme markers; (ii) sequence analysis of the mitochondrial cytochrome oxidase II (mtDNA *cox2*); (iii) direct sequencing of the elongation factor (EF1 α-1 nDNA) nuclear gene; and (iv) DNA microsatellites (Mattiucci *et al.*, 2019). It has recently been demonstrated (Mattiucci *et al.*, 2016) that sequencing of the internal transcribed spacer (ITS) region of rDNA and restriction fragment length polymorphisms (RFLPs)–polymerase chain reaction (PCR) of the same gene do not represent a fully diagnostic marker to discern specimens of *A. simplex* (s.s.), *A. pegreffii* and their hybrid categories. For higher resolution it should be combined with other genetic markers to avoid misidentifications (for a comprehensive review of the molecular systematics of *Anisakis* species, see Mattiucci *et al.*, 2018).

24.2.3 Particular pathological conditions caused by *Anisakis simplex* (s.l.)

Levsen and Berland (2012) reported the so-called 'stomach crater syndrome' in the stomach wall and mucosa of migrating Atlantic cod. It was characterized by more or less strong inflammatory reactions with cellular infiltration, as a result of numerous penetrating *A. simplex* (s.l.) larvae. The syndrome was apparently restricted to migrating cod caught at the spawning grounds off the Lofoten Islands in northern Norway during the 1970s. The reason for the sudden appearance of the syndrome in 1969 and its marked decrease 10–12 years later still remains unresolved.

Wild Atlantic salmon and sea trout returning to rivers in Scotland, England and Wales, as well as Norway, Iceland and Quebec, frequently showed bleeding, swollen and haemorrhagic vents, giving name to the condition now known as the 'red vent syndrome' or RVS. The condition is characterized by haemorrhages and moderate to severe inflammation around the vent caused by large numbers of unencapsulated *A. simplex* (s.l.) larvae in the surrounding tissue and urogenital papilla region (Beck *et al.*, 2008; Noguera *et al.*, 2009). However, besides these sometimes severe lesions, affected fish were generally in good overall condition and there were no signs of RVS-induced wild salmon mortality or any other infections. The causative reason for the 'red vent syndrome' was not fully elucidated. However, Noguera *et al.* (2009) hypothesized that climate-driven changes in the NE Atlantic pelagic ecosystem may have evoked physiological responses that eventually facilitated the development of this condition in space and time. This was supported by the findings of strong eosinophilic inflammatory responses predominantly in early-summer fish still in the pre-spawning phase during gonad growth and maturation.

Behavioural changes in smelt and eel induced by *Pseudoterranova* sp. infections in the fish muscle may reduce swimming performance which again can lead to increased mortality in the actual fish populations (Sprengel and Lüchtenberg, 1991; Rohlwing *et al.*, 1998). However, this strategy of increasing parasite transmission probability by impeding a given fish host's ability to avoid predation has not been investigated to occur in any *Anisakis* species.

24.2.4 Effects of larval ascaridoid nematodes on fish larvae and juveniles

Fish larvae and juveniles have generally poorer immune-defensive mechanisms to cope with parasite infections compared with adults since development of an adaptive immune defence system is dependent on fish age and size (Secombes and Wang, 2012). Wootten (1978) studied the occurrence of ascaridoids in several gadoid species (i.e. Atlantic cod, whiting (*Merlangius merlangus*), haddock (*Melanogrammus aeglefinus*), Norway pout (*Trisopterus esmarkii*) and poor cod (*Trisopterus minutus*)) from Scottish waters. Wootten (1978) found larval *Hysterothylacium aduncum* in fishes at about 4 months of age (first infection in whiting at 2–3 cm length), and first infection with *Anisakis* occurred at a larger size and age (first infection in haddock at 6–7 cm length). Skovgaard *et al.* (2011) studied the occurrence of larval *H. aduncum* in cod larvae from the north-eastern North Sea. They found first infection in 1 cm long cod larvae, with prevalence increasing with fish size. The majority of the infected fish harboured only one parasite (maximum of four nematodes per fish) which may suggest that intensities of more than one parasite can cause mortalities among the fish larvae.

Experimental infection studies suggest that mortality among small fishes/fish larvae can be caused by larval ascaridoids (e.g. *H. aduncum*, *Pseudoterranova decipiens* s.s.) migrating for example from the stomach into the viscera or muscle, potentially affecting vital organs on their path (McClelland, 1995; Balbuena *et al.*, 2000). It therefore appears that ascaridoid-related mortality in small fishes/fish larvae may possibly occur under natural conditions. Further research is needed to acquire a better understanding of this ascaridoid-induced mortality in early developmental stages of various fish species. The rise in ocean temperature due to climate change may also unbalance the 'parasite–fish equilibrium status', and again further research is required to assess its significance in the stock recruitment of fishes.

24.3 Expected Spread of *Anisakis* Species

As with many other marine metazoan parasites, *Anisakis* species have a complex life cycle which relies on a bottom-up transfer between successively higher marine trophic levels. Thus, among the most important drivers that shape the distribution of *Anisakis* species is the availability of suitable hosts at all trophic levels, as a prerequisite for successful completion of the life cycle. Important abiotic factors include water temperature, salinity and general oceanographic conditions such as the availability of nutrients at the primary production level (Mattiucci *et al.*, 2018). For instance, the marine phytoplankton production cycle is largely related to temperature, and any permanent change in water temperature, e.g. in Arctic seas, may directly affect the production and biomass of phytoplankton in space and time which again could create a mismatch for the timing of zooplankton production, subsequently affecting all successive trophic levels including fish, sea birds and marine mammals (Byrne, 2011; Post *et al.*, 2013). Thus, successful transmission of *Anisakis* species up the trophic chain largely depends on a synchrony in timing between the occurrence of free and infective larval stages and the availability of suitable planktonic hosts. Hence, climate-related rise in water temperature could result in a mismatch between development and hatching of *Anisakis* eggs and the availability of zooplankton and susceptible fish hosts in a given area.

Højgaard (1998) showed experimentally that hatching time of eggs, and the survival of free larvae of *A. simplex* (s.l.), was inversely related to water temperature (5–21°C) but did not vary significantly with salinity (0–28 psu). The hatching success of eggs was highest at 12°C and lowest at 21°C. Such optimal hatching conditions may consequently prevail over longer periods during the year and over larger geographical areas in North Atlantic and North Pacific waters, thus facilitating the presence of relatively higher numbers of free *Anisakis* larvae to be available for krill or copepods over a longer period of time in these waters. The optimal temperature range for hatching and survival of free larvae of *A. pegreffii* has not been elucidated yet, but the species seems to be distributed mainly in warmer waters (Fig. 24.2). However, the currently observed trend of generally elevated mean water temperatures at higher latitudes, such as in the northern North Sea and the Norwegian Sea, may imply that this species could expand its natural distribution area further to the north. Indeed, a few *A. pegreffii* larvae have recently been recorded in Atlantic mackerel caught in the northern North Sea and the southern Norwegian Sea (Levsen *et al.*, 2018b). Although the findings appear to be related to the migration of southwestern subpopulations of mackerel carrying the parasites along on their northward feeding migration, *A. pegreffii* is probably about to stretch its fingers into the Nordic Seas. The general northward spreading trend of this *Anisakis* species was further illustrated by a survey of Gay *et al.* (2018) who identified *A. pegreffii* in Atlantic cod from the northern North Sea (ICES division IVa). This finding could indicate that *A. pegreffii* may complete at least part of its life cycle in the North Sea, e.g. by transferring larvae through predation on other fish such as (migrating) mackerel between individual cods. While the cod population in the North Sea seems largely to be stationary, the recently extended feeding distribution of Atlantic mackerel up to the Svalbard Archipelago in the north, and beyond Iceland in the west, is positively correlated with water temperature (Trenkel *et al.*, 2014; Berge *et al.*, 2015). Thus, the availability of suitable plankton intermediate and cetacean definitive hosts, upon arrival of mackerel during summer in these areas, might permit *A. pegreffii* to survive in subarctic or Arctic waters and complete its life cycle, at least on a seasonal basis. This assumption is supported by the expectation that ice-free summers might give *Calanus finmarchicus*, one of the most abundant copepod species in the North Atlantic, the possibility to inhabit the Barents Sea and adjacent areas, thus additionally facilitating spread of *Anisakis* paratenic fish hosts such as Atlantic mackerel that feed on them (Stenevik and Sundby, 2007). The relatively high abundance of whales, such as the minke whale (*Balaenoptera acutorostrata*), in the area during summer (Ressler *et al.*, 2015) would represent another crucial link for further spread and dissemination of *A. simplex* (s.s.) in the Arctic, as well as the possible establishment of *A. pegreffii* on a more permanent basis in Nordic waters. A similar scenario may occur in the North Sea, where the rise in average water temperature may promote the invasion of new fish species such as anchovy and sardine from the south. Both fish species are known to carry *A. simplex* (s.s.) and *A. pegreffii* in their southern main distribution area in the NE Atlantic (Molina-Fernández *et al.*, 2015; Levsen *et al.*, 2018a). Thus, an extension of these fish species' distribution into the North Sea may contribute significantly to the further northward spread of *A. pegreffii*, as well.

Studies from South-West Pacific waters suggest that changing occurrence pattern and distribution range of marine organisms due to climate-related events may influence the occurrence and infection level of *Anisakis* species. For example, Bak *et al.* (2014) found that the prevalence of infection with *Anisakis* spp. larvae in chub mackerel fluctuated off the eastern coast of Korea. The higher infection level seemed to be related to better availability of invertebrates/plankton in the area, which, in turn, appeared to be influenced by local changes in water temperature. Similarly, recent El Niño (El Niño Southern Oscillation, ENSO) events caused a drastic fluctuation in ocean currents and sea water temperature. Liu and Zhang (2013) reported that the surface temperature around Taiwan increased by more than 2.7°C. In turn, this event has led to an increased number of fish species migrating into Taiwan Sea waters, resulting also in changing dynamics of *Anisakis* spp. infections (Bak *et al.*, 2014).

Many cetacean species are highly migratory, and they can travel long distances between breeding or feeding areas. Their distribution is controlled by a combination of demographic, ecological, evolutionary, habitat-related and man-made factors with prey availability being particularly critical (Learmonth *et al.*, 2006). Along these migratory routes, they also transport and eventually release certain stages of parasites. Moreover, during these journeys they feed and can acquire a more diverse parasite infracommunity, spreading them in the course of migration. Certain abiotic factors such as water temperature and salinity, together with biotic ones including availability of suitable intermediate hosts, then determine if these parasite species can survive locally and be able to close their life cycles.

There is some evidence that cetacean distributions and movements have changed recently in some areas of the NE Atlantic (Simmonds and Eliott, 2009; Evans and Bjørge, 2013). With a rise in water temperature, some animals may move further north in Arctic regions, remain there longer and compete with the endemic Arctic species (Moore and Huntington, 2008). Thus, climate change is expected to affect the range and migratory patterns of many marine mammals hosting several parasite species, which in turn could lead to a spread of organisms and pathogens to naïve populations through these vagrant hosts (Simmonds and Eliott, 2009). It should also be taken into account that some marine mammal species may simply be responding to regional variability in resource availability independent of climate change (Evans and Bjørge, 2013). In the last years, the status of several cetacean species in the NE Atlantic has changed somewhat (Evans and Bjørge, 2013). Several species of 'oceanic dolphins' (Delphinidae) and harbour porpoises, known to be among the most common definitive hosts for *A. simplex* (s.s.) and *A. pegreffii*, have shifted their distribution, mostly following their prey (Evans and Bjørge, 2013). With respect to these parasite/host associations, the consequence of changes in distribution could result in an overall epidemiological cascade effect on the fish paratenic hosts, showing increasing or decreasing infection levels. A different effect could involve extension of the sympatric regions of *A. simplex* (s.s.) and *A. pegreffii* (Fig. 24.2), shifting northwards by following the cetacean host movements and changing abiotic environmental conditions. The cetacean migrations, along with an overall tendency of generally warmer northern waters, could then result in a change of the geographical distribution of *A. pegreffii* towards Nordic Seas, while *A. simplex* (s.s.) could disappear from southern areas in which water temperatures could reach too high levels for the eggs to hatch and survive.

Warm-water vagrant species (e.g. Bryde's whale (*Balaenoptera edeni*), pygmy sperm whale (*Kogia breviceps*), dwarf sperm whale (*Kogia sima*), rough-toothed dolphin (*Steno bredanensis*), Atlantic spotted dolphin (*Stenella frontalis*)) could start crossing north-western Europe (Evans and Bjørge, 2013). In this direction, a recent finding by Klimpel *et al.* (2011) revealing the presence of *A. paggiae* in the meso- and bathypelagic fish *Anoplogaster cornuta* from the NE Atlantic Ocean could be linked to some unusual cetacean movements in northern waters. Kogiid whales (i.e. *K. breviceps* and *K. sima*) are common definitive hosts of this *Anisakis* species (Mattiucci *et al.*, 2005; Mattiucci and Nascetti, 2008) and Klimpel *et al.* (2011) suggested that the *A. paggiae* larvae in that fish host could have been introduced through migratory kogiid hosts.

24.4 Increase in Pathogen Mortality

In general, parasite transmission rates through the life cycle, and possibly their virulence, are expected to increase with rising water temperature. Additionally, increased temperature may affect the immunocompetence of aquatic animals (reviewed by Marcogliese, 2008). Other anthropogenic-derived stressors related to, or interacting with climate change, such as pollution (e.g. heavy metals and microplastics), stratification, acidification and eutrophication, will most likely also affect the complex parasite–host relationships. This illustrates the difficulties of making predictions on the potentially harmful effects on individual hosts and populations (reviewed by Marcogliese, 2008, 2016; Cable *et al.*, 2017).

However, temporarily or permanently prevailing unfavourable environmental conditions in the macrohabitat such as extreme temperature or salinity outside tolerable limits, at both ends of the scale, may directly induce mortality of released eggs or free larvae of *Anisakis* species, or may indirectly affect the survival of their hosts. For example, in certain enclosed seas such as the Baltic Sea there is a tendency towards lower salinity due to increased river runoff facilitated by enhanced rainfall during winter. Lowered salinity is thought to have a major influence on the Baltic Sea fauna, possibly inducing a shift in species composition from marine to freshwater species (Philippart *et al.* 2011). Since Baltic cod (*G. morhua*), basically a marine species, appears to be a major transport host for several anisakid species including *Contracaecum osculatum* (s.s.) and *P. decipiens* (s.l.) in the area (Mehrdana *et al.*, 2014; Zuo *et al.*, 2018), a gradual increase in population size of various freshwater fish may lower the availability of suitable transport hosts for these nematode species, thereby reducing transmission probability to the next host level. However, species of *Anisakis* seem not to be able to complete their life cycle in the Baltic Sea.

The survival of *Anisakis* (and any other endoparasite for that matter) is dependent on each host, at each level, providing suitable conditions for the parasite's survival, transmission and, eventually, reproduction. *A. simplex* (s.s.) and *A. pegreffii* are generalists at the fish host level, which implies that many phylogenetic distant fish species may act as paratenic or transport host for the parasites. Thus, different fish species represent different microhabitats inflicting different physiological responses on the larvae. There seem to exist comparatively large differences in the ability of different fish species to cope with *Anisakis* infections. While herring (*C. harengus*) and blue whiting seem to show only weak or moderate immunological reactions (A. Levsen, 2014, personal observations), Atlantic mackerel and saithe (*Pollachius virens*) appear to be able to launch strong responses, even inducing direct mortalities, since dead and partially disintegrated *A. simplex* (s.l.) larvae can frequently occur in both fish species (Priebe *et al.*, 1991; Levsen and Berland, 2012). Thus, any long-term rise in water temperature in certain oceanic areas may enhance the immunocompetence in some fish species to cope with *Anisakis* infections, while in others, pathologies may be induced due to e.g. within-host larval migrating behaviour causing lesions, which again may influence fish host survival, especially in immunosuppressed individuals due to other concurrent infections or environmental stressors.

As far as the cetacean definitive hosts are concerned, the efficiency in completing the reproductive step in the *Anisakis* life cycle including the production and shedding of fertilized eggs would also depend on the state of health and nutrition of the whale hosts. Increased mortality of cetacean hosts (dolphins/porpoises) due to emerging marine diseases (e.g. virus) facilitated by rising water temperature (Harvell *et al.*, 1999) could, in turn, negatively impact *Anisakis* egg production and recruitment. For example, stranded cetaceans debilitated by illness, or suffering from stomach occlusion caused by ingestion of plastic items, harboured just a few or no *Anisakis* specimens compared with healthy cetacean hosts that stranded for ethological or traumatic reasons, which showed comparatively high intensities of worms in the stomach or foregut (P. Cipriani, 2016, personal observations; Mazzariol *et al.*, 2018).

24.5 Control and/or Prevention

Although not known to be of any significance as a disease organism in maricultured fish species such as Atlantic salmon or rainbow trout (*Oncorhynchus mykiss*) (Skov *et al.*, 2009; Levsen and Maage, 2016), the sheer presence of *Anisakis* larvae in any products derived from aquaculture could seriously affect consumer confidence. Thus, in order to minimize the risk of anisakids occurring in cultured fish through ingestion of infected plankton or wild fish that occasionally may enter the cages, closed aquaculture systems equipped with appropriate filtering of the intake water should be used. Additionally, farmed fish must not be fed with fresh wild marine fish offal, as it may contain live parasitic nematodes that could be transferred into the production fish.

In cases where increased infection levels and/or pathological alterations in fish due to anisakids/*Anisakis* can be linked to increasing population size of definitive marine mammal hosts in certain areas, regulatory measures to reduce the actual host populations could be implemented (see also Buchmann and Mehrdana, 2016). However, such measures would be highly controversial among the general public, even more so since most marine mammal species are protected by international law. Nevertheless, some climate-related environmental stressors such as elevated water temperature may compromise the general condition and immunocompetence of actual seal or whale populations, thereby increasing their susceptibility to emerging marine diseases such as the phocine distemper virus (PDV) or the dolphin morbillivirus (DMV) (Harvell *et al.*, 1999), which again may call for culling measures in order to prevent further spread of the diseases.

Additionally, the basically natural but anthropogenically maintained or enhanced presence of *Anisakis* in wild fish hosts could also be controlled. The most obvious solution may be to stop or reduce the still ongoing practice on board fishing vessels of throwing fish offal carrying *Anisakis* back into the sea. Fish offal should be brought ashore and used (if possible) in by-products, or at least the parasites should be inactivated, e.g. by heat treatment, before any offal is discarded at sea (Cipriani *et al.*, 2018; González *et al.*, 2018).

24.6 Conclusions with Suggestions for Future Studies

It appears not likely that *Anisakis* species may gain importance as disease organisms able to induce mortalities in fish populations in the course of climate change. However, climate change could lead to a general increase in *Anisakis* biomass in certain areas/habitats or ecosystems, especially in Nordic and Arctic seas. Conversely, a decrease in *Anisakis* biomass might also be expected in unhealthy/unbalanced ecosystems, where the close interweaving between occurrence and abundance of definitive, intermediate and transport hosts in space and time could be disturbed.

Surveillance programmes and regular large-scale epidemiological studies of *Anisakis*/anisakid nematode occurrence in commercially and ecologically important key fish species could be particularly useful for monitoring quantitative variations of these parasite species, especially in those areas or fishing grounds that one expects to be comparatively strongly affected by climate-related changes such as the Nordic Seas and the Baltic Sea (Levsen *et al.*, 2018a). Moreover, *Anisakis* species may also prove useful as biological tags to track and follow changing migration routes of certain fish species including Atlantic mackerel and sardine. In this respect, biodiversity and distribution of larval *Anisakis* spp. are widely employed as biomarkers for stock distribution of fish hosts (Mattiucci *et al.*, 2014b). Thus, long-term investigations of the *Anisakis* species diversity across the entire distribution of some of their key fish host species could provide useful supplementary markers to detect changing migration routes and geographical origin of fish stock components, also as a consequence of climate-related changes.

Similarly, surveillance of occurrence and migration patterns of whale populations in actual areas could provide useful data and insight to better understand their parasite distribution. Moreover, international cooperative programmes could be initialized to check the *Anisakis*/anisakid burden in stranded whales or whales accidentally caught in fish nets. Thus, *Anisakis*/anisakid species, despite their zoonotic potential and undesired presence in fish products, may play a useful role, too, as bioindicators of alterations including climate changes that interfere with trophic interrelationships and dynamics in various marine habitats, or even entire ecosystems (see also review by Mattiucci *et al.*, 2018).

References

Bak, T.J., Jeon, C.H. and Kim, J.H. (2014) Occurrence of anisakid nematode larvae in chub mackerel (*Scomber japonicus*) caught off Korea. *International*

Journal of Food Microbiology 191, 149–156. https://doi.org/10.1016/j.ijfoodmicro.2014.09.002

Balbuena, J.A., Karlsbakk, E., Kvenseth, A.M., Saksvik, M. and Nylund, A. (2000) Growth and emigration of third-stage larvae of *Hysterothylacium aduncum* (Nematoda: Anisakidae) in larval herring *Clupea harengus*. *Journal of Parasitology* 86, 1271–1275. https://doi.org/10.1645/0022-3395(2000)086[1271:GAEOTS]2.0.CO;2

Bao, M., Mota, M., Nachón, D.J., Antunes, C., Cobo, F. *et al.* (2015) *Anisakis* infection in allis shad, *Alosa alosa* (Linnaeus, 1758), and twaite shad, *Alosa fallax* (Lacépède, 1803), from Western Iberian Peninsula Rivers: zoonotic and ecological implications. *Parasitology Research* 114, 2143–2154. https://doi.org/10.1007/s00436-015-4403-5

Bao, M., Strachan, N.J.C., Hastie, L.C., MacKenzie, K., Seton, H.C. and Pierce, G.J. (2017) Employing visual inspection and magnetic resonance imaging to investigate *Anisakis simplex* s.l. infection in herring viscera. *Food Control* 75, 40–47. https://doi.org/10.1016/j.foodcont.2016.12.030

Beck, M., Evans, R., Feist, S.W., Stebbing, P., Longshaw, M. and Harris, E. (2008) *Anisakis simplex sensu lato* associated with red vent syndrome in wild adult Atlantic salmon *Salmo salar* in England and Wales. *Diseases of Aquatic Organisms* 82, 61–65. https://doi.org/10.3354/dao01979

Berge, J., Heggland. K., Lonne, O.J., Cottier, F., Hop, H. *et al.* (2015) First records of Atlantic mackerel (*Scomber scombrus*) from the Svalbard Archipelago, Norway, with possible explanations for the extension of its distribution. *Arctic* 68, 54–61. https://doi.org/10.14430/arctic4455

Buchmann, K. and Mehrdana, F. (2016) Effects of anisakid nematodes *Anisakis simplex* (s.l.), *Pseudoterranova decipiens* (s.l.) and *Contracaecum osculatum* (s.l.) on fish and consumer health. *Food and Waterborne Parasitology* 4, 13–22. https://doi.org/10.1016/j.fawpar.2016.07.003

Byrne, M. (2011) Impact of ocean warming and ocean acidication on marine invertebrate life history stages: vulnerabilities and potential for persistence in a changing ocean. *Oceanography and Marine Biology: An Annual Review* 49, 1–42. https://doi.org/10.1201/b11009

Cable, J., Barber, I., Boag, B., Ellison, A.R., Morgan, E.R. *et al.* (2017) Global change, parasite transmission and disease control: lessons from ecology. *Philosophical Transactions of the Royal Society B* 372, 20160088. https://doi.org/10.1098/rstb.2016.0088

Cipriani, P., Smaldone, G., Acerra, V., D'Angelo, L., Anastasio, A. *et al.* (2015) Genetic identification and distribution of the parasitic larvae of *Anisakis pegreffii* and *Anisakis simplex* (s.s.) in European hake *Merluccius merluccius* from the Tyrrhenian Sea and Spanish Atlantic coast: implications for food safety. *International Journal of Food Microbiology* 198, 1–8. https://doi.org/10.1016/j.ijfoodmicro.2014.11.019

Cipriani, P., Acerra, V., Bellisario, B., Sbaraglia, G.L., Cheleschi, R. *et al.* (2016) Larval migration of the zoonotic parasite *Anisakis pegreffii* (Nematoda: Anisakidae) in European anchovy, *Engraulis encrasicolus*: implications to seafood safety. *Food Control* 59, 148–157. https://doi.org/10.1016/j.foodcont.2015.04.043

Cipriani, P., Sbaraglia, G., Palomba, L., Giulietti, L., Bellisario, B. *et al.* (2018) *Anisakis pegreffii* (Nematoda: Anisakidae) in European anchovy *Engraulis encrasicolus* from the Mediterranean Sea: considerations in relation to fishing ground as a driver for parasite distribution. *Fisheries Research* 202, 59–68. https://doi.org/10.1016/j.fishres.2017.03.020

Dezfuli, B.S., Pironi, F., Shinn, A.P., Manera, M. and Giari, L. (2007) Histopathology and ultrastructure of *Platichthys flesus* naturally infected with *Anisakis simplex* s.l. larvae (Nematoda: Anisakidae). *Journal of Parasitology* 93, 1416–1423. https://doi.org/10.1645/GE-1214.1

Evans, P.G.H. and Bjørge, A. (2013) Impacts of climate change on marine mammals. *MCCIP Science Review* 2013, 134–148.

Gay, M., Bao, M., MacKenzie, K., Pascual, S., Buchmann, K. *et al.* (2018) Infection levels and species diversity of ascaridoid nematodes in Atlantic cod, *Gadus morhua*, are correlated with geographic area and fish size. *Fisheries Research* 202, 90–102. https://doi.org/10.1016/j.fishres.2017.06.006

González, A.F., Gracia, J., Miniño, I., Romón, J., Larsson, C. *et al.* (2018) Approach to reduce the zoonotic parasite load in fish stocks: when science meets technology. *Fisheries Research* 202, 140–148. https://doi.org/10.1016/j.fishres.2017.08.016

Harvell, C., Kim, K., Burkholder, J., Colwell, R., Epstein, P. *et al.* (1999) Emerging marine diseases: climate links and anthropogenic factors. *Science* 285, 1505–1510. https://doi.org/10.1126/science.285.5433.1505

Højgaard, D.P. (1998) Impact of temperature, salinity and light on hatching of eggs of *Anisakis simplex* (Nematoda, Anisakidae), isolated by a new method, and some remarks on survival of larvae. *Sarsia North Atlantic Marine Science* 83, 21–28. https://doi.org/10.1080/00364827.1998.10413666

Klapper, R., Carballeda-Sangiao, N., Kuhn, T., Jensen, H.M., Buchmann, K. *et al.* (2018) Anisakid infection levels in fresh and canned cod liver: significant reduction through liver surface layer removal. *Food Control* 92, 17–24. https://doi.org/10.1016/j.foodcont.2018.04.029

Klimpel, S., Kuhn, T., Busch, M.W., Karl, H. and Palm, H.W. (2011) Deep-water life cycle of *Anisakis paggiae* (Nematoda: Anisakidae) in the Irminger Sea indicates kogiids distribution in the North Atlantic waters. *Polar Biology* 34, 899–906. https://doi.org/10.1007/s00300-010-0946-1

Køie, M., Berland, B. and Burt, M.D.B. (1995) Development to third-stage larvae occurs in the eggs of *Anisakis simplex* and *Pseudoterranova decipiens* (Nematoda, Ascaridoidea, Anisakidae). *Canadian Journal of Fisheries and Aquatic Sciences* 52, 134–139. https://doi.org/10.1139/f95-519

Learmonth, J.A., Macleod, C.D., Santos, M.B., Pierce, G.J., Crick, H.Q.P. and Robinson, R.A. (2006) Potential effects of climate change on marine mammals. *Oceanography and Marine Biology: An Annual Review* 44, 431–464.

Levsen, A. and Berland, B. (2012) *Anisakis* Species. In: Woo, P.T.K. and Buchmann, K. (eds) *Fish Parasites: Pathobiology and Protection*. CABI Publishing, Wallingford, UK, pp. 298–309.

Levsen, A. and Maage, A. (2016) Absence of parasitic nematodes in farmed, harvest quality Atlantic salmon (*Salmo salar*) in Norway – results from a large scale survey. *Food Control* 68, 25–29. https://doi.org/10.1016/j.foodcont.2016.03.020

Levsen, A., Svanevik, C.S., Cipriani, P., Mattiucci, S., Gay, M. *et al.* (2018a) A survey of zoonotic nematodes of commercial key fish species from major European fishing grounds – introducing the FP7 PARASITE exposure assessment study. *Fisheries Research* 202, 4–21. https://doi.org/10.1016/j.fishres.2017.09.009

Levsen, A., Cipriani, P., Mattiucci, S., Gay, M., Hastie, L.C. *et al.* (2018b) *Anisakis* species composition and infection characteristics in Atlantic mackerel, *Scomber scombrus*, from major European fishing grounds – reflecting changing fish host distribution and migration pattern. *Fisheries Research* 202, 112–121. https://doi.org/10.1016/j.fishres.2017.07.030

Liu, Q. and Zhang, Q. (2013) Analysis on long-term change of sea surface temperature in the China Seas. *Journal of Ocean University of China* 12, 295–300. https://doi.org/10.1007/s11802-013-2172-2

McClelland, G. (1995) Experimental infection of fish with larval sealworm, *Pseudoterranova decipiens* (Nematoda, Anisakinae), transmitted by amphipods. *Canadian Journal of Fisheries and Aquatic Sciences* 52, 140–155. https://doi.org/10.1139/f95-520

Marcogliese, D.J. (2008) The impact of climate change on the parasites and infectious diseases of aquatic animals. *Revue Scientifique et Technique (International Office of Epizootics)* 27, 467–484. https://doi.org/10.20506/rst.27.2.1820

Marcogliese, D.J. (2016) The distribution and abundance of parasites in aquatic ecosystems in a changing climate: more than just temperature. *Integrative and Comparative Biology* 56, 611–619. https://doi.org/10.1093/icb/icw036

Mattiucci, S. and Nascetti, G. (2008) Advances and trends in the molecular systematics of anisakid nematodes, with implications for their evolutionary ecology and host–parasite co-evolutionary processes. *Advances in Parasitology* 66, 47–148. https://doi.org/10.1016/S0065-308X(08)00202-9

Mattiucci, S., Nascetti, G., Dailey, M., Webb, S.C., Barros, N. *et al.* (2005) Evidence for a new species of *Anisakis* Dujardin, 1845: morphological description and genetic relationships between congeners (Nematoda: Anisakidae). *Systematic Parasitology* 61, 157–171. https://doi.org/10.1007/s11230-005-3158-2

Mattiucci, S., Cipriani, P., Webb, S.C., Paoletti, M., Marcer, F. *et al.* (2014a) Genetic and morphological approaches distinguishing the three sibling species of the *Anisakis simplex* species complex, with a species designation as *Anisakis berlandi* n. sp. for *A. simplex* sp. C (Nematoda: Anisakidae). *Journal of Parasitology* 15, 12–15. https://doi.org/10.1645/12-120.1

Mattiucci, S., Garcia, A., Cipriani, P., Santos, M.N., Nascetti, G. and Cimmaruta, R. (2014b) Metazoan parasite infection in the swordfish, *Xiphias gladius*, from the Mediterranean Sea and comparison with Atlantic populations: implications for its stock characterization. *Parasite* 21, 35. https://doi.org/10.1051/parasite/2014036

Mattiucci, S., Acerra, V., Paoletti, M., Cipriani, P., Levsen, A. *et al.* (2016) No more time to stay 'single' in the detection of *Anisakis pegreffii*, *A. simplex* (s.s.) and hybridization events between them: a multi-marker nuclear genotyping approach. *Parasitology* 143, 998–1011. https://doi.org/10.1017/S0031182016000330

Mattiucci, S., Cipriani, P., Levsen, A., Paoletti, M. and Nascetti, G. (2018) Molecular epidemiology of *Anisakis* and anisakiasis: an ecological and evolutionary road map. *Advances in Parasitology* 99, 93–263. https://doi.org/10.1016/bs.apar.2017.12.001

Mattiucci, S., Bello, E., Paoletti, M., Webb, S.C., Timi, J.T. *et al.* (2019) Novel polymorphic microsatellite loci in *Anisakis pegreffii* and *A. simplex* (s.s.) (Nematoda: Anisakidae): implications for species recognition and population genetic analysis. *Parasitology* 146, 1387–1403. https://doi.org/10.1017/S003118201900074X

Mazzariol, S., Centelleghe, C., Cozzi, B., Povinelli, M., Marcer, F. *et al.* (2018) Multidisciplinary studies on a sick-leader syndrome-associated mass stranding of sperm whales (*Physeter macrocephalus*) along the Adriatic coast of Italy. *Scientific Reports* 8, 11577. https://doi.org/10.1038/s41598-018-29966-7

Mehrdana, F., Bahlool, Q.Z.M., Skov, J., Marana, M.H., Sindberg, D. *et al.* (2014) Occurrence of zoonotic nematodes *Pseudoterranova decipiens*, *Contracaecum osculatum* and *Anisakis simplex* in cod (*Gadus morhua*) from the Baltic Sea. *Veterinary Parasitology* 205, 581–587. https://doi.org/10.1016/j.vetpar.2014.08.027

Molina-Fernández, D., Malagón, D., Gómez-Mateos, M., Benítez, R., Martín-Sánchez, J. and Adroher, F.J. (2015) Fishing area and fish size as risk factors of *Anisakis* infection in sardines (*Sardina pilchardus*) from Iberian waters, southwestern Europe. *International*

Journal of Food Microbiology 203, 27–34. https://doi.org/10.1016/j.ijfoodmicro.2015.02.024

Moore, S.E. and Huntington, H.P. (2008) Arctic marine mammals and climate change: impacts and resilience. *Ecological Applications* 18, S157–S165. https://doi.org/10.1890/06-0571.1

Murphy, T.M., Berzano, M., O'Keeffe, S.M., Cotter, D.M., McEvoy, S.E. *et al.* (2010) Anisakid larvae in Atlantic salmon (*Salmo salar* L.) grilse and post-smolts: molecular identification and histopathology. *Journal of Parasitology* 96, 77–82. https://doi.org/10.1645/GE-2194.1

Noguera, P., Collins, C., Bruno, D., Pert, C., Turnbull, A. *et al.* (2009) Red vent syndrome in wild Atlantic salmon *Salmo salar* in Scotland is associated with *Anisakis simplex* sensu stricto (Nematoda: Anisakidae). *Diseases of Aquatic Organisms* 87, 199–215. https://doi.org/10.3354/dao02141

Pascual, S., Rodríguez, H., Pierce, G.J., Hastie, L.C. and González, A.F. (2018) The NE Atlantic European hake: a neglected high exposure risk for zoonotic parasites in European fish markets. *Fisheries Research* 202, 69–78. https://doi.org/10.1016/j.fishres.2017.12.008

Philippart, C.J.M., Anadón, R., Danovaro, R., Dippner, J.W., Drinkwater, K.F. *et al.* (2011) Impacts of climate change on European marine ecosystems: observations, expectations and indicators. *Journal of Experimental Marine Biology and Ecology* 400, 52–69. https://doi.org/10.1016/j.jembe.2011.02.023

Post, E., Bhatt, U.S., Bitz, C.M., Brodie, J.F., Fulton, T.L. *et al.* (2013) Ecological consequences of sea-ice decline. *Science* 341, 519–524. https://doi.org/10.1126/science.1235225

Priebe, K., Huber, C., Märtlbauer, E. and Terplan, G. (1991) Detection of antibodies against the larva of *Anisakis simplex* in the pollock *Pollachius virens* using ELISA. *Journal of Veterinary Medicine, Series B* 38, 209–214.

Ressler, P.H., Dalpadado, P., Macaulay, G.J., Handegard, N. and Skern-Mauritzen, M. (2015) Acoustic surveys of euphausiids and models of baleen whale distribution in the Barents Sea. *Marine Ecology Progress Series* 527, 13–29. https://doi.org/10.3354/meps11257

Rohlwing, T., Palm, H.W. and Rosenthal, H. (1998) Parasitation with *Pseudoterranova decipiens* (Nematoda) influences the survival rate of the European smelt *Osmerus eperlanus* retained by a screen wall of a nuclear plant. *Diseases of Aquatic Organisms* 32, 233–236. https://doi.org/10.3354/dao032233

Secombes, C.J. and Wang, T. (2012) The innate and adaptive immune system of fish. In: Austin, B. (ed.) *Infectious Disease in Aquaculture: Prevention and Control*. Woodhead Publishing, Cambridge, UK, pp. 3–68. https://doi.org/10.1533/9780857095732.1.3

Simmonds, M.P. and Eliott, W.J. (2009) Climate change and cetaceans: concerns and recent developments. *Journal of the Marine Biological Association of the United Kingdom* 89, 203–210. https://doi.org/10.1017/S0025315408003196

Skov, J., Kania, P.W., Olsen, M.M., Lauridsen, J.H., Buchmann, K. (2009) Nematode infections of maricultured and wild fishes in Danish waters: a comparative study. *Aquaculture* 298, 24–28. https://doi.org/10.1016/j.aquaculture.2009.09.024

Skovgaard, A., Bahlool, Q.Z.M., Munk, P., Berge, T. and Buchmann, K. (2011) Infection of North Sea cod, *Gadus morhua* L., larvae with the parasitic nematode *Hysterothylacium aduncum* Rudolphi. *Journal of Plankton Research* 33, 1311–1316. https://doi.org/10.1093/plankt/fbr027

Sluiters, J. (1974) *Anisakis* sp. larvae in the stomachs of Herring (*Clupea harengus* L.). *Zeitschrift für Parasitenkunde* 44, 279–288. https://doi.org/10.1007/BF00366111

Smith, J.W. and Wootten, R. (1975) Experimental studies on the migration of *Anisakis* sp. larvae (Nematoda: Ascaridida) into the flesh of herring, *Clupea harengus* L. *International Journal for Parasitology* 5, 133–136. https://doi.org/10.1016/0020-7519(75)90019-3

Smith, J.W. and Wootten, R. (1978) *Anisakis* and anisakiasis. *Advances in Parasitology* 16, 93–163. https://doi.org/10.1016/S0065-308X(08)60573-4

Sprengel, G. and Lüchtenberg, H. (1991) Infection by endoparasites reduces swimming speed of European smelt *Osmerus eperlanus* and European eel *Anguilla anguilla*. *Diseases of Aquatic Organisms* 11, 31–35. https://doi.org/10.3354/dao011031

Stenevik, E.K. and Sundby, S. (2007) Impacts of climate change on commercial fish stocks in Norwegian waters. *Marine Policy* 31, 19–31. https://doi.org/10.1016/j.marpol.2006.05.001

Suzuki, J., Murata, R., Hosaka, M. and Araki, J. (2010) Risk factors for human *Anisakis* infection and association between the geographic origins of *Scomber japonicus* and anisakid nematodes. *International Journal of Food Microbiology* 137, 88–93. https://doi.org/10.1016/j.ijfoodmicro.2009.10.001

Tolonen, A. and Karlsbakk, E. (2003) The parasite fauna of the Norwegian spring spawning herring (*Clupea harengus* L.). *ICES Journal of Marine Science* 60, 77–84. https://doi.org/10.1006/jmsc.2002.1307

Trenkel, V.M., Huse, G., MacKenzie, B.R., Alvarez, P., Arrizabalaga, H. *et al.* (2014) Comparative ecology of widely distributed pelagic fish species in the North Atlantic: implications for modelling climate and fisheries impacts. *Progress in Oceanography* 129, Part B, 219–243. https://doi.org/10.1016/j.pocean.2014.04.030

Valentini, A., Mattiucci, S., Bondanelli, P., Webb, S.C., Mignucci-Giannone, A. and Nascetti, G. (2006)

Genetic relationships among *Anisakis* species (Nematoda: Anisakidae) inferred from mitochondrial *cox2* sequences and comparison with allozyme data. *Journal of Parasitology* 92, 156–166. https://doi.org/10.1645/GE-3504.1

Wootten, R. (1978) The occurrence of larval anisakid nematodes in small gadoids from Scottish waters. *Journal of the Marine Biological Association of the United Kingdom* 58, 347–356. https://doi.org/10.1017/S0025315400028022

Zuo, S., Kania, P.W., Mehrdana, F., Marana, M.H. and Buchmann, K. (2018) *Contracaecum osculatum* and other anisakid nematodes in grey seals and cod in the Baltic Sea: molecular and ecological links. *Journal of Helminthology* 92, 81–89. https://doi.org/10.1017/S0022149X17000025

25 Lepeophtheirosis (*Lepeophtheirus salmonis*)

Mark D. Fast[1]* and Sussie Dalvin[2]

[1]*University of Prince Edward Island, Charlottetown, Prince Edward Island, Canada;* [2]*Institute of Marine Research, Bergen, Norway*

25.1 Brief Introduction

Lepeophtheirus salmonis, the salmon louse, is an ectoparasitic copepod that infects wild and cultured fish (namely salmonids) over two oceans and presents very different concerns with respect to environmental and host challenges in each. *Lepeophtheirus salmonis* belongs to the family Caligidae, which comprises a series of marine species commonly referred to as 'sea lice'. Like most of the other 41 families within the Siphonostomatoida, the Caligidae are parasites of invertebrate and vertebrate hosts (Poley and Fast, 2020). Even within the well-known species of sea lice that have plagued salmon aquaculture for decades, variation in genus, species, subspecies and host interactions are unique to each ocean and coastal environment. Therefore, it is important to consider these other sea lice species in the context of this chapter on *L. salmonis*, as changing environments may tip the ecological balance in favour of one of these other species in the future. Within North Atlantic waters three species of sea lice, *Caligus elongatus*, *Caligus curtus* and *L. salmonis*, have been described from Atlantic salmon (*Salmo salar*) and are associated with aquaculture in the Bay of Fundy (BoF), Canada (Hogans and Trudeau, 1989), as well as in Norway and Scotland. In the North Pacific, *Caligus clemensi*, *L. salmonis* and *Lepeophtheirus cuneifer* have been described from wild and farmed salmonids. Yazawa *et al.* (2008) and Skern-Mauritzen *et al.* (2014) have further differentiated *L. salmonis* in the Atlantic and Pacific into two subspecies of *L. salmonis*. Here, they are referred to as *L. salmonis salmonis* for those found in the Atlantic and as *L. salmonis oncorhynchi* for those from the Pacific (Skern-Mauritzen *et al.*, 2014). This divergence of *L. salmonis* subspecies is consistent with independent co-evolution of the parasite with salmonid species in their respective oceans occurring over the last 2.5–11 million years (Yazawa *et al.*, 2008). These differences in genus, species and subspecies are very important in understanding sea lice biological differences and abundances within the field and industry, and the impact on salmonid hosts.

Major differences do exist across and within species in terms of host range/preference (Table 25.1). The various species within the genus *Caligus* tend to be generalists compared with *L. salmonis*, especially *C. elongatus*. *C. elongatus* has been described from over 80 species of teleosts and elasmobranchs (Kabata, 1979), and Oines and Heuch (2007) have shown two different genotypes of *C. elongatus* with associations to different hosts, wild/farmed fish, based on season. In the South Pacific, introduction of salmon aquaculture has resulted in a unique Atlantic salmon–sea louse interaction with *Caligus rogercresseyi*. Much less is known about its original host range relationship, but similar to *L. salmonis*, *C. rogercresseyi* has exhibited a high degree of host specificity for salmonids and populations show strong selection (i.e. resistance development) based on drug interventions and life on a cultured host (Aaen *et al.*, 2015; Jaramillo, 2018). *L. cuneifer* and *C. curtus* have also been described by several authors on elasmobranch, gadid and salmonid hosts (reviewed in Pike and Wadsworth, 1999). In comparison, *L. salmonis* is much more the specialist in both oceans. In the Atlantic it has

*mfast@upei.ca

Table 25.1. Host susceptibility to *Lepeophtheirus salmonis*, *Caligus elongatus*, and/or *Caligus clemensi* infection over different host size ranges. (Data from Johnson and Albright, 1992; Mustafa and MacKinnon, 1999; Bjørn *et al.*, 2001; Fast *et al.*, 2002; Beamish *et al.*, 2005; Jones *et al.*, 2006a, 2007, 2008b; Sutherland *et al.*, 2011; Braden *et al.*, 2015a,b.)

Host species	Size in seawater at first exposure			
	Juvenile, <0.7 g	Juvenile, 0.7–30 g	Juvenile, 30–400 g	Adult
Salmo salar (Atlantic salmon)	N/A	N/A	+ +	+ +
Salmo trutta (sea trout)	N/A	+ + +	+ + +	+ + +
Salvelinus alpinus (Arctic charr)	N/A	N/A	+ + +	+ + +
Oncorhynchus keta (chum salmon)	N/A	+ + +	+ + +	+ + +
Oncorhynchus nerka (sockeye salmon)	N/A	+ +	+ +	+ +
Oncorhynchus mykiss (rainbow trout)	N/A	N/A	+	+
Oncorhynchus tshawytscha (chinook salmon)	N/A	N/A	–	–
Oncorhynchus gorbuscha (pink salmon)	+ + +	– –	– –	+ +
Oncorhynchus kisutch (coho salmon)	N/A	– – –	– – –	– –

N/A indicates data do not exist; '+' denotes level of susceptibility (+ + +, highly susceptible); '–' denotes level of resistance (– – –, highly resistant).

been described only from salmonids (reviewed in Pike and Wadsworth, 1999) and hence acquired the common name of 'salmon louse', whereas the Pacific form has a greater number of salmonid hosts to choose from, but it also commonly infects the three-spined stickleback (*Gasterosteus aculeatus*) (Jones *et al.*, 2006a; Jones and Prosperi-Porta, 2011). This creates an interesting twist to the ecology of this parasite, since three-spined sticklebacks co-aggregate with juvenile pink and chum salmon in the nearshore environment and *L. salmonis* does not appear to develop past chalimus life stage on the stickleback (Jones *et al.*, 2006ab). *L. cuneifer* and *C. clemensi* are also found on sticklebacks but the life cycles for these species, and *C. curtus* (in the Atlantic), are not completed, which complicates a description of their ecological interactions. These three caligids are not expected to differ greatly from *C. elongatus* or *L. salmonis*, which have eight-stage life cycles (Piasecki and MacKinnon, 1995; Hamre *et al.*, 2014). *C. elongatus* and *L. salmonis* do differ, however, in the number of attached chalimus (four in *C. elongatus*; two in *L. salmonis*) and mobile pre-adult stages (zero in *C. elongatus*; two in *L. salmonis*). With respect to generation time, the original work on Atlantic *L. salmonis* from Wootten *et al.* (1982) reported a generation time of 42 days between 9 and 12°C, whereas Johnson and Albright (1991) reported 40 days for males and 52 days for females at 10°C in the Pacific, suggesting little difference in these traits across subspecies. Similarly, Piasecki and MacKinnon (1995) reported 43 days generation time for *C. elongatus* at 10°C. There are no data for generation time of *C. clemensi* or *C. curtus*, but there is no reason to expect these species would have considerably different generation times.

Reduced abundance on cultured salmon and an inability to maintain the *Caligus* spp. or *L. cuneifer* in the laboratory have hampered our understanding of these species, in comparison to both subspecies of *L. salmonis*, which have close to 100% prevalence on salmonids in both oceans and have been amenable to laboratory experimentation. Fast *et al.* (2003) provided the first evidence of differences between the Pacific and Atlantic *L. salmonis*, showing different protease secretions and responses to different host species. Saksida *et al.* (2007) also reported reduced disease and requirement for treatment on Pacific farms infected with *L. salmonis*, and Bricknell *et al.* (2006) observed different tolerance to low salinity in Atlantic *L. salmonis* copepodids to that reported by Johnson and Albright (1991) in Pacific *L. salmonis*. Interestingly, while some of these biological differences exist across species/subspecies of sea lice, the host responses appear to be relatively consistent against these caligid parasites. Coho salmon (*Oncorhynchus kisutch*) show low levels of infection of both subspecies of *L. salmonis* under laboratory and field conditions, as well as of *C. clemensi* (Johnson and Albright, 1992; Fast *et al.*, 2002; Beamish *et al.*, 2005), whereas Atlantic salmon appear to be uniquely susceptible to both subspecies (Johnson and Albright, 1992; Fast *et al.*, 2002), as well as to *C. clemensi* and *C. elongatus* (Hogans and Trudeau, 1989; Jones *et al.*, 2006a).

25.1.1 *Lepeophtheirus salmonis salmonis* (Atlantic Ocean) occurrence in North America and Norway

Sea lice have been observed and recorded on wild salmon in Atlantic Canada since the 1930s and 1940s (reviewed in Chang *et al.*, 2011); however, in 1994 a large outbreak of lice in the Letang area spread quickly through the BoF industry (Hogans, 1995). Although Atlantic salmon historically inhabited nearly 50 rivers in the inner and outer BoF, some being involved in the commercial fishery at the time, the total number of returning adults in BoF rivers in recent years is approximately 5000 per year. In comparison there are approximately 15 million farmed salmon in sea cages in southwestern New Brunswick (reviewed in Chang *et al.*, 2011). In Newfoundland and Labrador, recreational catch of Atlantic salmon has ranged between 50,000 and 60,000 per year (DFO, 2016), whereas culture production reached >25,000 tonnes in 2016, or >5 million farmed salmon. These numbers relate the likely minor contribution of wild sourced sea lice to their overall abundance in Atlantic Canada and USA.

On farms, the early years of production in the 1980s–1990s were mainly dominated by *C. elongatus* (97%) generally in low numbers, <2% *L. salmonis* and <1 % *C. curtus* (Hogans and Trudeau, 1989). However, following the outbreak in the mid-1990s in the Letang area, *L. salmonis* has been the main species of concern ever since, with only transient/seasonal infections of *C. elongatus* in low numbers (<1 per fish when present) (reviewed in Chang *et al.*, 2011) and few if any observations of *C. curtus*. In the mid-1990s and then again since 2009, the infection levels of *L. salmonis* in BoF have been quite high, averaging more than 20–30 lice per fish in some management zones during the summer and autumn seasons (reviewed in Chang *et al.*, 2011; ACFFA, 2016). Unfortunately, the data for Nova Scotia and Newfoundland prevalence and abundance of lice are not as well described, although expected to be dominated by *L. salmonis*.

In Norway, *L. salmonis* has commonly been observed on wild-caught salmon for hundreds of years (Berland and Margolis, 1983), where it was regarded as a positive sign indicating that the salmon had newly returned to the river and thus still had good flesh quality. The first scientific description of *L. salmonis* was published as early as 1838, from wild Atlantic salmon in Danish waters, under the name *Caligus salmonis* by Krøyer (1838). Similar to the east coast of Canada and Atlantic America, the introduction of aquaculture has led to a large increase of host fish and thus an increase in sea lice populations (Torrissen *et al.*, 2013; Taranger *et al.*, 2015). The number of farmed salmonids greatly exceeds the number of wild salmonids in the North Atlantic. Farmed salmonids amount to approximately 400 million individuals (Statistics Norway, 2019), whereas there are about 0.5 million wild Atlantic salmon returning annually (ICES, 2017). The prevalence of *C. elongatus* is lower than of *L. salmonis* and is most commonly found in the northern regions towards later summer (www.Barentswatch.no (accessed 8 April 2020)). The number of *L. salmonis* permitted on farmed fish is strictly regulated and all counts must be reported for this weekly. The current limit is a maximum of 0.5 adult female lice per fish (Kaur *et al.*, 2017).

25.1.2 *Lepeophtheirus salmonis oncorhynchi* (Pacific Ocean) occurrence in North America

Pacific North America is a completely different story in all aspects. While the salmon farming industry numbers have changed somewhat in the intervening years, in 2003/04 the annual production of Atlantic salmon was close to 80,000 tonnes, whereas the average annual biomass of adult Pacific salmon returning to coastal areas was 1.42 million tonnes from 1990 to 1999 (Beamish *et al.*, 2007). In 2004, a survey on wild adult Pacific salmon (n = 666) in the coastal waters of central British Columbia showed nearly all salmon had lice and pink (*Oncorhynchus gorbuscha*) and chum salmon (*Oncorhynchus keta*) had 100% prevalence across the locations sampled in Queen Charlotte Strait and Smith and Rivers inlets (Beamish *et al.*, 2005). Pink, chum and sockeye (*Oncorhynchus nerka*) salmon had average intensities of 41.5–52.0 sea lice per fish, whereas chinook (*Oncorhynchus tshawytscha*) and coho salmon had average intensities of 16.1–18.5 sea lice per fish. In that study, *L. salmonis* were about twice as abundant as *C. clemensi* and most of the latter were in the chalimus stage, whereas *L. salmonis* were mostly in pre-adult/adult stages (Beamish *et al.*, 2005). Beamish *et al.* (2009) also observed >60% prevalence of lice on juvenile Pacific salmon and herring in the Gulf Islands area, virtually all of which were *C. clemensi*. Evaluation of farm ($n = 20$) data from 2003 to 2005 showed interannual and seasonal variations, in which *L. salmonis* were the more abundant species and had greater numbers on older fish (Saksida *et al*., 2007). In 2003 when *L. salmonis* pre-adult/adult life stages were at their lowest (mean of 2.2 per fish), *C. clemensi* had its highest mean adult lice count of 1.6 per fish. A 2011–2015 audit verified industry data from over 3000 farm counts showing an average 1.6 *L. salmonis* pre-adult/adults per fish, 0.78 per fish were adult females, 0.79 per fish were chalimus and 0.43 per fish were *C. clemensi* (DFO, 2016). While interannual and seasonal variations occur in sea lice species' abundances, the number of lice on farmed salmon is generally greater in second-year fish and increases in autumn coincidently at the same time of the adult Pacific salmon migration (Jones *et al*., 2016).

25.2 Morphological Diagnosis of the Pathogen/Disease

Understanding the parasite life cycle (both planktonic and parasitic) and the effects of temperature and salinity on development are imperative for trying to model or project regional infestation pressure and climate change impacts. Furthermore, individual life stages have drastically different pathological impacts on the host and development of disease. As with all crustaceans, the sea louse's life cycle begins with a free-living nauplius (I) larva. This hatches from an egg within the adult female egg string and undergoes two moults (nauplius I to nauplius II; nauplius II to copepodid) before entering its first infectious life stage, the copepodid. All of the first three free-living stages are lecitotrophic (non-feeding) and depend on the yolk deposited into the egg by the mother. The yolk consists of lipids and protein (Tucker *et al*., 2000; Dalvin *et al*., 2011). The copepodid needs to locate a host fish before the energy reserves are depleted. Locating and successful attachment to a host have previously been reported to be significantly greater for recently moulted individuals (Tucker *et al*., 2000). However, more recent and more intensive investigation of age impacts on copepodid activity indicate that attachment to the host initially is very low, followed by a rapid increase to maximum infectivity and then a slower decrease falling to zero as unattached copepodids starve to death (Brooker *et al*., 2018).

Copepodids respond to a range of environmental cues including light and pressure waves generated by approaching fish to locate and attach to a host (Browman *et al*., 2004). Once the copepodid has attached to a fish, it will utilize chemosensory receptors to determine if it is a suitable host and leave the fish if not. Molecular manipulation of taste receptors in *L. salmonis* produces lice that will remain on non-host fish for extended periods, indicating that lice attached to non-host fish require a secondary cue or stimulation of these receptors to identify appropriate hosts (Komisarczuk *et al*., 2017). Salmonids express initial erratic behaviour as lice attach, with increasing jumping and rolling behaviour and changes in swim patterns (Bui *et al*., 2016). A few days post-infection, transient dark-coloured spots can be observed at the site of infection (Fig. 25.1). The cause and possible function of these spots is currently unknown; however, melanophores and melanin-based pigmentation have been shown to aggregate in salmonid skin/fin tissue following stress, toxic stimuli and bacterial pathogen exposure (Dukovcic *et al*., 2010; Kittilsen *et al*., 2012). Melanophores are the pigmented cells responsible for rapid and transient chromomotor colour changes in poikilothermic vertebrates (Thody and Shuster, 1989), and melanin-based skin pigmentation is negatively correlated with plasma cortisol and positively correlated with reduced levels of egg-bearing female *L. salmonis* (Kittilsen *et al*., 2012). These observations would concur with skin melanization playing a role in inhibition of parasite proliferation through the toxic effects of melanin, and possibly with its involvement in cellular and/or antibody-mediated immunity (reviewed in Côte *et al*., 2018).

Temperature is a major regulator of development in *L. salmonis*, with faster development at higher

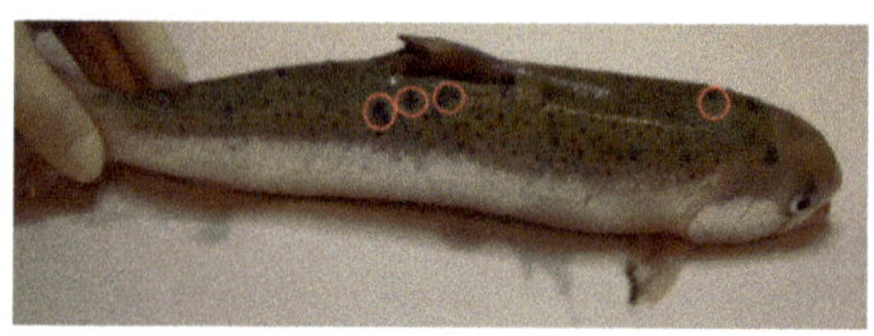

Fig. 25.1. Transient skin pigmentation at copepodid attachment site in Atlantic salmon appears within hours to days post-infection.

temperatures. The developmental pattern as a function of temperature has been studied in a number of publications and recently in two studies spanning a wide range of temperatures in free-living larvae (Samsing *et al.*, 2016) and on fish (Hamre *et al.*, 2019). Minimum and maximum thresholds for accomplishment of the lifecycle (development of adult females) did not deviate from the temperatures that allow for growth of Atlantic salmon, but the potential for new infections is significantly lowered below 5°C as development is slow, egg production low and infection success low (Samsing *et al.*, 2016; Brooker *et al.*, 2018; Hamre *et al.*, 2019; Table 25.2). At >21°C (i.e. at 24°C) infections do not develop through to adult life stages.

Following finding an appropriate host and initiation of feeding, the parasite moults into a chalimus I stage larvae and maintains its attachment to the host through a frontal filament (FF). Production of an FF, or the attachment structure in *L. salmonis*, is tightly linked to the moult cycles and is composed of an external FF and a basal plate, formed from a byssus thread-like material secretion that attaches to the fish skin. FF extrusion appears to be similar across caligid copepods; however, it appears that *L. salmonis* produces a new FF with each moult to a subsequent chalimus stage, whereas *C. elongatus* is reported to produce only one FF of which the length is increased at each moult through the addition of new material (Johnsons and Albright, 1991; Gonzalez-Alanis *et al.*, 2001). Pre-adult/adult stages of the *Caligus* spp. can be easily discerned from *L. salmonis* based on the presence of lunules, and from each other based on genital complex shape and numbers of setae on the distal margin of the exopod of the first leg (Jones and Johnson, 2014). Similarly, adult *C. clemensi* can be differentiated from pre-adult/adults of the two *Lepeophtheirus* spp. by the presence of lunules, and pre-adult/adults of the two *Lepeophtheirus* spp. can be distinguished from each other based on the position of the large spine on the exopod of the third leg (Jones and Johnson, 2014). Finally, *C. curtus* and *C. elongatus* can be distinguished at the adult stage by differences in the shape of the genital complex and abdomen, differences in the setae on the distal margin of the exopod of the first leg, as well as differences in the number of setae on the exopod of the fourth leg (Parker, 1968; Kabata, 1979; Margolis and Kabata, 1988; Piasecki, 1996; Jones and Johnson, 2014).

Table 25.2. Time of infection to first appearance of different life stages of *Lepeophtheirus salmonis* over 8°C temperature range. (Data presented with permission from Hamre *et al.*, 2019.)

	Time from infection to first appearance (days)			
Temperature (°C)	Chalimus 2	Pre-adults	Adult males	Adult females
4	30	47	77	98
12	8	12	20	25
20	4	6	9	12

25.3 Expected/Potential Impact on Sea Lice Demographics

A recent study by Klinger *et al.* (2017) suggests increases in global ocean temperatures of 2–5°C by the end of the century, based on IPCC (2014) models, that could be twice as high in some coastal environments than others (Pinsky and Byler, 2015; Saba *et al.*, 2016). This will likely cause increased production for salmon aquaculture towards polar and other regions but may also incentivize offshore aquaculture production in salmon. In eastern Canada and Maine, USA, which produces >50,000 tonnes of Atlantic salmon per annum and where wild Atlantic salmon populations are threatened (i.e. extirpated from the majority of their natal rivers and the Gulf of Maine distinct population segment has been listed as endangered since 2000), the highest scenario predicted warming over the next century (RCP8.5 for 2055–2099; USGCRP, 2017) nearly double (3.5–4°C) that of the Pacific coast of North America (2–2.5°C). On the west coast of South America, the opposite trend has been observed. In northern and south-central Chile, cooling of 0.2°C per decade has been observed since 1970 (Falvey and Garreaud, 2009), linked to intensification of coastal wind stress and upwelling of cool, nutrient-rich water from depth; and yet further south, an increase more similar to that in polar regions might be observed (Yanez *et al.*, 2018). In all cases, global climate change (GCC) appears likely to have different regional impacts and therefore will also have different impacts on aquaculture and fisheries populations that serve as hosts for *L. salmonis* and other sea lice species, as well as directly on the lice themselves. Of course, climate change impacts go beyond shifts in temperature, and for estuarine and marine organisms also include changes in winds and ocean currents, rain runoff, nutrient supply, salinity and other ocean chemistry such as pH.

25.3.1 Temperature effects on *Lepeophtheirus salmonis*

As discussed above (Section 25.1), *L. salmonis* are naturally occurring marine parasites and their prevalence, abundance and development are influenced by temperature, salinity and other environmental conditions. Because of this, even though major differences exist between Pacific and Atlantic salmon wild and farm populations, and the subspecies of *L. salmonis* as well, general trends in their abundance and prevalence can be similar. Historically, farms in the BoF, Atlantic Canada, begin to observe declines in *L. salmonis* abundance starting in November/December and this continues until approximately June (reviewed in Chang *et al.*, 2011; ACFFA, 2017). Furthermore, chalimus stages are least abundant in farmed salmon in this area between February and May, suggesting recruitment and development are hampered through these low-temperature months (<4°C; Fig. 25.2A and B). Chalimus stage abundance increases thereafter or is highest from June to September/October. In British Columbia, Pacific Canada, increases in lice abundance are generally observed to begin in the summer reaching their maximum in the autumn, and generally decrease from late autumn to early spring (reviewed in Saksida *et al.*, 2011; DFO, 2019; Fig. 25.3A and B). Average abundances tend to be much higher with *L. salmonis salmonis* compared with *L. salmonis oncorhynchi* (ACFFA, 2017; DFO, 2019) and seasonality of farm abundances is also impacted by management and intervention strategies that include anti-louse treatments (discussed in Section 25.5), especially where thresholds have been set for lice numbers to ensure they are maintained at low levels from 1 March to 30 June. These dates are concurrent with wild juvenile salmon outmigration in the Pacific (initially set for pink salmon, *O. gorbuscha*). As implied by these requirements for treatment in the Pacific, there is seasonal variation in the interactions between wild and farmed fish. Apart from the outmigration of smolts in the spring, adult wild salmon return to spawn in late summer and autumn, carrying high numbers of *L. salmonis oncorhynchi* and *C. clemensi* (Beamish *et al.*, 2005), at the same time that highest lice loads are observed on farms.

In Atlantic Canada, a portent of potential warming impacts on the BoF salmon-farming region (Fig. 25.2A and B) has already been observed over the last 6–7 years (2012–2018). Average water temperatures reaching record highs (16–18°C) in some bay management areas (BMAs) (i.e. BMA 1), and average temperatures (>14°C) in the region being maintained for extended periods in the summer (i.e. 2010, 2012, 2017, 2018), have been coupled with temperatures >8°C extending into December and minimum winter temperatures at or above 4°C; all in stark contrast to historic temperatures of the mid-to-late 1990s (ACFFA, 2017; Fig. 25.2C). These extended warm periods have been estimated to reduce the time to development of *L. salmonis* to the female adult stage by 60% (ACFFA, 2017). As identified by Hamre *et al.* (2019), maintaining minimum temperatures above 5°C throughout the year is an important physiological threshold for recruitment success in the initial infectious stage of *L. salmonis*, resulting in the potential for new infections to occur throughout the year without the historical 'winter fallow' afforded in some regions. As can be observed (Fig. 25.2B), since 2014/2015 the relative lice counts have not been dropping as low during the winter season as in previous years.

Reduced generation time and extended periods of infectious pressure will likely be accompanied by greater reproductive output from *L. salmonis* as well. The reproductive output from individual females is determined by the number of eggs and egg deposition frequency, both of which are regulated by temperature. Furthermore, the ability to survive and infect new fish will determine infection pressure. The number of eggs is relatively constant between 6 and 15°C, with the highest number at 6°C (600 eggs per female). At 3 and 18°C much lower production is observed (approximately 300 per female) (Samsing *et al.*, 2016). Heuch *et al.* (2000) showed that females produced new egg strings every 11 days at 7.2°C but that this was cut in half to 5 days at 12.2°C. Further studies of egg production at 6, 12 and 18°C showed a frequency of 17, 6 and 4 days to produce new egg strings, respectively (Hamre *et al.*, 2019). As adult stages do not

(A)

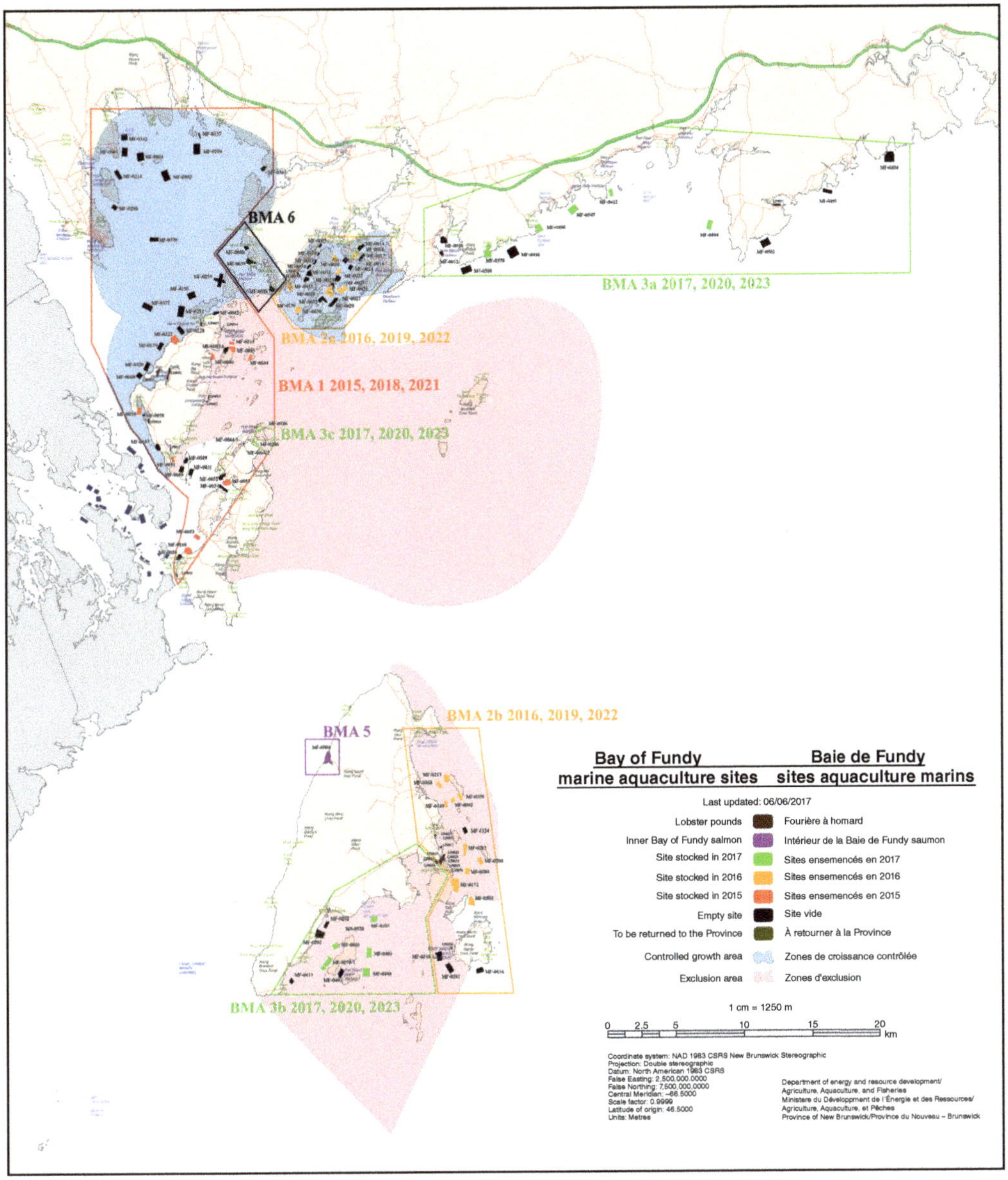

Fig. 25.2. (A) Aquaculture bay management areas (BMAs) of New Brunswick, Bay of Fundy, Canada, as of spring 2019. (B) Rolling average sea lice counts (avg_mob, average number of mobile lice (combination of pre-adult stages and adult males); avg_af, number of adult females) and average surface water temperatures in New Brunswick, Bay of Fundy, Canada, salmon farms from 2009 to 2018. (C) Average surface water temperatures for Bay of Fundy, Canada, from 1995 to 1998 (recreated using data from Chang *et al.*, 2011).

Continued

(B)

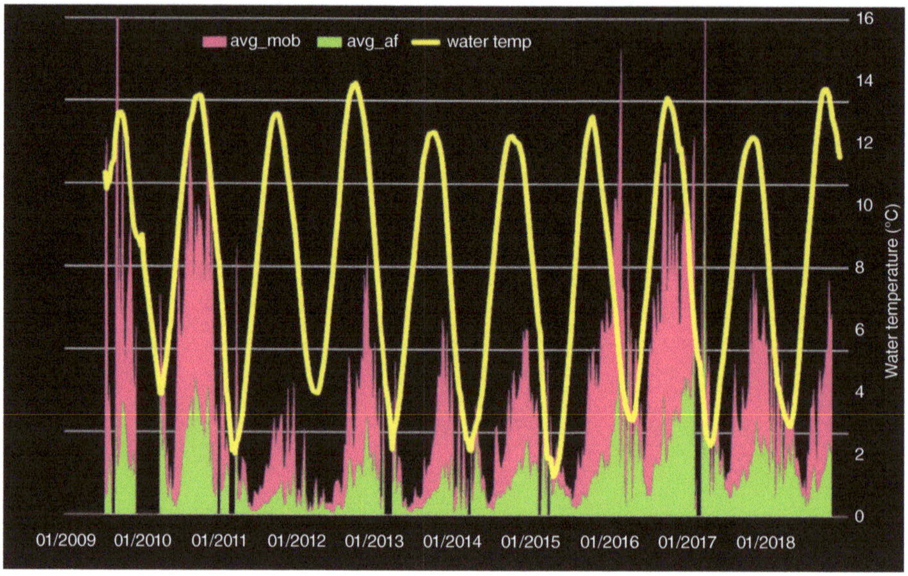

(C)

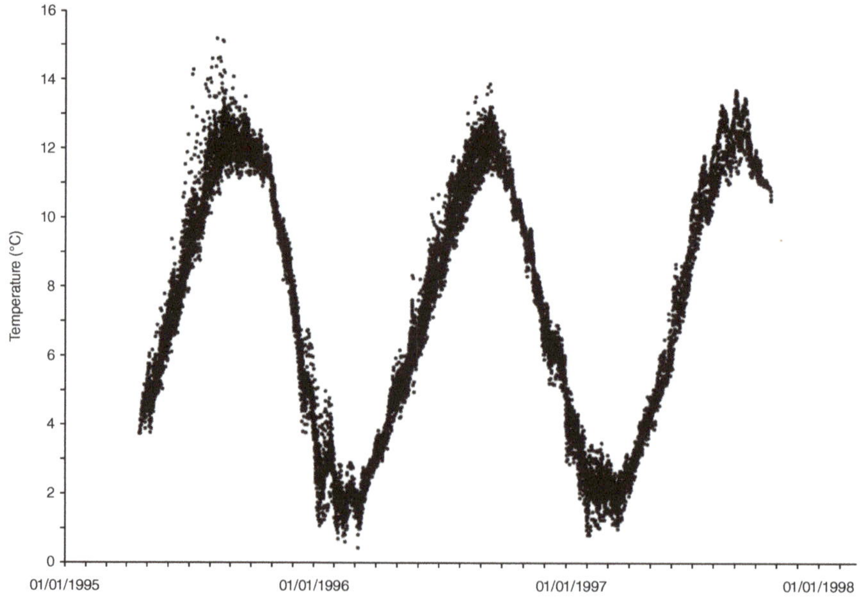

Fig. 25.2. Continued

(A)

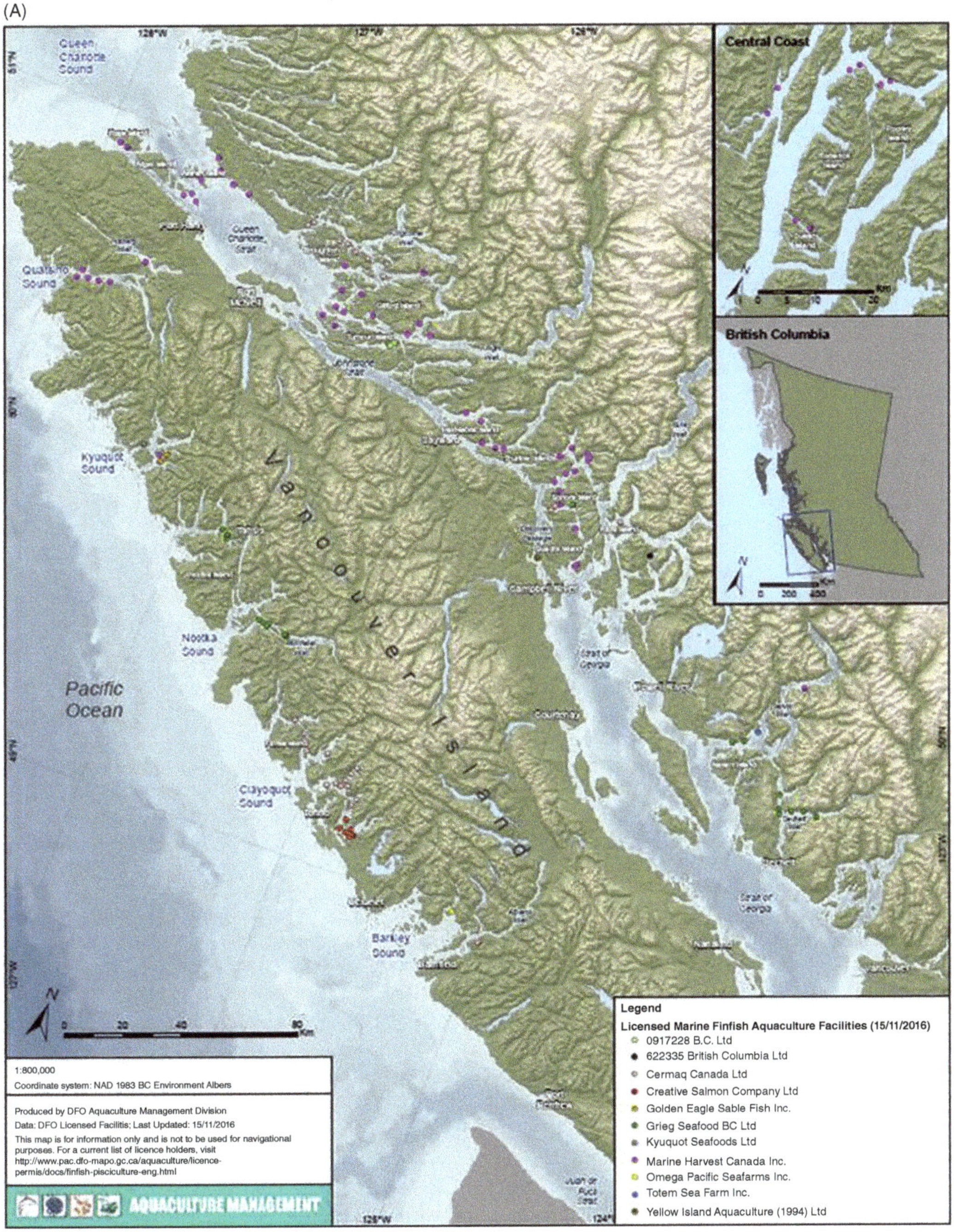

Fig. 25.3. (A) Marine finfish aquaculture sites in British Columbia, Canada, as of spring 2019. (B) Sea lice, *Lepeophtheirus salmonis*, abundance at British Columbia salmon farms in fish health zones 3–5, 2011 to 2019. (From DFO, 2019.)

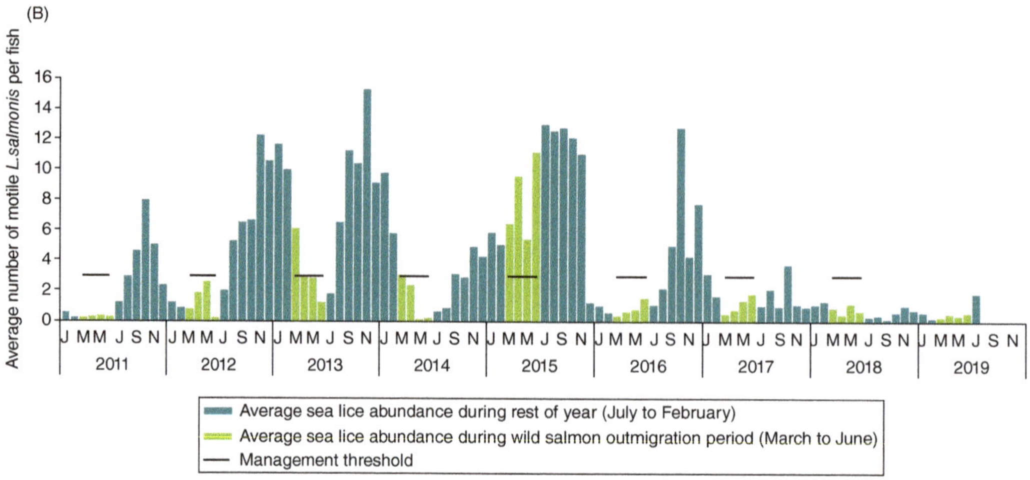

Fig. 25.3. Continued.

develop at 24°C in *L. salmonis*, we would expect that egg string development would likely begin to slow and eventually stop between 18 and 24°C, but there are currently no data on egg development that have been collected at these temperatures. The developmental time to copepodids and overall survival times decrease with lower temperatures, reducing the potential for long-distance spread of the infection. Development to copepodites was demonstrated only at temperatures above 3°C and the infection success is low at low temperatures (Samsing *et al.*, 2016; Brooker *et al.*, 2018; Hamre *et al.*, 2019).

Although the temperature increments may be different in other regions with wild and farmed salmonids, such as British Columbia, Scotland, Ireland, the Faroe Islands, Norway and southern Chile (*C. rogercresseyi*), increased water temperatures in coastal areas will lead to reduced generation time, increased fecundity and extended infectious pressure of *L. salmonis*. These changes would likely increase the economic strain caused by *L. salmonis* in Atlantic salmon-farming regions, as they would likely require moving further offshore to reduce exposure or increased therapeutic and mechanical interventions to remove the lice (Abolofia *et al.*, 2017; Section 25.6). As discussed below (Section 25.4), potential improvements in growth/performance of salmon in sea cages could reduce time at sea, balancing this equation. Unfortunately for wild hosts, these anthropogenic adjustments are not possible, and may lead to greater ecological strain on these populations at outmigration and return to natal rivers. Furthermore, increased temperature at northern latitudes may expand the range of *L. salmonis*' interaction with other salmonid populations and species such as ocean-going Arctic charr (*Salvelinus alpinus*). In potential areas of cooling, such as central and northern Chile, impacts on local species of sea lice, *C. rogercresseyi*, would have the opposite effect. Montory *et al.* (2018) showed that a reduction from 18 to 6°C in culture conditions significantly extended the incubation time of *C. rogercresseyi* by 50% and greater impacts were observed in the planktonic stages, increasing development times to 50 and 100 h for nauplii I and II, respectively, when temperature was decreased to 6°C.

Temperature variation may have major indirect impacts on *L. salmonis* spread and demographics as well. Hutchings *et al.* (2011) reviewed the impact of climate change on fisheries, aquaculture and marine biodiversity, identifying that species biodiversity will likely gain in lower latitudes and throughout the Arctic, as marine fish and invertebrates respond to increasing temperature by shifting northwards and also to greater depths (Cheung *et al.*, 2011). In particular, free-living copepod communities in the North Atlantic and European shelf seas have shown northward extension (1980s–2000s) of more than 10° latitude, whereas the colder/temperate subarctic and arctic species have declined (Beaugrand *et al.*,

2002). Increased diversity of fish species into the temperate climates exploited by *L. salmonis* would increase the likelihood of introducing ectoparasitic copepod species that can compete for host space/niches. Increased diversity of fish and invertebrates in these environments may also increase the potential predation of planktonic (i.e. juvenile salmon and many other fish are often planktivorous) and even parasitic stages (wrasse, lumpfish and other species will feed on adult sea lice stages of many species) of *L. salmonis*. As mentioned above, within the Pacific *C. clemensi* frequently co-occurs with *L. salmonis oncorhynchi*, and in the Atlantic *C. elongatus* co-occurs with *L. salmonis salmonis*. In both cases, these infections on salmon aquaculture sites and some wild populations are transient and often associated with migration/residency of mackerel and herring within the vicinity (Lees *et al.*, 2008; Beamish *et al.*, 2009; Saksida *et al.*, 2011). While competition between these species has yet to be quantified, in both Scottish and Canadian farms greater relative prevalence of *Caligus* spp. occurs in the first year of production with *L. salmonis* making up the majority of lice by the end of the first and during the entire second year of production. In the absence of aquaculture, large numbers of Pacific herring (*Clupea pallasii*) in the Strait of Georgia, British Columbia in April–June have also resulted in high prevalence of *C. clemensi* in Pacific salmon juveniles. Once the herring population moved out of the region, and the source of infection was removed, *C. clemensi* prevalence dropped in the salmon populations in the area. These migration and residency patterns are influenced by temperature. In the Norwegian Sea ecosystem, colder years are associated with better survival in adult herring (*Clupea harengus*), whereas warmer years have been associated with better juvenile survival, and due to difference in feeding grounds, both could potentially be linked to greater food abundance (planktonic copepod, *Calanus finmarchicus*) (Engelhard and Heino, 2006).

Louse species with greater dispersal in open ocean, possibly through adult life stages, like the cosmopolitan parasites *C. elongatus* and *C. clemensi*, may have a selective advantage in some areas over those that rely more heavily on dispersal through eggs, nauplii and copepodid stages (*L. salmonis* and perhaps *C. rogercresseyi*). Finally, increased Arctic Ocean temperatures may also facilitate species invasions from the Pacific into the North Atlantic. Invasions originally occurred around 3.5 million years ago but these populations were cut off from each other, leading to divergence of *L. salmonis* subspecies and their co-evolution with salmonid hosts in the two oceans. Warming temperatures could reintroduce them and lead to 're'-hybridization of the subspecies since they are reproductively compatible (Skern-Mauritzen *et al.*, 2014). As each species has co-evolved with different host species for extended periods, a hybrid subspecies may be able to exploit different susceptibilities across the different hosts.

25.3.2 Salinity effects on *Lepeophtheirus salmonis*

Salinity is another major determinant of development and survival in *L. salmonis*. Accompanying the increased temperatures expected from GCC is a decrease in rainfall and runoff and increased salinity in many of the coastal environments (west coast of Norway is an exception; Fig. 25.4) utilized by salmon aquaculture and exploited by *L. salmonis*. Other ectoparasitic copepods also show strong relationships to temperature and salinity based on their normal physiological ranges. The anchor worm, *Lernaea cyprinacea* (Copepoda: Cyclopoida),

Fig. 25.4. Norwegian aquaculture production zones as of spring 2019.

normally found in completely freshwater systems, exhibits increased infestation levels of cyprinid hosts at higher temperatures, within the physiological optimal range of 23–30°C for this parasite (Sanchez-Hernandez, 2017). Welicky *et al.* (2017) showed that hypersalinity under drought conditions was able to effectively eliminate *L. cyprinacea*. Tolerance to low salinity is an issue for *L. salmonis*, with the planktonic stages being more susceptible to salinity perturbations than the parasitic and especially adult life stages (Wright *et al.*, 2016). Adult *L. salmonis* have been recorded on wild salmon in freshwater rivers several weeks after river entry (reviewed in Wagner *et al.*, 2008) and showed the ability to survive low salinity (7–28 ppt) for up to 7 days in controlled aquaria (Connors *et al.*, 2008). Significant variability also exists within different family backgrounds of *L. salmonis salmonis* with respect to thermal and haline sensitivity (Ljungfeldt *et al.*, 2017). For these reasons, observations of louse behaviour suggest they may actively avoid waters with salinity <20 ppt (Heuch *et al.*, 1995; Crosbie *et al.*, 2019) and lead to modified host behaviour in which infected fish, such as sea trout (*Salmo trutta*), seek and reside for extended periods in low-salinity waters such as estuarine surface waters or river water (Gjelland *et al.*, 2014). Johnson and Albright (1991) showed no hatching of *L. salmonis oncorhynchi* eggs at 10 ppt (10°C), whereas at 15 and 20 ppt this increased to 70 and 78%, yet still <20% of nauplii were active. Increased hatching and activity were observed with increasing salinity at >25 ppt. Within the infectious copepodid stage of *L. salmonis oncorhynchi*, hyposalinity exposure resulted in large-scale gene expression changes in concordance with this salinity gradient, such that minimal changes occurred at 28–29 ppt and a threshold of response was identified at 27 ppt, with the largest response at 25 ppt (Sutherland *et al.*, 2012). Similarly, *L. salmonis salmonis* has also shown increased hatching success, from 3 to 80%, following a gradient of 0 to 30 ppt salinity, and negative effects on copepodids were manifested at salinities <27 ppt (Bricknell *et al.*, 2006). Furthermore, several hours at 26 ppt severely compromised survival and infection potential, and rapid mortality was achieved below 12 ppt (Bricknell *et al.*, 2006). These studies suggest that regardless of the subspecies, *L. salmonis* development and spread will be favoured by increased salinity associated with GCC. In *C. rogercresseyi*, the same appears to hold true in that salinities between 26 and 32 ppt result in a hatching success of 100%, whereas lower salinities (14 ppt) reduced hatching success by 60% (Montory *et al.*, 2018). Finally, the same is likely true regarding potential competition for *L. salmonis* as discussed with thermal increases, in that evidence suggests that cosmopolitan marine parasites like *C. clemensi* and *C. elongatus*, which have likely been naturally selected for a more marine lifestyle (>33 ppt) based on their host range, may have a lower tolerance than *L. salmonis* (i.e. specific to anadromous hosts) for low salinity. In Muchalat Inlet, British Columbia, Elmoslemany *et al.* (2015) observed very low *C. clemensi* numbers on salmon (>95% *L. salmonis oncorhynchi*), whereas other farming regions contained a more mixed infection, and associated this with low salinity of the area. *C. elongatus* has also displayed significant relationships based on temperature and salinity, showing significantly lower prevalence at both low temperature and low salinity (Heuch *et al.*, 2002). The lack of chalimus stages on sea trout and the general abundance of *C. elongatus* related to salinity are also findings recorded by others (Landsberg *et al.*, 1991; Mo and Heuch, 1998).

25.3.3 Acidification effects on *Lepeophtheirus salmonis*

There are numerous studies (listed above) that have contributed to our knowledge of *L. salmonis*, and related sea lice species, and their responses to temperature and salinity; however, the literature is lacking in terms of other environmental variables that will be impacted by GCC, most importantly ocean acidification (OA). To project impacts of OA on *L. salmonis*, we use other crustacean physiological studies as a model for our understanding. In some cases, OA has been studied in combination with or separate from increased water temperatures.

In the American lobster (*Homarus americanus*), thermal acclimatizaton allows for carbon dioxide (CO_2), pH and bicarbonate (HCO_3^-) to adjust within the haemolymph; however, acute exposure to elevated CO_2 and subsequently reduced pH can result in metabolic acidosis and impact cardiac performance (Qadri *et al.*, 2007). In the GCC-OA situation these environmental changes are expected to be felt chronically and over the long term. Ability to compensate for OA varies across crustaceans, but for the most part, compensation is better understood in adults and is affected by the ability to compensate for acid–base disturbances (Whiteley and Taylor, 2015). Through the elevation of HCO_3^- ions, crustaceans can buffer changes in pCO_2 and pH in the haemolymph. The inability to elevate HCO_3^-, in cases of reaching physiological limits, has been associated with acidosis, hypoxaemia and eventually death (Whiteley and Taylor, 2015). Predictions indicate a reduction in ocean pH, due to rising atmospheric CO_2 levels and the subsequent increase in ocean CO_2 absorption, to 7.7–7.8 by the end of the century (i.e. 2100), down from the current pH 8.1, and a further decrease to pH 7.4 by 2300. Whiteley's (2011) review on end-of-the-century pCO_2 predictions suggested little impact on populations and the ability of crustaceans to cope. However, while having little impact on egg development, predicted 2300 pH levels affected embryonic development and, when combined with elevated temperatures, showed adverse effects on calcification rate, growth rate and moulting frequency over the medium term (i.e. exposure of weeks). High-activity crustaceans such as those that make diel movements in order to feed or to encounter hosts, like *L. salmonis*, have higher levels of the non-bicarbonate buffer haemocyanin and greater ability to buffer these acid–base disturbances to cope with OA. The nauplii and copepodid stages in *L. salmonis* would likely be the most at risk of OA because these parasitic stages utilize the host for ion regulation and maintenance of acid–base balance as they are osmoconformers.

While we do not know the relative levels of haemocyanin in *L. salmonis* haemolymph, planktonic (naupliar) and infective copepodids, pre-adults and adult stages, all likely would be considered high-activity crustaceans and therefore more likely to be able to cope. Unfortunately, we have no long-term data at elevated temperature or reduced pH on crustacean species of importance (i.e. free living or parasitic), so even for those able to cope with medium-term chronic exposure, we must expect there will be some physiological limit or threshold at which long-term chronic exposure begins and some populations to experience major deleterious effects.

25.4 Expected/Potential Impacts on Host (Reservoir, Intermediate and Definitive) Populations

The work by Hutchings *et al.* (2011) and Klinger *et al.* (2017) suggests that both wild and cultured salmonids, the major host for *L. salmonis*, will observe expansions northward into new areas, and potentially deeper waters, due to the increased temperatures in the temperate, coastal marine environment resulting from climate change. Potential for expansion of reservoir and/or as yet undiscovered intermediate hosts for *L. salmonis* also exists. Moreover, these shifts will generally be favourable for the parasite's development and fecundity to a maximal temperature threshold (<24°C). How these factors will impact the host must be examined against a backdrop of current parasite effects on the host, the incremental effects that increased temperature and OA will have on host immunological competence, as well as any population-level improvements that might occur specifically against *L. salmonis* (i.e. selective breeding, anti-louse vaccines, etc.).

L. salmonis are found naturally on all ocean-migrating salmonids, with greatest intensities of infection and impacts of *L. salmonis salmonis* occurring on sea trout (*S. trutta*) and Atlantic salmon (*S. salar*). In these two salmonids, infection is characterized by absent, weak or delayed wound healing and immunological responsiveness (i.e. parasite-localized inflammation) and maintenance of high parasite burden. Although sea trout have been shown to harbour greater infection intensities under controlled laboratory exposure (Dawson *et al.*, 1997), the susceptible genotype/phenotype is best described and reviewed for Atlantic salmon infected with *L. salmonis salmonis* (Wagner *et al.*, 2008; Fast, 2014; Poley and Fast, 2020). Maintenance of high parasite burden results in development of skin lesions, focused just posterior and lateral to the head region and directly posterior to the dorsal fin. Initial discoloration/melanization (Fig. 25.1) of the attachment site at the copepodid generally disappears after a few days; and small focal areas of discoloration (white), sometimes observed at the late chalimus stage, begin to coalesce and deepen – in extreme cases extending into the dermis and underlying musculature – as the parasite moults through the pre-adult to the adult life stages. The inability of the host to mount significant inflammation and re-epithelialization of the wound at the attachment site during the chalimus-stage infection is epitomized by reduced pro-inflammatory signals and complete removal of epithelium at the attachment site, and worsening lesion development over time (Fig. 25.5A–C). Area, degree and depth of ulceration worsen as the host is unable maintain structural integrity of the skin

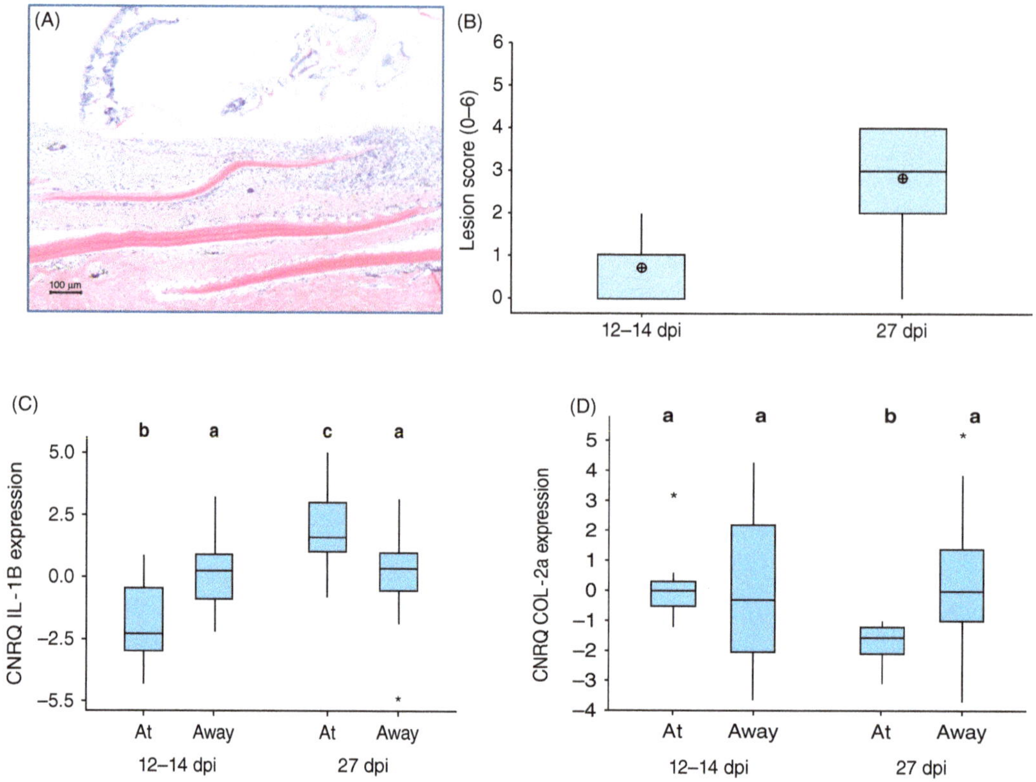

Fig. 25.5. (A) Chalimus-stage larvae of *Lepeophtheirus salmonis salmonis* infection site on Atlantic salmon skin. (B) Posterior to dorsal fin lesion development over the course of *L. salmonis salmonis* infection (12–14 days post-infection (dpi), chalimus II larvae; 27 dpi, pre-adult/adult transition) at 11°C (scoring system of Johnson *et al.*, 1996). (C) Atlantic salmon skin $\log_2$ calibration normalized relative expression (CNRQ) of the pro-inflammatory mediator interleukin-1β (IL-1β) 'at' the site of lice attachment and adjacent to that section ('away') over the course of infection. (D) Atlantic salmon skin $\log_2$ CNRQ of the extracellular matrix protein collagen 2a (COL-2a) 'at' the site of lice attachment and adjacent to that section ('away') over the course of infection. Results presented as box-and-whisker plots of median CNRQ, with 1st and 3rd quartiles represented within the box, whiskers representing the minimum and maximum values, and asterisks denoting outliers; letters (a, b, c) denote differences between salmon skin sites over time (two-way analysis of variance and Tukey's honest significant difference test, $p < 0.05$). (All figures presented with permission from MacDonald *et al.*, 2015.)

likely through the reduction of collagen production (Fig. 25.5D) and incorporation into the extracellular matrix.

The subspecies *L. salmonis oncorhynchi* has a greater host species range than subspecies *L. salmonis salmonis* potentially due to its greater diversity of salmonid hosts, but also the inclusion of a reservoir host, the three-spined stickleback (*G. aculeatus*). Atlantic salmon displays the susceptible genotype/phenotype against both subspecies of *L. salmonis*; however, studies examining Pacific salmonids are generally confined to infections with *L. salmonis oncorhynchi*. In this case sockeye salmon (*O. nerka*) and chum salmon (*O. keta*) also display susceptibility to *L. salmonis* infection, leading to lesion development and mortality in described for Atlantic salmon in extreme cases (Johnson *et al.*, 1996; Jones *et al.*, 2007; Braden *et al.*, 2015a,b). Rainbow trout (*O. mykiss*) and chinook salmon (*O. tshawytscha*) typically show a lower degree of susceptibility than those previously mentioned; however, pink salmon (*O. gorbuscha*) and especially coho salmon (*O. kisutch*) display a highly resistant phenotype characterized by well-developed inflammation at the attachment site, rapid parasite rejection and limited additional pathology (Johnson and Albright, 1992; Fast *et al.*, 2002; Jones *et al.*, 2007). In coho salmon, a strong inflammatory response, epithelial hyperplasia and eventual parasite encapsulation result in >90% reduction in lice loads in less than 2 weeks (Johnson and Albright, 1992). Pink salmon responses to early stages of *L. salmonis oncorhynchi* infection are also characterized by strong inflammatory signals, iron sequestration and C-type lectin pattern recognition induction (Fast, 2014).

Less is known on the impacts of, and responses towards, *L. salmonis* in potential reservoir hosts compared with the definitive hosts described above. To date, only the three-spined stickleback has been reported to harbour *L. salmonis* infections and this is described only for the *L. salmonis oncorhynchi* subspecies, despite extensive examination of three-spined sticklebacks in the North Atlantic (Jones *et al.*, 2006a; Jensen *et al.*, 2016). In this case, despite a high prevalence of infection in the field and significantly higher parasite abundances than on juvenile pink and chum salmon under controlled laboratory infections, three-spined sticklebacks do not appear to provide a suitable environment for completion of the life cycle or even development from attached chalimus to pre-adult life stages (Jones *et al.*, 2006a,b). This has raised the question of whether this species may actually act as a 'sink' or a 'dead-end' host (Losos, 2008).

Extended periods of high temperature favour parasite development and fecundity, albeit to a maximal temperature threshold (<24°C), and the same is generally true for salmon innate and adaptive immunity. Nikoskelainen *et al.* (2004) showed phagocytosis, respiratory burst, opsinization capacity and lytic activity of both total and alternative complement pathways all peaked or were enhanced in rainbow trout at 15 and 20°C compared with lower temperatures. Robledo *et al.* (2018) showed a positive correlation between complement-activating pathways and sea lice infection in resistant versus susceptible Atlantic salmon families. Furthermore, it is well described that poikilothermic vertebrates, more specifically coho salmon (Paterson, 1971), rainbow trout (Fryer *et al.*, 1976) and sockeye salmon (Alcorn *et al.*, 2002), have faster and more pronounced antibody production at higher temperatures within their physiological optimal ranges. The importance of complement activation and antibody-mediated immunity is well known in vertebrates, being shown specifically in vaccinated charr by Braden *et al.* (2019) and reviewed in Fast (2014) with respect to their association with sea lice and other ectoparasite infections. Finally, significant reductions in lice infection in vaccinated Atlantic salmon are associated with IgT^{+} cell trafficking (Y. Carpio, S.L. Purcell, F. Takizawa, Y. Leal, J. Velazquez *et al.*, 2020, unpublished results).

Temperature increase up to 20°C appears to favour both the parasite and the host's immune system intrinsically; yet despite our understanding of poikilotherm/salmonid immune systems, we have a major knowledge gap with respect to a breakpoint temperature at which either the parasite or the host seizes the advantage in the salmon–lice relationship. While this may be overly speculative in terms of the host or parasite gaining an advantage at the extreme temperatures predicted under GCC, other impacts of GCC would lead to faster growth in freshwater and seawater phases of the salmon life cycle, which could lead to earlier outmigration to sea in smolts, earlier returns of adults and reduced seawater grow-out times in aquaculture. The net impact of these on the host–parasite relationship is unknown, but some of the potential implications are discussed below.

The responsiveness of salmonids to altered environmental cues, such as earlier rises in temperature in the spring or different timing of snowmelt resulting in changes in estuarine/riverine salinity, will likely impact their residence times and migration routes, as discussed briefly in Nekouie *et al.* (2018). Changes in Arctic migrations have already begun to occur (see Section 25.3) and reduced polar ice/increased ocean temperatures of the last few years in the Arctic have been associated with pink salmon stocks of Kola Peninsula (Russia) origin migrating into Norwegian rivers. In 2017, more than 6000 fish were caught in 263 rivers (Hårdensson and Uglem, 2019). Successful spawning was also documented in several rivers. Pink salmon susceptibility/resistance to *L. salmonis salmonis* is unknown but despite the resistant phenotype to *L. salmonis oncorhynchi* in juveniles >2.5 g, returning mature pink salmon carry large numbers of *L. salmonis oncorhynchi* and *C. clemensi* into

Pacific coastal waters (Beamish *et al.*, 2005) and exhibit a susceptible phenotype (Braden *et al.*, 2015a). This could increase infection pressure from multiple species and subspecies of sea lice on already threatened wild populations of Atlantic salmon and sea trout.

Increased CO_2 leads to ocean as well as freshwater acidification, and both wild and hatchery populations of salmonids are more sensitive to acidic water during the parr–smolt transformation than at other stages (Rosseland and Skogheim, 1984; Staurnes *et al.*, 1993). Chronic low pH has been shown to impair seawater readiness (through impacts on Na^+, K^+ and ATPase) and reduce survival in Atlantic and Pacific salmonids (i.e. sockeye) (Staurnes *et al.*, 1993; Kennedy and Picard, 2012). Sub-lethal effects of low pH in the freshwater environment may reduce the ability of these already susceptible hosts to survive infection with *L. salmonis* and may also impact more-resistant host species. Juvenile pink salmon exposed to CO_2-induced acidification both during freshwater stages and at early seawater entry show dose-dependent reductions in growth, among other indirect stress responses (Ou *et al.*, 2015). Pink salmon has the highest specific growth rate of all Pacific salmonids (Parker, 1969) and early growth in pink salmon is strongly related to resistance to *L. salmonis oncorhynchus* infection, such that 0.3 g pink salmon show poor scale development, lack of cellular proliferation and high levels of mortality when exposed to copepodid-stage lice (Jones *et al.*, 2008b; Sutherland *et al.*, 2011). As the fish grow to 0.7 and 2.4 g in size, scale formation and cell-mediated immunity appear to mature and are associated with little to no mortality. Therefore, freshwater, estuarine and seawater acidification could have implications for the development of pink salmon smolts and their subsequent resistance to *L. salmonis* infection. Furthermore, as pink salmon smolts enter the marine environment they feed heavily on phytoplankton/zooplankton, whereby phytoplankton spring blooms are a strong predictor of pink salmon productivity (Malick *et al.*, 2015). These authors suggest climate change could cause latitudinal shifts in salmon productivity, as generally stated by Hutchings *et al.* (2011), but short-term GCC impacts on food availability could also impact pink salmon growth and potential anti-parasitic responses, as prior work on reduced feed rations in resistant pink salmon life stages showed delayed inflammatory responses to challenge (Jones *et al.*, 2008a). Finally, production of essential fatty acids within phytoplankton assemblages could be negatively impacted through the process of 'homeoviscous adaptation', or adaptation of membrane lipid composition to maintain fluidity (Hixson and Arts, 2016). Phytoplankton contribute greatly to long-chain polyunsaturated fatty acids (LCPUFAs) in aquatic organisms and warming temperatures have been shown to shift LCPUFAs from *n*-3 to *n*-6 fatty acids (Hixson and Arts, 2016). Again, this could impact inflammatory response/regulation, important in the resistance mechanism to lice, in these more sensitive life stages of salmon.

25.5 Control and Prevention of *Lepeophtheirus salmonis*

With the exception of a few field studies involving prophylactic treatment of released sea trout and salmon, control and prevention of *L. salmonis* infection are strictly managed in farmed fish in the hope of preventing cross-infection of wild populations. Section 25.4 discussed the inherent variability of susceptibility to infection across the salmonid lineage, however the introduction of resistance phenotypes to cultured Atlantic salmon has remained an unrealized goal for the industry. Prior studies involving controlled lice challenges in Atlantic salmon exhibited moderate heritability (15–25%) for the number of lice per fish (Gjerde *et al.*, 2011), suggesting potential for the approach, and led to the selection and production of salmon families showing reduced numbers of lice by the Norwegian company SalmoBreed AS (https://salmobreed.no/en/ (accessed 8 April 2020)). However, the traits for sea lice resistance are highly polygenic (Gharbi *et al.*, 2015; Tsai *et al.*, 2016) and still no clinical impacts have been reported in industrial application of these lines to date.

The inability to make gains in resistance through genetic manipulation, coupled with the minor advances in ectoparasite vaccinology (Raynard *et al.*, 2002; Carpio *et al.*, 2011; Y. Carpio, S.L. Purcell, F. Takizawa, Y. Leal, J. Velazquez *et al.*, 2020, unpublished results) and significantly lower relative protection afforded by anti-louse functional feeds, has – until recently – left the salmon farming industry with only chemotherapeutic intervention as an option in the face of management thresholds.

25.5.1 Chemical control methods of *Lepeophtheirus salmonis*

Treatment regimens

The industry production cycle for Atlantic salmon runs from 14 to 24 months (in the sea) with stocking increasing at multiple times of the year so that harvest can match market demand throughout the year. The development of management zones has occurred in all major salmon-farming regions based on multiple factors including oceanographic currents, a rotation system in stocking with a minimum mandatory fallowing per site and per BMA, and separation of successive year classes by BMA. Seasonal trends in lice abundance (winter lows and summer highs) in the 1990s were affected by the use of anti-louse drugs such as azamethiphos, hydrogen peroxide, ivermectin and permethrin. Unfortunately monitoring programmes on lice abundances over time were not maintained and the types, number and efficacy of these listed treatments are not available during this time frame. However, decreases in lice abundance from 1994/1995 to 1996 have been attributed to the use of azamethiphos in Canada (Chang *et al.* 2011), whereas use of organophosphates in Norway, dating back to the late 1970s, began to observe reduced efficacy leading to its termination by 1999. In 2000, emamectin benzoate, sold under the trade name SLICE® as an in-feed treatment, became available to farmers and was virtually the sole drug used in Atlantic Canada from 2000 to 2009 (the only product registered in Canada until 2009). Again, this was to treat infections i.e. with *L. salmonis*, and a few *C. elongatus* infections limited to Nova Scotia. In 2008, reduced efficacy of SLICE was observed, in which only half of the qualifying treatments (17/33) analysed by Jones *et al.* (2012) were established to be effective. An evaluation of SLICE efficacy in nearby Cobscook Bay from 2002 to 2005, using pre-treatment lice loads for reference, showed 68–100% efficacy and a duration of protection of 4–16 weeks (Gustafsson *et al.*, 2007). Conversely, maximum effectiveness in Jones *et al.* (2012) was 75.7% in 2008, or only a 24% reduction in pre-treatment lice load. By 2009 SLICE administration at the established treatment dose (50 μg/kg daily for 7 days) was virtually ineffective.

Emergency access to deltamethrin, AlphaMax®, was granted in BMA 2A in 2009 (using skirts) and in BMA 1 in 2010 (using well-boat administration). During this time, deltamethrin consistently reduced pre-adult and male adult *L. salmonis* in the range of 88–98% compared with pre-treatment levels (Whyte *et al.*, 2014). Cage-level reductions for adult female and chalimus lice stages varied considerably, often at 50% or less. *In vitro* bioassays conducted during this time generated EC_{50} values ranging from 0.20 (95% CI 0.14–0.28) ppb to 2.45 (95% CI 1.80–3.30) ppb, actually higher than those reported in Norway by Sevatdal *et al.* (2005) of 0.09 (95% CI 0.02–0.20) ppb to 1.03 (95% CI 0.57–1.82) ppb, where the drug had been used consistently for over a decade.

The benzoylurea teflubenzuron, sold under the trade name Calicide® as an in-feed treatment impacting non-adult lice during their moult, was also used at this time but only in small amounts. It was followed quickly in 2010 by the approval for use under emergency drug release of hydrogen peroxide (Paramove® 50) administered as a bath in well boats; its usage increased from 2011 to 2015, although high temperature and other factors may have resulted in a decline in 2016 (ACFFA, 2016). Azamethiphos (Salmosan®) was also granted use through emergency drug release in 2010, and usage has been steady since 2013. Gautam *et al.* (2017) evaluated the effectiveness in controlling *L. salmonis* pre-adults and adults using these two bath treatments in Atlantic Canada, using data from 1185 treatment events at 57 farms from 2010 to 2015. These data showed Salmosan administered by tarpaulin had the greatest effect in the summer, with loads above 10 sea lice per fish before treatment. In autumn performance varied greatly. However, most striking was the overall effectiveness regardless of tarpaulin/well-boat application for Salmosan/Paramove 50 treatment, where these baths rarely if ever reached 80% efficacy, regardless of assessment date (i.e. 1–7 days post-treatment), and generally resulted in parasite reductions of 65% or less. Salmosan bioassays conducted on lice during 2009–2012 by Whyte *et al.* (2016) showed increasing EC_{50} values from 2009 to 2012, even when the drug was rarely used at all (in 2011 and 2012).

Particular issues arising during the period of SLICE's lack of efficacy were warming winter trends and frequent infections of 10–20 adult females per fish being maintained in some zones over winter. This occurred again in 2016, causing bath treatments to extend into January 2017 (in BMA 3A and BMA 3B; ACFFA, 2016), and reduced the period of low lice numbers (i.e. break in infection pressure) experienced from January to April that was quite helpful in reducing treatments in the past. Despite the lack of efficacy of SLICE, avermectins/macrocyclic lactones continue to be used as an in-feed treatment, especially within the first year of production in Atlantic Canada (ACFFA, 2016). This involves either ivermectin or SLICE, with the dosage and timing of these treatments likely increased/extended. However, use of these and other chemotherapeutants has reduced markedly in the Norwegian salmon-farming industry since 2016 (Helgesen *et al.*, 2018).

Due to the reduced infection levels of *L. salmonis oncorhynchi* on farmed Atlantic salmon in Pacific Canada, the management and treatment interventions are quite different again from those in Atlantic Canada. A treatment trigger has been set in British Columbia, whereby a maximum threshold of 3 mobile lice per fish, between 1 March and 30 June, requires a treatment or harvest event on the farm whereas infection intensities above this threshold between 1 July and 28 February require elevated monitoring or can involve treatment or harvest. The 1 March–30 June time frame has been set to match the timing of outmigration of smolts from Pacific salmon-producing rivers and reduce the likelihood of lice infections passing from farmed to wild smolts. From 2001 to 2009, the number of treatments for the entire industry ranged from 4 (2007) to 19 (2004) per annum, and after 2004, >75% of these occurred between October and March (Peacock *et al.*, 2012). Since then, the average annual number of treatments has been in the range of 14–17 (DFO, 2016). Until 2014, SLICE was the only treatment used in Pacific Canada. In April 2014, hydrogen peroxide was used for the first time by the salmon production company MOWI (https://mowi.com/ (accessed 14 April 2020)). Since then, treatments have been split between SLICE and hydrogen peroxide.

Basis of Lepeophtheirus salmonis *pesticide resistance*

Several authors have studied the population genetics of *L. salmonis salmonis* in the North Atlantic, and through microsatellite and single nucleotide polymorphism (SNP) analysis reached the same conclusion that it consists of a single panmictic population (Todd *et al.*, 2004; Glover *et al.*, 2011; Besnier *et al.*, 2014). This is significant for connectivity between salmon-farming regions like North America, Scotland and Norway and wild populations off Greenland. More importantly, in the case of pesticide resistance, it provides the opportunity for rare mutations to spread throughout the aquatic environment quickly. Lees *et al.* (2008) were the first to report reduced efficacy of SLICE treatments, in Scotland, showing longer times to achieve efficacy and post-treatment counts increasing from 2003 to 2006. In 2006, abundances reached 35% of pre-treatment levels for days 56–62, whereas they remained above 40% for all other days. As mentioned above, treatment failures were observed in Atlantic Canada in 2008 and in Norway in 2008 (Aaen *et al.*, 2015). Besnier *et al.* (2014) genotyped over 500 *L. salmonis* from regions spread over the North Atlantic (2000–2010) and discovered a cluster of SNPs on linkage groups (LG) 1 and 5 that were under positive selection; their genome-wide association study showed a large region on LG5 that made a significant contribution to the variation in SLICE resistance. This *de novo* set of mutations occurred between 10 and 40 generations previously (in agreement with the treatment failure timelines) and these alleles conveying resistance were shown to spread throughout the North Atlantic extremely quickly (Besnier *et al.*, 2014). A gene coding for cytochrome b5 domain-containing protein was found in this broad LG5 cluster, and b5 is a known activator of cytochrome P450 (CYP450). While the CYP450 enzymes are important in phase I oxidation of toxic molecules and known to be involved in pyrethroid resistance, both a glutathione-*S*-transferase 1 isoform and a UDP–glucuronosyltransferase (2A3) gene, involved in phase II oxidation and overexpressed in houseflies resistant to another macrocyclic lactone (spinosad), were recently found to be overexpressed in resistant BoF male lice exposed to salmon receiving different dosages of emamectin benzoate or ivermectin (Højland and Kristensen, 2017; Whyte *et al.*, 2019). Male *L. salmonis*, which show much higher emamectin benzoate EC_{50} values than females, also show higher expression of the nicotinic acetylcholine receptor (nAChR) α7 and lower expression of nAChR α3 compared with females, as do populations resistant to emamectin benzoate relative to populations sensitive to emamectin benzoate. This profile forms a consensus with Scottish *L. salmonis* data and Pacific *L. salmonis oncorhynchi* data as well (Carmichael *et al.*, 2013; Sutherland *et al.*, 2014; Poley *et al.*, 2015).

Similar to *L. salmonis salmonis* (Atlantic), a 2011 microsatellite study on *L. salmonis oncorhynchi* also found no population structure. However, Saksida *et al.* (2013) reported a Pacific population with elevated EC_{50} to emamectin benzoate.

Like in the Besnier *et al.* (2014) study, Pacific lice samples (n = 478) were genotyped from farmed and wild locations (2005 to 2014), detecting the emergence of a rare genotype of 778 SNPs with a high frequency among lice with elevated EC_{50} values and from populations where SLICE treatment failures had occurred in British Columbia in 2013 and 2014 (Messmer *et al.*, 2018). The results suggested that the rare genotype was locally expanded in farms after SLICE treatment. Of these SNPs, 748 were further located on LG5, which was associated with resistance in Atlantic *L. salmonis* (Besnier *et al.*, 2014). The novel genotype emerged in *L. salmonis oncorhynchi* samples from the Klemtu region in 2013, where 485 of sampled lice made up this cluster and coincided with the first loss of efficacy of SLICE in Klemtu. In 2014, this genotype cluster was reduced in Klemtu and associated with a return of SLICE efficacy, whereas first observations of reduced efficacy were now observed in Quatsino and these coincided with the first presence of the novel genotype in this region. With removal of SLICE treatment and in-breeding of the local louse population with the dominant wild-type genotype, this novel genotype associated with reduced SLICE efficacy has not returned, nor has the phenotype of the reduced efficacy of SLICE. The importance of this genetic region in both subspecies of *L. salmonis* for conferring resistance to SLICE has been confirmed, even if the exact mechanism remains elusive and may still be polygenic.

Perhaps more importantly, the ability of novel mutations to fix in the population and result in continued treatment failures is a drastic contrast between *L. salmonis salmonis* and *L. salmonis oncorhynchi*. McEwan *et al.* (2015) used an agent-based model (ABM) to simulate the spread of a simple co-dominant allele linked to drug resistance through a salmon-farming region over time. Major findings of this work were the impact even relatively small refugia (i.e. population equal to the number of farmed salmon) could have on the development and spread of resistance, as well as the very low fitness cost the mutation would have on this development. The modelling outcomes of populations with refugia paralleled those of the Pacific salmon louse experience, whereby even with little to no fitness cost, these drug-resistant mutations would not fix in the population and numbers of treatments stayed low; whereas this happened rapidly in a region where there were very few to no refugia and treatment frequency escalated quickly, which would parallel the Atlantic Canada experience. Follow-up work by McEwan *et al.* (2016) using ABMs also highlighted the theoretical risk of these types of mutation fixing in the population: regardless of the intervention strategy being mosaic, rotation or responsive, a mutation without fitness cost accumulated over the same time frames. And really only the combination therapy delayed drug resistance in the population.

Perhaps an excellent field example of how these mutations can be difficult to remove from the population was in organophosphate resistance in *L. salmonis salmonis*. Kaur *et al.* (2016) showed the importance of the Phe362Tyr mutation in organophosphate resistance in Norway and its heterozygous presence in the population in 1998, likely from overuse of this treatment through the 1990s (Kaur *et al.*, 2015). The mutant allele, however, was maintained in the population despite a lack of selection pressure between 2000 and 2007 when organophosphates were not used, and spread widely to the point where it showed up in Canadian samples in 1999, 2002 and 2009, without the common usage of azamethiphos in this area (Kaur *et al.*, 2017).

The maintenance of resistant alleles and connectivity of populations across the North Atlantic, as exemplified by emamectin benzoate and organophosphate resistance, may also be a mechanism for the spread of resistant alleles related to other drugs, and explain the poorer than expected performance of deltamethtrin and hydrogen peroxide in Atlantic Canada. Potentially exacerbating the spread and emergence of these rare alleles is the occurrence of assortative mating in *L. salmonis*. Recently completing a high-resolution genetic map for this species, Danzmann *et al.* (2019) found examples of assortative mating, whereby females with relatively high recombination rates were choosing males also with relatively high recombination rates for reproduction. This was potentially reinforced by a reduction in the viability of progeny from parental matings of different intrinsic recombination levels.

25.5.2 Non-chemical control methods of *Lepeophtheirus salmonis*

With the development of resistance to available chemotherapeutants, alternative solutions to prevent or remove *L. salmonis* infections have been introduced. The use of these methods is relatively recent and under continuous development to improve efficacy and reduce negative impacts. Prevention strategies focus on keeping farmed fish away from the infective copepodids (Bui *et al.*, 2016) which, in full salinity, are mainly in the top 5 m of the water column. This can be accomplished by attracting the fish to depth using lights or underwater feeding (Frenzl *et al.*, 2014), by mechanical obstruction of the top part of the cage (snorkel cages; Stien *et al.*, 2016) or by keeping copepodids out of the cage using plankton skirts (Frank *et al.*, 2015). Removal of existing lice infections can be accomplished by the use of cleaner fish (lumpfish and wrasse) that feed mainly on the larger stages of lice from the surface of the fish. The main challenges to this method have been to modify the cage environment to improve the welfare and survival of the cleaner fish, as well as to establish breeding programmes of cleaner fish because the volume of cleaner fish needed makes wild-caught fish likely unsustainable (Skiftesvik *et al.*, 2013; Imsland *et al.*, 2016; Hvas *et al.*, 2018). Mechanical solutions in addition to the use of warm and/or fresh water have also become increasingly popular in Norway. The main problem with these types of treatment is fish welfare problems and mortality (Overton *et al.*, 2018).

A complicating factor with the use of freshwater and/or thermal baths as a treatment for *L. salmonis* is that this may enhance their rate of adaptation to global climate changes. In a laboratory-based pedigree experiment with *L. salmonis salmonis*, Ljungfeldt *et al.* (2017) demonstrated a significant degree of variation to thermal treatments especially and cautioned the potential for enhanced tolerance to this treatment. Finally, evidence from other crustaceans for the implications of thermal evolution may be even more striking. Zhang *et al.* (2018) showed that derived populations of *Daphnia magna* (i.e. under thermal evolutionary pressure) tolerated toxicity to zinc oxide nanoparticles at higher temperatures and suggested that evolution towards warming trends could shape responses to other anthropogenic stressors. In the case of the salmon louse, this would be most impactful in the area of chemical treatments to control the parasite, if thermal evolutionary pressures can enhance cross-resistance.

25.6 Conclusions and Suggestions for Future Studies

In summary, the current trends in global/ocean warming and their elevated impact on coastal areas that salmon and *L. salmonis* inhabit would suggest that continued increases in coastal ocean temperatures will likely lead to expansion into new regions for wild salmon and salmon aquaculture (i.e. northward). However, as has already been observed, increased temperature leads to reduced generation time, reduced overwintering fallow and increased fecundity in *L. salmonis* in many regions. These trends could lead to higher abundances of *L. salmonis* on farmed and wild salmon in most regions and could drive further reliance on chemotherapeutic and mechanical interventions. Development of 'lice avoidance' strategies for salmon aquaculture is likely to continue, and could certainly drive further innovation in cage design, but in the meantime the industry also needs to use more closed and semi-closed cages, move further offshore, and potentially even reduce the growing season of open-cage culture as a whole. The combination of longer land-based hatchery times, larger sizes of smolts entering seawater and semi-closed systems, alongside offshore culture, could eventually eliminate nearshore open-cage culture as we currently understand it. As cannot be stressed enough, the physiological optimal temperatures of this parasite and its salmonid hosts overlap to a significant degree, and we cannot currently predict how the parasite–host relationship will be affected as ocean temperatures increase above 20°C. Presently the host immune system, our ability to bolster it and our other intervention strategies are being outpaced by the ingenuity of this parasite, and there is no reason to expect that to cease. It is not surprising that *L. salmonis* having co-evolved with salmon for millions of years has developed a similar temperature profile; however, the origins of parasitism in this copepod species can be more clearly comprehended by examining not only the shared temporal and spatial existence, but also the importance of phytoplankton to juvenile salmonids – as well as the interactions between them – when fish first enter the ocean environment. It is an oddity that the same GCC mechanisms could enhance food availability to juvenile fish yet also positively select for the planktonic stages of this parasite. Further, that OA which could negatively affect planktonic primary production and reduce growth and/or condition in early-stage salmonids, could also be beneficial with respect to negative impacts on *L. salmonis* infective stages. In the end, this chapter highlights our knowledge gaps in these areas, and that we cannot truly understand the potential outcomes until we have a better understanding of the impacts predicted temperatures in concert with lowered pH will have on lice development and infection pressure in the coastal environment. Furthermore, our understanding of potentially competing ectoparasitic copepod species (*C. clemensi*, *C. elongatus*, etc.) that share these environments must also be examined to determine if future conditions may provide them a selective advantage. We must begin to address the threat of polar ice retreat, inter-ocean migration of hosts and parasites, 're'-hybridization of the two *L. salmonis* subspecies and salmonid hosts entering new environments, by examining the fitness of the host and hybridized parasites under these conditions/environments. And last but not least, we must examine the impact thermal evolution will have on *L. salmonis* selection with respect to drug tolerance because virulence is of major importance to wild and farmed salmonids alike.

Acknowledgements

The authors would like to thank Drs Sonja Saksida and Jordan Poley for their input and review of earlier drafts of this chapter. The authors would also like to thank the INTPART programme for the 'Cross Atlantic Sea Lice (CASL)' project funding; the Atlantic Canada Opportunities Agency (ACOA)-funded project 'Mitigating the impacts of climate change on salmon aquaculture (MICCSA)'; the Ocean Frontier Institute project funding which initiated work on salmon and sea lice interactions at elevated temperatures; and the strong collaborations between the University of Bergen, Institute of Marine Research (Norway) and the Atlantic Veterinary College, University of Prince Edward Island (Canada).

References

Aaen, S.M., Helgesen, K.O., Bakke, M.J., Kaur K. and Horsberg, T.E. (2015) Drug resistance in sea lice. A threat to salmonid aquaculture. *Trends in Parasitology* 31, 72–81.

Abolofia, J., Wilen, J.E. and Asche, F. (2017) The cost of lice: quantifying the impacts of parasitic sea lice on farmed salmon. *Marine Resource Economics* 32, 329–349.

ACFFA (Atlantic Canada Fish Farmers Association) (2016) *2016 New Brunswick Annual Sea Lice Management Report*. ACCFA, Letang, New Brunswick, Canada. Available at: https://static1.squarespace.com/static/56e827cb22482efe36420c65/t/58e3b26b893fc0130751d5ce/1491317360506/2016+Sea+Lice+Mgt+Report.pdf (accessed 9 April 2020).

ACFFA (Atlantic Canada Fish Farmers Association) (2017) *2017 New Brunswick Annual Sea Lice Management Report*. ACCFA, Letang, New Brunswick, Canada. Available at: https://static1.squarespace.com/static/56e827cb22482efe36420c65/t/5ada031e1ae6cf6be3d702fe/1524237088865/2017+Sea+Lice+Mgt+Report+.pdf (accessed 9 April 2020).

Alcorn, S.W., Murra, A.L. and Pascho, R.J. (2002) Effects of rearing temperature on immune functions in sockeye salmon (*Oncorhynchus nerka*). *Fish and Shellfish Immunology* 12, 303–334.

Beamish, R.J., Neville, C.M., Sweeting, R.M. and Ambers, N. (2005) Sea lice on adult Pacific salmon in the coastal waters of Central British Columbia, Canada. *Fisheries Research* 76, 198–208.

Beamish, R., Wade, J., Pennell, W., Gordon, E., Jones, S. *et al.* (2009) A large, natural infection of sea lice on juvenile Pacific salmon in the Gulf Islands area of British Columbia, Canada. *Aquaculture* 297, 31–37.

Beaugrand, G., Reid, P.C., Ibañez, F., Lindley, J.A. and Edwards, M. (2002) Reorganization of North Atlantic marine copepod biodiversity and climate. *Science* 296, 1692–1694.

Berland, B. and Margolis, L. (1983) The early history of 'lakselus' and some nomenclatural questions relating to copepod parasites of salmon. *Sarsia* 68, 281–288.

Besnier, F., Kent, M., Skern-Mauritzen, R., Lien, S., Malde, K. *et al.* (2014) Human-induced evolution caught in action: SNP-array reveals rapid amphi-atlantic spread of pesticide resistance in the salmon ecotoparasite *Lepeophtheirus salmonis*. *BMC Genomics* 15, 937.

Bjørn, P.A., Finstad, B. and Kristoffersen, R. (2001) Salmon lice infection of wild sea trout an Arctic char in marine and freshwaters: the effects of salmon farms. *Aquaculture Research* 32, 947–962.

Braden, L.M., Barker, D.E., Koop, B.F. and Jones, S.R.M. (2015a) Differential modulation of resistance biomarkers in skin of juvenile and mature pink salmon, *Oncorhynchus gorbuscha* by the salmon louse, *Lepeophtheirus salmonis*. *Fish and Shellfish Immunology* 47, 7–14.

Braden, L.M., Koop, B.F. and Jones, S.R.M. (2015b) Signatures of resistance to *Lepeophtheirus salmonis* include a TH2-type response at the louse–salmon interface. *Developmental and Comparative Immunology* 48, 178–191.

Braden, L.M., Shyte, S.K., Brown, A.B.J., Van Iderstine, C., Letendre, C. et al. (2019) Vaccine-induced protection against furunculosis involves pre-emptive priming of humoral immunity in Arctic charr. *Frontiers in Immunology* 10, 120.

Bricknell, I.R., Dalesman, S.J., O'Shea, B., Pert, C.C. and Luntz, A.J. (2006) Effect of environmental salinity on sea lice *Lepeophtheirus salmonis* settlement success. *Diseases of Aquatic Organisms* 71, 201–212.

Brooker, A.J., Skern-Mauritzen, R. and Bron, J.E. (2018) Production, mortality, and infectivity of planktonic larval sea lice, *Lepeophtheirus salmonis* (Krøyer, 1837): current knowledge and implications for epidemiological modelling. *ICES Journal of Marine Science* 75, 1214–1234.

Browman H., Boxaspen, K. and Kuhn, P. (2004) The effect of light on the settlement of the salmon louse, *Lepeophtheirus salmonis*, on Atlantic salmon, *Salmo salar* L. *Journal of Fish Diseases* 27, 701–708.

Bui, S., Oppedal, F. and Stien, L., Dempster, T. (2016) Sea lice infestation level alters salmon swimming depth in sea-cages. *Aquaculture Environment Interactions* 8, 429–435.

Carmichael, S.N., Bron, J.E., Taggart, J.B., Ireland, J.H., Bekaert, M. *et al.* (2013) Salmon lice (*Lepeophtheirus salmonis*) showing varying emamectin benzoate susceptibilities differ in neuronal acetylcholine receptor and GABA-gated chloride channel mRNA expression. *BMC Genomics* 14, 408.

Carpio, Y., Basabe, L., Acosta, J., Rodriguez, A., Mendoza, A. *et al.* (2011) Novel gene isolated from *Caligus rogercresseyi*: a promising target for vaccine development against sea lice. *Vaccine* 29, 2810–2820.

Chang, B.D., Page, F.H., Beattie, M.J. and Hill, B.W.H. (2011) Sea louse abundance on farmed salmon in Southwestern New Brunswick area of the Bay of Fundy. In: Jones, S.R.M. and Beamish, R. (eds) *Salmon Lice: An Integrated Approach to Understanding Parasite Abundance and Distribution*. Wiley, Chichester, UK, pp. 83–115.

Cheung, W.W.L., Dunne, J., Sariento, J.L. and Pauly, D. (2011) Integrating ecophysiology and plankton dynamics into projected maximum fisheries catch potential under climate change in the Northeast Atlantic. *ICES Journal of Marine Science* 68, 1008–1018.

Connors, B.M., Juarez-Clunga, E. and Dill, L.M. (2008) Effects of varying salinities on *Lepeophtheirus salmonis* survival on juvenile pink and chum salmon. *Journal of Fish Biology* 72, 1825–1830.

Côte, J., Boniface, A., Blanchet, S., Hendry, A.P., Gasparini, J. and Jacquin, L. (2018) Melanin-based coloration and host–parasite interactions under global change. *Proceedings of the Royal Society B: Biological Sciences* 285, 20180285.

Crosbie, W.D.W., Oppedal, F., Samsing, F. and Dempster, T. (2019) Effects of step salinity gradients on salmon lice larvae behaviour and dispersal. *Aquaculture Environment Interactions* 11, 181–190.

Dalvin, S., Frost, P., Loeffen, P., Skern-Mauritzen, R., Baban, J. *et al.* (2011) Characterisation of two vitellogenins in the salmon louse *Lepeophtheirus salmonis*: molecular, functional and evolutional analysis. *Diseases of Aquatic Organisms* 94, 211–224.

Danzmann, R.G., Norman, J.D., Rondeau, E.B., Messmer, A.M., Kent, M.P. *et al.* (2019) A genetic linkage map for the salmon louse (*Lepeophtheirus salmonis*): evidence for high male:female and inter-familial recombination rate differences. *Molecular Genetics and Genomics* 294, 343–363.

Dawson, L.H.J., Pike, A.W., Houlihan, D.F. and McVicar, A.H. (1997) Comparison of the susceptibility of sea trout (*Salmo trutta* L.) and Atlantic salmon (*Salmo salar* L.) to sea lice (*Lepeophtheirus salmonis* (Krøyer, 1837)) infections. *ICES Journal of Marine Science* 54, 1129–1139.

DFO (Fisheries and Oceans Canada) (2016) DFO sea lice audits of BC marine finfish aquaculture sites. DFO, Ottawa. Available at: https://open.canada.ca/data/en/dataset/5cfd93bd-b3ee-4b0b-8816-33d388f6811d (accessed 11 April 2020).

DFO (Fisheries and Oceans Canada) (2019) *Average number of lice per fish on BC salmon farms*. DFO, Ottawa. Available at: http://www.pac.dfo-mpo.gc.ca/aquaculture/reporting-rapports/lice-ab-pou/index-eng.html (accessed 9 April 2020).

Dukovcic, S.R., Hutchison, J.R. and Trempy, J.E. (2010) Potential of the melanophore pigment response for detection of bacterial toxicity. *Applied and Environmental Microbiology* 76, 8243–8246.

Elmoslemany, A., Review, C.R., Milligan, B., Stewardson, L. and Vanderstichel, R. (2015) Wild juvenile salmonids in Muchalat Inlet, British Columbia, Canada: factors associated with sea lice prevalence. *Diseases of Aquatic Organisms* 117, 107–120.

Engelhard, G.H. and Heino, M. (2006) Climate change and condition of herring (*Clupea harengus*) explain long-term trends in extent of skipped reproduction. *Oecologia* 149, 593–603.

Falvey, M. and Garreaud, R.D. (2009) Regional cooling in a warming world: recent temperature trends in the southeast Pacific and along the west coast of subtropical South America (1979–2006). *Journal of Geophysical Research* 114, D04102.

Fast, M.D. (2014) Fish immune responses to parasitic copepod (namely sea lice) infection. *Developmental and Comparative Immunology* 43, 300–312.

Fast, M.D., Ross, N.W., Mustafa, A., Sims, D.E., Johnson, S.C. *et al.* (2002) Susceptibility of rainbow trout *Oncorhynchus mykiss*, Atlantic salmon *Salmo salar* and coho salmon *Oncorhynchus kisutch* to experimental infection with sea lice *Lepeophtheirus salmonis*. *Diseases of Aquatic Organisms* 52, 57–68.

Fast, M.D., Burka, J.F., Johnson, S.C. and Ross, N.W. (2003) Enzymes released from *Lepeophtheirus salmonis* in response to mucus from different salmonids. *Journal of Parasitology* 89, 7–13.

Frank, K., Gansel, L.C., Lien, A.M. and Birkevold, J. (2015) Effects of a shielding skirt for prevention of sea lice on the flow past stocked salmon fish cages. *Journal of Offshore Mechanics and Arctic Engineering* 137, 011201.

Frenzl, B., Stien, L.H., Cockerill, D., Oppedal, F., Richards, R.H. et al. (2014) Manipulation of farmed Atlantic salmon swimming behaviour through the adjustment of lighting and feeding regimes as a tool for salmon lice control. *Aquaculture* 424–425, 183–188.

Fryer, J.L., Pilcher, K.S., Sanders, J.E., Rohovec, J.S., Zinn, J.K. *et al.* (1976) Temperature, infectious disease, and the immune responses in salmonid fish. *Ecological Research Series EPA-600/3-76-021*. US Environmental Protection Agency, Duluth, Minnesota.

Gautam, R., Vanderstichel, R., Boerlage, A.S., Revie, C.W. and Hammell, K.L. (2017) Evaluating bath treatment effectiveness in the control of sea lice burdens on Atlantic salmon in New Brunswick, Canada. *Journal of Fish Diseases* 40, 895–905.

Gharbi, K., Matthews, L., Bron, J., Roberts, R., Tinch, A. and Stear, M. (2015) The control of sea lice in Atlantic salmon by selective breeding. *Journal of the Royal Society Interface* 12, 0574.

Gjelland, K.O., Serra-Llinares, R.M., Hedger, R.D., Arechavala-Lopez, P., Nilsen, R. *et al.* (2014) Effects of salmon lice infection on the behaviour of sea trout in the marine phase. *Aquaculture Environment Interactions* 5, 221–223.

Gjerde, B., Ødegård, J. and Thorland, I. (2011) Estimates of genetic variation in the susceptibility of Atlantic salmon (*Salmo salar*) to the salmon louse *Lepeophtheirus salmonis*. *Aquaculture* 314, 66–72.

Glover, K.A., Stølen, Å.B., Messmer, A., Koop, B.F., Torrissen, O. and Nilsen, F. (2011) Population genetic structure of the parasitic copepod *Lepeophtheirus salmonis* throughout the Atlantic. *Marine Ecology Progress Series* 427, 161–172.

Gonzalez-Alanis, P., Wright, G.M., Johnson, S.C. and Burka, J.F. (2001) Frontal filament morphogenesis in the salmon louse, *Lepeophtheirus salmonis*. *Journal of Parasitology* 87, 561–574.

Gustafsson, L., Ellis, S., Robinson, T., Marenghi, F. and Endris, R. (2007) Efficacy of emamectin benzoate against sea lice infestations of Atlantic salmon, *Salmo salar* L.: evaluation in the absence of an untreated contemporary control. *Journal of Fish Diseases* 29, 621–627.

Hamre, L.A., Eichner, C., Caipang, C.M.A., Dalvin, S.T., Bron, J.E. *et al.* (2014) The salmon louse *Lepeophtheirus salmonis* (Copepoda: Caligidae) life cycle has only two chalimus stages. *PLoS ONE* 8, e73539.

Hamre, L.A., Bui, S., Oppedal, F., Skern-Mauritzen, R. and Dalvin, S. (2019) Development of the salmon louse *Lepeophtheirus salmonis* parasitic stages in temperatures ranging from 3 to 24°C. *Aquaculture Environment Interactions* 11, 429–433.

Hårdensson, B.H. and Uglem, I. (2019) *Pukkellaks i Norge 2017, version 1.0. Dataset/Occurrence. Norwegian Institute for Nature Research*, Trondheim, Norway. Available at: http://data.nina.no:8080/ipt/resource?r=pukkellaks2019&v=1.0 (accessed 9 April 2020).

Helgesen, K.O., Marin, S.L. and Fast, M.D. (2018) Sea lice resistance. In: *Technical Report: An Overview of Emerging Diseases in the Salmonid Farming Industry*. Elanco Animal Health and the Global Salmon Initiative. Available at: https://www.cahs-bc.ca/wp-content/uploads/2019/03/EMERGING_DISEASES_TECHNICAL_REPORT_JAN_17_2019.pdf (accessed 11 April 2020).

Heuch, P.A., Parsons, A. and Boxaspen, K. (1995) Diel vertical migration – a possible host-finding mechanism in salmon louse (*Lepeophtheirus salmonis*) copepodids. *Canadian Journal of Fisheries and Aquatic Sciences* 52, 681–689.

Heuch, P.A., Nordhagen, J.R. and Schram, T.A. (2000) Egg production in the salmon louse [*Lepeophtheirus salmonis* (Krøyer)] in relation to origin and water temperature. *Aquaculture Research* 31, 805–814.

Heuch, P.A., Knutsen, J.A., Knutsen, H. and Schram, T.A. (2002) Salinity and temperature effects on sea lice over-wintering on sea trout (*Salmo trutta*) in coastal areas of the Skagerrak. *Journal of the Marine Biology Association of the United Kingdom* 82, 887–892.

Hixson, S.M. and Arts, M.T. (2016) Climate warming is predicted to reduce omega-3, long-chain, polyunsaturated fatty acid production in phytoplankton. *Global Change Biology* 22, 2744–2755.

Hogans, W.E. (1995) Infection dynamics of sea lice, *Lepeophtheirus salmonis* (Copepoda: Caligidae) parasitic on Atlantic salmon (*Salmo salar*) cultured in marine waters of the lower Bay of Fundy. *Canadian Technical Report on Fisheries and Aquatic Sciences No. 2067*. Fisheries and Oceans Canada, Ottawa.

Hogans, W.E. and Trudeau, D.J. (1989) Preliminary studies on the biology of sea lice, *Caligus elongatus*, *Caligus curtus*, and *Lepeophtheirus salmonis* (Copepoda: Caligidae) parasitic on cage-cultured salmonids in the lower Bay of Fundy. *Canadian Technical Report on Fisheries and Aquatic Sciences No. 1715*. Fisheries and Oceans Canada, Ottawa.

Højland, D.H. and Kristensen, M. (2017) Analysis of differentially expressed genes related to resistance in spinosad- and neonicotinoid-resistant *Musca domestica* L. (Diptera: Muscidae) strains. *PLoS ONE* 12, e0170935.

Hutchings, J.A., Côté, I.M., Dodson, J.J., Fleming, I.A., Jennings, S. *et al.* (2011) Climate change, fisheries, and aquaculture: trends and consequences for Canadian marine biodiversity. *Environmental Reviews* 20, 220–311.

Hvas, M., Folkedal, O., Imsland, A. and Oppedal, F. (2018) Metabolic rates, swimming capabilities, thermal niche and stress response of the lumpfish, *Cyclopterus lumpus*. *Biology Open* 7, bio036079.

ICES (International Council for the Exploration of the Sea) (2017) Report of the Working Group on North Atlantic Salmon (WGNAS), 29 March–7 April 2017, Copenhagen, Denmark. *ICES CM 2017/ACOM:20*. ICES, Copenhagen. Available at: http://ices.dk/sites/pub/Publication%20Reports/Expert%20Group%20Report/acom/2017/WGNAS/wgnas_2017.pdf (accessed 14 April 2020).

Imsland, A.K., Reynolds, P., Nytro, A.V., Eliassen, G., Hangstad, T.A. *et al.* (2016) Effects of lumpfish size on foraging behaviour and co-existence with sea lice infected Atlantic salmon in sea cages. *Aquaculture* 465, 19–27.

IPCC (Intergovernmental Panel on Climate Change) (2014) *Climate Change 2014: Synthesis Report. Contribution of Working Groups I, II and III to the Fifth Assessment Report of the Intergovernmental Panel on Climate Change* [Core Writing Team, Pachauri, R.K. and Meyer, L.A. (eds)]. IPCC, Geneva, Switzerland.

Jaramillo, R. (2018) Drug resistance of sea lice *Caligus rogercresseyi* in Chile. *Approaches in Poultry, Dairy and Veterinary Sciences* 2, 1–2.

Jensen A.J., Zydlewski, G.B., Barker, S. and Pietrak, M. (2016) Sea lice infestation of a wild fish assemblage in the Northwest ocean. *Transactions of the American Fisheries Society* 145, 7–16.

Johnson, S.C. and Albright, L.J. (1991) The developmental stages of *Lepeophtheirus salmonis* (Krøyer, 1837) (Copepoda: Caligidae). *Canadian Journal of Zoology* 69, 929–950.

Johnson, S.C. and Albright, L.J. (1992) Comparative susceptibility and histopathology of the response of naive Atlantic, chinook and coho salmon to experimental infection with *Lepeophtheirus salmonis* (Copepoda: Caligidae). *Diseases of Aquatic Organisms* 14, 179–193.

Johnson, S.C., Blaylock, R.B., Elphick, J. and Hyatt, K.D. (1996) Disease induced by the sea louse (*Lepeophtheirus salmonis*) (Copepoda: Caligidae) in wild sockeye salmon (*Oncorhynchus nerka*) stocks of Alberni Inlet, British Columbia. *Canadian Journal of Fisheries and Aquatic Sciences* 53, 2888–2897.

Jones, P.G., Hammell, K.L., Dohoo, I.R. and Revie, C.W. (2012) Effectiveness of emamectin benzoate for treatment of *Lepeophtheirus salmonis* on farmed Atlantic salmon *Salmo salar* in the Bay of Fundy, Canada. *Diseases of Aquatic Organisms* 102, 53–64.

Jones, S.R.M. and Johnson, S.C. (2014) Biology of sea lice, *Lepeophtheirus salmonis* and *Caligus* spp., in western and eastern Canada. *Canadian Science Advisory Secretariat Research Document No. 2014/019*. Fisheries and Oceans Canada, Ottawa.

Jones, S.R.M. and Prosperi-Porta, G. (2011) The diversity of sea lice (Copepoda: Caligidae) parasitic on threespine stickleback (*Gasterosteus aculeatus*) in coastal British Columbia. *Journal of Parasitology* 97, 399–405.

Jones, S.R.M., Kim, E. and Dawe, S.C. (2006a) Experimental infections with *Lepeophtheirus salmonis* (Krøyer) on threespine sticklebacks, *Gasterosteus aculeatus* L., and juvenile Pacific salmon, *Oncorhynchus* spp. *Journal of Fish Diseases* 29, 489–495.

Jones, S.R.M., Prosperi-Porta, G., Kim, E., Callow, P. and Hargreaves, N.B. (2006b) The occurrence of *Lepeophtheirus salmonis* and *Caligus clemensi* (Copepoda: Caligidae) on threespine stickleback *Gasterosteus aculeatus* in coastal British Columbia. *Journal of Parasitology* 92, 473–480.

Jones, S.R.M., Fast, M.D., Johnson, S.C. and Groman, D.B. (2007) Differential rejection of salmon lice by pink and chum salmon: disease consequences and expression of proinflammatory genes. *Diseases of Aquatic Organisms* 75, 229–238.

Jones, S.R.M., Fast, M.D. and Johnson, S.C. (2008a) Influence of reduced feed ration on *Lepeophtheirus salmonis* infestation and inflammatory gene expression in juvenile pink salmon. *Journal of Aquatic Animal Health* 20, 103–109.

Jones, S., Kim, E. and Bennett, W. (2008b) Early development of resistance to the salmon louse, *Lepeophtheirus salmonis* (Krøyer), in juvenile pink salmon, *Oncorhynchus gorbuscha* (Walbaum). *Journal of Fish Diseases* 31, 591–600.

Jones, S.R.M., Keith, I., Saksida, S., Morrison, D., Milligan, B. and Hewison, T. (2016) The role of salmon lice data from salmon aquaculture in British Columbia in management, regulation and research. Presented at *11th International Sea Lice Conference*, Westport, *Ireland*, 26–28 September 2016.

Kabata, Z. (1979) *Parasitic Copepoda of British Fishes*. The Ray Society, London.

Kaur, K., Bakke, M.J., Nilsen, F. and Horsberg, T.E. (2015) Identification and molecular characterization of two acetylcholinesterases from the salmon louse, *Lepeophtheirus salmonis*. *PLoS ONE* 10, e0125362.

Kaur, K., Jansen, P.A., Aspehaug, V.T. and Horsberg, T.E. (2016) Phe362Tyr in AChE: a major factor responsible for azamethiphos resistance in *Lepeophtheirus salmonis* in Norway. *PLoS ONE* 11, e0149264.

Kaur, K., Besnier, F., Glover, K.A., Nilsen, F., Aspehaug, V.T. *et al.* (2017) The mechanism (Phe362Tyr mutation) behind resistance in *Lepeophtheirus salmonis* predates organophosphate use in salmon farming. *Scientific Reports* 7, 12349.

Kennedy, C.J. and Picard, C. (2012) Chronic low pH exposure affects the seawater readiness of juvenile Pacific sockeye salmon. *Fish Physiology and Biochemistry* 38, 1131–1143.

Kittilsen, S., Johansen, I.B., Braastad, B.O. and Overli, O. (2012) Pigments, parasites and personality: towards a unifying role for steroid hormones. *PLoS ONE* 7, e34281.

Klinger, D.H., Levin, S.A. and Watson, J.R. (2017) The growth of finfish in global open-ocean aquaculture under climate change. *Proceedings of the Royal Society B: Biological Sciences* 284, 20170834.

Komisarczuk, A.Z., Grotmol S. and Nilsen, F. (2017) Ionotropic receptors signal host recognition in the salmon louse (*Lepeophtheirus salmonis*, Copepoda). *PLoS ONE* 12, e0178812.

Krøyer, H. (1838) Om snyltekrebsene, især med hensyn til den danske fauna. *Naturhistorik Tidsskrift* 2, 131–157.

Landsberg, J.H., Vermeer, G.K., Richards, S.A. and Perry, N. (1991) Control of the parasitic copepod *Caligus elongatus* on pond-reared red drum. *Journal of Aquatic Animal Health* 3, 206–209.

Lees, F., Gettinby, G. and Revie, C.W. (2008) Changes in epidemiological patterns of sea lice infestation on farmed Atlantic salmon, *Salmo salar* L., in Scotland between 1996 and 2006. *Journal of Fish Diseases* 31, 259–268.

Ljungfeldt, L.E.R., Quintela, M., Besnier, F., Nilsen, F. and Glover, K.A. (2017) A pedigree-based experiment reveals variation in salinity and thermal tolerance in the salmon louse, *Lepeophtheirus salmonis*. *Evolutionary Applications* 10, 1007–1019.

Losos, C. (2008) Behavioural interactions of sea lice, threespine sticklebacks, and juvenile Pacific salmon. MSc thesis, Simon Fraser University, British Columbia, Canada.

MacDonald, M.M., Groman, D., Stryhn, H., Frasca, S. and Fast, M.D. (2015) Molecular and histopathological disruption of wound healing in an ectoparasitic model. In: Program of the 40th Annual Eastern Fish Health Workshop, Charleston, South Carolina, 2–6 March 2015, p. 49. Available at: https://scmss.coastal.edu/sites/default/files/student_experience/40th%20EFHW%20Program.pdf (accessed 14 April 2020).

McEwan, G.F., Groner, M.L., Fast, M.D., Gettinby, G. and Revie, C.W. (2015) Using agent-based modelling to predict the role of wild refugia in the evolution of resistance of sea lice to chemotherapeutants. *PLoS ONE* 10, e0139128.

McEwan, G.F., Groner, M.L., Burnette, D.L., Fast, M.D. and Revie, C.W. (2016) Managing aquatic parasites for reduced drug resistance: lessons from the land. *Journal of the Royal Society Interface* 13, 20160830.

Malick, M.J., Cox, S.P., Mueter, F.J. and Peterman, R.M. (2015) phenology to salmon productivity along a north-south gradient in the Northeast Pacific Ocean. *Canadian Journal of Fisheries and Aquatic Sciences* 72, 697–708.

Margolis, L. and Kabata, Z. (eds) (1988) Guide to the parasites of fishes of Canada. Part II – Crustacea. *Canadian Special Publication of Fisheries and Aquatic Sciences No. 101*. Department of Fisheries and Oceans, Ottawa.

Messmer, A.M., Leong, J.S., Rondeau, E.B., Mueller, A., Despins, C.A. *et al.* (2018) A 200K SNP chip reveals a novel Pacific salmon louse genotype linked to differential efficacy of emamectin benzoate. *Marine Genomics* 40, 45–57.

Mo, T.A. and Heuch, P.A. (1998) Occurrence of *Lepeophtheirus salmonis* (Copepoda: Caligidae) on sea trout (*Salmo trutta*) in the inner Oslo Fjord, south-eastern Norway. *ICES Journal of Marine Science* 55, 176–180.

Montory, J.A., Cumillaf, J.P., Cubillos, V.M., Paschke, K., Urbina, M.A. and Gebauer, P. (2018) Early development of the ectoparasite *Caligus rogercresseyi* under combined salinity and temperature gradients. *Aquaculture* 486, 68–74.

Mustafa, A. and McKinnon, B.M. (1999) Genetic variation in susceptibility of Atlantic salmon to the sea louse *Caligus elongatus* Nordmann, 1832. *Canadian Journal of Zoology* 77, 1332–1335.

Nekouie, O., Vanderstichel, R., Thakur, K., Arriagada, G., Patanasatienkul, T. *et al.* (2018) Association between sea lice (*Lepeophtheirus salmonis*) infestation on Atlantic salmon farms and wild Pacific salmon Muchalat Inlet, Canada. *Scientific Reports* 8, 4023.

Nikoskelainen S., Bylund, G. and Lilius, E.M. (2004) Effect of environmental temperature on rainbow trout (*Oncorhynchus mykiss*) innate immunity. *Developmental and Comparative Immunology* 28, 581–592.

Oines, O. and Heuch, P.A. (2007) *Caligus elongatus* Nordmann genotypes on wild and farmed fish. *Journal of Fish Diseases* 30, 81–91.

Ou, M., Hamilton, T.J., Eom, J., Lyall, E.M., Gallup, J. *et al.* (2015) Responses of pink salmon to CO_2-induced aquatic acidification. *Nature Climate Change* 5, 950–955.

Overton, K., Dempster, T., Oppedal, F., Kristiansen, T.S., Gismervik, K. and Stien, L.H. (2018) Salmon louse treatments and salmon mortality in Norwegian aquaculture: a review. *Reviews in Aquaculture* 11, 1398–1417.

Parker, R.R. (1968) Mortality of juvenile pink salmon in central British Columbia coastal waters. *Manuscript Report Series No. 956*. Fisheries Research Board of Canada, Nanaimo, Canada.

Parker, R.R. (1969) Predator–prey relationship among pink and chum salmon fry and coho smolts in a central

British Columbia inlet. *Manuscript Report Series No. 1019*. Fisheries Research Board of Canada, Nanaimo, Canada.

Paterson, W.D. (1971) The antibody response of juvenile coho salmon (Oncorhynchus kisutch) to Aeromonas salmonicida, the causative agent of furunculosis. PhD thesis, Oregon State University, Corvallis, Oregon.

Peacock, S.J., Krkošek, M., Proboszcz, S., Orr, C. and Mark, A. (2013) Cessation of a salmon decline with control of parasites. *Ecological Applications* 23, 606–620.

Piasecki, W. (1996) The developmental stages of *Caligus elongatus* von Nordmann, 1832 (Copepoda: Caligidae). *Canadian Journal of Zoology* 74, 1459–1478.

Piasecki, W. and MacKinnon, B.M. (1995) Life cycle of a sea louse, *Caligus elongatus* von Nordmannm 1832 (Copepoda, Siphonostomatoida, Caligidae). *Canadian Journal of Zoology* 73, 74–82.

Pike, A. and Wadsworth, S. (1999) Sea lice on salmonids: biology and control. *Advances in Parasitology* 44, 233–337.

Pinsky, M.L. and Byler, D. (2015) Fishing, fast growth and climate variability increase the risk of collapse. *Proceedings of the Royal Society B: Biological Sciences* 282, 20151053.

Poley, J.D. and Fast, M.D. (2020) Ectoparasitic arthropods of fish: orders Siphonostomatoida, Cyclopoida, Arguloida, Isopoda. In: Kibenge, F. and Freeman, M. (eds) *Aquaculture Parasitology: Pathogens of Fish, Crustaceans, and Molluscs*, 1st edn. Elsevier, Amsterdam (in press).

Poley, J.D., Igboeli, O.O. and Fast, M.D. (2015) Towards a consensus: multiple experiments provide evidence for constitutive expression differences among sexes and populations of sea lice (*Lepeophtheirus salmonis*) related to emamectin benzoate resistance. *Aquaculture* 448, 445–450.

Qadri, S.A., Camacho, J., Wang, H., Taylor, J.R., Grosell, M. and Worden, M.K. (2007) Temperature and acid–base balance in the American lobster *Homarus americanus*. *Journal of Experimental Biology* 210, 1245–1254.

Raynard, R.S., Bricknell, I.R., Billingsley, P.F., Nisbet, A.J., Vigneau, A. and Sommerville, C. (2002) Development of vaccines against sea lice. *Pest Management Science* 58, 569–575.

Robledo, D., Gutierrez, A.P., Barria, A., Yanez, J.M. and Houston, R.D. (2018) Gene expression response to sea lice in Atlantic salmon skin: RNA sequencing comparison between resistant and susceptible animals. *Frontiers in Genetics* 9, 287.

Rosseland, B.O. and Skogheim, O.K. (1984) A comparative study on salmonid fish species in acid aluminium-rich water. II Physiological stress and mortality of one- and two-year old fish. *Institution of Freshwater Research Drottingholm Report* 61, 186–194.

Saba, V.S., Griffies, S.M., Anderson, W.G., Winton, M., Alexander, M.A. *et al.* (2016) Enhanced warming of the Northwest Atlantic Ocean under climate change. *Journal of Geophysical Research: Oceans* 121, 118–132.

Saksida, S., Constantine, J., Karreman, G.A. and Donald, A. (2007) Evaluation of sea lice abundance levels on farmed Atlantic salmon (*Salmo salar* L.) located in the Broughton Archipelago of British Columbia from 2003 to 2005. *Aquaculture Research* 38, 219–231.

Saksida, S.M., Morrison, D., Sheppard, M. and Keith, I. (2011) Sea lice management on salmon farms in British Columbia, Canada. In: Jones, S.R.M. and Beamish, R. (eds) *Salmon Lice: An Integrated Approach to Understanding Parasite Abundance and Distribution*. Wiley, Chichester, UK, pp. 235–278.

Saksida, S.M., Morrison, D., McKenzie, P., Milligan, B., Downey, E. *et al.* (2013) Use of Atlantic salmon, *Salmo salar* L., farm treatment data and bioassays to assess for resistance of sea lice, *Lepeophtheirus salmonis*, to emamectin benzoate (SLICE®), in British Columbia, Canada. *Journal of Fish Diseases* 36, 515–520.

Samsing, F., Oppedal, F., Dalvin, S., Johnsen, I., Vagseth, T. and Dempster, T. (2016) Salmon lice (*Lepeophtheirus salmonis*) development times, body size, and reproductive outputs follow universal models of temperature dependence. *Canadian Journal of Fisheries and Aquatic Sciences* 73, 1841–1851.

Sanchez-Hernandez, J. (2017) *Lernaea cyprinacea* (Crustacea: Copepoda) in the Iberian Peninsula: climate implications on host–parasite interactions. *Knowledge and Management of Aquatic Ecosystems* 418, 11.

Sevatdal, S., Copley, L., Wallace, C., Jackson, D. and Horsberg, T.E. (2005) Monitoring of the sensitivity of sea lice (*Lepeophtheirus salmonis*) to pyrethroids in Norway, Ireland and Scotland using bioassays and probit modelling. *Aquaculture* 244, 19–27.

Skern-Mauritzen, R., Torissen, O. and Glover, K.A. (2014) Pacific and Atlantic *Lepeophtheirus salmonis* (Krøyer, 1838) are allopatric subspecies: *Lepeophtheirus salmonis salmonis* and *L. salmonis oncorhynchi* subspecies novo. *BMC Genetics* 15, 32.

Skiftesvik, A.B., Bjelland, R.M., Durif, C.M.F., Johansen, I.S. and Browman, H.I. (2013) Delousing of Atlantic salmon (*Salmo salar*) by cultured vs. wild ballan wrasse (*Labrus bergylta*). *Aquaculture* 402–403, 113–118.

Statistics Norway (2019) Aquaculture. Available at: https://www.ssb.no/en/fiskeoppdrett/ (accessed 14 April 2020).

Staurnes, M., Blix, P. and Reite, O.B. (1993) Effects of acid water and aluminum on parr–smolt transformation and seawater tolerance in Atlantic salmon, *Salmo salar*. *Canadian Journal of Fisheries and Aquatic Sciences* 50, 1816–1827.

Stien, L.H., Dempster, T., Bui, S., Glaropoulos, A., Fosseidengen, J.E. *et al*. (2016) 'Snorkel' sea lice barrier technology reduces sea lice loads on harvest-sized Atlantic salmon with minimal welfare impacts. *Aquaculture* 458, 29–37.

Sutherland, B.J.G., Jantzen, S.G., Sanderson, D.S., Koop, B.F. and Jones, S.R.M. (2011) Differentiating size-dependent responses of juvenile pink salmon (*Oncorhynchus gorbuscha*) to sea lice (*Lepeophtheirus salmonis*) infections. *Comparative Biochemistry and Physiology Part D: Genomics and Proteomics* 6, 213–223.

Sutherland, B.J.G., Jantzen, S.G., Yasuike, M., Sanderson, D.S., Koop, B.F. and Jones, S.R.M. (2012) Transcriptomics of coping strategies in free-swimming *Lepeophtheirus salmonis* (Copepoda) larvae responding to abiotic stress. *Molecular Ecology* 21, 6000–6014.

Sutherland, B.J., Koczka, K.W., Yasuike, M., Jantzen, S.G., Yazawa, R. *et al*. (2014) Comparative transcriptomics of Atlantic *Salmo salar*, chum *Oncorhynchus keta* and pink salmon *O. gorbuscha* during infections with salmon lice *Lepeophtheirus salmonis*. *BMC Genomics* 15, 200.

Taranger, G.L., Karlsen, O., Bannister, R.J., Glover, K.A., Husa, V. *et al*. (2015) Risk assessment of the environmental impact of Norwegian Atlantic salmon farming. *ICES Journal of Marine Science* 72, 997–1021.

Thody, A.J. and Shuster S. (1989) Melanophores, melanocytes and melanin: endocrinology and pharmacology. In: Greaves, M.W. and Shuster, S. (eds) *Pharmacology of the Skin I*. Springer, Berlin/Heidelberg, Germany, pp. 257–269.

Todd, C.D., Walker, A.M., Ritchie, M.G., Graves, J.A. and Walker, A.F. (2004) Population genetic differentiation of sea lice (*Lepeophtheirus salmonis*) parasitic on Atlantic and Pacific salmonids: analyses of microsatellite DNA variation among wild and farmed hosts. *Canadian Journal of Fisheries and Aquatic Sciences* 61, 1176–1190.

Torrissen, O., Jones, S., Asche, F., Guttormsen A., Skilbrei, O.T. *et al*. (2013) Salmon lice – impact on wild salmonids and salmon aquaculture. *Journal of Fish Diseases* 36, 171–194.

Tsai, H.-Y., Hamilton, A., Tinch, A.E., Guy, D.R., Bron, J.E. *et al*. (2016) Genomic prediction of host resistance to sea lice in farmed Atlantic salmon populations. *Genetics, Selection, Evolution* 48, 47.

Tucker, C.S., Sommerville, C. and Wootten, R. (2000) An investigation into the larval energetics and settlement of the sea louse, *Lepeophtheirus salmonis*, an ectoparasitic copepod of Atlantic salmon, *Salmo salar*. *Fish Pathology* 35, 137–143.

USGCRP (US Global Change Research Program) (2017) *Climate Science Special Report: Fourth National Climate Assessment*, Vol. I [Wuebbles, D.J., Fahey, D.W., Hibbard, K.A., Dokken, D.J., Stewart, B.C. and Maycock, T.K. (eds)]. USGCRP, Washington, DC. Available at: https://science2017.globalchange.gov (accessed 9 April 2020).

Wagner, G.N., Fast, M.D. and Johnson, S.C. (2008) Physiology and immunology of *Lepeophtheirus salmonis* infections of salmonids. *Trends in Parasitology* 24, 176–183.

Welicky, R.L., De Swardt, J., Gerber, R., Netherlands, E.C. and Smit, N.J. (2017) Drought-associated absence of alien invasive anchorworm, *Lernaea cyprinacea* (Copepoda: Lernaeidae), is related to changes in fish health. *International Journal of Parasitology: Parasites and Wildlife* 6, 430–438.

Whiteley, N. (2011) Physiological and ecological responses of crustaceans to ocean acidification. *Marine Ecology Progress Series* 430, 257–271.

Whiteley, N.M. and Taylor, E.W. (2015) Responses to environmental stresses: oxygen, temperature and pH. In: Chang, E.S. and Thiel, M. (eds) *The Natural History of the Crustacea, Growth, Moulting and Physiology*. Oxford University Press, Oxford, pp. 320–358.

Whyte, S.K., Westcott, J.D., Jiminez, D., Revie, C.W. and Hammell, K.L. (2014) Assessment of sea lice (*Lepeophtheirus salmonis*) management in New Brunswick, Canada, using deltamethrin (AlphaMax®) through clinical field treatment and laboratory bioassay responses. *Aquaculture* 422/423, 54–62.

Whyte, S.K., Westcott, J.D., Revie, C.W. and Hammell, K.L. (2016) Sensitivity of salmon lice (*Lepeophtheirus salmonis*) in New Brunswick, Canada, to the organophosphate Salmosan® (w/w 50% azamethiphos) using bioassays. *Aquaculture* 464, 593–600.

Whyte, S.K., Poley, J.D., Mueller, A., Van Iderstine, C., Fitzpatrick, K.E. *et al*. (2019) Avermectin treatment for *Lepeophtheirus salmonis*: impacts on host (*Salmo salar*) and parasite immunophysiology. *Aquaculture* 501, 488–501.

Wootten, R., Smith, J.W. and Needham, E.A. (1982) Aspects of the biology of the parasitic copepods *Lepeophtheirus salmonis* and *Caligus elongatus* on farmed salmonids and their treatment. *Proceedings of the Royal Society of Edinburgh* 81B, 185–197.

Wright, D.W., Oppedal, F. and Dempster, T. (2016) Early-stage sea lice recruits on Atlantic salmon are freshwater sensitive. *Journal of Fish Diseases* 39, 1179–1186.

Yanez, E., Lagos, N.A., Norambuena, R., Silva, C., Letelier, J. *et al*. (2018) Impacts of climate change on marine fisheries and aquaculture in Chile. In: Phillips, B.F. and Pérez-Ramírez, M. (eds) *Climate Change Impacts on Fisheries and Aquaculture: A Global Analysis*, Vol. 1. Wiley, Chichester, UK, pp. 239–332.

Yazawa, R., Yasuike, M., Leong, J., von Schalburg, K.R., Cooper, G.A. *et al.* (2008) EST and mitochondrial DNA sequences support a distinct Pacific form of salmon louse, *Lepeophtheirus salmonis*. *Marine Biotechechnology* 10, 741–749.

Zhang, C., Jansen, M., De Meester, L. and Stoks, R. (2018) Thermal evolution offsets the elevated toxicity of a contaminant under warming: a resurrection study in *Daphnia magna*. *Evolutionary Applications* 11, 1425–1436.

Index

Page numbers in **bold** type refer to figures and tables.

www.ingramcontent.com/pod-product-compliance
Lightning Source LLC
LaVergne TN
LVHW070903300926
849162LV00011B/1845